Natural constants in SI units

The numerical values are taken from CODATA.

Quantity	Symbol	Value	Error (ppm)
speed of light in vacuum	c	$2.99792458 \cdot 10^8$ m/s	exact
gravitational constant	G	$6.67259 \cdot 10^{-11}$ m^3/(kgs^2)	128
electron charge, elementary charge	e, e_0	$1.60217733 \cdot 10^{-19}$ C	0.30
Planck's quantum of action	h	$6.6260755 \cdot 10^{-34}$ J $\cdot$ s	0.60
Planck's constant	$\hbar = (2\pi)^{-1}h$	$1.05457266 \cdot 10^{-34}$ J $\cdot$ s	0.60
Avogadro's number	N_A	$6.0221367 \cdot 10^{23}$ mol^{-1}	0.59
Faraday constant	$F = N_A e_0$	$9.6485309 \cdot 10^4$ C/mol	0.30
electron mass	m_e	$9.1093897 \cdot 10^{-31}$ kg	0.59
		0.51099906 MeV	0.30
Rydbeg constant	$R_\infty = (2h)^{-1}m_e c\alpha^2$	$1.0973731534 \cdot 10^7$ m^{-1}	0.0012
fine-structure constant	$\alpha = e_0^2(2\varepsilon_0 hc)^{-1}$	$7.29735308 \cdot 10^{-3}$	0.045
	α^{-1}	137.0359895	0.045
electron radius	$r_e = \hbar(m_e c)^{-1}\alpha$	$2.81794092 \cdot 10^{-15}$ m	0.13
e^--Compton wavelength	$\lambda_C = h(m_e c)^{-1}$	$2.42631058 \cdot 10^{-12}$ m	0.089
Bohr radius	$a_0 = r_e\alpha^{-2}$	$5.29177249 \cdot 10^{-11}$ m	0.045
atomic mass unit	$u = \frac{1}{12}m(^{12}C)$	$1.6605402 \cdot 10^{-27}$ kg	0.59
proton mass	m_p	$1.6726231 \cdot 10^{-27}$ kg	0.59
		938.27231 MeV	0.30
neutron mass	m_n	$1.6749286 \cdot 10^{-27}$ kg	0.59
		939.56563 MeV	0.30
magnetic flux quantum	$\Phi_0 = h(2e_0)^{-1}$	$2.06783461 \cdot 10^{-15}$ Wb	0.30
specific electron charge	$-e_0 m_e^{-1}$	$-1.75881962 \cdot 10^{11}$ C/kg	0.30
Bohr magneton	$\mu_B = e_0\hbar(2m_e)^{-1}$	$9.2740154 \cdot 10^{-24}$ J/T	0.34
magnetic moment of electron	μ_e	$9.2847701 \cdot 10^{-24}$ J/T	0.34
nuclear magneton	$\mu_N = e_0\hbar(2m_p)^{-1}$	$5.0507866 \cdot 10^{-27}$ J/T	0.34
magnetic moment of proton	μ_p	$1.41060761 \cdot 10^{-26}$ J/T	0.34
gyromagnetic ratio	γ_p	$2.67522128 \cdot 10^8$ rad/sT	0.30
quantum Hall resistance	R_H	25812.8056 Ω	0.045
universal gas constant	R	8.314510 J/(mol K)	8.4
Boltzmann constant	$k, k_B = R N_A^{-1}$	$1.380658 \cdot 10^{-23}$ J/K	8.5
Stefan-Boltzmann constant	$\sigma = \pi^2 k_B^4(60\hbar^3 c^2)^{-1}$	$5.67051 \cdot 10^{-8}$ W/m^{2}K^4	34
Wien's constant	$b = \lambda_{max}T$	$2.897756 \cdot 10^{-3}$ m $\cdot$ K	8.4
permeability of free space	μ_0	$4\pi \cdot 10^{-7}$ Vs/(Am)	exact
permittivity constant of free space	$\varepsilon_0 = (\mu_0 c^2)^{-1}$	$8.85418781762 \cdot 10^{-12}$ As/(Vm)	exact

Handbook of Physics

Springer
New York
Berlin
Heidelberg
Barcelona
Hong Kong
London
Milan
Paris
Singapore
Tokyo

Walter Benenson   John W. Harris
Horst Stocker   Holger Lutz

Handbook of Physics

With 797 Illustrations

 Springer

Walter Benenson
Department of Physics
Cyclotron Laboratory
Michigan State University
East Lansing, MI 44844
USA
benenson@nscl.nscl.msu.edu

John W. Harris
Department of Physics
272 Whitney Avenue
Yale University
New Haven, CT 06520
USA
john.harris@yale.edu

Horst Stocker
Institut für Theoretische Physik
Johann Wolfgang Goethe-Universität
Robert Mayer Strasse 8-10
D-60054 Frankfurt am Main
Germany
stoecker@th.physik.uni-frankfurt.de

Holger Lutz
Fachhochschule Gießen-Friedberg
Fachbereich Elektrotechnik II
Wilhelm-Leuschner-Strasse 13
61169 Friedberg
Germany
holger.lutz@e2.fh-friedberg.de

Translated from the German *Tashenbuch der Physik* edited by Horst Stöcker, published by Verlag Harri Deutsch, Frankfurt am Main, Germany, Fourth Edition © 2000.

Library of Congress Cataloging-in-Publication Data
Benenson, W. (Walter).
    Handbook of physics/Walter Benenson, John W. Harris, Horst Stocker, Holger Lutz.
      p. cm.
    Includes bibliographical references and index.
    ISBN 0-387-95269-1 (alk. paper)
    1. Physics—Handbooks, manuals, etc. I. Harris, John W. II. Stocker, Horst.
III. Lutz, Holger. IV. Title.
QC61.H37 2001
530—dc21   2001020442

Printed on acid-free paper.

Production managed by Frank McGuckin; manufacturing supervised by Jeffrey Taub.
Typeset by Integre Technical Publishing Company, Inc., Alburquerque, NM.
Printed and bound by Hamilton Printing Co., Rensselaer, NY.
Printed in the United States of America.

9 8 7 6 5 4 3 2 1

ISBN 0-387-95269-1   SPIN 10831209

Springer-Verlag   New York Berlin Heidelberg
A member of BertelsmannSpringer Science+Business Media GmbH

Preface

Applications of physics can be found in a wider and wider range of disciplines in the sciences and engineering. It is therefore more and more important for students, practitioners, researchers, and teachers to have ready access to the facts and formulas of physics.

Compiled by professional scientists, engineers, and lecturers who are experts in the day-to-day use of physics, this *Handbook* covers topics from classical mechanics to elementary particles, electric circuits to error analysis.

This handbook provides a veritable toolbox for everyday use in problem solving, homework, examinations, and practical applications of physics, it provides quick and easy access to a wealth of information including not only the fundamental formulas of physics but also a wide variety of experimental methods used in practice.

Each chapter contains

> ➤ all the important concepts, formulas, rules and theorems
> ▲ numerous examples and practical applications
> ■ suggestions for problem solving, hints, and cross references
> M measurement techniques and important sources of errors
> as well as numerous tables of standard values and material properties.

Access to information is direct and swift through the user friendly layout, structured table of contents, and extensive index. Concepts and formulas are treated and presented in a uniform manner throughout: for each physical quantity defined in the *Handbook*, its characteristics, related quantities, measurement techniques, important formulas, SI-units, transformations, range of applicability, important relationships and laws, are all given a unified and compact presentation.

This *Handbook* is based on the third German edition of the *Taschenbuch der Physik* published by Verlag Harri Deutsch. Please send suggestions and comments to the Physics Editorial Department, Springer-Verlag, 175 Fifth Avenue, New York, NY 10010.

Walter Benenson, *East Lansing, Michigan, USA*
John W. Harris, *New Haven, Connecticut, USA*
Horst Stocker, *Frankfurt, Germany*
Holger Lutz, *Friedberg, Germany*

Contents

Part II   Vibrations and Waves   253

8   Vibrations   255

9   Waves   287

Contributors

Mechanics: Christoph Best (Universität Frankfurt), with Helmut Kutz (Mauserwerke AG, Oberndorf) and Rudolph Pitka, (Fachhochschule Frankfurt)

Oscillations and Waves, Acoustics, Optics: Kordt Griepenkerl (Universität Frankfurt), with Steffen Bohrmann (Fachhochschule Technik, Mannheim) and Klaus Horn (Fachhochschule Frankfurt)

Electricity, Magnetism: Christian Hofmann, (Technische Universität Dresden), with Klaus-Jürgen Lutz (Universität Frankfurt), Rudolph Taute (Fachhochschule der Telekom, Berlin), and Georg Terlecki, (Fachhochschule Rheinland-Pfalz, Kaiserslautern)

Thermodynamics: Christoph Hartnack (Ecole de Mines and Subatech, Nantes), with Jochen Gerber (Fachhochschule Frankfurt), and Ludwig Neise (Universität Heidelberg)

Quantum physics: Alexander Andreef (Technische Hochschule Dresden), with Markus Hofmann (Universität Frankfurt) and Christian Spieles (Universität Frankfurt)

With contributions by:

Hans Babovsky, Technische Hochschule Ilmenau

Heiner Heng, Physikalisches Institut, Universität Frankfurt

Andre Jahns, Universität Frankfurt

Karl-Heinz Kampert, Universität Karlsruhe

Ralf Rüdiger Kories, Fachhochschule der Telekom, Dieburg

Imke Krüger-Wiedorn, Naturwissenschaftliche-Technische Akademie Isny

Christiane Lesny, Universität Frankfurt

Monika Lutz, Fachhochschule Gießen-Friedberg

Raffaele Mattiello, Universität Frankfurt

Jörg Müller, University of Tennessee, Knoxville

Jürgen Müller, Denton Vacuum, Inc., and APD Cryogenics, Inc. Frankfurt

Gottfried Munzenberg, Universität Gießen and GSI Darmstadt

Helmut Oeschler, Technische Hochschule Darmstadt

Roland Reif, ehem. Technische Hochschule Dresden

Joachim Reinhardt, Universität Frankfurt

Hans-Georg Reusch, Universität Munster and IBM Wissenschaftliches Zentrum Heidelberg

Matthias Rosenstock, Universität Frankfurt

Wolfgang Schäfer, Telenorma (Bosch-Telekom) GmbH, Frankfurt

Alwin Schempp, Institut für Angewandte Physik, Universität Frankfurt

Heinz Schmidt-Walter, Fachhochschule der Telekom, Dieburg

Bernd Schürmann, Siemens, AG, München

Astrid Steidl, Naturwissenschaftliche-Technische Akademie, Isny

Jürgen Theis, Hoeschst, AG, Höchst

Thomas Weis, Universität Dortmund

Wolgang Wendt, Fachhochschule Technik, Esslingen

Michael Wiedorn, Gesamthochschule Essen und PSI Bern

Bernd Wolf, Physikalisches Institut, Universität Frankfurt

Dieter Zetsche, Mercedes-Benz AG, Stuttgart

We gratefully acknowledge numerous contributions from textbooks by:

Walter Greiner (Johann Wolfgang Goethe-Universität) and Werner Martienssen (Physikalisches Institut, Universität Frankfurt)

The second edition included contributions by:

G. Brecht, FH Lippe, and DIN committee AEF

H. Dirks, FH Darmstadt

E. Groth, FH Hamburg

K. Grupen, Uni Siegen

U. Gutsch, FH Hanover

S. Jordan, FH Schweinfurt

P. Kienle, TU München

U. Kreibig, Rheinisch-Westfälische Technische Hochschule, Aachen

J.L. Leichsenring, FH Köln

H. Löckenhoff, FH Dortmund

H. Merz, Uni Münster

J. Michele, FH Wilhemshaven

H.D. Motz, Gesamthochschule Wuppertal

H. Niedrig, TH Berlin

R. Nocker, FH Hanover

H.J. Oberg, FH Hamburg

A. Richter, TH Darmstadt

D. Riedel, FH Düsseldorf

W.-D. Ruf, FH Aalen

J.A. Sahm, TU Berlin

H. Schäfer, FH Schmalkalden

G. Zimmerer, Uni Hamburg

The third edition benefited from the efforts of:

G. Flach and N. Flach, who worked on format and illustrations

R. Reif (Dresden), who contributed to the sections on mechanics and nuclear physics

P. Ziesche (Dresden) and D. Lehmann (Dresden), who contributed to the sections on condensed-matter physics

J. Moisel (Ulm), who contributed to the sections on optics

R. Kories (Dieburg), who contributed to the sections on semiconductor physics

E. Fischer (Arau), who provided detailed suggestions and a thorough list of corrigenda for the second edition

H.-R. Kissener, who helped with the revisions of the entire book.

Part I
Mechanics

1
Kinematics

Kinematics, the theory of the motion of bodies. Kinematics deals with the mathematical description of motion without considering **the applied forces**. The quantities position, path, time, velocity and acceleration play central roles.

1.1 Description of motion

Motion, the change of the position of a body during a time interval. To describe the motion, numerical values (**coordinates**) are assigned to the **position** of the body in a **coordinate system**. The **time** variation of the coordinates characterizes the motion.

Uniform motion exists if the body moves equal distances in equal time intervals. Opposite: **non-uniform motion**.

1.1.1 Reference systems

1. Dimension of spaces

Dimension of a **space**, the number of numerical values that are needed to determine the position of a body in this space.

■ A straight line is one-dimensional, since **one** numerical value is needed to fix the position; an area is two-dimensional with **two** numerical values, and ordinary space is three-dimensional, since **three** numerical values are needed to fix the position.

■ Any point on Earth can be determined by specifying its longitude and latitude. The dimension of Earth's surface is 2.

■ The space in which we are moving is three-dimensional. Motion in a plane is two-dimensional. Motion along a rail is one-dimensional. Additional generalizations are a point, which has zero dimensions, and the four-dimensional space-time continuum (Minkowski space), the coordinates of which are the three space coordinates and one time coordinate.

➤ For constraints (e.g., guided motion along rail or on a plane), the space dimension is restricted.

2. Coordinate systems

Coordinate systems are used for the mathematical description of motion. They attach numerical values to the positions of a body. A motion can thereby be described as a mathematical function that gives the space coordinates of the body at any time.

There are various kinds of coordinate systems ($\vec{e}_i$: unit vector along i-direction):

a) Affine coordinate system, in the two-dimensional case, two straight lines passing through a point O (enclosed angle arbitrary) are the coordinate axes (**Fig. 1.1**); in the three-dimensional case, the coordinate axes are three different non-coplanar straight lines that pass through the coordinate origin O. The coordinates ξ, η, ζ of a point in space are obtained as projections parallel to the three coordinate planes that are spanned by any two coordinate axes onto the coordinate axes.

b) Cartesian coordinate system, special case of the affine coordinate system, consists of respectively perpendicular straight coordinate axes. The coordinates x, y, z of a space point P are the orthogonal projections of the position of P onto these axes (**Fig. 1.2**).

Line element: $\qquad\qquad\qquad\qquad d\vec{r} = dx\,\vec{e}_x + dy\,\vec{e}_y + dz\,\vec{e}_z.$
Areal element in the x, y–plane: $\quad dA = dx\,dy.$
Volume element: $\qquad\qquad\qquad dV = dx\,dy\,dz.$

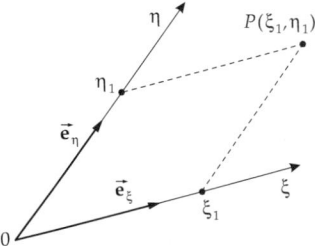

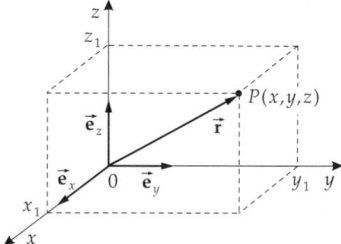

Figure 1.1: Affine coordinates in the plane, coordinates of the point P: ξ_1, η_1.

Figure 1.2: Cartesian coordinates in three-dimensional space, coordinates of the point P: x, y, z.

Right-handed system, special order of coordinate axes of a Cartesian coordinate system in three-dimensional (3D) space: The x-, y- and z-axes in a right-handed system point as thumb, forefinger and middle finger of the right hand (**Fig. 1.3**).

c) Polar coordinate system in the plane, Polar coordinates are the distance r from the origin and the angle φ between the position vector and a reference direction (positive x-axis) (**Fig. 1.4**).

Line element: $\quad d\vec{r} = dr\,\vec{e}_r + r\,d\varphi\,\vec{e}_\varphi.$
Areal element: $\quad dA = r\,dr\,d\varphi.$

d) Spherical coordinate system, generalization of the polar coordinates to 3D space. Spherical coordinates are the distance r from origin, the angle ϑ of the position vector relative to the z-axis, and the angle φ between the projection of the position vector onto the x-y-plane and the positive x-axis (**Fig. 1.5**).

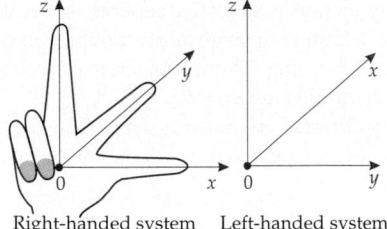

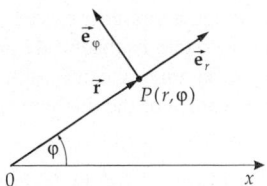

Figure 1.3: Right- and left-handed systems.   Figure 1.4: Polar coordinates in the plane.
Coordinates of the point P: r, ψ.

Line element: $\qquad\qquad \mathrm{d}\vec{\mathbf{r}} = \mathrm{d}r\,\vec{\mathbf{e}}_r + r\,\mathrm{d}\vartheta\,\vec{\mathbf{e}}_\vartheta + r\,\sin\vartheta\,\mathrm{d}\varphi\,\vec{\mathbf{e}}_\varphi.$

Volume element: $\qquad\quad \mathrm{d}V = r^2\,\sin\vartheta\,\mathrm{d}r\,\mathrm{d}\vartheta\,\mathrm{d}\varphi.$

Spherical angle element: $\quad \mathrm{d}\Omega = \sin\vartheta\,\mathrm{d}\vartheta\,\mathrm{d}\varphi.$

e) Cylindrical coordinate system, mixing of Cartesian and polar coordinates in 3D space. Cylindrical coordinates are the projection (z) of the position vector $\vec{\mathbf{r}}$ onto the z-axis, and the polar coordinates (ρ, φ) in the plane perpendicular to the z-axis, i.e., the length ρ of the perpendicular to the z-axis, and the angle between this perpendicular and the positive x-axis (**Fig. 1.6**).

Line element: $\qquad\quad \mathrm{d}\vec{\mathbf{r}} = \mathrm{d}\rho\,\vec{\mathbf{e}}_\rho + \rho\,\mathrm{d}\phi\,\vec{\mathbf{e}}_\phi + \mathrm{d}z\,\vec{\mathbf{e}}_z.$

Volume element: $\qquad \mathrm{d}V = \rho\,\mathrm{d}\rho\,\mathrm{d}\phi\,\mathrm{d}z.$

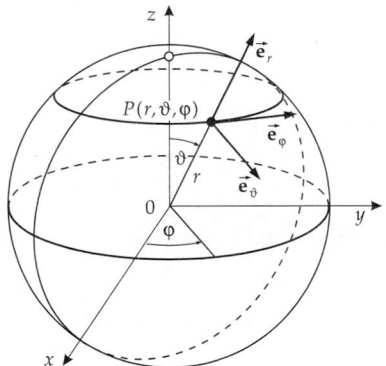

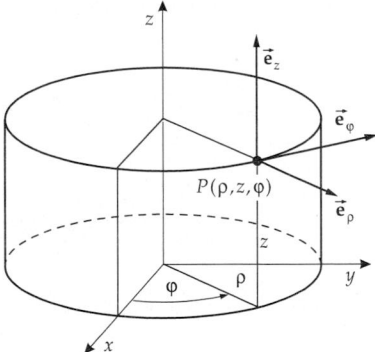

Figure 1.5: Spherical coordinates.   Figure 1.6: Cylindrical coordinates.

3. Reference system

A reference system consists of a system of **coordinates** relative to which the position of the mechanical system is given, and a **clock** indicating the time. The relation between the reference system and physical processes is established by **assignment**, i.e., by specification of reference points, reference directions, or both.

■ For a Cartesian coordinate system in two dimensions (2D), one has to specify the origin and the orientation of the x-axis. In three dimensions, the orientation of the y-axis must also be specified. Alternatively, one can specify two or three reference points.

▲   There is no absolute reference system. Any motion is a relative motion, i.e., it depends on the selected reference system. The definition of an **absolute** motion without specifying a reference system has no physical meaning. The specification of the reference system is **absolutely necessary** for describing any motion.

➤   Any given motion can be described in many different reference systems. The appropriate choice of the reference system is often a prerequisite for a simple treatment of the motion.

4. Position vector and position function

Position vector, $\vec{\mathbf{r}}$, vector from the coordinate origin to the space point (x, y, z). The position vector is written as a column vector with the spatial coordinates as components:

$$\vec{\mathbf{r}} = \begin{pmatrix} x \\ y \\ z \end{pmatrix}.$$

Position function, $\vec{\mathbf{r}}(t) = \begin{pmatrix} x(t) \\ y(t) \\ z(t) \end{pmatrix}$, specifies the position of a body at any time t. The motion is definitely and completely described by the position function.

5. Path

Path, the set of all space points (positions) that are traversed by the moving body.

■   The path of a point mass that is fixed on a rotating wheel of radius R at the distance $a < R$ from the rotation axis, is a circle. If the wheel rolls on a flat surface, the point moves on a shortened cycloid (**Fig. 1.7**).

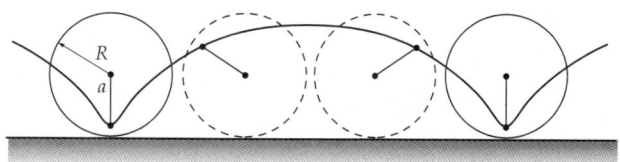

Figure 1.7: Shortened cycloid as superposition of rotation and translation.

6. Trajectory

Trajectory, representation of the path as function $\vec{\mathbf{r}}(p)$ of a parameter p, which may be for instance the elapsed time t or the path length s. With increasing parameter value, the point mass runs along the path in the positive direction (**Fig. 1.8**).

➤   Without knowledge of the time-dependent position function, the velocity of the point mass cannot be determined from the path alone.

a) Example: Circular motion of a point mass. Motion of a point mass on a circle of radius R in the x, y-plane of the 3D space. Parametrization of the trajectory by the rotation angle φ as function of time t:

●   in spherical coordinates: $r = R$, $\vartheta = \pi/2$, $\varphi = \varphi(t)$,
●   in Cartesian coordinates: $x(t) = R \cdot \cos \varphi(t)$, $y(t) = R \cdot \sin \varphi(t)$, $z(t) = 0$ (**Fig. 1.9**).

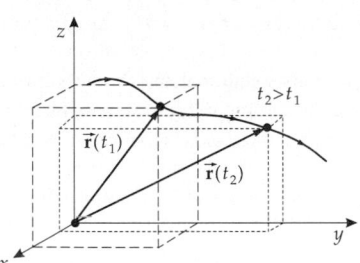

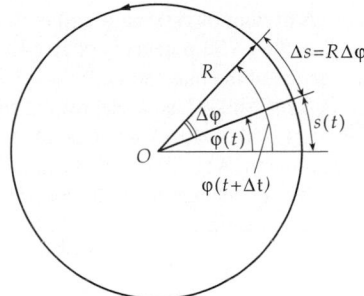

Figure 1.8: Trajectory $\vec{r}(t)$.

Figure 1.9: Motion on a circle of radius R. Element of rotation angle: $\Delta\varphi$, element of arc length: $\Delta s = R \cdot \Delta\varphi$.

b) Example: Point on rolling wheel. The trajectory of a point at the distance $a < R$ from the axis of a wheel (radius R) that rolls to the right with constant velocity is a shortened cycloid. The parameter representation of a shortened cycloid in Cartesian coordinates in terms of the rolling angle $\phi(t)$ (**Fig. 1.10**) reads:

$$x(t) = vt - a\sin\phi(t),$$
$$y(t) = R - a\cos\phi(t).$$

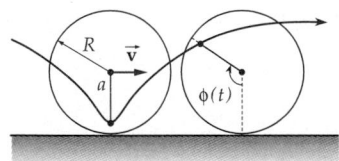

Figure 1.10: Parameter representation of the motion on a shortened cycloid by the rolling angle ϕ as function of time t.

7. Degrees of freedom

of a mechanical system, number of independent quantities that are needed to specify the position of a system definitely.

■   A point mass in 3D space has three translational degrees of freedom (displacements in three independent directions x, y, z). A free system of N mass points in 3D space has $3 \cdot N$ degrees of freedom.

If the motion within a system of N mass points is restricted by inner or external constraints, so that there are k auxiliary conditions between the coordinates $\vec{r}_1, \vec{r}_2, \ldots, \vec{r}_N$,

$$g_\alpha(\vec{r}_1, \vec{r}_2, \ldots, \vec{r}_N, t) = 0, \quad \alpha = 1, 2, \ldots, k,$$

there remain only $f = 3 \cdot N - k$ degrees of freedom with the system.

■   For a point mass that can move only in the x, y-plane (condition: $z = 0$), there remain two degrees of freedom. The point mass has only one degree of freedom if the motion is restricted to the x-axis (conditions: $y = 0$, $z = 0$).

A system of two mass points that are rigidly connected by a bar of length l has $f = 6 - 1 = 5$ degrees of freedom (condition: $(\vec{r}_1 - \vec{r}_2)^2 = l^2$, $\vec{r}_1, \vec{r}_2$: position vectors of the mass points).

A rigid body has six degrees of freedom: three translational and three rotational. If a rigid body is fixed in one point (gyroscope), there remain three degrees of freedom of rotation. A rigid body that can only rotate about a fixed axis is a physical pendulum with only one rotational degree of freedom.

A non-rigid continuous mass distribution (continuum model of a deformable body) has infinitely many degrees of freedom.

1.1.2 Time

1. Definition and measurement of time

Time, t, for quantification of processes varying with time.
  Periodic (recurring) processes in nature are used to fix the time unit.
  Time period, time interval, Δt, the time distance of two events.

M   Time measurement by means of **clocks** is based on periodic (pendulum, torsion vibra-
    tion) or steady (formerly used: burning of a candle, water clock) processes in nature.
    The pendulum has the advantage that its period T depends only on its length l (and
    the local gravitational acceleration g): $T = 2\pi\sqrt{l/g}$. Mechanical watches use the
    periodic torsional motion of the **balance spring** with the energy provided by a spiral
    spring. Modern methods employ electric circuits in which the frequency is stabilized
    by the resonance frequency of a quartz crystal, or by atomic processes.

    Stopwatch, for measuring time intervals, often connected to mechanical or electric
    devices for start and stop (switch, light barrier).

    Typical precisions of clocks range from minutes per day for mechanical clocks,
    over several tenths of seconds per day for quartz clocks, to 10^{-14} (one second in
    several million years) for atomic clocks.

2. Time units

Second, s, SI (International System of Units) unit of time. One of the basic units of the
SI, defined as 9,192,631,770 periods of the electromagnetic radiation from the transition
between the hyperfine structure levels of the ground state of Cesium 133 (relative accuracy:
10^{-14}). Originally defined as the fraction 86400^{-1} of a mean solar day, subdivided into
24 hours, each hour comprising 60 minutes, and each minute comprising 60 seconds. The
length of a day is not sufficiently constant to serve as a reference.

$$[t] = s = \text{second}$$

Additional units:

1 minute (min)	=	60 s
1 hour (h)	=	60 min = 3600 s
1 day (d)	=	24 h = 1440 min = 86400 s
1 year (a)	=	365.2425 d.

➤   The time standard is accessible by special radio broadcasts.
➤   The Gregorian **year** has 365.2425 days and differs by 0.0003 days from the tropical
    year.
Time is further divided into weeks (7 days each) and months (28 to 31 days) (Gregorian
calendar).

3. Calendar

Calendar, serves for further division of larger time periods. The calendar systems are related to the lunar cycle of ca. 28 days and to the solar cycle of ca. $365\frac{1}{4}$ days. Since these cycles are not commensurate with each other, intercalary days must be included.

Most of the world uses the **Gregorian calendar**, which was substituted for the former **Julian calendar** in 1582, at which time the intercalary rule was modified for full century years. Since then, the first day of spring falls on March 20 or 21.

➤   The Julian calendar was in use in eastern European countries until the October Revolution (1917) in Russia. It differed from the Gregorian one by about three weeks.

Intercalary day, inserted at the end of February in all years divisible by 4. Exception: full century years that are not divisible by 400 (2000 is leap year, 1900 is not).

Calendar week, subdivision of the year into 52 or 53 weeks. The first calendar week of a year is the week that includes the first Thursday of the year.

➤   The first weekday of the civil week is Monday, however it is Sunday according to Christian tradition.

Gregorian calendar years are numbered consecutively by a **date**. Years before the year 1 are denoted by "B.C." (before Christ) or B.C.E. (before the Common Era to Jews, Buddhists, and Muslims).

➤   There is no year Zero. The year 1 B.C. is directly followed by the year 1 A.D., or C.E. (Common Era)

➤   Julian numbering of days: time scale in astronomy.

Other calendar systems: Other **calendar systems** presently used are the calendar (**lunisolar calendar**, a mixture of solar and lunar calendar) that involves years and leap months of different lengths; years are counted beginning with 7 October 3761 B.C. ("creation of the world") and the year begins in September/October; the year 5759 began in 1998), and the Moslem calendar (purely lunar calendar with leap month; years are counted beginning with the flight of Mohammed from Mecca on July 16, 622 A.D.; the Moslem year 1419 began in the year 1998 of the Gregorian calendar).

1.1.3   Length, area, volume

1. Length

Length, l, the **distance** (shortest **connecting line**) between two points in space.

Meter, m, SI unit of length. One of the basic units of the SI, defined as the distance traveled by light in vacuum during 1/299792458 of a second (relative accuracy: 10^{-14}). The meter was originally defined as the 40-millionth fraction of the circumference of earth and is represented by a **primary standard** made of platinum-iridium that is deposited in the *Bureau International des Poids et Mesures* in Paris.

$$[l] = \mathrm{m} = \text{meter}.$$

Additional units see **Tab. 33.0/3**.

2. Length measurement

Length measurement was originally carried out by defining and copying the unit of length (e.g., primary meter, tape measure, yardstick, screw gauge, micrometer screw, often with a nonius scale for more accurate reading).

Interferometer: for precise optical measurement of length (see p. 383) in which the wavelength of monochromatic light is used as scale.

Sonar: for acoustical distance measurement by time-of-flight measurement of ultra-sound for ships; used for distance measurements with some cameras.

Radar: for distance measurement by means of time-of-flight measurement of electro-magnetic waves reflected by the object.

Lengths can be measured with a relative precision as good as 10^{-14}. Using micrometer screws, one can reach precisions in the range of 10^{-6} m.

Triangulation, a geometric procedure for surveying. The remaining two edges of a triangle can be evaluated if one edge and two angles are given. Starting from a known basis length, arbitrary distances can be measured by consecutive measurements of angles, using a **theodolite**.

Parallax, the difference of orientation for an object when it is seen from two different points (**Fig. 1.11**). Applied to distance measurement.

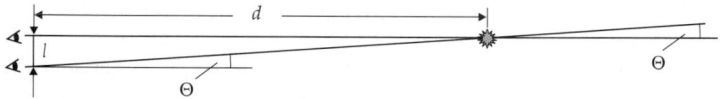

Figure 1.11: Parallax Θ for eyes separated by a distance l and the object at a distance d: $\tan \Theta = l/d$ or $\Theta \approx l/d$ for $d \gg l$.

3. Area and volume

Area A and **volume** V are quantities that are derived from length measurement.

Square meter, m^2, SI unit of area. A square meter is the area of a square with edge length of 1 m.

$$[A] = m^2 = \text{square meter.}$$

Cubic meter, m^3, SI unit of volume. A cubic meter is the volume of a cube with edge length 1 m.

$$[V] = m^3 = \text{cubic meter.}$$

$\boxed{M}$ Areas can be measured by subdivision into simple geometric figures (rectangles, triangles), the edges and angles of which are measured (e.g., by triangulation), and then calculated. Direct area measurement can be undertaken by counting the enclosed squares on a measuring grid.

Analogously, the volume of hollow spaces can be evaluated by filling them with geometric bodies (cubes, pyramids, ...).

For the measurement of the volume of fluids, one uses standard vessels with known volume. The volume of solids can be determined by submerging them in a fluid (see p. 182).

For a known density ρ of a homogeneous body, the volume V can be determined from the mass m, $V = \frac{m}{\rho}$.

➤ **Decimal prefixes for area and volume units**:
The decimal prefix refers only to the length unit, not to the area or volume unit:

$$1 \text{ cubic centimeter} = 1 \text{ cm}^3 = (1 \text{ cm})^3 = \left(1 \cdot 10^{-2} \text{ m}\right)^3 = 1 \cdot 10^{-6} \text{ m}^3.$$

1.1.4   Angle

1. Definition of angle

Angle, ϕ, a measure of the divergence between two straight lines in a plane. An angle is formed by two straight lines (**sides**) at their intersection point (**vertex.**) It is measured by marking on both straight lines a distance (radius) from the vertex, and determining the length of the arc of the circle connecting the endpoints of the two distances (**Fig. 1.12**).

angle and arc			**1**
	Symbol	Unit	Quantity
$\phi = \dfrac{l}{r}$	ϕ	rad	angle
	l	m	length of circular arc
	r	m	radius

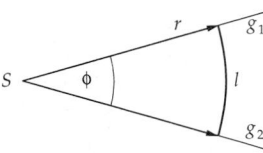

Figure 1.12: Determination of the angle ϕ between the straight lines g_1 and g_2 by measurement of the arc length l and radius r, $l = r \cdot \phi$. S: vertex

2. Angle units

a) Radian, rad, SI unit of plane angle. 1 rad is the angle for which the length of the circular arc connecting the endpoints of the sides just coincides with the length of a side. A full circle corresponds to the angle 2π rad.

➤   Radian (and degree) are supplementary SI units, i.e., they have unit dimensionality.

$$1 \text{ rad} = 1 \text{ m}/1 \text{ m}.$$

b) Degree, °, also an accepted unit for measurement of angles. A degree is defined as 1/360 of the angle of a complete circle. Conversion:

$$1 \text{ rad} = \frac{360°}{2\pi} = 57.3°,$$

$$1° = \frac{2\pi}{360°} = 0.0175 \text{ rad}.$$

Subdivisions are:

$$1 \text{ degree } (°) = 60 \text{ arc minutes } (') = 3600 \text{ arc seconds } ('').$$

c) Gon (formerly **new degree**), a common unit in surveying: 1 **gon**, 1/100 of a right angle.

$$1 \text{ gon} = 0.9° = 0.0157 \text{ rad}$$

$$1° = 1.11 \text{ gon}$$

$$1 \text{ rad} = 63.7 \text{ gon}$$

M   **Measurement of angles**: Measurement of angles is performed directly by means of an angle scale, or by measuring the chord of an angle and converting if the radius is known. When determining distances by triangulation, the **theodolite** (see p. 10) is used for angle measurement.

3. Solid angle

Solid angle, Ω, is determined by the area of a unit sphere that is cut out by a cone with the vertex in the center of the sphere (**Fig. 1.13**).

solid angle			
	Symbol	Unit	Quantity
$\Omega = \dfrac{A}{r^2}$	Ω	sr	solid angle
	A	m^2	area cut out by cone
	r	m	radius of sphere

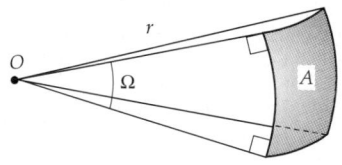

Figure 1.13: Determination of the solid angle Ω by measuring area A and radius r ($\Omega = A/r^2$).

Steradian, sr, SI unit of the solid angle.
1 steradian is the solid angle that cuts out a surface area of 1 m^2 on a sphere of radius 1 m (**Fig. 1.14**). This surface can be arbitrarily shaped and can also consist of disconnected parts.
▲   The full spherical angle is 4π sr.
➤   Radian and steradian are dimensionless.

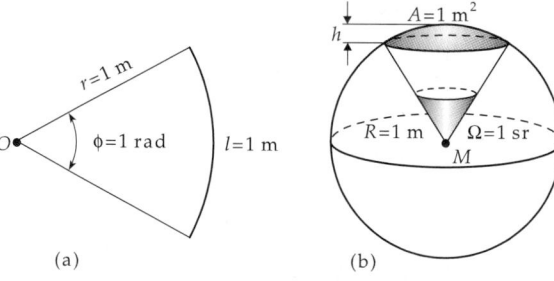

(a)

(b)

Figure 1.14: Definition of the angular units radian (rad) (a) and steradian (sr) (b). The (curved) area of the spherical segment A is given by $A = 2\pi R \cdot h$.

1.1.5   Mechanical systems

1. Point mass

Point mass, idealization of a body as a mathematical point with vanishing extension, but finite mass. A point mass has no rotational degrees of freedom. When treating the motion of a body, the model of point mass can be used if it is sufficient under the given physical conditions to study only the motion of the center of gravity of the body, without taking the spatial distribution of its mass into account.

➤ In the mathematical description of motion without rotation, every rigid body can be replaced by a point mass located in the **center of gravity** of the rigid body (see p. 94).

■ For the description of planetary motion in the solar system, it often suffices to consider the planets as points, since their extensions are very small compared with the typical distances between sun and planets.

2. System of point masses

System consisting of N individual point masses $1, 2, \ldots, N$. Its motion can be described by specifying the position vectors $\vec{r}_1, \vec{r}_2, \ldots, \vec{r}_N$ as a function of the time t: $\vec{r}_i(t)$, $i = 1, 2, \ldots N$ (**Fig. 1.15a**).

3. Forces in a system of point masses

a) Internal forces, forces acting between the particles of the system. Internal forces are in general two-body forces (pair forces) that depend on the distances (and possibly the velocities) of only two particles.

b) External forces, forces acting from the outside on the system. External forces originate from bodies that do not belong to the system.

c) Constraint reactions or reaction forces (external forces) result from constraining the system. The interaction between the system and the constraint is represented by reactions that act perpendicularly to the enforced path. Constraint reactions restrict the motion of the system.

■ Guided motion: Mass on string fixed at one end, mass on an inclined plane, point mass on a straight rail, bullet in a gun barrel.

4. Free and closed systems

Free point mass, free system of point masses, a point mass or a system of point masses can react to the applied forces without constraints.

   Closed system, a system that is not subject to external forces.

5. Rigid body

Rigid body, a body the material constituents of which are always the same distances from each other, hence rigidly connected to each other. For the distances of all points i, j of the rigid body: $|\vec{r}_i(t) - \vec{r}_j(t)| = r_{ij} = \text{const.}$ (**Fig. 1.15b**).

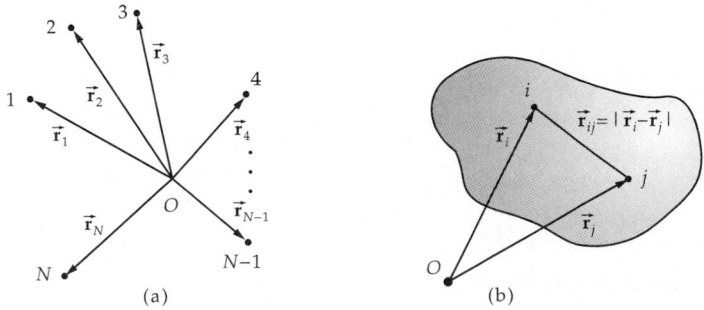

Figure 1.15: Mechanical systems. (a): system of N point masses, (b): rigid body.

6. Motion of rigid bodies

Any motion of a rigid body can be decomposed in two kinds of motion (**Fig. 1.16**):

a) Translation, all points of the body travel the same distance in the same direction; the body is shifted in a parallel fashion. The motion of the body can be described by the motion of a representative point of the body.

b) Rotation, when all points of the body rotate about a common axis. Any point on the body keeps its distance from the rotation axis and moves along a circular path.

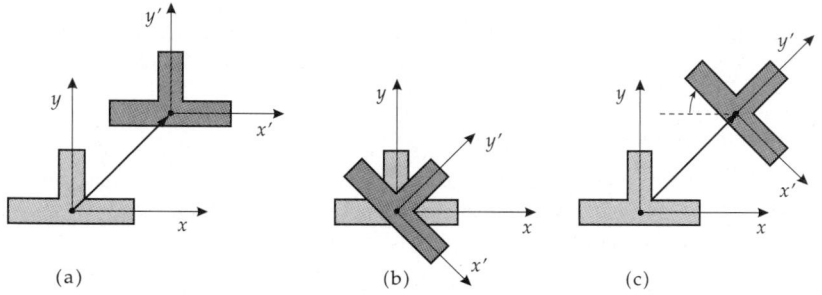

Figure 1.16: Translation and rotation of a rigid body. (a): translation, (b): rotation, (c): translation and rotation.

7. Deformable body

A deformable body can change its shape under the influence of forces. Described by
- many discrete point masses that are connected by forces, or
- a continuum model according to which the body occupies the space completely.

1.2   Motion in one dimension

We now consider motion along a straight-line path. The distance x of the body from a fixed point on the axis of motion is used as the coordinate. The sign of x indicates on which side of the axis the body is located. The choice of the positive x-axis is made by convention.

Position-time graph, graphical representation of the motion (**position function** $x(t)$) of a point mass in two dimensions. The horizontal axis shows the time t, the vertical axis the position x (coordinate).

1.2.1   Velocity

Velocity, a quantity that characterizes the motion of a point mass at any time point. One distinguishes between the mean velocity $\bar{v}_x$ and the instantaneous velocity v_x.

1.2.1.1   Mean velocity

1. Definition of mean velocity

Mean velocity, $\bar{v}_x$, over a time interval $\Delta t \neq 0$, gives the ratio of the path element Δx traveled during this time interval and the time Δt needed (**Fig. 1.17**).

mean velocity $= \dfrac{\text{path element}}{\text{time interval}}$			$\mathbf{LT^{-1}}$
$\begin{aligned} \bar{v}_x &= \dfrac{x_2 - x_1}{t_2 - t_1} \\[2mm] &= \dfrac{x(t_1 + \Delta t) - x(t_1)}{(t_1 + \Delta t) - t_1} \\[2mm] &= \dfrac{\Delta x}{\Delta t} \end{aligned}$	Symbol	Unit	Quantity
	$\bar{v}_x$	m/s	mean velocity
	x_1, x_2	m	position at time t_1, t_2, resp.
	$x(t)$	m	position function
	t_1, t_2	s	initial and final time point
	Δx	m	path element traveled
	Δt	s	time interval

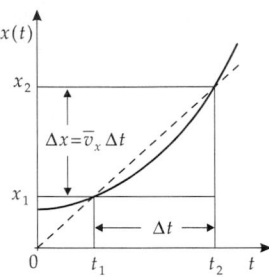

Figure 1.17: Mean velocity $\bar{v}_x$ of one-dimensional motion in a position vs. time graph.

2. Velocity unit

Meter per second, ms^{-1}, the SI unit of velocity.
1 m/s is the velocity of a body that travels one meter in one second.

■ A body that travels a distance of 100 m in one minute has the mean velocity

$$\bar{v}_x = \frac{\Delta x}{\Delta t} = \frac{100 \text{ m}}{60 \text{ s}} = 1.67 \text{ m/s}.$$

3. Measurement of velocity

Velocity measurement can be performed by time-of-flight measurement over a section of known length. Often it is done by converting the translational motion into a rotational one.

Speedometer, for measuring speeds of cars. The rotational motion of the wheels is transferred by a shaft into the measuring device where the pointer is moved by the centrifugal force arising by this rotation (centrifugal force tachometer).

In the **eddy-current speedometer**, the rotational motion is transferred to a magnet mounted in an aluminum drum on which the pointer is fixed, eddy currents create a torque that is balanced by a spring.

Electric speedometers are based on a pulse generator that yields pulse sequences of higher or lower frequency corresponding to the rotation velocity.

Velocity measurement by **Doppler effect** (see p. 300) is possible using radar (automobiles, airplanes, astronomy).

➤ The velocity $\bar{v}_x$ can have a positive or a negative sign, corresponding to motion in either the positive or negative coordinate direction.

➤ The mean velocity depends in general on the time interval of measurement Δt. Exception: motion with constant velocity.

1.2.1.2   Instantaneous velocity

1. Definition of instantaneous velocity

Instantaneous velocity, limit of the mean velocity for time intervals approaching zero.

instantaneous velocity			LT^{-1}
	Symbol	Unit	Quantity
$v_x(t) = \lim\limits_{\Delta t \to 0} \dfrac{\Delta x}{\Delta t} = \dfrac{d}{dt}x(t) = \dfrac{dx(t)}{dt} = \dot{x}(t)$	$v_x(t)$	m/s	instantaneous velocity
	$x(t)$	m	position at time t
	Δt	s	time interval
	Δx	m	path element

The function $x(t)$ represents the position coordinate x of the point at any time t. In the position-time graph, the instantaneous velocity $v_x(t)$ is the slope of the tangent of $x(t)$ at the point t (**Fig. 1.18**).

The following cases must be distinguished (the time interval Δt is always positive):

$v_x > 0$: $\Delta x > 0$ and hence $x(t + \Delta t) > x(t)$. The body moves along the positive coordinate axis, i.e., the x-t curve increases: the derivative of the curve $x(t)$ is positive.

$v_x = 0$: $\Delta x = 0$ and hence $x(t + \Delta t) = x(t)$, the distance Δx is constant (zero). In this coordinate system the body is at rest (possibly only briefly), i.e., v_x is the horizontal tangent to the x vs. t curve, and the derivative of the curve $x(t)$ vanishes.

$v_x < 0$: $\Delta x < 0$ and hence $x(t + \Delta t) < x(t)$. The body moves along the negative coordinate axis, i.e., the x-t curve decreases, the derivative of the curve $x(t)$ is negative.

2. Velocity vs. time graph

Velocity vs. time graph, graphical representation of the instantaneous velocity $v_x(t)$ as function of time t. To determine the position function $x(t)$ for a given velocity curve $v_x(t)$, the motion is subdivided into small intervals Δt (**Fig. 1.19**). If the interval from t_1 to t_2 is subdivided in N intervals of length $\Delta t = (t_2 - t_1)/N$, t_i is the beginning of the ith time interval and $\bar{v}_x(t_i)$ the mean velocity in this interval, then

$$x(t_2) = x(t_1) + \lim_{\Delta t \to 0} \sum_{i=1}^{N-1} \bar{v}_x(t_i) \cdot \Delta t = x(t_1) + \int_{t_1}^{t_2} v_x(t)\, dt.$$

path = definite integral of the velocity over the time			L
	Symbol	Unit	Quantity
$x(t) = x(t_1) + \displaystyle\int_{t_1}^{t} v(\tau)\, d\tau$	$x(t)$	m	curve of motion
	$v(t)$	m/s	velocity curve
$x(t_2) = x(t_1) + \displaystyle\int_{t_1}^{t_2} v(t)\, dt$	t_1, t_2	s	beginning and ending time points

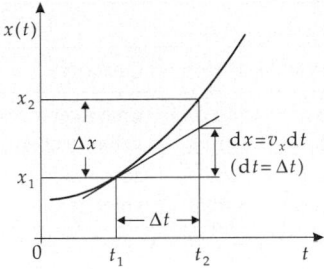

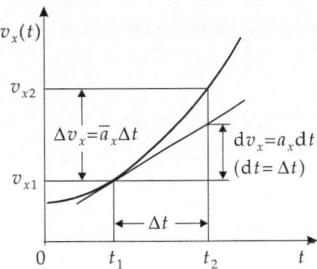

Figure 1.18: Instantaneous velocity v_x at time t_1 of one-dimensional motion in a position vs. time graph.

Figure 1.19: Velocity vs. time graph of one-dimensional motion. $\bar{a}_x$: mean acceleration, a_x: instantaneous acceleration at time t_1.

1.2.2 Acceleration

Acceleration, the description of non-uniform motion (motion in which the velocity varies). The acceleration, as well as the velocity, can be positive or negative.

➤ Both an increase (**positive acceleration**) and a decrease of velocity (**deceleration**, as result of a deceleration process, negative acceleration) are called acceleration.

1. Mean acceleration,

$\bar{a}_x$, change of velocity during a time interval divided by the length of the time interval:

acceleration $=\dfrac{\textbf{change of velocity}}{\textbf{time interval}}$			$\mathbf{LT^{-2}}$
	Symbol	Unit	Quantity
$\bar{a}_x = \dfrac{\Delta v_x}{\Delta t} = \dfrac{v_{x2} - v_{x1}}{t_2 - t_1}$	$\bar{a}_x$	m/s^2	mean acceleration
	Δv_x	m/s	velocity change
	Δt	s	time interval
	v_{x1}, v_{x2}	m/s	initial and final velocity
	t_1, t_2	s	initial and final time

Meter per second squared, m/s^2, SI unit of acceleration. 1 m/s^2 is the acceleration of a body that increases its velocity by 1 m/s per second.

If the mean acceleration and initial velocity are given, the final velocity reads

$$v_{x2} = v_{x1} + \bar{a}_x \cdot \Delta t.$$

The time needed to change from the velocity v_{x1} to the velocity v_{x2} for given mean acceleration is

$$\Delta t = \frac{v_{x2} - v_{x1}}{\bar{a}_x}.$$

2. Instantaneous acceleration

Instantaneous acceleration, limit of the mean acceleration for very small time intervals ($\Delta t \to 0$).

instantaneous acceleration			LT^{-2}
	Symbol	Unit	Quantity
$a_x(t) = \lim\limits_{\Delta t \to 0} \dfrac{\Delta v_x}{\Delta t} = \dfrac{dv_x}{dt} = \dfrac{d}{dt} v_x(t)$	Δt	s	time interval
	Δv_x	m/s	velocity change
	$a_x(t)$	m/s^2	acceleration
	$v_x(t)$	m/s	velocity

The instantaneous acceleration $a_x(t)$ is the first derivative of the velocity function $v_x(t)$, and hence the second derivative of the position function $x(t)$:

$$a_x(t) = \frac{dv_x(t)}{dt} = \dot{v}_x(t) = \frac{d}{dt}\frac{dx(t)}{dt} = \frac{d^2 x(t)}{dt^2} = \ddot{x}(t).$$

Graphically, it represents the slope of the tangent in the velocity-time diagram (**Fig. 1.20**). The following cases are to be distinguished:

$a_x > 0$: $\Delta v_x > 0$ and hence $v_{x2} > v_{x1}$. For $v_{x1} > 0$ the body moves with increasing velocity, i.e., in the v vs. t graph the curve is rising.

$a_x = 0$: $\Delta v_x = 0$ and hence $v_{x2} = v_{x1}$. The body does not change its velocity (possibly only briefly).

$a_x < 0$: $\Delta v_x < 0$ and hence $v_{x2} < v_{x1}$. For $v_{x1} > 0$ the body moves with decreasing velocity.

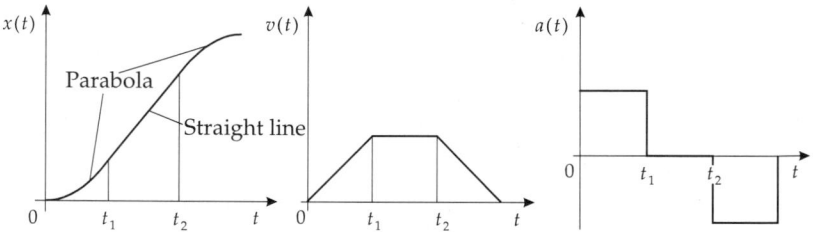

Figure 1.20: Graphs for position vs. time, velocity vs. time, and acceleration vs. time. Starting from the origin, the body is first uniformly accelerated, then moves with constant velocity, and thereafter is uniformly decelerated to rest.

3. Determination of velocity from acceleration

If the acceleration is given as function of time $a_x(t)$, the velocity is determined by integration:

velocity = integral of acceleration over time			LT^{-1}
$v_x(t) = v_x(t_1) + \displaystyle\int_{t_1}^{t} a(\tau)\, d\tau$	Symbol	Unit	Quantity
	$v_x(t)$	m/s	velocity curve
$v_x(t_2) = v_x(t_1) + \displaystyle\int_{t_1}^{t_2} a_x(t)\, dt$	$a_x(t)$	m/s^2	acceleration curve
	t_1, t_2	s	initial and final times

➤ If a body has velocity $v_{1x} < 0$ and undergoes a positive acceleration $a_x > 0$, the velocity decreases in absolute value.

1.2.3 Simple motion in one dimension

Here we discuss uniform and uniformly accelerated motion as the simplest forms of motion and discuss their physical description.

➤ For motion in one dimension, one can omit the index x and the vector arrow over the symbols for velocity v and acceleration a. One should note, however, that v and a can take positive and negative values and thus are components of vectors.

1. Uniform motion

Uniform motion, a motion in which the body does not change its velocity, $\bar{v}_x = v_x = $ const. (**Fig. 1.21**).

laws of uniform motion			
	Symbol	Unit	Quantity
$x(t) = x_0 + v_x t$	$x(t)$	m	position at time t
$v_x(t) = v_x = v_0$	x_0	m	initial position ($t = 0$)
$a_x(t) = 0$	v_x	m/s	uniform velocity
	v_0	m/s	initial velocity
	t	s	time

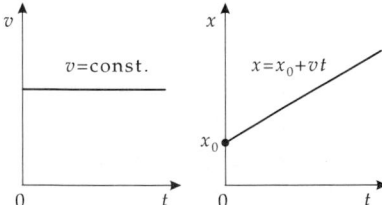

Figure 1.21: Uniform motion.

▲ Uniform motion arises if no force acts on the body.

➤ The curve of motion $x(t)$ is the integral of the velocity curve $v_x(t) = $ const. and is given by

$$x(t) = x_0 + \int_0^t v_x(t')\, dt' = x_0 + v_0 t.$$

Clearly, $v_x(t)$ is a straight line, and the integral corresponds to the area below the straight between the points 0 and t on the time axis.

2. Uniformly accelerated motion

Uniformly accelerated motion, a motion with constant acceleration. Then $\bar{a}_x = a_x = a$ and

$$v_x(t) = at + v_0,$$

if v_0 is the initial velocity (**Fig. 1.22**).

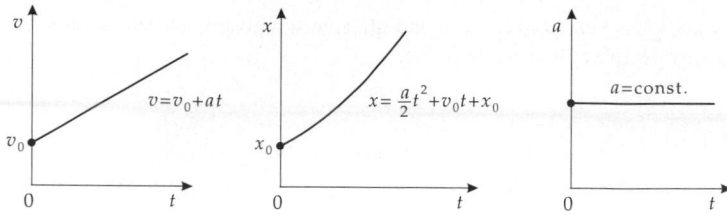

Figure 1.22: Uniformly accelerated motion.

It follows by integration that

$$x(t) = \int_0^t v_x(t')\,dt' + x_0 = \int_0^t (at' + v_0)\,dt' + x_0 = \frac{a}{2}t^2 + v_0 t + x_0.$$

This result can also be read from the velocity vs. time graph: the area below the curve is composed of a rectangle of area $v_0 \cdot t$ and a triangle of area $at^2/2$ (height at and basis t) (**Fig. 1.23**).

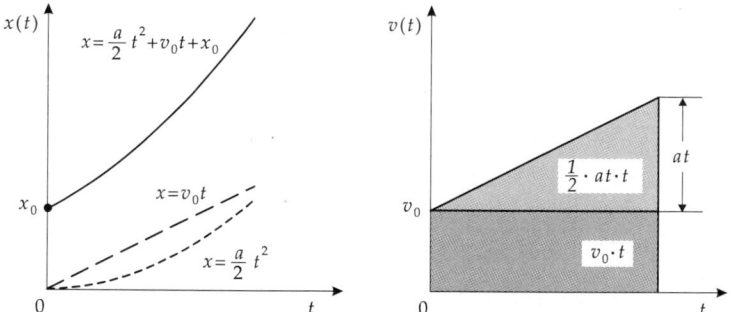

Figure 1.23: Graphs for uniformly accelerated motion.

uniformly accelerated motion			
	Symbol	Unit	Quantity
$x(t) = \dfrac{a}{2}t^2 + v_0 t + x_0$	$x(t)$	m	position at time t
	$v_x(t)$	m/s	velocity
$v_x(t) = at + v_0$	t	s	time
$a_x(t) = a$ = const.	a_x, a	m/s^2	acceleration
	v_0	m/s	initial velocity
	x_0	m	initial position

▲   A uniformly accelerated motion results if a constant force acts upon the body. By rearrangement, one gets:

- Initial and final velocity v_0 and $v_x(t)$ given, function of motion $x(t)$ wanted:

$$x(t) = \frac{v_0 + v_x(t)}{2} t + x_0.$$

- Initial velocity v_0 and position function $x(t)$ given, $x_0 = 0$, final velocity $v_x(t)$ wanted:

$$v_x(t) = \sqrt{v_0^2 + 2ax(t)}.$$

- Special case: start from rest ($v_0 = 0$, $x_0 = 0$):

$$v_x(t) = at = \sqrt{2ax(t)}, \qquad x(t) = \frac{v_x(t)t}{2} = \frac{at^2}{2}.$$

3. Deceleration

Uniform deceleration (see **Fig. 1.24**) is a special case of uniformly accelerated motion. During deceleration, the velocity and acceleration have opposite signs, hence the magnitude of the velocity is reduced, for example, until the instantaneous velocity reaches zero. The **braking distance** needed s_B to bring an object to rest can be determined from the initial velocity and the deceleration. The initial velocity can be determined when the braking distance s_B and deceleration are known.

uniform deceleration			
$t_B = \dfrac{\|v_0\|}{\|a\|} = -\dfrac{v_0}{a}$	Symbol	Unit	Quantity
$s_B = \dfrac{v_0^2}{2\|a\|}$	s_B	m	braking distance
	t_B	s	braking time
	$\|v_0\|$	m/s	magnitude of initial velocity
$v_0 = \sqrt{2\|a\|s_B}$	$\|a\|$	m/s^2	braking deceleration

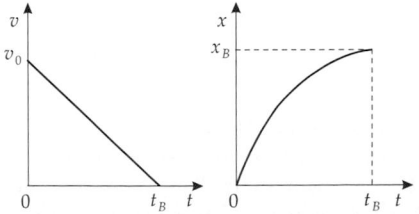

Figure 1.24: Velocity vs. time and position vs. time graphs for a uniform deceleration process. x_B: braking distance, t_B: braking time.

- ➤ Consideration of a deceleration process as a uniformly decelerated motion is an idealization. Braking is in general a non-uniform process.
- ■ For an automobile, one can assume a deceleration of about $|a| = 4$ m/s^2. For a velocity of 50 km/h $= 13.9$ m/s, there results a braking distance of

$$s_B = \frac{v_0^2}{2|a|} = \frac{(13.9 \text{ m/s})^2}{2 \cdot 4 \text{ m/s}^2} = 24 \text{ m}.$$

➤ For automobiles, the following estimate of the braking distance holds:

$$s_B \approx \left(\frac{v_0}{10 \text{ km/h}}\right)^2 \text{m} + 3 \cdot \frac{v_0}{10 \text{ km/h}} \text{ m}.$$

Here a response time of the driver of ca. 1 s is included.

1.3 Motion in several dimensions

Motion in several dimensions is usually represented in vector notation.

1. Trajectory in three-dimensional space

To fix the position of a point in 3D space, three coordinates must be specified. In a Cartesian coordinate system, these are referred to as the **position vector**, which has components x, y and z:

$$\vec{r}(t) = \left(\begin{array}{c} x(t) \\ y(t) \\ z(t) \end{array}\right).$$

The vector function $\vec{r}(t)$ describes the trajectory of a point or body in space (**Fig. 1.25**). The components of the position vector specify the x-, y- and z-coordinate of the point at time t.

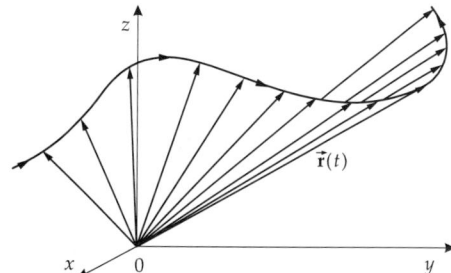

Figure 1.25: Trajectory in three dimensions.

2. Tangent and normal

Tangent to a curve at a point M, a straight line touching the curve at this point. Analytically, it results from taking the derivative of the curve with respect to the time at this point. Hence, it represents the velocity vector of a point mass. The positive direction of the tangent points along the instantaneous direction of motion. The **normal** to a curve at a point M is a straight line perpendicular to the tangent in this point. It is orthogonal to the instantaneous direction of motion (**Fig. 1.26**).

■ The tangent to a circle is orthogonal to the radius vector. The normal is parallel to the radius vector.

➤ In 3D space there is more than one normal to a given point of the space curve. All normals through the tangential point form the **normal plane**. The **osculating plane** is the limit position of a plane through M and two neighboring points on the curve as the two points tend towards M.

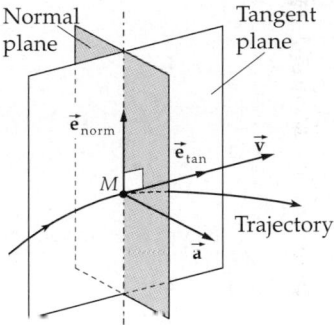

Figure 1.26: Tangent and normal plane of a trajectory. The tangent lies in the osculating plane, which is perpendicular to the normal plane.

1.3.1 Velocity vector

Velocity vector, $\vec{\mathbf{v}}$, specifies direction and magnitude of the velocity of the point mass.

1. Mean velocity

Mean velocity $\vec{\bar{\mathbf{v}}}$, in a time interval Δt, defined by (**Fig. 1.27**)

$$\vec{\bar{\mathbf{v}}} = \frac{\vec{\mathbf{r}}(t_2) - \vec{\mathbf{r}}(t_1)}{t_2 - t_1} = \frac{\Delta \vec{\mathbf{r}}}{\Delta t} = \begin{pmatrix} \frac{\Delta x}{\Delta t} \\ \frac{\Delta y}{\Delta t} \\ \frac{\Delta z}{\Delta t} \end{pmatrix}, \qquad \Delta \vec{\mathbf{r}} = \begin{pmatrix} \Delta x \\ \Delta y \\ \Delta z \end{pmatrix}.$$

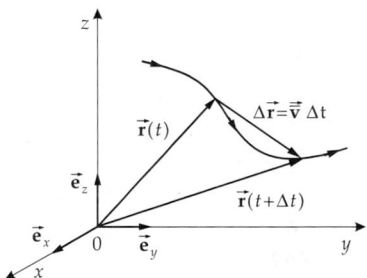

Figure 1.27: Mean velocity $\vec{\bar{\mathbf{v}}}$.

2. Instantaneous velocity

Instantaneous velocity is obtained by taking the limit as $\Delta t \to 0$ (**Fig. 1.28**):

instantaneous velocity				$\mathbf{LT^{-1}}$
$\vec{\mathbf{v}}(t) = \lim\limits_{\Delta t \to 0} \dfrac{\vec{\mathbf{r}}(t + \Delta t) - \vec{\mathbf{r}}(t)}{\Delta t}$ $= \dfrac{d\vec{\mathbf{r}}}{dt} = \dot{\vec{\mathbf{r}}}(t) = \begin{pmatrix} \dot{x}(t) \\ \dot{y}(t) \\ \dot{z}(t) \end{pmatrix}$	Symbol	Unit	Quantity	
	$\vec{\mathbf{v}}(t)$	m/s	velocity vector	
	Δt	s	time interval	
	t	s	time	
	$\vec{\mathbf{r}}(t)$	m	trajectory	
	$\dot{x}, \dot{y}, \dot{z}$	m/s	velocity components	

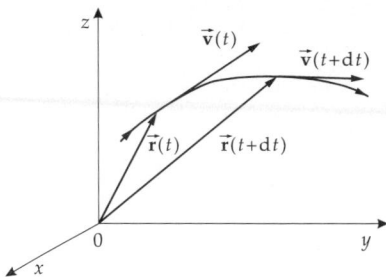

Figure 1.28: Instantaneous
velocity $\vec{\mathbf{v}}(t)$.

The components of the velocity vector $\vec{\mathbf{v}}$ are the derivatives of the coordinate functions $x(t)$, $y(t)$ and $z(t)$ with respect to time. They specify its projections onto the x-, y- and z-axes:

$$v_x = \dot{x}, \quad v_y = \dot{y}, \quad v_z = \dot{z}.$$

3. Properties of the velocity vector

The magnitude of the velocity vector, v, represents the path distance traveled per unit time.

▲ The velocity vector $\vec{\mathbf{v}}$ points along the direction of motion.

➤ The velocity vector $\vec{\mathbf{v}}(t)$ depends on the change of the position vector, $d\vec{\mathbf{r}} = \vec{\mathbf{v}}\, dt$. It is possible for the orientation of the position vector to change while its magnitude remains constant (circular motion). The variation of the distance from the origin in vector notation is found by means of the product and chain rules of differentiation to be:

$$\frac{d|\vec{\mathbf{r}}|}{dt} = \frac{d\sqrt{\vec{\mathbf{r}}^2}}{dt} = \frac{\vec{\mathbf{r}} \cdot \vec{\mathbf{v}}}{|\vec{\mathbf{r}}|}.$$

In particular, the distance remains constant if $\vec{\mathbf{r}} \cdot \vec{\mathbf{v}} = 0$, i.e., if the velocity vector is perpendicular to the radius vector. A motion for which the distance from the origin or another fixed point remains unchanged is a **circular motion**.

Tangent unit vector, $\vec{\mathbf{e}}_{tan}$, a vector of unit length that points along the positive tangent to a curve. The velocity can then be written as

$$\vec{\mathbf{v}} = v\,\vec{\mathbf{e}}_{tan}, \qquad \vec{\mathbf{e}}_{tan} = \frac{\vec{\mathbf{v}}}{v}.$$

4. Example: Circular motion in a plane

A circular motion in the x–y-plane with constant angular velocity ($\varphi(t) = \omega t$), and $\omega = \dfrac{d\varphi}{dt}$ is given by the position vector (**Fig. 1.29**)

$$\vec{\mathbf{r}}(t) = \begin{pmatrix} x(t) \\ y(t) \\ z(t) \end{pmatrix} = \begin{pmatrix} r\cos\omega t \\ r\sin\omega t \\ 0 \end{pmatrix}.$$

Unit of angular velocity: $[\omega] = \text{rad/s}$.

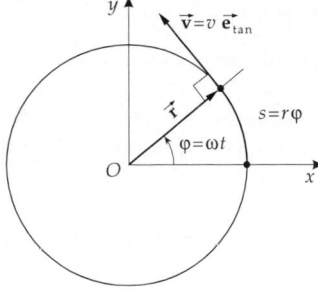

Figure 1.29: Circular motion. The magnitude of the velocity is denoted by v.

Hence, the velocity vector $\vec{\mathbf{v}}$ is

$$\vec{\mathbf{v}}(t) = \dot{\vec{\mathbf{r}}}(t) = \begin{pmatrix} \dot{x} \\ \dot{y} \\ \dot{z} \end{pmatrix} = \begin{pmatrix} -r\omega \sin \omega t \\ r\omega \cos \omega t \\ 0 \end{pmatrix}.$$

Its magnitude is $|\vec{\mathbf{v}}(t)| = \sqrt{\dot{x}^2 + \dot{y}^2 + \dot{z}^2} = r\omega$.

1.3.2 Acceleration vector

1. Acceleration vector

Acceleration vector, $\vec{\mathbf{a}}$, the time derivative of the velocity vector; it specifies the change of velocity per unit time (**Fig. 1.30**). As in the case of velocity, one can introduce a mean acceleration vector $\vec{\bar{\mathbf{a}}}$ over a time interval Δt,

$$\vec{\bar{\mathbf{a}}}(t) = \frac{\vec{\mathbf{v}}(t + \Delta t) - \vec{\mathbf{v}}(t)}{\Delta t},$$

and an instantaneous acceleration vector by the limit $\Delta t \to 0$:

$$\vec{\mathbf{a}}(t) = \begin{pmatrix} a_x(t) \\ a_y(t) \\ a_z(t) \end{pmatrix} = \lim_{\Delta t \to 0} \frac{\vec{\mathbf{v}}(t + \Delta t) - \vec{\mathbf{v}}(t)}{\Delta t} = \frac{d\vec{\mathbf{v}}(t)}{dt} = \begin{pmatrix} \dot{v}_x(t) \\ \dot{v}_y(t) \\ \dot{v}_z(t) \end{pmatrix} = \begin{pmatrix} \ddot{x}(t) \\ \ddot{y}(t) \\ \ddot{z}(t) \end{pmatrix}.$$

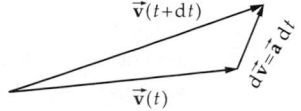

Figure 1.30: Acceleration vector.

The components of the acceleration vector are the second derivatives of the coordinate functions with respect to time:

$$a_x = \ddot{x}, \quad a_y = \ddot{y}, \quad a_z = \ddot{z}.$$

2. *Example: Acceleration vector for circular motion*

For circular motion with constant angular velocity ω, the acceleration vector is

$$\vec{a}(t) = \frac{d}{dt} \begin{pmatrix} -r\omega \sin \omega t \\ r\omega \cos \omega t \\ 0 \end{pmatrix} = \begin{pmatrix} -r\omega^2 \cos \omega t \\ -r\omega^2 \sin \omega t \\ 0 \end{pmatrix} = -\omega^2 \vec{r}(t).$$

Acceleration vector and radius vector are antiparallel, the acceleration vector points to the center.

The magnitude of the acceleration is

$$|\vec{a}(t)| = \sqrt{\ddot{x}^2 + \ddot{y}^2 + \ddot{z}^2} = r\omega^2 \sqrt{\cos^2 \omega t + \sin^2 \omega t + 0} = r\omega^2.$$

3. *Tangential and normal acceleration*

Tangential acceleration, $\vec{a}_{tan}$ and **normal acceleration**, $\vec{a}_{norm}$, the projections of the acceleration vector onto the tangent and the normal perpendicular to it, respectively (**Fig. 1.31**):

$$\vec{a} = \vec{a}_{tan} + \vec{a}_{norm}.$$

According to the product rule of differentiation:

$$\vec{a} = \frac{d(v\,\vec{e}_{tan})}{dt} = \frac{dv}{dt}\vec{e}_{tan} + v\frac{d\vec{e}_{tan}}{dt}.$$

The first term is the tangential acceleration,

$$\vec{a}_{tan} = \frac{dv}{dt}\vec{e}_{tan}, \qquad a_{tan} = \dot{v}.$$

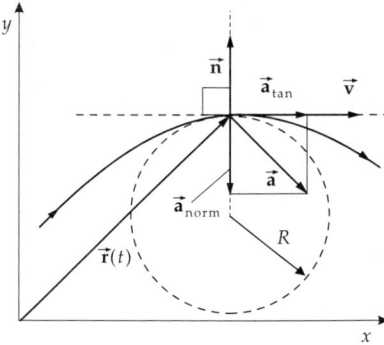

Figure 1.31: Tangential and normal acceleration $\vec{a}_{tan}$, $\vec{a}_{norm}$.

▲   The magnitude of the tangential component of the acceleration is the change of magnitude of the velocity with time.

The second term is the normal acceleration,

$$\vec{a}_{norm} = v\frac{d\vec{e}_{tan}}{dt}.$$

➤   Since the magnitude $|\vec{e}_{tan}|$ of the tangent unit vector invariably remains equal to unity,

$$\frac{d}{dt}(\vec{e}_{tan})^2 = 2\vec{e}_{tan} \cdot \frac{d\vec{e}_{tan}}{dt} = 0.$$

The time derivative of the tangent unit vector is orthogonal to the tangent unit vector. The second term represents the normal component of the acceleration. The plane defined by $\vec{e}_{tan}$ and $d\vec{e}_{tan}/dt$ is the **osculating plane** of the trajectory.

4. Example: Circular motion

For a circular motion with constant angular velocity,

$$\vec{a}(t) = \begin{pmatrix} -r\omega^2 \cos\omega t \\ -r\omega^2 \sin\omega t \\ 0 \end{pmatrix} = -\omega^2 \vec{r}(t),$$

i.e., the acceleration vector is antiparallel to the radius vector and thus to the normal vector, and points towards the center. Hence, the tangential component vanishes,

$$\vec{a}_{tan}(t) = 0,$$

and the normal component is

$$a_{norm}(t) = r\omega^2 = \frac{v^2}{r},$$

where $v = r\omega$ was inserted.

5. Curvature of trajectory and acceleration

The normal component of the acceleration vector is related to the curvature of the trajectory.

   Radius of curvature, R, in a point of a trajectory, the radius of a circle that has the same curvature as the trajectory at this point.

▲   The normal component of the acceleration vector is

$$a_{norm} = \frac{v^2}{R},$$

R being the radius of curvature of the trajectory.

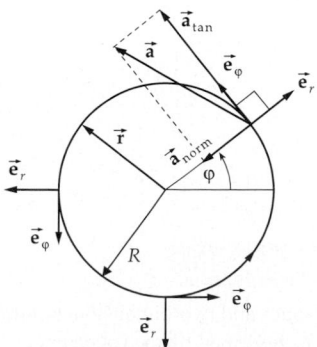

Figure 1.32: Non-uniform circular motion, $\vec{e}_{tan} = \vec{e}_\varphi$, $\vec{e}_{norm} = \vec{e}_r$.

➤ A straight line has a radius of curvature $R = \infty$. The normal acceleration vanishes for motion along a straight line.

➤ For a non-uniform circular motion (**Fig. 1.32**) both the normal acceleration (**centripetal acceleration**) a_r and the tangential acceleration a_φ differ from zero:

$$\vec{v}(t) = r\,\dot\varphi\,\vec{e}_\varphi, \qquad \vec{a}(t) = a_r\,\vec{e}_r + a_\varphi\,\vec{e}_\varphi,$$

$$a_r = -r\,\dot\varphi^2 = -r\,\omega^2, \qquad a_\varphi = r\,\ddot\varphi = r\,\dot\omega.$$

6. *Position, velocity and acceleration vectors in different coordinate systems*

a) *Cartesian coordinates:*

$$\vec{r}(t) = x(t)\,\vec{e}_x + y(t)\,\vec{e}_y + z(t)\,\vec{e}_z$$
$$\vec{v}(t) = \dot{x}(t)\,\vec{e}_x + \dot{y}(t)\,\vec{e}_y + \dot{z}(t)\,\vec{e}_z$$
$$\vec{a}(t) = \ddot{x}(t)\,\vec{e}_x + \ddot{y}(t)\,\vec{e}_y + \ddot{z}(t)\,\vec{e}_z$$

b) *Polar coordinates:*

$$\vec{r}(t) = r\,\vec{e}_r$$
$$\dot{\vec{e}}_r = \dot\varphi\,\vec{e}_\varphi, \qquad \dot{\vec{e}}_\varphi = -\dot\varphi\,\vec{e}_r$$
$$\vec{v}(t) = \dot{r}\,\vec{e}_r + r\,\dot\varphi\,\vec{e}_\varphi$$
$$\vec{a}(t) = (\ddot{r} - r\,\dot\varphi^2)\,\vec{e}_r + (r\,\ddot\varphi + 2\,\dot{r}\,\dot\varphi)\,\vec{e}_\varphi$$

c) *Spherical coordinates:*

$$\vec{r}(t) = r\,\vec{e}_r$$
$$\dot{\vec{e}}_r = \dot\vartheta\,\vec{e}_\vartheta + \sin\vartheta\,\dot\varphi\,\vec{e}_\varphi, \quad \dot{\vec{e}}_\vartheta = \dot\varphi\,\cos\vartheta\,\vec{e}_\varphi - \dot\vartheta\,\vec{e}_r, \quad \dot{\vec{e}}_\varphi = -\dot\varphi\,\cos\vartheta\,\vec{e}_\vartheta - \sin\vartheta\,\dot\varphi\,\vec{e}_r$$
$$\vec{v}(t) = \dot{r}\,\vec{e}_r + r\,\dot\vartheta\,\vec{e}_\vartheta + r\,\sin\vartheta\,\dot\varphi\,\vec{e}_\varphi$$
$$\vec{a}(t) = (\ddot{r} - r\,\dot\vartheta^2 - r\,\sin^2\vartheta\,\dot\varphi^2)\,\vec{e}_r + (r\,\ddot\vartheta + 2\,\dot{r}\,\dot\vartheta - r\,\sin\vartheta\,\cos\vartheta\,\dot\varphi^2)\,\vec{e}_\vartheta$$
$$+ (r\,\sin\vartheta\,\ddot\varphi + 2\,\sin\vartheta\,\dot{r}\,\dot\varphi + 2r\,\cos\vartheta\,\dot\vartheta\,\dot\varphi)\,\vec{e}_\varphi$$

d) *Cylindrical coordinates:*

$$\vec{r}(t) = \rho\,\vec{e}_\rho + z\,\vec{e}_z$$
$$\dot{\vec{e}}_\rho = \dot\phi\,\vec{e}_\phi, \quad \dot{\vec{e}}_\phi = -\dot\phi\,\vec{e}_\rho, \quad \dot{\vec{e}}_z = 0$$
$$\vec{v}(t) = \dot\rho\,\vec{e}_\rho + \rho\,\dot\phi\,\vec{e}_\phi + \dot{z}\,\vec{e}_z$$
$$\vec{a}(t) = (\ddot\rho - \rho\,\dot\phi^2)\,\vec{e}_\rho + (\rho\,\ddot\phi + 2\,\dot\rho\,\dot\phi)\,\vec{e}_\phi + \ddot{z}\,\vec{e}_z$$

1.3.3 *Free-fall and projectile motion*

Free-fall, **projectile motion**, refer respectively to one- and two-dimensional motion under the influence of Earth's gravitation. Such motion is described by the trajectory

$$\vec{\mathbf{r}}(t) = \begin{pmatrix} x(t) \\ y(t) \end{pmatrix}$$

and the velocity vector

$$\dot{\vec{\mathbf{r}}}(t) = \begin{pmatrix} v_x(t) \\ v_y(t) \end{pmatrix}.$$

The x-coordinate represents the horizontal distance from origin, the y-coordinate the height. In any case, the acceleration vector is the vector of gravitational acceleration $\vec{\mathbf{g}}$,

$$\ddot{\vec{\mathbf{r}}}(t) = \vec{\mathbf{g}} = \begin{pmatrix} 0 \\ -g \end{pmatrix}.$$

➤   The assumption of constant acceleration is only justified as long as the air friction is negligible, and the height of fall is small compared with the distance from Earth's center, so that the gravitational acceleration varies negligibly during the motion.

1. Free-fall

Let the body initially be at rest and move under the influence of gravity from a height h_0 downwards. If one ignores air friction, or assumes motion in vacuum, the motion is described by the position on the y-axis (instantaneous height) $y(t)$, the **velocity of fall** $v(t) = v_y(t)$, and the initial height h_0:

$$x(t) = 0, \qquad y(t) = h_0 - \frac{gt^2}{2},$$

$$v_x(t) = 0, \qquad v_y(t) = -gt.$$

Fall time t_F and **impact velocity** $v(t_F)$ are given by

$$t_F = \sqrt{\frac{2h_0}{g}}, \qquad v(t_F) = -\sqrt{2h_0 g}.$$

2. Vertical projectile motion upwards

The body is initially at height h_0 and gets a velocity v_0 upwards:

$$x(t) = 0, \qquad y(t) = h_0 + v_0 t - \frac{gt^2}{2},$$

$$v_x(t) = 0, \qquad v_y(t) = v_0 - gt.$$

The maximum height H is reached at time T_H when the velocity $v_y(t)$ reaches zero (**Fig. 1.33**):

$$H = h_0 + \frac{v_0^2}{2g}, \qquad T_\mathrm{H} = \frac{v_0}{g}, \qquad T = \text{flight time.}$$

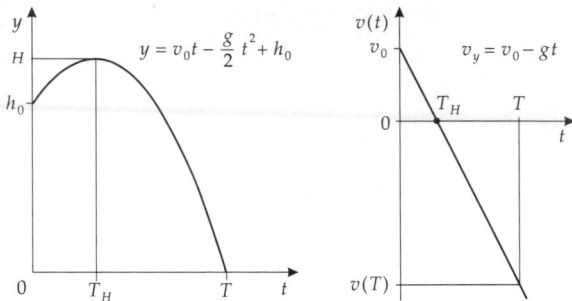

Figure 1.33: Vertical projectile motion upwards.

3. Inclined projectile motion

At the beginning, the body has not only a velocity component in y-direction (height), but also a component along the x-direction (horizontal). The horizontal motion is uniform because it is not affected by the gravitational force. Let the motion begin at $x = y = 0$; it is then described by

$$x(t) = v_{x0}t, \qquad y(t) = v_{y0}t - \frac{gt^2}{2},$$

$$v_x(t) = v_{x0}, \qquad v_y(t) = v_{y0} - gt.$$

The components of the initial velocity are specified by the launch angle α (**Fig. 1.34**):

$$\vec{v}_0 = \begin{pmatrix} v_{x0} \\ v_{y0} \end{pmatrix} = \begin{pmatrix} v_0 \cos \alpha \\ v_0 \sin \alpha \end{pmatrix}.$$

For $h_0 = 0$, the time of ascension until the peak of flight T_H, and the flight time T until the impact, are given by

$$T_H = \frac{T}{2}, \qquad T = \frac{2v_{y0}}{g} = \frac{2v_0 \sin \alpha}{g}.$$

The body has the same velocity at impact as at launch.

The trajectory of the inclined projectile motion is a **parabola**,

$$y(x) = x \tan \alpha - \frac{g}{2v_0^2 \cos^2 \alpha} x^2.$$

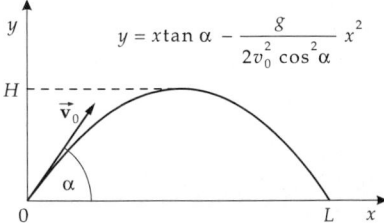

Figure 1.34: Inclined projectile motion upwards.

Projectile motion height H and **projectile motion range** L are given by

$$H = \frac{v_{y0}^2}{2g} = \frac{v_0^2 \sin^2 \alpha}{2g}, \qquad L = \frac{2v_0^2 \sin \alpha \cos \alpha}{g} = \frac{v_0^2 \sin 2\alpha}{g}.$$

➤ The maximum range $\left(\dfrac{dL}{d\alpha} = 0 \right)$ is reached for an angle α of $45°$. It amounts to

$$L_{\max} = \frac{v_0^2}{g}.$$

4. Real projectile motion

Actually, the trajectory of a projectile is modified by air friction. The velocity of fall cannot increase unlimitedly, but tends to a limit value $v_{\max}$ at which the friction force of air equals the gravitational force:

$$v_{\max} = \sqrt{\frac{2mg}{\rho c_W A}}$$

(m mass of body, ρ density of air, c_W air-resistance coefficient, A cross-sectional area of the body).

The trajectory of real projectile motion must be determined by solving a differential equation.

1.4   Rotational motion

Rotational motion, motion of a body in which the mutual distances between all points, and to a fixed rotation axis, remain constant. It is characterized by a rotation angle $\varphi(t)$ that specifies the position of the body at any time t.

Rotation, spatially periodic rotational motion in which the system performs full turns.

Circular motion, the motion of a mass point on a trajectory in a constant distance from a fixed rotation axis. It is the simplest example of rotational motion (**Fig. 1.35**).

The quantities angle, angular velocity and angular acceleration, needed for a description of rotational motion, correspond to the position, velocity and acceleration of translational motion, respectively.

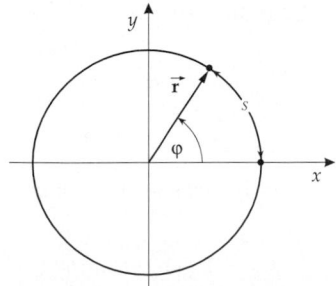

Figure 1.35: Circular motion of a point mass. Rotation angle: φ, path: $s = r\varphi$.

1.4.1   Angular velocity

1. Definition of angular velocity

Angular velocity, $\vec{\omega}$, a vector pointing along the rotation axis. Its magnitude gives the change of the rotation angle of a body per unit time, the orientation specifies the sense of rotation (**Fig. 1.36**). As in the case of the velocity of translational motion, one can introduce the mean angular velocity over the time interval Δt,

$$|\vec{\omega}| = \frac{\Delta\varphi}{\Delta t},$$

and in the limit $\Delta t \to 0$ the instantaneous angular velocity:

angular velocity = $\dfrac{\text{element of rotation angle}}{\text{time interval}}$			$\mathbf{T^{-1}}$
	Symbol	Unit	Quantity
$\|\vec{\omega}\| = \lim\limits_{\Delta t \to 0} \dfrac{\Delta\varphi}{\Delta t} = \dfrac{\mathrm{d}\varphi}{\mathrm{d}t} = \dot{\varphi}$	$\vec{\omega}$	rad/s	angular velocity
	φ	rad	rotation angle
	$\Delta\varphi$	rad	element of rotation angle
	Δt	s	time interval

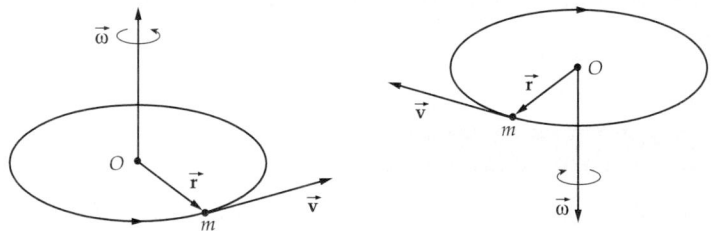

Figure 1.36: Angular velocity of circular motion.

2. Unit of angular velocity

Radian per second, rad/s, SI unit of angular velocity.
1 rad/s is the angular velocity of a body that changes its rotation angle in one second by one radian ($\approx 57.3°$).

■   Earth rotates once about its axis every 24 h. The angular velocity is

$$\omega = \frac{2\pi \text{ rad}}{24 \text{ h}} = \frac{2\pi \text{ rad}}{86400 \text{ s}} \approx 7.27 \cdot 10^{-5} \text{ rad/s}.$$

3. Rotational frequency and period

Rotational frequency, n, number of turns per unit time. The relation to the angular velocity is

$$\omega = 2\pi n, \qquad n = \frac{\omega}{2\pi}.$$

The rotational frequency can be given in r.p.s. (**revolutions per second**) or r.p.m. (**revolutions per minute**).

Period, T, the time for one revolution:

$$\omega = \frac{2\pi}{T}, \qquad T = \frac{1}{n} = \frac{2\pi}{\omega}.$$

■   The period of Earth's rotation is $T = 24$ h. Its rotational frequency is

$$n = \frac{1}{T} = \frac{1}{24\ \text{h}} = 1.157 \cdot 10^{-5}\ \text{s}^{-1}.$$

4. Right-hand rule

Right-hand rule specifies the orientation of the angular velocity vector $\vec{\omega}$ for a given sense of rotation (left or right rotation):

▲   The angular-velocity vector $\vec{\omega}$ is by definition oriented as follows: the thumb of the right hand indicates the orientation of $\vec{\omega}$ when the bent fingers indicate the sense of rotation (**Fig. 1.37**).

➤   Looking along the vector of angular velocity, the rotation is to the right, and thus **clockwise**.

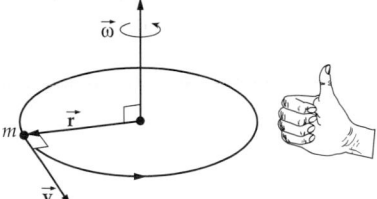

Figure 1.37: Relative orientation of angular velocity $\vec{\omega}$, radius vector $\vec{r}$ and orbital velocity $\vec{v}$ according to the right-hand rule.

➤   By convention, angular velocity, radius vector and orbital velocity are oriented with respect to each other just as thumb, forefinger and middle finger of the right hand.

➤   When using the left hand, the orientation of angular velocity would be just the opposite.

■   Because the Earth rotates eastwards, the angular velocity vector points from the south pole to the north pole.

5. Angular velocity as axial vector

Angular velocity is an **axial vector**, i.e., under a point reflection at the origin (**inversion**), $\vec{r} \rightarrow -\vec{r}$, it does not change its direction, contrary to a **polar vector** (like the velocity vector, or the acceleration vector):

$$\vec{r} \rightarrow -\vec{r}: \qquad \vec{v} \rightarrow -\vec{v}, \quad \vec{\omega} \rightarrow \vec{\omega}.$$

➤   The vector product of two polar vectors is an axial vector. The vector product of a polar and an axial vector is a polar vector.

1.4.2 Angular acceleration

Angular acceleration, $\vec{\alpha}$, change of angular velocity per unit time, an axial vector quantity. If the rotation axis remains fixed, the angular acceleration points parallel or antiparallel to the angular velocity. As in the case of acceleration in translational motion, one introduces the mean angular acceleration over the time interval Δt,

$$\bar{\vec{\alpha}} = \frac{\Delta \vec{\omega}}{\Delta t},$$

and by the limit $\Delta t \to 0$ the instantaneous angular acceleration:

angular acceleration =	change of angular velocity		T^{-2}
	time interval		

	Symbol	Unit	Quantity
$\vec{\alpha} = \lim\limits_{\Delta t \to 0} \dfrac{\Delta \vec{\omega}}{\Delta t} = \dfrac{d\vec{\omega}}{dt}$	$\vec{\alpha}$	rad/s^2	angular acceleration
	$\vec{\omega}(t)$	rad/s	angular velocity
	$\Delta \vec{\omega}$	rad/s	change of angular velocity
	Δt	s	time interval

Radian per square second, rad/s^2, SI unit of angular acceleration.
1 rad/s^2 is the angular acceleration if the angular velocity changes by 1 rad/s per second.
➤   If the rotation axis is fixed in space during the motion, the angular acceleration points along the rotation axis. The angular acceleration results only in an increase (angular acceleration and angular velocity parallel) or decrease of the rotational speed, or an inversion of the sense of rotation (angular acceleration and angular velocity antiparallel). In general, the angular acceleration expresses both the change of the rotational speed and also the change of orientation of the rotation axis.

1.4.3 Orbital velocity

1. Definition of orbital velocity

Orbital velocity, **tangential velocity**, $\vec{v}$, of a point mass on a circular orbit, the vector product of angular velocity $\vec{\omega}$ and the position vector $\vec{r}$ (**Fig. 1.38**):

orbital velocity = angular velocity × position vector			LT^{-1}

	Symbol	Unit	Quantity
$\vec{v} = \dfrac{d\vec{r}}{dt} = \vec{\omega} \times \vec{r}$	$\vec{v}$	m/s	orbital velocity
	$\vec{r}$	m	position vector
	$\vec{\omega}$	rad/s	angular velocity

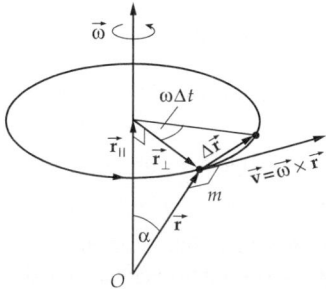

Figure 1.38: Orbital velocity $\vec{v}$ as vector product of angular velocity $\vec{\omega}$ and position vector $\vec{r}$.

▲   For circular motion, the orbital velocity vector is perpendicular to the position vector and perpendicular to the angular velocity vector if the rotation axis passes through the origin.

2. Decomposition of the orbital velocity vector

The vector $\vec{r}$ can be decomposed into two components: $\vec{r}_{\parallel}$ parallel to the angular velocity (rotation axis) $\vec{\omega}$ and $\vec{r}_{\perp}$ orthogonal to it. Then $\vec{\omega} \times \vec{r}_{\parallel} = 0$, and therefore

$$\vec{v} = \vec{\omega} \times \vec{r} = \vec{\omega} \times \vec{r}_{\perp}.$$

Hence, for the orbital velocity, only the **perpendicular distance** of the point mass from the rotation axis is relevant.

The magnitude of the orbital velocity is given by

$$|\vec{v}| = |\vec{\omega}|\,|\vec{r}_{\perp}| = |\vec{\omega}|\,|\vec{r}|\sin\alpha,$$

where α is the angle between the rotation axis and the position vector. The orbital velocity is proportional to the angular velocity and to the perpendicular distance from the rotation axis.

In particular, for the circumferential speed of a wheel of radius R:

$$v = R\omega = 2\pi Rn = \frac{2\pi R}{T},$$

where n is the rotational frequency and T the period of rotation.

3. Example: Orbital velocity of Earth

Earth has a radius R of 6380 km. The circumferential speed at the equator is

$$v = \omega R = \frac{2\pi\ \text{rad}}{24\ \text{h}} \cdot 6380\ \text{km} = 464\ \text{m/s} = 1670\ \text{km/h}.$$

The orbital velocity of a point at a latitude of $45°$, that has a perpendicular distance to Earth's axis at

$$R_{\perp} = R/\sqrt{2},$$

is $v = 1670/\sqrt{2}$ km/h $= 1180$ km/h.

2
Dynamics

Dynamics, the theory of the motion caused by forces. Dynamics describes how bodies move under the action of external forces. Unlike kinematics, it is concerned with the causes of the motion of a body. The concepts mass and force are introduced for the description of the dynamics of the motion.

2.1   Fundamental laws of dynamics

Forces are the cause of the change of the state of motion of bodies. Newton's laws establish a relation between the forces and the kinematical quantities velocity and acceleration.

2.1.1   Mass and momentum

2.1.1.1   Mass

1. Inertial and gravitational mass

Inertial mass, the resistance of a body to a change of motion.
   Gravitational mass, the strength of attraction on one body by another due to the gravitational force (e.g., in the gravitation field of earth).
▲   Inertial and gravitational masses of a body are equal.
➤   This equivalence is an empirical fact that has been established by high-precision experiments. This equality is a basic postulate of the general theory of relativity.
   Mass, m, elementary property ascribed to a body. **Point masses** have this property only; extended bodies (**rigid bodies**) are also characterized by their moments of inertia (see p. 111). The moment of inertia of a rigid body depends on the distribution of its mass and on the choice of the rotation axis.

2. Unit of mass

Kilogram, kg, SI unit of mass. One of the seven basic quantities of the SI.
1 kg is defined as the mass of the **primary kilogram**, a platinum-iridium cylinder stored in Paris. The relative accuracy of the mass standard is 10^{-9}.

$$[m] = \text{kg} = \text{kilogram}.$$

37

3. Measurement of mass

A mass can be measured by weighing, i.e., by comparison of the weight of the body with that of a body of known mass (**balance** according to the **rule of levers, balance scale** with movable counterweight). Weighing is one measurement that can be carried out with high accuracy by simple means.

The **spring balance** measures the weight of a body directly by the extension of a spring (dynamometer).

The mass of atomic particles can be measured by its inertia, e.g., by deflection in an electric field, magnetic field (**mass spectrometer, mass spectrograph**), or both.

➤ Mass and weight are different qualities. The weight depends on the acting gravitational force. A body of mass 1 kg has on the moon the mass of 1 kg, but it weighs only 1/6 of what it weighs on earth (see p. 53).

4. Density,

ρ, ratio of mass to volume of a homogeneous body:

density = $\dfrac{\textbf{mass}}{\textbf{volume}}$			$\mathbf{ML^{-3}}$
	Symbol	Unit	Quantity
$\rho = \dfrac{m}{V}$	ρ	kg/m^3	density
	m	kg	mass
	V	m^3	volume

5. Unit of density

Kilogram per cubic meter, SI unit of density.
One kilogram per cubic meter is the density of a homogeneous body having a volume of one cubic meter and a mass of one kilogram.

$$[\rho] = \text{kg/m}^3.$$

➤ The density is sometimes given in kilograms per cubic decimeter (kg/dm^3), or in grams per cubic centimeter (g/cm^3):

$$1 \text{ kg/dm}^3 = 1 \text{ g/cm}^3 = 10^3 \text{ kg/m}^3.$$

Water has a density of about 1 g/cm^3 at 20 °C, and metals have densities three (aluminum) to twenty (platinum) times that. Gasoline has a density of about 0.7 g/cm^3 (see **Tab. 7.1**).

➤ Density depends on the temperature of the body (**volume expansion coefficient**), and, particularly for gases, also on the pressure.

[M] The density of solids can be measured with a **Mohr's balance** which uses the buoyancy force of the body in a fluid (see p. 180).

6. Density of inhomogeneous bodies

In an **inhomogeneous** body with a continuous mass distribution, the density varies with the spatial coordinate $\vec{r}$, $\rho = \rho(\vec{r})$. Assume the body to be decomposed into volume elements ΔV in which the density is approximately constant. The mass in the volume element ΔV

at the point $\vec{r}$ is Δm (**Fig. 2.1**). For the density in the volume element ΔV: $\rho = \Delta m / \Delta V$. For a continuous mass distribution, one gets for the density at the point $\vec{r}$:

$$\rho(\vec{r}) = \lim_{\Delta V \to 0} \frac{\Delta m}{\Delta V} = \frac{dm}{dV}, \quad dm = \rho(\vec{r}) \cdot dV.$$

The total mass m of the body is given by a volume integral,

$$m = \int dm = \int \rho(\vec{r}) \, dV.$$

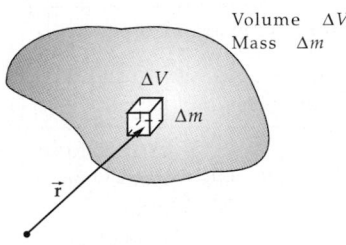

Volume ΔV
Mass Δm

ΔV

Δm

$\vec{r}$

Figure 2.1: Density $\rho(\vec{r})$ of an inhomogeneous body with a continuous mass distribution.

2.1.1.2   Momentum

1. Definition of momentum

Momentum, the quantity of motion of a body that is given by the product of its mass and velocity. The momentum is, like the velocity, a vector quantity; its orientation coincides with the direction of motion of the body. It specifies the state of motion of a body relative to a reference system.

momentum = mass · velocity			$\mathbf{MLT^{-1}}$
$\vec{\mathbf{p}} = m\vec{\mathbf{v}}$	Symbol	Unit	Quantity
	$\vec{\mathbf{p}}$	kg m/s	momentum of body
	m	kg	mass of body
	$\vec{\mathbf{v}}$	m/s	velocity of body

2. Unit of momentum

Kilogram meter per second, kg m/s, SI unit of momentum.
One kilogram meter per second is the momentum of a body with a 1 kg mass that moves with the velocity of 1 m/s.

$$[p] = \frac{\text{kg m}}{\text{s}} = \text{Ns}, \quad \text{N} = \text{Newton} = \text{kgm/s}^2 \quad \text{(see p. 41)}.$$

■   A body of 10 kg mass that moves with 3 m/s has a momentum of

$$p = mv = 10 \text{ kg} \cdot 3 \text{ m/s} = 30 \text{ Ns}.$$

A body with twice the mass (20 kg) has twice the momentum at the same velocity:

$$p = mv = 20 \text{ kg} \cdot 3 \text{ m/s} = 60 \text{ Ns}.$$

2.1.2   Newton's laws

Newton's laws establish a relation between force (for definition see p. 41) and change of momentum. Newton's first law expresses the principle of inertia, the second, the principle of action and the third, the principle of action and reaction.

2.1.2.1   Inertia (Newton's first law)

1. Newton's first law

(**Galileo's principle of inertia**), describes the inertial power or the **inertia** of bodies:

Newton's first law: A body that is not under external forces does not change its momentum.			
	Symbol	Unit	Quantity
$\vec{F} = 0 \Longrightarrow \vec{p} = $ const.	$\vec{F}$	N	external force
	$\vec{p}$	kg m/s	momentum
$m = $ const. $\Longrightarrow \vec{v} = $ const.	m	kg	mass
	$\vec{v}$	m/s	velocity

➤   Newton's first law holds even in the case of mass m that is not constant, e.g., for a rocket (recoil propulsion). The conclusion $m = $ const. $\Longrightarrow \vec{v} = $ const., then, no longer holds.

The notion of constant velocity is always relative to a special reference system.

■   A passenger sitting on a transatlantic flight moves with constant velocity $v = 0$ relative to the plane, but on a curve relative to a point on Earth's surface. Relative to a point outside of the Earth, one must add the Earth's rotation, and relative to the Sun, the rotation of the Earth about the Sun. The Sun in turn moves relative to the center of the Milky Way, which again moves relative to other galaxies.

2. Inertial systems,

reference systems in which Newton's first law holds. A reference system that moves uniformly on a straight line relative to an inertial system is also an inertial system (**Fig. 2.2**). Hence, there are arbitrarily many inertial systems in which the laws of physics hold in identical form (see p. 137).

■   A body that moves free of forces on a horizontal frictionless rail maintains a constant velocity. This is an idealized case, since neither the friction with the rail nor the air friction can be completely excluded. Motion in space far from large bodies comes closer to the idealization.

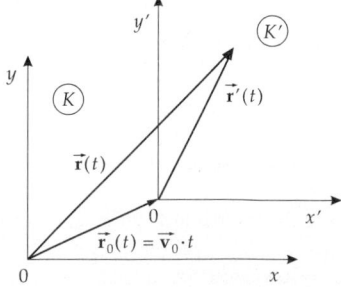

Figure 2.2: Relative motion of two inertial systems K and K' with relative velocity $\vec{v}_0$. $\vec{r}_0(t)$: position vector of the coordinate origin of K' in K at time t.

2.1.2.2 Fundamental law of dynamics (Newton's second law)

Newton's second law (**action principle**), describes how the state of motion of a body is changed by forces (for definition, see the next section) acting on it (**Fig. 2.3**):

Newton's second law: If a force acts upon a body, then the resulting change of momentum is proportional to the acting force. The change of momentum points along the direction of the force.	MLT^{-2}

$\dfrac{d\vec{\mathbf{p}}}{dt} = \dfrac{d(m\vec{\mathbf{v}})}{dt} = \vec{\mathbf{F}}$	Symbol	Unit	Quantity
	$\vec{\mathbf{v}}$	m/s	instantaneous velocity
	$\vec{\mathbf{p}}$	kg m/s	momentum
	$\vec{\mathbf{F}}$	N	force
	m	kg	mass

If the mass of the body can be considered constant during the dynamical process, then:

$$m\vec{\mathbf{a}} = \vec{\mathbf{F}},$$

where $\vec{\mathbf{a}}$ is the acceleration, $\vec{\mathbf{a}} = \dfrac{d\vec{\mathbf{v}}}{dt}$, with the SI unit

$$[\vec{\mathbf{a}}] = \mathrm{ms}^{-2}.$$

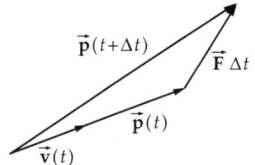

Figure 2.3: Force $\vec{\mathbf{F}}$ and momentum change $\vec{\mathbf{p}}(t + \Delta t) - \vec{\mathbf{p}}(t)$.

▲ Newton's second law is the **fundamental law of dynamics**.

■ If a force acts on a body with twice the mass of another body, then it gets only half of the acceleration.

➤ Newton's second law holds even when the mass of the body varies during the motion (as with a rocket). Corresponding to the product rule of differentiation it then has the form

$$\frac{d\vec{\mathbf{p}}}{dt} = \frac{dm}{dt}\vec{\mathbf{v}} + m\frac{d\vec{\mathbf{v}}}{dt} = \vec{\mathbf{F}}.$$

➤ If one considers length, time and mass to be the fundamental quantities of motion (as in the SI), Newton's second law leads to the unit of force. If, however, length, time and force were adopted as fundamental quantities, Newton's second law would define the mass.

2.1.2.3 Force

1. Definition of force

In the SI, the definition of force is based on Newton's second law:

  Force, the product of mass of a body, and of its acceleration caused by the force. The force is a vector quantity and points along the acceleration. It is thus defined by the table below:

force = mass · acceleration				MLT^{-2}
		Symbol	Unit	Quantity
$\vec{F} = m\vec{a}$	$\vec{F}$	N	applied force
	m	kg	mass
	$\vec{a}$	m/s^2	resulting acceleration

2. Unit of force

Newton, N, the SI unit of force:
1 newton is the force that accelerates a mass of 1 kg by 1 m/s^2.

$$[F] = \mathrm{N} = \text{newton} = \mathrm{kg\,m/s^2}.$$

Non-SI units are:

$$1 \text{ kilopond (kp)} = 9.80665 \text{ N},$$

$$1 \text{ Dyne (dyne)} = 10 \ \mu\mathrm{N}.$$

Mass, the proportionality factor of force and acceleration: The more mass a body has, the less it is accelerated by a force applied to it. This allows the determination of the mass as the ratio of applied force and resulting acceleration,

$$m = \frac{|\vec{F}|}{|\vec{a}|}.$$

3. Impulse of a force,

the product $\vec{F}\Delta t$. The impulse of a force gives the change $\Delta\vec{p} = \vec{p}_2 - \vec{p}_1$ of momentum (**Fig. 2.4**).

impulse of force = force · time interval for constant force				MLT^{-1}
		Symbol	Unit	Quantity
$\Delta\vec{p} = m(\vec{v}(t + \Delta t) - \vec{v}(t)) = \vec{F}\Delta t$		$\Delta\vec{p}$	kg m/s	change of momentum
		Δt	s	time interval
		$\vec{v}$	m/s	velocity
		$\vec{F}$	N	acting force

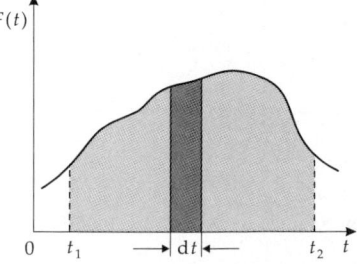

Figure 2.4: One-dimensional motion. The impulse of a force is the area below the curve $F(t)$ in a graph of force vs. time.

➤ If the force is not constant over the time interval Δt, the integral form must be used:

$$\Delta \vec{p} = \int_{t_1}^{t_2} \vec{F}\, dt.$$

2.1.2.4 Reaction principle (Newton's third law)

Newton's third law (reaction principle), states that, for each force $\vec{F}$ acting on a body 1, there exists a second force $\vec{F'}$ acting on another body 2 that has equal magnitude, but opposite direction (**Fig. 2.5**):

$$\vec{F} = -\vec{F'}, \qquad \textbf{action = reaction.}$$

Newton's third law: Two bodies exert forces equal in magnitude and opposite in direction on each other.		$\mathbf{MLT^{-2}}$	
$\vec{F} = -\vec{F'}$	Symbol	Unit	Quantity
	$\vec{F}$	N	force of 2 on 1
	$\vec{F'}$	N	force of 1 on 2

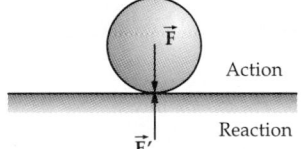

Action

Reaction

Figure 2.5: The principle of action and reaction.

■ Two people on a frictionless surface hold the ends of a rope. If one pulls on the rope and moves forward, the second also moves, but in the opposite direction.

➤ Although the forces F and F' are equal in magnitude, the body with the larger mass receives less acceleration than the body with the smaller mass.

2.1.2.5 Inertial forces

1. Definition of inertial forces

Inertial forces, virtual forces felt by an observer in a reference system that carries out an accelerated motion relative to an inertial system (**Fig. 2.6**). Contrary to the abovementioned

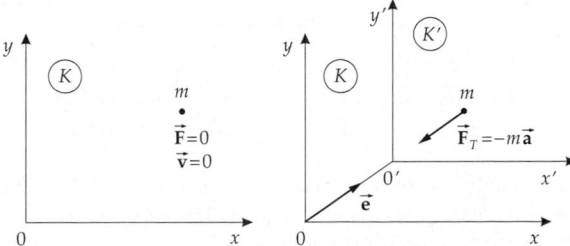

Figure 2.6: Inertial force in a reference system K' that is uniformly accelerated with respect to the reference system K.

forces, inertial forces are not a cause, but a consequence, of accelerated motion. In the case of accelerated translational motion of a reference system, the inertial forces point in a direction that is opposite to the acceleration vector.

The inertial force in accelerated translational motion of the reference system has the opposite orientation, but the same magnitude, as the force that causes the acceleration of the reference system.	MLT^{-2}		
	Symbol	Unit	Quantity
$\vec{F}_T = -m\vec{a}$	$\vec{F}_T$	N	inertial force
	m	kg	mass
	$\vec{a}$	m/s^2	acceleration

2. Examples of inertial forces

■ A point mass m is at rest in the reference system K ($\vec{F} = 0$). A second reference system K' is moving in the x, y-plane relative to K with velocity $\vec{v}$ and constant acceleration $\vec{a} = d\vec{v}/dt \neq 0$ in the direction of the vector $\vec{e}$. Under the influence of the inertial force $\vec{F}_T = -m\vec{a}$, one observes in K' an accelerated motion of the point mass that is antiparallel to the displacement vector $\vec{d}(t) = \overrightarrow{OO'}$.

■ A body of mass 1 kg is in a car that is being accelerated by 3 m/s^2. A measurement taken in the car yields a virtual force of

$$F_T = -ma = -1 \text{ kg} \cdot 3 \text{ m/s}^2 = -3 \text{ N}.$$

This is the magnitude of force needed to accelerate the body by 3 m/s^2.

3. Inertial forces in rotational motion

Other inertial forces arise in rotational motion (see p. 31).

■ An observer on a rotating disk feels a radial acceleration towards the outside. This virtual force is called the **centrifugal force**.

2.1.2.6   Principle of d'Alembert

Dynamical equilibrium exists if the sum of the applied force $\vec{F}$ and the opposite inertial force $\vec{F}_T$ vanishes (**d'Alembert's principle**).

Body in dynamical equilibrium			
	Symbol	Unit	Quantity
$\vec{F} + \vec{F}_T = 0$	$\vec{F}$	N	acting force
$\vec{F} - m\vec{a} = 0$	$\vec{F}_T$	N	inertial force
	$\vec{a}$	m/s^2	acceleration

➤ Unlike static equilibrium, the existence of dynamical equilibrium does not mean that the body does not change its state of motion. The appearance of inertial forces just implies that an acceleration is taking place.

➤ This rule allows a calculation of the motion of a body under the conditions that forces and inertial forces mutually compensate. Dynamical processes are thereby reduced to static-equilibrium problems.

2.1.2.7 Composition of forces

1. Resulting force,

$\vec{F}_R$, replaces two forces $\vec{F}_1$ and $\vec{F}_2$ acting on a point mass by a single force $\vec{F}_R$. Forces are added as vectors according to the **force parallelogram (Fig. 2.7)**.

resulting force = vector sum of individual forces			MLT^{-2}
	Symbol	Unit	Quantity
$\vec{F}_R = \vec{F}_1 + \vec{F}_2$	$\vec{F}_R$	N	resulting force
	$\vec{F}_1, \vec{F}_2$	N	force vectors
$F_{Rx} = F_1 \cos\alpha_1 + F_2 \cos\alpha_2$	F_{Rx}, F_{Ry}	N	components of resulting force
$F_{Ry} = F_1 \sin\alpha_1 + F_2 \sin\alpha_2$	φ	rad	angle between $\vec{F}_1$ and $\vec{F}_2$
$F_R = \sqrt{F_1^2 + F_2^2 + 2\,F_1\,F_2\,\cos\varphi}$	α_1	rad	angle between $\vec{F}_1$ and x-axis
$\alpha = \arctan \dfrac{F_1\,\sin\alpha_1 + F_2\,\sin\alpha_2}{F_1\,\cos\alpha_1 + F_2\,\cos\alpha_2}$	α_2	rad	angle between $\vec{F}_2$ and x-axis
	α	rad	angle between $\vec{F}_R$ and x-axis

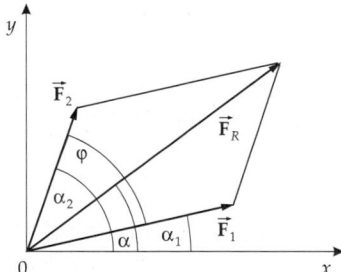

Figure 2.7: Adding forces. The force parallelogram.

2. Force polygon

By repeating this process, arbitrarily many forces acting at the same point can be replaced by a single resulting force:

▲ $\vec{F}_R = \vec{F}_1 + \vec{F}_2 + \vec{F}_3 + \cdots$.

This can be represented graphically by a **force polygon (force diagram)**. The force arrows are lined up by parallel shifts (conserving magnitude and orientation). The resulting vector is the force arrow from the beginning of the first force arrow to the end of the last one **(Fig. 2.8)**.

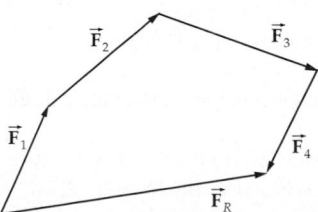

Figure 2.8: Force polygon.

3. Addition of components

The resulting force can also be calculated by summing components (see p. 1115):

$$\vec{F}_R = \begin{pmatrix} F_{Rx} \\ F_{Ry} \\ F_{Rz} \end{pmatrix} = \vec{F}_1 + \vec{F}_2 = \begin{pmatrix} F_{1x} + F_{2x} \\ F_{1y} + F_{2y} \\ F_{1z} + F_{2z} \end{pmatrix}.$$

➤ If two vectors point in the same direction ($\varphi = 0$), then

$$|\vec{F}_R| = |\vec{F}_1 + \vec{F}_2| = |\vec{F}_1| + |\vec{F}_2|.$$

If they point in opposite direction ($\varphi = \pi$), then

$$|\vec{F}_R| = |\vec{F}_1 + \vec{F}_2| = |\vec{F}_1| - |\vec{F}_2|.$$

If the forces are perpendicular ($\varphi = \pi/2$), then

$$|\vec{F}_R| = |\vec{F}_1 + \vec{F}_2| = \sqrt{|\vec{F}_1|^2 + |\vec{F}_2|^2}.$$

2.1.2.8 Decomposition of forces

1. General decomposition of forces

The decomposition of a force $\vec{F}$ into two forces $\vec{F}_1, \vec{F}_2$ pointing in given directions is accomplished by means of the **scalar product** using the force parallelogram.

decomposition of a force			
$F_1 = F \dfrac{\sin(\alpha_2 - \alpha)}{\sin(\alpha_2 - \alpha_1)}$	Symbol	Unit	Quantity
	$\vec{F}$	N	given force
$F_2 = F \dfrac{\sin(\alpha - \alpha_1)}{\sin(\alpha_2 - \alpha_1)}$	$\vec{F}_1, \vec{F}_2$	N	force vectors
	α	rad	angle between $\vec{F}$ and x-axis
$\alpha_1 = \alpha - \arccos \dfrac{F^2 + F_1^2 - F_2^2}{2\,F\,F_1}$	α_1	rad	angle between $\vec{F}_1$ and x-axis
$\alpha_2 = \alpha + \arccos \dfrac{F^2 + F_2^2 - F_1^2}{2\,F\,F_2}$	α_2	rad	angle between $\vec{F}_2$ and x-axis

2. Tangential and normal force

The decomposition of a force in the special case of two perpendicular directions can also be accomplished by means of the **scalar product (Fig. 2.9)**:
The component F_1 of the force $\vec{F}$ along the orientation given by the unit vector $\vec{e}_1$ is the scalar product of $\vec{F}$ and the unit vector $\vec{e}_1$:

$$F_x = \vec{F} \cdot \vec{e}_1 = F \cos\alpha, \quad \alpha \text{: angle between } \vec{F} \text{ and } \vec{e}_1.$$

The component F_2 of $\vec{F}$ along the orientation 2 perpendicular to orientation 1, $\vec{e}_1 \cdot \vec{e}_2 = 0$, is given by

$$F_2 = \vec{F} \cdot \vec{e}_2 = F \cos(\frac{\pi}{2} - \alpha) = F \sin\alpha.$$

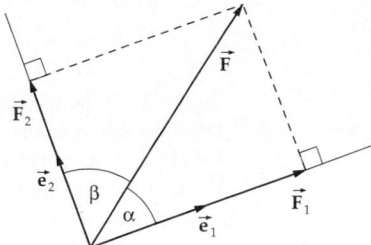

Figure 2.9: Decomposition of a force $\vec{F}$ into two perpendicular components $\vec{F}_1$ and $\vec{F}_2$.

Tangential force, force acting along the tangent to the trajectory. The tangential force causes a pure orbital acceleration (**tangential acceleration**), since it changes the magnitude v of the velocity, but not the direction, of $\vec{v}$:

$$\vec{F}_{\text{tan}} = m \cdot v \cdot \vec{e}_{\text{tan}}.$$

Normal force, force acting along the principal normal to the trajectory. The normal force causes a pure **normal acceleration** that does not affect the magnitude of the velocity, but changes only the orientation of $\vec{v}$:

$$\vec{F}_{\text{norm}} = m \cdot \frac{v^2}{R} \cdot \vec{e}_{\text{norm}},$$

R: radius of curvature of the trajectory, $\vec{e}_{\text{norm}}$: unit vector along the principal normal.

3. Centripetal force

In uniform circular motion with radius of curvature R, the tangential force vanishes. The normal force is called the **centripetal force**

$$\vec{F}_r = -m \frac{v^2}{R} \vec{e}_r.$$

It causes a uniform acceleration towards the center of the circle (**centripetal acceleration**). The centripetal force is a **central force**.

4. Application of decomposing a force

The force on a body is decomposed into components along and pependicular to the restraint when a body is supported in a definite manner. The support (fixed bearing, rail, supporting plane) provides a counterforce (**guiding force**, **constraint reaction**, **reaction force**) which, without consideration of friction, just equals the force acting in this direction. The guiding force is perpendicular to the curve or plane in space to which the mass is constrained.

5. Application to inclined planes

Inclined plane: One needs the components of the weight $\vec{F}_G$ perpendicular (**normal force** $\vec{F}_N$) and parallel to the inclined plane (**force along the** $\vec{F}_H$). The force along the slope accelerates the body while the normal force is counteracted by the plane (**Fig. 2.10**).
    One finds:

force along the slope:	$F_H = F_G \sin\alpha = mg \sin\alpha,$
normal force:	$F_N = F_G \cos\alpha = mg \cos\alpha.$

α is the angle of inclination of the plane with respect to the horizontal.

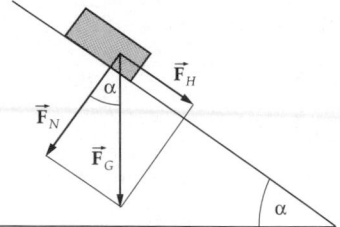

Figure 2.10: Inclined plane. Decomposition of the weight $\vec{F}_G$ into normal force $\vec{F}_N$ and force along the slope $\vec{F}_H$.

■ A body of $m = 2$ kg glides down on a plane with $\alpha = 30°$ inclination. The force along the slope is

$$F_H = F \sin \alpha = mg \sin \alpha = 2 \text{ kg} \cdot 9.81 \text{ m/s}^2 \cdot 0.5 = 9.81 \text{ N}.$$

The corresponding acceleration is

$$a = \frac{F_H}{m} = 4.91 \text{ m/s}^2 = \frac{1}{2}g,$$

i.e., half the gravitational acceleration. For an angle of $\alpha = 45°$, the reduction factor is $1/\sqrt{2} \approx 0.707$, for $\alpha = 60°$, $\sqrt{3}/2 \approx 0.866$. The fraction of the weight that is counteracted by the plane (the normal force) is $\sqrt{3}/2$, $1/\sqrt{2}$ and $1/2$, respectively.

➤ For an inclination of $45°$, the force along the slope and the normal force are equal:

$$F_H = F_N = \frac{1}{\sqrt{2}} F_G \approx 0.707 \, F_G.$$

➤ The tangent of α gives the ratio of height difference and horizontal distance; it is called the **slope**.

■ In order to overcome gravity, a train of mass 1000 t on a rise of $h/l = 1 : 150$ needs a force of

$$F_H = F_G \sin \alpha = mg \frac{h}{l} = 10^6 \text{ kg} \cdot 9.81 \text{ m/s}^2 \cdot \frac{1}{150} = 65.4 \text{ kN}.$$

2.1.3 Orbital angular momentum

1. Definition of orbital angular momentum

Angular momentum, orbital angular momentum, $\vec{l}$, the vector product of the position vector $\vec{r}$ and the momentum $\vec{p} = m\vec{v}$, $\vec{v}$ is the velocity of the point mass (**Fig. 2.11**).

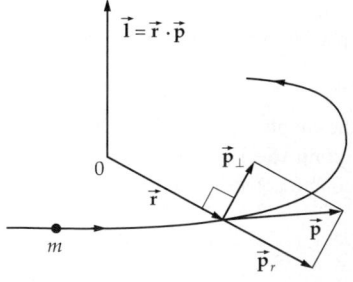

Figure 2.11: Orbital angular momentum $\vec{l}$ of a point mass m.

Radial momentum, $\vec{p}_r$, component of the momentum $\vec{p}$ of a point mass in the direction of the position vector $\vec{r}$:

$$\vec{p}_r = (\vec{p} \cdot \vec{e}_r)\,\vec{e}_r, \qquad \vec{e}_r : \text{ unit vector in } \vec{r}\text{-direction.}$$

The component of $\vec{p}$ that lies in the plane spanned by $\vec{r}$ and $\vec{p}$ and points perpendicular to the radial momentum is given by the vector $-\vec{e}_r \times (\vec{e}_r \times \vec{p})$. This component of $\vec{p}$ that is perpendicular to the position vector enters the **orbital angular momentum**.

orbital angular momentum = position vector x momentum			$\mathbf{ML^2T^{-1}}$
	Symbol	Unit	Quantity
$\vec{l} = \vec{r} \times \vec{p} = m\vec{r} \times \vec{v}$	$\vec{l}$	kg m^2/s	angular momentum
	$\vec{r}$	m	position vector
$l = r \cdot m \cdot v \cdot \sin\alpha$	$\vec{p}$	kg m/s	momentum
	$\vec{v}$	m/s	orbital velocity
	m	kg	mass
	α	rad	angle between $\vec{r}$ and $\vec{p}$

Kilogram times meters squared per second, kg m^2/s, SI unit of the angular momentum.

2. Properties of orbital angular momentum

➤ The orbital angular momentum of a point mass is a vector that is perpendicular to the direction of motion of the mass point, and perpendicular to the position vector. Its magnitude is given by $l = r \cdot p \cdot \sin\alpha$, with α being the angle between the position and momentum vectors, respectively.

➤ The orbital angular momentum depends on the choice of the reference point.

➤ The orbital angular momentum vanishes if the momentum vector has no component perpendicular to the position vector. Motion along a straight line through the coordinate origin as reference point corresponds to zero orbital angular momentum.

■ For circular motion, the orbital velocity $\vec{v}$ is the vector product of angular velocity $\vec{\omega}$ and position vector $\vec{r}$, $\vec{v} = \vec{\omega} \times \vec{r}$. Hence, the angular momentum of circular motion is

$$\vec{l} = m\vec{r} \times (\vec{\omega} \times \vec{r}) = mr^2\,\vec{\omega} = J \cdot \vec{\omega}.$$

The quantity $J = mr^2$ is denoted as the moment of inertia of a mass point.

➤ The angular momentum of a circular motion points along the angular velocity vector. Hence, it is perpendicular to the trajectory plane.

3. Moment of inertia of a point mass

For circular motion, the product of the mass m and the square of the perpendicular distance r from the rotation axis.

moment of inertia of a point mass			$\mathbf{ML^2}$
	Symbol	Unit	Quantity
$J = m \cdot r^2$	J	kg m^2	moment of inertia
	m	kg	mass
	r	m	distance from the rotation axis

➤ In rotational motion, the moment of inertia J and the angular momentum $\vec{l} = J \cdot \vec{\omega}$ correspond to the mass m and the momentum $\vec{p} = m \cdot \vec{v}$ of the translational motion.

2.1.4 Torque

1. Definition of torque

Torque, moment of a force, the vector product of the position vector $\vec{r}$ and the force $\vec{F}$ acting at the point $\vec{r}$ (**Fig. 2.12**).

torque = position vector × force				$\mathbf{ML^2T^{-2}}$
	Symbol	Unit	Quantity	
$\vec{\tau} = \vec{r} \times \vec{F}$	$\vec{\tau}$	Nm	torque	
	$\vec{r}$	m	position vector	
$\tau = r \cdot F \cdot \sin\alpha$	$\vec{F}$	N	force	
	α	rad	angle between the position and force vectors	

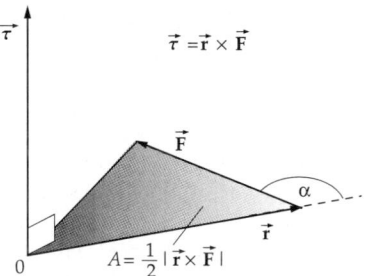

Figure 2.12: Torque $\vec{\tau}$ of a force $\vec{F}$.

$$\vec{\tau} = \vec{r} \times \vec{F}$$

$$A = \frac{1}{2} |\vec{r} \times \vec{F}|$$

Newton meter, the SI unit of the torque:
1 newton meter is the torque created by a force of 1 N that acts perpendicular to the position vector at a distance of 1 m from the center of rotation.

$$[\vec{\tau}] = \text{newton meter} = \text{Nm} = \text{N} \cdot \text{m}.$$

2. Properties of torque

▲ The torque vector is perpendicular to the plane A, which contains the position vector $\vec{r}$ and the force $\vec{F}$. The magnitude of the torque is the product of the distance of the point of application of the force from the reference point (coordinate origin) and the force component acting perpendicular to the position vector of the point at which the force is applied.

➤ The torque has its maximum value when $\vec{r}$ and $\vec{F}$ are perpendicular to each other ($\sin\alpha = 1$). Since the component of the force orthogonal to the position vector is the only one that contributes to the torque, a force pointing radially inwards or outwards relative to the force center, $\vec{F} \sim \vec{r}$, yields no torque. Such forces are called **central forces**. A motion that results from the action of a central force is called **central motion**.

▲ If one doubles the distance from the reference point to the point of application of a constant force, the torque also doubles. Application: wrench.

■ A force of $F = 5$ N acts at a distance of $d = 20$ cm from the rotation axis on a wrench. The torque is

$$\tau = F \cdot d = 5\,\text{N} \cdot 20\,\text{cm} = 1\,\text{Nm}.$$

3. Resulting torque

If several forces $\vec{\mathbf{F}}_i$, $i = 1, 2, \ldots$ act on a body, the individual torques $\vec{\tau}_i = \vec{\mathbf{r}}_i \times \vec{\mathbf{F}}_i$ can be summed as vectors to form the resulting torque vector.

composition of torques			$\mathbf{ML^2T^{-2}}$
$\vec{\tau}_R = \vec{\tau}_1 + \vec{\tau}_2 + \cdots$	Symbol	Unit	Quantity
	$\vec{\tau}_R$	Nm	resulting torque
	$\vec{\tau}_1, \vec{\tau}_2, \ldots$	Nm	individual torques

➤ For two opposite forces of equal magnitude (**couple**), $\vec{\mathbf{F}}_2 = -\vec{\mathbf{F}}_1$, the resulting force vanishes, $\vec{\mathbf{F}}_1 + \vec{\mathbf{F}}_2 = 0$. The resulting torque $\vec{\tau}_R$, however, does not vanish if the forces act at different points (**Fig. 2.13**):

$$\vec{\tau}_R = \vec{\mathbf{r}}_1 \times \vec{\mathbf{F}}_1 + \vec{\mathbf{r}}_2 \times \vec{\mathbf{F}}_2 = (\vec{\mathbf{r}}_1 - \vec{\mathbf{r}}_2) \times \vec{\mathbf{F}}_1.$$

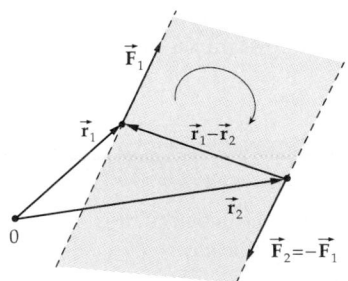

Figure 2.13: Torque of a couple (perpendicular to the plane including $\vec{\mathbf{r}}_1 - \vec{\mathbf{r}}_2$ and $\vec{\mathbf{F}}_1, \vec{\mathbf{F}}_2$).

4. Torque: change of angular momentum with time

The variation of the angular momentum $\vec{\mathbf{l}} = \vec{\mathbf{r}} \times \vec{\mathbf{p}}$ of the orbital motion of a point mass with time, according to the product rule of differentiation, is given by

$$\frac{d\vec{\mathbf{l}}}{dt} = \frac{d(\vec{\mathbf{r}} \times \vec{\mathbf{p}})}{dt} = \frac{d\vec{\mathbf{r}}}{dt} \times m\vec{\mathbf{v}} + \vec{\mathbf{r}} \times \frac{d\vec{\mathbf{p}}}{dt}.$$

The first term on the right side vanishes, since $d\vec{\mathbf{r}}/dt = \vec{\mathbf{v}}$ and the vector product of parallel vectors equals zero. According to Newton's second law, the change of momentum $d\vec{\mathbf{p}}/dt$ can be substituted for the force $\vec{\mathbf{F}}$.

▲ The variation of the angular momentum with time equals the torque of the applied force (see **Fig. 2.14**).

change of angular momentum = torque			$\mathbf{ML^2T^{-2}}$
$\dfrac{d\vec{\mathbf{l}}}{dt} = \vec{\mathbf{r}} \times \vec{\mathbf{F}} = \vec{\tau}$	Symbol	Unit	Quantity
	$\vec{\mathbf{l}}$	kg m^2/s	angular momentum
	$\vec{\mathbf{r}}$	m	position vector
	$\vec{\mathbf{F}}$	N	acting force
	$\vec{\tau}$	Nm	produced torque

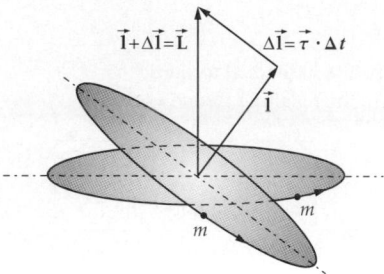

Figure 2.14: Torque $\vec{\tau}$ and change of angular momentum $\Delta\vec{\mathbf{l}}$. The magnitude of the angular velocity and the orientation of the rotation axis are changed.

➤ If the torque points parallel or antiparallel to the orbital angular momentum, only the magnitude of the orbital angular momentum changes while the orientation of the orbital plane remains fixed in space. If the torque is not parallel or antiparallel to the orbital angular momentum, the orientation of the angular velocity vector also changes, i.e., the orbital plane is tilted.

➤ If the point mass moves under the action of a **central force** that points along the position vector $\pm\vec{r}$, the **torque** vanishes. The **orbital angular momentum** is then a **conserved quantity of motion**, both in magnitude and orientation.

➤ The gravitational force is a central force. Kepler's second law (area rule) for planetary motion of elliptic trajectories around the Sun follows from the conservation of orbital angular momentum.

2.1.5   The fundamental law of rotational dynamics

For the rotational motion of a body with orbital angular momentum $\vec{\mathbf{L}} = J \cdot \vec{\omega}$ and a moment of inertia J that is constant in time, $dJ/dt = 0$, one has

$$\frac{d\vec{\mathbf{L}}}{dt} = J\,\frac{d\vec{\omega}}{dt} = J\,\vec{\alpha} = \vec{\tau}.$$

▲ The angular acceleration $\vec{\alpha} = \dot{\vec{\omega}}$ is proportional to the torque $\vec{\tau}$ of the force. The moment of inertia J enters as proportionality factor.

1. The fundamental law of rotational dynamics

This law governs all rotational motion:

torque = moment of inertia · angular acceleration			$\mathbf{ML^2T^{-2}}$
	Symbol	Unit	Quantity
$\vec{\tau} = J \cdot \dfrac{d\vec{\mathbf{L}}}{dt} = J \cdot \vec{\alpha}$	$\vec{\tau}$	N m	torque
	J	kg m^2	moment of inertia
	$\vec{\alpha}$	rad/s^2	angular acceleration
	$\vec{\mathbf{L}}$	kg m^2/s	angular momentum

By integration, one obtains

$$\int_{t_1}^{t_2} \vec{\tau}\,dt = \Delta\vec{\mathbf{L}}.$$

▲ The **impulse of a torque** (time integral over the torque) equals the change of angular momentum.

2. *Comparison of translational and rotational motion*

translation		rotation	
position	$\vec{r}$	angle	φ
path element	$d\vec{r}$	angle element	$d\varphi$
velocity	$\vec{v} = \dfrac{d\vec{r}}{dt}$	angular velocity	$\vec{\omega} = \dfrac{d\varphi}{dt}$
acceleration	$\vec{a} = \dfrac{d\vec{v}}{dt} = \dfrac{d^2\vec{r}}{dt^2}$	angular acceleration	$\vec{\alpha} = \dfrac{d\vec{\omega}}{dt} = \dfrac{d^2\vec{e}_\omega}{dt^2}$
mass	m	moment of inertia	$J = m\,r^2$
momentum	$\vec{p} = m\,\vec{v}$	angular momentum	$\vec{L} = J\,\vec{\omega}$
force	$\vec{F} = m\,\vec{a} = \dot{\vec{p}}$	torque	$\vec{\tau} = J\,\vec{\alpha} = \dot{\vec{L}}$
kinetic energy	$E_{\text{kin}} = \dfrac{1}{2}mv^2$	kinetic energy	$E_{\text{kin}} = \dfrac{1}{2}J\omega^2$
work	$dW = \vec{F}\,d\vec{r}$	work	$dW = \vec{\tau}\,d\vec{e}_\omega$
power	$P = \vec{F}\,\vec{v}$	power	$P = \vec{\tau}\,\vec{\omega}$

uniform motion	
$a = 0$ $v = v_0 = \text{const.}$ $x = v_0 t + x_0$	$\dot{\omega} = 0$ $\omega = \omega_0 = \text{const.}$ $\varphi = \omega_0 t + \varphi_0$
uniformly accelerated motion	
$a = a_0 = \text{const.}$ $v = a_0 t + v_0$ $x = \dfrac{a_0}{2}t^2 + v_0 t + x_0$	$\dot{\omega} = \dot{\omega}_0 = \text{const}$ $\omega = \dot{\omega}_0 t + \omega_0$ $\varphi = \dfrac{\dot{\omega}_0}{2}t^2 + \omega_0 t + \varphi_0$

2.2 Forces

Several kinds of forces are characterized below.

2.2.1 Weight

1. *Definition of weight*

Weight, the attractive force (gravitation) of Earth that affects all bodies. It is proportional to the mass of the body.

The proportionality constant is the **acceleration** of gravity g that, at a given point, is identical for all bodies independent of their mass.

weight = mass · acceleration of gravity			$\mathrm{MLT^{-2}}$
	Symbol	Unit	Quantity
$F_G = mg$	F_G	N	weight
	m	kg	mass of body
	g	m/s^2	acceleration of gravity

➤ The mass of bodies applicable here is the **gravitational mass**. It always equals the inertial mass. This statement has been experimentally demonstrated and serves as a postulate of the general theory of relativity.

The acceleration acting on a body of mass m in a gravitational field, according to the fundamental law of dynamics, is

$$a = \frac{F_G}{m} = g.$$

■ In a vacuum, a steel ball and a feather fall with equal velocity. The difference in their velocities in air is caused by the greater air resistance of the feather as compared to its smaller weight.

2. Acceleration of gravity

■ A body of mass 1 kg at Earth's surface experiences a force $F_G = 1\ \mathrm{kg} \cdot 9.81\ \mathrm{m/s^2} = 9.81$ N. Its acceleration is

$$a = \frac{F_G}{m} = \frac{9.81\ \mathrm{N}}{1\ \mathrm{kg}} = 9.81\ \mathrm{m/s^2}.$$

A body with twice the mass (2 kg) experiences twice the force 19.62 N, its acceleration due to gravity, however, is again 19.62 N/2 kg = 9.81 m/s^2.

➤ The acceleration of gravity is position-dependent. It depends on the height above sea level, on the latitude (due to the rotation and oblateness of Earth), and to a small extent on density fluctuations of Earth's crust. The **standard acceleration of gravity** is 9.80665 m/s^2.

The acceleration of gravity is different on every planet in the solar system.

▲ At a given position, all bodies experience the same gravitational acceleration.

2.2.2 Spring torsion forces

1. Hooke's law

Because of its **elasticity**, a stretched spring exerts a restoring force that, according to **Hooke's law**, is proportional to its elongation. The proportionality constant is called **spring constant** (**Fig. 2.15**).

Hooke's law: force ~ elongation			$\mathrm{MLT^{-2}}$
	Symbol	Unit	Quantity
$F_x = -k\,x$	F_x	N	spring force
	k	N/m	spring constant
	x	m	elongation from rest position

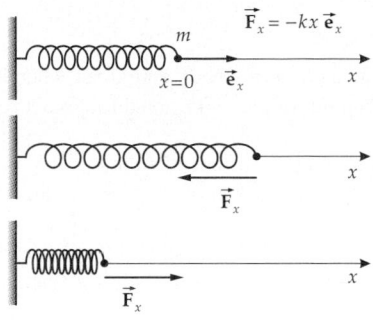

Figure 2.15: Spring forces

➤ Hooke's law holds only approximately, and only for small elongations from the rest position. For larger elongations nonlinearities arise, i.e., the force no longer increases linearly with the elongation; if it is extended enough, the spring breaks.

■ A weight of mass m = 1 kg hangs on a spring with spring constant k = 100 N/m. The elongation d of the spring is

$$d = \frac{|F_x|}{k} = \frac{mg}{k} = \frac{1 \text{ kg} \cdot 9.81 \text{ m/s}^2}{100 \text{ N/m}} = 0.0981 \text{ m} = 9.81 \text{ cm}.$$

2. Properties of springs

The following types of springs exist:
- **tension springs** produce a compressional force under elongation,
- **compression springs** produce an expansion force under compression,
- **torsion springs** oppose an external torque by the production of a counter torque.

If several springs are connected, the set can be replaced by a single **equivalent spring** with a **resulting spring constant**. Any network of springs can be decomposed into combinations of parallel and serial connections of springs:

Parallel connections of springs: the individual spring constants are added (**Fig. 2.16**),

$$k_{\text{res}} = k_1 + k_2 + \cdots.$$

Serial connections of springs: the reciprocal values of the individual springs are added (**Fig. 2.17**),

$$\frac{1}{k_{\text{res}}} = \frac{1}{k_1} + \frac{1}{k_2} + \cdots.$$

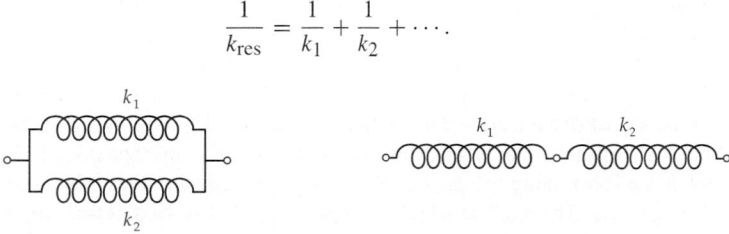

Figure 2.16: Parallel connection of springs.   Figure 2.17: Serial connection of springs.

M Springs are used for the measurement of forces (**dynamometer**). The spring is fixed at one end, and the force to be measured is applied at the other end. The elongation or compression of the spring is then proportional to the acting force. The calibration can be achieved by means of a body of known mass, and hence known weight.

2.2.3   Frictional forces

Frictional force, force which acts to oppose the motion of a body and arises when it is in contact with another body, or moving through a fluid (or gas). Frictional forces act parallel to the plane of contact.

Friction between solids, friction arising at the contact surface of solids.

➤   The friction between solids is approximately independent of the extent of the contact surface and the relative velocity.

➤   Frictional forces in viscous fluids or gases depend on the velocity of the moving body.

There are three types of friction between solids: **static friction**, **sliding friction** and **rolling friction**.

2.2.3.1   Static friction

1. Definition of static friction

Static friction, **rest friction**, a force caused by the coarseness of the contact surfaces. It appears as a resistance to motion. Static friction occurs only if the body is at rest with respect to the contact surface. If a force acts on the body, motion begins only when this force exceeds the static-frictional force F_H. The static-frictional force is proportional to the **normal force** that presses one body against the other:

$$F_H \leq F_{H,\max} = \mu_0 F_N.$$

The proportionality constant μ_0 that specifies the maximum value of the static-frictional force is called **coefficient of static friction** (**Fig. 2.18**).

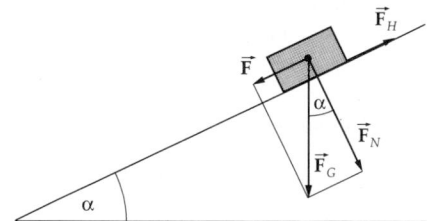

Figure 2.18: Static friction.

2. Properties of static friction

➤   Static friction is independent of the area of the contact surface.

➤   The coefficient of static friction depends on the surface material of the two bodies, and on their surface structure (coarseness) (see **Tab. 7.3/3**).

| M | The coefficient of static friction for the materials can be determined by setting a body of mass m of one material onto an inclined plane of the other material and increasing the angle of inclination α until the body just starts moving. The body begins to move when the force along the slope $F = mg \sin \alpha$ exceeds the force of static friction F_H, $F = F_{H,\max}$. The angle at which it happens is called **static-friction angle** φ. For the static-friction angle,

$$F = mg \sin \varphi = F_{H,\max} = \mu_0 F_N = \mu_0 mg \cos \varphi.$$

The coefficient of static friction is

$$\mu_0 = \tan \varphi.$$

▲   The coefficient of static friction μ_0 equals the tangent of the static-friction angle φ.

2.2.3.2   Sliding friction

Sliding friction arises when a body moves on the contact surface. The sliding-frictional force points opposite to the velocity of the body, its magnitude is proportional to the magnitude of the normal force (**Fig. 2.19**).

sliding-frictional force			MLT^{-2}
	Symbol	Unit	Quantity
$F_{GR} = \mu F_N$	F_{GR}	N	sliding-frictional force
	μ	1	coefficient of sliding friction
	F_N	N	normal force

The proportionality factor μ is called **coefficient of sliding friction** (see **Tab. 7.3/2**).

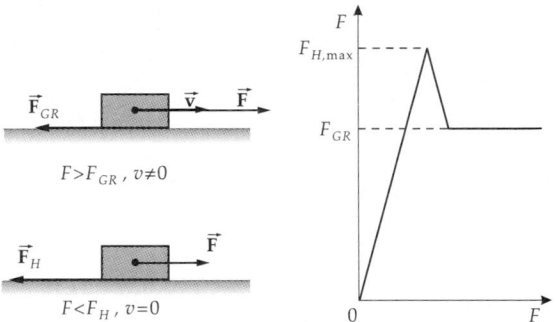

Figure 2.19: Solid friction. Static and sliding friction.

The coefficient of sliding friction is in general lower than the maximum value of the coefficient of static friction (see **Tab. 7.3/3**).

■   A metal block of 10 kg slides on a wood surface. The coefficient of static friction for metal on wood is $\mu_0 \approx 0.5$, the coefficient of sliding friction is $\mu \approx 0.4$. To put the resting block into motion, a force that exceeds the static friction must be applied:

$$F_{H,max} = \mu_0 F_N = \mu_0 mg = 0.5 \cdot 10 \text{ kg} \cdot 9.81 \text{ m/s}^2 = 49 \text{ N}.$$

As soon as the metal block moves, only the sliding friction acts:

$$F_{GR} = \mu F_N = \mu mg = 0.4 \cdot 10 \text{ kg} \cdot 9.81 \text{ m/s}^2 = 39 \text{ N}.$$

2.2.3.3   Rolling friction

1. Definition of rolling friction

Rolling friction arises when a body (e.g., a wheel) on a plane does not slide, but rolls. In a **rolling motion**, any point on the circumference line of the wheel (radius R) moves just as fast relative to the wheel center as the wheel moves forward as a whole (**Fig. 2.20**):

$$R\omega = v.$$

The velocity of a point on the circumference at the contact point with the floor equals zero because the circumference velocity $R\omega$ originating in the circular motion is just compensated by the linear motion v of the wheel. Rolling friction occurs because of the deformation of the wheel and support. A frictional force $\vec{F}_R$ that acts at the wheel's circumference and points opposite to the compressional force acting on the wheel axis causes the support force to act not at the point P_1 (the instantaneous rotation axis), but rather at the point P_2. The support force is the resultant of the normal force $\vec{F}_N$ and the force $\vec{F}_R$. The wheel rolls uniformly if the sum of normal force, support force, and compressional force vanishes (**Fig. 2.21**). The torque of the compressional force with respect to the instantaneous rotation axis through the point P_1 is

$$\tau = R \cdot F_R, \quad R: \text{ radius of the wheel.}$$

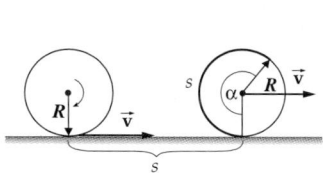

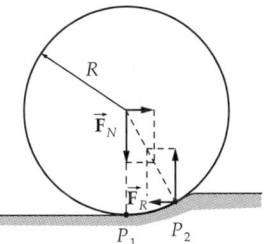

Figure 2.20: Rolling motion. The distance s traversed by the axis equals the length of the unwound circumference: $s = R\alpha$.

Figure 2.21: Rolling friction.

2. Coefficient of rolling friction,

f, expresses the proportionality between the support force F_N and the torque τ created by the frictional force:

$$\tau = f \cdot F_N.$$

It follows that

$$F_R = \frac{f}{R} F_N.$$

The rolling friction depends on the load, the wheel diameter, and the material of both wheel and support.

➤   The rolling-frictional force decreases as the wheel diameter increases.

➤   The coefficient of rolling friction has the dimension of a length. It is velocity-dependent. For steel on steel, it is between 0.01 cm at 4 m/s and 0.05 cm at 30 m/s (see **Tab. 7.3/1**).

2.2.3.4   Rope friction

1. Definition of rope friction

Rope friction, frictional force between rope (also belt or tape) and roller (pulley). During the lifting, the force F_2 compensates both for the load F_1 and for the frictional force $F_{GR} = F_2 - F_1$ (**Fig. 2.22**).

rope friction			MLT^{-2}
	Symbol	Unit	Quantity
$$F_{\mathrm{GR}} = F_1\left(e^{\mu_0\alpha} - 1\right)$$ $$= F_2\left(1 - e^{-\mu_0\alpha}\right)$$	F_{GR}	N	sliding frictional force
	F_1	N	load
	F_2	N	compressional force
	e	1	Euler number = 2.7183...
	μ_0	1	coefficient of static friction
	α	rad	angle made by the rope

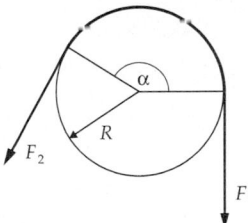

Figure 2.22: Lifting of a load F_1 by the force F_2. The rope friction depends on the angle α.

➤ When pulling a load up, F_1 is the load, F_2 is the lifting force. When lowering a load, F_2 is the load, F_1 is the lowering force:

$$F_{\mathrm{lift}} = e^{\mu\alpha} F_{\mathrm{load}},$$

$$F_{\mathrm{lower}} = e^{-\mu\alpha} F_{\mathrm{load}}.$$

➤ These formulas hold if the cylinder is at rest and the rope moves with uniform velocity, or if the rope is at rest and the cylinder rotates with uniform velocity.

2. Properties of rope friction

➤ In rope friction, the coefficient of sliding friction depends on the velocity of the rope and on the radius of the pulley radius.

▲ The rope is at rest if the compressional force is too small to lift the load, or too large to let it down:

$$F_{\mathrm{load}}\, e^{-\mu\alpha} < F < F_{\mathrm{load}}e^{\mu\alpha}.$$

▲ If the rope does not move when lifting a load F_1 by the force F_2, then:

$$F_2/F_1 \leq e^{\mu_0\alpha},$$

where μ_0 is the coefficient of static friction. For technical applications (traction belt) in which no sliding between rope and support occurs, one has correspondingly to apply the coefficient of static friction.

➤ For friction coefficients see **Tab. 7.3/2 to Tab. 7.3/3.**

2.3 Inertial forces in rotating reference systems

Inertial forces arise both in translational and rotational motions. The rotation is described as the **circular motion** of a point mass, and then the resulting inertial forces are determined. Circular motion is not a uniform straight-line motion, but involves acceleration, and hence must result from a force. The acceleration does not necessarily lead to an increase of velocity, but rather to a change of its direction.

Equation of motion of a point mass with mass m in a **non-inertial coordinate system** that moves with acceleration $\vec{a}_0$ and rotates with the angular velocity $\vec{\omega}$:

$$m\ddot{\vec{r}} = \vec{F} - m\vec{a}_0 - m\vec{\omega} \times (\vec{\omega} \times \vec{r}) - m\dot{\vec{\omega}} \times \vec{r} - 2m\vec{\omega} \times \dot{\vec{r}}.$$

Centrifugal force: $\vec{F}_c = -m\vec{\omega} \times (\vec{\omega} \times \vec{r}).$
Coriolis force: $\vec{F}_C = -2m\vec{\omega} \times \dot{\vec{r}}.$

2.3.1   Centripetal and centrifugal forces

The acceleration $\vec{a}$ of a point mass at the position $\vec{r}$ that moves with angular velocity $\vec{\omega}$ on a circular orbit is

$$\vec{a} = \frac{d\vec{v}}{dt} = \frac{d}{dt}(\vec{\omega} \times \vec{r}).$$

Differentiation of the vector product according to the chain rule yields

$$\vec{a} = \frac{d\vec{\omega}}{dt} \times \vec{r} + \vec{\omega} \times \frac{d\vec{r}}{dt}.$$

acceleration under rotation			$\mathbf{LT^{-2}}$
	Symbol	Unit	Quantity
$\vec{a} = \vec{\alpha} \times \vec{r} + \vec{\omega} \times (\vec{\omega} \times \vec{r})$	$\vec{a}$	m/s^2	acceleration
	$\vec{\alpha}$	rad/s^2	angular acceleration
	$\vec{r}$	m	distance from center
	$\vec{\omega}$	rad/s	angular velocity

The first term describes the contribution of the angular acceleration $\vec{\alpha}$ to the acceleration. The second term represents the central acceleration created by the force that keeps the body on its circular path.

1. Centripetal force

Centripetal acceleration, a_r, the radial acceleration in the motion of a point mass on a circular path. It points towards the center of the circle and has magnitude

$$a_r = |\vec{\omega} \times (\vec{\omega} \times \vec{r})| = \omega^2 \cdot r \cdot \sin\vartheta,$$

where ϑ specifies the angle between the position vector and the rotation axis. If $\vec{r}$ is perpendicular to the rotation axis,

$$a_r = \omega^2 \cdot r,$$

with r the **perpendicular distance** of the body from the rotation axis.

According to Newton's second law, the central acceleration is caused by a force:

Centripetal force, $\vec{F}_r$, force that causes the central acceleration, and hence keeps the body on the circular path:

centripetal force			$\mathbf{MLT^{-2}}$
	Symbol	Unit	Quantity
$F_r = m \cdot a_r = m \cdot \omega^2 \cdot r$ $= m \cdot \dfrac{v^2}{r}$	F_r	N	centripetal force
	m	kg	mass
	a_r	m/s^2	central acceleration
	ω	rad/s	angular velocity
	r	m	distance from rotation axis
	v	m/s	velocity

In vector notation, the centripetal force is given by

$$\vec{\mathbf{F}}_r = -F_r\,\vec{\mathbf{e}}_r = m\,\vec{\omega}\times(\vec{\omega}\times\vec{\mathbf{r}}).$$

The centripetal force points towards the center of the circle. Because of inertia, an observer rotating with the mass feels, however, a force that points outwards.

2. Centrifugal force

Centrifugal force, $\vec{\mathbf{F}}_c$, force felt by an observer moving on a circular path. It points from the center outwards, and its magnitude equals that of the centripetal force (**Fig. 2.23**):

$$\vec{\mathbf{F}}_c = F_r\,\vec{\mathbf{e}}_r = -m\,\vec{\omega}\times(\vec{\omega}\times\vec{\mathbf{r}}).$$

➤ The centrifugal force is an inertial force, i.e., it arises only in the accelerated reference system and is felt only by an observer in such a system.

■ A car with mass $m = 800$ kg moving on a curve with radius of curvature $r = 10$ m with the speed $v = 30$ km/h experiences a centrifugal force

$$F_c = \frac{mv^2}{r} \approx 5.5 \text{ kN}.$$

This force can be offset by banking the curved road. To take a curve without any frictional force, one needs a banked curve with a slope α of

$$\tan\alpha = \frac{F_c}{F_G} = \frac{v^2/r}{g} \approx 0.7 \Longrightarrow \alpha \approx 35°$$

(F_G gravitational force, g acceleration of gravity).

M **Centrifugal-force governor**, two pendulums mounted on an axis. When the axis rotates, the pendulums are pushed outwards by the centrifugal force. The force can be used for **controling the rate of rotation**.

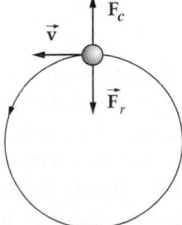

Figure 2.23: Centrifugal force. As viewed in the rotating coordinate system, circular motion is a balance between the centrifugal force $\vec{\mathbf{F}}_c$ and the centripetal force $\vec{\mathbf{F}}_r$ that keeps the body on the circular path.

2.3.2 Coriolis force

1. Definition of Coriolis force

Coriolis force $\vec{F}_C$, force felt by an observer who moves on a rotating coordinate system inwards or outwards from the axis. It acts perpendicular to the direction of motion of the observer and perpendicular to the rotation axis. The physical origin of the Coriolis force lies in the higher orbital velocity of the points that are farther from the rotation axis.

In vector notation, the Coriolis force can be written:

Coriolis force			MLT^{-2}
	Symbol	Unit	Quantity
$\vec{F}_C = -2m\vec{\omega} \times \vec{v}$	$\vec{F}_C$	N	Coriolis force
	m	kg	mass
	$\vec{\omega}$	rad/s	angular velocity of rotation
	$\vec{v}$	m/s	velocity of mass in the rotating system

2. Body on rotating bar

When a mass moves away from the center of rotation, its moment of inertia $J = mr^2$ increases continuously. This increase produces torque even if the angular velocity ω is constant (**Fig. 2.24**):

$$M = \frac{\mathrm{d}L}{\mathrm{d}t} = \omega \cdot \frac{\mathrm{d}J}{\mathrm{d}t},$$

$$= \omega \cdot m\frac{\mathrm{d}}{\mathrm{d}t}(r^2) = 2m \cdot \omega \cdot r \cdot v_r.$$

This torque must be supplied by the driving unit to maintain a constant rate of rotation. Hence, the mass experiences a force of magnitude

$$F_C = 2m \cdot \omega \cdot v_r.$$

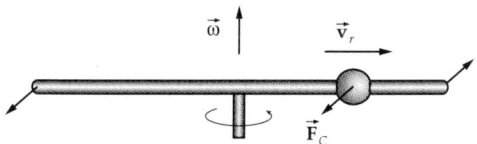

Figure 2.24: Orientation of the Coriolis force on a body that moves outwards on a rotating bar.

3. Coriolis force: examples

The trajectory of a body that moves uniformly in an inertial system appears as curved when projected onto a rotating system, for example a rotating disk (**Fig. 2.25**).

■   If a body on Earth's surface moves north, then, because of the Earth's rotation $\vec{\omega}$, it experiences a Coriolis force that drives it east on the Northern Hemisphere, west on the Southern Hemisphere. The obvious reason for the contrary deflections is: on the Northern Hemisphere, the body moving north runs into regions with continuously decreasing circumferential velocity of Earth, thus gets ahead of Earth's rotation due to its inertia. On the Southern Hemisphere, on the contrary, when moving north, the body runs into regions with continuously increasing circumferential velocity, and thus falls behind Earth's rotation (**Fig. 2.26**).

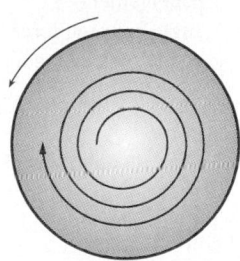

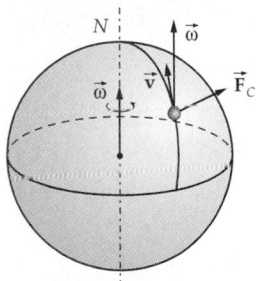

Figure 2.25: The trajectory of a body that moves uniformly in an inertial system appears as a spiral if seen from a rotating disk.

Figure 2.26: Coriolis force $\vec{F}_C$ on the Earth surface. A body moving on the Northern Hemisphere with velocity $\vec{v}$ north is deflected east (on the Southern Hemisphere, west). $\vec{\omega}$: angular velocity of Earth's rotation.

2.4   Work and energy

The concepts work and energy are fundamental for the description of physical processes. Energy is a conserved quantity. It occurs in various forms that can be converted into each other.

2.4.1   Work

1. Definition of work

Work, a force $\vec{F}$ that displaces a body along the path element $d\vec{r}$ performs work:

$$dW = \vec{F}(\vec{r}, t) \cdot d\vec{r} = F(\vec{r}, t) \cos \alpha \, dr,$$

where α is the angle between the force and the path element (**Fig. 2.27**).

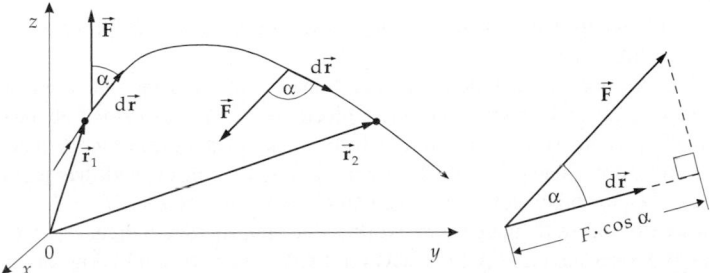

Figure 2.27: Work along the path from $\vec{r}_1$ to $\vec{r}_2$.

work = force · path			$\mathbf{ML^2T^{-2}}$
	Symbol	Unit	Quantity
$dW = \vec{\mathbf{F}} \cdot d\vec{\mathbf{r}}$ $= |\vec{\mathbf{F}}|\,|d\vec{\mathbf{r}}|\,\cos\alpha$	dW $\vec{\mathbf{F}}$ $d\vec{\mathbf{r}}$ α	$J = Nm$ N m rad	work force path element angle between force and path element

2. Unit of work

Joule, the SI unit of work: 1 joule is the work performed when a body is displaced by a force of 1 N over a distance of 1 m.

$$[W] = \text{joule} = J = N \cdot m = \frac{kg\, m^2}{s^2}$$

For additional units, see **Tab. 33.0/3 and 33.0/5**. Non-SI units:

$$1 \text{ kilopondmeter (kpm)} = 9.80665 \text{ J}$$

$$1 \text{ erg} = 10^{-7} \text{ J}$$

$$1 \text{ electron volt (eV)} = 1.602 \cdot 10^{-19} \text{ J}$$

3. Properties of work

➤ The sign of the work depends on the relative direction of the motion and the force.

$dW > 0$:   The displacement has a component along the direction of force ($\cos\alpha > 0$).

$dW < 0$:   The displacement has a component opposite to the direction of force ($\cos\alpha < 0$).

■ A body is displaced by a force $F = 10$ N by $s = 20$ cm along the direction of force. The work performed is in this case

$$W = Fs = 10 \text{ N} \cdot 0.2 \text{ m} = 2 \text{ J}.$$

If the body is displaced twice that distance, $s = 40$ cm, along the direction of force, twice the amount of work is performed:

$$W = Fs = 10 \text{ N} \cdot 0.4 \text{ m} = 4 \text{ J}.$$

The work would also be twice as large if twice the amount of the force acts along the original direction.

➤ If the force does not act along the direction of motion of the body, only the force component along the motion (i.e., the projection of the force vector onto the direction of motion) contributes to the work. A force acting perpendicular to the path element performs no mechanical work ($\cos\alpha = 0$). The amount of work has its maximum value when the body is displaced parallel to the force ($\cos\alpha = 1$).

➤ **Constraint forces** do not perform work, since they are perpendicular to the path.

■ A body moves on a rail. A force acts on it with an angle of 45°. The component of the force along the direction of motion is $F\cos 45° = \dfrac{1}{\sqrt{2}} F$.

4. Work as integral

The total work performed along the path from $\vec{r}_1$ to $\vec{r}_2$ is the path integral over the force.

work = integral of force along the path			ML^2T^{-2}
	Symbol	Unit	Quantity
$W = \displaystyle\int_{\vec{r}_1}^{\vec{r}_2} \vec{F}(\vec{r}) \cdot d\vec{r}$	W	$J = Nm$	work
	$\vec{F}(\vec{r})$	N	force vector at position $\vec{r}$
	$\vec{r}$	m	position vector
	$\vec{r}_1$	m	initial position
	$\vec{r}_2$	m	final position

Here $\vec{r}$ runs over all points on the path from $\vec{r}_1$ to $\vec{r}_2$.

➤ For a one-dimensional motion, the work is obtained as the area below the curve $F(x)$ (**Fig. 2.28**),

$$W = \int_{x_1}^{x_2} F(x)\, dx.$$

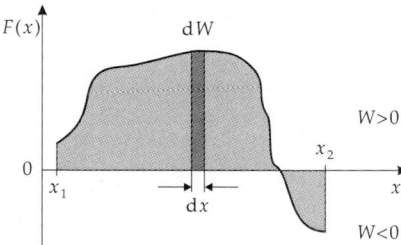

Figure 2.28: One-dimensional motion. Work is the area below the curve $F(x)$.

2.4.2   Energy

1. Definition and properties of energy

Energy, a quantity characterizing the state (position, state of motion, temperature, deformation, etc.) of a body. The energy increases when work is performed on the body, it decreases when work is performed by the body. The work thereby causes a change of the state of the body (displacement, acceleration, increase of temperature, change of shape, etc.).

▲ **Energy** measures how much work was put into the body, or was performed by it.
    Energy has the same SI unit as work: the **joule**.

➤ Energy is a quantity that depends on the choice of the system of reference. It can be specified only with respect to the reference system.

■ If a locomotive pulls a train up a mountain, then it increases its potential energy. If the train rolls down again, this energy can be released as heat of friction (by braking), or converted into energy of motion (kinetic energy).
    There are various forms of energy that can be converted into each other.

■ Electrical or chemical energy is converted by the locomotive into kinetic and potential energy of the train, which can in turn be converted to heat by braking. Heat is also a form of energy.

2. Energy conservation

Energy cannot be destroyed in physical processes, but various kinds of energy can be converted into each other.

Law of energy conservation: In a closed system, the total energy remains constant in all physical processes. Energy can only be converted into different forms, or be exchanged between partial systems.			ML^2T^{-2}
	Symbol	Unit	Quantity
$\sum E_i = E_{\text{pot}} + E_{\text{kin}} + \cdots = \text{const.}$	E_i	J	energy of kind i
	E_{pot}	J	potential energy
	E_{kin}	J	kinetic energy
	$\ldots$	J	other kinds of energy

3. Energy as state parameter

Energy is a property of a definite state of a system (e.g., of the position and the velocity in a gravitational field). The **energy difference** between two states must be put into the system if it changes from a state of lower energy to a state of higher energy.

▲   The **zero of energy** can be fixed arbitrarily, since only energy differences affect physical processes. One can thus add an arbitrary constant energy to the energy of every system without affecting the physical content.

Besides the mechanical energy forms, energy can be stored in electromagnetic fields. **Heat** is also an energy form. Energy of motion can be transformed into heat by friction. **Heat engines** convert heat into mechanical energy (**steam engine**, see p. 709).

2.4.3 Kinetic energy

1. Definition of kinetic energy

Work done during acceleration, the work performed on accelerating a mass m with the acceleration $\vec{a}$ against the inertial force $\vec{F}_T = -m\vec{a}$, $dW'_B = -m\vec{a}\,d\vec{r}$.

Kinetic energy, **energy of motion**, the energy of motion supplied to the body by the work done during acceleration. It can be released, e.g., by braking as heat of friction:

$$dW_B = -dW'_B = -\vec{F}_T\,d\vec{r} = m\,\frac{d\vec{v}}{dt}\,\vec{v}\,dt = m\,v\,dv = d\left(\frac{m}{2}v^2\right).$$

work done during acceleration			ML^2T^{-2}
	Symbol	Unit	Quantity
$dW_B = m\vec{a}\cdot d\vec{r}$	W_B	J	work done during acceleration
	m	kg	mass of body
$W_B = \frac{1}{2}m\left(v^2 - v_0^2\right)$	$\vec{a}$	m/s^2	acceleration
	$d\vec{r}$	m	path element
	v	m/s	final velocity
	v_0	m/s	initial velocity

The work done during acceleration depends on, besides the mass m, only the initial velocity v_0 and the final velocity v.

Kinetic energy of a point mass of mass m, the quantity

$$E_{\text{kin}} = \frac{1}{2}mv^2.$$

It specifies the work done during acceleration that was needed to accelerate the mass point from rest ($v_0 = 0$) to its instantaneous velocity v.

2. Kinetic energy and reference system

➤ The kinetic energy depends on the state of motion of the body, and thus on the reference system. This expresses the arbitrariness of the choice of the zero of energy. A body with the velocity $\vec{v}$ has in one reference system the kinetic energy

$$E_{\text{kin}} = \frac{1}{2}mv^2.$$

In another reference system that moves uniformly with the velocity $\vec{v}_0$ relative to the first one, its kinetic energy is

$$E'_{\text{kin}} = \frac{1}{2}m(v')^2 = \frac{1}{2}m\left(v^2 + 2\vec{v}\vec{v}_0 + v_0^2\right).$$

◼ A body of mass 5 kg at 2 m height above the floor has a potential energy of 98.1 J (see below). If it falls, the potential energy is converted into kinetic energy. When it reaches the floor, the total potential energy is transformed into kinetic energy. Its velocity is then

$$v = \sqrt{\frac{2E_{\text{kin}}}{m}} = \sqrt{\frac{2 \cdot 98.1 \text{ J}}{5 \text{ kg}}} = 6.26 \text{ m/s}.$$

2.4.4 Potential energy

Generally, the energy that depends only on the position of the body, but not on its velocity, is referred to as potential energy.

2.4.4.1 Lifting against the gravitational force

1. Lifting and potential energy

Work done in lifting in the gravitational field, the work performed in lifting a body against the constant gravitational force $F_G = mg$.

work done in lifting			$\mathbf{ML^2T^{-2}}$
	Symbol	Unit	Quantity
$W_H = F_G \Delta h = mg\,\Delta h$	W_H	J	work done in lifting
	F_G	N	gravitational force
	m	kg	mass of lifted body
	g	m/s^2	free acceleration of gravity (9.81 m/s^2)
	Δh	m	height difference

Potential energy, position energy, energy supplied to a body by work done in lifting. It depends on the position of the body (**Fig. 2.29**).

➤   This formula holds only if the gravitational force can be considered constant.

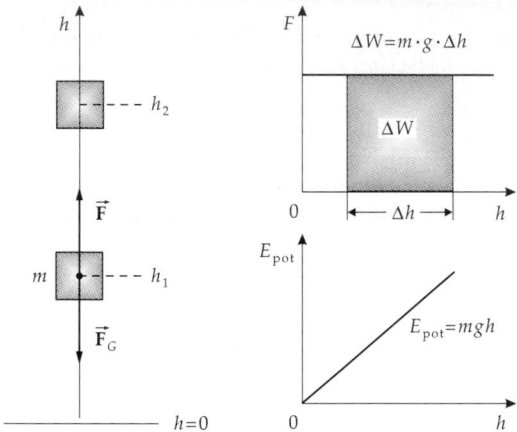

Figure 2.29: Work done in lifting.

2. Properties of potential energy

Potential energy, the quantity

$$E_{\text{pot}} = mgh.$$

The height h is measured from an arbitrarily chosen zero height.

➤   The potential energy depends on the selected zero height, but the difference of the potential energy between two points, and hence the work done in lifting, is independent of the choice of the zero height.

■   A body of 5 kg mass is lifted 2 m. The work done in lifting is

$$W_H = mgh = 5 \text{ kg} \cdot 9.81 \text{ m/s}^2 \cdot 2 \text{ m} = 98.1 \text{ J}.$$

If it is lifted to twice the height, or if the mass is twice as large, twice the amount of work must be done.

➤   A similar type of work is done when an electric charge is moved against the force of an electric field (see p. 447).

2.4.4.2   Work of deformation and tension energy of a spring

1. Work of deformation,

the work performed on deforming a body. The work of deformation occurs when a spring is stretched by the length x against the **restoring force** (spring force) $F_x = -kx$ (**Fig. 2.30**).

The spring force is not constant, unlike the gravitational force (in a restricted height interval), but is instead proportional to the elongation x for small spring displacements.

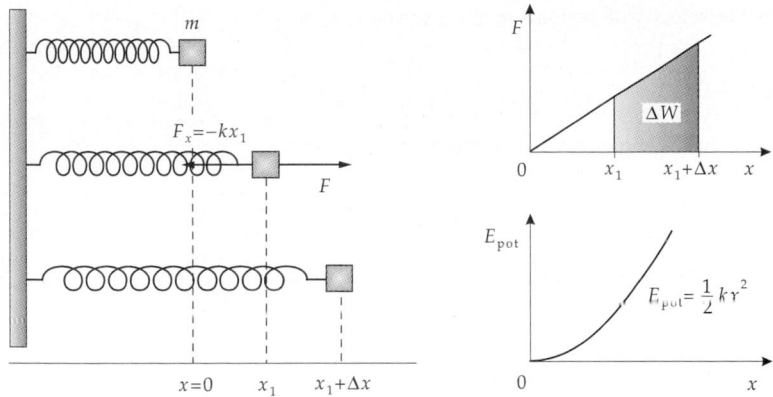

Figure 2.30: Work of deformation and tension energy of a spring.

The work done by an external force $F = -F_x$ in stretching the spring is therefore

$$W_F = \int_{x_{\min}}^{x_{\max}} F \, dx = \int_{x_{\min}}^{x_{\max}} kx \, dx,$$

if the spring is extended from $x_{\min}$ to $x_{\max}$. One finds:

work of deformation			$\mathbf{ML^2T^{-2}}$
$W_F = \dfrac{1}{2}k(x_{\max}^2 - x_{\min}^2)$	Symbol	Unit	Quantity
	W_F	J	work of deformation
	k	N/m	spring constant
	$x_{\min}$	m	initial elongation from rest position
	$x_{\max}$	m	final elongation from rest position

2. Tension energy,

the potential energy of an elastically deformed body, represents the work of deformation stored in the body. It depends on the state of deformation of the body, and is released when the body takes its original form again.

Tension energy E_F of a spring, the quantity

$$E_F = \frac{1}{2}kx^2.$$

It represents the work that was needed to deform the spring from the stress-free state ($x = 0$) up to the elongation x.

➤ Part of the work of deformation is always converted to heat by friction. Hence, the sum of kinetic and potential energy is only approximately conserved; the vibration is damped.

3. Example: Vibration of a spring

In the vibration of a spring, kinetic and potential energy are converted into each other during each cycle of the motion. When friction is neglected, the total energy E is

$$E = E_{\text{kin}} + E_{\text{pot}} = \frac{1}{2}mv^2 + \frac{1}{2}kx^2 = \text{const.}$$

Hence, the velocity of the mass m at a given elongation x is

$$v = \sqrt{\frac{2E}{m} - \frac{k}{m}x^2}.$$

The maximum elongation x_{max} is reached when $v = 0$:

$$x_{max} = \sqrt{\frac{2E}{k}}.$$

At the maximum elongation, the total energy is stored as potential energy. For $x = 0$, however, the total energy is kinetic energy:

$$E = \frac{1}{2}mv_{max}^2,$$

v_{max} the velocity at $x = 0$.

2.4.5   Frictional work

Frictional work, the work performed against the frictional force. The work supplied is transformed into heat.

➤   The energy converted into heat by frictional work cannot be completely converted back into mechanical energy by a heat engine.

For **sliding friction**, the frictional force F_R is approximately constant and proportional to the normal force (support force) of the body. It acts opposite to the direction of motion. For similar surfaces of the moving body, the friction force does not depend on the area of the support surface.

sliding-frictional work			$\mathbf{ML^2T^{-2}}$
	Symbol	Unit	Quantity
$dW_R = F_R\,dx$	dW_R	J	frictional work
	F_R	N	sliding-frictional force
$= \mu F_N\,dx$	dx	m	path element
	μ	1	coefficient of sliding friction
	F_N	N	normal force

Sliding friction on dry surfaces is to a first approximation independent of the velocity. For **gas** and **liquid friction**, the frictional force is velocity-dependent (see p. 198).

2.5   Power

Power, P, work done per unit time. It is useful for the characterization of machines.

power = $\dfrac{\text{work}}{\text{time}}$			$\mathbf{ML^2T^{-3}}$
	Symbol	Unit	Quantity
$P = \dfrac{\Delta W}{\Delta t}$	P	W	power
	ΔW	J	completed work
	Δt	s	needed time

Watt, W, SI unit of power.
1 watt is the power of a machine that performs 1 joule of work per second.

$$[P] = \text{watt} = \text{W} = \frac{\text{J}}{\text{s}} = \frac{\text{kg} \cdot \text{m}^2}{\text{s}^3}.$$

For additional units, see **Tab. 33.0/3**.
Non-SI unit:

$$1 \text{ horsepower (HP)} = 735.4988 \text{ W}$$

➤   If the power is time-dependent, the instantaneous power is

$$P = \frac{\text{d}W}{\text{d}t}.$$

■   An engine does work of 600 kJ per minute. Its power is

$$P = \frac{\Delta W}{\Delta t} = \frac{600 \text{ kJ}}{60 \text{ s}} = 10 \text{ kW}.$$

➤   In colloquial language, the term power often connotes the work done. In physics and engineering, however, power denotes the work delivered in a physical system per unit time.

2.5.1   Efficiency

Efficiency, η, the ratio of work released in an energy conversion (**effective power**) to the input work (**nominal power**). Since machines in general do work continuously, the efficiency is usually defined as ratio of output power to input power:

efficiency $= \dfrac{\textbf{useful work}}{\textbf{input work}} = \dfrac{\textbf{output power}}{\textbf{input power}}$			1
$\eta = \dfrac{P_{\text{out}}}{P_{\text{in}}}$	Symbol	Unit	Quantity
	η	1	efficiency
$= \dfrac{P_{\text{in}} - P_{\text{loss}}}{P_{\text{in}}}$	P_{out}	W	output power
	P_{in}	W	input power
$= 1 - \dfrac{P_{\text{loss}}}{P_{\text{in}}}$	P_{loss}	W	lost power

The efficiency has the dimension 1; it is often given as a percentage.
■   The output shaft of a gear unit provides a power of 40 kW. The needed input power on the drive shaft is 50 kW. The efficiency is

$$\eta = \frac{40 \text{ kW}}{50 \text{ kW}} = 0.8 = 80\%.$$

20 % of the input energy is lost as friction heat.

▲ An efficiency of $\eta = 1$ corresponds to a perfectly (loss free) working machine.
▲ Because of energy conservation and the inevitable losses, the efficiency is always less than unity,

$$\eta < 1.$$

Total efficiency of serially connected machines, obtained by multiplication of the individual efficiencies:

$$\eta_{\text{tot}} = \eta_1 \cdot \eta_2 \cdot \cdots .$$

The total efficiency lies therefore between zero and unity; it cannot be larger than the efficiency of any single machine.

2.6   Collision processes

Collisions, short-term interactions between two or more moving bodies that represent a closed system. Collisions are characterized by very large forces of short range. For the description of collisions precise knowledge of the interaction force is not needed; it suffices to calculate the exchange of energy and momentum between the particles.

1. Kinematic relations for two-body collisions

Two-body collisions, the collision of two bodies in which large forces of short range act over a short time interval. During the collision, energy and momentum are transfered between the collision partners; hence, the velocity, direction of motion and internal energy of the bodies may change. Outside of the interaction region, the collision partners move force-free (straight-line uniform motion).

Kinematic relations for two-body collisions:

Collision partners:	A, B
Mass of collision partners:	m_A, m_B
Velocities before collision:	$\vec{v}_A, \vec{v}_B$
Velocities after collision:	$\vec{u}_A, \vec{u}_B$
Momenta before collision:	$\vec{p}_A = m_A \vec{v}_A, \quad \vec{p}_B = m_B \vec{v}_B$
Momenta after collision:	$\vec{p}_A{}' = m_A \vec{u}_A, \quad \vec{p}_B{}' = m_B \vec{u}_B$
Kinetic energy before collision:	$E_{\text{kin}} = \dfrac{m_A}{2} v_A^2 + \dfrac{m_B}{2} v_B^2$
Kinetic energy after collision:	$E'_{\text{kin}} = \dfrac{m_A}{2} u_A^2 + \dfrac{m_B}{2} u_B^2$
Change of internal energy of the collision partners:	ΔW

2. Energy and momentum conservation

Momentum conservation:

$$m_A \vec{v}_A + m_B \vec{v}_B = m_A \vec{u}_A + m_B \vec{u}_B .$$

Energy conservation:

$$E_{\text{kin}} = E'_{\text{kin}} + \Delta W .$$

$\Delta W > 0 \, (E'_{kin} < E_{kin})$:   **Endothermal collision**. Kinetic energy is converted into internal energy of the collision partners (excitation of collision partners).

$\Delta W < 0 \, (E'_{kin} > E_{kin})$:   **Exothermal collision**. Intrinsic energy of the collision partners is converted into kinetic energy (de-excitation of collision partners).

According to conservation or non-conservation of mechanical energy, the collisions are classified as elastic or inelastic, respectively.

3. Elastic collision,

total mechanical energy and total momentum are conserved (**Fig. 2.31**):

$$\Delta W = 0, \quad E_{kin} = E'_{kin}.$$

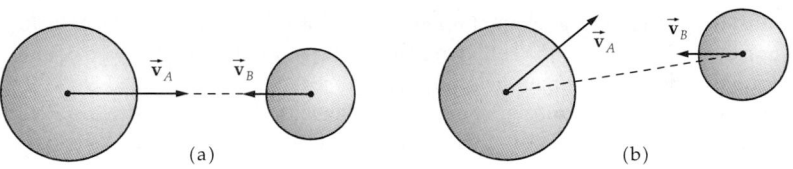

(a)   (b)

Figure 2.31: Elastic collisions. (a): central collision, (b): non-central collision.

■ The collision of two billiard balls is elastic to a very good approximation.
■ In atomic physics, collisions between electrons arise due to the Coulomb interaction. If the emission of electromagnetic waves is neglected, the collisions are elastic.

4. Inelastic collision,

during the collision process, a part of the mechanical energy is converted into other forms of energy (heat, deformation energy). The total energy is conserved only if, not only the kinetic energy of the collision partners is taken into account, but also the change of their intrinsic excitation energy ΔW (**Fig. 2.32**).

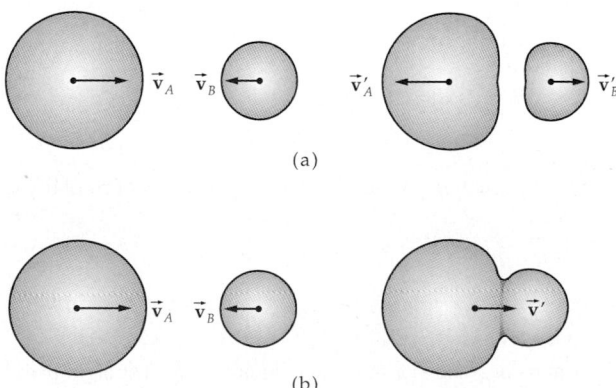

(a)

(b)

Figure 2.32: Inelastic collisions. (a): partly inelastic collision, (b): totally inelastic collision.

■   The bounce of a tennis ball on the floor is connected with an energy loss (friction), and hence is inelastic. The ball possesses a smaller magnitude of velocity just after the bounce than it did just before.

Totally inelastic collision, a collision in which both of the colliding bodies have the same velocity after the collision, i.e., they stick together.

■   Two snowballs collide totally inelastically and stick together. The lost energy is spent in the deformation of the balls.

5. Collision geometry

For motion in several dimensions, one distinguishes collisions by their geometry.

Straight-line collision, the centers of gravity of the colliding bodies move along their connecting lines before and after the collision. One coordinate (distance of centers of gravity) is sufficient to describe the collision.

Non-central collision, the centers of gravity of the colliding bodies move in different directions.

Collision normal, the direction of force transfer during the collision. The collision normal points perpendicular to the **collision plane**, the contact plane of the two bodies.

For rigid bodies one distinguishes collisions by the torque:

Central collision, the collision normal at the moment of collision points parallel to the connecting line of the centers of gravity. There is no torque ($\sin \phi = 0$, ϕ: angle between lever arm and orientation of force) (**Fig. 2.33 (a)**).

Off-center collision, the collision normal does not point along the connecting line of the centers of gravity, hence there is a torque. The bodies begin to rotate (**Fig. 2.33 (b)**).

➤   For point masses, there are only central collisions, since only extended bodies can rotate.

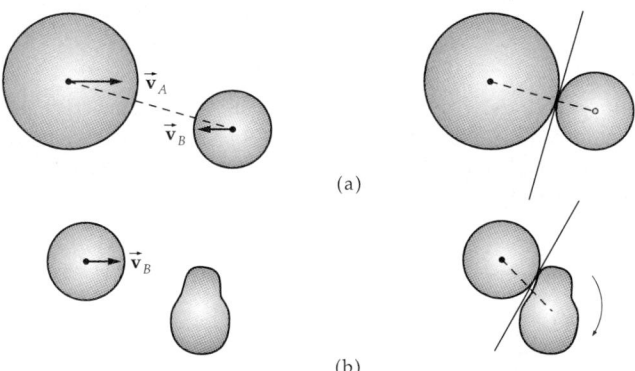

(a)

(b)

Figure 2.33: Central (a) and off-center collision (b) of rigid bodies.

2.6.1 Elastic straight-line central collisions

Let two bodies of mass m_A and m_B move in common along a straight path that coincides with the x-axis. The total energy and the total momentum along the path orientation are conserved quantities (**Fig. 2.34**).

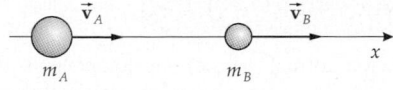

Figure 2.34: Elastic straight-line central collision.

Hence,

$$\frac{1}{2}m_A v_A^2 + \frac{1}{2}m_B v_B^2 = \frac{1}{2}m_A u_A^2 + \frac{1}{2}m_B u_B^2,$$

$$m_A v_A + m_B v_B = m_A u_A + m_B u_B.$$

Sorting the terms by their correspondence to the bodies A and B yields

$$m_A(v_A^2 - u_A^2) = m_B(u_B^2 - v_B^2),$$

$$m_A(v_A + u_A)(v_A - u_A) = m_B(u_B + v_B)(u_B - v_B),$$

$$m_A(v_A - u_A) = m_B(u_B - v_B).$$

Division of the last two equations yields

$$v_A + u_A = u_B + v_B.$$

This equation can be solved for u_B and inserted into the momentum conservation law:

$$u_B = v_A + u_A - v_B.$$

There remains only one unknown quantity in the momentum equation, the velocity u_A. Similarly, one finds the velocity of body B after the collision:

$$u_A = \frac{m_A - m_B}{m_A + m_B} v_A + \frac{2m_B}{m_A + m_B} v_B, \qquad u_B = \frac{2m_A}{m_A + m_B} v_A + \frac{m_B - m_A}{m_A + m_B} v_B.$$

1. Collision of two bodies with equal masses

If both bodies have equal mass, then

$$u_A = v_B, \qquad u_B = v_A.$$

The colliding bodies exchange their velocities.

2. Collision between a heavy and a light body

Let body A be very much heavier than body B: $m_A \gg m_B$. Then approximately

$$u_A \approx v_A, \qquad u_B \approx 2v_A - v_B.$$

The heavy body A remains almost unaffected. The relative velocity of the second body after the collision is just the negative of the relative velocity before the collision:

$$u_B - u_A \approx -(v_B - v_A).$$

Thus, the light body is reflected by the heavy body.

2.6.2   Elastic off-center central collisions

Momentum is exchanged only along the collision normal (y-axis); the components of the momenta perpendicular to the collision normal (x-axis) before and after the collision are equal (**Fig. 2.35**):

$$m_A v_{Ax} = m_A u_{Ax},$$

$$m_B v_{Bx} = m_B u_{Bx}.$$

Momentum conservation along the collision normal:

$$m_A v_{Ay} + m_B v_{By} = m_A u_{Ay} + m_B u_{By}.$$

Energy conservation:

$$\frac{m_A}{2}\left(v_{Ax}^2 + v_{Ay}^2\right) + \frac{m_B}{2}\left(v_{Bx}^2 + v_{By}^2\right) = \frac{m_A}{2}\left(u_{Ax}^2 + u_{Ay}^2\right) + \frac{m_B}{2}\left(u_{Bx}^2 + u_{By}^2\right).$$

Velocity components after the collision:

$$u_{Ax} = v_{Ax}, \quad u_{Bx} = v_{Bx},$$

$$u_{Ay} = \frac{m_A - m_B}{m_A + m_B} v_{Ay} + \frac{2m_B}{m_A + m_B} v_{By},$$

$$u_{By} = \frac{2m_A}{m_A + m_B} v_{Ay} + \frac{m_B - m_A}{m_A + m_B} v_{By}.$$

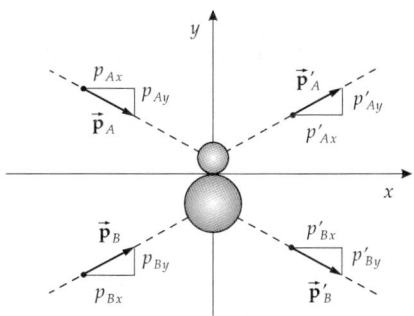

Figure 2.35: Elastic off-center central collision.

2.6.3   Elastic non-central collision with a body at rest

Body A with momentum $\vec{p}_A = m_A \vec{v}_A$ collides with body B at rest ($\vec{p}_B = 0$). After the collision body A moves with the momentum $\vec{p}_A{}' = m_A \vec{u}_A$, and body B has the recoil momentum $\vec{p}_B{}' = m_B \vec{u}_B$ (**Fig. 2.36**). The collision process is not completely fixed by the energy and momentum conservation laws: there are only 4 equations for calculating the 6 components of the final momenta. The end points of $\vec{p}_A{}'$ lie on the **momentum sphere** with the radius $p_A \cdot \dfrac{m_B}{m_A + m_B}$, where the center of this sphere divides the momentum $\vec{\mathbf{p}}_A$ according to the ratio of masses (**Fig. 2.37**),

$$\left(\vec{\mathbf{p}}_A{}' - \frac{m_A}{m_A + m_B}\vec{\mathbf{p}}_A\right)^2 = \left(\frac{m_B}{m_A + m_B}\vec{\mathbf{p}}_A\right)^2 .$$

There is rotational symmetry about the $\vec{\mathbf{p}}_A$-axis, and hence the collision process is characterized by the polar **scattering angle** ϑ.

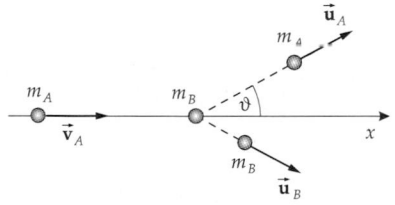

Figure 2.36: Elastic collision of the body A with a body B at rest.

One can distinguish the following cases.

$m_A > m_B$: There exists a maximum scattering angle ϑ_{max}, $\sin\vartheta_{max} = m_B/m_A$. Possible scattering angles lie in the interval $0 \leq \vartheta \leq \vartheta_{max}$.

$m_A = m_B$: The scattering angle lies in the interval $0 \leq \vartheta \leq \pi$. The momenta after the collision always include the angle $\pi/2$ (Thales' law).

$m_A < m_B$: All scattering angles between 0 and π are allowed: $0 \leq \vartheta \leq \pi$.

➤ In an inelastic collision, the radius of the momentum sphere changes while the center remains in place. The radius increases (decreases) for $\Delta W < 0$ ($\Delta W > 0$).

➤ As the radius of the momentum sphere vanishes, the inelastic collision approaches a totally inelastic collision.

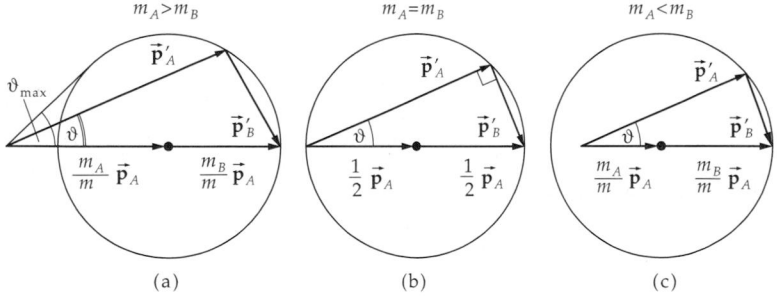

Figure 2.37: Momentum sphere ($m = m_A + m_B$). (a): $m_A > m_B$, (b): $m_A = m_B$, (c): $m_A < m_B$.

■ A body collides with a wall that is parallel to the y-direction. The direction of the collision is perpendicular to the wall, so that only the x-component of its momentum is changed. The process corresponds to an elastic collision with a very heavy body,

$$p'_x = -p_x .$$

The **reflection law** of the elastic collision follows from this example.

▲ If a body collides elastically with a fixed wall, its reflection angle ε' equals the incidence angle ε, and the magnitude of the momentum remains unchanged. The directions of motion before and after the collision are coplanar (**Fig. 2.38**).

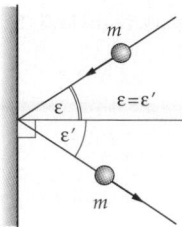

Figure 2.38: Reflection law for elastic collision on a wall. ε: angle of incidence, ε': angle of reflection.

2.6.4   Inelastic collisions

In inelastic collisions, part of the energy of motion is lost. It is used for permanent deformation of the collision partners and is converted into deformation heat.

2.6.4.1   Partly inelastic collisions

Energy loss ΔW, lies between the energy loss in a totally inelastic collision as a maximum value and zero:

$$0 < \Delta W < \frac{m_A m_B}{2(m_A + m_B)}(v_A - v_B)^2.$$

How large this fraction is depends on the inelastic deformability of the collision partners.

2.6.4.2   Totally inelastic collision

After the collision $u_a = u_b = u$. From the law of momentum conservation, it follows that

$$m_A \cdot v_A + m_B \cdot v_B = (m_A + m_B)u,$$

and therefore

$$u = \frac{m_A v_A + m_B v_B}{m_A + m_B}.$$

Kinetic energy before and after the collision:

$$E_{kin} = \frac{1}{2}m_A v_A^2 + \frac{1}{2}m_B v_B^2,$$

$$E'_{kin} = \frac{1}{2}(m_A + m_B)u^2 = \frac{1}{2}\frac{(m_A v_A + m_B v_B)^2}{m_A + m_B}.$$

Energy loss $\Delta W = E_{kin} - E'_{kin}$ in a totally inelastic collision:

$$\Delta W = \frac{m_A \cdot m_B}{2(m_A + m_B)}(v_A - v_B)^2.$$

If a body collides with a another **at rest** ($v_B = 0$), the ratio of kinetic energies before and after the collision depends only on the masses:

$$\frac{E'_{kin}}{E_{kin}} = \frac{m_A}{m_A + m_B} \leq 1.$$

The ratio of energy loss to initial kinetic energy $E_{\text{kin}}(t_0)$ is in this case ($v_B = 0$):

$$\frac{\Delta W}{E_{\text{kin}}(t_0)} = \frac{m_B}{m_A + m_B} \leq 1.$$

Equal masses $m_A = m_B$: half of the kinetic energy E_{kin} is lost. In macroscopic collision processes, this amount is converted into deformation and heat energy of the collision partners.

2.7　Rockets

Recoil principle, follows from the law of momentum conservation, applied to **rocket** propulsion. Unlike propulsion based on friction, rockets also work in a vacuum.

➤　Rockets are used for transportation into space and serve as carriers of payloads such as satellites (for information transmission, Earth and meteorological observation, research) and manned spaceships. Their significance on Earth is limited. Projectiles with jet propulsion are not rockets, since they do not carry their recoil mass (**reaction mass**), but rather suck it in as air.

2.7.1　Thrust

Rocket, continuously expels hot gases produced by combustion of the fuel by means of an oxidizer, also carried. The hot gases are emitted in the backward direction and push the rocket forward by their **recoil** (**Fig. 2.39**). The rocket mass therefore decreases during acceleration. Unlike the **jet engine**, that sucks in air and expels it in backward direction, a rocket can also be used in a vacuum.

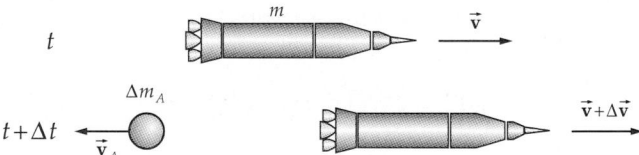

Figure 2.39: Rocket.

1. Acceleration of a rocket

To calculate the acceleration of the rocket, we consider a small time interval Δt, in which a mass Δm_A is ejected by the rocket with velocity $\vec{v}_A$, whereby the velocity of the rocket increases from $\vec{v}$ to $\vec{v} + \Delta\vec{v}$. For the momentum balance, the momentum $\Delta m_A \vec{v}_A$ of the ejected gas must be taken into account. The change of momentum of the system rocket plus ejected gas during this interval is

$$\Delta\vec{p} = [(m - \Delta m_A)(\vec{v} + \Delta\vec{v}) + \Delta m_A \vec{v}_A] - m\vec{v},$$

$$= m\Delta\vec{v} + \Delta m_A[\vec{v}_A - (\vec{v} + \Delta\vec{v})].$$

Introducing the escape velocity

$$\vec{v}_0 = \vec{v}_A - \vec{v}$$

of the ejected gas relative to the rocket, and neglecting the product of two small terms, $\Delta m_A \cdot \Delta \vec{v} \approx 0$, the momentum conservation law (in the absence of external forces) reads:

$$\Delta \vec{p} = m \Delta \vec{v} - \Delta m_A \vec{v}_0 = 0.$$

2. Recoil

The momentum difference is called **recoil**. A recoil arises always when one body pushes another body away. It expresses Newton's third law (action = reaction).

After division by Δt and letting $\Delta t \to 0$, one finds

$$\frac{d\vec{p}}{dt} = \lim_{\Delta t \to 0} \frac{\Delta \vec{p}}{\Delta t} = m \frac{d\vec{v}}{dt} - \frac{dm}{dt} \vec{v}_0 = 0.$$

3. Equation for rocket thrust

rocket thrust				$\mathbf{MLT^{-2}}$
		Symbol	Unit	Quantity
$\vec{F}_{\text{thrust}} = \dfrac{dm(t)}{dt} \vec{v}_0 = \dot{m}\,\vec{v}_0$		$\vec{F}_{\text{thrust}}$	N	thrust
		t	s	time
		$m(t)$	kg	mass at time t
		$\dot{m}$	kg/s	mass flow
		$\vec{v}_0$	m/s	escape velocity

If an additional external force F_a acts (e.g., Earth's gravitation), it enters on the right side of the equation for $\dfrac{d\vec{p}}{dt}$ in place of the zero. One writes:

$$m \frac{d\vec{v}}{dt} = \frac{dm}{dt} \vec{v}_0 + F_a = \dot{m}\,\vec{v}_0 + F_a$$

and calls the first term on the right side the thrust $\vec{F}_{\text{thrust}}$. The acceleration of the rocket $\vec{a}$ is obtained in the case of external forces $\vec{F}_a$ (gravitation, friction) given by:

$$\vec{a} = \frac{d\vec{v}}{dt} = \frac{1}{m(t)} (\vec{F}_{\text{thrust}} + \vec{F}_a).$$

■   A Saturn-V rocket has an initial mass $m_0 = 2.95 \cdot 10^6$ kg, a burning time of the first stage of $t_B = 130$ s, and an empty mass at the end of burning of the first stage of $m_{\text{empty}} = 1.0 \cdot 10^6$ kg. The mass flow is

$$\dot{m} = \frac{m_0 - m_{\text{empty}}}{t_B} = \frac{2.95 \cdot 10^6 \text{ kg} - 1.0 \cdot 10^6 \text{ kg}}{130 \text{ s}} = 1.50 \cdot 10^4 \text{ kg/s}.$$

For an escape velocity of $v_0 = 2220$ m/s, the thrust is

$$F_{\text{thrust}} = \dot{m} v_0 = 1.50 \cdot 10^4 \text{ kg/s} \cdot 2220 \text{ m/s} = 3.3 \cdot 10^7 \text{ N}.$$

2.7.2 Rocket equation

1. Final velocity and maximum altitude of a rocket

To calculate the final velocity of the rocket, the rocket acceleration must be integrated over time. This is relatively simple if the escape velocity v_0 and the mass flow $\dot{m}$ during the burning time t_B are constant. For the mass at time t then: $m(t) = m_0 - \dot{m}t$, where m_0 is the original mass of the rocket. If one includes as external force only a gravitational force with constant gravitational acceleration, $F_a = m(t)g$, the rocket acceleration is

$$a(t) = \frac{\dot{m}}{m_0 - \dot{m}t} v_0 - g.$$

By integration over the time, one finds for the velocity v at time t:

$$v(t) = v_0 \ln\left(\frac{m_0}{m_0 - \dot{m}t}\right) - gt.$$

An additional integration yields the height h at the time t:

$$h(t) = \frac{v_0(m_0 - \dot{m}t)}{\dot{m}}\left[\frac{m_0}{m_0 - \dot{m}t} - 1 - \ln\left(\frac{m_0}{m_0 - \dot{m}t}\right)\right] - \frac{1}{2}gt^2.$$

2. Form of rocket equation

At the end of the burning, the final velocity and height are:

rocket equation			
	Symbol	Unit	Quantity
$v_B = v_0 \ln\left(\dfrac{m_0}{m_{\text{empty}}}\right) - gt_B$	v_B	m/s	velocity at burning closure
$h_B = \dfrac{v_0\, m_{\text{empty}}}{\dot{m}}$	h_B	m	height at burning closure
	v_0	m/s	escape velocity
$\times\left[\dfrac{m_0}{m_{\text{empty}}} - 1 - \ln\left(\dfrac{m_0}{m_{\text{empty}}}\right)\right]$	m_0	kg	initial mass
	m_{empty}	kg	mass at burning closure
$-\dfrac{1}{2}gt_B^2$	$\dot{m}$	kg/s	mass flow
	g	m/s^2	free acceleration of gravity
$m_{\text{empty}} = m_0 - \dot{m}t_B$	t_B	s	burning time

3. Properties of the rocket equation

➤ This equation holds only under the assumption of a constant free-fall acceleration, i.e., if the rocket moves close to the Earth's surface. Air friction is also ignored.

➤ The final velocity and height that can be reached depend only on the escape velocity and the logarithm of the ratio m_0/m_{empty} of start mass m_0 to empty mass m_{empty}. Hence, the payload of a rocket is typically only 10 % of the initial mass.

➤ The chemical energy stored in a chemical fuel is not sufficient to lift the fuel into an orbit around Earth. However, a majority of the burnt fuel is released on Earth (or in the atmosphere) after transferring its energy to the rocket. Only due to this fact can rockets with chemical fuels work at all.

■ For the first stage of a Saturn-V rocket characterized above, the final velocity is

$$v_B = v_0 \ln\left(\frac{m_0}{m_{\text{empty}}}\right) - g t_B,$$

$$= 2.22 \cdot 10^3 \text{ m/s} \ln\left(\frac{2.95 \cdot 10^6 \text{ kg}}{1.0 \cdot 10^6 \text{ kg}}\right) - 9.81 \text{ m/s}^2 \cdot 130 \text{ s},$$

$$= 1\,126 \text{ m/s}.$$

The height of the first stage at the end of burning is $h_B = 45.6$ km.

2.8 Systems of point masses

System of point masses, system consisting of N individual point masses (particles) $1, \ldots, N$, their motion is described by specifying their position vectors $\vec{r}_1, \ldots, \vec{r}_N$ as function of time t: $\vec{r}_i(t)$, $i = 1, \ldots, N$.

Center of gravity, center of mass, point in a system of point masses (**Fig. 2.40**), the position vector $\vec{R}$ of which is calculated from the masses m_i and the position vectors $\vec{r}_i$ according to

$$\vec{R} = \frac{1}{M} \sum_{i=1}^{N} m_i \vec{r}_i, \qquad M = \sum_{i=1}^{N} m_i.$$

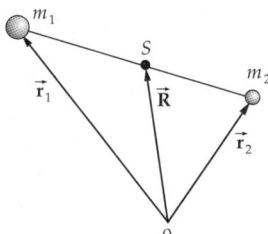

Figure 2.40: Center of mass S of a system of two point masses m_1, m_2.

2.8.1 Equations of motion

1. Forces in particle systems

Internal forces, forces acting between the particles of the system. Internal forces are in general a sum of two-body forces $\vec{F}_{ik}$ that depend on the positions (and possibly the velocities) of the pairs of particles (i, k).

According to Newton's third law (reaction principle), the force $\vec{F}_{ik}$ acting on the point mass i owing to the point mass k is opposite in direction and equal in magnitude to the force $\vec{F}_{ki}$ acting on the point mass k owing to the point mass i.

External forces, forces acting from outside the system. The external force $\vec{F}_i^{\text{ext}}$ on the point mass i does not depend on the coordinates of the other point masses (**Fig. 2.41**).

$$\vec{F}_{ik} = \vec{F}_{ik}(\vec{r}_i, \vec{r}_k), \quad \vec{F}_{ik} = -\vec{F}_{ki}, \quad \vec{F}_i^{\text{ext}} = \vec{F}_i^{\text{ext}}(\vec{r}_i)$$

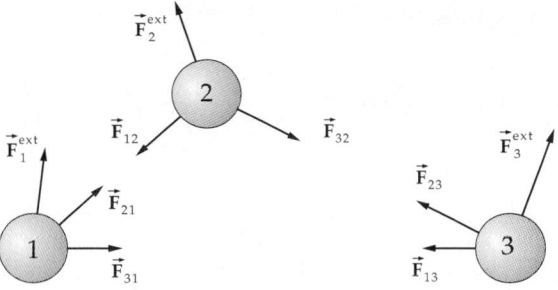

Figure 2.41: Internal and external forces in a system of point masses. The internal forces $\vec{\mathbf{F}}_{ik} = -\vec{\mathbf{F}}_{ki}$ cancel each other.

Constraint forces, reaction forces, originate from the support of the system. Constraint forces restrict the motion of the system.

Free system of point masses, a system of point masses that can follow the applied forces without constraints.

Closed system, a system of point masses free of external forces.

2. Dynamical law for systems of point masses

dynamical law for systems of point masses			
$m_i\,\ddot{\vec{\mathbf{r}}}_i = \vec{\mathbf{F}}_i, \quad i = 1, \dots, N$	Symbol	Unit	Quantity
$\displaystyle \vec{\mathbf{F}}_i = \sum_{k \neq i = 1}^{N} \vec{\mathbf{F}}_{ki} + \vec{\mathbf{F}}_i^{\text{ext}}$	m_i	kg	mass of point mass i
	$\vec{\mathbf{r}}_i$	m	position vector of point mass i
	$\vec{\mathbf{F}}_i$	N	force on point mass i
$\displaystyle \vec{\mathbf{F}}^{\text{ext}} = \sum_{i=1}^{N} \vec{\mathbf{F}}_i^{\text{ext}}$	$\vec{\mathbf{F}}_{ik}$	N	two-body force between i and k
	$\vec{\mathbf{F}}^{\text{ext}}$	N	total external force
	$\vec{\mathbf{F}}_i^{\text{ext}}$	N	external force on mass point i

The equations of motion of a system of point masses consist of a system of coupled differential equations of second order in time for the position vectors of the point masses. The equations are coupled through the spatial dependence of the forces. The general solution of the system involves $6N$ free parameters that must be determined in such a way that the given initial conditions for the positions and velocities of the point masses are fulfilled.

3. Momentum, angular momentum and energy of systems of mass points

Total momentum of the system:

$$\vec{\mathbf{p}} = \sum_{i=1}^{N} \vec{\mathbf{p}}_i = \sum_{i=1}^{N} m_i\,\dot{\vec{\mathbf{r}}}_i.$$

Total angular momentum of the system:

$$\vec{\mathbf{l}} = \sum_{i=1}^{N} \vec{\mathbf{l}}_i = \sum_{i=1}^{N} m_i\,(\vec{\mathbf{r}}_i \times \vec{\mathbf{p}}_i).$$

Total energy of the system:

$$E = E_{\text{kin}} + E_{\text{pot}}, \quad E_{\text{kin}} = \sum_{i=1}^{N} \frac{m_i}{2} \dot{\vec{r}}_i^2, \quad E_{\text{pot}} = \sum_{i<k=1}^{N} U_{ik}(|\vec{r}_i - \vec{r}_k|) + \sum_{i=1}^{N} U_i^{\text{ext}}(\vec{r}_i).$$

➤   The potential energy of the system is the sum of the potential energies owing to the internal and the external forces. The potential U_{ik} of the internal force $\vec{F}_{ik}$ can only depend on the distance $r_{ik} = |\vec{r}_i - \vec{r}_k|$ of the particles i, k to fulfil $\vec{F}_{ik} = -\vec{F}_{ki}$. The total potential of the internal forces is obtained by summing over all pairs (i, k). The potential U_i^{ext} of the external force $\vec{F}_i^{\text{ext}}$ depends only on the position of the particle.

2.8.2 Momentum conservation law

Because of the fundamental law of dynamics, the change of the total momentum $\vec{p}$ of the system per unit time equals the sum of applied forces. According to the reaction principle, the internal forces cancel each other, and therefore only the external forces contribute to the change of the total momentum.

1. Momentum conservation law

change of the total momentum per unit time = sum of external forces			MLT^{-2}
	Symbol	Unit	Quantity
$\dfrac{d\vec{p}}{dt} = \vec{F}^{\text{ext}}$	$\vec{p}$	Ns	total momentum
	$\vec{F}^{\text{ext}}$	N	external force

Momentum conservation law: If no external forces are applied, the total momentum is conserved.

The total momentum of a system of point masses that is free of external forces is constant.			MLT^{-1}
	Symbol	Unit	Quantity
$\vec{p} = \sum_i \vec{p}_i = \text{const.}$	$\vec{p}$	Ns	total momentum
	$\vec{p}_i$	Ns	momentum of point mass i

2. Center-of-mass law

The **center-of-mass law** corresponds to the momentum conservation law of the N-particle system:

	Symbol	Unit	Quantity
The center of mass of a system of point masses moves as if the total mass were rigidly connected to the center-of-mass, and were affected by the vector sum of the external forces.			
$$M\,\ddot{\vec{R}} = \vec{F}^{\text{ext}}$$ $$\vec{R} = \frac{1}{M}\sum_{i=1}^{N} m_i\,\vec{r}_i, \quad M = \sum_{i=1}^{N} m_i$$	m_i $\vec{r}_i$ M $\vec{R}$ $\vec{F}^{\text{ext}}$	kg m kg m N	mass of point mass i position vector point mass i total mass position vector center-of-mass external forces

2.8.3 Angular momentum conservation law

The time variation of the total angular momentum $\vec{l}$ of a system of point masses is given by

$$\frac{d\vec{l}}{dt} = \sum_{i=1}^{N}(\vec{r}_i \times \vec{F}_i) = \sum_{i=1}^{N}(\vec{r}_i \times \vec{F}_i^{\text{ext}}) = \sum_{i=1}^{N} \vec{\tau}_i^{\text{ext}}.$$

The vector $\vec{\tau}_i^{\text{ext}}$ is the torque exerted by the external force $\vec{F}_i^{\text{ext}}$ on the point mass i.

The internal forces do not change the total angular momentum because they act along the connecting lines of the point masses (**Fig. 2.42**):

$$\vec{r}_i \times \vec{F}_{ki} + \vec{r}_k \times \vec{F}_{ik} = (\vec{r}_i - \vec{r}_k) \times \vec{F}_{ki} = 0.$$

The time rate of change of the total angular momentum equals the sum of the torques of the external forces.

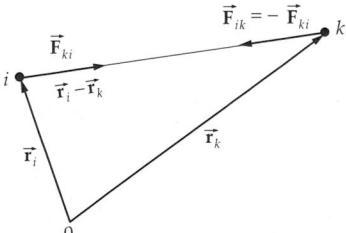

Figure 2.42: Vanishing torque of internal forces.

The total angular momentum of the system of point masses is conserved if the external forces vanish.

angular momentum conservation law: In a closed system of point masses, the total angular momentum is conserved.			$\mathbf{ML^2T^{-1}}$
	Symbol	Unit	Quantity
$$\vec{l} = \sum_{i=1}^{N} \vec{l}_i = \text{const.}$$	$\vec{l}$ $\vec{l}_i$	kg m²/s kg m²/s	total angular momentum angular momentum of point mass i

2.8.4 Energy conservation law

Conservative forces, forces that can be represented by a potential. Necessary and sufficient condition for the existence of a potential of the force $\vec{F}$: rot $\vec{F} = 0$. A conservative force does no work along a closed path:

$$\oint \vec{F}\,\mathrm{d}\vec{r} = 0.$$

Dissipative forces, forces having no potential.

Decomposition of the force $\vec{F}_i$ on the particle i into its conservative and dissipative part:

$$\vec{F}_i = \vec{F}_{i,\mathrm{cons}} + \vec{F}_{i,\mathrm{diss}}.$$

The change of the total energy of a system of point masses with time equals the power of the dissipative forces:

$$\frac{\mathrm{d}E}{\mathrm{d}t} = \frac{\mathrm{d}}{\mathrm{d}t}\left(E_{\mathrm{kin}} + E_{\mathrm{pot}}\right) = \sum_{i=1}^{N} \vec{F}_{i,\mathrm{diss}} \cdot \dot{\vec{r}}_i.$$

▲   For dissipative forces, the work for motion from position $\vec{r}_1$ to position $\vec{r}_2$ depends on the actual path between initial and final points.

➤   Frictional forces that are proportional to the velocity cause the system to release mechanical energy to the environment. Friction forces are dissipative forces. When moving a body from $\vec{r}_1$ to $\vec{r}_2$, the frictional work increases with the length of the selected path.

■   Damped vibration of a single point mass:
Equation of motion: $m\,\ddot{x} + k\,x + \mu\,\dot{x} = 0$.
Energy: $E = E_{\mathrm{kin}} + E_{\mathrm{pot}}$, $E_{\mathrm{kin}} = \dfrac{m}{2}\,\dot{x}^2$, $E_{\mathrm{pot}} = \dfrac{k}{2}\,x^2$.
Energy change: $\dfrac{\mathrm{d}}{\mathrm{d}t}E = \dfrac{\mathrm{d}}{\mathrm{d}t}\left(\dfrac{m}{2}\,\dot{x}^2 + \dfrac{k}{2}\,x^2\right) = -\mu\,\dot{x}^2 < 0$.
The sum of kinetic and potential energy of the pendulum decreases continuously because of the friction term ($\mu > 0$).

The total energy of the system of point masses is conserved if the dissipative forces vanish.

law of energy conservation: The total energy of a system of point masses is conserved if no dissipative forces arise.			ML^2T^{-2}
	Symbol	Unit	Quantity
$E = E_{\mathrm{kin}} + E_{\mathrm{pot}} = \mathrm{const.}$	E	J	total energy
	E_{kin}	J	total kinetic energy
	E_{pot}	J	total potential energy

2.9 Lagrange's and Hamilton's equations

2.9.1 Lagrange's equations and Hamilton's principle

1. Generalized mechanical quantities

Generalized coordinates, q_k, coordinates that are optimally adapted to the given mechanical system. Generalized coordinates can have different physical meanings (length, angle, etc.). The number of generalized coordinates equals the number of degrees of freedom of the system.

$$q_k(t), \qquad k = 1, \ldots, f, \qquad f\text{: number of degrees of freedom}$$

■ Generalized coordinates for
pendulum: angle φ of elongation from rest position.
point mass on a spherical surface: spherical coordinates θ, φ.
Generalized velocities, $\dot{q}_k$, first derivative of the generalized coordinates q_k with respect to time,

$$\dot{q}_k(t), \qquad k = 1, \ldots, f, \qquad f\text{: number of degrees of freedom.}$$

Generalized forces, Q_k, defined by the expressions

$$Q_k = \sum_{i=1}^{3N} F_i \frac{\partial x_i}{\partial q_k}, \qquad k = 1, \ldots, f.$$

x_i, $i = 1, \ldots, 3N$ are the Cartesion coordinates of a system of N mass points.

2. Lagrange's function,

difference between the kinetic energy $E_{\text{kin}} = T$ and the potential energy $E_{\text{pot}} = V$ as functions of the generalized coordinates q_k and generalized velocities $\dot{q}_k$,

$$L(q_k, \dot{q}_k, t) = T(q_k, \dot{q}_k) - V(q_k, t).$$

➤ The Lagrange function has the dimension of energy.
■ Lagrange function of simple mechanical systems:

Free point mass: $\qquad\qquad\qquad\qquad L = T = \dfrac{m}{2}\dot{\mathbf{r}}^2 = \dfrac{m}{2}(\dot{x}^2 + \dot{y}^2 + \dot{z}^2).$

Point mass in potential field $V(\vec{r})$: $\quad L = \dfrac{m}{2}\dot{\mathbf{r}}^2 - V(\vec{r}).$

Spring vibration, spring constant k: $\quad L = \dfrac{m}{2}\dot{x}^2 - \dfrac{k}{2}x^2.$

Pendulum, pendulum length l: $\qquad L = \dfrac{m}{2}l^2\dot{\varphi}^2 + mgl\,\cos\varphi.$

Physical pendulum: $\qquad\qquad\qquad L = \dfrac{J}{2}\dot{\varphi}^2 + mgl\,\cos\varphi.$

Distance from rotation axis to center of mass l, moment of inertia J.

3. Lagrange's equations,

system of f differential equations of second order with respect to time for determining the generalized coordinates q_k as functions of time:

$$\frac{\mathrm{d}}{\mathrm{d}t}\frac{\partial L}{\partial \dot{q}_k} - \frac{\partial L}{\partial q_k} = 0, \qquad k = 1, \ldots, f\;.$$

Constraint forces or auxiliary conditions no longer appear in the Lagrange equations. The solutions involve $2f$ integration constants.

▲ The Lagrange equations and Newton's second law are equivalent formulations of mechanics.

4. Examples of Lagrange's formalism

■ One-dimensional motion of a point mass in potential $V(x)$, Cartesian coordinate x:
Generalized coordinate: $q = x$.   Generalized velocity: $\dot{q} = \dot{x}$.
Lagrange function: $L = T - V = \dfrac{m}{2}\dot{x}^2 - V(x)$.
Lagrange equation:

$$\frac{\partial L}{\partial \dot{q}} = m\,\dot{x}, \qquad \frac{\mathrm{d}}{\mathrm{d}t}\frac{\partial L}{\partial \dot{q}} = m\,\ddot{x},$$

$$\frac{\partial L}{\partial q} = \frac{\partial V}{\partial x}, \qquad m\ddot{x} + \frac{\partial V}{\partial x} = 0.$$

Because $-\partial V/\partial x = F_x$, Newton's equations of motion follow from the Lagrange equation $m\,\ddot{x} = F_x$ for the motion of a point mass under the influence of the force F_x.

■ Motion in a central-symmetric potential $V(r)$:
Generalized coordinates: r, ϑ.   Generalized velocities: $\dot{r}, \dot{\vartheta}$.
Lagrange function: $L = T - V = \dfrac{m}{2}(\dot{r}^2 + r^2\,\dot{\vartheta}^2) - V(r)$.
Lagrange equations:

$$\frac{\partial L}{\partial \dot{r}} = m\,\dot{r}, \qquad \frac{\mathrm{d}}{\mathrm{d}t}\frac{\partial L}{\partial \dot{r}} = m\,\ddot{r}, \qquad \frac{\partial L}{\partial r} = m\,r\,\dot{\vartheta}^2 - \frac{\partial V}{\partial r},$$

$$\frac{\partial L}{\partial \dot{\vartheta}} = m\,r^2\,\dot{\vartheta}, \qquad \frac{\partial L}{\partial \vartheta} = 0.$$

Equations of motion:

$$m\,\ddot{r} = m\,r\,\dot{\vartheta}^2 - \frac{\partial V}{\partial r} = m\,r\,\dot{\vartheta}^2 + F(r), \qquad \frac{\mathrm{d}}{\mathrm{d}t}(m\,r^2\,\dot{\vartheta}) = 0.$$

$F(r)$ is the magnitude of the applied central force. The last equation implies the conservation of angular momentum $l = m\,r^2\,\dot{\vartheta}$.

5. Virtual displacement,

instantaneous infinitesimal displacement $\delta\vec{r}$ of a point mass, taking into account the restricting auxiliary conditions for the motion, without change in the time variable:

$$\vec{r} \longrightarrow \vec{r} + \delta\vec{r} \quad \text{for } \delta t = 0.$$

Virtual displacements are imaginary displacements that need not correspond to the actual course of the trajectory.

➤ When using generalized coordinates, virtual displacements may be made arbitrarily, without taking into account auxiliary conditions.

➤ The virtual displacement of a system of N point masses is composed of the virtual displacements of every individual point mass, $\delta\vec{r}_i$, $i = 1, \ldots, N$.

Virtual trajectory, trajectory $\hat{q}_k(t)$, between two fixed points $q_k(t_1)$, $q_k(t_2)$, that differs infinitesimally from the actual trajectory $q_k(t)$ by combining the virtual displacements δq_k at a fixed time t ($\delta t = 0$) (**Fig. 2.43**),

$$\hat{q}_k(t) = q_k(t) + \delta q_k(t).$$

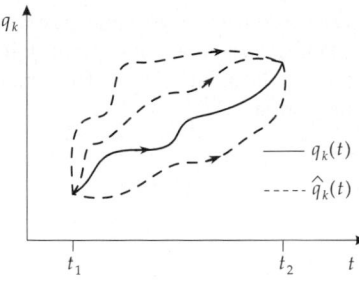

Figure 2.43: Virtual trajectories $\hat{q}_k(t)$. $q_k(t)$: actual trajectory.

6. Action function and Hamiltonian principle

Action function, action integral W, integral of the Lagrange function $L(q_k, \dot{q}_k, t)$ over time,

$$W = \int_{t_1}^{t_2} L(q_k(t), \dot{q}_k(t), t)\, dt.$$

➤ The action function has the dimension energy times time.

Principle of minimum action, Hamilton's principle, the trajectory of a mechanical system as a function of time is distinguished from all other virtual trajectories by the fact that the action integral takes an extremum value (usually a minimum):

$$W = \int_{t_1}^{t_2} L(q_k(t), \dot{q}_k(t), t)\, dt = \text{extremum}.$$

➤ Hamilton's principle does not depend on the choice of coordinates. An extremum principle is equivalent to the equations of motions of Newton or Lagrange.

➤ Extremum principles in other branches of physics: Fermat's principle of the shortest path in optics; Ritz's method for approximate calculation of energy eigenvalues in quantum mechanics.

2.9.2 Hamilton's equations

1. Generalized momentum,

p_k, defined as the derivative of the Lagrange function $L = T - V$ with respect to the generalized velocity $\dot{q}_k$:

$$p_k = \frac{\partial L}{\partial \dot{q}_k}, \qquad k = 1, \ldots, f, \qquad f: \text{ number of degrees of freedom.}$$

➤ The quantities q_k and p_k introduced this way are called **canonically conjugate**.

■ In circular motion, the rotation angle φ is the generalized coordinate. The canonically conjugate momentum is the angular momentum l.

2. Hamiltonian

H, is obtained if one eliminates the generalized velocities $\dot{q}_k$ from the theoretical description and uses instead the canonically conjugate momenta p_k:

$$H(q_k, p_k, t) = \sum_{k=1}^{f} \dot{q}_k p_k - L(q_k, \dot{q}_k, t).$$

➤   The Hamiltonian depends on the generalized coordinates, the canonically conjugate momenta, and possibly on the time. If the Hamiltonian is time-independent, it represents the total energy (sum of kinetic energy and potential energy). The total energy is a conserved quantity (or "integral") of the motion:

$$\frac{\partial H}{\partial t} = \frac{dH}{dt} = 0, \qquad H = T + V = E = \text{const.}$$

3. Legendre transformation

The transition from the Lagrange function $L(q_k, \dot{q}_k)$ to the Hamilton function $H(q_k, p_k)$ is called a **Legendre transformation**.

A function $f(x, y)$ of the two variables x, y can be transformed into an equivalent function h that depends on the variables x and $p = \partial f / \partial y$, by

$$h(x, p) = f(x, y) - y\, p.$$

Because of

$$\frac{\partial h}{\partial y} = \frac{\partial f}{\partial y} - p = 0,$$

the function h depends on x and p, but no longer on y.

The Legendre transformation is often applied in thermodynamics to transform state variables into other state variables. For instance, one obtains the free energy F as function of temperature T by replacing in the intrinsic energy $E(S, \ldots)$ the entropy variable S by the temperature variable $T = \partial U / \partial S$:

$$F(T, \ldots) = U(S, \ldots) - T\, S.$$

4. Hamilton's equations

Time derivative of the generalized coordinates and momenta,

$$\dot{q}_k = \frac{\partial H}{\partial p_k}, \quad \dot{p}_k = -\frac{\partial H}{\partial q_k}, \quad k = 1, \ldots, f, \quad f: \text{ number of degrees of freedom.}$$

Hamilton's equations are a system of $2f$ differential equations of first order with respect to time. The solutions contain $2f$ integration constants that can be freely chosen (e.g., the initial values of the coordinates and momenta). Hamilton's equations are equivalent to the Lagrange equations.

■   One-dimensional harmonic oscillator:

Lagrange function: $L = \dfrac{m}{2}\dot{x}^2 - \dfrac{k}{2}x^2.$

Generalized momentum: $p = \dfrac{\partial L}{\partial \dot{x}} = m\,\dot{x}.$

Hamiltonian: $H = p\,\dot{x} - L = \dfrac{p^2}{2m} + \dfrac{k}{2}x^2 = T(t) + V(t)$

$$= E = \text{const.}$$

Hamilton equations: $\dot{x} = \dfrac{\partial H}{\partial p} = \dfrac{p}{m}, \quad \dot{p} = -\dfrac{\partial H}{\partial x} = -kx.$

These equations lead to Newton's equation of motion, $m\ddot{x} = -kx$.

5. Phase space

Cyclic coordinate, generalized coordinate that does not enter the Lagrange function:

$$\frac{\partial L}{\partial \varphi} = 0 \implies \frac{d}{dt} \frac{\partial L}{\partial \dot\varphi} = \frac{d}{dt} p_\varphi = 0.$$

▲   The conjugate momentum that corresponds to a cyclic coordinate is an integral of the motion.

Configuration space, f-dimensional space of the generalized coordinates q_k. Trajectory in configuration space: $q_k(t)$, $k = 1, \ldots, f$.

Phase space, abstract space with $2f$ dimensions; the coordinates are the generalized coordinates q_k and the canonically conjugate momenta p_k. Trajectory of the system in phase space: $(q_k(t), p_k(t))$, $k = 1, \ldots, f$.

➤   For conservative systems, every trajectory in phase space is characterized by a definite value of the Hamiltonian (total energy). Spatially periodic motions correspond to closed trajectories in phase space.

■   In phase space, a one-dimensional harmonic oscillator makes an ellipse that is characterized by the energy $E = \dfrac{m}{2} A^2 \omega^2$ (A: amplitude, ω: angular frequency) (**Fig. 2.44**).

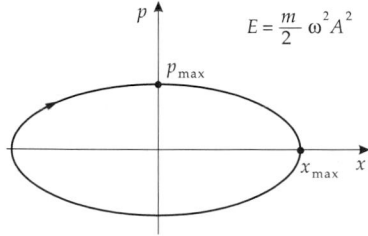

$$p_{max} = m\omega A, \quad x_{max} = A$$

Figure 2.44: Trajectory of a harmonic oscillator in phase space.

3
Rigid bodies

Rigid body, a body the constituents of which always keep the same distances between each other, i.e., are rigidly bound to each other. One may imagine the rigid body as composed of point masses (**Fig. 3.1**). The distances between all pairs of point masses i, j of the rigid body:

$$|\vec{r}_i(t) - \vec{r}_j(t)| = r_{ij} = \text{const.}$$

A rigid body cannot be deformed.

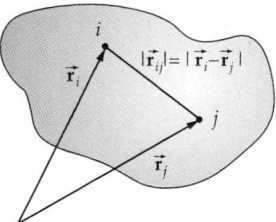

Figure 3.1: Rigid body.

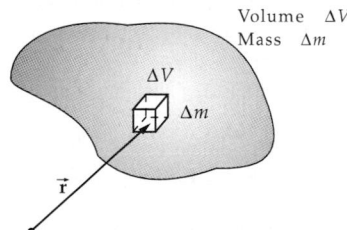

Figure 3.2: Density $\rho(\vec{r})$ of an inhomogeneous rigid body with a continuous mass distribution.

3.1 Kinematics

3.1.1 Density

Density ρ of a **homogeneous** body, the ratio of its mass m to its volume V,

$$\rho = \frac{m}{V}.$$

In an **inhomogeneous** body with a continuous mass distribution, the density varies with the spatial coordinate $\vec{r}$ (**Fig. 3.2**). One imagines the body to consist of volume elements ΔV with approximately constant density. The mass in the volume element ΔV at the point $\vec{r}$ is Δm. The density in the volume element ΔV is given by: $\rho = \Delta m / \Delta V$. For a continuous mass distribution, the density at point $\vec{r}$ is obtained by

$$\rho(\vec{r}) = \lim_{\Delta V \to 0} \frac{\Delta m}{\Delta V} = \frac{dm}{dV}, \quad dm = \rho(\vec{r}) \cdot dV.$$

The total mass m of the body is given by the volume integral:

$$m = \int dm = \int \rho(\vec{r}) \, dV.$$

3.1.2   Center of mass

1. Definition of the center of mass

Center of mass, **center of gravity**, a point at which all of the force from the **weights** of all elements of the body can be considered to act. The action of gravity on a rigid body can be represented by a single force of magnitude

$$F_G = mg$$

that acts on the center of mass, m being the total mass of the body.
▲   For a symmetric body of homogeneous density, the center of mass lies on the symmetry axis.
   To keep a body in equilibrium, one can
•   support the body at the center of mass;
•   support the body at several points in such a way that the resultant of the supporting forces lies at the center of mass.
▲   A rigid body under the action of its weight is in equilibrium if it is supported at the center of mass.
➤   The weight then has no torque with respect to the center of mass of the body.

2. Center-of-mass coordinates

The position vector $\vec{R}$ of the center of mass is given by:

center-of-mass coordinates			L
	Symbol	Unit	Quantity
$\vec{R} = \dfrac{\sum_i \vec{r}_i \, \Delta m_i}{m}$	$\vec{R}$	m	position vector of center of mass
	$\vec{r}_i$	m	coordinate of element i
$m = \sum_i \Delta m_i$	Δm_i	kg	mass of element i
	m	kg	total mass

Integral form for a continuous mass distribution:

$$\vec{R} = \frac{\int \vec{r} \, dm}{\int dm} = \frac{\int_V \vec{r} \, \rho(\vec{r}) \, dV}{\int_V \rho(\vec{r}) \, dV},$$

$\rho(\vec{r})$: density of the body, dV: volume element.

For a homogeneous body ($\rho = $ const.),

$$\vec{R} = \frac{1}{V} \int\limits_V \vec{r}\,dV \,.$$

3. Determination of the center of mass

M **Graphical determination** of the center of mass of an area: the area to be considered is divided into parts with known areas and centers of mass. One then attaches a force to the center of mass of each partial area having a magnitude proportional to the size of the partial area and pointing in an arbitrary identical direction. One then determines the resultant of all of these forces. The procedure is then repeated with another arbitrarily fixed orientation of the partial forces. The intersection point of the lines of action of the two resultants obtained this way is the center of mass.

Experimental determination of the center of mass of a plate (**Fig. 3.3**):

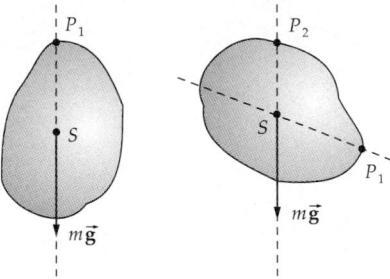

Figure 3.3: Determination of the center of mass of a plate.

- The plate is successively suspended at various points P_1, P_2, ... that do not lie on the same line, and in each case the line along which gravity acts is determined. The intersection point of the various lines so determined is the center of mass S.
➤ The center of mass of a body can lie outside of the body volume.

4. Center-of-mass rule

The motion of the center of mass is not affected by internal forces of the body. The center of mass moves as a point particle that carries the total mass of the entire body and is under the action of the resultant of all external forces.

■ Let two bodies with masses $m_1 = 1$ kg and $m_2 = 3$ kg be connected by a bar of length $l = 2$ m. The mass of the bar is negligible. If the coordinate system is chosen so that the first body lies at the origin, and the second one on the x-axis, the coordinates are

$$\vec{r}_1 = \begin{pmatrix} 0 \\ 0 \end{pmatrix}, \qquad \vec{r}_2 = \begin{pmatrix} l \\ 0 \end{pmatrix} \,.$$

The center of mass then has the coordinates

$$\vec{R} = \frac{m_1\vec{r}_1 + m_2\vec{r}_2}{m_1 + m_2} = \begin{pmatrix} 1.5 \text{ m} \\ 0 \end{pmatrix} ,$$

i.e., it is at a distance 1.5 m away from body 1, and thus 0.5 m away from body 2.

3.1.3 Basic kinematic quantities

1. Coordinate systems

Space-fixed coordinate system, K', coordinate system with the origin fixed in space and with space-fixed directions for the axes. Unit vectors along the axes: $\vec{e}_x{}', \vec{e}_y{}', \vec{e}_z{}'$.

Body-fixed coordinate system, K, an arbitrary point S (reference point) on the rigid body is selected as the coordinate origin. The coordinate axes are fixed to the body. Unit vectors along the axes: $\vec{e}_x(t), \vec{e}_y(t), \vec{e}_z(t)$. These unit vectors along the axes in general vary with time, as seen from the space-fixed coordinate system (**Fig. 3.4**).

➤   One may select the **center of mass** of the rigid body as the origin of the body-fixed coordinate system. For a gyroscope one uses the support point as the coordinate origin (**Fig. 3.5**).

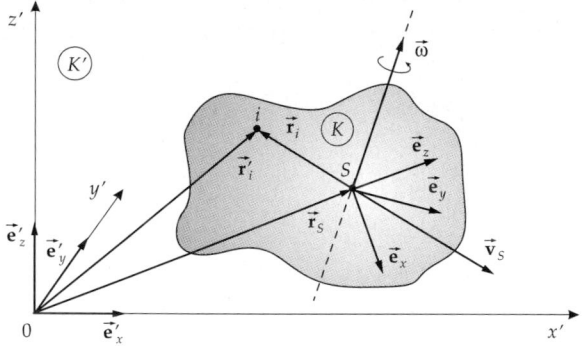

Figure 3.4: Body-fixed (K) and space-fixed coordinate systems (K').

Meaning of symbols in Fig. 3.4:

$\vec{r}_i{}'$:	position vector of point i in the space-fixed reference system K',
$\vec{r}_i$:	position vector of point i in the body-fixed reference system K,
$\vec{r}_S$:	position vector of the reference point in the space-fixed reference system K',
$\vec{v}_S$:	translational velocity of the reference point,
$\vec{v}_i{}'$:	velocity of point i in the space-fixed reference system K',
$\vec{v}_i$:	velocity of point i in the body-fixed reference system K,
$\vec{\omega}$:	angular-velocity vector for rotations about an axis through the reference point.

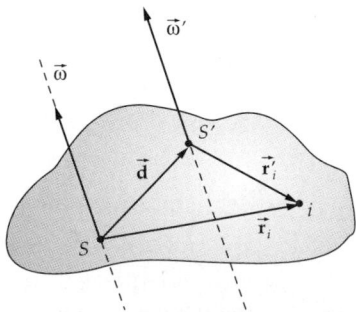

Figure 3.5: Shift of the reference point of the body-fixed coordinate system by $\vec{d}$ from S to S'.

2. Relations between basic kinematic quantities

The quantities in the reference systems K and K' are related as follows.

position vector:　$\vec{r}_i'(t) = \vec{r}_S(t) + \vec{r}_i(t)$,

velocity:　$\vec{v}_i'(t) = \vec{v}_S(t) + \vec{\omega} \times \vec{r}_i(t)$,

acceleration:　$\vec{a}_i'(t) = \vec{a}_S(t) - \dot{\omega} \times \vec{r}_i(t) - 2\vec{\omega} \times \dot{\vec{r}}_i(t) - \vec{\omega} \times (\vec{\omega} \times \vec{r}_i)$.

➤　The translational velocity $\vec{v}_S$ depends on the choice of the reference point S for the body-fixed coordinate system. The angular velocity $\vec{\omega}$ is independent of the choice of this reference point, i.e., body-fixed coordinate systems that refer to different reference points rotate with the same magnitude of angular velocity about axes that are parallel to each other.

3. General motion of a rigid body,

composed of the **translation** of the reference point S with velocity $\vec{v}_S(t)$, and of a **rotation** with angular velocity $\vec{\omega}(t)$ about an axis through S. Both the orientation of the rotation axis and the magnitude of the angular velocity can vary with time.

Fixed axis, axis fixed in the rigid body by external bearings.

Free axis, axis in the rigid body that does not change its orientation as long as there is no torque acting. A free axis is not stabilized by external bearings.

➤　For any rigid body, one can find three free axes that are perpendicular to each other. The axes with the largest and the smallest moment of inertia are always free axes. The third free axis points perpendicularly to the two previously specific axes.

➤　The principal axes of inertia of a rigid body are free axes.

4. Example: Motion of a dumbbell

The motion of a dumbbell can be decomposed into the rotation of the two masses about the center of mass, and the translational motion of the center of mass. If $\vec{\omega}$ is the angular velocity of rotation and $\vec{v}$ the translational velocity, $\vec{R}$ describes a translation of the center of mass:

$$\vec{R}(t) = \vec{R}_0 + \vec{v}t$$

($\vec{R}_0$: position of the center of mass at time $t = 0$). The relative coordinates $\Delta\vec{r}_i = \vec{r}_i - \vec{R}$ describe a rotation:

$$\Delta\vec{r}_1(t) = l_1 \begin{pmatrix} \cos\omega t \\ \sin\omega t \end{pmatrix}, \qquad \Delta\vec{r}_2(t) = -l_2 \begin{pmatrix} \cos\omega t \\ \sin\omega t \end{pmatrix},$$

l_1 and l_2 being the (constant) distance of each of the bodies from the center of mass; $\vec{l}_1$, $\vec{l}_2$ denote the vectors from the center of mass to the two dumbbell masses. The entire motion is then described by the equations

$$\vec{r}_1(t) = \vec{R}(t) + \Delta\vec{r}_1(t) = \vec{R}_0 + \vec{v}t + l_1 \begin{pmatrix} \cos\omega t \\ \sin\omega t \end{pmatrix},$$

$$\vec{r}_2(t) = \vec{R}(t) + \Delta\vec{r}_2(t) = \vec{R}_0 + \vec{v}t - l_2 \begin{pmatrix} \cos\omega t \\ \sin\omega t \end{pmatrix}.$$

3.2 Statics

Statics, theory of the equilibrium of forces on a rigid body. It serves in particular for evaluating forces that arise in trusses, bearings and beams (**architectural statics**).

3.2.1   Force vectors

1. Force vector and point of application

Forces acting on a rigid body are represented by **force vectors**. These differ from ordinary vectors because they also involve a point of application that specifies at what point the force acts.

Force vector, characterized by its magnitude (length), its direction (**action line**) and its **point of application**. The force vector is visualized by an arrow beginning at the point of application and pointing along the line of action; its length specifies the magnitude of the force (**Fig. 3.6**).

▲   A force acting on a rigid body can be arbitrarily shifted along its action line. (**Fig. 3.7**).

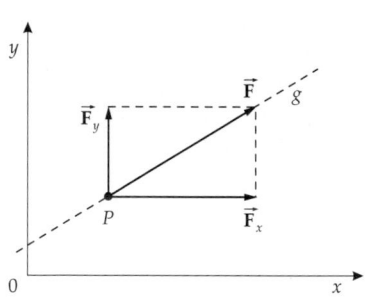

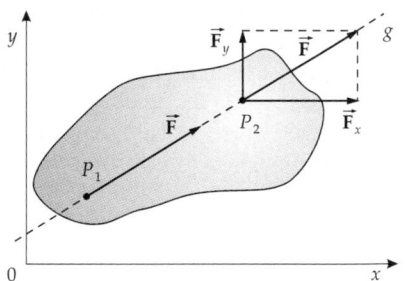

Figure 3.6: Force vector $\vec{F}$ with the action point P, the line of action g, the components $\vec{F}_x, \vec{F}_y$ and the magnitude $F = \sqrt{F_x^2 + F_y^2}$.

Figure 3.7: Shift of a force along its line of action g from P_1 to P_2.

2. Composition of plane forces

Plane system of forces, set of forces that all lie within a plane.

Resulting force, replaces two plane forces $\vec{F}_1$ and $\vec{F}_2$ at the same point of application by a single force $\vec{F}_R$. This is done by means of the **parallelogram of forces** (see p. 45). Here the second force vector is shifted parallel to the end of the first one. The connection line from the point of application of the first force vector to the end point of the second represents the resulting force $\vec{F}_R$ (**Fig. 3.8**).

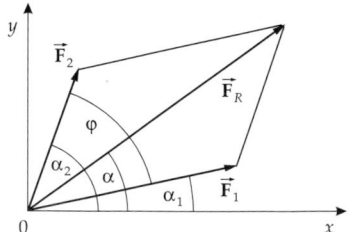

Figure 3.8: Parallelogram of forces. Addition of the forces $\vec{F}_1$ and $\vec{F}_2$ to form the resulting force $\vec{F}_R$.

The magnitude of the resulting force is obtained by the cosine rule:

resulting force = vector sum of individual forces			MLT^{-2}
	Symbol	Unit	Quantity
$\vec{F}_R = \vec{F}_1 + \vec{F}_2$ $|\vec{F}_R| = \sqrt{|\vec{F}_1|^2 + |\vec{F}_2|^2 + 2|\vec{F}_1||\vec{F}_2|\cos\varphi}$	$\vec{F}_R$ $\vec{F}_1, \vec{F}_2$ φ	N N rad	resulting force force vectors angle between $\vec{F}_1$ and $\vec{F}_2$

3. Polygon of forces

By repeating this procedure, arbitrarily many forces that act on the same point can be replaced by a single resulting force:

▲　$\vec{F}_R = \vec{F}_1 + \vec{F}_2 + \vec{F}_3 + \cdots$

This can be represented graphically by a **polygon of forces** (see p. 45): The force arrows are aligned by parallel shifting (i.e., keeping the magnitude and orientation fixed). The resultant is the force arrow from the beginning of the first to the end of the last force arrow (**Fig. 2.8**).

➤　The resulting force can also be evaluated by adding the components of the individual forces:

$$\vec{F}_R = \begin{pmatrix} F_{Rx} \\ F_{Ry} \\ F_{Rz} \end{pmatrix} = \vec{F}_1 + \vec{F}_2 = \begin{pmatrix} F_{1x} + F_{2x} \\ F_{1y} + F_{2y} \\ F_{1z} + F_{2z} \end{pmatrix}.$$

4. Parallel or opposite forces

If two vectors point in the same direction ($\varphi = 0$), then

$$|\vec{F}_R| = |\vec{F}_1 + \vec{F}_2| = |\vec{F}_1| + |\vec{F}_2|.$$

If they point in opposite directions ($\varphi = \pi$), then

$$|\vec{F}_R| = |\vec{F}_1 + \vec{F}_2| = ||\vec{F}_1| - |\vec{F}_2||.$$

If the forces are perpendicular to each other ($\varphi = \pi/2$), then

$$|\vec{F}_R| = |\vec{F}_1 + \vec{F}_2| = \sqrt{|\vec{F}_1|^2 + |\vec{F}_2|^2}.$$

To add two forces acting on two different points of a rigid body, shift them to the intersection point of their lines of action and add them there according to the parallelogram of forces (**Fig. 3.9**).

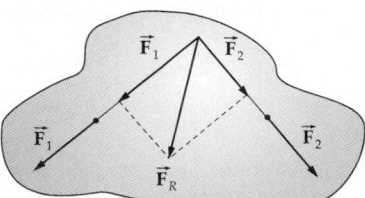

Figure 3.9: Addition of plane forces acting on a rigid body.

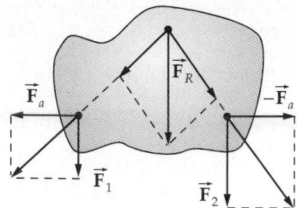

Figure 3.10: Addition of plane parallel forces on a rigid body.

If two forces point in the same direction, but along parallel lines of action with a finite perpendicular distance between them, there is no intersection point. One therefore adds to the forces $\vec{F}_1$ und $\vec{F}_2$ opposite auxiliary forces $\vec{F}_a$ and $-\vec{F}_a$ with the same line of action. The auxiliary vectors cancel when added, but allow the shift of the actual forces to a common point of application (**Fig. 3.10**).

3.2.2 Torque

1. Torque of an applied force

Torque, the product of the magnitude of the applied force and the length of the **lever arm** to a reference point where the body is mounted rotatably (**center of rotation**). Similar to a force that can cause a translational motion, a torque can put a freely movable rigid body into a **rotational motion** about the center of mass (**rotation**, see p. 31) (**Fig. 3.11**).

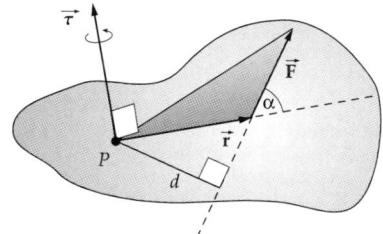

Figure 3.11: Torque $\vec{\tau}$ of the force $\vec{F}$ with respect to the center of rotation P.

magnitude of torque			$\mathbf{ML^2T^{-2}}$
	Symbol	Unit	Quantity
$\tau = F \cdot d$	τ	Nm	magnitude of torque
	F	N	applied force
	d	m	lever arm

Newtonmeter, Nm, SI unit of the torque. 1 Nm is the torque about the center of rotation generated by a force of 1 N on a lever arm of 1 m.

$$1 \ \text{Nm} = 1 \ \text{N} \cdot 1 \ \text{m}$$

➤ The lever arm is the vertical distance of the line of action of the force from the center of rotation.

➤ If the point of application of the force is given, the lever arm is

$$d = r \sin \alpha ,$$

where $\vec{r}$ is the vector from the center of rotation to the point of application of the force, and α is the angle between $\vec{r}$ and the force vector $\vec{F}$.

2. Properties of the torque

The torque is a vector pointing along the direction of the rotation that the torque would create:

$$\vec{\tau} = \vec{\mathbf{r}} \times \vec{\mathbf{F}}, \quad |\vec{\tau}| = |\vec{\mathbf{r}}|\,|\vec{\mathbf{F}}|\,\sin\alpha = d\,|\vec{\mathbf{F}}|.$$

The vector product

$$\vec{\tau} = \vec{\mathbf{r}} \times \vec{\mathbf{F}}$$

is also denoted the **moment** of the force $\vec{\mathbf{F}}$.

■   A force of $F = 5$ N acts in a distance of $d = 20$ cm on a screw. The applied torque is

$$\tau = F \cdot d = 5\,\text{N} \cdot 20\,\text{cm} = 1\,\text{Nm}.$$

➤   If the line of action of the force passes through the center of rotation, the lever arm equals zero and the torque vanishes.

▲   If one doubles the lever arm and keeps the force constant, the torque also doubles. Application: wrench.

3. Resulting torque

The torques produced by the forces $\vec{\mathbf{F}}_1, \vec{\mathbf{F}}_2, \ldots, \vec{\mathbf{F}}_n$ can be combined to form a resulting moment $\vec{\tau}_R$ (**Fig. 3.12**),

$$\vec{\tau}_R = \sum_{i=1}^{n} \vec{\mathbf{r}}_i \times \vec{\mathbf{F}}_i \,,$$

where $\vec{\mathbf{r}}_i$ is the position vector of the point of application of the force $\vec{\mathbf{F}}_i$.

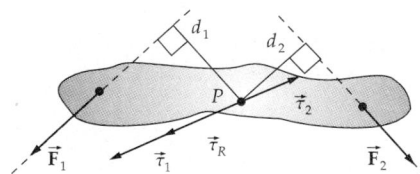

Figure 3.12: Addition of torques. The forces $\vec{\mathbf{F}}_1$, $\vec{\mathbf{F}}_2$ form a plane set of forces. The moments $\vec{\tau}_1$, $\vec{\tau}_2$ are perpendicular to the plane.

addition of torques			$\mathbf{ML^2T^{-2}}$
	Symbol	Unit	Quantity
$\vec{\tau}_R = \vec{\tau}_1 + \vec{\tau}_2 + \cdots$	$\vec{\tau}_R$	Nm	resulting torque
	$\vec{\tau}_1, \vec{\tau}_2, \ldots$	Nm	torques

3.2.3 Couples

1. Couple and torque of a couple

Couple, two antiparallel forces of equal magnitude, $\vec{\mathbf{F}}_1, \vec{\mathbf{F}}_2 = -\vec{\mathbf{F}}_1$, that act on different points of the rigid body so that their lines of action do not coincide. A couple cannot be reduced to a single force.

For a couple, the resulting force vanishes, $\vec{\mathbf{F}}_1 + \vec{\mathbf{F}}_2 = 0$, hence the translational state of the rigid body is not changed by a couple. The resulting torque, however, does not vanish.

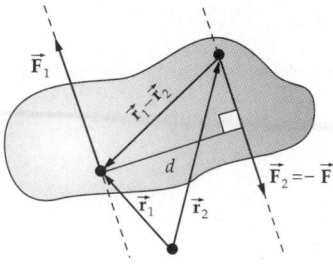

Figure 3.13: Torque of a couple $(\vec{F}_1, \vec{F}_2)$. d: perpendicular distance of the lines of action of the forces.

Torque of a couple depends only on the forces and the distance vector between the point of applications (**Fig. 3.13**):

$$\vec{\tau} = (\vec{r}_1 - \vec{r}_2) \times \vec{F}_1, \qquad \tau = F_1 \cdot d, \qquad d: \text{ distance of the lines of action.}$$

A couple can cause a rotation of the body. The sense of rotation is fixed by the definition of the vector product so that $\vec{r}_1 - \vec{r}_2, \vec{F}_1$ and $\vec{\tau}$ form a right-handed system. The torque of a couple is independent of the reference point. Unlike shifting a force vector off its line of action, the balance of torques remains unaffected when shifting a couple in its plane on the rigid body.

▲   A couple can be moved within its plane without changing its static influence on the rigid body. The vector of the torque of a couple is a **free vector**.

2. Reduction of a plane-force system

Every plane-force system acting on a rigid body can be reduced to a resulting single force and a couple. The point of application of the resultant can be freely chosen (**Fig. 3.15**).

Parallel shift of a force, a force $\vec{F}$ can be shifted parallel to its line of action from the point of application P to the action point P' if one introduces a couple $\vec{F}, -\vec{F}$ (**Fig. 3.14**).

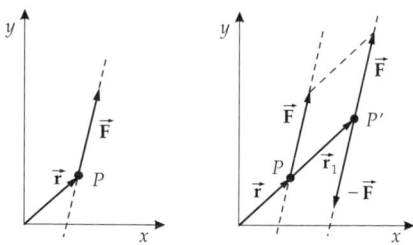

Figure 3.14: Parallel shift of a force $\vec{F}$ by introducing the shift moment $\vec{\tau}_1 = \vec{r}_1 \times \vec{F}$.

Shift moment, $\vec{\tau}_1$, compensates the change of the torque of force $\vec{F}$ due to the shift, $\vec{\tau}_1 = \vec{r}_1 \times \vec{F}$.

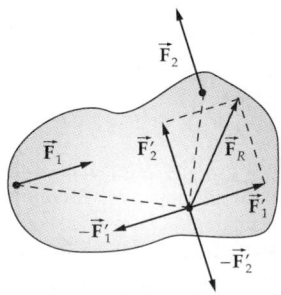

Figure 3.15: Reduction of a plane-force system $\vec{F}_1, \vec{F}_2$ to a single resultant force $\vec{F}_R$ and two couples $(\vec{F}_1, -\vec{F}'_1)$ and $(\vec{F}_2, -\vec{F}'_2)$. The torques of these can be combined to form a single torque.

3.2.4 Equilibrium conditions of statics

A body is at rest if the following conditions are fulfilled (**Fig. 3.16**):

The resultant of all applied forces vanishes. The sum of all torques vanishes. The first rule guarantees that the body is not put into a translational motion; the second rule guarantees that it does not perform a rotation.

$$\vec{F}_R = \vec{F}_1 + \vec{F}_2 + \cdots = 0$$

$$\vec{\tau}_R = \vec{\tau}_1 + \vec{\tau}_2 + \cdots = 0$$

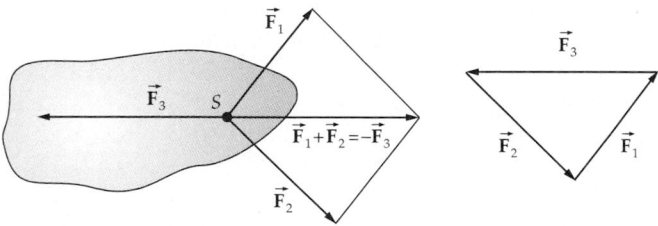

Figure 3.16: Equilibrium of a rigid body. S: center of mass.

➤ In component notation, these two vector equations correspond to the following six equations:

$$F_{1x} + F_{2x} + \cdots = \sum_i F_{ix} = 0 \qquad \tau_{1x} + \tau_{2x} + \cdots = \sum_i \tau_{ix} = 0$$

$$F_{1y} + F_{2y} + \cdots = \sum_i F_{iy} = 0 \qquad \tau_{1y} + \tau_{2y} + \cdots = \sum_i \tau_{iy} = 0$$

$$F_{1z} + F_{2z} + \cdots = \sum_i F_{iz} = 0 \qquad \tau_{1z} + \tau_{2z} + \cdots = \sum_i \tau_{iz} = 0$$

➤ If all forces act on the same point, the equilibrium condition reduces to

$$\vec{F}_1 + \vec{F}_2 + \cdots = 0 \,,$$

since the sum of the torques then also vanishes. If all forces are coplanar, the component equation for the coordinate perpendicular to the plane can be omitted.

Forces with lines of action intersecting at one point are in equilibrium if the force diagram forms a **closed polygon**.

The law of levers follows from the second rule: If two forces F_1 and F_2 act on a rigid body at the distances d_1 and d_2 from the center of rotation and are in equilibrium,

$$F_1 : F_2 = d_2 : d_1 \,.$$

1. Static stability

A body standing on a surface gets a **support force** that balances its weight. The support force is the resultant of forces applied where the body rests on the support. Hence, it can act only between the **edges**, i.e., the extremum points where the body is still being supported.

▲ A body is stable if the vertical line from the center of mass intersects the support plane within the edges (**Fig. 3.17**).

➤ If a support is added to the body as shown in the figure, the edge is shifted to the point of application of the bearing.

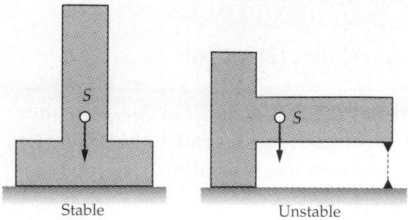

Figure 3.17: Tilting of a body. The body remains stable as long as a vertical line from the center of mass S intersects the support area between the extremum supporting points. Broken line: possible location of a support for stabilization.

A body is more stable if
- the horizontal distance of the center of mass from the edges is larger, i.e., the center of mass is closer to the middle;
- the center of mass is lower;
- the weight of the body is larger.

Tilting moment, the torque needed to tilt the body over:

tilting moment = distance from edge · weight force			$\mathbf{ML^2T^{-2}}$
	Symbol	Unit	Quantity
$\tau = d \cdot mg$	τ	Nm	tilting moment
	d	m	horizontal distance of center of mass from the edge
	mg	N	weight force

2. Spatial statics

Spatial statics, composition and resolution of forces in 3D space, where the lines of action in general do not intersect each other in space and are not parallel to each other. The addition of the forces and moments leads to a resulting force $\vec{F}$ and a resulting moment $\vec{\tau}$. The resulting moment can be represented by a couple $\vec{F}_1, \vec{F}_2$, which can be shifted until the force $\vec{F}_1$ coincides with the point of application of $\vec{F}$. The addition of $\vec{F}$ and $\vec{F}_1$ yields a single force $\vec{F}_{res}$. There remain two forces $\vec{F}_2 = -\vec{F}_1$ and $\vec{F}_{res}$ that cannot be further simplified.

3.2.5 Technical mechanics

3.2.5.1 Bearing reactions

Bearing, a point where a rigid body in static equilibrium under applied forces (e.g. weight) is supported.

Bearing reaction, force acting from the bearing on the body. It originates from the forces acting on the supported body (in general, the weight) that must be offset according to the equilibrium condition of statics.

1. Various types of bearings

One distinguishes:

Roller bearings, which support only loads perpendicular to the bearing (for example, a plate supported by a beam);

Thrust bearings and **Journal bearings**, which provide lateral and axial support, respectively, but permit rotations (for example, an axle for a rotating shaft);

Clamps, which prevent both displacements and rotations, thus supporting both forces and moments (for example, in a **vice**).

▲   At points where the body is not supported, no internal forces or moments may appear.

2. Connections between rigid bodies,

transfer of forces from one body to another.

One distinguishes (see **Fig. 3.18**):

- **Socket**, transmits longitudinal forces only;
- **Joint**, transmits forces along and perpendicular to the beam, but permits rotations;
- **Hinge**, transmits forces and moments parallel to the axis;
- **Rigid connection**, transmits all forces and moments.

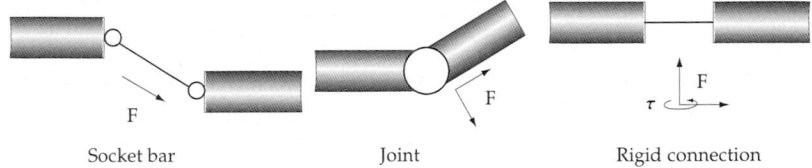

Socket bar   Joint   Rigid connection

Figure 3.18: Connections.

3.2.5.2   Trusses

Truss, construction for compensation and distribution of forces, in particular in buildings. A truss consists of straight **beams** or **rods** that are flexibly joined or clamped at their **junctions**. They transfer external forces, that in general are applied only at junctions, along the beam orientation.

Plane truss, truss with beams and all forces in a plane (**Fig. 3.19**). One has to calculate the forces on all beams if the external forces and the bearings are given. To have a determinate system:

plane truss			
	Symbol	Unit	Quantity
$2K = S + 3$	K	1	number of junctions
	S	1	number of rods

➤   For the forces acting on **beams**, see p. 153.

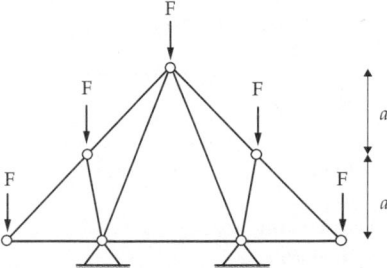

Figure 3.19: Plane truss.

3.2.6   Machines

▲   **Golden rule of mechanics**: A smaller force must be compensated by a longer path
(energy conservation).

3.2.6.1   Lever

1. Kinds of levers

Lever, a rigid body that is supported at one point, or can rotate about a fixed axis. Two
forces $\vec{F}_1$ (force) and $\vec{F}_2$ (load), the lines of action of which have perpendicular distances
d_1 and d_2 from the center of rotation, generate the torques $\vec{\tau}_1$, $\tau_1 = d_1 \cdot F_1$ and $\vec{\tau}_2$, $\tau_2 = d_2 \cdot F_2$. The lever is in equilibrium if the total torque $\vec{\tau} = \vec{\tau}_1 + \vec{\tau}_2$ vanishes,

$$\vec{\tau} = \vec{\tau}_1 + \vec{\tau}_2 = 0.$$

Lever arm, perpendicular distance of the center of rotation from the line of action of a
force acting on the lever.
Straight lever, a rod that can rotate about a point bearing.
One-armed straight lever, load and force act on the same side, as seen from the center of
rotation.
Two-armed straight lever, load and force act on different sides of the center of rotation.
Bent lever, the lever arms include an angle (**Fig. 3.20**).

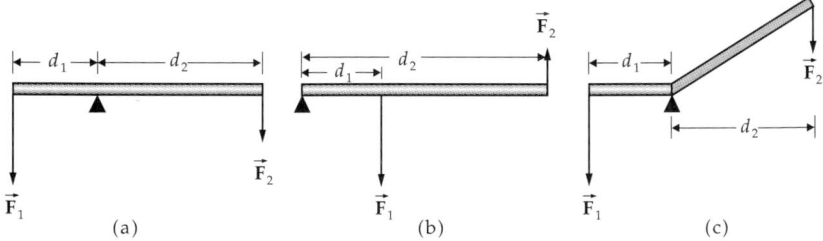

Figure 3.20: Lever. (a): Two-armed straight lever, (b): one-armed straight lever, (c): bent
lever.

Levers are applied to lift or shift loads or to reach a balance of forces.

2. Law of levers

law of levers: In equilibrium, the ratio of forces is the inverse of the ratio of the lever arms			ML^2T^{-2}
	Symbol	Unit	Quantity
$\vec{\tau}_1 = -\vec{\tau}_2$	F_1	N	applied force to keep the equilibrium
$F_1 d_1 = F_2 d_2$	F_2	N	load
$F_1 : F_2 = d_2 : d_1$	$\vec{\tau}$	Nm	torque
	d_1	m	force arm
	d_2	m	load arm

➤ The law of levers holds also for **bent levers**.

■ **Scale**, for measuring an unknown weight. The scale can be brought to balance either by changing or shifting the counterweight (**bridge scale**).

Wheelbarrow, one-armed lever, force arm is longer than load arm.

Catapult, the load arm is longer than the force arm, hence one can accelerate an object over a long path.

Pressure lever (**nutcracker**), two joined one-armed levers with the force arm longer than the load arm—for amplifying forces, as in **scissors** or **pliers**.

3.2.6.2 Wedges and screws

1. Wedge,

transforms the force $\vec{F}$ of hammer blows into two forces $\vec{F}_{N,1}, \vec{F}_{N,2}$ (normal forces) acting perpendicular to the sides of the wedge (**Fig. 3.21**). According to the law of vector decomposition,

$$F_{N,1} = F_{N,2} = \frac{F}{2\sin\alpha} ,$$

where α is the inclination angle (half of the wedge angle) of the surface with respect to the applied force.

2. Screw

analogous to an inclined plane wound around a cylinder. A screw is characterized by its pitch h (distance between subsequent screw turns) and its mean thread radius r. If an external force F_1 acts at a distance R from the screw axis, the point of application moves over a distance $b = 2\pi R$ each turn of the screw, and the screw moves forward by the amount h (**Fig. 3.22**). Hence, for the driving force F_2 exerted by the screw,

$$F_2 = F_1 \frac{2\pi R}{h} .$$

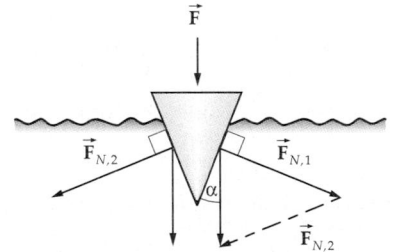

Figure 3.21: Wedge.

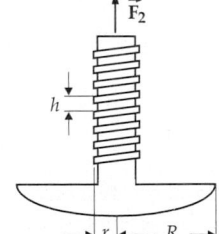

Figure 3.22: Screw.

Friction for screws, in contrast to the driving force exploited in drilling, screws are used for clamping one body to another. The forward driving force now acts as a support force that causes a correspondingly large friction force. This acts against the rotation and prevents the screw from loosening. The frictional force is about the same whether the screw is being driven forward or backward. For a tension force F_2 and a friction coefficient μ, the opposing force on a lever of length R is

$$F_1 = F_2 \frac{\mu h}{2\pi R} .$$

3.2.6.3 Pulleys

1. Pulley,

combined with **ropes**, **chains**, **gears** or **V-belts** for transmission and amplification of forces. In general, a device consisting of one or several (possibly different diameters) pulleys guiding a rope. An external force F_1 pulls at the rope, while the load (with the weight F_2) is fixed either at the other end of the rope or at the axis of one of the rollers. If the rollers of the device have different diameter (gear) or some of them are free rollers (pulley), the same torque causes different forces. To analyse this problem, one determines which force F_1 is needed according to the law of levers (equality of torques) to compensate the weight F_2.

2. Types of pulleys

Fixed pulley (**Fig. 3.23**), guides a rope. The force is transmitted and its magnitude remains unchanged. One pulls at one end of the rope, the other end carries the load. Static balance holds when

$$F_1 = F_2 .$$

Free pulley (**Fig. 3.24**), the rope is fixed at one end, the load is carried by the pulley. If one pulls the rope a distance d, the pulley—and thus the load—moves $d/2$. According to the law of levers, the equilibrium condition now reads

$$F_2 = \frac{F_1}{2} .$$

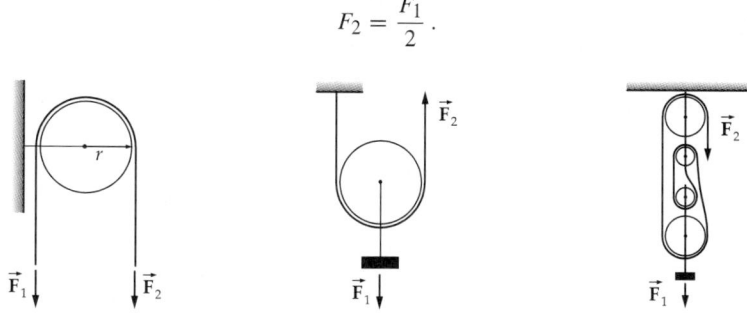

Figure 3.23: Fixed pulley.   Figure 3.24: Free pulley.   Figure 3.25: System of pulleys.

Systems of pulleys (**Fig. 3.25**), contains two groups of pulleys, with $2n$ pulleys in total which guide the rope. For static equilibrium

$$F_2 = \frac{F_1}{2n} .$$

n is the number of pulleys in each group, or the number of ropes that move in the middle of the pulley parallel at one side (with the same direction of motion).

➤   The diameter of the pulleys does not enter into the equation.

3. Gears,

devices for transmission and conversion of forces, in particular for converting torques. A gear is driven by a torque τ_1 on a **drive shaft** with rotation velocity ω_1 and transmits another torque τ_2 with another rotation velocity ω_2 on the **driven end**. In an ideal gear without friction, energy is conserved:

$$P_1 = P_2 \qquad \Longleftrightarrow \qquad \tau_1 \omega_1 = \tau_2 \omega_2 \,,$$

(P_1, P_2: powers, τ_1, τ_2: torques, ω_1, ω_2: angular velocities). Real gears lose energy to friction. The heat produced must be dissipated by cooling mechanisms. The losses can be reduced by **lubrication**.

4. Belt drives,

two pulleys are tightly coupled by a belt. Since the belt exerts the same force F on both pulleys (**Fig. 3.26**), the ratio of the torques is given by

$$\frac{\tau_1}{\tau_2} = \frac{Fr_1}{Fr_2} = \frac{r_1}{r_2} \,.$$

Let v be the velocity of the belt. The angular velocities of the pulleys are given by

$$\frac{\omega_1}{\omega_2} = \frac{v/r_1}{v/r_2} = \frac{r_2}{r_1} \,.$$

▲   The torques are proportional to the radii, the angular frequencies are inversely proportional to the radii.

■   V-belts in engines. Electric drives by small electric motors with low torque, but high rate of rotation. Chain gear for bicycles.

5. Gearboxes,

transmission of the force is not by a belt, but through direct contact of the gears. In particular: **toothed gears**. The efficiency is higher and the construction is more compact, but the requirements on the material are more stringent (**Fig. 3.27**).

■   Drive shafts of engines for vehicles. Machine tools. Clocks.

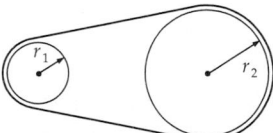

Figure 3.26: Belt drive.

Figure 3.27: Toothed gear.

6. Multistage gears,

result from the chaining of several simple gears. Used in particular as **transmissions** for automobiles, since the internal combustion engine works efficiently over only a small range of rotational frequency: By shifting the tooth wheels, one can select any of various combinations of gear ratios and thus produce a variety of ratios between the rotation speed of the crankshaft and the drive shaft. In modern transmissions, all gear wheels are spinning simultaneously and are connected as needed to the crankshaft. In a synchronized transmission, an additional friction coupling brings the gear and the crankshaft to the same rotational speed.

7. Automatic gears,

change automatically depending on the rate of rotation. One either uses conventional switch gears that are changed automatically by central force governors, or **planetary gears**. In the latter device, the planetary wheels run freely between a tooth wheel tightly connected to the drive shaft, and a gear rim. If the gear rim is fixed, the planetary wheels perform a rotation that is used as the driving gear. If the gear rim is free, however, it is driven instead of the planetary gear. Shifting gears is simply achieved by braking the gear rim.

8. Continuous gear,

can be realized by **hydraulics (liquid gears)**. The transmission of the force works by the viscous flow of a light oil: At low rotation rates it rotates almost freely, at higher turns the friction increases and thus the coupling becomes tighter. Application in automobiles with automatic transmissions.

Continuous mechanical gears, use cone-shaped pulleys: the drive radius, and thus the transmission, can be varied by changing the position of the V-belt depending on the torque.

9. Differential gears,

serve to distribute torque. These are gears in which the rotational speed and moment of the drive shafts are not uniquely determined. Torques are delivered to the shafts depending on the resistance in each of the shafts. Usually realized as a cone-gear differential in which four conical gears engage each other in a circle.

➤   In a broader sense, gears include screw gears (for transforming rotational motion into translation or vice versa) and hydraulic presses (see p. 173).

10. Crank mechanism,

for conversion of a (periodic) translational motion into a rotation and vice versa (e.g., driving a shaft with a piston). A **connecting rod** of length l is connected at one end to a rotating shaft by a joint at a distance r from the rotation axis. The other end slides on a rail back and forth between two **end points** (**Fig. 3.28**). The relation between rotation angle α and path s on the rail, both being measured from the upper end point, is

$$s = r\left(1 + \frac{\lambda}{2}\sin^2\alpha - \cos\alpha\right), \quad \text{for } \lambda^2 \ll 1$$

with a **connection-rod ratio** λ:

$$\lambda = \frac{r}{l}.$$

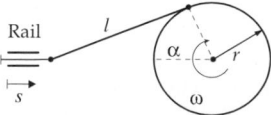

Figure 3.28: Crank mechanism.

3.3 Dynamics

Dynamics of rigid bodies, describes the motion of rigid bodies under the action of forces. The mechanical behavior of the rigid body follows from six differential equations that cover the translational motion of the center of mass $\vec{R}$ under the action of the force $\vec{F}$ and the time variation of the angular momentum $\vec{L}$ by the torque $\vec{\tau}$:

$$m\ddot{\vec{R}} = \vec{F}, \qquad \frac{d\vec{L}}{dt} = \vec{\tau}.$$

3.4 Moment of inertia and angular momentum

The concepts of torque, angular momentum and moment of inertia involved in the description of **rotational motions** are the analogs to the concepts of force, (linear) momentum and mass for linear motions. They are related to each other by the fundamental law of dynamics for rotational motion.

■   The simplest form of a rotation is the circular motion of a point mass about a fixed axis (**Fig. 3.29**).

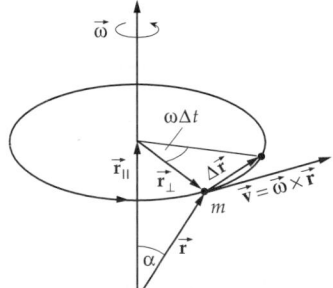

Figure 3.29: Circular motion of a point mass m with the orbital velocity $\vec{v}$ and angular velocity $\vec{\omega}$. The rotation axis and the sense of rotation are specified by the vector $\vec{\omega}$.

We now consider **rotation about fixed axes**. The **theory of the top** deals with the description of rotations about movable axes.

The rotation of a rigid body about a fixed axis can be described by analogy to linear motion. The angle ϕ, which describes the position of the body at a given time, is analogous to the coordinate x.

3.4.1 Moment of inertia

The moment of inertia describes the angular acceleration produced by an applied torque. It depends on the shape and mass distribution of the body, and on the orientation of the rotation axis. The moment of inertia plays the same role for rotation as the mass does for a translational motion; it describes the resistance of a body to a change of its state of motion (here: angular velocity).

1. Moment of inertia with respect to an axis

Moment of inertia, J_X with respect to an axis X, the proportionality constant between the torque τ_X about the axis X and the resulting angular acceleration α_X of rotation about the axis:

torque = moment of inertia · angular acceleration	$\mathbf{ML^2T^{-2}}$		
	Symbol	Unit	Quantity
$\tau_X = J_X \cdot \alpha_X$	τ_X	Nm	torque
	J_X	kg m^2	moment of inertia
	α_X	rad/s^2	angular acceleration

Kilogram times meter squared, kg m^2, SI unit of moment of inertia:
1 kilogram times 1 meter squared is the moment of inertia of a body that is given an angular acceleration of 1 rad/s^2 when a torque of 1 Nm is applied to it.
➤   This formula is analogous to presenting the force as "mass times acceleration".
➤   All quantities refer to the rotation axis X. The moment of inertia of a body depends on the choice of axis.
To calculate the moment of inertia of a rigid body, it is resolved into mass elements that move at a fixed distance from the rotation axis.

2. Moment of inertia of a point mass,

that moves with the angular velocity $\vec{\omega}$ along a circular orbit with radius r (**Fig. 3.30(a)**), follows from the fundamental law of dynamics:

$$\vec{\tau} = \vec{r} \times \vec{F} = \vec{r} \times m\frac{d\vec{v}}{dt}, \quad \vec{v} = \vec{\omega} \times \vec{r}, \quad |\vec{\tau}| = r \cdot mr\frac{d\omega}{dt} = mr^2\alpha.$$

One finds:

moment of inertia of a point mass	$\mathbf{ML^2}$		
	Symbol	Unit	Quantity
$J_X = m \cdot r^2$	J_X	kg m^2	moment of inertia with respect to axis X
	m	kg	mass
	r	m	distance from rotation axis X

▲   The moment of inertia of a point mass is the product of the mass m and the square of its perpendicular distance r from the rotation axis.

3. Moment of inertia of a rigid body,

obtained by resolving the body into mass elements Δm and summing up (**Fig. 3.30(b)**):

moment of inertia of a rigid body	$\mathbf{ML^2}$		
	Symbol	Unit	Quantity
	J_X	kg m^2	moment of inertia with respect to the axis X
$J_X = \sum_{i=1}^{N} \Delta m_i \, r_i^2 = \iiint r^2 \, dm$	Δm_i	kg	ith mass element
$dm = \rho \, dV$	ρ	kg/m^3	density
	dV	m^3	volume element
	r_i	m	distance of element i from rotation axis X

▲ The moment of inertia of a body depends on the choice of rotation axis.
➤ The moment of inertia of a rigid body is a **tensor quantity** (see p. 121).

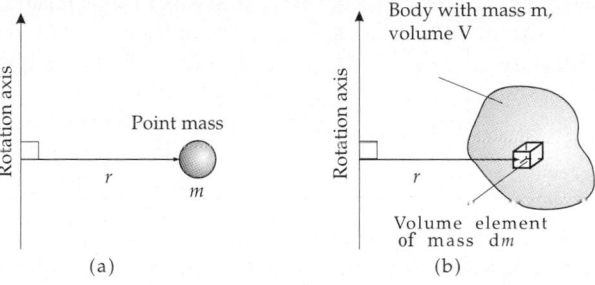

Figure 3.30: Moment of inertia. (a): point mass on a circular orbit, (b): rigid body.

4. Moment of inertia of planar bodies

Equatorial moments of inertia (Fig. 3.31(a)):

$$J_x = \int y^2 \, dA \,, \quad J_y = \int x^2 \, dA \,, \quad dA = dx \, dy \,.$$

Polar moment of inertia (Fig. 3.31(b)):

$$J_p = \int r^2 \, dA \,, \quad r^2 = x^2 + y^2 \,, \quad dA = dx \, dy \,.$$

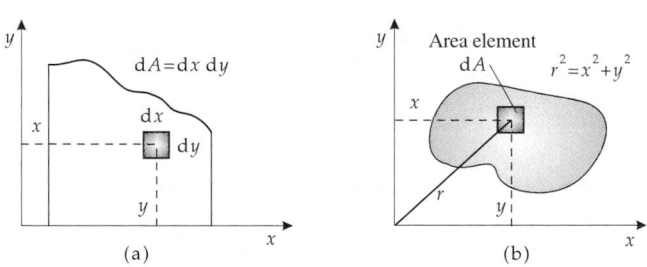

Figure 3.31: Plane moments of inertia. (a): equatorial, (b): polar.

Relation between equatorial and polar moments of inertia: $J_p = J_x + J_y$.

3.4.1.1 Steiner's rule

Steiner's rule establishes a relation between the moment of inertia with respect to an axis X_S through the center of mass, and the moment about an arbitrary parallel axis X (**Fig. 3.32**):

▲ **Steiner's rule**: The moment of inertia of a body with respect to an arbitrary axis X a distance r_S from the center of mass S:

Steiner's rule			**ML^2**
	Symbol	Unit	Quantity
$J_X = mr_S^2 + J_S$	J_X	kg m^2	moment of inertia relative to axis X
	m	kg	mass of body
	r_S	m	distance of the axis X from center of mass S
	J_S	kg m^2	moment of inertia relative to an axis through the center of mass parallel to axis X

➤ The rotation of a body about an arbitrary axis can thus be interpreted as a rotation about a center-of-mass axis parallel to the selected rotation axis (moment of inertia J_S), in addition to a rotation of the center of mass about the selected axis (moment of inertia mr_S^2) in which the total mass is thought to be concentrated in the center of mass.

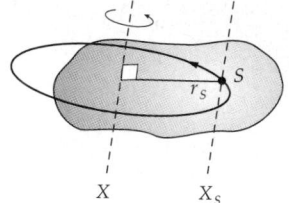

X X_S

Figure 3.32: Steiner's rule.

3.4.1.2 Moments of inertia of geometrical bodies

type	rotation axis	moment
thin rod (length l)	S	$\dfrac{1}{12}ml^2$
	S l	$\dfrac{1}{3}ml^2$
plate (edge lengths a, b, c)	b S c a	$\dfrac{1}{12}m(a^2 + b^2)$
	b S c a	$\dfrac{1}{12}ma^2$
thin circular disk (radius r)	S	$\dfrac{1}{2}mr^2$
	S	$\dfrac{1}{4}mr^2$

type	rotation axis	moment
thin circular ring (radius r)		mr^2
		$\frac{1}{2}mr^2$
cuboid (edge lengths a, b, c)		$\frac{1}{12}m(a^2 + b^2)$
		$\frac{1}{12}m(4a^2 + b^2)$
circular cylinder (radius r, height h)		$\frac{1}{2}mr^2$
		$\frac{1}{12}m(h^2 + 3r^2)$
circular cone (radius r, height h)		$\frac{3}{10}mr^2$
		$\frac{3}{80}m(h^2 + 4r^2)$
sphere (radius r)		$\frac{2}{5}mr^2$
		$\frac{7}{5}mr^2$
hollow sphere (inner radius r_i, outer radius r_a)		$\frac{2}{5}m\frac{r_a^5 - r_i^5}{r_a^3 - r_i^3}$
		$m\frac{7r_a^5 + 5r_a^2 r_i^3 - 2r_i^5}{5(r_a^3 - r_i^3)}$
ellipsoid (half axes a, b, c)		$\frac{1}{5}m(a^2 + b^2)$

3.4.2 Angular momentum

Similar to the momentum in the case of translational motion, one introduces an **angular momentum** for rotations of a rigid body about a fixed axis. Angular momentum is a vector quantity that points along the rotation axis.

1. Definition of the angular momentum of rigid bodies

Angular momentum, $\vec{L}$, the product of the moment of inertia J_X relative to the rotation axis X and the angular velocity $\vec{\omega}$:

angular momentum = moment of inertia · angular velocity			ML^2T^{-1}
	Symbol	Unit	Quantity
$\vec{L} = J_X \cdot \vec{\omega}$	$\vec{L}$	kg m^2/s	angular momentum
	J_X	kg m^2	moment of inertia
	$\vec{\omega}$	rad/s	angular velocity

Kilogram times meters squared per second, SI unit of angular momentum.
1 kg m^2/s is the angular momentum of a body with moment of inertia 1 kg m^2 that rotates with angular velocity 1 rad/s.
➤ This definition is the analog to the definition of (linear) momentum = mass times velocity.

2. Fundamental law of dynamics for rotational motion

For rotational motions: the torque equals the change of the angular momentum per unit time.

change of angular momentum per unit time = torque			ML^2T^{-2}
	Symbol	Unit	Quantity
$\dfrac{d\vec{L}}{dt} = \vec{r} \times \vec{F} = \vec{\tau}$	$\vec{L}$	kg m^2/s	angular momentum
	$\vec{r}$	m	distance from center of rotation
	$\vec{F}$	N	acting force
	$\vec{\tau}$	Nm	acting torque

➤ If the torque points parallel or antiparallel to the angular momentum, only the magnitude of the angular momentum (and thus the angular velocity) changes. If the torque and the angular velocity are not parallel or antiparallel, for a freely movable rigid body the orientation of the angular momentum—and thus the instantaneous rotation axis—also changes.

3. Angular momentum as a conserved quantity

▲ The angular momentum is an integral of motion if the torque vanishes:
$\tau = 0$, $L = $ const. (**Fig. 3.33**).

■ Two masses of 1 kg each rotate at the ends of a rod of length 100 cm about the center with a rate of rotation of 2/s. The moment of inertia is

$$J = J_1 + J_2 = 2 \cdot 1 \text{ kg} \cdot (50 \text{ cm})^2 = 0.5 \text{ kg m}^2.$$

The angular momentum of the system is

$$L = J\omega = 0.5 \text{ kg m}^2 \, 4\pi \text{ rad/s} = 6.28 \text{ kg m}^2/\text{s}.$$

Figure 3.33: On conservation of angular momentum. The vertical component of the angular momentum is a conserved quantity. This quantity vanishes in the figure on the left. When he changes the axis of the rotating wheel, the person standing on the rotary table is put into rotation; the corresponding angular momentum just compensates the change in the angular momentum of the wheel.

When the distance of the masses from the rotation center is cut in half to 25 cm, the angular momentum is conserved while the moment of inertia reduces to

$$J' = J'_1 + J'_2 = 2 \cdot 1 \text{ kg} \cdot (25 \text{ cm})^2 = 0.125 \text{ kg m}^2 = J/4.$$

To keep the angular momentum unchanged, the angular velocity now must be

$$\omega' = \frac{L}{J'} = \frac{6.28 \text{ kgm}^2/\text{s}}{0.125 \text{ kg m}^2} = 50.27 \text{ rad/s} = 4\,\omega$$

i.e., 8 rotations per second, or four times the initial rate of rotation.

3.4.2.1   Equilibrium for rotational motion

Similar to the equilibrium condition for translational motion, $\sum_i \vec{F}_i = 0$, there exists an equilibrium condition for rotational motion:

▲   A body rotates uniformly (special case: remains at rest) if the sum of all acting torques vanishes:

static equilibrium for rotations	$\mathbf{ML^2T^{-2}}$
$\sum_i \vec{\tau}_i = \sum_i (\vec{r}_i \times \vec{F}_i) = 0$	

➤   When the position vectors of the applied forces, $\vec{r}_i$, are represented as the sum of the center-of-mass vector, $\vec{R}_S$, and the distance vector of the ith point of application from the center of mass, $\Delta\vec{r}_i$,

$$\vec{r}_i = \vec{R}_S + \Delta\vec{r}_i \,,$$

the equilibrium condition reads

$$\sum_i (\vec{r}_i \times \vec{F}_i) = \vec{R}_S \times \sum_i \vec{F}_i + \sum_i (\Delta\vec{r}_i \times \vec{F}_i)\,.$$

If the sum of the external forces vanishes, $\sum_i \vec{F}_i = 0$, the equilibrium condition simplifies to

$$\sum_i (\vec{r}_i \times \vec{F}_i) = \sum_i (\Delta\vec{r}_i \times \vec{F}_i)\,.$$

Hence, it is sufficient that the sum of the torques with respect to the center of mass vanish.

3.5   Work, energy and power

If a force $\vec{\mathbf{F}}$ acts on the point $\vec{\mathbf{r}}$ of a rigid body, during a rotation by the angle element $\Delta\phi$ (rotation axis X) it does work

$$\Delta W = \vec{\mathbf{F}} \cdot \Delta\vec{\mathbf{r}} = F \sin\alpha \, r \, \Delta\phi = F_t \, r \, \Delta\phi,$$

where $\Delta r = r\Delta\phi$ is the distance traveled by the particle during the rotation by the angle $\Delta\phi$. The angle enclosed by $\vec{\mathbf{r}}$ and $\vec{\mathbf{F}}$ is α, so that $\vec{\mathbf{F}}_t$ is the component of the force in the direction of rotation (tangential component) (**Fig. 3.34**). Since the torque with respect to the rotation axis X is given by $\tau_X = F_t \, r$:

work = torque · angle element			$\mathbf{ML^2T^{-2}}$
	Symbol	Unit	Quantity
$\Delta W = \tau_X \cdot \Delta\phi$	ΔW	J	work done
	τ_X	Nm	torque with respect to axis X
	$\Delta\phi$	rad	angle element

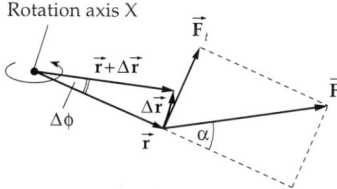

Figure 3.34: Work in rotation.

The mean power exerted by the applied torque in the time interval Δt is

$$\bar{P} = \tau_X \frac{\Delta\phi}{\Delta t} = \tau_X\bar{\omega},$$

where $\bar{\omega}$ is the mean angular velocity.

➤   Only the component of the torque along the rotation axis contributes to the work done. The perpendicular component causes only a change in orientation of the rotation axis, and does not contribute to the work.

The torque τ_X denotes the component along the rotation axis X, i.e., along $\vec{\omega}$. In vector notation therefore:

power = torque · angular velocity			$\mathbf{ML^2T^{-3}}$
	Symbol	Unit	Quantity
$P = \vec{\tau} \cdot \vec{\omega}$	P	W	instantaneous power
	$\vec{\tau}$	Nm	torque vector
	$\vec{\omega}$	rad/s	angular velocity

3.5.1 Kinetic energy

1. Kinetic energy of a rigid body

If the origin of the body-fixed coordinate system is established at the center of mass S, the kinetic energy of a rigid body is the sum of the kinetic energy of translation of the center of mass with velocity $\vec{v}$ and the kinetic energy of rotational motion with angular velocity $\vec{\omega}$ about an axis X_S through the center of mass:

$$E_{kin} = \frac{m}{2} v^2 + \sum_{i,k=1}^{3} J_{ik}\, \omega_i\, \omega_k, \qquad i, k = x, y, z.$$

m: total mass, J_{ik}: components of the tensor of inertia $\hat{J}$, ω_i: components of the angular velocity vector $\vec{\omega}$.

In matrix notation,

$$E_{kin} = \frac{1}{2} (\,\omega_x \quad \omega_y \quad \omega_z\,) \begin{pmatrix} J_{xx} & J_{xy} & J_{xz} \\ J_{yx} & J_{yy} & J_{yz} \\ J_{zx} & J_{zy} & J_{zz} \end{pmatrix} \begin{pmatrix} \omega_x \\ \omega_y \\ \omega_z \end{pmatrix} = \frac{1}{2} \vec{\omega}^T \hat{J} \vec{\omega}.$$

$\vec{\omega}^T$ is the (row) vector transposed to the column vector $\vec{\omega}$.

In the system of principal axes,

$$E_{kin} = \frac{1}{2}(J_x^2 \omega_x^2 + J_y^2 \omega_y^2 + J_z^2 \omega_z^2).$$

2. Kinetic energy for a fixed rotation axis

Kinetic energy of a rigid body rotating about a fixed axis X:

rotation energy			$\mathbf{ML^2T^{-2}}$
$E_{rot} = \dfrac{1}{2} J_X \cdot \omega^2$	Symbol	Unit	Quantity
	E_{rot}	J	rotation energy
	J_X	kg m^2	moment of inertia
	ω	rad/s	angular velocity

The kinetic energy of rotational motion is proportional to the square of the angular velocity.

3. Kinetic energy of a point mass

For a point mass moving on a circular orbit of radius r, the kinetic energy is

$$E_{rot} = \frac{1}{2} m \cdot v^2 = \frac{1}{2}(m \cdot r^2)\omega^2 = \frac{1}{2} J \cdot \omega^2,$$

J being the moment of inertia of the point mass (see p. 111).

➤   According to Steiner's rule, the moment of inertia with respect to any axis X with a perpendicular distance r_S from the center of mass is

$$J_X = m \cdot r_S^2 + J_S,$$

where J_S is the moment of inertia about an axis parallel to X through the center of mass. Hence, one obtains for the rotation energy

$$E_{\mathrm{rot}} = \frac{1}{2} J_S \cdot \omega^2 + \frac{1}{2} m r_S^2 \cdot \omega^2 \,.$$

The first term $\frac{1}{2} \cdot J_S \cdot \omega^2$ represents the kinetic energy of rotation about the axis through the center of mass; the second term $\frac{1}{2} \cdot m \cdot (r_S \cdot \omega)^2 = \frac{1}{2} m v^2$ gives the kinetic energy of the circular motion of the center of mass about the actual rotation axis of the system.

Steiner's rule allows a separation of the motion into a motion of the center of mass about the rotation axis and a rotation of the body about a center-of-mass axis. For a rigid body, both rotations have the same angular velocity ω.

The general motion of a rigid body is a translation of the center of mass with a super-imposed rotation about an axis through the center of mass. The total kinetic energy can therefore be separated into the **translation energy** $\frac{1}{2} m v_S^2$ of the center of mass, and the **rotation energy** $\frac{1}{2} J_S \cdot \omega^2$:

$$E_{\mathrm{total}} = E_{\mathrm{kin}} + E_{\mathrm{rot}} = \frac{1}{2} m v_S^2 + \frac{1}{2} J_S \omega^2 \,.$$

4. Potential energy of a rigid body,

energy of position of the center of mass,

$$E_{\mathrm{pot}} = m \, g \, h_S \,,$$

with m: total mass, g: acceleration of gravity, h_S: height of the center of mass above the reference level.

5. Energy conservation

In the absence of friction, the law of energy conservation holds provided the rotational energy is included. The sum of kinetic energy of translation, kinetic energy of rotation, and potential energy are constant if no dissipative forces are present:

▲   **Law of energy conservation**:

$$E_{\mathrm{kin}} + E_{\mathrm{rot}} + E_{\mathrm{pot}} = \mathrm{const.}$$

3.5.2   Torsional potential energy

Potential energy arises in rotations of **spiral springs.** When twisting the axis through the angle ϕ, a restoring torque τ is generated:

Hooke's law for spiral springs			$\mathbf{ML^2T^{-2}}$
	Symbol	Unit	Quantity
$\tau = \kappa \cdot \phi$	τ	Nm	torque
	κ	Nm	torsional constant
	ϕ	rad	twist angle from rest position

The quantity κ, the **torsional constant,** corresponds to the spring constant k of linear springs.

The **potential energy of a spiral spring** is thus

$$W_{\text{pot}} = \frac{1}{2}\kappa \cdot \phi^2 \,.$$

Similar to the case of a linear spring ($\frac{k}{2}x^2$) the potential energy is proportional to the square of the twist angle ϕ.

3.6 Theory of the gyroscope

Gyroscope, a rotating rigid body that is kept fixed at one point. The rotation axis, and thus the orientation of the angular velocity $\vec{\omega}$, of the gyroscope vary with time (**Fig. 3.35**).
    According to the **fundamental law of dynamics for rotational motion**,

$$\frac{d\vec{L}}{dt} = \vec{\tau},$$

the motion of the gyroscope results from the total applied torque $\vec{\tau}$. In this equation, the angular momentum $\vec{L}$ is a freely varying vector quantity.

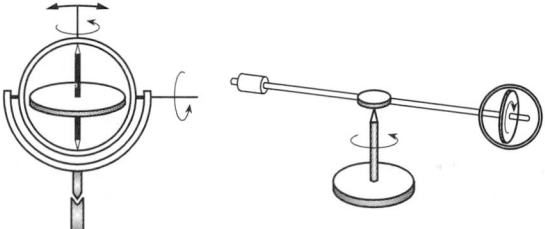

Figure 3.35: Gyroscope. To guarantee free rotation of the gyroscope axis, bearings (cardanic suspensions) with very low friction are used.

Bearing moment, the torque needed to keep the rotation axis in a definite orientation or plane. The bearing force results from the suppression of the motion of a free gyroscope, as discussed below.

3.6.1 Tensor of inertia

1. Definition of the tensor of inertia

Tensor of inertia, $\hat{J}$, a **tensor of second rank** that establishes the relation between the angular velocity $\vec{\omega}$ of a body and its angular momentum $\vec{L}$:

tensor of inertia			$\mathbf{ML}^2$
	Symbol	Unit	Quantity
$\vec{L} = \hat{J} \cdot \vec{\omega}$	$\vec{L}$	kg m^2/s	angular momentum
	$\hat{J}$	kg m^2	tensor of inertia
	$\vec{\omega}$	rad/s	angular velocity

The tensor of inertia has the same dimension as the moment of inertia; it differs from the latter quantity in that it is not related to a definite axis.

2. Inertia tensor in matrix notation

The inertia tensor can be represented in matrix notation:

$$\hat{J} = \begin{pmatrix} J_{xx} & J_{xy} & J_{xz} \\ J_{yx} & J_{yy} & J_{yz} \\ J_{zx} & J_{zy} & J_{zz} \end{pmatrix}.$$

The inertia tensor is a real symmetric tensor:

$$J_{xy} = J_{xy}^*, \qquad J_{xz} = J_{xz}^*, \qquad J_{yz} = J_{yz}^*,$$

$$J_{xy} = J_{yx}, \qquad J_{xz} = J_{zx}, \qquad J_{yz} = J_{zy}.$$

It can be characterized by only six independent elements.

In component representation, the relation between angular momentum and angular velocity is given by

$$L_x = J_{xx}\omega_x + J_{xy}\omega_y + J_{xz}\omega_z,$$

$$L_y = J_{yx}\omega_x + J_{yy}\omega_y + J_{yz}\omega_z,$$

$$L_z = J_{zx}\omega_x + J_{zy}\omega_y + J_{zz}\omega_z.$$

In compact notation:

$$L_i = J_{ij}\omega_j, \qquad i, j = 1, 2, 3,$$

where $i, j = 1, 2, 3$ stands for the x-, y- and z-direction, and the summation runs over the second index j (**Einstein sum convention**).

3. Calculation of the inertia tensor

To calculate the inertia tensor of an extended body, one starts with the inertia tensor of a point mass Δm, which has the form

$$\hat{J} = \Delta m \begin{pmatrix} y^2 + z^2 & -xy & -xz \\ -yx & x^2 + z^2 & -yz \\ -zx & -zy & x^2 + y^2 \end{pmatrix},$$

where x, y and z are the Cartesian coordinates of the point mass. The diagonal components (**moments of inertia**) involve the perpendicular distance from the corresponding axis, e.g.,

$$J_{xx} = \Delta m \cdot r_x^2 = \Delta m (y^2 + z^2).$$

r_x is the perpendicular distance of the x-axis. The off-diagonal elements are called the **products of inertia**.

The inertia tensor of an extended body is obtained by dividing the body into small mass elements Δm_i and summing or integrating:

$$\hat{J} = \sum_i \hat{J}_i = \sum_i \Delta m_i \begin{pmatrix} y_i^2 + z_i^2 & -x_i y_i & -x_i z_i \\ -y_i x_i & x_i^2 + z_i^2 & -y_i z_i \\ -z_i x_i & -z_i y_i & x_i^2 + y_i^2 \end{pmatrix},$$

x_i, y_i and z_i being the coordinates of the ith element.

For the components of the inertia tensor:

$$J_{kl} = \sum_i \Delta m_i \left(r_i^2 \, \delta_{kl} - x_{ik} x_{il} \right) .$$

Kronecker symbol: $\delta_{kl} = 1$ for $k = l$, otherwise zero. For a given coordinate system, the components of the inertia tensor are given by the mass distribution of the body.

➤   The summation over the mass elements Δm_i can be written as an integral,

$$J_{kl} = \iiint \left(r^2 \, \delta_{kl} - x_k x_l \right) dm .$$

4. Example: Inertia tensor of a cube

Inertia tensor of a cube with edge length a and mass m. The homogeneous mass density is given by ρ_0, $dm = \rho_0 \, dV = \rho_0 dx \, dy \, dz$, $m = \rho_0 V$. We take the lower left corner as a reference point (coordinate origin), i.e., the integration limits in the volume integral are 0 and a for all directions:

$$J_{11} = \rho_0 \int_0^a \int_0^a \int_0^a (x^2 + y^2) \, dx \, dy \, dz = \frac{2}{3} m a^2 ,$$

$$J_{12} = -\rho_0 \int_0^a \int_0^a \int_0^a xy \, dx \, dy \, dz = -\frac{1}{4} m a^2 .$$

One obtains:

$$\hat{J} = m a^2 \begin{pmatrix} 2/3 & -1/4 & -1/4 \\ -1/4 & 2/3 & -1/4 \\ -1/4 & -1/4 & 2/3 \end{pmatrix} .$$

5. System of principal axes

The form of the inertia tensor depends on the choice of coordinate system. However, one can always find a **system of principal axes** in which the tensor has a diagonal form:

$$\hat{J} = \begin{pmatrix} J_x & 0 & 0 \\ 0 & J_y & 0 \\ 0 & 0 & J_z \end{pmatrix} .$$

The axes of such a coordinate system are called **principal axes**. J_x, J_y and J_z specify the moments of inertia relative to the principal axes (**principal moments of inertia**).

6. Types of gyroscopes

One distinguishes:

Asymmetric gyroscope: $J_x \neq J_y \neq J_z$.

Symmetric gyroscope: $J_x = J_y \neq J_z$ or
$J_y = J_z \neq J_x$ or
$J_x = J_z \neq J_y$.

Spherical gyroscope: $J_x = J_y = J_z$.

➤   For bodies with axes of symmetry, these axes coincide with the principal axes.
■   For a sphere, any axis through its center is a principal axis.
      For a cube, the principal axes are perpendicular to the lateral faces.
      For a long cylinder, one principal axis points along the cylinder axis (smaller moment
      of inertia), the two other principal axes are perpendicular to the first one and pass
      through the cylinder center (larger moments of inertia).
In the system of principal axes:

$$L_x = J_x\,\omega_x\,, \quad L_y = J_y\,\omega_y\,, \quad L_z = J_z\,\omega_z\,.$$

Therefore, the angular-momentum and angular-velocity vectors are collinear if they are
parallel to a principal axis. If this is not the case, then the two vectors can have different
orientations, with the deviation depending on the differences between the principal moments of inertia J_x, J_y and J_z.

▲   The angular velocity $\vec{\omega}$ and angular momentum $\vec{L}$ are parallel only for rotation about
      a principal axis.
▲   A unilaterally suspended gyroscope always orients itself in such a way that it rotates
      about the principal axis with the largest moment of inertia (**Fig. 3.36**).

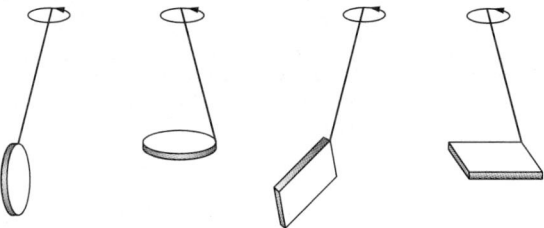

Figure 3.36: Unilaterally suspended gyroscopes orient themselves along a principal axis.

3.6.2   Nutation and precession

Symmetry axis, geometrically prominent symmetry axis of a symmetrical gyroscope.
      Instantaneous-rotation axis, direction of the angular velocity.

3.6.2.1   Nutation

Nutation, nodding motion, the motion of a gyroscope that is free of external forces. It
arises if the principal axis moments are not all equal and the rotation is not about a principal
axis.

1. Force-free symmetric gyroscope

Motion of a **force-free symmetric** gyroscope ($J_x = J_y \neq J_z$):
      Since no forces, and hence no torques, are acting, the angular-momentum vector has a
fixed orientation in space, $\vec{L} = $ const. The instantaneous-rotation axis, and thus the angular-
velocity vector $\vec{\omega}$ and the angular-momentum vector include a fixed angle, the value of
which results from the inertia tensor. The vector $\vec{\omega}$ rotates with constant angular velocity

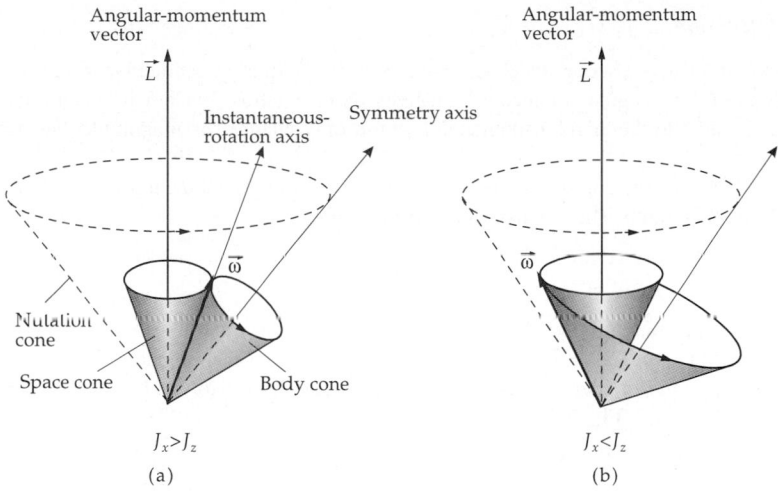

Figure 3.37: Axes of a force-free symmetric gyroscope ($J_x = J_y \neq J_z$). The angular-momentum vector $\vec{L}$ is space-fixed, the symmetry axis moves on the nutation cone about the direction of the angular momentum. The angular-velocity vector $\vec{\omega}$ (instantaneous-rotation axis) moves on the space cone (herpolhode) about the angular-momentum vector. The relative orientation of the axes is determined by the condition that the body cone (polhode) rolls with its outer surface ($J_x > J_z$) (a) or with its inner surface ($J_x < J_z$) (b) on the space cone.

about the angular-momentum vector, forming a circular cone, the **space cone** (**herpolhode**) that is space-fixed with the angular-momentum vector as symmetry axis (**Fig. 3.37**).

2. Body cone

The figure axis must not coincide with the rotation axis, but can include a fixed angle with it, and thus with the angular-velocity vector. As a result, another circular cone arises, the **body cone** (**polhode**), which has the symmetry axis as the central axis and rolls with its outer surface ($J_x > J_z$) or with its inner surface on the space cone ($J_x < J_y$). The two cones touch each other just along the instantaneous-rotation axis. Thus, the motion can be described by the rolling of two cones on each other; the cone tips lie in the support point of the gyroscope, and the symmetry axis moves on the **nutation cone** about the angular-momentum axis.

➤ A rotating body supported at its center of mass is a force-free gyroscope, since the total torque resulting from the weight vanishes.

➤ Because of friction effects, the gyroscope always orients itself along a principal axis. Therefore, nutation is observed only by pushing the gyroscope so that the angular-momentum vector moves away from the principal axis of inertia for a short time.

➤ For a non-symmetric gyroscope, the space cone, symmetry cone and nutation cone are not circular cones. The surfaces may not even be closed.

3.6.2.2 Precession

Gravity gyroscope, gyroscope with support point not coinciding with the center of mass, so that its weight introduces a torque on it.

1. Precession,

the motion of a gyroscope under an external torque acting perpendicular to the angular momentum. The angular momentum changes its orientation, but not its magnitude. (A torque parallel to the angular momentum would only change the magnitude, but not the orientation.)

The change of the angular momentum follows from the fundamental law of dynamics for rotational motion. The angular momentum vector,

$$\vec{\tau} = \vec{\mathbf{r}} \times \vec{\tau},$$

points perpendicular to $\vec{\mathbf{r}}$, and thus to the rotation axis. As a consequence, the change in the angular momentum $\Delta\vec{\mathbf{L}} = \vec{\tau}\Delta t$ is perpendicular to the angular momentum $\vec{\mathbf{L}}$, which leads to a rotation of the angular-momentum axis. The rotation proceeds in a plane perpendicular to the applied force (**Fig. 3.38**).

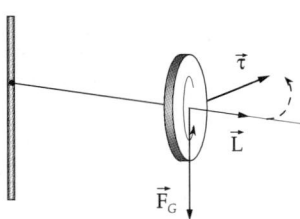

Figure 3.38: Precession of a rotating disk under the influence of Earth's gravitation $\vec{\mathbf{F}}_G$. The rotation axis specified by the angular momentum $\vec{\mathbf{L}}$ begins to rotate in the horizontal plane.

2. Precession velocity

The angular velocity of precession $\omega_p = \dfrac{\Delta\phi}{\Delta t}$ can be found by expressing the rotation angle $\Delta\phi$ of the angular-momentum axis by the change ΔL of the angular momentum,

$$\Delta\phi = \frac{\Delta L}{L} = \frac{\tau}{L}\Delta t,$$

and from that calculating the angular velocity ω_p (of rotation of the angular momentum, not that of the gyroscope):

precession velocity				$\mathbf{T^{-1}}$
	Symbol	Unit	Quantity	
$\omega_p = \dfrac{\tau}{L} = \dfrac{\tau}{J\omega}$	ω_p	rad/s	precession velocity	
	τ	Nm	torque	
	L	kg m²/s	angular momentum	
	J	kg m²	moment of inertia	
	ω	rad/s	rotation velocity of the gyroscope	

3. Precession rate

Instead of the precession velocity, one often adopts the **precession rate** f_p. The precession rate indicates how often the gyroscope axis rotates about the vertical per unit time:

$$f_p = \frac{\omega_p}{2\pi} = \frac{\tau}{2\pi J\omega}.$$

It becomes large when the applied torque τ increases, or the moment of inertia J of the gyroscope or its rotation velocity ω decreases.

4. Rotation direction

The direction of rotation of the gyroscope axis about the vertical depends on a number of factors.

▲ A gyroscope tends to adapt the orientation of its angular-momentum vector on the shortest path along the direction of an applied torque.

➤ The rotation of the gyroscope is assumed to occur about a principal axis. J is the moment of inertia relative to this axis. If this is not true, precession and nutation interact with each other.

■ **Non-symmetrically suspended gyroscope**, a gyroscope that is suspended only on one side of its horizontal-rotation axis. The weight does not act at the point of suspension, but provides a torque. The rotation axis does not turn downwards, but rotates in the horizontal plane.

3.6.2.3 Gyroscope moments

Gyroscope moment, the torque created by the **bearing forces** that must be compensated by the bearing of a tightly supported gyroscope if the rotation axis rotates. One finds:

gyroscope moment			$\mathbf{ML^2T^{-2}}$
	Symbol	Unit	Quantity
$\vec{\tau} = \vec{\mathbf{L}} \times \vec{\omega}_p$	$\vec{\tau}$	Nm	gyroscope moment
	$\vec{\mathbf{L}}$	kg m^2/s	angular momentum
	$\vec{\omega}_p$	rad/s	enforced precession velocity

■ The horizontal rotation axis of a rotating disk rotates about the vertical. $\vec{\omega}_p$ points vertically, $\vec{\mathbf{L}}$ horizontally. The bearings are under a force that tends to rotate the angular-momentum axis in the vertical direction. This force must be compensated by the bearings.

Bicycle, the wheels act as stabilizing gyroscopes. To make the bicycle fall over, a torque must act that rotates the orientation of the angular-momentum vector of the wheels; the faster the wheels rotate, the stronger must be the torque.

An additional stabilization stems from the precession torque at the front wheel that arises if the wheel turns sideways in a curve (rotation about the longitudinal axis). The resultant torque turns the front wheel in the curve's direction.

3.6.3 Applications of gyroscopes

1. Gyrocompass,

a gyroscope with a rotation axis freely movable in the horizontal plane, but with the vertical axis fixed by the suspension. The gyroscope thereby carries out a forced rotation with the Earth's rotation ω_E and tries to align its angular momentum parallel to it. The angular velocity of the Earth points permanently north, hence the gyrocompass always aligns to the north. In this way, it may supplement, or substitute for, a magnetic compass.

➤ The main problem with the gyrocompass is due to the slowness of the Earth's rotation, which makes the effect very small and difficult to protect against perturbations. One uses a gyroscope with a very large rate of rotation and as low a bearing friction as possible (e.g., in a liquid).

➤ On a moving ship there is another torque due to the motion along a meridian, which causes a deviation of the gyrocompass. Airplanes may move even faster than the local rotation velocity of the Earth, and hence the gyrocompass cannot be used.

➤ In the vicinity of the poles, the gyrocompass fails, just as the magnetic compass does, since the rotation axis of Earth points nearly normally to the surface, and hence the torque projected on the horizontal plane becomes very small.

2. *Gyroscope horizon,*

to determine the horizon position in an airplane, based on angular momentum conservation. A gyroscope is set into rotation on the ground. When low friction air bearings and cardanic suspensions are used, it keeps its original orientation.

3. *Gyroscope pendulum,*

improvement of the gyroscope horizon, where the gyroscope is brought to a slow precession. One exploits the fact that the precession always occurs about the vertical direction. The gyroscope pendulum is distinguished from the conventional **plumbline** or **pendulum** by its very low oscillation frequency, hence it does not respond to short-term accelerations in curved flight.

4. *Rate gyroscope*

For measuring the rate of turning of a vehicle by means of the moments of the gyroscopic motion induced by the turning. The gyroscopic moments are measured at the bearings with springs. The elongation of the spring at the top is proportional to the rotational velocity.

4
Gravitation and the theory of relativity

4.1 Gravitational field

4.1.1 Law of gravitation

1. Gravitation

The property of bodies to interact with each other through their masses is called **gravitation**. The electric force between bodies depends on the charge but not the mass. For the gravitational force only the mass enters, and the force is always attractive as opposed to the electric force, which depends on the sign of the charge. The gravitational force is always attractive and described by the universal law of gravitation:

law of gravitation			$\mathbf{MLT^{-2}}$
	Symbol	Unit	Quantity
$F_g = G \dfrac{m_1 m_2}{r_{12}^2}$	F_g	N	gravitational force
	G	N m^2/kg^2	gravitational constant
	m_1, m_2	kg	masses of bodies
	r_{12}	m	center-of-mass separation of the bodies

2. Properties of the gravitational force

The gravitational force always points towards the other body (**Fig. 4.1**). In vector notation:
The force acting on the body 2 is

$$\vec{F}_{g,2} = -G \frac{m_1 m_2}{r_{12}^2} \frac{\vec{r}_{12}}{|\vec{r}_{12}|},$$

where $\vec{r}_{12}$ represents the vector from the center of mass of body 1 to the center of mass of body 2. Potential theory states that, for the calculation of the gravitational force between

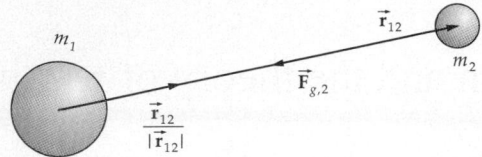

Figure 4.1: Gravitational force. The force acting on the body m_2 points opposite to the displacement vector from m_1 to m_2.

extended spherical homogeneous mass distributions, the bodies can be considered points, with the masses concentrated at the corresponding centers of mass.

➤   The expression $\vec{r}/|\vec{r}|$ (vector divided by its magnitude) represents the unit vector along the vector $\vec{r}$. The force acting on the body 2 points from body 2 to body 1 (notice the minus sign in the formula).

▲   The gravitational force is always an attractive force.

➤   The **gravitational constant** G is a natural constant. Its value is

$$G = 6.67259 \cdot 10^{-11} \; \mathrm{Nm^2/kg^2} \, .$$

➤   The formula gives both the magnitude of the force exerted by body 1 on body 2, and vice versa (2 on 1). The gravitational force always points towards the attracting body.

▲   The gravitational force between two bodies is proportional to the mass of each body and inversely proportional to the square of the distance between them.

Notice the similarity of this expression to Coulomb's law (see the section on Electricity). However, masses always attract each other, whereas the force between charges with the same sign is repulsive. The gravitational field strength is introduced by analogy to the electric field strength.

3. Gravitational field strength,

$\vec{E}_g$, a vector quantity which, for any point $\vec{r}$ in space, gives the force per unit mass that acts on a body due to gravitation:

$$\vec{E}_g = -G \, \frac{M}{r^2} \frac{\vec{r}}{|\vec{r}|} \, .$$

The gravitational field $\vec{E}_G$ depends only on the mass M of the attracting body, which is located at the coordinate origin and is considered to be the source of the gravitational field. The force on a test particle of mass m is $\vec{F} = m \, \vec{E}_g$. It points towards the attracting body and determines the acceleration of the test particle.

4. Gravitational potential,

Φ, potential of the gravitational field, describes the **work in the gravitational field**.

gravitational potential			$\mathbf{L^2 T^{-2}}$
	Symbol	Unit	Quantity
$\Phi = -G \, \dfrac{M}{r}$	Φ	J/kg = Nm/kg	potential of gravitational field
	G	N m²/kg²	gravitational constant
	M	kg	mass of the gravitating body
	r	m	distance between the test body and gravitating body

The gravitational force $\vec{F}$ is calculated from the potential Φ of the gravitational field as

$$\vec{F}_g(\vec{r}) = -m \operatorname{grad} \Phi(r).$$

➤ The potential of the gravitational force is $V(r) = m\,\Phi(r), \quad \vec{F}_g = -\operatorname{grad} V(r)$.
The potential energy of a test particle of mass m at the point $\vec{r}$ in the gravitational field of a body of mass M is

$$E_{\text{pot}}(\vec{r}) = m\,\Phi(\vec{r}).$$

The work needed to move a test particle of mass m from point $\vec{r}_1$ to point $\vec{r}_2$ against the gravitational force equals the difference of the potential energies at the points $\vec{r}_2$ and $\vec{r}_1$:

$$W_{12} = -\int_{\vec{r}_1}^{\vec{r}_2} \vec{F}_g \, d\vec{r} = E_{\text{pot}}(\vec{r}_2) - E_{\text{pot}}(\vec{r}_1) = GmM\left(\frac{1}{r_1} - \frac{1}{r_2}\right).$$

5. Attraction to Earth,

weight, the force exerted by Earth on a body at Earth's surface due to gravitation. It is specified by the law of gravitation, the mass and radius of Earth, and the mass of the test particle.

Acceleration of gravity g, nearly constant acceleration due to the attractive force of Earth that acts on all falling bodies: $g = 9.80665$ m/s^2 for mean sea level at about 45° geographical latitude.

➤ The acceleration of gravity is not the same everywhere on Earth's surface. It depends on the geographic latitude, as a result of the non-spherical shape of Earth, and the centrifugal force of Earth's rotation, and also depends on the height at which the measurement is made. Lastly, density fluctuations in Earth's crust lead to concentrations of mass that may modify both the magnitude and direction of Earth's attraction. The latter effect is exploited in searching for raw-material deposits.

➤ According to the law of gravitation, the ratio of the acceleration of gravity g_r at a distance $r > R$ from Earth's center, and g on the Earth's surface is

$$\frac{g_r}{g} = \frac{R^2}{r^2}, \qquad R: \text{ Earth's radius.}$$

➤ The hypothesis of a "fifth force," represented by a Yukawa term, with a strength parameter α and range parameter λ, as an additional term to the potential energy of the gravitational field,

$$V(r) = -G\frac{Mm}{r}\left(1 + \alpha e^{-r/\lambda}\right),$$

leads to an effective gravitational constant that depends on the distance r of the test particle from the gravitating mass M. This hypothesis has not be verified by experiment.

4.1.2 Planetary motion

Besides Earth's attraction, gravitation also manifests itself in the motion of the planets. Planetary motion was described empirically in 1609 by Johannes Kepler, as formulated in **Kepler's laws**. These laws can be derived from the law of gravitation and Newton's laws.

1. Kepler's first law

All planets move in elliptic orbits, with the Sun at one focal point.

➤ An ellipse is described by specifying its major semi-axis and either its minor semi-axis or its eccentricity. In our solar system, the planetary orbits are very close to circles.

Ecliptic, the plane of the Earth's orbit. It serves as an astronomical reference frame. **Perihelion,** the point of Earth's orbit with the minimum distance to the Sun. **Aphelion,** the point of the Earth's orbit with maximum distance to the Sun.

➤ The seasons on Earth are not caused by the difference of the distances to the Sun at the perihelion or aphelion, but by the inclination of Earth's equator with respect to the ecliptic. This inclination implies that sometimes the northern hemisphere is turned more towards the Sun, and at other times more away from the Sun.

2. Kepler's second law

A radius vector drawn from the Sun to a planet covers equal areas in equal time intervals (**Fig. 4.2**).

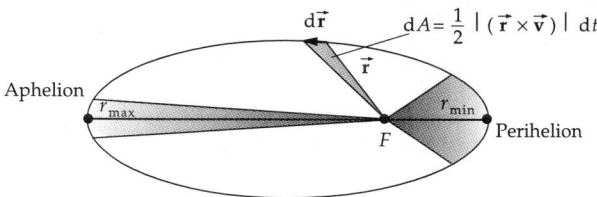

Figure 4.2: Kepler's second law. F: The focal point of the ellipse. The shadowed regions around r_{min} and r_{max} are of equal area.

➤ This statement follows from the conservation of angular momentum $\vec{l} = \vec{r} \times \vec{p}$: The areal element dA covered in the time interval dt is given by $2 \cdot dA = |\vec{r} \times d\vec{r}|$, hence $2m \cdot dA/dt = |\vec{r} \times \vec{p}| = |\vec{l}|$. If the angular momentum is a conserved quantity, $|\vec{l}| = $ const., the area dA covered per time interval dt is the same for all sections of the orbit. In particular, it follows that the orbital velocity at the perihelion, v_P, is higher than at the aphelion, v_A, since $l/m = r_{min} v_P = r_{max} v_A \implies v_P > v_A$.

3. Kepler's third law

The squares of the periods T_1 and T_2 of two planets are related as the cubes of the major semi-axes a_1 and a_2 of their orbits:

$$T_1^2 : T_2^2 = a_1^3 : a_2^3, \qquad \frac{T^2}{a^3} = \text{const}.$$

➤ Kepler's laws describe the planetary motion caused by the gravitational attraction by the Sun. They do not take the mutual attraction between the planets into account.

➤ According to the general theory of relativity deviations from the $\frac{1}{r^2}$-law arise near the Sun, as is manifested by the slow precession of the elliptic orbit of Mercury (rosette curve).

▲ Parabolas and hyperbolas are also possible orbits of celestial bodies. They pass, however, only once in the vicinity of the central stellar body; afterwards the celestial body leaves the planetary system (example: some comets).

4.1.3   Planetary system

4.1.3.1   Sun and planets

1. The Sun,

the central star of the **solar system** which consists of nine planets and the smaller celestial bodies (satellites, comets, asteroids). The nine planets of the solar system are partly earth-like in size and composition (Mercury, Venus, Mars), and partly much larger gaseous giants (Jupiter, Saturn, Uranus, Neptune).

Data on the Sun		
radius	696,000 km	= 109 Earth radii
mass	$1.99 \cdot 10^{30}$ kg	= 332,000 Earth masses
mean density	1,410 kg/m^3	
acceleration of gravity	273.7 m/s^2	= 27.9 times that on Earth

2. Planets and solar system

Planet, a non-self-luminous celestial body. Unlike **fixed stars**, planets are made visible by light reflected from them. Under the influence of the gravitational force of a central star the planets move in elliptic orbits around them. A star may have several planets revolving around it in different orbits (**planetary system**).

The solar system contains nine planets.

➤   It is not yet clear whether additional planets besides those currently known exist in the solar system. Since the sun light reflected by a possible further planet would be too small to be measured with present technology, one tries to determine the existence of additional planets via their gravitational force on other planets and the resulting distortions of their orbits.

➤   Indications of planets outside our solar system have been observed.

Basic data for planets of the solar system:

Planet	Major semi-axis of orbit (10^6 km)	Period of revolution (a)	Diameter (km)	Mass (in Earth masses)	Rotational period
Mercury	57.9	0.241	4,840	0.053	59 d
Venus	108.2	0.615	12,400	0.815	243 d
Earth	149.6	1.000	12,756	1.000	23 h 56 min
Mars	227.9	1.881	6,800	0.107	24 h 37 min
Jupiter	778	11.862	142,800	318.00	9 h 50 min
Saturn	1,427	29.458	120,800	95.22	10 h 14 min
Uranus	2,870	84.015	47,600	14.55	10 h 49 min
Neptune	4,496	164.79	44,600	17.23	15 h 40 min
Pluto	5,946	247.7	5,850	ca. 0.1	unknown

3. Basic data for Earth

Data on Earth	
equator radius	6378.163 km $= R_E$
polar radius	6356.777 km $= R_P$
flattening	$0.003356 = (R_E - R_P)/R_E$
mass	$5.977 \cdot 10^{24}$ kg
mean density	5517.0 kg/m^3
acceleration of gravity	9.80665 m/s^2
escape velocity	11.19 km/s

Escape velocity (parabolic velocity): The minimum velocity of a planet needed to leave the gravitational field of the central body.

➤   The rotation period of Earth is not exactly 24 hours, but is about 4 minutes less. These 4 minutes correspond to the angular distance the Earth travels in one day in its orbit around the Sun.

4. Titius-Bode relation

The radii a_n of the planetary orbits follow a geometrical series approximately:

$$a_n \approx a_{\text{Earth}} \, k^n \, , \qquad k \approx 1.85 \, ,$$

($n_{\text{Earth}} = 0, n_{\text{Venus}} = -1, n_{\text{Mercury}} = -2, n_{\text{Mars}} = 1, n_{\text{Jupiter}} = 3, n_{\text{Saturn}} = 4, \ldots$).

➤   The missing value $n = 2$ corresponds to the belt of asteroids between Mars and Jupiter.

➤   The origin of this relation is presumed to lie in the mutual perturbations of the planets and the resulting conditions for stable orbits.

5. Astronomical unit,

AE, the mean distance Earth–Sun,

$$1\text{AE} = 149.6 \cdot 10^6 \text{ km} \, .$$

Pluto, the outermost known planet, is about 40 AE distant from the Sun; Mercury, the innermost, ca. 0.4 AE. Hence, the solar system is very much smaller than the distance to the nearest star (Proxima Centauri, 4.3 ly $\approx$ 272,265 AE).

Light year, ly, the distance traversed by light in one year:

$$1 \text{ ly} = 9.4605 \cdot 10^{12} \text{ km} = 63{,}240 \text{ AE} \, .$$

Parsec, pc (parallax second), the distance at which the radius of Earth's orbit around the Sun is observed to subtend an angle of 1 arc second:

$$1 \text{ pc} = 3.262 \text{ ly} = 30.857 \cdot 10^{12} \text{ km} \, .$$

6. Measurement of astronomical quantities

M   **Parallax**, the virtual displacement of a star (e.g., with respect to other, more remote stars) in the sky in the course of one year, due to the motion of Earth on its orbit. The nearer a star, the larger its parallax.

Parallax range finding, measurement of the distance to a star by comparison of photographs taken in the course of one year. A star at a distance of 1 pc performs a parallax motion of 1 arc second. The method is applicable up to about 100 ly. For larger distances, indirect methods (luminosity, Doppler shift, . . .) are used.

7. Moon,

stellar body orbiting a planet. The diameter of **Earth's moon** is about one fourth of Earth's diameter. Many planets, in particular the larger planets Jupiter, Saturn and Uranus, have several moons with nearly the dimension of planets. The **rings of Saturn,** which consist of rocks and dust orbiting the planet, resemble moons.

Data on Earth's Moon		
diameter	3476.0 km	= 27 % of Earth's diameter
mass	$7.350 \cdot 10^{22}$ kg	= 1.2 % of Earth's mass
mean density	3 342 kg/m^3	= 61 % of Earth's density
acceleration of gravity	1.620 m/s^2	= 16.6 % of g on Earth
escape velocity	2.37 km/s	

8. Planet rotation

Planets (and moons) rotate about their own axes; Earth once in 24 hours, Earth's Moon once per month (ca. 28 days). Hence, Earth's Moon always turns the same face towards Earth; the other half of its surface remains permanently out of sight of Earth.

Equator, great circle in the plane of rotation of the planet. The inclination of this equatorial plane against the orbital plane determines the length of the day in the course of the year and is responsible for the occurrence of seasons.

9. Asteroids and comets

Asteroids, **small planets,** significantly smaller than any of the nine planets. Most of the asteroids are found in an **asteroid belt** between Mars and Jupiter. Their diameters range from a few kilometers up to 740 km (Ceres).

Comet, an object on a hyperbolic or highly eccentric elliptic orbit. The hyperbolic orbit approaches the Sun (or Earth) only once, the elliptic orbit in periodic intervals that may reach 200 years. The most famous comet is **Halley's comet** with a period of 76 years. When comets are remote from the Sun (i.e., not within the orbits of the nine planets) they are not observable. Comets typically have sizes between 1 km and 100 km. Frozen gases on the surface of the comets evaporate when they approach the Sun and become visible as a **comet tail**.

Meteor, a luminous phenomenon caused by **meteorites** that enter Earth's atmosphere and burn out due to the air friction. Their often metallic residues sometimes reach Earth's surface.

4.1.3.2   Satellites

Satellite, a body moving on an orbit in the gravitational field of another body, in general a planet. Originally, the term referred to moons; nowadays **artificial satellites** are also included.

▲   For satellites, Kepler's first law may be modified as follows: satellites move along conic sections, i.e., on circular, elliptic, parabolic or hyperbolic curves, depending on the satellite's initial velocity.

Satellites on parabolic and hyperbolic orbits escape the gravitational field of the central object.

1. First critical velocity

Circular orbit velocity, v_K, **first critical velocity**, the velocity that a body must have to move on a circular orbit near Earth's surface. It is the minimum velocity of a satellite to avoid impact on the surface of Earth. The circular orbit velocity follows from the balance between the centrifugal force and the gravitational force of Earth that provides the centripetal force to maintain the circular motion.

2. Second critical velocity

Parabolic orbit velocity, v_P, **second critical velocity** or **escape velocity**, the minimum velocity that a body must have to leave the gravitational field of Earth. The body then moves on a parabolic orbit arbitrarily far away from Earth.

For Earth, the critical velocities are (**Fig. 4.3**):

critical velocities				LT^{-1}
$v_K = \sqrt{\dfrac{GM}{R}} = 7912 \text{ m/s}$ $v_P = \sqrt{2}\, v_K = \sqrt{\dfrac{2GM}{R}}$ $= 11190 \text{ m/s}$	**Symbol**	**Unit**		**Quantity**
	v_K	m/s		circular orbit velocity
	v_P	m/s		parabolic orbit velocity
	G	N m^2/kg^2		gravitational constant
	M	kg		Earth's mass
	R	m		Earth's radius

➤   For velocities $v_K < v < v_P$, elliptic orbits result. Hyperbolic orbits arise for $v > v_P$.

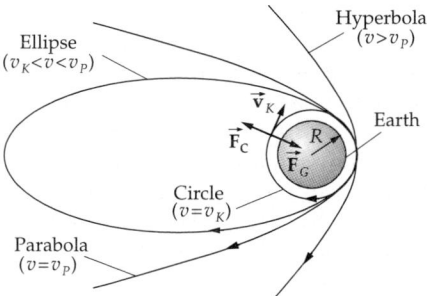

Figure 4.3: Satellite orbits. $\vec{F}_c$: centrifugal force, $\vec{F}_G$: weight (centripetal force), R: Earth's radius, v_K: first critical velocity, v_P: second critical velocity.

3. Third critical velocity

Third critical velocity, the minimum velocity that a body on Earth must have to leave the solar system. It follows from the same formula as the second critical velocity, but now the Sun's mass and the distance from the Sun have to be inserted:

$$v = \sqrt{\frac{2GM_{\text{Sun}}}{r_{\text{Sun}-\text{Earth}}}} = 42.1 \text{ km/s}.$$

➤   Using the relation $g = GM/R^2$, v_K and v_P can also be expressed in terms of the acceleration of gravity g at Earth's surface.

4.2   Special theory of relativity

1. Special theory of relativity,

developed by Albert Einstein (1905) to explain phenomena in motion at velocities near the speed of light.

The central concept of the special theory of relativity is the postulate that the laws of physics are the same in any uniformly moving reference frame, and the postulate of the **constancy of the speed of light in vacuum** in all inertial systems. This postulate leads to a new definition of the concepts of time and space in the framework of a **space–time continuum**.

2. General theory of relativity,

extension of the special theory of relativity, also developed by Einstein (1916), that also includes arbitrarily accelerated reference frames in the **relativity principle**.

■   The general theory of relativity leads to an equal treatment of gravitation and inertial forces by means of a curved space–time continuum, and constitutes the basis of modern **cosmology**.

3. Relativistic effects

Differences between the ordinary, non-relativistic physics and the special or general theory of relativity become important only for velocities close to the speed of light, and for motions in the vicinity of extremely massive objects, respectively. They are in general not observable in everyday life.

■   Physical applications of the theory of relativity are found in elementary-particle physics (particle accelerators), in atomic physics, and in astronomy and astronautics. Because of the increasing sensitivity of precision measurements, relativistic effects may also be demonstrated using highly sensitive instruments in macroscopic processes on Earth (time dilatation in airplanes).

4.2.1   Principle of relativity

1. Inertial system,

a frame in which Newton's laws hold, in particular the law of inertia. In such a frame, a body that is free of forces remains in its state of motion. Therefore, inertial systems are those frames that move with uniform speed relative to each other.

➤   The velocity of a system cannot be specified without reference to a system relative to which the velocity is being measured. Hence, an inertial system cannot be defined as a system that moves with uniform velocity without referring to another frame that is also an inertial system.

▲   A system that moves with uniform velocity $v =$ const. relative to an inertial system is also an inertial system.

Event, an incident that is fixed in a coordinate system by specifying its time coordinate t and its spatial coordinate x. Therefore, any physical event in a given reference frame is assigned to a **coordinate** (x, t) in the space–time continuum.

2. Galilean transformation,

transformation of the coordinates when changing from one inertial system to another inertial system *without accounting for the special theory of relativity*. Let x and x' denote the space coordinate, t and t' the time coordinate in the two frames, respectively. If the coordinate origins of both systems coincide at time $t = 0$, and their relative motion is in the x-direction with velocity v (**Fig. 4.4**), the Galilean transformation is then:

Galilean transformation			
	Symbol	Unit	Quantity
$x' = x - vt$	x, x'	m	space coordinates
$t' = t$	t, t'	s	time coordinates
	v	m/s	relative velocity of the reference frames

The second relation, $t' = t$, says that the time measurement (motion of a watch, pendulum motion, etc.) does not depend on the velocity of the spatial motion of the chronometer.

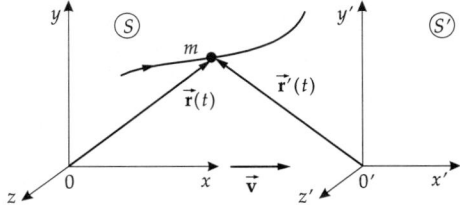

Figure 4.4: Galilean transformation. The coordinate origin O' of the frame S' moves relative to the coordinate origin O of the frame S on a straight line and uniformly with the velocity $\vec{v}$ along the x-axis. Trajectory in S: $\vec{r}(t) = \vec{r}'(t) + \vec{v}t$. For both frames, the same time scale is assumed.

➤ A system S' is denoted as a moving frame if it moves relative to the frame S of the observer with the velocity $\vec{v} \neq 0$. And vice versa, for an observer who is at rest in S', the frame S is moving with the velocity $-\vec{v}$.

3. Trajectory,

$x(t)$, characterizes the motion of a body m in a given frame. Its trajectory in S:

$$\vec{r}(t) = \vec{r}'(t) + \vec{v}t .$$

■ According to the Galilean transformation, the trajectory in a frame S' that moves with velocity v along the x-direction is given by

$$x'(t') = x'(t) = x(t) - vt .$$

■ A body moving uniformly with velocity u has the trajectory

$$x(t) = x_0 + ut , \qquad x_0 : \text{coordinate at time } t = 0 .$$

In a coordinate system moving with the velocity v the trajectory is given by

$$x'(t) = x(t) - vt = x_0 + ut - vt = x_0 + (u - v)t .$$

Under a Galilean transformation, the velocity u' in the moving frame S' is thus obtained by subtracting the original velocity u of the body and the relative velocity v of the moving system S':

$$u' = u - v , \quad u = u' + v .$$

4. Relativity principle in classical, non-relativistic mechanics

The laws of classical mechanics have the same form in any inertial system.
- ■ Transformation of Newton's second law:

Observer in S: $\quad \vec{F} = m\ddot{\vec{r}}$.

Observer in S': $\quad \dot{\vec{r}} = \dot{\vec{r}}' + \vec{v}, \quad \dot{\vec{v}} = 0$,

$\qquad\qquad\quad \ddot{\vec{r}} = \ddot{\vec{r}}', \quad \vec{F}' = m\ddot{\vec{r}}'$.

The force law has the same mathematical form for both observers.

5. Maxwell's equations,

describe the propagation of electromagnetic waves, do **not** follow this relativity principle:
- ▲ Electromagnetic waves (light) propagate in vacuum with the speed

$$c = 2.997\,924\,58 \cdot 10^8 \text{ m/s}.$$

> If this velocity were to transform according to the Galilean transformation, the above value would be valid only in a unique, and hence distinguished, reference frame. This contradicts experimental experience.

For the propagation of sound in gases, the sound velocity quoted in the literature holds for the reference frame in which the gas is at rest. A very rapidly moving source of sound may actually be faster than the sound emitted by the source, and in this way it may generate a shock wave.

This leads to the question of whether a source moving faster than the speed of light can pass the light emitted by itself.

6. Ether hypothesis,

analogy between light and sound propagation. According to this hypothesis, electromagnetic waves are carried by a medium called the **ether**. The reference frame in which the ether is at rest would constitute an absolute coordinate system.
- ■ The value of the speed of light would then hold just in the reference frame in which the ether is at rest.
- M In particular, the existence of an ether would imply that electromagnetic waves in a moving reference system propagate (analogous to sound propagation) with distinct velocities forward (i.e., direction of motion of the source) and sideways. This hypothesis was tested for the first time in the **Michelson-Morley experiment** (1887) by means of a **Michelson interferometer**. Here one observes with an interference setup whether the speed of light changes because of Earth's motion. The moving system in which the experiment was performed is Earth itself on its path around the Sun. The experiment proves that light propagates with equal velocity c along Earth's orbit and in the perpendicular direction, disproving the ether hypothesis.

7. Special relativity principle

All inertial systems are equivalent. In a vacuum, light propagates in any inertial system and in all directions with the same speed: the speed of light in vacuum c.
- ➤ Contrary to the ether hypothesis (which presupposes an absolute motion), according to the relativity principle there exists only **relative motion** in the selected reference frame; hence, the term **theory of relativity**.

4.2.2   Lorentz transformation

1. Introduction of the Lorentz transformation

The validity of the relativity principle is maintained only if the Galilean transformation is replaced by another transformation, the **Lorentz transformation**. Let the coordinates of an event in 3D space relative to a reference frame S be given by x, y, z and the time t. The coordinates x', y', z', t' of the same event in a coordinate system S' that moves uniformly with the speed v along the x-axis relative to the first system, are (**Fig. 4.5**):

Lorentz transformation			
$x' = \dfrac{x - vt}{\sqrt{1 - v^2/c^2}}$	Symbol	Unit	Quantity
	x, y, z	m	space coordinates in frame S
	t	s	time coordinate in frame S
$y' = y$	x', y', z'	m	space coordinates in frame S'
$z' = z$	t'	s	time coordinate in frame S'
	v	m/s	relative velocity of S' against S
$t' = \dfrac{\left(t - \dfrac{v}{c^2}x\right)}{\sqrt{1 - v^2/c^2}}$	c	m/s	speed of light

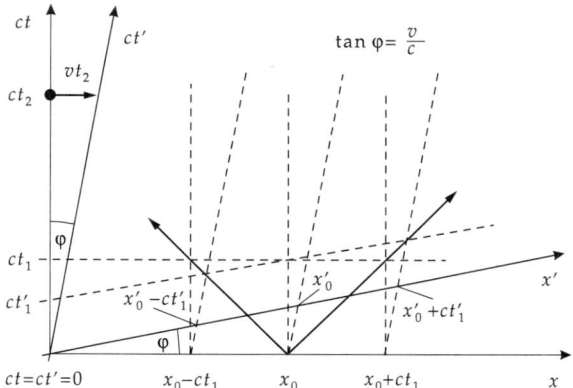

Figure 4.5: Lorentz transformation in the Minkowski graph. Besides the axes (x, ct), (x', ct') of the two frames, the world line (= trajectory in Minkowski space) of a light pulse emitted at $(x = x_0, t = 0)$ is plotted. The scale on the axes of system S' may be determined by recognizing that the light pulse propagates in both systems with the speed of light c.

➤   The inverse of the Lorentz transformation is obtained by changing the sign of velocity. The frame S moves with velocity $-v$ relative to the frame S'.

$$x = \frac{x' + vt'}{\sqrt{1 - v^2/c^2}}, \quad t = \frac{\left(t' + \dfrac{v}{c^2}x'\right)}{\sqrt{1 - v^2/c^2}}.$$

2. Relativistic factor,

γ, characteristic parameter of the Lorentz transformation:

$$\gamma = \frac{1}{\sqrt{1 - \dfrac{v^2}{c^2}}}\,.$$

For velocities much below the speed of light,

$$v \ll c \quad \Longrightarrow \quad \gamma \approx 1\,.$$

▲   For $v \ll c$, the Lorentz transformation becomes the Galilean transformation.
■   This guarantees that the Lorentz transformation does not contradict common experience, since relativistic effects become measurable only for large speeds beyond our everyday range of experience.

3. Minkowski diagram and world point,

serve for visualization of the Lorentz transformation. The position x and on the abscissa the time t (or ct) are plotted on the ordinate, so that to any event a **world point** (t,x) may be assigned in the graph (**Fig. 4.6**).

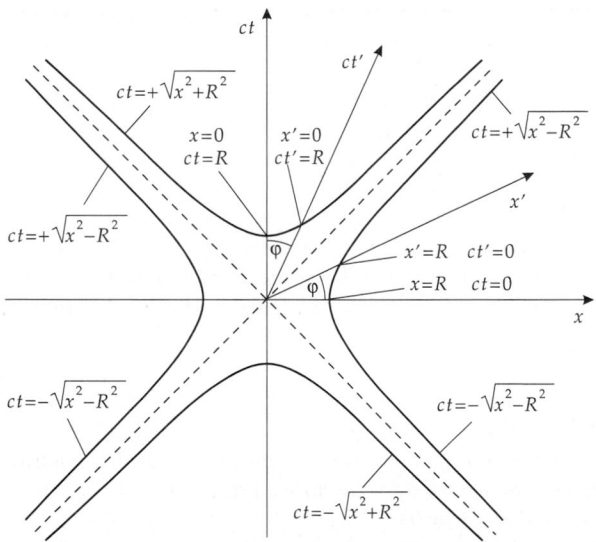

Figure 4.6: Lorentz transformation in a Minkowski diagram. The axes x, ct and x', ct' of the two frames, and the hyperbolas $ct = \pm\sqrt{x^2 \pm R^2}$ are plotted.

World line, the trajectory of a particle in a Minkowski diagram. For convenience, the units on the axes are taken so that a motion with the speed of light, $x(t) = ct$, appears as a straight line with a slope of $45°$, and therefore the distance is plotted in light seconds (ls) and the time in seconds. A **light second** is the distance passed by light in 1 second: $1 \text{ ls} \approx 3 \cdot 10^8$ m.

When making a Lorentz transformation, the coordinate axes of the moving frame are plotted in a Minkowski diagram. The coordinates of the origin $(t' = 0, \ x' = 0)$ are $(t = 0, \ x = 0)$, i.e., the origins of both coordinate systems lie at the same world point. The x'-axis of the frame S' is given by

$$t' = 0 \quad \Longrightarrow \quad \gamma\left(t - \frac{v}{c^2}x\right) = 0 \quad \Longrightarrow \quad t = \frac{v}{c^2}x \, .$$

This corresponds to a straight line enclosing the angle φ with the x-axis, with

$$\tan \varphi = \frac{v}{c} \, .$$

Correspondingly, one gets the same value, although counted in the opposite direction, for the angle between the ct'-axis and the ct-axis. Finally, the scales on the axes of frame S' have to be wider by the factor γ (> 1) than in the frame S (see p. 145).

➤   For an observer in the system S', the system S moves with the speed $-v$.

4. Comparison with the Galilean transformation

The most radical change in the Lorentz transformation as compared with the Galilean transformation is the statement that the time coordinate cannot be the same in both systems. This follows directly from the postulate of the constancy of the speed of light, and this consequence cannot be avoided.

■   Two events that occur simultaneously at distinct space points in one reference frame are not simultaneous in another reference frame. This **relativity of simultaneity** is a general phenomenon and is connected with the fact that the information on an event cannot propagate faster than the speed of light from one space point to another one.

▲   The largest propagation velocity of a physical phenomenon is the speed of light.
The relativistic factor γ is not defined for the velocity $v = c$ (division by zero), and becomes imaginary for velocities $v > c$. Therefore, a massive body cannot reach a velocity $v \geq c$ in vacuum. This experience is expressed by the **addition theorem of velocities**.

5. Tachyons,

hypothetical particles that move at or faster than the speed of light, but cannot go below it.

$\boxed{\text{M}}$ Tachyons would emit light in vacuum. Radiation arises if a massive particle moves in an optical medium with refractive index n faster than $c_{gr} = c/n$ (c: vacuum speed of light, c_{gr}: group velocity).

4.2.2.1   Addition of velocities

1. Addition of velocities under Lorentz transformation

Let a body move with the velocity $\vec{u}'$ in a reference frame S' that has a relative velocity $\vec{v}$ against the frame S. The velocity $\vec{u}$ of the body relative to the frame S does not follow by simple vector addition of $\vec{u}'$ and $\vec{v}$. According to the Lorentz transformation, it is given by the

addition theorem of velocities			
$u_x = \dfrac{u'_x + v}{1 + \dfrac{v}{c^2}u'_x}$	Symbol	Unit	Quantity
	u_x, u_y, u_z	m/s	velocity in S
	u'_x, u'_y, u'_z	m/s	velocity in S'
$u_y = \dfrac{u'_y}{\gamma\left(1 + \dfrac{v}{c^2}u'_x\right)}$	v	m/s	relative velocity of S' along the x-axis of S
	c	m/s	speed of light
	γ	1	relativistic factor
$u_z = \dfrac{u'_z}{\gamma\left(1 + \dfrac{v}{c^2}u'_x\right)}$			

➤   Inversion by changing the sign of the relative velocity, $v \to -v$.

2. Derivation of the addition theorem

The above expressions are obtained if the uniform motion of a particle in a moving coordinate frame S',

$$x' = u'_x t, \quad y' = u'_y t, \quad z' = u'_z t,$$

undergoes a Lorentz transformation, and one then considers the resulting expressions for $x(t)$, $y(t)$ and $z(t)$ in the (rest) frame S of the observer. For this purpose, it is suitable to consider the distance (dx, dy, dz) passed during a short time dt. According to the differentiation rules,

$$dx = \gamma\,dx' + \gamma v\,dt', \quad dy = dy', \quad dz = dz', \quad dt = \gamma\,dt' + \frac{v}{c^2}dx'.$$

In the moving frame S', another time interval dt' elapses as compared with the interval in the rest system S.

Velocity in the frame S:

$$u_x = \frac{dx}{dt} = \frac{\gamma\,dx' + \gamma v\,dt'}{\gamma\,dt' + \gamma\dfrac{v}{c^2}dx'} = \frac{\dfrac{dx'}{dt'} + v}{1 + \dfrac{v}{c^2}\dfrac{dx'}{dt'}}.$$

Similarly, one finds the velocities u_y and u_z.

3. Conclusions from the addition theorem

▲   For low velocities $v \ll c$, the relativistic addition of velocities reduces to the ordinary, non-relativistic vector addition of velocities, $u = u' + v$.

▲   For velocities close to the speed of light, one finds, however, $u < u' + v$, i.e., the velocity is smaller than the simple vector sum.

In particular, for $u'_x \approx c$ and $v \approx c$, the relativistic addition theorem leads to

$$u_x = \frac{u'_x + v}{1 + \dfrac{v}{c^2}u'_x} \approx \frac{c + c}{1 + \dfrac{c}{c^2}c} \approx c.$$

▲   The velocity of a body cannot exceed the speed of light.

4.2.3 Relativistic effects

Relativistic effects, effects predicted by means of the Lorentz transformation.

4.2.3.1 Length contraction

1. Distance in the moving system

The distance between two points on the x'-axis in the frame S' is given by

$$l' = x_2' - x_1'.$$

In the frame S, the length l is measured by determining the coordinates of the initial point and the endpoint x_1, x_2 **at the same time** t, $l = x_2 - x_1$. The Lorentz transformation then yields

$$x_1' = \gamma(x_1 - vt), \quad x_2' = \gamma(x_2 - vt),$$

or

$$l = \frac{1}{\gamma}l'.$$

In the frame S the **length** of the same distance appears to be **shortened by the factor** $1/\gamma$.

2. Length contraction

The length of a distance in a moving frame appears to an observer in his own rest frame to be contracted by the factor

$$\frac{1}{\gamma} = \sqrt{1 - \frac{v^2}{c^2}}.$$

➤   The relativity principle leads to the seeming paradox that, for an observer in the frame S', the length of a distance in the frame S appears **also to be contracted**: $l' = (1/\gamma)l$. This paradox is resolved by the relativity of simultaneity of the measurement in both systems.

4.2.3.2 Time dilatation

1. Time interval in a moving system

If in the moving frame S' two events occur at the positions x_1' and x_2' at the times t_1' and t_2', the time distance Δt between the events in the rest frame S is given by

$$\Delta t = t_2 - t_1 = \gamma\left[\left(t_2' + \frac{vx_2'}{c^2}\right) - \left(t_1' + \frac{vx_1'}{c^2}\right)\right],$$

$$= \gamma\left(\Delta t' + \frac{v}{c^2}(x_2' - x_1')\right).$$

If both events happen in the moving frame S' **at the same position** $(x_2' = x_1')$, then

$$\Delta t = \gamma\,\Delta t'.$$

2. Time dilation

The **time** between two events in a moving frame appears to an observer in the rest frame to be **increased** by the factor

$$\gamma = \frac{1}{\sqrt{1 - \dfrac{v^2}{c^2}}} \, .$$

➤   This statement holds also for an observer in the frame S': $\Delta t' = \gamma \, \Delta t$. The time interval in the other frame appears to any observer to be **increased**.

It further follows that two events that occur simultaneously ($\Delta t' = 0$) in a moving frame do not appear as simultaneous events in a rest frame if the events do not occur at the same position:

$$\Delta t = \gamma \frac{v}{c^2} (x_2' - x_1') \, .$$

3. Example: Cosmic radiation

Upon entering Earth's atmosphere, the primary cosmic radiation generates (by collisions with air molecules) a hard secondary radiation that consists of energetic particles. Muons created at a height of about 30 km have a lifetime of $2 \cdot 10^{-6}$ s in their rest frame. At a velocity of $v = 0.9995 \, c$ ($\gamma \approx 32$), these fast muons could (without relativistic effects) traverse a distance of only ≈ 600 m. Hence, they would not be observed at Earth's surface. When taking the time dilatation into account, a lifetime of $32 \cdot 2 \cdot 10^{-6}$ s $\approx 6 \cdot 10^{-5}$ s results. This time interval is sufficiently long to let the particles traverse the path from where they were created to Earth's surface. Hence, the muons created by cosmic radiation can be detected in laboratories on the ground.

4.2.4 Relativistic dynamics

Relativistic dynamics, generalization of dynamics for velocities that are not small compared with the speed of light. It takes the relativistic increase of mass into account and leads to the concept of the equivalence of mass and energy.

4.2.4.1 Relativistic increase of mass

1. Increase of mass

Because of the addition theorem of velocities, the law of momentum conservation, $\vec{p} = m\vec{v}$, can hold in relativistic dynamics only if the mass becomes velocity-dependent (**Fig. 4.7**).

relativistic increase of mass			**M**
	Symbol	Unit	Quantity
$m(v) = \dfrac{m_0}{\sqrt{1 - \dfrac{v^2}{c^2}}} = \gamma m_0$	$m(v)$	kg	mass at velocity v
	m_0	kg	rest mass
	v	m/s	velocity of the body
	c	m/s	speed of light
	γ	1	relativistic factor

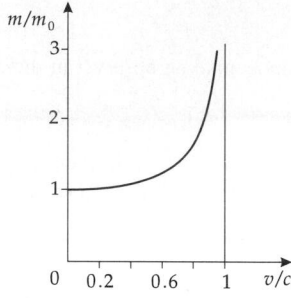

Figure 4.7: Relativistic increase of mass.

➤ The relativistic mass may become arbitrarily large as the velocity of the body approaches the speed of light. Therefore, it is impossible to accelerate a body by a force or by collisions to the speed of light, since this would require an infinite expense of energy.

2. *Relativistic momentum*,

$$\vec{p} = m(v)\vec{v} = \frac{m_0 \vec{v}}{\sqrt{1 - \dfrac{v^2}{c^2}}} = \gamma m_0 \vec{v}.$$

When this expression is inserted into the momentum-balance equation, the law of momentum conservation, and all relations derived from it, continue to hold without modification.

3. *Relativistic force*

For the **relativistic force**:

$$\vec{F} = \frac{d\vec{p}}{dt} = \frac{d}{dt}\left(\frac{m_0 \vec{v}}{\sqrt{1 - \dfrac{v^2}{c^2}}}\right).$$

There is a distinction made between forces acting parallel or perpendicular to the motion. Let $\vec{v}$ be parallel to the x-axis,

$$F_x = \frac{m_0 a_x}{\left(1 - v^2/c^2\right)^{3/2}} = m_0 \gamma^3 a_x,$$

$$F_y = \frac{m_0 a_y}{\sqrt{1 - v^2/c^2}} = m_0 \gamma a_y,$$

$$F_z = \frac{m_0 a_z}{\sqrt{1 - v^2/c^2}} = m_0 \gamma a_z.$$

$\vec{a}$ is the acceleration vector.

▲ To accelerate a body farther along its direction of motion, a force increased by a factor γ^3 is required as compared with the non-relativistic case. For an acceleration perpendicular to the motion, the corresponding factor is only γ.

4.2.4.2   Relativistic kinetic energy

1. Relativistic work,

the work performed on accelerating a body,

$$\Delta W = F \Delta s = m_0 \gamma^3 a \Delta s = m_0 \gamma^3 \frac{\Delta v}{\Delta t} v \Delta t = m_0 \gamma^3 v \Delta v \,,$$

F acting force, Δs distance covered, Δv velocity increase, Δt time interval.
  For acceleration from rest, $u = 0$, up to a velocity $u = v$ the integration yields

$$W = \int_0^v \frac{m_0 u}{\left(1 - \dfrac{u^2}{c^2}\right)^{3/2}} \, du = m_0 c^2 \left(\frac{1}{\sqrt{1 - v^2/c^2}} - 1 \right),$$

the expression for the relativistic kinetic energy.

2. Relativistic kinetic energy

relativistic kinetic energy			$\mathbf{ML^2 T^{-2}}$
$E_{\text{kin}} = m_0 c^2 \left[\dfrac{1}{\sqrt{1 - \dfrac{v^2}{c^2}}} - 1 \right]$ $= m_0 c^2 (\gamma - 1)$	**Symbol**	**Unit**	**Quantity**
	E_{kin}	J	kinetic energy
	m_0	kg	rest mass
	v	m/s	velocity
	c	m/s	speed of light
	γ	1	relativistic factor

➤   In the non-relativistic case,

$$\gamma = \frac{1}{\sqrt{1 - \dfrac{v^2}{c^2}}} \approx 1 + \frac{1}{2} \frac{v^2}{c^2} \,, \qquad E_{\text{kin}} \approx \frac{m_0}{2} v^2 \,.$$

This is the non-relativistic expression for the kinetic energy.

3. Equivalence of mass and energy

Since the zero level of energy can be set arbitrarily, one assigns to any body a relativistic total energy $E = mc^2$, with a **velocity-dependent mass** $m = \gamma m_0$.

▲   **Equivalence of mass and energy**:
  A body with mass m has **relativistic total energy** E,

$$E = mc^2 \,.$$

A body at rest has **rest energy (mass energy)**

$$E_0 = m_0 c^2 \,.$$

■   The mass energy can be released only by converting it to another form of energy.

Application of the theory of relativity to elementary particles (relativistic quantum field theory) leads to just such processes.

If particles and antiparticles closely approach each other, the mass energy $2m_0c^2$ of both particles may be converted to other kinds of energy, in particular to electromagnetic radiation (**pair annihilation**). Conversely, particle-antiparticle pairs may be created from radiation energy (**pair creation**).

4. Energy-momentum relation for relativistic particles

energy-momentum relation			
$\dfrac{E^2}{c^2} = p^2 + m_0^2 c^2$	**Symbol**	**Unit**	**Quantity**
	E	J	relativistic total energy
	p	kg m/s	momentum
	m_0	kg	rest mass
	c	m/s	speed of light

where for E the relativistic total energy mc^2 has to be inserted.

5. Center-of-mass energy,

E_{cm} (cm = center of mass), in a collision of two particles the total energy of both particles, measured in the center-of-mass system is

$$E_{cm} = \sqrt{m_1^2 c^4 + m_2^2 c^4 + 2E_1 E_2 \left(1 - \frac{v_1}{c}\frac{v_2}{c}\right)\cos\theta}$$

(E_1, E_2, relativistic energy of the particles 1 and 2 in an arbitrary system; v_1, v_2, their velocities in this system; θ, angle between the particles). If the particle 2 is at rest in the **laboratory system**, then

$$E_{cm} = \sqrt{m_1^2 c^4 + m_2^2 c^4 + 2E_{1lab}\, m_2\, c^2}\,.$$

The center-of-mass energy characterizes the total energy available in collisions of elementary particles. The velocity in the center-of-mass system is

$$\frac{\vec{v}_{cm}}{c} = \frac{\vec{p}_{1lab}\, c}{E_{1lab} + m_2 c^2}$$

($\vec{p}_{1lab}$—momentum in the laboratory system). The relativistic factor is

$$\gamma_{cm} = \frac{E_{1lab} + m_2 c^2}{E_{cm}}\,.$$

➤   In thermodynamics, the variables pressure and entropy are invariant against Lorentz transformations, whereas the temperature and the amount of heat depend on the state of motion of the system.

4.3   General theory of relativity and cosmology

General theory of relativity, extension of the special theory of relativity to arbitrary (non-inertial) systems. It deals in particular with **gravitation**, using the mathematical tool of a **curved four-dimensional space-time continuum.**

1. General relativity principle

An inertial system in a gravitational field is equivalent to a reference frame in a gravitation-free space that is uniformly accelerated (relative to an inertial frame). This means that an observer cannot distinguish by any experiment which of the systems he is in.

■   An astronaut in a falling elevator, slowed only by air friction, falls with 5/6 of the gravitational acceleration at Earth's surface. He feels only the remaining sixth part of the gravitational force and may therefore believe to be on the Moon, where the weight force is only 1/6 of that on Earth.

Curvature of space, arises as a consequence of the presence of masses and manifests itself by the gravitational force.

2. Test of the general theory of relativity (GTR)

• **Light deflection** in the gravitational field of the Sun. A beam of light from a remote star that passes close to the surface of the Sun is deflected by the space curvature by an angle of $1.75''$. The star then seems to change its position relative to neighboring stars. The phenomenon can be demonstrated during a solar eclipse. Light is also deflected according to Newton's theory, but only by half of the value predicted by GTR. Light deflection is thus no test of the GTR on its own, but the precise experimental value is such a test.

• **Rotation of the apse line** (the line connecting aphelion and perihelion) of the inner planets, due to a modification of Newton's law of gravitation in strong gravitational fields. After accounting for the influence of the other planets, GTR has predicted for Mercury an excessive rotation of $43''$ per century, which has been confirmed by experiment.

• **Red shift** of star light. According to GTR, light is affected by gravitation. The energy spent by the light to leave the gravitational field of a star causes a reduction of the radiation energy, i.e., a shift of the spectral lines towards the long-wave (infra-red) region. The red-shift of spectral lines is also predicted by Newton's theory (combined with the quantum-mechanical rule $E = h \cdot f$).

Black hole, a star with a very strong gravitational field, so that light cannot leave the space region.

3. Properties of the universe

GTR predicts that the universe is either infinite or finite, depending on the total mass of the universe. A finite universe can be compared with the surface of a sphere: it has no boundary, but nevertheless is finite.

Hubble effect, proof of the expansion of the universe. The spectra of very remote stars show a shift to the infra-red, the radiating objects thus move away from the observer. This Hubble shift (cosmologic red-shift) is to be interpreted only by imperfect analogy to the optical Doppler effect.

Hubble constant H specifies the increase of expansion velocity:

$$H = 50 \text{ to } 100 \text{ km/s per Mpc}$$

(1 Mpc = 1 Megaparsec = 3.26 Mill. light years). In a curved space, any observer may believe that all other points move away from him (like the points of the surface of a balloon being blown up).

It depends on the mass available in the universe whether the universe reaches a maximum extension and then collapses (**closed universe**), or whether it continues to expand (**open universe).** The majority of the mass of the universe seems to exist as **dark matter**, invisible to all types of telescopes and other devices. The investigation of the rotation of galaxies suggests that galaxies are enclosed by halos of dark matter.

Big bang, hypothesis that the universe developed ca. $1–2·10^{10}$ years ago from **one** point (**singularity**) of extremely high energy density. It then quickly expanded, and was cooled by that expansion.

3-Kelvin-background radiation, the observed strongly cooled, nearly isotropic thermal radiation in the universe, the remainder of the radiation from the first seconds after the big bang.

4.3.1 Stars and galaxies

1. Stars and their classification

Star, self-luminous stellar object. A star releases energy by a **nuclear-fusion process** that proceeds at very high temperature ($\approx 10^6$ K) in its interior.

Classification:

Stars are classified according to the wavelengths (colors) of the emitted light, and by their magnitude. The typical distances between stars in galaxies are light years, the distances between galaxies are millions of light years. About 5,000 to 10,000 stars are visible to the naked eye, with a small telescope, 100,000. In total, about 10 billion individual stars are accessible by astronomic instruments.

2. Star catalogs

Stars are classified according to **Sky maps** (star catalogs). The brightest stars have proper names from Arabic or Greek. Most of the stars visible with an unaided eye are denoted according to the sky mapping of Bayer (1603); the names consist of a Greek letter specifying the luminosity of the star in its constellation, and the name of the constellation. If the Greek alphabet is not sufficient, the name continues with Latin letters and numbers. Weaker stars are classified by catalog numbers.

■ The brightest star in the constellation Cassiopeia:

Proper name	Schedir
Bayer's Name	α Cassiopeiae (short: α Cas)
Bonn sky mapping	BD +55°139

3. Stellar brightnesses and spectral classes

Stellar brightness, specifies the apparent brightness of a star. Originally from 1^m to 6^m (m, *magnitudo*, Latin for size), today it ranges from the brightness of the Sun, -27^m, to the weakest recordable stars, 23^m. Smaller (more negative) numbers mean brighter stars; each class is $10^{0.4} = 2.512$ times brighter than the next following class.

stellar brightness	example
-27^m	Sun
-13^m	full Moon
-11^m	half Moon
-5^m to -1^m	close planets
up to -2^m	brightest stars (Sirius, Vega)
$+6^m$	observation limit of eye
$+14^m$	Pluto
$+23^m$	photographic observation limit

Spectral class, classifies the type of spectrum of the light emitted by a star.

Spectrum of the light from a star, consists of broad **emission bands**, overlayed by **absorption lines**. Spectral classes are denoted by a Latin capital letter and a number.

■   The Sun has the spectral type G 2.

▲   The spectral class of a star is closely related to its surface temperature.

4. Galaxy,

disk- or spiral-shaped ensemble (diameter 30,000 parsec) of stars. **Milky Way**, spiral-shaped galaxy with a total mass of about 200 billion Sun masses, the Sun being located in one spiral arm. The Milky Way is visible in the sky as a dim band of light. It is surrounded by spherical **stellar clusters**. Galaxies are combined into nebula groups and **nebula clusters** (with diameters of several million light years).

4.3.1.1   Star evolution

1. Energy source of the stars

Stars get their energy from nuclear-fusion processes that take place in the star interior at several million degrees Celsius. In these reactions, hydrogen fuses to helium, catalyzed by carbon and nitrogen (**Bethe-Weizsaecker cycle** or **carbon-nitrogen cycle**). This "hydrogen burning" proceeds relatively slowly.

■   The Sun has consumed only about 3 parts per thousand of its mass over the 4.5 billion years of its existence. In stars with larger mass the energy conversion proceeds very much faster.

When the hydrogen is burned, the energy production in the star decreases. As a consequence, the star contracts since the gravitational force dominates. During the contraction process, the pressure and temperature in the central region increase, so that higher-mass fusion processes up to carbon become feasible. The total energy production again rises steeply, and the contraction due to gravitation is stopped. Ultimately, a **red giant star** develops: the star explodes and reaches temperatures of up to 1 billion degrees Celsius in its interior.

■   The Sun will reach this stage probably in 3.5 billion years. Stars with large mass finally become unstable after consuming their fuel and first form pulsating stars, later novae and supernovae, and finally white dwarfs, neutron stars or black holes.

2. Special states of stars

Double star, a system of two stars rotating about each other due to gravitation.

Variable stars, stars with varying brightness. Periodic variables arise by shadowing of double stars, or by periodic instabilities of the fusion process.

Novae (exploding variable stars), stars which have an explosively expanding gas shell and grow in brightness within about one day by 7 to 10 stellar magnitudes, and then fade away again over months or years. Thereby, only a minor part of the star mass is expelled. Several novae occur periodically. In our Milky Way system, 166 novae have been observed so far.

Supernovae, explosive final stages in the evolution of massive stars. Supernovae occur much more rarely than novae but reach increases of brightness of up to 20 stellar magnitudes (increase of brightness by a factor 10^8). About 7 to 10 supernova explosions are supposed to have happened in the Milky Way system in the past two millennia; several of them have been recorded by ancient historians. After a supernova, the remnants of the star are mostly only expanding gas shells (**gas nebulae**), and possibly white dwarves.

Pulsar, radio source with periodically varying intensity. The periods are in the range of milliseconds to seconds. The pulse length is about 5 % of the period. Pulsars are most likely rapidly rotating neutron stars with extraordinarily strong magnetic fields.

Neutron star, remnants of a star after the supernova stage. Stars release the major part of their energy in a supernova and then collapse so strongly under their own gravitational force that they no longer consist of common matter (atomic nuclei + electron shells). They now consist of tightly packed neutrons, after absorption of the shell electrons by the nuclear protons (see p. 885). Neutron stars have masses of the order of the Sun's mass. Typical radii are ca. 10 km, densities ca. $3 \cdot 10^{17}$ kg/m^3 (density of nuclear matter). The radio radiation arises from plasma clouds accelerated in the gravitational field; the periodicity arises because of the rotation of the system. During a further contraction of a neutron star of sufficient mass, a **black hole** may arise.

5
Mechanics of continuous media

5.1 Theory of elasticity

The theory of elasticity deals with the effects of external, in general static, forces on the shape of rigid bodies.

Elastic deformation, a reversible deformation process in which the body returns to its original shape after the removal of the external force.

Plastic deformation, an irreversible deformation process in which the deformation of the body persists after the removal of the external force.

5.1.1 Stress

1. Definition and properties of stress

Stresses, internal forces within a body. The stresses existing within a body are described by decomposing the body into small volume elements onto which these forces act (**Fig. 5.1**). The stresses produce **deformations of shape** of the volume elements.

Stress, S, the quotient of the applied force and cross-sectional area element upon which the force is acting.

Normal stress, σ, acts perpendicular to the area element.

Shear stress, τ, acts parallel to the area.

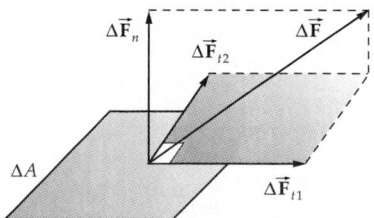

Figure 5.1: Decomposition of the force $\Delta \vec{F}$ acting on the area ΔA into a normal component $\Delta \vec{F}_n$ and two perpendicular tangential components $\Delta \vec{F}_{t1}$, $\Delta \vec{F}_{t2}$.

stress			ML^{-1}T^{-2}
	Symbol	Unit	Quantity
$\vec{S} = \dfrac{\Delta\vec{F}}{\Delta A}$	$\vec{S}$	N/m^2	stress vector
	$\vec{\sigma}$	N/m^2	normal-stress vector
$\vec{\sigma} = \dfrac{\Delta\vec{F}_n}{\Delta A}$	$\vec{\tau}$	N/m^2	shear-stress vector
	ΔA	m^2	area element
	$\Delta\vec{F}$	N	acting force
$\vec{\tau} = \dfrac{\Delta\vec{F}_t}{\Delta A}$	$\Delta\vec{F}_n$	N	normal component of $\vec{F}$
	$\Delta\vec{F}_t$	N	tangential component of $\vec{F}$

Newton per square meter, N/m^2, SI unit of stress:
1 N/m^2 is the stress on an area of 1 m^2 if a force of 1 N is acting on it.

➤ The typical order of magnitude of stress is MN/m^2 = N/mm^2.

➤ For a pressure load, the stress has a negative sign.

➤ It is assumed that the cross-section does not change under deformation.

■ A load of m = 1 kg is fixed to a wire of diameter d = 1 mm. The stress on the wire is

$$S = \frac{F}{A} = \frac{mg}{\pi(d/2)^2} = \frac{1\ \text{kg} \cdot 9.81\ \text{m/s}^2}{\pi \cdot (0.5\ \text{mm})^2} = 12.5\ \text{N/mm}^2 = 12.5\ \text{MN/m}^2 .$$

2. Stress tensor,

$\hat{\tau}$, describes the **state of stress** of a small cubic element of the body. The state of stress can be described in general by specifying **nine** quantities. For any face of the cube, three force components must be given (**Fig. 5.2**). If the cube is sufficiently small, the forces acting on opposite sides are equal, so that the state of stress may be described by the elements τ_{ij} of the **stress tensor**:

$$\hat{\tau} = \begin{pmatrix} \tau_{xx} = \sigma_x & \tau_{xy} & \tau_{xz} \\ \tau_{yx} & \tau_{yy} = \sigma_y & \tau_{yz} \\ \tau_{zx} & \tau_{zy} & \tau_{zz} = \sigma_z \end{pmatrix} .$$

■ The first index of the components of the stress tensor characterizes the area; the second index specifies the direction of force. For instance, the element τ_{xy} gives the force acting in y-direction onto the lateral area element with the normal perpendicular to the x-axis.

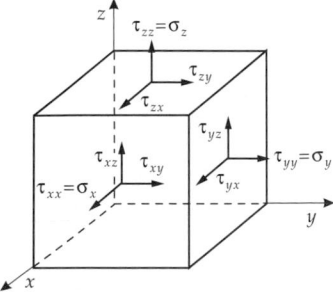

Figure 5.2: Components of the stress tensor.

The diagonal elements are the **normal stresses** (components of stress along the surface normal), the off-diagonal elements represent the **shear stresses** or **tangential stresses** (components of stress perpendicular to the surface normal). The stress tensor is symmetric:

$$\tau_{xy} = \tau_{yx}, \qquad \tau_{xz} = \tau_{zx}, \qquad \tau_{yz} = \tau_{zy}.$$

$\hat{\tau}$ therefore contains only six independent quantities: three normal stresses and three shear stresses.

5.1.1.1   Tension, bending, shear, torsion

The following definitions describe elementary types of loads.

Tension or **compression**, arise if the shear stresses vanish, and the force acts uniformly on the body. The body responds with **strain** and **transverse strain** (**Fig. 5.3** and **Fig. 5.4**). **Isotropic pressure (hydrostatic pressure)**, an equal pressure acts on all faces of the body (**Fig. 5.5**).

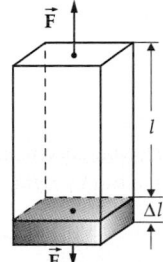

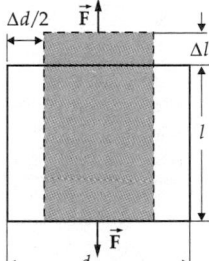

Figure 5.3: Strain.   Figure 5.4: Transverse strain.

Shear, occurs when the forces act parallel to the surface of the body. The body responds with a deformation that is also called shear. The angles between the edges of the body change (**Fig. 5.6**). **Bending**, the shear stresses vanish, but the pressure or the tension acts non-uniformly and causes a non-uniform deformation of the body; some parts of the body undergo a tensile load, others a compressional load (**Fig. 5.7**).

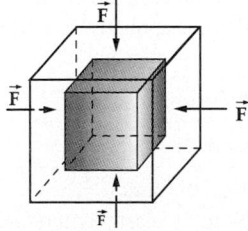

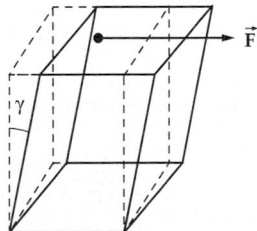

Figure 5.5: Isotropic compression.   Figure 5.6: Shear. Shear angle γ.

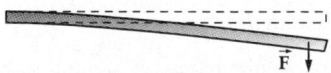

Figure 5.7: Bending.

Torsion, as in the case of shear only forces parallel to the surfaces of the body occur. They point, however, in different directions at different positions and hence generate a **torque**. This leads to a twist of the body axes.

Practical examples of load can be a mixture of these elementary examples.

5.1.2 Elastic deformation

Elastic deformation is described as the change in the geometry of a body under the action of external forces.

Method of finite elements: To describe the deformation of a body, one considers a small cubic element of the body and the deformation generated by the applied stress. The deformation of an extended body may then be calculated by summing the deformations of the elements.

Basically, there are two kinds of deformation of a cube:

Strain, ε, the length of one or several edges of the cube is changed, but a right angle is maintained:

$$\varepsilon = \frac{\Delta l}{l},$$

where l is the original length, and Δl is the change of length.

➤   Compressions are negative strains.

Shear, γ, a change of one or several angles of the cube without changing the edge lengths. Shear denotes the deviation of the corresponding angle from a right angle (in rad).

In practice, the following four cases arise:

- **strain**
- **transverse strain**
- **isotropic compression**
- **shear**

5.1.2.1   Strain

1. Properties of strain

Strain, due to an external tensile force; the body is stretched along the direction of the applied force, or contracts due to an external compression force. In the elastic region, the change of length follows **Hooke's law**, it is proportional to the applied stress (**Fig. 5.8**):

stress = elasticity modulus · strain (Hooke's law)		$\mathbf{ML^{-1}T^{-2}}$	
$\varepsilon = \dfrac{1}{E}\sigma$ $\sigma = E\varepsilon$	Symbol	Unit	Quantity
	ε	1	strain
	E	N/m^2	elasticity modulus
	σ	N/m^2	normal stress

2. Elastic modulus and coefficient of linear extension

Elastic modulus, Young's modulus, E, gives the required stress σ per unit strain (fractional change of length $\varepsilon = \Delta l/l$). E is a quantity which depends on the material. SI unit

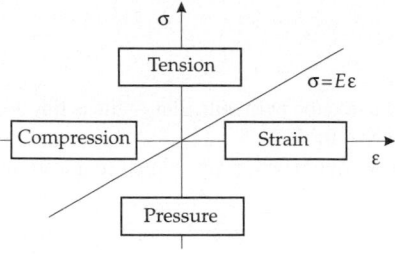

Figure 5.8: Hooke's law. The strain ε is proportional to the stress σ.

of E:

$$[E] = \frac{N}{m^2}.$$

The elastic modulus is usually given in units of N/mm^2 = MN/m^2 or GN/m^2.

Coefficient of linear extension α, the reciprocal value of the elastic modulus, gives the resulting strain per unit applied stress,

$$\alpha = \frac{1}{E}.$$

SI unit of the coefficient of linear extension α:

$$[\alpha] = \frac{m^2}{N}.$$

➤ Hooke's law holds only approximately for small strains. For higher strains, the relation between normal stress and strain is **nonlinear**. The elastic modulus is a parameter that also depends on the temperature. Typical values are in the range of 10^4 and 10^5 N/mm^2 (see **Tab. 7.2**).

■ The elastic modulus of gold is 81000 N/mm^2. In order to compress a cube of gold of edge length
$l = 10$ cm by 1% of its edge length ($\varepsilon = -0.001$), a stress of

$$\sigma = E\varepsilon = -81 \cdot 10^9 \text{ N/m}^2 \cdot 0.001 = -81 \text{ N/mm}^2$$

has to be applied, i.e., the mass that must be placed on its face is

$$m = \frac{F}{g} = \frac{A \cdot \sigma}{g} = \frac{l^2 \sigma}{g} = 82.6 \cdot 10^3 \text{ kg} = 82.6 \text{ t}.$$

In general, the strain ε of a cubic element of a body is a function $\varepsilon(\sigma)$ of the normal stress σ applied.

Elasticity modulus for a given normal stress, the change $d\sigma$ of the normal stress required for a change of strain by $d\varepsilon$:

$$E(\sigma) = \frac{d\sigma}{d\varepsilon}.$$

Hence, the elasticity modulus is the derivative of the function $\sigma(\varepsilon)$, or graphically, the slope of the curve of normal stress in the graph of stress versus strain.

5.1.2.2   Transverse strain

1. Definition of transverse strain

Transverse strain, the change of edge length of a cube perpendicular to the acting force.

▲   A tensile force stretches a body and makes it thinner.

Fractional change of thickness, (**transverse strain**) ε_q, proportional to the strain and to the normal stress:

transverse strain, transverse contraction			1
	Symbol	Unit	Quantity
$\varepsilon_q = \dfrac{\Delta d}{d}$	d	m	thickness
	Δd	m	change of thickness
$= -\nu \cdot \varepsilon = -\dfrac{1}{\mu}\varepsilon$	ε_q	1	transverse strain
	ε	1	strain
$= -\dfrac{\nu}{E}\sigma = -\dfrac{1}{\mu E}\sigma$	ν	1	coefficient of transverse strain
	μ	1	Poisson number
	E	N/m^2	elasticity modulus
	σ	N/m^2	normal stress

Coefficient of transverse strain, ν, proportionality factor between strain and transverse strain.

2. Poisson number,

Poisson coefficient, μ, reciprocal value of the coefficient of transverse strain ν, gives the ratio of fractional change of thickness $\Delta d/d$ and the fractional change of length $\Delta l/l$:

$$\mu = \frac{1}{\nu} = -\frac{\Delta d/d}{\Delta l/l}\,.$$

➤   The negative sign between ε_q and ε expresses the experimental fact that the diameter of a cylindrical wire is reduced under tension while its length increases.

➤   Typical values of the coefficient of transverse strain: $\nu \approx 0.3$ to 0.4, $\mu \approx 2$ to 3.

■   In the example given above of a cube of gold with the edge length $l = 10$ cm, which is compressed by a mass of 82.6 t by 1% ($\varepsilon = -0.001$), the cube becomes wider by

$$\varepsilon_q = -\nu\varepsilon = 0.42 \cdot 0.001 = 0.42\%.$$

3. Change of volume

Due to strain and transverse strain, the volume of a rod with a square cross-section is altered:

$$\Delta V = V' - V = (d + \Delta d)^2(l + \Delta l) - d^2 l\,.$$

V, V' volume without and with stress, respectively, ΔV change of volume, l, d length and diameter of the rod without stress, Δl change of length (along orientation of tension), Δd change of diameter (perpendicular to orientation of tension). For small changes, the terms quadratic in Δd and Δl may be ignored:

$$\Delta V = d^2 \Delta l + 2d \cdot l \Delta d\,.$$

The fractional change of volume is

$$\frac{\Delta V}{V} = \frac{\Delta l}{l} + 2\frac{\Delta d}{d} = \varepsilon(1 - 2\nu).$$

➤ For $\nu = 0.5$ the volume does not change, for $\nu < 0.5$ it increases. Values of $\nu > 0.5$ would mean a decrease of the volume under an applied tensile stress, a situation that does not occur physically.

◼ The cube of gold of 10 cm edge length changes its volume by

$$\frac{\Delta V}{V} = \varepsilon(1 - 2\nu) = -0.001(1 - 2 \cdot 0.42) = -0.16\%,$$

in absolute numbers:

$$\Delta V = -0.00016 \cdot V = -0.00016 \cdot 1000 \text{ cm}^3 = 0.16 \text{ cm}^3.$$

4. Strain tensor,

$\hat{\varepsilon}$, determines the general state of strain of the body if a point mass at a position $\vec{r} = (x_1, x_2, x_3)$ is shifted due to the strain by the displacement vector $\vec{s}(\vec{r})$ to $\vec{r} + \vec{s}(\vec{r})$:

$$dx_i \rightarrow dx_i + ds_i = dx_i + \sum_{k=1}^{3} \frac{\partial s_i}{\partial x_k} dx_k.$$

The components of the strain tensor $\hat{\varepsilon}$ are expressed as the partial derivatives of the components of the displacement vector $\vec{s}$ with respect to the coordinates x_i, $i = 1, 2, 3$:

$$\hat{\varepsilon} = \frac{1}{2}\begin{pmatrix} \varepsilon_1 & \gamma_{12} & \gamma_{13} \\ \gamma_{21} & \varepsilon_2 & \gamma_{23} \\ \gamma_{31} & \gamma_{32} & \varepsilon_3 \end{pmatrix}, \quad \varepsilon_i = 2\frac{\partial s_i}{\partial x_i}, \quad \gamma_{ik} = \gamma_{ki} = \frac{\partial s_k}{\partial x_i} + \frac{\partial s_i}{\partial x_k}.$$

The strain tensor is a symmetric tensor.

5.1.2.3 Isotropic compression

1. Properties of isotropic compression

Isotropic compression, the volume change of a body under a compression force acting with equal magnitude from any side, unlike strain and transverse strain, where the force acts only in one direction.

The fractional change of volume is

$$\frac{\Delta V}{V} = 3\varepsilon(1 - 2\nu),$$

where the factor 3 takes into account that three normal stresses are acting instead of one. The stress is written as

$$\sigma = -\Delta p,$$

where Δp denotes the pressure load, and using $\varepsilon = \sigma/E$, then

$$-\Delta p = \frac{\Delta V}{V}\frac{E}{3(1 - 2\nu)}.$$

By analogy to the elasticity modulus, one defines:

pressure = bulk modulus · fractional change of volume			$ML^{-1}T^{-2}$
	Symbol	Unit	Quantity
$-\Delta p = K \dfrac{\Delta V}{V}$	Δp	$Pa = N/m^2$	pressure
	K	N/m^2	bulk modulus
	ΔV	m^3	change of volume
	V	m^3	volume of the body

2. Bulk modulus,

K, gives the pressure required per fractional change of volume.
   Customary unit for K: $N/mm^2 = MN/m^2$ or GN/m^2.

➤   Typical values of the bulk modulus are between 100 and 200 GN/m^2,
    (ice: $K \approx 10\ GN/m^2$, lead: $K \approx 44\ GN/m^2$; see **Tab. 7.3/2**).

■   Copper has a bulk modulus of 126,000 N/mm^2. Under atmospheric pressure (about
    10^5 Pa), the volume of a block of copper changes by

$$\frac{\Delta V}{V} = -\frac{\Delta p}{K} = 7.9 \cdot 10^{-7} = 0.000079\,\% \,.$$

   Hence, the volume of a block of copper of 1 m^3 changes by about 0.8 cm^3.
   Bulk modulus K and elasticity modulus E are related by the coefficient of transverse
strain:

$$K = \frac{E}{3(1 - 2\nu)} \,.$$

In thermodynamics, when describing fluids and gases, it is customary to use the reciprocal
value of the bulk modulus K, the compressibility κ.

3. Compressibility,

κ, the reciprocal value of the bulk modulus (see **Tab. 7.3/4**):

$$\kappa = \frac{1}{K} = \frac{\Delta V/V}{-\Delta p} \,.$$

For gases

$$\kappa = \frac{A}{V(p + p_T)} \,.$$

A increases with temperature and is characteristic for the particular gas, volume V, external
pressure p, Van der Waals pressure p_T. For the ideal gas $A = 1$ and $p_T = 0$.

5.1.2.4   Bending of a rod (beam)

1. Definition of bending

Bending, occurs if a pointwise supported or mounted component of construction is under
load away from the supporting (pivoting) points. Here we consider only the case of a beam
that is assumed to be oriented along the z-axis and to have constant cross-section (x, y).
Let the loading force act perpendicular to the z-axis.

Cases of load in bending:

- cantilever beams, one end is tightly mounted (tangent horizontal), point load applied at the free end, or continuous load distributed along the z-axis;
- simple beams, tightly mounted at both ends, pointwise or continuous load;
- one end is tightly mounted, the other end is supported;
- both ends are supported.

In one part of the cross-section of the beam, there is a compression load, in the other part there is a tension load. Both regions are separated by the **neutral axis** that passes through the center of mass of the beam cross section (see **Fig. 5.9**).

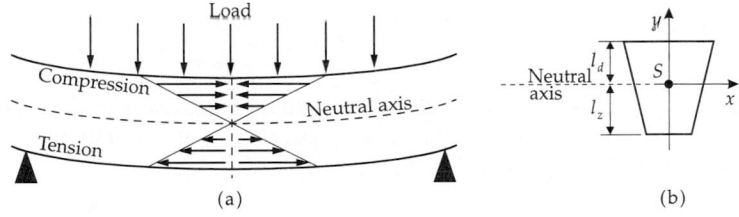

(a)   (b)

Figure 5.9: Schematic illustration of bending, compression and tension load distribution in a beam supported at both ends. The neutral axis passes through the center of mass S of the perpendicular cross-section. l_d, l_z: distances between the outermost axes of the compression and tension regions and the neutral axis, respectively. (a): longitudinal cross-section, (b): perpendicular cross-section.

2. Bending moment,

M_b, the product of force F and force arm l. For a cantilevered beam of length l tightly mounted at one end and loaded at the free end, the force arm is measured from the free end to the point of attachment. The bending moment on a cross-sectional area perpendicular to the beam axis z is zero at the free end; the maximum value arises at the fixed end, $M_{b,\mathrm{max}} = F \cdot l$.

For a beam fixed at one end loaded by several point loads (or by a continuous load), the bending moment on a selected cross-sectional area is the sum (or the integral) over the bending moments of the individual forces.

For a beam that is freely supported or tightly mounted at both ends and loaded by a single load, the maximum bending moment occurs at the load point.

For a beam that is supported or tightly mounted at both ends and supporting a constant continuous load (or by a sum of equidistant and equal point loads), the maximum bending moment occurs at the midpoint of the beam.

bending moment			$\mathbf{ML^2T^{-2}}$
	Symbol	Unit	Quantity
$M_b = \sum_i F_i \cdot l_i$	M_b	Nm	bending moment
	F_i	N	ith acting force
	l_i	m	ith force arm

If several forces are applied, then the bending moments must be added. Right-handed moments (clockwise) and left-handed moments enter the sum with opposite signs.

Plane area moment of inertia, J, characterizes the shape and magnitude of the cross-sectional area of the beam (see **Fig. 5.9 (b)**).

Axial plane area moment of inertia J_a, with respect to the neutral axis:

$$J_x = \int y^2 \, dA, \quad J_y = \int x^2 \, dA, \quad dA \text{ area element.}$$

Polar area moment of inertia J_p, with respect to the center of mass:

$$J_p = \int r^2 \, dA = \int (x^2 + y^2) \, dA = J_x + J_y.$$

Resistive moment, W_b:

$$W_{x,\text{tens}} = \frac{J_x}{e_{\text{tens}}}, \quad W_{x,\text{press}} = \frac{J_x}{e_{\text{press}}},$$

where e_{tens}, e_{press} are the distances between the outermost axes of the tension and pressure regions of the beam cross-section and the neutral axis, respectively (see **Fig. 5.9**).
    The **maximum bending stress** is given by

$$\sigma_b = \frac{M_b}{W_b}.$$

3. Deflection,

determined by the geometry of the support system, and by the ratio

$$\frac{F}{E J_a}$$

of the applied force F and the product of the elasticity modulus E and the axial **plane area moment of inertia** J_a of the perpendicular beam cross-section. The axial plane area moments for a circular cross-section of diameter d and for a rectangular cross-section (width b and height h) are:

$$J_{a,\text{circle}} = \frac{\pi}{64} d^4 \approx 0.049 \, d^4, \quad J_{a,\text{rectangle}} = \frac{bh^3}{12} \approx 0.083 \, bh^3.$$

The maximum load of a beam with rectangular cross-section is proportional to the width and to the third power of the height, but inversely proportional to the beam length.

4. Examples: bending moments and deflections for typical cases of load

- Cantilever beam, point load F at the free end (**Fig. 5.10 (a)**):

$$F_A = F, \quad s = \frac{l^3}{3} \frac{F}{E J_a}, \quad M_{b,\text{max}} = l F.$$

- Cantilever beam, uniform load, sum F (**Fig. 5.10 (b)**):

$$F_A = F, \quad s = \frac{l^3}{8} \frac{F}{E J_a}, \quad M_{b,\text{max}} = \frac{l}{2} F.$$

- Simple beam, point load F, asymmetrical (**Fig. 5.10 (c)**):

$$F_A = \frac{b}{l} F, \quad a + b = l, \quad F_B = \frac{a}{l} F$$

$$s = \frac{a^2 b^2}{3l} \frac{F}{E J_a}, \quad M_{b,\text{max}} = \frac{ab}{l} F.$$

- Simple beam, uniform load, sum F (**Fig. 5.10 (d)**):

$$F_A = F_B = F/2, \quad s \approx \frac{l^3}{77}\frac{F}{E J_a}, \quad M_{b,\max} = \frac{l}{8}F.$$

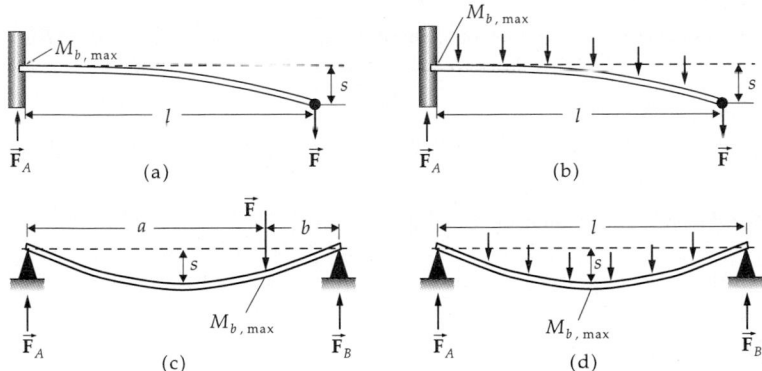

Figure 5.10: Bending contours (statically determined) of a beam. (a) point load, for a cantilever beam, (b) uniform load, for a cantilever beam, (c) point load, for a simple beam supported at both ends, (d) uniform load, for a simple beam supported at both ends.

5. Example: steel girder

A steel girder (elasticity modulus 200 GN/m^2) with a square cross-section of edge length 10 cm and a length of 2 m carries a load of 1000 kg mass. The plane area moment of inertia J_a is

$$J_a = J_{a,\text{rectangle}} = 0.083 \cdot (0.1 \text{ m}) \cdot (0.1 \text{ m})^3 = 8.3 \cdot 10^{-6} \text{ m}^4.$$

One obtains

$$\frac{F}{E J_a} = 5.9 \cdot 10^{-3} \text{ m}^{-2}.$$

For various cases of load, the deflections and normal stresses are as follows.

cantilever, uniform	$s = \dfrac{l^3}{8}\dfrac{F}{E J_a} = 5.9 \text{ mm}$	$M_b = \dfrac{l}{2}F = 9810 \text{ Nm}$
simple, uniform	$s = \dfrac{l^3}{77}\dfrac{F}{E J_a} = 0.6 \text{ mm}$	$M_b = \dfrac{l}{8}F = 2450 \text{ Nm}$
cantilever, load at the end	$s = \dfrac{l^3}{3}\dfrac{F}{E J_a} = 16 \text{ mm}$	$M_b = lF = 19620 \text{ Nm}$
simple, load in the middle	$s = \dfrac{(l/2)^2(l/2)^2}{3l}\dfrac{F}{E J_a}$ $= 1 \text{ mm}$	$M_b = \dfrac{(l/2)(l/2)}{l}F = 4900 \text{ Nm}$

- If the length of a girder is doubled, the deflection increases by a factor of eight, and the maximum normal stress increases by a factor of two.
- When the lengths of the sides of the cross-sectional area are halved, the area moment of inertia decreases by a factor of one sixteenth, hence the deflection increases by a factor of sixteen.

6. Bending stress,

σ_b, the stress generated in bending, quotient of bending moment M_b and **resistive moment** W_b:

$$\sigma_b = \frac{M_b}{W_b}, \quad W_{b,\text{circle}} = \frac{\pi d^3}{32} = 0.098\, d^3, \quad W_{b,\text{rectangle}} = \frac{bh^2}{6} = 0.167\, bh^2.$$

■ For the preceding example (steel girder), one finds

$$W_b = 1.67 \cdot 10^{-4} \text{ m}^3.$$

Hence, the maximum stresses are:

cantilever, uniform load $\sigma_b = 59 \text{ N/mm}^2$,
simple, uniform load $\sigma_b = 15 \text{ N/mm}^2$,
cantilever, load at the end $\sigma_b = 118 \text{ N/mm}^2$,
simple, load in the middle $\sigma_b = 3 \text{ N/mm}^2$.

The tensile strength of steel varies over the range of 400 to 1200 N/mm^2. When the edge length of the cross-section is halved, the resistive moment decreases to one eighth, and the stress increases by a factor of eight.

5.1.2.5   Shear

1. Properties of shear

Shear, deformation of a body in which the right angles in a small cubical element change by the **shear angle** γ. Shear occurs if forces act parallel to a face of the cube.

▲ For small shear strains, the shear angle is proportional to the **shear stress** τ.

shear stress = shear modulus · shear angle			$\mathbf{ML^{-1}T^{-2}}$
	Symbol	Unit	Quantity
$\tau = G\gamma$	τ	N/m^2	shear stress
	G	N/m^2	shear modulus
	γ	rad	shear angle

2. Shear modulus,

G, proportionality factor that gives the required shear stress per unit of shear angle.
SI unit of G:

$$[G] = \frac{\text{N}}{\text{m}^2} = 1 \text{ Pa}.$$

In general, the required shear stress τ is a function of the desired shear angle γ, and generally one defines the shear modulus by

$$G = \frac{d\tau}{d\gamma} \, .$$

▲ Shear modulus G and elasticity modulus E are connected by the coefficient of transverse strain ν:

$$G = \frac{E}{2(1 + \nu)} \, .$$

Since $0 \leq \nu \leq 0.5$, it follows that

$$\frac{E}{3} \leq G \leq \frac{E}{2} \, .$$

➤ In anisotropic materials that behave differently in different directions, a different modulus must be used for each different spatial direction.

5.1.2.6   Torsion

1. Torsion and torsion stress

Torsion, shear stresses act along different orientations so that there is a torque on the body.

Torsion stress, K, the ratio of applied torque τ_t to **resistive moment** W_t under torsion of the body:

$$K = \frac{\tau_t}{W_t} \, , \qquad [K] = \frac{N}{m^2} \, .$$

▲ The resistive moment W_t depends on the geometry of the body.
■ For a circular cross-section of diameter d:

$$W_t = \frac{\pi}{16} d^3 = 0.196 \, d^3, \qquad [W_t] = m^3 \, .$$

In the torsion of rods cross-sections are twisted by a torsion angle ϕ that depends on the position along the axis.

2. Twisting,

ψ, for a body that is a right circular cylinder, the torsion angle ϕ per unit length, $\psi = \phi/l$, or $\psi = d\phi/dl$. The twisting is proportional to the torque τ_t, but inversely proportional to the shear modulus G (**Fig. 5.11**):

twisting					$\mathbf{L^{-1}}$
		Symbol	Unit	Quantity	
		ψ	rad/m	twisting	
		ϕ	rad	torsion angle	
$\psi = \dfrac{d\phi}{dl} = \dfrac{W_t}{GJ_p} \quad K = \dfrac{\tau_t}{GJ_p}$		l	m	length of the body	
		W_t	m^3	resistive moment	
		J_p	m^4	polar area moment	
		G	N/m^2	shear modulus	
		K	N/m^2	torsion stress	
		τ_t	Nm	torque	

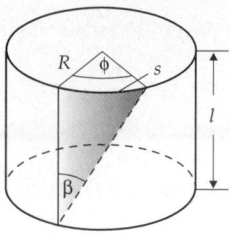

Figure 5.11: Torsion of a cylindrical rod, radius R, length l, torsion angle ϕ. Deflection s at the circumference of the end face: $s = R \cdot \phi = l \cdot \beta$.

3. Polar area moment of inertia,

J_p, the area moment of the cross-section with respect to its center of mass:

$$J_p = \int r^2 \, dA, \quad r^2 = x^2 + y^2, \quad dA = dx \, dy.$$

■   For a circular cross-section of diameter d:

$$J_p = \frac{\pi}{32} \cdot d^4 = 0.098 \, d^4, \qquad [J_p] = \text{m}^4.$$

For a circular ring with outer radius R_1 and the inner radius R_2:

$$J_p = \frac{\pi}{2} \left(R_1^4 - R_2^4 \right).$$

➤   If the body does not have a circular cross-section, one has to replace the polar area moment J_p in the formula by the torsion moment J_t ($J_t \leq J_p$).

5.1.2.7   Energy and work in deformations

1. Work of deformation

In an elastic deformation of a body, work is performed. If one considers only the strain ε, according to the definition of work, one obtains:

$$\Delta W = F \Delta l = \sigma A \cdot l \Delta \varepsilon = V \sigma \Delta \varepsilon.$$

In integral notation:

work of deformation			$\mathbf{ML^2T^{-2}}$
	Symbol	Unit	Quantity
$W = V \int \sigma(\varepsilon) \, d\varepsilon$	W	J	performed work
	V	m^3	volume of the body
	$\Delta \varepsilon$	1	change of strain
	σ	N/m^2	normal stress
	A	m^2	area
	l	m	extension of the body
	Δl	m	change of length

$\sigma(\varepsilon)$ is the normal stress applied in the deformation process. The integral ranges from the original value of the strain to the final one.

For a pressure load $\sigma < 0$ a compression ($\Delta\varepsilon < 0$) results. The work done is

$$\Delta W = -V\sigma\Delta\varepsilon > 0 .$$

Work is performed both in compression and expansion of a body.

2. Energy conservation law in elastic deformations

If a deformation is perfectly elastic, the work done to deform the body is released when the body relaxes.

➤ There are no perfectly elastic deformations. Part of the work expended is always lost as **dissipated heat**, for reasons discussed in thermodynamics.

■ In order to compress the cube of gold treated above with edge length 10 cm by 1%, the work

$$\Delta W = V\sigma\Delta\varepsilon = 1000 \text{ cm}^3 \cdot (-810 \text{ N/mm}^2) \cdot (-0.001) = 810 \text{ J}$$

is performed.

5.1.3   Plastic deformation

1. Properties of plastic deformation

Plastic deformation, the deformation is maintained partly or completely after the force is removed. Therefore, the work expended for deformation cannot be gained back completely.

This is expressed by the **hysteresis curve** of plastic deformation: the applied stress σ is plotted against the resulting strain ε for a load process with alternating tension and pressure phases (**Fig. 5.12**). **Stress-strain diagram** (σ-ε **diagram**): In a perfectly elastic deformation, the same curve is followed in the strain phase and in the compression phase. Plastic deformations are characterized by the occurrence of hysteresis, i.e., of two distinct branches of the curve traversed in different directions. Even for a vanishing stress σ, a **residual strain** ε_1 or a **residual compression** ε_2 persists.

2. Energy loss in plastic deformation

The work done in this process is proportional to the area enclosed by both curves:

energy lost in plastic deformation			$\mathbf{ML^2T^{-2}}$
	Symbol	Unit	Quantity
$W = V \oint \sigma \, d\varepsilon$	W	J	energy lost
	V	m^3	volume
	σ	N/m^2	normal stress
	ε	1	strain

➤ Plastic deformations play an important role in materials processing (pressing, rolling, bending, etc.).

5.1.3.1   Regions in tensile load

The behavior of materials under a tensile load is determined by specially designed machines and plotted in a **stress-strain diagram** (**Fig. 5.13**).

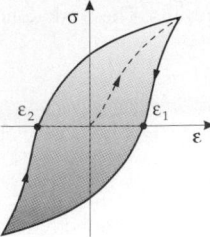

Figure 5.12: Hysteresis curve in plastic deformation. Dashed curve: first deformation. Shadowed area: energy loss in plastic deformation.

1. Regions in tensile load

The following regions may be distinguished:

a) elastic region, the strain follows Hooke's law and the deformation disappears completely if the stress is no longer acting;

b) plastic-elastic region, the deformation does not disappear completely after the decay of the stress, but Hooke's law still holds;

c) plastic region, the deformation is also maintained to a large extent without stress. Usually, the stress-strain curve becomes flat in this region. For large strains, the stress required decreases, since the internal structure of the body has already been altered significantly under the strain.

d) Break point, the strain at which the body undergoes rupture.

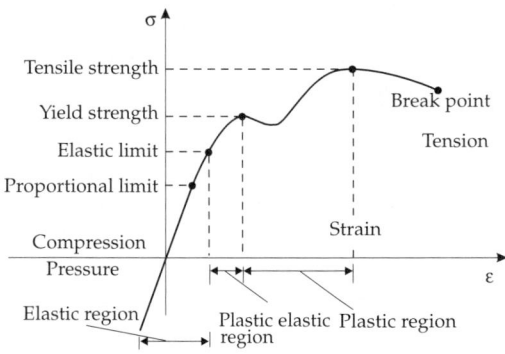

Figure 5.13: Stress-strain diagram.

2. Parameters and properties of tensile loads

M   The stress-strain diagram is determined by machine according to the ISO standard under fixed external conditions, such as temperature. A defined test course (speed of tension etc.) is followed.

➤   All material constants depend on the detailed composition of the material (in particular for alloys).

a) Hooke's straight line, tangent to the stress-strain curve at the origin. Its slope is the **elastic modulus** E of the body for small strains.

   The transition points between the regions of the stress-strain graph are described by **critical stresses**:

b) Yield point, R_p, or **yield strength**, σ_f, stress at which a certain deformation persists as a plastic deformation. It is customary to take the 0.2 %-yield point $R_{p\,0.2}$ obtained when plotting a parallel to Hooke's straight line that intersects the abscissa at $\varepsilon_r = 0.2\,\%$. The intersection point between this straight line and the stress-strain curve gives the yield point.

c) Tensile strength, R_m, or **rupture stress**, σ_B, the maximum stress occuring in the stress-strain diagram. If higher stress is applied to a body, the break point is reached, i.e., the body fractures.

➤ Typical values for metals are 10 to 20 N/mm^2; for ordinary steels, values of 400 to 1200 N/mm^2 may be reached. Special steels reach values up to 4500 N/mm^2.

d) Yield strength (**flow limit**), the point beyond which the tensile force no longer increases even for an additional extension. Some materials exhibit a non-monotonic transition between elastic and plastic or nonelastic region, i.e., at the end of the plastic region the stress decreases first and then increases again. In this case, one distinguishes an upper and lower tensile yield point corresponding to the local minima of the stress-strain curve.

e) Break point, ε_B, the value of strain where the body fractures.

➤ Typical values for the fracture strain are 0.02 (copper) through 0.45 (V2A steel) to 0.5 (aluminum and gold).

➤ Unlike elastic deformations, in plastic deformations there are no (or only very small) changes of volume. Correspondingly, the coefficient of transverse strain is $\nu = 0.5$.

5.1.3.2 Buckling

1. Buckling and buckling stress

Buckling, occurs when a rod under compressive stress moves sideways at its center (**Fig. 5.14**).

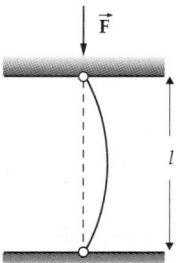

Figure 5.14: Buckling of a rod by a force $\vec{F}$. Due to the deformation of the rod, the compressive stress becomes a bending stress under which the rod gives way much more easily.

Buckling occurs if the applied compressive stress σ exceeds the buckling stress σ_k. **Euler formula** for buckling stress:

buckling stress: Euler formula			$\mathbf{ML^{-1}T^{-2}}$
	Symbol	Unit	Quantity
$\sigma_k = \pi^2 \dfrac{E}{\lambda^2}$	σ_k	N/m^2	buckling stress
	E	N/m^2	elasticity modulus
	λ	1	thinness ratio

➤ Safety factors of 5 to 10 must be included in the design of machine components.

2. Thinness and safety factor

Thinness ratio, λ, describes the thinness of a rod:

$$\lambda = l\sqrt{\frac{A}{J_a}}$$

(l length of rod, A cross-sectional area, J_a plane area moment of inertia).

◾ A circular rod of diameter 1 cm and length 1 m has a plane area moment

$$J_a = \frac{\pi}{64} \cdot d^4 = 0.049 \cdot (1\ \text{cm})^4 = 490\ \text{mm}^4,$$

and hence a thinness ratio

$$\lambda = 1\ \text{m} \cdot \sqrt{\frac{79\ \text{mm}^2}{490\ \text{mm}^4}} = 400.$$

For an elasticity modulus of 200 GN/m^2, one obtains a buckling stress

$$\sigma_k = \pi^2 \frac{200\ \text{GN/m}^2}{400^2} = 12.3\ \text{MN/m}^2.$$

This corresponds to a maximum load

$$F = \sigma_k \cdot A = 975\ \text{N}.$$

With a corresponding safety factor of 8, the rod can be loaded with 12 kg.

Safety factor, in structural design the ratio of a stress limit value (yield stress, rupture stress, buckling stress) and the actual stress.

5.1.3.3 Hardness

1. Definition of hardness

Hardness, the resistance of a body to the indentation of a small test body into its surface. In such a process, high stresses occur at a point on the body, which may lead to a local deformation.

The hardness of a material is determined by standardized methods of measurement and denoted by a number. All methods of measurement are based on a standardized indenter that is pressed with a certain force during a certain time into the surface (**Fig. 5.15**). From the applied force, the geometry of the indenter, and the deformation, the hardness number may be determined (see **Tab. 7.2**).

➤   The indenter must have a higher hardness than the specimen to be tested in order not to become deformed itself.

2. Brinell hardness,

HB, the indenter is a sphere. The Brinell hardness is the ratio of the applied force F and the area of indentation A, multiplied by a factor 0.102:

$$HB = 0.102\ \frac{F}{A}.$$

The factor 0.102 converts the SI unit N into the old unit kgf and guarantees that the old hardness values may also be used unchanged in SI.

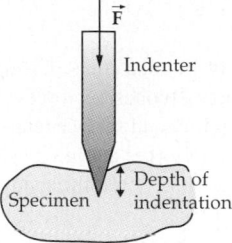

Figure 5.15: Measurement of hardness. One measures the depth of indentation (i.e., the area of indentation) of a prescribed indenter pressed with a fixed force $\vec{\mathbf{F}}$ for a certain time into a specimen.

➤ Since spherical surfaces do not penetrate easily into hard materials, this method can be applied to soft materials only.

➤ The hardness values are meaningful only if the diameter of the indentation is between 0.2 and 0.7 times the diameter of the test sphere.

3. Vickers hardness,

HV, the indenter is a diamond pyramid with a square base. Again, the ratio of the acting force and the surface of indentation is given when the latter quantity may be determined simply from the diagonal d of the square area of indentation:

$$HV = 0.102\frac{F}{A} = 0.189\frac{F}{d^2} \,.$$

➤ The Brinell hardness and the Vickers hardness have about the same numerical value. However, the Vickers method can also be applied to hard materials and therefore serves in general as a reference method.

➤ The relation between the Vickers hardness and the tensile strength R_m of steel:

$$R_m \approx 3.38 \, HV \,.$$

4. Rockwell hardness,

HR, with a standardized indenter (Rockwell-B: steel sphere of diameter 1.59 mm, Rockwell-C: diamond cone, vertex angle 120°) the depth of indentation is measured for a given force (Rockwell-B: 883 N, Rockwell-C: 1373 N). Each 2 μm depth of indentation corresponds to a unit of hardness. For a better comparison, an initial force of 98 N is introduced in both methods. Rockwell-B is used for moderately hard materials, Rockwell-C applies to very hard materials (hardened steels). The Rockwell method allows for an automatized hardness test, but is less precise.

➤ In some ranges, the hardness values obtained with different methods are similar.

5.2   Hydrostatics, aerostatics

Hydrostatics (**aerostatics**), the theory of the properties of liquids (and gases) at rest, in contrast to **hydrodynamics** (aerodynamics) dealing with the flow of liquids (and gases). In this context, one introduces the concepts of **pressure** and **buoyancy** as the forces of liquids on the bodies immersed.

5.2.1 Liquids and gases

Liquid, state of matter characterized by the mobility of molecules. Liquids may take an arbitrary shape, but there still are appreciable forces (cohesion forces) between the molecules, manifesting themselves in a low compressibility and surface tension.

Gas, a state of matter in which only weak, short-range forces act in collisions between molecules. Gases are characterized by a high compressibility (see thermodynamics) and by a lack of surface tension and cohesion. The flow of gases may also be described by hydrodynamics, but the high compressibility and the resulting density fluctuations must be taken into account.

5.2.2 Pressure

1. Definition of pressure

Pressure, force per unit area acting normally to a surface element within a fluid. Due to the high mobility of the molecules of the fluid, the force acting at one position propagates immediately and isotropically with the same magnitude through the entire volume of the fluid. Within a fluid at rest, the normal force exerted on a small test surface (e.g., part of the wall of the vessel or of the surface of a submerged body) has the same magnitude everywhere and is independent of the orientation of the test surface (**isotropic pressure**, **Fig. 5.16**). This holds only if the pressure due to gravity (see p. 174) can be ignored. Shear stresses do not exist in fluids.

pressure = $\dfrac{\text{force}}{\text{area}}$			$\mathbf{ML^{-1}T^{-2}}$
	Symbol	Unit	Quantity
$p = \dfrac{F_N}{A}$	p	Pa	pressure
	F_N	N	applied normal force
	A	m^2	area

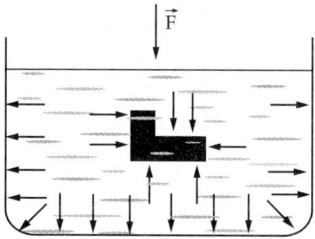

Figure 5.16: Isotropic pressure acts uniformly and isotropically; the direction of force is shown by arrows.

2. SI unit of pressure,

Pascal, Pa, SI unit of pressure.
1 pascal is the pressure exerted by a force of 1 N on an area of 1 m^2.

$$[p] = \text{Pa} = \text{pascal} = \text{N/m}^2$$

➤   Pressure is not a vector quantity. It acts in any direction with the same magnitude.
➤   Attention! The same symbol, p, is used for pressure and for linear momentum.
Atmospheric pressure, at sea level about 1 bar = 10^5 Pa.

3. Measurement of pressure

M **Autoclave**, pressure vessel for generating very high pressure (1000–10000 bar).

    Vacuum pump, generates very low pressure (presently down to 10^{-11} bar). Pressure is determined by measuring the force acting on a known area: in a **manometer** by springs, in an **aneroid barometer** by the deformation of an evacuated metal box, in **Bourdon's tube** by the deformation of a tube directly transmitted to a pointer. **Mercury barometer**, measures the pressure by comparing the unknown pressure to the known pressure due to gravity of a liquid column. Modern methods use **piezo-electric elements** (see chapter on electrotechnics) in which the force applied to a crystal generates an electric voltage.

5.2.2.1 Piston pressure

1. Definition of piston pressure

Piston pressure, the pressure generated within a liquid by pressing a movable piston into a cylinder in the liquid container (**Fig. 5.17**). In static equilibrium the pressure p of the liquid just compensates the external forces F_1 and F_2. Therefore,

$$F_1 = A_1 p , \qquad F_2 = A_2 p ,$$

and hence

$$p = \frac{F_1}{A_1} = \frac{F_2}{A_2} , \qquad \frac{F_1}{F_2} = \frac{A_1}{A_2} .$$

▲   The piston pressure is the same throughout the fluid.

2. Hydraulic press,

a device to amplify forces. A small external force F_1 acts on a small area A_1. At the large area A_2, a large force

$$F_2 = \frac{A_2}{A_1} F_1$$

is produced.

➤   From energy conservation, it follows that the piston stroke at the larger area is lower by a factor A_1/A_2 than the stroke at the smaller area. The same follows from the property of incompressibility of the medium.

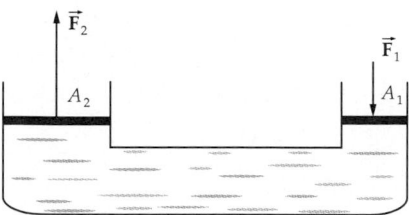

Figure 5.17: Piston pressure in a hydraulic press.

3. Hydraulics,

application of the piston principle to transmit and amplify forces in technological settings. Typical applications are the hydraulic brake, the hydraulic lift, and the pressure transducer.

A particular advantage is the possibility of changing the direction of a force without using mechanical elements such as levers or rollers.

Unlike liquids, gases are highly compressible. The **compressional work** done in compressing a gas volume is stored as **internal energy** (see thermodynamics) in the gas and may be released at any position and at any time. Compressed gases (compressed air) serve as energy-storage devices, and are used in machine controls (**pneumatics**).

5.2.2.2   Pressure due to gravity in liquids

1. Definition of the pressure due to gravity

Pressure due to gravity, the pressure generated within a liquid by its own weight. It results from the force exerted by a liquid column of height h and volume $V = hA$ on its base area A:

gravity pressure			$\mathbf{ML^{-1}T^{-2}}$
	Symbol	Unit	Quantity
	p	Pa	gravity pressure
	ρ	kg/m^3	density of liquid
$p = \dfrac{\rho V g}{A} = h\rho g$	V	m^3	volume of liquid column
	A	m^2	base area of liquid column
	h	m	height of liquid column
	g	m/s^2	gravitational acceleration = 9.81 m/s^2

■   A water column 10 m high generates a pressure of

$$p = h\rho g = 10 \text{ m} \cdot 1000 \text{ kg/m}^3 \cdot 9.81 \text{ m/s}^2 = 9.81 \cdot 10^4 \text{ Pa}$$

on the base area. A mercury column (density 13600 kg/m^3) producing the same pressure has a height of

$$h = \frac{p}{\rho g} = \frac{9.81 \cdot 10^4 \text{ Pa}}{13600 \text{ kg/m}^3 \cdot 9.81 \text{ m/s}^2} = 735 \text{ mm} \,.$$

The pressure due to gravity in a liquid depends on the depth. Hence, the isotropic pressure in the liquid is the same only at a given level because it depends on depth (**Fig. 5.18**).

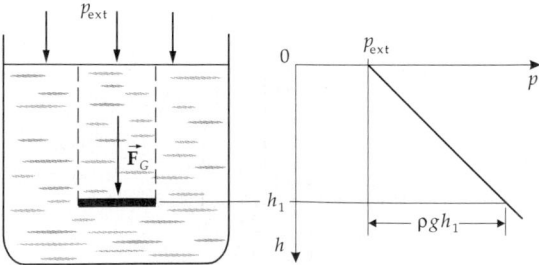

Figure 5.18: Pressure due to gravity in a liquid. p_{ext}: external pressure.

2. Hydrostatic paradox,

the pressure at the bottom of a vessel depends only on the density of the liquid and the height of the liquid column, but not on the shape of the vessel, and hence not on the quantity of liquid (**Fig. 5.19**).

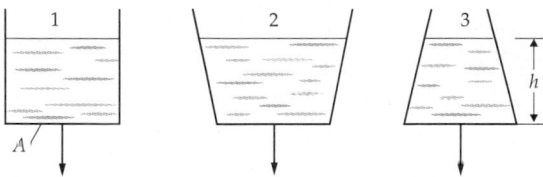

Figure 5.19: Hydrostatic paradox. For equal height h of the column, the pressure on the base area A is independent of the shape of the vessels 1, 2, 3.

3. Manometer

M   **Mercury gauge**, device for pressure measurement by comparison with the pressure due to gravity of a mercury column. On one side of the gauge, there is p, the pressure to be measured, and $\rho g h_1$ (ρ density, g acceleration of gravity, h_1 height), the pressure due to gravity; and at the other side, there is the pressure due to gravity of the liquid column $\rho g h_2$, and a reference pressure p_0. In equilibrium,

$$p - p_0 = \rho g(h_2 - h_1).$$

Hence, the difference in pressure is proportional to the difference of heights. The heavier the liquid, the greater the measurable pressure. That is why mercury is used to measure air pressure. In its simplest form, the gauge consists of a glass tube closed at the upper end with the lower end submerged in mercury. The reference pressure, i.e., the pressure in the cavity at the upper end, is the vapor pressure of mercury which is very low (vacuum). The corresponding device designed for measurements of the atmospheric pressure is called a **barometer**. **Fig. 5.20** shows the barometer according to Torricelli.

4. Connected vessels

In connected tubes, the liquid rises to equal height in each vessel if the same external pressure acts everywhere (**Fig. 5.21**). Capillary forces are ignored.

M   A manometer based on connected vessels is used for measurements of small pressure differences.

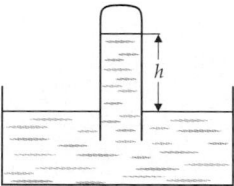
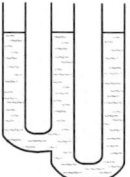

Figure 5.20: Simplest form of manometer: barometer to measure the atmospheric pressure, as invented by Torricelli. The height of the liquid in the glass tube is proportional to the atmospheric pressure.

Figure 5.21: Connected vessels.

5.2.2.3   Compressibility

1. Definition of compressibility

Compressibility, the change of a volume of liquid due to a change in pressure. It is defined as the ratio of the fractional change of volume to the change of pressure:

compressibility			$\mathbf{M^{-1}LT^2}$
	Symbol	Unit	Quantity
$\kappa = \dfrac{\Delta V}{V \Delta p}$	κ	$1/\text{Pa} = \text{m}^2/\text{N}$	compressibility
	ΔV	m^3	decrease of volume
	V	m^3	original volume
	Δp	Pa	increase of pressure

Typical compressibility values are in the range of 10^{-9} 1/Pa (see **Tab. 7.3/9**).

■   Under standard conditions (temperature 0 °C and pressure 101.325 kPa), water has compressibility $0.5 \cdot 10^{-9}$ 1/Pa. Under atmospheric pressure of 10^5 Pa, the volume of 1 m^3 water changes by

$$\Delta V = \kappa V \Delta p = 0.5 \cdot 10^{-9} \ 1/\text{Pa} \cdot 1 \ \text{m}^3 \cdot 10^5 \ \text{Pa} = 0.5 \cdot 10^{-4} \ \text{m}^3 = 50 \ \text{cm}^3 \ .$$

2. Coefficient of volume expansion,

γ, describes the expansion of a liquid as the temperature increases. The fractional expansion of a volume of liquid is proportional to the increase of temperature if it is small compared with the original temperature.

coefficient of volume expansion			1
	Symbol	Unit	Quantity
$\dfrac{\Delta V}{V} = \gamma \Delta \theta$	$\Delta V/V$	1	fractional change of volume
	γ	1/K	coefficient of volume expansion
	$\Delta \theta$	K	change of temperature

The coefficient of volume expansion has units 1/K. It depends on the temperature of the material, and usually the temperature is given. $\theta_0 = 0$ °C.

➤   The coefficient of volume expansion of water at 20 °C is $\gamma = 0.18 \cdot 10^{-3}$ 1/K. Other liquids reach a multiple of this value. For ideal gases at this temperature,

$$\gamma = \frac{1}{\theta_0} = 3.4 \cdot 10^{-3} \ 1/\text{K} \ .$$

5.2.2.4   Pressure due to gravity in gases

1. Calculation of pressure due to gravity in gases

In the calculation of the pressure due to gravity in gases, one must take into account the compressibility of the gas. The density ρ of a gas at pressure p is given by

$$\rho = \rho_0 \frac{p}{p_0} \ ,$$

where ρ_0 denotes the density at a reference pressure p_0. The change of pressure Δp for a change of height Δh above the base area of the gas column is

$$\Delta p = -\frac{\Delta m g}{A} = -\rho g \, \Delta h \, .$$

(A cross-sectional area of the gas column, Δm mass within the layer Δh, g gravitational acceleration.) This expression may be rewritten as

$$\int_{p_0}^{p_1} \frac{dp}{p} = -\int_0^{h_1} \frac{\rho_0 g}{p_0} \, dh$$

(p_0 pressure at the bottom, p_1 pressure at height h_1). Integration with $p = p_1$, $h = h_1$ gives

$$\ln\left(\frac{p}{p_0}\right) = -\frac{\rho_0 g}{p_0} h \, .$$

2. Barometric equation

The **barometric equation** (**Fig. 5.22**) from the preceding expression:

barometric equation			
	Symbol	Unit	Quantity
$p = p_0 e^{-Ch}$ $C = \dfrac{\rho_0 g}{p_0}$	p h C p_0 ρ_0 g	Pa m 1/m Pa kg/m^3 m/s^2	pressure at height h height constant pressure at ground level density at ground level acceleration of gravity

The pressure in a column of gas (in particular, in the atmosphere of Earth) decreases exponentially with height. The constant C for air has the value

$$C = 0.1256/\text{km}$$

for a pressure of $p_0 = 101.3$ kPa at ground level and a temperature of $0\,°C$.
▲   For each ca. 8 m increase of altitude near ground level, the air pressure decreases by 100 Pa = 1 mbar.

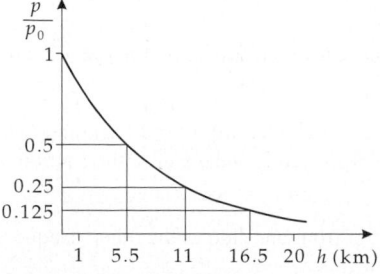

Figure 5.22: Solution of the barometric equation.

3. International barometric equation

The decrease of temperature with increasing altitude is not taken into account in the barometric equation. Inclusion of this variation of temperature leads to the **International barometric equation**:

$$p = \left(1 - \frac{0.00651/\text{m} \cdot \text{h}}{288}\right)^{5.255} \cdot 101.325 \text{ kPa} .$$

This equation is valid up to an altitude of 11 km. The density of air is given by

$$\rho = \left(1 - \frac{0.00651/\text{m} \cdot \text{h}}{288}\right)^{4.255} \cdot 1.2255 \, \frac{\text{kg}}{\text{m}^3} .$$

4. Standard atmosphere

The atmospheric pressure fluctuates by about 10 %, depending on the weather and temperature.

   Standard pressure and **standard density** of air at sea level and for 15 °C are

$$p_0 = 101.325 \text{ kPa} , \qquad \rho_0 = 1.293 \text{ kg/m}^3$$

(previously: 760 Torr, 1 atm = physical atmosphere). This is the **ISO standard atmosphere**.

5.2.2.5 Pumps

Pumps, machines to transport liquids and gases.

1. Types of pumps

a) Piston pump, a piston moving back and forth in a tube. In one stroke the material to be pumped is drawn in through a **suction valve**, in the reverse stroke it is expelled through a **pressure valve**. Used for engines (**Fig. 5.23**).

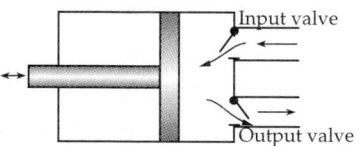

Figure 5.23: Principle of the piston pump. The moving piston alternatively draws fluid from one region and pushes it through an outlet into another.

b) Diaphragm pump, a membrane is used instead of a piston (e.g., for corrosive liquids, fuel pumps).

c) Vane pump, one or several vanes placed in a cylinder are moved back and forth, instead of the piston; the pressure valves are incorporated into the vanes, the suction valves are mounted in the inlet pipe.

d) Gear pump, meshing gears press the liquid from one side to the other (frequently as a pump for lubricants).

e) Rotary pump, also **turbine** or **centrifugal pump**, liquid enters the central region and is caught by rotating vanes, accelerated and pressed outward by the centrifugal force (high throughput water pumps driven by an electric motor such as a turbo-pump) (**Fig. 5.24**).

f) Water-jet pump, a jet of water flowing through a nozzle transports air outward (see suction effects of flowing fluids).

g) Vapor-ejector pump, an escaping jet of vapor transports water.

h) Diffusion pump, to generate high vacuum. A material such as oil or mercury is vaporized in the forevacuum. It rises, thereby conveying gas molecules to be pumped off via diffusion into the vapor beam, after condensation on the cooled walls, it is fed back (**mercury diffusion pump**) (**Fig. 5.25**).

i) Molecular pump, a turbine pump drives gas molecules into regions of higher pressure because of the friction in the collisions of particles with a rotating disk.

j) Getter pump, for ultra-high vacuum, based on the adsorption of residual gas molecules on a working substance (getter).

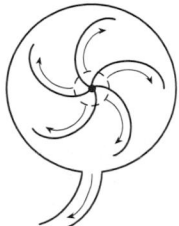

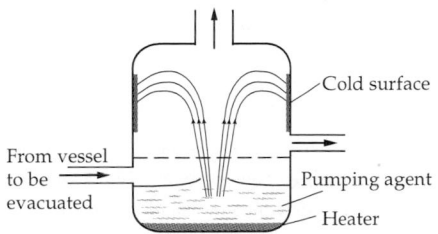

Figure 5.24: Rotary pump. The inlet pipe is connected axially.

Figure 5.25: Diffusion pump.

2. Properties and parameters of pumps

Pumping height, H, the maximum height up to which a liquid can be conveyed by a pump. This quantity is determined by the available pump pressure that can compensate the pressure of a water column of this height. The parameter H also limits the flow velocity that may be achieved in a pipeline; the pumping height is correlated with the pumping flow, depending on the detailed design.

  Pumping flow, Q, volume flow, the volume of liquid conveyed per unit time. It depends on the dimensions of the pump, and on the flow velocity achieved.

  Characteristic curve of a pump, a plot of the pumping height versus the pumping flow. In general, the characteristic curve turns down at higher pumping flow.

  Pumping capacity, P_Q, pumping power, the work per unit time that can be done by the pump against gravity, the product of gravitational force per volume ρg, volume flow Q and pumping height H:

$$P_Q = g\rho H Q .$$

Efficiency of a pump, the ratio of the pumping capacity achieved P_Q to the mechanical power supplied P_0:

$$\eta = \frac{P_Q}{P_0} .$$

3. Suction pumps and pressure pumps

Suction pumps, exploit the atmospheric pressure by generating a subpressure region (e.g., by volume expansion due to moving a piston). The suction effect then arises due to the pressure difference between atmospheric pressure and the subpressure value. Hence, the maximum pumping pressure is the atmospheric pressure, and the maximum suction height for water is about 10 m.

Pressure pumps operate directly in the medium, independent of the atmospheric pressure.

4. Turbines

Turbine, the inverse of a pump. In a turbine, the energy of flow is converted into mechanical energy (rotational energy) (e.g., to operate generators). In contrast to the piston engine, this does not happen by moving a piston, rather a shaft is driven directly by the flow.

Water wheel, oldest device to convert flow energy into mechanical energy. The water wheel may be driven by water falling onto the vanes, or by water flowing below the wheel and carrying the vanes along. Efficiency 80 to 85 %. The power for the former case is given by:

$$P = g\rho Qh$$

(g acceleration of gravity, ρ density of liquid, Q volume flow, h height of fall).

a) Water turbine, hydraulic engine that obtains energy from a water flow. In the **water jet turbine**, a jet of water hits vanes fixed to **runners**. In the **Kaplan turbine** and the **Francis turbine**, the water flows from outside through guiding vanes onto the moving vanes, releasing kinetic energy when moving inward, and is discharged near the wheel axle. Power: up to 250 MW.

b) Steam turbine, for production of energy in thermal power stations. First, the steam is expanded in fixed guide wheels (which can not occur in water turbines because of the incompressibility of water) and thereby accelerated to high velocity; then it drives one or several moving vanes. The various types are characterized by the relation between velocity and pressure in the turbine.

c) Gas turbine, driven by the combustable gases: combination of a proper turbine driven by hot combustion waste gases and a **compressor** preceding the combustion that presses air into the combustion chamber. Application for airplanes as **turboprop engine** involving a propeller on the shaft, and **jet engine** without the propeller; also for automobile generators, occasionally for land-based vehicles. The advantages are simple construction with few moving units, low weight per unit of power, high rate of rotation (up to 20 000 rev/min), an efficiency up to 35 % for multi-stage devices and cheap fuel.

5.2.3 Buoyancy

1. Buoyant force

Buoyancy, a force directed in a direction opposite to Earth's attraction and acting on all bodies submerged in a liquid (or gas). Buoyancy results from the difference in pressure on the upper and lower face of the body (**Fig. 5.26**). If the upper face of the body with an area A is at the depth h_1, and the lower face (of the same area) at the depth h_2, then

$$F_A = F_2 - F_1 = A(p_2 - p_1) = A\rho_{Fl}\, g\, (h_2 - h_1)$$

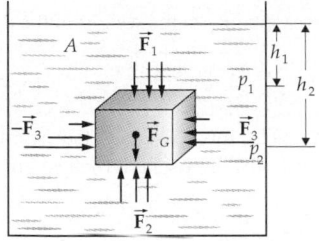

Figure 5.26: Buoyancy. The lateral forces $\vec{F}_3$ cancel each other; the force $\vec{F}_2$ (below) exceeds the force $\vec{F}_1$ (above).

(ρ_{Fl} density of the liquid, p_1, p_2 pressure at h_1 and h_2, F_1, F_2 force at the upper and lower face of the body, F_A buoyant force, g gravitational acceleration). The quantity $A(h_2 - h_1)$ is the volume V of liquid displaced by the body. Hence:

buoyant force			$\mathbf{MLT^{-2}}$
	Symbol	Unit	Quantity
$F_A = \rho_{Fl}\, g\, V$	F_A	N	buoyant force
	ρ_{Fl}	kg/m^3	density of liquid
$= m_{\text{disp}}\, g = F_{G,\text{disp}}$	g	m/s^2	gravit. acceleration (9.81 m/s^2)
	V	m^3	volume of the body
$= \dfrac{\rho_{Fl}}{\rho_K}\, F_G$	m_{disp}	kg	displaced mass of liquid
	$F_{G,\text{disp}}$	N	weight force of m_{disp}
	ρ_K	kg/m^3	density of the body
	F_G	N	weight force of the body

➤   The density stands for the mean density of the entire body, i.e., total mass divided by total volume.

2. Principle of Archimedes and properties of buoyancy

Principle of Archimedes, the buoyant force experienced by a body submerged in a liquid equals the weight of the displaced quantity of liquid.

➤   This rule holds also for partly submerged bodies.

There are three kinds of buoyant forces:

$F_A < F_G$:   The body sinks when its density is larger than the density of the liquid;

$F_A = F_G$:   The body remains suspended when its density equals the density of the liquid;

$F_A > F_G$:   The body floats and is only partly submerged when its density is less than the density of the liquid.

■   The density of iron is 7.8 times that of water. An iron body experiences a buoyant force

$$F_A = \frac{\rho}{\rho_{\text{body}}} F_G = \frac{1}{7.8} F_G = 0.13 F_G\,,$$

i.e., 13 % of its weight. The effective weight of iron is only 87 % of its true weight when submerged in water.

The effective weight of a submerged body is the real weight minus the buoyancy:

$$F_{\text{eff}} = F_G - F_A = \left(1 - \frac{\rho_{Fl}}{\rho_K}\right) F_G\,.$$

A body in air also experiences a buoyant force corresponding to the weight of the displaced air.

3. Balloon

flying object kept in the air by the buoyant force. The force is generated by filling the balloon with a gas having a density lower than that of the atmosphere (heated air, helium; in the past, hydrogen).

4. Measurement of densities by means of Mohr's balance

Buoyancy may be employed to measure the density of a solid ρ_K. One measures the force required to balance a scale when the body is suspended while submerged in a liquid, F_{Fl}, and in the air, F_G (**Mohr's balance, Fig. 5.27**). The difference equals the difference of the buoyant forces,

$$F_G - F_{Fl} = F_{A,Fl} - F_{A,\text{air}} = (\rho_{Fl} - \rho_{\text{air}})Vg \approx \rho_{Fl}Vg$$

($F_{A,Fl}$ buoyant force in the liquid, $F_{A,\text{air}}$ buoyant force in the air, ρ_{Fl} density of the liquid, ρ_{air} density of air, V volume of the body, g gravitational acceleration). In general, the density of the air may be ignored compared to the density of the liquid. If both sides are divided by

$$F_G = \rho_K V g = mg,$$

(m mass of the body) then

$$\rho_K = \frac{\rho_{Fl}}{1 - \dfrac{F_{Fl}}{mg}}.$$

➤   Measurement of density in this manner is only feasible if the body does not float, i.e., its density is greater than that of the liquid.

If the density of the body is less than that of the liquid, an auxiliary weight may be added to the body. The force in the liquid F_{Fl} is then replaced by the difference $F_H - F_{Fl}$ of the force for the auxiliary weight alone, F_H, and together with the body, F_{Fl}:

$$\rho_K = \frac{\rho_{Fl}}{1 - \dfrac{F_H - F_{Fl}}{mg}}.$$

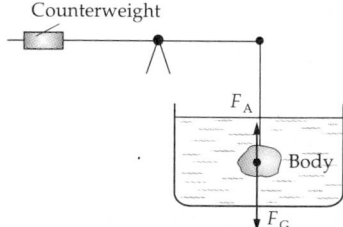

Counterweight

F_A

Body

F_G

Figure 5.27: Measurement of the density using Mohr's balance.

Conversely, the density of the liquid may be determined using a body of known density. By rearranging the above formula, one has

$$\rho_{Fl} = \rho_K \left(1 - \frac{F_{Fl}}{mg}\right).$$

By submerging the same body in two liquids of different densities ρ_1 and ρ_2, one can determine the ratio of the densities from the measured balancing forces $F_{Fl,1}$ and $F_{Fl,2}$:

$$\frac{\rho_1}{\rho_2} = \frac{1 - \dfrac{F_{Fl,1}}{mg}}{1 - \dfrac{F_{Fl,2}}{mg}} .$$

5. Determination of density from submersion depth

Another method to determine the density of a liquid is based on the submergence of a floating body. Let A be the (constant) cross-sectional area, H the height of the floating body, and h the submersion depth, the balance of force is

$$0 = F_A - F_G = h A \rho_{Fl} g - H A \rho_K g .$$

From there,

$$\rho_{Fl} = \frac{H}{h} \rho_K .$$

The density of a floating body may also be determined by

$$\rho_K = \frac{h}{H} \rho_{Fl} .$$

5.2.4 Cohesion, adhesion, surface tension

1. Cohesion,

the property of liquids and solids to link up and form non-disrupting filaments and layers. It arises because of attractive forces between the molecules. The attractive forces arise from the charge distribution (polarization) within the molecules and the resulting electrostatic attraction (see **Van der Waals forces**, p. 666). The cohesive forces in gases are much weaker than in liquids and have a noticeable effect only near the boiling temperature.

■   **Siphon (Fig. 5.28).** As soon as the liquid exceeds the highest point of the tube, it is pulled down into the other half of the tube by gravity. Cohesion prevents the liquid filament from breaking. Such phenomena do not occur for gases; rather, the density of the gas varies according to the barometric formula.

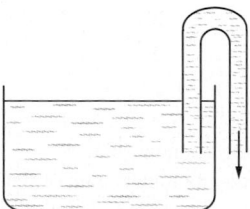

Figure 5.28: Siphon. The fluid is extracted from a vessel by the cohesive forces and the gravitational pressure.

2. Surface tension,

force on the surface of a liquid caused by the molecular forces within the liquid (**Fig. 5.29**). In the interior of the liquid, cohesive forces act isotropically with the same magnitude, since any molecule is surrounded in any direction by other molecules in the same way. At

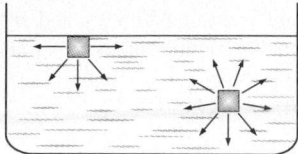

Figure 5.29: Surface tension. The cohesive forces compensate each other only in the interior of the liquid.

the surface, however, a resultant cohesive force arises towards the interior, which must be compensated by a pressure within the liquid.

Surface energy, the potential energy resulting from the surface tension.

The surface tension opposes an increase in the surface area. In order to enlarge the surface by an amount ΔA, an amount of work ΔW is required. The ratio of the work ΔW to the surface increase ΔA is called the surface tension σ:

surface tension			$\mathbf{MT^{-2}}$
	Symbol	Unit	Quantity
$\sigma = \dfrac{\Delta W}{\Delta A}$	σ	$J/m^2 = kg/s^2 = N/m$	surface tension
	ΔW	J	work performed
	ΔA	m^2	surface gained

➤ Typical values for the surface tension are 0.02 N/m for hydrocarbons, 0.07 N/m for strongly polarized molecules such as water or glycerine, and the extreme case of mercury 0.49 N/m. The surface tension depends on the temperature of the material. It can be very sensitive to contamination by certain substances (detergents).

3. Measurement of surface tension

M   Surface tension is measured by a wire frame of length d (**Fig. 5.30**) submerged in a liquid and pulled out by an amount Δs, thereby forming a thin liquid film of surface $\Delta A = 2d\,\Delta s$. If the frame is pulled out of the liquid with a force F, the work done to generate the liquid film is $\Delta W = F\,\Delta s$. Therefore,

$$\sigma = \frac{\Delta W}{\Delta A} = \frac{F\,\Delta s}{2d\,\Delta s} = \frac{F}{2d} \,.$$

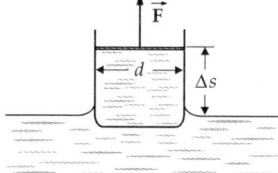

Figure 5.30: Measurement of surface tension. A liquid film is drawn with a wire frame, and the force $\vec{F}$ is measured.

4. Specific properties of surface tension

Surface tension represents a force per unit length of the boundary line.

▲   The force F_σ acting due to the surface tension on a boundary line of length l is

$$F_\sigma = l\sigma \,.$$

➤ A system always tends to approach the state of lowest potential energy. For this reason, the surface of a liquid is always a **minimum surface**.

■   The body with the minimum surface for a given volume is the sphere. If no other forces are present, a drop of liquid takes a spherical shape. Special case: soap-bubble.

5.2.4.1 Capillarity

1. Adhesion,

denotes the attractive forces between the molecules of two **distinct** materials, unlike cohesion, which is between molecules of the same material. Adhesion may occur between solid, liquid, or gaseous materials. In particular, in the contact of a liquid (drop) and a solid (supporting surface) the following cases must be distinguished, depending on the ratio of strengths of cohesive and adhesive forces (**Fig. 5.31**):

● the adhesive forces dominate: the liquid spreads over the entire supporting surface (**perfect wetting**),

● the cohesive forces dominate: the liquid contracts into drop-like objects (**no wetting**).

Rim angle, ϕ, the angle between the liquid surface and the supporting surface at the contact point. For wetting liquids, $0 \le \phi \le \pi/2$. For a non-wetting liquid, $\pi/2 < \phi \le \pi$.

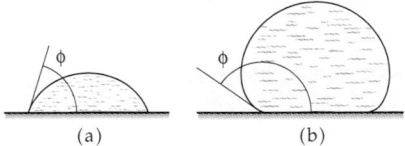

Figure 5.31: Contact of a liquid drop with a solid support area. (a): wetting, rim angle $\phi < \pi/2$, (b): no wetting, rim angle $\phi > \pi/2$.

2. Capillary action,

the phenomenon of the rising of a liquid in a thin tube (**capillary**) (**Fig. 5.32**). It is caused by the surface tension at the boundary line of the liquid, and the resulting force is $F_\sigma = \sigma l = \sigma \cdot 2\pi r$ (*l*: circumference). This force is compensated by the weight of the liquid column $F_G = mg = \rho \cdot h \cdot \pi r^2$ (*m* mass of the liquid column). From $F_G = F_\sigma$, one obtains:

capillary elevation height (capillary ascension)			L
	Symbol	Unit	Quantity
$h = \dfrac{2\sigma}{g\rho r}$	h	m	elevation height
	σ	N/m	surface tension
	ρ	kg/m^3	density of fluid
	g	m/s^2	gravitational acceleration
	r	m	inner radius of capillary

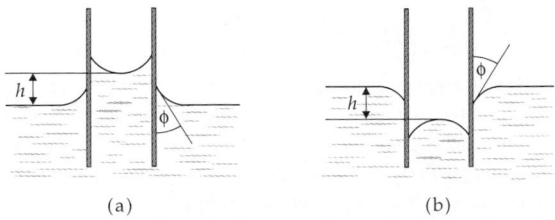

Figure 5.32: Capillarity. (a): capillary ascension, (b): capillary depression.

■   In a capillary of inner diameter 1 mm, water (surface tension 0.07 N/m, density 1000 kg/m^3) ascends to

$$h = \frac{2\sigma}{g\rho r} = \frac{2 \cdot 0.07 \text{ N/m}}{9.81 \text{ m/s}^2 \cdot 1000 \text{ kg/m}^3 \cdot 0.5 \text{ mm}} = 29 \text{ mm}.$$

➤   For a given substance, the elevation depends only on the radius of the capillary.
➤   The surface tension of the liquid may be determined from the capillary ascension (depression).

Wetting energy, E_{wetting}, a measure of adhesion strength. The wetting energy is released in wetting a surface of area A. It may be calculated from the wetting angle ϕ and the surface tension σ:

$$E_{\text{wetting}} = A\sigma(1 + \cos\phi).$$

5.3   Hydrodynamics, aerodynamics

Fluid mechanics, theory of flow in liquids (**hydrodynamics**) and gases (**aerodynamics**). It describes the transport of matter due to differences in pressure and external forces, taking into account the internal friction. Again, gases differ from liquids by their high compressibility. If, however, the flow velocity is significantly below (by about one third) the velocity of sound, gases behave practically like incompressible fluids.

The central concept of flow mechanics is the flow field.

5.3.1   Flow field

1. Definition of the flow field

At a given instant, every particle of a flowing fluid has a velocity defined by magnitude and direction. The basic assumption of hydrodynamics is that the mean velocity of the particles over a small volume is nearly constant. One therefore may assign to any point in the fluid a mean velocity $\vec{v}$ of the mass particles in a small volume element around this point. The velocity distribution in space and time arising in this way is called the **velocity field** $\vec{v}(x, y, z, t)$. Analogously, one introduces the **pressure field** $p(x, y, z, t)$, the **temperature field** $T(x, y, z, t)$ and the **density field** $\rho(x, y, z, t)$.

➤   This description holds only in the local thermodynamical equilibrium (see p. 691). Only then may the pressure and the temperature be defined meaningfully, and a relation to the density may be established via the equation of state. Flow that is not in local thermodynamic equilibrium is described by the **kinetic theory** (**transport theory**).

2. Properties of the velocity field

The velocity field is a vector field; its value $\vec{v}(x, y, z, t)$ gives the mean velocity of the particles that at the instant t are within a small volume element around the position (x, y, z). One distinguishes between time-independent (stationary) and time-dependent (non-stationary) flow, and also between space-dependent (non-uniform) and space-independent (uniform) flow. For stationary flows:

$$\vec{v} = \vec{v}(x, y, z), \quad \frac{\partial\vec{v}}{\partial t} = 0.$$

Streamlines and **pathlines** serve for visualization of the flow field (**Fig. 5.34**). **Stream-lines** follow the velocity vectors in a given instant, i.e., a tangent to a streamline gives the direction of flow at this point (**Fig. 5.33**). The streamlines must be distinguished from the **pathlines**, which describe the real motion of the material particles over a certain period.
▲   For steady flows, streamlines and pathlines coincide.
The mathematical description of flow is done with the tools of **vector analysis**.

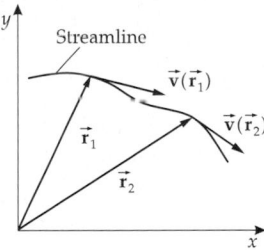

Figure 5.33: Streamline. The velocity vector $\vec{v}(\vec{r})$ corresponds to the tangent to the streamline at the point $\vec{r}$.

3. Examples of streamline plots

In a streamline plot, the finite density of lines n (n: number of streamlines intersecting a unit area) characterizes the flow velocity: $n \sim |\vec{v}|$.
   Stream tube, tube-like space region. The boundary lines of the tube coincide with streamlines (**Fig. 5.35**). In stationary flow, the liquid does not cross the boundary of the stream tube (**Fig. 5.36**).

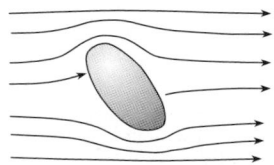

Figure 5.34: Flow field around a plate.

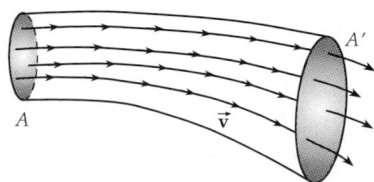

Figure 5.35: Velocity field. Streamlines in a stream tube with the cross-sectional areas A and A'.

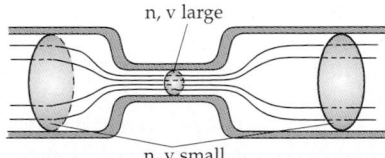

Figure 5.36: Streamline density n in a tube of variable cross-section.

5.3.2   Basic equations of ideal flow

Ideal liquid, liquid that is incompressible and does not exhibit friction. In an ideal liquid, no vortices can occur, rot $\vec{v} = 0$. As the name suggests, this idealization cannot be realized physically.
   Ideal flow, an incompressible flow without frictional forces.
➤   **Ideal gases** are gases with a compressibility that follows the law of ideal gases. The flow of real gases is not ideal flow.

5.3.2.1   Continuity equation

1. Setting up the continuity equation

Continuity equation, expresses the conservation of mass. One considers (**Fig. 5.37**) a tube with the cross-sectional area A through which a liquid is flowing. The mass Δm of all particles passing the area A in a time interval Δt is given by the product of area, time interval, density ρ and velocity $\vec{v}$ of the liquid:

$$\Delta m = \rho \, v \, A \, \Delta t \, .$$

At another position along the tube, where the cross-section is A' and the velocity is $\vec{v}'$, the same mass must pass the area per unit time because there is assumed to be no sources or sinks for the material. Then

$$\rho v A = \rho' v' A' \, .$$

Incompressible liquid: $\rho = \rho'$, and therefore:

continuity equation for incompressible fluids			$\mathbf{L^3 T^{-1}}$
$vA = v'A'$	Symbol	Unit	Quantity
	v, v'	m/s	velocities
	A, A'	m^2	cross-sectional areas

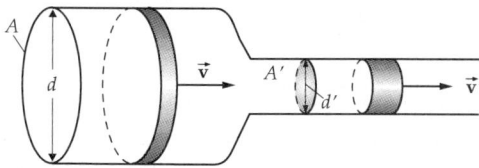

Figure 5.37: Flow in a tube of varying cross-section A.

▲   The smaller the cross-sectional area of a tube, the higher the velocity of the liquid passing through it.

Volume flux, volume flow, $Q = vA$, $[Q] = $ m^3/s. Volume of liquid that passes a tube of cross-sectional area A per unit time.

   Current density, mass current density, the vector $\vec{j} = \rho \, \vec{v}$.

➤   An analogous equation holds for the conservation of the electric charge for electric currents in electrodynamics. Generally, a continuity equation expresses the conservation of a physical quantity.

2. Continuity equation in differential form

▲   The volume of liquid flowing into a small cube fixed in space equals the volume flowing out of this cube in the same time interval.

The **differential formulation of the continuity equation** follows from this statement: The volume flow $Q_{in,x}$ through the face of a cuboid perpendicular to the x-direction is

$$Q_{in,x} = v_x(x) \cdot \Delta y \cdot \Delta z \, ,$$

Δx, Δy and Δz denote the edge lengths of the cuboid. The volume flow through the opposite face is

$$Q_{out,x} = v_x(x + \Delta x) \cdot \Delta y \cdot \Delta z.$$

According to Taylor's theorem,

$$v_x(x + \Delta x) \approx v_x(x) + \frac{\partial v_x(x)}{\partial x} \Delta x.$$

The same treatment for the y- and z-direction yields the excess of the volume flow through the cuboid,

$$\Delta Q = \left(\frac{\partial v_x}{\partial x} + \frac{\partial v_y}{\partial y} + \frac{\partial v_z}{\partial z} \right) \cdot \Delta x \cdot \Delta y \cdot \Delta z.$$

The quantity in parentheses is the **divergence** of the vector field $\vec{v}$:

$$\text{div } \vec{v} = \frac{\partial v_x}{\partial x} + \frac{\partial v_y}{\partial y} + \frac{\partial v_z}{\partial z}.$$

The differential formulation of the continuity equation reads:

continuity equation in differential form			LT^{-1}
	Symbol	Unit	Quantity
$\text{div } \vec{v} = 0$	$\vec{v}$	m/s	velocity field

3. Velocity potential, Laplace and Poisson equations

The continuity equation may be solved by the introduction of a **velocity** or **flow potential** Φ. The velocity potential is a scalar field. The streamlines are trajectories orthogonal to the equipotential surfaces $\Phi = \text{const.}$ The **gradient** of Φ is a vector field that, at any position, points along the steepest slope of Φ. The gradient of the velocity potential Φ is the velocity field $\vec{v}$:

$$\text{grad } \Phi = \left(\frac{\partial \Phi}{\partial x}, \frac{\partial \Phi}{\partial y}, \frac{\partial \Phi}{\partial z} \right) = \vec{v}.$$

After inserting Φ, the continuity equation reads

$$\left(\frac{\partial^2}{\partial x^2} + \frac{\partial^2}{\partial y^2} + \frac{\partial^2}{\partial z^2} \right) \Phi = 0.$$

The equation is called **Laplace's equation**. If on the right-hand side of the equation a finite source density q appears instead of zero, one has **Poisson's equation**:

$$\left(\frac{\partial^2}{\partial x^2} + \frac{\partial^2}{\partial y^2} + \frac{\partial^2}{\partial z^2} \right) \Phi = -4\pi q.$$

Laplace operator, Δ, scalar product of the **del** or **nabla operator** $\vec{\nabla}$ with itself, sum over all partial second derivatives,

$$\Delta = \vec{\nabla} \cdot \vec{\nabla} = \frac{\partial^2}{\partial x^2} + \frac{\partial^2}{\partial y^2} + \frac{\partial^2}{\partial z^2}.$$

An extensive body of analytical and numerical tools exists for the solution of Laplace's equation under given boundary conditions (**boundary-value problem**).

4. Helmholtz condition

A flow can only then be represented by a potential Φ if it is **irrotational** or **vortex-free**, i.e., if no closed streamlines occur. For a steady flow field, this means that no particle in the liquid follows a closed path. The vortex property of the flow field may be expressed in vector analysis in terms of the **curl**, $\mathrm{rot}\,\vec{\mathbf{v}}$ of the velocity field,

$$\mathrm{rot}\,\vec{\mathbf{v}} = \begin{pmatrix} \dfrac{\partial v_z}{\partial y} - \dfrac{\partial v_y}{\partial z} \\[2mm] \dfrac{\partial v_x}{\partial z} - \dfrac{\partial v_z}{\partial x} \\[2mm] \dfrac{\partial v_y}{\partial x} - \dfrac{\partial v_x}{\partial y} \end{pmatrix}.$$

If the curl vanishes everywhere, the flow is irrotational,

$$\mathrm{rot}\,\vec{\mathbf{v}} = 0.$$

This is the **Helmholtz condition**.

5. Sources and sinks

Source or **sink**, region of space where streamlines begin (source) or terminate (sink). The number of streamlines entering through a surface enclosing the source (sink) differs from the number of streamlines leaving the volume through this surface. For the divergence of the velocity field (**Fig. 5.38**):

$$\mathrm{div}\,\vec{\mathbf{v}} = q\,, \qquad q: \text{source density}, \qquad q > 0: \text{source}, \qquad q < 0: \text{sink}.$$

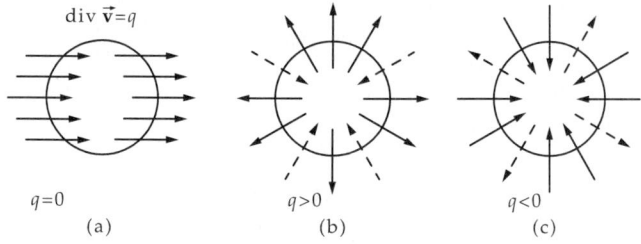

Figure 5.38: Divergence of the velocity field. (a): source-free flow, (b): source $q > 0$, (c): sink $q < 0$.

5.3.2.2 Euler's equation

Euler's equation, describes incompressible, non-viscous flow. It expresses Newton's second law:

$$\rho \cdot \left((\vec{\mathbf{v}} \cdot \mathrm{grad})\,\vec{\mathbf{v}} + \frac{\partial \vec{\mathbf{v}}}{\partial t} \right) \equiv \rho \cdot \frac{d\vec{\mathbf{v}}}{dt} = \vec{\mathbf{F}} - \mathrm{grad}\,p\,.$$

On the right-hand side of the equation appear the force per unit volume $\vec{\mathbf{F}}$ acting on the liquid, for example, the gravitational force, and the gradient of the pressure along which the pressure force acts. The left-hand side represents the **total** of the velocity field with respect to time,

$$\frac{d\vec{\mathbf{v}}}{dt} = \left(\vec{\mathbf{v}} \cdot \text{grad}\right)\vec{\mathbf{v}} + \frac{\partial \vec{\mathbf{v}}}{\partial t}.$$

It represents the change of the velocity of a small volume element in a reference frame moving with the liquid. Hence, the left-hand side of the equation is the acceleration, and the right hand side corresponds to the applied forces:

- the external force per unit volume $\vec{\mathbf{F}}$,
- the pressure force per unit volume along the pressure gradient $-\text{grad}\,p$.
- ➤ For a viscous flow, Euler's equation is extended to the Navier-Stokes equation (see p. 200).

5.3.2.3 Bernoulli's law

1. Bernoulli's law,

establishes a relation between the cross-sectional area of a tube and the pressure in the tube. One distinguishes:

- **static pressure**, which acts with equal magnitude perpendicular and parallel to the flow direction;
- **pressure due to gravity** (**geodesic pressure**), which corresponds to the hydrostatic pressure in a liquid column;
- **dynamic pressure**, which occurs because of the flow. The dynamic pressure depends on the flow velocity.
- ➤ In a flowing liquid, the pressure is not the same in different directions, it is **not isotropic**. The static pressure is just the isotropic component of the total pressure.
- ▲ **Bernoulli's law**:
  In steady flow, the sum of static and dynamic pressure is constant.

2. Derivation of Bernoulli's equation

Bernoulli's law follows from energy conservation. If a volume ΔV of a liquid has a kinetic energy $\frac{1}{2}\rho\Delta V v^2$ (ρ density, v velocity) at a point where the tube cross-section is A, and the kinetic energy $\frac{1}{2}\rho\Delta V v'^2$ at another point where the cross-section is A', then the difference

$$\Delta W_{\text{kin}} = \frac{1}{2}\rho\Delta V(v'^2 - v^2),$$

must originate from the pressure difference and the difference of the potential energies $\Delta V\rho g(h - h')$ (h, h' are the corresponding heights).

Pressure energy, W_p, the work to be expended to force the volume ΔV at a pressure p into the tube,

$$W_p = pA\Delta s = p\Delta V.$$

Then:

$$\Delta W_{kin} = \Delta V(p - p') + \Delta V\rho g(h - h'),$$

and therefore:

Bernoulli's equation			$ML^{-1}T^{-2}$
	Symbol	Unit	Quantity
$p + \dfrac{1}{2}\rho v^2 + \rho g h = \text{const.}$	p	Pa	static pressure
	ρ	kg/m^3	density
	v	m/s	flow velocity
	g	m/s^2	gravitational acceleration
	h	m	height

The first term is the static pressure, the second and third are the dynamic pressure and the pressure due to gravity (**Fig. 5.39**).

➤ Bernoulli's equation holds for steady, non-viscous flow, and is therefore an idealization.

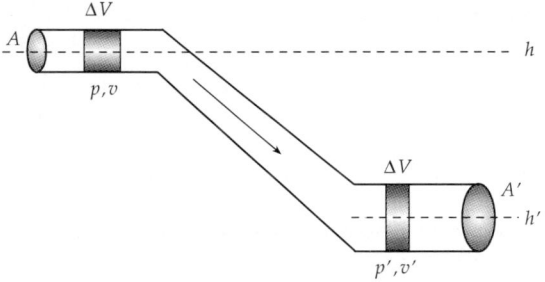

Figure 5.39: On Bernoulli's equation.

3. Methods of measurement based on Bernoulli's law

Flow nozzle, constriction of a pipe along the flow direction to increase the flow velocity. Basic tool to convert pressure energy into kinetic energy. Application in turbines, jet nozzles.

Diffuser, channel extension. Inverse of nozzle: kinetic energy of the flowing liquid is converted to pressure energy. Application in flow-type pumps.

The continuity equation and Bernoulli's equation form the basis of several methods of pressure measurement (**Fig. 5.40**):

M **Static-pressure tube**, for measurements of static pressure. The pressure is probed at an opening in the pipe and measured by a manometer.

M **Pitot tube**, for measurements of static and dynamic pressure. The pressure arises at the mouth of a tube in a direction opposite to the flow direction.

M **Prandtl's impact tube**, combines the Pitot tube and the static pressure tube for measurements of the dynamic pressure as the difference between the total and static pressure. For known density ρ of the liquid, one can evaluate the flow velocity v from the dynamic pressure p_S, $v = \sqrt{2 p_S / \rho}$.

4. Venturi tube,

(**nozzle device**), for measuring the volume flow Q according to the Venturi principle (see p. 195). The difference between the static pressure before and in a nozzle constriction is

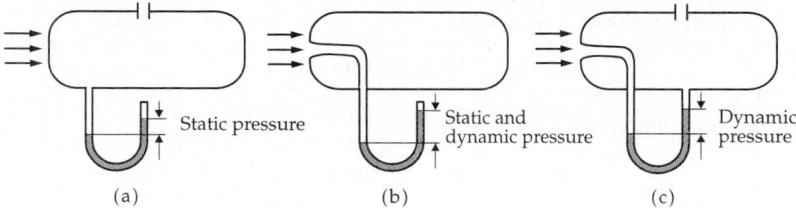

Figure 5.40: Methods for pressure measurements based on Bernoulli's law. (a): static-pressure tube, (b): Pitot tube (static pressure and dynamic pressure), (c): Prandtl's impact tube (dynamic pressure).

measured. The faster the liquid flows, the smaller the static pressure (**Fig. 5.41**):

$$Q = A_1 \cdot \sqrt{\frac{1}{(A_1/A_2)^2 - 1}} \cdot \sqrt{\frac{2\Delta p}{\rho}} ,$$

$$= A_1 \cdot \sqrt{\frac{1}{(A_1/A_2)^2 - 1}} \cdot \sqrt{2g\Delta h}$$

(A_1 cross-section of tube, A_2 constricted cross-section, Δp pressure difference, ρ density of fluid, g acceleration of gravity, Δh difference of heights in the ascension tube).

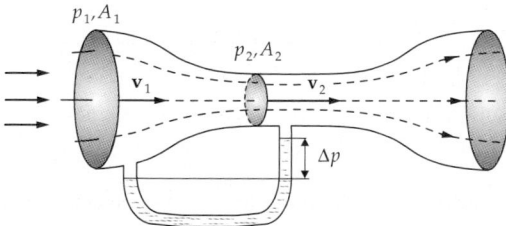

Figure 5.41: Venturi tube for measuring volume flow according to the Bernoulli equation.

➤   In real, viscous flows the friction must be taken into account. In practice, one applies correction factors determined by a calibration.

5.3.2.4   Torricelli's effluent formula

1. Effluent velocity

The effluent velocity of a liquid through a small aperture on the surface of a vessel under the influence of the weight is obtained from Bernoulli's equation. If one compares a small volume of liquid at an arbitrary point in the vessel (height h_1, at rest) with another volume at the effluent aperture (height h_2, velocity v), with atmospheric pressure p_0, one gets

$$\rho g h_1 + p_0 = \rho g h_2 + \frac{\rho}{2} v^2 + p_0 ,$$

and therefore:

effluent velocity $\sim \sqrt{\text{height}}$			$\mathbf{LT^{-1}}$
	Symbol	Unit	Quantity
$v = \sqrt{2gh}$	v	m/s	effluent velocity
	g	m/s^2	gravitational acceleration
	h	m	height of liquid column above the effluent aperture

2. Torricelli's effluent law

The effluent velocity in a liquid column of height h above the effluent aperture equals the velocity of free fall of a body from the height h (**Fig. 5.42**).

The horizontal distance L of the jet from the outlet of the vessel at the depth h_2 below the aperture is

$$L = 2\sqrt{h_1 \cdot h_2}\,.$$

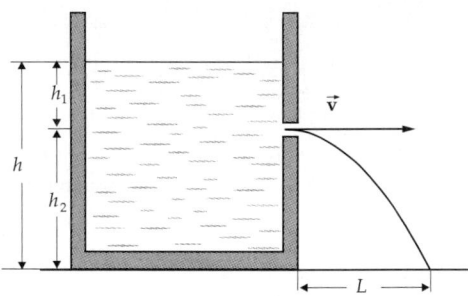

Figure 5.42: Torricelli's effluent law. The effluent velocity v depends on the height h_1 of the liquid column above the aperture.

For an aperture in the bottom of the vessel follows an effluent velocity of

$$v = \sqrt{2gh}\,.$$

If an additional pressure p_{ext} acts on the surface of the liquid, the effluent velocity is

$$v = \sqrt{2\left(gh + \frac{p_{\text{ext}}}{\rho}\right)}\,.$$

3. Effluent velocity

By the same consideration, one may find the **effluent velocity** from a pipe in which an overpressure p (compared with the exterior) exists:

effluent velocity			$\mathbf{LT^{-1}}$
	Symbol	Unit	Quantity
$v = \sqrt{\dfrac{2p}{\rho}}$	v	m/s	effluent velocity
	p	Pa	overpressure
	ρ	kg/m^3	density

➤   In the considerations above, the friction within the liquid (see viscosity, p. 198) has been ignored. Friction may be taken into account by multiplying the velocity by a **velocity coefficient** ϕ (water: $\phi \approx 0.97$). Moreover, a constriction of the jet arises when the liquid leaves the exit aperture; the effect may be taken into account by the **coefficient of contraction** α (sharp-edged exit: $\alpha \approx 0.61$). The product of both corrections is called **coefficient of discharge** μ, $\mu = \phi\alpha$. In order to take the friction and the influence of the exit aperture into account, the values for the effluent velocity v and the distance L calculated with the above formulae must be multiplied by the coefficient of discharge μ.

4. Dam,

effluent of a liquid over the top edge of a container, e.g., over locks in rivers (**Fig. 5.43**). The volume flow Q is

$$Q = \frac{2\kappa}{3} \cdot h \cdot b \cdot \sqrt{2gh}$$

(h height of flow over dam, b lateral width, g gravitational acceleration). The coefficient of contraction κ may be determined according to Swiss standards, as follows:

$$\kappa = 0.615 \cdot \left(1 + \frac{1}{1.6 + 1000\,h}\right)\left(1 + 0.5\frac{h^2}{H^2}\right), \quad h \text{ in m}.$$

The expression holds for fall height $H - h \geq 0.3$ m, level $H \geq 2h$ and height of flow $h = 0.025$ m $\ldots 0.8$ m.

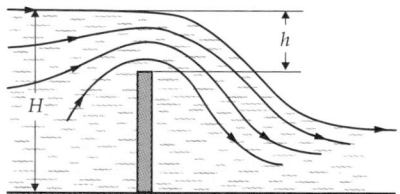

Figure 5.43: Dam flow of liquid over an edge.

5.3.2.5   Suction effects

According to Bernoulli's law, the static pressure in a flowing fluid is smaller than the static pressure in the liquid at rest. This causes **suction effects** in flows:

▲   **Venturi principle**, by reducing the cross-sectional area of a pipe and the resulting acceleration of the flow, the static pressure in the tube may fall below the atmospheric pressure in the vicinity; hence, another liquid may be sucked in.

a) Water-jet pump, suction of a gas by a liquid (**Fig. 5.44**). The liquid (water, mercury) flowing at high speed through a nozzle leads to a reduction of the static pressure, which causes suction of the gas from the vessel to be evacuated. **Mercury-vapor diffusion pumps** of this design are used in vacuum technology; such pumps reach pressure values of 1 Pa $= 10^{-5}$ bar. The pressure that can be reached is limited by the vapor pressure of the liquid.

b) Sprayer, for suction of a liquid into an air flow (**Fig. 5.45**). The top of the sprayer capillary is placed into an air flow. Owing to the reduced static pressure, as compared with the pressure on the liquid in the vessel, the liquid is sucked in.

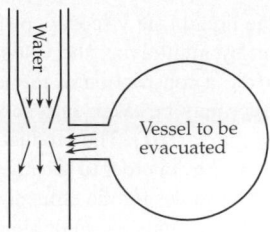

Figure 5.44: Water-jet pump.

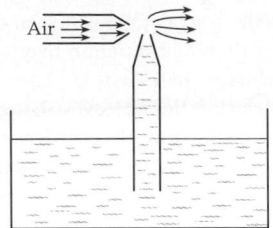

Figure 5.45: Sprayer.

c) Hydrodynamic paradox: A liquid or gas flowing out of a tube may attract a plate placed on the tube exit (**Fig. 5.46**). This happens if the effluent velocity becomes so large that the external pressure exceeds the remaining static pressure in the liquid flowing between the tube exit and the plate. For the same reason, two vehicles moving closely side by side are attracted to each other.

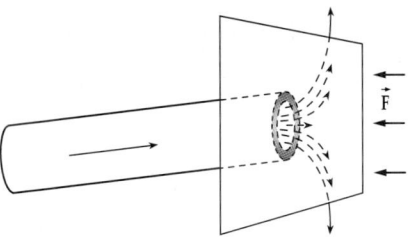

Figure 5.46: Hydrodynamic paradox. A plate is sucked in by a jet emerging from a pipe.

5.3.2.6 Buoyancy in flow around bodies

1. Buoyancy

Buoyancy on a body immersed in a flow, arises according to Bernoulli's law if the flow velocity at different faces of the body have different magnitudes. At the face with the higher velocity, there is an **underpressure**, on the other side, an **overpressure**.

   Magnus effect (**Fig. 5.47**), a cylinder rotating in a flowing liquid experiences a force perpendicular to the flow. Owing to the rotation, the flow on one side of the cylinder is diminished, on the other side it is increased. The net effect is a difference of static pressures, and thus a sideward acceleration.

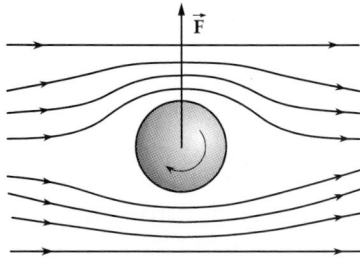

Figure 5.47: Magnus effect.

2. Buoyant force

Wing, a body in a flow formed in such a way that the speed of flow on the upper side is higher than that on the lower side. Because of the resulting pressure difference, the body experiences a **dynamic buoyant force**:

dynamic buoyant force			MLT^{-2}
	Symbol	Unit	Quantity
$F_A = c_A \dfrac{\rho}{2} A v^2$	F_A	N	dynamic buoyant force
	c_A	1	buoyancy coefficient
	ρ	kg/m^3	density of liquid
	v	m/s	flow velocity
	A	m^2	max. projected area

The buoyant force is proportional to the square of the flow velocity (cf. frictional force) and to a typical area extension. The latter quantity corresponds to the largest area when projecting the wing onto an arbitrary plane (length times width) (**Fig. 5.48**).

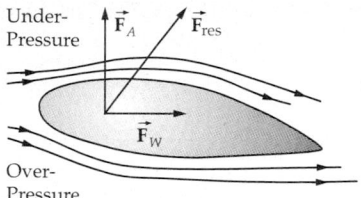

Under-Pressure $\vec{F}_A$ $\vec{F}_{res}$

$\vec{F}_W$

Over-Pressure

Figure 5.48: Dynamic buoyant force $\vec{F}_A$ in flow around a wing. Buoyant force $\vec{F}_A$ and drag force $\vec{F}_W$ of the wing sum to the resulting force $\vec{F}_{res}$.

➤ The buoyancy coefficient is determined in an aerodynamic tunnel. Typical values vary between 0.02 and 0.05.
➤ In calculations for airplanes, the **drag force** F_W must also be taken into account,

$$F_W = c_w \frac{\rho}{2} A v^2, \quad c_w: \text{drag coefficient.}$$

The resulting force F_{res} points upwards and backwards. Its action point is called **center of pressure**. It may be determined in an aerodynamic tunnel from the torque on the wing, which depends on the **angle of attack**. The backward component of the force is compensated by the propelling force of the engines.

5.3.3 Real flow

Real flow differs from ideal flow by the presence of friction. One distinguishes:
● **laminar flow**, which differs from the flow of an ideal liquid mainly by a modified speed,
● **turbulent flow**, which is no longer stationary and where, at a fixed space point, both the orientation and the velocity of a flowing liquid vary at random.

5.3.3.1 Internal friction

Internal friction, friction originating from cohesion forces between molecules of liquids or gases. The kinetic energy of the fluid is dissipated by friction, which manifests itself as an increase in temperature.

1. Laminar flow,

a flow in which individual films of liquid of finite thickness slide over each other, without notable mixing between the layers, as e.g., in the flow between two parallel plates moving with respect to each other (**Fig. 5.49**).

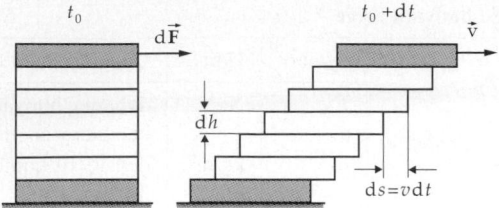

Figure 5.49: Layers of liquid in laminar flow between two plates moving with respect to each other.

The liquid moves in the same direction over the entire volume considered, but the individual layers move with different velocities. Frictional forces arise in this sliding and cause a uniform decrease of velocity across the flow profile (**Fig. 5.50**). The opposite is **turbulent flow**.

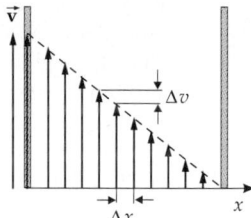

Figure 5.50: Velocity profile in laminar flow between two parallel plates moving with respect to each other.

Velocity gradient, dv/dx, the difference of velocities of two neighboring layers, referred to the thickness of a layer. A plot of the velocity of a layer versus its position shows the **velocity profile** $v(x)$; the first derivative dv/dx of the profile represents the velocity gradient.

2. Newtonian viscosity,

describes the strength of the frictional force between neighboring layers of a laminar flow. The force acting on such a layer is proportional to the area of the layer, and to the velocity gradient with respect to the neighboring layers:

Newtonian viscosity			MLT^{-2}
	Symbol	Unit	Quantity
$F_R = \eta A \dfrac{dv}{dx}$	F_R	N	frictional force
	η	$Pa \cdot s = N \cdot s/m^2$	dynamic viscosity
	A	m^2	area of layer
	dv/dx	$1/s$	velocity gradient

The proportionality constant η is called **dynamic viscosity**, or simply **viscosity**. The unit of viscosity is **Pascal second** (Pa · s). The higher the viscosity of a liquid, the greater the force required to move the layers against each other. A typical order of magnitude for η is 10^{-5} Pa · s for gases, 10^{-3} Pa · s for water and between 0.1 and 0.01 Pa · s (depending on temperature) for lubricating oils.

M   The viscosity manifests itself directly when one pulls a plate out of a narrow vessel. If the distance between the plate and the wall of the vessel is sufficiently small, viscosity shows up as a braking force.

Non-SI unit: Poise (named after the physicist Poiseuille)

$$1 \text{ Poise} = 0.1 \text{ Pa} \cdot \text{s}.$$

3. Fluidity and kinematic viscosity

Fluidity, ϕ, the reciprocal value of dynamic viscosity:

$$\phi = \frac{1}{\eta}, \quad [\phi] = \frac{m^2}{Ns}.$$

Kinematic viscosity, ν, the ratio of dynamic viscosity η and density ρ of the liquid:

$$\nu = \frac{\eta}{\rho}, \quad [\nu] = \frac{m^2}{s}.$$

Obsolete unit:

$$1 \textbf{ Stokes} = 1 \text{ St} = 10^{-4} \text{ m}^2/\text{s}.$$

Typical orders of magnitude of the kinematic viscosity are $10^{-6} \text{ m}^2/\text{s}$ for water, $10^{-4} \text{ m}^2/\text{s}$ for air, and from 1 to several hundred m^2/s for motor oils.

Although the dynamic viscosity gives the force acting on a layer of liquid, the kinematic viscosity takes into account the density of the liquid, and hence the mass $\Delta m = \rho \Delta V = \rho A \Delta x$ of the layer of liquid. The kinematic viscosity specifies the acceleration:

$$a = \frac{F_R}{\Delta m} = \frac{F_R}{\rho A \Delta x} = \nu \frac{\Delta v}{(\Delta x)^2}$$

(a acceleration, Δm mass of layer, F_R frictional force, A area, Δx thickness of layer, ν kinematic viscosity, Δv velocity difference).

➤ The viscosity is a constant that depends on the material; it is strongly temperature- and pressure-dependent. The dependence on the temperature is described approximately by

$$\eta = A \, e^{b/T}$$

with material-dependent constants A and b; hence, it decreases with increasing temperature. The viscosity and its temperature dependence is of particular importance for lubricants.

The dynamic viscosity of gases is much lower than that of liquids (air $1.7 \cdot 10^{-5}$ Pa $\cdot$ s, water $1.8 \cdot 10^{-3}$ Pa $\cdot$ s for 0 °C).

The viscosity of solutions and mixtures of fluids is strongly dependent on the concentration.

➤ **Non-Newtonian materials**, materials for which the Newtonian viscosity is not valid and/or the deformation of which is not plastic. Such materials are polymeric materials (**liquid plastics**) and **dispersions** (liquids containing solids or other liquids suspended as small spheres; also denoted **suspension** or **colloid**, depending on their dimension).

5.3.3.2   Navier-Stokes equation

1. Equation of motion of real flow

The continuity equation also holds for real flow. Euler's equation is extended to the **Navier-Stokes equation**:

$$\rho \cdot \left((\vec{\mathbf{v}} \cdot \mathrm{grad}) \, \vec{\mathbf{v}} + \frac{\partial \vec{\mathbf{v}}}{\partial t} \right) = \rho \cdot \frac{\mathrm{d}\vec{\mathbf{v}}}{\mathrm{d}t} = \vec{\mathbf{F}} - \mathrm{grad}\, p + \eta \cdot \Delta \vec{\mathbf{v}} .$$

The left-hand side represents the substantial derivative of the velocity field. Besides the external force per unit volume, $\vec{\mathbf{F}}$, and the pressure force per unit volume $-\mathrm{grad}\, p$, the right-hand side contains an additional force term

$$\eta \cdot \Delta \vec{\mathbf{v}} = \eta \cdot \left(\frac{\partial^2 \vec{\mathbf{v}}}{\partial x^2} + \frac{\partial^2 \vec{\mathbf{v}}}{\partial y^2} + \frac{\partial^2 \vec{\mathbf{v}}}{\partial z^2} \right) .$$

It depends on the curvature of the velocity distribution and gives the frictional force. Δ denotes the Laplace operator.

The Navier-Stokes equation is the basic equation of the hydrodynamics of viscous liquids. Together with the continuity equation, it describes any flow of an incompressible liquid, in particular turbulent flow. There are efficient numerical algorithms for solving the equation.

2. Special cases of real flow

The following special cases can be distinguished.

- Flow with negligible friction: $\eta \approx 0$. The Navier-Stokes equation then reduces to the Euler equation (see p. 733).
- Steady flow: the time derivative vanishes.
- Sluggish flow for very high viscosity: $\eta \to \infty$. The left-hand side of the Navier-Stokes equation may be ignored; the flow is determined by the balance of pressure gradient and friction.
- Rotational flow in turbulences. Instead of solving the equations directly, one expresses the change of the vortex strength in a volume element by the energy dissipation due to friction. In this way, turbulent flow may be described efficiently.

5.3.3.3   Laminar flow in a tube

1. Modelling of laminar flow in a tube

The laminar flow in a cylindrical pipe of inner radius R may be imagined as being composed of many hollow cylinders of thickness Δr in which liquid flows with equal speed. The outermost hollow cylinder adheres to the wall, and is at rest. The velocity of the other hollow cylinders results from the balance of the frictional forces F_R (described by Newton's viscosity formula) and the pressure force F_p. If one considers a hollow cylinder of radius r symmetric to the axis of the tube of length l, the pressure force acting on the cross-sectional area A is

$$F_p = pA = \pi p r^2 .$$

The opposing frictional force

$$F_R = -\eta A \frac{\Delta v}{\Delta r} = -\eta 2\pi r l \frac{\Delta v}{\Delta r}$$

in the equilibrium state equals the pressure force. Hence, the velocity gradient is

$$\frac{\Delta v}{\Delta r} = -\frac{pr}{2\eta l}.$$

The velocity gradient increases with increasing pressure, and decreases with increasing velocity and increasing tube length. It increases linearly with the distance from the tube axis.

2. Derivation of the Hagen-Poiseuille law

One goes from the difference quotient $\Delta v/\Delta r$ to the differential quotient dv/dr and separates the resulting differential equation. One obtains

$$r \, dr = -\frac{2\eta l}{p} \, dv.$$

Integration yields

$$r^2 = -\frac{4\eta l}{p} v + C,$$

with an integration constant C. The latter is specified by the requirement that at the wall $(r = R)$ the velocity vanishes, $v = 0$; hence $C = R^2$. Rewriting yields the law for laminar flow in a tube:

Hagen-Poiseuille law			LT^{-1}
	Symbol	Unit	Quantity
$v(r) = \dfrac{p}{4\eta l}(R^2 - r^2)$	$v(r)$	m/s	velocity profile
	r	m	distance from tube axis
	p	Pa	pressure
	η	Pa s	dynamic viscosity
	l	m	length of tube
	R	m	inner radius of tube

The velocity profile is a parabola (**Fig. 5.51**). The maximum velocity $v_0 = v(0)$ occurs at the tube axis; it is proportional to the pressure and to the square of the radius (and hence to the tube cross-section), and inversely proportional to the viscosity and the length of the tube.

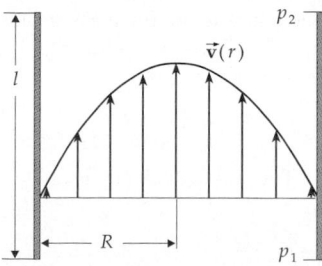

Figure 5.51: Law of Hagen-Poiseuille. Velocity profile of laminar flow in a tube.

3. *Properties of laminar flow in a tube*

The **decrease of pressure** p between the ends of the tube is proportional to the tube length l, the flow velocity v_0 on the axis and to the viscosity η, and inversely proportional to the tube cross-section:

$$p = \frac{4\eta l}{R^2} v_0 .$$

The volume flow $\Delta V / \Delta t$, i.e., the volume ΔV of liquid passing the tube per unit time Δt, is obtained by integration of the velocity profile $v(r)$ over the tube cross-section:

$$\frac{\Delta V}{\Delta t} = \frac{\pi R^4}{8\eta l} p .$$

➤ Hence, it is easier to increase the volume flow by enlarging the cross-section of the tube than by increasing the pressure.

➤ For a given volume flow, the decrease of pressure is

$$p = \frac{8\eta l}{\pi R^4} \frac{\Delta V}{\Delta t} .$$

| M | The viscosity may be measured in a way related to the relation between pressure and volume flow. One measures the time required for a definite quantity of liquid to flow through the opening of a funnel. The pressure results from the density of the liquid and the height of the liquid column above the funnel.

5.3.3.4 Flow around a sphere

A similar consideration yields the force acting on a sphere submerged in a laminar flow of liquid:

Stokes' law of friction			$\mathbf{MLT^{-2}}$
	Symbol	Unit	Quantity
$F_R = 6\pi \eta r v$	F_R	N	frictional force
	η	Pa s	dynamic viscosity
	r	m	radius of sphere
	v	m/s	flow velocity

The Stokes frictional force is thus proportional to the radius of the sphere (not to the cross-sectional area), to the flow velocity and to the dynamic viscosity of the liquid.

| M | **Höppler's sphere viscosimeter**, for measurements of the dynamic viscosity η based on Stokes' law, by determining the sinking speed v of a sphere of radius r. The sinking speed follows from the balance between the friction force F_R and the weight force F_G reduced by the buoyant force F_A:

$$F_R = 6\pi \eta r v = F_G - F_A = \frac{4}{3}\pi r^3 (\rho_K - \rho_{Fl})$$

(ρ_K density of sphere, ρ_{Fl} density of liquid). The sinking velocity is

$$v = \frac{2gr^2(\rho_K - \rho_{Fl})}{9\eta} ,$$

and for the dynamic viscosity one obtains

$$\eta = \frac{2gr^2(\rho_K - \rho_{Fl})}{9v}.$$

5.3.3.5   Bernoulli's equation

For real flow with friction, Bernoulli's law must be modified.

▲   **Law of Bernoulli**:
The sums of static and dynamic pressure, measured at two distinct positions of a tube, differ by the magnitude of the pressure decrease calculated according to the Hagen-Poiseuille law.

$$\left(p_1 + \frac{1}{2}\rho v_1^2 + \rho g h_1\right) - \left(p_2 + \frac{1}{2}\rho v_2^2 + \rho g h_2\right) = \Delta p,$$

where p_1, p_2 is the pressure, v_1, v_2 the velocity of the liquid, and h_1, h_2 the height at the two points of measurement; Δp denotes the pressure decrease. The latter quantity is positive if the first point of measurement lies upstream of the second one.

   Lost head, h_V, the height by which the point of inflow has to be lifted to compensate the friction:

$$h_V = \frac{\Delta p}{\rho g}.$$

It is determined by the coefficient of tube friction ρ.

5.3.4   Turbulent flow

1. Characterization of turbulent flow

Turbulent flow, a flow characterized by random variation in direction and speed at a fixed space point. It is no longer stationary. But when measuring over a period much longer than a period typical for the turbulent changes, one obtains a mean velocity distribution. If the distribution is time-independent, turbulent flow is treated like steady flow, and one tries to include the effects of turbulence by appropriate coefficients of friction.

2. Formation of vortices,

arises because of friction in the **detachment of liquid layers**. If an ideal liquid flows around a sphere the pressure takes the maximum value where the surface is perpendicular to the flow ("in front" and "backward") since the speed vanishes there; the pressure takes the lowest value (and the speed the highest value) where the spherical surface is parallel to the flow ("above" and "below"). Hence, liquid particles flowing around the sphere are first decelerated (dynamic pressure), then accelerated (according to Bernoulli's principle) and finally decelerated to again fit into the normal flow. The latter deceleration of the liquid elements is enforced by friction, so that the particles come to rest before reaching the symmetry axis. Hence, **vortices** are generated that occur pairwise because of conservation of angular momentum.

3. Reynolds number,

a nondimensional quantity, specifies the role of vortex formation. For higher Reynolds numbers, vortices develop spontaneously from small perturbations (**Fig. 5.52**). Turbulent flows are an example of nonlinear dynamics (see p. 211) of an extended system.

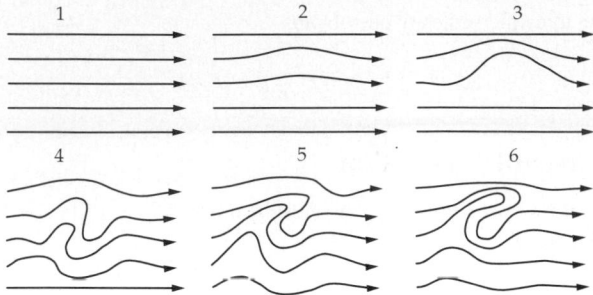

Figure 5.52: Transition from laminar flow to turbulent flow. Formation of vortices from a small perturbation.

Additional energy is drawn from the flowing liquid by the friction between the particles of the liquid in a vortex, which is represented by an additional frictional force.

▲   The frictional force in turbulent flow is larger than that in laminar flow.

5.3.4.1   Drag coefficient

1. Drag force

In turbulent flow, there are two drag forces acting on a body (**Fig. 5.53**):

- **surface-friction drag**, F_R, the force between the liquid and the surface of the body, described by the friction law of laminar flow;
- **pressure drag**, F_D, the difference of pressure onto the front and the back of a body, acting additionally in turbulent flow. The pressure difference originates in the formation of vortices at the back of the body. In the vortices, the liquid is moving very rapidly, hence the static pressure there is smaller than at the front face, according to Bernoulli's equation.

Both components added yield the drag force, $\vec{F}_W$,

$$\vec{F}_W = \vec{F}_R + \vec{F}_D \, .$$

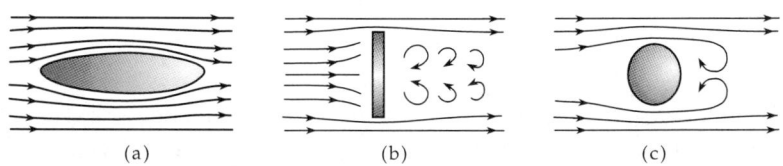

Figure 5.53: Drag in the flow around bodies. (a): frictional force in laminar flow, (b): drag in turbulent flow around a plate, (c): friction and pressure drag in the flow around a sphere.

2. Drag coefficient,

characterizes the magnitude of the drag force:

drag force			MLT^{-2}
	Symbol	Unit	Quantity
$F_W = c_w \dfrac{\rho}{2} A v^2$	F_W	N	drag force
	c_w	1	drag coefficient
	ρ	kg/m^3	density of liquid
	A	m^2	cross-sectional area of the body
	v	m/s	flow velocity

The drag coefficient is a dimensionless quantity. It depends significantly on the shape of the body.

▲   The drag force is proportional to the cross-sectional area of the body and to the square of the velocity.

➤   Typical values for the drag coefficient vary between 0.055 (streamlined body) and 1.1 up to 1.3 (plate).

| M | The drag coefficient is measured directly in an aerodynamic tunnel. The measurements may be taken with models scaled down in size using scaling laws.

3. Streamlined body,

a drop-like body with the lowest possible drag coefficient. The pressure decrease along a streamlined body proceeds so smoothly that no vortices are formed; this is achieved by a properly designed tail.

➤   The drag force on a body in the atmosphere is largely caused by vortex formation. One therefore tries to suppress vortices as far as possible by designing slots or guide vanes, and by keeping the flow laminar.

The power P needed for moving a body in a turbulent flow is (due to $P = F_W v$) equal to

$$P = c_w \frac{\rho}{2} A v^3 .$$

■   When doubling the velocity, the power must be raised by a factor of eight.

4. Wind load

on buildings, by pressure or suction (ripping off of roofs). **Beaufort degrees** (see **Tab. 33.0/6**).

■   The air pressure in the interior of a house in a strong wind is higher than the pressure above the roof (see p. 191).

▲   The wind pressure p_w increases with the square of the wind velocity:

$$p_w = c_p v^2 , \quad [p_w] = \text{Pa} = \text{pascal} .$$

The proportionality factor has the dimension kg/m^3. Typical numerical values are $c_p = 1.0$ kg/m^3.

Typical dynamic wind loads on buildings		
height above ground	wind velocity/(m/s)	dynamic pressure/(kPa)
up to 8 m	30	0.5
8 to 20 m	36	0.8
20 to 100 m	42	1.1
beyond 100 m	46	1.3

5.3.5 Scaling laws

1. Types of scaling

Scaling laws, set up a relation between fluid-mechanical properties of scaled-down models and those of the original bodies. The model must fulfill the following two conditions.

- **Geometric similarity**: The model must be a length-preserving, scaled-down representation of the original, both in the geometric measures and in the surface properties.
- **Hydrodynamic similarity**: Density, viscosity, velocity of the fluid and drag force in the model experiment must be in a certain ratio to those of the original situation.

2. Reynolds number,

Re, describes the hydrodynamic similarity.

Reynolds number			1
	Symbol	Unit	Quantity
$$\mathrm{Re} = \frac{L\rho v}{\eta} = \frac{Lv}{\nu}$$	Re	1	Reynolds number
	L	m	characteristic length
	ρ	kg/m^3	density of liquid
	v	m/s	flow velocity
	η	Pa s	dynamic viscosity
	ν	m^2/s	kinematic viscosity

The Reynolds number is a dimensionless quantity. L denotes a typical extension in the geometry considered, e.g., the diameter of a sphere or the edge length of a cube. The Reynolds number is a measure for the ratio of the inertial force of a volume of liquid to the drag force acting on it. The behavior of the flow is determined by the interplay of both quantities. The Reynolds number depends on the temperature.

▲ **Similarity laws**:
  The drag coefficients of geometrically similar bodies coincide if the Reynolds numbers for both cases coincide.

➤ This law is the foundation for the measurement of drag coefficients for models in **aerodynamic tunnels**.

▲ In order to get hydrodynamic similarity in scaling-down the model, either the velocity must be increased in a proportional relation, or the kinematic viscosity must be decreased correspondingly. The latter may be achieved by diminishing the dynamic viscosity, or by increasing the density.

3. Critical Reynolds number,

$\mathrm{Re_{crit}}$, gives a criterion for the transition from laminar flow to turbulent flow. If the Reynolds number of a flow exceeds the critical Reynolds number, $\mathrm{Re} > \mathrm{Re_{crit}}$, the flow becomes turbulent (**Fig. 5.54**).

➤ The critical Reynolds number depends sensitively on the geometry of the flow. For a smooth pipe, it varies between 1000 and 2500. The transition from laminar flow to turbulent flow does not happen suddenly; it also depends on the presence of disturbances in the flow.

  In particular, turbulence occurs only beyond a certain minimum velocity. Therefore, in the flow around a body, the vortices arise at its backside where the streamlines join again and the liquid thereby is accelerated, both in radial and axial direction. **Laminar boundary layer**, generated in the flow around a body submerged in a real liquid. In the boundary

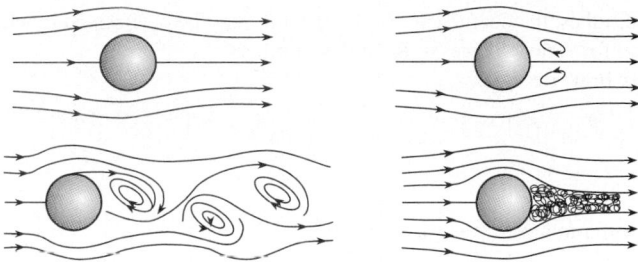

Figure 5.54: Formation of vortices and transition to turbulent flow for increasing Reynolds number.

layer, the flow velocity is low because of the friction at the surface of the body. In this situation, the Reynolds number is below the critical Reynolds number. The formation of vortices starts only beyond the boundary layer (**Fig. 5.55**).

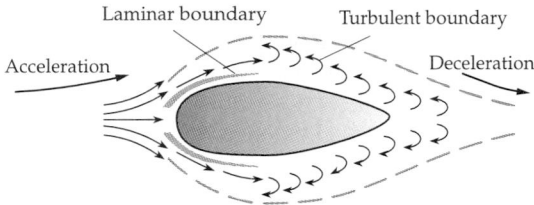

Figure 5.55: Laminar and turbulent boundary layers.

Froude number, Fr, another similarity number that takes into account the influence of gravitation: Dynamical similarity requires the same ratio of inertial force and gravitational force. The Froude number is of importance for the description of surface waves (e.g., in the flow around the hull of a ship):

$$\mathrm{Fr} = \frac{v}{\sqrt{Lg}}$$

(v flow velocity, L characteristic length, g gravitational acceleration). Ideally, in a model investigation, both the Froude numbers and the Reynolds numbers for model and original should coincide. This is, however, impossible because of their different dependences on L. In the investigation of flow in pipes, where the Earth's attraction has only a minor influence on the internal motion of a liquid, one uses the Reynolds number. In studies of the flow around a ship's hull, where the influence of surface waves is more important, or in effluent or jet problems, one uses the Froude number.

5.3.5.1   Tube friction

1. Law of tube friction,

the proportionality between the lost head and the length l of the tube:

$$h_V = \lambda \frac{l}{d} \frac{v^2}{2g}$$

(d diameter, l length of the tube, v velocity of flow, g gravitational acceleration). The proportionality constant λ is called **coefficient of tube friction**.

For smooth tubes, the coefficient of tube friction may be determined by empirical formulas holding for various ranges of Reynolds numbers:

- **laminar flow**: $\mathrm{Re} < \mathrm{Re}_{\mathrm{crit}}$,

$$\lambda = \frac{64}{\mathrm{Re}}.$$

- **Blasius formula**: $\mathrm{Re}_{\mathrm{crit}} \leq \mathrm{Re} \leq 10^5$,

$$\lambda = \frac{0.3164}{\sqrt[4]{\mathrm{Re}}}$$

- **Nikuradse formula**: $10^5 \leq \mathrm{Re} \leq 10^8$,

$$\lambda = 0.0032 + \frac{0.221}{\mathrm{Re}^{0.237}}$$

- **Kirschmer-Prandtl-Kármán formula**: $\mathrm{Re} > \mathrm{Re}_{\mathrm{crit}}$,

$$\frac{1}{\lambda} = \left(2 \cdot \log \frac{\sqrt{\lambda} \cdot \mathrm{Re}}{2.51} \right)^2.$$

The equation is a transcendental equation which must be solved numerically or graphically.

2. Roughness

For tubes with a rough surface, the coefficient of tube friction depends on the **mean height of roughness** k. This quantity specifies the typical size of elevations on the surface:

type of tube	height of roughness k
plastic tubes	≈ 0.007 mm
steel tubes	0.05 mm
rusted steel tubes	0.15 mm to 4 mm
cast iron tubes	0.1 mm to 0.6 mm
concrete channels	1 mm to 3 mm
built channels	3 mm to 5 mm

➤   Whether a given tube is smooth or rough depends on the **relative roughness**

$$k_{\mathrm{rel}} = \frac{k}{d}$$

(d tube diameter) of the tube, and on the Reynolds number. For

$$\mathrm{Re} \cdot \frac{k}{d} > 1300$$

the tube is rough, for values up to 65 it is smooth, with a mixed region in between.

5.3.6 Flow with density variation

Flow with density variation, occurs in gases. For liquids, the density variation is almost always negligible. The prevailing phenomena are the propagation of small density variations (sound) and large density variations (shock waves). Density variations have also to be taken into consideration for flows at high velocity (nozzles), and for atmospheric flow (meteorology).

The equation of motions for flow of compressible media employ the equation of state of the medium, which relates pressure, density, and temperature.

1. Sound,

the propagation of small pressure variations. It proceeds by sound waves (see p. 311) that propagate with a constant **sound velocity** c, which is dependent on the medium, the temperature and the pressure. For an ideal gas, the sound velocity is given by

$$c = \sqrt{\kappa RT/M}$$

(κ isentropic coefficient of gas, R universal gas constant, T temperature, M molar mass).
▲   In a homogeneous gas at rest, the propagation of sound proceeds via spherical waves emerging from the source and propagating uniformly (isotropically) with the velocity of sound.

If the source of sound moves relative to the observer, the motion of the source is superimposed on the propagation of the sound waves.

2. Mach cone,

propagation of sound from a source moving with a velocity v_q above the sound velocity c. The source of sound escapes from the sound waves emitted, $v_q t > ct$. Hence, the spherical waves emitted at different times superimpose in such a manner that a cone-shaped wave front arises, with a maximum of pressure increase on the cone surface (**Fig. 5.56**). An observer passed by this wave front registers a supersonic boom.

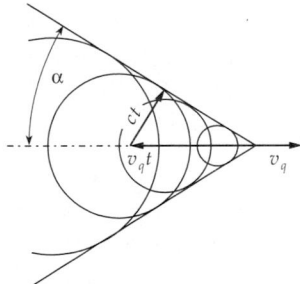

Figure 5.56: Mach cone. A source of sound moves at supersonic velocity $v_q > c$. α: Mach angle.

Mach angle, α, half the apex angle of the Mach cone:

$$\sin \alpha = \frac{c}{v_q} = \frac{1}{M}$$

(v_q velocity of the source of sound, c sound velocity, M Mach number). The Mach number M gives the velocity of the source of sound in units of the sound velocity.

3. Shock wave,

(**compression shock**), large discontinuous change of pressure that propagates with **supersonic velocity**. The pressure jump in such a wave is localized within distances of few molecular mean free-paths (in the range of micrometers). Continuous waves of large amplitudes are transformed into shock waves, since the sound propagates more rapidly in regions of high pressure (high temperature) than in regions of low pressure. Therefore, the continuous rise at the beginning of the wave is overtaken by the crest.

➤   Shock waves arise in detonations.

6
Nonlinear dynamics, chaos and fractals

Nonlinear dynamics, deals with the complex phenomena caused by nonlinear terms in the equations of motion, in particular with deterministic chaos.

1. Example: Oscillators and vibrations with nonlinear damping.

Oscillators with nonlinear damping, applied force, or both. Oscillators of this kind display a **broad spectrum of resonances** that vary with the amplitude (**amplitude dependence** of the resonance frequency) and possibly show **self-excitation**.

Vibrations in mechanical elements are approximately linear only for small amplitudes. For large displacements, **distortions** arise in the vibrations which, in extreme situations, may result in unexpected material breaks.

Electronic components almost always display some nonlinear characteristics. Hence, electronic amplifiers distort the input information at large modulation range (**distortion**).

2. Example: Forces between planets.

The forces between the planets depend nonlinearly on the coordinates (via the distances, which involve square roots). In the two-body case, the equations of motion can still be solved. For the multi-body problem, however, no general solution exists even for simple two-body forces.

In the planetary system, the attraction between Sun and planets dominates by orders of magnitude, but the mutual attraction between the planets causes perturbations of the orbits. Nonlinear dynamics investigates the **stability** of planetary orbits against these perturbation terms.

3. Turbulences

Turbulences in fluids and gases are examples of extended nonlinear processes. They typically occur only if a certain critical **parameter** (here the Reynolds number, see p. 206) becomes large enough (**bifurcation**).

Turbulences in the atmosphere govern the weather. They illustrate the **sensitive dependence of the dynamical evolution on the initial conditions**. Some turbulences (many 100 km diameter) are predictable only if the initial conditions are known exactly. Such systems are deterministic, but nevertheless not predictable (**deterministic chaos**).

4. Stadium billiards

A **billiards** is a space region bounded by reflecting walls where particles move freely otherwise. If the walls are curved (**stadium billiards**), the trajectory of a particle in general depends sensitively on the initial conditions. One then cannot predict whether, or when, the particle leaves the billards space through an opening (**Fig. 6.1**).

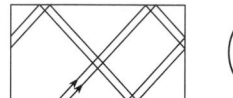 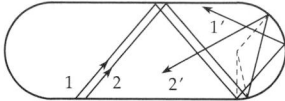

Figure 6.1: Rectangular and stadium billiards. In stadium billiards, the two originally close trajectories diverge more and more.

6.1   Dynamical systems and chaos

Dynamics generally deals with the **time evolution** of systems. The concept of a **dynamical system** plays a leading role. One distinguishes between **conservative** (energy-conserving) systems and **dissipative** (energy-losing) systems. Conservative systems serve for investigations of **integrability**, dissipative systems for studies of the **long-time behavior,** the existence of **attractors**, and the **sensitive dependence on the initial conditions** that leads to **strange attractors** and to **deterministic chaos**.

6.1.1   Dynamical systems

1. Dynamical system,

abstract method of description of a (physical, chemical, economical, ecological, ...) process. The state of a dynamical system is represented by a number of **variables** that describe the physical situation and are subject to a **time evolution**.

2. Examples of dynamical systems

- A mathematical pendulum (see p. 260) is described by its displacement from the rest position. The variable is the angle θ of displacement. The time evolution is determined by the differential equation of the pendulum:

$$\frac{d^2\theta}{dt^2} = -\omega^2 \sin\theta\,,$$

  $\omega = \sqrt{g/l}$ is the angular frequency of the vibration for small displacements, l the pendulum length, g the gravitation acceleration. The nonlinearity (**anharmonicity**) in this simple system consists of the appearance of higher powers of θ in the series expansion of the sine function.
- Other examples of dynamical systems: the motion of bodies in classical mechanics, the flow of currents in electric circuits, the course of chemical reactions, the evolution of economic variables, the population growth in biology.

3. Counter-example to the dynamical system

Contrast: **thermodynamical equilibrium**, which is considered in thermodynamics (see p. 626). It does not describe the time evolution, but gives information on the steady state of the system, depending on the environmental conditions. **Kinetic theory** establishes the connection between the dynamical system (molecular motion) and the criteria for equilibrium.

4. Deterministic system,

a system in which the time evolution can be determined for all future from the knowledge of the present (and possibly the past).
■   Any classical mechanical system is deterministic; the motion is determined by Newton's equations of motion. It suffices to know positions and momenta at some instant to fix the time evolution of the system for all time.

Stochastic systems, which are affected by influences of which only probability distributions are known, are non-deterministic: gas molecules in thermodynamics, kinetic theory, Brownian motion, also quantum systems and models in economy and biology, where stochastic terms (**noise**) simulate random variations.

5. Continuous system,

a system the variables of which are changing continuously so that, to any real value of time t, a state of the system can be assigned. Its time evolution may be described by a system of **differential equations** that state how rapidly any variable is changing for a given state of the system.
■   The motion of bodies in classical mechanics and the behavior of electric circuits are described by continuous variables (positions, currents).

6. Discrete system,

a system the variables of which change from one time step t_n to the next one t_{n+1}, without employing any state of the system between these instants of time. Its time evolution is determined by a **mapping** that specifies the values of the variables at the instant t_{n+1} if their values at the moment t_n and possibly at other, previous instants of time $t_{n-1}, t_{n-2}, \cdots$ are given.
■   Discrete systems occur in mathematical models, e.g., in modelling economical data (gross national product in different years) and in the description of continuous systems in terms of Poincaré cuts (see p. 216).

7. Linear system and superposition principle

Linear system, a system in which cause and effect are proportional to each other; it therefore can be represented by a linear equation.
■   Harmonic oscillator: the restoring force is proportional to the elongation x,

$$\ddot{x} = -\omega^2 x \,,$$

ω is the angular frequency of the vibration.
▲   **Superposition principle**:
If two solutions $x_1(t)$ and $x_2(t)$ of a linear system are known, then the **linear superposition** or **linear combination**

$$x(t) = \alpha\, x_1(t) + \beta\, x_2(t)$$

with arbitrary coefficients α, β is also a solution of the systems.

In particular, the properties of the system at larger values of the variables may be derived by scaling.

■   The resonance frequency of a harmonic oscillator does not depend on the amplitude.
■   The harmonic oscillator has two elementary solutions, e.g., the sine and cosine vibration that differ only in the phase. By linear combination of the elementary solutions, a solution with arbitrary amplitude and phase may be constructed.

Because of the superposition principle, it is sufficient to know only a few fundamental solutions of the equations for a linear system.

8. Nonlinear system

Cause and effect are not proportional to each other, the system cannot be described by a linear equation.

■   Nonlinear restoring forces and/or damping cause the properties of an oscillator to vary with the amplitude. Such oscillators can exhibit a large number of resonances with frequencies depending on the **amplitude** of the excitation.
■   Mathematical pendulum: For large elongations, the restoring force does not increase proportional to the angle, but only to the sine of the angle (i.e., it is weaker than in the linear case). At small elongations, the system carries out oscillations about the rest position; at large elongations, loops may occur.

6.1.1.1   Space of states and phase space

1. Configuration space,

the space spanned by the space variables of a physical system.

▲   The time evolution of a dynamical system is represented by specifying a **trajectory** in configuration space, i.e., to any time point t a point $x(t)$ in the configuration space is assigned.

a) Examples for trajectories

■   Trajectory of a point mass in classical mechanics; the configuration space is the three-dimensional space in which the motion happens.
■   **Fibonacci sequence**, as a dynamical system, defined by the prescription

$$x_n = x_{n-1} + x_{n-2}$$

with the initial conditions $x_0 = x_1 = 1$. The configuration space is the real axis, the trajectory is the sequence $(x_0, x_1, x_2, \ldots)$.

M x-t **graph**, used to represent the motion of a system in two dimensions. On the vertical axis one or several variables are represented, on the horizontal axis time is plotted.

b) Example: Mathematical pendulum θ, t. The x-t graph of a mathematical pendulum at small amplitudes is a sine function (**Fig. 6.2**).

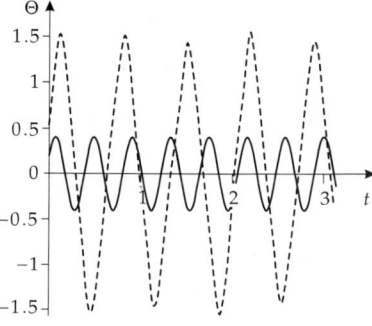

Figure 6.2: θ-t graph of the pendulum. For small amplitudes the pendulum vibrates harmonically, for larger amplitudes both shape and frequency of the vibration are altered.

2. Space of states

In order to calculate the further evolution of a system, it is in general not sufficient to know only the present state of the system; the present rates of change (time derivatives) of the variables are also needed.

■   In the case of the mathematical pendulum, the displacement and the velocity must both be known.

Space of states, the space spanned by all quantities that must be known at *one* moment t to calculate the further time evolution. Every point in the space of states uniquely characterizes the present and future states of the system.

a) Examples of spaces of states

■   In order to predict the additional members of the Fibonacci sequence, the present number x_n and the preceding number x_{n-1} must be known. Every point in the space of states is thus represented by **two** numbers.

■   The space of states for a system of classical mechanics is the **phase space** (see p. 91) spanned by the space variables and the related momentum variables.
The phase space of the mathematical pendulum is spanned by the variable θ and its time derivative $\dot{\theta}$.

b) Trajectory in phase space

➤   The concept of phase space is also often used for other systems; it then denotes the space of states.

Trajectory in phase space, the motion of a system in time through the phase space: any instant of time corresponds to a point in phase space (**Fig. 6.3**).

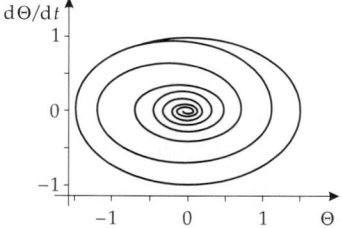

Figure 6.3: Phase-space trajectories of the harmonic oscillator (ellipse, periodic motion) and of the damped pendulum (spiral).

M x-y **graph**, represents the motion of a system in phase space: each axis corresponds to a phase-space coordinate.

c) Example: Mathematical pendulum.
For small elongations, the phase-space trajectory is an ellipse (**Fig. 6.4**).

d) Properties of phase-space trajectories

▲   Closed phase-space trajectories represent periodic motions.

In a deterministic system, the position of the system in phase space at any instant of time determines the future course of the trajectory, i.e., the entire future evolution of the system.

▲   Phase-space trajectories cannot intersect each other.

Otherwise, at the intersection point of two trajectories which path the system would follow would be undetermined.

➤   **Singularity**, a point in phase space into which many trajectories converge and the system remains, or which is reached only asymptotically. At such points the system loses information on the trajectory.

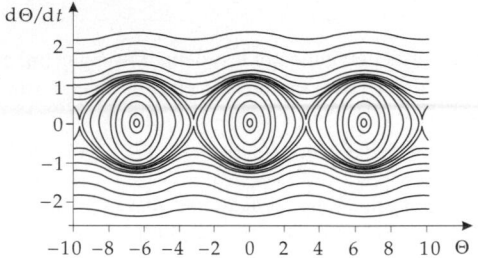

Figure 6.4: Phase-space trajectories of the mathematical pendulum. For small amplitudes Θ, the trajectory corresponds to that of the ideal (harmonic) oscillator, at larger elongations distortions arise, lastly loops occur.

3. Poincaré cut

a) Definition of the Poincaré cut: A simple way to visualize the behavior of a system is the Poincaré cut. Here, not the full phase space, but only a subspace (hypersurface) is considered. It is spanned by $n-1$ phase-space coordinates. Whenever all other phase-space coordinates take previously fixed values, the actual value of the phase-space coordinate just considered is marked by a point.

Poincaré cut, a subspace of phase space defined by the prescription that a phase-space coordinate takes a certain value. One then considers the intersection points of the cut with the phase-space trajectories.

➤   This procedure must be distinguished from a **phase-space projection** where the values of the phase-space coordinates considered are continuously plotted. In the cut, however, the system is considered only at those times when the selected phase-space coordinate takes a certain value.

b) Example: Poincaré cut of an anharmonic oscillator. The variable according to which the cut is made is the phase $\sin \omega t = 0$ of the external excitation. Technically, this can be achieved by means of a **stroboscope**. **Fig. 6.5** shows the corresponding phase-space trajectory.

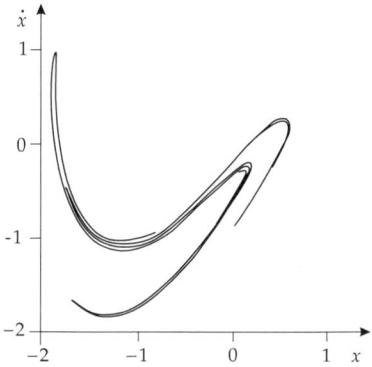

Figure 6.5: Poincaré cut of an anharmonic oscillator (Duffing oscillator).

c) Example: Pendulum. In the case of a pendulum, one can use the zero line $\theta = 0$ of the displacement as Poincaré cut. Every trajectory intersects this line at two distinct points: once when the pendulum moves from the left to the right ($\dot{\theta} > 0$), and once in the opposite direction ($\dot{\theta} < 0$). The Poincaré cut is a straight line on which $\dot{\theta}$ is plotted and these two points are marked.

Alternatively, one can take the zero line of the velocity $\dot{\theta} = 0$ as a cut. One then plots the value of θ when $\dot{\theta}$ equals zero. This just happens for the points $\pm\theta_{max}$, the maximum elongations.

d) Properties of the Poincaré cut

▲   Any point on a Poincaré cut corresponds to exactly one point in phase space.
➤   Contrast: in a projection the coordinate projected out can no longer be reconstructed. Therefore, a point on a Poincaré cut completely determines the trajectory passing it, and hence also the next intersection point of the trajectory with the Poincaré cut.

4. Poincaré mapping,

attaches to any point on the Poincaré cut the corresponding next following intersection point of the phase-space trajectory (**Fig. 6.6**). The mapping allows the reduction of the dynamics of the system to the question of at which point the phase-space trajectory intersects the Poincaré cut the next time.

▲   Poincaré mapping reduces a continuous dynamical system to a discrete dynamical system.

The Poincaré cut allows a classification of periodic systems:

- **Periodic** or **quasiperiodic** phase-space trajectory, intersects the Poincaré cut only in a finite number of points that are arranged on a curve.
- **Chaotic** phase-space trajectory, intersects the Poincaré cut at infinitely many, irregularly distributed points.

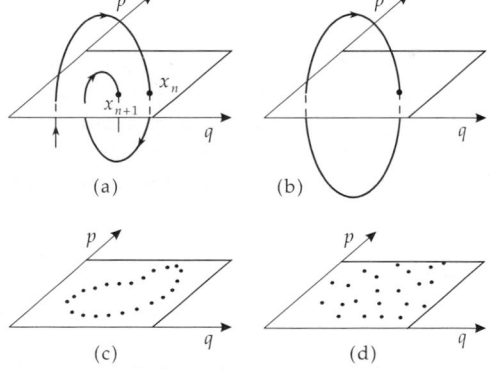

Figure 6.6: Visualization of the trajectory in phase space by a Poincaré mapping (schematic). One considers the sequence of points $x = (q, p)$ where the trajectory crosses the p, q plane vertically from up to down. (a): Poincaré mapping $x_n \to x_{n+1}$. (b): periodic trajectory. (c): regular trajectory. The intersection points lie on an invariant curve. (d): chaotic trajectory. The intersection points are irregularly distributed over the plane.

6.1.2   Conservative systems

Conservative system, a system the energy of which does not change with time. Such a system is characterized by the existence of an **energy function** that assigns an energy value to any point in phase space. The system then moves on the equipotential surfaces of this function.

■   Mechanical systems without friction represent conservative systems, as do electric circuits without resistances.

The motion of planets, taking into account the gravitational attraction of the planets to the Sun and to each other, is an example of a nonlinear conservative system. The two-body problem (Sun + one planet) may still be solved analytically, but the multi-body problem (Sun + several planets) can no longer be solved.

6.1.2.1   Liouville's theorem

The behavior of conservative systems in phase space is characterized by Liouville's theorem. One considers trajectories in phase space starting from several closely neighboring points.

➤   The phase space includes both the position variables and the linear momenta. Vicinity of points in phase space therefore means: similar positions and similar velocities.

▲   **Theorem of Liouville**:
The magnitude of an area occupied by an ensemble in phase space does not change in the course of the time evolution of the system (**Fig. 6.7**).

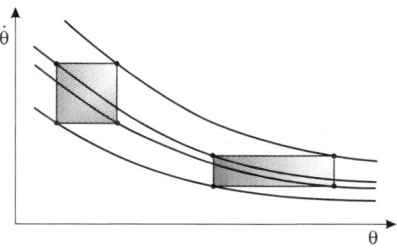

Figure 6.7: Theorem of Liouville: The size of a phase-space element of an ensemble does not change in the time evolution. The hatched areas have equal magnitude.

➤   Consider e.g., a square area in phase space. Liouville's theorem states that the points of this square during the time evolution of the system continue to cover an area of the same magnitude. Nothing is stated about the shape of this area. It may be an extended rectangle, may have a fully irregular form, or may represent a fractal (see p. 227).

Liouville's theorem represents a strong constraint on the dynamics of a conservative system.

6.1.2.2   Integrability

1. Example: Conservative system–harmonic oscillator.

Classical examples of a conservative system are the harmonic oscillator (see p. 90) and systems of coupled harmonic oscillators. Their solution is always a **quasi-periodic** motion, i.e., a motion that may be represented by a superposition of harmonic vibrations of various frequencies:

$$x(t) = c_1 \sin(\omega_1 t + \phi_1) + c_2 \sin(\omega_2 t + \phi_2) + \cdots$$

($c_1, c_2, \ldots$ constants, $\omega_1, \omega_2, \ldots$ vibration angular frequencies, $\phi_1, \phi_2, \ldots$ phases).

■   Coupled pendula may be described by a superposition of fundamental modes:

$$x_1(t) = A \sin\left(\frac{\omega_1 + \omega_2}{2} t\right) \cos\left(\frac{\omega_1 - \omega_2}{2} t\right),$$

$$x_2(t) = A \sin\left(\frac{\omega_1 - \omega_2}{2} t\right) \cos\left(\frac{\omega_1 + \omega_2}{2} t\right).$$

If one adopts the difference and the sum of x_1 and x_2 as variables z_1 and z_2 then:

$$z_1(t) = x_1(t) - x_2(t) = A_1 \sin(\omega_1 t + \phi_1),$$

$$z_2(t) = x_1(t) + x_2(t) = A_2 \sin(\omega_2 t + \phi_2).$$

2. Integration of dynamical systems

To integrate the equation of motion of a given dynamical system, one tries to find such co-ordinates in which the system carries out harmonic vibrations. The question arises whether any conservative system may be reduced by an appropriate **coordinate transformation** to one or several coupled harmonic oscillators.

a) Integrable system, a system that, for an appropriate choice of variables, may be written as a superposition of harmonic oscillators. It is characterized by the existence of **constants of motion (integrals of motion,)** i.e., quantities that do not change during the time evolution (as the energy and the vibration angular frequencies ω_i). Knowledge of all constants of motion completely characterizes the motion, except for specification of the phases ϕ_i.

■   All linear systems are integrable.
     The two-body problem (motion of a planet about the Sun) is integrable.

b) Non-integrable system, a system the motion of which is neither periodic nor quasi-periodic. Therefore, the system cannot be represented by a harmonic oscillator by means of a coordinate transformation.

Non-integrable systems may exhibit periodic behavior in one part of their phase space while behaving irregularly in another part. In particular, they show a **sensitive dependence on the initial conditions**, and thus chaotic behavior (see p. 221).

■   The multi-body problem (orbits of two or more planets around the Sun) is not integrable. There are always certain stable orbits; other orbits are unstable and lead to the escape of the planet from its orbit, to a breakdown of the system, or both.

6.1.3 Dissipative systems

1. Definition of a dissipative system

Dissipative system, a system that loses energy in the course of its time evolution.
■   A classical pendulum with damping, an electric circuit with resistance.
Liouville's theorem does not hold for dissipative systems.
▲   In a dissipative system, the size of the area in phase space covered by an ensemble decreases during the time evolution of the system.
Dissipative systems are characterized by the existence of attractors and limit cycles that govern the **long-term behavior**.

2. Fixed point and limit cycle

Fixed point, a point at which the system remains and no longer changes after reaching it. It may be the endpoint of one or several phase-space trajectories, or an isolated point.

   Limit cycle, a periodic motion reached by the system, after the **transients (transient oscillations)** faded away. A system that has a limit cycle will reach this cycle after a sufficiently long time for a large variety of initial conditions, and will not leave it. The information on the initial conditions is then largely lost.

➤   Because of Liouville's theorem, such behavior is not possible for conservative systems.

3. Attractors

Fixed points and limit cycles are the simplest examples of attractors.

Attractor, a region in phase space that cannot be left by the system, having once reached it.

Attraction pool of an attractor, all those points in phase-space trajectories that run into the attractor.

A dissipative system can be described in terms of the knowledge of its attractors and their attraction pools. The problem of nonlinear dynamics of dissipative systems is to find and characterize the attractors that determine the **long-term behavior** of the system.

6.1.3.1   Strange attractors, deterministic chaos

The simplest attractors are point attractors (the system reaches a definite state and remains there) and limit cycles (the system reaches a periodic motion). Knowledge of these attractors allows a complete statement on the behavior of the system after sufficiently long time. Other attractors exist, however, that allow only for a statement on what part of phase space the system will be found. They are characterized by the fact that the actual motion of the system is predictable only if the initial conditions are exactly known. Any uncertainty in the initial conditions amplifies in such a way that, after some time, nothing can be stated about the state of the system.

1. Sensitive dependence on the initial conditions,

a very small change in the initial conditions causes the system to reach a completely different state after sufficiently long time.

■   **Bernoulli mapping**, an iterative mapping according to the prescription:

$$x_{n+1} = \begin{cases} 2x_n, & \text{if} \quad x_n \text{ from the interval } [0;\ 0.5], \\ 2x_n - 1, & \text{if} \quad x_n \text{ from the interval } [0.5;\ 1]. \end{cases}$$

The real numbers x_n lie between zero and unity. If the initial value x_0 is not known precisely, one cannot predict whether a value x_n lies in the upper or lower half of the interval.

If x_0 is written in the binary system, $x_0 = 0.b_1 b_2 b_3 \ldots$, with binary digits $b_i = 0$ or 1, the Bernoulli mapping simply shifts the point to the right, i.e., $x_1 = 0.b_2 b_3 b_4 \ldots$ etc. If $b_2 = 0$, this number lies in the lower half of the interval. An irrational number has infinitely many, apparently random, binary digits b_i, hence the system behaves predictably only if the initial condition is known with complete accuracy.

2. Ergodicity,

the property of a motion in which, after some sufficiently long time, the trajectory approaches any given point of phase space with arbitrary precision. An ergodic motion covers the entire phase volume.

3. Lyapunov exponent,

λ, specifies the speed of increase of a small perturbation

$$|f(x + \Delta x, t) - f(x, t)| = \Delta x\, e^{-\lambda t},$$

for sufficiently long time t and sufficiently small distances Δx.

4. *Example: Duffing oscillator.*

Duffing oscillator, a nonlinear oscillator that is represented by the equation of motion

$$m\ddot{x} = -D_1 x - D_2 x^2 - D_3 x^3 - b\dot{x} + F\sin\omega t.$$

D_1 is the spring constant of the linear component of the system, while D_2 and D_3 describe nonlinear modifications of the spring force that manifest themselves at large elongations (**Fig. 6.8**). b represents a (linear) friction force. The nonlinear behavior of the system may be simulated with these four constants.

The term $F\sin\omega t$ describes a periodic external force of amplitude F and angular frequency ω that excites the oscillator.

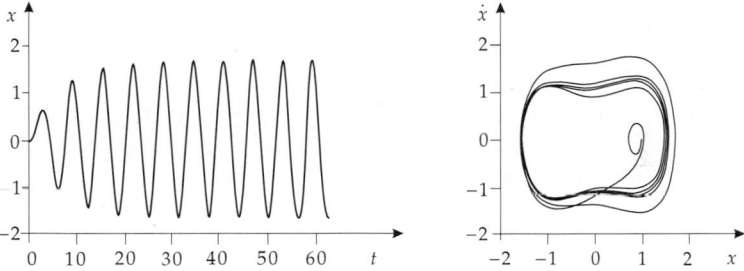

Figure 6.8: Duffing oscillator with specific choice of parameters. The vibration differs significantly from the harmonic form, the phase-space trajectory is deformed. The behavior nevertheless remains regular.

5. *Chaotic system,*

a system that displays a sensitive dependence on the initial conditions, but nevertheless covers only a restricted phase-space region.

➤ A sensitive dependence on the initial conditions may occur also in linear systems, e.g., for exponentially diverging trajectories. A chaotic system has the additional feature that its motion remains confined to a finite region of phase space.

6. *Deterministic chaos,*

the initial conditions being uncertain, the behavior of the system cannot be predicted for a long time, although it behaves strictly deterministically. The system is deterministic, but not predictable.

7. *Strange attractor,*

an attractor on which the system depends sensitively on the initial conditions. The system moves into the attractor, but its motion on the attractor is chaotic (**Fig. 6.9**).

6.2   Bifurcations

External parameter, a quantity characterizing bulk properties of the system. The parameter is set from outside, i.e., by the researcher.

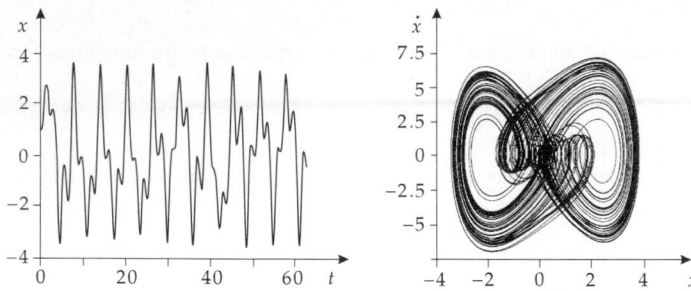

Figure 6.9: Duffing oscillator in the chaotic region. The x-t-diagram shows an irregular reversal behavior near the zero line. The phase-space trajectory covers an area, it remains in a strange attractor.

■   The masses of bodies in the multi-body problem, the spring constant and the damping of an oscillator. Parameters may in particular determine the degree of nonlinearity, e.g., by varying the characteristic of a spring in an oscillator.

The theory of chaos, among other questions, deals with which parameters lead to a chaotic system.

6.2.1   Logistic mapping

1. Definition of logistic mapping

Logistic mapping, a discrete dynamical system with a variable x that is determined by the mapping

$$x_{n+1} = r\, x_n\, (1 - x_n).$$

x_n, x_{n+1} are values of the variable in successive steps, r is a parameter. The logistic mapping is one of the simplest examples of a nonlinear discrete dynamical system.

2. Graph of logistic mapping

Fig. 6.10 shows the x_n–x_{n+1}-graph of the logistic mapping. On the horizontal axis the value x_n is plotted, on the vertical axis the corresponding subsequent value.

The x_n–x_{n+1}-graph may serve for visualization of the dynamics of the logistic mapping. One starts at a given point x_n on the horizontal axis, then moves up vertically to the curve, and from there to the left, where one finds the subsequent value x_{n+1}. From there, one again moves back horizontally to the plotted diagonal, and then vertically downwards until again reaching the horizontal axis, but now in the point x_{n+1}. Then one starts again.

When skipping the paths passed twice, it suffices to go vertically from the diagonal to the curve, and horizontally back to the diagonal again.

➤   The configuration space of the logistic mapping is the real axis along which the variable x is plotted. Since no further information than the actual value of x is needed to get the future values, the space of states is the real axis as well.

Fig. 6.10 shows several iterations for various values of the parameter r.

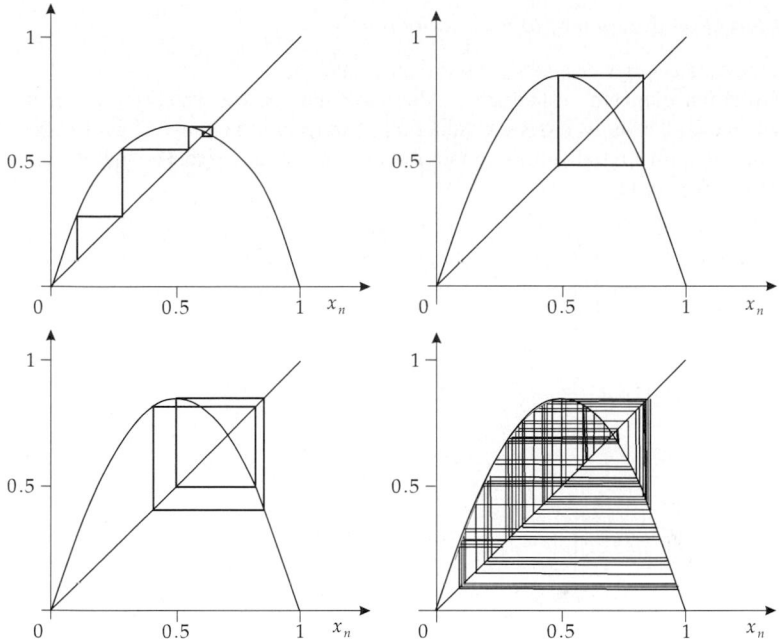

Figure 6.10: Iteration steps of the logistic equation at various values of the parameter r that determines the steepness of the parabola.

3. Properties of logistic mapping at various parameter values r

- Attractive **fixed point** as attractor. The system runs from most initial conditions towards a fixed point x obeying

$$x = r\,x\,(1 - x) \qquad \Longrightarrow \qquad x_{n+1} = x_n = x\,.$$

- For larger values of the parameter r, which determines the slope of the parabola, a limit cycle of period 2 occurs. After the transients have died out the system jumps back and forth between two values.
- If the steepness of the parabola is still increased, a limit cycle of period 4 arises: **period doubling**. After four steps, the system reaches the initial state.
- For even higher r, one finds limit cycles with longer and longer periods (8, 16, 32, ...). The distances between the various values r_n where the next higher period sets in become shorter and shorter.
- Starting from a certain critical value r_∞, this period becomes infinite, hence the system becomes aperiodic.

4. Trajectory of logistic mapping

The **trajectory** of the logistic mapping consists of all points x_n that are reached, starting from an initial value x_0. If one omits the transients by starting to mark e.g., from the 100th iteration, then the trajectory consists only of the fixed point itself (and possibly of its nearest neighborhood). For the limit cycle of period 2, it consists of two points (the two values taken by x), for the limit cycle of period n it consists of n distinct points. In the aperiodic region finally infinitely many points arise, and an entire section of the axis is blackened.

5. Bifurcation diagram of logistic mapping

From these trajectories, one gets the bifurcation diagram.

Bifurcation diagram, a diagram on the horizontal axis of which the parameter r is plotted, and on the vertical axis the values x_n as trajectory: For any value of r, one takes all values of x_n that result from a certain initial value at this r, where again the transients are omitted (**Fig. 6.11**).

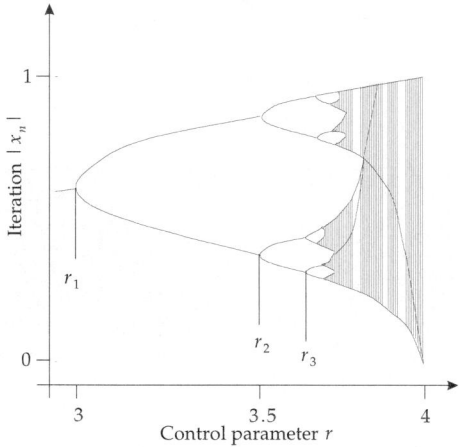

Figure 6.11: Bifurcation diagram of the logistic equation (schematically): For every value of the control parameter r, those x are blackened that correspond to any value x_n which arises at this parameter value (transients are not shown).

- $r < r_1$: The system has a fixed point, for any r only one point is blackened. The bifurcation diagram shows only a single branch.
- $r_1 < r < r_2$: The system is in a periodic limit cycle, only two points are blackened. On the bifurcation diagram, two branches arise.
- For further increasing parameter value r, there appear increasingly higher periods, and correspondingly many branches of the curve.
- Lastly, the dynamics become aperiodic, entire regions are blackened.

Hence, in the bifurcation diagram one sees on the left one curve that, at a certain value of r, bifurcates in each case into two branches. With increasing value of the control parameter, the bifurcations succeed more and more rapidly until at some critical value r_∞ the curve is split into infinitely many branches.

6. Bifurcation,

(Latin: branching), in general a qualitative change in the behavior of a system (here: transition from a fixed point to a period 2, then to period 4, etc., lastly to an aperiodic motion) for a small variation of a continuous parameter (here: r).

The particular role of the logistic mapping is due to the fact that the Poincaré mappings (see p. 217) of many dynamical systems have a similar structure and pass the same sequence of bifurcations. These systems are said to approach chaotic motion in the Feigenbaum scenario.

▲   Systems approaching chaos in the Feigenbaum scenario are characterized by a sequence of period doublings until chaos is reached.

➤   There are still other **routes to chaos**; not all systems follow the Feigenbaum scenario.

6.2.2 Universality

Universality, originates from the evidence that the Poincaré mappings of many systems have a form similar to that of the logistic mapping, so that these systems also pass through a sequence of period doublings.

In 1979 Feigenbaum succeeded in deriving universal properties of these systems:

▲ If r_n denotes the value of the parameter r at the nth period doubling and r_∞ its value where chaotic motion is reached, the distances $r_\infty - r_n$ form a geometric series:

$$r_\infty - r_n = C\,\delta^{-n}$$

(first **Feigenbaum law**). C is a constant depending on the system, but the number δ is universal: it has the same value for all systems following this scenario:

$$\delta = 4.669201\ldots\ ,\qquad \text{first } \textbf{Feigenbaum constant}.$$

The parameter values for which period doubling arises are thus connected by a simple relation that can be experimentally verified. Hence, chaotic motion does not mean at all that no statements on properties of the motion can be made.

➤ Two more Feigenbaum laws describe additional universal properties, in particular the position of the attractor elements x_n.

6.3 Fractals

1. Fractal dimension

D of a set, is determined by a scale for surveying the set. On a straight line the scale is a segment of fixed length l, on a plane the scale is a square with the side l, in space a cube of edge length l. One then counts how many times the scale is needed to cover the set completely. If the scale is reduced in size, this number $N(l)$ increases in a D-dimensional space with the power D:

$$\frac{N(l)}{N(l_0)} \sim \left(\frac{l_0}{l}\right)^D$$

(l_0 original scale, l new scale).

■ When halving the side length of the unit square in 2D space, one needs $2^2 = 4$ of the smaller squares to cover the same area. In three dimensions, one correspondingly needs $2^3 = 8$ times as many cubes.

2. Objects with broken fractal dimension

There are objects for which the number of required scales does not increase with an integer number D, but rather increases with a fractional number as exponent.

a) Cantor set, a subset of the interval between 0 and 1 in one dimension. One removes the middle third part of the interval; from the remaining two thirds, one again removes the medium third, etc. When covering this set with a scale, one finds for its dimension D:

$$D = \frac{\ln 2}{\ln 3} \approx 0.63\,.$$

Thus, when halving the size of the scale, one needs only ca. $2^{0.63} \approx 1.55$ times more scales than before.

b) Coast line: If one measures the coast length of a country on a low-resolution map, one finds a lower value than when using a high-resolution map, which shows more of the inlets and bays.

c) Koch curve and Koch's snowflake. **Koch curve**, obtained by the following construction principle: A section of length l is subdivided into three parts of equal lengths. The medium third is then replaced by two straight sections of equal length $l/3$ that enclose an angle of 60° (**Fig. 6.12**). The procedure is then repeated for each of the four straight sections, and so on. The dimension of the resulting Koch curve is

$$D = \frac{\ln 4}{\ln 3} \approx 1.262 .$$

Figure 6.12: First step of constructing a Koch curve.

Koch's snowflake, evolves from an equilateral triangle by dissolving its sides into Koch curves. In each iteration step, the circumference of the figure increases by the factor 4/3; the area, however, remains finite (**Fig. 6.13**).

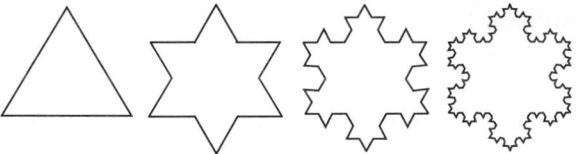

Figure 6.13: The first four steps of constructing a Koch snowflake.

d) Sierpinski triangle, results from an equilateral triangle by successive removal of the corresponding, by the (linear) factor 2 reduced triangles that join the side-midpoints of the triangles from the preceding iteration step. In each iteration step, the area decreases by the factor 3/4 (**Fig. 6.14**). The dimension of the Sierpinski triangle is

$$D = \frac{\ln 3}{\ln 2} \approx 1.585 .$$

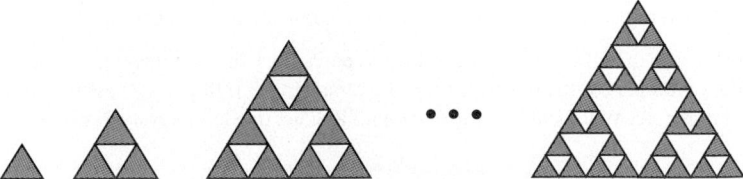

Figure 6.14: First steps of constructing the Sierpinski triangle. For convenience of representation, the scale varies.

3. Fractal,

object with a fractal dimension, in contrast to objects such as straight lines, areas, volumes that have integer dimensions.

Self-similarity, the property that a fractal set in the lateral magnification looks like the original set.

■   In the Cantor set, each third of an interval looks like the interval itself (**Fig. 6.15**).

Figure 6.15: Cantor set.

▲   All known strange attractors are fractals.

➤   The set of all points in phase space that belong to the attractor has a fractal dimension. This is the fundamental link between nonlinear dynamics and fractal geometry.

Other fractal sets arise if one considers for which values of its parameters a system displays chaotic behavior.

4. Mandelbrot set,

best known fractal object: the set of all points μ of the complex plane obeying the constraint that the iterated mapping

$$z \rightarrow z^2 - \mu$$

(starting at $z_0 = 0$) shall not diverge towards infinity. This set ("apple manikin," **Fig. 6.16**) is self-similar and is constrained by a fractal curve.

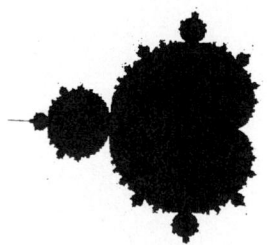

Figure 6.16: Mandelbrot set.

Formula symbols used in mechanics

symbol	unit	designation	symbol	unit	designation
α	rad/s^2	angular acceleration	f	$1/s$	frequency
α	1	extension number	f	m	coefficient of
γ	1	relativistic factor			rolling friction
ε	1	extension	Fr	1	Froude number
η	1	efficiency	G	$N\,m^2/kg^2$	gravitation constant
η	Pa s	dynamical viscosity	G	N/m^2	shear modulus
κ	$1/Pa$	compressibility	g	m/s^2	acceleration of gravity
μ	1	coefficient of sliding friction	HB	1	Brinell hardness
			HR	1	Rockwell hardness
μ	1	Poisson coefficient	HV	1	Vickers hardness
μ_0	1	coefficient of static friction	h, H	m	height
			J	m^4	areal moment of inertia
ν	1	transverse-extension number	$\hat{J}$	$kg\,m^2$	tensor of inertia
			J	$kg\,m^2$	moment of inertia
ν	m^2/s	kinematic viscosity	K	N/m^2	compression modulus
ρ	kg/m^3	density	k	N/m	spring constant
σ	N/m^2	normal tension	$\vec{L}$	$kg\,m^2/s$	angular momentum
σ	N/m	surface tension	$\vec{l}$	$kg\,m^2/s$	orbital angular momentum
τ	Nm	torque			
τ	N/m^2	shear stress	l	m	length
Φ	J/kg	gravitational potential	m	kg	mass
ϕ	rad	angle	P	W	power
ϕ	$m^2/(Ns)$	fluidity	p	Pa	pressure
ω	rad/s	angular velocity	$\vec{p}, p$	$kg\,m/s$	momentum
A	m^2	area	Q	m^3/s	volume flow
a	m/s^2	acceleration	Re	1	Reynolds number
c	m/s	speed of light	r	m	radius
c_a	1	coefficient of buoyancy	$\vec{r}$	m	position vector
			$\vec{r}(t)$	m	spatial function
c_w	1	air-resistance coefficient	S	N/m^2	tension
			s	m	path
D_m	Nm	directing moment	t	s	time
d	m	distance, lever arm	Δt	s	time interval
E	J	energy	V	m^3	volume
E	N/m^2	elasticity modulus	W	J	work
$\vec{e}$	1	unit vector	v	m/s	velocity
e	1	Euler's number	x, y, z	m	spatial coordinate
$\vec{F}$	N	force			

7
Tables on mechanics

7.1 Density

7.1.1 Solids

The density of solids is given at a temperature 293.15 K = 20 °C.

7.1/1 Simple metals

Material		Density ρ $(10^3$ kg/m$^3)$	Material		Density ρ $(10^3$ kg/m$^3)$
aluminum	Al	2.707	gadolinium	Gd	7.900
antimony	Sb	6.684	germanium	Ge	5.350
arsenic	As	5.727	gold	Au	19.320
barium	Ba	3.510	hafnium	Hf	13.300
beryllium	Be	1.850	holmium	Ho	8.795
bismuth	Bi	9.800	indium	In	7.28
cadmium	Cd	8.648	iridium	Ir	22.420
calcium	Ca	1.540	iron	Fe	7.897
cesium	Cs	1.878	lanthanum	La	6.145
cerium (cub.)	Ce	6.657	lead	Pb	11.373
cerium (hex.)		6.757	lithium	Li	0.530
chromium	Cr	7.190	lutetium	Lu	9.840
cobalt	Co	8.830	magnesium	Mg	1.746
copper	Cu	8.954	manganese	Mn	7.210
dysprosium	Dy	8.550	mercury (fluid)	Hg	13.546
erbium	Er	9.006	molybdenum	Mo	10.200
europium	Eu	5.243	neodymium	Nd	7.004
gallium	Ga	5.904	neptunium	Np	20.45

(continued)

7.1/1   Simple metals (*continued*)

Material		Density ρ $(10^3$ kg/m$^3)$	Material		Density ρ $(10^3$ kg/m$^3)$
nickel	Ni	8.906	tantalum	Ta	16.690
niobium	Nb	8.570	tantalum		
osmium	Os	22.480	(powder)		14.401
palladium	Pd	12.080	tellurium	Te	6.250
platinum	Pt	21.450	tellurium		
plutonium	Pu	19.84	(amorphous)		6.00
polonium	Po	9.320	terbium	Tb	8.229
potassium	K	0.851	thallium	Tl	11.860
praseodymium	Pr	6.773	thorium	Th	11.7
protactinium	Pa	15.37	thulium	Tm	9.321
radium	Ra	5.500	tin (grey)	Sn	5.75
rhenium	Re	20.530	tin (white)		7.304
rhodium	Rh	2.400	titanium	Ti	4.540
rubidium	Rb	1.520	tungsten	W	19.350
ruthenium	Ru	12.300	uranium	U	18.700
samarium	Sm	7.520	vanadium	V	5.960
scandium	Sc	2.989	ytterbium	Yb	6.965
selenium	Se	4.81	yttrium	Y	4.469
silver	Ag	10.500	zinc	Zn	7.144
sodium	Na	0.971	zirconium	Zr	6.520
strontium	Sr	2.630			

7.1.1.1   Metallic alloys

7.1/2   Construction materials

Material	Composition	ρ $(10^3$ kg/m$^3)$
aluminum alloys		
dural	Al (0.5 % Cu)	2.787
aluminum bronze	*	2.7
AlCuMg	*	2.8
AlMg	5 % Mg	2.6
cast aluminum(Si)	12 % Si	2.65
copper alloys		
delta metal	56 % Cu, 40 % Zn, 2 % Fe, 1 % Pb	8.6
brass (rolled)	30 % Zn	8.522
cast brass	*	8.4
phosphorus bronze	4.5 % Sn, 0.2 % P	8.91
bronze	25 % Sn	8.666
manganine	12 % Mn, 2 % Ni	8.5
new silver	15 % Ni, 22 % Zn	8.618

(*continued*)

7.1/2   Construction materials (*continued*)

Material	Composition	ρ $(10^3\text{kg}/\text{m}^3)$
iron alloys		
cast iron	Fe+0.4 % C	7.272
invar	36 % Ni	8.7
Steel		
	0.5 % C	7.833
	1.0 % C	7.801
	1.5 % C	7.753
St304, St316, St347		8.0
St410, St414		7.7
chromium steel	3 % Cr	7.7
tombac	6...20 % Sn	8.7...8.9
nickel alloys		
chromium nickel steel	24 % Fe, 16 % Cr	8.250
chromium nickel V	20 % Cr	8.410
monel	32 % Cu, 1 % Mn	8.9

7.1/3   Electric materials

Material	Composition	$\rho/(10^3\text{kg}/\text{m}^3)$
resistance alloys		
manganin	86 % Cu, 12 % Mn, 2 % Ni	8.5
isabellin	70 % Cu, 10 % Mn, 20 % Ni	8.0
constantan	55 % Cu, 1 % Mn, 44 % Ni	8.8
nickelin	67 % Cu, 3 % Mn, 30 % Ni	8.8
contact materials		
silver bronze	1...7 % Ag, 0.2 % Cd, remainder Cu	8.9...9.2
hard silver	3...4 % Cu, remainder Ag	10.4
silver-cadmium	5...20 % Cd, remainder Ag	10.1

7.1/4   Magnetic materials

Material	Composition	$\rho/(10^3\text{kg}/\text{m}^3)$
trafoperm	steel with 2.5...4.5 % Si	7.57...7.7
permenorm	steel with 36...40 % Ni	· 8.15
mu metal	Ni-Fe alloy with $\approx$ 50 % Ni	8.6
AlNiCo 9/5	11...13 % Al, <5 % Co, <1 % Ti, 2...4 % Cu, 21...28 % Ni, remainder Fe	6.8
AlNiCo 18/9	6...8 % Al, 24...34 % Co, 5...8 % Ti, 3...6 % Cu, 13...19 % Ni, remainder Fe	7.2
SECo 112/110	rare earth-cobalt alloy	8.1

7.1.1.2 Non-metals

7.1/5 Ferrites

Material	Composition	$\rho/(10^3\mathrm{kg/m^3})$
SIFERRITE DB	15 % BaO, 85 % Fe_2O_3	5
SIFERRITE DS	16 % SrO, 84 % Fe_2O_3	4.4...4.6
MAGNETOFLEX 35	52 % Co, 13 % V, 35 % Fe	8.1
SIFERRITE U 60	iron oxides, Ba,Co	4.8
SIFERRITE K	iron oxides, Ni, Zn	4.2...4.4
SIFERRITE M	iron oxides, Ni, Mn, Zn	4.5...4.6
SIFERRITE N	iron oxides, Ni, Mn, Zn	4.7...4.8

7.1/6 Glass

Material	$\rho/(10^3\mathrm{kg/m^3})$	Material	$\rho/(10^3\mathrm{kg/m^3})$
aluminum silicate glass	2.53	bottle glass	2.6
barite crown glass	2.90	flint glass (light)	2.5...3.2
(bright; optical)		flint glass (heavy)	3.5...5.9
barite crown glass	3.56	glass fiber (textiles)	2.46
(dark; optical)		glass fiber	2.53
lead glass	2.89	(fiber glass)	
boron silicate glass	2.23	quartz glass	2.2
window glass	2.48		

7.1/7 Ceramics

Material	$\rho/(10^3\mathrm{kg/m^3})$	Material	$\rho/(10^3\mathrm{kg/m^3})$
porcelain	2.3...2.6	steatite	2.7
rutile	3.7	barium titanate	5
corund	3.8	Al_2O_3	3.9
ZrO_2	5.5	SiC	3.2
Si_3N_4	3.2	diamond (sintered)	3.5

7.1/8 Synthetics

Material	Composition	$\rho/(10^3\,\mathrm{kg/m^3})$
thermosets		
phenoplasts bacelite bacelite	phenole aldehyde phenole aldehyde with wood powder phenole aldehyde with asbestos	1.27 ... 1.35 1.35 ... 1.45 1.7 ... 2.1
amino plasts	aniline urea with wood powder melamine with wood powder melamine with asbestos	1.2 ... 1.25 1.45 ... 1.5 1.45 ... 1.55 1.7 ... 2.0
polyester resins	with glass texture	1.7 ... 1.9
thermoplasts		
cellulose derivates	cellulose A, soft cellulose acetate A, medium cellulose acetate A, hard cellulose acetobutyrate cellulose nitrate ethyl cellulose benzyl cellulose	1.32 1.33 1.34 1.20 1.38 1.14 1.22
ethylene derivates	high-pressure polyethylene low-pressure polyethylene polypropylene polystyrole styrole/butadien mix polymeres styrole/acryl nitril polyacryl acid ester polyvinyl chloride (PVC)	0.92 0.94 0.90 ... 0.91 1.05 1.06 1.08 1.18 1.38
polycarbonate		1.2
proteins	polyurethane synthetic horn polyamide (ultramide A) polyamide (rilsan) polyamide (vestamide)	1.21 1.35 1.15 1.04 1.02
fluorine carbonates (teflon)	polychlorine trifluorine ethylene polytetrafluorine ethylene	2.1 ... 2.2 2.2
silicones	silicone rubber silicone resin	1.2 ... 2.3 1.65
elastomers		
neoprene buna S perbunan	polychlorine butadiene butadiene/styrol mix polymeres butadiene/acrylnitril mix polymeres	1.24 1.2 1.2

7.1/9   Semiconductors

Material		$\rho/(10^3\,\mathrm{kg/m^3})$	Material		$\rho/(10^3\,\mathrm{kg/m^3})$
elemental	Ge	5.32	$A_{IV}B_{VI}$	PbS	7.50
semiconductors	Si	2.33		PbSe	8.15
	Se	4.79		PbTe	8.16
	Te	6.24	$A_{III}B_V$	BN	2.25
$A_{II}B_{IV}$	ZnS	4.09		BP	2.97
	ZnSe	5.26		AlP	2.38
	ZnTe	5.70		AlAs	3.79
	CdS	4.84		AlSb	4.26
	CdSe	5.74		GaP	4.13
	CdTe	5.86		GaAs	5.32
	HgSe	8.26		GaSb	5.60
	HgTe	8.20		InP	4.78
$A_{IV}B_{IV}$	SiC	3.22		InAs	5.66
				InSb	5.77

7.1/10   Building materials

Remark: One distinguishes between packed density ρ_R and true density ρ. The packed density is defined by $\rho_R = \mathrm{mass/total\ volume}$. The true density takes the pore volume into account and is defined as follows: $\rho = \mathrm{mass/volume\ of\ solid\ material}$. The table lists the packed density.

Material	$\rho/(10^3\,\mathrm{kg/m^3})$	Material	$\rho/(10^3\,\mathrm{kg/m^3})$
bricks		**natural stones**	
full bricks	1.0...2.2	granites, syenites	2.6...2.8
clinkers	1.6...2.2	basalt, diabas	2.9...3.9
air bricks	0.8...2.0	marble, diorite	2.6...2.8
gas concrete bricks	0.5...0.8	sandstone	2.6...2.7
fireclay bricks	0.8...2.1	pumice stone	0.2...1.3
earthenware	2.0...2.5	shale	2.6...2.7
wood 15 weight-% moist		plaster stone	2.0...2.2
		asbestos	2.5...2.6
spruce, fir	0.43...0.49	quartz	2.65
pine	0.48...0.56	limestone	2.4...2.8
larch	0.55...0.63	grey wacke	2.6...2.7
oak	0.63...0.72	gneiss	2.6...2.9
beech	0.66...0.76		

7.1/11   Bulk goods

Remark: The table gives the bulk density for **loose** accumulations. It is defined as mass per unit volume, including the heap pores and the pores in the individual grains.

Bulk goods	$\rho/(10^3\text{kg/m}^3)$	Bulk goods	$\rho/(10^3\text{kg/m}^3)$
cotton wool (air-dried)	0.080	sand	1.2...1.6
peas	0.700	snow (fresh)	0.08...0.19
hay	0.050	snow (old)	0.2...0.4
lime	0.500	cement	0.9...1.2
potatoes	0.670	gravel	1.8
maize	0.750	polystyrol	0.015

7.1.2   Fluids

The density is temperature-dependent because of expansion. The table below lists the densities for the temperature 293.15 K = 20 °C. The density of the same phase at any other temperature T can be calculated from the relation $\rho_T = \dfrac{\rho}{1 + \gamma(T - 293.15\text{ K})}$.

7.1/12   Fluids under normal conditions

Material	$\rho/(10^3\text{kg/m}^3)$	Material	$\rho/(10^3\text{kg/m}^3)$
acetone	0.792	sodium hydroxide (40 %)	1.43
alcohols		pentane	0.626
pentanol	0.814	**acids**	
ethyl alcohol	0.789	acetic acid	1.049
butyl alcohol	0.810	nitric acid (50 %)	1.31
glycerol	1.260	nitric acid (100 %)	1.502
isobutyl alcohol	0.801	hydrochloric acid (40 %)	1.195
isopropyl alcohol	0.785	sulphuric acid (50 %)	1.40
methyl alcohol	0.793	sulphuric acid (100 %)	1.834
propyl alcohol	0.804	**oils**	
bromine ethane	1.430	petroleum	0.73...0.94
ethyl acetate	0.901	heating oil	0.95...1.08
iodine ethane	1.933	lubricating oil	0.90...0.92
petrol (vehicle)	0.68...0.72	olive oil	0.91
petrol (airplane)	0.72	paraffin oil	0.87...0.88
benzene	0.921	cooking oil	0.87
trichlorine methane	0.879	silicon oil	0.76
chlorine benzene	1.066	turpentine	0.86
di-ethyl ether	0.714	transformer oil	0.87
fluorine benzene	1.024	vaseline oil	0.8
glycerol	1.26	toluene	0.867
kerosene	0.82	tetrachlorine methane	1.595
xylene	0.88	water	1.003
sea water	1.01...1.05	heavy water	1.1
milk	1.03		

7.1/13   Density of several metals in the liquid state

Material	$T/(°C)$	$\rho/(10^3 \text{kg/m}^3)$	Material	$T/(°C)$	$\rho/(10^3 \text{kg/m}^3)$
Al	660	2.380	Na	100	0.928
	900	2.315		400	0.854
	1100	2.261		700	0.780
Bi	300	10.03	Sb	409	6.834
	600	9.66		574	6.729
	962	9.20		704	6.640
Fe	1530	7.23	Pb	400	10.51
Au	1100	17.24		600	10.27
	1200	17.12		1000	9.81
	1300	17.00	Ag	960.5	9.30
K	64	0.82		1092	9.20
Hg	100	12.875		1300	9.00

7.1.3   Gases

The density of gases is strongly temperature-dependent. This dependence is nonlinear for a real gas.

The table lists the density ρ_0 for $T_0 = 273.15$ K (and normal pressure $p_0 = 1.0132 \cdot 10^5$ Pa). If the gases behave as ideal gases, one can calculate the density ρ for other values of pressure or temperature according to $\rho = \rho_0 \cdot (p/p_0) \cdot (T_0/T)$.

Gas	$\rho_0/(\text{kg/m}^3)$	Gas	$\rho_0/(\text{kg/m}^3)$
ethane	1.355	krypton[*]	3.68
ethylene	1.2611	coal gas	≈ 0.58
ammonia	0.7708	air, dry	1.2928
argon[*]	1.783	methane	0.7167
acetylene	1.1715	neon[*]	0.900
butane	2.70	ozone	2.14
isobutane	2.67	propane	2.01
chlorine	3.17	radon[*]	9.73
hydrogen chloride	1.639	oxygen[*]	1.429
frigene	5.51	sulphur carbonide	3.40
blast-furnace gas	1.28	sulphur dioxide	2.931
helium[*]	0.1785	hydrogene sulphide	1.54
carbon dioxide[*]	1.9768	nitrogen [*]	1.2504
carbon monoxide[*]	1.2502	hydrogen[*]	0.08988
xenon[*]	5.85		

[*] These gases behave like ideal gases in the temperature region $T < 1000$ K

7.2   Elastic properties

The following table presents the following quantities: the elasticity modulus E, the shear modulus G, and the transversal-extension number ν, the yield stress σ_f, the rupture stress σ_B and the Brinell hardness HB. All these quantities are strongly dependent on the prehistory of the material under consideration. They are therefore to be considered only as approximate values.

7.2/1   Elastic properties

Material	$E/(10^{10}\text{Pa})$	$G/(10^{10}\text{Pa})$	ν
Ag (annealed)	8.05	2.59	0.38 . . . 0.407
Al (annealed)	6.85	2.45	0.359 . . . 0.369
Au (cast)	8.06	2.91	0.422
Bi (cast)	3.19	1.2	0.33
Cd (cast)	4.99	1.92	0.3
Co (annealed)	19.6 . . . 20.6	—	0.34
Cu (rolled)	11.2	4.15	0.358 . . . 0.378
Cr	27.9	11.5	—
Fe (cast)	10 . . . 13	3.5 . . . 5.3	0.23 . . . 0.31
Fe (welded)	21	7.7	0.28
In	5.2	—	—
Ir	5.2	—	0.44
Mg (cast)	15.6	0.35	0.31
Mn	15.7	—	—
Mo (cast)	30900	11810	0.324
Nb (annealed)	15.6	3.8	0.38
Ni (annealed)	20.2	7.7	0.300
Os	55.5	—	—
Pb (cast)	1.62	0.562	0.446
Pd (cast)	11.3	5.11	0.393
Pt (annealed)	14.7	6.09	0.387
Rh (annealed)	27.5	—	0.32
Ru (annealed)	42.2	—	—
Sb	7.8	—	0.33
Sn (cast)	12.7	1.8	0.33
Ta (annealed)	18.3	6.9	0.39
Ti	11.6	4.4	—
U	16.6	8.3	0.21
V (annealed)	14.8	—	—
W (annealed)	34.2 . . . 40	8.8 . . . 21.5	—
Zn (cast)	4.06 . . . 5.86	1.64 . . . 4.78	0.33
Zr	7.4	—	—

7.2/2   Critical stresses[*]

Material	$\sigma_f/(10^7\,\mathrm{Pa})$	$\sigma_B/(10^7\,\mathrm{Pa})$	$HB/(10^7\,\mathrm{Pa})$
Ag (annealed)	—	13.5	20.6
Al (annealed)	5.63 ... 6.44	8.96 ... 10.75	18.4
Au (cast)	—	12.4	18.9
Bi	—	—	7
Ca	—	6.0	41.6
Cd	—	6.3	19.6
Co (annealed)	—	48.6	129.1
Cr (annealed)	—	8	68.8
Cu (rolled)	6.85	20 ... 25	52
Fe (cast)	—	1.84 ... 22.5	—
In	3.0	5.05	0.98
Ir	—	22	212
La	—	13	40
Mg (cast)	11.2	29.4	4.4
Mo (cast)	29.4	30.8	134
Nb (annealed)	—	32.2 ... 40.6	73.5
Ni (annealed)	20.5	34.5 ... 56.1	90 ... 120
Os	—	—	348.7
Pb (cast)	0.49 ... 0.98	1.47 ... 1.76	3.75 ... 4.18
Pd (cast)	—	18.2	31
Pt (annealed)	—	14.0	29.9
Rh (annealed)	—	55	54
Ru (annealed)	—	—	179.5
Sn (cast)	—	2.94 ... 3.92	29.2 ... 44.1
Ta (annealed)	—	31 ... 44.7	44.1 ... 122.4
Ti (annealed)	7.5	29.6	102.8
U	—	38.6	—
V (annealed)	52.5	56.5	74.2
W (annealed)	10.8	69.9 ... 80.9	196 ... 245
Zn (cast)	1.17	1.47 ... 2.4	4.8 ... 5.2
Zr	11.3	24.7	33.3

7.2/3   Wires[*]

Material	$E/$ (GPa)	$\sigma_B/$ (GPa)
steel	196	3.4
Be	290	1.52
W	400	2.75

7.2/4   Whiskers[*]

Material	$E/$ (GPa)	$\sigma_B/$ (GPa)
graphite	980	20.5
Al_2O_3	410	1.08 ... 17.6
BeO	410	19
SiC	450	3.05
B_4C	450	9.8

[*] Instead of the yield stress σ_f, one often quotes the conventional tensile strength R_p, instead of the fracture stress σ_B the yield strength R_m.

7.2/5   Steel

The elasticity modulus $E = (195 \ldots 206)$ GPa, the shear modulus $G = (79 \ldots 89)$ GPa and the Poisson number $\nu = 0.23 \ldots 0.31$ are close to each other for all types of steel. The various steels differ in the fracture stress σ_B (or tension resistance R_m), in the yield stress σ_f (or yield limit R_p) and the hardness (for example, the Brinell hardness HB).

Steel sort	Composition (example)	$\sigma_B/(10^8\text{Pa})$	$\sigma_f/(10^8\text{Pa})$	$HB/(10^8\text{Pa})$
mass steel	≈ 0.25 % C	≈ 4.7	2.5	≈ 13
spring steel	≈ 0.47% C, ≈ 1.65 % Si, ≈ 0.65 % Mn	14	12.2	41
rail steel	0.55 % C, 0.2 % Si, 0.8 % Mn	≈ 7.5	≈ 4	20
piano-string wire	0.9 % C, 0.15 % Si, 0.4 % Mn	≤ 36	—	—
silver steel	0.9 % C, 0.33 % Si, 0.4 % Mn, 0.1 % W	9	4.5	25
file steel	1.3 % C, 0.25 % Si, 0.35 % Mn	6	—	17
V2A-steel	< 0.1 % C, 0.4 % Si, 0.3 % Mn, 18 % Cr, 8 % Ni	≈ 6.5	> 2.7	≈ 16.5
transformer sheets	0.07 % C, 3.7 % Si, 0.2 % Mn	≤ 12	—	—
cast steel	0.1 % C, 0.3 % Si, 0.4 % Mn	3.8	1.8	11
hard metal	6 % C, 88 % W, 6 % Co	—	—	160

7.2/6 Ceramic materials

σ_{bB} is the fracture stress for a bending load, E is the elastic modulus.

Material	Chemical formula	$\sigma_{bB}/(\text{MPa})$	$E/(\text{GPa})$
aluminum oxide	Al_2O_3	400	400
zirconium oxide	ZrO_2	600	240
silicon carbide	SiC	440	440
silicon nitride	Si_3N_4	700	210
diamond (sintered)	—	300	900

7.2/7 Synthetic materials

σ_B is the fracture stress (or tension resistance R_m).
σ_{dB} is the fracture stress for a pressure load and σ_{bB} the corresponding stress for a bending load. δ denotes the fracture extension in percent.

Material	E/GPa	σ_B/MPa	σ_{dB}/MPa	σ_{bB}/MPa	HB/GPa	$\delta/\%$
polyamides glass-fiber strengthened	$1.5\dots3.2$ $10\dots18$	$60\dots90$ $120\dots220$	$93\dots98$ 108	$93\dots98$ $122\dots147$	$147\dots176$ $274\dots294$	$6\dots12$ $4\dots6$
polycarbonates glass-fiber strengthened	$2\dots3.5$ $3.5\dots9.5$	$55\dots75$ $70\dots140$	$78\dots88$ 130	78 $171\dots219$	$147\dots157$ —	$5\dots7$ $2\dots5$
polystyrol glass-fiber strengthened	$3\dots3.6$ $5\dots10$	$45\dots65$ $96\dots117$	98 $103\dots130$	98 —	$137\dots147$ 3	$2\dots4$
polyethylene HD polyethylene LD	$0.4\dots1.5$ $0.15\dots0.6$	$20\dots35$ $8\dots20$	24.5 12.3	21.6 $11.8\dots16.7$	$44\dots57$ —	$12\dots20$ $8\dots11$
polypropylene glass-fiber strengthened	$0.65\dots1.4$ $2.5\dots6$	$18\dots38$ $40\dots75$	59 48	78 69	61.7 —	$10\dots20$ $7\dots70$
polyvinyl chloride (hard)	$2.9\dots3.6$	$50\dots80$	—	—	—	$3\dots4$
polyvinyl chloride (soft)	$0.45\dots0.6$	$15\dots30$	—	—	—	$50\dots300$
polytetrafluorene ethylene	$0.45\dots0.75$	$9\dots12$	—	—	—	$250\dots500$

7.2/8 Fiber materials

Material	$\sigma_B/(\text{MPa})$	$\delta/\%$	Material	$\sigma_B/(\text{MPa})$	$\delta/\%$
acetate silk	$176\dots215$	25	glass	2100	—
bamboo	345	—	silk	410	—
viscose	$265\dots440$	$15\dots24$	wool	$156\dots172$	—
nylon	$490\dots635$	$15\dots35$	SiO_2	$1380\dots1480$	—

7.3   Dynamical properties

7.3.1   Coefficients of friction

Sliding friction and static friction are strongly dependent on the adhesive properties of the surface of individual materials. Therefore, the data on coefficients of friction fluctuate within certain boundaries. The data quoted in the subsequent tables are to be understood as approximate guide values. Many values are mean values. For more accurate purposes, the coefficient of friction must be determined experimentally in each case.

7.3/1   Rolling friction

on		$f/(\text{cm})$
Material	Material	
rubber	asphalt	0.10
rubber	concrete	0.15
wood	wood	$0.5 \ldots 0.8$
steel	steel (hardened)	$0.005 \ldots 0.01$
steel	steel (soft)	0.05

7.3/2   Coefficient of sliding friction

on		Coefficient of sliding friction μ		
			lubricated with	
Material	Material	dry	H_2O	grease
bronze	bronze	0.20	0.10	0.06
	grey cast	0.18		0.08
	steel	0.18		0.07
oak	oak=*	$0.20 \ldots 0.40$	0.10	$0.05 \ldots 0.15$
	oak$\perp$*	$0.15 \ldots 0.35$	0.08	$0.04 \ldots 0.12$
grey-cast	grey cast		0.31	0.1
	copper	0.25		
	wood	0.35	0.25	
rubber	asphalt	0.5	0.3	0.2
	concrete	0.6	0.5	0.3
	grey cast	$0.4 \ldots 0.5$		
leather	oak	0.4		
belt	metal	0.28	0.25	0.12
steel	oak	$0.2 \ldots 0.5$	0.26	$0.02 \ldots 0.1$
	ice		0.014	
	steel	$0.1 \ldots 0.3$		$0.02 \ldots 0.08$
	brake lining	$0.5 \ldots 0.6$		
	polyethylene	$0.4 \ldots 0.5$		
	teflon	$0.03 \ldots 0.05$		
	polyamide	$0.3 \ldots 0.5$		0.1
	hostaflon	$0.35 \ldots 0.45$		

(continued)

7.3/2   Coefficient of sliding friction (*continued*)

on Material	Material	Coefficient of sliding friction μ dry	lubricated with H_2O	grease
polyethylene	polyethylene	0.5...0.7		
teflon	teflon	0.035...0.055		
polyamide	polyamide	0.4...0.5		

* = motion along grain, ⊥ motion perpendicular to grain.

7.3/3   Coefficient of static friction

on Material	Material	Static friction μ_0 dry	lubricated with H_2O	grease
bronze	bronze			0.11
	steel	0.19		0.10
oak	oak=*	0.40...0.60		0.18
	oak⊥*	0.50		
grey cast	grey cast			0.16
hemp rope	wood	0.5		
leather	oak	0.5		
belt	metal	0.6	0.25	0.62
	oak	0.5...0.6		0.11
steel	ice		0.03	
	steel	0.15...0.3		0.1

* = motion along grain, ⊥ motion perpendicular to grain.

7.3.2   Compressibility

The compressibility of a material is expressed by its compression modulus

$$\kappa = \left(\frac{1}{V}\right)\left(\frac{\Delta V}{\Delta p}\right).$$

ΔV is the change of volume under a change of pressure Δp. The compression modulus is dependent both on the temperature and on the pressure. For gases:

$$\kappa = \frac{A}{V(p + p_T)}.$$

A is a function increasing with temperature, p the external pressure, and p_T the Van der Waals pressure at temperature T.

7.3.2.1 Gases

The following tables give the compressibility of several gases as deviations from the behavior of an ideal gas, expressed by the quantity $\kappa + \dfrac{1}{p}$.

7.3/4 Helium

pressure/(MPa)	$\left(\dfrac{1}{V}\dfrac{\Delta V}{\Delta p} + \dfrac{1}{p}\right)/(10^3\ \mathrm{Pa^{-1}})$							
	$-253\ °C$	$-208\ °C$	$-183\ °C$	$-150\ °C$	$-100\ °C$	$-50\ °C$	$0\ °C$	$50\ °C$
0 – 0.1	0	10.34	8.97	6.57	4.67	3.62	2.47	2.1
0.1 – 1	−0.74	8.88	7.09	5.56	4.13	3.21	2.57	2.17
1 – 5	22.2	9.43	7.12	5.56	4.1	3.19	2.55	2.16
5 – 10	29.6	9.29	7.21	5.51	4.07	3.14	2.49	2.12

7.3/5 Nitrogen

pressure/(MPa)	$\left(\dfrac{1}{V}\dfrac{\Delta V}{\Delta p} + \dfrac{1}{p}\right)/(10^3\ \mathrm{Pa^{-1}})$							
	$-130\ °C$	$-100\ °C$	$-50\ °C$	$0\ °C$	$50\ °C$	$100\ °C$	$200\ °C$	$400\ °C$
0 – 0.1	−33.1	−17.9	−6.65	−2.47	0	1.08	1.71	1.80
0.1 – 1	−36.4	−18.5	−6.96	−2.14	0	1.12	1.96	2.11
1 – 2	−43	−18.9	−6.66	−1.84	0.21	1.22	2.04	2.11
2 – 4	−60.7	−20.7	−6.09	−2.1	0.5	1.4	2.08	2.12
4 – 6	−83.1	−20.7	−5.17	0	0.872	1.62	1.56	2.15
6 – 8	—	−17.4	−3.93	−0.05	1.22	1.84	2.84	2.17
8 – 10	—	−8.67	−2.29	0.7	1.58	2.07	2.33	2.17
10 – 20	—	—	2.87	2.41	2.59	2.29	2.69	2.29
20 – 40	—	—	6.73	4.36	3.83	3.15	2.85	2.17
40 – 60	—	—	5.94	5.15	3.95	3.41	2.72	2.03
60 – 80	—	—	4.7	4.7	3.53	3.12	2.54	1.93
80 – 100	—	—	3.78	3.43	3.07	2.78	2.34	1.81

7.3/6 Hydrogen

pressure/(MPa)	$\left(\dfrac{1}{V}\dfrac{\Delta V}{\Delta p} + \dfrac{1}{p}\right)/(10^3\ \mathrm{Pa^{-1}})$							
	$-208\ °C$	$-183\ °C$	$-150\ °C$	$-50\ °C$	$0\ °C$	$50\ °C$	$100\ °C$	$200\ °C$
0 – 0.1	−33.2	−4.49	1.09	3.11	3.63	2.96	2.92	2.53
0.1 – 1	−15	−3	1.7	3.28	3.28	3.06	2.82	2.48
1 – 2	−15.2	−1.96	2.07	7.14	3.29	3.08	2.81	2.51
2 – 4	−11.7	−0.28	2.76	1.63	3.38	3.10	2.77	2.47
4 – 6	−0.93	1.96	3.52	3.72	3.45	3.09	2.74	2.45
6 – 8	6.87	4.24	4.31	3.96	3.51	3.12	2.71	2.46
8 – 10	—	6.41	10.2	4.51	3.58	3.1	2.7	2.45

7.3/7   Methane

pressure/(MPa)	$\left(\dfrac{1}{V}\dfrac{\Delta V}{\Delta p}+\dfrac{1}{p}\right)/(10^3\ \mathrm{Pa}^{-1})$						
	−70 °C	−50 °C	−25 °C	0 °C	25 °C	50 °C	100 °C
0 – 0.1	−29.9	−23.6	−16.8	−11.8	−9.03	−5.83	−2.88
0.1 – 2	−35.2	−25.1	−17.3	−12.2	−8.75	−6.32	−3.36
2 – 4	−51.8	−30.1	−18.7	−12.5	−8.56	−6.05	−2.94
4 – 6.1	−107	−40.8	−20.6	−12.8	−8.36	−5.75	−2.60
6.1 – 8.1	−67.4	−46.2	−21.0	−12.3	−7.88	−4.97	−2.06
8.1 – 10.1	23.0	−29.0	−113	−10.8	−6.54	−4.15	−1.51
10.1 – 12.1	30.5	0.60	84.0	−8.32	−5.36	−3.27	−2.09
12.1 – 14.1	26.4	11.7	−3.38	−4.93	−3.27	−2.13	1.94
14.1 – 16.2	25.1	16.6	3.80	−0.99	−1.38	−0.95	−0.19
16.2 – 18.2	22.2	−17.2	7.83	1.99	0.27	0.24	0.47
18.2 – 20.2	20.4	50.6	9.55	4.91	2.47	1.66	1.33
20.2 – 30.4	16.0	14.1	10.8	7.66	5.32	3.91	2.72
30.4 – 40.5	11.7	10.8	9.51	8.15	6.59	5.45	3.92
40.5 – 50.6	9.18	8.64	7.88	6.99	6.27	5.54	4.32
50.6 – 60.8	7.48	7.19	6.72	6.20	5.70	5.11	4.15
60.8 – 81.1	5.93	5.74	5.44	3.22	4.77	4.49	3.86
81.1 – 101.3	4.63	4.47	4.29	8.9	4.05	3.73	3.35

7.3/8   Nitrogen monoxide

pressure/(MPa)	$\left(\dfrac{1}{V}\dfrac{\Delta V}{\Delta p}+\dfrac{1}{p}\right)/(10^3\ \mathrm{Pa}^{-1})$							
	−70 °C	−50 °C	−25 °C	0 °C	25 °C	50 °C	100 °C	150 °C
0 – 0.1	−6.64	−6.04	−5.43	−3.45	0	0	0	0
0.1 – 2.5	−11.4	−6.66	−3.19	−2.27	−0.94	−0.35	1.2	2.64
2.5 – 5	−11.3	−7.31	−3.79	−2.01	0.17	1.29	1.5	
5 – 7.5	−9.75	−6.05	−3.18	−1.21	0	0.83	1.56	1.99
7.5 – 10	−5.38	−3.5	−0.92	−0.20	0.18	1.16	1.55	2.09
10 – 15	0.64	0.54	0.80	1.51	2.16	1.96	2.29	2.35
15 – 20	6.77	4.75	4.02	2.76	2.64	2.95	2.71	2.65
20 – 30	9	6.67	5.53	4.54	3.99	3.63	3.26	2.99
30 – 40	8.34	7.82	6.02	5.41	4.65	4.19	3.49	3
40 – 61	6.69	6.17	5.53	5.03	4.45	4.09	3.51	3.11
61 – 81	5.09	4.85	4.51	4.18	4.98	3.63	3.16	2.86
81 – 101	4.08	1.15	3.71	3.51	2.32	3.09	2.82	2.58

7.3/9   Carbon dioxide

pressure/(MPa)	$\left(\dfrac{1}{V}\dfrac{\Delta V}{\Delta p}+\dfrac{1}{p}\right)/(10^3\ \mathrm{Pa^{-1}})$							
	0 °C	10 °C	20 °C	30 °C	40 °C	50 °C	60 °C	80 °C
0 – 5	−160	−158	−44.9	−35.7	−30.0	−25.3	−21.8	−16.9
5 – 7.5	73.4	68.2	−230	−221	−61.8	−41	−30.9	−20.5
7.5 – 10	54.5	52.5	47.3	30	−132	−47.3	−24.6	
10 – 15	36.9	36.3	34.5	29.9	19.6	−15.6	−30.3	−24.3
15 – 20	26.1	25.6	24.6	23.6	21.3	17.4	11.1	−3.09
20 – 30	18.3	17.8	17.4	17	16	14.8	13.2	8.85
30 – 40	12.9	12.7	12.4	12	11.7	11.3	10.8	9.38
40 – 50	15.1	9.8	9.64	9.43	9.09	8.9	8.66	7.97
50 – 60	2.85	7.95	7.84	7.79	7.68	7.42	7.16	6.79
60 – 71	6.82	6.81	6.65	6.57	6.46	6.34	6.22	5.9
71 – 81	5.85	5.84	5.83	5.73	5.64	5.52	5.43	5.15
81 – 91	5.2	5.13	5.02	5.93	4.88	4.82	4.75	4.58
91 – 101	4.58	4.47	4.42	4.25	4.25	4.23	4.12	4.01

7.3.2.2   Liquids and solids

7.3/10   Temperature dependence of compressibility

	$\kappa\ /\mathrm{MPa^{-1}}$						
$T/$ °C	acetone	tetrachlorine methane	benzene	trichlorine methane	ethyl alcohol	methyl alcohol	water
0	82	89.8	80.9	86.6	98.7	107	50
10	110	97	87	91.8	104	114	47.8
20	125	103.5	94.5	100	111	121.5	45.8
30	133.4	112.8	102	109	118.5	129.5	44.6
40	150	122	110	118.5	126.5	138.5	44.1
50	160	132.6	118.5	129.5	136	147.6	44

7.3/11   Compressibility of fluids under normal conditions

Material	$\kappa/(\mathrm{MPa^{-1}})$
olive oil	63
paraffin oil	62.67
mercury	4
kerosene	69.6

7.3/12   Compressibility of solids at 0 °C

Material	$\kappa/(\mathrm{MPa^{-1}})$	Material	$\kappa/(\mathrm{MPa^{-1}})$
Al	1.38	Si	0.324
Au	0.617	Mo	0.47
Cd	2.13	Cu	0.74
Fe	0.597	Pl	0.385

7.3.3   Viscosity

7.3/13   Viscosity of fluids at normal pressure and $20\,°C$

Material	$\eta/(\mu Pa \cdot s)$	Material	$\eta/(\mu Pa \cdot s)$
acetone	330	terpentine	1490
ethyl alcohol	1192	o-xylene	807
methyl alcohol	591	m-xylene	615
benzene	649	p-xylene	643
carbon disulfide	367	mercury	1550
ether	234	kerosene	1460
glycerol	$141.2 \cdot 10^4$	toluene	585
nitric acid	1770	pitch	$3 \cdot 10^{13}$
sulphuric acid	$22 \cdot 10^3$	heavy water	1260

7.3/14   Viscosity of cryogenic fluids at saturation pressure

hydrogen		nitrogen		oxygen		argon	
T/K	$\eta/(\mu Pa \cdot s)$	T/K	$\eta/(\mu Pa \cdot s)$	T/K	$\eta/(\mu Pa \cdot s)$	T/K	$\eta/(\mu Pa \cdot s)$
15	217	60		60	5800	85	2720
16	197	70	2200	70	3580	90	2300
17	178	80	1410	80	2500	95	1970
18	161	90	1040	90	1890	100	1970
19	147	100	850	100	1520	105	1540
20	134	110	760	110	1280	110	1410

7.3/15   Viscosity of water dissolutions of glycerol in $(mPa \cdot s)$ at normal pressure

glycerol	temperature $/°C$					
(mass %)	0	20	40	60	80	100
20	2.44	1.76	1.07	0.731	0.635	...
40	8.25	3.72	2.07	1.3	0.918	0.668
60	29.9	10.8	5.08	2.85	1.84	1.28
80	255	60.1	20.8	9.42	5.13	3.18
90	1,310	219	60.0	22.5	11.0	6.00
95	3,690	523	121	39.9	17.5	9.08
100	12,070	1412	284	81.3	31.9	14.8

7.3/16   Viscosity of water at various temperatures

$T/°C$	$\eta/(\mu Pa \cdot s)$	$T/°C$	$\eta/(\mu Pa \cdot s)$
0	1793	60	469
10	1309	70	406
20	1006	80	357
30	800	90	315
40	657	100	284
50	550		

7.3/17 Viscosity vs temperature at normal pressure

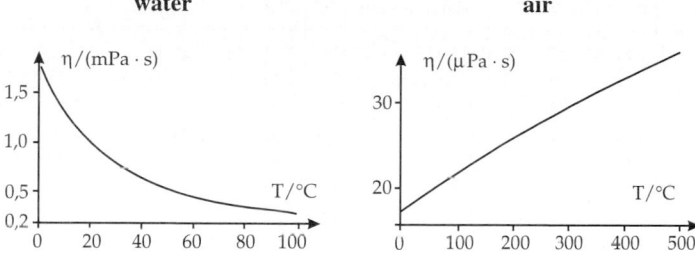

water   **air**

7.3/18 Viscosity of gases at normal pressure and 20 °C

Material	$\eta/(\mathrm{Pa} \cdot \mathrm{s})$	Material	$\eta/(\mu\mathrm{Pa} \cdot \mathrm{s})$
air	18.1	chlorine	14.7
ammonia	10.8	methane	12
carbon monoxide	18.4	nitrogen monoxide	18.6
carbon dioxide	16	nitrogen	18.4
hydrogen	9.5	oxygen	20.9
hydrogen sulphide	13	sulphur dioxyde	13.8

7.3/19 Viscosity of gases at normal pressure and $T_0 = 273.15$ K

Material	$\eta/(\mu\mathrm{Pa} \cdot \mathrm{s})$	Material	$\eta/(\mu\mathrm{Pa} \cdot \mathrm{s})$	Material	$\eta/(\mu\mathrm{Pa} \cdot \mathrm{s})$	Material	$\eta/(\mu\mathrm{Pa} \cdot \mathrm{s})$
N_2	16.65	C_5H_{10}	6.65	CO_2	13.67	C_3H_6	7.84
NO	18.00	C_4H_{10}	6.89	C_2H_6	12.23	C_3H_7OH	7.15
NH	9.35	C_5H_{12}	6.38	C_2H_4	8.55	H_2S	11.79
Ar	20.85	C_3H_7OH	7.20	$C_3H_6O_2$	6.85	CS_2	9.20
H_2	8.40	C_3H_4	8.08	C_2H_2	9.55	SH_4	10.76
H_2O	8.83	C_5H_{10}	6.65	C_6H_6	6.93	C_5H_{10}	6.39
(vapor)		CH_3Br	12.32	Br_2	13.90	CCl_4	9.06
air	17.08	CH_2Cl_2	9.16	C_3H_{10}	6.82	C_2N_2	9.33
He	18.60	CH_3OH	8.70	C_4H_{10}	6.90	HCN	6.72
O_2	19.10	CH_3Cl	10.84	HBr	17.10	C_6H_{12}	6.53
Kr	23.30	NOCl	9.89	HI	17.00	C_3H_6	8.08
Xe	21.10	CO	11.32	HCl	13.20	Cl_2	12.45
CH_4	10.28	C_5H_{10}	6.23	PH_3	10.72	$CHCl_3$	9.33
Ne	29.75	C_3H_8	7.50	C_6H_{14}	6.00	$C_4H_8O_2$	9.60
SO_2	11.58	$C_5H_{10}O_2$	7.40	$(CH_3)_2O$	8.70	C_2H_5OH	7.75
CO	16.62			$(CH_5)_2O$	6.80	C_2H_5Cl	9.11

7.3/20 Temperature correction factor

For gases, the dependence of viscosity on the absolute temperature can be represented by the formula

$$\eta = \eta_{T_0} \sqrt{\frac{T}{T_0}} \, \frac{1 + \frac{C}{T_0}}{1 + \frac{C}{T}}.$$

The temperature correction factor C is only weakly temperature-dependent.

7.3/20   Temperature correction factor (*continued*)

Material	$C/°C$	$\vartheta/°C$	Material	$C/°C$	$\vartheta/°C$	Material	$C/°C$	$\vartheta/°C$
N_2	103.9	25 – 280	$(C_2H_5)_2O$	404	122 – 309	C_3H_6	312.6	20 – 120
NO	128	20 – 250	C_5H_{10}	368	20 – 120	C_3H_7OH	515.6	122 – 273
NH	503	20 – 300	C_4H_{10}	368	20 – 120	SO_2	306	300 – 825
Ar	142	20 – 827	C_3H_7OH	459.9	119 – 308	H_2S	331	0 – 100
C_2H_2	198.2	20 – 120	I_2	568	106 – 523	CS_2	499.5	114 – 310
$C_3H_6O_2$	541.5	119 – 306	HI	390	0 – 100	C_4H_4S	467	20 – 245
C_6H_6	447.5	130 – 313	O_2	126.6	20 – 280	PH_3	290	0 – 100
Br_2	533	190 – 600		125	15 – 630	CO_2	254	25 – 280
HBr	375	0 – 100	Kr	188	0 – 100		213	300 – 824
C_3H_{10}	377.4	20 – 120	Xe	252	0 – 100	CO	101.2	22 – 277
air	106.8	20 – 280	CH_4	162	20 – 500	CCl_4	335	128 – 315
	111	16 – 825	CH_3Br	276	20 – 120		365.4	128 – 315
H_2	73	20 – 200	CH_3OH	486.9	111 – 312	Cl_2	351	20 – 250
	86	100 – 200	CH_2Cl_2	425	22 – 309	HCl	360	0 – 250
	105	200 – 250	CH_3Cl	441	20 – 308	$CHCl_3$	373	121 – 308
	234	713 – 822	H_3AS	300	0 – 100	C_2H_2	330	0 – 100
water vapor	673	100 – 350	Ne	61	20 – 100	HCN	901	20 – 330
He	83	100 – 200	C_5H_{10}	382.8	122 – 306	C_3H_6	372	20 – 120
	95	200 – 250	C_3H_8	278	20 – 250	C_6H_{12}	350.9	122 – 306
	173	682 – 815		290	25 – 280	C_2H_6	252	20 – 250
			C_2H_4	225	20 – 250	$C_4H_8O_2$	504	128 – 314

7.3.4   Flow resistance

7.3/21   Resistance coefficient

Body shape		c_W	Body shape		c_W
		1.1		$R : r = 2$	1.22
	$a : b = 1$ $a : b = 4$ $a : b = 10$ $a : b = 18$	1.1 1.19 1.29 1.4		$l : d = 2$ $l : d = 5$ $l : d = 10$ $l : d = 20$	0.2 0.06 0.083 0.094
	without bottom (parachute)	1.33		with bottom	1.17
	without bottom	0.34		with bottom	0.4
	$Re < 2 \cdot 10^5$ $Re = 10^6$	0.45 0.13	with bottom	$\alpha = 60°$ $\alpha = 30°$	0.51 0.34

(*continued*)

7.3/21   Resistance coefficient (*continued*)

Body shape		c_W	Body shape		c_W
	$\mathrm{Re} > 10^5$ $l : d = 1.8$ $\mathrm{Re} < 4, 5 \cdot 10^5$ $l : d = 0.75$ $\mathrm{Re} > 5.5 \cdot 10^5$ $l : d = 0.45$	0.1 0.6 0.2		$\mathrm{Re} \approx 8 \cdot 10^4$ $h : d = 1$ $l : d = 2$ $l : d = 5$ $l : d = 10$	0.63 0.68 0.74 0.82
	$\mathrm{Re} \approx 5 \cdot 10^5$ $l : d = 30$	0.78		$\mathrm{Re} \approx 10^6$ $l : d = 5$ $l : d = 8$ $l : d = 18$	0.08 0.1 0.2
		0.4 ⋮ 0.55			0.3 ⋮ 0.4
		0.23			0.6 ⋮ 0.7

7.3.5   Surface tension

7.3/22   Surface tension of fluids and dissolutions

Fluid	$\sigma/$ $(10^{-3}\mathrm{Nm}^{-1})$	Fluid	$\sigma/$ $(10^{-3}\mathrm{Nm}^{-1})$
acetone	23.7	olive oil	33
ethyl alcohol	22.3	paraffin oil	26
methyl alcohol	22.6	terpentine	27
aniline	43	water	
benzol	28.9		
chloroform	27.2	water at $5\,^\circ\mathrm{C}$	74.92
glycerine	64	water at $10\,^\circ\mathrm{C}$	74.22
mercury	475	water at $20\,^\circ\mathrm{C}$	72.75
		water at $30\,^\circ\mathrm{C}$	71.18
Dissolutions			
sulphuric acid (conc.)	55	nitric acid	41
Per 1 weight-% the following value must be added to that of pure water			
calcium chloride	0.29	KOH	0.32
copper sulphate	0.11	sodium chloride	0.28
potassium chloride	0.19	NaOH	0.5

Part II
Vibrations and Waves

Vibration, a change of the state of a system periodic in time (**oscillator**) and occurring when
- a system is displaced from its mechanical, electrical or thermal equilibrium by an external perturbation, and
- forces arise that drive the system back towards equilibrium.

Vibrations occur in almost all physical systems.

Wave (see p. 287), change of the state of a system periodic in space and time and occurring when
- a system consists of subsystems that all are oscillatory,
- the subsystems may interact with each other, hence, energy may be transferred from one subsystem to another neighboring subsystem, and
- at least one of the subsystems is driven from its mechanical, electrical or thermal equilibrium by an external perturbation.

Energy is transfered from one subsystem to other subsystems without any mass transport being involved.

■ Sound consists of density waves which occur in media. Light consists of electromagnetic waves within a certain frequency interval.

8
Vibrations

1. Periodic processes,

processes or configurations that repeat regularly. If a process repeats continously in fixed time intervals, then it is called **periodic in time**. If a spatial configuration repeats in fixed space intervals, then this arrangement is called **spatially periodic**.

2. Period,

period, T, the smallest time interval for the repetition of phenomena that are periodic in time:

$$u(t + T) = u(t).$$

The SI unit of period is the second s. The period is determined by the system parameters.

Frequency, f, the number of repetitions per second of a phenomenon that is periodic in time, $f = 1/T$.

Hertz, Hz, SI unit of frequency. 1 Hz = 1/s.

■   The frequency 1 Hz means that a process repeats itself once per second. In North America the line voltage has a frequency of 60 Hz; it changes its direction 120 times per second.

3. Oscillator,

oscillator, a system in which vibrations may occur.

■   A pendulum is a mechanical oscillator, e.g., a mass hanging on a string. An oscillator circuit is an example of an electrical oscillator.

Equilibrium position, the state of a system capable of oscillation before it experiences an external perturbation, i.e., the mechanical, electrical or thermal **equilibrium state**.

4. Harmonic vibration,

a periodic process in which the change of state is described by a sine or cosine function (**Fig. 8.1**). These functions differ by a phase shift of $\pi/2$ radians.

harmonic vibration			
	Symbol	Unit	Quantity
$u(t) = A\cos(2\pi ft + \phi)$	u		state of system
	A		amplitude
$= A\sin\left(2\pi ft + \phi + \dfrac{\pi}{2}\right)$	f	Hz	frequency
	t	s	time
	ϕ	rad	phase shift

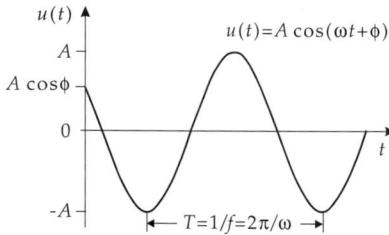

Figure 8.1: Harmonic vibration.

Here $u(t)$ describes the state of the system at time t. The physical meaning of u depends on the system under consideration (linear or angular coordinate, stress, electric or magnetic field strength, etc.).

■   For a mass on a spring, $u(t)$ is the displacement, and for an oscillator circuit u may be the electric voltage, current or charge. The dimension of u depends on the system considered.

5. Phase and amplitude

Phase, phase angle, argument of the sine or cosine function, $2\pi ft + \phi$, specifies the instantaneous state of vibration.

Phase constant, initial phase, ϕ, value of the phase at $t = 0$, describes the state of the system at the initial instant.

Amplitude A, maximum value of the function $u(t)$.

6. Frequency, angular frequency and period

angular frequency = 2π · frequency			$\mathbf{T^{-1}}$
	Symbol	Unit	Quantity
$\omega = 2\pi f$	f	Hz	frequency
	ω	rad/s	angular frequency

The harmonic vibration has the form

$$u(t) = A\cos(\omega t + \phi).$$

The relation between period, frequency and angular frequency is:

period = reciprocal value of frequency			$\mathbf{T}$
	Symbol	Unit	Quantity
$T = 1/f = 2\pi/\omega$	T	s	period
	f	Hz	frequency
	ω	rad/s	angular frequency

➤ There are always frictional forces in nature. Therefore, bodies come to rest if the energy lost by friction is not compensated by a supply of energy from outside. For this reason, no process is exactly described by a harmonic vibration that continues to infinity.

➤ The sine function describes a vibration that also occurred throughout the past ($t \rightarrow -\infty$). In nature an oscillation begins only if an oscillatory system is supplied with energy, e.g., when an impulse force is applied to a pendulum. The state of the system at which the vibration starts is the same state to which the system returns when the energy initially supplied has been dissipated by friction.

Natural frequency, frequency depending only on the system parameters; an oscillator vibrates with this frequency in the absence of external forces (free vibration).

7. Types of vibrations

Vibrations are subdivided into:

- **Free vibrations**, the vibration is excited once and proceeds without further external influence. The frequency is constant and is uniquely determined by the system parameters.
- **Damped vibrations**, friction is present. The oscillator continually loses energy.
- **Forced vibrations**, the oscillator is driven by an external periodic force. If the oscillator vibrates at the frequency of the external force, the vibrating system is called a **resonator**.

Combination of the last two cases: **forced damped vibrations**. A periodic external force drives a damped oscillator; the vibration does not decay since the external excitation permanently supplies energy to the oscillator.

Vibration	Free	Forced
Undamped	no friction no external excitation energy constant	no friction external excitation energy supply resonance catastrophe
Damped	friction no external excitation energy loss	friction external excitation energy supply and energy loss resonance

■ An antenna transmitting radio waves is an example of a forced electromagnetic vibration in which energy is lost due to release of electromagnetic energy.

8.1 Free undamped vibrations

Free undamped vibration, vibration without external excitation and without friction, is exactly described by a harmonic time dependence. Amplitude and frequency are time-independent.

8.1.1 Mass on a spring

1. Definition of mass on a spring

Spring and mass system, **mass on a spring**, a body fixed to a cylindrical helical spring.
  Oscillator, name of the body fixed to the spring.
■   Cart attached to a spring fixed at one end, moving without friction on a horizontal plane (**Fig. 8.2**).

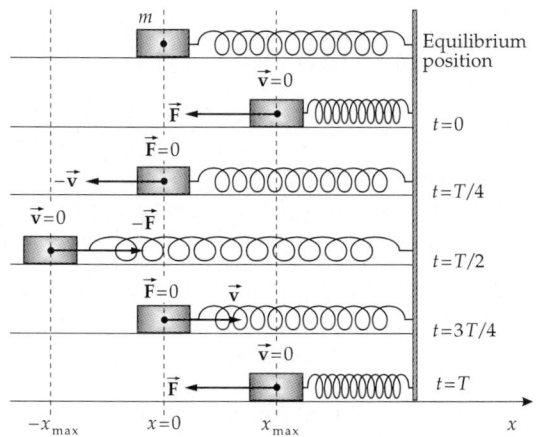

Figure 8.2: Mass on a spring. Restoring force $\vec{F}$ and velocity $\vec{v}$.

  Equilibrium position: top figure, the spring is relaxed.
  Perturbation: external force compresses (or stretches) the spring by the length x; the system leaves the mechanical equilibrium position.
  Displacement, x, specifies how much the system is shifted out of equilibrium, i.e., how much the spring is compressed or stretched.
➤   The description of the system is simplest if the origin of the coordinate frame is the rest position of the mass. In the following, the coordinate frame is always chosen in this way.

2. Restoring force

Restoring force, force driving the system back towards the equilibrium position.
  Linear force law, **Hooke's law**, the restoring force is proportional to the displacement and points towards the rest position, the essential condition for **harmonic** vibrations:

restoring force = −spring constant · displacement			MLT^{-2}
	Symbol	Unit	Quantity
$F = -kx$	F	kg m/s^2	restoring force
	k	kg/s^2	spring constant
	x	m	displacement

➤   The spring force is proportional to the displacement only within certain limits. Hence, the following equations are valid only within these limits.

If the cart is released at a point displaced from the equilibrium position, it is accelerated by the restoring force. Due to its inertia, it rolls beyond the equilibrium point and compresses or stretches the spring. Again the spring force acts on the cart, but now in the opposite direction.

3. Equation of motion of the mass on a spring

Equation of motion, follows with the formulation for the restoring force and Newton's equation, $F = ma = m\ddot{x}$,

equation of motion and solution of the spring-mass system			
$\ddot{x} = -\dfrac{k}{m}x$ $x(t) = A\cos(\omega t + \phi)$ $\dot{x}(t) = -A\omega\sin(\omega t + \phi)$ $\ddot{x}(t) = -A\omega^2\cos(\omega t + \phi)$ $\omega = \sqrt{\dfrac{k}{m}}, \quad f = \dfrac{1}{2\pi}\sqrt{\dfrac{k}{m}}$ $T = 2\pi\sqrt{\dfrac{m}{k}}$	Symbol	Unit	Quantity
	x	m	displacement
	$\dot{x}$	m/s	velocity
	$\ddot{x}$	m/s^2	acceleration
	k	kg/s^2	spring constant
	m	kg	mass of oscillator
	t	s	time
	A	m	amplitude
	ω	rad/s	angular frequency
	f	Hz	frequency
	ϕ	rad	phase constant
	T	s	period

Fig. 8.3 illustrates the time evolution of the quantities $x(t)$, $\dot{x}(t)$ and $\ddot{x}(t)$.

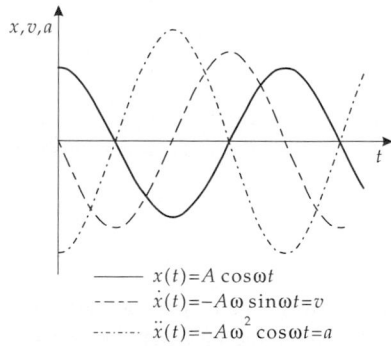

$$\text{———} \quad x(t)=A\cos\omega t$$
$$\text{- - - -} \quad \dot{x}(t)=-A\omega\sin\omega t=v$$
$$\text{·······} \quad \ddot{x}(t)=-A\omega^2\cos\omega t=a$$

Figure 8.3: Displacement $x(t)$, velocity $v = \dot{x}(t)$ and acceleration $a = \ddot{x}(t)$ of a mass on a spring.

➤ The vibrating body reaches its maximum velocity $|v_{max}| = A\omega$ when passing through the equilibrium position. The acceleration takes its maximum values at the turning points, $|a_{max}| = A\omega^2$.

The experimental setup shown in **Fig. 8.2** corresponds to a horizontally vibrating oscillator. For a hanging oscillator, one has to take into account that in the equilibrium state the spring is already pre-stretched by gravity. The origin of the coordinate frame should coincide with the balance point of the spring under gravity for the solution to take the form above.

4. Energy of the mass and spring system

The energy of the mass and spring system is the sum of the kinetic and potential energies (**Fig. 8.4**):

- Kinetic energy E_{kin}, the energy of motion of the oscillator.
- Potential energy E_{pot}, deformation energy stored in the stretched or compressed spring.

energy of the mass and spring system				$\mathbf{ML^2T^{-2}}$
$E_{\text{kin}}(t) = \dfrac{m\dot{x}^2}{2} = \dfrac{mA^2\omega^2 \sin^2(\omega t + \phi)}{2}$	Symbol	Unit	Quantity	
	W_{kin}	J	kinetic energy	
$E_{\text{pot}}(t) = \dfrac{kx^2}{2} = \dfrac{mA^2\omega^2 \cos^2(\omega t + \phi)}{2}$	W_{pot}	J	potential energy	
	m	kg	mass of oscillator	
	x	m	displacement	
$E_{\text{kin}}(t) + E_{\text{pot}}(t) = \dfrac{mA^2\omega^2}{2}$	A	m	amplitude	
	ω	rad/s	angular frequency	
	t	s	time	
$= \dfrac{kA^2}{2} = \text{const.}$	ϕ	rad	phase constant	
	k	kg/s^2	spring constant	

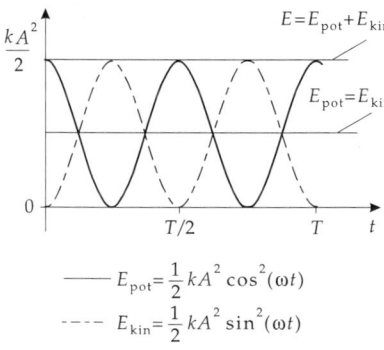

$$E = E_{\text{pot}} + E_{\text{kin}}$$

$$E_{\text{pot}} = E_{\text{kin}}$$

$$\text{———} \quad E_{\text{pot}} = \frac{1}{2}kA^2\cos^2(\omega t)$$

$$\text{- - - -} \quad E_{\text{kin}} = \frac{1}{2}kA^2\sin^2(\omega t)$$

Figure 8.4: Kinetic energy $E_{\text{kin}}(t)$, potential energy $E_{\text{pot}}(t)$ and total energy E of the mass on a spring.

Both the kinetic and the potential energy of the system are time-dependent. The total energy is constant; for a given spring constant, it is determined by the square of the amplitude.

8.1.2 Standard pendulum

Standard pendulum, a body hanging freely on a string in a gravitational field. The body is displaced and then released. Let the coordinate origin lie on a vertical line with the suspension point of the pendulum.

1. Mathematical pendulum and quantities of description

Mathematical pendulum, idealized standard pendulum with the following assumptions:
- non-stretchable string of negligible mass
- frictionless suspension of the pendulum
- point-like mass of the pendulum body

The description of motion involves the length of the cord l, the mass m of the pendulum, and the angle of displacement $\alpha(t)$ between the vertical and the displaced pendulum at the time t or the **horizontal displacement** $x(t)$ of the pendulum body at the instant t:

$$x(t) = l \sin \alpha(t).$$

Restoring force, F, accelerates the pendulum towards the rest position (**Fig. 8.5**):

restoring force of the standard pendulum			MLT^{-2}
	Symbol	Unit	Quantity
$F = -mg \sin \alpha$	F	N	restoring force
	m	kg	mass of pendulum body
	g	m/s^2	gravitational acceleration
	α	rad	angle of displacement

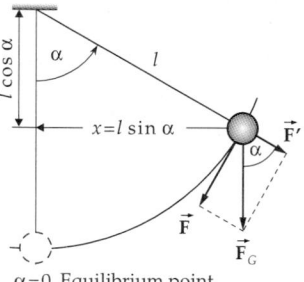

Figure 8.5: Standard pendulum of length l. α: angle of displacement, x: horizontal displacement, $\vec{F}_G$: weight force, $F_G = mg$, $\vec{F}$: restoring force, $F = mg \sin \alpha$, $\vec{F}'$: force along the string of pendulum.

2. Linearization of the equation of motion

The equation of motion can be linearized by restricting it to small displacements α, by approximating $\sin \alpha$ with the angle α itself:

equation of motion of the linearized mathematical pendulum			
	Symbol	Unit	Quantity
$x = l\alpha$	F	N	force
	m	kg	mass
$v = \dot{x} = l\dot{\alpha}$	l	m	length of cord
	x	m	horizontal displacement
$a = \dot{v} = l\ddot{\alpha}$	v	m/s	velocity
	a	m/s^2	acceleration
$F = ma = ml\ddot{\alpha} = -mg\alpha$	α	rad	elongation angle
	$\dot{\alpha}$	rad/s	angular velocity
$\ddot{\alpha} = -\dfrac{g}{l}\alpha$	$\ddot{\alpha}$	rad/s^2	angular acceleration
	g	m/s^2	acceleration of gravity

➤ An approximation that replaces the sine function by the first term of the series expansion is made in the description of many oscillations. Only then can the problem in general be solved analytically.

➤ When approximating $x \approx l\alpha$, note that the unit of the angle α is rad and not degree.

■ $3°$ corresponds to $3° \cdot (2\pi/360°) = 0.052$ rad. For a cord length of 0.5 m, the horizontal displacement $x \approx l\alpha = 0.5$ m $\cdot$ 0.052 rad $= 0.026$ m.

3. Solution of the linearized equation of motion of the mathematical pendulum

vibration solution of the linearized mathematical pendulum			
$x(t) = A\cos(\omega t + \phi)$	Symbol	Unit	Quantity
	$x(t)$	m	displacement
$\omega = \sqrt{\dfrac{g}{l}}$	t	s	time
	A	m	amplitude, maximum displacement
	l	m	cord length
$f = \dfrac{1}{2\pi}\sqrt{\dfrac{g}{l}}$	ω	rad/s	angular frequency
	g	m/s^2	acceleration of gravity
	f	Hz	frequency
$T = 2\pi\sqrt{\dfrac{l}{g}}$	ϕ	rad	phase constant
	T	s	period

▲   For small displacements, the period of the standard pendulum depends on the string length and the gravitational acceleration; it is independent of the mass and the vibration amplitude.

➤   For larger displacements of the pendulum, the period T must be multiplied by correction factors (see **Tab. 12.1/1**).

➤   All harmonic systems that carry out free vibrations obey a differential equation of the form $\ddot{x} = -\omega^2 x$. The constant ω^2 is determined by the system parameters.

8.1.2.1   Vibration and circular motion

▲   Periodic motion is closely related to circular motion: the parallel projection of circular motion yields a harmonic space-time function.

If a radius vector of length R rotating with constant angular velocity ω in the x-y plane needs a time T for one revolution, the projection of the radius vector onto the y-axis (x-axis) displays a sine (cosine) dependence on time t,

$$y(t) = R\sin(\omega t + \phi), \quad \omega = \frac{2\pi}{T},$$

$$x(t) = R\cos(\omega t + \phi), \quad \omega = \frac{2\pi}{T}.$$

Here ϕ is the angle between the radius vector and the x-axis at the instant $t = 0$ (**Fig. 8.6**).

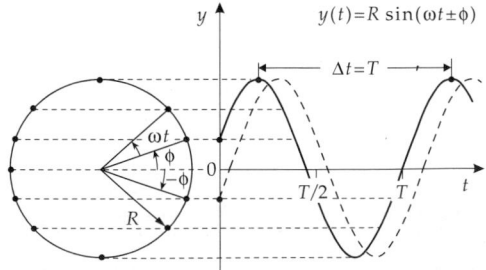

Figure 8.6: Parallel projection of a circular motion onto the y-axis.

It is often convenient to represent vibrations or rotations by a complex radius vector (**Fig. 8.7**):

$$x(t) + \mathrm{j}y(t) = R(\cos(\omega t + \phi) + \mathrm{j}\sin(\omega t + \phi)) = R\mathrm{e}^{\mathrm{j}(\omega t + \phi)} \qquad (\mathrm{j}^2 = -1).$$

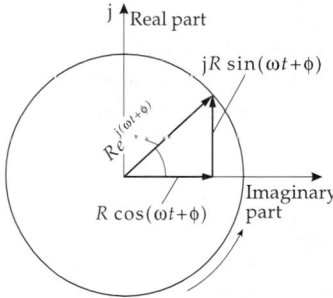

Figure 8.7: Complex representation of the circular motion of a radius vector.

Conversely, one often adopts a complex formulation for the solution of the equation of motion of an oscillator. This is possible because the real part and the imaginary part, if taken separately, are independent solutions to a linear differential equation.

8.1.3 Physical pendulum

1. Definition of the physical pendulum

Physical pendulum, gravitational pendulum, a rigid body that, under the action of gravity, carries out rotational motions about a fixed axis A, which does not pass the center of mass of the body.

■ Rod pendulum: A hanging rod with a pivot at the upper end (**Fig. 8.8**).

The torque $\tilde{\mathbf{M}}$ and the angular momentum $\tilde{\mathbf{L}}$ of the pendulum are normal to the plane of vibration.

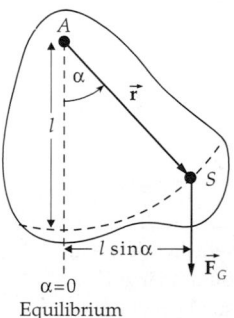

Figure 8.8: Physical pendulum. S: center of gravity, $\vec{\mathbf{F}}_G$: weight force.

2. Equation of motion of the physical pendulum

Equation of motion: according to the **basic law of rotational motion**, the torque of the weight with respect to the rotation axis A equals the product of the moment of inertia J_A and the angular acceleration $\ddot{\alpha}$.

angular momentum and torque about the axis A			
	Symbol	Unit	Quantity
$L = J_A \dot{\alpha}$	L	Nm s	angular momentum
	τ	Nm	torque
$\tau = \dot{L} = J_A \ddot{\alpha}$	l	m	distance axis–center of gravity
	m	kg	mass of pendulum
$\tau = -lmg \sin\alpha$	g	m/s^2	acceleration of gravity
	α	rad	angle of displacement
	J_A	kg m^2	moment of inertia about axis A

For small angles α ($\sin\alpha \approx \alpha$):

equation of motion for the physical pendulum			
	Symbol	Unit	Quantity
$\ddot{\alpha} = -\dfrac{lmg}{J_A}\alpha$	α	rad	angle of displacement
	$\ddot{\alpha}$	rad/s^2	angular acceleration
$\alpha(t) = \alpha_{max}\cos(\omega t + \phi)$	l	m	distance axis–center of gravity
	m	kg	mass of pendulum
$\omega = \sqrt{\dfrac{mgl}{J_A}}$	g	m/s^2	gravitational acceleration
	J_A	kg m^2	moment of inertia of pendulum about axis A
$f = \dfrac{1}{2\pi}\sqrt{\dfrac{mgl}{J_A}}$	α_{max}	rad	maximum amplitude
	ω	rad/s	angular frequency
	f	Hz	frequency
$T = 2\pi\sqrt{\dfrac{J_A}{mgl}}$	t	s	time
	ϕ	rad	phase constant
	T	s	period

M **Moments of inertia** J_A of arbitrary rigid bodies may be determined by measuring m, l and T and using the above equation.

3. Reduced pendulum length

of a physical pendulum, the string length of an equivalent mathematical pendulum with the same period as that of the physical pendulum.

reduced pendulum length				**L**
	Symbol	Unit	Quantity	
$l' = \dfrac{J_A}{ml}$	l	m	distance axis–center of gravity	
	l'	m	reduced pendulum length	
	m	kg	mass of physical pendulum	
	J_A	kg m^2	moment of inertia of pendulum about axis A	

➤ According to Steiner's law, the moment of inertia J_A for rotation about the axis A can be replaced by

$$J_A = J_S + ml^2 .$$

J_S is the moment of inertia for rotation about an axis parallel to the axis A and passing through the center of gravity S. One may replace the moment of inertia J_A in the expression for the reduced pendulum length by the moment of inertia J_S about the center of gravity:

$$l' = \frac{J_S}{ml} + l .$$

4. Example: Homogeneous rod pendulum

The center of gravity of a homogeneous rod pendulum of mass m and length L lies half way down the rod, $l = L/2$. The moment of inertia of the rod with respect to a rotation axis through one end point is

$$J_A = \frac{1}{3} m L^2 .$$

For the reduced pendulum length l', one finds

$$l' = \frac{1}{3} m L^2 \frac{2}{mL} = \frac{2}{3} L .$$

The moment of inertia of the rod with respect to the rotation axis through the center of gravity is

$$J_S = \frac{1}{12} m L^2 .$$

One obtains the same value for the reduced pendulum length:

$$l' = \frac{L}{6} + \frac{L}{2} = \frac{2}{3} L .$$

8.1.4 Torsional vibration

1. Definition of torsional vibration

Torsion (see p. 165), the twisting of a body, causes a torque τ proportional but opposite to the torque producing the torsion. For small torsion angles α, $\tau = -D^*\alpha$.

Torsional constant, D^*, proportionality factor between τ and α.

Torsional vibration, **rotary vibration**, results if a body is twisted by external torques, i.e., is driven out of its mechanical equilibrium and then vibrates about a longitudinal axis (**Fig. 8.9**).

Torsional oscillator, system performing torsional vibrations.

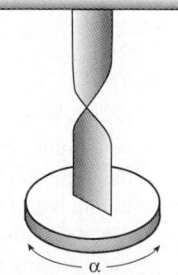

Figure 8.9: Rotary or torsional vibration. A disk of mass m is suspended by a metal strip, which is twisted.

2. Equation of motion of torsional vibration

Equation of motion, follows from Newton's law $\tau = J_A \ddot{\alpha}$ (τ torque, $\ddot{\alpha}$ angular acceleration):

equation of motion and solution for torsional vibration			
$\ddot{\alpha} = -\dfrac{D^*}{J_A}\alpha$	Symbol	Unit	Quantity
	α	rad	torsional angle
$\alpha(t) = \alpha_{max} \cos(\omega t + \phi)$	$\ddot{\alpha}$	rad/s^2	angular acceleration
	D^*	Nm/rad	torsional constant
$\omega = \sqrt{\dfrac{D^*}{J_A}}$	J_A	kg m^2	moment of inertia
	α_{max}	rad	amplitude
	ω	rad/s	angular frequency
$f = \dfrac{1}{2\pi}\sqrt{\dfrac{D^*}{J_A}}$	f	Hz	frequency
	t	s	time
$T = 2\pi\sqrt{\dfrac{J_A}{D^*}}$	ϕ	rad	phase constant
	T	s	period

3. Kinetic and potential energy of the torsional pendulum

$$E_{kin} = \frac{1}{2} J_A \cdot \dot{\alpha}^2, \qquad E_{pot} = \frac{1}{2} D^* \alpha^2 .$$

M   The **Moment of inertia** may be determined by measuring the period, using the relation

$$J_A = -D^* \frac{\alpha}{\ddot{\alpha}} = \frac{T^2}{4\pi^2} D^* .$$

The torsional constant D^* may be determined by measuring the torsional angle α and the corresponding torque τ, or by measuring the period T of torsional vibration of a body of known moment of inertia (e.g., circular disk about center).

8.1.5   Liquid pendulum

1. Definition of liquid pendulum

Liquid pendulum, a liquid column in a U-shaped pipe out of equilibrium and vibrating about the rest (equilibrium) position (**Fig. 8.10**). In the rest position, the columns in both legs have equal height.

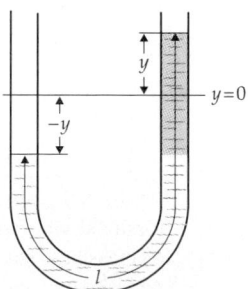

Figure 8.10: Liquid pendulum. The weight of the excessive liquid volume in one leg of the U-shaped pipe (shadowed) provides the restoring force.

Restoring force, results from the weight of the excess liquid column. If the levels are displaced by $\pm y$ from the rest position, one has a:

restoring force for liquid pendulum			MLT^{-2}
	Symbol	Unit	Quantity
$F = -2yA\rho g$	F	N	restoring force
	y	m	displacement of liquid column
	A	m^2	cross-sectional area of pipe
	ρ	kg/m^3	density of liquid
	g	m/s^2	acceleration of gravity

2. Equation of motion of the liquid pendulum

equation of motion and solution for the liquid pendulum			
	Symbol	Unit	Quantity
$m\ddot{y} = -2A\rho gy$	y	m	displacement
$y(t) = B\cos(\omega t + \phi)$	$\ddot{y}$	m/s^2	acceleration
	m	kg	mass of liquid
$\omega = \sqrt{\dfrac{2A\rho g}{m}} = \sqrt{\dfrac{2g}{l}}$	A	m^2	cross-sectional area
	ρ	kg/m^3	density of liquid
	g	m/s^2	gravitational acceleration
$f = \dfrac{1}{2\pi}\sqrt{\dfrac{2g}{l}}$	B	m	amplitude
	ω	rad/s	angular frequency
$T = 2\pi\sqrt{\dfrac{l}{2g}}$	f	Hz	frequency
	ϕ	rad	phase constant
	t	s	time
$m = lA\rho$	T	s	period
	l	m	length of liquid column

➤ The quantity m is the total mass of liquid in both legs, i.e., $m = lA\rho$, where l is the total length of the liquid column. The quantity y then describes the motion of the level in one leg, while the level in the other leg is given by $-y$.

8.1.6   Electric circuit

Oscillator circuit, a combination of inductor and capacitor connected to a circuit (**Fig. 8.11**).

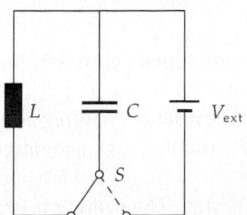

Figure 8.11: Parallel oscillator circuit with inductor (inductance L) and capacitor (capacity C). V_{ext}: voltage applied by the switch S for exciting the initial state.

The displacement out of the rest position of a pendulum here corresponds to charging a capacitor by an external voltage V_{ext}. In the initial state (maximum electrostatic energy, analogous to the potential energy), the capacitor voltage takes the maximum value; no current flows through the inductor. When discharging the capacitor, the current generates a magnetic field in the inductor (analogous to the kinetic energy). If the capacitor is discharged (analogous to the zero passage of a pendulum; maximum magnetic energy), the magnetic field decreases, thereby inducing a current that charges the capacitor, but with reversed voltage. In the time evolution, the total energy of the system oscillates back and forth between the capacitor and the inductor. The restoring force, which corresponds to the gravitational force on a pendulum, is inversely proportional to the capacitance.

Oscillation equation: Since the circuit is closed, the voltages of the inductor V_L and the capacitor V_C must sum to zero:

undamped electric oscillator circuit			
$0 = V_L + V_C$	Symbol	Unit	Quantity
$V_L = L\,\dot{I}$	Q	C	charge of capacitor
$V_C = Q/C$	V_L	V	voltage at inductor
$I = \dot{Q}$	V_C	V	voltage at capacitor
	t	s	time
$0 = L\ddot{Q} + \dfrac{Q}{C}$	A	C	amplitude, max. charge of capacitor
$Q(t) = A\cos(\omega t + \phi)$	ω	rad/s	angular frequency
	f	Hz	frequency
$\omega = \sqrt{\dfrac{1}{LC}}$	ϕ	rad	phase constant
	L	Vs/A	inductivity of inductor
$f = \dfrac{1}{2\pi}\sqrt{\dfrac{1}{LC}}$	C	As/V	capacity
	T	s	period
$T = 2\pi\sqrt{LC}$	I	A	current

Electrostatic energy E_{el} and **magnetic energy** E_{magn} of oscillator circuit:

$$E_{el} = \frac{1}{2C}\,Q^2 = \frac{1}{2}CV_C^2, \qquad E_{magn} = \frac{1}{2}\,LI^2\,.$$

➤   The oscillator circuit is an important basic element in electrical engineering. It is used, for example, to generate electromagnetic oscillations in transmitter antennas.

8.2 Damped vibrations

Damped vibration, the energy of the oscillator does not remain constant, but is released into the environment.

■   Mechanical oscillators lose energy by friction because of coupling to the environment. Frictional forces oppose the motion of the oscillator. Hence, after some time the oscillation ceases (**Fig. 8.12**).

■   A pendulum can never pivot perfectly without friction. The bearing is heated by friction, and a fraction of the energy leaves the system as heat.

Oscillation equation with additional friction force F_R:

oscillation equation with friction			MLT^{-2}
	Symbol	Unit	Quantity
	F	N	total force
	m	kg	mass
$m\ddot{x} = F = -kx + F_R$	x	m	displacement
	$\ddot{x}$	m/s^2	acceleration
	k	kg/s^2	restoring-force constant
	F_R	N	frictional force

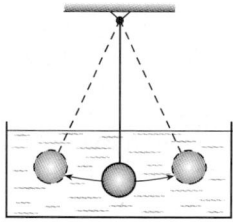

Figure 8.12: Damped vibration. Pendulum in an oil bath.

8.2.1 Friction

The form of the oscillation equation depends on the type of friction (see p. 56). The equation of motion may be solved analytically for only a few types of friction.

8.2.1.1   Sliding friction and rolling friction

1. Coulomb friction

Coulomb friction, solid friction, F_R, friction independent of the magnitude of the velocity and opposite to its direction (see p. 56). For motion along the x-direction,

$$F_R = -\mathrm{sgn}(v_x)\mu F_N \,,$$

where $\mathrm{sgn}(x)$ is $+1$ if $x > 0$, -1 if $x < 0$, and 0 if $x = 0$.

Normal force, F_N, force with which the body is pressed onto the supporting surface. If no other forces than gravity act on the body, F_N is the normal component of the weight $F_G = mg$.

Oscillation equation:

$$m\ddot{x} + kx + \mathrm{sgn}(v_x)\mu F_N = 0 \,.$$

2. Properties of the solution of the oscillation equation for sliding friction

- The frequency (and hence the period) remains constant and equals that of the undamped oscillation.
- The amplitude decreases **linearly** with time.
- The oscillation may terminate at a displacement different from zero.
- The period is **finite**.
- The vibration amplitude decreases per period T by $4x_0$. The amplitudes form an arithmetic series.
- The rest position alternates each half period between x_0 and $-x_0$.
- The vibration comes to rest if the displacement is smaller than x_0 after a half vibration.

The solution cannot be given in closed form (i.e., analytically) as a function of time, but only for a certain time interval. For instance (**Fig. 8.13**):

vibration with sliding friction			L
$x(t) = -(\Delta x - x_0) \cdot \sin\left(\omega t + \dfrac{\pi}{2}\right) - x_0$ $\text{for } 0 \le t \le \dfrac{T}{2}$ $x(t) = -(\Delta x - 3x_0) \cdot \sin\left(\omega t + \dfrac{\pi}{2}\right) + x_0$ $\text{for } \dfrac{T}{2} \le t \le T$ $x_0 = \dfrac{\mu F_N}{k}$	**Symbol**	**Unit**	**Quantity**
	x	m	displacement
	Δx	m	initial displacement
	x_0	m	final displacement
	ω	rad/s	angular frequency
	t	s	time
	T	s	period
	k	kg/s^2	force constant
	F_N	N	normal force
	μ	1	friction coefficient

x_0 is the displacement for which the restoring force equals the frictional force. Hence, the system has to be displaced by more than x_0 to initiate the oscillation.

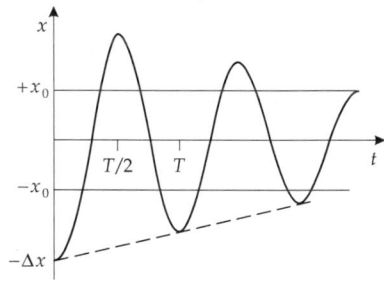

Figure 8.13: Damped vibration for velocity-independent friction. The maximum displacement decreases linearly with the time.

8.2.1.2 Viscous friction

1. Oscillation equation for viscous friction

Viscous friction, Stokes' friction, proportional to the magnitude of velocity and opposed to it:

$$F_R = -bv = -b\dot{x} \, .$$

Damping constant, damping coefficient, b, proportionality constant between the viscous-friction force and the velocity.

Oscillation equation:

oscillation equation for viscous friction			MLT^{-2}
	Symbol	**Unit**	**Quantity**
	m	kg	mass
	x	m	displacement
$m\ddot{x} + b\dot{x} + kx = 0$	$\dot{x}$	m/s	velocity
	$\ddot{x}$	m/s^2	acceleration
	k	kg/s^2	force constant
	b	kg/s	damping constant

2. Solution of the oscillation equation for viscous friction

vibration for viscous friction			
$x(t) = Ae^{-\delta t}e^{\pm j\sqrt{\omega_0^2 - \delta^2}\, t}$ $= Ae^{-\delta t}e^{\pm j\omega_0\sqrt{1 - D^2}\, t}$ $\delta = \dfrac{b}{2m}$ $D = \dfrac{\delta}{\omega_0} = \dfrac{b}{2m\omega_0}$ $\omega_0 = \sqrt{\dfrac{k}{m}}$	Symbol	Unit	Quantity
	x	m	displacement
	A	m	initial amplitude
	ω_0	1/s	angular frequency
	t	s	time
	δ	1/s	decay constant
	D	1	degree of damping
	b	kg/s	damping constant
	m	kg	mass

The eigenfrequency of the undamped vibration is determined by the mass of the oscillator m and the restoring force constant k:

$$\omega_0 = \sqrt{k/m}\,.$$

Dissipation factor, d, twice the value of the degree of damping:

$$d = 2\,D = b/\sqrt{mk}\,.$$

Quality factor, Q, reciprocal value of the dissipation factor d :

$$Q = \frac{1}{d} = \frac{\sqrt{mk}}{b}\,.$$

3. Damped torsional vibration with viscous friction

damped torsional vibration with viscous friction			
	Symbol	Unit	Quantity
$J_A\,\ddot{\alpha} + b\,\dot{\alpha} + D^*\,\alpha = 0$	α	rad	torsional angle
	b	kg m^2/s	friction constant
	J_A	kg m^2	moment of inertia, axis A
	D^*	kg m^2/s^2	angular restoring-force coefficient

4. Cases of degrees of damping

Case distinction with respect to the degree of damping D (**Fig. 8.14**):
- **Underdamped case**, $D < 1$ ($\omega_0 > \delta$), weak damping:

$$\omega' = \sqrt{\omega_0^2 - \delta^2} = \omega_0\sqrt{1 - D^2}\,, \quad \omega' < \omega_0 \text{ real}\,,$$

$$x(t) = Ae^{-\delta t}\cos\left(\sqrt{\omega_0^2 - \delta^2}\,t + \phi\right)\,.$$

The angular frequency ω' of the damped vibration is smaller than the angular frequency ω_0 of the undamped vibration. The vibration amplitude decreases exponentially, the period remains constant. The envelope of the oscillation curve $x(t)$ is an exponential function.

- **Overdamped case,** $D > 1$ $(\omega_0 < \delta)$:

$$\text{Damping frequency:} \quad \omega' = j\sqrt{\delta^2 - \omega_0^2}, \quad \omega' \text{ imaginary.}$$

$$x(t) = A_1 e^{(-\delta + \sqrt{\delta^2 - \omega_0^2})t} + A_2 e^{(-\delta - \sqrt{\delta^2 - \omega_0^2})t}.$$

Here the system no longer oscillates. When driven out of the rest position, the system returns exponentially to equilibrium, but slower than in the critical damping limit.

- **Critical Damping,** $D = 1$ $(\omega_0 = \delta)$:

$$\omega' = \omega_0 = \delta, \quad x(t) = (A_1 + A_2 t)e^{-\delta t}.$$

The solutions in the overdamped and critical cases are not vibrations in the proper sense because, after a displacement away from the rest position, the system does not pass through the rest position again.

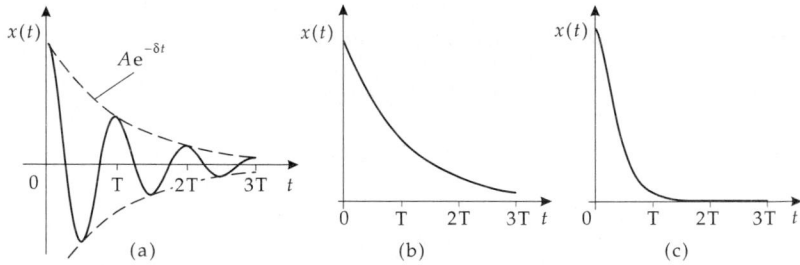

Figure 8.14: Damped vibration. (a): underdamped, (b): overdamped, (c): critical damping.

M The case of aperiodic motion is important for practice, since in this case, after a disturbance of the system, the equilibrium state is reached most rapidly. Measuring and indicating instruments are adjusted in this manner, for example, the **ballistic galvanometer**.

5. *Characteristic quantities of vibrations with viscous damping*

characteristic quantities of vibration with viscous damping			
$\omega_0 = \sqrt{\dfrac{k}{m}}$	Symbol	Unit	Quantity
	D	1	degree of damping
$\omega' = \sqrt{\dfrac{k}{m} - \left(\dfrac{b}{2m}\right)^2}$	δ	1/s	decay coefficient
	ω_0	1/s	angular frequency of undamped vibration
$\delta = \dfrac{b}{2m}$	ω'	1/s	angular frequency of damped vibration
$D = \dfrac{\delta}{\omega_0} = \dfrac{b}{2}\dfrac{1}{\sqrt{mk}}$	d	1	dissipation factor
$d = 2D = \dfrac{2\delta}{\omega_0} = \dfrac{b}{\sqrt{mk}}$	Q	1	quality factor
	b	kg/s	damping constant
$Q = \dfrac{1}{d} = \dfrac{1}{2D}$	Λ	1	logarithmic decrement
	m	kg	mass
$\Lambda = \ln(x(t)/x(t+T)) = \delta T$	k	kg/s^2	force constant

Logarithmic decrement, Λ, logarithm of the ratio of two amplitudes separated by one period,

$$\Lambda = \ln\left(\frac{x(t)}{x(t+T)}\right) = \delta T \,.$$

8.2.1.3 Newton friction

The friction force F_R proposed by Newton is proportional to the square of the velocity,

$$F_R = -b\,v^2\,.$$

This type of friction arises in viscous media if the body moves at a speed below a certain limit that depends on the viscosity of the medium. A **nonlinear differential equation** in x arises that in general cannot be solved analytically:

$$m\ddot{x} + kx + b\dot{x}^2 = 0\,.$$

8.2.2 *Damped electric oscillator circuit*

1. *Damped electric oscillator circuit,*

contains, besides capacitor C and inductor L, an ohmic resistor R (**Fig. 8.15**).

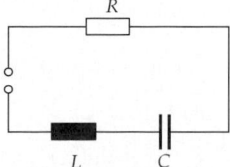

Figure 8.15: Damped electric oscillator circuit with capacitor C, inductor L and ohmic resistor R.

damped electric oscillator circuit

$0 = V_L + V_C + V_R$	Symbol	Unit	Quantity
$V_L = L\,\dot{I}$	Q	C	charge of capacitor
$V_C = Q/C$	V_L	V	voltage at inductor
$V_R = R\,I$	V_C	V	voltage at capacitor
$I = \dot{Q}$	V_R	V	voltage at resistor
	I	A	current
$0 = L\,\ddot{Q} + R\,\dot{Q} + \dfrac{Q}{C}$	t	s	time
	R	Ω	resistance
$\omega_0 = \sqrt{\dfrac{1}{LC}}$	L	Vs/A	inductance
	C	As/V	capacitance
$\omega' = \sqrt{\dfrac{1}{LC} - \left(\dfrac{R}{2L}\right)^2}$	ω_0	rad/s	angular frequency of undamped vibration
$\delta = \dfrac{R}{2L}$	ω'	rad/s	angular frequency of damped vibration
$D = \dfrac{\delta}{\omega_0} = \dfrac{R}{2}\sqrt{\dfrac{C}{L}}$	δ	1/s	decay constant
	D	1	degree of damping

2. Analogies between mechanical and electromagnetic damped vibrations

Characteristics	mechanical vibration	electromagnetic vibration
oscillation equation	$m\,\ddot{x} + b\,\dot{x} + k\,x = 0$	$L\,\ddot{I} + R\,\dot{I} + \dfrac{1}{C}\,I = 0$
undamped angular frequency ω_0	$\sqrt{\dfrac{k}{m}}$	$\sqrt{\dfrac{1}{LC}}$
damped angular frequency ω'	$\sqrt{\dfrac{k}{m} - \left(\dfrac{b}{2m}\right)^2}$	$\sqrt{\dfrac{1}{LC} - \left(\dfrac{R}{2L}\right)^2}$
decay constant δ	$\dfrac{b}{2m}$	$\dfrac{R}{2L}$
degree of damping $D = \delta/\omega_0$	$\dfrac{b}{2}\sqrt{\dfrac{1}{mk}}$	$\dfrac{R}{2}\sqrt{\dfrac{C}{L}}$
quality factor Q	$\dfrac{\sqrt{mk}}{b}$	$\dfrac{1}{R}\sqrt{\dfrac{L}{C}}$

m: mass, L: inductance, k: restoring-force coefficient (spring etc.), C: capacitance, b: damping constant, R: ohmic resistance.

8.3 Forced vibrations

1. Definition of a forced vibration

Forced vibration, vibration under the influence of an external force F_{ext} on the oscillator. After a transient oscillation, the vibrator follows the frequency imposed by the external force.

Oscillation equation for $F_{ext} = B \cos(\omega_{ext}t)$ and viscous friction:

oscillation equation for a forced damped vibration		Symbol	Unit	Quantity
$F = m\ddot{x}$ $= -kx - b\dot{x} + B\cos(\omega_{ext}t)$ $x(t) = A(\omega_{ext})\sin(\omega_{ext}t + \phi(\omega_{ext}))$ $A(\omega_{ext}) = \dfrac{B}{\sqrt{(m\omega_{ext}^2 - k)^2 + b^2\omega_{ext}^2}}$ $\phi(\omega_{ext}) = \arctan\dfrac{-b\omega_{ext}}{k - m\omega_{ext}^2}$		F m $\ddot{x}$ x A k b B ω_{ext} ϕ_{ext} t	N kg m/s^2 m m kg/s^2 kg/s N rad/s rad s	force mass of oscillator acceleration displacement amplitude force constant damping constant exciting amplitude exciting angular frequency phase shift time

2. Properties of the solution

The solution is a superposition of the general solution to the homogeneous equation (without the inhomogeneity F_{ext}; this corresponds to free damped vibrations with the angular frequency $\omega_0 = \sqrt{k/m}$), and a particular solution to the inhomogeneous equation; it is a sine function with angular frequency ω_{ext} and amplitude and phase that depend on ω_{ext}.

▲ The amplitude of the vibration is proportional to the maximum value of the driving force and depends on its frequency. For large frequencies, the amplitude approaches zero, independent of the friction, $A \to 0$ for $\omega_{ext} \to \infty$ (see **Fig. 8.16**).

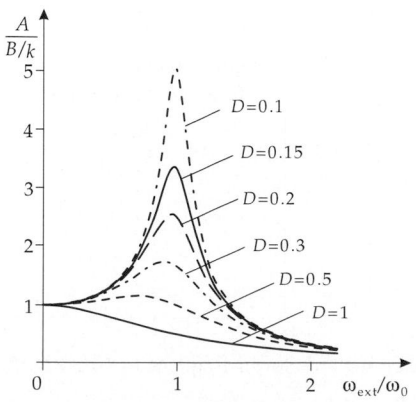

Figure 8.16: Forced vibration. Normalized amplitude $\dfrac{A(\omega_{res})}{B/k}$ as a function of ω_{ext}/ω_0 for various degrees of damping D.

Resonance frequency, ω_{res}, angular frequency of the external excitation at which the resulting amplitude reaches the maximum value. It is obtained from the minimum of the denominator of $A(\omega_{ext})$ for positive values of ω_{ext}.

Resonance amplitude, A_{max}, amplitude of the vibration at the resonance frequency. It follows from the substitution of ω_{res} for ω_{ext} in $A(\omega_{ext})$.

Phase shift, ϕ, phase difference between response and excitation of the oscillator (see **Fig. 8.17**).

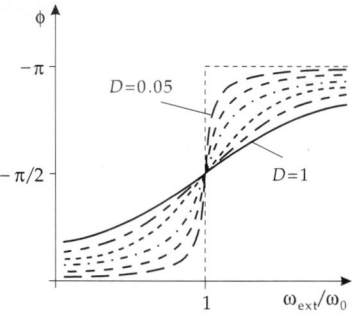

Figure 8.17: Forced vibration. Phase shift ϕ as function of ω_{ext}/ω_0 for various degrees of damping D.

▲   For $\omega_{ext} = \omega_0$ one finds $\phi = -\pi/2$.
This property may also be used to define the resonance.

3. Resonance in forced vibrations

characteristic quantities of the resonance			
	Symbol	Unit	Quantity
$\omega_{res} = \sqrt{\dfrac{k}{m} - \dfrac{b^2}{4m^2}}$	ω_{res}	rad/s	resonance angular frequency
$= \omega_0\sqrt{1 - 2D^2}$	A		resonance amplitude
	B		excitation amplitude
$A_{max} = \dfrac{B}{b\sqrt{\dfrac{k}{m} - \left(\dfrac{b}{2m}\right)^2}}$	k	kg/s^2	restoring-force coefficient
	m	kg	mass
$= \dfrac{B}{2kD\sqrt{1 - D^2}}$	b	kg/s	damping constant
	δ	1/s	decay coefficient
$\phi = \arctan\left(\dfrac{-2\delta\omega_{ext}}{\omega_0^2 - \omega_{ext}^2}\right)$	D	1	degree of damping
$D = \dfrac{b}{2m\omega_0}$	ω_0	rad/s	angular frequency of free vibration
$\omega_0 = \sqrt{\dfrac{k}{m}}$			
$\delta = \dfrac{b}{2m}$			

▲   The maximum of the resonance is shifted towards lower frequencies (see **Fig. 8.16**) with increased damping of the vibration.

4. Characteristics of the resonance

a) Resonance selectivity, value of the renormalized amplitude curve at the resonance frequency, renormalized resonance amplitude (**Fig. 8.18**),

$$\frac{A(\omega_{res})}{B/k} = \frac{A_{max}}{B/k} .$$

b) Half-width, width of the resonance, region of excitation angular frequency $\Delta\omega_{ext}$ between the angular frequencies with amplitude $A_{max}/\sqrt{2}$ (**Fig. 8.18**),

$$\Delta\omega_{ext}/\omega_0 ,$$

c) Resonance catastrophy, arises for vanishing friction, $b = 0$, in the limit $\omega_{ext} \to \omega_0$: the vibration amplitude tends to infinity.

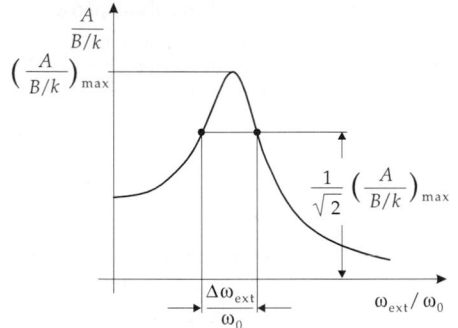

Figure 8.18: Forced vibration. Half-width $\Delta\omega_{ext}/\omega_0$ and resonance selectivity $\dfrac{A_{max}}{B/k}$.

➤   In technical applications, resonances are often highly undesirable because they may lead to damage to the oscillating system at large amplitudes. To prevent the occurrence of resonances, a device must work at frequencies far below the resonance frequency, or the resonance frequency must be crossed rapidly for the machine to work above ω_0. Moreover, in the construction of bridges and buildings in earthquake regions, resonances must be eliminated as far as possible, or need to be damped sufficiently.

8.4   Superposition of vibrations

Superposition law, holds because of the **linearity of the equations of motion** for harmonic vibrations:

▲   Harmonic vibrations can be superposed without influencing each other.

If a system carries out several vibrations simultaneously, then the corresponding oscillation equation may be solved for each vibration separately. The instantaneous displacement of the oscillator is obtained as the sum of the displacements of the individual oscillations.

8.4.1   Superposition of vibrations of equal frequency

From the two harmonic vibrations

$$x_1(t) = A_1 \cos(\omega t + \phi_1) , \quad x_2(t) = A_2 \cos(\omega t + \phi_2) , \quad \Delta\phi = \phi_1 - \phi_2$$

and, with the addition theorems for trigonometric functions, one obtains a resulting harmonic vibration with the same frequency as that of the original vibrations (**Fig. 8.19**):

superposition of vibrations of equal frequency			
$x_1(t) = A_1 \cos(\omega t + \phi_1)$	Symbol	Unit	Quantity
$x_2(t) = A_2 \cos(\omega t + \phi_2)$	$x_1(t), x_2(t)$		vibrations 1, 2
$\Delta\phi = \phi_1 - \phi_2$	$x_{1+2}(t)$		resulting vibration
	A_1, A_2		amplitudes 1, 2
$x_{1+2}(t) = x_1(t) + x_2(t)$	A_{1+2}		resulting amplitude
$\quad\quad = A_{1+2} \cos(\omega t + \phi_{1+2})$	ω	rad/s	angular frequency
	t	s	time
$A_{1+2} = \sqrt{A_1^2 + A_2^2 + 2A_1 A_2 \cos \Delta\phi}$	ϕ_1, ϕ_2	rad	phase constants 1, 2
	$\Delta\phi$	rad	phase difference
$\tan \phi_{1+2} = \dfrac{A_1 \sin\phi_1 + A_2 \sin\phi_2}{A_1 \cos\phi_1 + A_2 \cos\phi_2}$	ϕ_{1+2}	rad	resulting phase constant

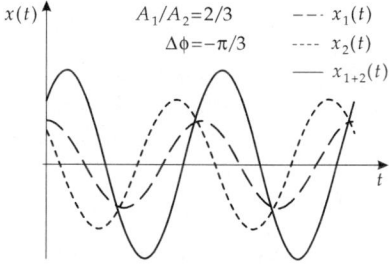

Figure 8.19: Superposition of vibrations $x_1(t)$, $x_2(t)$ of equal frequency ω and the phase difference $\Delta\phi$ for special values of A_1/A_2 and $\Delta\phi$.

Maximum enhancement: $\Delta\phi = 0$, $\quad A_{1+2} = A_1 + A_2$.
Superposition of vibrations of equal amplitude ($A_1 = A_2 = A$):

$$A_{1+2} = 2A \cos(\Delta\phi/2) \quad\quad \phi_{1+2} = \frac{\phi_1 - \phi_2}{2}.$$

- Maximum enhancement: $\Delta\phi = 0$, $A_{1+2} = 2A$.
- Cancellation: $\Delta\phi = \pi$, $A_{1+2} = 0$ (**Fig. 8.20**).

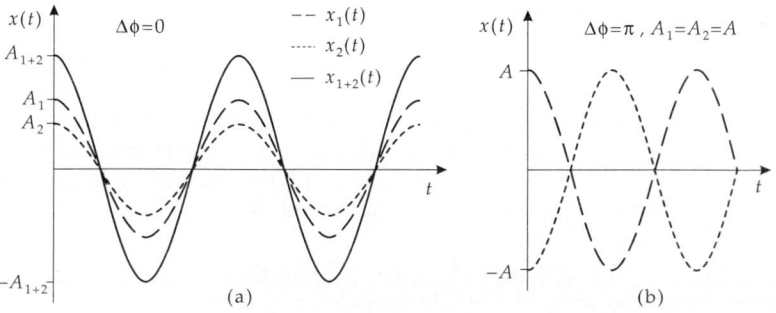

Figure 8.20: Superposition of vibrations $x_1(t)$, $x_2(t)$ of equal frequency ω.
(a): maximum enhancement ($\Delta\phi = 0$), (b): cancellation ($\Delta\phi = \pi$).

8.4.2 Superposition of vibrations of different frequencies

Vibrations:

$$x_1(t) = A_1 \cos(\omega_1 t + \phi_1), \quad x_2(t) = A_2 \cos(\omega_2 t + \phi_2).$$

With the simplifying assumption $\phi_1 = \phi_2 = 0$, $A_1 = A_2 = A$ and with the addition theorems for trigonometric functions:

superposition of vibrations of distinct frequencies			
$x_1(t) = A \cos \omega_1 t$	Symbol	Unit	Quantity
$x_2(t) = A \cos \omega_2 t$	x_1, x_2		displacements 1, 2
$x_{1+2}(t) = 2A \cos\left(\dfrac{\omega_1 - \omega_2}{2}t\right)$	x_{1+2}		resulting displacement
	A		amplitude
$\times \cos\left(\dfrac{\omega_1 + \omega_2}{2}t\right)$	ω_1, ω_2	rad/s	angular frequencies 1, 2
	t	s	time

1. Beats,

occur when the difference between the superposed frequencies is small compared with the frequencies themselves, $\omega_2 = \omega_1 + \Delta\omega$, $|\Delta\omega| \ll \omega_1$. The result may be interpreted as a vibration with the angular frequency $(\omega_1 + \omega_2)/2$ and an amplitude that varies slowly and periodically with the frequency $|(\omega_1 - \omega_2)|/2$ (**Fig. 8.21**).

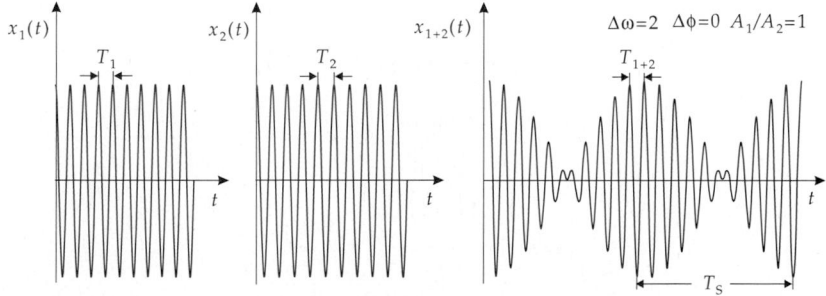

Figure 8.21: Beat. Superposition of vibrations $x_1(t)$, $x_2(t)$ with a small difference of frequencies $\Delta\omega$. T_1, T_2: periods of the individual vibrations, T_S: beat period, T_{1+2}: period of the resulting vibration.

2. Frequency and period of beats

Beat period, T_S, defined as the time interval between two successive zero passages of the beat amplitude, obtained from $\pi = |(\omega_1 - \omega_2)|T_S/2$ as $T_S = 2\pi/|(\omega_1 - \omega_2)|$,

frequency and period of beat

	Symbol	Unit	Quantity
$f_S = \|f_1 - f_2\|$	f_S	s^{-1}	beat frequency
$\dfrac{1}{T_S} = \left\|\dfrac{1}{T_1} - \dfrac{1}{T_2}\right\|$	f_1, f_2	s^{-1}	frequencies of vibration 1, 2
	T_S	s	beat period
	T_1, T_2	s	periods of vibration 1, 2

■   In general, beats are unwanted in music. They arise if two sounds with fundamental tones differing only slightly in frequency are heard simultaneously. The ear registers the resulting sound as dissonant and recognizes the oscillating amplitude of the beats. Beats offer a very effective way of tuning a musical instrument against a tuning fork or standard oscillator.

3. Frequency and period in the general case

frequency and period of the resulting vibration

	Symbol	Unit	Quantity
$\omega_{1+2} = \dfrac{\omega_1 + \omega_2}{2}$	ω_{1+2}	rad/s	angular frequency of resulting vibration
$f_{1+2} = \dfrac{f_1 + f_2}{2}$	ω_1, ω_2	rad/s	angular frequencies of vibrations 1, 2
$T_{1+2} = \dfrac{2\pi}{\omega_{1+2}} = \dfrac{4\pi}{\omega_1 + \omega_2}$	f_{1+2}	s^{-1}	frequency of resulting vibration
$= 2\dfrac{T_1 T_2}{T_1 + T_2}$	f_1, f_2	s^{-1}	frequencies of vibrations 1, 2
	T_1, T_2	s	periods of vibrations 1, 2
	T_{1+2}	s	period of resulting vibration

➤   For large frequency differences $\Delta\omega$ of the superimposed vibrations, the time evolution of the resulting vibration is in general not harmonic (**Fig. 8.22**).

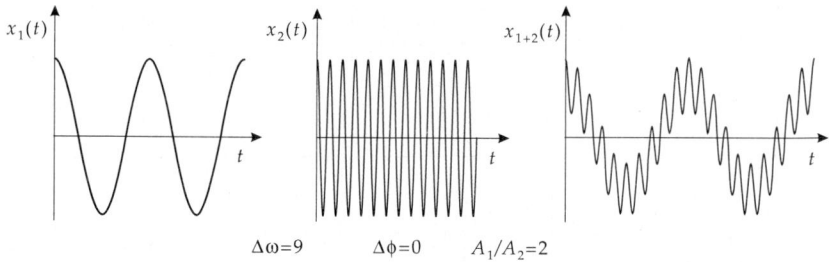

$$\Delta\omega = 9 \qquad \Delta\phi = 0 \qquad A_1/A_2 = 2$$

Figure 8.22: Superposition of vibrations $x_1(t)$, $x_2(t)$ with large frequency difference $\Delta\omega$.

8.4.3   *Superposition of vibrations in different directions and with different frequencies*

1. Lissajous patterns

In order to describe the superposition of two vibrations in different directions (e.g., in x- and y-direction), it is convenient to start from a representation of the individual vibrations in a Cartesian coordinate frame:

$$x(t) = A_x \sin(\omega_x t + \phi_x),$$

$$y(t) = A_y \sin(\omega_y t + \phi_y).$$

Representation of the resulting vibration in polar coordinates:

$$r(t) = \sqrt{x(t)^2 + y(t)^2}, \quad \alpha(t) = \arctan\frac{y(t)}{x(t)},$$

where r is the length of the resulting vector, and α is the angle between the positive x-axis and the vector, with counterclockwise taken as the positive direction.
The vector $\vec{r} = (x(t), y(t))$ describes **Lissajous patterns**, the shapes of which are determined by the ratio of A_x and A_y, the ratio of ω_x and ω_y, and the difference of phase angles $\Delta\phi = \phi_x - \phi_y$ (**Fig. 8.23**).

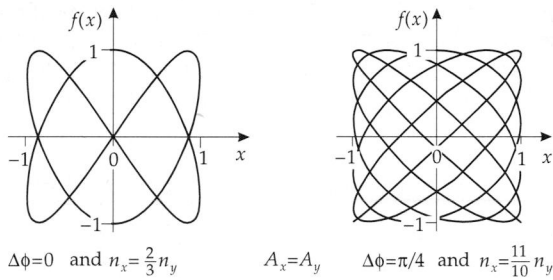

$$\Delta\phi=0 \quad \text{and} \quad n_x=\tfrac{2}{3}n_y \qquad\qquad A_x=A_y \quad \Delta\phi=\pi/4 \quad \text{and} \quad n_x=\tfrac{11}{10}n_y$$

Figure 8.23: Lissajous patterns.

Properties: periodic structure; a curve in 2D space is traversed periodically.

- $x(t) = A\cos(\omega t), \quad y(t) = A\sin(\omega t)$:
  $\vec{r}$ describes a circle.
- $x(t) = A_x\cos(\omega t), \quad y(t) = A_y\cos(\omega t)$:
  One obtains the equation of a straight line: $y(t) = \dfrac{A_y}{A_x}x(t)$.

➤ Lissajous patterns may be visualized by means of an oscilloscope by controlling the beam displacements in x- and y-direction according to the actual frequencies and amplitudes. Due to the finite persistence of the oscilloscope screen and the relatively high oscilloscope frequencies available, the entire pattern can be made to appear as a standing curve.

2. Two-dimensional harmonic oscillator

Equations of motion of the two-dimensional oscillator:

$$m\ddot{x} = -kx, \quad m\ddot{y} = -ky, \quad \omega = \sqrt{\frac{k}{m}}.$$

Solution:

$$x(t) = A_x\cos(\omega t + \phi_1), \quad y(t) = A_y\cos(\omega t + \phi_2).$$

The amplitudes A_x, A_y and the phase angles ϕ_1, ϕ_2 are determined by the initial conditions.

The trajectory is obtained by eliminating the time coordinate t:

$$\frac{x^2}{A_x^2} - xy \frac{2 \cos \phi}{A_x A_y} + \frac{y^2}{A_y^2} = \sin^2 \phi , \quad \phi = \phi_2 - \phi_1 .$$

The trajectory is an ellipse. For $\phi = \pi/2$, the principal axes coincide with the coordinate axes:

$$\frac{x^2}{A_x^2} + \frac{y^2}{A_y^2} = 1 .$$

8.4.4 Fourier analysis, decomposition into harmonics

The superposition of sine or cosine functions is an oscillatory function itself, i.e., a periodic phenomenon. Conversely, arbitrary periodic phenomena may be represented as a superposition of pure sine and cosine oscillations. This is a statement of Fourier's theorem.

▲ **Fourier decomposition**: Any periodic function may be represented by a (possibly infinite) sum over sine and cosine functions of different frequencies and amplitudes. The Fourier frequencies are integer multiples of a fundamental frequency.

1. Fourier series,

mathematical representation of a periodic function $x(t)$ of period T by a superposition of sine and cosine oscillations,

$$x(t) = \frac{a_0}{2} + \sum_{k=1}^{\infty} (a_k \cdot \cos(k \cdot \omega t) + b_k \cdot \sin(k \cdot \omega t)) ,$$

with the **Fourier coefficients**

$$a_k = \frac{2}{T} \int_0^T x(t) \cdot \cos(k\omega t) dt , \qquad k = 0, 1, 2, 3, \ldots$$

and

$$b_k = \frac{2}{T} \int_0^T x(t) \cdot \sin(k\omega t) dt , \qquad k = 1, 2, 3, \ldots ,$$

where $\omega = 2\pi/T$. The Fourier amplitudes specify the weights of the individual frequency components in the periodic function $x(t)$.

$k = 1$:   **fundamental oscillation (first harmonic)**
$k = 2$:   **first overtone (second harmonic)**
$k = 3$:   **second overtone (third harmonic)**

2. Fourier analysis,

investigation of the frequencies and amplitudes of the harmonic components occurring in the decomposition of a given periodic function.

Fourier spectrum, representation of the result of a Fourier analysis by a frequency-amplitude plot, showing the amplitudes of the Fourier terms of the sum as vertical lines against the corresponding frequencies (**Fig. 8.24**).

Fourier synthesis, construction of a complex time signal out of several sine and cosine functions of different frequencies and amplitudes.

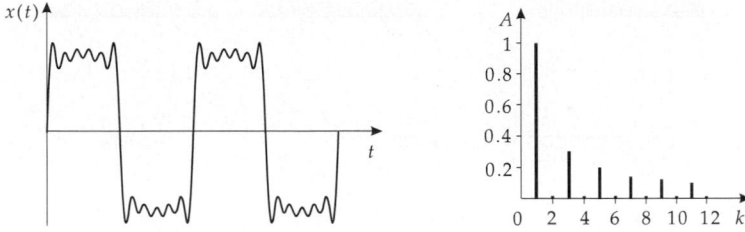

Figure 8.24: Fourier analysis of a periodic function.

3. Complex representation of a Fourier series

The representation of the Fourier series by complex functions reads:

$$x(t) = \sum_{k=-\infty}^{\infty} c_k \cdot e^{j\omega \cdot k \cdot t} \, ,$$

with the coefficients

$$c_k = \frac{1}{T} \int_{-T/2}^{T/2} x(t) \cdot e^{-j\omega \cdot k \cdot t} \, dt \, , \quad k = ..., -2, -1, 0, 1, 2, ... $$

T is the period of the analyzed signal.

➤   Relation between the coefficients a_n, b_n and c_n:

$n = 0$: $a_0 = 2c_0$.

$n > 0$: $a_n = c_n + c_{-n}$, $b_n = j(c_n - c_{-n})$.

In acoustics, sound waves are analyzed by Fourier decomposition. Tones involving only one Fourier term sound "synthetic." The musical impression made by a sound is determined by the type and amplitude of an admixture of additional terms.

In "synthetic" music (synthesizer), any instrument or voice may be "Fourier"-synthesized by computer.

8.5   Coupled vibrations

1. Vibrations of coupled oscillating systems

Coupled vibrations, vibrations originating in systems consisting of several self-oscillating subsystems that affect each other. The subsystems may exchange energy among their individual elements.

In the following, coupling is considered for the example of two pendula connected by a helical spring (**Fig. 8.25**) as an example of coupled vibrations.

Assumption:
- Both pendula have equal mass m and equal string length l, hence equal restoring force coefficient k and period T.
- **Weak coupling**, the coupling between the oscillators is much weaker than the restoring force of the vibrators themselves.

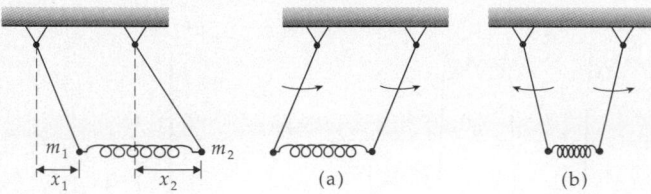

Figure 8.25: Coupled pendula. (a): equal-phase vibration, (b): opposite-phase vibration.

➤ Systems with only one oscillator have one fixed frequency at which the free oscillator vibrates. For several oscillators coupled to each other, different kinds of vibration (**vibration modes**) generally arise.

2. Fundamental vibrations,

the vibration modes of a coupled system in which the oscillators do not exchange energy. Fundamental vibrations of two coupled pendula:

- **equidirectional** or **equal-phase vibration** in which both pendula synchronously carry out the same motion,
- **opposite-directional** or **opposite-phase vibration**, in which both pendula synchronously vibrate against each other.

  A displacement of only one oscillator leads to a decay of its vibration while the second oscillator starts vibration. Then the vibration of the second oscillator fades away and the first one again starts to vibrate. The total energy of the system is continuously exchanged between the two pendula.

3. Equations of motion for coupled identical oscillators

equations of motion of two coupled pendula			
$m\ddot{x}_1(t) = -kx_1 - k_{12}(x_1 - x_2)$	Symbol	Unit	Quantity
$m\ddot{x}_2(t) = -kx_2 - k_{12}(x_2 - x_1)$	m	kg	mass of pendulum
$m(\ddot{x}_1 - \ddot{x}_2) = -(k + 2k_{12})(x_1 - x_2)$	k	kg/s^2	restoring-force coefficient of single pendulum
$m(\ddot{x}_1 + \ddot{x}_2) = -k(x_1 + x_2)$			
$x_1(t) = A \sin\left(\dfrac{\omega_I + \omega_{II}}{2}t\right)$	k_{12}	kg/s^2	restoring-force coefficient of coupling spring
$\cdot \cos\left(\dfrac{\omega_I - \omega_{II}}{2}t\right)$	$x_{1,2}$	m	displacements of pendulum 1, 2
$x_2(t) = A \sin\left(\dfrac{\omega_I - \omega_{II}}{2}t\right)$	$\ddot{x}_{1,2}$	m/s^2	accelerations of pendulum 1, 2
$\cdot \cos\left(\dfrac{\omega_I + \omega_{II}}{2}t\right)$	A	m	amplitude
$\omega_I = \sqrt{\dfrac{k}{m}} = \omega$	$\omega_{I,II}$	1/s	angular frequency of fundamental vibrations
$\omega_{II} = \sqrt{\dfrac{k + 2k_{12}}{m}}$	t	s	time

Here k_{12} is the restoring-force coefficient of the coupling spring between the pendula. The solution given above is obtained by forming $(F_1 - F_2)/m$ and $(F_1 + F_2)/m$, then solving the equation for the new variables $x_1 - x_2$ and $x_1 + x_2$, and from there calculating again x_1 and x_2.

➤   Each of the two oscillators carries out a beat-like oscillation with a time delay between them.

The fundamental vibrations are contained in the solutions $x_1(t)$ and $x_2(t)$:

• Equal-phase fundamental mode, $x_1(t) = x_2(t)$:
  The equations for F_1 and F_2 reduce to two decoupled equations for a simple pendulum. The solution to the differential equation yields the angular frequency $\omega_I = \omega = \sqrt{k/m}$ of the free vibration of the uncoupled oscillators.

• Opposite-phase fundamental mode, $x_1(t) = -x_2(t)$:
  The solution yields the angular frequency $\omega_{II} = \sqrt{(k + 2k_{12})/m}$ for each of the two oscillators.

4. Angular frequencies of the fundamental modes

The angular frequency ω_I of the **equidirectional**, or synchronous, fundamental mode equals the angular frequency of the individual pendula ω, since the coupling is not in effect and hence does not influence the pendulum vibration,

$$\omega_I = \omega = \sqrt{k/m}.$$

For the **opposite-directional**, or antisynchronous, fundamental mode, the restoring force differs from that without coupling. The approximate description of the vibration process by a linear formulation for the restoring force of the individual oscillator with a modified restoring force coefficient,

$$F = k'x \quad \text{with} \quad k' \neq k = \frac{mg}{l},$$

yields

$$\omega_{II} = \sqrt{\frac{k + 2k_{12}}{m}}.$$

Hence, the restoring force of the opposite-phase fundamental mode corresponds to a restoring force coefficient $k' = k + 2k_{12}$.

Beat, occurs for both pendula if the system does not perform one of the fundamental oscillations. Example: Only one pendulum is moved out of the rest position and then released. It then transfers its energy completely to the other pendulum and forces it to oscillate. Then the second pendulum transfers the energy back to the first one, and so forth.

This motion is a superposition of the two fundamental modes.

Coupling coefficient, K, of two identical self-oscillatory systems, defined as

$$K = \frac{\omega_I^2 - \omega_{II}^2}{\omega_I^2 + \omega_{II}^2}$$

with the fundamental frequencies ω_I and ω_{II}. For weak coupling $K \ll 1$, $\omega_I \approx \omega_{II}$.

➤   The principle of coupled oscillators is employed in the ballast tanks of ships to reduce rolling motion at sea. The rolling motion of the ship is transfered to water in a tank at the bottom of the ship. The flow of water is strongly damped and hence the vibrational energy of the ship is in the end converted into heat.

9
Waves

Waves, propagation of a vibrational state, periodic in space and time, in which energy is transported without simultaneous mass transport.

Systems in which waves arise may be envisaged as being composed of infinitely many mutually coupled oscillators. The vibrational state of the individual oscillator depends on space and time. The energy of the system is permanently redistributed among the oscillators.

Free waves arise if no external force acts on the system, and no energy is lost (e.g., by friction). The wave progagates because of the coupling between neighboring oscillators.

Mechanical model of a wave, e.g., by a finite number of pendula weakly coupled to the nearest neighbors by springs. Apart from the displacement of the masses, all pendula remain at their position, only the energy is transferred from pendulum to pendulum (see p. 283).

Waves are described by a function of the form $f(\vec{r}, t)$ where f represents the displacement of the oscillator at the space point $\vec{r}$ and time t.

➤ **Shock waves**, non-periodic waves of large amplitude that may be connected with mass transport. The speed of propagation depends on the amplitude (**nonlinear wave**). The superposition principle does not hold for shock waves.

9.1 Basic features of waves

1. Description of waves by the wave equation

Wave equation, linear partial differential equation of second order in space and time for the function $f(\vec{r}, t)$. The wave equation governs the propagation of the wave in space and time:

$$\triangle f(\vec{r}, t) - \frac{1}{c^2} \frac{\partial^2 f(\vec{r}, t)}{\partial t^2} = \frac{\partial^2 f(\vec{r}, t)}{\partial x^2} + \frac{\partial^2 f(\vec{r}, t)}{\partial y^2} + \frac{\partial^2 f(\vec{r}, t)}{\partial z^2} - \frac{1}{c^2} \frac{\partial^2 f(\vec{r}, t)}{\partial t^2} = 0.$$

The most general solution to the wave equation is a superposition of waves propagating with the same velocity c in any direction $\vec{e}_i$,

$$F(\vec{r}, t) = \sum_i f_i \left(t - \frac{\vec{e}_i \cdot \vec{r}}{c} \right) .$$

➤   It is in general easier to consider infinitely extended waves for the mathematical description of wave phenomena. But in nature, as a rule, only spatially confined waves occur. This limitation manifests itself in the shape of the solutions of the wave equation, which has to be solved with the corresponding boundary conditions.

2. Phase and wave front of a wave

Phase of a wave, the argument of the solution function f, written in the form $\omega t - \vec{k} \cdot \vec{r} + \phi$, a quantity which describes the vibrational state of the wave.

   Wave front, **normal surface**, the set of points $\vec{r}$ where f at a time t has the same phase (**Fig. 9.1**).

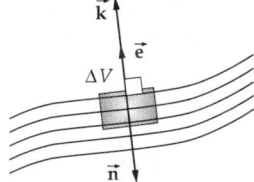

Figure 9.1: Wave front, wave normal $\vec{n}$, wave vector $\vec{k}$, propagation vector $\vec{e}$, ΔV volume element.

➤   Since the wave is periodic in space, there are always infinitely many wave fronts. According to the shape of the wave front one distinguishes:
- plane waves,
- cylindrical waves,
- spherical waves.

➤   Any wave front may be considered a plane in sufficiently small spatial regions ΔV.
Normal, normal to the wave front.

3. Wave vector and wave number

Wave vector, $\vec{k}$, constant vector appearing in the solution to the wave equation. Its meaning may be understood by considering the function $f(\omega t - \vec{k} \cdot \vec{r} + \phi)$ for the case $t = 0$. Then f has the same value for all points $\vec{r}$ with $\vec{k} \cdot \vec{r} = $ const., i.e., for points $\vec{r}$ lying on planes perpendicular to $\vec{k}$. The planes of equal phase move parallel to each other with velocity c along the direction of $\vec{k}$. Hence, the vector $\vec{k} = k \cdot \vec{e}$ represents the **direction of wave propagation** (**Fig. 9.1**),

$$\vec{k} = k \cdot \vec{n} , \quad \vec{n}: \text{ unit vector of the normal to the wave or wave normal} .$$

A wave propagating in the opposite direction has a wave vector $-\vec{k}$.

   Propagation vector, $\vec{e}$, $\hat{\vec{k}}$, wave vector normalized to 1,

$$\vec{e} = \hat{\vec{k}} = \vec{k}/k .$$

Wave number, k, magnitude of the wave vector $|\vec{k}|$.

4. Phase velocity, frequency and wavelength of waves

Phase velocity, c, the velocity of the moving wave fronts. For sound, c is the speed of sound, for light c is the velocity of light in the corresponding medium.

Period, T, the time after which at a fixed space point the vibrational motion repeats itself (**Fig. 9.2**, left).

Frequency, f, number of repetitions per second of a defined vibrational state at a fixed space point (**Fig. 9.2**, right).

Angular frequency, ω, analogous to the definition for vibrations: $\omega = 2\pi f$.

Wavelength, λ, distance between two successive wave fronts of equal phase, characteristic quantity of spatial periodicity. The relation between wave number and wavelength is

$$k = \frac{2\pi}{\lambda}, \quad \lambda = \frac{2\pi}{k}.$$

Periodicity in time: $\omega \cdot T = 2\pi$, **periodicity in space:** $k \cdot \lambda = 2\pi$.

The phase velocity (**Fig. 9.3**) is given by:

phase velocity			$\mathbf{LT^{-1}}$
	Symbol	Unit	Quantity
$c = \dfrac{\omega}{\|\vec{k}\|} = \dfrac{\lambda}{T} = \lambda f$	c	m/s	phase velocity
	ω	rad/s	angular frequency
	$\vec{k}$	1/m	wave vector
	T	s	period
	f	Hz	frequency
	λ	m	wavelength

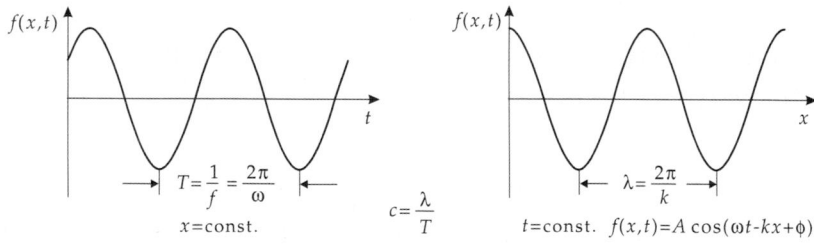

Figure 9.2: Frequency f and wavelength λ of a harmonic wave. c: phase velocity.

5. Phase velocity of various waves

a) Longitudinal waves in liquids:

$$c = \sqrt{K/\rho}, \quad K \text{ compression modulus}, \quad \rho \text{ density}.$$

b) Longitudinal waves in gases:

$$c = \sqrt{\kappa\, p/\rho}, \quad \kappa \text{ isentropic exponent}, \quad p \text{ pressure}, \quad \rho \text{ density}.$$

c) Torsional waves in rods (circular cross-section):

$$c = \sqrt{G/\rho}, \quad G \text{ shear modulus}, \quad \rho \text{ density}.$$

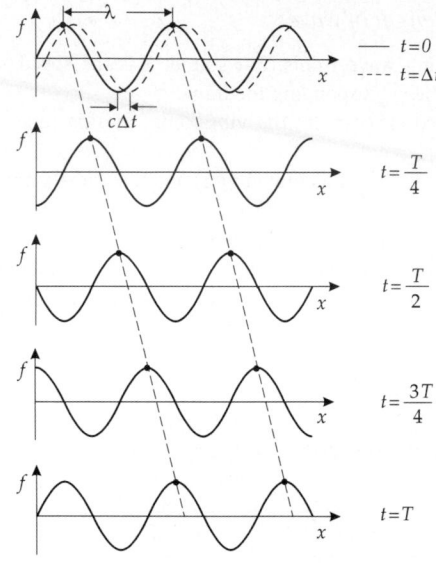

Figure 9.3: Propagation of a harmonic wave.

d) *Transverse waves on a string:*

$$c = \sqrt{\frac{F}{A\,\rho}}, \quad F \text{ tension force}, \quad A \text{ string cross section}, \quad \rho \text{ density}.$$

e) *Electromagnetic waves in a vacuum:*

$$c = \frac{1}{\sqrt{\varepsilon_0 \cdot \mu_0}},$$

ε_0 electric field constant, μ_0 magnetic field constant (see p. 454 and 469).

f) *Electromagnetic waves in a medium:*

$$c = \frac{1}{\sqrt{\varepsilon_r \cdot \varepsilon_0 \cdot \mu_r \cdot \mu_0}},$$

ε_0 electric field constant, μ_0 magnetic field constant, ε_r relative permittivity, μ_r relative permeability.

6. *Plane and spherical waves as special solutions of the wave equation*

a) *Plane wave,* the wave fronts are planes perpendicular to the propagation vector (see **Fig. 9.4**).

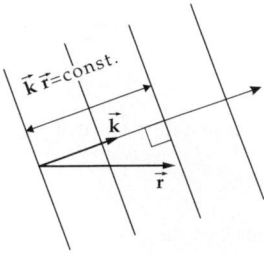

Figure 9.4: Wave fronts of a plane wave.

plane wave

	Symbol	Unit	Quantity
$f(\vec{\mathbf{r}}, t) = A\cos(\omega t - \vec{\mathbf{k}}\vec{\mathbf{r}} + \phi)$	$f(\vec{\mathbf{r}}, t)$		displacement at position $\vec{\mathbf{r}}$ at instant t
	A		amplitude
$\vec{\mathbf{k}}^2 = \dfrac{\omega^2}{c^2}$	ω	rad/s	angular frequency
	t	s	time
	$\vec{\mathbf{k}}$	1/m	wave vector
$\lambda = \dfrac{2\pi}{|\vec{\mathbf{k}}|}$	$\vec{\mathbf{r}}$	m	position
	ϕ	rad	phase shift
	c	m/s	phase velocity
	λ	m	wavelength

b) Spherical wave, spherically symmetric solution to the wave equation. The wave fronts are surfaces of concentric spheres around the source at $r = 0$ ($\vec{\mathbf{k}} \cdot \vec{\mathbf{r}} = kr$) (**Fig. 9.5**):

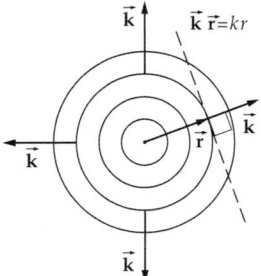

Figure 9.5: Wave fronts of a spherical wave.

spherical wave

	Symbol	Unit	Quantity
$f(\vec{\mathbf{r}}, t) = \dfrac{A}{|\vec{\mathbf{r}}|}\cos(k|\vec{\mathbf{r}}| - \omega t + \phi)$	$f(\vec{\mathbf{r}}, t)$		local displacement
	A		amplitude
	ω	rad/s	angular frequency
	t	s	time
	$\vec{\mathbf{k}}$	1/m	wave vector
	$\vec{\mathbf{r}}$	m	position
	ϕ	rad	phase shift
	c	m/s	phase velocity
	λ	m	wavelength

7. Complex representation of waves

Plane wave:

$$f(\vec{\mathbf{r}}, t) = e^{-j(\omega t - \vec{\mathbf{k}}\vec{\mathbf{r}})}.$$

Spherical wave:
- outgoing from point $r = 0$

$$f(\vec{r}, t) = e^{-j(\omega t - kr)},$$

- converging to point $r = 0$

$$f(\vec{r}, t) = e^{-j(\omega t + kr)}.$$

8. Superposition principle and the Huygens principle

Superposition principle, linear waves overlay each other without mutual interaction. The resulting displacement at position $\vec{r}$ at instant t is the sum of the displacements of all individual waves.

➤ The superposition principle does not hold for nonlinear waves (**shock waves**, **gravity waves**).

Huygens principle, principle for constructing wave fronts in wave propagation (**Fig. 9.6**).

▲ Any point of a wave front serves as starting point of an **elementary wave**. The wave front at a later instant is obtained as the envelope of the superposition of all elementary waves emerging from a given wave front.

▲ **Elementary waves** are outgoing **spherical waves**. The wave front of an elementary wave emitted at instant t has after the time Δt the radius $r = c \cdot \Delta t$. Except for the direction of the normal to the total wave front, the elementary waves mutually cancel each other by interference.

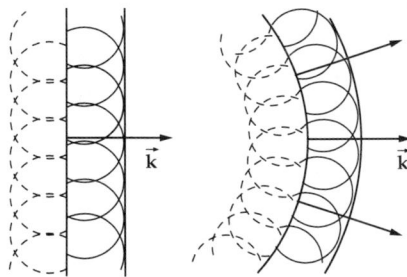

Figure 9.6: Propagation of a wave front according to the Huygens principle. (a): plane wave, (b): spherical wave.

9. Vector waves

Many physical quantities, e.g., magnetic or electric field strengths, are vector quantities and are described by a vector-wave equation,

$$\Delta \vec{g}(\vec{r}, t) - \frac{1}{c^2} \frac{\partial^2 \vec{g}(\vec{r}, t)}{\partial t^2} = \frac{\partial^2 \vec{g}(\vec{r}, t)}{\partial x^2} + \frac{\partial^2 \vec{g}(\vec{r}, t)}{\partial y^2} + \frac{\partial^2 \vec{g}(\vec{r}, t)}{\partial z^2} - \frac{1}{c^2} \frac{\partial^2 \vec{g}(\vec{r}, t)}{\partial t^2} = 0,$$

with the vector quantity $\vec{g}(\vec{r}, t)$. The function $\vec{g}$ (and hence the wave equation) may be decomposed into the Cartesian components $g_x(\vec{r}, t)$, $g_y(\vec{r}, t)$, $g_z(\vec{r}, t)$, which are the solutions to the scalar-wave equations. Special solutions are given above.

■ In electrodynamics, $\vec{g}$ stands, e.g., for the vectors of the magnetic flux density $\vec{B}$ or the electric field strength $\vec{E}$ that obey vector-wave equations.

Vector wave, solution of the vector-wave equation, e.g., a plane wave,

$$\vec{g}(\vec{r}, t) = \vec{A} \cos(\omega t - \vec{k}\vec{r} + \phi), \quad \vec{k}^2 - \frac{\omega^2}{c^2} = 0.$$

The vector $\vec{A}$ specifies both the **wave amplitude**, $|\vec{A}|$, and the **orientation of displacement** of the oscillators, $\hat{\vec{A}} = \vec{A}/A$.

➤   The vectors $\vec{g}(\vec{r}, t)$ and $\vec{A}$ may be decomposed into their components g_x, g_y, g_z and A_x, A_y, A_z with respect to a Cartesian coordinate frame. These components are solutions to the corresponding scalar-wave equation.

10. *Longitudinal wave,*

wave in which the propagation vector $\hat{\vec{k}}$ and the displacement of the individual oscillators $\vec{A}$ are parallel to each other (**Fig. 9.7**),

$$\vec{A} = |\vec{A}| \hat{\vec{k}}.$$

■   A helical spring lying on a support plane is given an impulse along its longitudinal axis. The induced local compression propagates along the spring; the individual sections of the spring vibrate along the spring axis, which thus defines the propagation vector of the compression wave.

➤   Sound is an example of a longitudinal wave in which pressure variations, hence compression waves, propagate through the medium.

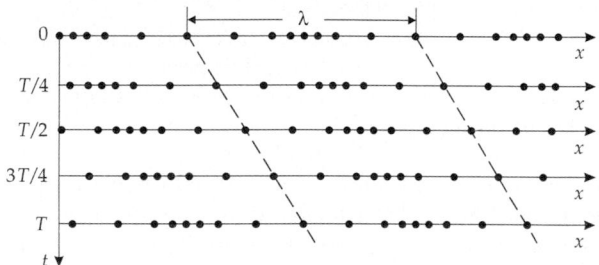

Figure 9.7: Propagation of a longitudinal wave.

11. *Transverse wave,*

wave in which the oscillators vibrate perpendicularly to the wave-propagation vector,

$$\hat{\vec{k}} \cdot \vec{A} = 0.$$

■   If the end of an extended rope is quickly moved up and down, crests and troughs run along the rope. The individual sections of the rope are displaced perpendicularly to the rope axis while the wave itself travels along the rope.

➤   Electromagnetic waves are transverse waves, with the electric- and magnetic-field vectors pointing perpendicularly to the wave-progagation vector.

9.2 Polarization

Polarization, orientation of the wave vector $\vec{k}$ with respect to the wave displacement vector $\hat{\vec{A}}$.

Longitudinal polarization, the wave number vector is parallel to the local wave displacement vector.

Transverse polarization, the wave-number vector points perpendicularly to the wave-displacement vector.

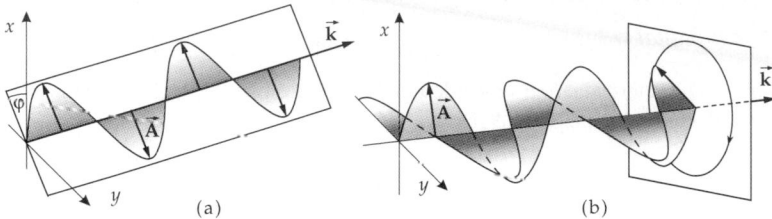

Figure 9.8: Polarization of transverse waves. (a): linear polarization, (b): circular polarization.

Distinction between transverse waves according to the behavior of the displacement vector:

- **Linear polarization**, the displacement vector $\vec{A}$ does not change its orientation in a plane perpendicular to $\vec{k}$ (**Fig. 9.8a**).
- **Elliptic polarization**, the displacement vector $\vec{A}$ rotates in the plane perpendicular to $\vec{k}$. The end point of $\vec{A}$ describes an ellipse in this plane.
- **Circular polarization**, the displacement vector $\vec{A}$ rotates in the plane perpendicular to $\vec{k}$. The end point of $\vec{A}$ describes a circle, special case of elliptic polarization (**Fig. 9.8b**).

If $\vec{k}$ points in z-direction of a coordinate frame, the displacement vector $\vec{A}$ lies in the x-y plane. The rotation of $\vec{A}$ may be represented by a superposition of two linearly polarized vibrations along the x- and y-axis, respectively: $x(t) = A\sin(\omega t - \phi_x)$ and $y(t) = B\sin(\omega t - \phi_y)$.

9.3   Interference

Interference, notation of phenomena occurring in the superposition of waves. In the more restricted sense interference means superposition of **coherent** waves.

9.3.1   Coherence

Coherent waves: Two waves are coherent if their phase difference does not vary with time.
■   A laser generates **coherent** monochromatic light.
An extended conventional light source can generate coherent waves through reflection of a focused beam by a half transparent mirror.
■   **Coherent** waves may be generated with an extended conventional source of light by reflection of a pencil-like ray by a semitransparent mirror, or by a plane-parallel plate.
■   Two waves without an established phase relation are called **incoherent**.
Wave train, a wave confined in time and space that is composed of a superposition of infinitely many waves of different frequencies and phase shifts.

Coherence, the property of wave trains of being able to interfere. The effects resulting from superposition can be experimentally demonstrated in the time average.

For wave trains, interference may be detected if the waves are superimposed in the region of observation, and if the intensity maxima and minima do not permanently vary in position.

Coherence length, l, largest path difference of two wave trains for which an interference may just still take effect. If a wave train is generated (for light: emitted) within the time interval τ:

coherence length			**L**
	Symbol	Unit	Quantity
$l = c\tau$	l	m	coherence length
	c	m/s	propagation speed
	τ	s	time interval

9.3.2 Interference

For linear waves, the superposition principle holds.

▲   The instantaneous displacement of the resulting wave at a given spatial position is obtained by adding the instantaneous displacements of all partial waves at this position (**Fig. 9.9**).

1. Examples of interference

Superposition of two waves of identical amplitude A and angular frequency ω, but different phase angles ϕ, propagating in the same direction. First wave:

$$y_1(x, t) = A \cos(\omega t - kx + \phi_1).$$

Second wave:

$$y_2(x, t) = A \cos(\omega t - kx + \phi_2).$$

Using the addition theorem for the cosine function, one gets for the resulting wave

$$y_{\text{res}}(x, t) = y_1(x, t) + y_2(x, t)$$

$$= 2A \cos\left(\omega t - kx + \frac{\phi_1 + \phi_2}{2}\right) \cdot \cos\left(\frac{\Delta\phi}{2}\right)$$

with the **phase difference**

$$\Delta\phi = \phi_1 - \phi_2.$$

2. Path difference and intensity in interference

Path difference, δ, for a given phase difference $\Delta\phi$ defined as follows:

path difference			**L**
	Symbol	Unit	Quantity
$\delta = \dfrac{\Delta\phi}{2\pi}\lambda$	δ	m	path difference
	$\Delta\phi$	rad	phase difference
	λ	m	wavelength

Intensity, I, notation for the square of the wave amplitude.
The resulting wave described above has the intensity

$$I = 2A^2 \left(1 + \cos(\Phi_1 - \Phi_2)\right), \quad \Phi_1 = \omega_1 t + \phi_1, \quad \Phi_2 = \omega_2 t + \phi_2.$$

If two waves of frequencies f_1 and f_2 are superposed, the intensity of the resulting wave has a period T (see p. 279):

$$T = \left| \frac{1}{f_1} - \frac{1}{f_2} \right|.$$

If the time of observation is essentially larger than T, only the mean value of the intensity can be measured:

$$\bar{I} = 2A^2 = I_1 + I_2,$$

i.e., the **interference term** $2A \cos(\Phi_1 - \Phi_2)$ drops out. The same holds in general for the superposition of incoherent waves and of wave trains:

▲   In the superposition of incoherent waves, there is no interference; the intensities of the individual waves simply sum.

3. Special cases of interference

- **Constructive interference**, **enhancement**, $\delta = n\lambda$, n integer. Superposing waves of equal amplitude may lead to maximum enhancement, the amplitude of the resulting wave being then twice the amplitude of the initial waves.
- **Destructive interference**, **cancellation**, $\delta = (2n + 1)\lambda/2$, n integer. The waves cancel each other, the resulting wave has zero amplitude.
- $\delta = (n + 1/4)\lambda$, n integer. The resulting amplitude is $\sqrt{2}A$, the phase of the resulting wave is shifted so that its zero passages are between those of the original waves.

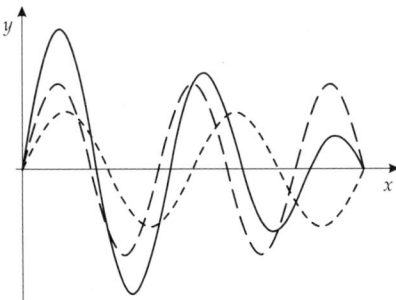

Figure 9.9: Interference. Superposition of two waves y_1, y_2 (dashed lines) of different frequencies and amplitudes at a fixed time as function of the position x. The resulting wave y is shown by the solid line.

9.3.3 Standing waves

Standing waves, are generated by superposition of two waves of equal frequency, amplitude and phase, but with opposite directions of propagation (**Fig. 9.10**). The wave numbers $(\vec{k}, -\vec{k})$ of both waves have the same magnitude, but are antiparallel.

Mathematical description:

$$y_1(\vec{r}, t) = A \cos(\vec{k} \cdot \vec{r} - \omega t),$$

$$y_2(\vec{r}, t) = A \cos(-\vec{k} \cdot \vec{r} - \omega t),$$

$$y(\vec{r}, t) = y_1(\vec{r}, t) + y_2(\vec{r}, t) = -2A \cos(\omega t) \cos(\vec{k} \cdot \vec{r}).$$

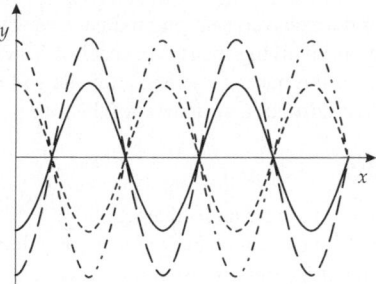

Figure 9.10: Standing wave. Displacement y at the positions x for different instants of time.

▲   The minima and maxima of the standing wave are fixed in space.
Node, notation for a space-fixed minimum of a standing wave.
   Antinode, notation for a space-fixed maximum of a standing wave.

9.3.3.1   Standing waves in rods tightly mounted at one end

If a density wave travels along a rod of length l, it is reflected at its ends. The end at which the rod is attached forms a fixed end (**Fig. 9.11**). Standing waves arise in the rod if the wavelength λ_n is:

standing wave: one free, one fixed end			L
	Symbol	Unit	Quantity
$\lambda_n = \dfrac{4l}{2n+1}$	λ_n	m	wavelength
	l	m	rod length
	n	1	number of nodes

These standing waves are called the **natural vibrations of the rod**. Waves of the same type also arise for half-closed pipes. The node number n (≥ 0) corresponds to the number of nodal points of the standing wave; the node at the fixed end is not counted.

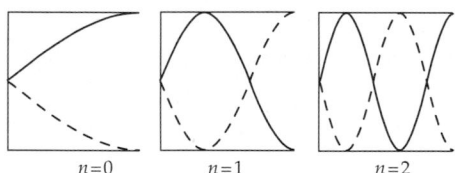

$$n=0 \qquad\qquad n=1 \qquad\qquad n=2$$

Figure 9.11: Natural vibrations of a rod with one free and one fixed end.

Fundamental vibration, standing wave with $n = 0$. Its wavelength is

$$\lambda_0 = 4l \,.$$

Fundamental frequency, f_0, frequency of the fundamental vibration,

$$f_0 = \frac{c}{\lambda_0} = \frac{c}{4l} \,,$$

where c is the phase velocity of the density wave in the rod.
   Harmonic, standing wave with a node number n different from zero.

M   A rod is excited into oscillation by giving it a transverse or longitudinal impulse. The impulse generates a complicated excitation involving many frequencies. Vibrations with frequencies that do not correspond to the natural vibrations of the rod decay much more rapidly than vibrations with the **natural frequencies** of the rod.

9.3.3.2   Standing waves on strings

String, elastic object having a length considerably larger than its diameter.

If a string is attached at both ends, transverse waves may be excited that are reflected at the fixed ends. For suitable values of the wavelength, there arise standing waves called the **natural vibrations** of the string (**Fig. 9.12**).

Condition for the wavelength λ_n:

standing wave: two fixed ends			**L**
$\lambda_n = \dfrac{2l}{n+1}$	Symbol	Unit	Quantity
	λ_n	m	wavelength
	l	m	string length
	n	1	node number (≥ 0)

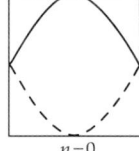 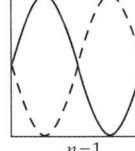 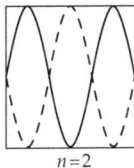

$$n=0 \qquad\qquad n=1 \qquad\qquad n=2$$

Figure 9.12: Natural vibrations of a string with two fixed ends.

Fundamental vibration, standing wave for the case $n = 0$, wavelength

$$\lambda_0 = 2l .$$

Fundamental frequency, f_0, frequency of the fundamental vibration,

$$f_0 = \frac{c}{\lambda_0} = \frac{c}{2l} ,$$

where c is the phase velocity of the wave on the string.

The pitch (frequency of the fundamental mode) decreases with increasing diameter of the string (e.g., piano strings).

9.3.3.3   Standing waves in Kundt's tube

M   **Kundt's tube**, device to make longitudinal standing waves visible in air. It consists of a glass tube, closed at one end by a vibrating membrane (e.g., loudspeaker), and a movable piston that closes the other end. The bottom of the (horizontal) tube is covered with cork powder.

The membrane excites the air column in the tube into vibration. The length of the column may be controlled by the position of the piston. The waves are reflected at the piston surface (fixed end), hence standing waves may arise for appropriate tube lengths. At the positions of vibrational nodes, the cork powder remains at rest, while it spreads perpendicular to the tube axis at the positions of antinodes.

➤ By shifting the piston, the length of the air column may be varied, and thus the resonance condition for formation of standing waves may be observed.

➤ The pressure distribution along the air column may be visualized by a similar device, the flame tube invented by Rubens.

Resonance condition for the wavelength λ_n in the case of two free ends (**Fig. 9.13**):

standing wave: two free ends			L
$\lambda_n = \dfrac{2l}{n+1}$	Symbol	Unit	Quantity
	λ_n	m	wavelength
	l	m	tube length
	n	1	integer, non-negative

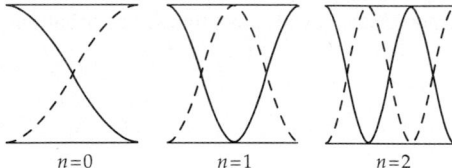

$$n=0 \qquad\qquad n=1 \qquad\qquad n=2$$

Figure 9.13: Normal vibrations in Kundt's tube with two free ends.

Fundamental vibration, standing wave for the case $n = 0$, wavelength

$$\lambda_0 = 2l .$$

Fundamental frequency, f_0, frequency of the fundamental vibration,

$$f_0 = \frac{c}{\lambda_0} = \frac{c}{2l} ,$$

where c is the phase velocity of the wave in air (sound velocity).

9.3.4 Waves with different frequencies

1. Superposition of two harmonic waves

Two harmonic waves

$$y_1(x, t) = A \cos(\omega_1 t - k_1 x) ,$$
$$y_2(x, t) = A \cos(\omega_2 t - k_2 x)$$

with **different** frequencies and wave numbers, but **equal** amplitude sum to form

$$y(x, t) = y_1(x, t) + y_2(x, t) = 2A \cos(\omega t - kx) \cos(\Delta\omega t - \Delta kx)$$

with

$$\omega = \frac{\omega_1 + \omega_2}{2} , \quad k = \frac{k_1 + k_2}{2} , \quad \Delta\omega = \frac{\omega_1 - \omega_2}{2} , \quad \Delta k = \frac{k_1 - k_2}{2} .$$

This corresponds to a wave of angular frequency ω whose amplitude is modulated with the frequency $\Delta\omega$.

Envelope of the wave:

$$\cos(\Delta\omega t - \Delta k x).$$

Group velocity, v_{gr}, velocity of motion of the wave envelope,

$$v_{gr} = \frac{\Delta\omega}{\Delta k} = \frac{\omega_1 - \omega_2}{k_1 - k_2}.$$

2. Wave packet,

wave group, spatially confined (**localized**) wave that may be generated by superposition of infinitely many harmonic waves with a continuous distribution $c(\vec{k})$ of wave vectors (**Fourier synthesis**):

$$f(\vec{r}, t) = \int c(\vec{k}) \cos(\omega t - \vec{k}\vec{r}) \, d^3\vec{k}, \quad \vec{k} = k(\omega)\vec{e}, \quad \vec{e}: \text{propagation vector}.$$

The wave packet may be generated for any envelope by an appropriate choice of the distribution $c(\vec{k})$.

Group velocity of a wave packet in a medium, v_{gr}, defined as $\dfrac{d\omega}{d\vec{k}}$.

group and phase velocity in one-dimensional medium			LT^{-1}
	Symbol	Unit	Quantity
$v_{gr} = v - \lambda \dfrac{dv}{d\lambda}$	v_{gr}	m/s	group velocity in medium
	v	m/s	phase velocity in medium
$v_{gr} = \dfrac{d\omega}{dk}$	λ	m	wavelength
	k	1/m	wave number
	ω	1/s	angular frequency

Group velocity and phase velocity differ if the propagation speed of waves in a medium depends on the wavelength, i.e., if dispersion occurs (see p. 305).

The transport of energy (more general: information) by a wave packet proceeds with the group velocity.

9.4 Doppler effect

Doppler effect, frequency and wavelength registered by an observer depend on the relative speed of the source of wave and the observer.

■   The tone of a horn of a car moving towards an observer seems higher than the tone from the car at rest.

The number of wave fronts reaching the observer within a certain time interval changes if the source of waves moves towards or away from the observer (**Fig. 9.14**).

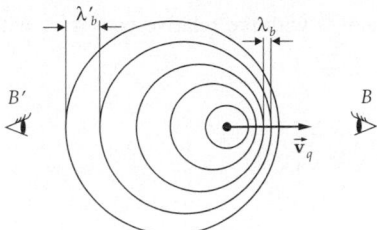

Figure 9.14: Doppler effect. Wave fronts of a source moving with velocity $\vec{v}_q$ in the rest frame of the observers B, B'.
λ_b, λ'_b: wavelength measured by the observer.

1. Case distinction for Doppler effect in a medium

For the Doppler effect in a medium, the relation between the frequency f_q and the wavelength λ_q in the rest frame of the wave source and the frequency f_b and wavelength λ_b in the rest frame of the observer depends on whether source, observer, or both, are moving:

Doppler effect: moving source, observer at rest			
$f_b = \dfrac{f_q}{\left(1 \pm \frac{v_q}{c}\right)}$ $\lambda_b = \left(\dfrac{c}{f_q}\right)\left(1 \pm \dfrac{v_q}{c}\right)$	Symbol	Unit	Quantity
	f_b	Hz	frequency in the observer rest frame
	f_q	Hz	frequency in the source rest frame
	λ_b	m	wavelength in the observer rest frame
	λ_q	m	wavelength in the source rest frame
	v_q	m/s	velocity of source in medium
	c	m/s	phase velocity in medium

➤ In the formulas given above, the plus (minus) sign holds when the source moves away from (towards) the observer.

Doppler effect: source at rest, moving observer			
$f_b = f_q\left(1 \pm \dfrac{v_b}{c}\right)$ $\lambda_b = \dfrac{c}{f_q}\dfrac{1}{\left(1 \pm \frac{v_b}{c}\right)}$	Symbol	Unit	Quantity
	f_b	Hz	frequency in the observer rest frame
	f_q	Hz	frequency in the source rest frame
	λ_b	m	wavelength in the observer rest frame
	λ_q	m	wavelength in the source rest frame
	v_b	m/s	velocity of observer in medium
	c	m/s	phase velocity in medium

➤ In the formulas given above, the plus (minus) sign holds if the observer moves towards (away from) the source.

2. Doppler effect for electromagnetic waves without dispersion,

frequency f' in the **moving reference frame** for:

a) transverse Doppler effect: observer moves with the **relative velocity** v with respect to the source, perpendicular to the propagation vector of the electromagnetic wave:

$$f' = f\sqrt{1 - (v/c)^2}\,;$$

b) longitudinal Doppler effect: observer moves with the **relative velocity** v with respect to the source:

- source moves away from the observer,

$$f' = f \sqrt{\frac{c - v}{c + v}},$$

- source approaches the observer,

$$f' = f \sqrt{\frac{c + v}{c - v}}.$$

$\boxed{\text{M}}$ The Doppler effect is employed in **radar speed measurement,** in which electromagnetic waves are reflected, e.g., by moving cars.

9.4.1   Mach waves and Mach shock waves

Mach wave, conical wave front with the source as apex. Mach waves arise when a source traverses a medium with a velocity v_q that exceeds the propagation speed c of the waves in the medium. The half angle of the apex α of the **Mach cone** may be calculated with the **Mach formula,**

Mach angle α			1
	Symbol	Unit	Quantity
$\sin \alpha = \dfrac{c}{v_q}$	α	rad	half apex angle of Mach cone
	c	m/s	sound velocity in the medium
	v_q	m/s	velocity of source in the medium

- ■ **Supersonic bang** for sound waves.
- ■ **Čerenkov radiation** for electromagnetic waves, generated by charged particles moving with a velocity $v > c/n$, where $n > 1$ is the refractive index of the material through which the wave is passing.

Mach number M, quotient of the velocity of source v_q and the sound velocity c.

- ■ Commercial airliners typically fly at $v < 1000\,$km/h, $M < 1$. The Concorde reaches $M > 2$.

Mach shock waves, arise if the sound velocity of the medium traversed by the source depends on the density of the medium. In general, the sound velocity increases with increasing density. The sound velocity is largest near the source because its motion causes a compression of the medium. The fronts of maximum density therefore deviate from the conical shape by a typical curvature. The Mach formula still holds in the *following sense:* If the tangent to the wave front is shifted along the curved front, the tangent coincides with the cone calculated via the Mach formula at the position of the source. The Mach shock front then lies within this cone.

9.5   Refraction

1. Definition of refraction

Refraction, change of the direction of wave propagation at the interface of two media with different propagation speeds.

Refraction may be interpreted by means of Huygens' concept of elementary waves: Any point of the interface reached by the incident wave front serves as source of an elementary wave with the propagation speed of the medium. The elementary waves then generate a new wave front (**Fig. 9.15**).

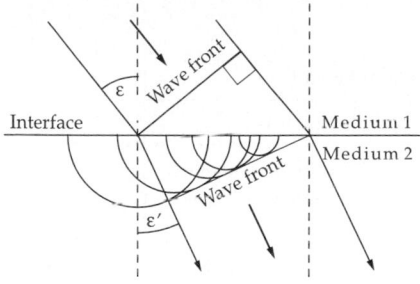

Figure 9.15: Refraction in the wave picture. ε, ε': angles between the propagation vectors before and after refraction, and the normal of the interface.

➤ Refraction can also be explained by Fermat's principle, which says that light propagation between two points proceeds along the optical path for which the minimum transit time is required. One must take into account that the propagation speed of light depends on the local (possibly position-dependent) refractive index of the medium traversed. Hence, the determination of the optical path is a typical **variational problem**.

2. Refraction law

▲ If the propagation speed of the wave in the media is c_1 and c_2, respectively, then for an incidence angle ε, the refraction angle ε' is (**Fig. 9.16**):

refraction law (Snell's law)			
	Symbol	Unit	Quantity
$\dfrac{\sin \varepsilon}{\sin \varepsilon'} = \dfrac{c_1}{c_2}$	ε	rad	incidence angle
	ε'	rad	refraction angle
	c_1, c_2	m/s	wave velocities in medium 1, 2

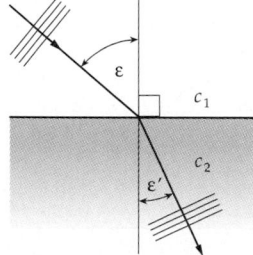

Figure 9.16: Refraction law. ε: angle of incidence, ε': refraction angle. c_1, c_2: wave velocities in medium 1, 2, respectively.

➤ The elementary waves excited at the interface also propagate back into the medium from which the primary wave is incident. Hence, if a wave strikes an interface, it is partly refracted and partly reflected.

9.6 Reflection

Law of reflection (Fig. 9.17):
- The angle of incidence equals the angle of reflection.
- The reflected ray lies in the **plane of incidence**, which includes the incident ray and the normal of the reflecting surface.

angle of incidence = angle of reflection			
	Symbol	Unit	Quantity
$\varepsilon = \varepsilon_r$	ε	rad	angle of incidence
	ε_r	rad	angle of reflection

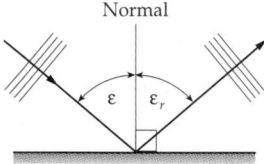

Figure 9.17: Law of reflection. ε: angle of incidence, ε_r: angle of reflection.

9.6.1 Phase relations

1. Phase shift under reflection

The phase of the wave changes under reflection, depending on the type of reflecting interface:
- ▲ If a wave is reflected by an interface behind which the wave velocity is higher than in the medium before the interface, the phase of the reflected wave remains unchanged.
- ▲ If a wave is reflected by an interface behind which the wave velocity is lower than in the medium before the interface, the phase of the reflected wave changes by π.
- ■ If light propagates in a vacuum and hits a plate of glass, the light is reflected with a phase shift of π.

2. Phase relations for mechanical waves

- **Reflection at a free end**, the point where reflection occurs is free to move: no phase shift arises.
- **Reflection at a fixed end**, the point where reflection occurs is less movable than the rest of the system: a phase shift of π radians arises.

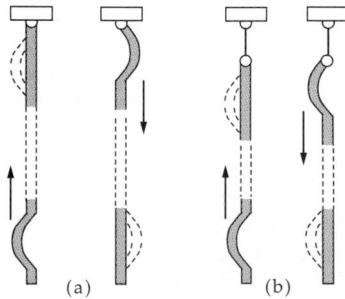

(a)   (b)

Figure 9.18: Phase shift under reflection. (a): reflection at a fixed end, (b): reflection at a free end.

■ If one end of a helical spring is tightly fixed to a rigid wall, and longitudinal or transverse waves run along the spring, the attached end is a fixed end (reflection with phase shift).

If the end of a helical spring is attached to a wall by a thin long cord, this is a free end (no phase shift) (see **Fig. 9.18**).

9.7 Dispersion

Dispersion, dependence of the phase velocity of a wave on its wavelength:

a) *normal dispersion:* the phase velocity v increases with increasing wavelength λ,

$$\frac{\mathrm{d}v}{\mathrm{d}\lambda} > 0, \quad v_{\mathrm{gr}} < v.$$

The group velocity $v_{\mathrm{gr}} = v - \lambda \dfrac{\mathrm{d}v}{\mathrm{d}\lambda}$ (see p. 300) is smaller than the phase velocity v.

b) *anomalous dispersion:* the phase velocity v decreases with increasing wavelength λ,

$$\frac{\mathrm{d}v}{\mathrm{d}\lambda} < 0, \quad v_{\mathrm{gr}} > v.$$

The group velocity $v_{\mathrm{gr}} = v - \lambda \dfrac{\mathrm{d}v}{\mathrm{d}\lambda}$ is larger than the phase velocity v.

c) *no dispersion:* the phase velocity v does not depend on the wavelength λ,

$$\frac{\mathrm{d}v}{\mathrm{d}\lambda} = 0, \quad v_{\mathrm{gr}} = v.$$

The group velocity v_{gr} equals the phase velocity v.

9.8 Diffraction

Diffraction, deviation from the straight propagation of waves. Explanation by Huygens' elementary waves emitted from any point of an object reached by the wave.

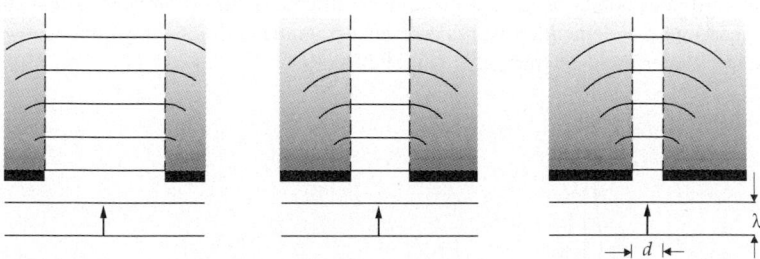

Figure 9.19: Propagation of a plane wave of wavelength λ into the shadow region behind a slit of width d. The diffraction effect increases with decreasing ratio of slit width to wavelength.

Shadow, region behind an object which in the sense of geometric optics is not accessible to rays emitted by the source. The wave intrudes by diffraction into the geometric shadow region behind the object (**Fig. 9.19**). The details of the diffraction pattern are determined by the ratio of the wavelength to the geometric extension of the object.

9.8.1   Diffraction by a slit

1. Diffraction of a plane wave by a slit

Let a plane wave be incident perpendicularly on a long rectangular slit of width d. The wave fronts are then parallel to the aperture plane. Any point in the plane of the slit acts as an emitter of Huygens' elementary waves (**Fig. 9.20**). On a screen far behind the aperture plane, the intensity pattern I_α depends on the diffraction angle α, i.e., the angle through which the wave-propagation vector is deflected:

intensity distribution for diffraction by a slit				1
$$I_\alpha = I_0 \dfrac{\sin^2\left(\dfrac{\pi d \sin\alpha}{\lambda}\right)}{\left(\dfrac{\pi d \sin\alpha}{\lambda}\right)^2}$$	Symbol	Unit	Quantity	
	α	rad	diffraction angle	
	I_α	1	intensity at α	
	I_0	1	intensity at $\alpha = 0$	
	d	m	slit width	
	λ	m	wavelength	

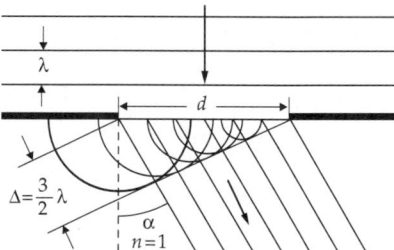

Figure 9.20: Diffraction by a slit according to the wave picture. λ: wavelength, d: slit width, Δ: path difference, α: diffraction angle.

➤   This formula holds only if the distance between the screen and the aperture plane is very large compared with the slit width.

This form of intensity distribution is explained by the fact that the elementary waves emitted from different points of the slit plane have different path differences Δ that depend on α. Therefore the elementary waves emerging from the half-slits may constructively or destructively interfere with each other (**Fig. 9.21**).

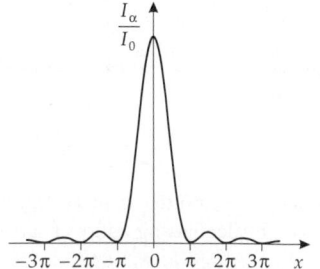

Figure 9.21: Diffraction by a slit of width d. Intensity distribution as function of $x = \pi d \sin(\alpha)/\lambda$.

2. Intensity maxima and minima in diffraction by a slit

Position of intensity minima: angles α_n satisfying the condition

$$\sin \alpha_n = \pm n \frac{\lambda}{d}, \quad n = 1, 2, 3, \ldots.$$

Position of intensity maxima: angles α_n satisfying the condition

$$\sin \alpha_n = \pm \left(n + \frac{1}{2} \right) \frac{\lambda}{d}, \quad n = 1, 2, 3, \ldots.$$

➤   The **main maximum** is the dominant zeroth intensity, or diffraction maximum at $\alpha_n = 0$.

▲   In diffraction by a circular diaphragm of diameter d, the first interference minimum occurs at

$$\sin \alpha = 1.22 \frac{\lambda}{d}.$$

Because of diffraction by circular diaphragms (lens apertures, etc.), optical instruments may image two distinct points only if the points subtend an angle ε,

$$\varepsilon \geq 1.22 \frac{\lambda}{d}.$$

This limitation is called the **resolving power**.

9.8.2 Diffraction by a grating

Let a plane wave be incident on a grating of slit width d and distance g between the slits. Let the number of slits be q (**Fig. 9.22**).

Grating constant g, notation for the distance between the slits ("grooves") of a ruled grating.

The intensity distribution on a screen far behind the plane grating may be explained by the superposition of Huygens' elementary waves emerging from the grating grooves (**Fig. 9.23**). The elementary waves generated by different grooves superpose with path differences depending on the diffraction angle α, as follows:

intensity pattern for a diffraction grating				1
	Symbol	Unit	Quantity	
$I_\alpha = I_0 \dfrac{\sin^2\left(\dfrac{\pi d \sin \alpha}{\lambda}\right)}{\left(\dfrac{\pi d \sin \alpha}{\lambda}\right)^2} \cdot \dfrac{\sin^2\left(\dfrac{q\pi g \sin \alpha}{\lambda}\right)}{\sin^2\left(\dfrac{\pi g \sin \alpha}{\lambda}\right)}$	α	rad	diffraction angle	
	I_α	1	intensity at α	
	I_0	1	intensity at $\alpha = 0$	
	d	m	slit width	
	g	m	grating constant	
	q	1	number of slits	
	λ	m	wavelength	

Position of the intensity maxima at angles α_n satisfying the condition

$$\sin \alpha_n = \pm n \frac{\lambda}{g}, \quad n = 0, 1, 2, \ldots.$$

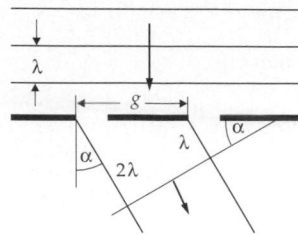

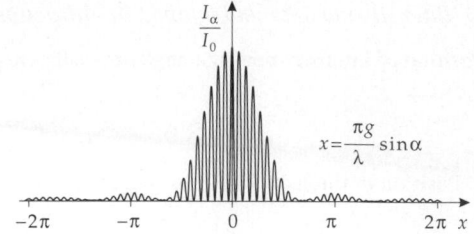

Figure 9.22: Diffraction at a grating with grating constant g. λ: wavelength, α: diffraction angle.

Figure 9.23: Intensity distribution for diffraction at a double slit (grating with $q = 2$) as a function of $x = \pi g \sin(\alpha)/\lambda$. λ: wavelength, g: grating constant, α: diffraction angle.

9.9   Modulation of waves

Waves may be used as carriers of signals. The information is impressed on them during their generation, and this information may be obtained upon receiving the waves.

Modulation, process of impressing information onto a wave in sending.

Demodulation, process of picking up the information when receiving the wave.

Addressing, selection of the receiver of a signal, mostly by selecting a particular frequency of the carrier wave conveying the signal.

Transmitting signals by modulation of electromagnetic waves is an extremely important technology (radio, television, cellular phones, etc.).

1. Amplitude modulation

Amplitude modulation (AM), variation of the amplitude of a high-frequency carrier wave in the sequence of the low-frequency signal to be transmitted. Modulating signal: $\Delta A \sin(\Omega t)$.

The time dependence of the displacement y of an amplitude-modulated wave at a fixed position is then

$$y(t) = (A + \Delta A \sin(\Omega t)) \sin(\omega t),$$

where ω is the angular frequency of the carrier wave, and Ω is the angular frequency of the signal (**Fig. 9.24**).

■   Amplitude modulation is employed in AM radio broadcasting for the long-wave, medium-wave and short-wave ranges.

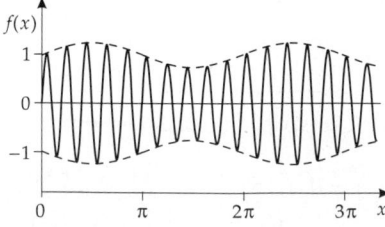

Figure 9.24: Amplitude modulation. Example: $\delta A = 0.25 A, \ \Omega = 0.1 \omega.$

2. Frequency modulation

Frequency modulation (FM), variation of the frequency of a high-frequency carrier wave in the sequence of the low-frequency signal. Modulating signal: $\frac{\Delta\omega}{\Omega}\sin(\Omega t)$.

The dependence of the displacement y of a frequency-modulated wave at a fixed position is then given by

$$f(t) = A\sin\left(\omega t - \frac{\Delta\omega}{\Omega}\cos(\Omega t)\right),$$

where ω is the angular frequency of the carrier wave, Ω the angular frequency of the signal (**Fig. 9.25**).

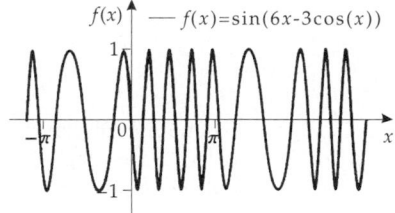

Figure 9.25: Frequency modulation with $\Delta\omega = 0.5\omega$ and $\Omega = \omega/6$.

■ Very-high-frequency (VHF) radio transmission, for example in television, uses frequency modulation (FM) of electromagnetic waves.

Phase modulation, variation of the phase angle of a carrier wave by the signal.

➤ Phase modulation and frequency modulation are identical if the modulation is performed with a sine oscillation.

3. Pulse modulation

Pulse modulation, (**Fig. 9.26**), variation of
- the amplitude, frequency or phase of a pulse function,
- the duration of a pulse.

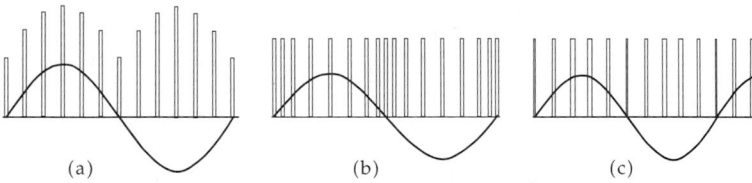

Figure 9.26: Methods of pulse modulation. (a): pulse amplitude modulation, (b): pulse frequency modulation, (c): pulse duration modulation.

9.10 Surface waves and gravity waves

Surface waves, boundary waves at the free surface of a liquid.

▲ Surface waves are neither purely longitudinal, nor purely transverse, waves.

The particles of the liquid carry out complicated ellipse-like motions. Gravity waves at the liquid–gas boundary exhibit a dependence of the propagation velocity on the wavelength (**dispersion, Fig. 9.27**):

phase velocity of surface waves			LT^{-1}
$$v_0 = \sqrt{\dfrac{g\lambda}{2\pi} + \dfrac{2\pi\sigma}{\lambda\rho}}$$	Symbol	Unit	Quantity
	v_0	m/s	phase velocity
	g	m/s²	gravitational acceleration
	λ	m	wavelength
	σ	N/m	surface tension
	ρ	kg/m³	density of liquid

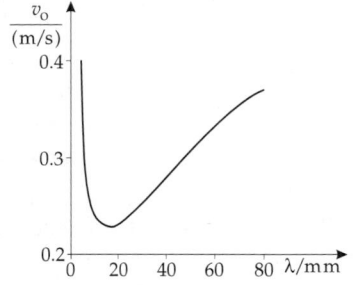

Figure 9.27: Dispersion of surface waves ($h > 0.5\,\lambda$). h: depth of liquid, λ: wavelength.

The density of the liquid must be large relative to the density of the gas. The depth of the liquid h must be larger than 0.5λ.

For low depth of liquid $h < 0.5\lambda$:

$$v_0 = \sqrt{gh}\,.$$

The following table lists various kinds of water waves:

Name	Period	Cause
capillary waves	up to 1 s	wind
ordinary gravity waves	1 … ≈ 12s	wind
ocean swells,	0.5 … 5 min	ordinary gravity
infra gravity waves		waves, wind
tsunamis	5 min up to several hours	earthquakes, wind and changes of air pressure
tidal waves	12, 24 hours	Moon, Sun
trans-tidal waves	>24 h	Moon, Sun, storms

10
Acoustics

Acoustics, the science of vibrations and waves in elastic media. In the more narrow sense, it deals with the audible region of frequencies between 16 Hz and 20 kHz. Physiological and psychological aspects of hearing are also part of the field of acoustics.

■ Elastic media comprise air and water, and solid bodies such as metals, concrete and wood.

10.1 Sound waves

Sound waves, propagation of pressure variations in elastic media.

▲ Both **longitudinal waves** and **transverse waves** occur in **solid** elastic media.

▲ In longitudinal waves, the particles oscillate parallel to the direction of wave propagation.

▲ In gases, and to a large extent in liquids, there is no shear viscosity. Therefore, only **non-polarizable** longitudinal waves arise.

Longitudinal waves propagate in elastic media as rarefaction and compression fronts.

Rarefaction front, ensemble of neighboring points with minimum pressure.

Compression front, ensemble of neighboring points with maximum pressure.

▲ There is **no** sound in a vacuum.

10.1.1 Sound velocity

In a three-dimensional homogeneous medium, the sound from an ideal point source propagates in the form of spherical waves.

Sound velocity, c, speed of propagation of sound waves in a medium.

Meter/second, m/s, SI unit of sound velocity. The sound velocity has the value 1 m/s if sound propagates 1 m in 1 s.

$[c] = $ m/s.

The sound velocity depends on the properties of the medium.

➤ For large amplitudes, the sound velocity depends on the amplitude.

311

1. Velocity of sound in gases

Velocity of sound in gases, depends on the adiabatic coefficient κ (see p. 690) and the temperature T or the pressure p of the gas:

velocity of sound in gases			LT^{-1}
	Symbol	Unit	Quantity
$c_G = \sqrt{\dfrac{p \cdot \kappa}{\rho}}$ $= \sqrt{\kappa R_s T}$	c_G	m/s	velocity of sound
	p	Pa	pressure
	κ	1	adiabatic coefficient
	ρ	kg/m^3	density of gas
	T	K	temperature
	R_s	J/(K kg)	specific gas constant

■   The velocity of sound of many commonly used gases is in the range of $c \approx 200 - 1300$ m/s, i.e., in the range of the mean molecular velocities.

▲   The velocity of sound in gases depends strongly on the temperature.

The temperature dependence of the velocity of sound in air in the range between $-20\,°C$ and $40\,°C$ may be approximated linearly:

$$c_L = (331.5 + 0.6 \cdot T) \text{ m/s}, \qquad T \text{ in } °C$$

(see **Tab. 12.1/2** and **12.1/3**, velocity of sound of various gases).

2. Velocity of sound in liquids

Velocity of sound in liquids, depends on the compression modulus K (see p. 160) and the density ρ of the liquid:

velocity of sound in liquids			LT^{-1}
	Symbol	Unit	Quantity
$c_{Fl} = \sqrt{\dfrac{K}{\rho}}$	c_{Fl}	m/s	velocity of sound
	ρ	kg/m^3	density
	K	N/m^2	compression modulus

■ c_{Fl} is in the range of $1100 - 2000$ m/s (water at $20\,°C$: $c_W = 1480$ m/s).

■   Sound velocities in liquids: water ($20\,°C$) 1480 m/s, benzene ($20\,°C$) 1330 m/s, methyl alcohol ($20\,°C$) 1156 m/s, naphta ($25\,°C$) 1295 m/s, transformer oil ($32.5\,°C$) 1425 m/s (see **Tab. 12.1/6** and **12.1/7**).

3. Velocity of sound in solids

Velocity of sound in solids, depends on the elasticity modulus E (see p. 168) and the density ρ of the solid:

velocity of sound in solids (rods)			LT^{-1}
	Symbol	Unit	Quantity
$c_{So} = \sqrt{\dfrac{E}{\rho}}$	c_{So}	m/s	velocity of sound
	E	N/m^2	elasticity modulus
	ρ	kg/m^3	density

➤ Sound waves in solids may be longitudinal waves or transverse waves.
➤ For non-isotropic solids, the velocity of sound depends on the direction of propagation.
➤ In applications of ultrasound, the wave is transversely confined to a small region of the body. The velocity of sound is then given by

$$c = \sqrt{\frac{E(1-v)}{\rho(1+v)(1-2v)}} \; ;$$

v is the coefficient of transverse contraction.

■ c_{So} is in the range of 1200–6000 m/s (concrete: $c = 3100$ m/s, iron: $c = 5000$ m/s).
■ Velocity of sound in solids: iron 5000 m/s, lead 1200 m/s, tin 2490 m/s, PVC (soft) 80 m/s, PVC (hard) 1700 m/s, concrete 3100 m/s, beech wood 3300 m/s, cork 500 m/s (see **Tab. 12.1/9, 12.1/10, 12.1/11**).

10.1.2 Parameters of sound

1. Sound pressure,

p, superposed on the static equilibrium pressure p_0 (e.g., air pressure) and connected with the compressions and rarefactions of the medium. The pressure p has a sinusoidal dependence on time and space. For an excitation frequency f, the pressure p in **one dimension** is given by:

harmonic sound pressure			$\mathbf{ML^{-1}T^{-2}}$
	Symbol	Unit	Quantity
$p(x,t) = \hat{p} \cos\left[2\pi f\left(t - \dfrac{x}{c}\right)\right]$	p	Pa	pressure
	p_0	Pa	static pressure
	p_{tot}	Pa	total pressure
	$\hat{p}$	Pa	pressure amplitude
$p_{tot} = p_0 + p(x,t)$	f	1/s	frequency
	t	s	time
	x	m	position
	c	m/s	velocity of sound

Effective sound pressure, p_{eff}, analogous to the effective value of electric alternating currents:

$$p_{eff} = \frac{\hat{p}}{\sqrt{2}} \; .$$

➤ In the three-dimensional case, the sound pressure decreases with increasing distance from the source as follows (**Fig. 10.1**),

$$\hat{p} = \hat{p}(r_0)\,\frac{r_0}{r}: \qquad \text{point source,}$$

$$\hat{p} = \hat{p}(r_0)\,\sqrt{\frac{r_0}{r}}: \qquad \text{linear source.}$$

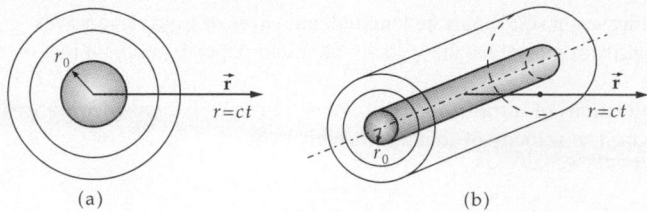

Figure 10.1: The sound pressure amplitude depends on the distance from the source. (a): point-like source, (b): linear source. r_0: reference distance from the source of sound.

2. Wavelength of sound,

wavelength of sound, λ, distance between two zero crossings of the cosine or sine curve at the same time instant and in the same direction. The wavelength is proportional to the reciprocal value of the frequency:

wavelength = velocity of sound/frequency			L
	Symbol	Unit	Quantity
$\lambda = \dfrac{c}{f}$	λ	m	wavelength of sound
	c	m/s	velocity of sound
	f	Hz	sound frequency

■ For an exciting frequency of $f = 300$ Hz, the wavelength in air is $\lambda \approx 1$ m.

3. Sound frequencies

Frequency bands of sound:
  Infrasound, sound of frequencies $f < 16$ Hz,
  Audible sound, sound within the **audible range**, $16\,\text{Hz} < f < 20\,\text{kHz}$,
  Ultrasound, sound with frequencies $f > 20$ kHz.
■ Bats emit sound in the ultrasonic range.
■ **Galton whistle**, pipe of variable length. It generates sound up to the ultrasonic range (< 30 kHz).
⎡M⎤ Ultrasound is employed for range finding and signal transmission, as well as for materials testing, cleaning, and underwater sound ranging (sonar).
Hypersound, sound of frequencies $f > 10$ GHz, generated by piezoelectric excitation of quartz crystals.
⎡M⎤ Application of hypersound in phonon spectroscopy and molecular dynamics.
Debye frequency, upper limit of frequency for sound vibrations. The limit corresponds to a wavelength in the range of twice the molecular distance.
■ In iron the interatomic distance is $2.9 \cdot 10^{-10}$ m. With a sound velocity of $c \approx 5 \cdot 10^3$ m/s, one obtains from $f = c/\lambda$ a Debye frequency of $\approx 10^{13}$ Hz.

10.1.2.1 Sound displacement

Sound displacement, **displacement** $y(x, t)$, displacement of the vibrating particles of the medium from their rest position:

$$y(x, t) = \frac{1}{2\pi f} \frac{1}{\rho c} \hat{p} \sin \left\{ 2\pi f \left(t - \frac{x}{c} \right) \right\} .$$

▲   For plane waves, the phase shift of the sound displacement y with respect to the sound pressure p is $\pi/2$.

10.1.2.2   Sound particle velocity and wave resistance

1. Sound particle velocity,

v, the velocity of the vibrating particles of the medium in a sound wave:

$$v(x, t) = \frac{dy(x, t)}{dt} \, .$$

The space and time dependence of the sound particle velocity $v(x, t)$ is given by

$$v(x, t) = \frac{\hat{p}}{\rho \cdot c} \cos\left[2\pi f \left(t - \frac{x}{c}\right)\right] \, .$$

The amplitude of the sound particle velocity $\hat{v}$ is proportional to the pressure amplitude $\hat{p}$. The inverse of the proportionality coefficient is the characteristic acoustic impedance Z:

amplitude of sound particle velocity			LT^{-1}
	Symbol	Unit	Quantity
$\hat{v} = \dfrac{1}{Z}\,\hat{p}$ $= \dfrac{1}{\rho \cdot c}\,\hat{p}$	$\hat{v}$ $\hat{p}$ ρ c Z	m/s Pa kg/m^3 m/s kg/(m^{2}s)	amplitude of sound particle velocity pressure amplitude density velocity of sound characteristic acoustic impedance

➤   In practice, the effective value of the amplitude of sound particle velocity, $v_{\text{eff}} = \hat{v}/\sqrt{2}$, is usually given instead of the amplitude $\hat{v}$ itself.

2. Acoustic radiation resistance,

Characteristic acoustic impedance, Z, characteristics of the medium with respect to wave propagation: product of the density of the medium ρ and the velocity of sound c. The characteristic acoustic impedances are material constants:

$$Z = \frac{\hat{p}}{\hat{v}} = \rho \cdot c \, .$$

$[Z] = \text{kg/(m}^2\text{s)}$, SI unit of the characteristic acoustic impedance Z.

■   Characteristic acoustic impedances (in kg/(m^{2}s)) of several media under standard conditions (p_n, T_n): air 427, water $1.4 \cdot 10^6$, concrete $8 \cdot 10^6$, glass $13 \cdot 10^6$, steel $39 \cdot 10^6$.

➤   If two media have identical characteristic acoustic impedances, no reflection occurs at their interfaces (see p. 304).

Hearing threshold, lower limit of audibility at $f = 1000$ Hz, i.e., the minimum value of sound volume that still may be registered by a human.

   Reference-sound pressure $p_{\text{eff},0}$, effective sound pressure at the **minimum hearing threshold**, according to DIN 45630

$$p_{\text{eff},0} = 2 \cdot 10^{-5} \, \text{Pa} \, .$$

10.1.2.3   Energy density

Energy density of a sound wave, w, the transported sound energy ΔW per volume element ΔV:

$$w = \frac{dW}{dV} = \lim_{\Delta V \to 0} \frac{\Delta W}{\Delta V} \, .$$

For a sound wave, the energy density w is proportional to the square of the amplitude of the sound particle velocity $\hat{v}$, or to the square of the amplitude of the sound pressure $\hat{p}$:

energy density of a sound wave			$\mathbf{MT^{-2}L^{-1}}$
	Symbol	Unit	Quantity
$w = \dfrac{1}{2}\dfrac{\hat{p}^2}{\rho c^2}$ $= \dfrac{1}{2}\hat{v}^2 \cdot \rho$	w	J/m³	energy density
	$\hat{p}$	Pa	pressure amplitude
	ρ	kg/m³	density
	c	m/s	velocity of sound
	$\hat{v}$	m/s	amplitude of sound particle velocity

The sound energy W in the volume V is obtained by integrating the energy density w over the volume V:

$$W = \int_V w \, dV \, .$$

10.1.2.4   Sound intensity and sound power

1. Sound intensity,

Sound intensity, I, the energy W of the sound wave passing through an area A per unit time, product of the energy density w and the velocity of sound c:

sound intensity			$\mathbf{MT^{-3}}$
	Symbol	Unit	Quantity
$I = \dfrac{1}{A}\dfrac{dW}{dt}$ $= w \cdot c$	I	W/m²	sound intensity
	w	J/m³	energy density
	c	m/s	velocity of sound
	W	J	energy
	t	s	time
	A	m²	area

Watt/square meter, W/m², SI unit of the sound intensity I.

$[I] = \text{W/m}^2.$

The sound intensity, expressed by the amplitudes of sound particle velocity $\hat{v}$ and sound pressure $\hat{p}$, is given by

$$I = \frac{1}{2}\hat{v}\hat{p} = \frac{1}{2}\rho c \hat{v}^2 = \frac{1}{2}\frac{\hat{p}^2}{\rho c} \, .$$

Sound intensity expressed by the effective values of sound pressure and sound particle velocity:

$$I = p_{\text{eff}} \cdot v_{\text{eff}} = \frac{p_{\text{eff}}^2}{Z} .$$

■   Sound characteristics for air at 20 °C at a distance $r = 3$ m from a source of sound that for a sound power of $P = 1 \cdot 10^{-3}$ W emits a tone of frequency $f = 440$ Hz: sound intensity $I = P/4\pi r^2 = 8.85 \cdot 10^{-6}$ W/m^2, $\rho c = 408$ kg/(m^{2}s), sound particle velocity $\hat{v} = \sqrt{2I/(\rho c)} = 2.08 \cdot 10^{-4}$ m/s, sound displacement $\hat{y} = \hat{v}/(2\pi f) = 0.75 \cdot 10^{-7}$ m, sound pressure $\hat{p} = \sqrt{2I\rho c} = 0.85 \cdot 10^{-?}$ Pa, relative pressure variation $\hat{p}/p_0 = 10^{-7}$.

2. Sound power,

P, of a source of sound, sound intensity I integrated over a closed surface O about the source:

sound power = sound intensity · area			$\mathbf{ML^2T^{-3}}$
	Symbol	Unit	Quantity
$P = \oint_O I \, dA$	P	W	sound power
	I	W/m^2	sound intensity
	dA	m^2	areal element
	O	m^2	closed surface

Watt, W, SI unit of sound power P.

$[P] = $ W.

■   Sound power of several sources of sound: conversation: 10^{-5} W, trumpet: 0.1 W, cry: 10^{-3} W, organ: up to 10 W.

10.1.3 Relative quantities

1. Definition of relative quantities

In acoustics and telecommunication technology, one often uses **dimensionless relative quantities**:
- **factors** denote ratios of linear quantities, e.g., reflection **factor**,
- **degrees** denote ratios of quadratic quantities, e.g., (**degree of**) efficiency,
- **measures** or **levels** denote the logarithm of ratios, e.g., transmission **level**, sound pressure **level**.

Decibel, abbreviation **dB**, for dimensionless quantities M proportional to the base 10 logarithm of the quotient of two physical quantities X_0, X_1 of the same dimension.

- For ratios of **linear** quantities x_1, x_2:

$$M = 20 \log \frac{x_1}{x_2} \text{ dB} .$$

- For ratios of **quadratic** quantities X_1, X_2:

$$M = 10 \log \frac{X_1}{X_2} \text{ dB} .$$

2. Sound-relative quantities

Sound pressure level, SPL, L_p, logarithmic scale for relative **sound pressures**:

$$L_\mathrm{p} = 10 \log \frac{p_\mathrm{eff}^2}{p_\mathrm{eff,0}^2} \ \mathrm{dB} = 20 \log \frac{p_\mathrm{eff}}{p_\mathrm{eff,0}} \ \mathrm{dB}.$$

Reference-sound pressure:

$$p_\mathrm{eff,0} = 2 \cdot 10^{-5} \ \mathrm{Pa}.$$

Sound power level, L_w, logarithmic scale for relative **sound power**:

$$L_\mathrm{w} = 10 \log \frac{P}{P_0} \ \mathrm{dB}.$$

Reference-sound power:

$$P_0 = 10^{-12} \ \mathrm{W}.$$

Sound intensity level:

$$L_\mathrm{I} = 10 \log \frac{I}{I_0} \ \mathrm{dB}.$$

Reference-sound intensity:

$$I_0 = 10^{-12} \ \mathrm{W/m^2}.$$

■ An effective sound pressure of $p_\mathrm{eff} = 3 \cdot 10^{-3}$ Pa corresponds to a sound pressure level of

$$L_\mathrm{p} = 20 \log \frac{3 \cdot 10^{-3} \ \mathrm{Pa}}{2 \cdot 10^{-5} \ \mathrm{Pa}} \ \mathrm{dB} = 20 \log \left(1.5 \cdot 10^2 \right) \ \mathrm{dB} = 43.5 \ \mathrm{dB}.$$

3. Addition of sound levels

The relative sound intensities of n sources of sound may be added to the relative total sound intensity,

$$\frac{I}{I_0} = \sum_{k=1}^{n} \frac{I_k}{I_0} = \sum_{k=1}^{n} 10^{0.1 \, L_{Ik}},$$

L_{Ik}: sound intensity level of the source of sound k.
The **total sound level** L_G is given by

$$L_\mathrm{G} = 10 \log \frac{I}{I_0} \ \mathrm{dB} = 10 \log \left(\sum_{i=1}^{n} 10^{L_i/10} \right) \ \mathrm{dB}.$$

▲ Two sound levels are **not** added linearly.

■   For $L_1 = 70$ dB and $L_2 = 80$ dB

$$L_G = 10\log(10^7 + 10^8)\text{ dB} = 80.4\text{ dB}.$$

$L_1 = 0$ dB and $L_2 = 0$ dB yields $L_G = 3$ dB.

■   The sound of a moving truck can completely hide the chirp of birds.

■   Two equal sources of 100 dB each have a sound level raised by only 3 dB compared with one source: $L_G = 103$ dB. Two equal sources of zero dB each have together $0\text{ dB} + 0\text{ dB} = 3\text{ dB}$.

▲   **Sound levels** are added with a **sound level excess** L_Z and the **sound level difference** ΔL,

$$\Delta L = L_1 - L_2,$$

successively—term by term, always starting from the higher level L_1, i.e.,

$$L_G = L_1 + L_Z.$$

$\Delta L = 0$ dB	$L_Z = 3$ dB
$\Delta L = 3$ dB	$L_Z = 1.8$ dB
$\Delta L = 5$ dB	$L_Z = 1.2$ dB
$\Delta L = 7$ dB	$L_Z = 0.8$ dB
$\Delta L = 10$ dB	$L_Z = 0.4$ dB
$\Delta L \geq 20$ dB	$L_Z = 0$ dB

10.2   Sources and receivers of sound

Source of sound, body vibrating in a medium and periodically emitting compression and rarefaction fronts, i.e., waves.

10.2.1   Mechanical sound emitters

1. Strings

Rods and **strings**, linear sources of sound.

Natural vibrations arise by exciting standing waves, with frequencies determined by the sizes of the oscillatory object.

■   String instruments (piano, violin, guitar).

For **two fixed ends**, the wavelength λ_n of the natural vibration of a rod or a string of length l is:

wavelength for natural string vibrations (2 ends fixed)			L
	Symbol	Unit	Quantity
$\lambda_n = \dfrac{c}{f_n} = \dfrac{2l}{n+1}$	λ	m	wavelength
	n	1	number of nodes
$n = 0, 1, 2, \ldots$	f_n	Hz	frequency
	c	m/s	velocity of sound

The sound pitch of the string of fixed length depends on the longitudinal tension (tuning of instruments).

Fundamental vibration (1st harmonic), f_0, for $n = 0$.
Overtone, higher harmonic, f_n, for $n > 0$.
1st overtone (2nd harmonic): $f_1 = 2f_0$.
2nd overtone (3rd harmonic): $f_2 = 3f_0$.

2. Membranes

Membrane, mostly circular surfaces fixed only at the boundary, two-dimensional analog of the string fixed at both ends.

The natural vibrations of a membrane are labeled by two integers (n, m).

■   Drum, kettledrum.

Wavelengths $\lambda_{m,n}$ of the natural vibrations of a **circular** membrane of radius R:

wavelength of natural vibrations of a circular membrane			L
	Symbol	Unit	Quantity
$\lambda_{m,n} = \dfrac{2\pi Rc}{B_{m,n}} \sqrt{\dfrac{\rho}{\sigma_F}}$	λ	m	wavelength
	R	m	radius of membrane
	c	m/s	velocity of sound
	$B_{m,n}$	1	zeros of Bessel functions
	σ_F	N/m^2	tension of membrane surface
	ρ	kg/m^3	density of membrane

➤ σ_F must be measured with the membrane at rest.

Fundamental vibration: the entire membrane vibrates in phase.

Overtone: formation of nodal lines on the membrane, corresponding to the nodes of the string. Out-of-phase oscillation of the segments of the membrane confined by the nodal lines.

Subdivision of the shapes of vibration depending on the position of nodes:
- The nodal lines coincide with the diameters of the membrane.
- Circular nodal lines with the centers at the center of the membrane.
- Nodal lines combined out of both cases cited above.

Plate and **bell**, two- and three-dimensional analogs of vibrating rod. The vibrational shapes are similar to those of the membrane.

| M | **Chladni's acoustic figures**, patterns (analogous to **Kundt's dust figures** in the sound tube) formed if a membrane is covered with cork powder and then excited to vibration. The dust then accumulates along the nodal lines, thus visualizing the **vibration mode** of the membrane.

10.2.1.1   Vibrating air columns

Siren, consists of a rotating circular disk with series of holes arranged concentrically, and a nozzle blowing air onto the disk. Thus, a periodic release and interruption of the airstream is generated. The periodic vibrations of air pressure are perceived as a tone. The frequency of the tone increases with the rotational speed of the disk.

Edge-tone generator, consists of a sharp edge or a thin wire blown on by an air stream. Vortices arise periodically at the edge and thereby generate periodic pressure variations.

The frequency f of the tone generated depends on the distance d between the nozzle and the edge or the wire, and on the speed v_S of the airflow.

Edge-tone frequency			T^{-1}
	Symbol	Unit	Quantity
$f = \gamma \dfrac{v_S}{d}$	f	Hz	frequency
	γ	1	proportionality constant
	v_S	m/s	flow speed
	d	m	distance nozzle–edge

■ The **propeller noise** of helicopters is due to edge-tone generation.

■ Wind "whistling" is edge-tone generation at the corners of buildings, projections, etc. If the edge or the wire is coupled to a resonator, the frequency of vortex formation is determined by the resonant frequency of the resonator. The resonator is usually a tube in which standing waves are generated.

■ Standing waves may be formed by blowing on a bottle top, which then "whistles."

■ Pipes and flutes make sound with edge-tone formation.

Horns

■ Including all brass instruments.

The lips are closed, the air pressure generated by the abdominal muscles increases in the mouth cavity until it exceeds the lip tension.The lips are opening, and air is released, and the pressure in the mouth cavity drops. The lips are closing again because of their tension. The process repeats periodically and leads to periodic pressure variations in the instrument. Standing waves in the air column arise in the case of resonance, i.e., if the lip tension fits properly to the length of the instrument.

Woodwind instruments—except for flutes and pipes—involve an elastic reed set to vibration by the airflow and then modulates the airflow and the air pressure.

10.2.2 Electro-acoustic transducers

Sound transducer, device that converts electric energy into sound energy and vice versa.

Sound emitter, a mechanical system set into vibration by mechanical, electric or magnetic forces.

■ **Loudspeaker**, consists mostly of a sound membrane in the field of a permanent magnet. Applying an alternating voltage causes a forced vibration of the membrane, which then generates sound waves.

1. Electrically driven sound emitter

Electromagnetically driven emitter, metallic membrane in the field of a permanent magnet.

■ Loudspeaker, horn, telephone receiver (electromagnetic).

Electrodynamically driven emitter, vibrating coil with membrane.

■ Loudspeaker: the distortion factor is appreciably smaller than that for an electromagnetic system. A higher output power may be radiated without distortion. Higher efficiency due to an exponential horn.

Piezoelectric sound emitter, contains a **piezoelectric** element. The size varies with the applied electric voltage. When an alternating voltage is amplified, the surface vibrates and generates sound waves. Application: mostly in the ultrasonic range.

➤ **Piezoelectric crystals** (quartz, Seignette salt) perform motions when the electric charges on two layers on parallel surfaces (cut according to a preference orientation) are varying, and vice versa.

■ Crystal tone pickup, crystal microphone, high-frequency loudspeaker.

Thermal sound generation, conversion of heat into sound energy.

■   Spark sound waves, thermophone, singing electric arc.
Magnetostriction emitter
■   Generation of ultrasound.
Condenser microphone

2. Electro-acoustic transmission factor

for sound emitter, B_S, quantity specifying the frequency range that may be transmitted by a reversible sound transducer. For a sound generator (e.g., a loudspeaker), the electro-acoustic transmission factor B_S is the ratio of the emitted sound pressure p_r at a distance of 1 m and the voltage U applied to the transducer.

electro-acoustic transmission factor				$L^{-3}TI$
	Symbol	Unit	Quantity	
$B_S = \dfrac{p_r}{U}$	B_S	Pa/V	electro-acoustic transmission-factor emitter	
	p_r	Pa	sound pressure at a distance of 1 m	
	U	V	voltage	

Reference transmission factor for sound sources, B_{S0}, defined as $B_{S0} = 0.1$ Pa/V.

3. Measure of electro-acoustic transmission for a transmitter,

G_S, quantity given frequently instead of the transmission factor B_S, proportional to the base 10 logarithm of the ratio of the transmission factor B_S to a reference transmission factor B_{S0}:

measure of electro-acoustic transmission				1
	Symbol	Unit	Quantity	
$G_S = 20 \cdot \log \dfrac{B_S}{B_{S0}}$	G_S	dB	electro-acoustic transmission measure	
	B_S	Pa/V	electro-acoustic transmission factor	
	B_{S0}	Pa/V	reference transmission factor	

G_S is given in dB ($B_{S0} = 0.1$ Pa/V).

4. Loudspeaker sensitivity,

$\overline{E}_k$, quantity introduced to characterize a loudspeaker, the product of the transmission factor $\overline{B}_S$ averaged over the frequency range $f = 0.25 - 4$ kHz, the square root of the impedance Z of the loudspeaker, and the ratio of the distance r from the loudspeaker to a reference distance r_0 of 1 m:

characteristic loudspeaker sensitivity				
	Symbol	Unit	Quantity	
	$\overline{E}_k$	Pa/$\sqrt{VA}$	characteristic loudspeaker sensitivity	
$\overline{E}_k = \overline{B}_S \cdot \sqrt{Z} \cdot \dfrac{r}{r_0}$	$\overline{B}_S$	Pa/V	mean value transmission factor	
	Z	Ω	impedance	
	r	m	distance from loudspeaker	
	r_0	m	reference distance	

5. Range of loudspeakers,

r, defined as the product of the characteristic loudspeaker sensitivity $\overline{E_k}$ and the square root of the electric power received P, divided by the desired sound pressure p:

range of loudspeakers			L
	Symbol	Unit	Quantity
$r = \dfrac{\overline{E_k}}{\overline{B_S} \cdot \sqrt{Z}} \cdot r_0$ $= \dfrac{\overline{E_k} \cdot \sqrt{P}}{p} \cdot r_0$	r $\overline{E_k}$ $\overline{B_S}$ Z P p r_0	m Pa/$\sqrt{\text{VA}}$ Pa/V Ω VA Pa m	range characteristic sensitivity transmission factor impedance appearent power sound pressure reference distance

10.2.2.1 Sound receivers or microphones

Sound receivers or **microphones**, convert sound energy into electric energy.

1. Kinds of microphones

Piezoelectric transducer, inversion of the piezoelectric sound source. It consists of a piezoelectric element with a surface that responds to the pressure variations generated by the incident sound wave. In the piezoelectric element, a voltage is formed that is proportional to the sound pressure.
■   Application in **microphones**.

Piezoresistive transducer, based on the change of the resistance in a piezoelectric element owing to pressure variations. Modulation of the current via the change of the resistance.
■   Application in **telephones**.

Magnetostrictive transducer, consists of a ferromagnetic material that changes its length as a function of the applied magnetic field. Hence, sound waves may be generated by alternating magnetic fields.
■   Application in **ultrasonic experiments**.

Electrostatic transducer, capacitor with one plate formed as metallic membrane. Sound causes a deformation of the membrane, and thus a variation of the capacitance and a corresponding variation of the electric voltage.
■   Application in condenser microphones in studios, and in hand microphones. Also used for sound generation, mainly in headphones.

Electrodynamic transducer, sound pressure deforms a membrane. The membrane moves a coil in the field of a permanent magnet, thereby inducing an electric current in the coil.
■   Application in small, portable microphones and headphones.

Bio-acoustic transducer, sound energy induces biological processes. Most important example is the human sense of hearing, which converts sound into neural currents via a series of mechanical and chemical processes.

2. Electro-acoustic transmission factor for sound receiver,

B_E, ratio of received sound pressure p to generated electric voltage V.

electro-acoustic transmission factor			$L^3T^{-1}I^{-1}$
	Symbol	Unit	Quantity
$B_E = \dfrac{V}{p}$	B_E	V/Pa	electro-acoustic transmission-factor receiver
	p	Pa	received sound pressure
	V	V	electric voltage

Reference transmission factor for acoustic sound receiver, B_{E0}, defined as $B_{E0} = 10$ V/Pa.

3. Electro-acoustic transmission measure for sound receivers

Electro-acoustic transmission measure			1
	Symbol	Unit	Quantity
$G_E = 20 \cdot \log \dfrac{B_E}{B_{E0}}$	G_E	1	electro-acoustic transmission measure receiver
	B_E	V/Pa	electro-acoustic transmission factor
	B_{E0}	V/Pa	reference transmission factor receiver

4. Microphone sensitivity,

E_M, analogous to loudspeaker sensitivity:

microphone sensitivity			dB
	Symbol	Unit	Quantity
$E_M = \dfrac{\sqrt{P}}{p}$	E_M	$\sqrt{VA}$/Pa	microphone sensitivity
	P	VA	received electric power
	p	Pa	sound pressure

5. Stereo signals

difference signal: $D = L - R$
sum signal: $S = L + R$
left-signal: $S + D = L + R + L - R = 2L$
right-signal: $S - D = L + R - L + R = 2R$

Frequencies for **stereo radio**:

main-signal frequency (sum, mono): $f_M = 30$ Hz $\ldots$ 15 kHz
subcarrier frequency: $f_H = 38$ kHz
stereo additive frequency: $f_S = f_H \pm f_M$
upper sideband: $f_{Su} = f_H + f_M = 38.03 \ldots 53$ kHz
lower sideband: $f_{Sl} = f_H - f_M = 23 \ldots 37.97$ kHz

10.2.3 Sound absorption

1. Distortions of sound propagation,

occur by:
- sound reflection,
- sound diffraction,

- sound refraction,
- sound interference,
- sound absorption.

2. Sound absorption,

Sound damping, energy loss in the propagation of a sound wave because of
- internal friction,
- isentropic compression,
- excitation of intrinsic degrees of freedom (like rotation of molecules) of the medium transmitting the sound wave.

Exponential decrease of the sound intensity I by sound absorption with increasing distance r from the source of sound according to

$$I(r) = I(r_0) \cdot e^{-\alpha(r - r_0)}.$$

$I(r_0)$ is the sound intensity at a reference distance r_0 from the source.

3. Sound-damping coefficient

α, depends on the frequency of the sound source and on absorption properties of the medium (see **Tab. 12.1/4** and **12.1/8**).

$[\alpha] = \text{m}^{-1}$, SI unit of the sound-damping coefficient α.

■ Sound-damping coefficient in cm^{-1} for various frequencies in some liquids at 20 °C: water 23.28 (307 MHz), 55.3 (482 MHz), 172 (843 MHz), benzene 711.5 (307 kHz), 1150 (482 kHz), tetrahedral chlorine methane 492 (307 kHz), 1115.2 (482 kHz), 3269 (843 kHz); see **Tab. 12.1/8**.

Sound-absorbing material, material acting as sound insulator.

   Technological realization:
- homogeneous or porous material, conversion of sound into heat by deformation of the material or friction.
- resonators, convert sound with frequencies near the resonance frequency into heat, due to loss of flow or frictional loss.

4. Degree of sound reflection,

ρ, ratio of reflected sound intensity to perpendicularly incident sound intensity:

degree of sound reflection			1
	Symbol	Unit	Quantity
$\rho = \dfrac{I_r}{I_e}$	ρ	1	degree of sound reflection
	I_e	W/m^2	sound intensity of incident wave
	I_r	W/m^2	sound intensity of reflected wave

5. Degree of sound absorption,

α, $[\alpha] = 1$, dimensionless quantity for the absorptivity of a body. The quantity α gives the normalized difference of incident and reflected sound intensity:

degree of sound absorption			1
	Symbol	Unit	Quantity
$\alpha = \dfrac{I_e - I_r}{I_e}$	α	1	degree of sound absorption
	I_e	W/m^2	sound intensity of incident wave
	I_r	W/m^2	sound intensity of reflected wave

➤ The degree of sound absorption α, $[\alpha] = 1$, should **not** be confused with the sound-damping coefficient α, $[\alpha] = 1/m$ quoted above!

■ Degrees of sound absorption for several building materials at various frequencies: light concrete 0.07 (125 Hz), 0.22 (500 Hz), 0.10 (2000 Hz), wooden doors 0.14 (125 Hz), 0.06 (500 Hz), 0.10 (2000 Hz), wooden panels 0.25 (125 Hz), 0.25 (500 Hz), 0.08 (2000 Hz) (see **Tab. 12.1/16**).

| M | Measurement of the degree of sound absorption by means of Kundt's tube.

6. Sound-transmission degree,

τ, the ratio of transmitted sound intensity I_d to incident sound intensity I_e,

$$\tau = \frac{I_d}{I_e} = \frac{p_d^2}{p_e^2} .$$

Degree of sound dissipation, δ, ratio of sound intensity absorbed in the wall I_a to incident sound intensity I_e,

$$\delta = \frac{I_a}{I_e} = \frac{I_e - I_r - I_d}{I_e} = \alpha - \tau = 1 - \rho - \tau .$$

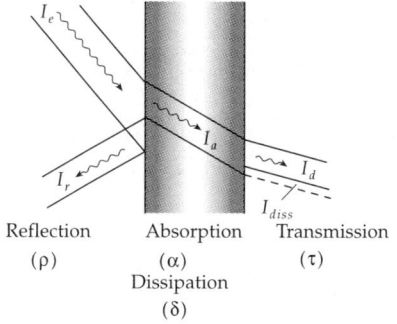

Reflection (ρ) Absorption (α) Dissipation (δ) Transmission (τ)

Figure 10.2: Reflection, absorption, dissipation and transmission of sound waves.

▲ Energy conservation holds for the sound energy at the interface of two media:

$$\rho + \tau + \delta = 1 , \qquad \rho + \alpha = 1 .$$

Degree of sound reflection ρ, degree of sound absorption α, and degree of sound transmission τ for a sound wave at perpendicular incidence onto the interface may be expressed by the characteristic sound impedance values Z_1, Z_2 of the two media (**Fig. 10.2**):

$$\rho = \left(\frac{Z_2 - Z_1}{Z_2 + Z_1}\right)^2 , \qquad \tau = \frac{4Z_1 Z_2}{(Z_1 + Z_2)^2} , \qquad \alpha = 1 - \left(\frac{Z_2 - Z_1}{Z_2 + Z_1}\right)^2 .$$

Matching, the reflected wave vanishes for $Z_1 = Z_2$.

10.2.4 Sound attenuation

Sound attenuation, hindrance of sound propagation by reflection at barriers, in particular reflections by interfaces between media of different sound propagation properties.

Sound-reflection factor, r, dimensionless quantity, ratio of the pressure amplitude of the reflected wave $\hat{p}_r$ to the pressure amplitude of the incident wave $\hat{p}_e$.

sound-reflection factor			1
	Symbol	Unit	Quantity
$r = \dfrac{\hat{p}_r}{\hat{p}_e}$ $= \dfrac{Z_2 - Z_1}{Z_1 + Z_2}$	r	1	sound reflection factor
	Z_1, Z_2	kg/(m²s)	characteristic sound impedance of medium 1, 2
	$\hat{p}_e$	kg/(ms²)	pressure amplitude of incident sound
	$\hat{p}_r$	kg/(ms²)	pressure amplitude of reflected sound

$r = 0$: no reflection, $r = \pm 1$: complete reflection.
Relation between **degree of sound reflection** ρ and **sound-reflection factor** r:

$$\rho = \frac{I_r}{I_e} = \frac{\hat{p}_r^2}{\hat{p}_e^2} = r^2 .$$

▲ Maximum attenuation is achieved when one uses a material for reflection material having a characteristic sound impedance differing as much as possible from that of the medium wherein the incident wave propagates.

attenuation measure R of a wall			1
	Symbol	Unit	Quantity
$R = 10 \log \dfrac{I_e}{I_\tau}$ $= L_1 - L_2$	R	dB	attenuation measure
	I_e	W/m²	sound intensity in front of the wall
	I_τ	W/m²	sound intensity behind the wall
	L_1	dB	sound level in front of the wall
	L_2	dB	sound level behind the wall

Technical realization:
- **Sound in air** is usually attenuated by a separating wall of a material as heavy and hard as possible.
- **Sound in solids**, optimum attenuation by soft sound-insulating layers with low characteristic sound impedance.
- **Footstep sound**, sound in buildings caused by footsteps. Footstep sound propagates through ceilings. Attenuation by a **floating floor** that is not put directly on the concrete floor, but rather on a soft intermediate layer, or by suspended ceilings.

10.2.4.1 Reverberation

Reverberation, usually an exponential decay of the sound field after switching off the acoustic excitation.

Reverberation time, T_N, time interval after which the sound energy drops by 60 dB, i.e., to 1 ppm $= 10^{-6}$ of the original value (**Fig. 10.3**).

reverberation time (Sabine's law)			T
	Symbol	Unit	Quantity
$T_N = 0.163 \dfrac{V}{\alpha A}$	T_N	s	reverberation time
	V	m³	volume of room
	A	m²	absorption areas
	α	1	degree of sound absorption

■   A hall of $V = 500$ m³ has a typical reverberation time of $T_N = 1$ s.

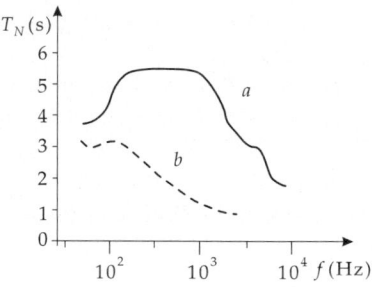

Figure 10.3: Reverberation time in the Dresden Cathedral depending on the frequency (determined from records). (a): empty, (b): occupied with 4000 persons.

10.2.5   Flow noise

Liquids generate a broad-band **flow noise** when flowing out of pipes, or around obstacles and curvatures. It is caused by pressure variations in the vortex field of the flow.

Avoided by:

- encasement of the pipes,
- water sound attenuators,
- acoustic filter chains (low-pass filters) in sound attenuators of ventilation pipes or exhausts of vehicles.

Additional flow noise corresponds to narrow-band blown tones, broad-band noise owing to implosion of steam bubbles, and very-broad-band free-beam noise that arises in a stream of gas flows into a gas at rest.

10.3   Ultrasound

1. Properties of ultrasound

Ultrasound, frequencies $f > 20$ kHz.
  Hypersound, frequencies $f > 10$ GHz $= 10^{10}$ Hz.
  Wavelength of ultrasound in air at a mean velocity of sound of $c \approx 330$ m/s:

$$\lambda_{\text{air}} < 1.5 \text{ cm}.$$

Ultrasonic waves can be focused, and parallel rays may be formed that propagate along straight lines with weak diffraction effects.
  Generation of ultrasound, by magnetostriction.

| M | Measurement of velocity and attenuation of ultrasound:

pulse-echo method,

reverberation method.

2. *Application of ultrasound*

Ultrasonic diagnostics in medicine, therapy, microsurgery.
    Materials testing of solids:
determination of elastic properties.
    Ultrasound in electronics and microelectronics:
ultrasonic delay line,
ultrasonic surface-wave filter,
ultrasonic microscope,
ultrasonic welding device.
    Hydroacoustics:
underwater sound position finding, SONAR (sound navigation and ranging),
echo depth determination,
underwater communication.
    Control of production processes by means of ultrasound:
level measuring,
hydrometry,
tracing of chemical processes,
determination of concentrations,
quality control (materials testing with a precision up to 10^{-4} m).
    Power ultrasound in the range of $\approx 20 \ldots 40$ kHz:
Ultrasonic cavitation in solid surfaces:
ultrasonic drilling,
ultrasonic cleaning,
ultrasonic welding.

10.4   Physiological acoustics and hearing

Hearing, human sensory faculty that detects sound waves and analyzes loudness levels and frequencies. Example of a **bio-acoustic transducer**.
    External ear, flat horn collecting the sound and channeling it into the auditory canal.
    Auditory canal, passage connecting the external ear and the eardrum.
    Eardrum, horn-like membrane of about 0.5 cm^2 area forced to vibrate by the sound waves.
    Hammer, **anvil** and **stirrup**, three **auditory ossicles** onto which the vibration of the membrane is transferred. They act as a system of levers that match the distinct characteristic impedances of the external ear (air) and the internal ear (essentially water).
    Oval window and **round window**, two membranes between the middle ear and the internal ear located behind the middle ear. The stirrup transfers the vibrations to these membranes, which amplify the pressure variations by an additional factor of 20 up to 30.
    Internal ear, bisected space behind the middle ear filled with an incompressible liquid rich in sodium ions. Proper bio-acoustic transducer.
    Cochlea, bisects the internal ear, filled with a liquid rich in potassium ions. Hence, there is an electric potential difference between the liquids in the cochlea and in the internal ear.
    Basilar membrane, membrane at the cochlea that is deformed mechanically by the vibrations of the round and the oval window via the liquid in the internal ear.

Capillary cells of Corti's organ, attached to the basilar membrane. The motions of the basilar membrane generate electric-potential variations in these cells, resulting in exciting currents in the auditory nerve. These then give rise to a perception of sound in the brain.

10.4.1 Perception of sound

1. Frequency range of hearing

Frequency range of hearing, frequency range between 16 Hz and 20 000 Hz, the range of vibrations and waves of elastic media that may be perceived by the human ear (**Fig. 10.4**),
- **Frequency range** of **speech**: $\approx$ 10 Hz. . . 10 kHz.
- **Intelligible speech**: $\approx$ 300 Hz. . . 3 kHz.

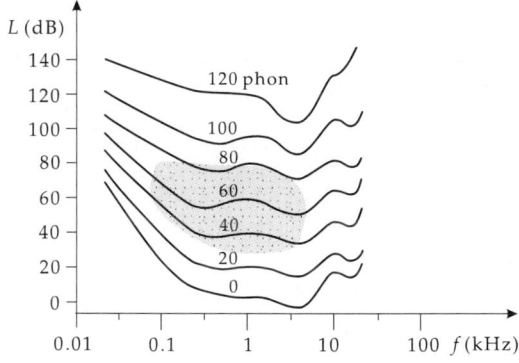

Figure 10.4: Curves of equal sound intensity. Emphasized: range of hearing of man.

➤   The frequency range of hearing reduces with increasing age. Moreover, entire frequency regions may drop out permanently because of overload (lumping of capillaries).

▲   Equal sound-intensity levels at different frequencies are perceived as having different loudness.

2. Loudness level

Weber-Fechner law: The change of the perception of sound ΔL is proportional to the logarithm of the ratio of the sound intensities,

$$\Delta L \sim \log I_2/I_1 \ .$$

Loudness level, L_S, measure of the subjective perception of sound intensity by the ear, frequency-dependent. It is chosen in such a way that, at a sound frequency of 1 kHz, the value of the loudness level equals the sound-pressure level:

$$L_S = 10 \log \left(\frac{I}{I_0} \right) = 20 \log \frac{P}{P_0} \ \text{dB} \ .$$

Reference-sound intensity:

$$I_0 = 10^{-12} \ \text{W/m}^2 \ ,$$

corresponds to the hearing threshold of the ear at 1 kHz.

Phon, the unit of the loudness level L_S. Phon is a dimensionless quantity.

▲   The **hearing threshold** is 4 phons (corresponds to $I_0 = 10^{-12}$ W/m^2).
The human ear has an extremely large **dynamical region**: the **ability of hearing** covers 12 **orders of magnitude in intensity** with displacement amplitudes between 10^{-11} m (1/10 of atomic radius) and 10 micrometers.

➤   Zero phon does **not** correspond to the frequency-dependent **standard hearing threshold**!

➤   A difference of $\Delta L_S = 1$ phon is just perceivable for the human ear.

▲   The **pain threshold** is 120 phons (corresponding to $I \approx 1$ W/m^2).

▲   For $f = 1000$ Hz the sound pressure level is equal to the loudness level,

$$L_p - L_S \quad \text{for} \quad f = 1 \text{ kHz} .$$

■   Hence, for $f = 1000$ Hz: **sound-pressure levels** of 40, 80, 120 dB correspond to **loudness levels** of 40, 80, 120 phons.

▲   Sound **intensities** I are added, $I_G = I_1 + I_2 + I_3$.

▲   Sound **levels** are added to a sound level excess (see p. 319).

10.4.2   Evaluated sound levels

Evaluation curve A, takes into account the complex relation between the physical sound-level spectrum and the human perception of sound.

| M | A frequency-dependent evaluation factor Δ_i^* (in dB) is added to the measured frequency-dependent sound levels L_i.

A-evaluated sound level:

$$L_A = 10 \log \left(\sum_{i=1}^{n} 10^{(L_i + \Delta_i^*)/10} \right) \text{dB} .$$

f/Hz	90	220	400	1000	3000	60000
Δ_i^*/dB	-20	-10	-5	0	$+2$	0

Loudness, S, physiological quantity for a subjective comparison of sources of sound. The loudness is defined in such a way that a doubling of its value corresponds to a doubling of the **subjectively perceived** loudness:

$$S = 2^{0.1(L_S - 40)} \text{ sone} .$$

Sone, dimensionless quantity, unit of loudness.

➤   A doubling of the loudness corresponds to a change of the loudness level of $\Delta L_S = 8 - 10$ phons.

■   The loudness of $S = 1$ sone corresponds to a loudness level of $L_S = 40$ dB.

10.5   Musical acoustics

The human ear judges the sound according to the **loudness** and the **frequency spectrum**.
Sound may always be represented by superposition of sinusoidal pressure variations with different frequencies and amplitudes (**Fourier representation**).

Frequency range of music: ≈ 16 Hz – 16 kHz.

Classification of hearing impressions:

- **Tone**, a purely sinusoidal pressure variation, contains a single frequency (harmonic vibration, **Fig. 10.5 (a)**). A pure sine oscillation cannot be generated by customary musical instruments, but may be generated electronically.
- **Sound**, superposition of tones with different frequencies and amplitudes for which the frequencies have integral ratios with respect to each other (**Fig. 10.5 (b)**).
- **Noise**, superposition of tones with a continuous spectrum of frequencies. Noise is not a periodic vibration.
- **Bang**, superposition of tones with a continuous spectrum and nearly constant intensities, i.e., the contributions with different frequencies all have about the same amplitude (**Fig. 10.5 (c)**).

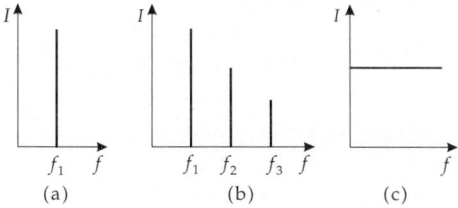

Figure 10.5: Frequency spectra (schematically). (a): tone, (b): sound, (c): bang.

1. Diatonic tone scale

Tone scale, stepwise arrangement of tones within an octave.

Sounds are classified as follows:

- **Consonance**: the frequency ratio of tones f_2/f_1 may be expressed by integers N_1, N_2 smaller than or equal to eight.
- **Dissonance**, if this is impossible.
- ➤ This definition of consonance and dissonance is a purely subjective one and corresponds to the Western perception of sound.

Interval, notation for the frequency ratio of two tones.

Table of interval notations:

Frequency ratio	Interval	Perception
1:1	prime	consonant
16:15	minor second	dissonant
10:9; 9:8	major second	dissonant
6:5	minor third	consonant
5:4	major third	consonant
4:3	quart	consonant
3:2	quint	consonant
8:5	minor sext	consonant
5:3	major sext	consonant
9:5; 16:9	minor septim	dissonant
15:8	major septim	dissonant
2:1	octave	consonant

▲   The octave corresponds to a frequency doubling.
■ $A = 110$ Hz, $a = 220$ Hz, $a^1 = 440$ Hz, $a^2 = 880$ Hz, $a^3 = 1760$Hz.
▲   The octave is subdivided into 12 **half-tones** (minor seconds).
Whole tone, notation for a major second.
   Concert pitch, a^1, normalized to $f = 440$ Hz.

2. Chromatic scale

Chromatic scale: 12 half-tone steps in **well-tempered (tempered)** scale.
▲   Frequency ratio per **half-tone interval**:

$$1 : {}^{12}\sqrt{2} = 1 : 1.059463 \,.$$

▲   Frequency ratio per **whole-tone interval**:

$$1 : {}^{6}\sqrt{2} = 1 : 1.1222462 \,.$$

Pitch, designation for the frequency of a tone.
   Sound intensity, designation for the intensity amplitude of a tone.
   Fundamental tone: customary musical instruments do not generate pure sinusoidal tones, but a superposition of sine waves, with a mixing ratio depending on the kind of instruments and on their pitch. The lowest frequency of a given superposition is the tone of the instrument. As a rule, the fundamental tone has the largest amplitude.
   Overtones, the tones in a sound that have a higher frequency than the fundamental tone.
   Harmonic vibrations:

fundamental vibration	f_1	first harmonic
first overtone	$f_2 = 2f_1$	second harmonic
second overtone	$f_3 = 3f_1$	third harmonic

   Timbre of sound, designation for the mixing ratio of the amplitudes of the various tones involved in a sound.
■   Musical instruments are distinguished by their timbres of sound.
   Tonal range of an instrument, range of frequencies between the **fundamental tones** of the highest and lowest pitch that can be generated by an instrument, further characteristics of an instrument besides the timbre (**Fig. 10.6**).

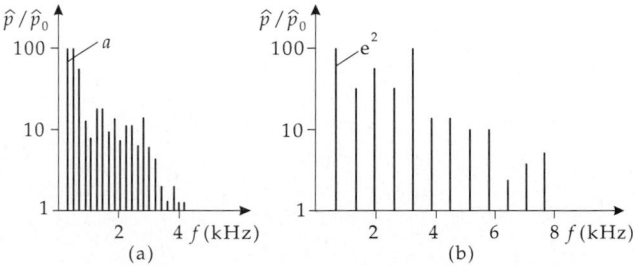

Figure 10.6: Frequency spectra of string instruments. (a): cello, (b): violin.

■   Since the tonal range of many instruments, and also of human voices, depends strongly on the ability of the musician, in a restrictive way, the tonal range puts a requirement on a voice or an instrument in classical music.

Tonal ranges of various instruments and voices:

Instrument or voice	Lowest frequency (Hz)	Highest frequency (Hz)
violin	200	3000
piano	30	4000
flute	250	2500
cello	70	800
contrabass	40	300
tuba	50	400
trumpet	200	1000
organ	16	1600
bass	100	350
baritone	150	400
tenor	150	500
alto	200	800
soprano	250	1200

11
Optics

Optics, the science of light, i.e., the range of wavelengths of electromagnetic radiation that may be perceived by the human eye. This range covers the wavelengths between $\lambda = 380$ nm and $\lambda = 780$ nm (1 nm $= 10^{-9}$ m). Electromagnetic radiation close to, but outside, the visible range is also included.

Optics deals with processes that occur when light interacts with media.

1. Main characteristics of light

Propagation velocity of light, depends on the medium in which the light propagates.

Speed of light in vacuum, fundamental universal constant with the value

$$c = 299\,792\,458 \text{ m/s}.$$

▲ In all media the speed of light is less than it is in vacuum.

■ The speed of light in water is $2.24 \cdot 10^8$ m/s, in glass $(1.85 \pm 0.25) \cdot 10^8$ m/s, in diamond $1.22 \cdot 10^8$ m/s.

Wavelength λ and **frequency** f are related to the speed of propagation c as follows:

light speed = frequency · wavelength			
$c = f\lambda$	Symbol	Unit	Quantity
$k = \dfrac{2\pi}{\lambda}$	k	1/m	wave number
	λ	m	wavelength
	ω	rad/s	angular frequency
$f = \dfrac{1}{T}$	c	m/s	light speed
	f	1/s	frequency
$\omega = \dfrac{2\pi}{T} = 2\pi f = ck$	T	s	period

335

2. Subdivision of electromagnetic waves

Frequency f (Hz)	Wavelength λ (m)	Notation
$> 10^{19}$	$< 3 \cdot 10^{-11}$	γ-radiation
$> 10^{17}$	$< 3 \cdot 10^{-9}$	X-rays
$10^{15} \dots 10^{17}$	$3 \cdot 10^{-7} \dots 3 \cdot 10^{-9}$	ultraviolet radiation
$\sim 0.5 \cdot 10^{15}$	$\sim 6 \cdot 10^{-7}$	visible light
$10^{13} \dots 10^{14}$	$3 \cdot 10^{-5} \dots 3 \cdot 10^{-6}$	infrared radiation
$10^{9} \dots 10^{13}$	$0.3 \dots 3 \cdot 10^{-5}$	microwaves
$\sim 10^{8}$	3	ultra-short radio waves
$\sim 10^{7}$	30	short radio waves
$\sim 10^{6}$	300	medium radio waves
$\sim 10^{5}$	3000	long radio waves

For subdivisions of the ultraviolet range, see **Tab. 12.2/8**.

3. Spectral colors and regions

Spectral color, sensory perception of the eye for various ranges of wavelengths of the spectrum.

Ranges of spectral colors:

Color	Frequency (10^{12} Hz)	Wavelength (10^{-9} m)
violet	659...769	455...390
blue	610...659	492...455
green	520...610	577...492
yellow	503...520	597...577
orange	482...503	622...597
red	384...482	780...622

4. Theoretical models of light

Wave theory, model for optical phenomena in which light is considered a wave phenomenon.

Corpuscular theory, model for optical phenomena in which light is considered to consist of **corpuscles** (Latin word for particle) that move along straight lines in the absence of interaction with matter.

Wave-particle duality, certain experiments can be explained only within wave theory, other experiments only within corpuscular theory. The need for two contradictory models to explain the phenomena in full is called wave-particle dualism.

➤   Classical wave theory fails when it tries to explain experiments in which light interacts with atomic particles. Examples are the photoelectric effect (photo effect) and the Compton effect. Nor is wave theory sufficient for the explanation of the phenomena of heat radiation (Planck's radiation law).

5. Subfields of optics

•   **Classical optics**, describes the phenomena of optics with the models of classical physics.

- **Geometric optics** or **ray optics**, a branch of classical optics. It describes the interaction of light with objects with dimensions appreciably larger than the wavelength of light.
- **Wave optics**, branch of classical optics. It describes the interaction of light with objects with dimensions of the same order of magnitude as the wavelength of light.
- **Quantum optics**, describes optical processes by the methods of quantum mechanics. This approach leads to a particle picture in particular when describing the interaction of light and matter.

Electron optics, ion optics, generation of images by means of electron (ion) beams by particle deflection in combinations of inhomogeneous electric and magnetic fields that act analogously to the refractive components in light optics.

11.1   Geometric optics

Geometric optics or **ray optics**, describes the interaction of light with objects with dimensions appreciably larger than the wavelength of light.

■   Interaction of light with lenses, mirrors, prisms, and apertures.

Light path, **optical path length**, λ, product of geometric path length l of the ray and the refractive index n of the medium traversed by the ray,

$$\lambda = l \cdot n .$$

1. Fermat's principle,

extremum principle from which ray optics can be derived:

▲   Light propagates in such a way that the light path takes an extremum value, usually a minimum.

Light follows the **shortest path in time** in traveling between two points. Since the speed of light depends on the medium, the light path between two points in different media is not necessarily the shortest geometric distance.

2. Properties of light rays

Fermat's principle is based on the concept of **light rays** (**Fig. 11.1**):

- Light propagation can be described by single rays. A ray of light in a homogeneous medium follows a straight line, similar to particle motion in a force-free space. In an inhomogeneous medium, light rays may be curved.
- Rays are always perpendicular to the wave front of the corresponding wave.
- Rays may intersect each other and do not influence each other.
- The direction of motion of the rays may be reversed.
- The direction of a ray of light changes in general at the interface between two media in which light propagates with different speeds.
➤   The rule that rays do not influence each other corresponds to the **superposition principle** of linear wave theory.

3. Types of rays

- **Bundle of rays**, spatial set of light rays.
- **Pencil of rays**, plane set of rays. Partial set of a bundle, obtained e.g., by collimating the bundle with a slot.
- **Divergent rays**, rays starting from a point (as for an outgoing spherical wave, see **Fig. 11.2 (a)**).

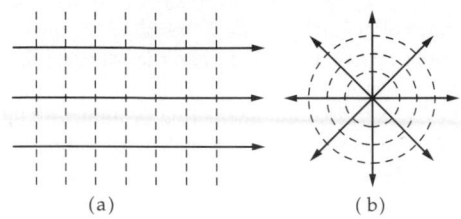

(a)   (b)

Figure 11.1: Wave fronts
(dashed lines) and rays
(arrows). (a): plane wave,
(b): spherical wave.

- **Convergent rays**, rays converging into a point (as for an incoming spherical wave, see **Fig. 11.2 (b)**).
- **Parallel rays**, all rays point parallel to each other. This pattern corresponds to a plane wave (see **Fig. 11.2 (c)**).
- **Homocentric rays**, generic term for diverging, converging, and parallel rays.
- **Diffuse rays**, the rays are randomly oriented (see **Fig. 11.2 (d)**), contrary to homocentric rays. Diffuse rays arise, e.g., by reflection of parallel rays by a rough surface.

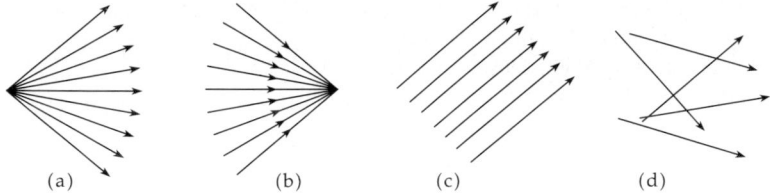

(a)   (b)   (c)   (d)

Figure 11.2: Bundle of rays. (a): divergent bundle, (b): convergent bundle, (c): parallel bundle of rays, (d): diffuse bundle.

11.1.1   Optical imaging—fundamental concepts

Optical image, conversion of a homocentric bundle of rays leaving an **object point** into another homocentric bundle with its center at the **image point**.

Object point, O, **source point**, G, any point from which light emerges.

Image point, B, center of the bundle of rays originating from one object point.

1. Real and virtual images

Real image, bundles of rays belonging to image points converge (**Fig. 11.3 (e)**).

Virtual image, the bundles of rays belonging to image points are divergent; the rays themselves do not intersect each other, but their backward extensions do.

Virtual image point, B', intersection point of the extended rays in virtual imaging (**Fig. 11.3 (a) – (d)**).

2. Optical elements and their characteristics

Optical elements: lenses, mirrors, apertures, plates, prisms, etc. and their combinations in functional groups.

■   Lens, eye piece, condenser and image inversion system.

Optical axis, symmetry axis of optical elements with respect to rotations, e.g., the connecting line through the centers of curvature of the refracting surfaces of an optical system.

Centered system, a system for which the optical axes of all optical elements coincide.

■   The centers of curvature of all optically refracting surfaces lie on a straight line, the optical axis.

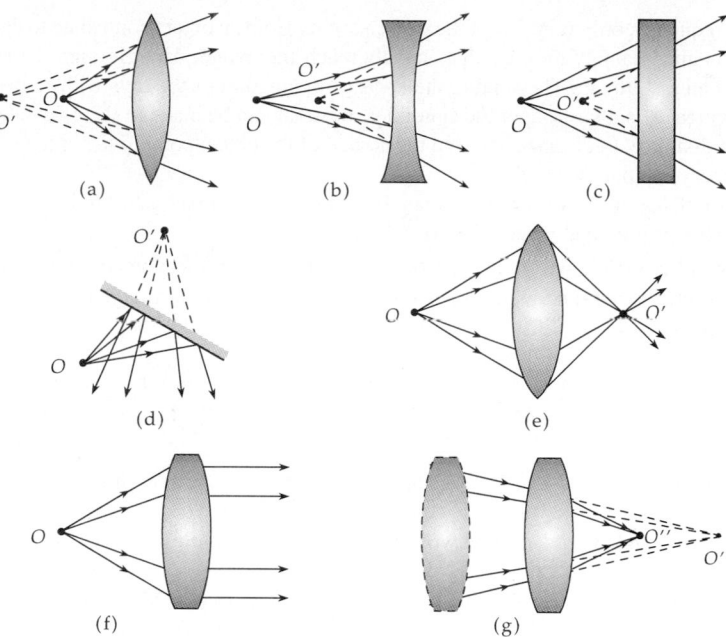

Figure 11.3: Optical imaging (schematically). (a) – (d): virtual image, (e): real image, (f): imaging to infinity, (g): real imaging of a virtual object point O' to O''.

3. Focal points of optical elements

Object focus, $\bar{F}$, point from which rays emerge when they are parallel to the optical axis behind the optical system.

  Image focus, F', point at which the rays incident parallel to the optical axis intersect.

  Principal planes: To construct the image, one introduces plane surfaces at which the change of the direction of the rays will proceed, instead of the mostly curved surfaces of the optical elements. These auxiliary planes are perpendicular to the optical axis; their positions have to be determined so that the image constructed with their help coincides with the real image generated by the actual (curved) surfaces of the optical elements.

➤   The principal planes are an auxiliary concept to simplify the calculation of the imaging and the graphic approximation of the leading rays in imaging. The actual change of directions occurs, of course, at the confining surfaces of lenses, prisms or mirrors.

  Principal points, intersection points of the principal planes with the optical axis.

  For lenses there are two surfaces where refraction occurs. Correspondingly, two principal planes and two principal points are introduced:

- **object principal point**, H, principal point located closer to the object,
- **image principal point**, H', principal point located closer to the image.

4. Focal lengths and object distances

Object focal length, **object-side focal length**, $\bar{f}$, distance between object principal point and object focal point.

  Image focal length, **image-sided focal length**, f', distance between image principal point and image focal point.

➤ Frequently, only very few of the rays emerging from an object contribute to the optical imaging, namely the rays that actually reach the image plane through the apertures of an instrument. The smaller the angle of inclination of the rays with respect to the optical axis, the stronger the simplifications that can be made in the calculations.

Object distance, a, distance between the normal of the object point to the optical axis, and the object principal plane, $a = \overline{HO}$.

Image distance, a', distance between the normal of the image point to the optical axis, and the image principal plane, $a' = \overline{H'O'}$.

Focus-object distance, z, distance of the object plane from the object focus, $z = \overline{FO}$.

Focus-image distance, z', distance of the image plane from the image focus, $z' = \overline{F'O'}$.

Relations:

$$z = a - \bar{f}, \qquad z' = a' - f'.$$

Intercept distances, $\bar{s}$ and s', distance of object or image measured from the corresponding lens apex.

Imaging equation, relation between the conjugated quantities (object distance, image distance) of imaging.

Object size, y, lateral size of the object (perpendicular to optical axis).

Image size, y', lateral size of the real image (perpendicular to optical axis).

Paraxial region, space region near the axis where the angle α between rays and optical axis is so small that $\sin \alpha$ and $\tan \alpha$ can be replaced with sufficient accuracy by the angle α. The image equations then simplify significantly.

The paraxial region cannot be defined generally; it depends on the actual accuracy required.

Gaussian optics, notation for optics in the paraxial region.

➤ Gaussian optics is also a first approximation for analysis **outside** the paraxial region in order to determine the basic properties of an optical system.

In the following, we shall mainly treat centered systems in the paraxial region.

5. Sign conventions

* **Direction of light** from left to right.
* Use of oriented segments. Distances are measured as follows: from the reference point towards the right (along the direction of light) as positive, towards the left as negative.
* The y-direction upwards is positive.
* The curvature radius (lens, mirror) is positive if the center of curvature (C) lies to the right of the apex (S), and negative if C lies to the left of S.
* Conjugated quantities (quantities corresponding to each other in the image space and the object space) are quantities that may be imaged into each other; they get the same letters. The quantities in the image space are labeled by an upper prime at the right.
* For quantities that occur pairwise, but are not related by imaging, the quantity on the object side is specified by a bar, e.g., $\bar{F}$ (object side) and F' (image side).
* A reference leg is fixed for angle measurement. An angle is positive if the other leg has to be turned counterclockwise to coincide with the reference leg; otherwise, it is negative. Arrows in angles point away from the reference leg.
* ➤ In the subsequent figures, a reference is often made to the sign of a quantity by adding it in brackets in front of the symbol. Hence, $(-)f$ means that the value of f is negative.

6. Notations in formulae and figures

◼   The object size is mapped to the image size; these quantities are conjugated to each other and yield the notations y and y'.

Symbol	Meaning
$\bar{C}, C'$	centers of curvature
S, S'	apex points
d	lens thickness
n	refractive index
a (also g), a' (also b)	object distance, image distance
$\bar{f}, f'$	object focal length, image focal length
y (also G)	object size
y' (also B)	image size
$\bar{F}, F'$	object focus, image focus
H, H'	object and image principal points
O	point on the optical axis
s, s'	intercept distances (from apex)
i	distance between principal planes
β'	linear magnification
Γ'	magnification

11.1.2   Reflection

Mirror, plane or curved surface, with a roughness small compared with the wavelength of the incident radiation.

In order to describe the reflection of a light ray geometrically, the normal to the mirror surface at the point hit by the ray is required (**Fig. 11.4**).

Perpendicular, **perpendicular of incidence**, the normal to the surface at the point hit by the ray.

Angle of incidence, ε, angle between the perpendicular to the surface and the incident ray.

Angle of reflection, ε_r, angle between the perpendicular to the surface and the reflected ray.

Reflection law (see p. 304):

▲   The angle of incidence equals the angle of reflection,

$$\varepsilon = \varepsilon_r .$$

▲   The incident ray, the perpendicular, and the reflected ray always lie on a plane.

➤   The reflection is independent of the wavelength (color). Therefore, reflection causes no **chromatic aberration**, contrary to refraction (imaging by lenses).

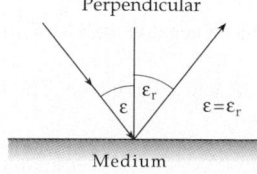

Figure 11.4: Reflection law. Perpendicular, angle of incidence ε and angle of reflection ε_r.

11.1.2.1 Plane mirror

Relation between image point and object point:
▲ The virtual image point and the object point are equidistant from the mirror and lie on the same perpendicular.

The virtual image arising for a plane mirror is erect and laterally (left-right) inverted. The image size equals the object size (**Fig. 11.5**).

➤ Since the rays are only an auxiliary tool of representation, an arbitrary number of rays may be drawn in any direction from any object point. All rays striking the mirror yield the same virtual image point (no imaging error).

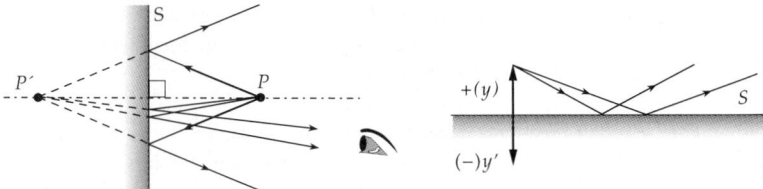

Figure 11.5: Image formation at the plane mirror. P: object point, P': virtual image point, y: object, y': image. Homocentric bundles generate the image according to the law of reflection. Depending on the position, the eye selects a certain fraction of the rays, which leads to the perception of a virtual image.

11.1.2.2 Concave mirror

Concave mirror, general designation for mirrors that collect parallel incident rays into a convergent bundle (**Fig. 11.6**).

■ Most concave mirrors resemble a spherical dish (**spherical mirror**), or a rotational paraboloid (**parabolic mirror**), or other rotational conics with axial symmetry.

1. Characteristics of concave mirrors

Apex, S, of a mirror, intersection point of the optical axis and the mirror surface.

Focal point of concave mirror, by definition the point where the rays incident very close and parallel to the optical axis intersect each other.

Focal length, $\bar{f}$, distance between the focal point and the apex.

➤ For the mirror the principal planes H and H' coincide with the osculating plane at the apex S.

focal length of spherical concave mirror = half of sphere radius			L
	Symbol	Unit	Quantity
$(-)\bar{f} = \dfrac{(-)r}{2}$	$\bar{f}$	m	focal length
	r	m	radius of mirror

According to the sign convention, a concave mirror has a negative radius of curvature and **negative focal length**.

▲ For spherical concave mirrors, the focal points $\bar{F}$ and F' coincide. The image focal length equals the object focal length,

$$\bar{f} = f'.$$

➤   Actually, the reflection does not occur at the principal plane, but at the surface of the mirror. In the paraxial region (see p. 340), the difference may be ignored.

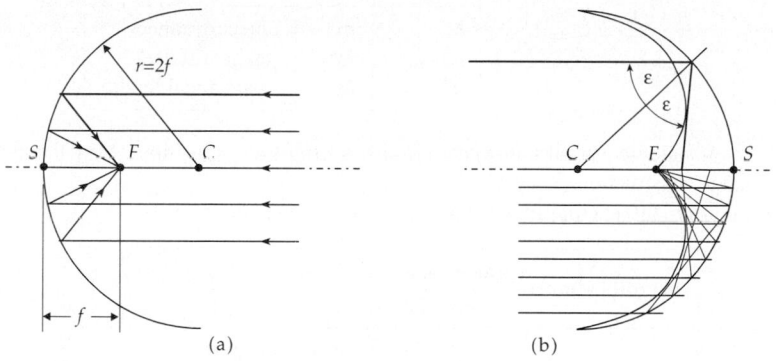

Figure 11.6: Spherical concave mirror. C: center, S: apex, F: focal point. (a): curvature radius r and focal length f, (b): catacaustic line (envelope of reflected rays) and aperture aberration.

2. Image construction for concave mirror

▲   **Image construction** with two preferential rays (**Fig. 11.7**):
       Focal ray, strikes the mirror through the focal point and is reflected parallel to the optical axis.
       Parallel ray, is incident parallel to the optical axis and is reflected through the focal point.
Rays passing the center of curvature of a spherical concave mirror (**central rays**) are reflected into themselves.

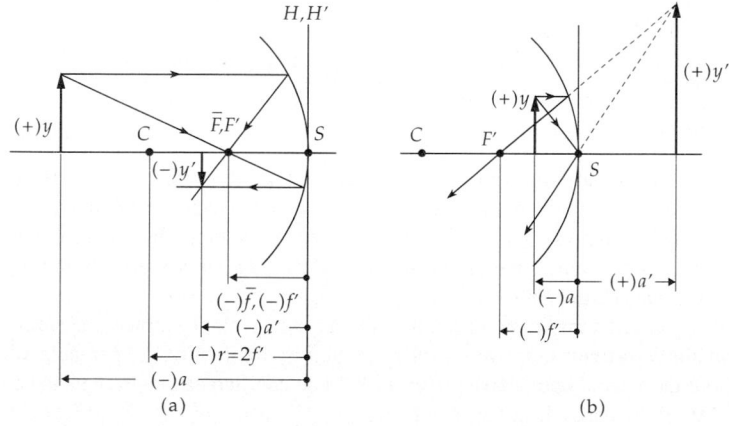

Figure 11.7: Image construction for spherical concave mirror. (a): object beyond twice the focal length, (b): object within the focal length.

3. Imaging equation and magnification of the concave mirror

Imaging equation for the concave mirror:

imaging equation for concave mirror			L^{-1}
$\dfrac{1}{a'} + \dfrac{1}{a} = \dfrac{1}{f'}$	Symbol	Unit	Quantity
	a	m	object distance
	a'	m	image distance
	f'	m	image focal length

➤   The imaging equation immediately follows from the application of ray theorems to image construction.

Magnification, **lateral magnification**, β', of the concave mirror:

magnification = $\dfrac{\text{image size}}{\text{object size}}$			1
$\beta' = \dfrac{y'}{y} = -\dfrac{a'}{a}$	Symbol	Unit	Quantity
	β	1	lateral magnification
	y'	m	image size
	y	m	object size
	a	m	object distance
	a'	m	image distance

➤   According to the sign conventions, the lateral magnification is positive (negative) if the image is erect (inverted).

Images of the spherical concave mirror, depending on the object distance a:

object distance a	image distance a'	image	lateral magnification β
$-\infty < a < 2f'$	$2f' < a' \leq f'$	real, reduced, inverted	$-1 < \beta < 0$
$2f'$	$2f'$	real, equal size, inverted	-1
$2f' < a < f'$	$-\infty < a' \leq 2f'$	real, enlarged, inverted	$-\infty < \beta < -1$
$f' < a < 0$	$0 < a' < \infty$	virtual, enlarged, erect	$1 < \beta < \infty$

4. Non-paraxial cases

• **Spherical concave mirror** or **spherical mirror**, the larger the distance of the parallel incident rays from the optical axis, the larger is the distance of the intersection point of the reflected rays on the optical axis from the focal point. In the sense of Gaussian optics, this phenomenon is an imaging defect called **aperture aberration** (**spherical aberration**) (**Fig. 11.6 (b)**).

  The reflected rays have a continuously curved envelope surface, the **catacaustic**.

• **Parabolic mirror**, convave mirror generated by rotation of the parabola $y^2 = 2cx$ about the x-axis (optical axis) (**Fig. 11.8**). The **coefficient** c represents the curvature radius of the parabola at the apex point.

➤   In the sense of Gaussian optics, the parabolic mirror with parabolic coefficient c and the spherical mirror with radius $r = c$ are equivalent. In particular, the same imaging equations hold.

  All rays parallel to the optical axis intersect in the focal point of the parabolic mirror, which thus has a vanishing aperture aberration. However, the imaging defects of the parabolic mirror are very strong, even for parallel rays that are only slightly inclined with respect to the optical axis (coma).

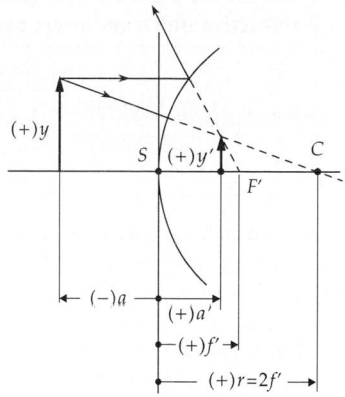

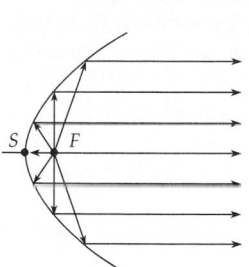

Figure 11.8: Generation of a parallel bundle of rays by a parabolic mirror.

Figure 11.9: Image construction for a convex mirror by means of central ray and focal ray.

11.1.2.3   Convex mirror

Convex mirror, spherical or other rotational surface reflecting at the outer side surface (**Fig. 11.9**).
- Parallel incident light is diverging after reflection.
- According to the sign convention, the convex mirror has a positive radius of curvature and a **positive focal length**,

$$(+)f = \frac{(+)r}{2}.$$

- A convex mirror always creates virtual, reduced and erect images.

11.1.3   Refraction

Refraction (see p. 302), change of the direction of a ray when passing the interface between two media.
➤ The light does not enter the second medium entirely; a certain fraction is reflected.

11.1.3.1   Refractive index

Coefficient of refraction, **refractive index** n, material constant, characterizes the refractive behavior of the medium in the transition of light from vacuum into this medium.
▲ If two media border on each other, the medium with higher refractive index is said to have **higher optical density**, the medium with lower refractive index is said to have **lower optical density** than the corresponding other medium.
■ The coefficient of refraction for a vacuum is 1, the refractive coefficients for air, water and diamond are 1.0003, 1.333 and 2.417, respectively. The coefficient of refractions for glasses are in the range of 1.4 to 1.9 (e.g., quartz glass 1.46, optic boron crown 1.51, optic flint 1.61, heavy optic flint 1.76).
For additional values of refractive index see **Tab. 12.2/2**.

refractive index and propagation speed			1
$n_{medium} = \dfrac{c_{vacuum}}{c_{medium}}$	Symbol	Unit	Quantity
	n_{medium}	1	refractive index
	c_{vacuum}	m/s	phase velocity in vacuum
	c_{medium}	m/s	phase velocity in medium

In general, the index of refraction depends on the wavelength (see p. 305).

➤   In optical technology, one introduces the quantity $n' = c_{air}/c_{medium}$. The coefficient of refractions n and n' differ only slightly. For dry air under standard conditions, $n' = 1$ and $n = 1.0003$.

Angle of incidence, ε, angle between incident ray and the perpendicular. **Angle of refraction**, ε', angle between refracted ray and perpendicular.

11.1.3.2   Law of refraction

Law of refraction, Snell's law, describes the relation between the angles of incidence and refraction (**Fig. 11.10**):

Snell's law of refraction			
$\dfrac{\sin \varepsilon}{\sin \varepsilon'} = \dfrac{n_2}{n_1} = \dfrac{c_1}{c_2}$	Symbol	Unit	Quantity
	ε	rad	angle of incidence
	ε'	rad	angle of refraction
	n_1, n_2	1	refractive indices of medium 1, 2
	c_1, c_2	m/s	phase velocities of medium 1, 2

▲   The ratio of the sine of the angle of incidence and the sine of the angle of refraction is a constant that depends only on the material properties of both media.

▲   The incident ray, the perpendicular, and the refracted ray lie in one plane; the reflected ray lies in the same plane.

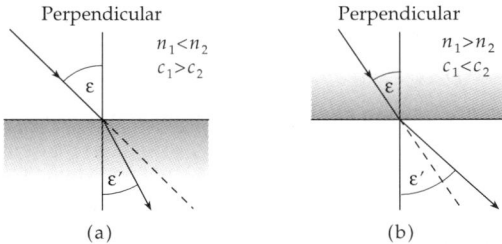

Figure 11.10: Snell's law of refraction. ε: angle of incidence, ε': angle of refraction. (a): $n_1 < n_2$, $c_1 > c_2$, refraction towards the perpendicular, (b): $n_1 > n_2$, $c_1 < c_2$, refraction away from perpendicular.

▲   When entering the optically more dense medium ($n_1 < n_2$, $c_1 > c_2$), the light ray is refracted towards the perpendicular; when entering an optically less dense medium ($n_1 > n_2$, $c_1 < c_2$), the ray is refracted away from the perpendicular.

■   In a transition from air to glass, the ray is refracted towards the perpendicular. For light of wavelength $\lambda = 632.8$ nm, the angles of incidence $\varepsilon = 10°, 30°, 60°, 80°$ correspond to the angles of refraction $\varepsilon' = 6.5°, 19.0°, 35.0°, 40.0°$.

Relative refractive index, notation for the ratio of the refractive indices of two media, $n = n_2/n_1$.

11.1.3.3   Fresnel's formulas

In reflection of light, the intensity of the reflected ray is less than that of the incident light (except for the case of total reflection, see below):

- in reflection by metallic layers because of weak absorption in the layer (see **Fig. 11.11**),
- at the interface between media of different refractive indices only a fraction of the intensity is reflected.

1. General Fresnel formulas for the intensity of light

Fresnel's formulas, quantitative statements on the splitting of intensity between the reflected and the transmitted ray under reflection, as a function of the state of polarization of light (**Fig. 11.11**). These formulas follow from Maxwell's equations of electrodynamics:

Fresnel's formulas for intensities of light			1
	Symbol	Unit	Quantity
$R_\parallel = \dfrac{\tan^2(\theta_i - \theta_t)}{\tan^2(\theta_i + \theta_t)}$	θ_i	rad	angle of incidence
	θ_t	rad	angle of emergence
	$R_\parallel$	1	reflection coefficient/fraction with polarization $\parallel$ incidence plane
$R_\perp = \dfrac{\sin^2(\theta_i - \theta_t)}{\sin^2(\theta_i + \theta_t)}$	$R_\perp$	1	reflection coefficient/fraction with polarization $\perp$ incidence plane

The transmitted fractions are

$$T_\parallel = 1 - R_\parallel, \quad T_\perp = 1 - R_\perp$$

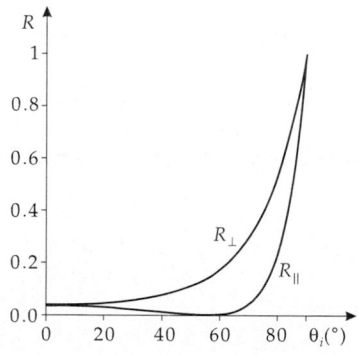

Figure 11.11: Dependence of reflection coefficients on state of polarization and incidence angle for a glass-air interface.

2. Fresnel's formulas for perpendicular incidence of light,

specify the reflected and transmitted fraction of intensity for the incidence angle $\theta_i = 0$.

$$R = \left(\frac{n-1}{n+1}\right)^2, \quad T = \frac{4n}{(n+1)^2}, \quad n = n_2/n_1.$$

➤ Because of the flatness of the curve in **Fig. 11.11**, it is often sufficient to adopt these simplified formulas.

■ At any air-glass interface at least 4 % of the intensity is reflected. Therefore, lenses in optical devices always have to be coated with antireflective material. Example: in an objective consisting of three groups of lenses (six interfaces) without coating, about 25 % of the light intensity would be lost.

➤ For a certain angle, Brewster's angle θ_B, the fraction $R_\| = 0$. The reflected fraction of the unpolarized light incident at this angle is linearly polarized (see p. 293).

Brewster's angle: $\tan\theta_B = n_2/n_1$.

11.1.3.4   Rainbow

Rainbow, atmospheric-optical phenomenon caused by the refraction and reflection of light in water droplets. The rainbow is part of a circle with the center on the connecting line of Sun and observer on the side opposed to the Sun. For an m-fold reflection in the droplet interior, the deflection angle δ is

$$\delta = 2(\varepsilon - \varepsilon') + m(\pi - 2\varepsilon').$$

Here ε is the angle of incidence and ε' the angle of refraction on entrance of the light ray into a water droplet (**Fig. 11.12**). The minimum deflection for a refractive index n is

$$\frac{\partial\delta}{\partial\varepsilon} = 0, \quad \cos\varepsilon_{min} = \sqrt{\frac{n^2 - 1}{n + 2m}}.$$

Main rainbow, has a radius of 42.5° and a width of 1.5°. It arises under two-fold refraction and single reflection of light by a drop of water. Dispersion causes color spreading with the sequence red, orange, yellow, green, indigo and violet from inside to outside.

　Secondary rainbow, has a radius of 52° and a width of 3°. It arises by two-fold refraction, two-fold reflection and dispersion in the water drop. The sequence of colors is reversed from the main rainbow.

➤ Rainbow formation is connected with interference phenomena, which depend on the size of the droplets. These interferences manifest themselves by alternating bright and dark rings and by the irregular sequence of colors in the secondary rainbow.

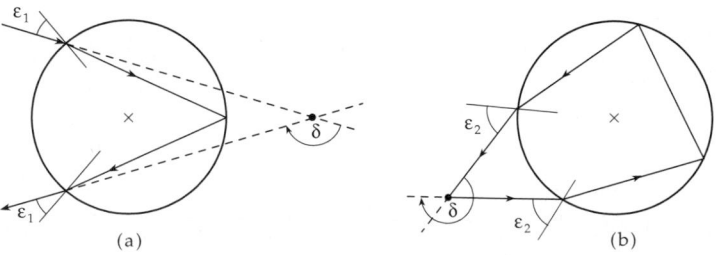

Figure 11.12: Path of rays in the rainbow. (a): main rainbow, (b): secondary rainbow.

11.1.3.5 Total reflection

Total reflection, occurs when light is incident from a medium of higher optical density, with an angle larger than or equal to the critical angle of total reflection, onto the interface with a medium of lower optical density.

  Critical angle of total reflection, ε_g, the angle of incidence for which the angle of emergence equals $\pi/2$ when the ray travels from the medium of higher optical density to the medium of lower optical density (**Fig. 11.13**).

critical angle of total reflection			1
$\sin \varepsilon_g = \dfrac{n_2}{n_1}$	Symbol	Unit	Quantity
	ε_g	rad	critical angle of total reflection
	n_1, n_2	1	refractive index of medium 1, 2

Critical angle of total reflection for several media; the surrounding medium is air:

Substance	ε_g	Substance	ε_g
diamond	23°	light crown glass	40°
heavy flint glass	34°	glycerol	43°
carbon sulfide	38°	water	49°

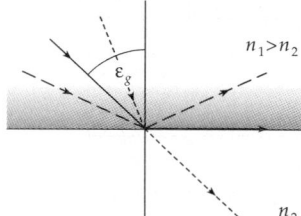

Figure 11.13: Critical angle of total reflection ε_g ($n_1 > n_2$). Refraction for $\varepsilon < \varepsilon_g$ (dotted line), critical case for $\varepsilon = \varepsilon_g$ (full line), total reflection for $\varepsilon > \varepsilon_g$ (dashed line).

  Critical angles of other substances may be calculated from the tabulated refractive indices (**Tab. 12.2/2**).

■   The refractive index of air is 1.0003, that of ice 1.310. If a ray passes through ice and hits an interface to air, the critical angle of total reflection is

$$\varepsilon_g = \arcsin \frac{n_{\text{air}}}{n_{\text{ice}}} = \arcsin \frac{1.0003}{1.310} = 0.868851 \,\text{rad} = 49.78°.$$

Rays hitting the interface at angles of incidence $\varepsilon > 49.78°$ are totally reflected.

➤   Total reflection is used in prisms for inverting the direction of rays.

Porro prism set, set of prisms for image inversion by four-fold total reflection (**Fig. 11.14**).

11.1.3.6 Light wave guide

Light wave guide or **pipe**, arrangement of mirrors or totally reflecting interfaces that confine light propagation in a definite direction (along the symmetry axis of the arrangement).

■   Tube with mirror-coated inner surface.

■   Most important application: Glass fibers for optical communication.

➤   There is a direct analogy to wave guides in microwave technology.

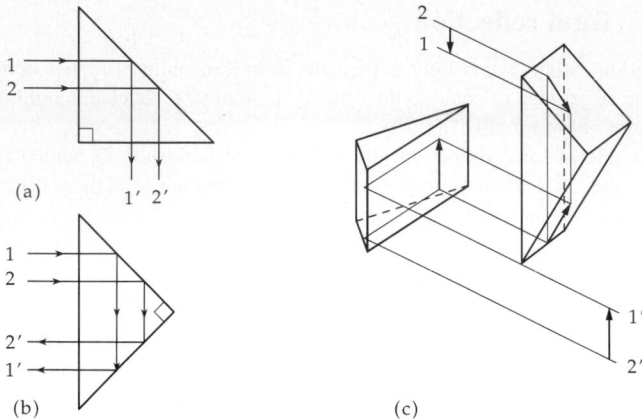

Figure 11.14: Ray inversion by totally reflecting prisms. (a): deflection by $\pi/2$, (b): deflection by π, (c): Porro prism set (image inversion).

1. Operational mode of light wave guides

a) Structure and properties of light wave guides: **Light wave guide**, consists of a **core** of refractive index n_1, and a **coating** of refractive index $n_2 < n_1$ (**Fig. 11.15**).

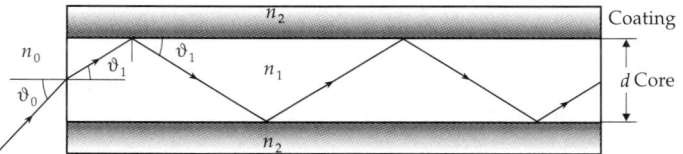

Figure 11.15: Structure of a light wave guide.

If the angle ϑ_1 is small enough, the light ray is **totally reflected** at the interface between core and coating and may leave the core of the light wave guide only through the end face.

Critical angle of total reflection ϑ_1 at the interface between core and coating is given by the equation

$$n_1 \sin(90° - \vartheta_1) = n_2 \quad \Rightarrow \quad \cos \vartheta_1 = n_2/n_1.$$

At the entrance face, the law of refraction holds, $n_0 \sin \vartheta_0 = n_1 \sin \vartheta_1$. Both relations combined yield the **numerical aperture** *NA* of the wave guide:

$$n_0 \sin \vartheta_0 = n_1 \sqrt{1 - \cos^2 \vartheta_1} = n_1 \sqrt{1 - (n_2/n_1)^2}$$

$$= \sqrt{n_1^2 - n_2^2} = \sqrt{n_{\text{core}}^2 - n_{\text{coating}}^2} = NA.$$

➤   Only light rays for which $n_0 \sin \vartheta_0$ is smaller than or equal to *NA* will be transmitted by the wave guide. For larger angles of incidence ϑ_0 the condition of total reflection is not fulfilled.

➤   This consideration holds only in the approximation of ray optics, i.e., for lengths of the wave guide that are considerably larger than the wavelength of the light used. In the important case of the *single mode* wave guide, this condition is not fulfilled; here a specification of a numeric aperture makes no sense and should be replaced by the specification of the characteristics of the **natural mode** of the wave guide (e.g., $1/e$ width in the Gaussian approximation).

■   The principle of the wave guide was demonstrated in 1870 by John Tyndall in London using a water jet (core, $n = 1.33$) in air (coating, $n = 1.00$).

▲   If a wave guide is bent, the condition for total reflection is violated for a fraction of the rays that go astray ("decouple"). The larger the difference of the refractive indices of core and coating, the less sensitive the wave guide with respect to such losses due to curvature.

b) Wave-optical boundary condition: If the coherence length of the light used is larger than the thickness d of the core of the wave guide, an additional **wave-optical boundary condition** has to be satisfied:

▲   After two-fold reflection at the interfaces, the wave front must constructively interfere with its not-yet-reflected fractions (**Fig. 11.16**), i.e., the optical path difference must be an integer multiple $m \cdot \lambda$ of the wavelength λ (polarization and phase shift are ignored).

Then

$$n_1 \sin \vartheta_1 = \frac{m\lambda}{2d} \leq NA = \sqrt{n_1^2 - n_2^2} \quad \Rightarrow \quad m \leq \frac{2d}{\lambda}\sqrt{n_1^2 - n_2^2}.$$

The largest integer number N to fulfill this condition is the number of allowed ray orientations. If this is fulfilled only for $N = 0$, it is a **single-mode wave guide**.

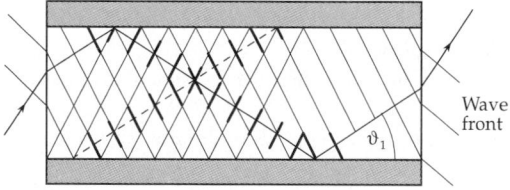

Figure 11.16: Quantization of propagation angles in the wave picture: The reflected fraction has to interfere constructively with the not-yet-reflected fraction of the wave.

▲   Wave guides with a small number of allowed propagation angles, and in particular *single-mode* wave guides, are quantitatively described by Maxwell's equations with the given boundary conditions. Instead of ray-optical propagation angles, one gets the **(natural) modes** of the wave guide. These are the allowed distributions of the electric and magnetic fields. There is a close analogy to the probability functions of a quantum-mechanical particle in a potential well.

➤   The intensity distribution in a *single-mode* wave guide calculated with Maxwell's equations can be described in good approximation by a Gaussian curve (**Fig. 11.17**).

2. Application of light wave guides

■   Simplest light wave guide: pipe with mirror coating on the inner surface, with a diameter of several millimeters. Applied to guide UV light to difficult-to-access positions where it is needed for hardening of UV cement (dentist).

Important applications: endoscopes, alternating-sign signaling systems, fiber-optics plates for electron-image tubes, and for reduction of the field of view of monitors (cash dispenser), and fiber-optical sensors (see below), fibers for **optical communication**.

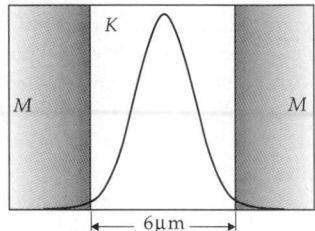

Figure 11.17: Cross-section and intensity distribution for a typical single-mode fiber. M: coating ($n = 1.455$), K: core ($n = 1.46$).

a) Optical communication: Main field of application for glass fibers: substitute for electric connections. Advantages:
- higher transfer capacity and lower attentuation
- very low failure rate
- insensitivity to EMI (electromagnetic interference)
- ➤ Since 1988 underwater cables for transatlantic (TAT-8) and transpacific (TPC-3) communication are glass fibers.

Signal dispersion, important characteristics that specifies how much a signal is spread out in time. The smaller the dispersion, the faster subsequent pulses can be sent. It is caused by both material and mode dispersion.

Material dispersion, describes the dependence of the refractive index, and thus of the light speed in the medium, on the wavelength.
- ➤ Material dispersion can **not** be avoided by using a monochromatic light source. Since the energy-time uncertainty relation connects the line width and the coherence length, a truly monochromatic source of light would imply a coherence length of ∞. For data transfer in the GHz- (or GBit/s) range one needs, however, pulses of lengths ≤ 1ns, which correspond to wave trains of a maximum length of 20 cm (for $n = 1.5$), and the corresponding frequency spread.

For this reason one tries to reduce the material dispersion of the fibers by special geometries and materials (*dispersion-shifted fiber, dispersion-flattened fiber*).

Mode dispersion, arises because light rays of different propagation angles pass through the fibers with different transit times.
- ▲ Mode dispersion is the most important distinguishing feature between different types of glass fibers.

Multi-mode fibers are used exclusively for short connections, since mode dispersion quickly reaches unacceptable values as the length increases.

Single-mode (mono-mode) fibers basically have no mode dispersion. Their transfer capacity is limited by material dispersion. However, they require a very high effort in installation (calibration with sub-μm-accuracy) because of their small core diameter.

b) Gradient-index fibers (GRIN fibers, GI fibers), contain a core with a refractive index that decreases continuously with increasing radius. They correspond to a series of gradient-index lenses (rod lenses, SELFOC^{TM} lenses) of very low diameter (**Fig. 11.18**).

The **optical path differences** take a minimum value for a radial dependence of the refractive index as follows:

$$n(r) = n_1(1 - \alpha r^2).$$

For this reason, gradient-index fibers take (with respect to their transmission capacity) an intermediate position between multi-mode and single-mode step-index fibers.

c) Fiber-optical sensors, level-measuring sensor, a light guide that uses air as low-refractive coating is positioned in a container (**Fig. 11.19**). If light is coupled in, the step in

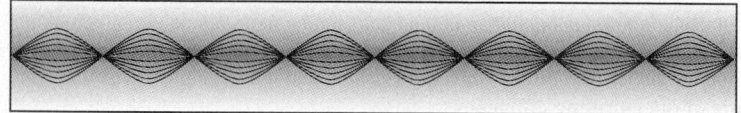

Figure 11.18: Gradient-index fiber as a sequence of gradient-index lenses.

the refractive index is so large for an empty container that almost no light is lost, despite the curvature, and the ray hits a detector (photo diode) at the remote end of the core. If the container is filled up to a level such that the core is immersed, the liquid serves as coating material. Because of the reduced step in the refractive index, much more light is lost, and the signal from the photo diode is changed.

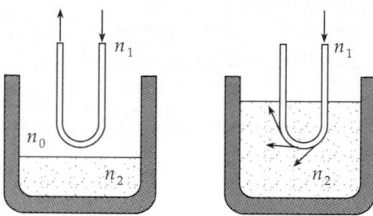

Figure 11.19: Simplest example of a fiber-optical system: level-measuring sensor.

By appropriate structuring of the interface (Bragg grating) between the core and the coating, the sensivity of such a sensor may be increased considerably. One may then demonstrate, for instance, the attachment of certain molecules. Further, by using these gratings, one may establish the length of a fiber with high precision (tension, pressure, temperature). The measured quantities then are not the total transmission, but phase and polarization changes, as well as absorption or reflection in very narrow ranges of wavelength.

d) Coupling of light into wave guides: **Coupling efficiency**, ratio of the light power coupled in and the emitted power of the source of light.
Fig. 11.20 shows that for maximum coupling:

$$B = D\frac{b}{g} < d, \qquad \text{and} \qquad \vartheta_{\text{WL}} < NA.$$

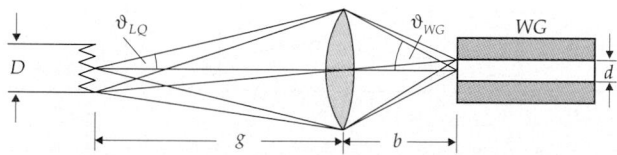

Figure 11.20: Coupling of the light of a halogen bulb LQ into a wave guide WG.

For a light source radiating isotropically (halogen bulb, arc lamp, light diode), this condition cannot be fulfilled. Moreover, the product of the size of the radiating area and the solid angle covered by the lens is a constant of optical imaging. It follows that, in a case of optical imaging that **reduces** the radiating area, the solid angle **increases**.

▲   For this reason, high-efficiency coupling may be achieved exclusively by lasers, since laser light fills the minimum phase space volume.

➤ For laser diodes, a matching to fibers with refractive index 1.5 is achieved by **enlarging** the radiating area and **reducing** the solid angle. In addition, an **anamorphotic** imaging system is used to convert an elliptic beam profile into a circular one.

Estimate of the coupling efficiency:

$$\eta = \left(\frac{d \cdot \vartheta_{WG}}{D \cdot \vartheta_{LQ}} \right)^2$$

For a more detailed consideration, the overlap integral of the natural mode $A(x, y)$ of the wave guide and the complex amplitude $B(x, y)$ of the light hitting the end face has to be evaluated:

$$\eta = \frac{\left[\int\int A(x, y)B^*(x, y)\mathrm{d}x\mathrm{d}y \right]^2}{\int\int A(x, y)A^*(x, y)\mathrm{d}x\mathrm{d}y \int\int B(x, y)B^*(x, y)\,\mathrm{d}x\,\mathrm{d}y}.$$

➤ For most practical applications, the functions $A(x, y)$ and $B(x, y)$ may be approximated by Gauss functions $\exp\left(\dfrac{-x^2}{\sigma^2} \right)$.

Estimate of the order of magnitude of the coupling efficiencies of various combinations of light sources and wave guides:

	halogen bulb	short-arc lamp	light diode	laser
wave guide 10 mm	1	1	1	1
plastic fiber 1 mm	0.001	0.01	1	1
multi-mode fiber 50 μm	10^{-5}	10^{-4}	0.01	1
single-mode fiber 6 μm	10^{-8}	10^{-7}	10^{-5}	1

e) Integrated optics, wave-guide structures with definite functions such as splitting, joining and switching of light (**Fig. 11.21**).

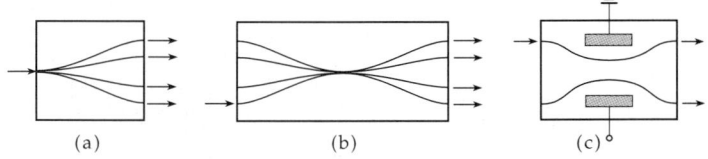

(a)   (b)   (c)

Figure 11.21: Examples of integrated optical elements: (a) brancher 1 × 4, (b) star coupler 4 × 4, (c) electro-optical switch.

▲ The wave guides are produced on wafers (cf. integrated circuits of microelectronics) by lithographic methods. Basic materials are e.g., glass, lithium niobate and polymers.

   In addition, electrodes, heating elements, etc. may be attached.

■ Switching of light between two wave guides by varying the refractive index (by means of electric fields or temperature).

Discrimination between:
   active components such as switches, modulators;
   passive components such as star coupler, brancher, etc.

11.1.3.7   Refraction by a prism

Prism, unit formed of transparent materials, with at least two plane surfaces enclosing an angle; the intersecting line is called **refracting edge**.

In the case of a triangular prism (**Fig. 11.22**), light strikes two interfaces. A ray is thus refracted twice. Let the refractive index of the prism be n_1, and that of the surrounding medium, n_2, where $n_2 < n_1$.

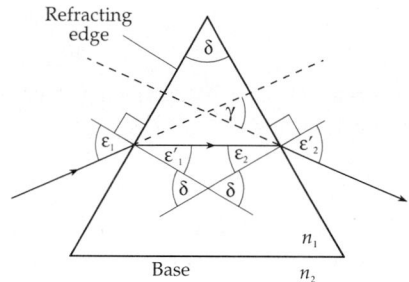

Figure 11.22: Refraction by a triangular prism for symmetric path of rays. δ: vertex angle of the prism, ε_1: angle of incidence, ε_2': angle of emergence, γ: deflection angle.

1. Deflection angle,

γ, of the emerging ray relative to the incident ray: $\gamma = \varepsilon_1 + \varepsilon_2' - \delta$.

deflection angle for prism				**1**
	Symbol	Unit	Quantity	
$\gamma = \varepsilon_1 - \delta$	γ	rad	deflection angle	
$\quad + \arcsin\left\{ \sin\delta\sqrt{\left(\dfrac{n_1}{n_2}\right)^2 - \sin^2\varepsilon_1}\right.$	γ_{min}	rad	minimum deflection angle	
$\quad\quad\left. - \cos\delta\sin\varepsilon_1 \right\}$	ε_1	rad	angle of incidence	
	n_1	1	refractive index, prism	
$\gamma_{min} = 2\arcsin\left(\dfrac{n_1}{n_2}\sin\dfrac{\delta}{2}\right) - \delta$	n_2	1	refractive index, medium	
	δ	rad	vertex angle, prism	

▲   The deflection angle γ takes a minimum value for a symmetric light path, $\varepsilon_1 = \varepsilon_2'$, $\varepsilon_1' = \varepsilon_2$.

If the dependence of the refractive index n_1 on the wavelength is taken into account, the deflection angle γ also depends on the wavelength (see p. 305): light is dispersed by the prism into a spectrum (**Fig. 11.23**).

<u>**M**</u>   Refractive indices may be determined by measuring the minimum deflection angle γ_{min},

$$n_2 = \frac{n_1\sin(\delta/2)}{\sin\left(\dfrac{\gamma_{min} + \delta}{2}\right)}.$$

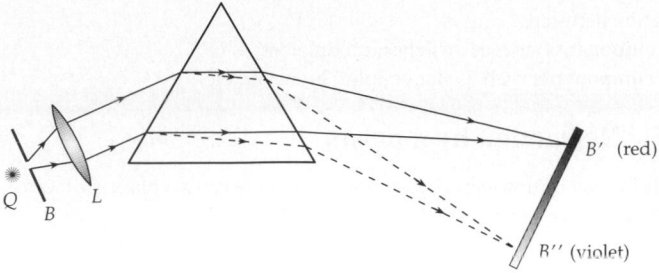

Figure 11.23: Spectral decomposition of light by refracting prism. Q: light source, B: diaphragm, L: lens, B', B'': images of diaphragm.

Since this method allows a determination of the refractive index with high precision, it is appropriate to determine its frequency dependence (see p. 305).

2. *Fraunhofer lines,*

absorption lines in the spectrum of the Sun, caused by absorption by various elements in the photosphere (and in few cases in Earth's atmosphere). The strongest lines are labeled by Latin capital letters (see **Tab. 12.2/9**).

Since the Fraunhofer lines arise by absorption, they appear as black lines in a spectrum of the Sun's light because the energy of the corresponding wavelengths is transferred to the absorbing elements. There are several hundred Fraunhofer lines.

Abbe number, ν_e, quantity characterizing the dispersion of an optical material,

$$\nu_e = \frac{n_e - 1}{n_{F'} - n_{C'}} \, ,$$

where n_e is the refractive index at the frequency of the mercury-e-line ($\lambda = 546.07$ nm), and n'_F and n'_C are the refractive indices at the cadmium lines F' ($\lambda = 480.0$ nm) and C' ($\lambda = 643.8$ nm) (see **Tab. 12.2/9**). The Abbe number ν_λ for the wavelength λ is obtained by replacing the refractive index n_e in the above formula by the corresponding value $n(\lambda)$ for the spectral line with wavelength λ.

11.1.3.8   Refraction by plane parallel glasses

▲   If a ray passes through a plane parallel glass of thickness d, the incident ray and the outgoing ray after two-fold refraction are displaced parallel by a distance δ (**Fig. 11.24**).

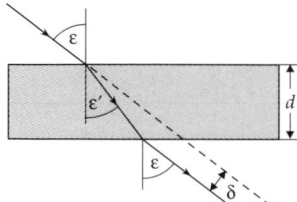

Figure 11.24: Lateral displacement δ of a ray by a plane parallel glass of thickness d.

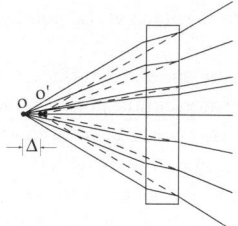

Figure 11.25: Axial displacement Δ of a pencil of rays by a plane parallel glass.

▲ If a pencil of rays passes the plane plate, the center of the pencil is axially displaced by Δ. An observer perceives an object as displaced by the corresponding amount (**Fig. 11.25**).

parallel displacement of ray by plane parallel glass			
$\delta = d\,\dfrac{\sin(\varepsilon - \varepsilon')}{\cos \varepsilon'}$	Symbol	Unit	Quantity
	ε	rad	angle of incidence
$\dfrac{\sin \varepsilon}{\sin \varepsilon'} = \dfrac{n_2}{n_1}$	ε'	rad	angle of refraction
	δ	m	lateral displacement of ray
	d	m	thickness of plate
$\Delta = d\left(1 - \dfrac{n_1}{n_2}\right) \quad$ for small ε	n_1	1	refractive index of air
	n_2	1	refractive index of plate
	Δ	m	displacement of vertex

In order to calculate δ, one first calculates the angle of refraction ε' from the given angle of incidence ε according to the law of refraction and inserts these values into the above formula.

➤ The quantity Δ is important for inversion prisms and must be taken into account in the construction of the image.

11.1.3.9   Refraction by spherical surfaces

Most lenses have spherical surfaces. Therefore, refraction by a spherical surface is of fundamental importance.

Let the spherical surface have a radius R and center C. Let the refractive index be n_2 inside, and n_1 outside, the sphere.

We consider an incident ray from an arbitrary point O outside the sphere to an arbitrary point A on the surface, and trace the refracted ray to the intersection point O' with the optical axis $\overline{OC}$ (**Fig. 11.26**).

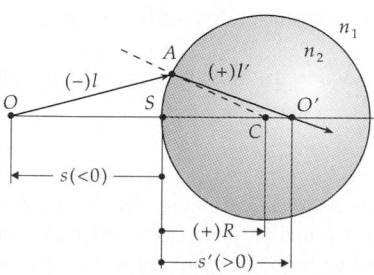

Figure 11.26: Refraction at the surface of a sphere of radius R.

Perpendicular $\overline{AC}$, the perpendicular to the tangential plane at the point where the ray reaches the surface.

Apex, S, intersection point of the optical axis with the spherical surface.

Intersection distances, s and s', distances of the intersection points O and O' of the incident ray and the refracted ray with the optical axis, measured from the apex S. They are counted positively from the apex to the right, and negatively to the left.

l and l' are the distances of the intersection points O and O', respectively, from the point A where the ray hits the spherical surface. The distance from the spherical surface is counted negatively towards the left, and positively towards the right

relation between the intersection distances			1
	Symbol	Unit	Quantity
	n_1	1	refractive index of medium outside the sphere
	n_2	1	refractive index of medium inside the sphere
$n_1 \dfrac{s - R}{l} = n_2 \dfrac{s' - R}{l'}$	R	m	radius of sphere
	s	m	distance $\overline{SO}$
	s'	m	distance $\overline{SO'}$
	l	m	distance $\overline{AO}$
	l'	m	distance $\overline{AO'}$

11.2   Lenses

Lens, transparent body with two interfaces; in general, at least one of them is curved. A lens generates an optical image.

Spherical lens, a lens bounded by two spherical surfaces.

■   Special cases: plane parallel glass, meniscus (lens with a convex and a concave side).
    Other shapes of lenses: aspherical lens, **cylindrical lens, Fresnel lens**, correction plate for Schmidt mirror.

In general, lenses consist of material of higher optical density than that of the surrounding medium (mostly air). Then:

•   convergent lenses are thicker in the middle than at the edge,
•   divergent lenses are thinner in the middle than at the edge.

■   Eyeglasses for near-sighted people are convex-concave divergent lenses **(Fig. 11.27 (f))**.

➤   In order to minimize imaging defects, the shape and orientation of the convergent or divergent lens has to be chosen so that the rays are refracted at both interfaces by about the same amount.

▲   The surface with larger curvature should point in the direction of the parallel ray.

11.2.1   Thick lenses

Thick lens, lens with refractive properties in the paraxial region that may be described by the refraction at two principal planes, the object principal plane and the image principal plane. In this construction, the ray is assumed to propagate parallel to the optical axis between the two principal planes.

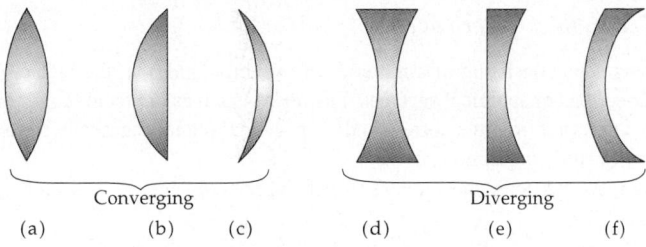

$$\underbrace{\hspace{4cm}}_{\text{Converging}} \qquad \underbrace{\hspace{4cm}}_{\text{Diverging}}$$

(a)   (b)   (c)   (d)   (e)   (f)

Figure 11.27: Lens shapes. (a): biconvex lens ($r_1 > 0, r_2 < 0, f' > 0$), (b): plane-convex lens ($r_1 = \infty, r_2 < 0, f' > 0$), (c): concave-convex lens ($r_1 < r_2 < 0, f' > 0$), (d): biconcave lens ($r_1 < 0, r_2 > 0, f' < 0$), (e): plane-concave lens ($r_1 = \infty, r_2 > 0, f' < 0$), (f): convex-concave lens ($0 < r_2 < r_1, \ f' < 0$). (a)–(c): convergent lenses, (d)–(e): divergent lenses.

1. Characteristic quantities of thick lenses

Object distance, a, the distance between the object principal plane and the object.

Image distance, a', the distance between the image principal plane and the image, counted positively along the direction of the incident rays, negatively in the opposite direction.

Meridional section, a section through an optical system containing the optical axis and an object point off the axis (see **Fig. 11.28**).

Meridional rays, rays propagating within the meridional section.

Sagittal section, plane perpendicular to the meridional section, containing the off-axis object point and a reference ray in the meridional section. The optical axis is inclined with respect to the sagittal section (see **Fig. 11.28**).

Sagittal rays, rays propagating within the sagittal section.

Focal point, F' or $\bar{F}$, point on the optical axis into which the rays which are incident parallel to the optical axis are being focused.

Focal length, $\bar{f}$ or f', the distance between the object principal point or the image principal point and the object focal point or the image focal point.

Focal planes, planes perpendicular to the optical axis that contain the image focal point or the object focal point, respectively.

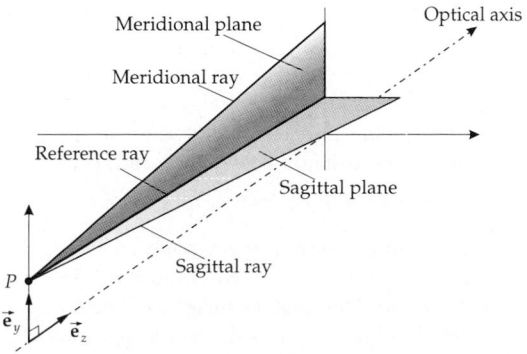

Figure 11.28: Meridional and sagittal planes.

2. Special case: thick spherical lens

Spherical lens, lens consisting of a material of refractive index n, the refracting surfaces of which are sections of spherical surfaces. Distances (such as the focal length or the object distance) are measured from the corresponding principal plane, negatively towards the left, and positively towards the right.

■   For the biconvex lens, the radii r_1, r_2 of the spheres are $r_1 > 0$ and $r_2 < 0$ (see **Fig. 11.27 (a)**).

▲   The focal lengths $\bar{f}$ and f', the image distance a', and the object distance a are measured from the corresponding nearest principal plane.

Mid-thickness, d, the distance between the apex points of a lens.

▲   The magnitudes of the two focal lengths $\bar{f}$ and f' differ when the refractive indices of the media on the two sides of the lens are different,

$$\frac{\bar{f}}{f'} = -\frac{n}{n'}.$$

If the lens is surrounded on both sides by the same medium, then the image focal length f' has the same magnitude as the object focal length $\bar{f}$,

$$\bar{f} = -f'.$$

3. Lens formula for thick lenses

For thick (spherical) lenses surrounded by air, the formulae for the focal length f', the distance s_H of the object principal point H from the object apex S, the distance $s'_{H'}$ of the image principal point H' from the image apex S', and the distance i between the principal planes (see **Fig. 11.29**) read as follows:

lens formulae for thick lens				L
$f' = \dfrac{nr_1r_2}{(n-1)\left[n(r_2-r_1)+(n-1)d\right]}$	Symbol	Unit	Quantity	
	f'	m	focal length	
$s_H = \dfrac{r_1d}{n(r_1-r_2)-(n-1)d}$	n	1	refractive index of lens	
	r_1, r_2	m	radius of spheres 1, 2	
	d	m	mid-thickness	
$s'_{H'} = \dfrac{r_2d}{n(r_1-r_2)-(n-1)d}$	s_H	m	distance $\overline{SH}$	
	$s'_{H'}$	m	distance $\overline{S'H'}$	
$i = \dfrac{(r_1-r_2-d)(n-1)d}{n(r_1-r_2)-(n-1)d}$	i	m	distance $\overline{HH'}$	

M   For common glass lenses in air, the distance between the principal planes is about one third of the lens thickness (distance between the apex points).

4. Construction of the image for a thick lens

Image point, to determine its position, three rays are used:

•   **Parallel ray**, runs parallel to the optical axis from the object point to the image principal plane, and then passes through the image focal point.

•   **Central ray**, runs from the object point to the intersection point of the object principal plane and the optical axis (**principal point**), and then runs parallel to the optical axis up to the image principal plane. Then it is continued parallel to the ray between the object point and the object principal point.

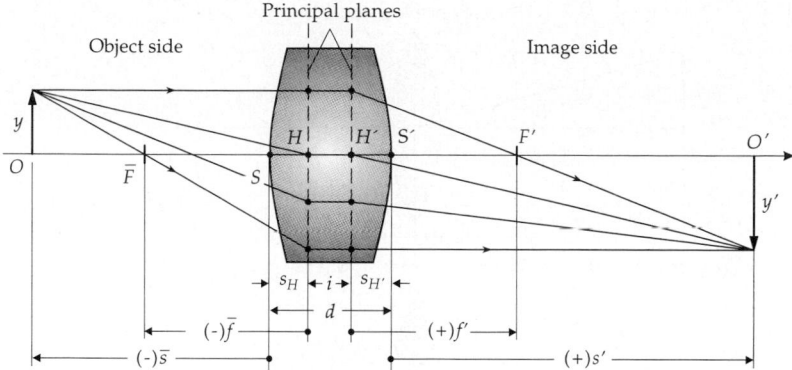

Figure 11.29: Construction of the image for a thick converging lens with two principal planes.

- **Focal ray**, a line is drawn from the object point through the object focus to the object principal plane. Afterwards, the focal ray runs parallel to the optical axis.

These three rays originate in an object point and meet again at an image point.

5. *Imaging equation and refractive power of a thick lens*

The imaging equation establishes a relation between the focal length f', the object distance a (OH), and the image distance a' $(H'O')$:

imaging equation			$\mathbf{L^{-1}}$
	Symbol	Unit	Quantity
$\dfrac{1}{f'} = \dfrac{1}{a'} + \dfrac{1}{a}$	f'	m	focal length
	a	m	object distance
	a'	m	image distance

Another formulation:

Newton's imaging equation, object distance and image distance with respect to the principal planes are replaced by the corresponding quantities related to the focal points, $z = a - \bar{f}$, $z' = a' - f'$:

$$z \cdot z' = \bar{f} f',$$

or with $\bar{f} = -f'$,

$$z \cdot z' = -f'^2.$$

Refractive power, B, of a lens or a system of lenses, defined by

refractive power = $\dfrac{1}{\textbf{focal length}}$			$\mathbf{L^{-1}}$
	Symbol	Unit	Quantity
$B = \dfrac{1}{f'}$	B	1/m	refractive power
	f'	m	focal length

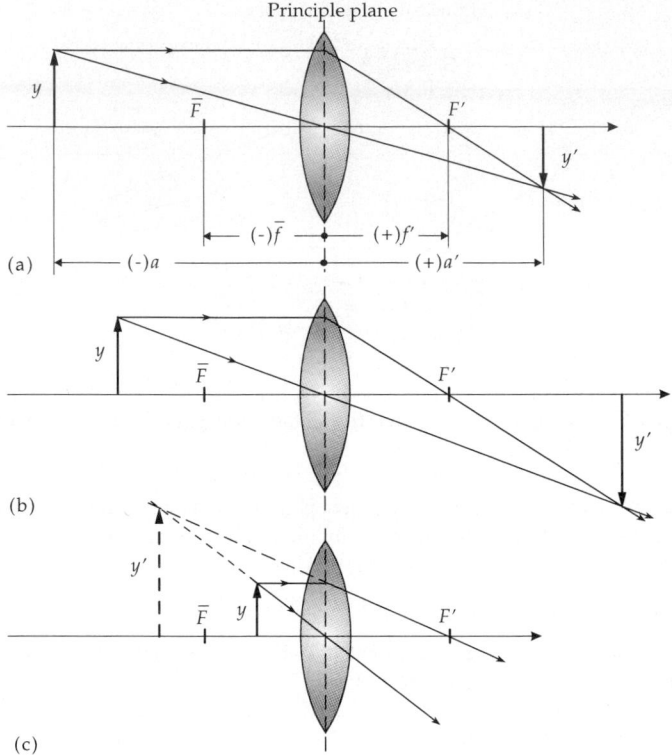

Figure 11.30: Image construction for a converging lens. (a): object outside of twice the focal length, image reduced, inverted, real; (b): object between twice the focal length and the focal length, image enlarged, inverted, real; (c): object within the focal length, image enlarged, erect, virtual.

Diopters, dpt, conventional unit of refractive power. 1 dpt = 1/m.

6. Converging lens

Fig. 11.30 displays a lens with the following properties:

- Rays striking a thin converging lens parallel to the optical axis converge at the **real image focus** F.
- Rays starting from the object focus point leave the lens parallel to the optical axis (reversal of the path).
- Parallel bundles of rays within the paraxial region that make an angle with respect to the optical axis intersect each other at a point on the focal plane.
- The image focal length f' is positive.

A converging lens generates various kinds of images, depending on the object distance a:

- $a > f$, the object position lies between the principal plane and the focus. The resulting image is enlarged, virtual and erect. The **magnifying glass** works within this range of object distances.
- $2f < a < f$, the object position lies between the focal length and twice the focal length. The image is real, inverted, and enlarged. The **slide projector** and the **overhead projector** work within this range of object distances.

- $a < 2f$, the distance between the object and the principal plane is larger than twice the focal length. The image is real, inverted, and reduced. The **telescope** works within this range of object distances.
- ➤ The maximum possible magnification is determined by practical limits, since the focal length of a lens cannot be reduced arbitrarily.

7. Diverging lens,

lens with the following properties (**Fig. 11.31**):
- Rays running parallel to the optical axis and striking a thin diverging lens are refracted in such a way that the deflected rays seem to originate from a point, the **virtual image focus** F'.
- Rays pointing towards the focus leave the lens parallel to the optical axis (reversal of the path of rays).
- The imaging equation is the same as that for converging lenses, but the image focal length f' is negative. Hence, the refractive power of a diverging lens is negative.

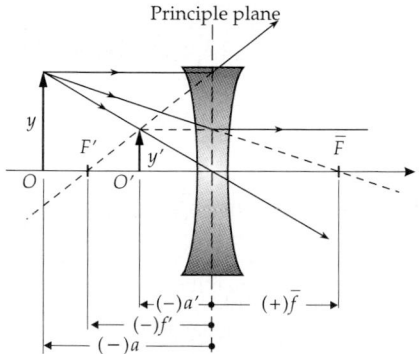

Figure 11.31: Imaging by a thin diverging lens. Reduced, erect, virtual image.

8. Bending: Several lenses of equal refractive power

Bending, notation for a series of lenses of different shape, but equal refractive power. The concept of bending is illustrated by a calculation: If the focal length f' and the refractive index n of a lens are given, then, for an arbitrarily chosen radius of curvature r_1, one can always find values of the radius of curvature r_2 and the mid-thickness d to satisfy the demands.

| M | The method of bending is used to minimize image defects by an appropriate choice of the values (r_1, r_2, d) for given values of the focal length and the refractive index. Two sets of bent converging and diverging lenses, each with identical focal lengths, are illustrated by **Fig. 11.32**.

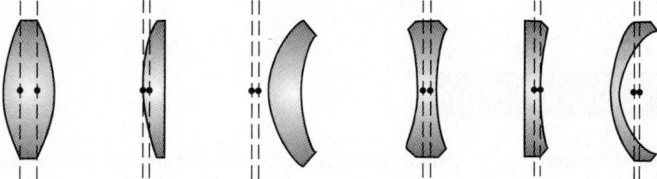

Figure 11.32: Bent lenses. Because of bending, the principal planes may move out of the lens, whereas the focal length remains unchanged.

9. Summary of properties of thick lenses

Image distance and magnification of lenses depending on the object distance:

object distance a	image distance a'	image	magnification β
\multicolumn{4}{c}{converging lens}			
$-\infty < a < 2\bar{f}$	$f' \leq a' < 2f'$	real, reduced, inverted	$0 < \beta < 1$
$2\bar{f} < a \leq \bar{f}$	$2f' < a' \leq \infty$	real, enlarged, inverted	$1 < \beta < \infty$
$f < a < 0$	$-\infty < a' < 0$	virtual, enlarged, erect	$-\infty < \beta < -1$
\multicolumn{4}{c}{diverging lens}			
$-\infty \leq a < 0$	$f' \leq a' < 0$	virtual, reduced, erect	$-1 \leq \beta < 0$

11.2.2   Thin lenses

Thin lens, lens with a thickness d small compared with the radii r_1 and r_2, hence:

$$n(r_2 - r_1) + (n - 1)d \approx n(r_2 - r_1), \quad r_1 - r_2 - d \approx r_1 - r_2.$$

The lens formulae then simplify as follows:

lens formulae for thin lens			L
$f' = \dfrac{r_1 r_2}{(n-1)(r_2 - r_1)}$	Symbol	Unit	Quantity
	f'	m	focal length
$s_H = \dfrac{r_1 d}{n(r_1 - r_2)}$	n	1	refractive index lens
	r_1, r_2	m	radii of spheres 1, 2
	d	m	mid-thickness
$s'_{H'} = \dfrac{r_2 d}{n(r_1 - r_2)}$	s_H	m	distance $\overline{SH}$
	$s'_{H'}$	m	distance $\overline{S'H'}$
$i = \dfrac{n-1}{n} d$	i	m	distance $\overline{HH'}$

Infinitely thin lens, the thickness d is neglected.
    The lens formulae then further simplify:

$$f' = \frac{r_1 r_2}{(n-1)(r_2 - r_1)}, \quad s_H = s'_{H'} = i = 0.$$

11.3   Lens systems

Lens system, arrangement of several lenses with a common optical axis, mostly used to correct image defects found in single lenses.

▲   An optical image may be constructed for a lens system if the positions of the principal planes of the individual lenses and the total focus are known. If there are only two principal planes, the construction of the image is the same as for a thick lens.

For a system of two lenses with the focal lengths f_1 and f_2 (or the refractive powers B_1 and B_2) and the distance d between the two middle principal planes H_{12} and H_{21}, the total focal length f', the refractive power B and the position of the principal planes H_1 and H_2 of the total system are related according to (see **Fig. 11.33**):

calculation of the total focal length			$\mathbf{L^{-1}}$
$\dfrac{1}{f'} = \dfrac{1}{f_1'} + \dfrac{1}{f_2'} - \dfrac{d}{f_1' f_2'}$	Symbol	Unit	Quantity
	f'	m	total focal length
$B = B_1 + B_2 - d\,B_1 B_2$	f_1'	m	focal length lens 1
	f_2'	m	focal length lens 2
$\overline{H_{11} H_1} = \dfrac{f'd}{f_2'}$	d	m	distance middle principal planes
	B	1/m	refractive power
$\overline{H_{22} H_2} = -\dfrac{f'd}{f_1'}$	B_1, B_2	1/m	refractive power lens 1, 2

For the case of closely spaced principal planes (d small), the last term may be ignored. The refractive powers of two lenses of the same material then simply sum, $B = B_1 + B_2$.

➤ Systems consisting of more than two lenses may be treated analogously, by successive reduction of two lenses to a single lens.

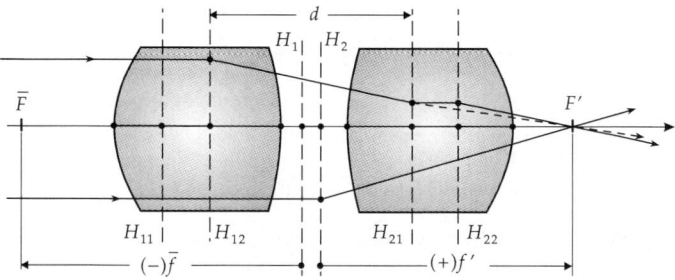

Figure 11.33: Construction of the image for a system of two thick lenses.

11.3.1 Lenses with diaphragms

Diaphragm, collimation of a bundle of light.

Aperture diaphragm, limit of rays generating an image.

Pupils, generally the images of diaphragms.

Entrance pupil, image of the aperture diaphragm of an optical system as seen from the object side.

Exit pupil, image of the aperture diaphragm of an optical system as seen from the image side.

➤ In the sense of technical optics, the eye's pupil is an aperture diaphragm.

➤ The aperture diaphragm of an optical system must be chosen so that the exit pupil coincides in size and position with the eye's pupil.

➤ There are special oculars for eyeglass wearers (for microscopes, telescopes, etc.) with the exit pupil shifted backwards.

Note:

▲ All rays passing through the entrance pupil must also pass through the exit pupil.

- In real lens systems, there is always at least one diaphragm, i.e., the rim of the lens. The lens diameter specifies what fraction of rays leaving the object contributes to image formation.

▲ The dimension of the diaphragm determines the brightness of the image.

Field diaphragm, determines the dimension of the image.

- A real image is usually projected on a screen. The dimension and the frame of the screen then determine the size of the image; the frame represents the field diaphragm.

11.3.2   Image defects

Image defects, or **aberrations**, are deviations of the rays from the ideal paths.

a) Aperture aberration, or **spherical aberration**, occurs when the rays strike the lens system parallel to the optical axis, but not close to the axis. Then these rays no longer converge at the ideal focus (**Fig. 11.34 (a)**).

Consequence: If paraxial rays and rays far off the axis are involved simultaneously, the focus is broadened to a finite size.

Correction: For a converging lens, correction is achieved by combining it with a diverging lens, and vice versa. However, the correction works perfectly only for a given object distance.

b) Astigmatism occurs in imaging of non-axial object points, since the refractive power of a spherical surface in the meridional section differs from that in the perpendicular sagittal section (**Fig. 11.34 (b)**).

Consequence: The image point is ovally distorted, the image is blurred.

Correction: change of the positions of the diaphragms and combination of different shapes of lenses of different materials.

Anastigmat, optical system displaying no astigmatism.

c) Coma or **asymmetry defect**, occurs in imaging of a point off the optical axis where the incident bundle of rays oblique to the optical axis is limited by a diaphragm. The image point has an oval shape with a comet-tail distortion. The defect depends sensitively on the position and shape of the diaphragm.

Consequence: blurred image.

Correction: appropriate positioning of the diaphragm, addition of more lenses.

d) Chromatic aberration, occurs when the light used for imaging is composed of different frequencies and the lens system displays dispersion, i.e., a frequency dependence of refraction (**Fig. 11.34 (c)**).

Consequence: Each color converges at its own focus. The image is blurred and has colored edges.

Correction: A converging lens is combined with a diverging lens of a material with different dispersion behavior (e.g., crown glass and flint glass).

The correction is nearly perfect when several types of glasses and multiple-lens systems are used.

e) Field curvature, the image of a plane object is not generated on a plane perpendicular to the optical axis, but on a curved surface. The defect occurs in imaging of extended objects. As a rule, the spacing between the curved surface and the plane increases with the distance from the optical axis (**Fig. 11.34 (d)**).

Consequence: An image incident on a plane screen becomes more and more blurred with increasing distance from the optical axis.

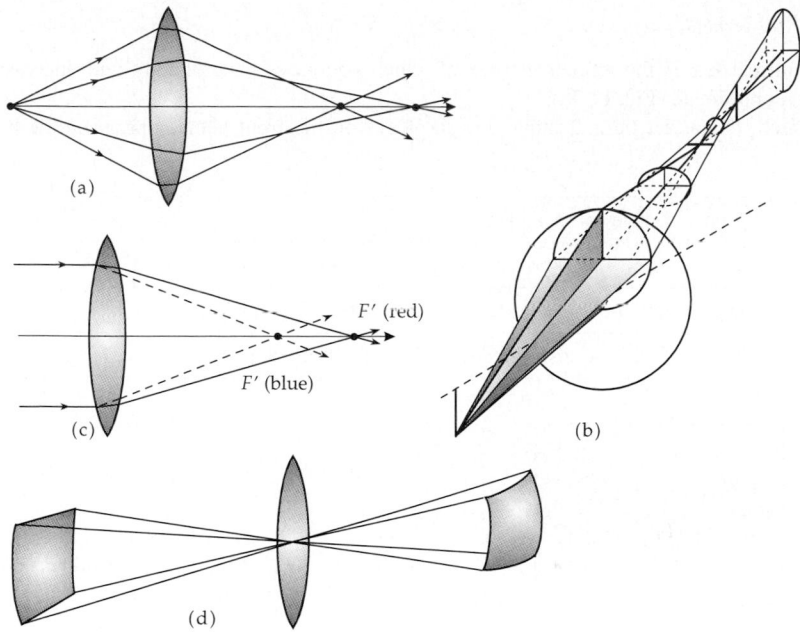

Figure 11.34: Image defects. (a): spherical aberration, (b): astigmatism in the propagation of an oblique bundle through a lens, (c): chromatic aberration, (d): field curvature.

Correction: change of the position of the diaphragm, and combination of different types of lenses of different material, or bending of the image surface (e.g., the film).

f) Distortion, occurs when the position of the diaphragm is not appropriate.
Consequence: object and image are no longer geometrically similar.
Correction: positioning of the diaphragm (or pupil) in the lens plane, bending of lenses.

g) Stray light, scattering of light by impurities in the lens material.
Consequence: the image becomes blurred.
Correction: use of cleaner types of glass.

11.3.2.1 Gradient-index lenses

Gradient-index lenses, lenses with a continuous deflection of the light rays caused by a variation of the refractive index.

Refractive index gradients may be easily generated in gases (pressure and temperature differences) and in liquids (temperature and concentration differences). The light is always deflected towards the region with higher refractive index.

- Streaks in rising warm air.
- Streaks in heating or mixing of liquids.
- At sunset, the Sun may still be observed although it has sunk geometrically below the horizon. The variation of the density of the atmosphere bends the light rays, and the Sun's shape appears deformed.
- Differences of pressure and temperature in the atmosphere limit the resolving power of astronomic telescopes.

1. Rod lenses,

cylindrical lenses the refractive index of which decreases parabolically with increasing radial coordinate (**Fig. 11.35**).

Pitch, characteristic indicating how an object on the front plane appears on the back plane:

Pitch	Image on back plane
1	true-sided (corresponding to two-fold imaging by normal lenses)
0.5	laterally inverted (corresponding to single imaging)
0.25	Fourier-transformed (imaging of the object to ∞)

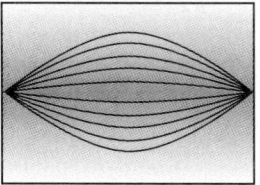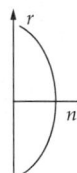

Figure 11.35: Rod lens with pitch $= 0.5$.

GRIN rod or **SELFOC lenses** (manufacturer names), mainly used in photocopiers, scanners, telefax devices and in optical telecommunication.

➤ The numerical aperture specified for rod lenses refers to the center of the entrance face, it decreases towards the cylinder surface.

2. Luneburg lens, Maxwell's fish eye,

gradient-index lenses with special variation of the refractive index. These lenses are of theoretical interest, since they represent the ideal solutions for two basic problems of optics. Their index distributions are difficult to realize in particular for the three-dimensional case.

Maxwell's fish eye, imaging of a point to a point:

$$n(r) = \frac{n_0}{1 + (r/r_0)^2}, \quad n_0, \ r_0 \text{ so that } n(r) \geq 1.$$

Luneburg lens, focusing of a parallel bundle to a point:

$$n(r) = \sqrt{2 - (r/r_0)^2}, \quad r_0 \text{ so that } n(r) \geq 1.$$

11.4 Optical instruments

Optical glass, a non-crystalline, and to a large extent homogeneous, substance free of streaks and bubbles, obtained from the melt of an inorganic mixture. Optical glasses are characterized by a refractive index and a dispersion formula. They have a high internal transmission factor in the visible range of wavelengths.

Abbe number, ν_e, defined by

$$\nu_e = \frac{n_e - 1}{n_{F'} - n_{C'}}.$$

Principal refractive index, n_e, refractive index for the e-line of mercury (λ = 546.07 nm, yellowish-green).

Principal dispersion, $n_{F'} - n_{C'}$, difference of the refractive indices for the cadmium lines F' (λ = 480.0 nm, blue) and C' (λ = 643.8 nm, red).

Crown glasses, glasses with $\nu_e > 55$.

Flint glasses, glasses with $\nu_e < 55$.

➤ The Abbe number ν_λ for the wavelength λ is obtained by replacing the refractive index n_e by the corresponding value $n(\lambda)$ for the spectral line of wavelength λ.

➤ Optical glasses do not have sufficient transparency in the UV and IR ranges. In these spectral ranges, synthetic monocrystals are used as optical components.

➤ For standard optics without high demands on precision, synthetic materials like polystyrol (corresponding to flint glass) and polymethylmethacrylate (corresponding to crown glass) are suitable for imaging components. Optical elements that consist of organic glasses are inexpensive but have high thermal expansion coefficients and low hardness.

11.4.1 Pinhole camera

Pinhole camera (**camera obscura**), archetype of the camera (**Fig. 11.36**), consisting of
- a box with a ground-glass screen as back,
- a small hole (pinhole diaphragm) or converging lens in the front side of the box.

Rays from an object incident through the pinhole or the lens generate an inverted real image on the ground-glass screen. If rays emerging from different object points reach the same image point, the image can become blurred.

The small opening of a pinhole camera guarantees than only rays from a small object region may reach a given image point.

Disadvantage: the smaller the aperture, the lower the illumination of the image.

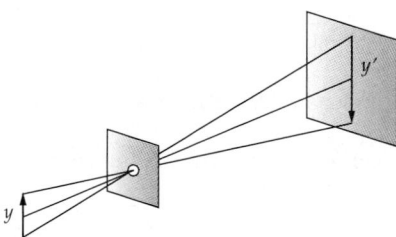

Figure 11.36: Principle of the pinhole camera.

11.4.2 Camera

Camera and **video camera**, optical instruments for recording images according to the principle of the pinhole camera.

In the camera, a converging lens is used for imaging. Modern cameras and video cameras involve additional lenses to correct for image defects. The image in a camera is recorded by a light-sensitive film, the image in a video camera is recorded by an electronic light sensor. However, present-day digital cameras also use electronic light sensors (e.g., CCD's).

The camera is normally adapted to various object distances by varying the distance between lens and film.

The linear magnification may be modified by changing the focal length:
- discontinuously: wide-angle lens, standard lens, telephoto lens;
- continuously: **zoom objective**.

Iris diaphragm, diaphragm for controlling the incident luminous flux.

Relative aperture, measure of the incident quantity of light, defined as the ratio of the diameter of the entrance pupil D_{EP} to the focal length f of the camera.

Focal ratio, k, characteristic parameter for the incident quantity of light, frequently used in practice, defined as the reciprocal value of the relative aperture:

focal ratio = $\dfrac{\text{focal length}}{\text{diameter of entrance pupil}}$				1
$k = \dfrac{f'}{D_{EP}}$	Symbol	Unit	Quantity	
	k	1	focal ratio	
	f'	m	focal length	
	D_{EP}	m	diameter of entrance pupil	

11.4.3   Eye

Eye, organ of man and animal for the perception of light.

1. Camera-like eye,

most powerful eye occurring in nature, to be found in vertebrates (including man) and cephalopods (e.g., octopus). The vertebrate eye (**Fig. 11.37**) consists essentially of:

- **sclera**, the stable skin enveloping the eye;
- **cornea**, the transparent part of the sclera, placed in front of the crystalline lens and therefore visible from outside, elastic, with a refractive index of $n \approx 1.38$;
- **crystalline lens**, deformable bi-convex lens, composed of several layers of distinct refractive indices;
- **ciliary muscle**, annular muscle to which the crystalline lens is fixed. A contraction causes the crystalline lens to become more spherical, thus its refractive power increases;
- **pupil**, circular diaphragm in front of the crystalline lens. The aperture may be varied between 2 mm and 8 mm;
- **retina**, light-sensitive sensory cells that convert light signals into current variations transmitted via the nerves to the brain.

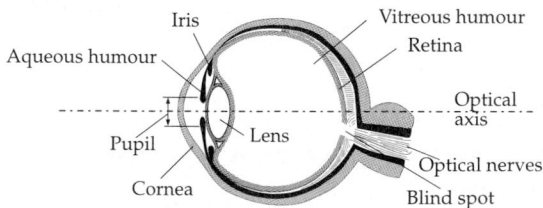

Figure 11.37: Structure of the human eye.

2. Properties of the normal-vision eye

Eye at rest: ciliary muscle fully relaxed, crystalline lens maximally stretched, maximum radii of the spherical surfaces, minimum refractive power of the lens. Rays from infinitely remote points are focussed onto the retina.

Reference distance of vision, a_B, smallest distance to which the eye can adapt for long times without eye strain. This distance is an average value for people with normal vision and is fixed at $a_B = 25$ cm.

Accommodation, adaptation of the refractive power of the crystalline lens for imaging objects at finite distance, by contraction of the ciliary muscle. This causes a compression of the crystalline lens perpendicular to the optical axis and thus an increase of the lens power, i.e., a reduction of the focal length.

Adaptation, response of the eye to the external light conditions by adapting the pupil diameter.

➤   The main contribution to the refraction is at the interface between air and the cornea. For this reason, man does not have sharp vision under water without additional devices, since the range of accommodation of the eye lens is exceeded.

➤   The eye views only real images on the retina. When using magnifying glasses or mirrors, the virtual intermediate image becomes a real image on the retina.

Visual angle, ε, angle of vision with the vertex in the eye and the legs including the object.

Magnification v of an optical instrument, ratio of the tangent of the visual angle ε of an object as seen with the instrument, where the distance object to eye is 25 cm (reference distance of vision), to the tangent of the visual angle ε_0 of the same object as seen by the unaided eye.

magnification of an optical instrument			1
	Symbol	Unit	Quantity
$$v = \frac{\tan \varepsilon}{\tan \varepsilon_0} \approx \frac{\varepsilon}{\varepsilon_0}$$	v	1	magnification
	ε	rad	visual angle with optical instrument
	ε_0	rad	visual angle without optical instrument

The tangent may be replaced by the angle itself only for small angles ε and ε_0.

3. Defects in vision and corrections for the human eye

The most frequent deficiencies of vision of the human eye are:

*   **Near-sightedness**, the refractive power of the eye is too high. Infinitely distant objects are blurred because their images occur in front of the retina.

    Correction: spectacles with diverging lenses reduce the total refractive power of the system.

*   **Far-sightedness**, the refractive power of the eye is too small for near objects. The focus lies behind the retina.

    Correction: spectacles with converging lenses increase the total refractive power.

*   **Age-related far-sightedness**, owing to weakening of the ciliary muscle and hardening of the crystalline lens, the lens can no longer be curved sufficiently to accommodate near objects.

*   **Astigmatism**, the refractive power of the eye is different along the meridional section and the sagittal section.

    Correction: spectacles with lenses curved distinctly in different directions.

Reference visual range, a_B, distance of 25 cm from eye at which normal-sighted persons may view objects sharply without effort (reading distance), $a_B = -25$ cm.

Near point, smallest distance at which the eye may still make a sharp image of an object. It is about 10 cm for children and young persons and increases with age.

11.4.4 Eye and optical instruments

How large an object is perceived to be depends on its visual angle, and hence on its distance from the eye. Maximum light magnitude and simultaneous sharp imaging is achieved when the object is at the near point. A further magnification may be achieved only by means of optical instruments such as magnifying glasses or microscopes. If the distance to a remote object cannot be changed essentially, e.g., when observing planets, telescopes are used.

11.4.4.1   Magnifying glass

Magnifying glass, a converging lens of at least three-fold magnification.

Reader's lens, a converging lens of less than three-fold magnification.

Magnifying glass and reader's lens yield virtual, erect and enlarged images (see **Fig. 11.38**).

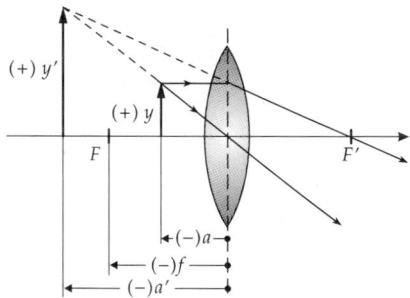

Figure 11.38: Construction of the image for a magnifying glass.

Standard magnification of magnifying glass, Γ'_L, defined as lateral magnification of the magnifying glass for the case that the object is placed in the focal plane of the magnifier, and the eye is accommodated to an infinite distance. Then one has:

magnification, magnifying glass $= -\dfrac{\text{reference range of vision}}{\text{focal length}}$				1
$\Gamma'_L = -\dfrac{a_B}{f'}$	Symbol	Unit	Quantity	
	Γ'_L	1	standard magnification magnifying glass	
	a_B	m	reference range of vision	
	f'	m	focal length of magnifying glass	

The reference range of vision is $a_B = -25$ cm.

11.4.4.2   Microscope

1. Construction of the microscope

Microscope, exceeds the maximum magnification that may be technically achieved by a magnifying glass, by a suitable combination of lenses. It provides a virtual, enlarged, inverted image of the object (see **Fig. 11.39**).

It consists of:

- **objective lens**, lens system of the type of a converging lens of very short focal length, oriented towards the object; generates an enlarged, inverted, real intermediate image;

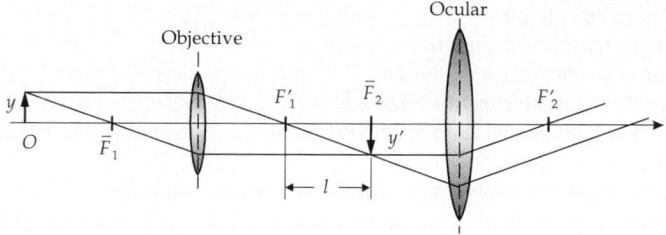

Figure 11.39: Path of rays in the microscope.

- **ocular lens**, lens system of the type of a converging lens, oriented towards the eye, used like a magnifying glass for viewing the intermediate image created by the objective lens;
- **illuminating device** or **condenser**, illuminates the object viewed;

ocular, consists in most cases of

- **field lens**, lens at the position of the real intermediate image that deflects rays incident from the side into the center of the ocular; the change in direction of the bundle causes an extension of the image region without affecting the magnification;
- **eye lens**, converging lens enlarging the image created by the objective lens and the field lens.

Optical tube length, l, distance $\overline{F_1' \bar{F}_2}$ between the neighboring focal planes of objective and ocular lens.

2. Magnification of the microscope

The total magnification Γ'_M of the microscope is

total magnification of microscope			1
	Symbol	Unit	Quantity
$\Gamma'_M = \Gamma'_{Ob} \cdot \Gamma'_{Oc}$ $= \dfrac{l}{f'_{Ob}} \dfrac{a_B}{f'_{Oc}}$	Γ'_M	1	total magnification of microscope
	Γ'_{Ob}	1	magnification of objective lens
	Γ'_{Oc}	1	magnification of ocular lens
	f'_{Ob}	m	focal length of objective lens
	f'_{Oc}	m	focal length of ocular lens
	l	m	optical tube length
	a_B	m	distance of normal vision $= 0.25$ m

Near field, light at a distance from the emitting object that is smaller than one wavelength λ.

Near-field microscope, generation of images by means of the near field in an optical scanning microscope. A screen with a diaphragm of diameter less than λ is positioned above the object to be scanned at a distance less than the wavelength. Light enters through this diaphragm. If the diaphragm is moved over the entire object, and the light reflected by the object is collimated in a conventional microscope, structures with an size less than one wavelength may be resolved, since the signal depends on object regions smaller than one wavelength. The optical near-field microscope can achieve resolutions below 50 nm (about $\lambda/10$). The shape of and distances between individual molecules may be observed.

11.4.4.3 Telescope

Telescope, optical instrument to increase the visual angle of very distant objects.

It consists essentially of:
- **objective lens**, lens nearest to the object;
- **ocular lens**, lens nearest to the eye.

Characteristic parameters of the telescope:
- **visual field**, object field imaged by the telescope. Specification in radians, or as a segment at a distance of 1000 m.
- **effective diameter of objective**, defines the entrance pupil D_{EP}. The quantity determines the amount of light entering the telescope, and thus limits the brightness of the image.
- **aperture ratio**, ratio of the objective diameter to the focal length of the objective.
- **luminosity**, ratio of the objective diameter to the magnification of the telescope.
- **magnification**, v_F. If the image is viewed with a relaxed eye, then

magnification of the telescope			1
	Symbol	Unit	Quantity
$v_F = -\dfrac{f'_{Ob}}{f'_{Oc}}$ $= \dfrac{D_{EP}}{D_{AP}}$	v_F	1	magnification of telescope
	f'_{Ob}	m	focal length of objective
	f'_{Oc}	m	focal length of ocular
	D_{EP}	m	diameter of entrance pupil
	D_{AP}	m	diameter of exit pupil

The magnification corresponds to the ratio of the tangent of the aperture angles with and without the telescope.
- **twilight number**, Z, measure of the twilight efficacy of the telescope,

$$Z = \sqrt{|v_F| D_{EP}}.$$

■   The 7×50 binocular has a twilight number $Z = \sqrt{7 \cdot 50} = 18.7$.

1. Astronomical telescope

Astronomical telescope, **Kepler's telescope**, yields an inverted, laterally inverted image (see **Fig. 11.40**). It consists of:
- **objective lens**, converging lens nearest to the object, generates a real intermediate image of the remote object in the focal plane F'_{Ob};
- **ocular lens**, converging lens nearest to the eye, with the focal plane $\overline{F}_{Oc}$ at the position of the image focal plane of the objective, for viewing the real intermediate image generated by the objective as by a magnifying glass.

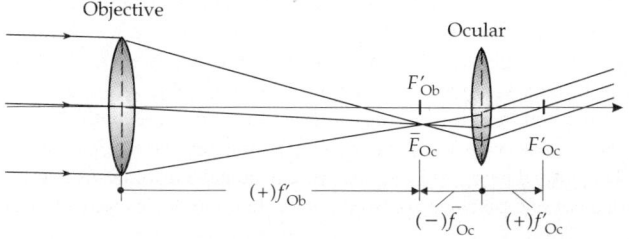

Figure 11.40: Path of rays in the astronomical, or Kepler, telescope.

The magnification of Kepler's telescope is negative. Its length L corresponds to the **sum of focal lengths** of the objective and ocular lenses,

$$L = f'_{Ob} + |\bar{f}_{Oc}|.$$

Kepler's telescope is used in astronomy.

2. Terrestrial telescope

Terrestrial telescope, astronomical telescope with an additional converging lens (erecting lens) between objective and ocular which inverts the laterally inverted intermediate image. The final image is erect and not laterally inverted (**Fig. 11.41**).

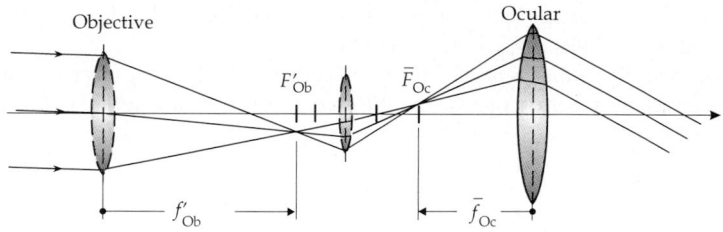

Figure 11.41: Path of rays in the terrestrial telescope.

➤ The image reversal may also be achieved by an inverting prism (**Porro prism system** in the prism binocular).

Mirror telescope, astronomical telescope with the objective replaced by a parabolic concave mirror. Advantages over lens combinations: larger aperture angles, no chromatic aberration. In Cassegrain's version the focal length of the main mirror is extended by a convex collector mirror. The image arises behind the main mirror and is observed through a diaphragm by a ocular lens (**Fig. 11.42**).

Schmidt mirror, mirror telescope with spherical concave mirror with focal length equal to half the curvature radius, and a thin plate of glass with aspherical surface for correction of the image defects for rays far off the axis. The correction plate is positioned in the center of curvature of the concave mirror. The image arises on a spherical surface in the middle between the correction plate and the mirror. The Schmidt telescope generates images of large fields of stars without coma and astigmatism.

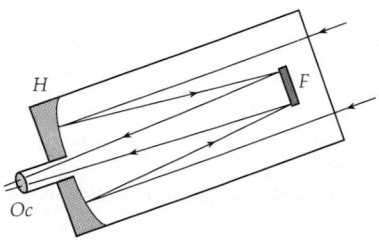

Figure 11.42: Mirror telescope according to Cassegrain. Oc: ocular lens, H: parabolic concave mirror, F: convex collector mirror.

3. Dutch telescope

Dutch telescope, **Huygens' telescope**, **Galilei's telescope**, generates an erect non-laterally inverted image (see **Fig. 11.43**).

It consists of:

• **objective lens**, converging lens nearest to the object;

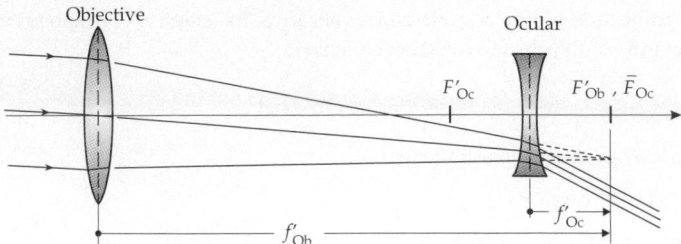

Figure 11.43: Path of rays in the Dutch or Huygens-Galilei telescope.

- **ocular lens**, diverging lens nearest to the eye, whose focal plane $\overline{F}_{Oc}$ coincides with the image focal plane F'_{Ob} of the objective.

There is no real intermediate image. The magnification of Galilei's telescope is positive. Its construction length L corresponds to the **difference of the focal lengths** of the objective and ocular lenses,

$$L = f'_{Ob} - \bar{f}_{Oc} .$$

The low construction length is an advantage of Galilei's telescope. Application: mainly in opera glasses.

11.5 Wave optics

Wave optics, explains optical phenomena related to diffraction, interference and polarization, based on the concept that light is a transverse electromagnetic wave.

11.5.1 Scattering

Diffuse scattering, occurs when light strikes a rough surface that consists of many area elements with different orientations. Refraction and reflection then occur in many different directions. Due to diffuse scattering, a bundle of parallel rays becomes a bundle of diffuse rays (**stray light**).

Scattering center, in Huygens' wave picture a single point emitting spherical waves that represent the stray light.

Rayleigh scattering, scattering of light by spherical particles the radii of which are very small compared with the wavelength of light. The intensity of the scattered radiation increases in proportion to the fourth power of the frequency, i.e., the fraction of the radiation scattered by the particles increases with decreasing wavelength.

■ The sky appears to be blue because blue light has the shortest wavelength in the visible range, and hence is scattered most intensely by the molecules and atoms in the air.

Human perception of objects depends on how much light is scattered or reflected by them. There are various approaches in computer graphics to stimulate preception (virtual reality):

Radiosity approach, method of **graphical data processing** with the aim of realistic computer representation of an interior scene. For this purpose, the surfaces in the scene are assumed to reflect diffusely because their view is then independent of the position of the observer. Hence, there is no need to generate a fully new picture when changing the observer's position in a virtual-reality application.

Ray tracing, an alternative method of representing realistic pictures in which the surfaces are assumed to be specular reflectors. This approach requires a new computation of the picture for any change of the position of the observer, and thus needs much more computational effort than the radiosity approach.

11.5.2 Diffraction and limitation of resolution

Diffraction (see p. 305), change of propagation direction of a wave striking an obstacle. Light enters also into the geometrical shadow region of the barrier and causes diffraction patterns on a screen by interference. The phenomenon may be explained by the concept of Huygens' elementary waves starting from any point of the obstacle hit by the wave and interfering.

1. Types of diffraction

Fraunhofer diffraction, diffraction phenomenon caused by parallel light (**Fig. 11.44**).
Fresnel diffraction, diffraction phenomenon caused by divergent light.

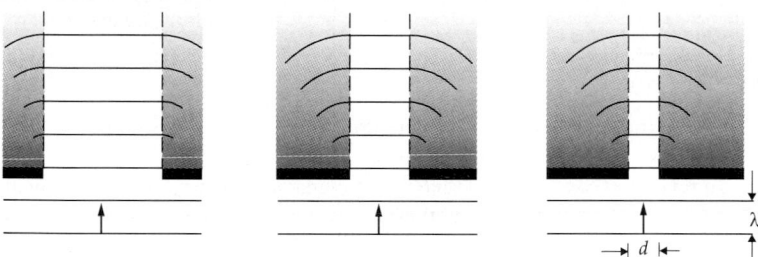

Figure 11.44: Fraunhofer diffraction.

Diffraction by a slit (Fig. 11.45, Fig. 11.46):

Intensity:
$$I_\alpha = I_0 \frac{\sin^2\left(\frac{\pi d \sin\alpha}{\lambda}\right)}{\left(\frac{\pi d \sin\alpha}{\lambda}\right)^2}.$$

Intensity minima:
$$\sin\alpha_n = \pm n\frac{\lambda}{d}, \quad n = 1, 2, 3, \ldots.$$

Intensity maxima:
$$\sin\alpha_n = \pm\left(n + \frac{1}{2}\right)\frac{\lambda}{d}, \quad n = 1, 2, 3, \ldots.$$

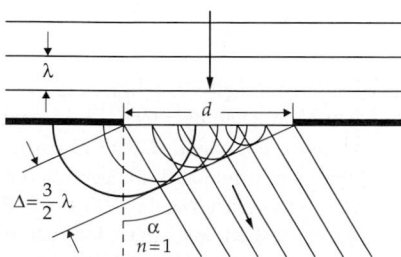

Figure 11.45: Diffraction by a slit. λ: wavelength, d: slit width, α: diffraction angle.

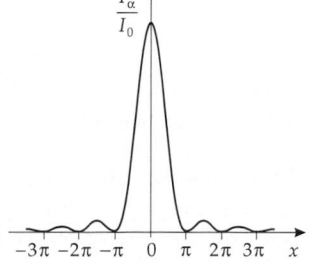

Figure 11.46: Intensity distribution for diffraction by a slit as a function of $x = \pi d \sin(\alpha)/\lambda$.

Diffraction by a grating (Fig. 11.47):

Intensity: $I_\alpha = I_0 \dfrac{\sin^2\left(\frac{\pi d \sin\alpha}{\lambda}\right)}{\left(\frac{\pi d \sin\alpha}{\lambda}\right)^2} \cdot \dfrac{\sin^2\left(\frac{q\pi g \sin\alpha}{\lambda}\right)}{\sin^2\left(\frac{\pi g \sin\alpha}{\lambda}\right)}$.

Intensity maxima: $\sin\alpha_n = \pm n \dfrac{\lambda}{g}, \quad n = 1, 2, 3, \ldots.$

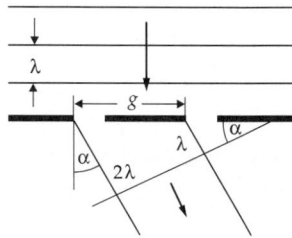

Figure 11.47: Diffraction by a grating. Notations: λ: wavelength, d: slit width, α: diffraction angle, g: grating constant, I_0: intensity at $\alpha = 0$, q: number of grating slits, I_α: intensity at α.

2. Diffraction by a crystal grating

Diffraction of X-rays by crystals may be interpreted as selective reflection by sets of lattice planes that are occupied by the components of crystalline structure.

Bragg's reflection condition, interference maxima occur when the angle of incidence **(grazing angle)** ϑ satisfies the condition

$$2d \sin\vartheta = k \cdot \lambda, \quad k = 1, 2, 3, \ldots.$$

d is the distance of the lattice planes, λ is the wavelength of the X-rays (**Fig. 11.48**). The optical path difference between two rays reflected by neighboring lattice planes is $\Delta = 2d \sin\vartheta$.

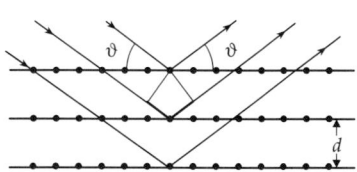

Figure 11.48: Bragg's reflection condition for diffraction by a crystal grating. d: distance of the lattice planes, Δ: optical path difference between two neighboring rays, ϑ: angle of incidence.

3. Influence of diffraction on optical imaging

For any optical imaging, the outer edge of a lens or any diaphragm or aperture represents an obstacle for electromagnetic waves. So, when a point is imaged by a telescope or other optical device, the resulting image is not a point, as assumed in ray optics, but a diffraction pattern. The pattern consists of a maximum of brightness (a bright disk the center corresponding to the image point of ray optics) and several subsidiary maxima (**Fig. 11.46**). When two closely spaced points are imaged, the two corresponding diffraction patterns overlap. If the object points are too close to each other, the maxima of the diffraction patterns are no longer perceived as being separated.

Diffractional disk, the smearing-out of an image point in an optical imaging caused by diffraction.

4. Resolving power,

smallest distance between two object points at which they still may be imaged by an optical instrument as separated points.

There is no objective criterion under which conditions the two diffractional disks are perceived as separated entities. The **Rayleigh criterion** is frequently used:

Rayleigh criterion: two object points are considered to be resolved when the central diffraction maximum of the first object coincides with the first diffraction minimum of the second object. The minimum visual angle δ between the two resolved objects is given by:

Rayleigh criterion			1
	Symbol	Unit	Quantity
$\sin \delta \geq 1.22 \dfrac{\lambda}{b}$	δ	rad	visual angle
	λ	m	wavelength
	b	m	aperture diameter

For small angles δ, the sine may be replaced by the angle in radians.

Spectral resolving power of a prism, the product of base length b and dispersion $|\mathrm{d}n(\lambda)/\mathrm{d}\lambda|$,

$$\frac{\lambda}{\Delta\lambda} = b \cdot \left| \frac{\mathrm{d}n(\lambda)}{\mathrm{d}\lambda} \right| .$$

■ A prism of flint glass ($|\mathrm{d}n/\mathrm{d}\lambda| = 1500 \text{ mm}^{-1}$) with a base of $b = 1$ cm enables the resolution of the sodium lines $\lambda_{D1} = 589.6$ nm and $\lambda_{D2} = 589.0$ nm. A crown glass prism with the same base ($|\mathrm{d}n/\mathrm{d}\lambda| \approx 55 \text{ mm}^{-1}$) does not attain the needed resolving power.

Spectral resolving power of a grating, the product of the order of maximum k and the number N of grating grooves,

$$\frac{\lambda}{\Delta\lambda} = k \cdot N .$$

■ A grating spectral apparatus with $N = 10^5$ grooves allows the separation in first diffraction order wavelengths which differ by only $\Delta\lambda = 10^{-5}\lambda$.

Resolving power of a microscope, defined by the minimum distance $x_{\min}$ between two object points that are still perceived as separated image points,

$$x_{\min} = \frac{\lambda}{A}, \quad A = n \cdot \sin\alpha .$$

n is the refractive index of the medium in front of the objective, α is the half of the aperture angle of the light cone emerging from an object point that is just covered by the objective lens. The quantity A is denoted as the **numerical aperture** of the objective lens.

11.5.3 Refraction in the wave picture

Refraction, change of the direction of propagation of waves at the interface of two media with different wave propagation speeds. In this type of transition, the wave frequency remains constant, but the wavelength varies.

This phenomenon may be represented and understood in terms of Huygens' elementary waves (**Fig. 11.49**). If a wave front strikes an interface between materials with different refractive indices at an angle different from 90°, every point of the interface becomes a source of Huygens' elementary waves (spherical waves). Every elementary wave now propagates through both half-spaces in front of and behind the interface, respectively. (The reflected fraction is not plotted.) Since the wave front reaches different points of the interface at different times, the elementary waves also arise at different times. The figure illustrates a snapshot showing both the maxima of individual wavelets and the plane wave fronts resulting from their interference (see p. 287).

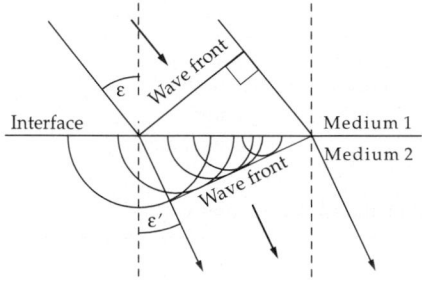

Figure 11.49: Refraction in the wave picture.

11.5.4 Interference

In order to produce interference between electromagnetic waves, the waves superimposed must be **coherent** (see p. 294), i.e., they must originate from the same region of a light source. Coherent light rays may be obtained by splitting a ray with mirrors or partially transmitting plates (beam splitter).

If the superposed waves are not coherent in space and time, the interference phenomena are not visible, since at a fixed point cancellation and reinforcement are randomly changing.

➤   A **laser** generates coherent light.

For thermal sources of light, the individual surface elements emit wave trains without a fixed phase relation. The phase differences change randomly. For such sources of light, interference patterns can therefore be made visible only when the superimposed wave trains are generated by a restricted areal element, for example with an aperture.

1. Coherence condition

In order to produce interference (see **Fig. 11.50**), the aperture angle α of an areal element of a light source of extension b must satisfy the condition

$$n \sin \alpha \ll \frac{\lambda}{2b} .$$

λ is the wavelength of radiation, n is the refractive index of the medium.

Coherence length, l, mean length of the individual wave trains.

Coherence time, τ, time needed for traversing the coherence length,

$$l = c \cdot \tau .$$

Coherence in time exists if, for the half-width Δf of a spectral line of frequency f,

$$\tau \approx \frac{1}{\Delta f}, \quad l \approx \frac{c}{\Delta f}.$$

■   Spectroscopic lamps have coherence lengths of the order of magnitude $l = 1 \cdot 10^{-1}$ m for a frequency band width of $\Delta f = 1$ GHz. HeNe-lasers reach values of $l = 150$ m and $\Delta f = 2$ MHz.

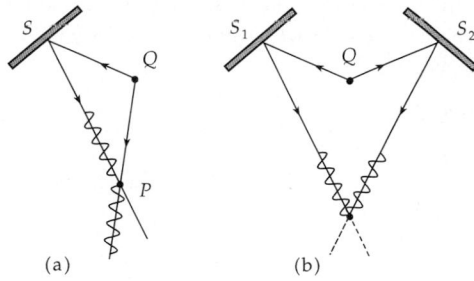

Figure 11.50: On coherence of light. Q: light source, S, S_1, S_2: mirrors. (a): no coherence at point P, (b): coherence in time at point P.

2. Interference in thin films

Interference in thin films occurs when
- light hits a layer with a refractive index that differs from the refractive index of the original medium,
- a fraction of the incident light is reflected by the interface between the layer and the surrounding medium, while another fraction enters into the layer.

In each impact of the ray with one of the two interfaces between the layer and the surrounding medium, the ray is split into two parts, one being reflected, the other penetrating into the medium behind the interface (**Fig. 11.51**).

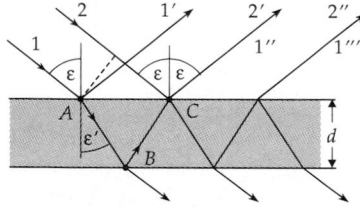

Figure 11.51: Interference at a plane-parallel plate of thickness d.

- Ray 1 hits the interface at point A and is partly reflected, which yields the ray $1'$.
- Another part penetrates at point A into the layer and is then partly reflected at point B. Let this reflected ray leave the layer at point C, yielding the ray $1''$. This ray is coherent with ray $1'$.

The remaining rays in the figure correspond to multiple reflections within the layer, and to the light leaving the layer at the back.

3. Path difference in interference at thin layers

For light of wavelength λ incident on a layer of thickness d and refractive index n, the rays $1''$ and 2 have, due to the different refractive indices of the layer and the surrounding medium (air), at the point C a **path difference** Δ of:

path difference in interference at thin layers			**L**
	Symbol	Unit	Quantity
$\Delta = 2d\sqrt{n^2 - \sin^2 \varepsilon} - \dfrac{\lambda}{2}$	Δ	m	path difference
	d	m	thickness of layer
	n	1	refractive index of layer
	ε	rad	angle of incidence
	λ	m	wavelength

The term $-\lambda/2$ corresponds to reflection by an optically more dense medium behind the thin layer. If the rays 1 and 2 interfere, one observes **constructive interference** (brightness) or **destructive interference** (darkness), depending on Δ.

4. Condition for constructive interference

For amplification (constructive interference):

condition for constructive interference			
	Symbol	Unit	Quantity
$2d\sqrt{n^2 - \sin^2 \varepsilon} = \left(m + \dfrac{1}{2}\right)\lambda$	d	m	thickness of layer
	n	1	refractive index of layer
$m = 0, 1, 2, \ldots$	ε	rad	angle of incidence
	λ	m	wavelength

For perpendicular incidence ($\sin \varepsilon = 0$) amplification occurs for

$$\lambda = \frac{2dn}{m + \frac{1}{2}} .$$

5. Condition for destructive interference

For cancellation (destructive interference):

condition for destructive interference			
	Symbol	Unit	Quantity
$2d\sqrt{n^2 - \sin^2 \varepsilon} = (m + 1)\lambda$	d	m	thickness of layer
	n	1	refractive index of layer
$m = 0, 1, 2, \ldots$	ε	rad	angle of incidence
	λ	m	wavelength

For perpendicular incidence, cancellation occurs for

$$\lambda = \frac{2nd}{m + 1} .$$

The interferences observed at plane-parallel plates correspond to fixed angles of incidence (**interferences of equal inclination**).

■ Oil films on water appear colored. For interference by a thin oil film a certain wavelength (color) is most positively reinforced, depending on the varying thickness, whereas other wavelengths interfere with varying degrees of cancellation or reinforcement.

6. Antireflection coatings,

method for reducing reflection by a surface, based on thin-film interference. The surface of a material of refractive index n_1 is covered by a layer of refractive index $n_2 < n_1$. The refractive index n_2 and the thickness d of the layer are chosen so that the reflected waves cancel each other at a chosen value of wavelength λ.

The cancellation is not limited to a single sharp wavelength, but extends over a certain range. Hence, one may, e.g., cover the green range of the visible spectrum. Thin layers are used for **optical coating** of lenses; these show weak purple reflections, since red and violet cannot be compensated completely when the layer is set for cancellation of light in the middle of the visible spectrum.

Fizeau fringes, equally spaced interference fringes occuring at two plane areas tilted with respect to each other with a wedge of air in between. The observed fringes correspond to positions at which the wedge thickness (**Fig. 11.52**) is $(n/2)\lambda$.

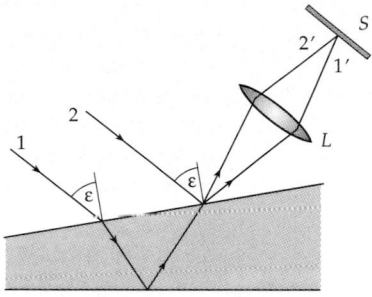

Figure 11.52: Interference by a wedge-like layer. L: lens, S: screen.

7. Newton's rings,

concentric circular bright and dark Fizeau fringes arising when the layer of air is confined by a plane area and a spherical area of curvature radius R.

Distance of the dark rings from the contact point:

$$r_{\min} = \sqrt{R\lambda k}, \quad k = 1, 2, 3, \ldots.$$

Distance of the bright rings from the contact point:

$$r_{\max} = \sqrt{R\lambda\left(k + \frac{1}{2}\right)}, \quad k = 0, 1, 2, 3, \ldots.$$

8. Interferometry,

branch of precision-measurement technology that exploits the interference of waves for measurement of physical quantities.

Michelson interferometer, optical device that may be considered the basic type of an interferometer (see **Fig. 11.53**). Light from a source Q is split by a partially transmitting plate P_1 into a reflected ray 2 and a transmitted ray 1, which are then reflected by two plane mirrors S_1 and S_2. After an additional splitting by the plate P_1, the reflected rays are superimposed in an observation telescope F. (In order to symmetrize the path of rays, i.e., to let both rays pass through a splitting plate the same number of times, one places an additional plate P_2 into the path of ray 2.) When combining the rays one observes interferences of equal displacement (concentric rings). If one mirror is tilted, one obtains interferences of equal separation (Fizeau fringes).

If the rays traverse the geometric paths s_1 and s_2, respectively, neighboring interference maxima are distinguished by a path difference of

$$n(s_2 - s_1) = k \cdot \lambda,$$

if the refractive index along both light paths is n. If the mirror S_1 is shifted by a distance $d = \lambda/2$ to the position $S_1{'}$, exactly one interference fringe passes the visual field of the observer. On this basis, one may measure differences of length by means of the Michelson interferometer with high precision.

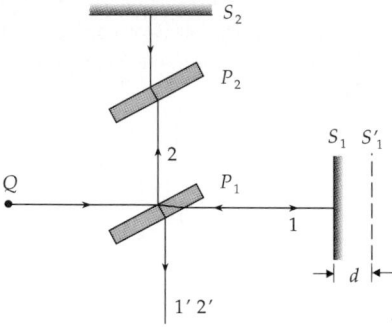

Figure 11.53: Principle of the Michelson interferometer. Q: light source, P_1, P_2: plates, S_1, S_2: mirrors, d: displacement of mirror S_1.

11.5.5   Diffractive optical elements

Diffractive optical elements (DOE), operate on the basis of the **diffraction** of light waves by fine structures. The description of their operation involves wave optics. To be contrasted with **refractive optical elements**, which are described by refraction of light rays.

■   Diffraction gratings, holograms, Fresnel-zone plates are diffractive optical elements.
➤   The "classical" (geometric optics) optical elements are either refractive (lenses, prisms) or reflective (mirrors).
➤   A more thorough consideration of refractive elements shows that diffraction effects also occur. For example, the edge of a lens acts as an aperture at which diffraction occurs. This diffraction effect limits the resolving power of optical instruments.

Diffraction effects become dominant when the typical structure dimension of the optical element is of the same order of magnitude as the wavelength used.

The structural dimensions of DOE are therefore only a few micrometers (10^{-6} m). The production of DOE, which are more complicated than simple diffraction gratings, has become possible only since the middle of the twentieth century.

11.5.5.1   Diffraction gratings

Diffraction gratings, decompose light into its spectral components (grating spectrograph) or deflect monochromatic radiation into one or several directions. The related formulae may be found in the chapter on waves.

■   The compact disc is a type of reflective diffraction grating.

11.5.5.2   Fresnel-zone plate

Fresnel-zone plate, arrangement for the focusing of coherent light which uses concentric transparent and nontransparent rings of widths which decrease with radius (**Fig. 11.54**).

Radius of the transparent rings is chosen so that the path of light from neighboring zones to the focal point differs by just one wavelength, i.e., the light interferes constructively at the focus.

Mean radius r_1 of the first transparent ring:

$$r_1 = \sqrt{(f + \lambda)^2 - f^2} = \sqrt{2f\lambda + \lambda^2} \approx \sqrt{2f\lambda} \quad \text{for } \lambda \ll f.$$

Radius r_n of the nth transparent ring:

$$r_n \approx \sqrt{2n\lambda f}.$$

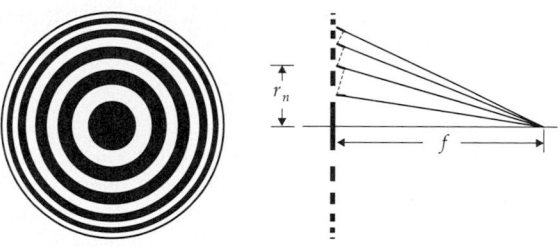

Figure 11.54: Fresnel-zone plate.

Such an element is focusing, but has low light efficiency (50 % loss by the dark fringes, and loss into higher-order diffraction patterns). Nevertheless, it is of importance for ranges of wavelength for which there are no refracting materials for normal lenses (X-ray microsope).

11.5.5.3   Fresnel-zone lens

Fresnel-zone lens (FZL), focusing element. The surface shape corresponds to a lens from which all superfluous glass has been removed.

A FZL is really a diffractive element. But refraction effects can be important, since it rather covers the transition range between refractive and diffractive elements (**Fig. 11.55, Fig. 11.56**).

Figure 11.55: Transition from normal lens to Fresnel-zone lens (FZL).

For incoherent light, the image quality of FZL can be poor. FZL are nevertheless applied when the weight or thickness of the lens has to be reduced.

■   FZL are used as collimators for beacons, etc. For movable lenses with diameter of up to one meter, the reduction of weight is of particular importance.

■   Most frequent application nowadays are the collimators for overhead projectors and "fish eyes" for rear windows of motor vehicles.

For coherent light, the gradation is chosen so that the optical paths of neighboring zones differ by just one wavelength and the partial waves therefore interfere constructively. With this additional wave-optical condition, a FZL becomes a diffractive element (**Fig. 11.57**);

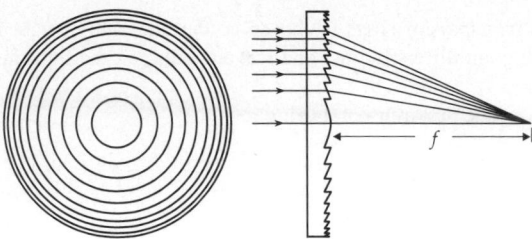

Figure 11.56: Fresnel-zone lens, based on refraction of light rays; it is, strictly speaking, not a diffractve optical element.

the height of the structure h is then in the μm range. In this case, one may reach the same imaging quality as with a normal lens.

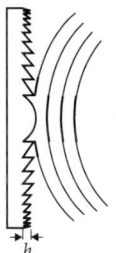

Figure 11.57: Fresnel-zone lens in which the optical paths differ step by step by an integer multiple of the wavelength. h: height of structure.

➤   For incoherent illumination, the resolving power of a FZL is determined by the diameter of the innermost zone, for coherent illumination, by the entire lens diameter.

11.5.5.4   Holograms

Hologram, optical element that stores not only an intensity distribution (as for a photograph), but also the relative phase distribution.

▲   Photographic recording methods (films, TV tubes, CCDs) use only the intensity, i.e., the square of the magnitude of the complex amplitude; the phase information contained in the wave field is lost. When coherent light is employed, the phase information may also be recorded, but in an indirect way. For this purpose, the light emerging from an object interferes with a reference wave. The original wave field may be extracted (**Fig. 11.58**) from the recorded interferogram.

Let $o(x, y) = |o(x, y)|e^{i\phi(x,y)}$ be the complex amplitude of the light emerging from the object, $r(x, y)$ be the complex amplitude of the reference wave in the hologram plane (x, y).

Without the reference wave, one records

$$oo^* = |o|e^{i\phi}|o|e^{-i\phi} = |o|^2.$$

With the reference wave, one records

$$(o + r)(o + r)^* = oo^* + rr^* + or^* + ro^* = |o|^2 + |r|^2 + or^* + ro^* .$$

If this picture is illuminated again by the reference wave r, one obtains

$$|o + r|^2 \cdot r = |o|^2 r + |r|^2 r + o|r|^2 + o^* r^2.$$

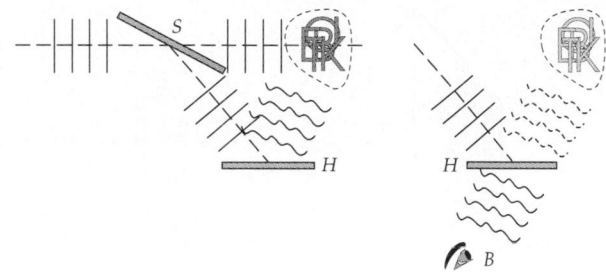

Figure 11.58: Recording (left) and viewing (right) of a hologram. (S): partially transmitting mirror, (H): hologram, (B): observer.

When a plane wave is used as the reference wave r, $|r|^2 =$ becomes const., and one obtains the wave front o emerging originally from the object. An observer views a virtual image of the object, by changing his position he may perceive it from different directions of view.

- One may obtain a normal image from the reconstructed wave o.
- If a hologram is subdivided, each fraction provides an image of the entire object, although viewed from different positions.
- A particularly fine-grained film is needed (grains of the order of magnitude of the wavelength) for the recording of holograms, e.g., dichromate gelatine, as well as a laser of sufficient power and coherence length.
- ➤ This simple description holds only for transmission holograms and coherent illumination in the reconstruction process. Reflection holograms may also be generated; the conditions concerning the coherence of the light source in the reconstruction then may be eased.
- ■ The holograms on credit cards are reflection holograms. A change of the angle of vision is perceived only when moving one's head horizontally. The other direction (up-down) is used for decomposing the light into its spectral components. This may be better recognized when the coherence length of the adopted light source is longer (the smaller and more remote the source). Well suited are, e.g., low-voltage halogen lamps. These reflection holograms are, however, not produced photographically, but are computer-generated.

11.5.5.5   Computer-generated holograms

Computer-generated holograms (CGH), holograms with a structure calculated to generate a definite image, and produced by means of microstructuring technology (lithography). Lithographic methods enable generation of units in the range of the light wavelength.

- ■ Holograms as antiforgery devices on credit cards, banknotes and seals.
  Beam formation for material processing by a laser.

Computational basis for CGH, calculation of the propagation of light waves by **Fourier transformation**.

Fraunhofer diffraction, approximate computation of diffraction patterns in the far field (distance between diffraction object and diffraction image $\gg$ wavelength of light). **Huygens' principle** (see p. 380): any point of a wave front may be considered a source of a spherical wave. The propagation of the wave field is described by the superposition of these spherical waves. The geometry of the diffraction problem is represented in **Fig. 11.59**.

Let the diffraction object be flat and positioned in the plane $z_1 = 0$ and be illuminated from the left by a plane wave $u(x_1, y_1)$, e.g., a laterally extended laser beam. Let its light

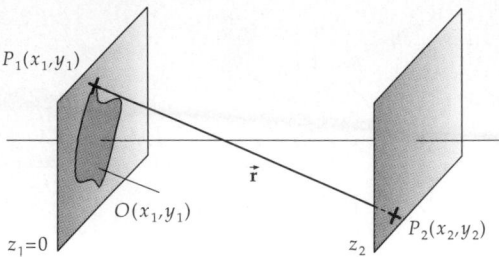

Figure 11.59: Geometry of the diffraction problem.

transmittance be described by the function $O(x_1, y_1)$; for a slit diaphragm $O(x_1, y_1) = 1$ over the slit aperture and $O(x_1, y_1) = 0$ elsewhere.

Complex amplitude in the plane of the diffraction pattern:

$$u(x_2, y_2) = \int dx_1 \int O(x_1, y_1) \frac{1}{r} \exp[i\vec{\mathbf{k}} \cdot \vec{\mathbf{r}}] \, dy_1.$$

Paraxial approximation: $(x_2 - x_1)^2 + (y_2 - y_1)^2 \ll z_2^2$, leads to the Fraunhofer approximation. One obtains:

$$u(x_2, y_2) \approx \int dx_1 \int O(x_1, y_1) \exp\left[\frac{-2\pi i}{\lambda z_2}(x_2 x_1 + y_2 y_1)\right] dy_1.$$

This is a two-dimensional, complex Fourier transformation.

The diffraction pattern is determined by the square of magnitude of the amplitude, $|u|^2 = uu^*$:

▲   Within the range of application of the Fraunhofer approximation, the diffraction pattern equals the square of magnitude of the complex Fourier transform of the diffraction object, within constant factors.

This statement is very important, since:

- The Fourier transformation may be inverted. By inverse transformation, one may calculate the diffraction object from the diffraction pattern.

- The mathematical description by means of Fourier transformation allows the use of corresponding theorems, in particular the convolution theorem.

- There are very efficient algorithms available for implementing the Fourier transformation on computers (FFT - fast Fourier transformation).

➤   Actually, the process is difficult, since for technical reasons pure amplitude holograms (phase part of $O(x_1, y_1)$ constant) or pure phase holograms (magnitude $|O(x_1, y_1)|$ constant) are usually produced, with limited resolving power. This leads to additional boundary conditions. On the other hand, the phase in the plane of the diffraction pattern may be freely chosen. To meet this condition, iterative algorithms are used (e.g., *Gerchberg-Saxton* algorithm).

➤   Huygens' principle is equivalent to choosing spherical waves as **Green functions** for solutions of the **Helmholtz equation**. This approach is still treated in many textbooks for historical reasons (*Fresnel-Kirchhoff diffraction integral, Rayleigh-Sommerfeld diffraction integral*). An expansion in terms of plane waves, on the contrary, allows a much simpler formulation and avoids superfluous approximations. Therefore, this approach is becoming increasingly significant in modern optics (**Fourier optics**).

11.5.6   Dispersion

Dispersion (see p. 305), dependence of the phase velocity on the wavelength (or frequency).

Since the refractive index of a medium is defined as the ratio of the propagation speed of a wave in a vacuum to the propagation speed in the medium, the wavelength dependence of the refractive index is the inverse of the wavelength dependence of the propagation speed (see p. 287).

- **Normal dispersion:**

$$\frac{dn}{d\lambda} < 0.$$

The refractive index of the medium decreases with increasing wavelength λ; the refraction angle also decreases with increasing wavelength (prism spectroscopic apparatus).

- **Anomalous dispersion:**

$$\frac{dn}{d\lambda} > 0.$$

The refractive index of the medium increases with increasing wavelength λ; the refraction angle increases with increasing wavelength.

- **No dispersion:**

$$\frac{dn}{d\lambda} = 0.$$

Example: electromagnetic waves in a vacuum.

➤ With a few exceptions, all media occuring in nature exhibit normal or no dispersion.

Fig. 11.60 shows dispersion curves of several optical materials.

Visible white light is a superposition of electromagnetic waves of various wavelengths that individually are perceived by an observer as distinct colors.

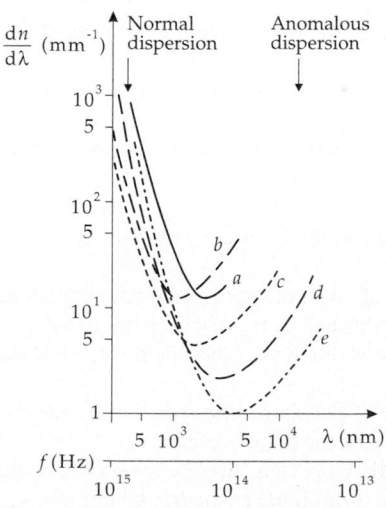

Figure 11.60: Dispersion of several optical materials. (a): Flint glass, (b): quartz, (c): fluorite, (d): NaCl, (e): KBr .

Spectral colors, the colors contained in white light, in the sequence of decreasing wavelengths: red, orange, yellow, green, blue, indigo, violet.

Spectrum, set of spectral colors, ordered by the wavelength.

Spectral decomposition, separation of the components of different wavelengths of a radiation.

M   Prisms are frequently used for spatial separation of the components of white light. The refraction of the components with different wavelengths by the interfaces of the prism proceeds with different refraction angles, due to the non-zero dispersion $dn(\lambda)/d\lambda$. The spectral resolving power of a prism of base length b is given by

$$\frac{\lambda}{\Delta\lambda} = b \left| \frac{dn(\lambda)}{d\lambda} \right| .$$

Achromatic prism, special prism for which to first order only refraction, but no dispersion, occurs. The incident light is deflected, but not decomposed, by colors. It consists of two prisms of flint glass and crown glass cemented together.

Dispersion cannot be compensated simultaneously over the entire spectrum without reducing the refractive power of the system.

Achromat, lens system consisting of a converging and a diverging lens in which chromatic aberrations are compensated for **two** wavelengths.

Apochromat, lens system of three lenses with particular choice of glasses, for which chromatic aberrations are compensated for **three** wavelengths.

11.5.7 Spectroscopic apparatus

Spectrum analysis, analysis of an emission or absorption spectrum to determine the qualitative and quantitative composition of substances.

Spectroscopic apparatus, optical device for spectral decomposition of polychromatic electromagnetic radiation:

spectroscope, visual observation of a spectrum,

spectrometer, determination of the wavelength of spectral lines by comparison with a calibrated scale of wavelengths,

spectrograph, complete registration of a spectrum by a photographic plate, and comparison with a calibration spectrum,

monochromator, selection of a narrow range of wavelengths out of a broad spectral range, to generate nearly monochromatic radiation,

spectroscopic photometer, combination of a spectroscopic apparatus and a photometer (determination of spectral material parameters).

Optical devices that generate an image of the entrance slit employ concave mirrors and lenses.

Requirements for spectroscopic apparatus:

- high luminosity: determines the brightness of the spectrum, important for sources of low intensity,
- high resolving power: determines the smallest wavelength difference between neighboring spectral lines that may still be established by the device as separated,
- broad range of dispersion: determines the width of the wavelength range that may be covered in a single session.

Prism spectroscope, spectral decomposition of polychromatic radiation by means of a prism, based on the variation of the refractive index with the wavelength.

Optical grating, regular arrangement of diffracting elements (grating grooves), characterized by their distance (grating constant) and their profile (Echelette grating).

Transmission grating, consists of parallel, nontransparent scratches on a glass plate.

Reflection grating, consists of parallel grooves scratched into the surface of a glass plate. The diffracted light may be concentrated to a large extent into one diffraction order by appropriate shaping of the grooves.

Grating spectroscope, spectral decomposition of polychromatic radiation by means of a grating, based on the variation of the position of intensity maxima with the wavelength.

➤ Prism spectroscopes in general have a broader range of dispersion and a lower resolving power than grating spectroscopes. Grating spectroscopes with a reflection grating reach a higher luminosity than analogous devices that employ a transmission grating.

11.5.8 Polarization of light

1. Types of polarization

Since electromagnetic waves are transverse, light may exhibit the polarization phenomena known from the theory of waves (see p. 293):

- **Linearly polarized light**, the electric field vector $\vec{E}$ and the propagation vector of the wave span a plane of vibration fixed in space.
- **Circularly polarized light**, the electrical field vector $\vec{E}$ runs on a helical path about the propagation vector. In the projection plane perpendicular to the propagation vector, the electric field vector $\vec{E}$ describes a circle. When looking in the opposite direction to the propagation direction, the light is called **right(left)-circularly** polarized if the field vector circulates clockwise (counterclockwise).
- **Elliptically polarized light**, the electric field vector $\vec{E}$ travels on an elliptical helix about the propagation vector. In the projection plane perpendicular to the propagation vector the electric field vector $\vec{E}$ describes an ellipse. When looking in a direction opposite to the propagation direction, the light is **right(left)-elliptically** polarized if the field vector circulates clockwise (counterclockwise).

2. Causes of polarization

For natural light emitted by the Sun, the electric field vector $\vec{E}$ oscillates in a plane perpendicular to the propagation direction of the wave without preference for a direction of vibration. All possible vibration directions occur in the light beam with the same statistical weight. Natural light is **unpolarized**. Light is **partly polarized** if a specific oscillation direction occurs preferably. If in the beam only a single oscillation direction occurs, then the light is completely **linearly polarized**. The preferred oscillation direction is denoted as the **polarization direction**. Linearly polarized light may be decomposed into two components of equal frequency and equal propagation direction that vibrate perpendicularly to each other. Other amplitude and phase relations lead to **right-** or **left-circularly polarized** light (equal amplitude and phase difference $\pi/2$ of the components), or **right-** or **left-elliptically polarized** light (phase difference $(2n + 1) \cdot \pi/2$, $n = 1, 2, \ldots$ and different amplitudes).

▲ Two light waves polarized perpendicularly to each other cannot interfere to zero intensity.

3. Polarizer,

a device that selects only the components of unpolarized light that vibrate linearly along a given direction perpendicular to the propagation vector.

Analyzer, polarization filter positioned in such a way that its transmission direction is perpendicular to the transmission direction of the polarizer. The analyzer then lets no light pass unless the polarization plane of the light is rotated between polarizer and analyzer.

If the oscillation direction of polarizer and analyzer subtend an angle ϕ, the analyzer lets pass only the component that is aligned along its oscillation direction. The amplitude of the transmitted wave is thereby reduced by the factor $\cos \phi$.

4. Optical activity,

the property of a substance to rotate the polarization plane of linearly polarized light, in which the rotation angle depends on the thickness of the layer of the substance. One distinguishes between right-rotating and left-rotating substances. Optical activity is observed both for isotropic and anisotropic materials.

- Quartz is optically active. The effect may be observed when polarized light passes the crystal along the optical axis, since then no double refraction arises.
- ➤ The **liquid-crystal display** (**LCD**) based on the rotation of the polarization plane by a nematic liquid crystal.

Faraday effect, magnetorotation, optically active substances rotate the polarization direction when they are penetrated by a magnetic field strength $\vec{H}$ that is aligned parallel to the propagation vector $\vec{k}$. The rotation angle α is given by

$$\alpha = V \, l \, H \, .$$

l denotes the thickness of the layer transmitted, V is the **Verdet constant**, a material-specific parameter that depends on the wavelength, H is the magnitude of the magnetic field strength. The rotation angle changes sign when the magnetic field is reversed.

- ▲ Light reflected or refracted by a medium is partially polarized.

11.5.8.1   Polarization by reflection

Brewster's angle, polarization angle α_p, angle of incidence at which the light reflected by a surface is completely linearly polarized perpendicular to the plane of incidence (**Fig. 11.61**). α_p obeys the condition that the refracted ray and the reflected ray be mutually perpendicular (Brewster's law):

$$\sin \alpha_p = n \sin(\pi/2 - \alpha_p) = n \cos \alpha_p \, .$$

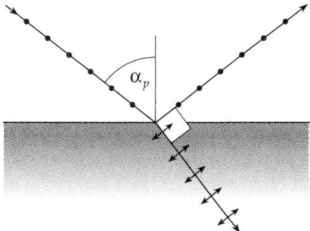

Figure 11.61: Polarization of light by reflection and transmission by an incidence angle equal to Brewster's angle α_p.

Reflection polarizers, polarizers based on Brewster's law for generating polarized light.

- Polarizers are also used in cameras to avoid disturbing reflections, e.g., by panes of glass. One exploits the fact that the reflected light is partly polarized and thus may be filtered out by a polarizer. The polarizer, strictly speaking, acts as an analyzer in this case.

11.5.8.2 Polarization by refraction

1. Double refraction

On entering certain crystals, a light beam is split into two fractions because of the dependence of the phase velocity of electromagnetic waves on the propagation and polarization directions (see **Fig. 11.62**).

Ordinary ray, obeys Snell's law of refraction. The refractive index n_o for the ordinary ray is independent of the propagation direction in the crystal.

Extraordinary ray, the refractive index n_{ao} depends on the propagation direction in the medium.

Double refraction occurs in crystals with an **anisotropic** structure. Such an **anisotropy** may also be generated artificially by external deformation, i.e., by mechanical load, by applying electric voltages or electromagnetic fields. In liquids, double refraction may be generated by flow (**flow bi-refringence**).

Optical axis in a crystal, preferred orientation of symmetry defined by the crystalline structure along which the waves propagate as in an isotropic medium. Along the optical axis $n_o = n_{ao}$, perpendicular to the optical axis $|n_o - n_{ao}|$ becomes a maximum.

2. Optical crystals

Optically uniaxial crystals, crystals with one optical axis (monoclinic, triclinic or rhombic crystals).

Optically biaxial crystals, crystals with two optical axes (tetragonal, hexagonal or rhomboedric crystals).

Principal section, plane in the crystal containing the light ray and the optical axis.

Kinds of double refraction:

- **Linear double refraction,** the phase velocities of mutually perpendicular components of linearly polarized waves differ.
- **Circular double refraction,** the phase velocities of opposite circularly polarized waves differ.

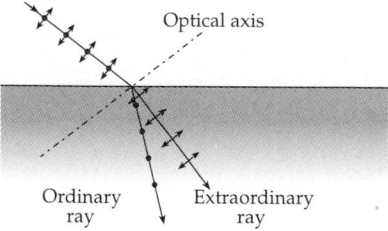

Figure 11.62: Double refraction in optically uniaxial crystals.

- ■ Doubly refracting crystals: Iceland spar, quartz, turmaline.
- ■ Refractive indices for ordinary and extraordinary ray for Iceland spar: $n_o = 1.66$, $n_{ao} = 1.49$.

3. Propagation of polarized rays in the crystal

The wave vector of the ordinary ray oscillates perpendicular to the principal section; the wave vector of the extraordinary ray oscillates parallel to the principal section. The ordinary ray propagates in all crystallographic orientations with the same velocity; the wave surfaces of the elementary waves are spherical surfaces. The propagation velocity of the extraordinary ray is dependent on the orientation; the wave surfaces of the elementary waves are surfaces of rotationally symmetric ellipsoids. Along the optical axis, the propagation velocities of ordinary and extraordinary rays coincide; sphere and rotation ellipsoid osculate along the optical axis (see **Fig. 11.63**).

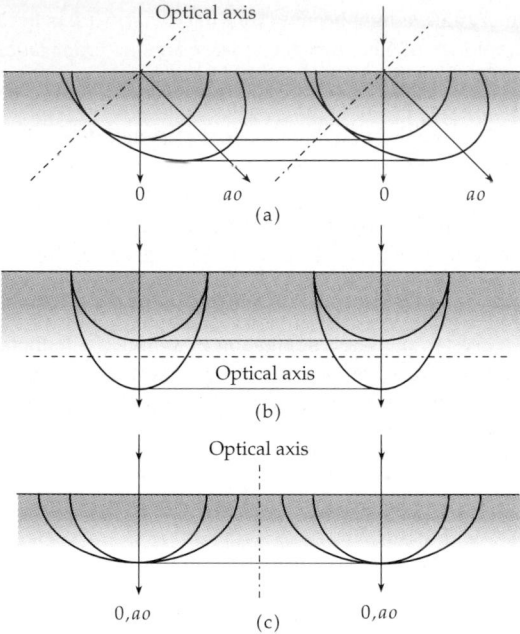

Figure 11.63: Path of polarized rays for perpendicularly incident radiation according to Huygens' principle. (a): optical axis (dashed-dotted line) making an angle with respect to the crystal plane. The extraordinary ray is not perpendicular to the incident wave front. (b): optical axis in the crystal plane. No splitting of the rays, but different propagation velocities for ordinary and extraordinary rays. (c): optical axis perpendicular to crystal plane. Ordinary and extraordinary rays cannot be distinguished.

Positively uniaxial crystals: the ordinary ray propagates faster than the extraordinary ray (**Fig. 11.64**). The sphere encloses the rotational ellipsoid, $c_o \geq c_{ao}$, $n_o \leq n_{ao}$.

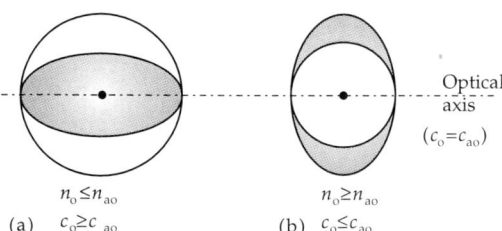

Figure 11.64: Wave surface of the elementary wave. (a): positively uniaxial crystals, (b): negatively uniaxial crystals.

Dichroism, the absorption maximum of the ordinary ray arises at a different wavelength than the absorption maximum of the extraordinary ray. When illuminating the crystal by linearly polarized light, it appears in different colors, depending on the polarization direction.

Negatively uniaxial crystals: the ordinary ray propagates slower than the extraordinary ray. The rotational ellipsoid encloses the sphere, $c_o \leq c_{ao}$, $n_o \geq n_{ao}$.

▲ Double refraction generates linearly polarized light. The polarization directions of ordinary and extraordinary rays are perpendicular to each other.

4. Nicol prism,

polarizer for the generation of linearly polarized light by double refraction in an appropriately cut Iceland spar cystal cemented by special glue ($n = 1.54$) (**Fig. 11.65**). The ordinary ray is separated by total reflection at the interface; for the ordinary ray, the glue is an optically thinner medium. The extraordinary ray penetrates through the interface and leaves the prism as completely linearly polarized light. The polarization direction is on the ray plane.

By selecting an appropriate cut in the spar rhombohedron, one may achieve a situation in which the incident beam is perpendicular to the front face of the crystal (**Glan-Thompson prism**).

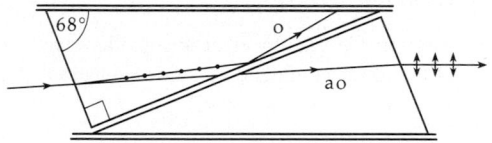

Figure 11.65: Generation of polarized light by a Nicol prism.

5. Photoelasticity,

application of double refraction to investigate the stresses on a loaded body. A model of an object is produced in plexiglass, e.g., a hook loaded like a real hook. Then, depending on the local stress, light is differently polarized at different positions on the model. This polarization may be detected by an analyzer, hence the positions of largest strain may be localized.

Pockels effect, in an electric field E piezoelectric crystals without center of symmetry (potassium dihydrogen phosphate, lithium niobate) become bi-refringent. The difference of the indices of refraction of the ordinary ray (n_o) and the extraordinary ray (n_{ao}) is proportional to the applied electric field strength,

$$|n_{ao} - n_o| \sim E \,.$$

Kerr effect, in a transverse electric field $E \approx 10^6$ V/m an optically isotropic substance (carbon sulfid, benzene) becomes bi-refringent. The difference of the refractive indices for the ordinary ray (n_o) and the extraordinary ray (n_{ao}) is proportional to the square of the applied electric field strength,

$$|n_{ao} - n_o| \sim E^2 \,.$$

➤ **Kerr cells** are used for delay-free intensity modulation of light.

11.6 Photometry

Photometry, measurement of light, measurement of the **photometric quantities** basic for vision and light technology.

Distinction:

- **Objective photometry**, measurement of photometric quantities by instruments that do not take into account the specific properties of human perception of light. The symbols for quantities measured in this manner are labeled by the index e (for energetic).
- **Subjective photometry**, measurement of photometric quantities, taking into account the subjective perception of the human eye, e.g., in the comparison of brightness.

    The symbols for quantities measured in this manner are labeled by the index v (for visual).

11.6.1 Photometric quantities

1. Radiant energy and energy density

Radiant energy, Q_e, energy transported by electromagnetic waves.

Energy density, w, of the electromagnetic radiation, radiation energy per volume element, given by:

energy density of electromagnetic waves			$MT^{-2}L^{-1}$
	Symbol	Unit	Quantity
$w = \dfrac{1}{2}(\vec{E} \cdot \vec{D} + \vec{H} \cdot \vec{B})$	w	J/m^3	energy density
	$\vec{E}$	V/m	electric field strength
	$\vec{D}$	C/m^2	dielectric displacement
	$\vec{B}$	T	magnetic induction
	$\vec{H}$	A/m	magnetic field strength

The quantity of energy within a region of space is obtained as a volume integral over the energy density.

2. Measurement of radiant energy

The radiant energy is measured by conversion into other forms of energy, e.g.:

Thermocouple, generation of an electric voltage by irradiation. The energy is calculated from the measured voltage. In particular, infrared radiation is measured with thermocouples.

Bolometer, semiconductor or electrolytically black-coated platinum wires or platinum foils. One measures the change of resistance due to heating by absorption of radiation. Bolometers mainly respond to infrared radiation, i.e., heat radiation.

Semiconductor, the resistance is changed under irradiation, due to the internal photo effect.

Photo diode, the electric current is measured during irradiation.

Photo emulsion, a surface is coated by a light-sensitive chemical. Incident light changes the color of the layer; the radiation energy is directly converted into chemical energy.

3. Radiant power and radiant flux

Radiant power, radiant flux, Φ_e, radiant energy transported per unit time into a region of space by the electromagnetic wave:

radiant power			$\mathbf{ML^2T^{-3}}$
$\Phi_e = \dfrac{dQ_e}{dt}$	Symbol	Unit	Quantity
	Φ_e	W	radiant power
	Q_e	J	radiant energy
	t	s	time

The radiant power displayed by a measuring instrument for a given radiation source depends on

- area of the receiver of the measuring instrument,
- distance of the receiver from the transmitter, the source of electromagnetic radiation,
- orientation of the area of receiver with respect to the transmitter,
- spectral sensitivity of the receiver.

Extended, arbitrarily shaped bodies may be considered point-like if the distance from them is large enough. Otherwise, one considers sufficiently small area elements on the body surface that again satisfy the point approximation. The measured quantity is then summed over these elements.

The area of the receiver is usually a plane; it does not correspond to a spherical shell about the transmitter. If the distance between receiver and transmitter is large enough, one may to a good approximation insert the (mostly plane) area of the receiver for the section of the spherical shell. It is presumed, however, that the receiver area points towards the transmitter.

4. Photometric limiting distance,

minimum distance beyond which, according to the DIN standard, the approximation given above may be regarded as satisfied: the distance between transmitter and receiver must be at least 10 times the largest transverse dimension of the receiver or the transmitter, respectively. If this condition is fulfilled, replacing the section of the spherical shell by a plane area causes an error of less than 2 %.

▲   The radiant power received by the receiver is proportional to the solid angle corresponding to its area if the radiation is homogeneously distributed over the area.

5. Radiant intensity,

I_e, proportionality factor between solid angle and radiant power:

radiant intensity $= \dfrac{\text{radiant power}}{\text{solid angle}}$			$\mathbf{ML^2T^{-3}}$
$d\Phi_e = I_e \, d\Omega \qquad I_e = \dfrac{d\Phi_e}{d\Omega}$	Symbol	Unit	Quantity
	Φ_e	W	radiant power
	I_e	W/sr	radiant intensity
	Ω	sr	efficient solid angle

The radiant flux into the solid angle Ω is given by

$$\Phi_e = \int_\Omega I_e \, d\Omega .$$

11.6.1.1   Radiation source

For transmitters that are not point-like, the measured radiant intensity depends on
- the area of the transmitter A_S,
- the relative orientation of transmitter area to the receiver area.

1. Radiation pattern

of a light source, $g(\alpha)$, a function specifying the dependence of the radiant intensity on the angle α at which the transmitter is seen:

angular dependence of radiant intensity			$\mathbf{ML^2T^{-3}}$
	Symbol	Unit	Quantity
$I_e(\alpha) = L_e(\alpha)A_S g(\alpha)$	$I_e(\alpha)$	W/sr	radiant intensity
	$g(\alpha)$	1	radiation pattern
	α	rad	angle between normal of transmitter and receiver area
	$L_e(\alpha)$	W/(m^2 sr)	radiance
	A_S	m^2	area of transmitter

Radiance, L_e, characteristic quantity for the properties of a transmitter. It depends on its material, its surface properties and its temperature, among other qualities.

2. Lambert source,

Lambertian source, source with a radiation pattern $g(\alpha) = \cos(\alpha)$. A Lambert source appears as equally bright for all observation angles α, since $A_S \cos(\alpha)$ is just the projection of the area in the direction of observation. Hence, the ratio of radiant intensity to the effective area A_{eff} at the angle α is constant,

$$\frac{I_e(\alpha)}{A_{\text{eff}}} = \frac{L_e(\alpha)A_S \cos(\alpha)}{A_S \cos(\alpha)} = L_e(\alpha).$$

Most thermal light sources are approximately Lambertian.
Conditions for a Lambertian source:
- No fixed phase relations of wave fields radiated by neighboring area elements of the transmitter.
- The material of the transmitter must be **optically dense**, i.e., it must be able itself to absorb the radiation emitted by the transmitter surface.

Lambert law, the radiation pattern given above,

$$g(\alpha) = \cos(\alpha).$$

3. Gaussian pattern and irradiance

Gaussian pattern, radiation pattern of the form

$$g(\alpha) = e^{-\alpha^2/\gamma^2}.$$

Here γ is a constant characterizing the radiation source. For decreasing values of γ the distribution g becomes more narrow, i.e., the radiation is increasingly concentrated in one direction. The Gaussian pattern is realized for a **laser**.

Specific radiant emittance, M_e, characteristic parameter of a transmitter, defined by:

specific radiant emittance = $\dfrac{\textbf{radiant flux}}{\textbf{transmitter area}}$			$\mathbf{MT^{-3}}$
$M_e = \dfrac{d\Phi_e}{dA_S}$	Symbol	Unit	Quantity
	M_e	W/m^2	specific radiant emittance
	Φ_e	W	radiant flux
	A_S	m^2	transmitter area

Irradiance, E_e, the radiant flux incident on the receiver area A_E:

irradiance = $\dfrac{\textbf{radiant flux}}{\textbf{receiver area}}$			$\mathbf{MT^{-3}}$
$E_e = \dfrac{d\Phi_e}{dA_E}$	Symbol	Unit	Quantity
	E_e	W/m^2	irradiance
	Φ_e	W	radiant flux
	A_E	m^2	effective receiver area

The effective receiver area A_E is obtained by projecting the actual receiver area A onto the connecting line between transmitter and receiver,

$$A_E = A \cos \beta \,.$$

β is the angle between the connecting line between transmitter and receiver and the perpendicular onto A.

4. Photometric inverse-square law,

gives the dependence of the irradiance E_e on the distance r from the transmitter, is valid only for spherical symmetry, without account for reflection and absorption:

photometric inverse-square law			$\mathbf{MT^{-3}}$
$E_e = \dfrac{I_e(\alpha)}{r^2} \cos \beta \, \Omega_0$	Symbol	Unit	Quantity
	E_e	W/m^2	irradiance
	r	m	distance transmitter-receiver

Irradiation, H_e, the energy incident per unit area in a given time interval between t_1 and t_2. It is obtained by integrating the irradiance over the time:

irradiation			$\mathbf{MT^{-2}}$
$H_e = \displaystyle\int_{t_1}^{t_2} E_e(t)\,dt$	Symbol	Unit	Quantity
	H_e	J/m^2	irradiation
	E_e	W/m^2	irradiance
	t	s	time

11.6.1.2   Spectral quantities

Spectral filters, change the spectral energy of transmitted radiation. Their action is based on absorption, interference, total reflection, etc., as represented by the degree of spectral transmission as a function of the wavelength (**filter curve**). Filters are classified according to the trend of the filter curve into edge filters (high-pass or low-pass filters), band-pass filters, narrow-band or line filters.

If the radiation consists of waves of different wavelengths, the contribution of the individual components to photometric quantities may be investigated by selecting single wavelength regions and measuring the corresponding photometric quantity for that fraction of the radiation.

➤   While a UV suppression filter removes UV radiation, a blue filter is transparent for blue light only, and a red filter is transparent for red light only. This is the customary nomenclature, as a matter of convention.

▲   The contribution of radiation from a wavelength range $d\lambda$ to a photometric quantity X_e is given by

$$\frac{\partial X_e}{\partial \lambda}\, d\lambda\,.$$

Spectral quantity, designation for the derivative of a photometric quantity with respect to the wavelength. Spectral quantities are specified by the index λ.

■   The derivative of the radiance with respect to the wavelength,

$$L_{e,\lambda} = \frac{\partial I_e}{\partial \lambda}\,,$$

is called **spectral radiance**.

Conversely, the radiance is calculated from the spectral radiance by integrating over the wavelength,

$$L_e = \int L_{e,\lambda}\, d\lambda\,.$$

11.6.1.3   Reflection, absorption, transmission

When electromagnetic radiation hits a layer, one observes the phenomena reflection, absorption and transmission. Only a fraction of the incident radiant flux Φ_e can be detected behind the layer as transmitted radiant flux Φ_t. Reflection and absorption depend on the material of the layer, and on the wavelength λ of the radiation (**Fig. 11.66**).

1. Spectral reflectance and absorptance

Spectral reflectance, $\rho(\lambda)$, ratio of the total reflected radiant flux Φ_r to the incident radiant flux Φ_e,

$$\rho(\lambda) = \frac{\Phi_r(\lambda)}{\Phi_e(\lambda)}\,.$$

The total reflected radiant flux may originate, as for a plate, by reflection at several surfaces. The reflectance depends significantly on the surface properties of the material.

■   The reflectance of snow is 0.93, of aluminum 0.69, and of black paper 0.05.

Spectral absorbance, spectral absorptive power, $\alpha(\lambda)$, ratio of the total absorbed radiant flux Φ_a to the incident radiant flux Φ_e,

$$\alpha(\lambda) = \frac{\Phi_a(\lambda)}{\Phi_e(\lambda)}\,.$$

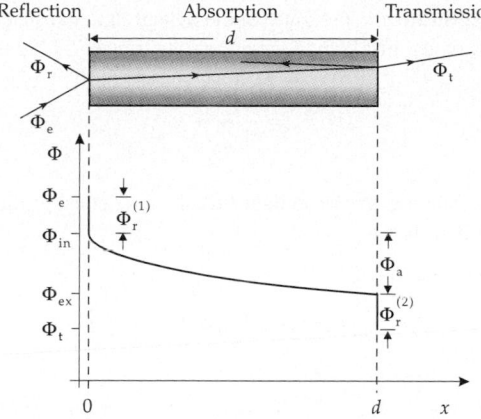

Figure 11.66: Reflection, absorption and transmission of electromagnetic radiation incident on a plate of thickness d.

The absorptance depends weakly on the temperature T of the material, $\alpha(\lambda) = \alpha(\lambda, T)$.

Absorption law, the radiant flux in the interior of the layer decreases exponentially with increasing penetration depth x,

$$\Phi(x) = e^{-a(\lambda)\, x} \ .$$

Absorption coefficient, $a(\lambda)$, unit: m^{-1}, characterizes the absorbing material.

2. Mean range and transmittance

Mean range of radiation, x_m, the penetration depth at which the radiant flux decreases to the fraction $1/e$ of the incident radiant flux,

$$x_m = \frac{1}{a} \ .$$

Spectral transmittance, $\tau(\lambda)$, ratio of the transmitted radiant flux Φ_t to the incident radiant flux Φ_e,

$$\tau(\lambda) = \frac{\Phi_t(\lambda)}{\Phi_e(\lambda)} \ .$$

The transmittance is a measure for the transparency of a layer for radiation.
According to the energy law,

$$\rho(\lambda) + \alpha(\lambda) + \tau(\lambda) = 1 \ .$$

Spectral pure absorptance, $\alpha_i(\lambda)$, the radiant flux absorbed in the layer, $\Phi_{in} - \Phi_{ex}$, is not related to the incident radiant flux, but to the radiant flux $\Phi_{in} = \Phi_e - \Phi_r^{(1)}$ just behind the entrance surface,

$$\alpha_i(\lambda) = \frac{\Phi_{in}(\lambda) - \Phi_{ex}(\lambda)}{\Phi_{in}(\lambda)} \ .$$

If the reflection is negligible, then $\alpha_i(\lambda) = \alpha(\lambda)$.

Spectral pure transmittance, $\tau_i(\lambda)$, ratio of the radiant flux Φ_{ex} just in front of the exit surface to the radiant flux Φ_{in} just behind the entrance surface,

$$\tau_i(\lambda) = \frac{\Phi_{ex}(\lambda)}{\Phi_{in}(\lambda)}.$$

The radiant flux Φ_{ex} subdivides in the radiant flux $\Phi_r^{(2)}$ reflected at the exit surface, and the transmitted radiant flux Φ_t.
  One has

$$\alpha_i(\lambda) + \tau_i(\lambda) = 1.$$

3. Black body,

a body with absorbance 1 over the entire range of wavelengths of electromagnetic radiation. There is no material having exactly this property; nevertheless, the concept of the black body is of central importance in the theory of heat radiation.

▲   **Kirchhoff's law**: The spectral radiance $L_{e,\lambda}$ of an arbitrary body of temperature T at wavelength λ equals the product of the absorbance of the body at this temperature and wavelength, and the spectral radiance $L_{e,\lambda}^{black}$ of a black body at the same temperature and wavelength.

Kirchhoff's law			$\mathbf{ML^{-1}T^{-3}}$
	Symbol	Unit	Quantity
$L_{e,\lambda} = \alpha(\lambda, T) \cdot L_{e,\lambda}^{black}$	$L_{e,\lambda}$	W/(m^3 sr)	spectral radiance
	$\alpha(\lambda, T)$	1	absorbance
	$L_{e,\lambda}^{black}$	W/(m^3 sr)	spectral radiance black body

Kirchhoff's law traces the spectral radiance of an arbitrary body back to the spectral radiance of a black body $L_{e,\lambda}^{black}$ (**Planck's radiation law, Fig. 11.67**):

$$L_{e,\lambda}^{black} = \frac{2hc^2}{\lambda^5} \frac{1}{e^{hc/\lambda kT} - 1},$$

c: vacuum speed of light, h: Planck's constant, k: Boltzmann constant.

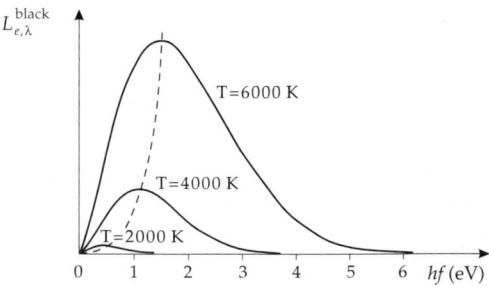

Figure 11.67: Spectral radiance of a black body at various temperatures versus the radiation energy in eV (f: frequency of radiation).

11.6.2 Photometric quantities

Photometric quantities, are based on an evaluation of radiation by the human eye. They describe radiation in such a way that the sensation of brightness is given, and thus they are important for **lighting engineering**.

▲   A **photometric** quantity Y generally results by evaluation of an **energetic** quantity X_e for the eye.

1. Relative and absolute sensitivity

In order to be able to describe the evaluation of an energetic quantity by an arbitrary receiver (and thus also by the eye) and to include the dependence of the sensitivity on the wavelength λ of light, the following quantities are introduced:

relative and absolute sensitivity			
	Symbol	Unit	Quantity
$s(\lambda) = \dfrac{\mathrm{d}Y}{\mathrm{d}X_e} = \dfrac{Y_\lambda}{X_{e\lambda}}$ $s_{\mathrm{rel}}(\lambda) = \dfrac{s(\lambda)}{s(\lambda_0)}$	λ	m	arbitrarily chosen wavelength
	λ_0	m	wavelength
	$s(\lambda)$		absolute spectral sensitivity
	$s_{\mathrm{rel}}(\lambda)$	1	relative spectral sensitivity
	X_e		energetic input quantity
	$X_{e\lambda}$		spectral energetic input quantity
	Y		output quantity
	Y_λ		spectral output quantity

■   If a radiant flux $\mathrm{d}\Phi_e = \Phi_{e\lambda} \cdot \mathrm{d}\lambda$ hits a receiver and thereby generates the current $\mathrm{d}J$, then Φ_e corresponds to the energetic input quantity X_e, and J corresponds to the output quantity Y. $\Phi_{e\lambda}$ is the corresponding spectral energetic input quantity $X_{e\lambda}$, J_λ the spectral output quantity Y_λ.

Radiation may be evaluated by means of these quantities even if it consists of a superposition of light of distinct wavelengths from an interval $[\lambda_1, \lambda_2]$. The evaluated output quantity Y is then obtained by the convolution of the spectral energy input quantity with the spectral sensitivity. The ratio of the output quantity obtained this way to the input quantity then yields the **absolute sensitivity**:

absolute sensitivity			
	Symbol	Unit	Quantity
$Y = \displaystyle\int_{\lambda_1}^{\lambda_2} X_{e\lambda} \cdot s(\lambda) \, \mathrm{d}\lambda$ $= s(\lambda_0) \displaystyle\int_{\lambda_1}^{\lambda_2} X_{e\lambda} \cdot s_{\mathrm{rel}}(\lambda) \, \mathrm{d}\lambda$ $s = \dfrac{Y}{X}$	Y		output quantity
	λ	m	wavelength
	λ_0	m	wavelength
	λ_1	m	lower-limit wavelength
	λ_2	m	upper-limit wavelength
	$X_{e\lambda}$		spectral energetic input quantity
	$s(\lambda)$		absolute spectral sensitivity
	$s_{\mathrm{rel}}(\lambda)$	1	relative spectral sensitivity
	s		absolute sensitivity

➤   In the definition of s, X is written instead of X_e, since the formula given above holds also for non-energetic quantities.

Spectral degree of brightness, relative spectral sensitivity of the eye. In the evaluation, one takes:

- for $X_{e\lambda}$ the spectral radiant flux $\Phi_{e\lambda}$,
- for $s_{rel}(\lambda)$ the **spectral relative luminosity** $V(\lambda)$ for daylight vision (see **Fig. 11.68**),
- for $s(\lambda_0)$ the absolute spectral sensitivity of the eye at $\lambda_0 = 555$ nm.

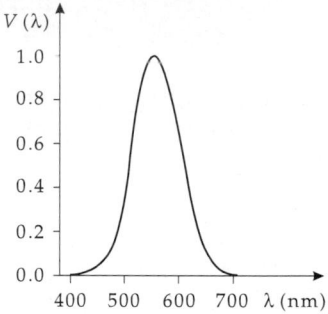

Figure 11.68: Spectral luminosity of the eye for daylight vision $V(\lambda)$.

2. *Luminous flux,*

Φ, determined by integration, owing to the dependence of the spectral luminosity on the wavelength:

definition of luminous flux			**J**
	Symbol	Unit	Quantity
$$\Phi = V(\lambda_0) \int_{380 \text{ nm}}^{780 \text{ nm}} \Phi_{e\lambda} V(\lambda)\, d\lambda$$	Φ	lm	luminous flux
	V	1	spectral luminosity
	λ	m	wavelength
	$\Phi_{e\lambda}$	cd/m	spectral radiant flux

Lumen, lm, SI unit of the luminous flux Φ.

■ Luminous flux of several light sources: mercury lamp 125 000 lm, fluorescent lamp 2300 lm, electric-light bulb 730 lm, light diode 0.01 lm.

Luminous intensity, I, the luminous flux $d\Phi$ emitted by a light source into a solid angle element $d\Omega$.

Candela, cd, SI unit of the luminous intensity. The candela is a basic quantity in the SI system (cf. kg, m, s, A), and hence cannot be expressed by other SI quantities.

▲ 1 candela is the luminous intensity of a source of radiation that emits a monochromatic radiation of frequency $f = 540$ THz ($\lambda = 555$ nm) and radiates a power of $(1/683)$ W/sr.

One has: 1 lm = 1 cd · sr.

➤ Formerly, the unit cd was defined through the luminance of a black body at the freezing point of platinum.

Luminance, L, contribution of the surface element dA of a source of light making an angle α with respect to the luminous intensity,

$$L = \frac{dI}{dA_S \cos\alpha}.$$

Illuminance, E, defined as ratio of the element of luminous flux to the illuminated area element,

$$E = \frac{\mathrm{d}\Phi}{\mathrm{d}A}.$$

Lux, lx, SI unit of illuminance, $1\ \mathrm{lx} = 1\ \mathrm{lm/m^2}$.
■ Illuminances: Sun (summer) 70 000 lx, Sun (winter) 5500 lx, daylight (covered sky) 1000 – 2000 lx, full Moon 0.25 lx, limit of color perception 3 lx.

Comparison of radiation-physical and photometric quantities		
radiation-physical		
radiant flux	Φ_e	W
radiant intensity	$I_e = \dfrac{\mathrm{d}\Phi_e}{\mathrm{d}\Omega}$	W/sr
radiance	$L_e = \dfrac{\mathrm{d}I_e}{\mathrm{d}A_S\,\cos\alpha}$	$\mathrm{W/(m^2 sr)}$
irradiance	$E_e = \dfrac{\mathrm{d}\Phi_e}{\mathrm{d}A_E}$	$\mathrm{W/m^2}$
photometric		
luminous flux	Φ	lm (cd·sr)
luminous intensity	$I = \dfrac{\mathrm{d}\Phi}{\mathrm{d}\Omega}$	cd
luminance	$L = \dfrac{\mathrm{d}I}{\mathrm{d}A_S\,\cos\alpha}$	$\mathrm{cd/m^2}$
illuminance	$E = \dfrac{\mathrm{d}\Phi}{\mathrm{d}A_E}$	$\mathrm{lx\ (lm/m^2)}$

Symbols used in formulae on vibrations, waves, acoustics and optics

Symbol	Unit	Designation	Symbol	Unit	Designation
α	rad	angular displacement	α	1	degree of sound absorption
$\dot{\alpha}$	rad/s	angular velocity	κ	1	adiabatic exponent
$\ddot{\alpha}$	rad/s^2	angular acceleration	ω	rad/s	angular velocity
δ	1/s	decay constant	B_S	Pa/V	electro-acoustic
$\Delta\phi$	rad	phase difference			transmission factor
λ	m	wavelength	c_{Fk}	m/s	sound velocity
Λ	1	logarithmic decrement			in solids
μ	1	friction coefficient	c_{Fl}	m/s	sound velocity
ϕ	rad	phase angle			in liquids
ω	rad/s	angular frequency	c_G	m/s	sound velocity
a	m/s^2	acceleration			in gases
A		amplitude	E	N/m^2	elasticity modulus
b	kg/s	damping constant	E_k	Pa/$\sqrt{\text{VA}}$	characteristic sensitivity
c	m/s	phase velocity	E_M	Pa/$\sqrt{\text{VA}}$	loudspeaker sensitivity
c	m/s	sound velocity	G_S	dB	electro-acoustic
D	1	degree of damping			transmission measure
d	1	loss factor	J	W/m^2	sound intensity
f	Hz	frequency	K	N/m^2	compression modulus
F	kg m/s^2	restoring force	p	Pa	sound pressure
F_N	N	normal force	P	W	sound power
F_R	N	friction force	p_0	Pa	static pressure
g	m/s^2	gravitational acceleration	r	1	reflectance
k	kg/s^2	restoring force coefficient	R	dB	measure of attenuation
k	1/m	wave number	R_i	J/(K kg)	specific gas constant
$\bar{\mathbf{k}}$	1/m	wave vector	T	K	temperature
m	kg	mass	T	s	reverberation time
Q	1	quality factor	v	cm/s	sound particle velocity
v	m/s	phase velocity	w	J/m^3	energy density
v_{gr}	m/s	group velocity	Z	kg/(m^{2}s)	characteristic acoustic
T	s	period			impedance
T_S	s	beat period	Z	Ω	impedance

Symbol	Unit	Designation	Symbol	Unit	Designation
α_g	rad	critical angle of total reflection	M_e	W/m^2	specific radiant emittance
			Φ	cd	luminous flux
β	1	linear magnification	Φ_e	W	radiant power
a	m	object distance	$\Phi_{e\lambda}$	cd/m	spectral radiant flux
a'	m	image distance			
a_B	m	near limit	Q_e	J	radiant energy
$A(\lambda, T)$	1	absorptance	s		absolute sensitivity
B	1/m	refractive power	$s(\lambda)$		absolute spectral sensitivity
E_e	W/m^2	irradiance			
$\bar{f}$	m	object focal length	$s_{\mathrm{rel}}(\lambda)$	1	relative spectral sensitivity
f'	m	image focal length			
H_e	J/m^2	irradiation	v	1	magnification
I_e	W/sr	radiant intensity	V	1	spectral relative luminosity
k	1	focal ratio			
L_e	$W/(m^2\,sr)$	radiance	y	m	object size
$L_{e,\lambda}$	$W/(m^3\,sr)$	spectral radiance	y'	m	image size

12
Tables on vibrations, waves, acoustics and optics

12.1 Tables on vibrations and acoustics

12.1/1 Correction factors for pendulum period at large displacement

Angle (°)	Angle (rad)	Correction factor
1	0.017453	1.00002
5	0.087266	1.00048
10	0.174533	1.00191
30	0.523599	1.01741
45	0.785398	1.03997

12.1/2 Sound velocity in gases

Gas	$c/(ms^{-1})$ at 0 °C	at 20 °C	Gas	$c/(ms^{-1})$ at 0 °C	at 20 °C
ammonia	415	428	argon	319	321
carbon dioxide	259	258	city gas	453	450
chlorine	206	—	oxygen	316	324
nitrogen	334	348	hydrogen	1284	1300
helium	965	1020	ethylene	317	329
methane	430	—	neon	435	453

12.1/3 Sound velocity in air

Gas	$c/(ms^{-1})$ 0 °C	10 °C	20 °C	30 °C
air	332	338	344	350

12.1/4   Damping coefficient for sound in gases

Gas	$T/(°C)$	f/kHz	p/MPa	α/cm^{-1}
nitrogen	19.9	598.9	0.097	0.0484
hydrogen	19.9	598.9	0.1	1.284
helium	17.5	598.9	0.099	1.061
nitrogen monoxide	16.3	598.9	0.095	0.656
carbon dioxide	18.7	304.4	0.085	2.073
oxygen	19.6	598.9	0.099	0.602

12.1/5   Sound field quantities in air at 20 °C

Sound pressure/ (Pa)	Sound particle velocity/ (cm·s^{-1})	Sound intensity/ (μW/cm^2)
0.01	$2.42 \cdot 10^{-5}$	$2.42 \cdot 10^{-9}$
0.05	$1.21 \cdot 10^{-4}$	$6.05 \cdot 10^{-8}$
0.10	$2.42 \cdot 10^{-4}$	$1.42 \cdot 10^{-7}$
0.50	$1.21 \cdot 10^{-3}$	$6.05 \cdot 10^{-6}$
1.00	$2.42 \cdot 10^{-3}$	$2.42 \cdot 10^{-5}$

12.1/6   Sound velocity in oil and mineral-oil products

Substance	$T/°C$	$c/(\text{ms}^{-1})$	Substance	$T/°C$	$c/(\text{ms}^{-1})$
petrol	25	1295	kerosene	34	1295
linseed oil	31.5	1772	olive oil	32.5	1381
paraffin oil	33.5	1420	pine oil	31	1468
turpentine oil	27	1280	transformer oil	32.5	1425
eucalyptus oil	29.5	1276	mustard oil	31.5	1825

12.1/7   Sound velocity in liquids at 20 °C

Liquid	$c/(\text{ms}^{-1})$	Liquid	$c/(\text{ms}^{-1})$
benzene	1330	glycerol	1920
water	1480	sea water	1470
heavy water	1399	ethyl alcohol	1165
kerosene	1451	mercury	1460
aniline	1656	acetone	1192
		methyl alcohol	1156

12.1/8   Sound-damping coefficients for liquids

Liquid	$T/°C$	f/MHz	α/cm^{-1}
acetone	20	307	25.6
	20	482	56
	20	843	167.7
water	20	307	23.28
	20	482	55.3
	20	843	172
toluene	20	307	71.9
	20	482	182.4
	20	843	575.6
glycerol	32.8	30	12.69
olive oil	21	1	0.0125
benzene	20	307	711.5
	20	482	1150
petrol	—	1	0.0096
trichloromethane	20	307	344
	20	482	720.2
	20	843	1748
tetrachloromethane	20	307	492
	20	482	1115.2
	20	843	3269
linseed oil	20.5	3.1	0.141
castor oil	21.4	15.7	5.18

12.1/9   Sound velocity in metals

Substance	$c/(\text{ms}^{-1})$
aluminum	5200
lead	1200
iron	5000
iridium	4900
copper	3500
brass	3400
nickel	4973
steel	5050
zinc	2680
tin	2490
silver	3650
titanium	6070

12.1/10   Sound velocity in synthetic materials and glasses (thin rods)

Substance	$c/(\mathrm{ms}^{-1})$
polystyrene	1800
PVC, soft	80
PVC, hard	1700
polycarbonate	1400
polyethylene	540
nylon	1800
plexiglass	1840
flint glass	3720
borate glass	4540
crown glass	5300
quartz glass	5400
porcelain	4880

12.1/11   Sound velocity in construction materials

Substance	$c/(\mathrm{ms}^{-1})$
concrete	3100
marble	3810
granite	3950
pine	3600
fir	3320
brick	3600
oak	4100
cork	500
beech	3300
brickwork	3500...4000

12.1/12   Acoustical-attenuation coefficients for building materials (mean values) and demands for construction

Building material	dB	Building structure	Required attenuation coefficient
single window	15	brickwork	50
double window (12 cm air)	<30	window	25
single wooden door	20	doors	30
double door (12 cm air)	<40	internal walls in apartments	40
straw mat, 5 cm	38	internal walls in schools	42
concrete wall, 10 cm	42	outer walls	48
brickwork, plastered, 12 cm	45	ceilings	52

12.1/13   Sound attenuation in air in dB/(100 m) for standard pressure

$T/°C$	Relative moisture/%	Frequency/Hz					
		125	250	500	1000	2000	4000
30	10	0.09	0.19	0.35	0.82	2.6	8.8
	20	0.06	0.18	0.37	0.64	1.4	4.4
	30	0.04	0.15	0.38	0.68	1.2	3.2
	50	0.03	0.10	0.33	0.75	1.3	2.5
	90	0.02	0.06	0.24	0.70	1.5	2.6
20	10	0.08	0.15	0.38	1.21	4.0	2.5
	20	0.07	0.15	0.27	0.62	1.9	6.7
	30	0.05	0.14	0.27	0.51	1.3	4.4
	50	0.04	0.12	0.28	0.50	1.0	2.8
	90	0.02	0.08	0.26	0.56	0.99	2.1
10	10	0.07	0.19	0.61	1.9	4.5	7.0
	20	0.06	0.11	0.29	0.94	3.2	9.0
	30	0.05	0.11	0.22	0.61	2.1	7.0
	50	0.04	0.11	0.20	0.41	1.2	4.2
	90	0.03	0.10	0.21	0.38	0.81	2.5
0	10	0.10	0.30	0.89	1.8	2.3	2.6
	20	0.05	0.15	0.50	1.6	3.7	5.7
	30	0.04	0.10	0.31	1.08	3.3	7.4
	50	0.04	0.08	0.19	0.60	2.1	6.7
	90	0.03	0.08	0.15	0.36	1.1	4.1

12.1/14   Loudness levels in dB

lower threshold of hearing	0	typewriter	50...70
ticking of pocket watch	10	loud street noise	70
rustle of leaves	20	crying	80
whispering	20	loud horn	90
muted conversation	40	motorbike	70...100
muted radio music	40	rock music	105
tearing of paper	40	riveting hammers	110
speech	40...50	pain threshold	130

12.1/15   Noise injurious to health

Response	Sound Intensity level /(dB)
psychic (displeasure, irritability)	> 30
vegetative (weak concentration, decreasing performance)	> 65
auditory damage (damage of inner ear, incurable)	> 80
mechanical damage (deafness)	> 120

12.1/16   Sound absorptance

Sound absorptance α of various building materials			
	α at		
Material	125 Hz	500 Hz	2000 Hz
plastering on brickwork	0.02	0.02	0.03
lime plastering	0.03	0.03	0.04
lightweight concrete	0.07	0.22	0.10
rough plastering	0.03	0.03	0.07
acoustical lightweight building boards, 2.5 cm thick			
at 3 cm distance	0.25	0.23	0.74
directly on massive wall	0.15	0.23	0.73
insulating plates, 2 cm thick			
directly on massive wall	0.13	0.19	0.24
at 3 cm distance	0.15	0.23	0.23
at 3 cm distance and glass wool	0.33	0.44	0.37
wooden doors	0.14	0.06	0.10
parquet	0.05	0.06	0.10
plywood, 3 mm, distance 2 cm	0.07	0.22	0.10
plywood, 3 mm, directly on wall	0.07	0.05	0.10
wooden panels	0.25	0.25	0.08

12.2   Tables on optics

12.2/1   The most important fiber types for optical telecommunication

Material	Plastic	Glass	Glass	Glass
type [a]	MM, SI	MM, SI	MM, GI	SM, SI
core diameter (μm)	200 – 600	50, 62.5, 200, …	50, 62.5, 85, …	4 – 10
mantle diameter (μm)	500 – 1000	125, 900	125	125
numerical aperture	≈ 0.5	0.15 – 0.5	0.2 – 0.3	
damping (dB/km)	50 – 1000 (650 nm)	5 (850 nm) 0.5 (1300 nm)	5 (850 nm) 0.5 (1300 nm)	0.4 (1300 nm) 0.2 (1550 nm)
transfer capacity	1 – 10 MHz · km [b]	10 MHz · km	1 GHz · km	10 – 100 GHz · km [c]

[a] MM: multi-mode, SM: single-mode, SI: step index, GI: gradient index

[b] transfer range limited to several meters because of high damping

[c] by principle not limited by mode dispersion. The possible transfer capacity follows from the material dispersion and the line width of the adopted light source. Actual transfer capacities achieved are in the range of 10 – 50 GHz·km.

12.2/2 Refractive indices n_d at $\lambda = 589.3$ nm (yellow sodium line)

Material	n_d	Material	n_d
gases at 0 °C and 1013 hPa		solid materials at 20 °C	
air	1.00029	diamond	2.417
nitrogen	1.00030	saphir (Al_2O_3)	1.769
oxygen	1.00027	lithium chloride	1.662
carbon dioxide	1.00045	sodium chloride	1.544
ammonia	1.00038	potassium chloride	1.490
hydrogen	1.00014	lithium fluoride	1.392
helium	1.000035	lithium bromide	1.784
neon	1.000067	lithium iodide	1.955
argon	1.000283	fluorite (CaF_2)	1.434
crypton	1.000429	ice (at 0 °C)	1.310
xenon	1.00071	quartz glass	1.459
liquids at 20 °C		SCHOTT BK1	1.51009
water	1.333	SCHOTT BK7	1.51680
methanol	1.329	SCHOTT F2	1.62004
ethanol	1.362	SCHOTT SF6	1.80518
acetone	1.359	SCHOTT FK3	1.46450
glycerol	1.455	window glass	≈ 1.51
benzene	1.501	plexiglass (PMMA)	≈ 1.49
carbon disulphide	1.628	polystyrene (PS)	≈ 1.59
bromonaphthaline	1.658	polycarbonate (PC)	≈ 1.59
linseed oil	1.486	several bi-refringent materials	
cedar oil	1.505	quartz (SiO_2)	1.544/1.553
		Iceland spar ($CaCO_3$)	1.658/1.486
		magnesium fluoride	1.389/1.377
		several infrared-transmitting materials and semiconductors at various wavelengths	
		zinc sulphide ($\lambda = 3$ μm/10.6 μm)	2.27/2.19
		germanium ($\lambda = 3$ μm/10.6 μm)	4.05/4.00
		silicon ($\lambda = 3$ μm/10.6 μm)	3.43/3.42
		gallium arsenide ($\lambda = 3$ μm/10.6 μm)	3.32/3.28

12.2/3 The most important types of lasers

Type of laser	Most important lines (nm)
helium-neon	632.8, 543, 594, 612 and others in the IR range
helium-cadmium	442, 325
argon ions	488, 514
carbon dioxide	$10.6\ \mu m$
excimer (XeF, KrF, ArF)	351, 248, 193
dye	tunable UV – IR
Nd:YAG	1064 (532 with frequency doubling)
semiconductors (e.g., InGaAs)	tunable ca. 660 – 1550

12.2/4 Coherence lengths of several light sources

Light source	Coherence length (order of magnitude)
Sun (visible spectral range)	$1\ \mu m$
light-emitting diode	$20\ \mu m$
mercury lamp	0.5 mm
laser diodes	mm – cm
He-Ne laser (single)	0.2 m
stabilized lasers	> 100 m

12.2/5 Illuminances

Light source	Illuminance /(lx)
Sun, summer time	70000
Sun, winter time	5500
daylight, overcast sky	1000 – 2000
full Moon	0.25
stars, clear/without Moon	0.001
workplace illumination	1000
living-room illumination	120
threshold of color perception	3
street illumination	1 – 16

12.2/6   Luminous fluxes

Light source	Luminous flux /(lm)
light-emitting diode	0.01
electric light bulb 60 W	730
electric light bulb 100 W	1380
fluorescent tube	2300
mercury lamp 60 W	5400
mercury lamp 100 W	125000

12.2/7   Relative luminosity

λ/nm	V/1	λ/nm	V/1	λ/nm	V/1	λ/nm	V/1
380	0	490	0.208	590	0.757	700	0.0041
390	0.0001	500	0.323	600	0.631	710	0.0021
400	0.0004	510	0.503	610	0.503	720	0.105
410	0.0012	520	0.710	620	0.381	730	0.000052
420	0.0040	530	0.862	630	0.265	740	0.000025
430	0.0116	540	0.954	640	0.175	750	0.000012
440	0.023	550	0.995	650	0.107	760	0.000006
450	0.038	555	1	660	0.061	770	0.000003
460	0.060	560	0.995	670	0.032	780	0.0000015
470	0.091	570	0.952	680	0.017		
480	0.139	580	0.870	690	0.0082		

12.2/8   Ultraviolet spectral range

Wavelength $\lambda/(10^{-7}\text{m})$	Designation	Effect
3.80 ... 3.15	long-wave UV	sudden pigmentation
3.15 ... 2.80	medium-wave UV	formation of erythema
2.80 ... 2.00	short-wave UV	bactericidal
< 2.00	vacuum UV	ozonization

12.2/9   Fraunhofer lines

Designation	Element	Wavelength/ (nm)	Designation	Element	Wavelength/ (nm)
A	O_2	759.3	F	H	486.1
B	O_2	686.7	f	H	434.0
C	H	659.3	G	Fe, Ti	430.8
D_1	Na	589.6	h	H	410.2
D_2	Na	589.0	H	Ca^+	396.8
E	Ca, Fe	527.0	K	Ca^+	393.3

Part III
Electricity

Electricity, deals with stationary and moving electric charges, the actions of force between them, and the electric and magnetic fields generated by them.

There are applications of electricity in these fields:

- electrical engineering, e.g., in direct-current, alternating-current and three-phase-current engineering, in calculations of circuits, and in the construction of generators and motors;
- electrochemistry, e.g., in charge transport in electrolytes, and in the production of batteries;
- electronics, e.g., in development and application of electronic components in analog and digital electronics, and in the development of computers;
- plasma physics, e.g., in generation of light, materials processing, energy production and formation of ion beams from ion sources;
- accelerator physics, in the transport and acceleration of ions and electrons;
- telecommunications engineering, information processing, signal processing.

Furthermore, electricity is of basic importance for the fields:

- atomic physics, and
- solid-state physics.

13
Charges and currents

Electric charges are bound to matter. Charged bodies may interact over large distances due to their electric field. The interaction of two point charges is described by Coulomb's law.

Electric currents occur when electric charges move. Currents may also interact over large distances via their magnetic fields. The interaction of two thin current-carrying wires is described by Ampere's law.

13.1   Electric charge

Electric charge, Q, property of bodies of exerting forces on each other by electric fields. Charge is bound to matter.

Coulomb, C, SI unit of electric charge Q. 1 coulomb is the charge transported by a steady electric current of 1 ampere for 1 second,

$$[Q] = 1 \text{ C} = 1 \text{ As}.$$

1. Negative and positive charges

There are two kinds of electric charges:

Negative charges, the sinks of the electric field.

■   **Electrons,**
   anions, negative ions, i.e., atoms that received additional electrons, negatively charged **elementary particles**.

Positive charges, sources of the electric field.

■   **Cations**, positive ions, i.e., atoms that delivered electrons.
   Holes in semiconductors, missing electrons in lattices of solids. Holes should not be confused with positrons.
   Positively charged **elementary particles**, like **protons** (H^+-ions) and **positrons** (the antiparticles of electrons).

▲   Like charges repel each other, unlike charges attract each other.

421

2. Elementary charge and charge conservation

The electric charge is quantized. Charge occurs as a multiple of the elementary charge.
   Elementary charge, the smallest quantity of free electric charge in nature:

elementary charge			TI
$e_0 = 1.602\,177\,33 \cdot 10^{-19}$ C	Symbol	Unit	Quantity
	e_0	C	elementary charge

▲   **Conservation of charge**, in a closed system the total charge is conserved; the sum of
   the positive and negative electric charges remains constant,

$$\sum_i Q_i = \text{const.}$$

■   A proton carries the charge e_0, an electron the charge $-e_0$. A uranium nucleus carries
   the charge $92\,e_0$. The unit of charge 1 C corresponds to about $6.24 \cdot 10^{18}$ elementary
   charges.

3. Conductors and insulators

Electric conductor, material in which freely movable charge carriers are present. Conductors have a low electric resistance (see p. 431).
   Electric nonconductor, insulator, material containing no freely movable charge carriers. Nonconductors possess a very large resistance against an electric current (see p. 431).
➤   In nonconductors, charges may be displaced distances in the atomic range by an
   electric field.

4. Electrostatic induction and polarization

Electrostatic induction, the displacement of electric charges within a body when it is put
into an electric field.
   Polarization, the formation of electric dipoles within a nonconductor due to the displacement of charges in the molecules or atoms of the insulator.
   Charge separation, arises within a conductor by electrostatic induction, hence an excess of positive charge or negative charge emerges in some regions. Altogether, the conductor itself remains electrically neutral.
➤   Owing to polarization, charges may also exert forces on insulators.

5. Measurement of charges

M   Charge may be measured by the force it causes, by the difference of the potential, or
   by the pulse of current generated by the flow of charge.
       Measurement of the voltage V between conductors for known capacitance C of
   the arrangement of conductors according to

$$Q = CU .$$

Measurement of the deflection of a **ballistic galvanometer** caused by the pulse of
current during the flow of charge through the galvanometer.

$$Q = \int_0^T I(t)\mathrm{d}t \quad I(t) \text{ current at time } t.$$

The duration of the current pulse should be less than 1 % of the galvanometer period.

M   **Millikan's oil-drop experiment**, measures the elementary charge. Charged oil droplets are positioned between the horizontal plates of a capacitor. The capacitor voltage is then varied until the gravitational force acting on the droplets is just compensated by the action of the electric force field in the capacitor. The charge of the droplet may then be determined from the capacitor voltage. This charge has always been found to be an integral multiple of a definite minimum amount of charge: the elementary charge.

By similar, more expensive, methods physicists have tried (unsuccessfully so far) to demonstrate the existence of fractions of the elementary charge in free matter.

13.1.1   Coulomb's law

1. Force between point charges

Coulomb's law, describes the force acting between two point charges:

▲   The force $\vec{F}_{12}$ between two point charges Q_1 and Q_2 is proportional to the product of the charges and decreases with the square of the distance r_{12} between the charges. It is a central force, i.e., the force is acting along the connecting line of the charges (**Fig. 13.1**).

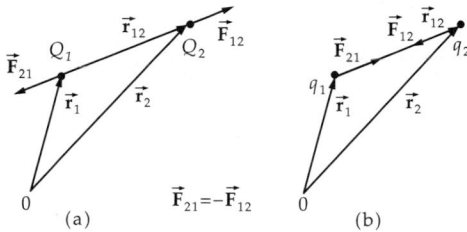

Figure 13.1: Coulomb's law. (a): charges Q_1 and Q_2 of the same sign, (b): charges q_1 and q_2 of opposite sign.

The force exerted by the charge Q_1 on the charge Q_2 is given by:

Coulomb's law			$\mathbf{MLT^{-2}}$
$\vec{F}_{12} = \dfrac{1}{4\pi\varepsilon_0}\dfrac{Q_1 Q_2}{r_{12}^2}\cdot\dfrac{\vec{r}_{12}}{r_{12}}$	Symbol	Unit	Quantity
	$\vec{F}_{12}, \vec{F}_{21}$	N	force between the charges
$\vec{F}_{12} = -\vec{F}_{21}$	Q_1, Q_2	C	charge 1, 2
	$\vec{r}_{12}$	m	distance vector of the charges
$\vec{r}_{12} = \vec{r}_2 - \vec{r}_1$	r_{12}	m	distance of the charges
	ε_0	C/(Vm)	permittivity of free space

The proportionality factor contains the **electric permittivity of free space**:

permittivity of free space			$\mathbf{L^{-3}T^4M^{-1}I^2}$
$\varepsilon_0 = 8.854\,187\,82\cdot 10^{-12}\ \dfrac{C}{Vm}$	Symbol	Unit	Quantity
	ε_0	C/(Vm)	permittivity of free space

2. Examples of Coulomb's law

■ A charge $Q = 10^{-5}$ C is repelled by another charge $Q = 5 \cdot 10^{-5}$ C at a distance $r = 1$ m with a force

$$F = \frac{1}{4\pi\varepsilon_0} \frac{10^{-5} \text{ C} \cdot 5 \cdot 10^{-5} \text{ C}}{1 \text{ m}^2} = 4.49 \text{ N} \,.$$

■ In the classical picture of the hydrogen atom, the force exerted by the proton on the electron is given by

$$\vec{F} = -\frac{1}{4\pi\varepsilon_0} \frac{e_0^2}{r^2} \cdot \vec{e}, \quad F = 8.24 \cdot 10^{-8} \text{ N} \,.$$

Here e_0 is the elementary charge (Q(proton) $= e_0$, Q(electron) $= -e_0$), and $r = 0.529 \cdot 10^{-10}$ m is the Bohr radius of the classical circular orbit that corresponds to the ground state of the hydrogen atom. The unit vector $\vec{e}$ points from the proton to the electron. The negative sign of the force vector indicates that the Coulomb force is attractive.

13.2 Electric charge density

Electric charge density, for describing spatial charge distributions.

The quantity Q states only that a certain amount of charge is localized in a restricted space region. The charge density, on the contrary, specifies the amount of charge within a small volume at any point in space, and hence yields more information than the integral. The charge density is a scalar function of the position.

1. Electric space charge density

Electric space charge density ρ, represents the ratio of the electric charge ΔQ within the space region ΔV at position $\vec{r}$ to the size of this space region (**Fig. 13.2 (a)**). If the charge density is position-dependent, the volume ΔV is to be reduced until the charge within this volume can be considered uniformly distributed. This corresponds to forming the limit:

electric charge density = $\dfrac{\text{charge}}{\text{volume element}}$			$\mathbf{L^{-3}TI}$
	Symbol	Unit	Quantity
$\rho(\vec{r}) = \lim\limits_{\Delta V \to 0} \dfrac{\Delta Q}{\Delta V} = \dfrac{dQ}{dV}$	ρ	C/m^3	space charge density
	dQ	C	charge within volume dV
	$\vec{r}$	m	position vector
	dV	m^3	volume element at position $\vec{r}$

Coulomb/meter^3, SI unit of the electric space charge density ρ,

$$[\rho] = \text{C/m}^3 \,.$$

Charge density for a uniform distribution of the charge Q over the volume V:

$$\rho = \frac{Q}{V} \,.$$

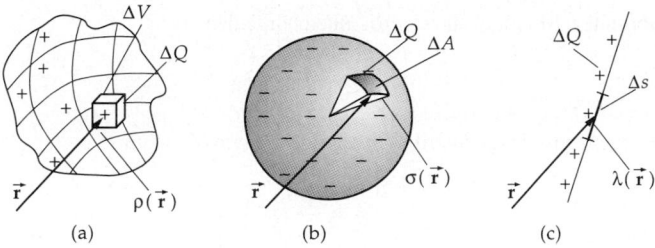

Figure 13.2: Electric charge density. (a): space charge density ρ, (b) surface charge density σ, (c): line charge density λ.

2. Electric surface charge density

Electric surface charge density σ, describes the charge distribution over a surface (**Fig. 13.2 (b)**).

Ratio of the electric charge ΔQ on the area ΔA at position $\vec{r}$ to the size of the area to be diminished until the charge ΔQ can be regarded as uniformly distributed. This corresponds to forming the limit:

electric surface charge density $=$ charge / area element			$L^{-2}TI$

	Symbol	Unit	Quantity
$\sigma(\vec{r}) = \lim\limits_{\Delta A \to 0} \dfrac{\Delta Q}{\Delta A} = \dfrac{dQ}{dA}$	σ	C/m^2	surface charge density
	dQ	C	charge on the area dA
	$\vec{r}$	m	position vector
	dA	m^2	area element at position $\vec{r}$

Coulomb/meter^2, SI unit of the electric surface charge density,
$[\sigma] = C/m^2$.
Surface charge density for a homogeneous charge distribution on the area A:

$$\sigma = \frac{Q}{A}.$$

3. Electric line charge density

Electric line charge density λ, describes the charge distribution along a wire-like conductor (**Fig. 13.2 (c)**). Ratio of the electric charge ΔQ on a line element Δs at position $\vec{r}$ to the length of the line element. Here, the line element Δs is to be diminished until the charge ΔQ can be considered uniformly distributed. This corresponds to forming the limit:

electric line charge density $=$ charge / line element			$L^{-1}TI$

	Symbol	Unit	Quantity
$\lambda(\vec{r}) = \lim\limits_{\Delta s \to 0} \dfrac{\Delta Q}{\Delta s} = \dfrac{dQ}{ds}$	λ	C/m	line charge density
	dQ	C	charge along the element ds
	$\vec{r}$	m	position vector
	ds	m	line element at position $\vec{r}$

Coulomb/meter, SI unit of the electric line charge density,

$$[\lambda] = C/m.$$

For a homogeneous charge distribution along the wire of length s,

$$\lambda = \frac{Q}{s}.$$

4. Mean charge density

Mean charge density, defined by:

mean space charge density $\bar{\rho} = \dfrac{Q}{V} = \dfrac{1}{V} \displaystyle\int_V \rho(\vec{r})\,dV,$

mean surface charge density $\bar{\sigma} = \dfrac{Q}{A} = \dfrac{1}{A} \displaystyle\int_A \sigma(\vec{r})\,dA,$

mean line charge density $\bar{\lambda} = \dfrac{Q}{s} = \dfrac{1}{s} \displaystyle\int_s \lambda(\vec{r})\,ds.$

Here V is the volume, not the potential difference.

13.3 Electric current

1. Electric current,

characterizes the motion of electrically charged particles within conductors. An electric current may cause heating of matter, electrochemical processes, and magnetization.
- ◼ A resistor in a circuit is heated by the current flowing through it.
- ◼ In a chemical solution, materials precipitate at the electrodes due to the exchange of charges.
- ◼ A coil carrying a current is surrounded by a magnetic field. A piece of iron placed into the coil is magnetized.

2. Current,

I, the quantity of charge ΔQ flowing through a cross-sectional area A per time interval Δt (**Fig. 13.3**). If the current varies during the time interval Δt, the interval is reduced until the current can be considered as constant.

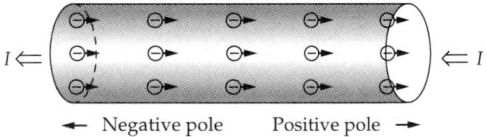

Figure 13.3: Current as a motion of carriers of charge, and the definition of the direction of current.

Current intensity at time t of a time-dependent electric current in a conductor: the amount of charge dQ flowing in an infinitesimally small time interval dt through a cross-sectional area of a conductor:

electric current = $\dfrac{\text{charge}}{\text{unit time}}$				**I**
$I = \lim\limits_{\Delta t \to 0} \dfrac{\Delta Q}{\Delta t} = \dfrac{dQ}{dt}$	Symbol	Unit	Quantity	
	I	A	current at time t	
	dQ	C = As	charge transported	
	dt	s	time interval	

For steady charge transport,

$$Q = I \cdot t.$$

3. SI unit of electric current,

Ampere, A, SI basis unit, SI unit of the electric current I. A current of 1 A in a conductor means that an amount of charge $\Delta Q = 1$ C flows through a cross-sectional area of the conductor during a time interval $\Delta t = 1$ s.

$$[I] = A = C/s.$$

▲ **Definition of the unit of current, ampere**: The current I has the value 1 A when two rectilinear, infinitely long conductors of negligibly small wire cross section arranged parallel to each other at a distance of $r = 1$ m and carrying equal, time-independent current I exert a force $F = 2 \cdot 10^{-7}$ N per 1 m of conductor length on each other (**Fig. 13.4**).

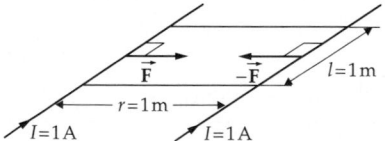

Figure 13.4: Definition of the current unit ampere.

▲ **Definition of the direction of current**, corresponds to the direction of motion of positive charges. In a metallic conductor, the direction of current is opposite to the direction of motion of the negative charge carriers, i.e., the electrons (**Fig. 13.3**).
In a circuit, the electrons move from the negative pole of the voltage source to the positive pole. Thus, the direction of current points from the positive pole (+) of the voltage source to its negative one (−).

4. Direct current,

direction and intensity I of the current are constant in time. The amount of charge ΔQ passing through a cross-sectional area during the time interval Δt is proportional to Δt:

$$I = \frac{\Delta Q}{\Delta t} = \text{const.}$$

5. Alternating current,

direction and intensity I of the current change periodically with time.
The effects of electric currents are listed in **Tab. 18.3/7** for direct current and alternating current, respectively.

■ If in a conductor a charge $\Delta Q = 3$ C passes a given cross-section during the time interval $\Delta t = 60$ s, the current is

$$I = \frac{\Delta Q}{\Delta t} = \frac{3 \text{ C}}{60 \text{ s}} = 50 \text{ mA} .$$

6. Measurement of current

The electric current is measured by its effects:

Current balance (mechanical action of force): Conductors carrying currents exert forces on each other by the magnetic field. This force may be compared with a weight by means of a balance.

Hot-wire ammeter (action of heat): A wire carrying a current is heated and thereby expands. The expansion can be measured.

Electrolysis (chemical action): The quantity of material precipitated by electrolysis per unit time is proportional to the current. The method used to serve for definition of the unit of current, the ampere.

Rotating-coil instrument: A current-carrying coil is deflected in a magnetic field. The deflection increases with the increase of current in the coil.

13.3.1   Ampere's law

Ampere's law, current-carrying conductors generate magnetic fields by which they exert forces on each other:

▲ The mutual force of two current-carrying conductors is proportional to the product of the currents I_1 and I_2 in the conductors and to the length l of the conductor, and inversely proportional to the distance r between the conductors (**Fig. 13.4**).

Ampere's law			$\mathbf{LT^{-2}M}$
	Symbol	Unit	Quantity
$F = \dfrac{\mu_0}{2\pi} \dfrac{I_1 I_2 l}{r}$	F	N	force
	I_1, I_2	A	current 1, 2
$\mu_0 = 4\pi \cdot 10^{-7}$ Vs/(Am)	r	m	distance
	l	m	length of conductor
	μ_0	Vs/(Am)	free-space permeability

➤ Ampere's law is used for the current definition of the unit of current.

13.4   Electric current density

1. Definition of electric current density

Electric current density, $\vec{J}$, describes the current distribution in extended conductors. The electric current density is a vector quantity the direction of which coincides with the direction of motion of positive charge carriers. The magnitude is calculated from the current ΔI passing through a cross-sectional area $\Delta A_\perp$ perpendicular to the direction of motion of the charge carriers, divided by this area (**Fig. 13.5**). If the current depends on the position, then the current density J is defined by the differential.

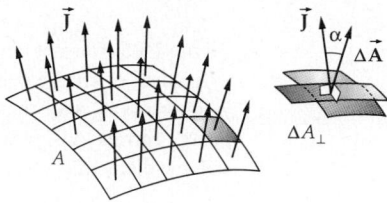

Figure 13.5: Definition of
the current density $\vec{J}$.

current density = $\dfrac{\text{current}}{\text{area element}}$			$L^{-2}I$

	Symbol	Unit	Quantity
$J = \lim\limits_{\Delta A_\perp \to 0} \dfrac{\Delta I}{\Delta A_\perp} = \dfrac{dI}{dA_\perp}$	J	A/m^2	current density
	$\Delta A_\perp$	m^2	area element
	ΔI	A	current through $\Delta A_\perp$
	$dA_\perp$	m^2	infinitesimal area element
	dI	A	current through $dA_\perp$

Ampere/meter^2, A/m^2, SI unit of the current density J. A current density of 1 A/m^2 corresponds to an electric current of intensity $I = 1$ A passing through a perpendicular surface $A_\perp = 1$ m^2,

$$[J] = A/m^2.$$

2. Properties of the current density

Whereas the electric current is a measure of the quantity of charge transported through a given cross-sectional area, the electric current density gives the direction of the charge transport and the magnitude of the transported charge at any point in space.

If the current I through an area $A_\perp$ is the same at any point of the area, the current density is

$$J = \frac{I}{A_\perp}.$$

■   A current of $I = 2$ A flowing in a metallic wire of cross-sectional area $A = 2.5$ mm^2 corresponds to a current density of

$$J = \frac{I}{A} = \frac{2\,\text{A}}{2.5\,\text{mm}^2} = \frac{2\,\text{A}}{2.5 \cdot 10^{-6}\,\text{m}^2} = 8 \cdot 10^5\,\text{A/m}^2.$$

The current density vector $\vec{J}$ points along the wire opposite to the direction of motion of the electrons, i.e., along the direction of positive current.

3. Product representation of the current density

The current density is the product of the space charge density ρ and the local mean velocity $\vec{v}$ of the charge carriers (**Fig. 13.6**),

$$\vec{J} = \rho \cdot \vec{v}.$$

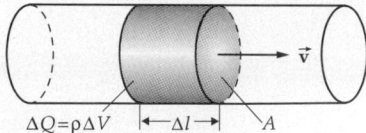

Figure 13.6: Current density as product of charge density ρ and velocity $\vec{v}$ of the charge carrier. $\Delta V = \Delta l \cdot A$: volume element.

4. Current as integral over current density and area

The current is given as the product of the component of the current density $J \cos \alpha$ perpendicular to the area ΔA and the area ΔA,

$$I = J \cos \alpha \cdot \Delta A \,.$$

If the current is not constant across the area ΔA, then one has to use the differential form

$$dI = J \cdot dA \cdot \cos \alpha = \vec{J} \cdot d\vec{A} \,.$$

The current through an arbitrary surface A follows by integration:

current = integral of current density over surface			I
	Symbol	Unit	Quantity
$I = \displaystyle\int_A \vec{J} \cdot d\vec{A}$	I	A	current through total area A
	$\vec{J}$	A/m^2	current density
	$d\vec{A}$	m^2	infinitesimal area element
	A	m^2	total area

The vector $d\vec{A}$ points along the surface normal, and its magnitude is equal to the area element dA.

5. Kirchhoff's first law

▲   The sum of all currents passing through a closed surface vanishes because of conservation of the electric charge:

$$\oint_A \vec{J} \cdot d\vec{A} = 0 \,, \qquad \text{Kirchhoff's first law.}$$

13.4.1 Electric current flow field

1. Electric current flow field and stream lines

Electric flow field, specifies the electric current density at any space point.

If the electric flow field does not change with time, it is denoted as a steady-state electric flow field. The current density is then constant in time, but may vary with the position. In a steady-state electric flow field, the quantity of charge flowing per unit time through a surface is constant.

Stream lines, serve for visualization of the electric current density.

The following conventions hold for stream lines:

- Stream lines correspond to the paths of motion of the positive charge carriers.
- The tangent to a stream line at a given point coincides with the orientation of the current density vector at this point.

2. Properties of stream lines

- The density of stream lines is a measure of the current.
- Stream lines may not intersect each other, since the direction of motion of the charge carriers is uniquely given at any space point.
- ■ The stream lines in a long straight wire run parallel to the wire axis.
- ■ The stream lines of a point-like source of current in an extended conducting medium run in radial direction outward. The current density decreases as the square of the distance from the source.
- ■ The stream lines of a metallic cylinder in an extended conducting medium are perpendicular to the axis of the conductor and point radially outward.
- ■ The current lines within a circularly bent conductor are concentric circles in the plane of conductor, parallel to the middle axis of the conducting loop.

13.5   Electric resistance and conductance

13.5.1   Electric resistance

1. Definition of electric resistance

Electric resistance of a conductor, determines the amount of current flow through the conductor for a given voltage at the ends of the conductor. The resistance R is the ratio of voltage V to current I:

resistance = $\dfrac{\text{voltage}}{\text{current}}$			$\mathbf{L^2 T^{-3} M I^{-2}}$
$R = \dfrac{V}{I}$	Symbol	Unit	Quantity
	R	$\Omega = \text{V/A}$	electric resistance
	V	V	voltage
	I	A	current

Ohm, Ω, SI unit of the electric resistance R. $1\,\Omega$ is the resistance of a conductor when, for a voltage $V = 1$ V at its ends, a current $I = 1$ A flows through the conductor,

$$[R] = \text{V/A}.$$

2. Ohm's law

In an ohmic conductor, the voltage V is proportional to the current I. The proportionality factor is the **ohmic resistance** R:

voltage = resistance · current (Ohm's law)			$\mathbf{L^2 T^{-3} M I^{-1}}$
$V = R \cdot I$	Symbol	Unit	Quantity
	V	V	voltage
	R	$\Omega = \text{V/A}$	electric resistance
	I	A	current

3. Current-voltage characteristic,

graphical representation of the relation between current and voltage.

Linear resistance, ohmic resistance, resistance with linear current-voltage characteristic (**Fig. 13.7 (a)**).

Nonlinear resistance, the relation between the current through the conductor and the voltage drop is nonlinear (**Fig. 13.7 (b)**).

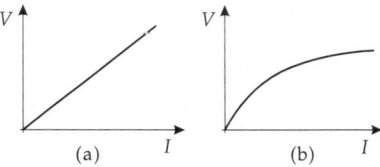

Figure 13.7: Current-voltage characteristics. (a): linear resistance, (b): nonlinear resistance.

Metallic conductors at constant temperature exhibit a linear current-voltage characteristic. A current flowing through a metallic conductor heats the conductor. For higher currents, the relation between current and voltage becomes nonlinear.

■ The current-voltage characteristic of a diode is nonlinear.

13.5.2 Electric conductance

Conductance, σ, reciprocal value of the electric resistance, quotient of current I and voltage V:

conductance $= \dfrac{1}{\text{resistance}} = \dfrac{\text{current}}{\text{voltage}}$		$L^{-2}T^3M^{-1}I^2$
	Symbol	Unit	Quantity

$\sigma = \dfrac{1}{R} = \dfrac{I}{V}$	Symbol	Unit	Quantity
	σ	S = A/V	conductance
	R	Ω	electric resistance
	I	A	current
	V	V	voltage

Siemens, S, SI unit of the conductance σ. If the electric resistance of a conductor is $R = 1\,\Omega$, the electric conductance is $\sigma = 1$ S,

$$[\sigma] = S = 1/\Omega = A/V.$$

13.5.3 Resistivity and conductivity

Resistivity, ρ (specific resistance), material-dependent quantity, independent of the geometry of the conductor.

Conductivity, κ (specific conductance), reciprocal value of the specific resistance.

1. Resistance of a wire,

R, proportional to the length of wire l and inversely proportional to the wire cross-section A. The proportionality constant is the specific resistance ρ (**Fig. 13.8**).

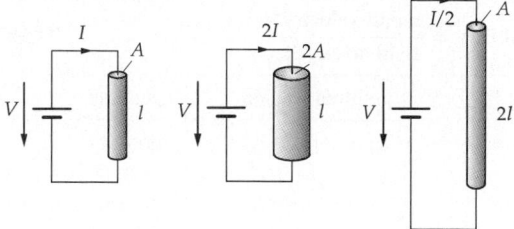

Figure 13.8: Resistance of a wire depending on the cross-section A and length l.

resistance = specific resistance · $\dfrac{\text{length}}{\text{cross-section}}$			$\mathbf{L^2T^{-3}MI^{-2}}$
	Symbol	Unit	Quantity
$R = \rho \cdot \dfrac{l}{A}$	R	Ω	resistance
	ρ	Ωm	specific resistance
$= \dfrac{1}{\kappa} \cdot \dfrac{l}{A}$	κ	S/m	conductivity
	l	m	wire length
	A	m^2	wire cross-section

2. SI units of the specific resistance and of the conductivity

Ohm·meter, Ωm, SI unit of the specific resistance ρ,

$$[\rho] = \Omega\text{m}.$$

➤ **Resistivity and space charge density are denoted by the same symbol ρ.**
➤ **The specific resistance is not related to the mass, but is rather a characteristic material quantity, contrary to the terminology adopted in thermodynamics.**
Siemens/meter, S/m, SI unit of the conductivity κ,

$$[\kappa] = \text{S/m}.$$

■ Resistivity of gold $2.04 \cdot 10^{-2}\ \Omega \cdot \text{mm}^2/\text{m}$, of platinum-rhodium (20 %) alloy $20 \cdot 10^{-2}\ \Omega \cdot \text{mm}^2/\text{m}$, graphite $800 \cdot 10^{-2}\ \Omega \cdot \text{mm}^2/\text{m}$.
The resistivity of metals is compiled in **Tab. 18.1/1**, of several alloys in **Tab. 18.1/4** and of several resistance alloys in **Tab. 18.3/1**. The resistivity of insulating substances is listed in **Tab. 18.2/5 and 18.2/6**.
■ A copper wire of length $l = 2$ m and cross-section $A = 1$ mm^2 has the resistivity $\rho = 0.0178\ \Omega\text{mm}^2/\text{m}$. The resistance of this wire is

$$R = \rho \cdot \frac{l}{A} = 0.0178\ \Omega\text{mm}^2/\text{m} \cdot \frac{2\ \text{m}}{1\ \text{mm}^2} = 0.0356\ \Omega\,.$$

13.5.4 Mobility of charge carriers

1. Mobility of charge carriers,

b, specifies the mean drift velocity $\bar{v}$ of charge carriers in an electric field of field strength E:

mobility = $\dfrac{\text{mean velocity}}{\text{field strength}}$			$\mathbf{T^2M^{-1}I}$
	Symbol	Unit	Quantity
$b = \dfrac{\bar{v}}{E}$	b	m²/(Vs)	mobility
	$\bar{v}$	m/s	mean drift velocity
$= \dfrac{\bar{v} \cdot l}{V}$	E	V/m	electric field strength
	l	m	distance
	V	V	voltage drop

Meter²/volt·second, m²/(Vs), SI unit of mobility b,

$$[b] = \text{m}^2/(\text{Vs}).$$

For a linear resistor, the mean drift velocity is proportional to the electric field strength.

The conductivity κ is the product of the density ρ and the mobility b of the charge carriers,

$$\kappa = \rho \cdot b.$$

■   Conductivity of refined gold: 45.7 Ω^{-1}m · mm^{-2}.
The conductivity of several contact materials is listed in **Tab. 18.3/3**.

2. Example of mobility of electrons

Let the voltage at the ends of a metallic wire of length 1 m be $V = 5$ V, and mean drift velocity of electrons in the wire be $\bar{v} = 50\ \mu\text{m/s} = 5 \cdot 10^{-5}$ m/s. The mobility of the electrons is then

$$b = \frac{\bar{v} \cdot l}{V} = \frac{5 \cdot 10^{-5}\ \text{m/s} \cdot 1\ \text{m}}{5\ \text{V}} = 10^{-5}\ \text{m}^2/\text{Vs}.$$

The charge density of electrons in the metal is $\rho = 1.36 \cdot 10^{10}$ C/m³. The conductivity of the metallic wire is given by

$$\kappa = \rho \cdot b = 1.36 \cdot 10^{10}\ \text{C/m}^3 \cdot 10^{-5}\ \text{m}^2/\text{Vs} = 1.36 \cdot 10^5\ \text{S/m}.$$

The resistivity of the wire is

$$\rho = \frac{1}{\kappa} = 7.35 \cdot 10^{-6}\ \Omega\text{m}.$$

13.5.5   Temperature dependence of the resistance

The resistivity ρ, and hence the electric resistance R, of a conductor are temperature-dependent. In many cases, one may assume the resistance to vary linearly with temperature. Then it is sufficient to give the resistance for a certain temperature (mostly room temperature $T_0 = 293.15$ K) and a temperature coefficient.

1. Temperature coefficient,

proportionality constant specifying the relative change of resistance $\Delta R / R$ for a change of temperature by $\Delta T = 1$ K:

resistance as function of temperature			$\mathbf{L^2T^{-3}MI^{-2}}$
	Symbol	Unit	Quantity
$R(T) = R_0(1 + \alpha \Delta T)$	R, R_0	Ω	resistance at temperature T, T_0
$\rho(T) = \rho_0(1 + \alpha \Delta T)$	ρ, ρ_0	Ωm	resistivity at temperature T, T_0
	ΔT	K	change of temperature
	α	1/K	temperature coefficient

1/kelvin, SI unit of temperature coefficient,

$$[\alpha] = 1/K.$$

2. Properties of the temperature coefficient

For many conductors, the temperature coefficient is in the range 10^{-3} 1/K, e.g., for gold $\alpha = 4 \cdot 10^{-3}$ 1/K.

The temperature coefficient is given for various conductors in **Tab. 18.1/1**, for alloys in **Tab. 18.1/4** and for resistance alloys in **Tab. 18.3/1**.

If the resistance varies nonlinearly with temperature, one adopts a power-series expansion

$$R = R_0 \cdot (1 + \sum_i \alpha_i (\Delta T)^i)$$

and introduces a corresponding number of coefficients α_i, $i = 1, \ldots, n$ in order to describe the variation of the resistance with temperature.

Cold conductor, **PTC** (positive temperature coefficient), the resistance strongly increases with increasing temperature, the temperature coefficient is positive. Metallic wires are PTC. They are used as thermostats, temperature sensors and current stabilizers.

Thermistor, **NTC** (negative temperature coefficient), the resistance decreases with increasing temperature, the temperature coefficient is negative. Semiconducting oxide ceramics are NTC. They are used as temperature sensors and voltage stabilizers (**Fig. 13.9**).

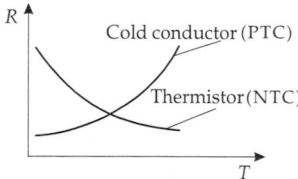

Figure 13.9: Characteristics of a thermistor ($\alpha < 0$) and a cold conductor ($\alpha > 0$). T: temperature.

➤ The electric resistance of metals may also depend on the pressure. One then introduces a pressure coefficient $(1/\rho)\mathrm{d}\rho/\mathrm{d}p$, analogously to the temperature coefficient. The pressure coefficient is given for several metals in **Tab. 18.1/2**.

13.5.6 Variable resistors

Variable-resistor units, change their resistance depending on external inductions.

Besides temperature-dependent and pressure-dependent resistors, the following resistor components are available:

- **Adjustable resistor, potentiometer**, changes its resistance by manual action. Linearly adjustable resistors are used as voltage dividers; logarithmically adjustable resistors control volume.
- **Photoresistor, LDR** (light-dependent resistor), its resistance value depends on the intensity of the incident light, used in exposure meters.
- **Voltage-dependent resistor, VDR, varistor**, the resistance value depends on the voltage applied, used for voltage stabilization.

13.5.7 Connection of resistors

1. Series connection of n resistors

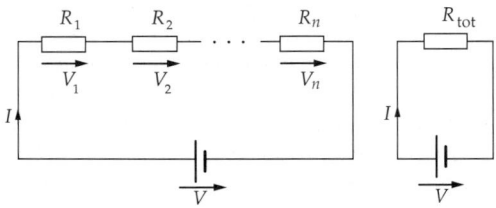

Figure 13.10: Series connection of resistors.

The current I is the same in all resistors. The total drop of voltage V is the sum of the partial voltages $V_i = R_i \cdot I$ at the resistors R_i and may be expressed by a total resistance R_{tot} (**Fig. 13.10**):

$$V = V_1 + V_2 + V_3 + \cdots + V_n \,,$$

$$V = R_{\text{tot}} \cdot I \,,$$

$$R_{\text{tot}} = R_1 + R_2 + R_3 + \cdots + R_n \,.$$

The total conductance G_{tot} is

$$\frac{1}{G_{\text{tot}}} = \frac{1}{G_1} + \frac{1}{G_2} + \frac{1}{G_3} + \cdots + \frac{1}{G_n} \,.$$

2. Parallel connection of n resistors

The voltage V is the same in all branches (**Fig. 13.11**). The partial currents $I_i = V/R_i$ in the branches sum to the total current I,

$$I = I_1 + I_2 + I_3 + \cdots + I_n \,,$$

$$\frac{1}{R_{\text{tot}}} = \frac{1}{R_1} + \frac{1}{R_2} + \frac{1}{R_3} + \cdots + \frac{1}{R_n} \,.$$

The total resistance R_{tot} is smaller than any of the single resistance R_i. The total conductance G_{tot} is the sum of the individual conductance values G_i,

$$G_{\text{tot}} = G_1 + G_2 + G_3 + \cdots + G_n \,.$$

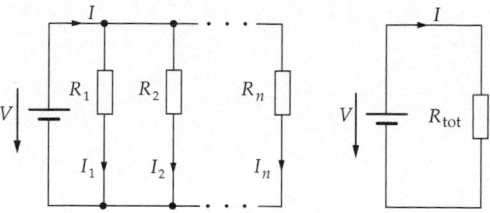

Figure 13.11: Parallel connection of resistors.

3. Potentiometer circuit,

to divide the total voltage V into smaller partial voltages (**Fig. 13.12**). The circuit is loaded by an external resistor R_a. The tapped partial voltage V_a is

$$V_a = V \frac{R_2 R_a}{R_1 R_2 + R_a(R_1 + R_2)} \, .$$

In the case without external load ($R_a \gg R_1 R_2/(R_1 + R_2)$, i.e., the current through the external resistor R_a may be ignored), the formula simplifies to

$$V_a = V \frac{R_2}{R_1 + R_2} \, .$$

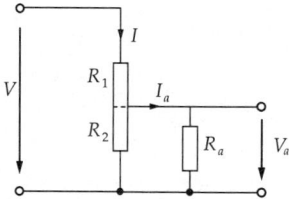

Figure 13.12: Potentiometer circuit.

14
Electric and magnetic fields

Electric fields are caused by electric charges, time-varying magnetic fields, or both.

Magnetic fields are created by permanent magnets or by currents, i.e., by moving electric charges.

A moving electric charge is surrounded by an electric field and a magnetic field. In its rest frame, an electric charge generates an electric field, but no magnetic field.

Electric and magnetic fields are vector fields.

Vector field, $\vec{V}(\vec{r})$, a function assigning a vector to any point in space with the coordinates $\vec{r} = (x, y, z)$:

$$\vec{V} = \vec{V}(\vec{r}).$$

Scalar field, $f(\vec{r})$, a function assigning a scalar to any point in space with the coordinates $\vec{r} = (x, y, z)$:

$$f = f(\vec{r}).$$

■ The electric field is a vector field. The electric field strength $\vec{E}(\vec{r})$ exists at any point in space $\vec{r}$.

The electric potential is a scalar field. A scalar, the potential $\varphi(\vec{r})$, is assigned to any point in space $\vec{r}$.

➤ The argument $\vec{r}$ is often omitted although a space dependence usually occurs, as in the case of the electric field of a point charge.

14.1 Electric field

Electric field, property of the space in the vicinity of electric charges. The electric field is a vector field. An electric field strength may be assigned to any point in space; it is proportional to the local force acting on electric charges.

1. Electric field strength,

$\vec{E}$, a vector with magnitude E that specifies the strength of the electric field and the direction of acceleration of a positive test charge. The electric field strength is determined by the force $\vec{F}$ experienced by a test charge Q in an electric field, divided by the test charge:

electric field = $\dfrac{\text{force}}{\text{test charge}}$			$\mathbf{LT^{-3}MI^{-1}}$
	Symbol	Unit	Quantity
$\vec{F} = Q\,\vec{E} \qquad \vec{E} = \dfrac{\vec{F}}{Q}$	$\vec{E}$	V/m	electric field strength
	$\vec{F}$	N	force on test charge
	Q	C	test charge

Volt/meter, V/m, SI unit of electric field strength $\vec{E}$. The field strength at a point in space $\vec{r}$ is $E = 1$ V/m if a charge $Q = 1$ C at the position $\vec{r}$ experiences a force $F = 1$ N,

$$[\vec{E}] = \text{V/m} = \text{N/C}\,.$$

➤   Since both positive and negative charges exist, the shielding of electric fields is possible. In contrast, the gravitational field cannot be shielded.

2. Test charge,

a charge placed into an electric field to measure its magnitude and direction. The charge should be so small that it does not significantly disturb the original field to be measured. In theoretical considerations, one may let the test charge be infinitesimally small although there is a physical lower limit (the elementary charge).

■   If a test charge of magnitude $Q = -10^{-9}$ C experiences a force $F = 10^{-5}$ N, the electric field strength at the position of the test charge is $E = 10^4$ V/m. The electric field strength points opposite to the direction of the force.

3. Uniform electric field,

the field strength is constant, both in magnitude and direction, at any point in the region of space considered. A test charge Q experiences the same force $\vec{F}$ at any point in the region of space:

$$\vec{E} = \frac{\vec{F}}{Q}\,.$$

■   The electric field in a parallel-plate capacitor is uniform if the separation of the plates is small compared with their extension except near the edges of the plates.

14.2   Electrostatic induction

1. Electric conductor,

a material containing freely movable charges.

■   Metals are conductors; the movable charges are the conduction electrons.
    Salt solutions are conductors; the movable charges are the positive and negative ions.
    A plasma is a conductor; the movable charges are the electrons and the positive ions.
  Like charges repel each other. Therefore, uncompensated charges move in a conductor until they have reached the largest possible mutual separation.

▲   The electric charge of a charged conductor is located on its surface. The interior of a metallic conductor is field-free. Otherwise, the free charge carriers would be displaced by the field forces acting upon them.

2. Electrostatic induction,

the displacement of movable charges in a conductor when it is placed into an electric field.

■   When a metal is placed between the plates of a charged capacitor, the conduction electrons move towards the positively charged capacitor plate. An electric field is then built up between the remaining (positively charged) atoms and the displaced (negatively charged) electrons. This field points opposite to the original capacitor field. The motion of the electrons ends when these two electric fields just compensate each other.

➤   For nonconductors, the charge separation in the atoms or molecules (formation of electric dipoles) is referred to as **polarization**.

14.2.1 Electric field lines

1. Field lines,

serve for visualization of the action of force in the spatial electric field.
The following conventions are normally employed:
- The direction of the field lines at a point corresponds to the direction of the electric field strength, i.e., to the direction of force acting on a positive charge at this point.
- Field lines emerge from a positive point charge (source), and point towards a negative point charge (sink).

Hence:
- In electrostatics, there are no closed field lines. The electrostatic field is **irrotational**.
- Field lines may not intersect each other: at any point, the direction of the electric field strength is unique.
- The greater the density of field lines, the greater the field strength.
- ▲ The field lines emerging from a charged metallic conductor are perpendicular to its surface.

If there were a component of the electric field tangential to the surface of the conductor, the charge carriers would move until a balance of forces were reached, i.e., the tangential component of the field strength vanishes.

2. Faraday cage,

when a charge-free region in an electric field is enclosed by a metallic cover, no electric field arises within the cover (screening).

■   During a thunderstorm, a car (acting as a Faraday cage) protects the passengers from lightning if they are within the car and are not touching the outer metallic skin.

3. Field lines of various charge distributions

a) Point charge, charge of infinitesimally small spatial extension. Electric field lines of a positive point charge by definition point radially outward (**Fig. 14.1 (a)**), electric field lines of a negative point charge point radially inward (**Fig. 14.1 (b)**). The electric field about a point charge is isotropic.
Fig. 14.1 (c) and (d) show the field lines of a system of two point charges.

b) Point charge near a conducting plate: **Fig. 14.2** shows the field lines of a point charge placed in front of a conducting plate.

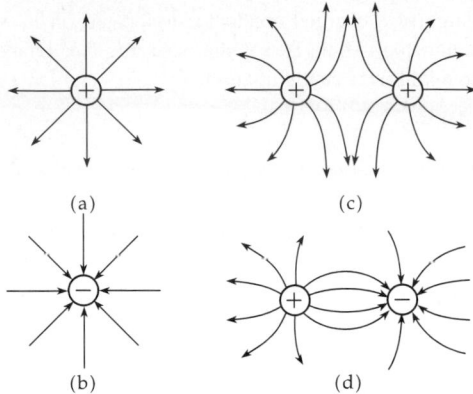

(a)   (c)

(b)   (d)

Figure 14.1: Electric field lines. (a): positive point charge, (b): negative point charge, (c): two charges of equal magnitude and sign, (d): two charges of equal magnitude but opposite sign.

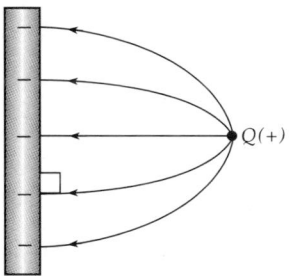

$Q(+)$

Figure 14.2: Electric field lines of a point charge near a conducting plate.

c) *Parallel-plate capacitor,* two oppositely charged conducting parallel plates at fixed distance. The electric field lines between the plates are parallel and perpendicular to the surfaces of the plates, except for the edge region (**Fig. 14.3 (a)**). The electric field within the parallel-plate capacitor is uniform.

Fig. 14.3 (b) shows the field lines of a spherical capacitor.

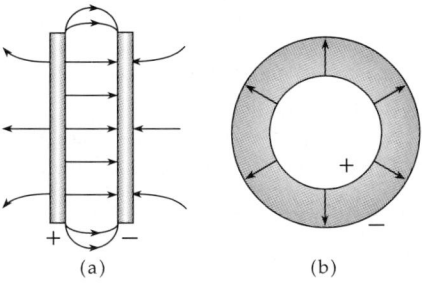

+   −

+

−

(a)   (b)

Figure 14.3: Electric field lines. (a): parallel-plate capacitor, (b): spherical capacitor.

4. Electric dipole,

two point charges $+Q$ and $-Q$ at a distance d. The positive charge is located at the position $\vec{r}_+$, the negative charge at the position $\vec{r}_-$.

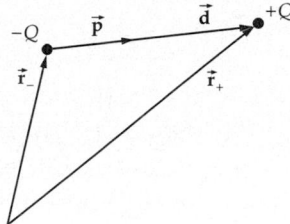

Figure 14.4: Electric dipole.
$\vec{\mathbf{p}}$: dipole moment.

Electric dipole moment, $\vec{\mathbf{p}}$, product of charge Q and distance vector $\vec{\mathbf{d}}$ of the charges,

$$\vec{\mathbf{p}} = Q(\vec{\mathbf{r}}_+ - \vec{\mathbf{r}}_-) = Q\vec{\mathbf{d}}.$$

The two point charges are denoted as poles. The connecting line between the poles is the dipole axis. The dipole moment $\vec{\mathbf{p}}$ is a vector along the dipole axis, which by definition points from the negative to the positive charge (**Fig. 14.4**).

5. Dipole in an electric field

▲   From the outside, a dipole appears as electrically neutral.
The potential energy E_{pot} of a dipole in an electric field $\vec{\mathbf{E}}$ is

$$E_{\text{pot}} = -\vec{\mathbf{p}} \cdot \vec{\mathbf{E}}.$$

In a uniform electric field $\vec{\mathbf{E}}$, a torque $\vec{\tau}$ acts on the dipole (**Fig. 14.5 (a)**),

$$\vec{\tau} = \vec{\mathbf{p}} \times \vec{\mathbf{E}} = Q \cdot (\vec{\mathbf{d}} \times \vec{\mathbf{E}}).$$

The torque turns the dipole into the direction of the electric field.
In an inhomogeneous field $\vec{\mathbf{E}}$, a dipole experiences a force $\vec{\mathbf{F}}$ pulling it into the region of higher field strength (**Fig. 14.5 (b)**),

$$\vec{\mathbf{F}} = \left(\vec{\mathbf{p}} \cdot \frac{\partial}{\partial \vec{\mathbf{r}}} \right) \vec{\mathbf{E}}.$$

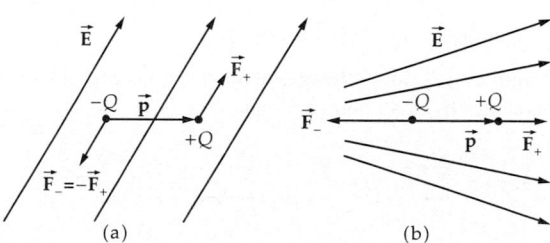

Figure 14.5: Dipole in an electric field $\vec{\mathbf{E}}$. (a): couple and torque in the uniform electric field, (b): force acting on an electric dipole in an inhomogeneous electric field ($F_- > F_+$).

■   A water molecule H_2O has a permanent electric dipole moment of $6.17 \cdot 10^{-30}$ C · m.

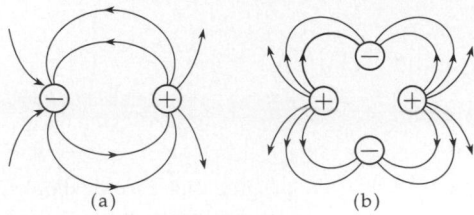

Figure 14.6: (a): electric field of a dipole, (b): electric field of a quadrupole.

(a)   (b)

6. Electric field at large distance from the dipole

The electric field of a dipole at a large distance from the charges is indicated in **Fig. 14.6 (a)**.

A charge distribution that exhibits no dipole moment may have a non-vanishing **quadrupole moment**. The electric field of a quadrupole is indicated in **Fig. 14.6 (b)**.

14.2.2   Electric field strength of point charges

1. Electric field strength of a point charge,

$\vec{E}$, a vector quantity. The magnitude gives the electric field strength of a point charge Q at a distance r from this point charge. The field intensity points radially outward from a positive charge, and inwards for a negative charge. It decreases in proportion to the inverse square of the distance:

field strength $\sim \dfrac{\textbf{charge}}{\textbf{square of distance}}$			$\mathbf{LT^{-3}MI^{-1}}$
	Symbol	Unit	Quantity
$\vec{E} = \dfrac{Q}{4\pi\varepsilon_0 r^2}\dfrac{\vec{r}}{r}$	$\vec{E}$	N/C = V/m	electric field strength of charge Q
	Q	C	charge generating the field
	$\vec{r}$	m	distance vector
	ε_0	C/(Vm)	permittivity of free space

■   A charge $Q = 10^{-6}$ C at the distance $r = 1$ m generates an electric field strength

$$E = \frac{Q}{4\pi\varepsilon_0 r^2} = \frac{10^{-6}\text{ C}}{4\pi\varepsilon_0 \cdot (1\text{ m})^2} = 8988\text{ V/m}.$$

The electric field vector points radially away from the point charge.

2. Electric field strength of many point charges

Electric field strength $\vec{E}$ of N point charges at the positions $\vec{r}_i$ is obtained by superposing the electric field strengths $\vec{E}_i$ of all point charges:

$$\vec{E}(\vec{r}) = \sum_{i=1}^{N} \vec{E}_i(\vec{r}_i) = \frac{1}{4\pi\varepsilon_0}\sum_{i=1}^{N}\frac{Q_i}{|\vec{r}-\vec{r}_i|^2}\frac{\vec{r}-\vec{r}_i}{|\vec{r}-\vec{r}_i|}.$$

3. Electric field strength of charge distributions

Electric field strength $\vec{E}$ of a spatial charge distribution $\rho(\vec{r}')$ is obtained by integrating

$$\vec{E}(\vec{r}) = \frac{1}{4\pi\varepsilon_0}\int_{V'}\frac{\rho(\vec{r}')}{|\vec{r}-\vec{r}'|^2}\frac{\vec{r}-\vec{r}'}{|\vec{r}-\vec{r}'|}dV'.$$

14.3 Force

The force on an electric test charge Q in an electric field is given by the product of the charge Q and the electric field strength $\vec{E}$. The force is a vector quantity. It points along the electric field vector for a positive charge Q, and opposite to the field for a negative charge Q:

force = test charge · electric field strength			$LT^{-2}M$
	Symbol	Unit	Quantity
$\vec{F} = Q \cdot \vec{E}$	$\vec{F}$	N	force on electric charge
	Q	C	electric charge
	$\vec{E}$	N/C = V/m	electric field strength

■ A negative charge $Q = -10^{-6}$ C in an electric field of strength $E = 200$ V/m experiences a force

$$F = 10^{-6} \text{ C} \cdot 200 \text{ V/m} = 2 \cdot 10^{-4} \text{ N}.$$

The force $\vec{F}$ points opposite in direction to the electric field strength $\vec{E}$.

14.4 Electric voltage

1. Definition of voltage

Electric voltage, V, between two points A and B, the work done by the force $\vec{F} = Q\vec{E}$ when displacing a test charge Q along a path s from point A to point B, divided by the test charge Q (**Fig. 14.7**).

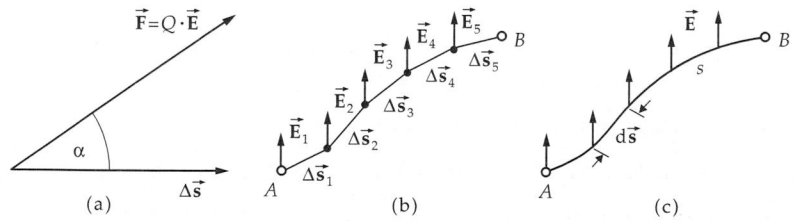

Figure 14.7: Displacement of a charge in an electric field. (a): displacement along a path element Δs, (b): displacement along a broken line from A to B, (c): displacement along a smooth path s from A to B.

If the force $\vec{F}$ along a path element $\Delta\vec{s}$ is constant, the voltage, i.e., the work ΔW per test charge Q, is

$$V = \frac{\Delta W}{Q} = \frac{F \Delta s \cos \alpha}{Q} = \vec{E} \cdot \Delta\vec{s}.$$

α is the angle between the direction of force and the direction of the path element (**Fig. 14.7**).

2. Representation of the voltage as an integral

For an arbitrary path s from point A to point B, the curve is subdivided into straight path elements $\Delta \vec{s}_i$. Then the electric voltage V_{AB} between the points A and B is obtained as the sum over the contributions ΔV_i from the individual path elements:

$$V_{AB} = \sum_i \Delta V_i = \sum_i \vec{E} \cdot \Delta \vec{s}_i .$$

As the subdivision gets more refined, the sum becomes an integral:

voltage = $\dfrac{\text{work}}{\text{test charge}}$			$\mathbf{L^2 T^{-3} M I^{-1}}$
$V_{AB} = \dfrac{W_{AB}}{Q}$ $= \displaystyle\int_A^B \vec{E} \cdot d\vec{s}$	Symbol	Unit	Quantity
	V_{AB}	$V = Nm/C$	voltage between A and B
	W_{AB}	$J = Nm$	work done
	Q	C	test charge
	$\vec{E}$	V/m	electric field strength
	$d\vec{s}$	m	path element

Volt, V, SI unit of the electric voltage V. The voltage is 1 V if work $W = 1\,J$ is done in displacing a charge of $Q = 1\,C$,

$$[V] = V = J/C .$$

▲   The integral of the electric field strength $\vec{E}$ along a closed path s equals zero,

$$\oint_s \vec{E} \cdot d\vec{s} = 0 .$$

This statement corresponds to energy conservation. The mesh rule, the second of Kirchhoff's laws, results from this principle.

3. Electric voltage between capacitor plates,

product of electric field strength E and distance d between the capacitor plates:

voltage = field strength · distance between plates			$\mathbf{L^2 T^{-3} M I^{-1}}$
$V = Ed$	Symbol	Unit	Quantity
	V	$V = Nm/C$	electric voltage
	E	$N/C = V/m$	electric field strength
	d	m	distance of plates

The electric force between the plates is constant. The electric field $\vec{E}$ is uniform. The electric field strength is

$$|\vec{E}| = \frac{V}{d} .$$

14.5   Electric potential

1. Definition and properties of the potential

Electric potential, φ_A, of a point A in the electric field, voltage between the point A and a fixed reference point P. The electric potential φ_A specifies the work W'_A to be done by the force $\vec{F}' = -Q\vec{E}$ in order to shift the charge Q from point P to the point A.

Usually the reference point P, where the potential is set to zero, is taken to be infinity, $\varphi_P = \varphi_\infty = 0$.

▲   The potential then depends only on the point A. Hence, the potential is a scalar function of the position.

The work W'_A defines the potential energy $E_{pot}(A)$ of a charge Q at the point A of the electric field $\vec{E}$,

$$E_{pot}(A) = Q \cdot \varphi_A , \quad E_{pot}(\infty) = 0 .$$

potential = $\dfrac{\textbf{work}}{\textbf{test charge}}$			$\mathbf{L^2 T^{-3} M I^{-1}}$
$\varphi_A = \dfrac{W'_A}{Q} = \dfrac{E_{pot}(A)}{Q}$ $= -\displaystyle\int_\infty^A \vec{E} \cdot d\vec{s}$	Symbol	Unit	Quantity
	φ_A	V = Nm/C	potential at point A
	W'_A	J = Nm	work on displacing Q
	Q	C	test charge
	$\vec{E}$	N/C = V/m	electric field strength
	$d\vec{s}$	m	infinitesimal path element
	E_{pot}	J	potential energy

2. Potential and field strength

Potential difference $\varphi_A - \varphi_B$, voltage between two points A and B:

$$\varphi_A - \varphi_B = -\int_\infty^A \vec{E} \cdot d\vec{s} - \left(-\int_\infty^B \vec{E} \cdot d\vec{s} \right)$$

$$= \int_A^B \vec{E} \cdot d\vec{s} = V_{AB} .$$

The component of the electric field strength $\vec{E}$ in x-, y-, z-direction is obtained by taking the derivative of the potential with respect to the corresponding direction:

$$E_x = -\frac{d\varphi}{dx} , \quad E_y = -\frac{d\varphi}{dy} , \quad E_z = -\frac{d\varphi}{dz} .$$

In three dimensions, one obtains the field strength $\vec{E}$ from the electric potential φ by means of the gradient:

$$\vec{E} = -\text{grad}\,\varphi = -\left(\frac{\partial \varphi}{\partial x}\vec{e}_x + \frac{\partial \varphi}{\partial y}\vec{e}_y + \frac{\partial \varphi}{\partial z}\vec{e}_z \right) .$$

$\vec{e}_x$, $\vec{e}_y$, $\vec{e}_z$ are unit vectors in x-, y-, z-direction.

➤ The electric field strength is independent of the choice of the reference point.

3. Potential equation

The potential of a charge distribution $\rho(\vec{r})$ is given by

$$\varphi(\vec{r}) = \frac{1}{4\pi\varepsilon_0} \int \frac{\rho(\vec{r}')}{|\vec{r} - \vec{r}'|} \, dV' \, .$$

Poisson equation, potential equation, differential equation for calculating the electric potential φ from the charge density $\rho(\vec{r})$,

$$\Delta\varphi = \left(\frac{\partial^2}{\partial x^2} + \frac{\partial^2}{\partial y^2} + \frac{\partial^2}{\partial z^2} \right) \varphi = -\frac{\rho}{\varepsilon_0} \, .$$

14.5.1 Equipotential surfaces

Equipotential surfaces, surfaces of equal electric potential. Equipotential surfaces cannot intersect or touch each other. The electric field strength is always perpendicular to the equipotential surfaces (**Fig. 14.8**). Equipotential surfaces correspond to the contour lines of maps; the direction of the steepest ascent is the normal to the contour lines.

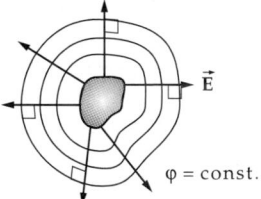

$\varphi = \text{const.}$

Figure 14.8: Equipotential surfaces $\varphi = \text{const.}$ and electric field strength $\vec{E}$ of a charge distribution.

The surfaces of conductors are equipotential surfaces. Otherwise, there would exist a component of the electric field strength along the surface that would cause a displacement of the charge along the conductor surface.

14.5.2 Field strength and potential of various charge distributions

1. Point charge

The potential φ of a point charge in 3D space is inversely proportional to the distance r from the charge. The electric field strength decreases as r^{-2}:

field strength and potential of a point charge			
	Symbol	Unit	Quantity
$\vec{E} = \dfrac{Q}{4\pi\varepsilon_0\, r^2} \dfrac{\vec{r}}{|\vec{r}|}$	$\vec{E}$	V/m	field strength at position $\vec{r}$
	φ	V	potential at position $\vec{r}$
$\varphi = \dfrac{Q}{4\pi\varepsilon_0\, r}$	Q	C	point charge
	$\vec{r}$	m	position vector
	ε_0	C/(Vm)	permittivity of free space

2. Dipole

At a large distance from the dipole ($|\vec{r}| \gg |\vec{d}|$), the potential of a dipole decreases as r^{-2} towards zero. For smaller distances, there are admixtures of potential fields of higher multipolarity that decay faster with increasing distance from the dipole. Hence, at large distance only the dipole field remains:

field strength and potential of a dipole

$$\vec{E} = \frac{1}{4\pi\varepsilon_0}\left(\frac{3(\vec{p}\cdot\vec{r})\vec{r}}{r^5} - \frac{\vec{p}}{r^3}\right)$$

$$\varphi = \frac{1}{4\pi\varepsilon_0}\frac{\vec{p}\cdot\vec{r}}{r^3}$$

$$\vec{p} = Q\vec{d}$$

$$\vec{d} = \vec{r}_+ - \vec{r}_-$$

Symbol	Unit	Quantity
$\vec{E}$	V/m	field strength at position $\vec{r}$
φ	V	potential at position $\vec{r}$
Q	C	charge
$\vec{p}$	Cm	dipole moment
$\vec{r}_+$	m	position vector positive pole
$\vec{r}_-$	m	position vector negative pole
$\vec{d}$	m	distance vector
ε_0	C/(Vm)	permittivity of free space

3. Charged hollow sphere

The electric field within a uniformly charged hollow sphere of radius R vanishes. The electric potential is constant over this space region. The potential φ outside the sphere ($r > R$) is inversely proportional to the distance r from the center of the sphere. The electric field strength decreases with the distance r as r^{-2}:

field strength and potential outside of a hollow sphere

$$\vec{E} = \frac{Q}{4\pi\varepsilon_0}\frac{1}{r^2}\frac{\vec{r}}{|\vec{r}|}$$

$$\varphi = \frac{Q}{4\pi\varepsilon_0}\frac{1}{r}$$

Symbol	Unit	Quantity
$\vec{E}$	V/m	field strength at distance r
φ	V	potential at distance r
Q	C	charge of the hollow sphere
r	m	distance from center
ε_0	C/(Vm)	permittivity of free space

4. Uniformly charged sphere

The electric field E within the sphere increases linearly with the distance r from the center of the sphere. The potential φ within the sphere increases with r^2:

field strength and potential within a sphere

$$\vec{E} = \frac{Q}{4\pi\varepsilon_0}\frac{r}{R^3}\frac{\vec{r}}{|\vec{r}|}$$

$$\varphi = \frac{Q}{8\pi\varepsilon_0}\left(\frac{3}{R} - \frac{r^2}{R^3}\right)$$

Symbol	Unit	Quantity
$\vec{E}$	V/m	field strength at distance r
φ	V	potential at distance r
Q	C	charge of sphere
r	m	distance from center
R	m	radius of sphere
ε_0	C/(Vm)	permittivity of free space

The electric field strength E outside of the sphere decreases as r^{-2} with increasing distance r from the center of the sphere. The potential φ decreases as r^{-1}:

field strength and potential outside of a sphere			
	Symbol	Unit	Quantity
$\vec{E} = \dfrac{Q}{4\pi\varepsilon_0}\dfrac{1}{r^2}\dfrac{\vec{r}}{\lvert\vec{r}\rvert}$	$\vec{E}$	V/m	field strength at distance r
	φ	V	potential at distance r
$\varphi = \dfrac{Q}{4\pi\varepsilon_0}\dfrac{1}{r}$	Q	C	charge of sphere
	r	m	distance from center
	ε_0	C/(Vm)	permittivity of free space

5. Charged hollow cylinder

In the interior of a thin-walled hollow cylinder of radius R and constant surface charge density σ, the electric field strength vanishes. The potential is constant in the interior. The electric field strength E outside a long hollow cylinder decreases hyperbolically as r^{-1} with increasing distance r from the cylinder axis. The potential decreases logarithmically with the distance from the axis:

field strength and potential outside of a hollow cylinder			
	Symbol	Unit	Quantity
$\vec{E} = \dfrac{\sigma}{\varepsilon_0}\dfrac{R}{r}\,\vec{e}_\rho$	$\vec{E}$	V/m	field strength at position $\vec{r}$
	φ	V	potential at position $\vec{r}$
	σ	C/m^2	surface charge density of hollow cylinder
$\varphi = -\dfrac{\sigma}{\varepsilon_0}\,R\,\ln\left(\dfrac{r}{R}\right)$	R	m	radius of hollow cylinder
	$\vec{r}$	m	position vector
	ε_0	C/(Vm)	permittivity of free space

$\vec{e}_\rho$ is a unit vector along the cylinder radius.

6. Uniformly charged cylinder

The electric field strength E in the interior of a cylinder with constant space charge density ρ increases linearly with the distance r from the cylinder axis. The potential φ increases with r^2:

field strength and potential within a uniformly charged cylinder			
	Symbol	Unit	Quantity
$\vec{E} = \dfrac{\rho}{2\pi\varepsilon_0}\,r\,\vec{e}_\rho$	$\vec{E}$	V/m	field strength at position $\vec{r}$
	φ	V	potential at position $\vec{r}$
	ρ	C/m^3	space charge density
$\varphi = -\dfrac{\rho}{4\pi\varepsilon_0}\,R^2\left[1+\left(\dfrac{r}{R}\right)^2\right]$	R	m	radius of cylinder
	$\vec{r}$	m	position vector
	ε_0	C/(Vm)	permittivity of free space

The electric field strength E outside of the cylinder decreases as $1/r$. The potential φ decreases logarithmically:

field strength and potential outside of a uniformly charged cylinder			
	Symbol	Unit	Quantity
$\vec{E} = \dfrac{\rho}{2\pi\varepsilon_0}\,\dfrac{R^2}{r}\,\vec{e}_\rho$	$\vec{E}$	V/m	field strength at position $\vec{r}$
	φ	V	potential at position $\vec{r}$
	ρ	C/m^3	space charge density
$\varphi = -\dfrac{Q}{2\pi\varepsilon_0}\ln\dfrac{r}{R}$	R	m	radius of cylinder
	$\vec{r}$	m	position vector
	ε_0	C/(Vm)	permittivity of free space

$\vec{e}_\rho$ is a unit vector along the cylinder radius.

7. Uniformly charged infinitely extended plate

For small distances from a plate (placed in the plane $x = 0$) the field is uniform: field strength and potential are proportional to the surface charge density $\sigma = Q/A$. The potential φ is proportional to the perpendicular distance x from the plate. The electric field strength is constant.

field strength and potential of an extended plate			
	Symbol	Unit	Quantity
$\vec{E} = \pm\dfrac{\sigma}{2\varepsilon_0}\,\vec{e}_x$	$\vec{E}$	V/m	field strength
	φ	V	potential at distance x
$\varphi = \mp\dfrac{\sigma}{2\varepsilon_0}\cdot x$	σ	C/m^2	surface charge density
	x	m	perpendicular distance from plate
	ε_0	C/(Vm)	permittivity of free space

$\vec{e}_x$ is a unit vector in the positive x-direction, normal to the plate. The upper (lower) sign holds for $x > 0$ ($x < 0$), respectively.

14.5.3 Electric flux

1. Definition of the electric flux

Let a square area ΔA be placed in a uniform electric field of field strength $\vec{E}$.

Electric flux or **displacement flux**, ψ, a measure of the total electric field penetrating the area ΔA. The displacement flux $\Delta\psi$ is the product of the electric field strength $\vec{E}$, the area ΔA, and the cosine of the angle α between the field direction and the surface normal,

$$\Delta\psi = \vec{E}\cdot\Delta\vec{A} = E\cdot\Delta A\cdot\cos\alpha\,.$$

For an arbitrary surface A in an inhomogeneous electric field $\vec{E}$, the area is subdivided into plane partial areas such that the field strength over any partial area may be considered as constant with respect to direction and magnitude. The resulting displacement fluxes are summed (**Fig. 14.9**), which corresponds to integration over the surface.

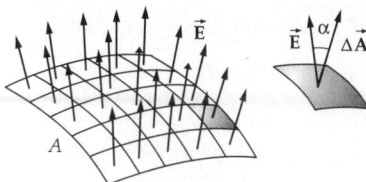

Figure 14.9: Electric flux ψ through a surface A and through an oriented surface element $\Delta\vec{A}$.

flux = integral of field strength over surface			$\mathbf{L^3T^{-3}MI^{-1}}$
	Symbol	Unit	Quantity
$\psi = \displaystyle\int_A \vec{E} \cdot d\vec{A}$	ψ	Vm	displacement flux
	$\vec{E}$	V/m	electric field strength
	$d\vec{A}$	m^2	oriented surface element
	A	m^2	total surface

Volt·meter, Vm, SI unit of the displacement flux ψ. 1 Vm is the electric flux of a uniform electric field of field strength $E = 1$ V/m that penetrates a plane surface $A = 1$ m^2 placed perpendicular to the field,
$$[\psi] = \text{Vm}.$$

2. Properties of the displacement flux

The displacement flux depends on the orientation of the surface A. When permuting the upper and lower sides of the surface, the displacement flux changes its sign.

■   A square surface of magnitude $A = 1$ dm^2 is placed into a uniform electric field of field strength $E = 100$ V/m, with the surface normal tilted by $\alpha = 30°$ relative to the field orientation. The electric flux through the surface is

$$\psi = E \cdot A \cdot \cos\alpha = 100 \text{ V/m} \cdot 0.01 \text{ m}^2 \cdot \cos 30° = 0.866 \text{ Vm}.$$

The displacement flux through a spherical surface A enclosing a point charge of magnitude Q equals the displacement flux of an arbitrary charge distribution of the same total charge Q,

$$\psi = \oint_A \vec{E} \cdot d\vec{A} = \frac{Q}{4\pi\varepsilon_0 r^2} \cdot 4\pi r^2 = \frac{Q}{\varepsilon_0}.$$

For an arbitrary closed surface in the electric field:

▲   The displacement flux through a closed surface is proportional to the charge enclosed. The proportionality factor is $1/\varepsilon_0$,

$$\psi = \oint_A \vec{E} \cdot d\vec{A} = \frac{Q}{\varepsilon_0}.$$

▲   If a surface A encloses a charge-free space region in an electric field, the electric fluxes through the partial surfaces may differ from each other. The total flux, however, vanishes, since no charges are present within the surface A.

■   The flux ψ through a spherical surface about a point charge $Q = 10^{-6}$ C is

$$\psi = \frac{Q}{\varepsilon_0} = \frac{10^{-6} \text{ C}}{8.854 \cdot 10^{-12} \text{ C/(Vm)}} = 1.13 \cdot 10^5 \text{ Vm}.$$

14.5.4 Electric displacement in a vacuum

1. Charge separation by electrostatic induction

Two rectangular conducting plates of equal size and area ΔA are placed near each other congruently in a uniform electric field such that the electric field intensity is perpendicular to the plates. If the plates are then separated and removed from the electric field, they are found to be charged, characterized by the surface charge density σ (**Fig. 14.10**). Owing to electrostatic induction, charges have been separated and shifted to a plate.

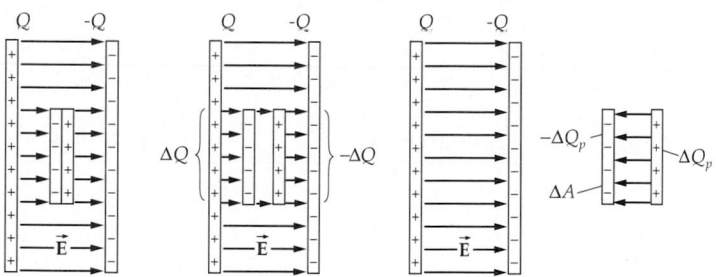

Figure 14.10: Charge separation by electrostatic induction.

2. Displacement,

$\vec{D}$, a vector quantity that measures the quantity of charge ΔQ per area element ΔA displaced by electrostatic induction. The magnitude of the displacement equals the surface charge density σ. In a vacuum, the orientation of the displacement vector coincides with the direction of the electric field strength.

If the displaced charge ΔQ is not constant over the area element ΔA, e.g., for curved surfaces or for insulators, one has to use the differential quotient:

displacement $= \dfrac{\textbf{quantity of charge}}{\textbf{area}}$			$\mathbf{L^{-2}TI}$
	Symbol	Unit	Quantity
$D = \lim\limits_{\Delta A \to 0} \dfrac{\Delta Q}{\Delta A} = \dfrac{\mathrm{d}Q}{\mathrm{d}A} = \sigma$	D	C/m^2	magnitude of displacement
	$\mathrm{d}Q$	C	charge on the area element $\mathrm{d}A_\perp$
	$\mathrm{d}A$	m^2	infinitesimal area element
	σ	C/m^2	surface charge density

Coulomb/square meter, C/m^2, SI unit of the electric displacement $\vec{D}$. 1 C/m^2 is the displacement if a quantity of charge $Q = 1$ C is shifted through an area $A = 1$ m^2 placed perpendicularly to the electric field lines,

$$[\vec{D}] = C/m^2 \,.$$

3. Properties of displacement

The displacement depends on the orientation of the area relative to the electric field. The displaced quantity of charge is proportional to the cosine of the angle between surface normal and electric field vector.

➤ If the surface normal is perpendicular to the electric field strength, the displacement vanishes.

▲ The integral of the displacement over a closed surface A equals the charge Q enclosed by this surface,

$$\oint_A \vec{D} \cdot d\vec{A} = Q .$$

4. Proportionality between displacement and field strength

▲ The displacement is proportional to the electric field strength of an external electric field.

Permittivity of free space ε_0, proportionality factor between the displacement and the field strength in a vacuum. At any position in a uniform or non-uniform field:

displacement ∼ field strength			$\mathsf{L^{-2}TI}$
	Symbol	Unit	Quantity
$\vec{D} = \varepsilon_0 \cdot \vec{E}$	$\vec{D}$	C/m^2	displacement
	ε_0	C/(Vm)	permittivity of free space
	$\vec{E}$	V/m	electric field strength

In matter, the relation between displacement and electric field is more complicated. Then a material-dependent quantity arises that may vary with the frequency, the temperature and other physical quantities. In particular, the dependence of this material parameter on the field strength of the external electric field may lead to nonlinear effects. Moreover, the orientations of the displacement and the field strength may differ (see p. 454).

■ A uniform electric field $E = 400$ V/m causes a displacement

$$D = \varepsilon_0 \cdot E = 8.854 \cdot 10^{-12} \text{ C/(Vm)} \cdot 400 \text{ V/m} = 3.54 \cdot 10^{-9} \text{ C/m}^2 .$$

Two metallic plates of area $A = 1$ cm^2 are placed near each other and in a field perpendicular to the orientation of the field strength ($\alpha = 0°$). The magnitude of the charge of the plates is then

$$Q = D \cdot A \cdot \cos\alpha = 3.54 \cdot 10^{-9} \text{ C/m}^2 \cdot 10^{-4} \text{ m}^2 \cos 0° = 3.54 \cdot 10^{-13} \text{ C} .$$

14.6   Electric polarization

1. Polarization of a dielectric

If a nonconductor is placed between the plates of a capacitor, then, for a fixed voltage, the amount of charge on the capacitor plates, and thus the capacitance of the capacitor, may change. This phenomenon is related to the change in the electric field between the plates. The material inserted becomes polarized. Because of this polarization, an electric field $\vec{E}_{pol}$ opposite to the original electric field $\vec{E}$ is built up. Therefore, the electric field strength in the capacitor is reduced.

The following types of polarization can be distinguished:

Displacement polarization, a displacement of electric charges in neutral atoms or molecules against each other. The electric field induces electric dipole moments (**Fig. 14.11 (a)**).

Orientation polarization, the permanent dipole moments already present in the material are aligned along the electric field (**Fig. 14.11 (b)**).

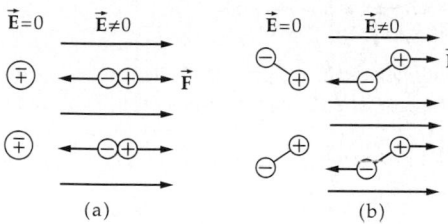

Figure 14.11: Polarization. (a): displacement polarization, (b): orientation polarization.

On the front faces of a partial volume $dV = d \cdot dA$ of the dielectric between the plates of a plate capacitor, the polarization charges $\pm dQ_{\mathrm{pol}}$ are generated. They cause an electric dipole moment

$$|\,d\vec{\mathbf{p}}\,| = d \cdot \sigma_p \, dA = \sigma_p \, dV \,.$$

σ_p is the surface charge density of the polarization charges.

2. Polarization vector

Polarization, $\vec{\mathbf{P}}$, electric dipole density per unit volume in the dielectric, characterizes the density of the polarization charges on the surface of the dielectric. The polarization $\vec{\mathbf{P}}$ is a vector along the dipole moment of the polarization charges. It points from the negative to the positive polarization charges. The magnitude of $\vec{\mathbf{P}}$ represents the surface density σ_p of the polarization charges,

$$\vec{\mathbf{P}} = \frac{d\vec{\mathbf{p}}}{dV} \,, \quad |\,\vec{\mathbf{P}}\,| = \sigma_p \,.$$

The polarization $\vec{\mathbf{P}}$ and the electric field $\vec{\mathbf{E}}$ are collinear. The field lines of the electric field $\vec{\mathbf{E}}_{\mathrm{pol}}$ generated by the polarization charges run from the positive to the negative surface charges of the dielectric, opposite to the field lines of the field $\vec{\mathbf{E}}$ (**Fig. 14.12**).

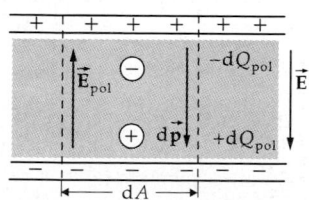

Figure 14.12: Polarization of a dielectric. $\pm dQ_{\mathrm{pol}}$: polarization charges, $d\vec{\mathbf{p}}$: electric dipole moment of the polarization charges, $\vec{\mathbf{E}}_{\mathrm{pol}}$: electric field of the polarization charges, $\vec{\mathbf{E}}$: original electric field.

The displacement polarization is given by

$$\vec{\mathbf{P}} = n\alpha \, \vec{\mathbf{E}} \,,$$

where n is the number per unit volume and α the **electric polarizability** of the atoms or molecules in the insulator. The polarizability is a molecular parameter.

14.6.1 Dielectric

Dielectric, insulator placed into an electric field.

1. Relative permittivity,

also **relative dielectric constant**, ε_r, a nondimensional material-dependent quantity that specifies the decrease of the electric field strength when a material (dielectric) is placed in an electric field,

$$[\varepsilon_r] = 1 .$$

The relative permittivity of vacuum is $\varepsilon_r = 1$. The relative permittivity of air is well approximated by unity.

For most dielectrics, ε_r varies in the range 1 to 100. There are dielectrics with ε_r up to 10 000.

■　The relative permittivity of water, cellulose and polystyrene are 81, 4.5 and 2.5, respectively.

2. Permittivity,

ε, product of permittivity of free space and relative permittivity,

$$\varepsilon = \varepsilon_0 \cdot \varepsilon_r .$$

Coulomb/volt·meter, C/(Vm), SI unit of permittivity ε,

$$[\varepsilon] = C/(Vm).$$

Electric polarization $\vec{P}$ in dielectric, given by

$$\vec{P} = (\varepsilon_r - 1)\, \varepsilon_0\, \vec{E} = \chi_e\, \varepsilon_0\, \vec{E} .$$

Electric susceptibility χ_e, defined by

$$\chi_e = \varepsilon_r - 1 .$$

3. Displacement in dielectric,

$\vec{D}$, given by the equation:

displacement = permittivity · field strength			$L^{-2}TI$
	Symbol	Unit	Quantity
$\vec{D} = \varepsilon_r\varepsilon_0\vec{E} = \varepsilon\vec{E}$	$\vec{D}$	C/m^2	displacement
	$\vec{E}$	V/m	electric field strength
$\vec{D} = \varepsilon_0\vec{E} + \vec{P}$	ε	C/(Vm)	permittivity
	ε_r	1	relative permittivity
$\vec{P} = (\varepsilon_r - 1)\, \varepsilon_0\, \vec{E} = \chi_e\varepsilon_0\vec{E}$	ε_0	C/(Vm)	permittivity of free space
	$\vec{P}$	C/m^2	electric polarization
	χ_e	1	electric susceptibility

■　The relative permittivity of pure water is $\varepsilon_r = 81$. If water is brought into a uniform electric field, the electric field intensity is reduced to $1/81$ of its original value due to the polarization charges generated in the water (**Fig. 14.13**).

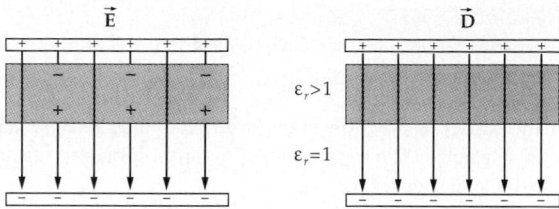

Figure 14.13: $\vec{D}$ and $\vec{E}$ field in a plate capacitor partly filled by a dielectric.

■   Relative permittivity (relative dielectric constant) of several substances: helium 1.0055, sulphur 3.5, capacitor paper 4 to 6, glycerol 43, ceramics 10 to 200.
The relative permittivity (relative dielectric constant) ε_r of various substances is listed in **Tab. 18.2/1 and 18.2/6.** For ceramics, ε_r is given in **Tab. 18.2/2,** for glasses in **Tab. 18.2/3** and for polymers in **Tab. 18.2/4.**

Electrostriction, change of shape and volume of a dielectric in an electric field. The phenomenon is observed in all aggregation states of matter. For solid insulators, the variation of length and volume (contractions) are in general proportional to the square of the electric field strength,

$$\left| \frac{\Delta V}{V} \right| \sim \varepsilon E^2 ,$$

$\Delta V / V$ relative change of volume, ε permittivity, E electric field strength.

14.7 Capacitance

Capacitance, C, of an arrangement of conductors, a scalar quantity specifying the quantity of electric charge that may be stored by this arrangement for given voltage V between the conductors.

Capacitor, arrangement of two conductors insulated against each other and charged to different values of potential.

capacitance = $\dfrac{\text{charge}}{\text{voltage}}$			$\mathbf{L^{-2}T^4M^{-1}I^2}$
$C = \dfrac{Q}{V}$	Symbol	Unit	Quantity
	C	F	capacitance of capacitor
	Q	C	charge of capacitor
	V	V	applied voltage

Farad, F, SI unit of capacitance C. A capacitor has the capacitance $C = 1$ F if for a voltage $V = 1$ V at the capacitor plates the charge $Q = 1$ C may be stored,

$$[C] = \text{F} = \text{C/V}.$$

➤   1 F is a very large unit. Typical capacitance values range from 1 pF to 1 mF. Capacitors with capacitances as large as 10 F are available for low voltages.

14.7.1   Parallel-plate capacitor

1. Properties of parallel-plate capacitors

The extension of the capacitor plates must be large compared to their separation so that edge effects may be neglected. The capacitance C is proportional to the area of the plates A and decreases with increasing separation,

$$C = \frac{\varepsilon_0 A}{d}, \varepsilon_r = 1 .$$

➤   The capacitance of a capacitor is increased by placing a dielectric between the capacitor plates. The capacitance of a dielectric of permittivity ε is

$$C = \frac{\varepsilon A}{d} = \frac{\varepsilon_0 \varepsilon_r A}{d} .$$

■   A capacitor with plates that are sheets of area $A = 10$ cm^2 with a separation of $d = 0.1$ mm has the capacitance

$$C = \frac{\varepsilon_0 A}{d} = \frac{8.854 \cdot 10^{-12} \text{ F/m} \cdot 10^{-3} \text{ m}^2}{10^{-4} \text{ m}} = 8.854 \cdot 10^{-11} \text{ F} \approx 90 \text{ pF} .$$

If there is capacitor paper of relative permittivity $\varepsilon_r = 4$ between the foils, the capacity is four times higher:

$$C \approx 360 \text{ pF} .$$

➤   Applying too high a voltage leads to breakdowns, and thus causes destruction of the capacitor.
➤   A charged capacitor discharges after some time, since the dielectric between the capacitor plates has a finite electric resistance.

2. Applications and special types of capacitors

Application of capacitors:
- separation of direct current and alternating current; smoothing of wavy direct current,
- in time-delay circuits as component of RC units,
- storing of charges,
- tuning of oscillating circuits in radio receivers.
  Special shapes of capacitors:
- **Electrolytic capacitor.** In applications, one has to observe the correct polarity of the voltage. High capacitance value. Applied for storing charge, e.g., in flash units or lasers.
- **Tunable capacitors, variable-disk capacitors** or **trimmer capacitors**. A set of plates is fixed (**stator**), the second set may be moved (**rotor**). Variable-disk capacitors are used for tuning of oscillator circuits.

14.7.2   Parallel connection of capacitors

Parallel connection of n capacitors, all capacitors are supplied by the same voltage, but the capacitor surfaces sum (**Fig. 14.14**). The total capacitance of a parallel connection of capacitors is thus the sum of the individual capacitances,

$$C_{\text{tot}} = C_1 + C_2 + \cdots + C_n .$$

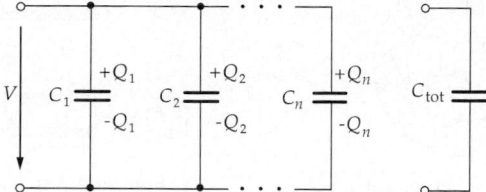

Figure 14.14: Parallel connection of n capacitors.

14.7.3 Series connection of capacitors

Series connection of n capacitors, the quantity of charge is the same on all capacitor plates (**Fig. 14.15**). Hence, the inverse value of the total capacitance equals the sum of the inverse values of individual capacitances,

$$\frac{1}{C_{\text{tot}}} = \frac{1}{C_1} + \frac{1}{C_2} + \cdots + \frac{1}{C_n}.$$

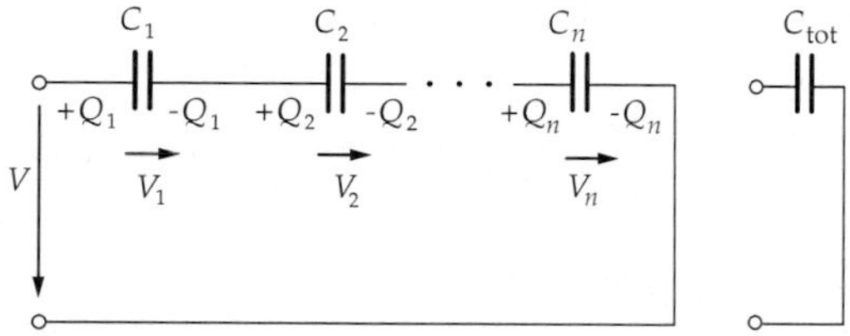

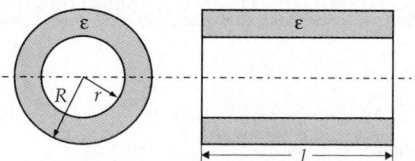

Figure 14.15: Series connection of n capacitors.

14.7.4 Capacitance of simple arrangements of conductors

1. Cylindrical capacitor

The capacitance is proportional to the length l of the cylindrical capacitor and inversely proportional to the logarithm of the ratio of the radii of the outer cylinder R and the inner cylinder r (**Fig. 14.16**):

$$C = 2\pi\varepsilon \, \frac{l}{\ln(R/r)}.$$

Figure 14.16: Cylindrical capacitor.

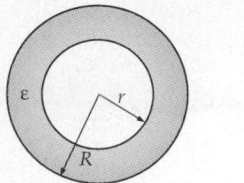

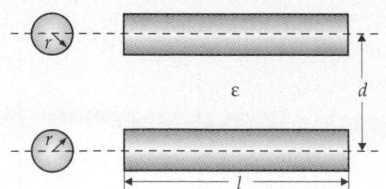

Figure 14.17: Spherical capacitor.   Figure 14.18: Two-wire lines.

2. Spherical capacitor, two concentric hollow spheres

The capacitance is proportional to the product of outer and inner radius R, r and inversely proportional to the difference between the outer and inner radius (**Fig. 14.17**):

$$C = 4\pi\varepsilon \, \frac{R \cdot r}{R - r} \, .$$

For $R \to \infty$ one gets the capacitance of a single sphere against an infinitely remote electrode, $C = 4\pi\varepsilon r$.

3. Two-wire lines

The capacitance is proportional to the conductor length l and inversely proportional to the logarithm of the ratio of the separation of the conductors d to the conductor radius r (**Fig. 14.18**):

$$C = \pi\varepsilon \, \frac{l}{\ln(d/r)} \qquad (d \gg r) \, .$$

4. Two spheres of equal radius

The capacitance of two spheres of equal radius r at a distance d between the centers is given by (**Fig. 14.19**)

$$C \approx 2\pi\varepsilon r \left[1 + \frac{r(d^2 + dr - r^2)}{d(d^2 - r^2)} \right] , d \gg r \, .$$

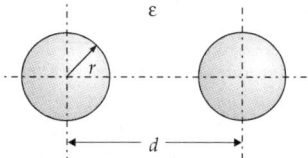

Figure 14.19: Capacitor consisting of two identical spheres.

14.8   Energy and energy density of the electric field

1. Energy density of the electric field,

w_e, the electric energy ΔW_e per volume ΔV. If the energy distribution is position-dependent, the energy density is given by

$$w_e = \lim_{\Delta V \to 0} \frac{\Delta W_e}{\Delta V} = \frac{\mathrm{d}W_e}{\mathrm{d}V} , \qquad \Delta W_e = w_e \cdot \Delta V \, .$$

energy density of the electric field			$L^{-1}T^{-2}M$
$w_e = \dfrac{1}{2}\vec{E}\cdot\vec{D}$	Symbol	Unit	Quantity
	w_e	J/m^3	energy density
	$\vec{D}$	C/m^2	displacement
	$\vec{E}$	V/m	field strength

2. Energy of the electric field,

W_e in the volume V, is obtained by integration of the energy density over the volume V,

$$ W_e = \int_V w_e \, dV = \frac{1}{2} \int_V \vec{E} \cdot \vec{D} \, dV . $$

The energy W_e of a charged parallel-plate capacitor is proportional to the square of the voltage between the capacitor plates:

energy of parallel-plate capacitor			$L^2T^{-2}M$
$W_e = \dfrac{1}{2}CV^2$	Symbol	Unit	Quantity
	W_e	J	energy
$= \dfrac{1}{2}\dfrac{Q^2}{C}$	Q	C	charge
	C	F	capacitance
	V	V	voltage

The energy W_e of a uniformly charged sphere is proportional to the square of the charge Q and inversely proportional to the sphere radius R:

energy of a uniformly charged sphere			$L^2T^{-2}M$
$W_e = \dfrac{1}{4\pi\varepsilon_0}\dfrac{3}{5}\dfrac{Q^2}{R}$	Symbol	Unit	Quantity
	W_e	J	energy
	Q	C	charge
	R	m	radius of sphere
	ε_0	C/(Vm)	permittivity of free space

14.9   Electric field at interfaces

When moving from a medium of permittivity ε_1 to a medium of permittivity ε_2, the electric field strength and the electric displacement change at the interface.

1. Change of the electric field strength

The tangential component of the electric field strength does not change in the transition (**Fig. 14.20**):

$$ E_{t1} = E_{t2}, \qquad \text{or} \qquad \frac{D_{t1}}{\varepsilon_1} = \frac{D_{t2}}{\varepsilon_2} . $$

The normal component of the electric field strength changes discontinuously.

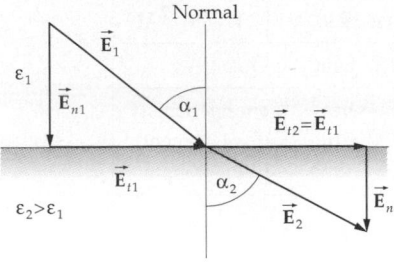

Figure 14.20: Electric field strength $\vec{E}$ at an interface between two media.

2. Change of the electric displacement

The normal component of the electric displacement does not change in the transition (**Fig. 14.21**):

$$D_{n1} = D_{n2}, \qquad \text{or} \qquad E_{n1} \cdot \varepsilon_1 = E_{n2} \cdot \varepsilon_2.$$

The tangential component of the electric displacement changes discontinuously.

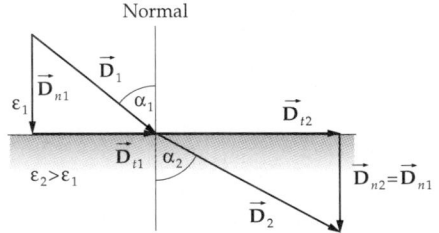

Figure 14.21: Electric displacement $\vec{D}$ at the interface between two media.

3. Angular relations of the electric fields at an interface

If α_1 denotes the angle between the perpendicular (normal to interface) and the direction of the electric field strength in the first medium, and α_2 the angle between the normal and the field strength in the second medium, the tangent values of the angles are related by the permittivity values:

electric field at interface			
	Symbol	Unit	Quantity
$\dfrac{\tan \alpha_1}{\tan \alpha_2} = \dfrac{\varepsilon_1}{\varepsilon_2} = \dfrac{\varepsilon_{r1}}{\varepsilon_{r2}}$	α_1, α_2	1	angle to normal in medium 1, 2
	$\varepsilon_1, \varepsilon_2$	C/(Vm)	permittivity of medium 1, 2
	$\varepsilon_{r1}, \varepsilon_{r2}$	1	relative permittivity of medium 1, 2

■ In a transition from a medium of low permittivity to a medium of higher permittivity, the electric field strength changes its direction away from the perpendicular.
■ In a transition from a medium of high permittivity to a medium of lower permittivity, the electric field strength changes its direction towards the perpendicular.

14.10 Magnetic field

Magnetostatics, treats magnetic fields constant in time, and magnetic phenomena caused by permanent magnets, or by steady currents.

The magnetic field of permanent magnets may be traced back to magnetic moments of the atomic constituents.

Current-carrying conductors are surrounded by a magnetic field that exerts forces on other current-carrying conductors. The magnetic field has a certain energy content.

Materials may be distinguished according to their behavior in a magnetic field.

Magnetic fields varying with time, occur when conductors carry time-dependent currents. The magnetic fields about a conductor induce a voltage in this conductor and other conductors. The conductors are characterized by their inductance. In order to generate magnetic fields, a certain expenditure of energy is necessary, which is stored in the fields as magnetic field energy.

Many applications exist, e.g., in alternating-current technology, in the construction of motors and generators, in three-phase-current technology, and in the construction of transformers.

■ A simple magnetic unit to be treated in this context is the coil.

14.11 Magnetism

1. Magnets

Permanent magnets, consist of magnetic iron or other magnetic materials. They exert forces on each other as well as on iron, nickel, cobalt and various alloys.

■ Materials for permanent magnets: AlNiCo alloys, sintered bodies such as Sr- and Ba-ferrites, CoPt- and FePt-alloys with ordered structure.

Electromagnets, consist of current-carrying coils with an iron core.

As in the case of electric dipoles, magnets have two poles denoted
magnetic north pole, and
magnetic south pole.

▲ Any subdivision of a permanent magnet yields two magnets, both having north and south poles.

2. Magnetic dipoles

There are no magnetic monopoles. Any elementary magnet is a **magnetic dipole**. The dipole axis is the connecting line of north and south poles. The magnetic dipole moment $\vec{m}$ is a vector along the dipole axis and pointing to the north pole. The magnetic moment of a body is determined by the torque on the body caused by an external magnetic field.

As in the case of electric dipoles, one finds:

▲ Like poles of two magnets repel each other, unlike poles attract each other.

Magnetic forces act over large distances even if the magnets are in a vacuum.

Magnetic field, range of the action of force of a magnet, or of a current-carrying conductor, on other magnets.

14.11.1 Magnetic field lines

1. Magnetic field lines,

serve for visualization of magnetic fields, as do the electric field lines of electric fields. Conventions:

• In the external region, the direction of magnetic field lines is defined as the direction from the north pole to the south pole of the magnet (**Fig. 14.22**).

• A test magnet would align along the tangent to the field lines of a field.

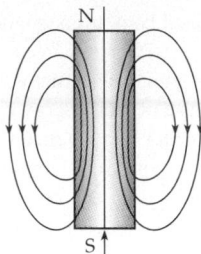

Figure 14.22: Magnetic field of a bar magnet.

The field lines have the following features:
- The field lines are always solenoidal. There are no magnetic charges (magnetic monopoles).
- The density of the magnetic field lines measures the magnetic flux density.
- Field lines about a current-carrying straight conductor are concentric circles. Their orientation is given by the right-hand rule.
- In a uniform magnetic field, the field lines are parallel lines.
- ➤ The magnetic field may be visualized by iron filings. The iron particles align to form chains, and thus map the magnetic field lines.

2. Geomagnetic field

The geomagnetic field displays quasi-periodic and partly aperiodic short-term fluctuations, from seconds up to days. These are caused by processes in the ionosphere and on the Sun. Moreover, one observes long-term polar motions. Since the magnetization of Earth's crust varies, the geomagnetic field may vary locally.

A compass needle aligns along the direction of the geomagnetic field tangential to Earth's surface. A magnet suspended in the geomagnetic field points with its north pole towards North, and with its south pole towards South. Since unlike poles are attracting each other, the geomagnetic south pole is close to the geographic north pole, whereas the geomagnetic north pole is close to the geographic south pole (**Fig. 14.23**).

Declination, deviation of the orientation of the geomagnetic field from the North-South axis. Magnetic field declinations in the United States vary from 20 degrees east to 20 degrees west.

Isogons, lines connecting points of equal declination on Earth's surface.

Inclination, angle between the horizontal and the orientation of the geomagnetic field.

Isoclines, lines connecting points of equal inclination on Earth's surface.

➤ A compass needle may also be used to determine the direction of the magnetic field about a current distribution.

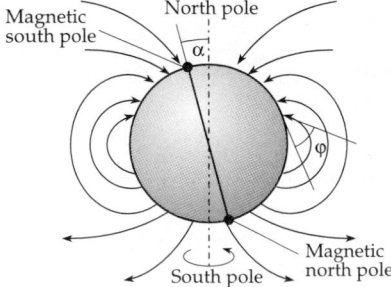

Figure 14.23: Geomagnetic field.

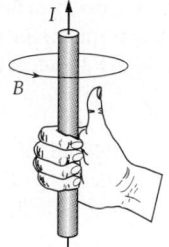

Figure 14.24: Right-hand rule: orientation of the magnetic field of a current. The right hand holds the wire, the thumb is aligned in the direction of the current, the remaining fingers point along the magnetic field lines.

3. Magnetic field of a current-carrying wire

The magnetic field lines are concentric circles about the current thread. The orientation of the magnetic field follows from the **cork-screw** or **right-hand rule**:

▲   If the thumb points along the direction of current, the remaining fingers point along the magnetic flux density vector, i.e., along the field lines (**Fig. 14.24**).

14.12   Magnetic flux density

1. Magnetic flux density,

magnetic induction, $\vec{B}$, a vector quantity. The magnitude B specifies the magnetic field intensity. The orientation of the magnetic flux density may be seen from the alignment of a test magnet: it points from the south pole of the test magnet to its north pole. A moving charge experiences a force proportional to the magnetic flux density.

Tesla, T, SI unit of the magnetic flux density $\vec{B}$,

$$[\vec{B}] = \mathrm{T} = \mathrm{Vs/m}^2 .$$

M   The measurement of the magnetic flux density can be reduced to the measurement of the magnetic flux, which may be determined by an induction coil (see for measurement of magnetic flux).

Hall effect (see p. 1003), a voltage V_H, the **Hall voltage**, is generated across a current-carrying conductor placed into a transverse magnetic field $\vec{B}$ oriented perpendicular to the conductor plane. V_H is proportional to the magnetic flux density B_z (**Fig. 14.25**),

$$V_H = \frac{I_x \cdot B_z}{n \cdot e_0 \cdot d} = \frac{b}{n \cdot e_0} J_x \cdot B_z .$$

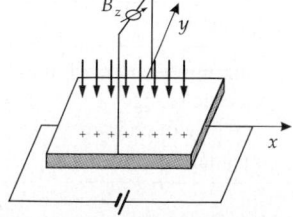

Figure 14.25: Hall effect.

$I_x = J_x \cdot b \cdot d$ is the current (in x-direction) passing the conductor of thickness d and width b. J_x is the current density, n the density of charge carriers in the conductor, and e_0 the elementary charge. In **Hall probes**, semiconductor materials are used because the density of charge carriers n is low, and hence the Hall voltage is high.

2. *Lorentz force,*

force acting on a charge moving in a magnetic field. The magnitude of force F is determined by the velocity v of the charge, the magnitude of the charge Q, the magnetic flux density B and the angle between the velocity vector $\vec{v}$ and the magnetic flux density $\vec{B}$. The Lorentz force $\vec{F}$ acts perpendicular to both $\vec{v}$ and $\vec{B}$.

The force vector is given by the vector product:

Lorentz force			$LT^{-2}M$
	Symbol	Unit	Quantity
$\vec{F} = Q\,(\vec{v} \times \vec{B})$ $F = Q \cdot v \cdot B \cdot \sin\alpha$	$\vec{F}$ Q $\vec{v}$ $\vec{B}$ α	N C m/s $T = Vs/m^2$ 1	Lorentz force electric charge velocity of charge magnetic flux density angle between $\vec{v}$ and $\vec{B}$

▲   **Three-fingers rule**: If the thumb of the right hand points along the direction of motion of the positive charge carriers, and the forefinger points along the magnetic flux density, then the middle finger shows the direction of the force on the charge carriers (**Fig. 14.26**).

➤   Force acting on negative charges: Use the left hand!

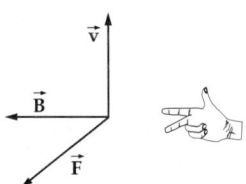

Figure 14.26: Three-fingers rule: the thumb points along the direction of motion of a positive charge, the forefinger along the magnetic flux density; the force is acting along the middle finger.

The maximum force F_{max} on the charge Q results when the velocity v is perpendicular to the magnetic flux density B,

$$F_{max} = Q \cdot v \cdot B\,.$$

From the maximum force, one finds the magnetic flux density:

magnetic flux density = $\dfrac{\text{maximum force}}{\text{charge} \cdot \text{velocity}}$			$T^{-2}MI^{-1}$
	Symbol	Unit	Quantity
$B = \dfrac{F_{max}}{Q \cdot v}$	F_{max} Q v B	N C m/s $T = Vs/m^2$	maximum Lorentz force electric charge velocity of charge magnetic flux density

3. Properties of the Lorentz force

The Lorentz force changes only the orientation of the velocity $\vec{v}$, but not its magnitude. In a uniform magnetic field $\vec{B}$ pointing perpendicular to the orbital plane, a particle of mass m and charge Q moves on a circular orbit of radius R,

$$R = \frac{mv}{QB}\,.$$

The orbital radius is inversely proportional to the magnetic flux density, and proportional to the linear particle momentum.

If the particle is moving in an additional electric field $\vec{E}$, the total force acting on the charge is given by

$$\vec{F} = Q \cdot \vec{E} + Q\,(\vec{v} \times \vec{B})\,.$$

If the electric and magnetic fields are aligned parallel to each other, the path of the particle becomes a helical line about the field direction, with the pitch depending on the position.

The force on a straight conductor of length l carrying a current I in the magnetic field $\vec{B}$ is

$$\vec{F} = I\,(\vec{l} \times \vec{B})\,.$$

$\vec{l}$ is a vector of magnitude l along the current flow. The force $\vec{F}$ points normally to the plane containing $\vec{l}$ and $\vec{B}$. The magnitude of the force is

$$F = I \cdot l \cdot B \cdot \sin\alpha\,,$$

where α is the angle enclosed by $\vec{l}$ and $\vec{B}$. The force reaches the maximum value when the current and the magnetic flux density are perpendicular.

14.13   Magnetic flux

1. Magnetic flux,

Φ, scalar quantity, a measure for the magnetic flux density (induction) through a surface in a magnetic field. For a plane area in a uniform magnetic field, the magnetic flux Φ equals the product of the magnetic flux density B, the area ΔA and the cosine of the angle between $\vec{B}$ and $\Delta\vec{A}$ (**Fig. 14.27**). If the normal to the area is perpendicular to the magnetic flux density, the magnetic flux vanishes.

$$\Phi = B \cdot \Delta A \cdot \cos\alpha = \vec{B} \cdot \Delta\vec{A}\,.$$

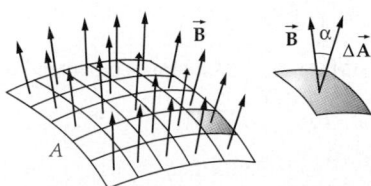

Figure 14.27: Magnetic flux through a surface.

For an arbitrarily shaped surface A in a nonuniform magnetic field, the surface is subdivided into plane area elements ΔA_i, so that the magnetic flux density through these area elements may be regarded as constant. The total flux Φ through the surface A is obtained by summing the individual fluxes:

$$\Phi = \sum_i \Delta \Phi_i = \sum_i \vec{B}_i \cdot \Delta \vec{A}_i \,.$$

2. Flux as an integral over the flux density

For more refined subdivisions of the total surface, the magnetic flux corresponds to the surface integral over the flux density:

flux = integral of flux density over surface			$\mathbf{ML^2T^{-1}Q^{-1}}$
$\Phi = \int\limits_A \vec{B} \cdot d\vec{A}$	Symbol	Unit	Quantity
	Φ	Wb = Vs	magnetic flux
	$\vec{B}$	T = Vs/m^2	magnetic flux density through A
	$d\vec{A}$	m^2	infinitesimal area element
	A	m^2	total area

Weber, Wb, SI unit of the magnetic flux Φ. 1 Wb is the intensity of the magnetic flux through a surface $A = 1$ m^2 if the magnetic flux density is $B = 1$ Vs/m^2,

$$[\Phi] = \text{Wb} = \text{Vs}\,.$$

3. Solenoidal property of the magnetic flux

The magnetic flux through a closed surface A vanishes,

$$\Phi = \oint\limits_A \vec{B} \cdot d\vec{A} = 0\,, \quad \text{div}\,\vec{B} = 0\,.$$

The magnetic field lines are closed; there are no magnetic charges (magnetic monopoles). This relation constitutes one of **Maxwell's equations** (see p. 496).

4. Determination of the flux density

The magnitude of the magnetic flux density is obtained from the magnetic flux $\Delta \Phi$ flowing through a surface $\Delta A_\perp$ placed perpendicularly to the flux. If the magnetic flux is position-dependent, ΔA is reduced in size until the magnetic flux may be considered as being uniformly distributed over the surface element. This approach corresponds to the limit

$$B = \lim_{\Delta A_\perp \to 0} \frac{\Delta \Phi}{\Delta A_\perp} = \frac{d\Phi}{dA_\perp}\,.$$

■   A magnetic flux $\Phi = 0.2$ Wb passing a surface $A = 6$ cm^2 placed perpendicularly to the flux corresponds to a magnetic flux density of

$$B = \frac{\Phi}{A} = \frac{0.2\,Wb}{6 \cdot 10^{-4}\,\text{m}^2} = 333.3\,\text{T}\,.$$

M   The magnetic flux may be measured by means of an induction coil of known number of turns n. Placing the coil into a magnetic field B results in an induced voltage pulse $\int V \, dt$ which is proportional to the enclosed flux:

$$\int_0^T V_{\text{ind}} \, dt = n\Phi = n \cdot B \cdot A \, .$$

If the area A of the induction coil is known, the magnetic flux density may also be determined. If the induction coil is removed from the magnetic field, a voltage pulse of opposite polarity is induced.

14.14   Magnetic field strength

1. Magnetic field strength,

$\vec{H}$, vector quantity, synonymously used for magnetic field. In isotropic magnetic materials, $\vec{H}$ is proportional to $\vec{B}$:

magnetic field strength = $\dfrac{\text{magnetic flux density}}{\text{permeability of free space}}$			$\mathbf{L^{-1}I}$
	Symbol	Unit	Quantity
$\vec{H} = \dfrac{\vec{B}}{\mu_0}$	H	A/m	magnetic field strength
	B	$T = Vs/m^2$	magnetic flux density
	μ_0	Vs/(Am)	permeability of free space

Ampere/meter, A/m, SI unit of the magnetic field strength $\vec{H}$,

$$[\vec{H}] = \text{A/m}.$$

➤   Notice the units:
   magnetic field strength, related to current: A/m .
   electric field strength, related to voltage: V/m .

2. Permeability of free space

permeability of free space			$\mathbf{LT^{-2}MI^{-2}}$
$\mu_0 = 4\pi \cdot 10^{-7} \dfrac{Vs}{Am}$	Symbol	Unit	Quantity
$= 1.257 \cdot 10^{-6} \dfrac{Vs}{Am}$	μ_0	Vs/(Am)	permeability of free space

➤   One might assume that, analogously to the electric field strength $\vec{E}$, the magnetic field strength $\vec{H}$ is the fundamental field concept, and that the magnetic flux density

$\vec{B}$, like the electric displacement density $\vec{D}$, follows from the field strength, i.e., is a deduced quantity. One should notice, however, that electric and magnetic fields are demonstrated through their forces on (moving) charges. The formula for this force involves the *electric field strength* and the *magnetic flux density*, but not the magnetic field strength.

3. Vector potential,

$\vec{A}$, vector quantity for calculation of the magnetic flux density $\vec{B}$. From the solenoidal property of the magnetic field, it follows that the magnetic flux density may be written as the rotation (curl) of a vector field $\vec{A}$,

$$\operatorname{div}\vec{B} = 0, \quad \vec{B} = \operatorname{rot}\vec{A}.$$

The vector potential $\vec{A}$ may be calculated from the current density distribution $\vec{J}(\vec{r})$ as solution of the differential equation

$$\Delta\vec{A} = -\mu_0\vec{J}(\vec{r}),$$

$$\vec{A} = \frac{\mu_0}{4\pi} \int \frac{\vec{J}(\vec{r}')}{|\vec{r}-\vec{r}'|} \, dV'.$$

For the magnetic flux density, one finds:

$$\vec{B} = \frac{\mu_0}{4\pi} \int \frac{\vec{J}(\vec{r}') \times (\vec{r}-\vec{r}')}{|\vec{r}-\vec{r}'|^3} \, dV'.$$

The potentials φ and $\vec{A}$ may be determined from two coupled differential equations, if the spatial charge density ρ and the current density $\vec{J} = \rho\vec{v}$ are specified as functions of position $\vec{r}$ and time t:

$$\Delta\vec{A}(\vec{r},t) - \mu_0\varepsilon_0\frac{\partial^2\vec{A}(\vec{r},t)}{\partial t^2} = -\mu_0\vec{J}(\vec{r},t), \quad \Delta\varphi(\vec{r},t) - \mu_0\varepsilon_0\frac{\partial^2\varphi(\vec{r},t)}{\partial t^2} = -\frac{\rho(\vec{r},t)}{\varepsilon_0}.$$

➤ For a particle of mass m, charge Q and momentum $\vec{p} = m\vec{v}$ in an electromagnetic field, the Lagrange function L and Hamilton function H read

$$L = \frac{m}{2}v^2 + Q\vec{v}\vec{A} - Q\varphi, \quad H = \frac{(\vec{p} - Q\vec{A})^2}{2m} + Q\varphi.$$

14.15   Magnetic potential difference and magnetic circuits

1. Magnetic potential difference,

V_{AB}, between two points A and B, the line integral of the magnetic field strength $\vec{H}$ along the path s:

magnetic potential difference = integral of magnetic field strength along path			**I**
$$V_{AB} = \int_A^B \vec{\mathbf{H}} \cdot d\vec{\mathbf{s}}$$	Symbol	Unit	Quantity
	V_{AB}	A	magnetic potential
	$\vec{\mathbf{H}}$	A/m	magnetic field strength
	$d\vec{\mathbf{s}}$	m	path element

Ampere, A, SI unit of the magnetic potential difference V,

$$[V] = 1 \text{ A}.$$

M   **Rogowski coil**, a flexible, long, thin coil for measurements of magnetic potential differences. The coil is placed in a magnetic field. Switching the magnetic field on or off generates a voltage pulse in the coil proportional to the magnetic potential difference between the end points of the coil.

2. Magnetic circuit and magnetic resistance

Magnetic circuit, the magnetic flux traverses a series of materials of different magnetic reluctance.

Magnetic reluctance, R_m, ratio of magnetic potential difference V to magnetic flux Φ in a medium:

magnetic reluctance = $\dfrac{\text{magnetic potential difference}}{\text{magnetic flux}}$			$\mathbf{L^{-2}T^2M^{-1}I^2}$
$$R_m = \frac{V}{\Phi}$$	Symbol	Unit	Quantity
	R_m	A/Wb	magnetic reluctance
	V	A	magnetic potential difference
	Φ	Wb	magnetic flux

Ampere/weber, A/Wb, SI unit of the magnetic reluctance R_m,

$$[R_m] = \text{A/Wb} = \text{A/(Vs)}.$$

3. Mesh rule and vertex rule in the magnetic circuit

Analogously to Kirchhoff's laws for current circuits, the following relations hold for magnetic circuits:

Mesh rule in a magnetic circuit, the sum over all magnetic potential differences of a mesh in a magnetic circuit equals the total current flow Θ,

$$V_{\text{tot}} = V_1 + V_2 + \cdots V_n = \Theta.$$

Vertex rule in a magnetic circuit, the sum over all magnetic fluxes at a vertex in a magnetic circuit equals the total flux,

$$\Phi_{\text{tot}} = \Phi_1 + \Phi_2 + \cdots + \Phi_n.$$

Thus, there are similar rules for series connection and parallel connection of magnetic reluctances, as compared with electric resistors:

4. Series connection of magnetic reluctances

The magnetic flux passes through a series of different materials with magnetic reluctance $R_{m1}, \ldots, R_{mn}$ (**Fig. 14.28**). The total reluctance is

$$R_{\text{tot}} = R_{m1} + \cdots + R_{mn}.$$

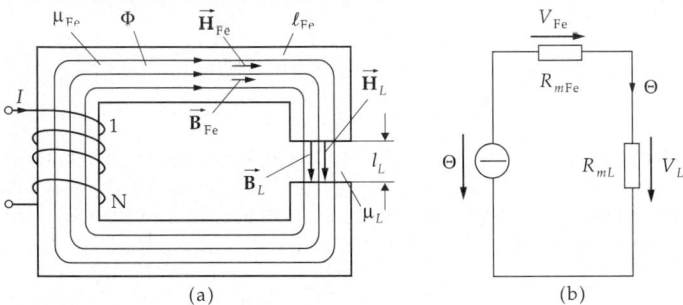

(a)   (b)

Figure 14.28: Total magnetic reluctance of a series connection of magnetic reluctances. (a): iron core with pole gap, (b): equivalent circuit.

5. Parallel connection of magnetic reluctances

The magnetic flux in a magnetic circuit separates into several branches with the magnetic reluctances $R_{m1}, \ldots, R_{mn}$. The reciprocal values of the magnetic reluctances sum to the reciprocal value of the total reluctance:

$$\frac{1}{R_{\text{tot}}} = \frac{1}{R_{m1}} + \cdots + \frac{1}{R_{mn}}.$$

6. Calculation of magnetic circuits

The rules described in the paragraphs 3, 4 and 5 are applied in technology for calculating magnetic circuits in which the magnetic flux successively passes distinct materials.

■   The total reluctance of a magnetic circuit containing an iron core with a pole gap (**Fig. 14.28**) is

$$R_m(\text{iron}) + R_m(\text{polegap}) = R_m(\text{total}).$$

14.15.1 Ampere's law

1. Current flow,

Θ, the current through a surface enclosed by a path s as line integral of the magnetic field strength $\vec{H}$ along the closed path s.

In order to evaluate the current flow Θ, the path s is subdivided into rectilinear path elements Δs. One then calculates the product of the magnetic field component tangential to the path element, and the length of the path element,

$$H \cdot \Delta s \cdot \cos \alpha = \vec{H} \cdot \Delta \vec{s}.$$

The orientation of the vectorial path element $\Delta\vec{s}$ corresponds to the direction of the path s that was traversed. Summing over all path elements yields the current flow

$$\sum H_i\,\Delta s_i = \Theta\,,$$

H_i being the component of $\vec{H}$ along $\Delta\vec{s}_i$.

For an arbitrarily shaped path in a nonuniform magnetic field, the path is subdivided until the path elements may be regarded as straight segments and the magnetic field along the path element as being uniform. One then obtains **Ampere's law**:

▲   The line integral of the magnetic field strength along a closed path equals the current flow through the surface enclosed by the path:

current flow = integral of magnetic field strength along path			I
	Symbol	Unit	Quantity
$\displaystyle \Theta = \oint_s \vec{H}\cdot d\vec{s}$ $\displaystyle = \int_A \vec{J}\cdot d\vec{A}$	Θ	A	current flow
	$\vec{H}$	A/m	magnetic field strength
	$d\vec{s}$	m	infinitesimal path element
	s	m	total path
	$\vec{J}$	A/m^2	current density
	$d\vec{A}$	m^2	infinitesimal surface element
	A	m^2	surface enclosed by path s

Ampere, A, SI unit of the current flow Θ,

$$[\Theta] = A.$$

2. Consequences of Ampere's law

If the path encloses the current completely, the line integral is independent of the shape of the path.

If the path encloses a current-carrying wire, the current flow Θ equals the current I flowing in the conductor,

$$\Theta = I.$$

If the path encloses a set of N current-carrying wires, the current flow Θ equals the sum of the currents I_n of the individual conductors,

$$\Theta = \sum_{n=1}^{N} I_n\,.$$

If the path encloses N turns of a coil, the current flow Θ equals the coil current I multiplied by the number N of enclosed turns,

$$\Theta = N \cdot I\,.$$

If the path encloses a current distribution characterized by the current density $\vec{J}$, the current flow equals the flux of the current density through the surface A enclosed by the path,

$$\Theta = \int_A \vec{J}\cdot d\vec{A}\,.$$

The last statement may also be written in differential form:

$$\mathrm{rot}\,\vec{\mathbf{H}} = \vec{\mathbf{J}}.$$

➤ Ampere's law allows the calculation of magnetic fields generated by simple current distributions.

14.15.2   Biot-Savart's law

1. Biot-Savart's law,

allows the calculation of the magnetic field strength of wire-shaped conductors of arbitrary geometry.

The contribution of the conductor element $\mathrm{d}\vec{\mathbf{s}}$ to the magnetic field strength is proportional to the current I and inversely proportional to the square of the distance r. The orientation of the magnetic field strength generated by the conductor element $\mathrm{d}\vec{\mathbf{s}}$ is given by the vector product of the distance vector $\vec{\mathbf{r}}$ and the orientation of the conductor element $\mathrm{d}\vec{\mathbf{s}}$ (**Fig. 14.29**).

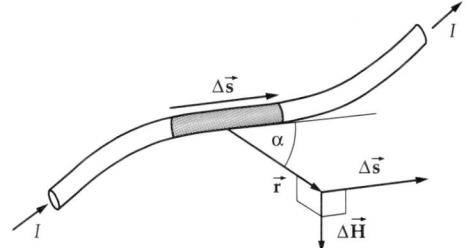

Figure 14.29: Biot-Savart's law.

The total field strength is obtained by summing over all contributions of the individual conductor elements (integral over $\mathrm{d}\vec{\mathbf{s}}$):

Biot-Savart's law			$\mathbf{L^{-1}I}$
	Symbol	Unit	Quantity
$\displaystyle \vec{\mathbf{H}} = \int_{s} \frac{I\,\mathrm{d}\vec{\mathbf{s}} \times \vec{\mathbf{r}}}{4\pi r^{3}}$	$\vec{\mathbf{H}}$	A/m	magnetic field strength
	I	A	current through conductor
	$\mathrm{d}\vec{\mathbf{s}}$	m	conductor element
	$\vec{\mathbf{r}}$	m	distance vector
	r	m	magnitude of distance vector

2. Magnetic moment of a steady current density distribution,

$\vec{\mathbf{J}}(\vec{\mathbf{r}})$, defined by

$$\vec{\mathbf{m}} = \frac{1}{2} \int \vec{\mathbf{r}} \times \vec{\mathbf{J}}(\vec{\mathbf{r}})\,\mathrm{d}V .$$

The magnetic field is in first order given by

$$\vec{\mathbf{B}} = \frac{\mu_0}{4\pi} \left[\frac{3(\vec{\mathbf{m}} \cdot \vec{\mathbf{r}})\vec{\mathbf{r}}}{r^5} - \frac{\vec{\mathbf{m}}}{r^3} \right] .$$

▲   The magnetic field of a steady current density distribution is in first order equivalent to the electric field of an electric dipole.

3. Examples on Biot-Savart's law

a) Closed conductor loop: In a closed conductor loop in a plane enclosing an area A, a current I flows. In a magnetic field $\vec{\mathbf{B}}$, the current loop experiences a torque $\vec{\tau}$:

$$\vec{\tau} = I\,(\vec{\mathbf{A}} \times \vec{\mathbf{B}})\,.$$

Since the magnetic moment $\vec{\mathbf{m}}$ is defined by $\vec{\tau} = \vec{\mathbf{m}} \times \vec{\mathbf{B}}$, one gets for the magnetic moment of a closed current loop

$$\vec{\mathbf{m}} = I \cdot \vec{\mathbf{A}}\,.$$

b) Charge on a circular path: A particle of mass m and charge Q moving with a linear momentum $\vec{\mathbf{p}}$ along a circular path corresponds to a circular current with the magnetic moment,

$$\vec{\mathbf{m}} = \frac{Q}{2m}\,\vec{\mathbf{l}}, \quad \vec{\mathbf{l}} = \vec{\mathbf{r}} \times \vec{\mathbf{p}}\,.$$

The magnetic moment $\vec{\mathbf{m}}$ is proportional to the orbital angular momentum $\vec{\mathbf{l}}$.

4. Force and energy of magnetic moments

A body with a magnetic moment $\vec{\mathbf{m}}$ in a uniform magnetic field $\vec{\mathbf{B}}$ experiences a torque $\vec{\tau}$,

$$\vec{\tau} = \vec{\mathbf{m}} \times \vec{\mathbf{B}}\,.$$

A body with a magnetic moment $\vec{\mathbf{m}}$ in a nonuniform magnetic field $\vec{\mathbf{B}}$ experiences a force $\vec{\mathbf{F}}$,

$$\vec{\mathbf{F}} = \left(\vec{\mathbf{m}} \cdot \frac{\partial}{\partial \vec{\mathbf{r}}}\right) \vec{\mathbf{B}}\,.$$

The potential energy E_{pot} of a body with a magnetic moment $\vec{\mathbf{m}}$ is given by

$$E_{\text{pot}} = -\vec{\mathbf{m}} \cdot \vec{\mathbf{B}}\,.$$

5. Types of magnetic moments

Magnetic moment of a bar magnet, defined by the product of magnetic flux Φ (pole intensity) and fictitious pole distance d,

$$\vec{\mathbf{m}} = \Phi \cdot \vec{\mathbf{d}}\,.$$

The vector $\vec{\mathbf{d}}$ points from the south pole to the north pole.
    Coulomb's magnetic moment, defined by $\vec{\mathbf{m}}_C = \Phi\,\vec{\mathbf{d}}$.
    Ampere's magnetic moment, defined by $\vec{\mathbf{m}}_A = \vec{\mathbf{m}}_C/\mu_0 = \Phi\,\vec{\mathbf{d}}/\mu_0$. Mainly used in atomic physics.

14.15.3   Magnetic field of a rectilinear conductor

1. Magnetic field strength of a current-carrying conductor

The magnetic field of a rectilinear conductor is proportional to the current in the conductor and inversely proportional to the distance from the conductor. The direction of the magnetic field follows concentric circles about the conductor (right-hand rule, see **Fig. 14.24**):

magnetic field strength = $\dfrac{\text{current}}{2\pi \cdot \text{distance}}$				$\mathbf{L^{-1}I}$
	Symbol	Unit	Quantity	
$H = \dfrac{I}{2\pi r}$	H	A/m	field strength at distance r from conductor	
	I	A	current through conductor	
	r	m	distance from conductor	

■ A rectilinear conductor carries the current 4 A. The magnetic field at a distance of $r = 1$ m is

$$H = \frac{4\,A}{2\pi \cdot 1\ m} \approx 0.64\ \text{A/m}.$$

If the distance r is doubled, one gets half the field strength, $H \approx 0.32$ A/m.

2. Force acting on conductors carrying current

A magnetic field generates a force on conductors carrying current. This force is used for the definition of the unit of current, the ampere (see **Fig. 13.4**).

The force F between two rectilinear parallel conductors of equal length l at distance a, with currents I_1 and I_2, respectively, is given by

$$F = \mu_0 \frac{l}{2\pi a} I_1 I_2 .$$

If both currents have the same orientation, the conductors attract each other; for opposite current directions, they repel each other (**Fig. 14.30**).

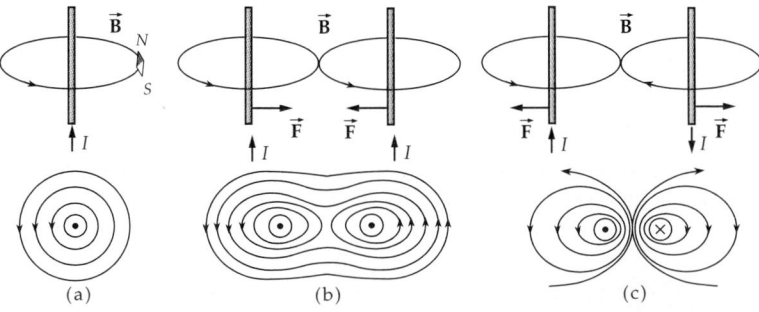

Figure 14.30: Magnetic field lines and force for current-carrying rectilinear conductors. (a): single conductor, (b): parallel double-wire conductor with parallel currents, (c): parallel double-wire conductor, antiparallel currents.

14.15.4 *Magnetic fields of various current distributions*

Biot-Savart's law and Ampere's law allow a calculation of the magnetic fields of several simple current distributions.

1. Magnetic field of a wire

The magnetic field strength H of a long rectilinear conductor of circular cross-sectional area (radius R) carrying a current I is given by:

external region of conductor ($r \geq R$)			
$H(r) = \dfrac{I}{2\pi r}$	Symbol	Unit	Quantity
	H	A/m	magn. field strength in external region
	I	A	current through conductor
	r	m	perpendicular distance from axis

internal region of conductor ($r \leq R$)			
$H(r) = \dfrac{I}{2\pi R^2}\, r$	Symbol	Unit	Quantity
	H	A/m	magn. field strength in internal region
	I	A	current through conductor
	r	m	perpendicular distance from axis
	R	m	radius of conductor

In the interior region, the magnetic field strength increases linearly with r up to the conductor radius R, in the external region it decreases as $1/r$ to zero.

2. Magnetic field strength in the center of a plane circular conductor loop

The magnetic field strength at the center of a circular current loop of radius R is given by the quotient of the current I and the loop diameter $2R$:

field strength $= \dfrac{\text{current}}{2 \cdot \text{radius}}$			$\mathbf{L^{-1}I}$
$H = \dfrac{I}{2R}$	Symbol	Unit	Quantity
	H	A/m	magnetic field strength
	I	A	circular current
	R	m	loop radius

For a large distance r from the plane of the circular current ($r \gg R$), the field strength on the symmetry axis is given by

$$H = \frac{AI}{2\pi r^3},$$

where A is the area enclosed by the conductor. The formula holds for arbitrarily shaped plane current loops because, at large distances, the detailed shape of the conductor becomes irrelevant.

3. Magnetic field of a long cylindrical coil

The magnetic field strength of a long cylindrical coil (solenoid) ($R \rightarrow 0, l \gg R$) is the product of coil current I and number of turns n, divided by the coil length l. The magnetic field is uniform within the coil, and strongly nonuniform in the external region, where it resembles the field of a bar magnet:

field strength =	$\dfrac{\text{winding number} \cdot \text{current}}{\text{length}}$		$L^{-1}I$
	Symbol	Unit	Quantity
$H = \dfrac{nI}{l}$	H	A/m	magnetic field within coil
	I	A	coil current
	l	m	length of coil
	n	1	number of turns

4. Magnetic axial field strength of a short cylindrical coil

The magnetic field strength in the center of a short cylindrical coil (length $l \rightarrow 0$) equals the product of coil current I and number of turns n, divided by the coil diameter $2R$, i.e., n times the field strength of a single circular loop:

field strength =	$\dfrac{\text{winding number} \cdot \text{current}}{2 \cdot \text{coil radius}}$		$L^{-1}I$
	Symbol	Unit	Quantity
$H = \dfrac{nI}{2R}$	H	A/m	magnetic field strength
	n	1	number of turns
	I	A	coil current
	R	m	coil radius

Magnetic field strength on the axis of a cylindrical coil of radius R and length l in the internal region:

$$H = \frac{nI}{\sqrt{l^2 + 4R^2}} \cdot$$

14.16 Matter in magnetic fields

If matter is placed into a magnetic field of strength $\vec{H}$, the magnetic flux density $\vec{B}$ is modified due to the interaction of the magnetic field with the electrons of matter. The change of the magnetic flux density depends on the material inserted.

1. Relative permeability,

μ_r, ratio of the magnetic flux density B in matter to the magnetic flux density B_0 in a vacuum at the same magnetic field strength H,

$$\mu_r = \frac{B}{B_0} \cdot$$

μ_r is listed in **Tab. 18.4/3** for several magnetic alloys.

Permeability, μ, product of the permeability of free space μ_0 and the relative permeability μ_r,

$$\mu = \mu_0 \cdot \mu_r \,.$$

▲ For isotropic magnetic materials, the magnetic flux density $\vec{B}$ in matter is proportional to the magnetic field strength $\vec{H}$. The proportionality factor is the permeability.

magnetic flux density = permeability · magnetic field strength			$\mathbf{T^{-2}MI^{-1}}$
	Symbol	Unit	Quantity
$\vec{B} = \mu\vec{H}$	$\vec{B}$	Vs/m^2	magnetic flux density
	μ	Vs/(Am)	permeability
$= \mu_r \cdot \mu_0 \cdot \vec{H}$	μ_0	Vs/(Am)	permeability of free space
	μ_r	1	relative permeability
	$\vec{H}$	A/m	magnetic field strength

2. Magnetic susceptibility

of a material, χ_m, difference between the relative permeability μ_r of matter and the relative permeability of a vacuum $\mu_r = 1$,

$$\chi_m = \mu_r - 1 \,.$$

χ_m is nondimensional: $[\chi] = 1$.
■ Magnetic susceptibility for
diamagnetics: Cu $-1 \cdot 10^{-5}$, Bi $-1.5 \cdot 10^{-4}$, H_2O $-7 \cdot 10^{-6}$;
paramagnetics: Al $2.4 \cdot 10^{-5}$, O_2 (gaseous) $3.6 \cdot 10^{-3}$;
ferromagnetics: Fe 10^4, AlNiCo alloys 3, ferrites (hard) 0.3.
Tab. 18.4/1 lists the molar magnetic susceptibility for the elements, **Tab. 18.4/2** the analogous quantities for several inorganic compounds.

3. Magnetic polarization,

$\vec{J}_m$, difference of the magnetic flux density $\vec{B}_m$ with matter and the magnetic flux density in a vacuum $\vec{B}_0$, given by the product of the magnetic susceptibility χ_m and the magnetic flux density $\vec{B}_0$ of the vacuum,

$$\vec{J}_m = \vec{B}_m - \vec{B}_0 = (\mu_r - 1) \cdot \vec{B}_0 = \chi_m \cdot \vec{B}_0 = \chi_m \mu_0 \cdot \vec{H} \,.$$

Volt · second/meter^2, Vs/m^2, SI unit of the magnetic polarization,

$$[\vec{J}] = \mathrm{Vs/m^2} \,.$$

4. Magnetization,

$\vec{M}$, product of magnetic susceptibility χ_m and magnetic field strength $\vec{H}$,

$$\vec{M} = \frac{\vec{B}_m}{\mu_0} - \vec{H} = (\mu_r - 1) \cdot \vec{H} = \chi_m \cdot \vec{H} \,.$$

Ampere/meter, A/m, SI unit of the magnetization $\vec{M}$,

$$[\vec{M}] = \mathrm{A/m} \,.$$

For many substances, the magnetization $\vec{M}$ is proportional to the magnetic field strength $\vec{H}$.

Magnetization curves, graphic representation of the variation of the magnetic flux density B versus the magnetic field strength H.

14.16.1   Diamagnetism

Diamagnetism, property of all substances. Diamagnetic behavior can only then be observed if it is not overwhelmed by the other types of magnetism.

▲   If a diamagnetic substance is placed in a nonuniform magnetic field, it feels a force towards regions of low magnetic field strength.

Diamagnetic behavior occurs for elements with closed electron shells. If a diamagnetic substance is placed in a magnetic field, intra-atomic currents are induced that, according to Lenz's rule, are opposed to the external magnetic field (see p. 485). Magnetic dipoles are induced in the substance, with the north pole pointing to the north pole of the external field, and the south pole pointing to the external south pole. Thus, the magnetic field is thereby weakened and the substance feels a force out of the field.

▲   The relative permeability of diamagnetic substances is smaller than unity, the magnetic susceptibility is negative,

$$\mu_r < 1, \quad \chi_m < 0 \, (-10^{-4} < \chi_m < -10^{-9}) \,.$$

➤   The field vectors $\vec{H}$ and $\vec{M}$ point opposite to each other. The density of field lines of $\vec{B}$ is lower in the interior of the material than in the external region.

▲   Diamagnetism is nearly independent of temperature.

■   Substances with diamagnetic behavior: Cu, Bi, Au, Ag, H_2.

14.16.2   Paramagnetism

Paramagnetism, occurs if there are noncompensated magnetic moments of electrons. This happens for atoms with only partially occupied electron shells. The originally randomly oriented atomic magnetic moments are aligned by an external magnetic field (**Fig. 14.31 (a)**).

▲   The relative permeability of paramagnetic substances is larger than unity, the magnetic susceptibility is positive,

$$\mu_r > 1, \quad \chi_m > 0 \, (10^{-6} < \chi_m < 10^{-4}) \,.$$

➤   The field vectors $\vec{H}$ and $\vec{M}$ are parallel. The density of the field lines $\vec{B}$ in matter is larger than in the external region.

Curie's law, describes the variation of the magnetic susceptibility χ_m with the absolute temperature T for paramagnetic matter,

$$\chi_m = \frac{C}{T} \,.$$

C is a material-dependent parameter.

■   Substances with paramagnetic behavior: Al, O_2, W, Pt, Sn.

14.16.3   Ferromagnetism

1. Ferromagnetism,

generated by the alignment of the magnetization direction of the Weiss domains along the direction of the external field. The magnetization curve of ferromagnetic substances is nonlinear (**Fig. 14.31 (b)**).

Weiss domains, crystal regions of equal magnetization, extension $10 \, \mu m$ to 1 mm, which in the nonmagnetized state are oriented at random.

Bloch walls, transition region between the Weiss domains where the magnetization is changing. Magnetization of a ferromagnetic substance proceeds by reversible and irreversible displacements of Bloch walls.

▲   The relative permeability of ferromagnetic matter depends on the magnetic field strength and is much larger than unity, the magnetic susceptibility is positive,

$$\mu_r \gg 1, \quad \chi_m > 0.$$

➤   The field vectors $\vec{H}$ and $\vec{M}$ are parallel. The density of $\vec{B}$ field lines in matter is larger than in the external region.

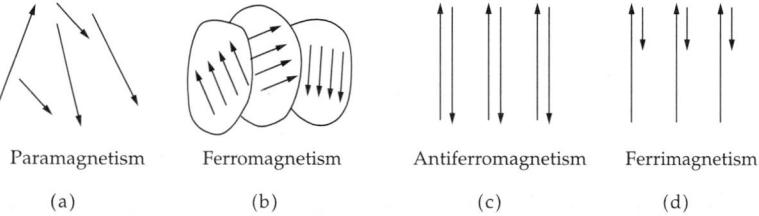

| Paramagnetism | Ferromagnetism | Antiferromagnetism | Ferrimagnetism |
| (a) | (b) | (c) | (d) |

Figure 14.31: Structure of magnetic substances. (a): paramagnetism, random orientation of the magnetic moments, (b): ferromagnetism, alignment of magnetic moments within Weiss domains separated by Bloch walls, (c): antiferromagnetism, two sublattices with equal, but oppositely oriented, magnetic moments, (d): ferrimagnetism, two sublattices with distinct and oppositely oriented magnetic moments.

2. Hysteresis curve,

magnetization curve of ferromagnetic materials. The area enclosed by the hysteresis curve is a measure of the magnetization energy needed to align the Weiss domains. The magnetization curve depends on the initial magnetic state of the ferromagnetic material. The hysteresis curve is symmetric against reflection about the origin of the coordinate system. This corresponds to a symmetry under inversion of the orientation of the magnetic field.

Magnetically hard material, ferromagnetic substance with a wide hysteresis curve. A large amount of work is needed for the remagnetization (**Fig. 14.32 (b)**).

Magnetically soft substance, ferromagnetic material with a narrow hysteresis curve. Only a small amount of energy is needed for the remagnetization (**Fig. 14.32 (a)**).

■   Magnetically hard substances are well suited for producing magnets, since they preserve an imposed magnetic field for a long time against disturbance (e.g., by other magnetic fields). Application: storage media.

Magnetically soft materials are used for transformer cores, since the flux density is high, but the energy loss due to remagnetization is low (recorder heads).

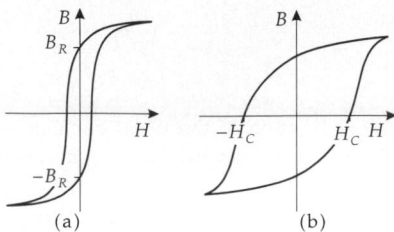

Figure 14.32: Hysteresis curves. (a): magnetically soft material, (b): magnetically hard material.

3. Virgin curve of magnetization and saturation flux

Virgin curve of magnetization, the magnetic flux density B of the non-magnetized substance in absence of a magnetic field H vanishes. If the magnetic field strength is increased, the virgin curve in the B-H diagram is followed (**Fig. 14.33**).

➤ If the material has been magnetized at least once, the virgin curve can no longer be reproduced.

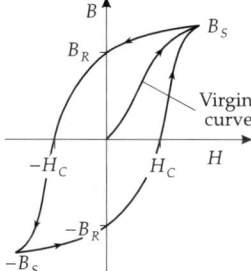

Figure 14.33: Hysteresis curve and virgin curve.

Saturation induction, B_S, the flux density at which all magnetic moments of the ferromagnetic material are aligned along the field direction. If the field strength is increased beyond this point, the flux density varies linearly with the field strength.

4. Remanence and coercitive field strength

Remanence, remanence flux density, B_R, the magnetic flux density remaining in the material after switching off the external magnetic field.

Coercitive field strength, H_C, a counter field that has to be applied in order to demagnetize the ferromagnetic material. For magnetically soft matter, H_C varies between 0.1 A/m and 10^3 A/m; for magnetically hard matter, between 10^3 A/m and 10^7 A/m.

■ Remanence and coercitive field strength for chromium steel: $R_B = 1.1$ T, $H_C = 5200$ A/m.

Remanence and coercitive field strength for several magnetic alloys are listed in **Tab. 18.4/3**.

5. Temperature dependence of ferromagnetism

Ferromagnetism decreases with increasing temperature. The ferromagnetic substance then becomes paramagnetic.

Curie-Weiss law, describes the temperature variation of the susceptibility χ_m of ferromagnetic substances (**Fig. 14.34**),

$$\chi_m = \frac{C}{T - T_C}.$$

T_C is the **ferromagnetic Curie temperature**, C is a material parameter.

■   Ferromagnetic Curie temperatures: Fe 1042 K, Co 1400 K, Ni 631 K, Dy 87 K.
▲   Above the Curie temperature, the substance is paramagnetic.
Tab. 18.5/1, 18.5/2 and 18.5/3 list the Curie temperature for several ferromagnetic elements and for binary iron- and nickel alloys.
■   Substances with ferromagnetic behavior: Fe, Co, Ni, Gd.

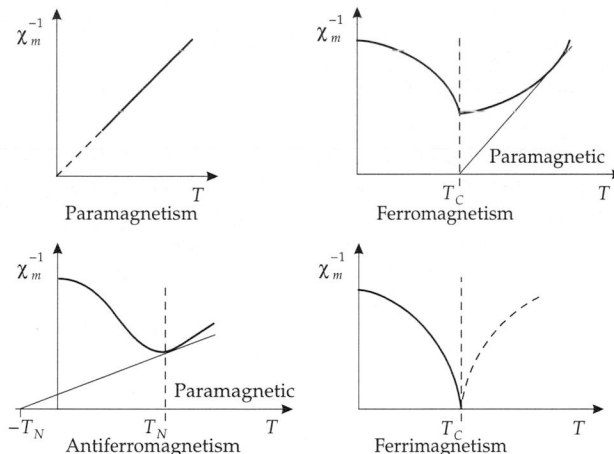

Figure 14.34: Temperature dependence of the magnetic susceptibility. T_C: Curie temperature, T_N: Néel temperature.

6. Magnetostriction,

elastic change of shape of ferromagnetic substances in magnetic fields due to displacements and turning of the Bloch walls; both positive and negative relative changes of length may occur.

Volume magnetostriction, change of volume, but with conservation of shape.

Joule magnetostriction, change of shape, but with volume conservation. The Joule magnetostriction is in general much larger than the volume magnetostriction.

Inverse magnetostriction, change of magnetization by mechanical stress.

14.16.4   Antiferromagnetism

Antiferromagnetism, if there are two sublattices in a crystal, with identical magnetic moments aligned antiparallel (see **Fig. 14.31 (c)**).
▲   The relative permeability of antiferromagnetic substances is larger than unity,

$$\mu_r > 1 \,.$$

Néel's law, describes the temperature dependence of the susceptibility of antiferromagnetic substances:

$$\chi_m = \frac{C}{T + T_N} \,.$$

Néel temperature is denoted T_N. C is a material parameter.
- Néel temperatures for several antiferromagnets: FeO 198 K, NiF$_2$ 73.2 K, CoUO$_4$ 12 K, CoO 328 K.

Tab. 18.7/1 lists the Néel temperature and the molar magnetic susceptibility for several antiferromagnets.
- Substances with antiferromagnetic behavior: CoO, NiCo, FeO, CoF$_3$, FeF$_3$.

14.16.5   Ferrimagnetism

Ferrimagnetism, occurs if in a crystal there are two sublattices with magnetic moments of different magnitude that generate a resulting magnetic moment (**Fig. 14.31 (d)**). Ferromagnetic properties such as hysteresis and antiferromagnetic properties occur, depending on the relative orientation of the moments of the sublattices.

Ferrites, ferrimagnetic materials, ion crystals. They are almost free of eddy currents because of their high specific reluctance. Ferrites are ceramic materials used as coil cores for high frequencies, e.g., as ferrite antennas.

The magnetic properties of several ferrites are listed in **Tab. 18.6/1**.
- Substances with ferrimagnetic behavior: NiFe$_2$O$_3$, CoFe$_2$O$_3$, hexagonal ferrites BaO· 6Fe$_2$O$_3$, PbO· Fe$_2$O$_3$, garnets 3Ce$_2$O$_3$·5Fe$_2$O$_3$, 3Sm$_2$O$_3$·5Fe$_2$O$_3$.

14.17   Magnetic fields at interfaces

When passing from one medium of permeability μ_1 to another medium of permeability μ_2, separated by an interface that itself carries no current, both the magnetic field strength and the flux density change at the interface.

1. Change of the magnetic field strength

The tangential component of the magnetic field strength H_t does not change in the transition,

$$H_{t1} = H_{t2} .$$

The normal component of the magnetic field strength changes discontinuously (**Fig. 14.35**).

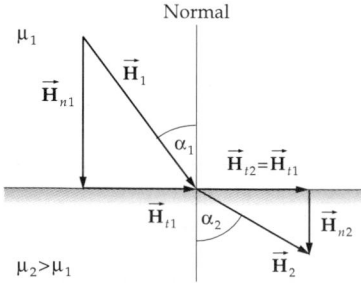

Figure 14.35: Magnetic field strength $\vec{H}$ at the interface between two media.

2. Change of the magnetic flux density

The normal component of the magnetic flux density B_n does not change in the transition,

$$B_{n1} = B_{n2}.$$

The tangential component of the magnetic flux density changes discontinuously (**Fig. 14.36**).

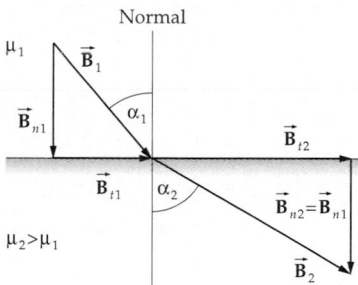

Figure 14.36: Magnetic flux density $\vec{B}$ at the interface between two media.

3. Angular relations of the magnetic field strengths at the interface

Let α denote the angle between the perpendicular (normal to interface) and the orientation of the magnetic field strength. The tangent values of the angles α_1 in the first medium and α_2 in the second medium are related as the permeabilities μ_1 and μ_2, or as the relative permeabilities μ_{r1} and μ_{r2}, respectively,

$$\frac{\tan \alpha_1}{\tan \alpha_2} = \frac{\mu_1}{\mu_2} = \frac{\mu_{r1}}{\mu_{r2}}.$$

■   In the transition from a medium of permeability μ_1 to a medium of higher permeability μ_2,

$$\alpha_1 < \alpha_2.$$

In the transition, the magnetic field strength bends away from the perpendicular.
   In the transition from a medium of permeability μ_1 to a medium of lower permeability μ_2,

$$\alpha_1 > \alpha_2.$$

In the transition, the magnetic field strength bends towards the perpendicular.

14.18 Induction

Induction, generation of voltages at the ends of a conductor or a conducting loop by changing the magnetic flux through the conductor or the conducting loop.

▲   The induced voltage V_{ind} equals the product of the time rate of change of the magnetic flux Φ, and the number n of conductors or number of turns of the conducting loop, respectively.

induced voltage = number of turns · $\dfrac{\text{change of flux}}{\text{time interval}}$			$\mathbf{L^2T^{-3}MI^{-1}}$
$V_{\text{ind}} = -n\,\dfrac{d\Phi}{dt}$	Symbol	Unit	Quantity
	V_{ind}	V	induced voltage
	$d\Phi$	Vs	change of magnetic flux
	dt	s	infinitesimal time interval
	n	1	number of turns of conductor

One distinguishes:
- motional induction, and
- transformer induction.

14.18.1 Faraday's law of induction

1. Faraday's law, motional induction,

induction of voltages in a conductor by moving the conductor in a constant magnetic field $\vec{B}$. The change of the magnetic flux is then determined by the area ΔA covered by the conductor,

$$\Delta\Phi = \vec{B} \cdot \Delta\vec{A}\,,$$

and the induced voltage is:

voltage $\sim \dfrac{\text{change of area}}{\text{time interval}}$ · flux density			$\mathbf{L^2T^{-3}MI^{-1}}$
$V_{\text{ind}} = -n\,\dfrac{d\vec{A}}{dt}\cdot\vec{B}$	Symbol	Unit	Quantity
	V_{ind}	V	induced voltage
	$d\vec{A}$	m^2	change of area
	dt	s	infinitesimal time interval
	$\vec{B}$	Vs/m^2	magnetic flux density
	n	1	number of turns of conductor

■   The magnetic flux in a conducting loop in a uniform magnetic field is proportional to the cosine of the angle α between the orientation of the magnetic field $\vec{B}$ and the normal to the area A (**Fig. 14.37**). If the conducting loop rotates with constant angular

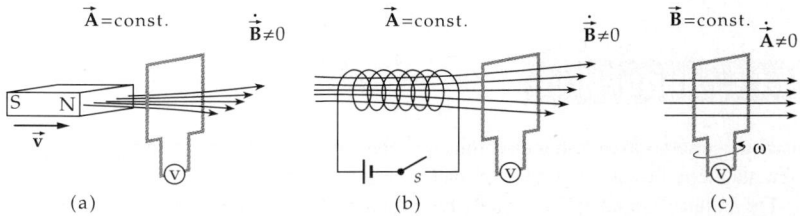

Figure 14.37: Motional induction in a conducting loop.

velocity ω, an alternating voltage of frequency $f = \omega/(2\pi)$ is generated at the ends of the loop. The induced voltage is given by

$$V_{\text{ind}}(t) = A \cdot B \cdot \omega \sin \omega t = \hat{v}_{\text{ind}} \cdot \sin \omega t \,,$$

$\hat{v}_{\text{ind}}$ is the amplitude of the alternating voltage.

➤ The operation of generators is based on motional induction.

2. Eddy current,

induced current in an extended conductor caused by a time-varying magnetic field. The lines of current form closed vortices.

■ **Eddy-current brake**: eddy currents are generated in a conductor when it moves in a magnetic field. In the magnetic field, a force acts on these currents that opposes the motion of the conductor. A rotating metallic disk is slowed when a transverse magnetic field is switched on.

3. Skin effect,

high-frequency alternating currents ($f > 10^7$ Hz) do not flow through the entire conducting cross-section, but only in a thin surface layer (current displacement). The time-dependent magnetic field induces a voltage in the interior of the conductor that is opposite to the applied external voltage and decreases towards the border. Hence, the current density J increases towards the surface:

$$J(r,t) = \hat{J}(r) \cos(2\pi\, ft + \phi(r)) \,, \quad \hat{J} = \hat{J}(R)\, e^{h(r)} \,,$$

with

$$h(r) = -\frac{\delta^2 R^4}{4} \left[1 - \left(\frac{r}{R} \right)^4 \right] \,,$$

$$\phi(r) = \phi(R) + \frac{\delta^2 R^4}{4} \left[1 - \left(\frac{r}{R} \right)^2 \right] \,,$$

$$\delta = \mu_0\, \kappa\, \omega \,,$$

where R denotes the conductor radius, r the distance from the conductor axis and κ the conductivity of the conductor.

14.18.2 Transformer induction

Transformer induction, induction of voltages in a conductor by a change of the surrounding magnetic field. The change of the magnetic flux $\Delta\Phi$ is determined by the change of the magnetic field ΔB,

$$\Delta\Phi = \Delta B \cdot A \cos\alpha \,,$$

where α is the angle between the flux density vector and the normal to the plane of the conductor loop.

voltage $\sim$ $\dfrac{\text{change of flux density}}{\text{time interval}} \cdot$ area			$L^2T^{-3}MI^{-1}$
	Symbol	Unit	Quantity
$V_{\text{ind}} = -n\dfrac{d\vec{B}}{dt} \cdot \vec{A}$	V_{ind}	V	induced voltage
	$d\vec{B}$	Vs/m^2	change of magnetic flux density
	dt	s	infinitesimal time interval
	$\vec{A}$	m^2	area
	n	1	number of current loops

Transformer induction is applied in transformers.

■   A test coil is placed in a current-carrying coil. Switching off the current, and thus the magnetic field, causes induction of a voltage pulse in the test coil.

Eddy-current losses, arise in transformers when the flux through the iron core changes. The eddy-current losses are reduced if the iron core is composed of metallic strips that are electrically isolated from each other by a varnish layer.

Switching on (off) the coil current causes high voltage peaks.

▲   **Lenz's law**, the magnetic field of an induced current opposes the change of the external magnetic field.

14.19   Self-induction

1. Self-induction,

a change of current I in a coil of n turns causes a change of the magnetic flux through this coil, and hence induces a voltage in the coil. The induced voltage is proportional to the change of current per unit time.

Self-inductance, **inductance**, L, property of the coil, proportionality factor between induced voltage and change of current per unit time.

▲   According to Lenz's rule, the induced voltage opposes the applied voltage:

induced voltage = winding number $\cdot$ $\dfrac{\text{change of flux}}{\text{time interval}}$			$L^2T^{-3}MI^{-1}$
	Symbol	Unit	Quantity
$v_{\text{ind}} = -L \cdot \dfrac{dI}{dt}$	v_{ind}	V	induced voltage
	dI	A	change of current
	dt	s	infinitesimal time interval
	L	H = Vs/A	inductance

Henry, H, SI unit of the inductance L,

$$[L] = H = Vs/A.$$

1 H is a very large unit. Usual inductances are in the range between $1\,\mu H = 10^{-6}$ H and 1 H.

▲   The inductance of a coil equals the product of the square of the number of turns n and the magnetic conductance Λ_m,

$$L = n^2 \cdot \Lambda_m.$$

2. Induction flux,

ψ, through a coil, product of the magnetic flux Φ and the number n of turns of the coil. The induction flux is proportional to the coil current I. The proportionality factor is the inductance L.

induction flux = inductance · current			$\mathbf{L^2T^{-2}MI^{-1}}$
	Symbol	Unit	Quantity
$\psi = L \cdot I$	ψ	Wb = Vs	induction flux
	L	H = Vs/A	inductance of coil
$= n \cdot \Phi$	I	A	current through coil
	n	1	number of turns of coil
	Φ	Wb = Vs	magnetic flux through coil

Weber, Wb, SI unit of the induction flux ψ,

$$[\psi] = \text{Wb} = \text{Vs}.$$

3. Series connection of inductances

Series connection of inductances, the total inductance L_{tot} of a series connection of inductances equals the sum of the individual inductances $L_1, \ldots, L_N$ (**Fig. 14.38**):

$$L_{\text{tot}} = L_1 + L_2 + \cdots + L_N .$$

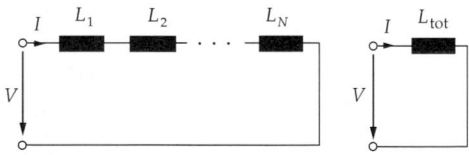

Figure 14.38: Series connection of inductances.

4. Parallel connection of inductances

Parallel connection of inductances, the reciprocal value of the total inductance L_{tot} of a parallel connection of inductances equals the sum of the reciprocal values of the individual inductances $L_1, \ldots, L_N$ (**Fig. 14.39**):

$$\frac{1}{L_{\text{tot}}} = \frac{1}{L_1} + \frac{1}{L_2} + \cdots + \frac{1}{L_N} .$$

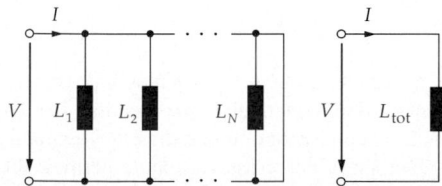

Figure 14.39: Parallel connection of inductances.

14.19.1 Inductances of geometric arrangements of conductors

a) Single line (**Fig. 14.40 (a)**):

$$L = \frac{\mu l}{2\pi} \left[\ln\left(\frac{2l}{r}\right) - \frac{3}{4} \right], \qquad r \text{ wire radius, } l \text{ wire length, } \mu \text{ permeability.}$$

b) Two-wire line, circular cross-section (**Fig. 14.40 (b)**):

$$L = \frac{\mu l}{\pi} \left[\ln\left(\frac{d}{r}\right) + \frac{1}{4} \right], \qquad r \text{ wire radius, } d \text{ wire distance, } l \text{ wire length.}$$

c) Two-wire line, rectangular cross-section (**Fig. 14.40 (c)**):

$$L = \frac{\mu l}{\pi} \frac{2a}{a+b}, \quad a \ll b, \ d \ll b$$

$$L = \frac{2\mu l}{\pi} \ln\left(1 + \frac{a}{a+b}\right), \quad d \ll a, \ d \ll b, \quad a, b \text{ edge lengths, } l \text{ wire length, } d \text{ wire distance.}$$

d) Ring conductor:

$$L = \mu R \left[\ln\left(\frac{R}{r}\right) + \frac{1}{4} \right], \qquad r \text{ conductor radius, } R \text{ ring radius.}$$

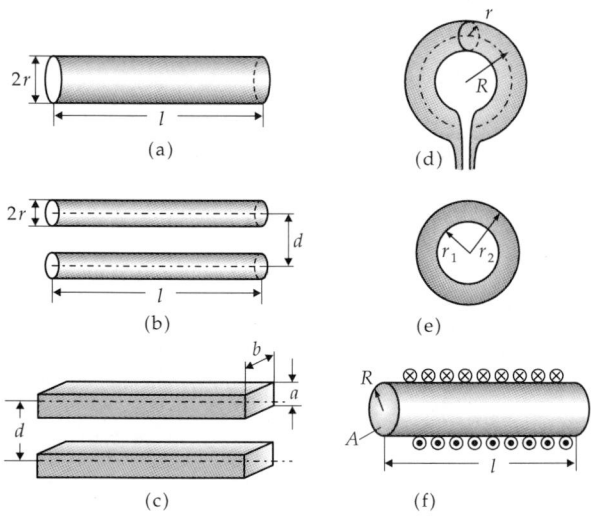

Figure 14.40: Inductances of various arrangements of conductors. (a): single line; (b): two-wire line, circular cross-section; (c): double-wire line, rectangular cross-section; (d): ring coil; (e): coaxial conductor; (f): cylindrical coil. r: conductor or coil radius, R: ring radius, l: conductor or coil length, d: distance of conductors, A: coil cross-section, μ: permeability.

e) Coaxial conductor (**Fig. 14.40 (e)**):

$$L = \frac{\mu l}{2\pi} \ln\left(\frac{r_2}{r_1}\right),$$
r_1 radius inner conductor, r_2 radius outer conductor,
l conductor length.

f) Long cylindrical coil, ring coil $l \gg r$ (**Fig. 14.40 (d)**):

$$L \approx \frac{\mu}{l} A n^2,$$
l cylinder length (mean ring circumference), A coil area, n turn number.

g) Short coil, one layer of windings

$$L = f \frac{\mu}{l} A n^2, \quad f \approx \frac{1}{1 + r/l}$$
l coil length, r coil radius, f coil form factor,
A coil area, n turn number.

14.19.2 Magnetic conductance

Magnetic conductance, Λ_m, a quantity that depends on the geometry and permeability of the magnetic circuit. The magnetic conductance of coil cores is specified by the manufacturer.

Henry, H, SI unit of the magnetic conductance Λ_m,

$$[\Lambda_m] = H = Vs/A.$$

The magnetic conductance of a **toroidal coil without iron** is obtained from the cross-sectional area A penetrated by the magnetic field, the mean length l of the magnetic field lines, and the magnetic free-space permeability constant μ_0,

$$\Lambda_m = \mu_0 \cdot \frac{A}{l}.$$

The magnetic conductance of a **toroidal coil with iron** is obtained from the cross-sectional area A penetrated by the magnetic flux, the mean length of the field lines, the permeability of free space μ_0 and the relative permeability μ_r of iron,

$$\Lambda_m = \mu_0 \cdot \mu_r \cdot \frac{A}{l}.$$

Magnetic reluctance R_m, reciprocal value of the magnetic conductance,

$$R_m = \frac{1}{\Lambda_m}.$$

The concept of magnetic reluctance is used in calculations of magnetic circuits.

■   A coil with an iron core of magnetic conductance $\Lambda_m = 5$ μH carries 40 windings. The inductance of this coil is

$$L = n^2 \cdot \Lambda_m = 40^2 \cdot 5 \cdot 10^{-6} \text{ H} = 8 \cdot 10^{-3} \text{ H} = 8 \text{ mH}.$$

14.20 Mutual induction

1. Magnetic coupling

of two coils occurs if both coils are traversed by the same magnetic flux Φ (**Fig. 14.41** and **14.42**).

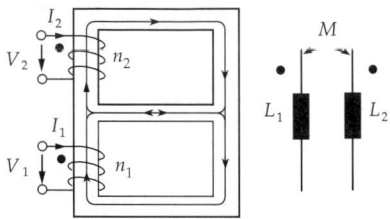

Figure 14.41: Magnetic coupling of two coils with windings in the same direction.

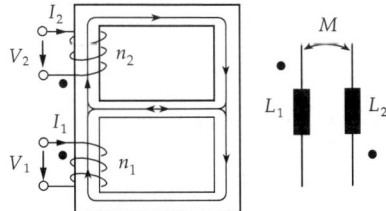

Figure 14.42: Magnetic coupling of two coils with windings in the opposite direction.

When the magnetic flux through one of the magnetically coupled coils is changing, a voltage pulse is induced in the other coil.

Starting from the first coil with a coil current I_1 generating the magnetic flux Φ_1, the following notations are introduced:

Useful flux Φ_N, the fraction of the magnetic flux that traverses the second coil:

$$\Phi_N = k_1 \cdot \Phi_1.$$

Coupling coefficient, k_1, denotes the fraction of the magnetic flux traversing the second coil.

Stray flux, Φ_S, the fraction of the magnetic flux that is lost:

$$\Phi_S = \Phi_1 - \Phi_N = (1 - k_1) \cdot \Phi_1.$$

■ In a real transformer, part of the magnetic flux is lost as stray flux.

2. Mutual inductance,

M, gives the induction flux through the second coil that is caused by the current I_1 in the windings of the first coil. The mutual inductance is proportional to the product of the number of turns n_1 and n_2 of the two coils, the magnetic inductance of the first coil, and the coupling coefficient k_1.

mutual inductance			$\mathbf{L^2T^{-2}MI^{-2}}$
	Symbol	Unit	Quantity
$M = k_1 \Lambda_1 n_1 n_2$	M	H	mutual inductance
	Λ_1	H	magnetic conductance
	n_1, n_2	1	numbers of turns
	k_1	1	coupling coefficient

Henry, H, SI unit of the mutual inductance M,

$$[M] = \text{H}.$$

▲   Assuming a constant permeability, the mutual inductance of two coupled coils is identical.

14.20.1   Transformer

1. Transformer,

converts low voltages to higher voltages, or vice versa. A transformer consists of a primary coil and a secondary coil, both traversed by the same magnetic flux (**Fig. 14.43**).

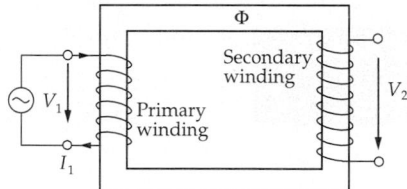

Figure 14.43: Transformer.

Primary winding, denotes the coil supplied by the (primary) voltage to be transformed.
Secondary winding, denotes the coil delivering the transformed (secondary) voltage.
Ideal transformer, a transformer without loss of power.
Efficiency of real transformers, better than 95 % for good transformers.
▲   If an alternating voltage is applied to the primary coil, then the magnetic flux through this coil varies, and thereby a voltage is induced in the secondary coil.

2. Transmission ratio,

u, gives the ratio of the voltage at the primary side to the voltage at the secondary side. If u is greater than unity, the voltage is transformed down; if u is less than unity, the voltage is transformed up. The phase shift between the voltages is 180° (Lenz's law).
    For ideal transformers, the ratio of voltages is

$$\frac{V_1}{V_2} = u = \frac{n_1}{n_2},$$

and the ratio of currents is

$$\frac{I_1}{I_2} = \frac{n_2}{n_1}.$$

▲   The ratio of the voltages at the primary coil and the secondary coil equals the reciprocal value of the ratio of the corresponding currents.
➤   If the voltage to be transformed contains a direct-current component, then this part is not transmitted: the voltage induced at the secondary side is a purely alternating voltage. Hence, the transformer may also be used for separating the alternating-current component from the direct-current component. This principle is used, e.g., in amplifier circuits.

3. Example: transformer

The primary coil of a transformer has $n_1 = 100$ turns, the secondary coil has $n_2 = 250$ turns. Let the voltage at the primary coil be $V_1 = 12$ V. Then, the voltage at the secondary side is

$$V_2 = \frac{n_2}{n_1} \cdot V_1 = \frac{250}{100} \cdot 12 \text{ V} = 30 \text{ V}.$$

If the secondary coil is connected to a load resistance $R = 300 \ \Omega$, the current is $I_2 = 0.1$ A. The current in the primary coil is

$$I_1 = \frac{n_2}{n_1} \cdot I_2 = \frac{250}{100} \cdot 0.1 \text{ A} = 0.25 \text{ A}.$$

14.21   Energy and energy density of the magnetic field

1. Energy density of the magnetic field,

magnetic energy ΔW_m per volume ΔV. If the energy is not uniformly distributed over the volume, ΔV is reduced until the energy in ΔV can be considered to be spatially uniform:

$$w_m = \lim_{\Delta V \to 0} \frac{\Delta W_m}{\Delta V} = \frac{\mathrm{d} W_m}{\mathrm{d} V} .$$

Generally, the energy density is the integral of the field strength $\vec{H}$ over the magnetic flux density $\vec{B}$:

$$w_m = \int\limits_0^{B_{\max}} \vec{H} \cdot \mathrm{d}\vec{B} .$$

If the magnetization characteristics is linear, i.e., the magnetic induction B varies linearly with the magnetic field strength H, the energy density w_m is proportional to the product of B and H:

magnetic energy density $= \dfrac{\text{magnetic flux density·magnetic field strength}}{2}$			$\mathbf{ML^{-1}T^{-2}}$
	Symbol	Unit	Quantity
$w_m = \dfrac{1}{2}\vec{B} \cdot \vec{H}$	w_m	J/m^3	magnetic energy density
	$\vec{B}$	Vs/m^2	magnetic flux density
	$\vec{H}$	A/m	magnetic field strength

The energy density is then proportional to the shadowed area in **Fig. 14.44**.

Hysteresis losses, the energy put in during magnetization is larger than the energy released in demagnetization. The energy difference is released as heat. The area enclosed by the hysteresis curve is a measure of the energy loss per magnetization cycle.

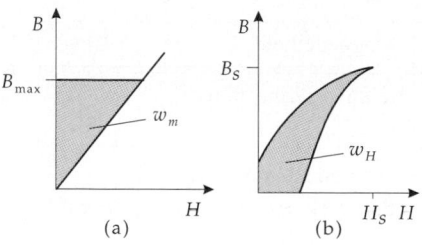

Figure 14.44: Energy density w_m of a magnetic field. (a): linear magnetization curve, (b): magnetization work of a hysteresis curve, B_S, H_S: saturation induction and saturation field strength, respectively.

2. Energy of magnetic field,

W_m, is obtained by integrating the energy density over the volume V occupied by the field. The energy of the magnetic field in a material with linear magnetization curve is given by

$$W_m = \int_V w_m \, dV = \frac{1}{2} \int_V \vec{H} \cdot \vec{B} \, dV \,.$$

Field energy of a coil, W_m, proportional to the square of the coil current I:

energy ~ inductance · current^2			$\mathbf{L^2 T^{-2} M}$
	Symbol	Unit	Quantity
$W_m = \dfrac{1}{2} L I^2$	W_m	J	magnetic energy
	L	H	inductance
	I	A	coil current

3. Analogy between electric and magnetic quantities

Electric field	Unit	Magnetic field	Unit
permittivity of free space $\varepsilon_0 = 1/(\mu_0 c^2)$	As/(Vm)	permeability of free space $\mu_0 = 1/(\varepsilon_0 c^2)$	Vs/(Am)
electric field strength $E = -\dfrac{dV}{ds}$	V/m	magnetic field strength $H = \dfrac{dI}{dl}$	A/m
electric voltage $V_{AB} = -\int_A^B \vec{E} \, d\vec{s}$	V	magnetic potential difference $V_{AB} = \int_A^B \vec{H} \, d\vec{s}$	A
electric current $I = \dfrac{dQ}{dt}$	A	induced voltage $V = -n \dfrac{d\Phi}{dt}$	V
electric charge $Q = \int I(t) \, dt$	As	magnetic flux $\Phi = B \, A$	Vs
permittivity $\varepsilon = \varepsilon_0 \varepsilon_r$	As/(Vm)	permeability $\mu = \mu_0 \mu_r$	Vs/(Am)

(*continued*)

3. Analogy between electric and magnetic quantities (continued)

Electric field	Unit	Magnetic field	Unit
relative permittivity ε_r	1	relative permeability μ_r	1
displacement density $\vec{D} = \varepsilon \vec{E}$	As/m^2	magnetic flux density $\vec{B} = \mu \vec{H}$	Vs/m^2
electric force $\vec{F} = Q\vec{E}$	N	magnetic force $\vec{F} = Q(\vec{v} \times \vec{B})$	N
electric dipole moment $\vec{p} = Q\vec{l}$	As m	magnetic dipole moment $\vec{m} = \Phi \cdot \vec{l}$	Vs m
capacitance $C = \dfrac{Q}{V}$	F	inductance $L = -\dfrac{V}{dI/dt}$	H
electric energy density $w_e = \dfrac{1}{2}\vec{D}\vec{E} = \dfrac{1}{2}\varepsilon\vec{E}^2$	Ws/m^3	magnetic energy density $w_m = \dfrac{1}{2}\vec{B}\vec{H} = \dfrac{1}{2}\mu\vec{H}^2$	Ws/m^3
electric energy of a capacitor $W_e = \dfrac{1}{2}CV^2$	J	magnetic energy of a coil $W_m = \dfrac{1}{2}LI^2$	J

14.22   Maxwell's equations

There are four **Maxwell equations**.

1. It follows from electrostatics that the electric field is a source field. The electric flux through a closed surface A is equal to the charge in the enclosed volume:

$$Q = \int_V \rho\, dV = \varepsilon_0 \oint_A \vec{E} \cdot d\vec{A} = \oint_A \vec{D} \cdot d\vec{A}\,.$$

2. The fact that no magnetic monopoles have been found suggests that the magnetic field is source-free. The total magnetic flux through a closed surface A vanishes:

$$\oint_A \vec{B} \cdot d\vec{A} = 0\,.$$

➤   This equation would need to be changed if magnetic monopoles were shown to exist. By analogy to the electric charge, the integral over the magnetic charge density would appear on the right-hand side.

3. It follows from the induction theorem that a change of the magnetic flux through a conductor loop would cause a voltage at the ends of the conductor. If the ends of

the conductor are connected, a current flows in the conductor. The induction theorem may be written in general form as follows:

$$\oint_s \vec{E} \cdot d\vec{s} = -\int_A \frac{d\vec{B}}{dt} \cdot d\vec{A}.$$

The time rate of variation of the magnetic flux density $\vec{B}$ integrated over a surface $\vec{A}$ equals the line integral of the electric field strength $\vec{E}$ along the closed path s about this surface.

▲ Any time-varying magnetic field generates a circulating electric field (**Fig. 14.45 (b)**).

4. The last of Maxwell's equations is obtained by introducing the displacement current:

$$I + \int_A \frac{d\vec{D}}{dt} \cdot d\vec{A} = \int_A \left(\vec{J} + \frac{d\vec{D}}{dt}\right) \cdot d\vec{A} = \oint_s \vec{H} \cdot d\vec{s}.$$

▲ Any time-varying electric field generates a circulating magnetic field (**Fig. 14.45 (a)**).

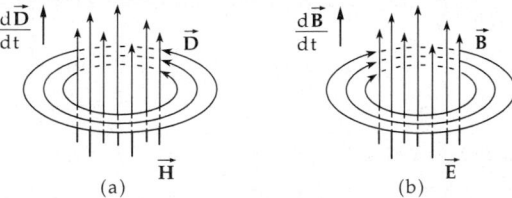

(a)   (b)

Figure 14.45: (a): time-dependent electric fields generate a circulating magnetic field, (b): time-dependent magnetic fields generate a circulating electric field.

14.22.1 Displacement current

It follows from magnetostatics that the magnetic field is always a circulating field. The magnetic field $\vec{H}$ summed along a closed path s equals the current I enclosed by the path:

$$I = \int_A \vec{J} \cdot d\vec{A} = \oint_s \vec{H} \cdot d\vec{s}.$$

The current I is the integral of the current density $\vec{J}$ over the surface A enclosed by the path.

1. Displacement current,

corresponds to the time rate of variation of the electric displacement density $\vec{D}$. In a circuit with a capacitor, a current flows until the capacitor is charged. The current is surrounded by a magnetic field. While the capacitor is charged, the electric field strength between the capacitor plates is changing. If there is a dielectric medium between the capacitor plates, the charges in the dielectric are displaced (polarization). This displacement of charges in turn generates a magnetic field.

Taking the displacement current into account, one arrives at the last of Maxwell's equations:

$$I + \int_A \frac{d\vec{D}}{dt} \cdot d\vec{A} = \int_A \left(\vec{J} + \frac{d\vec{D}}{dt} \right) \cdot d\vec{A} = \oint_s \vec{H} \cdot d\vec{s}.$$

➤   The system of Maxwell's equations is not complete without the displacement current.

2. Maxwell's equations in integral and differential form

Besides the integral form, the Maxwell equations may also be written in differential form.

Maxwell's equations in differential form		
meaning	integral form	differential form
solenoidality of magnetic field	$\oint_O \vec{B} \cdot d\vec{A} = 0$	$\operatorname{div} \vec{B} = 0$
displacement flux through surface equals the enclosed electric charge	$\oint_O \vec{D} \cdot d\vec{A} = Q$	$\operatorname{div} \vec{D} = \rho$
Faraday's induction law: time-varying magnetic fields generate an electric field	$\oint_s \vec{E} \cdot d\vec{s} = -\frac{\partial}{\partial t} \int \vec{B} \cdot d\vec{A}$	$\operatorname{rot} \vec{E} = -\frac{\partial \vec{B}}{\partial t}$
Ampere's law with Maxwell's supplement: time-varying electric fields generate a magnetic field	$\oint_s \vec{H} \cdot d\vec{s} = \frac{\partial}{\partial t} \int \vec{D} \cdot d\vec{A} + I$	$\operatorname{rot} \vec{H} = \frac{\partial \vec{D}}{\partial t} + \vec{J}$

14.22.2   Electromagnetic waves

From Maxwell's equations, it follows that a conductor in which charges are oscillating is surrounded alternating by electric and magnetic fields. The time-varying electric fields generate magnetic fields, the time-varying magnetic fields induce electric fields.

1. Electromagnetic waves,

propagation of electric and magnetic fields in space. Electromagnetic waves are propagating solutions of Maxwell's equations. Electromagnetic waves are transmitting energy. The spectrum is ranging from long-wave radio waves up to light waves and γ-quanta from the decay of atomic nuclei, or from energetic cosmic radiation (see table on p. 542).

➤   Electromagnetic waves in the range of radio waves may be generated by oscillator circuits.

2. Wave equation and its solution

Wave equations (see p. 287) for the fields $\vec{E}$ and $\vec{H}$ in a vacuum ($\rho = 0$, $\vec{J} = 0$):

$$\Delta \vec{E} - \mu_0 \varepsilon_0 \frac{\partial^2 \vec{E}}{\partial t^2} = 0, \quad \Delta \vec{H} - \mu_0 \varepsilon_0 \frac{\partial^2 \vec{H}}{\partial t^2} = 0.$$

Monochromatic solutions:

- plane wave moving along the direction $\vec{\mathbf{k}}$,

$$\vec{\mathbf{E}} = \vec{\mathbf{E}}_0\, e^{-j(\omega t - \vec{\mathbf{k}}\vec{\mathbf{r}})}, \quad \vec{\mathbf{H}} = \vec{\mathbf{H}}_0\, e^{-j(\omega t - \vec{\mathbf{k}}\vec{\mathbf{r}})},$$

- spherical waves emerging from the point $\vec{\mathbf{r}} = 0$ (upper sign) or converging to the point $\vec{\mathbf{r}} = 0$ (lower sign),

$$\vec{\mathbf{E}} = \vec{\mathbf{E}}_0\, e^{-j(\omega t \mp kr)}, \quad \vec{\mathbf{H}} = \vec{\mathbf{H}}_0\, e^{-j(\omega t + kr)}.$$

The vectors $\vec{\mathbf{E}}_0$, $\vec{\mathbf{H}}_0$ specify the intensity and polarization direction of the electromagnetic wave.

3. Speed of light in a vacuum,

c_0, propagation velocity of electromagnetic waves in a vacuum, a natural constant. The speed of light in a vacuum connects the electric free-space permittivity constant ε_0 and the magnetic free-space permeability constant μ_0:

speed of light in a vacuum			LT^{-1}
$c_0 = 299\,792\,458\ \text{m/s}$ $c_0 = \dfrac{1}{\sqrt{\varepsilon_0 \cdot \mu_0}}$	Symbol	Unit	Quantity
	c_0	m/s	vacuum speed of light
	ε_0	As/(Vm)	electric free-space permittivity constant
	μ_0	Vs/(Am)	permeability of free space

4. Speed of light in matter,

c, propagation velocity of electromagnetic waves in matter. The electric free-space permittivity constant is to be replaced by the permittivity $\varepsilon = \varepsilon_r \cdot \varepsilon_0$, and the magnetic free-space permeability constant by the permeability $\mu = \mu_r \cdot \mu_0$ of matter, respectively:

speed of light in matter			LT^{-1}
$c = \dfrac{1}{\sqrt{\varepsilon \cdot \mu}} = \dfrac{1}{\sqrt{\varepsilon_r \cdot \mu_r}} \cdot c_0$	Symbol	Unit	Quantity
	c	m/s	speed of light in matter
	ε	As/(Vm)	permittivity
	μ	Vs/(Am)	permeability
	ε_r	1	relative permittivity
	μ_r	1	relative permeability
	c_0	m/s	vacuum speed of light

5. Energy law of electrodynamics

One may derive the energy law of electrodynamics from Maxwell's equations:

$$\frac{\partial}{\partial t}\left(\frac{\vec{\mathbf{E}}\vec{\mathbf{D}} + \vec{\mathbf{H}}\vec{\mathbf{B}}}{2}\right) + \operatorname{div}(\vec{\mathbf{E}} \times \vec{\mathbf{H}}) = -\vec{\mathbf{J}}\vec{\mathbf{E}}.$$

The first term on the left-hand side describes the time rate of variation of the energy density w of the electromagnetic field,

$$w = w_e + w_m = \frac{\vec{\mathbf{E}}\vec{\mathbf{D}} + \vec{\mathbf{H}}\vec{\mathbf{B}}}{2}.$$

The second term on the left-hand side is the divergence of the energy flux density $\vec{S}$ (Poynting vector) of the electromagnetic field,

$$\vec{S} = \vec{E} \times \vec{H}.$$

The expression on the right-hand side of the equation represents the conversion of electromagnetic energy into other kinds of energy per unit time and unit volume.

14.22.3   Poynting vector

Poynting vector, $\vec{S}$, specifies magnitude and direction of the energy transport in electromagnetic fields. The Poynting vector at a given space point is obtained by the vector product of the electric field strength $\vec{E}$ and the magnetic field strength $\vec{H}$ at this space point. The Poynting vector has the dimension of an energy flux density.

Poynting vector = electric field strength × magnetic field strength			$T^{-3}M$
	Symbol	Unit	Quantity
$\vec{S} = \vec{E} \times \vec{H}$	$\vec{S}$	W/m^2	Poynting vector
	$\vec{E}$	V/m	electric field strength
	$\vec{H}$	A/m	magnetic field strength

Watt/square meter, W/m^2, SI unit of the Poynting vector $\vec{S}$. 1 watt/square meter is the magnitude of the Poynting vector at a space point where the electric field strength is $E = 1$ V/m and the magnetic field strength is $H = 1$ A/m and the field strength vectors are perpendicular to each other,

$$[\vec{S}] = \text{W/m}^2.$$

The energy W transported per unit time dt through a surface A is given by the integral of the Poynting vector over the surface:

$$\frac{dW}{dt} = \int_A \vec{S} \cdot d\vec{A}.$$

For free electromagnetic waves:
▲   The magnitude of the Poynting vector equals half of the product of the energy density of the electromagnetic wave and the speed of light,

$$S = \frac{c}{2}(w_e + w_m), \quad w_e = \frac{1}{2}\vec{E}\vec{D}, \quad w_m = \frac{1}{2}\vec{B}\vec{H}.$$

15
Applications in electrical engineering

1. Electric circuit,

consists of **source** and **load**, connected to each other so that an electric current may flow.

In a circuit, an electric field is generated by the source. The current flows through lines and loads from higher potential to lower potential.

Generally, electric circuits are treated in **network theory**.

In network theory, the sources and loads are generalized to **network elements** denoted two-terminal, four-terminal, etc., according to the number of external connection lines.

Two-terminal network, a network element with two external connections.

Active two-terminal network, a two-terminal network capable of releasing energy.

Passive two-terminal network, a two-terminal network that does not release energy.

■ An ohmic resistor is a passive two-terminal network (**Fig. 15.1 (a)**).

Sources of currents and voltages are active two-terminal networks.

Capacitors and inductors are mostly passive two-terminal networks. During discharging, a capacitor behaves as a source of voltage, and a coil behaves as a source of current after the current is switched off (**Fig. 15.1 (c), (b)**).

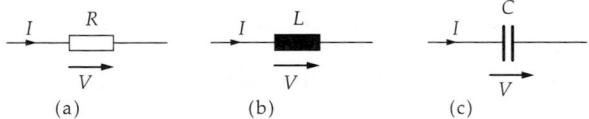

Figure 15.1: Circuit symbols. (a): resistor, (b): inductor, (c): capacitor.

Two-port network, network element with four external connections, one pair of input terminals and one pair of output terminals.

2. Voltage and current sources

Voltage sources and current sources are classified as ideal or real sources. **Ideal voltage source**, supplies a voltage that is independent of the extracted current.

▲ The internal resistance of an ideal voltage source is equal to zero.

Ideal current source, supplies a current that is independent of the applied voltage.
▲   The internal resistance of an ideal current source is infinite.
   In general, the assumption of an ideal current source or voltage source is always an approximation. If the finite internal resistance has to be taken into account, then real current and voltage sources have to be used in the considerations.
   Circuit symbols for ideal voltage and current sources are shown in **Fig. 15.2 (a)**.
   Direct-voltage source, supplies a voltage constant in time (**direct voltage**). Circuit symbols are shown in **Fig. 15.2 (b)**.
   Alternating-current source, supplies a voltage varying with time (**alternating voltage**). The circuit symbol is shown in **Fig. 15.2 (c)**.

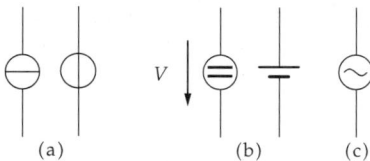

(a)   (b)   (c)

Figure 15.2: Circuit symbols. (a): ideal voltage and current source, (b): direct-voltage source, (c): alternating-voltage source.

   Depending on the type of source, one distinguishes **direct-current circuits** and **alternating-current circuits**.

15.1 Direct-current circuit

1. Direct voltage and direct current

Direct voltage, electric voltage constant in time with respect to magnitude and direction.
   To be distinguished from **rectified alternating voltage**, a voltage of time-independent polarity, but of magnitude varying in time in a wavelike manner. Rectification is achieved by a special circuit that usually includes diodes.
   Direct current, an electric current constant in magnitude and direction.
   Purely direct voltage is produced in electrochemical reactions, e.g., in accumulators and galvanic elements.
▲   The voltage V is symbolized by an arrow pointing from the higher potential value to the lower potential value.
▲   The current in the conductor flows from the positive pole to the negative pole of the voltage source (definition of the current direction).
Load current definition, usual convention in electric engineering. The direction of current and voltage are identical in the load.
   Therefore, the power released in the load is positive,

$$P_V = V \cdot I > 0 \,.$$

Within the voltage source, the current flow is opposed to the voltage.
   The power of the voltage source is therefore negative,

$$P_Q = V \cdot I < 0 \,.$$

2. Open-circuit voltage and terminal voltage

Open-circuit voltage, electromotive force, V_Q, denotes the voltage of an ideal voltage source.

Terminal voltage, V_C, specifies the voltage tapped at the voltage source by the external load. It is lower than the open-circuit voltage V_Q because of the finite internal resistance of the source,

$$V_C < V_Q.$$

Ideal voltage source, a voltage source with a terminal voltage that is independent of the load. Its internal resistance is equal to zero.

▲ The terminal voltage of an ideal voltage source is equal to the open-circuit voltage.

3. Network,

interconnection of electric components (**Fig. 15.3**) consisting of:
- **branch point**, a point connecting at least three feed lines,
- **branch**, an interconnection of components between two branch points,
- **mesh**, a closed chain of branches.

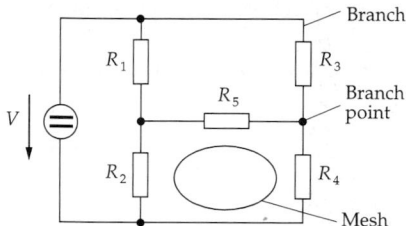

Figure 15.3: Network. Mesh, branch and branch point.

15.1.1 Kirchhoff's laws for direct-current circuit

Kirchhoff's laws enable the calculation of direct-current circuits.

1. Kirchhoff's first law, or branch-point rule

▲ The sum of all currents at a branch point is equal to zero:

$$I_1 + I_2 + I_3 + \cdots + I_N = 0.$$

The currents flowing out (in) are taken to be positive (negative).

➤ The branch-point rule follows from the law of the conservation of electric charge.

2. Kirchhoff's second law, or mesh rule

▲ The sum of all voltages around a mesh is equal to zero:

$$V_1 + V_2 + V_3 + \cdots + V_N = 0.$$

Voltages along (against) the circulation direction of the mesh are taken to have a positive (negative) sign.

➤ The partial voltages represent the work per test charge required to move the test charge through the corresponding sections of the mesh. The mesh rule follows from energy conservation (see p. 446).

15.1.2 Resistors in a direct-current circuit

Kirchhoff's laws may be used to calculate the total resistance of series connections or parallel connections of resistors.

1. Series connection of resistors

In a **series connection** of N resistors (see **Fig. 15.4**), each individual resistor carries the same current I. According to the mesh rule, the voltages V_i over the individual resistors sum to the total voltage V.

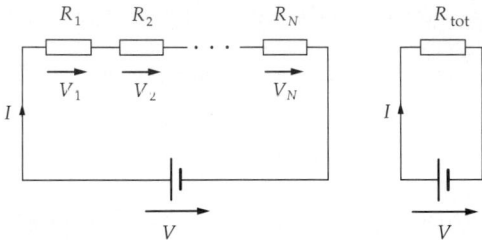

Figure 15.4: Series connection of resistors and equivalent circuit diagram.

▲   The total resistance of a series connection of resistors is equal to the sum of the individual resistances.

series connection of resistors			$\mathbf{L^2 T^{-3} M I^{-2}}$
	Symbol	Unit	Quantity
$R_{\text{tot}} = R_1 + R_2 + \cdots + R_N$	R_{tot}	Ω	total resistance
	R_i	Ω	individual resistances

The N resistors R_i may be replaced by a total resistor R_{tot}.

Voltage divider rules

▲   The ratio of the partial voltage V_i across a single resistor to the total voltage V across the total resistor is equal to the ratio of the single resistance R_i to the total resistance R_{tot}:

$$\frac{V_i}{V} = \frac{R_i}{R_{\text{tot}}} .$$

▲   Two partial voltages V_i and V_j are related to each other as the partial resistances R_i and R_j corresponding to the partial voltages:

$$\frac{V_i}{V_j} = \frac{R_i}{R_j} .$$

Voltage divider, a series connection of ohmic resistors supplied with a total voltage V. The resistors are chosen so that the desired voltage V_i can be tapped from the resistor chain.

2. Parallel connection of resistors

In a **parallel connection** of N resistors (see **Fig. 15.5**), each of the resistors is supplied by the same voltage V. According to the branch-point rule, the currents through the individual resistors sum to the total current I. The N resistors R_i may be replaced by a single resistor of total resistance R_{tot}.

▲   The reciprocal value of the total resistance of a parallel connection of resistors equals the sum of the reciprocal values of the individual resistances.

parallel connection of resistors			$\mathbf{L^{-2}T^3M^{-1}I^2}$
$\dfrac{1}{R_{\text{tot}}} = \dfrac{1}{R_1} + \dfrac{1}{R_2} + \cdots + \dfrac{1}{R_N}$ $G_{\text{tot}} = G_1 + G_2 + \cdots + G_N$	Symbol	Unit	Quantity
	R_{tot}	Ω	total resistance
	R_i	Ω	individual resistances
	G_{tot}	S	total conductance
	G_i	S	individual conductances

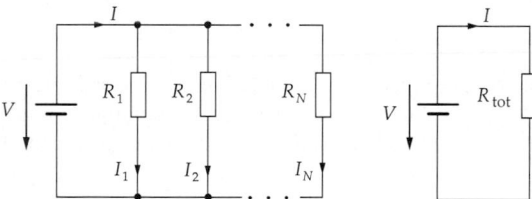

Figure 15.5: Parallel connection of resistors and equivalent circuit diagram.

Expressed in terms of conductances (reciprocal values of the resistances):

▲ **The total conductance of a parallel connection of resistors is equal to the sum of the individual conductances.**

For a parallel connection of two resistors R_1 and R_2, the total resistance is given by

$$R_{\text{tot}} = \frac{R_1 \cdot R_2}{R_1 + R_2} .$$

Current division rules

▲ The ratio of the partial current I_i through a single resistor R_i to the total current I is equal to the ratio of the conductance G_i of the single resistor to the total conductance G_{tot}:

$$\frac{I_i}{I} = \frac{G_i}{G_{\text{tot}}} = \frac{R_{\text{tot}}}{R_i} .$$

▲ Two partial currents I_i and I_j are related to each other as the individual conductances G_i and G_j:

$$\frac{I_i}{I_j} = \frac{G_i}{G_j} = \frac{R_j}{R_i} .$$

15.1.3 Real voltage source

1. Real voltage source,

has a finite internal resistance $R_i \neq 0$ (**Fig. 15.6**).

The magnitude of the current in the circuit is determined by the load resistance R_a and the internal resistance R_i of the current source:

$$I = \frac{V_Q}{R_a + R_i} .$$

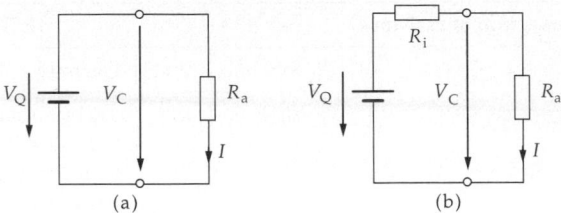

Figure 15.6: Load connected to voltage source. (a): ideal voltage source, (b): real voltage source.

▲ The terminal voltages of real voltage sources depend on the external load. The terminal voltage V_C is equal to the source voltage V_Q multiplied by the ratio of the load resistance R_a to the sum of load resistance and internal resistance R_i:

$$V_C = \frac{R_a}{R_a + R_i} V_Q.$$

2. Short-circuit current and open-circuit voltage

Short-circuit current, I_K, the current flowing if the external resistance R_a equals zero (**Fig. 15.7 (a)**). It is given by the ratio of source voltage V_Q to internal resistance R_i:

$$I_K = \frac{V_Q}{R_i}.$$

For a given source voltage, the short-circuit current depends only on the internal resistance of the voltage source.

Open-circuit voltage, V_L, obtained when no external load is connected to the voltage source (**Fig. 15.7 (b)**). The external resistance is infinite, the current vanishes:

$$V_L = V_Q.$$

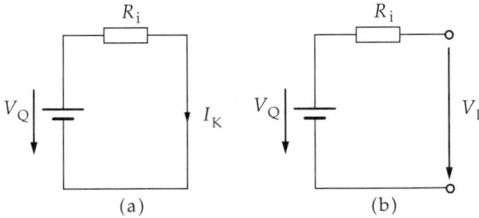

Figure 15.7: Real voltage source. (a): shorted, (b): unloaded.

M The internal resistance of a real voltage source may be determined by measuring the short-circuit current and the open-circuit voltage if R_i is independent of the current. Since measuring instruments have a finite resistance, both the short-circuit current and the open-circuit voltage can be determined only approximately.

15.1.4 Power and energy in the direct-current circuit

1. Power in the direct-current circuit

A load in a direct-current circuit supplied by a voltage V and a current I receives the power P:

power = voltage · current		$L^2T^{-3}M$
	Symbol	Unit	Quantity
$P = V \cdot I$	P	W	power
	V	V	voltage
	I	A	current

If R is the ohmic resistance of the load, Ohm's law (see p. 431) yields the relations

$$P = R \cdot I^2 = \frac{1}{R} \cdot V^2 .$$

2. Energy in the direct-current circuit

The energy W generated or consumed in a time interval Δt is proportional to the power P and to the length of the time interval Δt.

energy = power · time interval		$L^2T^{-2}M$
	Symbol	Unit	Quantity
$W = P \cdot \Delta t = V \cdot I \cdot \Delta t$	W	J	energy
	P	W	power
	Δt	s	time interval
	V	V	voltage
	I	A	current

The energy may be expressed by Ohm's law (see p. 431):

$$W = R \cdot I^2 \cdot \Delta t = \frac{1}{R} \cdot V^2 \cdot \Delta t .$$

In an ohmic resistance, the power is released as heat.

➤ Resistor components may be destroyed by a too-high thermal load. Therefore, in most cases the load capacity is indicated on the resistor by a color code.

If a load is connected to a voltage source, power is extracted from the source. Part of this source power is consumed by the load, another fraction is lost as dissipative power within the source itself (**Fig. 15.8**).

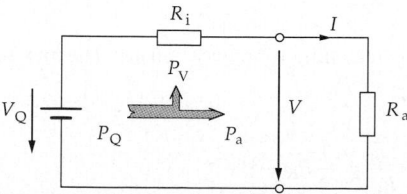

Figure 15.8: The power of a voltage source is partly consumed by the load (P_a), partly lost in the finite internal resistance as dissipative power (P_V).

3. Effective power, dissipative power and short-circuit power

Effective power, load power, P_a, the power received by the load:

$$P_a = \frac{R_a}{(R_a + R_i)^2}\, V_Q^2 .$$

Dissipative power, P_V, converted by the internal resistance R_i of the voltage source:

$$P_V = \frac{R_i}{(R_a + R_i)^2}\, V_Q^2 .$$

Power balance, the power of the voltage source P_Q is equal to the sum of the dissipative power P_V and the effective power P_a:

$$P_Q = P_a + P_V .$$

Short-circuit power, P_K, occurs if the external resistance R_a is equal to zero, i.e., the terminals of the voltage source are shorted.

▲   The short-circuit power is the maximum power the voltage source can supply. The short-circuit power is exclusively dissipative power. The consumed energy is released as heat.

4. Efficiency,

η, the ratio of effective power P_a to the power of the voltage source P_Q:

efficiency = $\dfrac{\textbf{effective power}}{\textbf{power of voltage source}}$				1
$\eta = \dfrac{P_a}{P_Q}$	Symbol	Unit	Quantity	
	η	1	efficiency	
$= \dfrac{P_a}{P_a + P_V}$	P_a	W	effective power	
	P_Q	W	source power	
	P_V	W	dissipative power	
$= \dfrac{R_a}{R_a + R_i}$	R_a	Ω	load resistance	
	R_i	Ω	internal resistance of voltage source	

15.1.5   Matching for power transfer

Matching for power transfer: source and load in the direct-current circuit are chosen so that the load receives the maximum of power from the voltage source. This happens if the load resistance R_a equals the internal resistance R_i of the source:

$$R_a = R_i .$$

Maximum load power, $P_{a,\max}$, is reached by matching for power transfer. The maximum load power is a quarter of the short-circuit power P_K:

$$P_{a,\max} = \frac{1}{4}\frac{V_Q^2}{R_a} = \frac{1}{4}P_K .$$

In the matching for load power, the efficiency is

$$\eta = \frac{R_a}{2R_a} = 50 \ \%.$$

15.1.6 Measurement of current and voltage

15.1.6.1 Current measurement

| M | **Current-measuring instruments**, or **ammeters**, are connected serially in the circuit (**series connection**). In order to avoid disturbance of the current flow, the internal resistance R_i of the instrument should be as small as possible compared with the remaining resistances in the circuit.

Range extension: If one wished to measure a current I that is outside the range of the measuring instrument, the range may be extended by a parallel connection of a resistor R_n, the **shunt resistor**. This resistor is chosen so that the current I_i through the ammeter is still within its range of the ammeter. The current I may be calculated from the shunt resistance R_n and the internal resistance R_i (see p. 505):

range extension for current measurement			I
$I = \left(1 + \dfrac{R_i}{R_n}\right) \cdot I_i$	Symbol	Unit	Quantity
	I	A	current
	I_i	A	current through ammeter
	R_i	Ω	internal resistance of ammeter
	R_n	Ω	shunt resistance

15.1.6.2 Voltage measurement

| M | **Voltage-measuring instruments**, or **voltmeters**, are connected parallel to the two-terminal network (**parallel connection**) at which the voltage has to be measured. The internal resistance of the measuring instrument should be as high as possible compared with the resistance of the two-terminal network.

Range extension, if one wished to measure a voltage V that is outside the range of the voltmeter, a resistor R_n is connected to the measuring instrument (series connection). The dropping resistor is chosen so that the voltage V_i across the voltmeter still keeps within the designed range of the instrument. The actual voltage V can be calculated from the dropping resistance R_n and the internal resistance R_i (see p. 504):

range extension for voltage measurement			$\mathbf{L^2 T^{-3} M I^{-1}}$
$V = \left(1 + \dfrac{R_n}{R_i}\right) \cdot V_i$	Symbol	Unit	Quantity
	V	V	voltage
	V_i	V	voltage across voltmeter
	R_i	Ω	internal resistance of voltmeter
	R_n	Ω	dropping resistance

15.1.6.3 Power measurement

For **power measurement** as well as for **resistance measurement**, and for plotting **current-voltage characteristics** by means of ammeters and voltmeters, the following choices for connections exist:

Voltage connection, the ammeter is in series with the parallel connection of voltmeter and resistor. Since part of the current ΔI flows through the voltmeter, the current I through the ammeter is higher than the current through the resistor. Therefore, voltmeters with a high internal resistance value should be used. The voltage V is measured correctly (**Fig. 15.9**).

Current connection, the voltmeter measures the voltage drop across resistor and ammeter. Since the ammeter has a (small) internal resistance where the voltage ΔV is dropping, the voltmeter measures a voltage V that is higher than the actual voltage across the resistor. Therefore, ammeters with a very low internal resistance should be used. The current I is measured correctly (**Fig. 15.10**).

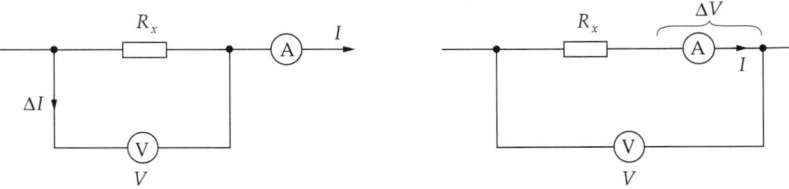

Figure 15.9: Voltage connection of ammeter and voltmeter.

Figure 15.10: Current connection of ammeter and voltmeter.

15.1.7 Resistance measurement by means of the compensation method

Besides the **resistance measurement** by ammeters and voltmeters, one can also employ the compensation method.

Compensation method, the resistance R_x is determined by comparison with a known resistance R_N by means of a bridge circuit (**Fig. 15.11**).

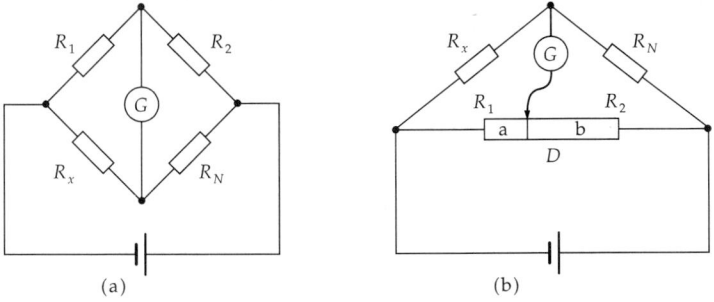

Figure 15.11: Wheatstone's bridge circuit. (a): circuit diagram, (b): realization by a resistance wire.

M **Wheatstone's bridge**: The variable resistance R_N is chosen so that no current flows through the galvanometer: a **zero balancing** is carried out. The bridge (G) is then current-free (balanced).

The unknown resistance R_x then follows from the known resistance values R_1, R_2 and R_N:

resistance measurement by Wheatstone's bridge circuit			$L^2T^{-3}MI^{-2}$
$R_x = \dfrac{R_1}{R_2} \cdot R_N$	Symbol	Unit	Quantity
	R_x	Ω	unknown resistance
	R_N	Ω	reference resistance
	R_1, R_2	Ω	known resistances

The reference resistors are precision resistors of low tolerance that may be combined to arbitrary resistance values.

M R_N may also be chosen as fixed, and the ratio of R_1 to R_2 may be varied. In practice, a resistance wire D is used. A slider is moved along the wire until the bridge becomes current-less (balanced). Since for a uniform wire of constant thickness the resistances R_1, R_2 of the intercepts are related as the corresponding lengths a, b. The unknown resistance R_x follows from the measured lengths as

$$R_x = \frac{R_1}{R_2} \cdot R_N = \frac{a}{b} \cdot R_N \, .$$

➤   For a precision measurement, the value of the reference resistance R_N should not differ too much from the unknown value R_x.

15.1.8 Charging and discharging of capacitors

Voltage at a capacitor, proportional to the time integral of the charging or discharging current $I(t)$:

$$V_C(t) = \frac{Q}{C} = \frac{1}{C} \int I(t)\mathrm{d}t \, .$$

Time constant, τ, the time interval needed to reduce the capacitor voltage to $1/e \approx 1/3$ of the original value. The time constant is the product of the capacitance C of the capacitor and the resistance value R of the resistor through which the capacitor is being charged or discharged:

$$\tau = R \cdot C \, .$$

■   A capacitor of capacitance $C = 1$ mF is discharged through a resistor of resistance $R = 1$ kΩ. The time constant is

$$\tau = 1 \text{ k}\Omega \cdot 1 \text{ mF} = 1 \text{ s} \, .$$

1. Charging of capacitor

A capacitor of capacitance C is connected to the voltage source V_0 through a resistor R. According to the mesh rule, the voltages across the resistor $V_R(t)$ and the capacitor $V_C(t)$ sum to the voltage V_0:

$$V_0 = V_C(t) + V_R(t) = \frac{1}{C} \int I(t)\mathrm{d}t + I(t) \cdot R \, .$$

From there follows the differential equation for the charging current:

$$\frac{dI(t)}{dt} = -\frac{1}{\tau} I(t) , \qquad I(0) = \frac{V_0}{R} = I_0 .$$

The **charging current**, $I(t)$, and the **capacitor voltage**, $V_C(t)$, are given by (**Fig. 15.12 (a)**):

$$I(t) = I_0 \cdot e^{-t/\tau} , \qquad V_C(t) = V_0 \cdot \left(1 - e^{-t/\tau}\right) , \qquad \tau = R \cdot C .$$

▲   The charging current decays exponentially from the initial value $I_0 = V_0/R$ with the time constant τ.

▲   The voltage at the capacitor increases with the same time constant τ up to the source voltage V_0.

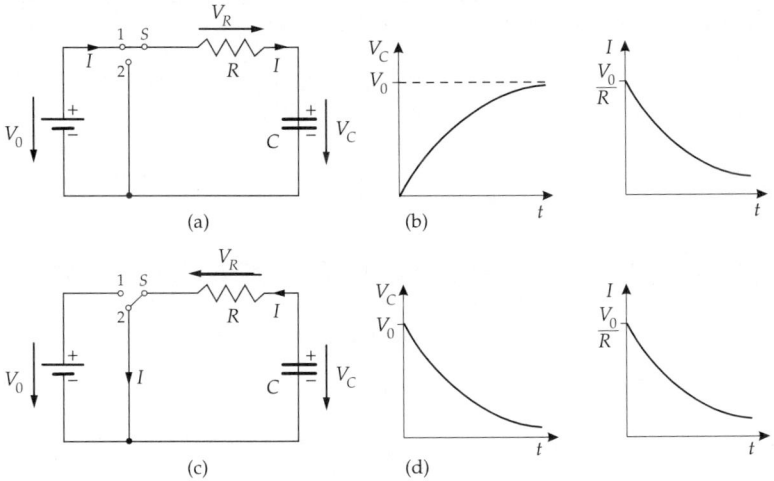

Figure 15.12: Charging and discharging of a capacitor. (a): circuit diagram, S: switch, (b): time dependence of voltage and current on charging, (c): circuit diagram, switch set to discharge, (d): time dependence of voltage and current on discharging. The time constant is $\tau = R \cdot C$.

2. Discharging of capacitor

A capacitor of capacitance C is discharged through a resistor R (**Fig. 15.12 (b)**). According to the mesh rule, the voltages across the capacitor $V_C(t)$ and the resistor $V_R(t)$ sum to zero:

$$0 = V_C(t) - V_R(t) = -\frac{1}{C} \int I(t)dt - I(t) \cdot R .$$

This leads to the differential equation for the discharge current:

$$\frac{dI(t)}{dt} = -\frac{1}{\tau} I(t) , \qquad I(0) = \frac{V_0}{R} = I_0 .$$

The **discharge current**, $I(t)$, and the **capacitor voltage**, $V_C(t)$, are then (**Fig. 15.12 (b)**):

$$I(t) = I_0 \cdot e^{-t/\tau} , \qquad V_C(t) = V_0 \cdot e^{-t/\tau} , \qquad \tau = R \cdot C .$$

▲   The discharge current decreases exponentially from the initial value $I_0 = V_0/R$ to zero, with the time constant τ.

▲   The voltage across the capacitor decreases exponentially with the same time constant τ from the initial value V_0 to zero.

15.1.9   Switching the current on and off in a RL-circuit

The **voltage across the coil** is proportional to the time variation of the current through it $I(t)$:

$$V_L(t) = L\,\frac{\mathrm{d}I(t)}{\mathrm{d}t}\,.$$

Time constant, τ, the time interval during which the coil current decreases to $1/e \approx 1/3$ of the initial value. The time constant is the quotient of the coil inductance L and the resistance R of the resistor passed by the starting current or breaking current:

$$\tau = \frac{L}{R}\,.$$

■   A coil of inductance $L = 100$ mH is shorted through a resistor of resistance $R = 10\ \Omega$. The time constant is then:

$$\tau = \frac{L}{R} = \frac{100\ \text{mH}}{10\ \Omega} = 0.01\ \text{s}\,.$$

1. Switching the current on

A coil of inductance L in series with a resistor R is connected to a voltage source V_0. According to the mesh rule, the negative coil voltage $V_L(t)$ and the voltage across the resistor $V_R(t)$ sum to the voltage V_0:

$$V_0 = V_L(t) + V_R(t) = L\,\frac{\mathrm{d}I(t)}{\mathrm{d}t} + R \cdot I(t)\,.$$

From there follows the differential equation

$$\frac{\mathrm{d}I(t)}{\mathrm{d}t} = -\frac{1}{\tau}\,I(t) + \frac{V_0}{L}\,,\qquad I(0) = 0\,.$$

Hence, the **coil current** $I(t)$ and the **coil voltage** $V_L(t)$ are

$$I(t) = I_0 \cdot \left(1 - e^{-t/\tau}\right),\qquad V_L(t) = V_0 \cdot e^{-t/\tau}\,,\qquad \tau = \frac{L}{R}\,.$$

▲   The coil current $I(t)$ increases asymptotically with the time constant τ up to the magnitude $I_0 = V_0/R$.

▲   The coil voltage $V_L(t)$ decreases exponentially with the same time constant τ from its initial value V_0 to zero.

2. Switching the current off

After switching off the voltage source, a coil of inductance L is shorted through a resistance R. According to the mesh rule, the coil voltage $V_L(t)$ equals the voltage at the resistor $V_R(t)$:

$$0 = V_L(t) + V_R(t) = L\frac{dI(t)}{dt} + R \cdot I(t).$$

Hence, the coil current $I(t)$ obeys the differential equation

$$\frac{dI(t)}{dt} = -\frac{1}{\tau}I(t), \qquad I(0) = I_0.$$

For the **coil current** $I(t)$ and the **coil voltage** $V_L(t)$ then follows:

$$I(t) = I_0 \cdot e^{-t/\tau}, \qquad V_L(t) = V_0 \cdot e^{-t/\tau}, \qquad \tau = \frac{L}{R}.$$

▲ The coil current decreases exponentially with a time constant τ from its original value $I_0 = V_0/R$ to zero.

▲ The coil voltage decreases with the same time constant τ from its original value V_0 to zero.

➤ When switching off a coil current, the occurrence of high voltages may lead to sparking at the switch contacts and may destroy electronic switching elements.

15.2   Alternating-current circuit

Alternating-current engineering deals with the behavior of resistors, capacitors and inductors when an alternating current is flowing through them or an alternating voltage is applied to them.

Alternating quantities may be represented by complex numbers that facilitate the calculation of physical quantities in alternating-current circuits. They may be represented by phasors in the complex plane: the phasor diagram.

15.2.1 Alternating quantities

Alternating quantity, a quantity with a time dependence given by a periodic function.

1. Characteristics of alternating quantities

Instantaneous value, **momentary value**, the value of an alternating quantity at an arbitrary time t.

Period, T, the time interval in which the alternating quantity x repeats all values in the same time sequence. The period is T if for any time t,

$$x(t + T) = x(t).$$

Frequency, f, reciprocal value of the period T,

$$f = \frac{1}{T}.$$

The simplest periodic functions are the **sine function** and the **cosine function**.

2. Sinusoidal alternating quantities

An alternating quantity with a sinusoidal time dependence is completely described by specifying:

- **amplitude, peak value,** $\hat{x}$, maximum value the alternating quantity x may take.
- **angular frequency, angular velocity,** 2π times the frequency: $\omega = 2\pi f$.
- **zero phase angle,** phase angle at the time $t = 0$: φ_0.

3. Alternating voltage and alternating current

Alternating voltage $v(t)$, described by

$$v(t) = \hat{v} \sin(\omega t + \varphi_v).$$

$\hat{v}$ denotes the amplitude, φ_v the zero phase angle of the alternating voltage (**Fig. 15.13**).
Alternating current $i(t)$, described by

$$i(t) = \hat{i} \sin(\omega t + \varphi_i).$$

$\hat{i}$ denotes the amplitude, φ_i the zero phase angle of the alternating current.

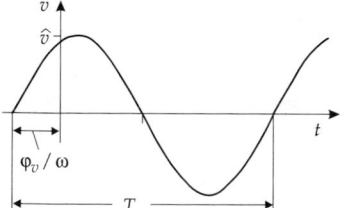

Figure 15.13: Period and zero phase angle of the sine function.

More complicated periodic functions may be constructed by superposition (linear combination) of sine and cosine functions (Fourier series).

15.2.1.1 Time average of periodic functions

1. Mean values of alternating quantities

Mean value, characterizes an alternating quantity $x(t)$ by a value, without specifying the detailed time behavior.

Several possibilities of averaging:

mean value or **arithmetic mean value**	$\bar{x} = \dfrac{1}{T} \displaystyle\int_0^T x(t)\, dt$
rectified value or **absolute mean value**	$\overline{|x|} = \dfrac{1}{T} \displaystyle\int_0^T |x(t)|\, dt$
effective value or **root-mean-square value**	$X = \sqrt{\dfrac{1}{T} \displaystyle\int_0^T x(t)^2\, dt}$

Crest factor, ratio of amplitude to effective value:

$$k_s = \frac{\hat{x}}{X} .$$

Form factor, ratio of effective value to absolute mean value:

$$k_f = \frac{X}{\overline{|x|}} .$$

2. Mean values of sinusoidal alternating quantities

For a sinusoidal alternating quantity, the mean values, crest factor and form factor are given by:

Mean value	Absolute mean value	Effective value	Crest factor	Form factor
$\overline{x} = 0$	$\overline{|x|} = \dfrac{2}{\pi}\hat{x}$ $\approx 0.637\,\hat{x}$	$X = \dfrac{1}{\sqrt{2}}\hat{x}$ $\approx 0.707\,\hat{x}$	$k_s = 1.414$	$k_f = 1.111$

Distortion factor, specifies the deviation of an alternating quantity from the sinusoidal shape.

3. Thermal load of ohmic components

In order to calculate the **thermal load of ohmic components** for a sinusoidal alternating voltage, the effective values of voltage and current have to be taken into account. At an ohmic resistor, current and voltage are in phase. The fraction with frequency zero in the Fourier expansion of the absorbed power is

$$P = \frac{\hat{v} \cdot \hat{i}}{2} = \frac{\hat{v}^2}{2R} = \frac{V^2}{R} = I^2 \cdot R .$$

The power consumption corresponds to that of a resistor in a direct-current circuit if the applied direct voltage corresponds to the effective value of the alternating voltage, and the equivalent direct current corresponds to the effective value of the alternating current.

4. Measurement of alternating voltage and alternating current

| M | Alternating current and alternating voltage can be measured by means of moving-coil instruments with a preconnected rectifier. Usually the measuring instrument is calibrated to display the effective value of a sinusoidal quantity. For non-sinusoidal alternating quantities, the displayed value has to be converted to the effective value by correction factors.

■ The voltage common in domestic appliances in the U.S. is measured by means of a moving-coil voltmeter and is equal to $V = 117$ V. The amplitude of the alternating voltage is

$$\hat{v} = \sqrt{2} \cdot V = \sqrt{2} \cdot 117\ \text{V} = 165\ \text{V} .$$

15.2.2   Representation of sinusoidal quantities in a phasor diagram

▲   Alternating quantities with a sinusoidal time dependence may be represented in a phasor diagram.

If a point P moves in a x-y coordinate system along a circle of radius r about the origin in the mathematically positive sense and with a constant angular velocity, the projection of this point onto the y-axis has a sine-like time behavior, the projection onto the x-axis a cosine-like time behavior (**Fig. 15.14**).

Mathematically positive sense of rotation, counterclockwise rotation.

1. Phasor,

vector from coordinate origin to the point P in the complex plane, position vector of P. The phasor is fully determined by giving its coordinates a with respect to the x-axis and b with respect to the y-axis of the reference system.

▲   The phasor is represented by a complex number in the complex number plane.

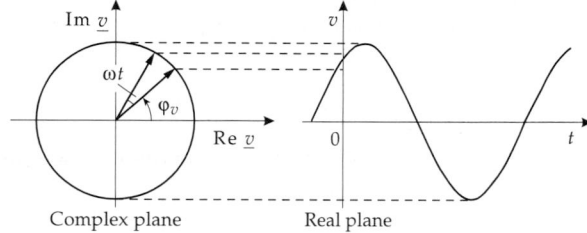

Figure 15.14: Relation between rotating phasor in the complex plane and sine function.

2. Cartesian representation of a complex number,

a pair of real numbers a and b written in the form

$$\underline{z} = a + jb$$

(**Fig. 15.15 (a)**); j denotes the **imaginary unit**, often also denoted as i. Real and imaginary parts of the complex number may be considered Cartesian coordinates of a point P in the

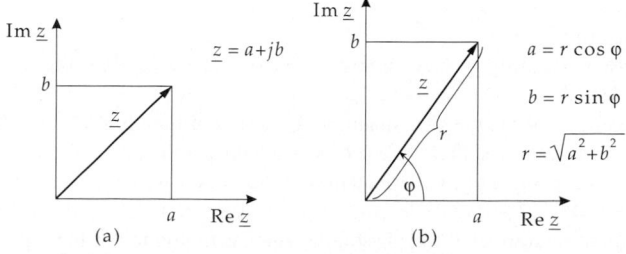

Figure 15.15: Representations of a complex number. (a): Cartesian representation in the complex number plane, (b): exponential representation.

(x, y)-plane, $P = P(a, b)$. A real number x is a complex number $\underline{z}$ with the imaginary part zero,

$$\underline{z} = x + j \cdot 0.$$

■   The complex number $3 + j \cdot 4$ has real part $a = 3$ and imaginary part $b = 4$.
A complex number may be viewed as a vector in the two-dimensional plane. The product of two complex numbers is again a complex number. The division operation may be done with complex numbers.

3. Exponential representation of complex numbers

Complex exponential function, represented by **Euler's formula**:

Euler's formula		
	Symbol	Quantity
$e^{j\varphi} = \cos\varphi + j\sin\varphi$	φ	phase
	$j = \sqrt{-1}$	imaginary unit

➤   The complex exponential function is $2\pi j$-periodic.
Exponential representation of complex numbers (**Fig. 15.15 (b)**):

$$\underline{z} = r \cdot e^{j\varphi}.$$

r gives the length of the phasor, the phase φ is the angle between the positive x-axis and the phasor, in the positive sense of rotation.

4. Conversion between the representations of complex numbers

The following relations hold for the conversion between the representations of complex numbers:

$$a = r\cos\varphi, \qquad b = r\sin\varphi,$$

and

$$r = \sqrt{a^2 + b^2}, \qquad \varphi = \arctan\left(\frac{b}{a}\right).$$

5. Phasor diagram and phasor characteristics

Phasor diagram, representation of phasors in the complex plane.
▲   The length of a phasor represents the amplitude of the corresponding alternating quantity.
   Phasor, determined by:
•   the physical quantity represented by the phasor. The formula symbol is written next to the phasor;
•   the magnitude of the physical quantity, the length of the phasor. One selects a representation in terms of either peak values or effective values;
•   the zero phase angle φ_0, the orientation of the phasor with respect to the real axis at the time $t = 0$;
•   the angular velocity ω of the phasor, the angular frequency of the represented quantity.
➤   Phasor quantities are represented by underlined symbols.
■   A current phasor $\underline{i}(t)$ is assigned to an alternating current $i(t)$.

6. Transformation of an alternating quantity into a phasor

Transformation between an alternating quantity and its phasor in the complex plane: the sine function describing the alternating quantity is viewed as the imaginary part of a complex number whose real part is a cosine function of equal phase and equal amplitude:

$$x(t) = \hat{x} \sin(\omega t + \varphi_0) \longrightarrow \underline{x}(t) = \hat{x} \cos(\omega t + \varphi_0) + j\hat{x} \sin(\omega t + \varphi_0)$$

$$= \hat{x} e^{j(\omega t + \varphi_0)}.$$

■ An alternating current

$$i(t) = \hat{i} \sin(\omega t + \varphi_i)$$

is mapped onto the phasor quantity

$$\underline{i}(t) = \hat{i} e^{j(\omega t + \varphi_i)}.$$

■ An alternating voltage

$$v(t) = \hat{v} \sin(\omega t + \varphi_v)$$

is mapped onto the phasor quantity

$$\underline{v}(t) = \hat{v} e^{j(\omega t + \varphi_v)}.$$

A calculation using complex phasors is often simpler and more transparent than the treatment with angular functions (see below).

15.2.3 Calculation rules for phasor quantities

1. Addition of phasor quantities

Phasor addition, corresponds to the **addition of complex numbers**. The real parts and imaginary parts of the phasors are added separately:

addition of two complex numbers		
$\underline{z}_1 + \underline{z}_2 = (a_1 + j \cdot b_1) + (a_2 + j \cdot b_2)$	Symbol	Quantity
$= (a_1 + a_2) + j \cdot (b_1 + b_2)$	$\underline{z}_1 = a_1 + j \cdot b_1$	first summand
	$\underline{z}_2 = a_2 + j \cdot b_2$	second summand
$= \underline{z}$	$\underline{z}$	sum

■ The sum of two phasors represented by the complex numbers $\underline{z}_1 = 3 + j \cdot 4$ and $\underline{z}_2 = 2 + j \cdot 5$, is

$$\underline{z} = \underline{z}_1 + \underline{z}_2 = (3 + j \cdot 4) + (2 + j \cdot 5) = (3 + 2) + j \cdot (4 + 5) = 5 + j \cdot 9.$$

The resultant phasor in the complex domain has real part $a = 5$ and imaginary part $b = 9$.

2. Subtraction of phasor quantities

Phasor subtraction, corresponds to the **subtraction of complex numbers**. The subtraction is done by component for the real part and the imaginary part, respectively:

subtraction of two complex numbers		
$\underline{z}_1 - \underline{z}_2 = (a_1 + j \cdot b_1) - (a_2 + j \cdot b_2)$	Symbol	Quantity
$\quad = (a_1 - a_2) + j \cdot (b_1 - b_2)$	$\underline{z}_1 = a_1 + j \cdot b_1$	minuend
	$\underline{z}_2 = a_2 + j \cdot b_2$	subtrahend
$\quad = \underline{z}$	$\underline{z}$	difference

■ From a phasor represented by the complex number $\underline{z}_1 = 3 + j \cdot 4$, the phasor $\underline{z}_2 = 2 + j \cdot 5$ is subtracted. The resulting phasor $\underline{z}$ is represented by

$$\underline{z} = \underline{z}_1 - \underline{z}_2 = 3 + j \cdot 4 - (2 + j \cdot 5) = (3 - 2) + j \cdot (4 - 5) = 1 - j.$$

The resulting phasor has real part $a = 1$ and imaginary part $b = -1$.

3. Multiplication of phasor quantities

Phasor multiplication, corresponds to the **multiplication of complex numbers**. The multiplication of two phasor quantities may be done more easily in the exponential representation. Here the phases of the complex numbers are summed, whereas the magnitudes are multiplied:

multiplication of two complex numbers		
$\underline{z}_1 \cdot \underline{z}_2 = r_1 \, e^{j\varphi_1} \cdot r_2 \, e^{j\varphi_2}$	Symbol	Quantity
$\quad = r_1 \cdot r_2 \, e^{j(\varphi_1 + \varphi_2)}$	$\underline{z}_1 = r_1 \, e^{j\varphi_1}$	first factor
	$\underline{z}_2 = r_2 \, e^{j\varphi_2}$	second factor
$\quad = \underline{z}$	$\underline{z}$	product

4. Division of phasor quantities

Phasor division, corresponds to the **division of complex numbers**.

As with the multiplication, the division of two phasor quantities may be done more easily in the exponential representation. The phases of the complex numbers are subtracted from each other, and the magnitudes are divided by each other:

division of two complex numbers		
$\dfrac{\underline{z}_1}{\underline{z}_2} = \dfrac{r_1 \, e^{j\varphi_1}}{r_2 \, e^{j\varphi_2}}$	Symbol	Quantity
$\quad = \dfrac{r_1}{r_2} \, e^{j(\varphi_1 - \varphi_2)}$	$\underline{z}_1 = r_1 \, e^{j\varphi_1}$	dividend
	$\underline{z}_2 = r_2 \, e^{j\varphi_2}$	divisor
$\quad = \underline{z}$	$\underline{z}$	quotient

5. Complex conjugation of a phasor quantity

Complex-conjugate phasor $\underline{z}^*$ of a phasor $\underline{z}$, has the same magnitude, but the opposite phase:

$$\underline{z}^* = |\underline{z}| \cdot e^{-j\varphi} \quad \text{for} \quad \underline{z} = |\underline{z}| \cdot e^{j\varphi}.$$

In the Cartesian representation, the complex-conjugate phasor reads

$$\underline{z}^* = a - jb \quad \text{for} \quad \underline{z} = a + jb.$$

▲ The complex-conjugate phasor is obtained by reflection of the original phasor at the real axis.

Complex conjugation applied twice to a phasor quantity (reflection of the reflected image) yields the original phasor quantity:

$$(\underline{z}^*)^* = \underline{z}.$$

6. Inversion of a phasor quantity

Inversion, special case of complex division. If the original phasor $\underline{z}$ has the length $|\underline{z}|$, then the inverted phasor $1/\underline{z}$ has the length $1/|\underline{z}|$. As in complex conjugation, the sign of phase changes:

$$\frac{1}{\underline{z}} = \frac{\underline{z}^*}{|\underline{z}|^2}.$$

In exponential representation,

$$\frac{1}{\underline{z}} = \frac{1}{r} e^{-j\varphi} \quad \text{for} \quad \underline{z} = r\, e^{j\varphi}.$$

In Cartesian representation,

$$\frac{1}{\underline{z}} = \frac{a - jb}{a^2 + b^2} \quad \text{for} \quad \underline{z} = a + jb.$$

■ If the complex resistance $\underline{Z}$ is given, the complex conductance $\underline{Y}$ is obtained by inversion,

$$\underline{Y} = \frac{1}{\underline{Z}} = \frac{\underline{Z}^*}{|\underline{Z}|^2},$$

and vice versa.

7. Differentiation of phasor quantities

Differentiation of phasors corresponds to the differentiation of complex functions. The differentiation is done with respect to the time.

Let a phasor quantity $\underline{z}$ be given by the magnitude z, the zero phase angle φ, and the angular frequency ω:

$$\underline{z}(t) = z\, e^{j(\omega t + \varphi)}.$$

Then, the time derivative reads

$$\frac{\mathrm{d}\underline{z}}{\mathrm{d}t} = \mathrm{j}\omega z\, \mathrm{e}^{\mathrm{j}(\omega t + \varphi)} = \mathrm{j}\omega\underline{z} = \omega\, z\, \mathrm{e}^{\mathrm{j}(\omega t + \varphi + \frac{\pi}{2})}\,.$$

▲   The time derivative corresponds to a rotation combined with stretching. In the complex phasor plane, the rotation proceeds counterclockwise by an angle $\pi/2$.

8. Integration of phasor quantities

Integration of phasors, corresponds to the integration of complex functions. The integration is done with respect to the time.

Let a phasor quantity $\underline{z}$ be given by the magnitude z, the zero phase angle φ, and the angular frequency ω:

$$\underline{z}(t) = z\, \mathrm{e}^{\mathrm{j}(\omega t + \varphi)}\,.$$

Then, the integral over the time reads

$$\int \underline{z}(t)\mathrm{d}t = \frac{1}{\mathrm{j}\omega}\cdot z\, \mathrm{e}^{\mathrm{j}(\omega t + \varphi)} = \frac{1}{\mathrm{j}\omega}\underline{z} = \frac{1}{\omega} z\, \mathrm{e}^{\mathrm{j}(\omega t + \varphi - \frac{\pi}{2})}\,.$$

▲   The integration corresponds to a rotation-stretching. The rotation proceeds in the complex phasor plane by the angle $-\pi/2$.

15.2.4   Basics of alternating-current engineering

15.2.4.1   Complex resistance

1. Definition of the complex resistance

Complex resistance, $\underline{Z}$, determined by:
- the ratio of the amplitudes of voltage and current, or the ratio of the effective values of voltage and current, respectively, and
- the phase shift of the voltage against the current.

complex resistance = $\dfrac{\text{complex voltage}}{\text{complex current}}$		$\mathbf{L^2T^{-3}MI^{-2}}$
$\underline{Z} = \dfrac{\underline{v}(t)}{\underline{i}(t)}$	**Symbol** **Unit**	**Quantity**
	$\underline{Z}$ Ω	complex resistance
	$\underline{v}(t)$ V	complex voltage
	$\underline{i}(t)$ A	complex current

Ohm, Ω, SI unit of the complex resistance $\underline{Z}$,

$$[\underline{Z}] = \Omega\,.$$

▲   If the voltage and the current have the same time dependence, then the complex resistance is time-independent.

2. *Cartesian form of the complex resistance,*

$\underline{Z}$, composed of:
- **resistance**, **resistive part of impedance**, R, the real part of the complex resistance,
- **reactive impedance**, **reactance**, X, the imaginary part of the complex resistance.

complex resistance, Cartesian form			$\mathbf{L^2T^{-3}MI^{-2}}$
	Symbol	Unit	Quantity
$\underline{Z} = R + jX$	$\underline{Z}$	Ω	complex resistance
	R	Ω	resistance
	X	Ω	reactive impedance

▲ The resistance equals the ohmic resistance of the circuit or two-terminal network.

3. *Exponential form of the complex resistance,*

$\underline{Z}$, expressed by:
- **Apparent resistance**, **impedance**, Z, absolute value of the complex resistance,

$$Z = |\underline{Z}| = \sqrt{R^2 + X^2}\,.$$

- **Phase angle**, φ_Z, arc-tangent of the ratio of reactive resistance X to resistance R,

$$\varphi_Z = \arctan\frac{X}{R}\,.$$

complex resistance, exponential form			$\mathbf{L^2T^{-3}MI^{-2}}$
	Symbol	Unit	Quantity
$\underline{Z} = Z \cdot e^{j\varphi_Z}$	$\underline{Z}$	Ω	complex resistance
	Z	Ω	impedance
	φ_Z	1	phase angle

The impedance Z gives the ratio of the voltage amplitude $\hat{v}$ and the current amplitude $\hat{i}$ (or the ratio of the effective values V and I), without taking into account the phase shift:

$$Z = \frac{\hat{v}}{\hat{i}} = \frac{V}{I}\,.$$

The phase angle φ_Z is the difference between the zero phase angles of the voltage, φ_v, and the current, φ_i:

$$\varphi_Z = \varphi_v - \varphi_i\,.$$

4. *Resistance phasor,*

representation of the complex resistance in the complex **resistance plane** (**Fig. 15.16**).
- ■ Let the complex resistance have the value $\underline{Z} = (50 + j \cdot 22)$ Ω. Then, the impedance is

$$Z = \sqrt{R^2 + X^2} = \sqrt{50^2 + 22^2}\ \Omega = 54.6\ \Omega\,,$$

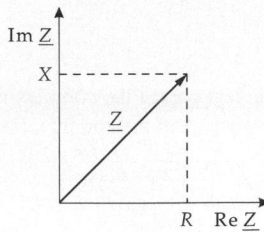

Figure 15.16: Representation of the complex resistance. R: resistance, X: reactive impedance, Z: impedance.

and the phase angle is

$$\varphi_Z = \arctan \frac{X}{R} = \arctan \frac{22\ \Omega}{50\ \Omega} = 23.7° \,.$$

15.2.4.2   Ohm's law in the complex domain

▲   The complex current $\underline{i}$ through an ohmic resistor is proportional to the complex voltage $\underline{v}$. The proportionality factor is the ohmic resistance (resistance) R.

complex voltage = ohmic resistance · complex current			$\mathbf{L^2 T^{-3} M I^{-1}}$
	Symbol	Unit	Quantity
$\underline{v}(t) = R \cdot \underline{i}(t)$	$\underline{v}(t)$	V	complex voltage
	R	Ω	ohmic resistance
	$\underline{i}(t)$	A	complex current

▲   For an ohmic resistor, current and voltage are in phase:

$$\varphi_Z = \varphi_v - \varphi_i = 0 \,.$$

15.2.4.3   Complex conductance

1. Complex conductance,

$\underline{Y}$, determined by:
- the ratio of the amplitudes of current and voltage (or the ratio of the effective values), and
- the phase shift between current and voltage.

The complex conductance is the reciprocal value of the complex resistance,

$$\underline{Y} = \frac{1}{\underline{Z}} = \frac{\underline{Z}^*}{\underline{Z}^2} = \frac{\underline{i}(t)}{\underline{v}(t)} \,.$$

$\underline{Z}^*$ is the complex-conjugate of the complex resistance.

2. Cartesian form of the complex conductance,

$\underline{Y}$, composed of:
- **conductive part of admittance**, **conductance**, G, real part of the complex conductance, and
- **susceptive part of admittance**, **susceptance**, B, imaginary part of the complex conductance.

complex conductance, Cartesian form			$L^{-2}T^3M^{-1}I^2$
	Symbol	Unit	Quantity
$\underline{Y} = G + jB$	$\underline{Y}$	S	complex conductance
	G	S	conductance
	B	S	susceptance

Siemens, S, SI unit of the complex conductance $\underline{Y}$,

$$[\underline{Y}] = S.$$

3. Exponential form of the complex conductance,

consists of:

- **admittance**, Y, magnitude of the complex conductivity:

$$Y = |\underline{Y}| = \sqrt{G^2 + B^2},$$

- **phase angle**, φ_Y, arc-tangent of the ratio of the susceptance B and the conductance G:

$$\varphi_Y = \arctan \frac{B}{G}.$$

complex conductance, exponential form			$L^{-2}T^3M^{-1}I^2$
	Symbol	Unit	Quantity
$\underline{Y} = Y \cdot e^{j\varphi_Y}$	$\underline{Y}$	S	complex conductance
	Y	S	admittance
	φ_Y	1	phase angle

The admittance Y gives the ratio of the current amplitude $\hat{i}$ and the voltage amplitude $\hat{v}$ (or the ratio of the effective values of current I and voltage V), without taking into account the phase shift:

$$Y = \frac{\hat{i}}{\hat{v}} = \frac{I}{V}.$$

Phase shift, φ_Y, the difference of the zero phase angles of the current φ_i and the voltage φ_v:

$$\varphi_Y = \varphi_i - \varphi_v.$$

4. Phasor of the conductance,

representation in the **complex conductance plane** (see **Fig. 15.17**).

The complex conductance is the reciprocal quantity of the complex resistance. From there follows:

▲ A positive susceptance B corresponds to a negative reactance X.

▲ The phase of the complex conductance equals the negative value of the phase of the complex resistance.

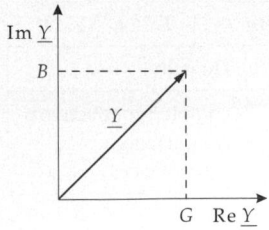

Figure 15.17: Representation
of the complex conductance.

▲ The admittance Y is the reciprocal value of the impedance Z.

■ Let the complex conductance be $\underline{Y} = (12 + j \cdot 27)$ S. Then, the admittance is

$$Y = \sqrt{G^2 + B^2} = \sqrt{12^2 + 27^2} \text{ S} = 29.5 \text{ S} ,$$

and the phase shift is

$$\varphi_Y = \varphi_i - \varphi_v = \arctan \frac{B}{G} = \arctan \frac{27 \text{ S}}{12 \text{ S}} = 66° .$$

Hence, the current leads the voltage by 66°.

15.2.4.4   Power in the alternating-current circuit

1. Power in the alternating-current circuit;

$p(t)$, product of the current $i(t)$ and the voltage $v(t)$:

power = current · voltage			$\mathbf{L^2 T^{-3} M}$
	Symbol	Unit	Quantity
$p(t) = i(t) \cdot v(t)$	$p(t)$	W	power
	$i(t)$	A	current
	$v(t)$	V	voltage

▲ In general, the power in the alternating-current circuit is time-dependent.
For a sinusoidal current of angular frequency ω,

$$i(t) = \hat{i} \sin \omega t$$

and a sinusoidal voltage out of phase by φ,

$$v(t) = \hat{v} \sin(\omega t + \varphi)$$

the power consists of a time-independent part and a time-dependent part pulsating with twice the angular frequency:

$$p(t) = V \cdot I \cdot \cos \varphi - V \cdot I \cdot \cos(2\omega t + \varphi) ,$$

with the effective values of voltage and current, $V = \hat{v}/\sqrt{2}$; $I = \hat{i}/\sqrt{2}$, respectively.

2. Real power and reactive power

Real power, P, denotes the time-independent part of the power for sinusoidal currents and voltages:

$$P = V \cdot I \cos\varphi.$$

Power factor, $\cos\varphi$, cosine of the phase shift φ of current and voltage.
■   For an ohmic resistance, $\cos\varphi = 1$; the real power is

$$P = V \cdot I.$$

For a pure inductance or capacitance, $\cos\varphi = 0$; the real power vanishes:

$$P = 0.$$

Reactive power, Q, the time-dependent part of the power. For sinusoidal currents and voltages, one has:

$$Q = V \cdot I \cdot \sin\varphi.$$

Reactive power factor, $\sin\varphi$, sine of the phase shift φ of current and voltage.
   Apparent power, S, product of the effective values of current I and voltage V:

$$S = V \cdot I = \sqrt{P^2 + Q^2}.$$

15.2.4.5   Complex power

1. Complex power,

$\underline{S}$, product of the complex voltage $\underline{v}$ and the complex-conjugate $\underline{i}^*$ of the complex current $\underline{i}$:

$$\underline{S} = \underline{v} \cdot \underline{i}^*.$$

The complex power consists of:
- **real power**, P, real part of the complex power,
- **reactive power**, Q, imaginary part of the complex power.

2. Cartesian form of the complex power

complex power, Cartesian form			$\mathbf{L^2 T^{-3} M}$
	Symbol	Unit	Quantity
$\underline{S} = P + jQ$	$\underline{S}$	W	complex power
	P	W	real power
	Q	W = var	reactive power

Watt, W, SI unit of the complex power $\underline{S}$,

$$[\underline{S}] = \text{W}.$$

➤ Although in the SI system the watt is assigned to power, the following units are also used:

voltampere-reactance, var, unit of the reactive power Q,

$$[Q] = \text{var} = V \cdot A,$$

voltampere, VA, unit of the apparent power S,

$$[S] = VA = V \cdot A.$$

3. Exponential form of the complex power

complex power, exponential form			$L^2T^{-3}M$
	Symbol	Unit	Quantity
$\underline{S} = S\,e^{j\varphi_S}$	$\underline{S}$	W	complex power
	S	W = VA	apparent power
	φ_S	1	phase angle

Power factor, $\cos\varphi_S$, denotes the ratio of the real power P to the apparent power S:

$$\cos\varphi_S = \frac{P}{S}.$$

■ For electromotors, the phase angle φ_S, and thus the power factor $\cos\varphi_S$, depend on the load. The power factor specified by the manufacturer holds only for full load.

15.2.4.6   Kirchhoff's laws for alternating-current circuits

1. Branch-point rule in the complex domain,

the sum of all complex currents $\underline{i}_1 \ldots \underline{i}_n$ flowing to and from a branch point is equal to zero:

$$\underline{i}_1 + \underline{i}_2 + \underline{i}_3 + \cdots + \underline{i}_n = 0.$$

If the current phasors are added at a branch point like two-dimensional vectors, one gets a closed polygon loop in the phasor diagram (**Fig. 15.18 (b)**).

2. Mesh rule in the complex domain,

the sum of all complex voltages along a network mesh is equal to zero:

$$\underline{v}_1 + \underline{v}_2 + \underline{v}_3 + \cdots + \underline{v}_n = 0.$$

If the phasors for the voltages along a mesh are added like vectors, the result is a closed polygon loop in the phasor diagram (**Fig. 15.18 (a)**).

15.2.4.7   Series connection of complex resistances

The same current flows through all circuit components (**Fig. 15.19**).

▲ The complex total resistance is equal to the sum of the complex individual resistances:

$$\underline{Z} = \underline{Z}_1 + \underline{Z}_2 + \underline{Z}_3 + \cdots + \underline{Z}_n.$$

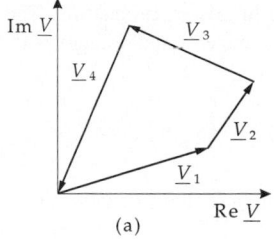 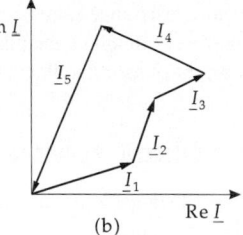

Figure 15.18: Addition of phasors. (a): voltage polygon for a mesh, (b): current polygon for a branch point.

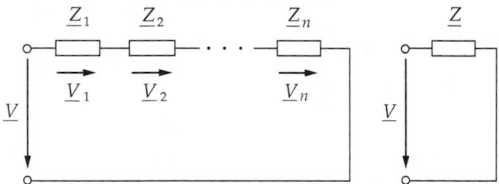

Figure 15.19: Series connection of complex resistances.

15.2.4.8   Parallel connection of complex resistances

The same voltage is applied to all circuit elements (**Fig. 15.20**).

▲   The complex conductance is equal to the sum of the complex individual conductances:

$$\underline{Y} = \underline{Y}_1 + \underline{Y}_2 + \underline{Y}_3 + \cdots + \underline{Y}_n \, .$$

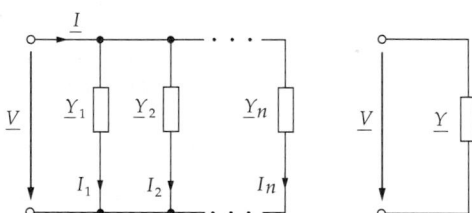

Figure 15.20: Parallel connection of complex resistances.

15.2.5   Basic components in the alternating-current circuit

The **two-terminal networks** (see p. 501) **resistor, capacitor and inductor** show a characteristic dependence of the complex resistance in an alternating-current circuit as function of the frequency. The complex resistances of the capacitor and the inductor depend on the frequency of the alternating voltage.

Locus, represents the dependence of a complex quantity on the frequency in the complex plane.

The complex resistance may be easily calculated in this representation. The phasor diagram immediately displays the phase shift between current and voltage as the angle between the current phasor and the voltage phasor.

15.2.5.1   Ohmic resistor

The complex resistance $\underline{Z}$ of an ohmic resistor is real and independent of the frequency of the alternating current.

The locus of the complex resistance is restricted to one point in the complex resistance plane corresponding to the value of the ohmic resistance R (**Fig. 15.21 (b)**).

▲   The reactance X vanishes.

Complex resistance	Impedance	Resistive part	Reactive part
$\underline{Z} = R$	$Z = R$	R	$X = 0$

The complex conductance $\underline{Y}$ of an ohmic resistor is also real and independent of the frequency.

▲   The susceptance B vanishes.

Complex conductance	Admittance	Conductive part	Susceptive part
$\underline{Y} = \dfrac{1}{R}$	$Y = \dfrac{1}{R}$	$G = \dfrac{1}{R}$	$B = 0$

▲   For an ohmic resistor, current and voltage are in phase. The phase of the complex resistance is zero:

$$\varphi_Z = \varphi_v - \varphi_i = 0.$$

In the phasor diagram (**Fig.15.21 (c)**), the current phasor and the voltage phasor are pointing in the same direction.

▲   The complex power of an ohmic resistor is real.

Complex power	Apparent power	Real power	Reactive power
$\underline{S} = V \cdot I$	$S = V \cdot I$	$P = V \cdot I$	$Q = 0$

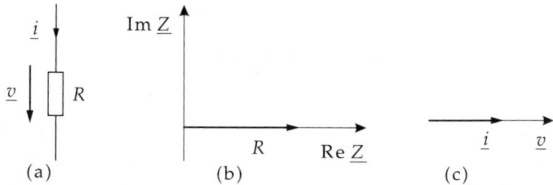

Figure 15.21: Ohmic resistor in an alternating-current circuit. (a): circuit symbol, (b): locus of the complex resistance, (c): phasor diagram for current and voltage.

15.2.5.2 Capacitor

Capacitor in the alternating-current circuit, the voltage $v(t)$ is the time integral of the current $i(t)$ flowing into the capacitor, divided by the capacitance C:

$$\underline{v}(t) = \frac{1}{C} \int_0^t \underline{i}(t')\, dt' \,.$$

1. Complex resistance of a capacitor

The complex resistance $\underline{Z}_C$ of a capacitor of capacitance C (**Fig. 15.22 (b)**) is purely imaginary and inversely proportional to the frequency $f = \omega/(2\pi)$.
▲ The reactance is negative.
▲ The resistance R vanishes.
The impedance is inversely proportional to the frequency f and the capacitance C; for low frequencies, it tends to infinity.

Complex resistance	Impedance	Resistance	Reactance
$\underline{Z} = -\dfrac{j}{\omega C}$	$Z = \dfrac{1}{\omega C}$	$R = 0$	$X = -\dfrac{1}{\omega C}$

■ A capacitor connected to a direct voltage, i.e., an alternating voltage of frequency $f = 0\,\text{Hz}$, has an infinitely high resistance.
➤ A capacitor connected to a high-frequency alternating voltage behaves like a short-circuit.
In the phasor diagram (**Fig. 15.22 (c)**), the current phasor and the voltage phasor are perpendicular to each other.
▲ The current leads the voltage by 90°:

$$\varphi_Z = \varphi_v - \varphi_i = -90° \,.$$

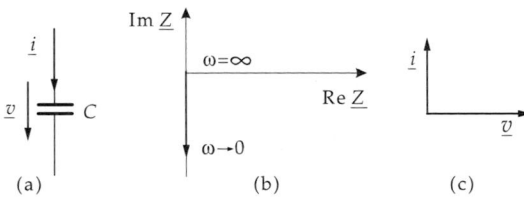

Figure 15.22: Capacitor in an alternating-current circuit. (a): circuit symbol, (b): locus of the complex resistance, (c): phasor diagram for current and voltage.

2. Complex conductance of a capacitor

The complex conductance $\underline{Y}_C$ is purely imaginary and proportional to the frequency $f = \omega/(2\pi)$ and to the capacitance C.
▲ The conductance vanishes.
▲ The susceptance is positive.

The admittance is proportional to the frequency f and the capacitance C.

Complex conductance	Admittance	Conductance	Susceptance
$\underline{Y} = j\omega C$	$Y = \omega C$	$G = 0$	$B = \omega C$

▲ The complex power at the capacitor is purely imaginary.
▲ The reactive power is negative.

Complex power	Apparent power	Real power	Reactive power
$\underline{S} = -jV \cdot I$	$S = V \cdot I$	$P = 0$	$Q = -V \cdot I$

15.2.5.3   Inductor

Inductor in an alternating-current circuit, the voltage $v(t)$ is equal to the product of the inductance L and the time derivative of the current $i(t)$:

$$\underline{v}(t) = L \frac{di(t)}{dt}.$$

1. Complex resistance of an inductor

The complex resistance $\underline{Z}_L$ (**Fig. 15.23 (b)**) of an inductor of inductance L is purely imaginary and depends on the frequency.
▲ The resistance R vanishes.
▲ The reactance X is positive.
  The impedance Z_L is proportional to the frequency f and vanishes for $f = 0$ Hz.

Complex resistance	Impedance	Resistance	Reactance
$\underline{Z} = j\omega L$	$Z = \omega L$	$R = 0$	$X = \omega L$

■ An ideal coil ($R = 0$) connected to a direct voltage is a short-circuit.
➤ A coil connected to a high-frequency alternating voltage has an infinitely high impedance.
In the phasor diagram (**Fig. 15.23 (c)**), the current phasor and the voltage phasor are perpendicular to each other.
▲ The voltage leads the current by $90°$:

$$\varphi_Z = \varphi_v - \varphi_i = 90° .$$

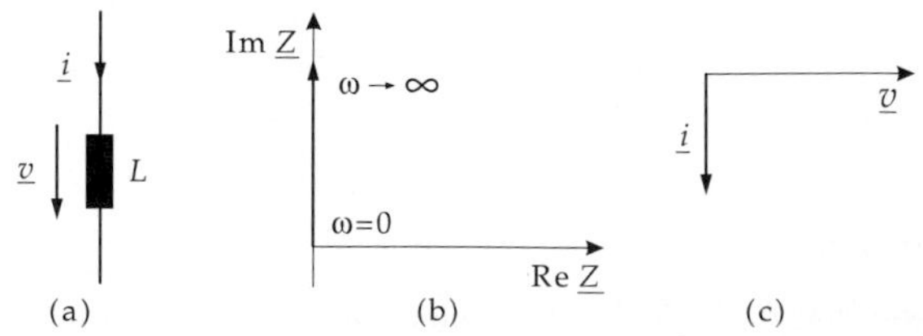

Figure 15.23: Inductor in an alternating-current circuit. (a): circuit symbol, (b): locus of the complex resistance, (c): phasor diagram for current and voltage.

2. Complex conductance of an inductor

The conductance $\underline{Y}_L$ is purely imaginary and inversely proportional to the frequency $f = \omega/(2\pi)$, as well as to the inductance L:
▲   The conductance vanishes.
▲   The susceptance is negative.
The admittance is inversely proportional to the frequency $f = \omega/(2\pi)$ and to the inductance L:

Complex conductance	Admittance	Conductance	Susceptance
$\underline{Y} = -\dfrac{j}{\omega L}$	$Y = \dfrac{1}{\omega L}$	$G = 0$	$B = -\dfrac{1}{\omega L}$

▲   The complex power at the inductor is purely imaginary.
▲   The reactive power is positive.

Complex power	Apparent power	Real power	Reactive power
$\underline{S} = jV \cdot I$	$S = V \cdot I$	$P = 0$	$Q = V \cdot I$

15.2.5.4   Complex resistances of the simplest two-terminal networks

Quantity	Resistor R	Capacitor C	Inductor L
$\underline{Z} = R + jX$	R	$-j\dfrac{1}{\omega C}$	$j\omega L$
R	R	0	0
X	0	$-\dfrac{1}{\omega C}$	ωL
$Z = \sqrt{R^2 + X^2}$	R	$\dfrac{1}{\omega C}$	ωL
$\phi_Z = \arctan(X/R)$	0	$-\pi/2$	$\pi/2$
$\underline{Y} = G + jB$	$\dfrac{1}{R}$	$j\omega C$	$-j\dfrac{1}{\omega L}$
G	$\dfrac{1}{R}$	0	0
B	0	ωC	$-\dfrac{1}{\omega L}$
$Y = \sqrt{G^2 + B^2}$	$\dfrac{1}{R}$	ωC	$\dfrac{1}{\omega L}$
$\phi_Y = \arctan(B/G)$	0	$\pi/2$	$-\pi/2$

15.2.6   Series connection of resistor and capacitor

Series connection of a resistor and a capacitor (Fig. 15.24).

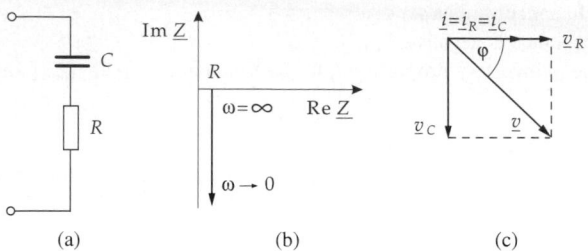

(a)   (b)   (c)

Figure 15.24: Series connection of a resistor and a capacitor in an alternating-current circuit. (a): circuit symbol, (b): locus of the complex resistance, (c): current-voltage phasor diagram.

The complex resistance is calculated according to the mesh rule: the complex total voltage equals the sum of the complex individual voltages of the resistor and capacitor.

The complex total resistance $\underline{Z}$ is (see **Fig. 15.24 (b)**):

resistance for series connection of R and C			$\mathbf{L^2T^{-3}MI^{-2}}$
	Symbol	Unit	Quantity
$\underline{Z} = R - \dfrac{\mathrm{j}}{\omega C}$	$\underline{Z}$	Ω	complex resistance
	R	Ω	ohmic resistance
	ω	s^{-1}	angular frequency
	C	F	capacitance
	j	1	imaginary unit

The resistive part of the impedance equals the ohmic resistance value R.

The reactive part of the impedance equals the reactive part of the capacitance C:

$$X = -\frac{1}{\omega C}.$$

The impedance of the series connection is

$$Z = \sqrt{R^2 + \left(\frac{1}{\omega C}\right)^2}.$$

▲   The phase shift (see **Fig. 15.24 (c)**) is between $0°$ and $-90°$. For high frequencies, it approaches $0°$, and for low frequencies it approaches $-90°$,

$$\varphi_Z = \varphi_v - \varphi_i = -\arctan\frac{1}{\omega RC}.$$

15.2.7   Parallel connection of a resistor and a capacitor

Parallel connection of a resistor and a capacitor (Fig. 15.25).

The complex resistance is calculated according to the branch-point rule: the complex total current equals the sum of the complex individual currents through the resistor and capacitor.

The complex total conductance $\underline{Y}$ is (see **Fig. 15.25 (b)**):

conductance for parallel connection of R and C			$\mathbf{L^{-2}T^3M^{-1}I^2}$
	Symbol	Unit	Quantity
$\underline{Y} = \dfrac{1}{R} + j\omega C$	$\underline{Y}$	S	complex conductance
	R	Ω	ohmic resistance
	ω	s^{-1}	angular frequency
	C	F	capacitance
	j	1	imaginary unit

The conductive part of the admittance equals the reciprocal value of the ohmic resistance R:

$$G = \frac{1}{R}.$$

The susceptive part of the admittance equals the susceptance of the capacitor C:

$$B = \omega C.$$

The admittance of the parallel connection is

$$Y = \sqrt{\frac{1}{R^2} + (\omega C)^2}.$$

▲   The phase shift (see **Fig. 15.25 (c)**) is between $0°$ and $-90°$. For high frequencies, it approaches $-90°$, for low frequencies, $0°$,

$$\varphi = \varphi_v - \varphi_i = -\arctan \omega RC.$$

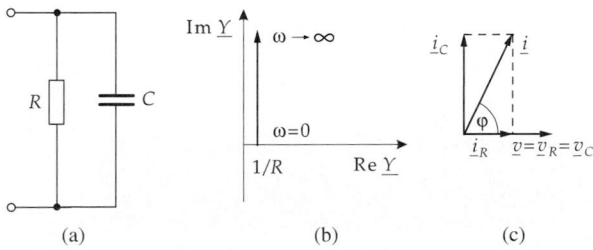

Figure 15.25: Parallel connection of resistor and capacitor. (a): circuit symbol, (b): locus of the complex conductance, (c): current-voltage phasor diagram.

15.2.8 Parallel connection of a resistor and an inductor

Parallel connection of a resistor and an inductor (Fig. 15.26).

The complex resistance is calculated according to the branch-point rule: the complex total current equals the sum of the complex individual currents through the resistor and the inductor.

The complex total conductance $\underline{Y}$ (see **Fig. 15.26 (b)**) is:

complex conductance for parallel connection of R and L			$\mathbf{L^{-2}T^3M^{-1}I^2}$
	Symbol	Unit	Quantity
$\underline{Y} = \dfrac{1}{R} - \dfrac{j}{\omega L}$	$\underline{Y}$	S	complex conductance
	R	Ω	ohmic resistance
	ω	s^{-1}	angular frequency
	L	H	inductance
	j	1	imaginary unit

The conductive part of the admittance (conductance) equals the reciprocal value of the ohmic resistance R:

$$G = \frac{1}{R}.$$

The susceptive part of the admittance (susceptance) equals the susceptance of the inductor L:

$$B = -\frac{1}{\omega L}.$$

The admittance is

$$Y = \sqrt{\frac{1}{R^2} + \left(\frac{1}{\omega L}\right)^2}.$$

▲ The phase shift (see **Fig. 15.26 (c)**) is between $0°$ and $90°$. For high frequencies, it approaches $0°$, for low frequencies, $90°$:

$$\varphi = \varphi_v - \varphi_i = \arctan\frac{R}{\omega L}.$$

15.2.9 Series connection of a resistor and an inductor

Series connection of a resistor and an inductor (Fig. 15.27).

The complex resistance is calculated according to the mesh rule: the complex total voltage equals the sum of the complex individual voltages of resistor and inductor.

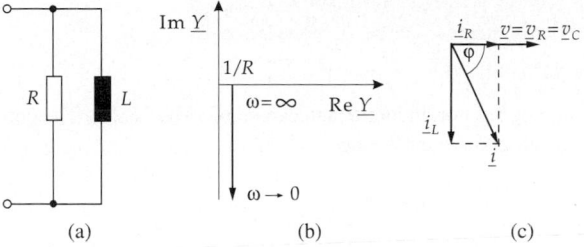

Figure 15.26: Parallel connection of a resistor and an inductor. (a): circuit symbol, (b): locus of the complex conductance, (c): current-voltage phasor diagram.

The complex total resistance $\underline{Z}$ (see **Fig. 15.27 (b)**) is:

resistance for series connection of R and L			$\mathbf{L^2T^{-3}MI^{-2}}$
	Symbol	Unit	Quantity
$\underline{Z} = R + j\omega L$	$\underline{Z}$	Ω	complex resistance
	R	Ω	ohmic resistance
	ω	s^{-1}	angular frequency
	L	H	inductance
	j	1	imaginary unit

The resistance is equal to the value of the ohmic resistance R.
The reactance is equal to the reactance of the inductor L:

$$X = \omega L .$$

The impedance is

$$Z = \sqrt{R^2 + (\omega L)^2} .$$

▲ The phase shift (see **Fig. 15.27 (c)**) is between $0°$ and $90°$. For low frequencies, it approches $0°$, and for high frequencies it approaches $90°$:

$$\varphi = \varphi_v - \varphi_i = \arctan \frac{\omega L}{R} .$$

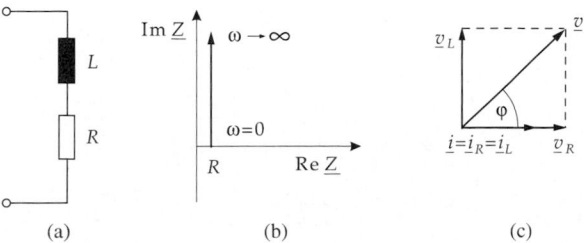

Figure 15.27: Series connection of a resistor and an inductor. (a): circuit symbol, (b): locus of the complex resistance, (c): current-voltage phasor diagram.

15.2.10 Series-resonant circuit

1. Series-resonant circuit,

a series connection of resistor, inductor, and capacitor (**Fig. 15.28**). The complex resistance is calculated according to the mesh rule.

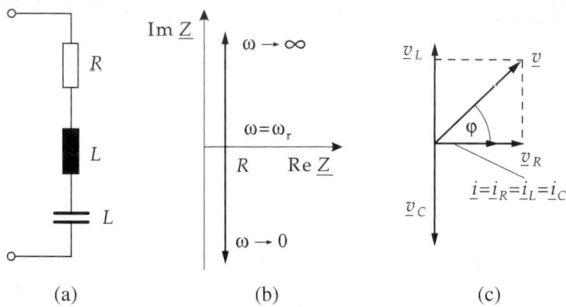

Figure 15.28: Series-resonant circuit. (a): circuit symbol, (b): locus of the complex resistance, (c): current-voltage phasor diagram.

The complex resistance (see **Fig. 15.28 (b)**) is

series-resonant circuit			$\mathbf{L^2T^{-3}MI^{-2}}$
	Symbol	Unit	Quantity
$$\underline{Z} = R + \mathrm{j}\left(\omega L - \dfrac{1}{\omega C}\right)$$	$\underline{Z}$	Ω	complex total resistance
	R	Ω	ohmic resistance
	ω	s^{-1}	angular frequency
	L	H	inductance
	C	F	capacitance
	j	1	imaginary unit

The resistance equals the ohmic resistance R.
The reactance equals the sum of the reactances of capacitor and inductor:

$$X = \omega L - \frac{1}{\omega C} \, .$$

The reactance depends on the frequency $f = \omega/(2\pi)$. It vanishes at the resonance frequency (see below).
The impedance is

$$Z = \sqrt{R^2 + \left(\omega L - \frac{1}{\omega C}\right)^2} \, .$$

The phase angle is

$$\varphi_Z = \arctan\left(\frac{\omega L - 1/(\omega C)}{R}\right) \, .$$

2. Resonance,

occurs when the capacitive reactance and the inductive reactance cancel each other. The total resistance is then real and equal to the ohmic resistance. The current takes a maximum value for a given applied voltage.

Series resonance, denotes the resonance in the series-resonant circuit.

Resonance frequency, for given inductance L and capacitance C, is

$$f_r = \frac{1}{2\pi} \frac{1}{\sqrt{LC}} .$$

The current in the series-resonant circuit takes a maximum value at the resonance frequency, and the phase angle changes by 180° (**Fig. 15.29**).

▲ At resonance the total resistance is real and takes a minimum value.

▲ Below the resonance frequency, the total current leads the total voltage, above the resonance frequency, the total voltage leads the total current.

▲ At the resonance frequency, the total current and the total voltage are in phase.

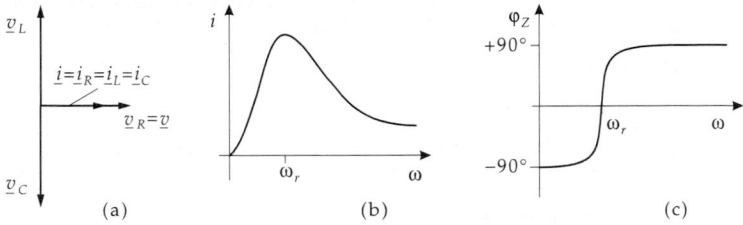

Figure 15.29: Series-resonant circuit. (a): current-voltage phasor diagram for resonance, (b): current amplitude, (c): phase angle for finite quality.

Quality of the series-resonant circuit, Q_R, the ratio of inductive or capacitive reactance at resonance $X_0 = X_C = X_L$ to the resistance R of the series connection:

$$Q_R = \frac{X_0}{R} .$$

▲ The lower the quality, the faster the oscillation in the circuit dies away; for low quality, the oscillation is damped more strongly, and the resonance curve $i(\omega)$ shows a broader maximum.

Damping factor of the series-resonant circuit, d_R, reciprocal value of the quality Q_R,

$$d_R = \frac{1}{Q_R} .$$

15.2.11   Parallel-resonant circuit

1. Parallel-resonant circuit,

a parallel connection of resistor, inductor and capacitor (**Fig. 15.30**). The complex conductance is calculated according to the branch-point rule.

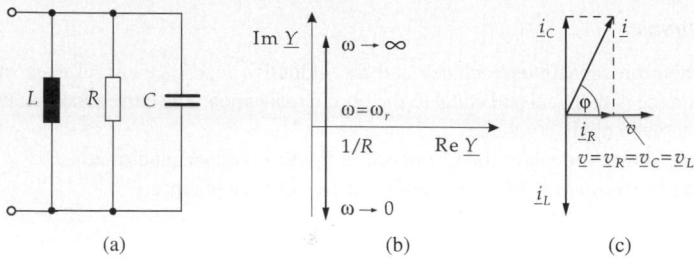

(a)   (b)   (c)

Figure 15.30: Parallel-resonant circuit. (a): circuit symbol, (b): locus of the complex conductance, (c): current-voltage phasor diagram.

The complex conductance is:

parallel-resonant circuit			$\mathbf{L^{-2}T^3M^{-1}I^2}$
	Symbol	Unit	Quantity
$\underline{Y} = \dfrac{1}{R} + j\left(\omega C - \dfrac{1}{\omega L}\right)$	$\underline{Y}$	S	complex total conductance
	R	Ω	ohmic resistance
	ω	s^{-1}	angular frequency
	L	H	inductance
	C	F	capacitance
	j	1	imaginary unit

The conductance is equal to the conductance of the ohmic resistor:

$$G = \frac{1}{R}.$$

The susceptance is equal to the sum of the susceptances of capacitor and inductor:

$$B = \left(\omega C - \frac{1}{\omega L}\right).$$

The susceptance depends on the frequency $f = \omega/(2\pi)$; it vanishes at the resonance frequency (see below).

The admittance is

$$Y = \sqrt{\frac{1}{R^2} + \left(\omega C - \frac{1}{\omega L}\right)^2}.$$

The phase angle of the complex conductance reads

$$\varphi_Y = \arctan\left(\omega RC - \frac{R}{\omega L}\right).$$

2. Resonance,

occurs when the susceptances of the inductor and the capacitor cancel each other.
▲   At resonance, the total conductance is real and equal to the reciprocal value of the ohmic resistance.

Parallel resonance, denotes the resonance in the parallel-resonant circuit.
Resonance frequency, f_r, the frequency at which the resonance occurs,

$$f_r = \frac{1}{2\pi} \frac{1}{\sqrt{LC}}.$$

The current takes a minimum value at the resonance frequency of the parallel-resonant circuit, and the phase angle changes by $180°$ (**Fig. 15.31**).
▲ At resonance, the total resistance is real and takes a maximum value.
➤ The parallel-resonant circuit acts as a **rejector circuit**.
▲ Below the resonance frequency, the voltage leads the current. Above the resonance frequency, the current leads the voltage.
▲ At resonance, the current and the voltage are in phase.

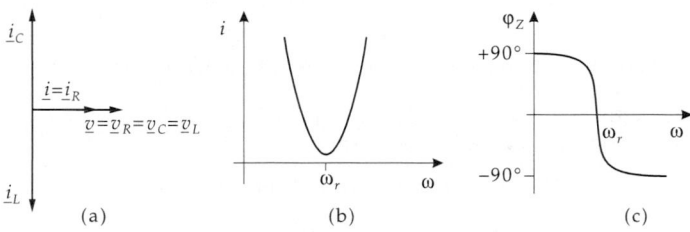

Figure 15.31: Parallel-resonant circuit. (a): current-voltage phasor diagram at resonance, (b): current amplitude, (c): phase angle for finite quality.

Quality of the parallel-resonant circuit, Q_P, ratio of the inductive or capacitive suscep-tance at resonance $Y_0 = Y_C = Y_L$ to the conductance G of the parallel connection:

$$Q_P = \frac{Y_0}{G}.$$

The lower the quality, the faster the oscillation in the circuit dies out; for low quality, the oscillation is damped more strongly, and the resonance curve $i(\omega)$ shows a broader minimum.

Damping factor of the parallel-resonant circuit, d_P, the reciprocal value of the quality Q_P,

$$d_P = \frac{1}{Q_P}.$$

15.2.12 Equivalence of series and parallel connections

1. Equivalent conversions

A series connection consisting of an ohmic resistor and a reactance (inductor or capac-itor) may be represented—for a definite angular frequency ω—by a parallel connection (**Fig. 15.32**).
▲ A parallel connection and a series connection of an ohmic resistance and a reactance have the same response if the complex resistances of both connections are the same.

▲   The equivalence can hold only for a definite frequency ω. For other frequencies, the complex resistances for series and parallel connection behave differently.

▲   The equivalence holds only for sinusoidal voltages and currents.

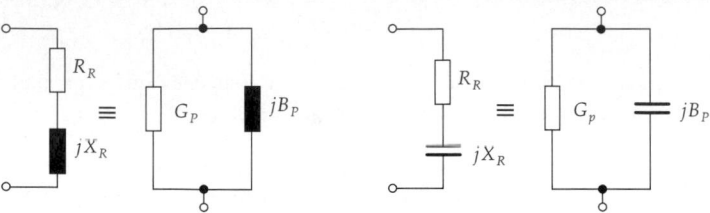

Figure 15.32: Equivalent conversions of series and parallel connections for a fixed frequency ω.

2. Transformation of a parallel connection to an equivalent series connection

parallel connection $\Longrightarrow$ series connection			$\mathbf{L^2T^{-3}MI^{-2}}$
$R_\mathrm{R} = \dfrac{G_\mathrm{P}}{G_\mathrm{P}^2 + B_\mathrm{P}^2}$ $X_\mathrm{R} = \dfrac{B_\mathrm{P}}{G_\mathrm{P}^2 + B_\mathrm{P}^2}$	Symbol	Unit	Quantity
	R_R	Ω	resistance of the series connection
	X_R	Ω	reactance of the series connection
	G_P	S	conductance of the parallel connection
	B_P	S	susceptance of the parallel connection

3. Transformation of a series connection to an equivalent parallel connection

series connection $\Longrightarrow$ parallel connection			$\mathbf{L^{-2}T^3M^{-1}I^2}$
$G_\mathrm{P} = \dfrac{R_\mathrm{R}}{R_\mathrm{R}^2 + X_\mathrm{R}^2}$ $B_\mathrm{P} = \dfrac{X_\mathrm{R}}{R_\mathrm{R}^2 + X_\mathrm{R}^2}$	Symbol	Unit	Quantity
	G_P	S	conductance of the parallel connection
	B_P	S	susceptance of the parallel connection
	R_R	Ω	resistance of the series connection
	X_R	Ω	reactance of the series connection

15.2.13 Radio waves

1. Generation and reception of electromagnetic waves

Resonant circuits are used for generating and receiving **electromagnetic waves**. Both the emission and the reception are carried out with antennas.

Mode of operation:

Linear oscillator, also **Hertz oscillator** or **Hertz dipole**, oscillating charge distribution that is surrounded by electromagnetic fields. The separation and propagation (see **Fig. 15.33**) of these fields is described by Maxwell's equations (see p. 496). At a distance of only a few wavelengths from the oscillating dipole, this field is already a **transverse wave**.

➤ The Hertz dipole may be represented by a resonant circuit with the inductance of a *coil* with a single loop, and the capacitance of a *capacitor* consisting of two wires extended to a linear dipole.

Electromagnetic oscillations at high frequencies are damped, in particular by **electromagnetic radiative losses**. The losses may be compensated by feeding the circuit by electric energy in the rhythm of the oscillations.

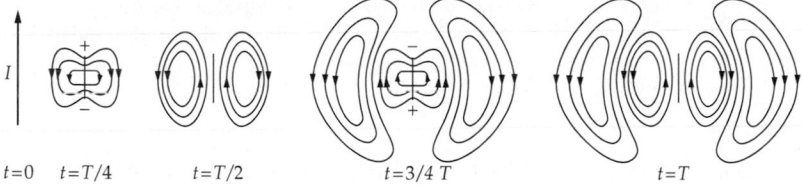

$t=0$ $t=T/4$ $t=T/2$ $t=3/4\,T$ $t=T$

Figure 15.33: Separation of the electromagnetic field generated by an oscillating Hertz dipole.

Resonance frequency of the linear oscillator, inversely proportional to the conductor length l of the oscillator:

resonance frequency $\sim \dfrac{1}{\text{conductor length}}$			$\mathbf{T^{-1}}$
$f = \dfrac{c}{2l}$	Symbol	Unit	Quantity
	f	Hz=1/s	resonance frequency
	c	m/s	speed of light
	l	m	length of conductor

2. Electromagnetic waves: propagation and applications

The propagation properties, hence the application, of electromagnetic waves depend strongly on the wavelength.

Wavelength	Frequency	Notation, use
high frequency		
30 km $\cdots$ 2 km	10 kHz $\cdots$ 150 kHz	ultra-long waves, VLF (very low frequency) underwater radio
2000 m $\cdots$ 600 m	150 kHz $\cdots$ 500 kHz	long waves, LW radio broadcasting
600 m $\cdots$ 200 m	500 kHz $\cdots$ 1.5 MHz	medium waves, MW radio broadcasting
100 m $\cdots$ 10 m	3 MHz $\cdots$ 30 MHz	short waves, SW radio broadcasting, amateur radio
10 m $\cdots$ 1 m	30 MHz $\cdots$ 300 MHz	ultra-short waves, USW, VHF (very high frequency) radio broadcasting, television police radio, air navigation

Wavelength	Frequency	Notation, use
high frequency (*continued*)		
1 m ⋯ 10 cm	300 MHz ⋯ 3 GHz	decimeter waves, UHF (ultra-high frequency) television, line-of-sight radio
10 cm ⋯ 1 cm	3 GHz ⋯ 30 GHz	centimeter waves line-of-sight radio, radar
10 mm ⋯ 1 mm	30 GHz ⋯ 300 GHz	millimeter waves
light waves		
1 mm ⋯ 1 μm	$3 \cdot 10^{11}$ Hz ⋯ $3 \cdot 10^{14}$ Hz	infrared, thermal radiation
760 nm	$3.95 \cdot 10^{14}$ Hz	red
589 nm	$5.09 \cdot 10^{14}$ Hz	yellow
527 nm	$5.70 \cdot 10^{14}$ Hz	green
486 nm	$7.65 \cdot 10^{14}$ Hz	violet
100 nm ⋯ 10 nm	$3 \cdot 10^{15}$ Hz ⋯ $3 \cdot 10^{16}$ Hz	ultra-violet
x-rays		
1 nm ⋯ 100 pm	$3 \cdot 10^{17}$ Hz ⋯ $3 \cdot 10^{19}$ Hz	
gamma rays		
100 pm ⋯ 0.1 pm	$3 \cdot 10^{19}$ Hz ⋯ $3 \cdot 10^{22}$ Hz	

➤ The wavelength ranges of x-rays and gamma radiation overlap. x-rays and gamma radiation differ by their generation mode (transitions between energy levels in atoms and in atomic nuclei, respectively).

15.3   Electric machines

Electric machines serve for the conversion of one form of energy into another one. The law of induction and the Lorentz force are used to operate generators and motors.

A **motor** receives electric energy and converts it into rotational energy.

A **generator** receives rotational energy and converts it into electric energy.

▲ In principle, any electric machine can work in the motor mode or in the generator mode, depending on the direction of energy flow.

➤ The energy conversion by electric machines has the advantage that the losses are particularly small. Efficiencies beyond 99 % may be achieved.

15.3.1   Fundamental functional principle

1. Moving conductor loop in a magnetic field

When a conducting loop is moved in a magnetic field, a voltage V_{ind} is induced in the loop. If this voltage can drive a current through the conductor, the conductor is under the action of a force F (Lorentz force) that opposes the direction of motion of the conductor.

Load current, I, flows in the conductor loop when the ends of the wire are connected through a resistor.

■ Technical application: Conductor loops and magnets are suitably arranged to rotate against each other. Either the magnets can rotate within a fixed arrangement of conductors, or the conductor loops can rotate between fixed magnets.

Stator, the fixed part of the machine.

Rotor or **runner**, movable, rotating part of the machine.

Armature, the part of the machine that carries the winding for the load current, depending on the design of the machine.

➤ Generators are usually constructed as **revolving-field machines** because of the ease of connecting to the current output, so that armature and stator are identical.

The magnetic flux Φ_E is guided in iron with possibly narrow air gaps between stator and rotor, and is generated by a coil carrying the **field current** I_E.

2. Induced voltage and torque

Induced voltage, V_{ind}, directly proportional to the exciting flow Φ_E and the turn speed n:

induced voltage ~ exciting flow · turn speed			$L^2T^{-3}MI^{-1}$
	Symbol	Unit	Quantity
$V_{ind} = k_1 \cdot \Phi_E \cdot n$	V_{ind}	V	induced voltage
	k_1	1	machine constant
	Φ_E	Wb = Vs	exciting flow
	n	min^{-1}	turn speed

The machine constant k_1 includes all constructive features of the machine.

The Lorentz force acts on the coil windings of the rotor, thereby generating a torque.

The **torque** is directly proportional to the load current I_1 and to the exciting flux Φ_E. The proportionality factor k_2 is another machine constant.

torque ~ load current · exciting flux			$L^2T^{-2}M$
	Symbol	Unit	Quantity
$\tau = k_2 \cdot I \cdot \Phi_E$	τ	Nm	torque
	k_2	1	machine constant
	I	A	load current
	Φ_E	Wb = Vs	exciting flux

➤ These two machine equations may be transferred accordingly to various types of machines.

15.3.2 Direct-current machine

Fig. 15.34 shows a section view of a quadrupole **direct-current machine** in the motor mode.

The outer stator carries the main poles of the exciting winding and four smaller commutating poles. The armature winding moves within the poles, which are arranged in slots on the rotor. If a coil of the armature winding leaves the range of action of a main pole and moves into the range of the opposite pole, the current direction in the armature must be inverted by the commutator.

Commutator, **collector** or **current reverser**, contact foils are assigned to each coil, insulated against each other and rotating with the rotor shaft. In the neutral zone between

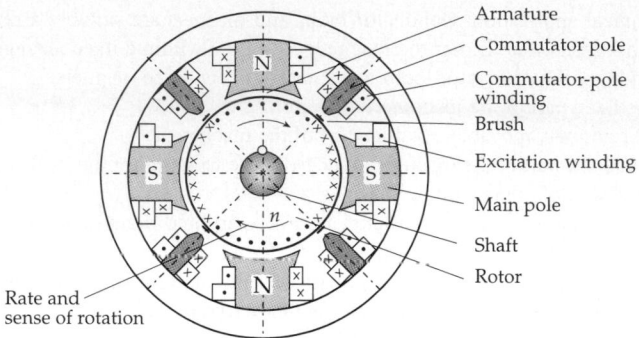

Armature
Commutator pole
Commutator-pole winding
Brush
Excitation winding
Main pole
Shaft
Rotor
Rate and sense of rotation

Figure 15.34: Configuration of a direct-current machine in motor mode.

the poles, the current of one direction is supplied or extracted by spatially fixed carbon brushes that are pushed against the commutator.

Commutating poles are excited by the armature current, in a direction contrary to the main poles. They facilitate reversal of the current direction under the brush.

Speed control for the direct-current motor, the rate of rotation n may be increased by increasing the terminal voltage V_T or by decreasing the exciting flux Φ_E:

turn speed of the direct-current motor			$\mathbf{T^{-1}}$
	Symbol	Unit	Quantity
$n = \dfrac{V_T - I_A \cdot R_A}{k_1 \cdot \Phi_E}$	n	min^{-1}	turn speed
	V_T	V	terminal voltage
	I_A	A	armature current
	R_A	Ω	internal resistance of armature
	k_1	1	machine constant
	Φ_E	Wb=Vs	exciting flux

➤   There is risk of destruction at too high a rate of rotation if the exciting flux vanishes! This is called **run-away of the motor**.

1. Starting a direct-current motor

To **start a direct-current motor**, the exciting current I_E has to be adjusted to the maximum allowed value in order to generate a sufficiently high starting torque. Since the armature circuit has only a very low internal resistance, a short-circuit current would flow when switching on the terminal voltage. In order to restrict the transient current pulse, a starting resistance is preconnected to the armature circuit; it thereafter must be shorted out after the motor reaches speed.

2. Connection of direct-current motors

Direct-current motors exhibit different performance characteristics that depend on how they are connected:

a) Shunt motor, the exciting circuit (E_1, E_2) and the armature circuit (A_1, B_2) are connected in parallel to each other and to the supply voltage (V_{supply}) (**Fig. 15.35 (a)**).

b) Separate-excited motor, the exciting coil (F_1, F_2) is fed by a separate voltage source (V_E) (**Fig. 15.35 (b)**).

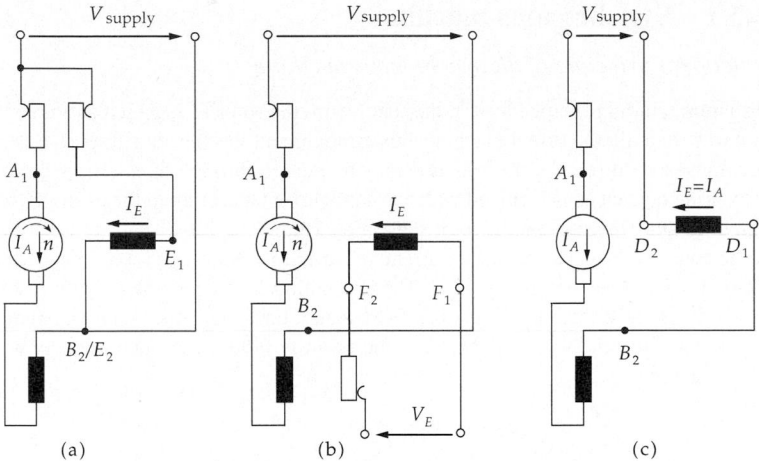

Figure 15.35: Connection of direct-current motors. (a): shunt motor, (b): separate-excited motor, (c): series motor.

A shunt motor and a separate-excited motor exhibit similar performance characteristics. Their rate of rotation may be varied by a **field rheostat** for the exciting field. For the adjusted value, the rate of rotation decreases only slightly under load.

c) Series motor, the exciting coil (D_1, D_2) is connected in series to the armature circuit (A_1, B_2) (**Fig. 15.35 (c)**). Hence, the exciting field and the torque are kept constant with increasing load, the turn speed, however, decreases.

➤ Therefore, a series motor may be used only if the operational conditions exclude running without a load!

■ Application for vehicle starters, drives of railroads and motors for cranes.

d) Compound motor, combination of shunt motor and series motor, separate exciting windings in shunt connection and series connection.

Therefore, changes of load cause only minor changes in rate of rotation, smaller than for a series motor. A run-away under no-load conditions is excluded by the fixed no-load speed of the series connection.

3. Reversal of rotation sense

The rotation direction of a direct-current machine may be reversed by inverting the field or current direction of the exciting winding. Since in a series machine the field winding and the armature winding are in series, a change of polarity acts simultaneously on both field directions, and therefore there is no change of the rotation sense.

➤ This suggests the operation of the series machine by single-phase alternating current, which leads to the principle of the **single-phase alternating current motor** or **universal motor**.

15.3.3 Three-phase machine

Three-phase machines, subdivided into **synchronous machines** and **asynchronous machines** (**three-phase induction machine**), depending on whether the armature runs synchronously or asynchronously with the main frequency.

15.3.3.1   Synchronous machine

1. Functional principle of the synchronous machine

The armature winding is embedded in the stator iron, the rotor is designed as a pole wheel. Compared with a direct-current machine, this arrangement has the advantage that the comparable high load currents of the armature may be guided through fixed terminals, and the lower exciting current can be guided with constant current direction in the exciting winding through only two slip rings on the rotor shaft (**Fig. 15.36**).

For the connection to a three-phase current, the armature winding is split into three phase windings staggered electrically by 120°. The alternating fields generated by the individual phase windings superpose to a **rotating field** that rotates with the main frequency and causes the rotor to run in coincidence with the rotating field, i.e., synchronously with the mains frequency.

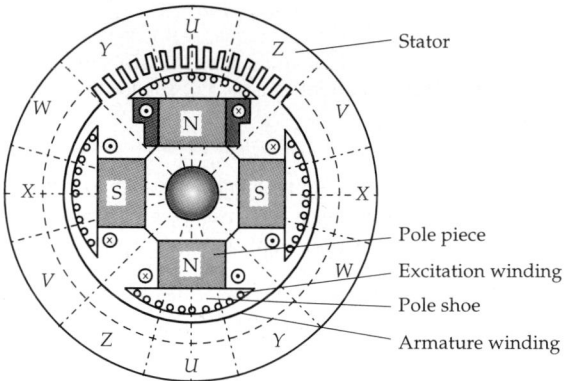

Figure 15.36: Configuration of a synchronous machine.

2. Rate of rotation equation

A synchronous engine has only one fixed rate of rotation n_{sync}, which is determined by the main frequency f divided by the number of pole pairs p of the engine:

turn speed = $\dfrac{\text{frequency}}{\text{number of pole pairs}}$			$\mathbf{T^{-1}}$
	Symbol	Unit	Quantity
$n_{sync} = \dfrac{f \cdot 60}{p}$	n_{sync}	min^{-1}	turn speed
	f	Hz	mains frequency
	p	1	number of pole pairs

■ The synchronous engine (**Fig. 15.36**) has two pole pairs and runs for a main frequency $f = 50$ Hz with a rate of rotation of $n_{sync} = 1500$ min^{-1}.

➤ Notice that the basic configuration of the phase windings of the stator has to correspond to the number of pole pairs of the rotor. The rate of rotation equation also shows that—contrary to the direct-current machine—a change of the exciting current cannot influence the turn speed.

Overexcitation, excitation of the synchronous machine beyond its demand of magnetization. Overexcitation causes the armature current to lead the terminal voltage, i.e., a capacitive action for the electric network.

Underexcitation, causes a lagging of the armature current, i.e., an inductive action. For strong underexcitation and simultaneous load, the pole wheel cannot keep step with the applied frequency, the machine drops out of synchronization.

3. Operation of the synchronous machine

▲ As **generator** a synchronous machine may only then be connected to the main supply or parallel to other generators if the following three **synchronizing conditions** are fulfilled:

- equal voltage,
- equal frequency,
- equal phase relation.

▲ As **motor**, a synchronous machine must be started only with a shorted exciting winding to avoid mechanical damage due to a sudden acceleration or high voltages induced in the armature winding.

➤ The **asynchronous run-up** is facilitated by a squirrel-cage winding in the pole faces.

15.3.3.2   Asynchronous machine

1. Functional principle of the asynchronous machine

The stator of an asynchronous machine (see **Fig. 15.37** for a quadrupole machine) is constructed like the stator of the synchronous machine, with three phase windings connected to the three-phase main voltage.

The three-phase windings generate a field that rotates with the main frequency and penetrates the rotor. The rotor is drum-like and arranged in layers of iron sheets, and carries a **squirrel-cage winding**. The rotating field induces voltages in the winding bars of the rotor causing a current and generating a field imposed to the rotor. This rotor field lags the rotational field by 90° and accelerates the rotor to follow the rotational field.

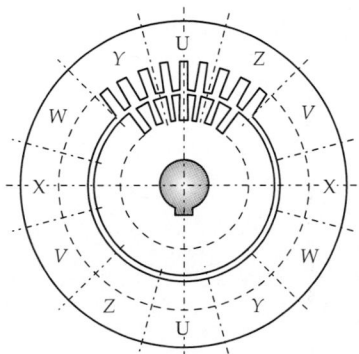

Figure 15.37: Configuration of an asynchronous machine.

If the rotor approaches the rate of rotation corresponding to the main frequency, the induction by the rotating field, and thus the acceleration of the rotor, no longer acts.

➤ Therefore, the rotor runs slightly slower than the rotating field, i.e., asynchronously, with the actual rate of rotational n.

2. Synchronous speed,

or **rotating-field speed**, n_{sync}, determined by the number of pole pairs p of the stator winding, as for the synchronous machine:

synchronous speed of asynchronous machine			$\mathbf{T^{-1}}$
$n_{sync} - \dfrac{f \cdot 60}{p}$	Symbol	Unit	Quantity
	n_{sync}	min^{-1}	synchronous speed
	f	Hz	mains frequency
	p	1	number of pole pairs

3. Slippage,

s, a measure for the load-dependent delay of the rotor against the rotating field.
- ■ For a nominal load, the slippage lies between 0.5 % and 10 %.
- ▲ The slippage s is given by the difference between the actual rate of rotation n and the rate of rotation of the rotating field n_{sync}, related to n_{sync}:

slippage of the asynchronous motor			1
$s = \dfrac{n_{sync} - n}{n_{sync}}$	Symbol	Unit	Quantity
	s	1	slippage
$= 1 - \dfrac{n}{n_{sync}}$	n	min^{-1}	actual rate of rotation
	n_{sync}	min^{-1}	synchronous rate of rotation

- ➤ Since the rotor of the asynchronous machine is designed to be drum-like without pronounced poles, the nominal speed is fixed only by the choice of the stator winding. A pole changing of the stator winding following **Dahlander** allows for speed changing between two fixed speed values.

In order to restrict the high starting current of an asynchronous machine, the stator winding is frequently switched on through a star-delta starter.

- ▲ The starting torque of the asynchronous machine is lower than the nominal torque. Therefore, for larger machines the start is provided by current-displacement rotors, or by start resistors connected into the rotor circuit by slip rings.
- ▲ Advantage of the asynchronous motor: It is not very sensitive and is nearly maintenance-free.
- ■ Frequently used as driving motor.

16
Current conduction in liquids, gases and vacuum

In liquids and gases, electric current is not transported only by electrons as it is in solids, but also by positive and negative ions. Also, the electric current in liquids can cause their decomposition.

16.1 Electrolysis

16.1.1 Amount of substance

Amount of substance, n, measure of the number of particles in a quantity of equal particles (atoms, molecules or ions), independent of their mass.

amount of substance = particle number/Avogadro constant			**N**
$n = \dfrac{N}{N_A}$ $N_A = 6.022\,136\,7 \cdot 10^{23}\ \text{mol}^{-1}$	Symbol	Unit	Quantity
	n	mol	amount of substance
	N	1	particle number
	N_A	1/mol	Avogadro constant

Mole, mol, SI unit of the amount of substance. 1 mol is the quantity of substance that contains just as many particles as 0.012 kg ^{12}C.

Avogadro constant, **Avogadro's number**, also Loschmidt number, N_A, the number of particles in 1 mol of substance.

16.1.2   Ions

1. Ionization and ions

Ionization, removal of one or several electrons from an atom, or addition of one or several electrons to an atom, so that the resulting object is electrically charged.

   Ions, atoms or molecules, which in total are not electrically neutral.
- Ions may be charged positively or negatively.
- Ions may carry one or several electric elementary charges.
- Ions are generated by the transfer of electrons, e.g., in the fragmentation of polar molecules ($H_2O \rightarrow OH^- + H^+$: **dissociation**).
- Positively charged ions often are ionized metal atoms.
- Negatively charged ions often are non-metallic molecule groups.
- ■ **Cations**, positive ions: Na^+, Ca^{++} (metals); H^+ (non-metal).
     Anions, negative ions: SO_4^{--}, Cl^-, NO_3^-.

2. Properties of ions

If a voltage is applied, anions move towards the anode, cations move towards the cathode.
➤ Frequently, salts are composed of ionic crystals. If they are dissolved in water, they divide into single ions ($NaCl \rightarrow Na^+ + Cl^-$).
   Ionic valence, z, the excess of positive charge over negative charge in an ion.

■
$$
\begin{aligned}
Li^+ & \quad z = 1 \\
Li^{2+} & \quad z = 2 \\
Cl^- & \quad z = -1 \\
U^{92+} & \quad z = 92
\end{aligned}
$$

➤ The ionic valence z should not be confused with the **atomic charge** Z that gives the number of protons in the nucleus, independent of the actual number of electrons in the shell.

16.1.3   Electrodes

Electrode, part of a solid conductor supplying a liquid, a gas, a vacuum or a solid with electric current.
   Electrochemical electrode, two-phase system consisting of a combination of an element (e.g., a copper rod) and solutions of its ions (e.g., copper-sulphate solution).
   Standard hydrogen electrode, platinum electrode in electrolyte bathed in hydrogen.
   Anode, positive electrode.
   Cathode, negative electrode.
▲ Anodes receive electrons, cathodes deliver electrons.
▲ Anions discharge at the anode, cations at the cathode.

16.1.4   Electrolytes

Electrolyte, a liquid that conducts the current. To a large extent, it consists of mobile ions.
➤ Pure water is a poor conductor because it dissociates only to a very small extent. The conductance may be strongly increased by adding a small portion of salts.
Hydratization, non-dense enclosure of dissolved ions in a cloud of polar solvent particles, such as water by electrostatic ion-dipole interaction.

Electronegativity, the tendency of an atom to bind electrons. In the periodic table of elements, fluorine and oxygen have the highest electronegativity, rubidium and caesium have the lowest electronegativity.

➤ Because of the special form of the molecule and the distinct electronegativities of hydrogen and oxygen, water has a static dipole moment, i.e., it is a **polar molecule**.

16.1.4.1   Electric conductance of an electrolyte

1. Ion motion in electrolytes

If an external electric field $\vec{E}$ is applied to an electrolyte, then the ions **drift** through the electrolyte at constant velocity (**Fig. 16.1**).

Drift velocity of ions v_{dr} in electrolytes, mean velocity of ions in an electrolyte in an external electric field $\vec{E}$.

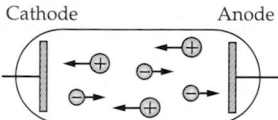

Figure 16.1: Motion of ions of an electrolyte in an external electric field.

drift velocity of ions in an electrolyte			$\mathbf{LT^{-1}}$
	Symbol	Unit	Quantity
$v_{dr} = \mu E$	v_{dr}	m/s	drift velocity
	μ	m^2/(Vs)	ion mobility
	E	V/m	electric field strength

▲ The ions drift parallel or antiparallel to the local electric field, depending on the sign of the ion charge.

The ion mobility depends both on the ion sort and on the medium the ions are drifting in.

The ion mobility in electrolytes is by about four orders of magnitude lower than the ion mobility in gases.

The energy gain in the external electric field is counteracted by the energy losses by collisions between the drifting ions and the surrounding molecules of the electrolyte. As a result, a mean drift velocity is observed.

2. Electric conductance of an electrolyte,

γ, conductance per unit length in an electrolyte (**Fig. 16.2**):

electric conductance of an electrolyte			$\mathbf{L^{-3}T^3M^{-1}I^2}$
	Symbol	Unit	Quantity
	γ	S/m	electric conductance
$\gamma = ze_0(\mu_+ n_+ + \mu_- n_-)$	z	1	ionic-charge number
	e_0	C	elementary charge
	$\mu_\pm$	m^2/(Vs)	ion mobility
	$n_\pm$	1/m^3	ionic density

Both positive and negative ions contribute to the electric current. Their mobilities are different, depending on the ionic charge and ionic radius.

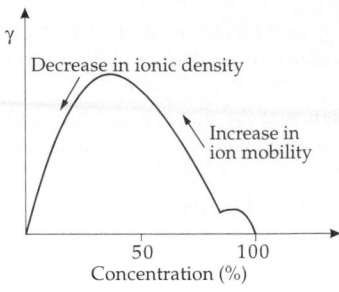

Figure 16.2: Electric
conductance of H_2SO_4 in
water (schematic)

Equivalent conductance, Λ, defined by

$$\Lambda = \frac{\gamma}{c},$$

where c is the concentration of the amount of substance, i.e., the number of moles of substance per unit volume.

➤ An electric current in an electrolyte leads to chemical reactions in the medium and at the electrodes, which may cause dissociation of the electrolyte.

Electrolysis, the decomposition of a substance by applying an electric voltage.

■ Water is decomposed into gaseous hydrogen and gaseous oxygen by applying a voltage between the two electrodes.

16.1.4.2 Faraday's laws

The quantitative relation between the electric current through the electrolyte and the quantities of substance precipitated at the electrodes is formulated in **Faraday's laws**.

1. Faraday's first law,

the mass precipitated is proportional to the transported quantity of charge only:

precipitated mass = charge/Faraday constant			**M**
	Symbol	Unit	Quantity
$m = \dfrac{MQ}{zF}$	m	kg	precipitated mass
	M	kg/mol	molar mass
	Q	C	transported charge
	z	1	charge number per molecule
	F	C/mol	Faraday constant

2. Faraday constant,

the proportionality constant between the transported quantity of a substance and the transported charge, product of two universal constants e_0 and N_A, the electron charge and Avogadro constant:

Faraday constant			**ITN^{-1}**
	Symbol	Unit	Quantity
$F = e_0 N_A$	F	C/mol	Faraday constant
$= 9.648\,530\,9 \cdot 10^4$ C/mol	e_0	C	elementary charge
	N_A	1/mol	Avogadro constant

■ Faraday constants of several substances (in C/mol): hydrogen 96 364, oxygen 96 486, nickel 96 515, tin 96 482.

3. Mass transport and mass deposition

The transported material is deposited at the electrodes as gas or metal.

▲ The amount of transported material is independent of the geometry of the cathode and the concentration of the electrolyte.

➤ The independence of the deposited mass on the external conditions served as a definition of the charge unit Coulomb (1 C = 1 As) until 1948.

■ In a silver nitrate solution ($AgNO_3$), a current $I = 1$ A flowing for 1 s deposits $n = 1/96 485$ mol $= 1.036 \times 10^{-5}$ mol of silver. This corresponds to a quantity of silver of 1.118 mg.

■ Electrolytic baths are frequently used for the production of pure metals, e.g., for **electrolytic copper**.

■ **Micromechanics**, microscopic mechanical elements that may control mechanical devices of smallest dimension:

Micromotors, microactors, microsensors. Micromechanical elements may be produced by galvanic methods (**LIGA methods**).

4. Electrochemical equivalent,

E, the mass of the electrolyte deposited for a given charge:

electrochemical equivalent			$\mathbf{MT^{-1}I^{-1}}$
	Symbol	Unit	Quantity
$E = \dfrac{m}{Q}$	E	kg/C	electrochemical equivalent
	m	kg	deposited mass
	Q	C	transported charge

An equivalent definition in terms of the molar mass is

$$E = \frac{M}{zF}.$$

■ Electrochemical equivalents (in 10^{-3} g/C): hydrogen 0.010 46, oxygen 0.082 91, nickel 0.304 15, platinum 0.505 88, silver 1.118 17.

5. Faraday's second law,

the masses deposited by the same quantity of electric charge are related like the electrochemical equivalents:

ratio of masses deposited by equal quantities of charge			1
	Symbol	Unit	Quantity
$\dfrac{m_1}{m_2} = \dfrac{E_1}{E_2}$	m_i	kg	deposited masses
	E_i	C/kg	electrochemical equivalents

16.1.4.3 Electric double layer

Electric double layer, arises at contact interfaces between materials with different concentrations of charge carriers (**Fig. 16.3 (a)**). Electric double layers locally compensate the potential differences caused by the difference in concentrations.

■ Electric double layers arise in the contact between solids (frictional electricity), between metals and electrolytes, and also between electrolytes with different ion concentrations.

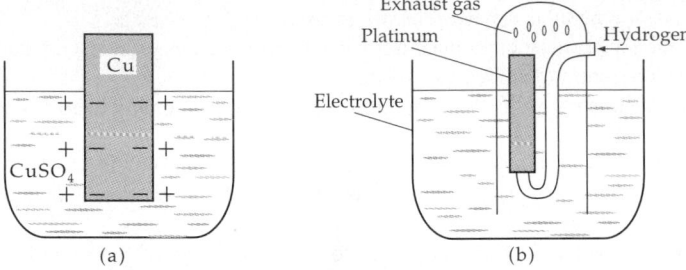

Figure 16.3: (a): Electric double layer at an electrode-electrolyte interface, (b): standard hydrogen electrode.

16.1.4.4 Nernst equation

The potential jump at an interface between electrolytes containing an ion species in different concentrations is proportional to the logarithm of the ratio of concentrations:

Nernst equation: potential jump			$\mathbf{L^2 T^{-3} M I^{-1}}$
	Symbol	Unit	Quantity
$\Delta U = -\dfrac{kT}{e_0} \ln \dfrac{c_1}{c_2}$	ΔU	V	potential difference
	k	J/K	Boltzmann constant
	T	K	temperature
	e_0	C	elementary charge
	c_i	mol/kg	ion concentrations

1. *Electromotive force,*

the equilibrium voltage between a metal and an electrolyte that contains a 1-normal concentration of the metal ions.

M The measurement of the electromotive force requires a second electrode; only the difference between the electromotive force of the two electrodes can be determined by the measurement.

2. *Standard hydrogen electrode,*

reference electrode for voltage measurement. It consists of a platinum sheet in a 1-normal H_3O^+ ionic solution bathed in gaseous hydrogen (**Fig. 16.3 (b)**).

By definition, the potential zero is assigned to the standard hydrogen electrode.

➤ Electromotive force is given analogously for non-metals and molecules as well.

3. *Electrochemical potential series,*

list of electromotive force of metals in an acid solution. Negative voltages imply the release of electrons; positive voltages imply the reception of electrons.

■ Elements of the electrochemical potential series (voltage in V): Li/Li^+ −3.02, Mg/Mg^{2+} −2.38, Zn/Zn^{2+} −0.76, Pb/Pb^{2+} −0.126, Cu/Cu^+ +0.35, Pt/Pt^{2+} +1.2.

➤ An atom may have different electromotive forces for different ions.

■ Au has an electromotive force of $+1.42$ V with respect to Au^+, but an electromotive force of $+1.5$ V with respect to Au^{3+}.

16.1.5 Galvanic cells

If two distinct metals are brought into contact with the same electrolyte, then a potential difference is established between them corresponding to the difference of the electromotive force (**Fig. 16.4**).

■ A copper electrode and a zinc electrode are immersed in an acid solution. A voltage V is established between them:

$$V = (0.35 - (-0.76)) \text{ V} = 1.11 \text{ V}.$$

If both electrodes are connected by an electric conductor, then a current flows.

More noble metal, stands at a lower position in the electrochemical potential series. In a galvanic cell it receives electrons from a metal at a higher position (less noble metal). It forms the anode, the metal standing higher forms the cathode.

In an electrolyte the current circuit is closed by the flow of ions.

If both electrodes are metallic, the cathode dissolves during the course of time while the more noble metal is deposited at the anode as long as it is present as an ion in the solution. This deposition permits the galvanic production of metals in purest form.

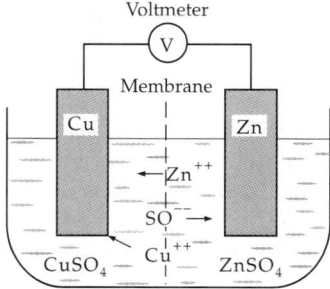

Figure 16.4: Galvanic cell.

Primary cells, galvanic cells carrying out an irreversible conversion of chemical energy into electrical energy.

In the course of time, the voltage decreases due to chemical changes in the electrodes.

Capacitance, K, of a galvanic cell, measured in ampere hours (Ah), a measure of the time (in h) a galvanic cell may deliver a current (in A).

16.1.5.1 Electrolytic polarization

Electrolytic polarization, the decrease of the voltage in a galvanic cell due to the formation of secondary galvanic cells at the electrodes.

■ An external voltage is applied to two platinum electrodes in a water solution; Pt-O_2 and Pt-H_2 double layers are formed by the electrolysis of water. These double layers are galvanic cells in themselves, diminishing the voltage between the electrodes.

Electrokinetic potential, V, the potential difference between the two parts of the double layer. The electrolytic reaction products at the electrodes may be dissolved chemically.

Constant galvanic cells, galvanic cells the voltage of which is kept nearly constant by preventing electrolytic polarization using chemical reactions.

Dry-cell battery, a constant galvanic cell with a non-liquid electrolyte.

■ **Zinc-carbon battery**, dry-cell battery consisting of a carbon rod and a cylindric zinc cover filled with electrolyte paste (**Fig. 16.5**). A layer of manganese oxide (MnO_2) is placed around the carbon rod, which oxidizes the hydrogen formed at the carbon and thus removes it. The voltage decreases only when the zinc is consumed.

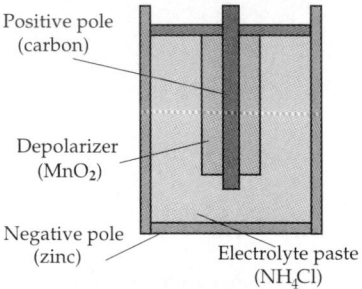

Positive pole (carbon)

Depolarizer (MnO_2)

Negative pole (zinc)

Electrolyte paste (NH_4Cl)

Figure 16.5: Zinc-carbon battery.

Leakage of a battery, destruction of a zinc-carbon battery by electrolytic dissociation of the zinc. The electrolyte released may cause corrosion.

16.1.5.2   Fuel elements

Fuel elements, **fuel cell**, galvanic cells in which the reaction energy from oxidation of the fuel (hydrogen, carbon) is continuously converted directly into electric energy by oxygen or air (**Fig. 16.6**). Water is produced as a byproduct of the combustion. A fuel cell consists of a porous anode at which the supplied fuel (H_2) is reduced ($H_2 \longrightarrow 2H^+ + 2e^-$), and a porous cathode at which the supplied oxidizer (O_2) is oxidized ($2H^+ + 2e^- + \frac{1}{2}O_2 \longrightarrow H_2O$). The electrodes are separated by an electrolyte, which permits the transport of ions (H^+) from the anode to the cathode, but stops the flow of electrons. The electrons are guided to the cathode as a load current through an external current circuit. Without current flow, a cell voltage of about 1 V is reached. Fuel cells are distinguished by a favorable current-voltage characteristic, a high power per mass unit, and a good energy efficiency.

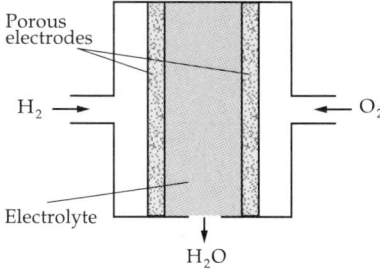

Porous electrodes

$H_2 \longrightarrow$

$\longleftarrow O_2$

Electrolyte

H_2O

Figure 16.6: Fuel cell.

M   Two platinum electrodes, which are bathed in hydrogen and oxygen and are electrically connected, are immersed in dilute sulfuric acid. At the hydrogen electrode, hydrogen is catalytically ionized to hydrogen ions. The electrons are passing through the conductor to the other electrode, where they burn cold with the hydrogen ions transported through the electrolyte and the oxygen present:

$$O_2 + 4H^+ + 4e^- \rightarrow 2H_2O.$$

The released energy of 286.2 kJ/mol can be used as electric energy.

➤ Currently, the efficiency of this direct conversion of chemical energy to electrical energy is about 60 %. The only byproduct of this ecologically sound technology is pure water.

16.1.5.3 Accumulators

Secondary cells, rechargeable galvanic cells in which the electrolytic polarization is exploited for storing electric energy.

■ **Lead-acid accumulator**, secondary cell consisting of lead electrodes immersed in sulphuric acid. The lead electrodes become covered with a layer of lead sulphate ($PbSO_4$). In the **charging** process, (PbO_2) is formed at the anode and metallic lead is formed at the cathode.
Anode:

$$PbSO_4 + 2OH^- \rightarrow PbO_2 + H_2SO_4 + 2e^-;$$

Cathode:

$$PbSO_4 + 2H^+ + 2e^- \rightarrow Pb + H_2SO_4.$$

A galvanic cell obtained in this way yields a voltage of 2.02 V. When extracting current from the cell, both reactions proceed inversely until the original state is nearly re-established.

➤ About 75 % of the stored chemical energy may be converted to electric energy.

16.1.5.4 Connection of galvanic cells

Parallel connection, the cathodes of the individual cells are connected to each other. The same is done with the anodes.

▲ The voltage of the parallel connection is the same as the voltage of the individual cell, but the capacitance K is the sum of the capacitances K_e of the individual cells.

voltage and capacitance of a parallel connection			
	Symbol	Unit	Quantity
$V = V_e$	V	V	voltage of parallel connection
	K	Ah	capacitance of parallel connection
$K = nK_e$	n	1	number of cells
	V_e	V	voltage of single cell
	K_e	Ah	capacitance of single cell

■ This is applied in a **starting assist**.

Series connection, the anode of a cell is connected to the cathode of the subsequent cell.

▲ The total voltage is the sum of the voltages of the individual cells.

voltage and capacitance of a series connection			
	Symbol	Unit	Quantity
$V = nV_e$	V	V	voltage of series connection
	K	Ah	capacitance of series connection
$K = K_e$	n	1	number of cells
	V_e	V	voltage of single cell
	K_e	Ah	capacitance of single cell

16.1.6   Electrokinetic effects

In a liquid in an external electric field, charged particles feel a force, which can cause them to move. The particles may either be charged from the beginning, or the charges may be induced by an electric double layer.

16.1.6.1   Electrophoresis

Electrophoresis, directed motion of suspended charged particles in a non-conductive liquid under the action of an external electric field.

The charge of the suspended particles induces a cloud of oppositely charged ions that surround the particles. Hence, the force acting on a particle does not depend only on its charge, but also on the ion concentration of the suspending agent.

■   This phenomenon is exploited in technology for **dehydration of the walls of build-ings**.

 Paper electrophoresis, electrophoresis of a molecular suspension on a paper carrier to which a direct voltage of several kV is applied. The various components of the suspension are separated because of their different drift velocities.

16.1.6.2   Electro-osmosis

Electro-osmosis, the motion of a liquid in a porous solid under the action of an external electric field. Double electric layers are formed at the liquid-solid interfaces, the liquid part of which separates and starts to move in the electric field. Because of internal friction, the whole liquid starts to move.

➤   In electro-osmosis only charges of one sign move, whereas in electrophoresis charges of both sign move.

16.1.6.3   Diaphragm electricity

Diaphragm electricity, the inverse effect of electro-osmosis. If a liquid is pressed through a porous solid, a current along the flow direction is observed that is due to the removal of part of the electric double layer.

16.2   Current conduction in gases

A rarefied neutral gas does not conduct current like an ideal vacuum does. The rarefied gas may become conducting only by the input of charge carriers. Both electrons and ions may serve as charge carriers. Denser gases usually are also insulators, as are liquids. But a certain number of ions are always generated by cosmic radiation and by natural radioactivity.

Gas discharge, current conduction in gases mainly at low pressure.

16.2.1   Non-self-sustained discharge

Non-self-sustained discharge, a gaseous discharge in which the charge carriers are produced from outside.

Sources for the production of charge carriers are:
•   hot gases in flames,
•   heated metallic surfaces,
•   cosmic radiation,

- ion sources,
- electron guns,
- short-wave electromagnetic radiation (e.g., UV or X-rays),
- radiation of radioactive nuclei.

Dark discharge, a gaseous discharge at very low current densities and low discharge voltages insufficient to ignite a self-sustained discharge. Mainly it may be sustained only by external ionization due to an external source of radiation.

Dark discharges generate only a faint glow of light in the gas. They arise at current densities $J < 10^{-9}$ A/m^2.

In an external electric field, the ions in a gas move uniformly because the energy gain by the external field is compensated by collisions between the molecules.

16.2.1.1 Drift velocity of ions in gases

Drift velocity, v_{dr}, directed velocity of ions through a gas in an external electric field $\vec{E}$.

drift velocity of ions in a gas			$\mathbf{LT^{-1}}$
	Symbol	Unit	Quantity
$v_{dr} = \mu E$	v_{dr}	m/s	drift velocity
	μ	m^2/(Vs)	ion mobility
	E	V/m	electric field strength

▲ Depending on the sign of the ionic charge, the ions move along or against the direction of the electric field.

The ion mobility depends on the type of ion and the medium.

The ion mobility in gases is by about four orders of magnitude higher than the ion mobility in electrolytes. The drift velocity is usually very small compared with the thermal velocity of the ions.

■ Ion mobility μ in air under standard conditions: hydrogen $5.7 \cdot 10^{-2}$ m^2/(Vs) for positive ions and $8.6 \cdot 10^{-2}$ m^2/(Vs) for negative ions, nitrogen $1.29 \cdot 10^{-2}$ m^2/(Vs) for positive ions and $1.82 \cdot 10^{-2}$ m^2/(Vs) for negative ions.

■ In an electric field $E = 1$ kV/m an H_2^+ ion in air under standard conditions is drifting with a velocity $v_{dr} = 5.7 \cdot 10^{-2}$ m^2/Vs $\cdot$ 1000 V/m $= 57$ m/s towards the cathode.

16.2.1.2 Electric conductance of gases

Electric conductance of a gas, γ, the conductance per unit length of a column of gas:

electric conductance of a gas			$\mathbf{L^{-3}T^3M^{-1}I^2}$
	Symbol	Unit	Quantity
	γ	S/m	electric conductance
$\gamma = ze_0(\mu_+ n_+ + \mu_- n_-)$	z	1	ionic valence
	e_0	C	elementary charge
	$\mu_\pm$	m^2/(Vs)	ion mobility
	$n_\pm$	1/m^3	ion density

Positive and negative ions are contributing to the electric current through a gas; their mobilities are distinct, however.

■ Air has an electric conductance of $\gamma \approx 1 \cdot 10^{-14}$ S/m near the ground.

16.2.1.3   Recombination

Recombination, the inverse process to ionization, i.e., ions and electrons agglomerate to form neutral atoms and molecules.

Recombination in gases proceeds mainly by thermal collisions.

Ion lifetime, τ, mean lifetime of an ion in the vicinity of other ions.

Recombination coefficient, α_i, proportionality factor between the reciprocal values of the mean lifetimes of the ions and their number density. The recombination coefficient depends mainly on temperature, pressure and ion species.

ion lifetime			T
	Symbol	Unit	Quantity
$\tau = \dfrac{1}{\alpha_i n_0}$	τ	s	ion lifetime
	α_i	m^3/s	recombination coefficient
	n_0	$1/m^3$	ion density at $t = 0$

The equation holds only if no new ions are formed during the decay.

16.2.1.4   Current-voltage characteristic of a gas

Ohm's law holds in gases only for small applied voltages. One distinguishes three ranges of voltages (ordered according to increasing applied voltage, **Fig. 16.7**).

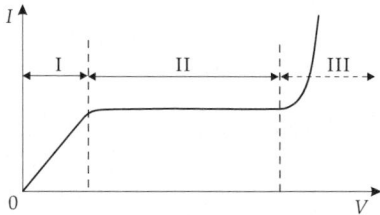

Figure 16.7: Current-voltage characteristic of a non-self-sustained gas discharge. I: recombination range, II: saturation range, III: proportional range, after that transition to a self-sustained gas discharge.

1. Characterization of the voltage ranges

Recombination range, the range of voltage in which the current increases proportionally with the voltage applied. Ohm's law is valid in the form:

current in the recombination range			I
	Symbol	Unit	Quantity
$I = \gamma \dfrac{A}{d} V$	I	A	current
	γ	S/m	electric conductance
	A	m^2	cross-sectional area of electrode
	d	m	distance of electrodes
	V	V	voltage

Saturation range, range of voltage in which nearly all ions reach the electrode. The current I is nearly independent of the voltage V.

In the saturation range, the recombination losses are negligible.

Proportional range, the range of voltage in which the energy of the ions and electrons is sufficiently high to ionize neutral atoms and molecules. The ionization current I increases linearly with the voltage V.

2. Application for the measurement of ionizing radiation

M | **Ionization chamber**, device for the measurement of the intensity of ionizing radiation. Two insulated electrodes are arranged in a gas-filled vessel. The voltage is chosen such that ions generated within the volume of the device contribute directly to the measurable current. Ionization chambers are operated in the saturation range.

Dead time, time that is needed after a registrated event until an ionization detector is again ready for registrating additional events. This time interval, during which the detector is insensitive to ionizing radiation, is determined by the drift velocity of the ions. The dead time specifies the time resolution of the detector.

M | **Geiger-Müller counter**, **trigger counter**, measuring device for the detection of single ionizing particles. Single ionizing particles generate ion avalanches by impact ionization in a gas-filled vessel (gas amplification). The avalanches are measured as discharge pulses (**Fig. 16.8, range II**). The dead time of a trigger counter is several hundred milliseconds.

Proportional counter, the voltage is chosen such that the counter may operate in the proportional range. The number of secondary charge carriers is proportional to the number of the primary charge carriers (**Fig. 16.8, range I**). The discharge pulse is also proportional to the energy loss ΔE of the particle. The proportional counter has a high time resolution, and is thus appropriate for the measurement of high pulse rates.

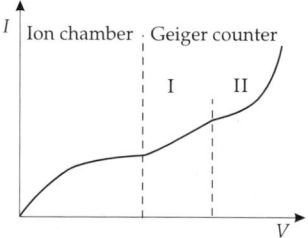

Figure 16.8: Working ranges of ionization detectors. V: detector voltage, I: ionization current, I: proportional range, II: trigger range.

16.2.2 Self-sustained gaseous discharge

Self-sustained gaseous discharge, a gaseous discharge in which the charge carriers are released by the applied voltage itself (**Fig. 16.9**).

16.2.2.1   Types of self-sustained gaseous discharges

1. Glow discharge,

a luminous discharge at mean current densities ($10^{-9}\,\text{A/m}^2 < J < 10^{-4}\,\text{A/m}^2$). The ions striking the cathode liberate electrons, which flow towards the anode.

Owing to the different mobility of the positive and negative charge carriers, zones of different space charges build up in the region between the electrodes. Therefore, the gas between cathode and anode does not glow uniformly.

■ **Fluorescent lamps**, lamps that reach a high luminous efficiency by gaseous discharges in the filling gases at low pressure. The UV radiation that arises is converted to visible light by appropriate layers. Radiation similar to daylight may be obtained by special luminescent layers on the inner surface of the tube.

2. Arc discharge,

bright luminous discharge at current densities $J > 10^{-4}$ A/m^2. The cathode is heated by the incident current and emits more electrons by thermo-ionic emission or field emission.

- **Carbon arc lamp**, a lamp in which an electric arc burns between two carbon electrodes. The **light spot** lies at the cathode.
- **Mercury-vapor lamp**, a lamp for high luminous fluxes. An arc discharge in a mercury gas under high pressure is burned between two metallic electrodes.

3. Spark discharge,

self-terminating arc discharge.

The **ignition voltage** of the spark discharge depends on the shape and the distance of the electrodes, as well as on the pressure of the gas between the electrodes.

➤ The light emission of the various gaseous discharges arises from impact excitation of the gas atoms in collisions with electrons.

Corona discharge, luminous discharge at high pressure and high electric fields. It surrounds high-voltage cables, or manifests itself as **Saint Elmo's fire**.

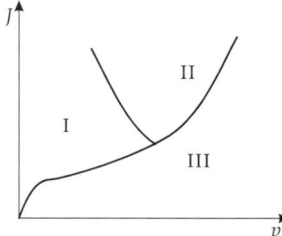

Figure 16.9: Types of gaseous discharge. I: glow discharge, II: arc discharge, III: dark discharge.

16.2.2.2   Current-voltage characteristic of a gaseous discharge

Ignition voltage, voltage at which a **dark discharge** turns into a **glow discharge**.

Self-sustained gaseous discharges have decreasing **resistance characteristics** (**Fig. 16.10**), or even a negative differential resistance dV/dI. Therefore, a dropping resistor (**current limiter**) is indispensable—for alternating current, an induction coil may be used.

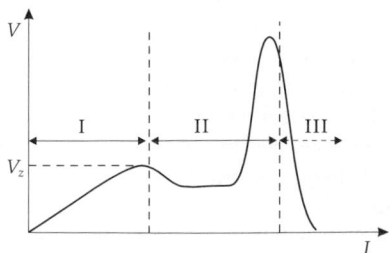

Figure 16.10: Current-voltage characteristic of a gaseous discharge. V_z: ignition voltage. I: dark discharge, II: glow discharge, III: arc discharge.

16.3   Electron emission

The emission of electrons from metals is the basis of various technical devices such as vacuum tubes and photomultipliers. By supplying energy externally, electrons from metals or other solids are released.

Electronic work function, W_A, the energy to be supplied to a conduction electron in a metal in order to free it from the metal to vacuum.

The electronic work function is between 1 eV and 5 eV. It depends on the type of metal and is particularly small for alkali metals.

At room temperature, the thermal energy of the conduction electrons is of the order of magnitude of 1 % of the work function W_A. But some of the electrons exceed this threshold.

16.3.1   Thermo-ionic emission

Thermo-ionic emission, the emission of electrons from a metal heated to point of glowing. The fraction of the electron gas in the metal at the upper end of the velocity distribution with energies exceeding the work function W_A increases with the temperature T proportional to $T^2 e^{-W_A/(k_B T)}$.

▲   The current density J of the emitted electrons as a function of the temperature T and the work function W_A is described by the **Richardson equation**:

Richardson equation			$\mathbf{IL^{-2}}$
	Symbol	Unit	Quantity
$J = AT^2 e^{-\frac{W_A}{k_B T}}$	J	A/m^2	current density of electrons
	A	A/(m^{2}K^2)	Richardson constant
	W_A	J	work function
	k_B	J/K	Boltzmann constant
	T	K	temperature

Richardson constant, proportionality factor in the Richardson equation:

$$A \approx 6 \cdot 10^{-3} \; \mathrm{Am^{-2}K^{-2}}.$$

▲   The Richardson constant is equal for all pure metals with a uniform emitting surface.

M   **Glow cathode**, an electrode consisting of a directly or indirectly heated carrier metal covered by BaO and alkaline metal admixtures in order to reduce the **work function** W_A. It is used as the cathode in vacuum tubes.

16.3.2   Photo emission

Photo emission, the release of electrons by light quanta of sufficient energy (see p. 820).

Einstein equation, gives the kinetic energy of the emitted electrons as a function of the frequency of the incident radiation and the work function (**Fig. 16.11**):

Einstein equation			ML^2T^{-2}
$E_{kin} = h\,f - W_A$	Symbol	Unit	Quantity
	E_{kin}	J	kinetic energy of emitted electrons
	h	Js	Planck's quantum of action
	f	1/s	photon frequency
	W_A	J	work function

The energy of the photoelectrons is independent of the intensity of the incident radiation. The radiation intensity determines only the magnitude of the photocurrent, i.e., the number of electrons released per unit time.

External photoeffect, emission of electrons from the illuminated surface into free space.

| M | **Photoelectric cell**, device to measure the illuminance. The photoelectric cell involves two electrodes, one of them illuminated. The electrons released by this electrode may be registered as current between both electrodes.

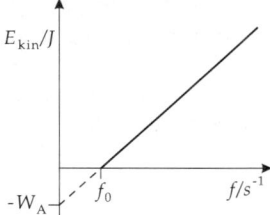

Figure 16.11: Dependence of the kinetic energy of photoelectrons on the frequency of the incident radiation.

Internal photoeffect, release of electrons inside the material. In a semiconductor, the effect causes a change of the electric conductance.

■   **Photovoltaic cell**, resistance depending on the illumination.

16.3.3 Field emission

Field emission, the emission of electrons from materials into vacuum under the influence of strong external electric fields.

Field emission requires field strengths of the order of 10^9 V/m. These values may be reached at sharp points.

| M | **Field emission microscope**, electron microscope for the magnification of atomic structures at sharp points.

In a vacuum tube, a sharp point serves as counter electrode for a metallic ring. A voltage of several kV generates a high field strength at the strongly curved point that accelerates the electrons from the point through the anode ring onto a luminescent screen. The atomic structures of the tip, and also the atoms of filling gases, may be made visible. The maximum magnification is 10^6.

| M | **Scanning tunneling microscope**, a microscope for the magnification of atomic structures on surfaces.

A tunnel current flows between the surface and a fine needle electrode. The current value strongly depends on the distance between them. By keeping the distance constant, the surface is scanned by the electrode with a discrimination of 10^{-11} m.

Single atoms may be observed with a scanning tunneling microscope.

16.3.4   Secondary electron emission

Secondary electron emission, the emission of electrons from a material due to the impact of fast charged particles. By the collisions, material molecules are ionized and electrons are released that can be separated from the material molecules by electric or magnetic fields.

After acceleration to sufficiently high energy, the released electrons may ionize more molecules by repeated collisions and in this way generate an electron avalanche.

| M | **Secondary electron multiplier**, a device to amplify weak electron currents. Electrons striking the first electrode liberate several electrons by impact ionization, and they are then accelerated by an electric field towards further electrodes, the **dynodes**. There, each electron liberates a number of secondary electrons. Hence, a series of dynodes may amplify the current by several orders of magnitude.

Photo multiplier, a device for the measurement of the lowest light intensities. A photoelectrode is connected to a secondary electron multiplier. It responds to incident photons by releasing a primary current, which may be subsequently amplified.

16.4   Vacuum tubes

Vacuum tube, evacuated glass bulb with inserted electrodes that control the flow of electrons by their electric potentials.

1. Cathode and anode in vacuum tubes

Cathode, negative electrode in the tube that releases electrons by thermo-ionic emission. It is heated either directly or indirectly.

Usually, the cathodes are covered with alkaline-earth oxides in order to reduce the work function and to increase the electron yield.

Anode, positive electrode opposed to the cathode.

➤   Vacuum tubes are evacuated in order to reduce as far as possible the collisions of electrons with gas molecules and to prevent oxidation of the hot cathode. The vacuum degrades with increasing age, however, due to evaporation of cathode material.

Anode potential, V_a, voltage between anode and cathode.

Anode current, I_a, current between anode and cathode.

➤   More complex vacuum tubes contain other electrodes besides the anode and cathode.

2. Plate resistance and characteristics

Plate resistance, R_i, the internal electric resistance of a vacuum tube.

By analogy to ohmic resistance, one defines:

plate resistance			$L^2T^{-3}MI^{-2}$
	Symbol	Unit	Quantity
$R_i = \dfrac{V_a}{I_a}$	R_i	Ω	plate resistance
	V_a	V	anode potential
	I_a	A	anode current

➤   In general, the plate resistance depends on the operating conditions of the tube.

Characteristics, diagrams of the electric properties of vacuum tubes.

To an increasing extent, vacuum tubes are being replaced by semiconductor components.

Current applications of vacuum tubes: special tubes (television tubes, x-ray tubes), tubes for high electric and mechanical loads, tubes for high power like transmitting vacuum tubes.

➤   Unlike semiconductors, tubes are rather insensitive to overvoltages and particle radiation.

16.4.1   Vacuum-tube diode

Vacuum-tube diode, simplest type of vacuum tube consisting of cathode and anode. Since electrons can flow only from cathode to anode, the vacuum-tube diode serves as rectifier.

Residual current, the current flowing in a vacuum-tube diode without an applied external voltage (**Fig. 16.12**).

➤   The electrons released by heating of the cathode provide a current between cathode and anode even without an external voltage (**Fig. 16.13**). The current stops only when a sufficiently high counter-voltage is applied.

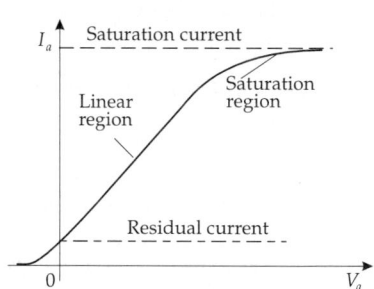
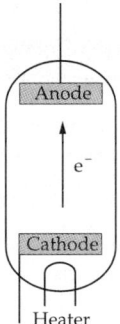

Figure 16.12: Characteristic of a vacuum-tube diode.

Figure 16.13: Vacuum-tube diode.

16.4.2   Vacuum-tube triode

Vacuum-tube triode, more complex vacuum tube for voltage amplification. The vacuum-tube triode contains a grid between anode and cathode (**Fig. 16.15**). The magnitude of the anode current may be controlled by applying a potential difference between grid and cathode (**Fig. 16.14**). The grid remains almost free of current, hence the current control works without power consumption.

The voltage signal applied to the grid is amplified by the triode.

➤   Vacuum tubes with additional grids (tetrode, pentode, . . .) exhibit a behavior qualitatively similar to the triode.

16.4.2.1   Vacuum-tube parameters

1. Grid voltage and slope conductance

Grid voltage, V_g, the voltage applied to the grid in order to control the anode current.

Slope conductance of the characteristic, S, the slope of the current-voltage characteristic for constant anode potential.

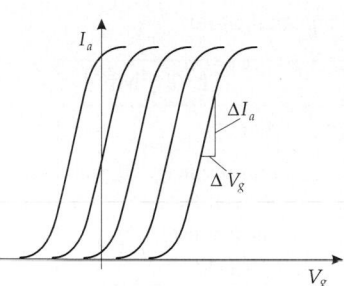

Figure 16.14: Characteristics of a vacuum-tube triode for varying direct voltage.

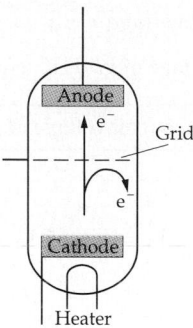

Figure 16.15: Vacuum-tube triode; for negative grid voltage, the electrons are decelerated.

slope of characteristic			$\mathbf{L^{-2}T^3M^{-1}I^2}$
$S = \dfrac{\Delta I_a}{\Delta V_g}$	Symbol	Unit	Quantity
	S	A/V	slope of characteristic
	I_a	A	anode current
	V_g	V	grid voltage

The slope of the characteristics is given for constant anode potential V_a. The formula holds only in the linear range. In general,

$$S = \frac{\partial I_a}{\partial V_g}, \quad V_a = \text{const.}$$

➤ For a large signal amplification of the tube, the slope conductance S should be as high as possible.

2. *Internal resistance and grid transparency*

Internal resistance of a triode, R_i, generalization of the plate resistance:

$$R_i = \frac{\partial V_a}{\partial I_a}, \quad V_g = \text{const.}$$

Transparency of the vacuum tube, D, the reaction of the anode potential V_a to the grid voltage V_g.

transparency of the triode			1
$D = \dfrac{\partial V_g}{\partial V_a}, \quad$ at $\quad I_a = \text{const.}$	Symbol	Unit	Quantity
	D	1	transparency
	V_g	V	grid voltage
	V_a	V	anode voltage
	I_a	A	anode current

3. Control voltage

Control voltage of grid, V_s, effectively acting voltage at the grid:

control voltage of grid			$\mathbf{L^2T^{-3}MI^{-1}}$
	Symbol	Unit	Quantity
$V_s = V_g + DV_a$	V_s	V	control voltage of grid
	V_g	V	grid voltage
	D	1	transparency
	V_a	V	anode potential

➤ The control voltage V_s and the anode current I_a are related by

$$I_a = SV_s .$$

4. Barkhausen equation,

relation between slope conductance S, transparency D and internal resistance R_i.

Barkhausen equation			1
	Symbol	Unit	Quantity
$SDR_i = 1$	S	A/V	slope conductance
	D	1	transparency
	R_i	Ω	internal resistance

5. Amplification factor of a vacuum tube,

A, ratio of the anode alternating voltage V_a to the grid alternating voltage V_g.

amplification factor of a vacuum tube			1
	Symbol	Unit	Quantity
$A = \left\| \dfrac{V_a}{V_g} \right\|$	A	1	amplification factor of a vacuum tube
	V_a	V	anode alternating voltage
	V_g	V	grid alternating voltage

The amplification factor of a vacuum tube V depends on the load resistance R_a in the anode circuit:

amplification factor of a vacuum tube			1
	Symbol	Unit	Quantity
$A = -S \dfrac{R_a R_i}{R_a + R_i}$	A	1	amplification factor of a vacuum tube
	S	A/V	slope conductance
	R_a	Ω	resistance in anode circuit
	R_i	Ω	internal resistance of vacuum tube

In order to reach a high amplification factor, the characteristic should be as steep as possible.

16.4.3　Tetrode

Tetrode, complex vacuum tube with two grids between anode and cathode. One distinguishes two types:

- **Screen-grid vacuum tube**, tetrode with an additional grid between anode and control grid; this grid reduces the transparency D and increases the amplification.
- **Space-grid vacuum tube**, tetrode with an additional grid between cathode and control grid; this grid increases the slope conductance S of the characteristic curve.

16.4.4　Cathode rays

Cathode rays, electron beams in evacuated vacuum tubes which, after being accelerated by the voltage between cathode and anode, leave the acceleration region through a hole in the anode plate. Cathode rays propagate along straight paths in the field-free space, or they may be deflected by electric and magnetic fields.

Cathode rays cause certain glasses, minerals and special fluorescent dyes to fluoresce.

- ■　**Braun's tube**, device in which a cathode beam may be guided over a luminous screen by electric or magnetic fields. Application: as display in television sets and oscilloscopes.

The velocity of the electrons in the cathode ray is determined by the accelerating field between cathode and anode:

velocity of cathode rays			$\mathbf{LT^{-1}}$
	Symbol	Unit	Quantity
$v = \sqrt{\dfrac{2e_0 V}{m_e}}$	v	m/s	velocity of cathode rays
	e_0	C	elementary charge
	V	V	voltage between anode and cathode
	m_e	kg	electron mass

- ➤　The equation holds only for $v \ll c$.
- ■　For a voltage between anode and cathode of $V = 50\,\text{V}$, one obtains $v = 4.19 \cdot 10^6$ m/s. This corresponds to 1.4 % of the speed of light.

16.4.5　Channel rays

Channel rays, rays of positively charged gas ions accelerated by the electric field towards the cathode and passing through it in channels (**Fig. 16.16**).

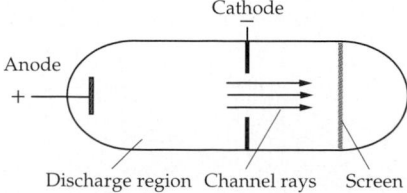

Discharge region　Channel rays　Screen

Figure 16.16: Generation of channel rays in a gaseous discharge.

velocity of channel rays			LT^{-1}
	Symbol	Unit	Quantity
$v = \sqrt{2ze_0 V/m_I}$	v	m/s	velocity of channel rays
	z	1	ionic valence
	e_0	C	elementary charge
	V	V	voltage between anode and cathode
	m_I	kg	ion mass

■ For a voltage between anode and cathode of $V = 50$ V, one obtains $v = 9.78 \cdot 10^4$ m/s for protons and $v = 1.85 \cdot 10^4$ m/s for N_2^+ ions. Because of the large ion mass, this velocity is very small compared with the velocity of the electrons of the same acceleration energy.

17
Plasma physics

Plasma, gaseous mixture of free electrons, ions and electrically neutral particles—atoms, molecules and free radicals. All components of the mixture have a high kinetic energy, but they are not necessarily in thermal equilibrium with each other. The electromagnetic interaction between the individual particles contributes significantly to the behavior of the system.

■ Much of the visible matter of the universe is in a plasma state, e.g., the Sun.

17.1 Properties of a plasma

Apart from the usual thermodynamic properties of a gas, such as temperature and pressure, a plasma also has peculiarities originating in its character as a mixture of partly charged and partly uncharged particles in different states of excitation.

Quasi-neutrality, fundamental property of a plasma: plasmas are electrically neutral in macroscopic regions, both in the spatial and temporal averages. Any volume element contains about the same quantity of positive and negative charge carriers.

➤ The kinetic energy of the plasma particles is large compared with the potential energy, which is caused by a local charge.

17.1.1 Plasma parameters

Owing to the large number of interacting species of particles, a large variety of quantities is required for a description of a plasma.

17.1.1.1 Degree of ionization

Degree of ionization, x_r, the fraction of ions of nuclear charge Z in a plasma of atoms and positively charged ions that are ionized at least r times:

degree of ionization			1
	Symbol	Unit	Quantity
$$x_r = \frac{\sum_{i=r}^{Z} n_i}{\sum_{i=0}^{Z} n_i} \leq 1$$	x_r	1	degree of ionization of the rth ionization step
	n_i	mol/m^3	concentration of i-times charged ions
	Z	1	nuclear-charge number

➤   Frequently, the degree of ionization x_1 of the first ionization step, $r = 1$, is denoted simply degree of ionization x.

Owing to the neutrality condition, the electron concentration is given by

$$n_e = \sum_{i=1}^{Z} i \, n_i .$$

If negatively charged ions are present, the equation has to include the corresponding negative terms.

Plasmas are classified according to their degree of ionization x_1:

- **weakly ionized plasmas**: degree of ionization $x_1 \ll 1$.
- **highly** or **fully ionized plasmas**: degree of ionization $x_1 \approx 1$.

➤   Plasmas may also be classified by the ratio of the charge-carrier density to the screening length, or by the ratio of kinetic and potential energy of the particles.

Plasmas with temperatures $T < 10^5$ K ($T > 10^6$ K) are refered to as cold (hot) plasmas. Nuclear fusion processes are possible only in plasmas of temperature $T > 10^8$ K.

17.1.1.2   Distribution functions of the plasma

The energy content of the plasma may be distributed over the usual excitations of a gas (rotational and vibrational excitations) and, to a large extent, electronic excitations.

1. Complete thermodynamic equilibrium,

ideal state of a plasma:

▲   In complete thermodynamic equilibrium, all distribution functions are determined by a single state variable, the temperature T.

▲   **Principle of detailed balance**: Every process occurs at the same rate as the inverse process.

In particular:

- the same number of atomic electrons are excited and de-excited,
- the same number of atoms that are ionized as ions are recombining with electrons to become neutral atoms,
- all possible chemical reactions are in equilibrium according to the **law of mass action**,
- direct reactions and inverse reactions occur at an equal rate (e.g., thermal dissociations).

2. Distribution functions of a plasma in complete thermodynamic equilibrium,

a) Electromagnetic radiation of plasma, corresponds to the **cavity radiation** (see p. 818) of a black radiator.

Planck's radiation law, the distribution of photons of energy hf at the radiation temperature T:

spectral radiation energy distribution of a plasma			MT^{-3}
	Symbol	Unit	Quantity
	$L_{e,f}(T)$	W/m^2	emitted spectral radiation intensity
$L_{e,f}(T) = \dfrac{2hf^3}{c^2} \dfrac{1}{e^{(hf)/(k_B T)} - 1}$	h	Js	Planck's quantum of action
	f	$1/s$	frequency of radiation
	k_B	J/K	Boltzmann constant
	T	K	plasma temperature
	c	m/s	vacuum speed of light

b) Maxwellian velocity distribution of the particles (ions and electrons at the same temperature T):

velocity distribution of a plasma in complete thermodynamic equilibrium			1
	Symbol	Unit	Quantity
	f	1	particle number in velocity range v, $v+dv$
$f(v) = \dfrac{4}{\sqrt{\pi}} v^2 \left(\dfrac{m}{2k_B T}\right)^{3/2} e^{-mv^2/(2k_B T)}$	v	m/s	particle velocity
	m	kg	particle mass
	k_B	J/K	Boltzmann constant
	T	K	plasma temperature

➤ Different particle species at the same temperature have different velocity distributions because of their different particle masses.

➤ Often the quantum mechanical degeneracy of the electrons cannot be ignored, so that, for the electron plasma in metals or in cold stars (white dwarfs), the **Fermi-Dirac distribution** must be used.

c) Boltzmann distribution, specifies the occupation of excited electronic states:

distribution of electronic excitations of a plasma			1
	Symbol	Unit	Quantity
	n_j	1	particle number in jth excited state
	n	1	total number of particles
$\dfrac{n_j}{n} = \dfrac{g_j}{g_0} e^{-E_j/(k_B T)}$	g_j	1	statistical weight of excited state j
	g_0	1	statistical weight of the ground state
	E_j	J	excitation energy of jth excited state
	k_B	J/K	Boltzmann constant
	T	K	plasma temperature

The partition function appearing in the denominator has been approximated by its first term g_0. A separate distribution holds for each of the individual degrees of ionization.

3. Saha equation,

describes ionization-recombination equilibrium:

Saha equation			1
	Symbol	Unit	Quantity
$$\frac{x^2}{1-x} = 2\frac{(2\pi m_e)^{3/2}}{h^3}\frac{g_1}{g_0 p}$$ $$\cdot (k_B T)^{5/2} e^{-E_I/(k_B T)}$$	x	1	ionization degree
	m_e	kg	electron mass
	h	Js	Planck's quantum of action
	g_i	1	statistical weight in ith ionized state
	p	N/m^2	plasma gas pressure
	E_I	J	ionization energy
	k_B	J/K	Boltzmann constant
	T	K	temperature

The Saha equation holds only in this simple form for equilibrium between the ground state and the first ionized state. To take into account further ionized states, one has to solve a system of Saha equations simultaneously. The partition functions have been replaced by their first term. The decrease of the ionization energy produced by the plasma has been ignored.

4. Real plasmas

In most cases, real plasmas deviate from complete thermodynamic equilibrium. But some of the statements referring to completely thermodynamic equilibrium may still be valid, depending on which of the partial equilibria are no longer valid.

➤ In chemically reactive plasmas, the equilibrium of the chemical reactions has to be taken into account as well. In completely thermodynamic equilibrium, the chemical reactions obey the law of mass action separately.

Local thermal equilibrium, partial equilibrium in which radiation equilibrium is no longer valid. For a sufficiently high electron concentration ($n_e > 10^{23}$ m^{-3}), the collision processes exceed the absorption and emission processes, so that the particle balances remain unaffected.

In local thermal equilibrium, the plasma is described by two state variables, a matter temperature T_m and a radiation temperature T_s.

Deviations from equilibrium, require the introduction of different temperatures for different elementary processes and for different particle species.

17.1.1.3   Energy content of the plasma

In the plasma, different forms of energy are continually converted into each other by the various interactions among the particles:

- energy of the electric and magnetic fields,
- ionization energy,
- translational energy of the neutral particles and charge carriers,
- dissociation energy and chemical binding energy,
- energy of electronic excitations,
- energy of rotational and vibrational excitations,
- radiation energy,
- energy of collective motions (plasma oscillations and plasma waves).

The establishment of thermal equilibrium among the various kinds of energy is determined by the coupling between them.

➤   The mean kinetic energies of the atoms and ions are rapidly equilibrated by collisions between particles of similar mass. The equilibration between ions and electrons proceeds more slowly, since only a small amount of kinetic energy is transferred in collisions involving very different masses.

17.1.1.4   Electric conductivity of plasmas

1. Charge-carrier drift of plasma particles in an external field

In an external electric field, the charge carriers of the plasma drift at constant velocity along the field lines. The drift velocity of ions is lower than that of electrons, hence the electric conductivity is dominated by the electronic transport (**Fig. 17.1 (a)**).

Coulomb logarithm, characteristic plasma parameter for describing the ratio of plasma temperature to electron density:

Coulomb logarithm			1
	Symbol	Unit	Quantity
$\ln \Lambda = \ln\left(\dfrac{aT^{3/2}}{\sqrt{n_e}}\right)$	$\ln \Lambda$	1	Coulomb logarithm
	a	$(Km)^{-3/2}$	proportionality constant
	T	K	temperature
	n_e	$1/m^3$	electron-number density

The proportionality factor a has the value $a \approx 10^7 \, (Km)^{-3/2}$. For most plasmas $\ln \Lambda \approx 15 \ldots 20$.

electric conductivity of plasma			$I^2 L^{-3} T^3 M^{-1}$
	Symbol	Unit	Quantity
$\sigma = \dfrac{e_0^2 n_e \tau_e}{m_e}$	σ	S/m	electric conductivity of plasma
	e_0	C	elementary charge
	n_e	$1/m^3$	electron-number density
	τ_e	s	mean time of flight between two collisions
	m_e	kg	electron mass

2. Properties of the electric conductivity of a plasma

The electric conductivity is governed by different processes depending on the degree of ionization:

Weakly ionized plasmas, the mean time of flight is limited by the collisions between electrons and neutral particles; τ_e is independent of the electron density, and $\sigma \sim n_e$.

Completely ionized plasmas, the collisions between charged particles are crucial. Then $\tau_e \sim 1/n_e$, and σ is independent of n_e.

Spitzer formula, gives the electric conductivity in completely ionized thermal plasmas, taking into account the electron-ion collisions. If the electron-electron collisions are included, the value of σ is halved.

Spitzer formula			$I^2L^{-3}M^{-1}T^3$
	Symbol	Unit	Quantity
$\sigma = \dfrac{64\sqrt{2\pi}\,\varepsilon_0^2}{e_0^2\sqrt{m_e}}\dfrac{(k_B T)^{3/2}}{\ln\Lambda}$	σ	S/m	electric conductivity
	ε_0	$C^2\,J^{-1}\,m^{-1}$	free-space permittivity constant
	e_0	C	elementary charge
	m_e	kg	electron mass
	k_B	J/K	Boltzmann constant
	T	K	plasma temperature
	$\ln\Lambda$	1	Coulomb logarithm

■   A nitrogen plasma at $T = 10^4$ K has an experimentally measured electric conductivity of $\sigma = 3000$ S/m.

17.1.1.5   Heat conductivity of a plasma

The transport of heat energy in a plasma may proceed in two ways:
- transport by transfer of translational energy of the particles available,
- **reaction heat conduction**, energy transport by transfer of excitation energy, dissociation energy and ionization energy.

The mechanism of heat conduction due to reactions means that, in regions of high temperature, heat energy is used for excitation or dissociation. The reaction products diffuse to cooler regions, and there release the heat energy by inverse processes (**Fig. 17.1 (b)**).

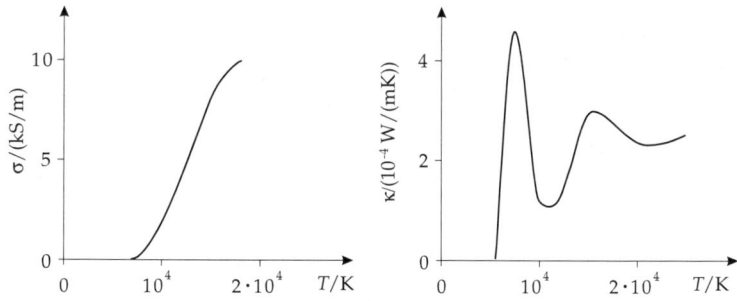

Figure 17.1: (a): electric conductivity σ and (b): heat conductivity κ of a nitrogen plasma.

17.1.1.6   Screening and Debye length

1. Potential about charged particle in a plasma

In a plasma, the potential about a charged particle differs significantly from that in a vacuum. A cloud of negative particles is formed around a positive particle that appreciably diminishes the range of the potential. Therefore, a screening potential is superimposed on the usual Coulomb potential:

screened electric potential			$\mathbf{I^{-1}ML^2T^{-3}}$
	Symbol	Unit	Quantity
$\varphi(r) = \dfrac{1}{4\pi\varepsilon_0}\dfrac{e_0}{r}\mathrm{e}^{-r/\lambda_\mathrm{D}}$	φ	V	electric potential
	r	m	distance from charge carrier
	ε_0	$C^2\,J^{-1}\,m^{-1}$	free-space permittivity constant
	e_0	C	elementary charge
	λ_D	m	Debye length

The potential given above holds for a plasma with $Z = 1$ in which $e_0 V \ll k_\mathrm{B}T$ everywhere.

2. Debye length,

λ_D, characteristic length that describes the screening of a potential. Within a Debye length, the potential drops to $1/e$ of the original value:

Debye length			**L**
	Symbol	Unit	Quantity
$\lambda_\mathrm{D} = \sqrt{\dfrac{\varepsilon_0 k_\mathrm{B} T}{2e_0^2 n_\mathrm{e}}}$	λ_D	m	Debye length
	ε_0	$C^2\,J^{-1}\,m^{-1}$	free-space permittivity constant
	k_B	J/K	Boltzmann constant
	T	K	temperature
	e_0	C	elementary charge
	n_e	$1/m^3$	electron-number density

■   For a Hydrogen plasma at $T = 10^4$ K and $n_\mathrm{e} = 10^{23}$ cm^{-3}, $\lambda_\mathrm{D} \approx 2 \times 10^{-5}$m.

3. Plasma classification by the Debye length

The Debye length may be used to classify a plasma:
- **Ideal plasma**, a plasma containing many charge carriers within a sphere of radius of one Debye length. The potential electric energy is significantly lower than the thermal energy.
- **Non-ideal plasma**, a plasma containing only few charge carriers within one Debye length of another charge carrier. Non-ideal plasmas exhibit characteristic anomalies (phase transitions, anomalous electric conductivities).

■   Dense plasmas are usually non-ideal plasmas.

17.1.1.7 Plasma oscillation frequency

Plasma oscillations, collective motion of a plasma caused by space charge fluctuations. The restoring force is due to the space charge field arising from the displacement of charge carriers.

Langmuir frequency, ω_Pe, fundamental frequency of plasma oscillations:

Langmuir frequency of electron oscillations			T^{-1}
	Symbol	Unit	Quantity
$\omega_{Pe} = \sqrt{\dfrac{e_0^2 n_e}{\varepsilon_0 m_e}}$	ω_{Pe}	rad s^{-1}	Langmuir frequency
	e_0	C	elementary charge
	n_e	1/m^3	electron-number density
	ε_0	C^2 J^{-1} m^{-1}	free-space permittivity constant
	m_e	kg	electron mass

➤ For ion oscillations, which also occur, the electron mass m_e must be replaced by the ion mass m_i.

17.1.2 Plasma radiation

1. Radiation from a plasma

Due to the high kinetic energy of the particles and the large number of excited atoms and ions, plasmas emit electromagnetic radiation in the range from microwaves up to hard x-rays (for highly ionized metal atoms).

Radiation from plasmas may originate from different kinds of transitions:

- discrete transitions between bound states,
- free-free transitions in the continuum (bremsstrahlung), i.e., transitions between unbound states,
- free-bound transitions in electron-ion recombination.
- bound-free transitions with dissociation in the lower state.
- ➤ The last three kinds of transitions yield continuous-emission spectra.

2. Characteristic quantities of plasma radiation

The radiation emitted by a plasma corresponds to spontaneous and stimulated emission, as well as to absorption in the plasma interior.

Spectral-radiation density, L_f, a quantity describing the radiation energy per frequency interval df emitted by a volume element.

Emission coefficient, ε_f, coefficient specifying the radiation energy emitted per unit volume and unit time within a frequency interval.

➤ The emission coefficient includes the spontaneous, but not the stimulated, emission. It is independent of the spectral-energy density at this position, but is itself frequency-dependent.

Effective-absorption coefficient, κ', a coefficient describing absorption, scattering and stimulated emission in a medium.

Optical depth, τ, a quantity that specifies the transparency of a column of matter for radiation. It is given by the integral of the effective-absorption coefficient along the column:

optical depth			1
	Symbol	Unit	Quantity
$\tau = \displaystyle\int_0^l \kappa'(x)\mathrm{d}x$	τ	1	optical depth
	l	m	length of column of matter
	κ'	m^{-1}	effective-absorption coefficient
	x	m	position along the column

After passing through a layer of matter with an optical depth $\tau = 1$, the radiation density is reduced to 1/e of the original value.

17.1.3 Plasmas in magnetic fields

17.1.3.1 Motion of charged particles in external fields

1. Force of an external field on plasma particles

In order to analyze the behavior of plasmas in external fields, one must consider the motion of individual particles.

A particle of charge q and velocity $\vec{v}$ in an electric field $\vec{E}$ and a magnetic field $\vec{B}$ experiences the Lorentz force

$$\vec{F}_L = q \left(\vec{v} \times \vec{B} \right) + \vec{F} .$$

$\vec{F}$ includes all external forces, including also the force $q\vec{E}$ due to the electric field. The entire motion may be separated into two distinct motions:

- **gyration**, rotation along a circular orbit about the direction of the (local) magnetic field,
- displacement of the center of the circle with the **drag velocity** $\vec{v}_F$.

Gyration radius r_G and gyration frequency $\vec{\omega}_G$ are given by:

gyration radius and gyration frequency			
	Symbol	Unit	Quantity
$r_G = \dfrac{mv_\perp}{qB}$	r_G	m	gyration radius
	ω_G	1/s	gyration frequency
	m	kg	particle mass
$\vec{\omega}_G = -\dfrac{q}{m}\,\vec{B}$	$v_\perp$	m/s	particle velocity perpendicular to magnetic field axis
	q	C	particle charge
	$\vec{B}$	T	magnetic flux density

The magnetic moment of the rotation remains constant; it is given by

$$\vec{\mu} = -m \frac{v_\perp}{2} \frac{\vec{B}}{B^2} .$$

➤ The particle motion has exactly this form only in a uniform, time-independent magnetic field and for a vanishing external force $\vec{F} = 0$.

2. Special cases of external fields

- $\vec{B}$ is constant in time and space, $\vec{F} = 0$.

  A particle rotates on helical orbits about the magnetic field lines. The drag velocity corresponds to the particle velocity along the magnetic field.

  For increasing magnetic flux density $\vec{B}$, the gyration radius r_G becomes ever smaller, i.e., the particles are bound more tightly to the magnetic field lines.

- $\vec{B}$ is constant in time and space, $\vec{F} \neq 0$.

  Besides the motion in helical orbits, there is an additional **transverse drag** perpendicular to $\vec{B}$ and to the force component $\vec{F}_\perp$ perpendicular to $\vec{B}$. The drag velocity is given by

$$\vec{v}_F = \frac{\vec{F}_\perp \times \vec{B}}{qB^2} .$$

- $\vec{B}$ is constant in time but not in space, $\vec{F} = 0$.

  Gradient-B drift, a drift motion in a non-uniform magnetic field whose gradient is perpendicular to the field. The drag velocity obeys

$$\vec{v}_F = \frac{v_\perp r_G}{2B} \operatorname{grad}_\perp B \,.$$

In a non-uniform field, the gradient of which is parallel to the magnetic flux direction, longitudinal kinetic energy is converted into rotational energy.

  Mirror effect, inversion of the direction of the drag velocity in a non-uniform magnetic field, the gradient of which is parallel to the magnetic field axis.

  Ions may be confined in a cylindrical, non-uniform magnetic field (magnetic bottle) by the mirror effect.

- $\vec{B}$ is not constant in either time or space.

  In a field increasing with time, the gyration radius r_G decreases; in a field decreasing with time, r_G increases.

  The magnetic flux enclosed by the particle during the gyration cycle is nearly constant.

17.1.3.2 Motion of charged particles in a magnetic field including collisions

Charged particles stop circulating around a magnetic field line because of collisions and are transferred to another field line. This corresponds to a drift motion across the magnetic field.

  The collisions act randomly and may be treated by adding an effective-stochastic-force term, which acts like a friction force.

  Langevin equation, equation describing the motion in a magnetic field including collisions and additional external forces:

Langevin equation				MLT^{-2}
	Symbol	Unit	Quantity	
	m	kg	particle mass	
	$\vec{v}$	m/s	particle velocity	
	t	s	time	
$m\dfrac{d\vec{v}}{dt} = q\left(\vec{v} \times \vec{B}\right) + \vec{F} - mf\vec{v}_m$	q	C	particle charge	
	$\vec{B}$	T	magnetic flux density	
	$\vec{F}$	N	external forces	
	f	1/s	collision frequency	
	$\vec{v}_m$	m/s	mean velocity	

17.1.3.3 Drift motion in an external electric field

In an external, time-independent electric field, the drift motion due to coupled fields may be determined by the averaged Langevin equation. Let the magnetic field point along the z-direction and the x-direction be chosen such that $E_y = 0$, then:

- E_z generates a drift motion along the magnetic field that is not affected by the magnetic field;
- The component E_x perpendicular to $\vec{B}$ generates a drift motion along the x-direction, but with the reduced mobility

$$\mu_x = \frac{1}{1 + \left(\omega_{Ge}^2 \big/ f_e^2 \right)} \mu_z, \quad \omega_{Ge} \text{ gyration frequency of the electrons,}$$

f_e collision frequency of the electrons.
- E_x excites a drift in y-direction, although $E_y = 0$.

17.1.3.4   Continuum theories

With increasing interaction between the particles, the model of individual particles has to be replaced by the model of a continuous medium. There are two approaches:
- **magnetohydrodynamics**, a combination of hydrodynamics and electrodynamics,
- **plasma dynamics**, hydrodynamics using different liquids for electrons, ions and neutral particles.

Analogous quantities arise for the hydrodynamic variables.

Magnetic pressure, additional plasma pressure arising because of the interaction between plasma and magnetic field. The magnetic pressure for a time-independent field is

$$p_m = \frac{B^2}{2\mu_0}, \quad \mu_0 \text{ permeability of free space.}$$

17.1.4   Plasma waves

The various interactions in a plasma, in particular in a medium far from equilibrium, cause a parse variety of possible wave excitations. The following quantities may display wave-like fluctuations:
- electric field strength E,
- electric space charge density ρ,
- magnetic flux density B,
- particle concentrations of charge carriers and neutral particles,
- temperatures of ions and electrons,
- drift velocities of the particles.
- ➤ The treatment of plasma waves requires the simultaneous treatment of Maxwell's equations and the transport equations for the charge carriers.

17.1.4.1   Plasma-acoustic waves in plasmas

1. Electron plasma waves

Electron plasma waves, **Langmuir waves**, longitudinal wave motion, connected with Langmuir oscillations of the electron density.

Electron plasma waves do not occur in cold plasmas. They are not affected by magnetic fields directed along the propagation direction of the wave.

dispersion relation of Langmuir waves			T^{-2}
	Symbol	Unit	Quantity
$<v_e>^2 k^2 - \omega^2 + \omega_{Pe}^2 = 0$	$<v_e>$	m/s	mean electron velocity
	k	1/m	wave number
	ω	1/s	angular frequency of wave
	ω_{Pe}	1/s	Langmuir angular frequency of electrons

2. Ion plasma waves,

additional longitudinal waves arising at low frequencies ($\omega \ll \omega_e$), since in this range ion density fluctuations also contribute to the wave motion in addition to the fluctuations of the electron density. Ion plasma waves are free of dispersion.

Ion sound velocity, propagation of ion plasma waves.

ion sound velocity			LT^{-1}
	Symbol	Unit	Quantity
$c_S = <v_i> \sqrt{1 + \dfrac{T_e}{T_i}}$	c_S	m/s	ion sound velocity
	$<v_i>$	m/s	averaged ion velocity
	T_e	K	electron temperature
	T_i	K	ion temperature

The ion sound velocity is affected by the electron and ion temperatures.

17.1.4.2   Magnetohydrodynamic waves

Magnetohydrodynamic waves, mixed hydrodynamic-electromagnetic waves which are strongly affected by the motion of the charge-carrier background.

Alfven waves, magnetohydrodynamic waves in a magnetic field parallel to the propagation direction. They are free of dispersion, and their phase velocity c_{ph} is

phase velocity of Alfven waves			LT^{-1}
	Symbol	Unit	Quantity
$c_{ph} = \dfrac{c_0}{\sqrt{1 + \dfrac{\mu_0 c_0^2 \rho_m}{B_a^2}}}$	c_{ph}	m/s	phase velocity
	c_0	m/s	vacuum speed of light
	μ_0	Vs/Am	free-space permeability constant
	ρ_m	kg/m^3	mass density of plasma
	B_a	T	external magnetic field

Alfven waves may be interpreted as electromagnetic waves propagating through a medium of increased relative permittivity:

$$\varepsilon_r = \sqrt{1 + \frac{\mu_0 c_0^2 \rho_m}{B_a^2}} \; .$$

17.1.4.3   Electromagnetic waves in plasmas

The propagation of electromagnetic waves in a plasma is modified (compared with the propagation in a vacuum) by the presence of free charge carriers. For $\omega \to \infty$, the waves behave like vacuum waves because no charge carriers may be dragged along. For $\omega \approx \omega_{Pe}$ and $\omega \approx \omega_{Ge}$ (ω_{Ge} gyration frequency of electrons), strong deviations occur. For circularly polarized waves propagating in a magnetic field parallel to the field axis, simple dispersion relations may be given:

Ordinary wave, circularly polarized electromagnetic wave with the $\vec{E}$ vector rotating against the gyration of electrons.

Extraordinary wave, circularly polarized electromagnetic wave with the $\vec{E}$ vector rotating in the same sense as the gyration of electrons.

dispersion relation of electromagnetic plasma waves			T^{-2}
	Symbol	Unit	Quantity
$$c_0^2 k^2 - \omega^2 + \dfrac{\omega_{Pe}^2}{1 \pm \dfrac{\omega_{Ge}}{\omega}} = 0$$	c_0	m/s	vacuum light speed
	k	m^{-1}	wave number
	ω	s^{-1}	angular frequency of wave
	ω_{Pe}	s^{-1}	Langmuir angular frequency of electrons
	ω_{Ge}	s^{-1}	gyrational angular frequency of electrons

The positive sign holds for ordinary waves, the negative sign for extraordinary waves.

In a plasma in which no external magnetic field is acting, the refractive index n obeys the **Eccles relation**:

Eccles relation for the refractive index			1
	Symbol	Unit	Quantity
$$n = \sqrt{1 - \dfrac{\omega_{Pe}}{\omega}}$$	n	1	refractive index of plasma
	ω_{Pe}	s^{-1}	Langmuir angular frequency of electrons
	ω	s^{-1}	angular frequency of wave

➤   Waves with $\omega = \omega_{Pe}$ are reflected when entering the plasma.

17.1.4.4   Landau damping

Besides the usual damping due to collisions between the plasma particles, energy of plasma motion is also transferred into electromagnetic waves.

Landau damping, damping of plasma waves by energy transfer in the convected wave field. Particles with velocities higher than the phase velocity of the wave are decelerated, particles with lower velocities are accelerated. If the velocity distributions of the plasma particles are Maxwellian, then the damping part (on the decreasing side), is predominant, so that altogether the wave releases energy.

➤   For an appropriate velocity distribution, the wave may also be amplified.

17.2   Generation of plasmas

In order to generate plasmas, sufficient energy has to be provided from outside to supply the minimum energy to the atoms and molecules that is needed for ionization. There are two mechanisms available:

- Increase of the energy content by heat supply. The energy supplied is distributed over the available degrees of freedom; the ionization proceeds by collisions, or by photo absorption. Mostly, these plasmas are close to thermal equilibrium.
- Increase of the energy content by calculated energy supply (radiation or electric current) without a significant increase of temperature. The ionization proceeds directly by the transfer of the energy supplied from outside to atoms and molecules. The resulting plasmas are far from thermal equilibrium ($T_e \gg T_i$).

17.2.1   Thermal generation of plasma

Plasma oven, a device for the heating of gas by contact with hot walls. The plasmas in plasma ovens are in equilibrium and satisfy the Saha equation. But the degree of ionization is limited by the maximum temperature achieved ($T \leq 3500$ K).

Q-machine, generates thermal plasmas with an increased degree of ionization reached by contact ionization of gas atoms by electrons leaving electrodes. The ionization energy of the gas must be lower than the electron work function of the electrode materials. The plasma cylinder is confined by a longitudinal magnetic field; the degree of ionization may reach 50 %.

➤ Besides mechanical heating, energy from chemical or nuclear reactions may also be used for heating plasmas. Heating in flames or explosions leads to plasmas of low temperature ($T < 10^4$ K), fusion plasmas with $T \approx 10^9$ K may be ignited by nuclear reactions.

■ The ignition of the plasma in a hydrogen bomb proceeds by the explosion of a nuclear-fission bomb in the center of the hydrogen vessel.

17.2.2   Generation of plasma by compression

By adiabatic compression of gases, the temperature may be increased so much that ionization starts and a plasma is generated. The compression may proceed by external forces (pistons, shock waves), or by **magnetic self-compression** of a conducting gas or plasma.

Shock tube, cylindrical tube in which a shock wave is initiated by destroying a membrane between a high-pressure region and a low-pressure region. Ionization occurs because of the strong heating of the gas when the shock wave passes (**Fig. 17.2**).

Shock waves may also be generated by the rapid heating of an amount of gas by pulse discharges, or by magnetic fields increasing in time (inductive hydrodynamic shock tube **Fig. 17.3**). In an electric-pulse discharge, a shock wave is released exclusively by sudden heating. When using magnetic fields, one exploits the magnetohydrodynamic properties of the plasma arising during the process to increase the temperature and the degree of ionization.

17.2.2.1   Pinch effect

Pinch effect, compression of charged liquids and gases in a magnetic field. The compression is generated by the passage of a large current or magnetic field (depending on the geometry) through the liquid or gas. The temperature of the plasma is increased by the compression.

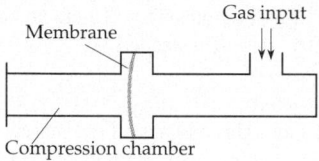

Figure 17.2: Mechanical shock tube.

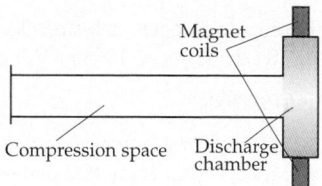

Figure 17.3: Inductive-hydrodynamic shock tube.

1. z-pinch,

a pinch in which the current flows axially through the plasma column. Because of the discharge between two electrodes, a current flows along the pinch axis and generates an azimuthal magnetic field $\vec{\mathbf{B}}_\theta$. In this field, a force density pointing radially inward

$$\vec{\mathbf{f}} = \vec{\mathbf{J}}_z \times \vec{\mathbf{B}}_\theta$$

acts on the charge carriers (**Fig. 17.4**). $\vec{\mathbf{J}}_z$ is the current density along the z-direction. The force pointing towards the pinch axis is just the **Lorentz force** of electrodynamics.

For sufficiently high current density, the force density exceeds the plasma pressure and compresses the plasma column, which thereby constricts (pinch effect) and separates from the walls of the vessel.

Bennett equation, equation giving the current needed for compressing a plasma column in a z-pinch:

Bennett equation			I^2
	Symbol	Unit	Quantity
$I^2 = \dfrac{8\pi}{\mu_0} N k_B T$	I	A	discharge current
	μ_0	Vs/Am	free-space permeability constant
	N	1/m	charge carrier density per unit length
	k_B	J/K	Boltzmann constant
	T	K	plasma temperature

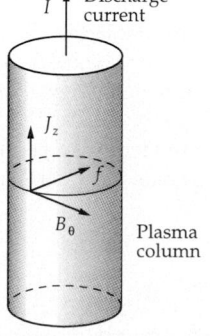

Figure 17.4: z-pinch.

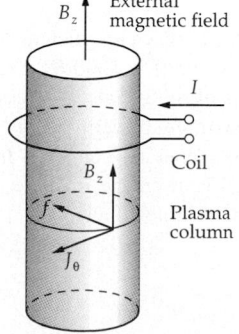

Figure 17.5: Theta-pinch.

■   In order to compress a fusion plasma ($T \approx 10^9$ K) of diameter $r = 15$ cm and charge carrier density $n = 10^{22}$ m^{-3}, a current $I = 2.1 \cdot 10^7$ A is needed.

2. Theta-pinch,

θ-**pinch**, a pinch in which an external coil generates an axial magnetic field increasing in time and inducing an azimuthal current in the plasma column that analogously leads to a radial-force density directed inward (**Fig. 17.5**).

The lifetime of pinch plasmas is limited ($\tau \approx 10 \ \mu$s) by plasma instabilities. Longer confinement times require other geometries, such as toroidal plasma columns.

17.3   Energy production with plasmas

Plasmas, like electrically conducting gases, can be confined by magnetic fields and thus kept away from contact with solid surfaces in the vicinity. This may be exploited in different ways:

*   Heat engines can be operated at higher maximum temperatures than allowed technically by the material of the combustion chamber (MHD generators).
*   Fusion plasmas can be confined without contact with the environment so that fusion reactions may take place in a reactor chamber for controlled energy production.

17.3.1   MHD generator

MHD generator, continuously running heat engine combining the functions of a turbine and a generator with a single working medium (**Fig. 17.6**).

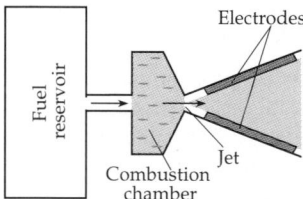

Figure 17.6: MHD generator.

A plasma flows from a combustion chamber through a nozzle into a space region traversed by an external magnetic field pointing perpendicular to the efflux axis. The resulting Lorentz force causes a charge separation of ions and electrons, which may be drawn off by electrodes.

➤   In order to reach a sufficient degree of ionization at typical combustion temperatures of $T = 2000 \ldots 3000 \, K$, alkali atoms must be added to the combustion gas. This doping of the gas also reduces the internal resistance of the generator, which limits the maximum power that may be obtained.

The MHD generator combines the operation modes of a turbine and a generator of a conventional heat engine into one operation step. The maximum potential efficiency is limited by the Carnot efficiency:

$$\eta = \frac{T_h - T_k}{T_h} \ .$$

Because of the high accessible values of T_h, which due to the magnetic confinement are not limited by the material of the walls, theory predicts significantly higher efficiencies for MHD generators than for conventional heat engines. The technical problems remain unsolved, however.

17.3.2 Nuclear fusion reactors

In the fusion of light nuclei, an amount of energy in the order of magnitude of 10 MeV per fusion reaction is released. But, in order to start the fusion reaction, sufficient thermal energy has to be supplied to the reaction partners to overcome the Coulomb repulsion. In a fusion power plant, part of the fusion energy delivered must be used to start additional reactions.

1. Nuclear reactions for nuclear fusion

The following nuclear reactions are possible candidates for a fusion reactor:

$$\,^2_1D + \,^3_1T \rightarrow \,^4_2He + \,^1_0n + 17.60 \text{ MeV}$$

$$\,^2_1D + \,^2_1D \rightarrow \,^3_2He + \,^1_0n + 3.27 \text{ MeV}$$

$$\,^2_1D + \,^2_1D \rightarrow \,^3_1T + \,^1_1H + 4.04 \text{ MeV}$$

$$\,^6_3Li + \,^2_1D \rightarrow 2\,^4_2He + 22.4 \text{ MeV}$$

$$\,^{11}_5B + \,^1_1H \rightarrow 3\,^4_2He + 8.47 \text{ MeV}.$$

The energy released is distributed uniformly over the reaction products.

➤ The reactions listed above are arranged according to decreasing cross-section for a given temperature. The D-T reaction requires the lowest technical effort, but the hard neutron radiation and the need to use radioactive tritium make its safety difficult to ensure.

2. Power density

The power density that may be reached by fusion reactions is given by:

power density from fusion reactions			$L^{-1}MT^{-3}$
	Symbol	Unit	Quantity
$p = n_1 n_2 \langle v\sigma \rangle \varepsilon$	p	W/m^3	power density
	n_i	1/m^3	number densities of reaction partners
	$\langle v\sigma \rangle$	m^3/s	velocity-averaged reaction rates
	ε	J	reaction energy

3. Confinement time,

τ, the time during which a fuel mixture maintains integrity, e.g., by external magnetic fields.

➤ Because of the high kinetic particle energy and the additional radiation pressure, fusion plasmas exert enormous pressures that may be balanced, depending on the density, for only a few ns.

4. Lawson criterion,

a criterion that connects the plasma fuel density required with the confinement time.

In order to produce a self-sustaining chain reaction in a reactor, the released fusion energy must be at least as large as the required thermal plasma energy. For a plasma composed of one particle type:

$$\frac{1}{4}n^2\langle v\sigma\rangle\varepsilon\tau > 3nk_BT .$$

Hence, one obtains a minimum value for the product $n\tau$.

Lawson criterion				$\mathbf{L^{-3}T}$
	Symbol	Unit	Quantity	
$n\tau > \dfrac{12k_BT}{\langle v\sigma\rangle\varepsilon}$	n	$1/m^3$	fuel density	
	τ	s	confinement time	
	k_B	J/K	Boltzmann constant	
	T	K	plasma temperature	
	$\langle v\sigma\rangle$	m^3/s	velocity-averaged reaction rates	
	ε	J	reaction energy	

■   For D-T reactions $n\tau > 5\cdot10^{19}$ s m^{-3}, for D-D reactions analogously $n\tau > 10^{21}$ s m^{-3}.

5. Energy losses in fusion plasmas,

leakage of energy, which has to be compensated by the fusion energy delivered:
- bremsstrahlung,
- discrete radiation by impurity atoms—particularly critical for impurities of high atomic number,
- synchrotron radiation (for toroidal confinement),
- heat conduction,
- particle loss.

6. Confinement techniques

In order to fulfill the Lawson criterion, there are two approaches for confining plasmas:
- **Magnetic confinement**: in a magnetic field, a plasma of low density holds together for a relatively long time. It is heated inductively from outside, in order to gain the necessary thermal energy.
- **Inertial confinement**: the fuel is compressed by energy input so that it holds together for a short time due to its own inertia. Thereby, a high density is reached.

17.3.3   Fusion with magnetic confinement

1. Variants of magnetic confinement

In order to confine a plasma of low density completely in a magnetic field, there are two choices:
- **Mirror machine**, linear θ-pinch, at the ends the magnetic field increases such that particles moving towards the ends are reflected. However, because of ion-ion collisions in the plasma, the required temperature is increased so much that an application in reactors is still questionable.

     ■   With the mirror machine 2XIIB at Lawrence Livermore National Laboratory (California, USA), ion temperatures of $T_i = 9$ keV (for $k_B = 1$) at a density of 10^{20} m^{-3} and a confinement time of $t = 1$ ms have been reached.

- **Toroidal plasma confinement**, a θ-pinch bent to a torus.

A simple θ-pinch bent to a torus does not provide a stable plasma confinement, since a resultant force component pointing outward acts on the charge carriers in the plasma. An additional meridional magnetic field may, however, force the charge carriers to move along spiral-like orbits about the torus axis.

2. Versions of generating the magnetic field

The meridional magnetic field may be generated by various choices:
- **Tokamak**: a transformer induces a current in the plasma, which itself generates the meridional magnetic field. Since this current occurs in a pulse, difficulties arise in continuous performance.

  - ■ Tokamak **JET**, the most advanced test device for magnetic fusion in Great Britain.

- **Stellarator**: an asymmetric coil geometry generates a combined azimuthal-meridional magnetic field. Diffusion losses are limited by the arrangement of the coils so that a stationary operation is possible in principle.

  - ■ Stellarator **Wendelstein**, test device for magnetic fusion at the Max Planck Institute for plasma physics, Munich, Germany.

The heating of the plasma to temperatures above 10^8 K proceeds by induction, or by injection of high-energetic particles. Besides the energy losses by radiation, the losses by plasma diffusion, i.e., motion perpendicular to the axis of the magnetic field, have also to be taken into account. The collisions between charged particles are not pure two-particle collisions, but may affect several particles due to the long range of the Coulomb interaction. Such collisions strongly reduce the lifetime of even a plasma in mechanical equilibrium.

17.3.4   Fusion with inertial confinement

In fusion under inertial confinement, a small amount of fuel enclosed in a spherical pellet is compressed by implosion to a multiple of the solid-state density after external irradiation. The symmetric compression of the fuel leads to a strong increase of its temperature; hence, in the center a fusion reaction is ignited. A thermonuclear burning wave then propagates outward. A plasma confinement by technical means is not needed, since the internal plasma is kept together by the external layers of the pellet during the time of burning (several hundred ps).

1. Structure of the fuel pellet

Fuel pellet, hollow sphere composed of several layers. The innermost layer consists of fuel, a frozen deuterium-tritium mixture. In the surrounding absorber, the energy input is deposited such that the external part (tamper) evaporates outward, while the inner part (pusher) is driven radially inward by the recoil. The fuel is thereby compressed into the hollow region (**Fig. 17.7**).

2. Methods of compression

In order to ignite the fuel pellet, energy injected from outside is deposited in a possibly symmetrical manner. For this purpose, there are several choices:
- Bombardment with **laser beams**. Laser beams may be well focussed with simultaneous high energy density. However, the coupling is not very efficient because of the sudden formation of a plasma layer outside the solid pellet surface, which absorbs the laser radiation. Furthermore, the laser efficiency is low.
- ➤ Laser irradiation generates extremely hot electrons, which may penetrate the entire pellet. This pre-heating of the fuel strongly increases the required compression work.

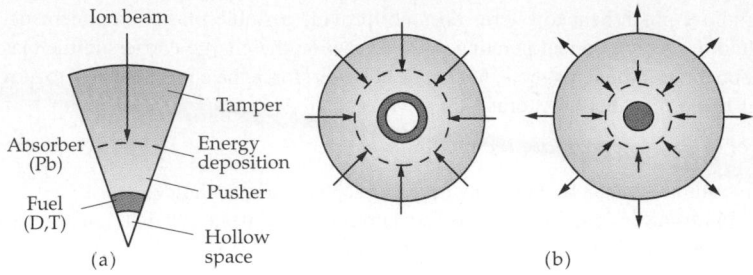

Figure 17.7: Fusion under inertial confinement. (a): structure of the fuel pellet, (b): pellet implosion.

- Bombardment with **ion beams**. Ion beams offer a strongly localized and efficient coupling of the beam to the pellet (due to the Bragg peak in the energy deposition profile). The focussing of the beam, in particular for heavy-ion beams, is technically difficult, and there is a lower limit for the size of the system.

Currently, the direct bombardment of pellets by ion beams appears insufficient for an effective ignition of the fuel because asymmetries of the energy deposition and hydro-dynamic instabilities (Rayleigh-Taylor instabilities at the interface of two substances of different density accelerated against each other) limit the maximum compression that can be achieved.

Indirectly driven pellet, a method for the compensation of the asymmetries in the irra-diation. The ion or laser beam is not guided directly onto the pellet, but hits gold radiation converters that convert a high percentage of the incoming radiation into weak x-rays, which are absorbed in a cavity and then re-emitted towards the pellet. A much better symmetriza-tion is expected; however, the required energy to ignite the fuel is increased by two orders of magnitude.

Symbols used in formulae on electricity and plasma physics

symbol	unit	designation
α	1/K	temperature coefficient
α_i	m^3/s	recombination coefficient
γ, σ	S/m	conductance
ε	C/(Vm)	permittivity
ε	J	reaction energy
ε_0	C/(Vm)	free-space electric-permittivity constant
ε_r	1	relative permittivity
κ	S/m	conductance
λ	C/m	line charge density
λ_D	m	Debye length
Λ_m	Vs/A	magnetic conductance
μ	Vs/(Am)	permeability
μ	$m^2/(Vs)$	ion mobility
μ_0	Vs/(Am)	free-space magnetic-permeability constant
μ_r	1	relative permeability
ρ	Ωm	specific resistance
ρ	C/m^3	space charge density
σ	C/m^2	surface charge density
τ	1	optical depth
Θ	A	magnetomotive force
φ	V	electric potential
Φ	Wb	magnetic flux
χ_m	1	magnetic susceptibility
ψ	Vm	electric flux
ψ	Wb	induction flux
$\vec{\omega}_G$	1/s	gyration frequency
ω_{Ge}	1/s	gyration frequency of electrons
ω_{Pe}	1/s	Langmuir frequency of electrons
ω_{Pe}	rad/s	Langmuir frequency
A	1	amplification factor of valve
A	$A/(m^2K^2)$	Richardson constant
b	$m^2/(Vs)$	mobility
B	S	susceptance
B	J/m^2	emitted radiant energy
$\vec{B}$	T	magnetic flux density
c	m/s	speed of light in matter
c_0	m/s	speed of light in vacuum
c_i	mol/kg	ionic concentration
c_S	m/s	ion sound velocity
C	F	capacitance
d_p	1	damping factor
D	1	grid transparency
$\vec{D}$	C/m^2	electric displacement density

(*continued*)

symbol	unit	designation
e_0	C	elementary charge
E	kg/C	electrochemical equivalent
$\vec{\mathbf{E}}$	V/m	electric field strength
f	1/s	radiation frequency
f	1/s	photon frequency
F	C/mol	Faraday constant
G	S	conductance
h	Js	Planck's quantum of action
$\vec{\mathbf{H}}$	A/m	magnetic field strength
I	A	current
I_a	A	anode current
$\vec{\mathbf{J}}$	A/m^2	current density
k_B	J/K	Boltzmann constant
$\ln \Lambda$	1	Coulomb logarithm
L	H	inductance
m_e	kg	electron mass
M	H	mutual inductance
M	kg/mol	molar mass
$\vec{\mathbf{M}}$	A/m	magnetization
n	mol	quantity of substance
N	1	particle number
N_A	1/mol	Avogadro constant
Q	C	charge
Q	W	reactive power
Q_p	1	quality
r_G	m	gyration radius
R	Ω	electric resistance
R_a	Ω	resistance in anode circuit
R_i	Ω	valve resistance
R_m	A/Wb	magnetic resistance
s	1	slippage
S	W	apparent power
S	A/V	transconductance
T	K	temperature
u	1	transmission ratio
V	V	electric voltage
V_a	V	anode voltage
V_g	V	grid voltage
V_s	V	control voltage of grid
$\langle v\sigma \rangle$	m^3/s	velocity-averaged reaction rates
$\langle v_e \rangle$	m/s	mean electron velocity
v_{dr}	m/s	drift velocity
W	J	binding energy of electron
W_A	J	work function
x	1	fractional ionization
X	Ω	reactance
Y	S	admittance
z	1	ionic valence
Z	Ω	impedance

18
Tables on electricity

18.1 Metals and alloys

18.1.1 Specific electric resistance

18.1/1 Metals at room temperature

Element	T/K	$\rho/10^{-8}\,\Omega\,\mathrm{m}$	$\dfrac{1}{\rho}\dfrac{\mathrm{d}\rho}{\mathrm{d}T}\Big/$ $10^{-3}\mathrm{K}^{-1}$
antimony	273	39.0	
bismuth	273	107	4.45
cadmium	273	6.8	4.26
cerium	290 – 300	82.8	
cobalt	273	5.6	6.58
dysprosium	290 – 300	92.6	
erbium	290 – 300	86.0	
europium	290 – 300	90.0	
gadolinium	290 – 300	131.0	
gallium	273	13.6	
holmium	290 – 300	81.4	
indium	273	8.0	5.1
iridium	273	4.7	4.9
lanthanum	290 – 300	61.5	
lutetium	290 – 300	58.2	
mercury	273	94.1	0.89
neodymium	290 – 300	64.3	
niobium	273	15.2	2.28

(continued)

18.1/1 Metals at room temperature (*continued*)

Element	T/K	$\rho/10^{-8}\,\Omega\,\mathrm{m}$	$\dfrac{1}{\rho}\dfrac{\mathrm{d}\rho}{\mathrm{d}T}\Big/$ $10^{-3}\mathrm{K}^{-1}$
osmium	273	8.1	4.2
polonium	273	40	
praseodymium	290 – 300	70.0	
promethium	290 – 300	75.0	
protactinium	273	17.7	
rhenium	273	17.2	3.1
rhodium	273	4.3	4.57
ruthenium	273	7.1	3.59
samarium	290 – 300	94.0	
scandium	290 – 300	56.2	
terbium	290 – 300	115	
thallium	273	15	5.2
thorium	273	14.7	3.3
thulium	290 – 300	67.6	
tin	273	11.5	4.63
titanium	273	39	5.5
uranium	273	28	3.4
ytterbium	290 – 300	25.0	
yttrium	290 – 300	59.6	
gold	273	2.06	4.5
platinum	273	9.81	3.93

18.1/2 Pressure dependence

As a rule, the electric conductance of metals increases when applying an external hydrostatic pressure. A measure for the magnitude of this change is the pressure coefficient $(1/\rho)(\mathrm{d}\rho/\mathrm{d}p)$ of the specific electric resistance.

Metal	T/K	Pressure $/10^2$ MPa		
		0	10	30
		$\dfrac{1}{\rho}\dfrac{\mathrm{d}\rho}{\mathrm{d}p}\,/10^{-5}\mathrm{MPa}^{-1}$		
lithium	303	−7.00	−7.52	−9.0
beryllium	298	1.77	1.63	1.46
sodium	303	58.8	23.6	4.04
magnesium	298	5.40	4.67	3.81
aluminum	301	4.29	4.06	3.6
potassium	303	134.4	30	0.88
calcium	303	−9.48	−12.2	−20.7
titanium	296	1.19	1.12	1.02
chromium	298	22.2	17.3	8.96
iron	303	2.42	2.26	1.90
cobalt	297	0.96	0.90	0.80

(*continued*)

18.1/2   Pressure dependence (*continued*)

Metal	T/K	Pressure $/10^2$ MPa		
		0	10	30
		$\dfrac{1}{\rho}\dfrac{\mathrm{d}\rho}{\mathrm{d}p}$ $/10^{-5}\mathrm{MPa}^{-1}$		
nickel	298	1.77	1.82	1.73
copper	303	1.92	1.80	2.42
zinc $\parallel$ c	303	9.68	8.76	6.72
zinc $\perp$ c	303	5.28	4.40	2.84
rubidium	303	157.0	14.4	-28.8
strontium	303	-45.3	-59.0	-118.8
zirconium	299	0.32	0.33	0.22
niobium	297	1.40	1.37	1.30
molybdenum	300	1.31	1.29	1.24
rhodium	299	1.65	1.62	1.56
palladium	299	2.10	2.04	1.93
silver	303	3.48	3.28	2.60
indium	296	1.25	1.09	0.85
tin $\parallel$ c	303	10.0	9.0	6.1
tin $\perp$ c	303	9.24	8.26	5.61
antimony	303	-9.84	-14.8	-2.80
barium	303	7.2	1.2	-13.6
cerium	297	-4.1	—	1.6
praseodymium	297	1.36	1.20	1.02
neodymium	297	1.57	1.32	1.03
tantalum	302	1.62	1.62	1.55
tungsten	302	1.33	1.31	1.25
iridium	296	1.39	1.37	1.33
platinum	296	1.92	1.88	1.78
gold	303	3.02	2.84	2.44
mercury (liqu.)	303	23.1	17.0	—
lead	303	13.7	11.6	6.96
bismuth	303	-14.8	-18.5	—
uranium	293	4.88	4.56	4.10

18.1/3   Relative change at the melting point

Metal	T_{melt}/K	$\rho_{\mathrm{liqu.}}/\rho_{\mathrm{solid}}$	Metal	T_{melt}/K	$\rho_{\mathrm{liqu.}}/\rho_{\mathrm{solid}}$
lithium	453	1.68	cadmium	594	1.89
sodium	370	1.46	indium	388	2.12
magnesium	924	1.63	tin	505	2.11
aluminum	934	1.82	antimony	904	0.71
potassium	337	1.55	tellurium	722	2.00
iron	1808	1.09	caesium	303	1.66
copper	1357	2.07	gold	1336	2.28
zinc	693	2.11	mercury	234	3.36
gallium	303	0.47	lead	601	1.98
rubidium	312	1.61	bismuth	544	0.47
silver	1234	1.9			

18.1/4 Alloys

Alloy	$\rho/10^{-6}$ Ωm	$\dfrac{1}{\rho}\dfrac{d\rho}{dT}10^{-3}$K
gold-chromium	0.33	0.001
graphite	8.00	−0.2
isabellin	0.50	0.02
commutator coal	40	—
constantan	0.50	0.03
manganin	0.43	0.02
chromium-nickel (80 Ni, 20 Cr)	1012	0.2
nickelin	0.43	0.2
novoconstantan	0.45	0.04
platinum-iridium (20 %)	0.32	2.0
platinum-rhodium (10 %)	0.20	1.7
resistin	0.51	0.008
red brass	0.127	1.5
nickel brass	0.30	0.4

18.1.2 Electrochemical potential series

18.1/5 Electrochemical potential series

The given values of the electromotive force V_0 refer to hydrogen as reference electrode and hold for a 1-n solution.

Material	Valence	V_0/V	Material	Valence	V_0/V
fluorine	1	+2.87	cadmium	2	−0.40
gold	1	+1.69	iron	2	−0.45
chlorine	1	+1.35	sulphur	2	−0.48
gold	3	+1.40	gallium	3	−0.55
bromium	1	+1.07	chromium	2	−0.91
platinum	2	+1.18	zinc	2	−0.76
mercury	2	+0.80	tellurium	2	−1.14
silver	1	+0.80	manganese	2	−1.19
graphite	2	+0.75	aluminum	3	−1.66
iodine	1	+0.54	uranium	3	−1.80
copper	1	+0.52	magnesium	2	−2.37
polonium	4	+0.76	beryllium	2	−1.85
oxygen	2	+0.39	sodium	1	−2.71
copper	2	+0.34	calcium	2	−2.87
arsenic	3	+0.23	strontium	2	−2.90
bismuth	3	+0.31	barium	2	−2.91
antimony	3	−0.51	potassium	1	−2.93
tin	4	+0.02	rubidium	1	−2.98
hydrogen	1	±0.00	lithium	1	−3.04
iron	3	−0.04	steel (galvanized)		−0.53 ⋯ −0.72
lead	2	−0.13	ingot iron		−0.21 ⋯ −0.48
tin	2	−0.14	cast iron		−0.18 ⋯ −0.42
nickel	2	−0.26	brass		+0.26 ⋯ +0.05
cobalt	2	−0.28	bronze		+0.36 ⋯ +0.03
indium	3	−0.34	chromium-nickel		+0.75 ⋯ −0.05

18.1/6   Thermoelectric potential series

The given values of the thermal electromotive force V_0 hold for platinum or copper as second metal and a temperature difference of 100 K.

Material	$V_0/(\text{mV}/100\,\text{K})$ platinum	copper	Material	$V_0/(\text{mV}/100\,\text{K})$ platinum	copper
tellurium	+50	+49	caesium	+0.5	—
silicon	+44.8	+44	lead	+0.44	−0.31
antimony	+4.75	+4.0	tin	+0.42	−0.33
chromium-nickel	+2.2	+1.45	magnesium	+0.42	−0.33
iron	+1.88	+1.08	tantalum	+0.41	−0.34
molybdenum	+1.2	−0.45	aluminum	+0.39	−0.36
brass	+1.1	+0.35	coal	+0.30	−0.45
cadmium	+0.9	+0.15	graphite	+0.22	−0.53
tungsten	+0.8	+0.05	mercury	±0	−0.75
V2A-steel	+0.8	+0.05	platinum	±0	−0.75
copper	+0.75	±0	thorium	−0.1	−0.85
silver	+0.73	−0.02	sodium	−0.2	−0.95
gold	+0.7	−0.05	palladium	−0.5	−1.25
zinc	+0.7	−0.05	nickel	−1.5	−2.25
manganese	+0.7	−0.05	cobalt	−1.7	−2.45
iridium	+0.66	−0.09	constantan	−3.3	−4.05
rhodium	+0.65	−0.10	bismuth	−6.5	−7.25

18.1/7   Thermoelectric voltage of common thermocouples

Reference temperature 0 °C

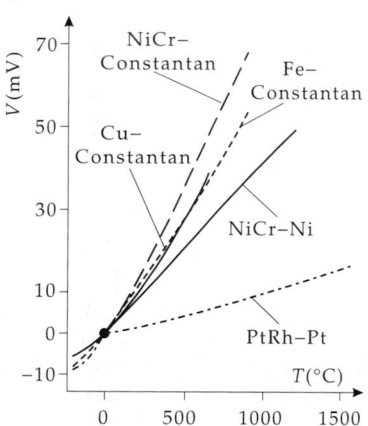

18.1/8   Common thermocouples

Temperature range	Thermocouple
−200 °C – 600 °C	Cu-constantan
−200 °C – 800 °C	Fe-constantan
0 °C – 1200 °C	NiCr-Ni
0 °C – 1600 °C	PtRh-Pt

18.1/9   Peltier coefficient P for various metals

The arrow indicates the flow direction of the electric current.

Metal couple	$T/°C$	$P/(\mu J/K)$
As → Pb	20	3.81
Bi$_\parallel$ → Bi$_\perp$	20	15.03
Cd$_\parallel$ → Cd$_\perp$	20	0.85
Cd → Ni	15	6.40
Cu → Ag	0	0.0703
Cu → Al	14	1.70
Cu → Au	0	0.3403
Cu → Bi	18	16.12
Cu → Ni	0	7.95
	14.4	5.80
Cu → Pd	0	0.588
Cu → Pt	0	0.238
Cu → constantan	15.5	2.436
Fe → Cu	0	0.664
Fe → Hg	18.4	1.1644
	99.64	1.388
	182.3	1.511
Fe → Ni	15	2.288
Fe → constantan	0	3.10
Pb → Bi	20	5.16
Pb → constantan	0	7.95
	100	11.43
	200	15.07
	300	18.42
Sb → Bi	20	44.79
Sb → Pb	20	0.78
Zn → Ni	15	6.42
Zn$_\parallel$ → Zn$_\perp$	20	0.53
graphite → Cu	20	2.94

18.2   Dielectrics

In the following tables, ε_r denotes the relative permittivity, δ the dielectric loss angle, and V_d the breakdown voltage.

18.2/1   Relative permittivity ε_r

The given values hold for room temperature.

Substance	Molecular formula	Frequency/MHz	ε_r
aluminum oxide	Al_2O_3	1	10
ammonium bromide	NH_4Br	100	7.1
ammonium chloride	NH_4Cl	100	7.0
apatite ($\perp$ optical axis)		300	9.5
apatite ($\parallel$ optical axis)		300	7.41
asphalt		< 1	2.68
barium chloride	$BaCl_2$	60	11.4
barium chloride (2 H_2O)		60	9.4
barium nitrate	$Ba(NH_3)_2$	60	5.9
barium sulphate	$BaSO_4$	100	11.4
beryl ($\perp$ optical axis)	$Be_3Al_2Si_6O_{18}$	0.01	7.02
beryl ($\parallel$ optical axis)	$Be_3Al_2Si_6O_{18}$	0.01	6.08
calcite ($\perp$ optical axis)	$CaCO_3$	0.01	8.5
calcite ($\parallel$ optical axis)	$CaCO_3$	0.01	6.08
Acetic amide	C_2H_5NO	400	4.0
acetic acid (2 °C)	$C_2H_4O_2$	400	4.1
calcium carbonate	$CaCO_3$	1	6.14
calcium fluoride	CaF_2	0.01	7.36
calcium sulphate (2 H_2O)	$CaSO_4$	0.01	5.66
cassiterite ($\perp$ optical axis)	SnO_2	10^6	23.4
cassiterite ($\parallel$ optical axis)	SnO_2	10^6	24
copper oxide	Cu_2O	100	18.1
copper oleate	$Cu(C_{18}H_{33}O_2)_2$	400	2.8
copper sulphate	$CuSO_4$	60	10.3
copper sulphate (2 H_2O)		60	7.8
diamond	C	100	5.5
dolomite ($\perp$ optical axis)	$CaMg(CO_3)_2$	100	8.0
dolomite ($\parallel$ optical axis)	$CaMg(CO_3)_2$	100	6.8
iron oxyde	Fe_3O_4	100	14.2
lead acetate	$Pb(C_2H_3O_2)_2$	1	2.6
lead carbonate	$PbCO_3$	100	18.6
lead chloride	$PbCl_2$	1	4.2
lead monoxide	PbO	100	25.9
lead nitrate	$Pb(NO_3)_2$	60	37.7
lead oleate	$Pb(C_{18}H_{32}O_2)_2$	400	3.27
lead sulphate	$PbSO_4$	1	14.3
lead sulphide	PbS	1	17.9
malachite	$Cu_2(OH)_2(CO_3)$	10^6	7.2

(continued)

18.2/1   Relative permittivity ε_r (*continued*)

Substance	Molecular formula	Frequency/MHz	ε_r
mercury chloride	Hg_2Cl_2	1	3.2
	$HgCl_2$	1	9.4
naphthalene	$C_{10}H_8$	400	2.52
phenol (10 °C)	C_6H_6O	400	4.3
red phosphorus	P_4	100	3.6
kalinite	$KAl(SO_4)_2 \cdot 12H_2O$	1	3.8
potassium carbonate	$KHCO_3$	100	5.6
potassium chlorate	$KClO_3$	60	5.1
potassium chloride	KCl	0.01	5.03
potassium chromate	K_2CrO_4	60	7.3
potassium iodide	KI	60	5.6
potassium nitrate	KNO_3	60	5.0
potassium sulphate	K_2SO_4	60	5.9
quartz ($\perp$ optical axis)	SiO_2	30	4.34
quartz ($\parallel$ optical axis)		30	4.27
rutile ($\perp$ optical axis)	TiO_2	100	86
rutile ($\parallel$ optical axis)		100	170
selenium	Se	100	6.6
silver bromide	$AgBr$	1	12.2
silver chloride	$AgCl$	1	11.2
silver cyanide	$AgCN$	1	5.6
zinc carbonate ($\perp$ optical axis)	$ZnCO_3$	10^6	9.3
zinc carbonate ($\parallel$ optical axis)		10^4	9.4
sodium carbonate	Na_2CO_3	60	8.4
sodium carbonate (10 H_2O)		60	5.3
sodium chloride	$NaCl$	0.01	6.12
sodium oleate	$NaC_{18}H_{38}O_2$	400	2.75
sodium perchlorate	$NaClO_4$	60	5.4
sugar		300	3.32
sulphur	S	—	4.0
thallium chloride	$TlCl$	1	46.9
tourmaline ($\perp$ optical axis)		0.01	7.10
tourmaline ($\parallel$ optical axis)		0.01	6.3
zirconium	Zr	100	12

18.2/2   Ceramics

Substance	ε_r	$\tan \delta$	$V_d/(kV/mm)$
porcelain	6...7	0.035	20...28
steatite	6...6.5	0.002	20...25
capacitor ceramics			
$ZrTiO_4$	28...30	$2.5...5.5 \cdot 10^{-4}$	32
TiO_2	78...88	$4...5.5 \cdot 10^{-4}$	27
$CaTiO_3$	150...165	$2...4 \cdot 10^{-4}$	22
$(SrBi)TiO_3$	900...1000	$5...10 \cdot 10^{-4}$	28
$(BaTiO_3)_{0.9} \cdot (BaZrO_3)_{0.075}$	2700...3000	$1...2 \cdot 10^{-2}$	13

18.2/3   Glasses

Type of glass	ε_r	$10^{-4}\tan\delta$
pyrex glass	4.1...4.6	45...130
quartz glass	3.75	1...2
corning glass	4.0	6

18.2/4   Electric properties of polymers

Property	Polyethylene	Teflon	Polyvinyl chloride	Polystyrene	Polymethyl methacrylate	Epoxy resin
thermal stability/$°C$	100	260	60–70	65–96	68–88	140
$\rho\,/\,\Omega\text{m}$	10^{15}–10^{17}	10^{15}–10^{16}	10^{14}–10^{16}	10^{17}–10^{18}	10^{14}–10^{16}	10^{13}–10^{14}
ε_r (1 MHz)	2.3	2	3–5	2.45–2.65	3.5–4.5	3.7
$\tan\delta$(1 MHz)	$2\cdot10^{-4}$	$2\cdot10^{-4}$	0.03–0.08	$(1\text{–}4)\cdot10^{-4}$	0.04–0.06	0.019
$V_d\,/(\text{kV}/\text{mm})$	18–20	20–30	14–20	20–35	18–35	18

18.2/5   Specific electric resistance of insulating materials

Insulating material	$\rho/\,\Omega\text{m}$	Insulating material	$\rho/\,\Omega\text{m}$
bakelite	10^{14}	plexiglass	10^{13}
benzene	10^{15}	polyethylene	$10^{10}\ldots10^{13}$
amber	$>10^{16}$	polystyrene	$10^{15}\ldots10^{16}$
celluloid	$10^{8}\ldots10^{10}$	polyvinyl chloride	up to 10^{13}
ivory	$2\cdot10^{6}$	porcelain	$5\cdot10^{12}$
earth, wet	$>10^{6}$	quartz glass	$5\cdot10^{16}$
flint glass	$3\cdot10^{8}$	shellac	10^{14}
galalith	$\approx10^{14}$	slate	10^{6}
glass	$>10^{11}$	sealing wax	$8\cdot10^{13}$
mica	$10^{13}\ldots10^{15}$	silicon	$8\cdot10^{7}$
gutta-percha	$\approx4\cdot10^{7}$	silicon oil	10^{13}
hard rubber	$10^{13}\ldots10^{16}$	transformer oil	$10^{10}\ldots10^{13}$
wood, dry	$10^{9}\ldots10^{13}$	vaseline	$10^{10}\ldots10^{13}$
marble	$10^{7}\ldots10^{8}$	vulcanized fibre	$10^{10}\ldots10^{13}$
rubber	$6\cdot10^{14}$	water, distilled	$(1\ldots4)\cdot10^{4}$
colophony	$5\cdot10^{14}$	river water	$10\ldots100$
paper	$10^{15}\ldots10^{16}$	sea water	0.3
paraffin	$10^{14}\ldots10^{16}$	soft rubber	$(2\ldots14)\cdot10^{11}$
paraffin oil	10^{14}	polyester resin	$(8\ldots14)\cdot10^{11}$
kerosene	$10^{10}\ldots10^{12}$		

18.2/6   Electric properties of insulating materials

Materials		$\rho/\Omega m$	ε (50 Hz)	ε (800 Hz)	$10^{-3}\tan\delta$ (50 Hz)	$10^{-3}\tan\delta$ (800 Hz)	$V_d/(\text{kV/mm})$
phenoplasts							
	pure casting resin	$10^9\ldots10^{14}$	—	8	—	75	10
	pure moulding resin	10^{13}	—	4.3	—	47	8
	mineral powder	$10^9\ldots10^{11}$	—	10	—	100	$5\ldots10$
phenol	asbestos fibre	$10^9\ldots10^{11}$	12	$6\ldots20$	—	$30\ldots300$	$5\ldots15$
form	wood flour	$10^{10}\ldots10^{12}$	—	9	—	70	$15\ldots20$
aldehyde	paper shred	$10^9\ldots10^{12}$	—	$6\ldots10$	—	$40\ldots100$	$8\ldots15$
	paper breadths	10^{19}	—	6	—	100	$1.5\ldots5.2$
	textile breadths	10^9	$5\ldots7$	—	$50\ldots600$	100	—
phenol	minerals	$10^9\ldots10^{11}$	—	4.8	—	$40\ldots150$	$1.6\ldots2.4$
furfural	wood flour	$10^{10}\ldots10^{12}$	—	$4.5\ldots80$	—	$100\ldots150$	$1\ldots2$
	textile	$10^9\ldots10^{11}$	—	$405\ldots6$	—	$80\ldots200$	$1\ldots2$
aminoplasts							
urea	wood flour	$10^{13}\ldots10^{14}$	6.6	—	$20\ldots34$	$20\ldots30$	$2.8\ldots2.9$
melanine	cellulose	$10^{12}\ldots10^{14}$	$6.2\ldots7.6$	$6.2\ldots7.5$	$32\ldots60$	$13\ldots100$	10
melanine	asbestos	10^{11}	$6.4\ldots10.2$	9	$70\ldots117$	70	—
aniline	moulding resin	10^{12}	$3\ldots4$	—	$10\ldots20$	—	1
cellulose derivates							
cellulose, soft		10^{15}	—	5.5	—	21	17
cellulose acetate, medium		10^{15}	—	5.4	—	23	17
cellulose acetate, hard		10^{15}	—	5.3	—	22	18
cellulose acetate, higher		10^{16}	—	4.3	—	20	19
cellulose acetobutyrate		10^{16}	—	3.5	—	10	21
cellulose nitrate		$10^{12}\ldots10^{13}$	—	$4\ldots9$	—	10	30
ethyl cellulose		$10^{13}\ldots10^{14}$	—	$2.5\ldots3.5$	—	$5\ldots25$	$60\ldots100$
benzyl cellulose		10^{14}	—	3.5	—	50	40
ethylene derivates							
high-pressure polyethylene		10^{16}	2.3	2.3	0.4	0.4	60
low-pressure polyethylene		10^{16}	—	2.3	—	$0.5\ldots1$	60
polypropylene		10^{13}	—	2.3	—	0.5	70
polystyrene		$10^{16}\ldots10^{17}$	—	2.5	—	$0.2\ldots0.7$	$50\ldots55$
polystyrene (styrene)		10^{14}	—	2.8	—	4	40
polystyrene (acrylonitrile)		10^{14}	3	—	10	40	
polymethacrylic ester		10^{15}	$3.5\ldots4.5$	$3.5\ldots3.5$	$40\ldots60$	$30\ldots50$	15
polyacrylic ester		10^{15}	—	3.5	—	40	15
polyvinyl chloride		4	3.4	$20\ldots40$	$20\ldots40$	50	
polycarbonate		10^{15}	3.5	3.2	0.5	1.65	100

(*continued*)

18.2/6   Electric properties of insulating materials (*continued*)

Materials	$\rho/\Omega m$	ε (50 Hz)	ε (800 Hz)	$10^{-3}\tan\delta$ (50 Hz)	$10^{-3}\tan\delta$ (800 Hz)	$V_d/(kV/mm)$
proteins						
polyurethane type U_g	10^{14}	4	3.3	10	10	—
poly type U_{30}	10^{14}	—	4.1	—	37	—
polyamide 6	10^{12}	—	6	300	20	1.14
polyamide 6 + GV	10^{12}	—	6.8	—	220	25
polyamide 66	10^{12}	—	5.5	—	200	28
polyamide 66 + GV	10^{12}	—	5.6	—	160	28
polyamide 11	10^{14}	3.7	3.7	50	50	20
polyamide 11 + GV	10^{15}	3.8	3.8	30	30	20
polyamide 12	10^{13}	4.2	4.2	90	90	31
polyamide 12 + GV	10^{12}	4.2	4.2	120	120	31
artificial horn	10^{5}	—	6	—	140	1...5
fluorocarbones						
polyfluoromonochloroethylene	10^{16}	2.3	2.8	15	24	20...30
polytetrafluoroethylene	10^{15}	2	2	0.2...0.5	0.2...0.5	20...60
silicons						
silicon resin	10^{15}	3	3	0.5...1	—	20...70
silicon rubber	10^{14}	2.5	2.5	20	—	20...30
elastomers						
neoprene	10^{5}	—	7.5	—	19	14
buna S	10^{3}	—	4...5	—	5	25
perbunan	10^{3}	—	18	—	17	—
modified natural materials						
vulcanized fibre	10^{8}	4	4	80	80	6
hard rubber	10^{12}	2.5...5	2.8...5	50	50	3

18.2/7   Electric properties of transformer oil

Property	Transformer oil	Castor oil
$\rho/\Omega m$	$10^{14}...10^{15}$	$5\cdot10^{10}...5\cdot10^{12}$
$\varepsilon_r(1\text{ MHz})$	2.1...2.3	4.0...4.4
$\tan\delta(1\text{ MHz})$	0.002...0.005	0.01...0.03
$V_d/(kV/mm)$	20	14...16

18.2/8   Some properties of electrets

Composition	$NaKC_4H_4O_6\cdot4H_2O$	KH_2PO_4	$NH_4H_2PO_4$
Curie point	$T_{C1}=258K; T_{C2}=295.5K$	123 K	147.9 K
melting point /°C	58	252.6	190
density /(g/cm^3)	1.775	2.34	2.311
spontaneous polarization /(μC/cm^2)	0.25	4.7	4.8

18.2/8 Some properties of electrets (*continued*)

Composition	KH_2AsO_4	$NH_4H_2AsO_4$
Curie point /K	95.6 T_C	216.1
melting point /°C	288	300
density /(g/cm^3)	2.85	1.803

Composition	$(CN_2H_6)AL(SO_4)_2 \cdot 12H_2O$	$(CH_2NH_2COOH)_3H_2SO_4$
Curie point	473	320...323
spontaneous polarization /($\mu C/cm^2$)	0.35	—

18.2/9 Ferroelectrics with oxygen octahedron structure

Compound	Formula	Structure	T_C/°C	ε_r
barium titanate	$BaTiO_3$	perovskite	120	1700...2000 (at T_C 8...$10 \cdot 10^3$)
lithium tantalate	$LiTaO_3$	ilmenite	> 450	
sodium niobate	$NaNbO_3$	perovskite	640; 518; 480; 360	anti-ferroelectric; 350
lead hafniate	$PbHfO_3$	perovskite	215; 163	anti-ferroelectric; 100; at 215 °C: 1000
lead niobate	$PbNb_2O_3$	cubic	570	280;
lead tantalate	$PbTaO_6$	cubic	260	300$\cdots$400; at 260 °C: 1100
lead titanate	$PbTiO_3$	perovskite	500	200; at 500 °C: 3500
lead zirconate	$PbZrO_3$	perovskite	235	anti-ferroelectric; 250; at 235 °C: 3750
strontium titanate	$SrTiO_3$	perovskite	−250	

18.3 Practical tables of electric engineering

18.3/1 Resistance alloys

Alloy	$\rho/\Omega\,mm^2m^{-1}$	α/K^{-1}	Max. working temperature/ °C
nickelin (67 % Cu, 30 % Ni, 3 % Mn)	0.4	0.0003	300
manganin (86 % Cu, 12 % Mn, 2 % Ni)	0.43	0.00001	300
constantan (54 % Cu, 45 % Ni, 1 % Mn)	0.5	±0.00003	400
chromium-nickel	1.0...1.2	0.00003	1000
megapyr (65 % Fe, 30 % Cr, 5 % Al)	1.4	−0.00006	1300
kanthal	1.45	0.00006	1300

18.3/2   Voltage of Weston standard elements

Temperature/°C	Voltage /V	Temperature/°C	Voltage /V
11	1.01874	20	1.01830
12	1.01868	21	1.01826
13	1.01863	22	1.01822
14	1.01858	23	1.01817
15	1.01853	24	1.01812
16	1.01848	25	1.01807
17	1.01843	26	1.01802
18	1.01839	27	1.01797
19	1.01834	28	1.01792

18.3/3   Contact materials

Material	Conductance $/\mathrm{m}\,\Omega^{-1} \cdot \mathrm{mm}^{-2}$	Melting temperature/°C	Properties
E-copper	56	1085	electric arc creates a badly conducting oxide layer; cheap
fine silver	60	960	conducting oxide layer; low hardness; unstable against sulphur; low transition resistance
fine gold	45.7	1063	chemically resistive; soft; contacts agglutinate easily
tungsten	18.2	3370	low burn-off; very hard
mercury	1.04	-38.9	maintenance-free; long serviceable life; chemically resistive; toxic!
coal	$0.03 \ldots 12$	—	no oxide layer, non-welding, self-lubricating, applicable up to 400 °C
silver bronze	$30 \ldots 50$	$700 \ldots 1100$	good spring properties
hard silver	$52 \ldots 56$	920	electric-arc resistive, hard
silver-cadmium	16	880	Cd acts electric-arc-quenching

18.3/4   Voltage ranges in electric engineering

Designation	Voltage range /V	Application
small voltage	$0 < V \leq 42$	electromechanical toy
low voltage	$0 < V \leq 1000$	operation networks of all kinds
medium voltage	$1000 < V \leq 30000$	high-voltage open wires
high voltage	$1000 < V \leq 110000$	high-voltage open wires
extra-high voltage	$110000 < V \leq 5 \cdot 10^6$	extra-high-voltage open wires

18.3/5   Guide values of some voltages

	V /V		V /V
antenna voltage	$(5 \ldots 40) \cdot 10^{-6}$	trolley, rapid transit	$500 \ldots 800$
nerval potential	$(0.5 \ldots 5) \cdot 10^{-2}$	spark plug	$(5 \ldots 15) \cdot 10^3$
lead accumulator	2	wire voltage of railway	$15 \cdot 10^3$
bicycle dynamo	6	x-ray tubes	up to $2 \cdot 10^5$
mains voltage	120 or 240	belt-type generators	up to $5 \cdot 10^6$

18.3/6   Gas transmittance κ of various quartz glasses

The gas-transmission coefficient specifies the quantity of gas in cm^3 passing at standard pressure and a pressure difference of $1.33 \cdot 10^2$ Pa through an area of $1\ cm^2$ per second for a glass thickness of 1 mm.

Helium		Hydrogen	
$T/°C$	κ	$T/°C$	κ
-78	$2 \cdot 10^{-13}$	200	$2 \cdot 10^{-12}$
0	$6 \cdot 10^{-12}$	300	10^{-11}
100	$6 \cdot 10^{-11}$	400	$3.7 \cdot 10^{11}$
200	$2 \cdot 10^{-10}$	500	$1.25 \cdot 10^{-10}$
400	10^{-9}	700	$2.52 \cdot 10^{-10}$
800	$5 \cdot 10^{-9}$	900	$6.4 \cdot 10^{-10}$

Neon		Nitrogen	
$T/°C$	κ	$T/°C$	κ
500	$1.4 \cdot 10^{-11}$	600	$6.5 \cdot 10^{-12}$
600	$2.8 \cdot 10^{-11}$	700	$1.32 \cdot 10^{-11}$
700	$4.2 \cdot 10^{-11}$	800	$4.3 \cdot 10^{-11}$
900	$1.18 \cdot 10^{-10}$	900	$1.19 \cdot 10^{-10}$

Argon	
$T/°C$	κ
800	$1.6 \cdot 10^{-12}$
900	$5.8 \cdot 10^{-11}$

18.3/7   Effect of electric current on the human body

Range	Response		Alternating current 15...200Hz effective value	Direct current
I	increase of blood pressure	minor muscle contraction in the fingers	0.4...4mA	1...20mA
	no influence on heart-beat frequency	nervous shock up to the forearm	0.8...4.5mA	25...40mA
	no influence on the stimulus-conducting system	letting go the electrode still possible	6...22mA	40...60mA
		letting go the electrode no longer possible	8.5...30mA	60...90mA
II	no unconsciousness yet, increase of blood pressure, irregular heartbeat;	reversible cardiac arrest at higher current intensities, partly already unconsciousness	25...80mA	80...300mA
III	flickering of ventricles, unconsciousness		80mA...8A	250mA...8A
IV	as in range II, arythmics, cardiac arrest, increase of blood pressure;	pulmonary swell out, burns, unconsciousness	> 3A	> 3A

18.4   Magnetic properties

18.4/1   Magnetic susceptibility of elements

The table lists the molar magnetic susceptibility $\chi_m = \chi \cdot M$ in SI units.
M is the molecular weight of the substance. These values hold under standard conditions.

Element	χ_m	Element	χ_m
Ag	−19.5	Br$_2$	−56.4
Al	+16.5	Cd	−19.8
Am	+1000	Ca	+40.0
Ar	−19.6	C(diam.)	−5.9
As(α)	−5.5	C(graph.)	−6.0
As(β)	−23.7	Ce(β)	+2500
As(γ)	−230	Ce(γ)	+2270
Au	−28.0	Cs	+29.0
Ba	+20.6	Cl$_2$	−40.5
Be	−9.0	Cr	+180
Bi	−280.1	Cu	−5.46
B	−6.7		

(continued)

18.4/1   Magnetic susceptibility of elements (*continued*)

Element	$\chi_m 10^{-9}$	Element	$\chi_m 10^{-9}$
Dy	+98 000	P(black)	−26.6
Er	+48 000	Pr	+5530
Eu	+30 900	Pt	+201.9
Gd	+185 000	Pu	+610.0
Ga	−21.6	Re	+67.6
Ge	−76.84	Rb	+17.0
Hf	+75.0	Rh	+111.0
He	−1.88	Ru	+43.2
Hg	−33.44	Sb	−99.0
Ho	+72 900	Se	−25.0
H_2	−3.98	Sc	+315
In	−107.0	Si	−3.9
I_2	−88.7	Sm	+1860.0
Ir	25.6	Sn(white)	+3.1
K	+20.8	Sn(grey)	−37.0
Kr	−28.8	Sr	+92.0
La	95.9	$S(\alpha)$	−14.9
Pb	−23.0	$S(\beta)$	−15.4
Li	+14.2	Ta	+154.0
Lu	> 0.0	Tc	+270.0
Mg	13.1	Te	−39.5
$Mn(\alpha)$	+529.0	Tb	+170 000
$Mn(\beta)$	+483.0	Tl	−50.9
Mo	+89.0	Th	+132
Na	+16.0	Tm	+24 700
Nd	+5930	Ti	+153.0
Ne	−6.74	W	+59.0
Nb	+195	U	+409.0
N_2	−12.0	V	+255.0
Os	+9.9	Xe	−43.9
O_2	+3449.0	Yb	+67
O_3	+6.7	Y	+187.7
Pd	+567.4	Zn	−11.4
P(red)	−20.8	Zr	−122.0

18.4/2   Magnetic susceptibility of inorganic compounds

Compound	$\chi_m\ 10^{-9}$	Compound	$\chi_m\ 10^{-9}$
Al_2O_3	-37.0	$CdBr_2$	-87.3
$Al_2(SO_4)_3$	-93.0	$CdCO_3$	-46.7
NH_3	-18.0	$CdCl_2$	-68.7
$NH_4C_2H_3O_2$	-41.1	$CdCrO_4$	-16.8
$(NH_4)_2SO_4$	-67.0	CdF_2	-40.6
$BaCO_3$	-58.9	CdO	-30.0
$Ba(BrO_3)_2$	-105.8	CdS	-50.0
BaO	-29.1	$CaCO_3$	-38.2
BaO_2	-40.6	$CaCl_2$	-54.7
$BeCl_2$	-26.5	CaF_2	-28.0
$Be(OH)_2$	-23.1	$Ca(OH)_2$	-22.0
BeO	-11.9	CaO	-15.0
Bi_2O_3	-83.0	CaO_2	-23.8

Compound	$\chi_m\ 10^{-9}$	Compound	$\chi_m\ 10^{-9}$
$CsBr$	-67.2	$CuCl$	-40.0
$CsBrO_2$	-75.1	$CuCl_2$	$+1080$
Cs_2CO_3	-103.6	Cu_2O	-20.0
CsO_2	$+1534.0$	CuO	$+238.9$
Cs_2S	-104.0	Cu_3P	-33.0
$Cr(C_2H_3O_2)_3$	$+5104$	CuP_2	-35.0
$CrCl_2$	$+7230$	$CuSO_4$	$+1330$
$CrCl_3$	$+6890$	Dy_2O_3	$+89\ 600$
Cr_2O_3	$+1960$	$Dy_2(SO_4)_3$	$+91\ 400$
CrO_3	$+40.0$	Dy_2S_3	$+95\ 200$
$Cr_2(SO_4)_3$	$+11\ 800$	Er_2O_3	$+73\ 920$
$Co(C_2H_3O_2)_2$	$+11\ 000$	Er_2S_3	$+77\ 200$
$CoBr_2$	$+13\ 000$	Eu_2O_3	$+10\ 100$

Compound	$\chi_m\ 10^{-9}$	Compound	$\chi_m\ 10^{-9}$
$BiCl_3$	-26.5	CO_2	-21.0
$Bi_2(CrO_4)_2$	$+154.0$	CO	-9.8
$Bi_2(SO_4)_3$	-199.0	$CeCl_3$	$+2490$
$BiPO_4$	-77.0	CeO_2	$+26.0$
$GaCl_3$	-63.0	$MgCl_2$	-47.4
Ga_2O	-34.0	MgO	-10.2
Ga_2S	-36.0	$MgSO_4$	-50.0
GaS	-23.0	$MnBr_2$	$+13\ 900$
Ga_2S_3	-80.0	$MnCO_3$	$+11\ 400$
$GeCl_4$	-72.0	MnO	$+4850$
GeO	-28.8	Mn_2O_3	$+14\ 100$
GeO_2	-34.3	Mn_3O_4	$+12\ 400$
GeS	-40.9	$MnSO_4$	$+13\ 660$
GeS_2	-53.3	Hg_2O	-76.3
$AuCl_3$	-112.0	Hg_2SO_4	-123.0
AuF_3	$+74.0$	$MoBr_3$	$+525.3$

(continued)

18.4/2   Magnetic susceptibility of inorganic compounds (*continued*)

Compound	$\chi_m\ 10^{-9}$	Compound	$\chi_m\ 10^{-9}$
AuP_3	-107.0	$MoBr_4$	$+520.0$
HfO_2	-23.0	Mo_3Br_6	-46.0
Ho_2O_3	$+88\ 100$	Mo_2O_3	-42.0
$Ho_2(SO_4)_3$	$91\ 700$	Mo_3O_8	$+42.0$
HCl	-22.6	Nd_2O_3	$+10\ 200$
$InBr_3$	-107.0	$Nd_2(SO_4)_3$	$+9990$
In_2O	-47.0	$NiCl_2$	$+6145.0$
In_2O_3	-56.0	NiO	$+660.0$
In_2S	-50.0	$NiSO_4$	$+4005.0$
InS	-28.0	NiS	$+190.0$
In_2S_3	-98.0	N_2O	-18.9
$IrCl_3$	-14.4	NO	$+1460$
IrO_2	$+224.0$	$OsCl_2$	$+41.3$
$FeBr_2$	$+13\ 600$	$PdCl_2$	-38.0
$FeCO_3$	$+11\ 300$	PdH	$+1077$
$FeCl_2$	$+14\ 750$	Pd_4H	$+2353$
FeO	$+7200$	Pt_2O_3	-37.70
$FePO_4$	$+11\ 500$	PuF_4	$+1760.0$
$FeSO_4$	$+10\ 200$	PuF_6	$+173.0$
La_2O_3	-78.0	PuO_2	$+730.0$
$Pb(C_2H_3O_2)_2$	-89.1	K_2CO_3	-59.0
$PbCO_3$	-61.2	KCl	-39.0
$PbCl_2$	-73.8	$K_3Fe(CN)_6$	$+2290.0$
PbO	-42.0	$K_4Fe(CN)_6$	-130.0
PbS	-84.0	KO_2	$+3230.0$
$LiC_2H_3O_2$	-34.0	KO_3	$+1185$
Li_2CO_3	-61.2	K_4MnO_4	$+20.0$
LiH	-10.1	PrO_2	$+1930.0$
$MgBr_2$	-72.0	ReO_2	$+44.0$
$MgCO_3$	-32.4	ReO_3	$+16.0$
Compound	$\chi_m\ 10^{-9}$	**Compound**	$\chi_m\ 10^{-9}$
$CoCl_2$	$+12\ 660$	$EuSO_4$	$+25\ 730$
Co_2O_3	$+4560$	EuS	$+23\ 800$
Co_3O_4	$+7380$	Gd_2O_3	$+53\ 200$
$Co_3(PO_4)_2$	$28\ 110$	Gd_2S_3	$+55\ 500$
$RbBr$	-56.4	Tl_3PO_4	-145.2
Rb_2CO_3	-75.0	Tl_2SO_4	-112.6
$RbCl$	-76.0	$Th(NO_3)_4$	-108.0
RbO_2	$+1527.0$	ThO_2	-16.0
Rb_2SO_4	-88.4	Tm_2O_3	$+51\ 444$
$RuCl_3$	$+1998.0$	$SnCl_4$	-115.0
RuO_2	$+162.0$	SnO	-19.0
Sm_2O_3	$+1988.0$	SnO_2	-41.0
Se_2Br_2	-113.0	TiC	$+8.0$
Se_2Cl_2	-94.0	$TiCl_2$	$+570.0$

(*continued*)

18.4/2   Magnetic susceptibility of inorganic compounds (*continued*)

Compound	$\chi_m\ 10^{-9}$	Compound	$\chi_m\ 10^{-9}$
SeO_2	−29.6	$TiCl_3$	+1110.0
SiC	−12.8	$TiCl_4$	−54.0
SiO_2	−29.6	Ti_2O_3	+125.6
AgBr	−59.7	TiS	+432.0
Ag_2CO_3	−80.9	WC	+10.0
AgCl	−49.0	WO_2	+57.0
Ag_2O	−134.0	WO_3	−15.8
$AgMnO_4$	−63.0	UF_4	+3530.0
Ag_3PO_4	−120.0	UF_6	+43.0
NaBr	−41.0	UO	+1600.0
Na_2CO_3	−41.0	UO_2	+2360.0
NaCl	−30.3	UO_3	+128.0
NAOH	−16.0	VCl_2	+2410.0
Na_2O	−14.5	VCl_3	+3030.0
Na_2O_2	−28.1	VO_2	+270.0
Na_2HPO_4	−56.6	V_2O_3	+1976.0
Na_2SO_4	−52.0	V_2O_5	+128.0
SrBr	−86.6	VS	+600.0
$SrCO_3$	−47.0	H_2O	−12.97
$SrCl_2$	−63.0	H_2O(ice)	−12.65
SrO	−35.0	D_2O	−12.76
SrO_2	32.3	D_2O(ice)	−12.54
$SrSO_4$	−15.5	Yb_2S_3	+18 300
SO_2	−39.8	Y_2O_3	+44.4
H_2SO_4	−39.8	$ZnCO_3$	−34.0
Ta_2O_5	−32.0	$ZnCl_2$	−65.0
Tb_2O_3	+78 340	ZnO	−46.0
TlBr	−63.9	$ZnSO_4$	−45.0
Tl_2CO_3	−101.6	ZnS	−25.0
TlCl	−57.8	ZrC	−26.0
TlCN	−49.0	$Zr(NO_3)_4 \cdot 5H_2O$	−77.0
Tl_2O_3	+76.0	ZrO_2	−13.8

18.4/3   Magnetic alloys of technical relevance

Material	Composition without iron component	Remanence B_r/T	Coercitive force $H_c/(\mathrm{A/m})$	Relative permeability μ_r
Magnetically hard metals				
carbon steel	1 % C	1...2	4000	—
chromium steel	5.8 % Cr; 1.1 % C	0.992	5200	—
tungsten steel	6 % W	1.1	4800	—
cobalt steel	36 % Co; 4.8 % Cr	0.93	18160	—
vicalloy	3.5 % Mn; 1.1 % C; 30–40 % Co; 14 % V	0.97	24000	—
KS-magnetic steel	9 % W; 1.5–3 % Cr; 0.4–0.8 % C	1	19200	—
tromalite	25 % Ni; 13 % Al	0.4	60000	—
Magnetically soft metals				
E-iron (1× annealed)	—	1.08	30.4	14600
E-iron (2× annealed)	—	0.085	12	4900
E-iron	3.5 % Si; vacuum molten	0.3	7.68	19400
permalloy	78.5 % Ni; 3 % Mo	—	< 8	−100000
nicalloy	40 % Ni	1.4	24	10000
hyperm 50	50 % Ni	1.5	6.8	28000
mu-metal	76 % Ni; 5 % Cu; 2 % Co	0.8	5	100000

18.5   Ferromagnetic properties

The notation in the following tables is:

T_C Curie temperature

σ_S specific saturation magnetization referred to the mass unit at room temperature (20 °C)

σ_0 specific saturation magnetization, extrapolated to $T = 0$ K

n_B effective number of magnetons, defined by $n_\mathrm{B} = \dfrac{\sigma_0 M_0}{N_\mathrm{A}\mu_\mathrm{B}}$

(M_0 is the molecular weight, N_A the Avogadro constant and μ_B the Bohr magneton.)

18.5/1 Ferromagnetic elements

Z		$T_C/°C$	$\sigma_S/(10^{-7}\text{Tm}^3\text{kg}^{-1})$	$\sigma_0/(10^{-7}\text{Tm}^3\text{kg}^{-1})$	n_B
26	Fe	770	218.0	221.9	2.219
27	Co	1120	161	162.5	1.715
28	Ni	358	54.39	57.5	0.604
64	Gd	20	0	253.5	7.55
65	Tb	−50	0	173.5	9.24
66	Dy	−186	0	235	10.20
67	Ho	−253	0	290	10.34
68	Er	−253	—	—	8.0
69	Tm	−235	—	—	7.0

18.5/2 Binary iron alloys

Element	Conc. atom%	$T_C/°C$	$\sigma_S/(10^{-7}\text{Tm}^3\text{kg}^{-1})$	n_B/atom
Al	7.1	756	207	2.05
	19.7	664	164	1.74
	24.9	441	134	1.29
	26.0	494	149	1.40
Au	6.2	767	174	2.08
	10.5	768	154	202
Co	20	950	236	2.42
	33	970	238	2.52
	50	980	233	2.42
	75	870	203	2.14
	80	910	184	1.94
Cr	17.7	678	196	1.70
	47.5	483	90	0.98
	68.8	268	35	0.53
Ir	4.0	750	200	2.25
	15.0	—	120	1.67
Ni	10	750	217	2.26
	20	720	209	2.22
	40	330	152	1.82
	60	560	136	1.45
	80	560	98	1.04
Os	8.1	—	158	1.97
	12.5	—	50	0.69
Pd	5.5	754	203	2.19
	40.0	—	129	1.89
	74.8	≈ 250	45	0.97

(continued)

18.5/2   Binary iron alloys (*continued*)

Element	Conc. atom%	$T_C/°C$	$\sigma_S/(10^{-7}Tm^3kg^{-1})$	$n_B/atom$
Pt	8.1	—	191	2.36
	12.4	—	177	2.43
	24.8	164	104	2.23
	50.0	—	32	0.75
Rh	10.0	—	209	2.32
	25.0	714	192	2.39
	40	624	161	2.26
Ru	7.0	660	200	2.18
	12.5	—	105	1.17
Sn	2.3	768	208	2.18
	6.0	768	197	2.16
Si	8.3	720	204	2.00
	15.9	653	174	1.67
	23.5	587	141	1.32
V	5.9	815	204	2.09
	10.6	805	184	1.91

18.5/3   Binary nickel alloys

Element	Conc. atom%	$T_C/°C$	$\sigma_S/(10^{-7}Tm^3kg^{-1})$	$n_B/atom$
Al	2.0	293	47.1	0.54
Au	3.4	321	46.0	0.58
Cr	1.7	298	49.8(−123 °C)	0.53
	6.7	72	25.4(−123 °C)	0.30
Mo	1.9	266	42.3	0.51
	4.2	120	23.1	0.37
Mn	25[1]	470	90	1.02
Pd	12.1	330	—	0.60
	45.2	217	—	0.57
	91.3	−116	—	—
Pt	9.1	245	37.7	0.55
	25.0	86	16.4	0.44
	45.0	−71	—	0.25
Sb	7.5	23	12.6	0.24

[1] (amorphous)   (*continued*)

18.5/3   Binary nickel alloys (*continued*)

Element	Conc. atom%	$T_C/°C$	$\sigma_S/(10^{-7}Tm^3kg^{-1})$	n_B/atom
Si	3.7	234	40.3	0.48
	6.8	117	23.7	0.36
	8.8	19	—	0.28
Sn	2.7	234	401	0.49
	9.0	225	9.9	0.30
Ta	3.6	—	—	0.41
	6.3	—	—	0.28
Ti	4.8	207	34.5	0.43
	10.3	30	—	0.22
W	2.1	270	39.2	0.49
	3.9	150	19.9	0.34
Y	5.5	67	15.3	0.29
Zn	4.1	300	45.3	0.52
	10.8	157	25.4	0.37

18.5.1   *Magnetic anisotropy*

The magnetic anisotropy is determined by the magnetization work. It differs in the various crystallographic orientations. The axis of easiest magnetization is determined by the minimum of magnetization work. For the most important crystallographic systems, the magnetization work reads as follows.

a) Cubic crystals:

$$E_a = K_1(\alpha_1^2\alpha_2^2 + \alpha_2^2\alpha_3^2 + \alpha_3^2\alpha_1^2) + K_2\alpha_1^2\alpha_2^2\alpha_3^2 + K_3(\alpha_1^2\alpha_2^2 + \alpha_2^2\alpha_3^2 + \alpha_3^2\alpha_1^2)^2 + \cdots;$$

$\alpha_1, \alpha_2, \alpha_3$ are the direction cosines referred to the axes of the elementary cell.

b) Hexagonal crystals:

$$E_a = K_1 \sin^2\phi + K_2 \sin^4\phi + K_3 \sin^6\phi + K_4 \sin^6\phi \sin^6\psi + \cdots;$$

ϕ is the angle between the magnetization direction and the [001]-axis.
ψ is the angle between the magnetization axis and the c-axis.

c) Tetragonal crystals:

$$E_a = K_1 \sin^2\vartheta + K_2 \sin^4\vartheta + K_3 \cos^2\alpha \cos^2\beta + \cdots;$$

ϑ is the angle between the magnetization axis and the tetragonal [001]-axis.
α and β are the angles between the magnetization axis and the tetragonal axes [100] and [010], respectively.

The anisotropy coefficients are temperature-dependent.

18.5/4   Anisotropy coefficients K_1 and K_2 of Fe-Co, Fe-Ni and Fe-Co-Ni alloys

Composition			20 °C		200 °C		300 °C		380 °C	
Atom%			K_1	K_2	K_1	K_2	K_1	K_2	K_1	K_2
Fe	Co	Ni	10^2 J/m^3							
100			420	150	300	22				
70	30		102	160						
60	40		45	−110						
50	50		−68	−390						
30	70		−433	50						
50		50	33	−180	25	−82	18	−7		
35		65	15	−70	12	−40	10	−32		
30		70	7	−17	2	−4	0	0		
10		90	−7	−23	−2	−10	0	−8		
		100	−34	53	5	20				
	65	35	−258	150						
	40	60	−108	−40						
	20	80	−4	8						
	10	90	16	−40						
	3	97	−10	9						
50	10	40	61	−160	19	4			7	−60
25	25	50	4	16	4	2			−3	22
20	15	65	9	−110	−1	−18			−3	−2
15	25	60	−26	34	−10	−45			−3	−15
10	40	50	−72	−4	−54	41			−9	−102
10	30	60	−38	−80	−17	−50			−12	−37
10	20	70	−29	17	−25	70			−14	29
10	10	80	−2	−39	−2	−20			−2	6

18.5/5   Directions of easy, medium and difficult magnetization in cubic crystals

K_1 K_2	+ $-\frac{9}{4}K_1$ $\vdots$ $+\infty$	+ $-9K_1$ $\vdots$ $-\frac{9}{4}K_1$	+ $-\infty$ $\vdots$ $-9K_1$	− $-\infty$ $\vdots$ $\frac{9}{4}|K_1|$	− $\frac{9}{4}|K_1|$ $\vdots$ $9|K_1|$	− $9|K_1|$ $\vdots$ $+\infty$
easy	[100]	[100]	[111]	[111]	[110]	[110]
medium	[110]	[111]	[100]	[110]	[111]	[100]
difficult	[111]	[110]	[110]	[100]	[100]	[111]

18.6   Ferrites

18.6/1   Magnetic properties of some ferrites with spinel structure

Parameter	Fe_3O_4	$MgFe_2O_4$	$MnFe_2O_4$	$CuFe_2O_4$
x-ray density $/(g/cm^3)$	5.24	4.52	5.0	5.25
Curie temperature $/°C$	585	440	300	455
magnetic moment/molecule $/(\mu_B)$	4.1	1.1	4.6	2.3 (cub.)
				1.3 (tetrag.)
spec. saturation magnetization $/(10^{-7} Tm^3/kg)$	92	27	80	25
anisotropy constant K_1 $/(10^2 J/m^3)$	-10.7	-2.5	-2.8	-6.3
anisotropy constant K_2 $/(10^2 J/m^3)$	-2.8	—	-0.2	—

	$CoFe_2O_4$	$NiFe_2O_4$	$Li_{0.5}Fe_{2.5}O_4$	
x-ray density $/(g/cm^3)$	5.29	5.37	4.75	
Curie temperature $/°C$	520	585	670	
magnetic moment/molecule $/(\mu_B)$	3.94	2.3	2.6	
spec. saturation magnetization $/(10^{-7} Tm^3/kg)$	80	50	65	
anisotropy constant K_1 $/(10^2 J/m^3)$	290	-6.2	-8.4	
anisotropy constant K_2 $/(10^2 J/m^3)$	—	-3	-0.2	

18.7   Antiferromagnets

18.7/1   Properties of some antiferromagnets

The following table lists the Néel temperature of the phase transition, the temperature T_C of the Curie-Weiss law and the molar magnetic susceptibility of several antiferromagnetic compounds.

Substance	T_N /K	T_C /K	$\chi_M\ 10^{-3}$ $/(cm^3/mol)$	Substance	T_N /K	T_C /K	$\chi_M\ 10^{-3}$ $/(cm^3/mol)$
Ti_2O_3	248	2000	0.24	$\alpha - VSe$	163	2570	0.62
VO_2	343	13.60	0.66	$ZnCr_2Se_4$	22	115	340
MnO	120	610	6	$MnSe$	247	740	19
MnS_2	48;20	592	7.1	$MnTe_2$	≈ 80	528	6.8
MnF_2	72	113	25	$MnAu_3$	145	-200	77.5
$MnCO_3$	32	64.5	43	$KMnF_3$	88.3	238	17.7
$RbMnF_3$	54	190	17.7	Mn_2SiO_4	50	163	18.8
FeO	198;186	190	8	FeF_2	78	15.9	117
$FeCl_2$	23	-48	320	FeI_2	10	23	85
FeP_2	250	17	1.18	αVS	1040	3000	0.066
$FeTiO_3$	68;56	-17	61	$LaCrO_3$	295	600	1.9
$FeSO_4$	≈ 22	30.5	78.5	$\beta - MnS$	165;110	528	6
CoF_2	37.7	52.7	50	$MnSe_2$	75	483	6.6
$\beta CoSO_4$	12	52	62	MnF_3	47	-8	75
NiF_2	73.2	100	20	$LaMnO_3$	131;100	-40	48.4
$NiSO_4$	37	82	15	$MnUO_4$	12	8	200
GdP	15	2	480	FeS	≈ 597	917	2.2

(continued)

18.7/1   Properties of some antiferromagnets (*continued*)

Substance	T_N /K	T_C /K	$\chi_M \, 10^{-3}$ /(cm^3/mol)
FeBr$_2$	11	−6	160
FeSn$_2$	380	230	1.95
YFeO$_3$	643	—	2.2
Fe$_2$SiO$_4$	65	150	20.4
CoCl$_2$	24.9	−20	60
CoUO$_4$	12	52	62
NiCl$_2$	52	−67	110
CuSO$_4$	34.5	77.5	12
GdAg	145	82	40
FeCO$_3$	35;20	14	17
LaFeO$_3$	738	480	12
CoO	328;291	280	5.3
KCoF$_3$	114;109	125	8.5
Nb$_2$Co$_4$O$_9$	30;27	10	133
NaTiO$_3$	23	55	23.4
EuTe	11;9.7	7	440
GdIn	28	66	73.5

18.8   Ion mobility

18.8/1   Ion mobility μ in air at $18\,^\circ$C and standard pressure

Gas	μ in 10^{-2}m^2/Vs	
	positive ions	negative ions
hydrogen	5.7	8.6
helium	5.1	6.3
argon	1.37	1.7
oxygen	1.33	1.8
nitrogen	1.29	1.82
ethyne	0.71	0.86
benzene	0.18	0.21

Part IV
Thermodynamics

19
Equilibrium and state variables

Thermodynamics describes macroscopic properties of matter in terms of appropriate physical quantities, and establishes universal relations between these quantities.

19.1 Systems, phases and equilibrium

19.1.1 Systems

Thermodynamic system, an arbitrary assembly of matter with properties that can be described uniquely and completely in terms of specific **state variables** (volume, energy, particle number, ...).

➤ In general, this matter is separated from an outside environment by walls. Other modes of confinement are also possible, e.g., the confinement of hot plasmas in strong magnetic fields.

19.1.1.1 Isolated systems

Isolated system, a system having no interaction with the environment. The container (walls) is impenetrable by any kind of energy and matter (**Fig. 19.1 (a)**).

➤ This cannot be realized entirely: any wall, for example, is heat-conducting. The magnetic plasma confinement in a vacuum also allows heat transport by radiation.

▲ In an isolated system, the total energy E (mechanical, electric, ...) is constant.

➤ Energy and particle number are conserved quantities: **microcanonical ensemble**.

Besides the energy, the particle number N and the volume V are quantities specifying an isolated system.

Dewar flasks, double-walled mirrored vessels with an intermediate vacuum layer that approach the requirements for containers of isolated systems.

■ Thermos bottles are constructed according to this principle.

➤ For experiments at low temperatures, several containers nested within each other may serve for keeping liquid coolant.

19.1.1.2   Closed systems

Closed system, a system that may exchange only energy with the environment, but cannot exchange matter (**Fig. 19.1 (b)**).

➤   Energy is not a conserved quantity, but the particle number is conserved: **canonical ensemble**.

The actual energy of the system fluctuates due to the exchange of energy with the environment. But in the equilibrium of the closed system with its environment a certain average value of the energy is established which may be related to a temperature of the system or the environment.

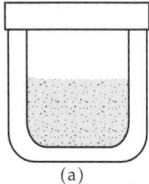
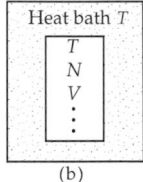

Figure 19.1: Thermodynamic systems. (a): closed system, (b): closed system in a heat bath.

To specify the macroscopic state, besides the particle number N and the volume V, one can use the **temperature**.

19.1.1.3   Open systems

Open system, may exchange energy as well as matter with its environment (**Fig. 19.2**). Neither energy nor particle number are conserved quantities.

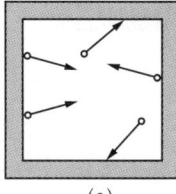
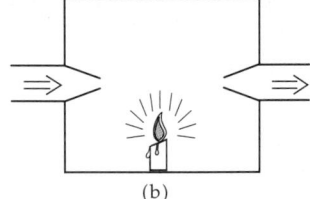

Figure 19.2: Open thermodynamic systems. (a): a particle reservoir, (b): a flowthrough system.

➤   If the open system is in equilibrium with its environment, then certain mean values of energy and particle number are established: **macrocanonical** (or **grand canonical**) ensemble.

Analogous to the relation between mean energy and temperature, the mean particle number may be related to a quantity denoted as the **chemical potential** μ.

Temperature T and **chemical potential** μ may be used to characterize an open system.

19.1.2   Phases

1. Homogeneous and heterogeneous systems

Homogeneous system, a system with the same properties throughout.

■   A container with (dry) air under standard conditions is a homogeneous system.

Heterogeneous system, a system whose properties may change discontinuously at certain interfaces.

■   A vessel containing water, water vapor and air is a heterogeneous system.

2. Phases and phase interfaces

Phase, a homogeneous part of a heterogeneous system.
  Phase interface, the boundary surface separating two phases.
■   For a closed pot with water, water vapor and air the surface of the water is a phase interface. There is a gaseous phase (vapor and air) and a liquid phase (water).
➤   In some cases, the macroscopic properties of the system depend on the size (and the shape) of the phase interfaces.
■   Pot with water, water vapor, and air.
      The system exhibits distinct macroscopic properties depending on whether the water is condensed at the bottom or distributed as droplets (fog).

3. Interface tension

tension arising at the interface between two phases that tends to reduce the interface area. It originates in different intermolecular interactions at the interface and in the interior of a phase. The interface tension of liquids against the gaseous phase is denoted surface tension.

4. Random surfaces

interfaces of two-phase systems, with very low or vanishing interface tension, that strongly fluctuate in shape. The behavior of random surfaces is determined by the elastic bending energy and the shear stiffness of the material.
➤   The statistics of random surfaces is significant for the thermodynamic description of micro-emulsions and of the thermal motion of cell membranes.

19.1.3 Equilibrium

1. Equilibrium state

the macroscopic state of an isolated system that evolves by itself after a sufficiently long waiting time (**Fig. 19.3 (a)**).
▲   In equilibrium, the macroscopic state variables no longer vary with time.
➤   Thermodynamic state variables can be defined and measured only in equilibrium.
➤   Frequently, it is meaningful to speak of thermodynamic equilibrium even if the thermodynamic state variables vary quite slowly.
■   The Sun continually loses energy by radiation and, therefore, is not in equilibrium. Nevertheless the use of thermodynamic state variables makes sense since the changes proceed very slowly.
  Global equilibrium, requires that the thermodynamic state variables not vary in time for all phases of the system.
  Local thermal equilibrium, a system that is not in global equilibrium, but in partial volumes behaves like an equilibrium system. In this case, the intensive variables are defined only locally.
■   Stars whose different zones are at distinct temperatures;
      Earth's atmosphere with different weather zones.

2. Steady state,

a state in which the macroscopic thermodynamic properties do not vary in time but an energy flow occurs (**Fig. 19.3 (b)**). A steady state system is not closed, but energy flows in and out. This is not the case for equilibrium states.

■   A pot placed upon an electric heating plate. After some time a steady state is reached in which the temperature of the food no longer changes. But energy must be supplied continuously, in order to prevent cooling of the pot, which continues to deliver energy (heat) to the environment.

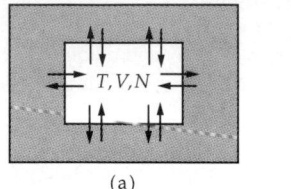   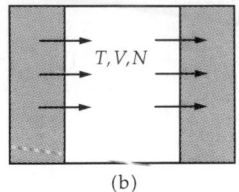

(a)   (b)

Figure 19.3: Thermodynamic systems. (a): system in equilibrium, (b): steady state.

3. *Thermal equilibrium,*

is established between two subsystems of an isolated system if they are brought in contact energetically (without exchange of matter) and a sufficiently long period elapses such that the net energy exchange ceases. The thermodynamic properties change until equilibrium is reached, and the temperatures in the two subsystems are equal.

4. *Zeroth law of thermodynamics,*

an empirical theorem on thermal equilibrium: All systems which are in thermal equilibrium with a given third system (the thermometer) are also in mutual thermal equilibrium.
➤   This law is the basis for defining the concept of temperature.

5. *Mechanical equilibrium,*

arises for systems with fixed boundaries if the forces of both systems acting on the boundaries are of equal magnitude. Hence, the pressures of both systems are equal,

$$p_1 = p_2 .$$

If the systems are not in mechanical equilibrium, then the volumes of the systems change until a balance of pressure is reached.

6. *Chemical equilibrium,*

for systems with variable particle number: the number of particles entering the system equals the number of particles leaving the system.

As for thermal equilibrium, the chemical equilibrium must be distinguished conceptually from a steady state, e.g., in a system with particle flow.

In chemical equilibrium, the chemical potentials of the systems are equal,

$$\mu_1 = \mu_2 .$$

Frequently, the conditions for chemical and mechanical equilibrium are related due to the partial pressure.
■   If a system consisting of carbon dioxide and water is under pressure, carbon dioxide is dissolved in water until the vapor pressure of the dissolved carbon dioxide is equal to the pressure of the gaseous carbon dioxide. While the particle numbers equalize, there is also an equalization of pressure.

19.1.3.1   Conditions for equilibrium

Different types of equilibrium result from different special conditions:

isolated isochoric equilibrium states ⇔ **maximum of entropy** S,

isothermal-isobaric equilibrium states ⇔ **minimum of free enthalpy** G,

isothermal-isochoric equilibrium states ⇔ **minimum of free energy** F,

adiabatic-isobaric equilibrium states ⇔ **minimum of enthalpy** H.

thermodynamic potentials			$\mathbf{ML^2T^{-2}}$
	Symbol	Unit	Quantity
$U(S, V, N) = TS - pV + \mu N$	U	J	internal energy
	F	J	free energy
	H	J	enthalpy
$F(T, V, N) = U - TS$	G	J	free enthalpy
	p	Pa	pressure
$H(S, p, N) = U + pV$	V	m^3	volume
	T	K	temperature
$G(T, p, N) = U + pV - TS$	S	J/K	entropy
	μ	J	chemical potential
	N	1	particle number

19.2   State variables

19.2.1   State property definitions

1. State property,

a physical quantity that specifies a macroscopic property as uniquely as possible.

■   Temperature, pressure, chemical potential, charge, dipole moment, refractive index, viscosity, chemical composition, size of phase interfaces, etc.

Microscopic properties, such as the positions or the momenta of the particles, are not thermodynamic state properties.

▲   Thermodynamic properties may be defined and measured only in equilibrium.

2. Equation of state,

a functional law connecting various thermodynamic state properties.

In thermodynamics equations of state must be determined empirically. Often one uses polynomials in the state variables; the virial coefficients of the variables then must be determined experimentally. Such empirically determined equations of state generally agree with experimental findings only within a very restricted range of values of the state variables.

■   The equation of state of an ideal gas (see p. 650) can give reliable results for real gases only at very low density. For higher densities, modified relations, such as the Van der Waals equation or the virial expansion, are used.

3. State variable,

a thermodynamic property of a system that may vary in time.

▲   In order to fix a thermodynamic state uniquely, only the state variables are needed. The remaining thermodynamic properties then take values that depend on the selected state variables.

➤   The number of the required state variables is related to the number of phases (see p. 734) of a system.

Generally, one distinguishes two categories of state properties: extensive and intensive quantities.

19.2.1.1   Extensive thermodynamic properties

Extensive thermodynamic properties, quantity proportional to the quantity of material in a system.

■   Volume, total energy, total mass are extensive thermodynamic properties.

▲   If the quantity of material is multiplied, then all extensive quantities are multiplied.

A thermodynamic property is also an extensive one if it is proportional to all other properties known to be extensive. The proportionality holds only as far as all non-extensive properties remain constant.

Heterogeneous total systems: the extensive properties of the total system are composed additively from the corresponding properties of the individual phases.

■   The volume of a pot of water, vapor and air is obtained from the volumes of the liquid phase, and the gaseous phase.

19.2.1.2   Intensive thermodynamic properties

Intensive thermodynamic property, property independent of the quantity of material and not additive for the various phases of the system. Intensive thermodynamic properties may take different values for the various phases, but not necessarily.

■   Density, pressure, temperature, refractive index are intensive properties.

Products of two intensive quantities are again intensive quantities. Quotients of two extensive quantities are intensive quantities.

■   The density is the quotient of the total mass and the volume.

The product of an extensive property and an intensive property is an extensive property.

■   The total charge is the product of the charge density (intensive) and the volume (extensive).

Intensive properties may be defined locally, i.e., they may vary in space.

■   The density of Earth's atmosphere decreases continuously with the height above its surface.

The water pressure in the ocean increases with depth.

The determination of the spatial dependence of intensive state variables either requires additional conditional equations (e.g., from hydrodynamics), or must be added in terms of additional equations of state.

19.2.1.3   Specific and molar properties

1. Specific quantity,

an intensive property of state, g, defined by the quotient of an extensive property G and the mass m,

$$g = G/m.$$

■   The specific heat q is the amount of heat per kilogram.

➤   In many textbooks on chemistry and physics, the concept of a specific quantity means the quotient of the property of state and the number of moles. This definition corresponds to the definition of a molar property quoted below.

▲   In technology, specific quantities are denoted by lowercase letters.

Most of the extensive quantities are specified by capital letters so that the corresponding specific quantity is characterized by the corresponding lowercase letter.

Extensive quantity		Specific quantity	
quantity of heat	Q	specific heat	q
heat capacity	C	specific-heat capacity	c
entropy	S	specific entropy	s
volume	V	specific volume	v
enthalpy	H	specific enthalpy	h

2. Molar quantity,

a quantity of state G_{mol} defined by the quotient of an extensive quantity G and the number of moles n,

$$G_{mol} = G/n .$$

■   The molar heat capacity c_{mol} is the heat capacity per mole.

In this book, molar quantities are specified by mol.

Relation between molar and specific quantities:

$$G_{mol} = g \cdot \frac{m}{n} = g \cdot M, \quad M = \frac{m}{n} : \quad \text{molar mass.}$$

➤   In technical textbooks, a subscript m or M is frequently used for molar quantities.

19.2.2 Temperature

Temperature, T, SI unit K (kelvin), a common intensive property of systems that are in mutual thermal equilibrium. Systems not in mutual thermal equilibrium may have different temperatures.

The temperature is related to the mean kinetic energy available for the individual particles.

■   In gases, the mean velocity of gas particles is directly related to the temperature. In solids the amplitude of oscillations of the particles about their lattice sites depends on the temperature.

➤   The oscillations of electrons cause, e.g., **thermal noise** and restrict the efficiency of sensitive measuring devices.

➤   The concept of temperature may be extended to systems that are not in equilibrium as an entity. This is possible as far as the total system may be decomposed in partial systems to which a **local** (position-dependent) **temperature** may be assigned.

19.2.2.1   Temperature units

The **symbol of temperature** is T in physical use.

➤   In technology, the temperature measured in Kelvin according to ISO is specified by the symbol T, the temperature measured in Celsius by t or ϑ.

a) Kelvin, the physical unit of temperature. Symbol: 1 kelvin = 1 K.

▲   One kelvin is the fraction 1/273.16 of the temperature difference between the triple point of water and absolute zero $T_0 \overset{\text{def}}{=} 0$ K.

b) Celsius, symbol °C, more frequently used unit of temperature for common use. It is based on the melting point (0 °C) and the boiling point (100 °C) of water under standard pressure (1013.25 hPa).

The Celsius scale is shifted with respect to the Kelvin scale by 273.15 degrees.

➤   The triple point of water is at 0.01 °C.

conversion kelvin–celsius				Θ
$\vartheta/°C = T/K - 273.15$ $T/K = \vartheta/°C + 273.15$	Symbol	Unit	Quantity	
	ϑ	°C	temperature in celsius degrees	
	T	K	temperature in kelvin	

▲   **Temperature differences** are identical in the Celsius and Kelvin scales:

$$(\vartheta_1 - \vartheta_2)/°C = (T_1 - T_2)/K.$$

c) Réaumur, symbol °R, subdivides the temperature difference between the melting point and the boiling point of water (under standard pressure) into 80 units (T(melting point) = 0 °R, T(boiling point) = 80 °R):

$$\vartheta/°C = T/K - 273.15 = 1.25\, T/°R, \qquad T/°R = 0.8\, \vartheta/°C.$$

d) Fahrenheit, Symbol °F, still in use in some English-speaking countries, in particular in USA. The limiting points of a freezing mixture (0 °F ≈ −17.8 °C) and the temperature of human blood (100 °F ≈ 37.8 °C):

$$T/°F = \frac{9}{5}\, \vartheta/°C + 32, \qquad \vartheta/°C = \frac{5}{9}\, T/°F - 17,\bar{7},$$

$$T/°F = \frac{9}{5}\, T/K - 459.67, \qquad T/K = \frac{5}{9}\, T/°F + 255.37\bar{2}.$$

e) Rankine, Symbol R, a Fahrenheit scale with the zero point shifted to absolute zero, analogous to the Kelvin scale:

$$T/R = \frac{9}{5}\, T/K = \frac{9}{5}\, \vartheta/°C + 491.67, \qquad T/K = \frac{5}{9}\, T/R.$$

➤   In atomic and nuclear physics, the Boltzmann constant is often set to $k = 1$, and the temperature is given in **electron volts** eV. Then:

$$1\, \text{eV} = 11604\, \text{K} \cdot k, \qquad 1\, \text{K} = 8.617 \cdot 10^{-5}\, \text{eV}/k.$$

19.2.2.2   Calibration points

Calibration points of temperature, points to fix the temperature scale. They are defined by temperature-dependent properties of materials (triple point, boiling point or solidification point for definite pressure).

IPTS fixed points, the fixed points of the **International Practical Temperature Scale** (IPTS-90) passed by the General Conference for Measures and Weights. They are listed in **Tab. 19.1**.

The boiling and solidification points refer to the standard pressure of 1013.25 hPa (except for the boiling point of hydrogen, marked by *).

Other characteristic temperatures that may be used as calibration points can be found in **Tab. 22.1/1**.

Standard temperature, fixing the temperature to

$$T_n = 273.15 \text{ K} = 0 \,^\circ\text{C}.$$

Standard conditions, fixing the temperature to the standard temperature and the pressure to the standard pressure 1013.25 hPa,

$$T_n = 273.15 \text{ K} = 0 \,^\circ\text{C}, \quad p_n = 1013.25 \text{ hPa} = 1.01325 \text{ bar}.$$

Table 19.1: IPTS-90 fixed points.

fixed point	substance	T/K	$\vartheta/^\circ C$
triple point	hydrogen	13.81	−259.34
boiling point*	hydrogen	17.042	−256.11
boiling point	hydrogen	20.28	−252.87
boiling point	neon	27.10	−246.05
triple point	oxygen	54.36	−218.79
boiling point	oxygen	90.19	−182.96
triple point	water	273.16	0.01
boiling point	water	373.15	100.00
solidification point	zinc	692.73	419.58
solidification point	silver	1235.08	961.93
solidification point	gold	1337.58	1064.43

* at a pressure of 333.306 hPa

19.2.2.3   Measurement of temperature,

1. Methods of measuring temperature,

are based on bringing a system whose thermal equilibrium state is connected uniquely to an easily observable thermodynamic variable, into thermal equilibrium with the system to be measured.

Thermometer, apparatus with which a property correlated with temperatures can be measured.

➤ The method of measuring the temperature is connected with an equation of state, namely the dependence of the observed state variable on the temperature.

Possible properties observed:

- the volume of a liquid (**liquid thermometer, Fig. 19.4 (a)**),
- the volume of a gas (**gas thermometer**),
- distinct extension of two metallic strips (**bimetal, Fig. 19.4 (c)**),
- extension of ceramic rods, e.g., control rods in muffle furnaces.
- deformation of ceramic cones in metallurgy (**Seger cone**),
- in the millikelvin range: the alignment of the nuclear spins of ^{60}Co in the monocrystal, and thus the anisotropy of gamma radiation,

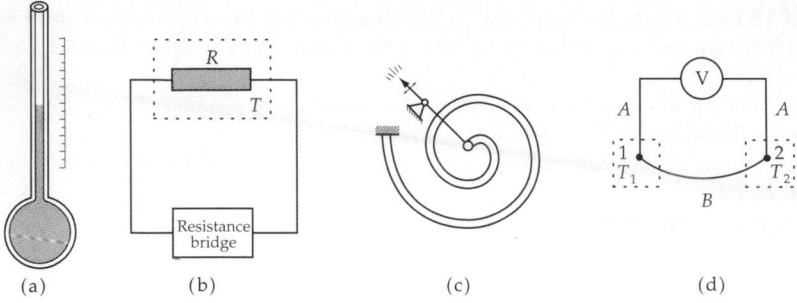

(a)   (b)   (c)   (d)

Figure 19.4: Schematic representation of various types of thermometers. (a): liquid thermometer (volume change of liquid), (b): resistance thermometer (temperature-dependent conductance), (c): bimetals (distinct linear expansion of metals), (d): thermocouples (distinct voltages at the contact points).

- the voltage at the junction of a pair of wires made from different metals (**thermocouple Fig. 19.4 (d)**),
- the color of light emitted by a solid or a gas (**pyrometer**),
- the resistance of certain conductors (**resistance thermometer Fig. 19.4 (b)**) with positive temperature coefficient (PTC) or negative temperature coefficient (NTC), see chapter on electricity, temperature dependence of resistance.

2. Operation range of thermometers

The schematic representation in **Fig. 19.5** shows the operation ranges (temperature along the abscissa) of various thermometers, arranged along the ordinate according to the function principle:

a) mechanical contact thermometer,
b) special forms of mechanical contact thermometer,
c) electric contact thermometer,
d) special forms of electric contact thermometer,
e) radiation thermometer.

3. Calibration of thermometers

For the calibration of thermometers in between the fixed points, the following devices are assigned:

Platinum resistance thermometer with special specifications for the temperature range 13.81 to 903.89 K.

The range is subdivided into five subranges for which special interpolation polynomials are used to calculate the temperature from the values of resistance.

Rhodium/platinum thermocouple with a platinum and a rhodium(10 %)-platinum compound as thermocouple in the temperature range 903.89 to 1337.58 K.

The relation between the temperature and the thermovoltage is interpolated by means of a quadratic equation.

Spectral pyrometer above 1337.58 K. Here, Planck's radiation law is used.

19.2.2.4   Kelvin scale and absolute zero

Rarefied gases show a very similar connection between temperature and volume expansion. The volume of a definite quantity of such a gas at definite pressure may be used as a measure of temperature, and other thermometers may be calibrated correspondingly.

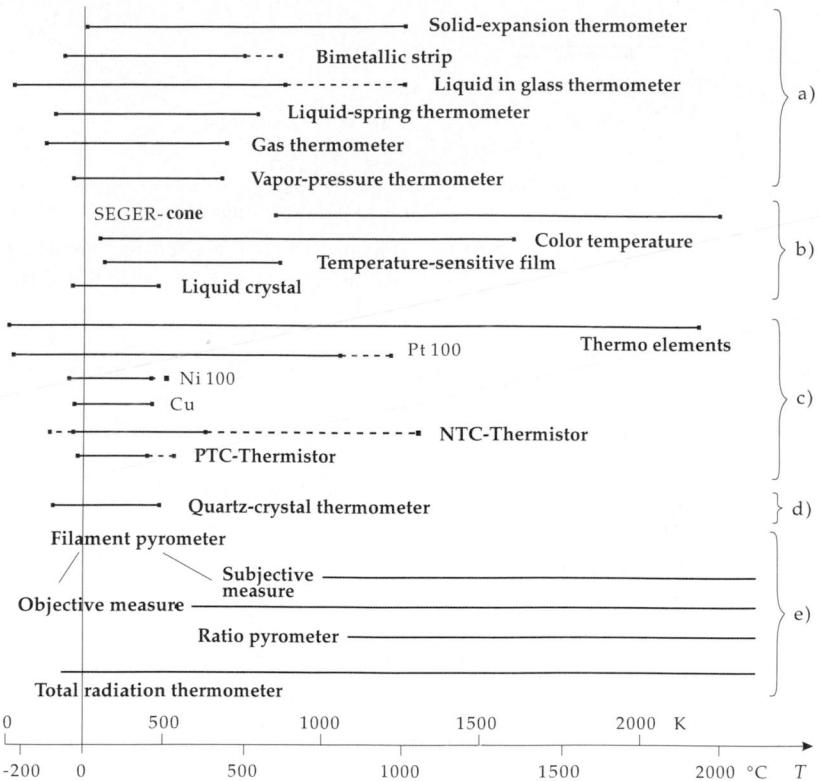

Figure 19.5: Operation ranges of thermometers.

1. Thermodynamic temperature,

T, determined through the volume of a rarefied gas (see p. 661) (**Fig. 19.6**),

$$T = T_0 \frac{V}{V_0} \, .$$

Pressure and particle number must remain constant.

2. Kelvin scale,

the temperature scale for which the triple point of water serves as fixed point. The pressure at the triple point is 619.6 Pa, and the temperature is defined as 273.16 K. The subdivision into degrees closely follows the **Celsius scale**, which was established earlier.

Conversion between the Kelvin and Celsius scales:

$$T/\text{K} = \vartheta/^\circ\text{C} + 273.15 \, .$$

3. Absolute zero,

extrapolation of the temperature-volume relation to the volume $V = 0$ (**Fig. 19.7**). The assumption of a gas whose volume may be diminished arbitrarily is of importance for the discussion of the ideal gas. In practice, at very low temperatures the volume of a gas can no longer be measured because of nascent liquefaction.

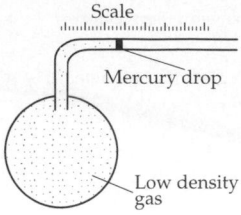

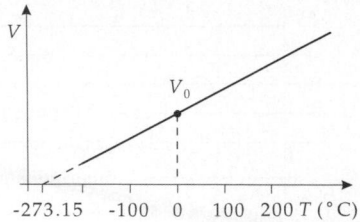

Figure 19.6: Gas thermometer (schemati-
cal).

Figure 19.7: V-T diagram of a rarefied gas.
Air liquefies at 80 K, H_2 at 20 K and He at
4.2 K.

At absolute zero, all motion of atoms and molecules ceases. The value of the temperature
is $T = 0\,\text{K} = -273.15\,°\text{C}$.

▲   Absolute zero cannot be attained. It is impossible to establish a system with exactly
$T = 0\,\text{K}$.

➤   This is a formulation of the third law of thermodynamics (see p. 702).

19.2.3 Pressure

Pressure, p, SI unit Pa (pascal), the magnitude of a force acting perpendicularly on a
measuring area A divided by the area (see p. 172):

pressure = $\dfrac{\textbf{normal component of force}}{\textbf{area}}$			$\textbf{ML}^{-1}\textbf{T}^{-2}$
	Symbol	Unit	Quantity
$p = \dfrac{F_\perp}{A}$	p	Pa	pressure
	$F_\perp$	N	normal force component
	A	m^2	area

Strictly speaking, the pressure is the component of the force vector $\vec{F}$ normal to the surface,
i.e., the scalar product of the force vector $\vec{F}$ and the normal vector $\vec{n}_A$ of the surface A,
divided by the area,

$$p = \frac{\vec{F} \cdot \vec{n}_A}{A}.$$

Microscopically, the pressure occurs because particles strike the surface, where they are
reflected and thereby transfer a certain momentum. The pressure is the average momentum
transferred to the wall per unit time and unit area.

Macroscopically, the pressure is related linearly to the density and thus inversely proportional to the volume occupied (see p. 651).

| M | **McLeod pressure gauge**, a mercury manometer for measuring low gas pressures
(**Fig. 19.8**) that operates according to the principle of volume measurement. A small
amount of gas of the system to be measured is confined. Its volume is then diminished
and the pressure difference between the reduced volume and the original system is
measured.

To the vacuum
system

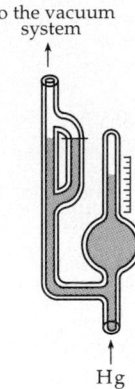

Hg

Figure 19.8: McLeod
pressure gauge.

19.2.3.1   Units of pressure

1. SI units of pressure

Pascal, abbreviation Pa, SI unit of pressure (see p. 172):

$$1\,\mathrm{Pa} = 1\,\frac{\mathrm{N}}{\mathrm{m}^2} = 1\,\frac{\mathrm{kg}}{\mathrm{m\,s}^2}\,.$$

In practice, pressures frequently are of the order 10^5 Pa (about normal air pressure); therefore, the more convenient unit bar is introduced.

Bar, 10^5 Pa.

➤   Formerly, the millibar was used frequently in meteorology. Currently, the identical SI unit hectopascal is used:

$$1\,\mathrm{Pa} = 10^{-5}\,\mathrm{bar}\,,\qquad 1\,\mathrm{bar} = 10^5\,\mathrm{Pa} = 10\,\frac{\mathrm{N}}{\mathrm{cm}^2}\,.$$

2. Pressure and energy density

The pressure has the same dimension as the energy density.

▲   Often, the pressure is related to the energy density in a simple manner.

■   In an ideal gas, the pressure is directly proportional to the mean kinetic energy density $e = \rho_N W_{\mathrm{kin}}$, ρ_N being the particle density and W_{kin} the mean kinetic energy:

$$p = \frac{2}{3}e\,.$$

3. Other units of pressure

The following units are not in the SI-system, but are found in many older technical books and in everyday settings (e.g., psi).

Technical atmosphere, at, corresponds to the pressure exerted by a mass of 1 kg at standard gravity

$$g = 9.806\,65\,\mathrm{m/s}^2$$

onto a square centimeter.

➤ The now obsolete force unit kilopond (kp) represents the weight of a mass of 1 kg at standard gravity.

$$1 \text{ at} = 1 \frac{\text{kp}}{\text{cm}^2} = 1 \frac{\text{kg}}{\text{cm}^2} g = 98\,066.5 \text{ Pa} = 0.980\,665 \text{ bar}, \qquad 1 \text{ bar} = 1.02 \text{ at}.$$

➤ 1 at corresponds to the pressure of a water column of a height of 10 m.
Atmospheric excess pressure, ate, excess pressure in atmospheres:

$$p_{\text{ate}} = p/\text{at} - 1.$$

Millimeters of water, mm WS, specifies the height of a column of water with a gravity pressure equivalent to the given pressure:

$$1 \text{ mm WS} = 10^{-4} \text{ at} = 9.80665 \text{ Pa}.$$

Standard atmosphere, atm, adjusted to the mean air pressure at the earth's surface.
   Torr, the raise of mercury in mm in an evacuated closed vertical glass tube immersed in an open mercury vessel, equivalent to air pressure.

$$1 \text{ atm} = 760 \text{ Torr} = 101325 \text{ Pa}$$
$$1 \text{ Torr} = 133.32 \text{ Pa} \,\hat{=}\, 1 \text{ mm Hg}$$
$$1 \text{ bar} = 0.987 \text{ atm} = 750.06 \text{ Torr}$$

4. Standard pressure and standard conditions

Standard pressure, **norm pressure**, reference value of pressure for specifying material properties.
▲ The standard pressure is one standard atmosphere:

$$p_{\text{n}} = 101\,325 \text{ Pa} = 1 \text{ atm} = 760 \text{ Torr}.$$

◼ In general, melting and boiling points are given for standard pressure.
Standard conditions, fixing the temperature to the standard temperature ($T = 273.15 \text{ K} = 0 °\text{C}$) and the pressure to the standard pressure $p_{\text{n}} = 1013.25$ hPa.

19.2.3.2 Measurement of pressure

1. Pressure gauges

Measurement of pressure is in general done by determining the force acting on a surface of known area (**Fig. 19.9**).
   Pressure balance and **piston manometer**, measure the force acting on a piston in a hollow cylinder. The counterforce is provided by weights or springs.
   Liquid manometers are used preferentially for measuring low pressures. Confining liquids are, for example, alcohol, water, mercury, or special liquids with possibly low vapor pressure, possibly temperature-independent density, and possibly favorable capillary properties.
   Thermovac tubes and **Penning tubes**, employ the **thermal** or **electric conductance** of gases for measuring pressure in the vacuum region.

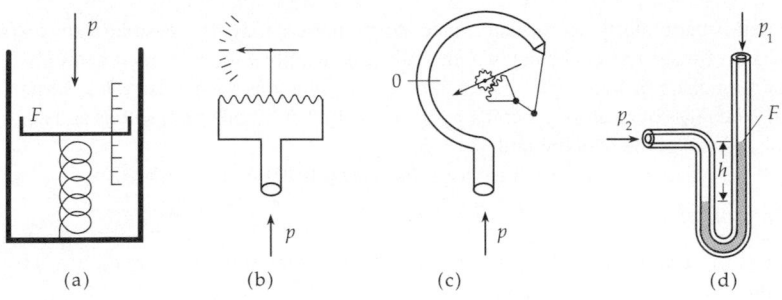

(a)   (b)   (c)   (d)

Figure 19.9: Schematic representation of the principle of operation of pressure gauges: the pressure acting on a fixed piston area is compensated by a counterforce. The magnitude of the counterforce may be determined by: the deflection of a Bourdon pressure gauge (a), the deflection of a pointer compensated by a spiral spring (b), the extension of a curled pressure hose (c) or (for known counterpressure) the height of ascent of a liquid column (d).

2. Local pressure

Pressure may also be defined locally, i.e., within a small subsystem.

> **M** For measurements of local pressure a small test surface is connected to the system. The net force F on the surface is measured, and the pressure is determined from the difference between F/A, where A is the area of the surface, and a calibrated pressure on the other side of the surface.

19.2.4 Particle number, amount of substance and Avogadro number

1. Particle number,

N, non-dimensional quantity, describes the number of particles present in the system.

➤ According to ISO, the symbol X may also be used, in particular if mixtures of different kinds of particles are being considered.

➤ Since N takes very large numbers for macroscopic systems, multiples of the Avogadro number are used.

2. Avogadro number,

Avogadro constant, N_A, used to determine the number of atoms or molecules in a sample of a substance.

▲ The Avogadro number is just the number of atoms or molecules in one mole of a substance.

$$N_A = 6.022\,136\,7 \cdot 10^{23} \text{ mol}^{-1}.$$

3. Atomic mass unit,

u (formerly amu), particularly suitable to specify the mass of individual particles (atoms, molecules) (see p. 912); it is defined as one twelfth of the mass of an atom of the carbon isotope ^{12}C:

$$1\,\text{u} = \frac{1}{12}m_{^{12}\text{C}} = 1.661 \cdot 10^{-27} \text{ kg}.$$

This unit is particularly appropriate, since atomic masses may be measured very precisely by means of mass spectrometers that are calibrated readily with carbon compounds.

For normal applications, e.g., in stoichometric calculations in chemistry, it is sufficient to specify the mass of an atom generally by its mass number (number of protons and neutrons; it is also given in the periodic table).

■   An oxygen molecule has the mass $m(O_2) = m(2\ ^{16}O) = 2 \cdot 16\ u = 32\ u$.

4. Amount of substance and molar volume

Amount of substance, n, SI unit mol, description of the number of particles, given as multiples of the Avogadro number.

Mol, SI unit for N_A particles (atoms, molecules) of a certain element (or a compound).

▲   The mole (mol) is the basic unit for the amount of substance: 1 mol is the amount of substance of a system which contains as many molecules as there are atoms in 0.012 kg of the carbon isotope ^{12}C.

Molar volume, the volume of one mole of a substance at **standard temperature and pressure (STP)** (0 °C temperature and 1.01325 bar pressure).

■   At STP, an ideal gas has a molar volume of about 22.4 liters.

5. Loschmidt constant,

the number of particles of an ideal gas at STP per molar volume:

Loschmidt constant $= \dfrac{\textbf{Avogadro number}}{\textbf{molar volume}}$			L^{-3}
$N_L = \dfrac{N_A}{V_m}$ $= 2.686\,75 \cdot 10^{25}\ \text{m}^{-3}$	**Symbol**	**Unit**	**Quantity**
	N_L	m^{-3}	Loschmidt constant
	N_A	mol^{-1}	Avogadro number
	V_m	m^3/mol	molar volume

6. Molar mass,

the mass of one mole of a substance:

molar mass $=$ Avogadro number $\cdot$ particle mass			MN^{-1}
$M = N_A \cdot m_N$ $= \dfrac{m}{n}$	**Symbol**	**Unit**	**Quantity**
	M	kg/mol	molar mass
	N_A	mol^{-1}	Avogadro number
	m_N	kg	particle mass
	m	kg	total mass
	n	mol	amount of substance

a) Molar mass of a mixture, mass of one mole of a mixture,

$$M_{\text{mixture}} = \frac{m_{\text{mixture}}}{n_{\text{mixture}}} = \frac{m_1 + m_2 + m_3 + \cdots}{n_1 + n_2 + n_3 + \cdots}.$$

It may be calculated from the molar masses M_i of the components i:

$$M_{\text{mixture}} = \frac{n_1 M_1 + n_2 M_2 + n_3 M_3 + \cdots}{n_1 + n_2 + n_3 + \cdots} = x_1 M_1 + x_2 M_2 + x_3 M_3 + \cdots.$$

(The mole fraction x_i is explained below.)

■   The molar mass of air is $M_{air} = 28.96 \, g/mol$. Its main constituents are nitrogen $M_{N_2} \approx 28 \, g/mol$ and oxygen $M_{O_2} \approx 32 \, g/mol$.

b) Molar mass of an element, see Periodic Table.

■   Molar masses of several elements (in g/mol): hydrogen 1.00797, oxygen 15.9994, nickel 58.71, silver 107.87, platinum 195.09.

➤   Rule of thumb: number of neutrons and protons $\approx$ molar mass in gram.

c) Molar mass of a compound, can be obtained additively from the components (atoms),

$$M(A_a B_b C_c) = a M(A) + b M(B) + c M(C) \,.$$

■   The molar mass of sulphuric acid (H_2SO_4) is approximately given by

$$M = 2 \cdot 1 \, g/mol + 32 \, g/mol + 4 \cdot 16 \, g/mol = 98 \, g/mol \,.$$

Molar masses of several gases see **Tab. 22.2/2.**

7. Amount of substance,

the number of moles of a substance:

amount of substance = $\dfrac{\text{particle number}}{\text{Avogadro number}}$			**N**
$n = \dfrac{N}{N_A}$ $n = \dfrac{m}{M}$ $N_A = 6.022\,136\,7 \cdot 10^{23} \, \text{mol}^{-1}$	Symbol	Unit	Quantity
	n	mol	amount of substance
	N	1	particle number
	M	kg/mol	molar mass
	N_A	mol^{-1}	Avogadro number
	m	kg	total mass

8. Universal gas constant,

the product of Avogadro number and Boltzmann constant:

gas constant = Boltzmann constant · Avogadro number			$ML^2T^{-2}\Theta^{-1}N^{-1}$
$R = k \cdot N_A$ $k = 1.380\,66 \cdot 10^{-23} \, J/K$	Symbol	Unit	Quantity
	R	J/(K mol)	gas constant
	N_A	mol^{-1}	Avogadro number
	k	J/K	Boltzmann constant

The value of R is

$$R = 8.314 \, J/(\text{mol K}) \,.$$

9. Mole fraction,

x_i, non-dimensional quantity, fraction of particles of kind i of the total number of particles:

$$x_i = \frac{N_i}{N_1 + N_2 + \cdots + N_n} \,, \qquad \sum_i x_i = 1 \,.$$

▲   The sum of all mole fractions always yields unity.

➤   According to ISO, a lowercase letter x_i is used for the mole fraction while X_i represents the total particle number of a species.

The mole fraction is an intensive variable and may take distinct values in different phases.

10. Mass fraction,

ξ_i, non-dimensional quantity, ratio of the total mass of a kind of particle to the total mass of all particles. It is equal to the product of the mole fraction and the ratio of the molar mass of the kind of particle i to the molar mass of the total system:

$$\xi_i = \frac{m_i}{m_{\text{tot}}} = x_i \frac{M_i}{M_{\text{tot}}} \, .$$

19.2.5   Entropy

1. Entropy as an extensive state function

Entropy, S, SI unit joule per kelvin, an extensive state function describing the disorder in the system (see p. 642).

The entropy change may be defined (for small temperature variations) via the reduced heat (see p. 702):

entropy change = $\dfrac{\text{heat change}}{\text{temperature}}$			$\mathbf{ML^2T^{-2}\Theta^{-1}}$
$\begin{aligned} \Delta S &= \frac{\Delta Q}{T} \\[2mm] &= \frac{C\,\Delta T}{T} \end{aligned}$	Symbol	Unit	Quantity
	S	J/K	entropy
	Q	J	amount of heat
	T	K	temperature
	C	J/K	heat capacity

Here only entropy differences are defined, not an absolute value of the entropy.

The absolute normalization is given by the third law of thermodynamics (see p. 702):

▲   The entropy at absolute zero is equal to zero,

$$S_{T=0} = 0 \, .$$

2. Microscopic consideration

Macrostate, the state characterized by the bulk properties of the system.

  Microstate, the state determined by the properties of the individual particles.

■   If a certain number of spheres is distributed over two containers, then the macrostate is specified by the number of spheres in each container, whereas the microstate is specified by identifying individual spheres in each of the containers.

Every thermodynamic macrostate may be realized by a large number of microscopically possible states (microstates).

■   In a system of three particles with three fixed distinct velocities, the state in which particle 1 has the highest velocity and particle 3 has the lowest velocity, and the state in which particle 1 has the lowest velocity and particle 2 has the highest velocity, are microscopically different. Macroscopically, both states are identical.

▲   The state with the most realizable possibilities is the most probable state.

■   Let a box be filled with gas. If one investigates whether a particle is in the left or right half of the box, the state with all particles in the left section is energetically allowed, but has only one microscopic realization. A uniform distribution of all particles to the left and right section has a much larger number of microscopic realizations and is therefore the most probable state.

▲   The **equilibrium state** is the state with the maximum number of microscopic realizations.

➤   Since the entropy increases with the number of realization possibilities, the entropy of the equilibrium state has a maximum.

3. Connection between entropy and number of microstates

entropy = Boltzmann constant · ln (number of realizations)	$\mathbf{ML^2T^{-2}\Theta^{-1}}$		
	Symbol	Unit	Quantity
$S = k \ln \Omega$	S	J/K	entropy
$k = 1.380\,66 \cdot 10^{-23}$ J/K	k	J/K	Boltzmann constant
	Ω	1	number of microstates

19.3   Thermodynamic potentials

19.3.1   Principle of maximum entropy—principle of minimum energy

Closed systems evolve to an equilibrium state characterized by a **maximum of entropy**. This state has the maximum number of microscopic realizations.

➤   This statement is a consequence of the second law of thermodynamics (see p. 701).

▲   In a closed system, all (irreversible) processes evolving by themselves increase the entropy until its maximum is reached in the equilibrium state.

In mechanics and electrodynamics, non-isolated systems tend to reduce their energy.

■   Mechanical systems tend to a local minimum of the potential energy.

▲   A non-isolated system of constant entropy evolves to a minimum of energy.

Both principles are connected via the laws of thermodynamics.

19.3.2   Internal energy as a potential

Internal energy, U, SI unit joule (J), extensive variable, describes the total energy contained in the system. In an isolated system, it is a central variable.

The internal energy is written as a function of the natural **extensive variables** entropy S, volume V and particle number N. If the dependence of the internal energy $U(S, V, N, \ldots)$ on the other variables is known, a complete knowledge of all thermodynamic quantities is guaranteed.

Differential representation of the internal energy:

$$dU = T\,dS - p\,dV + \mu\,dN + \cdots.$$

The intensive variables temperature T, pressure p and chemical potential μ may be described as functions of the natural extensive variables.

The intensive variables are described by a partial derivative with respect to an extensive variable (the other variables are assumed to be constant).

temperature, pressure and chemical potential as derivatives of U			
$T = \left.\dfrac{\partial U}{\partial S}\right\|_{V,N,\ldots}$	Symbol	Unit	Quantity
	U	J	internal energy
	T	K	temperature
$-p = \left.\dfrac{\partial U}{\partial V}\right\|_{S,N,\ldots}$	S	J/K	entropy
	p	Pa	pressure
	V	m^3	volume
$\mu = \left.\dfrac{\partial U}{\partial N}\right\|_{S,V,\ldots}$	μ	J	chemical potential
	N	1	particle number

▲   For isochoric adiabatic systems the internal energy U has a minimum, $dU \leq 0$ for $V = $ const., $S = $ const.

19.3.2.1   Internal energy in an ideal gas

In the ideal gas without rotational degrees of freedom, the following holds:

$$U = \frac{3}{2}NkT .$$

For isochoric changes of state, the following holds:

internal energy $\sim$ temperature			$\mathbf{ML^2T^{-2}}$
	Symbol	Unit	Quantity
	U	J	internal energy
$U = c_V m T$	c_V	J/(K kg)	specific-heat capacity at constant volume
	m	kg	mass
	T	K	temperature

19.3.3   Entropy as a thermodynamic potential

Entropy, S, SI unit joule per kelvin, in a closed system is a central variable, as is the internal energy. It describes the number of possible microstates in the system.

Differential representation of the entropy:

$$dS = \frac{1}{T}\,dU + \frac{p}{T}\,dV - \frac{\mu}{T}\,dN - \cdots .$$

If the dependence of the entropy $S(U, V, N, \ldots)$ on the variables $U, V, N, \ldots$ is known, then full knowledge of all thermodynamic quantities is guaranteed.

internal energy, pressure and chemical potential as derivatives of S			
$\dfrac{1}{T} = \dfrac{\partial S}{\partial U}\bigg|_{V,N,\dots}$	Symbol	Unit	Quantity
	U	J	internal energy
	T	K	temperature
$\dfrac{p}{T} = \dfrac{\partial S}{\partial V}\bigg|_{U,N,\dots}$	S	J/K	entropy
	p	Pa	pressure
	V	m^3	volume
$-\dfrac{\mu}{T} = \dfrac{\partial S}{\partial N}\bigg|_{U,V,\dots}$	μ	J	chemical potential
	N	1	particle number

▲ For isochoric systems with constant internal energy, the entropy S has a maximum, $dS = 0$ for $V = $ const., $U = $ const.

19.3.3.1 Entropy of the ideal gas

Entropy of an ideal gas without rotational degrees of freedom:

$$S(T, p) = Nk\left\{s_0(T_0, p_0) + \ln\left[\left(\frac{T}{T_0}\right)^{5/2}\left(\frac{p_0}{p}\right)\right]\right\}.$$

Rewritten in terms of N, V, U:

$$S(N, V, U) = Nk\left\{s_0(N_0, V_0, U_0) + \ln\left[\left(\frac{N_0}{N}\right)^{5/2}\left(\frac{U}{U_0}\right)^{3/2}\left(\frac{V}{V_0}\right)\right]\right\}.$$

From this equation, all equations of state of the ideal gas can be obtained by partial differentiation.

■ Differentiating with respect to the internal energy yields:

$$\frac{\partial S}{\partial U}\bigg|_{N,V} = \frac{1}{T} = \frac{3}{2}Nk\frac{1}{U} \quad \Rightarrow \quad U = \frac{3}{2}NkT;$$

Differentiating with respect to the volume yields the equation of state,

$$\frac{\partial S}{\partial V}\bigg|_{N,U} = \frac{p}{T} = Nk\frac{1}{V} \quad \Rightarrow \quad pV = NkT.$$

19.3.4 Free energy

1. Free energy,

also **Helmholtz potential**, F, SI unit joule (J), is of importance, in particular, for the description of processes running at constant temperature (isothermal).

The free energy is the difference between internal energy and the product of temperature and entropy,

$$F = U - TS.$$

This corresponds to a Legendre transformation from a function of entropy (internal energy) to a function of temperature (free energy).

The total differential of F is

$$dF = -S\,dT - p\,dV + \mu\,dN + \cdots.$$

Change of the free energy:

$$F = W + \int_{T_1}^{T_2} S\,dT.$$

2. Free energy as a function of the state variables

The free energy is written as a function of temperature, volume and particle number. If the dependence of the free energy $F(T, V, N, \ldots)$ on the other variables is known, then full knowledge of all thermodynamic quantities is guaranteed.

The remaining variables may be obtained by partial differentiation.

entropy, pressure and chemical potential as derivatives of F			
$-S = \left.\dfrac{\partial F}{\partial T}\right|_{V,N,\ldots}$	Symbol	Unit	Quantity
	F	J	free energy
	T	K	temperature
$-p = \left.\dfrac{\partial F}{\partial V}\right|_{T,N,\ldots}$	S	J/K	entropy
	p	Pa	pressure
	V	m^3	volume
$\mu = \left.\dfrac{\partial F}{\partial N}\right|_{T,V,\ldots}$	μ	J	chemical potential
	N	1	particle number

3. Free energy and isothermal processes

The change of the free energy dF_{sys} of a system at constant temperature (isothermal processes) represents the work delivered by (or supplied to) the system if the process is reversible.

Isothermal processes during which the system exchanges only heat but no work with the environment tend to a minimum of free energy, i.e., simultaneously to minimum internal energy and to maximum entropy.

▲   Isothermal and isochoric processes spontaneously evolve in the direction in which the free energy decreases.

Isothermal processes that actually increase the internal energy may occur spontaneously provided that energy is supplied from a heat bath. In order for this to occur, the gain of energy $T\,dS$ must exceed dU, the energy supplied. If there is no energy supplied, the process would run spontaneously in the opposite direction.

19.3.5 Enthalpy

1. Enthalpy,

H, SI unit joule (J), is of importance for the description of processes proceeding at constant pressure (isobaric).

■   In practice, chemical processes often proceed at constant pressure.
Displacement work, the product of pressure and volume.
    Enthalpy is the sum of internal energy and displacement work,

$$H = U + pV = TS + \mu N .$$

The enthalpy is written as a function of entropy, pressure and particle number. If the dependence of the enthalpy $H(S, p, N, \ldots)$ on the other variables is known, then full knowledge of all thermodynamic quantities is guaranteed.
    The total differential of the enthalpy is:

$$dH = dU + p\,dV + V\,dp ,$$

$$= T\,dS + V\,dp + \mu\,dN .$$

▲   For adiabatic, isobaric systems ($\Delta Q = 0$, $p = $ const.) the enthalpy H tends to a minimum, $dH = 0$.

2. Determination of the properties of state from the enthalpy

If the enthalpy $H(S, p, N, \ldots)$, is known, then all remaining properties can be obtained by partial differentiation.

temperature, volume and chemical potential as derivatives of H			
$T = \dfrac{\partial H}{\partial S}\Big|_{p,N,\ldots}$	Symbol	Unit	Quantity
	H	J	enthalpy
	T	K	temperature
$V = \dfrac{\partial H}{\partial p}\Big|_{S,N,\ldots}$	S	J/K	entropy
	p	Pa	pressure
	V	m^3	volume
$\mu = \dfrac{\partial H}{\partial N}\Big|_{S,p,\ldots}$	μ	J	chemical potential
	N	1	particle number

3. Isobaric and adiabatic processes

In principle, the **enthalpy** may be given for any system. It is particularly suitable for isobaric ($p = $ const., $dp = 0$) and adiabatic systems ($\Delta Q = 0$). Such systems do not exchange heat with the environment ($\Delta Q = 0$), but in an expansion they may do volume work against the constant external pressure (**Fig. 19.10**). For isobaric changes of state, the change of enthalpy is just the quantity of heat exchanged with the environment and other work (which does not contain the volume work against the external pressure).
    Technical work, W_t, SI unit joule (J), the total amount of work a machine may do in theory.

$$W_t = \int_{p_1}^{p_2} V\,dp .$$

▲   Irreversible processes will cause an isobaric adiabatic system to proceed on its own until an equlibrium state of minimum enthalpy is reached.

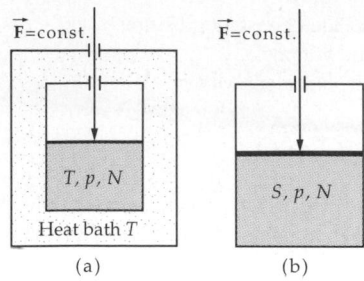

Figure 19.10: Thermo-dynamic systems. (a): isothermal-isobaric system, (b): adiabatic-isobaric system.

19.3.5.1 Enthalpy of the ideal gas

The enthalpy is the sum of the internal energy and the displacement work.

enthalpy $\sim$ temperature				$\mathbf{ML^2T^{-2}}$
	Symbol	Unit	Quantity	
$H = c_p m T$	H	J	enthalpy	
	c_p	J/(K kg)	spec. heat capacity at constant pressure	
	m	kg	mass	
	T	K	temperature	

Enthalpy of the ideal gas, microscopically (without rotational energy):

$$H(T, p, N) = \frac{5}{2} NkT \,.$$

19.3.5.2 Enthalpy and phase transitions

In phase transitions proceeding at constant pressure (isobaric) and at constant temperature (isothermal), the change of enthalpy of the substance is equal to the latent heat received (in melting, sublimating and boiling) or delivered (in solidifying, de-sublimating and conden-sation):

$$H_{\mathrm{fl}} = H_{\mathrm{solid}} + \Delta H_S \,.$$

Melting enthalpy, ΔH_S, the enthalpy spent in melting.
  Solidification enthalpy, $-\Delta H_S$, the enthalpy released in solidification. Analogously, the **evaporation enthalpy**, ΔH_V, is related to the **condensation enthalpy**, $-\Delta H_V$, and the **sublimation enthalpy**, $\Delta H_{\mathrm{sub}} = \Delta H_S + \Delta H_V$, to the **desublimation enthalpy**, $-\Delta H_{\mathrm{sub}}$.
  Mollier diagram, graph in which the entropy per mass unit is plotted against the en-thalpy per mass unit (**h, s diagram**).
  Analogously, graphs of other quantities, such as concentration against enthalpy (**h, x diagram**), may be used.

19.3.5.3 Reaction enthalpy and theorem of Hess

Reaction enthalpy, the enthalpy delivered or spent in a chemical reaction.
➤   Many chemical reactions proceed in open vessels at constant pressure.

▲   The question whether a chemical reaction proceeds spontaneously without external energy may be answered using the balance of enthalpy,

$$\Delta H = H_{\text{products}} - H_{\text{educts}} \, .$$

If the balance is negative, $\Delta H \leq 0$, the reaction proceeds spontaneously and exothermically.

▲   **Theorem of Hess**: The total enthalpy difference between products and reactants is independent of the reaction path.

Usually, the enthalpy balance depends strongly on the environmental pressure and the environmental temperature. Frequently, activation energy must be expended to start the reaction.

■   At room temperature, hydrogen and oxygen may be mixed. Despite the negative formation energy of water, the reaction does not proceed spontaneously. With a catalyst or open fire the reaction proceeds explosively (**oxyhydrogen reaction**).

Catalyst, a substance that allows for, or at least enhances, the reaction of other substances without being consumed itself.

■   Metallic platinum is a good catalyst for many reactions.

Exothermic reaction, reaction delivering enthalpy.

Endothermic reaction, reaction consuming enthalpy.

M   In chemistry, reaction enthalpies are measured simply by measuring the quantity of heat produced in the reaction in a **calorimeter** according to

$$\Delta H = \Delta Q|_p \, .$$

19.3.6   Free enthalpy

1. Free enthalpy,

also **Gibbs potential**, G, SI unit joule (J), a quantity introduced by J. W. Gibbs (1875) that is particularly suited to systems at given temperature and given pressure:

$$G = U - TS + pV \, .$$

▲   The free enthalpy per particle coincides with the chemical potential for systems of one kind of particle that do not exchange another kind of energy (e.g., electrical energy) with the environment.

The total differential of the free enthalpy reads:

$$dG = -S \, dT + V \, dp + \mu \, dN \, .$$

▲   For an isobaric-isothermal system ($p = \text{const.}$, $T = \text{const.}$), the free enthalpy is a minimum, $dG \leq 0$.

2. Free enthalpy as function of state variables

The free enthalpy is written as a function of temperature, pressure and particle number. If the dependence of the free enthalpy $G(T, p, N, \ldots)$ on the other variables is known, then full knowledge of all thermodynamic quantities is guaranteed.

If the function $G(T, p, N)$ is known, then all other quantities can be obtained by partial differentiation.

entropy, volume and chemical potential as derivatives of G			
$-S = \dfrac{\partial G}{\partial T}\Big|_{p,N,\dots}$	Symbol	Unit	Quantity
	G	J	free enthalpy
	T	K	temperature
$V = \dfrac{\partial G}{\partial p}\Big|_{T,N,\dots}$	S	J/K	entropy
	p	Pa	pressure
	V	m^3	volume
$\mu = \dfrac{\partial G}{\partial N}\Big|_{T,p,\dots}$	μ	J	chemical potential
	N	1	particle number

The change of the free enthalpy is just the work converted by the system in isothermal and isobaric **reversible** processes, without the volume work against the constant external pressure.

▲   In an isothermal isobaric system, irreversible processes proceed until a minimum of free enthalpy is reached.

19.3.6.1   Chemical reactions

The free enthalpy is of importance for reactions proceeding slowly.

Exoergic reactions, reactions in which free enthalpy is released.

Endoergic reactions, reactions in which free enthalpy is consumed.

Law of mass action, determines the conversion ratio between the products and reactants of a chemical reaction:

$$a_1 A_1 + a_2 A_2 + \cdots \rightleftharpoons b_1 B_1 + b_2 B_2 + \cdots ,$$

$$\frac{\left(x_{B_1}\right)^{b_1} \left(x_{B_2}\right)^{b_2} \cdots}{\left(x_{A_1}\right)^{a_1} \left(x_{A_2}\right)^{a_2} \cdots} = e^{\left(-\frac{\Delta G^0(p,T)}{kT}\right)} = K(p,T).$$

The quantity ΔG^0 is a constant characterizing the reaction. The equilibrium constant $K(p,T)$ is determined by the difference of the free enthalpies ΔG^0.

▲   For $K > 1$ the equilibrium lies on the side of the products, for $K < 1$ the concentration of the reactants is predominant.

For the most important reactions, acid-base reactions, and dissociations, the equilibrium constants are given in chemical tables.

19.3.6.2   Principle of Le Chatelier

Principle of Le Chatelier, statement on the change of an equilibrium state under external conditions.

▲   If a constraint is imposed or changed (change of temperature, pressure or concentration) an equilibrium state is shifted such that the constraint is relaxed.

■   Under external pressure, a system of water and steam will partly condense and thereby reduce its total volume.

19.3.7   Maxwell relations

Maxwell relations, relations connecting the partial derivatives of different thermodynamic potentials:

$$\left.\frac{\partial T}{\partial V}\right|_{S,N} = -\left.\frac{\partial p}{\partial S}\right|_{V,N} , \qquad \left.\frac{\partial S}{\partial V}\right|_{T,N} = \left.\frac{\partial p}{\partial T}\right|_{V,N} ,$$

$$\left.\frac{\partial T}{\partial p}\right|_{S,N} = \left.\frac{\partial V}{\partial S}\right|_{p,N} , \qquad -\left.\frac{\partial S}{\partial p}\right|_{T,N} = \left.\frac{\partial V}{\partial T}\right|_{p,N} .$$

➤ Usually, only systems of constant particle number ($dN = 0$) are considered so that the number of relations is reduced appreciably. But, if there are additional state variables, for example, a magnetic field and a **magnetic dipole moment** (see p. 463), other relations must be added.

Thermodynamic quadrangle, simple mnemonic aid giving a quick overview on the potentials and their variables and allows a quick reading of the Maxwell relations (**Fig. 19.11**).

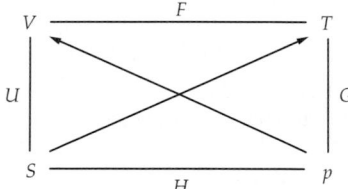

Figure 19.11: Thermodynamic quadrangle for $N = $ const.

It is designed especially for systems of constant particle number and without other state variables.

The variables V, T, p, S form the vertices of the quadrangle represented in **Fig. 19.11**. The edges represent the potentials depending on the variables at the corresponding vertices, e.g., $F(V, T)$.

▲ The derivative of a potential with respect to a variable (vertex) is given by the variable at the diagonal-opposite site. The two arrows in the diagonals indicate the sign.

■ The differentiation of F with respect to V yields a minus sign (see arrow) p: $\partial F/\partial V = -p$.

The Maxwell relations may be read as follows: The derivatives involving the variables along an edge of the quadrangle (e.g., $\partial V/\partial S$), with the variable in the diagonal-opposite vertex (here p) kept constant, are just equal to the corresponding derivative on the opposite edge (here $-\left(\dfrac{\partial T}{\partial p}\right)_S$). The signs depend on the sense in which the diagonals are passed.

19.3.8 Thermodynamic stability

1. Various kinds of equilibrium states

Equilibrium states, distinguished by a maximum of entropy, or by a minimum in the various thermodynamic potentials.

Closed isochoric states, characterized in equilibrium by a maximum of entropy S.

Isothermal-isobaric states, characterized in equilibrium by a minimum of the free enthalpy $G = U + pV - TS$.

Isothermal-isochoric states, characterized in equilibrium by a minimum of the free energy $F = U - TS$.

Adiabatic-isobaric states, characterized in equilibrium by a minimum of the enthalpy $H = U + pV$.

Adiabatic-isochoric states, characterized in equilibrium by a minimum of the internal energy U.

differential representation of thermodynamic potentials			
	Symbol	Unit	Quantity
$dU = -pdV + TdS$	U	J	internal energy
	F	J	free energy
$dF = -pdV - SdT$	H	J	enthalpy
$dH = Vdp + TdS$	G	J	free enthalpy
	p	Pa	pressure
$dG = Vdp - SdT$	V	m^3	volume
	T	K	temperature
	S	J/K	entropy

2. Survey of equilibrium criteria

System is ...	Isothermal	Isobaric	Isochoric	Adiabatic	Closed
entropy S maximum			$dV = 0$		$dU = 0$
internal energy U minimum			$dV = 0$	$\Delta Q = 0$	
free energy F minimum	$dT = 0$		$dV = 0$		
enthalpy H minimum		$dp = 0$		$\Delta Q = 0$	
free enthalpy G minimum	$dT = 0$	$dp = 0$			

▲   If a system is in stable equilibrium, then all spontaneous changes of the variables must initiate processes that bring the system back to equilibrium, i.e., counteract these spontaneous changes.

➤   This statement follows from the principle of Le Chatelier.

19.4 Ideal gas

Ideal gas, the gas particles may be treated like point particles of classical mechanics without any interaction. The ideal gas is a simple model of a real gas, assuming that the particles are of negligible size and have few mutual interactions. The approximation improves the more the gas is rarefied.

■   Under standard conditions air, hydrogen, and noble gases may be described quite well by an ideal gas.

▲   When describing real gases, one must take into account:

•   the internal volume of the particles,

•   the interaction between the gas particles.

19.4.1 Boyle-Mariotte law

Boyle-Mariotte law, general relation between pressure and volume of a gas at constant temperature, described in 1664 by **R. Boyle**, and independently by **E. Mariotte** in 1676.

▲ For constant temperature the product of pressure and volume is constant.

pressure · volume = constant				
		Symbol	Unit	Quantity
$pV = p_0 V_0,$ $\quad$ $T = \text{const.}$		p	Pa	pressure
		V	m^3	volume
$\dfrac{p}{p_0} = \dfrac{V_0}{V} = \dfrac{\rho}{\rho_0},$ $\quad$ $T, N = \text{const.}$		T	K	temperature
		ρ	kg/m^3	density
		N	1	particle number

■ If for constant temperature the volume of a cylinder is reduced to one half, the pressure of the gas is doubled (**Fig. 19.12**).

p–V diagram, diagram representing the pressure as a function of the volume, important for describing changes of state and thermodynamic machines.

If pressure and volume at fixed temperature are plotted against each other (**Fig. 19.13**), for the ideal gas one obtains hyperbolas.

$$p \sim \frac{1}{V}.$$

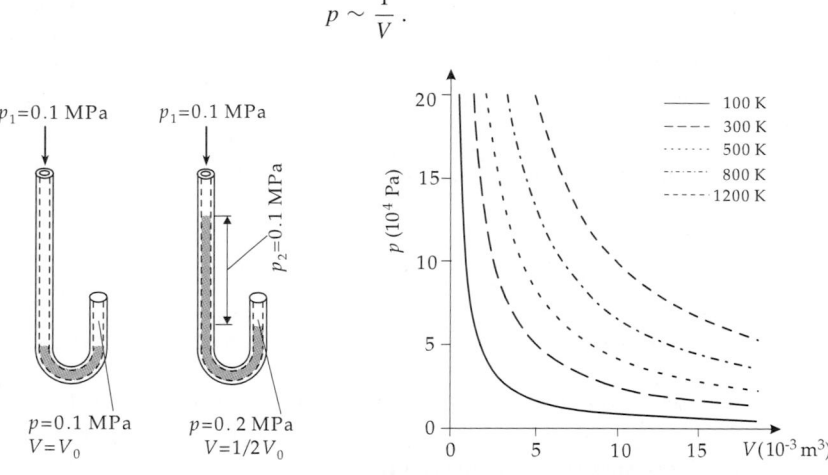

Figure 19.12: Relation between pressure and volume. Addition of liquid increases the pressure in the leg with closed end, the gas volume is reduced.

Figure 19.13: p–V diagram for 1 mole of an ideal gas. The isotherms are hyperbolas.

■ The **McLeod pressure gauge** is an application of Boyle-Mariotte's law.

19.4.2 Law of Gay-Lussac

1. Law of Gay-Lussac

dependence of the volume of a gas on the temperature, formulated in 1802 by **Gay-Lussac**:

$$V(\vartheta) = V_0(1 + \gamma \vartheta), \qquad V_0 : \text{volume at } \vartheta_0 = 0\,°C.$$

▲   If the absolute temperature of the gas in a cylinder is changed, the volume at constant
    pressure changes proportional to the temperature.

relative change of volume $\sim$ change of temperature			$\mathbf{L^3}$
	Symbol	Unit	Quantity
$\dfrac{\Delta V}{V_0} = \gamma \Delta T$	V	m³	volume
	γ	1/K	volume-expansion coefficient
$p = \text{const.}$	T	K	temperature
	p	Pa	pressure

In a notation in terms of temperature differences, the temperature difference may also be
given in celsius instead of kelvin.

Volume-expansion coefficient, γ, SI unit 1/kelvin, describes the relative change of volume for varying temperature.

The volume-expansion coefficient has almost the same value for all rarefied gases. For
the ideal gas

$$\gamma = 0.003661 \ \text{K}^{-1} = \frac{1}{273.15} \ \text{K}^{-1}, \ \text{referred to the volume at } 0°\text{C}.$$

The corresponding equation is identical to the definition of absolute temperature.

2. Rewriting the law of Gay-Lussac

For constant pressure, the volume of an ideal gas varies proportional to the temperature:

volume ratio = temperature ratio			
	Symbol	Unit	Quantity
$\dfrac{V}{V_0} = \dfrac{T}{T_0}, \quad p = \text{const.}$	V	m³	volume
	T	K	temperature
	p	Pa	pressure

For constant volume, the pressure of the ideal gas varies proportional to the temperature:

pressure ratio = temperature ratio			
	Symbol	Unit	Quantity
$\dfrac{p}{p_0} = \dfrac{T}{T_0}, \quad V = \text{const.}$	p	Pa	pressure
	T	K	temperature
	V	m³	volume

19.4.3   Equation of state

Equation of state of the ideal gas, describes the connection between the quantities
p_0, V_0, T_0 (pressure, volume and temperature) of an arbitrary initial state and the same
quantities p, V, T of a final state (see p. 627).

equation of state of ideal gas			
$\dfrac{pV}{T} = \dfrac{p_0 V_0}{T_0} = Nk$ $pV = NkT$ $k = 1.380\,66 \cdot 10^{-23}$ J/K	Symbol	Unit	Quantity
	p	Pa	pressure
	V	m^3	volume
	T	K	temperature
	N	1	particle number
	k	J/K	Boltzmann constant

This equation is obtained if two processes are carried out successively and the gas laws of Boyle and Gay-Lussac are applied.

Alternative notations may be found in the section on equations of state (see p. 661).

19.5 Kinetic theory of the ideal gas

Every particle in the gas has a definite velocity vector $\vec{v}$.

Velocity distribution, the distribution function of the particle velocities in a system.

▲   In the equilibrium state, the velocity distribution of a system does not change. The velocity of individual particles may change, of course.

19.5.1 Pressure and temperature

1. Microscopic interpretation of pressure,

the pressure is described as the momentum transferred per unit time and unit area onto the container walls by collisions of the gas particles (**Fig. 19.14**).

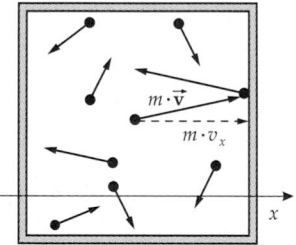

Figure 19.14: Scheme to calculate the pressure. Only the component of the momentum normal to the wall is taken into account.

Principal equation of gas theory, describes the relation between the pressure and the total kinetic energy.

pressure · volume = $\dfrac{2}{3}$ total kinetic energy			ML^{2}T^{-2}
$pV = \dfrac{2}{3} W_{\text{kin}}$	Symbol	Unit	Quantity
	p	Pa	pressure
	V	m^3	volume
	W_{kin}	J	total kinetic energy

2. Mean kinetic energy

Mean kinetic energy per particle, ε_{kin}, SI unit joule, total kinetic energy (see p. 66) divided by the particle number,

$$\varepsilon_{\text{kin}} = \frac{W_{\text{kin}}}{N} = \frac{1}{N} \sum_{i=1}^{N} \frac{m_i v_i^2}{2}.$$

Using the ideal gas law, the temperature dependence of the total energy and the mean kinetic energy are given by:

kinetic energy ~ particle number · temperature			$\mathbf{ML^2T^{-2}}$
$W_{\text{kin}} = \dfrac{3}{2} NkT$	Symbol	Unit	Quantity
	W_{kin}	J	total kinetic energy
	N	1	particle number
$\varepsilon_{\text{kin}} = \dfrac{3}{2} kT$	k	J/K	Boltzmann constant
	T	K	temperature
$k = 1.38066 \cdot 10^{-23}$ J/K	ε_{kin}	J	mean kinetic energy per particle

▲   The mean kinetic energy of the particles is proportional to the temperature. Microscopic interpretation of the temperature as a measure of the mean energy in a system.

19.5.1.1 Root-mean-square velocity

1. Root-mean-square velocity,

$\sqrt{\overline{v^2}}$, the root of the average value of the squares of velocity.
➤   The symbol v_{rms} (root mean square) is also used.
Assuming equal particle masses, the mean value of the squares of the velocity is twice the mean kinetic energy divided by the particle mass m,

$$\sqrt{\overline{v^2}} = \sqrt{\frac{2\varepsilon_{\text{kin}}}{m}}.$$

For an **ideal gas**, the following holds:

mean-square velocity in ideal gas			$\mathbf{LT^{-1}}$
$\sqrt{\overline{v^2}} = \sqrt{\dfrac{3kT}{m}}$	Symbol	Unit	Quantity
	$\sqrt{\overline{v^2}}$	m/s	mean-square velocity
$= \sqrt{3R_s T}$	k	J/K	Boltzmann constant
	T	K	temperature
$k = 1.38066 \cdot 10^{-23}$ J/K	m	kg	particle mass
	R_s	J/(K kg)	specific gas constant

2. Average velocity,

or **mean velocity**, $\overline{v}$, arithmetic mean of the magnitudes of velocity (without taking velocity directions into account).
➤   The mean velocity depends on the velocity distribution assumed (see p. 655).

3. Mean-velocity vector,

$\langle \vec{\mathbf{v}} \rangle$, a vector whose components are the mean values of the velocity components of the particles,

$$\langle \vec{\mathbf{v}} \rangle = \begin{pmatrix} \overline{v_x} \\ \overline{v_y} \\ \overline{v_z} \end{pmatrix}.$$

If there is no flow, the magnitude of the mean-velocity vector is zero, since all directions occur with equal probability.

▲   The root-mean-square velocity, the mean velocity, and the magnitude of the mean-velocity vector are three completely different quantities.

19.5.2 Maxwell–Boltzmann distribution

1. Velocity distribution,

a distribution function specifying the relative probability of a certain velocity in a system. The relative probability of velocities in the range v_1 to v_2 is given by the integral

$$h(v_1 \leq v \leq v_2) = \frac{N(v_1 \leq v \leq v_2)}{N(\text{total})} = \int_{v_1}^{v_2} f(\vec{\mathbf{v}}) \mathrm{d}^3 v \,.$$

The integral over all velocities yields unity,

$$\int_{v_1=0}^{v_2=\infty} f(\vec{\mathbf{v}}) \mathrm{d}^3 v = 1 \,.$$

2. Maxwell–Boltzmann distribution,

velocity distribution of an ideal gas (**Fig. 19.15**):

$$f(\vec{\mathbf{v}}) = \frac{1}{N} \frac{\mathrm{d}N}{\mathrm{d}v} = 4\pi v^2 \left(\frac{m}{2\pi kT} \right)^{3/2} \mathrm{e}^{\left(-\frac{\frac{1}{2}mv^2}{kT} \right)} \,.$$

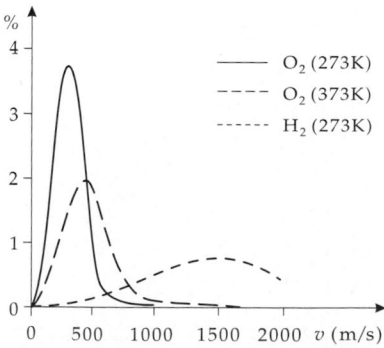

Figure 19.15: Maxwell–Boltzmann velocity distribution for various gases and various temperatures. Ordinate: % of molecules with v in the range of 10 m/s about the given velocity.

The term $4\pi v^2$ is from the assumption that the velocity distribution does not depend on the direction, $f(\vec{v}) \to f(v)$. Then:

$$\int f(v)\, dv_x dv_y dv_z = \int f(v) \cdot v^2 \sin\vartheta\, d\vartheta\, d\phi\, dv = \int 4\pi v^2 f(v) dv .$$

The term $(m/2\pi kT)^{3/2}$ is due to the normalization of the function to unity,

$$\int f(\vec{v}) d^3 v - 1$$

3. Boltzmann factor,

denotation for the exponential term. The term in the numerator of the exponential function is the kinetic energy,

$$e^{-\frac{E}{kT}} = e^{-\frac{mv^2}{2kT}} .$$

Generally, the Boltzmann factor is given by the exponential term with negative exponent, with the energy in the numerator and the temperature (multiplied by the Boltzmann constant) in the denominator.

▲   The velocity distribution depends on **temperature** and **particle mass**.

■   At the same temperature, oxygen molecules have a lower mean velocity than the lighter hydrogen molecules (see **Fig. 19.15**).

4. Most-probable velocity and mean velocity

Most-probable velocity, v_{max} or v_w, the velocity with the highest probability. v_w is the velocity at the **maximum** of the distribution function.

most-probable velocity $\sim \sqrt{\text{temperature}}$			
$v_w = \sqrt{\dfrac{2kT}{m}}$ $k = 1.380\,66 \cdot 10^{-23}$ J/K	Symbol	Unit	Quantity
	v_w	m/s	most-probable velocity
	k	J/K	Boltzmann constant
	T	K	temperature
	m	kg	particle mass

The **mean velocity**, $\bar{v}$, for a Maxwell-Boltzmann distribution is

$$\bar{v} = \sqrt{\frac{8kT}{\pi m}} = \sqrt{\frac{8}{3\pi}}\sqrt{\overline{v^2}} = \sqrt{\frac{8}{3\pi}} v_{rms} .$$

Its value is between v_w and $\sqrt{\overline{v^2}}$:

$$v_w = \sqrt{\frac{2}{3}}\sqrt{\overline{v^2}} = \sqrt{\frac{2}{3}} v_{rms} ,$$

$$\bar{v} = \sqrt{\frac{8}{3\pi}}\sqrt{\overline{v^2}} = \sqrt{\frac{8}{3\pi}} v_{rms} .$$

19.5.3 Degrees of freedom

Degree of freedom of a particle, description of the possibilities to take energy and convert it into some kind of motion. Hereby translational motion, rotational motion or vibrational motion may occur.

Number of degrees of freedom, f, non-dimensional quantity specifying the number of independent kinds of motion.

■ There are three degrees of freedom of translational motion, corresponding to motion along the x-axis, the y-axis and the z-axis.

a) Monatomic particles have only the three translational degrees of freedom.
■ All noble gases (He, Ne, Ar, Kr, Xe, Rn) are monatomic.

b) Diatomic particles have five degrees of freedom, three of translation, and two of rotation about two different axes perpendicular to the connecting line (**Fig. 19.16**).

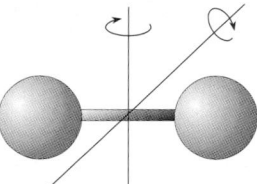

Figure 19.16: Rotational degrees of freedom of a diatomic molecule. The rotational axes are perpendicular to the connecting line.

The rotation about the molecular axis does not count as a degree of freedom, since the associated **moment of inertia** J is very small (for an ideal gas, even exactly zero) so that very high energies would be required to excite these rotations ($E = L^2/(2J)$, L: angular momentum).

■ Most molecules of air, such as N_2 and O_2, are diatomic.

c) Polyatomic molecules in most cases have three rotational axes, hence six degrees of freedom.

■ Sulphur dioxide (SO_2), ammonia (NH_3) and many hydrocarbon gases (methane $CH_4, \ldots$) belong to this category.

Vibrational degrees of freedom are excited in gases mostly at very high temperatures. The number of degrees of freedom therefore depends strongly on the temperature over a wide range of temperature.

In solids, the translational degrees of freedom ($f = 3$) and the vibrational degrees of freedom about the lattice sites ($f = 3$) yield six degrees of freedom in total.

19.5.4 Equipartition law

Equipartition law, **equipartition theorem**, thermal energy is apportioned equally among the degrees of freedom of a system.

▲ The thermal energy is distributed equally to each degree of freedom. On average, each degree of freedom carries the same energy.

The mean energy per gas particle (molecule) is:

mean energy ~ degrees of freedom · temperature			ML^2T^{-2}
	Symbol	Unit	Quantity
$\overline{W} = \dfrac{f}{2}kT$	$\overline{W}$	J	mean particle energy
	f	1	number of degrees of freedom
$k = 1.38066 \cdot 10^{-23}$ J/K	k	J/K	Boltzmann constant
	T	K	temperature

Mono-atomic gases thus have a mean energy per particle of

$$\overline{W} = \frac{3}{2}kT \ .$$

Diatomic gases correspondingly have a mean energy per particle of

$$\overline{W} = \frac{5}{2}kT \ .$$

Tri- and polyatomic molecules in general have a mean energy per particle of

$$\overline{W} = 3kT \ .$$

19.5.5 Transport processes

In real gases the particles mutually interact via molecular potentials. The gas particles collide, exchange momentum and energy, and fly apart with altered velocities. These collision processes are of great importance for the transport of energy and matter.

1. Characteristics of collision processes in gases

Mean free path l, often denoted by λ, SI unit meter, gives the length of the path of a particle (atom, molecule or—in metals—electron) between two collisions with other particles.

  Mean collision time τ, SI unit second, the mean time interval between two collisions.

  Collision frequency f, SI unit 1/second, the mean frequency of collisions per unit time.

■   At the temperature of 293 K and a pressure of $1.0 \cdot 10^5$ Pa, the molecules of air have a mean free path of $l = 6.4 \cdot 10^{-8}$ m. The mean free path increases with decreasing pressure. For a pressure of 100 Pa, $l = 6.4 \cdot 10^{-5}$ m.

The collision time and the collision frequency are related to the mean velocity $\overline{v}$ of the particles known from the velocity distribution and to their mean free path as follows:

collision time $= \dfrac{1}{\text{collision frequency}} = \dfrac{\text{mean free path}}{\text{mean velocity}}$			T
	Symbol	Unit	Quantity
$\tau = \dfrac{1}{f} = \dfrac{l}{\overline{v}}$	τ	s	collision time
	f	Hz	collision frequency
	l	m	mean free path
	$\overline{v}$	m/s	mean velocity

2. Cross-section,

σ, may be interpreted as the knock-on area of the colliding particles (**Fig. 19.17**).

  The mean free path is related to the cross-section as follows:

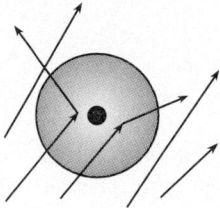

Figure 19.17: Cross-section in geometric interpretation. Particles passing through the grey area undergo a collision.

cross-section $= \dfrac{1}{\text{mean free path} \cdot \text{density}}$			$\mathbf{L^2}$
	Symbol	Unit	Quantity
$\sigma = \dfrac{1}{l\rho_N}$	σ	m^2	cross-section
	l	m	mean free path
	ρ_N	$1/m^3$	particle density

Barn, encoding word during Second World War, a unit of cross-section used in atomic and nuclear physics:

$$1\,b = 10^{-28}\ m^2.$$

3. Heat conductivity,

λ, SI unit watt per kelvin and meter, energy transport capability of a system. It is of importance for heat conduction.

heat conductivity (microscopic)			$\mathbf{MLT^{-3}\Theta^{-1}}$
	Symbol	Unit	Quantity
$\lambda = \dfrac{1}{3}\overline{v}l\rho c_V$	λ	$W/(K\,m)$	heat conductivity
	$\overline{v}$	m/s	mean velocity
	l	m	mean free path
	ρ	kg/m^3	density
	c_V	$J/(K\,kg)$	spec. heat capacity for const. volume

➤ Instead of density times specific heat capacity, one may also use the product of molar density and molar heat capacity, or the product of particle density and specific heat per particle.

4. Heat conductivity of monatomic gases

For monatomic gases, the following holds:

heat conduction (monatomic gas)			$\mathbf{MLT^{-3}\Theta^{-1}}$
	Symbol	Unit	Quantity
$\lambda = \dfrac{1}{2}k\overline{v}l\rho_N$	λ	$W/(K\,m)$	heat conductivity
	k	J/K	Boltzmann constant
	$\overline{v}$	m/s	mean velocity
	l	m	mean free path
	ρ_N	$1/m^3$	particle density

5. Heat conductivity of various materials

The heat conductivity of numerous materials may be found in **Tab. 22.3**.

■ Heat conductivity λ of several metals (in $W \cdot cm^{-1} \cdot K^{-1}$): copper 4.01, gold 3.17, lead 0.353, titanium 0.219.

Heat conductivity λ of several liquids and gases (in $W \cdot m^{-1} \cdot K^{-1}$): water 0.60, benzene 0.13, air 0.025, hydrogen 0.171, steam 0.016, chlorine 0.0081.

Heat conductivity λ of several materials (in $W \cdot m^{-1} \cdot K^{-1}$): cast iron 58, brass 113, sandstone 2.3, spruce 0.14, window glass 0.81, glass wool 0.04, PVC 0.16.

6. Diffusion constant,

D, SI unit square meter per second, describes the transport of matter (see nonequilibrium processes—diffusion).

diffusion constant (microscopic)			$\mathbf{L^2T^{-1}}$
	Symbol	Unit	Quantity
$D = \dfrac{1}{3}\bar{v}l$	D	m^2/s	diffusion constant
	$\bar{v}$	m/s	mean velocity
	l	m	mean free path

■ Diffusion constant D of various gas-gas systems (in cm^2/s): H-He 2.35, H-H_2 0.184, He-O_2 0.45, Ar-O_2 0.167, Kr-Xe 0.081.

7. Dynamic viscosity,

η, SI unit 1/(second meter), describes the internal friction.

viscosity (microscopic)			$\mathbf{L^{-1}T^{-1}}$
	Symbol	Unit	Quantity
	η	$1/(m\,s)$	viscosity
$\eta = \dfrac{1}{3}\bar{v}l\rho_N$	$\bar{v}$	m/s	mean velocity
	l	m	mean free path
	ρ_N	$1/m^3$	particle density

Dynamic viscosities of various substances may be found in **Tab. 22.3.1**.
The ratio

$$\frac{\lambda}{\eta} = \frac{C_{mV}}{M} = \frac{3}{2}k = \text{const.}$$

is experimentally confirmed to good approximation (C_{mV} is the specific molar heat at constant volume).

19.6   Equations of state

19.6.1   Equation of state of the ideal gas

Equation of state of the ideal gas, the relation between the quantities p_0, V_0, T_0 (pressure, volume and temperature) of an arbitrary initial state and the corresponding quantities p, V, T of a final state:

$$\frac{pV}{T} = \frac{p_0 V_0}{T_0} = \text{const.} = Nk \,.$$

Pressure and temperature are intensive properties, the volume is an extensive property. The product of an extensive and an intensive quantity is an extensive quantity (see p. 628), and is therefore proportional to the particle number N.

Definition via the particle density of the gas:

▲   The pressure is the product of the particle density $\rho_N = N/V$, the temperature, and the dimensional factor k, the Boltzmann constant.

pressure $\sim$ density $\cdot$ temperature			$\mathbf{ML^{-1}T^{-2}}$
	Symbol	Unit	Quantity
$p = \rho_N k T$	p	Pa	pressure
$k = 1.380\,66 \cdot 10^{-23}$ J/K	ρ_N	m^{-3}	particle density
	T	K	temperature
	k	J/K	Boltzmann constant

This definition no longer involves extensive variables.

19.6.1.1   Gas constants

1. Boltzmann constant and universal gas constant

Boltzmann constant, k, the proportionality factor of the ideal gas law,

$$k = 1.380\,66 \pm 0.000\,10 \cdot 10^{-23} \text{ J/K} \,.$$

Universal gas constant, **general gas constant** R, the product of Avogadro number and Boltzmann constant.

gas constant = Boltzmann constant $\cdot$ Avogadro number			
	Symbol	Unit	Quantity
$R = k \cdot N_A$	R	J/(K mol)	universal gas constant
$N_A = 6.022\,136\,7 \cdot 10^{23}$ mol^{-1}	N_A	mol^{-1}	Avogadro number
	k	J/K	Boltzmann constant

The value of the universal gas constant R is

$$R = 8.314 \text{ J/(mol K)} \,.$$

2. Equation of state of an ideal gas

The gas law reads:

ideal gas equation of state (universal gas constant)			
	Symbol	Unit	Quantity
$\dfrac{pV}{T} = nR = Nk$ $pV = nRT = NkT$ $k = 1.38066 \cdot 10^{-23}$ J/K	p	Pa	pressure
	V	m^3	volume
	T	K	temperature
	n	mol	quantity of substance
	R	J/(K mol)	universal gas constant
	N	1	particle number
	k	J/K	Boltzmann constant

When applying this equation of state, it is a disadvantage that, in general, the number of moles cannot be determined directly.

In technical thermodynamics, the following equation of state is frequently used:

ideal gas equation of state (specific gas constant)			
	Symbol	Unit	Quantity
$\dfrac{pV}{T} = mR_s$ $pV = mR_s T$	p	Pa	pressure
	V	m^3	volume
	T	K	temperature
	m	kg	gas mass
	R_s	J/(K kg)	specific gas constant

3. Specific gas constant,

R_s, or **individual gas constant**, R_i, material-dependent proportionality constant of the equation of state frequently used in technical thermodynamics.

➤ In technical thermodynamics, the specific gas constant is mostly denoted simply by R. The index s has been added here in order to avoid confusion with the universal gas constant. For distinct substances, one frequently attaches a material index $R_1, R_2 \ldots$ to the specific gas constant.

The specific gas constant is material-dependent.

specific gas constant = $\dfrac{\text{universal gas constant}}{\text{molar mass}}$			$\mathbf{L^2 T^{-2} \Theta^{-1}}$
	Symbol	Unit	Quantity
$R_s = \dfrac{R}{M} = \dfrac{nR}{m}$ $R = 8.314$ J/(K mol)	R_s	J/(K kg)	specific gas constant
	R	J/(K mol)	universal gas constant
	M	kg/mol	molar mass
	n	mol	amount of substance
	m	kg	mass

For specific gas constants of various gases, see **Tab. 22.2/2**.

4. Representation of the pressure by the specific gas constant

Representation by the gas density:

▲ The pressure is the product of density $\rho = m/V$, temperature and specific gas constant.

pressure = density · specific gas constant · temperature			$ML^{-1}T^{-2}$
	Symbol	Unit	Quantity
$p = \rho R_s T$	p	Pa	pressure
	ρ	kg/m^3	density
	T	K	temperature
	R_s	J/(K kg)	specific gas constant

19.6.1.2   Gas mixtures

Gas mixture, system of several kinds of particles, with $N_1, N_2, \ldots, N_n$ particles of type $i = 1, \ldots, n$.

Mole fraction, x_i, the fraction of particles of one kind in the total amount,

$$x_i = \frac{N_i}{N_1 + N_2 + \cdots + N_n}, \quad \sum_i x_i = 1.$$

The mole fraction gives the percentage composition of the system. It is an intensive variable and may take distinct values in different phases.

The **specific gas constant of a gas mixture**, R_G, can be written as

$$R_G = \frac{R_1 m_1 + R_2 m_2 + \cdots}{m_1 + m_2 + \cdots} = \frac{\sum_i R_i m_i}{\sum_i m_i}.$$

19.6.1.3   Calculation of quantities from the gas law

The following conversion formulae are based on the following definitions, in addition to the ideal gas law:

definitions for conversion			
$R = N_A k$	Symbol	Unit	Quantity
$R = M R_s$	R	J/(K mol)	universal gas constant
$m = \rho V$	R_s	J/(K kg)	specific gas constant
$m = nM$	k	J/K	Boltzmann constant
$N = n N_A$	M	kg/mol	molar mass
$N = \rho_N V$	m	kg	mass of gas
$n = \rho_m V$	n	mol	amount of substance
$\rho = \rho_m M$	N_A	mol^{-1}	Avogadro number
$k = 1.380\,66 \cdot 10^{-23}$ J/K	N	1	particle number
$R = 8.314$ J/(K mol)	ρ	kg/m^3	density
$N_A = 6.022\,136\,7 \cdot 10^{23}$ mol^{-1}	ρ_N	m^{-3}	particle density
	ρ_m	mol/m^3	molar density

Pressure in the ideal gas:

$$p = \frac{nRT}{V} = \frac{NkT}{V} = \frac{mR_sT}{V},$$

$$= \rho_m RT = \rho_N kT = \rho R_s T.$$

Volume of an ideal gas:

$$V = \frac{nRT}{p} = \frac{NkT}{p} = \frac{mR_sT}{p},$$

$$= \frac{n}{\rho_m} = \frac{N}{\rho_N} = \frac{m}{\rho}.$$

Temperature of an ideal gas:

$$T = \frac{pV}{nR} = \frac{pV}{Nk} = \frac{pV}{mR_s},$$

$$= \frac{p}{\rho_m R} = \frac{p}{\rho_N R} = \frac{p}{\rho R_s}.$$

Density of an ideal gas:

$$\rho = \frac{p}{R_s T} = \frac{pM}{RT} = \frac{pM}{N_A kT},$$

$$= \rho_N \frac{M}{N_A} = \rho_m M = \rho_N \frac{M}{N_A}.$$

19.6.1.4   Barometric formula

Barometric formula describes how the barometric pressure varies as a function of altitude above Earth's surface, assuming a constant gravitational acceleration.

The idea of the barometric formula is that the weight of a volume of gas is compensated by the pressure differential at upper and lower face of the volume: Then, one obtains the differential equation

$$\frac{\mathrm{d}p}{\mathrm{d}z} = -\frac{mg}{kT}p.$$

The solution of such a differential equation is an exponential function:

barometric formula			
	Symbol	Unit	Quantity
$p(z) = p_0\, e^{-\frac{mgz}{kT}}$ $k = 1.380\,66 \cdot 10^{-23}$ J/K	p	Pa	pressure
	p_0	Pa	pressure at $z = 0$
	z	m	altitude
	m	kg	mass of gas particle
	g	m/s^2	gravitational acceleration
	k	J/K	Boltzmann constant
	T	K	temperature

▲   The pressure in Earth's atmosphere decreases exponentially with the altitude. Here, it is assumed that the temperature is constant over the volume (isothermal atmosphere).

Using the equation of state $p = \rho_N k T$, with the specific density $\rho = m \cdot \rho_N$ one obtains by substituting the values p_0 and ρ_0:

$$p(z) = p_0 \cdot e^{-\frac{\rho_0 g z}{p_0}} = p_0 \cdot e^{-\frac{z}{z_0}}, z_0 = \frac{p_0}{\rho_0 g}.$$

Pressure correction factors may be found in **Tab. 22.5.1**.

19.6.2 Equation of state of real gases

The **equation of state of ideal gases** holds only in the limit of very low density. For real gases one must take into account the following properties in addition:
- The particles have a finite volume.
- The particles mutually interact.

19.6.2.1 Virial expansion of the real gas

Virial expansion, extension of the equation of state by inclusion of additional terms. In general, one uses a polynomial expansion in the pressure (or the density) with temperature-dependent coefficients. The standard representation of the virial expansion is as follows:

virial expansion of the equation of state of real gases			
	Symbol	Unit	Quantity
$p V_{\text{mol}} = RT \left(1 + \dfrac{B(T)}{V_{\text{mol}}} + \dfrac{C(T)}{V_{\text{mol}}^2} + \cdots \right)$ $V_{\text{mol}} = \dfrac{V}{n}$ $R = 8.314 \, \text{J/(K mol)}$	p	Pa	pressure
	V_{mol}	m³/mol	molar volume
	V	m³	volume
	n	mol	amount of substance
	R	J/(K mol)	universal gas constant
	T	K	temperature
	$B(T)$	mol/m³	second virial coefficient
	$C(T)$	mol²/m⁶	third virial coefficient

Virial coefficient, temperature-dependent coefficient in front of the power of an intensive quantity in the virial expansion (**Fig. 19.18**).

The virial coefficients depend on the substance, they may be taken from tables. Visually an expansion up to the second term ($B(T)$) is sufficient.

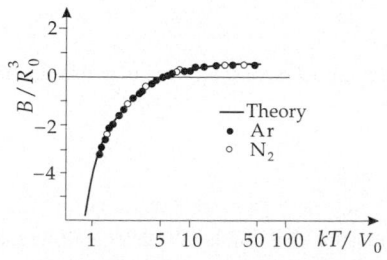

Figure 19.18: Virial coefficients of various gases, experiment (points) and theory (line).

19.6.2.2   Van der Waals equation

1. Assumptions for deriving the Van der Waals equation

Van der Waals equation, an equation of state for real gases, set up by Van der Waals (1873), with the following additions:

- Only the freely accessible volume is taken into account. The internal volume of the gas particles is subtracted from the total gas volume.
- The predominantly attractive interaction of the particles leads to a contraction of the gas. Hence, the pressure on the confining walls is reduced.

2. Internal volume and internal pressure

Internal volume, Nb', the volume occupied by the N particles. It is subtracted from the volume of the vessel. The volume correction is proportional to the particle number,

$$V \mapsto V - Nb'.$$

Internal pressure, a force per unit area acting inward. It originates from the mutually attractive force between the particles that cancels in the interior of the gas, but remains active for gas particles at the boundary (**Fig. 19.19**).

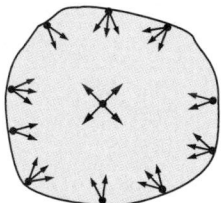

Figure 19.19: Illustration of internal pressure. Particles at the boundary are subject to the intermolecular forces only from a hemisphere.

3. Derivation of the Van der Waals equation

In general, the attractive force is assumed to be a dipole interaction. The potential is then proportional to the negative sixth power of the distance. The reduction of the pressure depends on the number of particles at the surface (proportional to the particle density) and the mean distance between the gas particles (also approximately proportional to the particle density).

The magnitude of the pressure reduction must be added to the real (measured) pressure to obtain the (hypothetical) pressure of the ideal gas,

$$p \mapsto p + a' \left(\frac{N}{V}\right)^2.$$

Frequently, the Avogadro number is incorporated into the constant. One then uses the molar density:

$$p \mapsto p + a\rho_{\text{mol}}^2.$$

After inserting internal volume and internal pressure, the Van der Waals equation reads:

Van der Waals equation			
	Symbol	Unit	Quantity
$\left(p + \left(\dfrac{n}{V}\right)^2 a\right)(V - nb) = nRT$ $R = 8.314 \text{ J}/(\text{K mol})$	p	Pa	pressure
	n	mol	amount of substance
	V	m^3	volume
	a	Nm^4/mol^2	constant of internal pressure
	b	m^3/mol	constant of internal volume
	R	J/(K mol)	universal gas constant
	T	K	temperature

The constants a and b for internal pressure and internal volume, respectively, are material parameters, see **Tab. 22.2/3**.

4. Van der Waals equation in technical thermodynamics

In technical thermodynamics, one often calculates with gas masses. The constants a and b are correspondingly redefined.

conversion of molar to specific constants			
	Symbol	Unit	Quantity
$a_S = \dfrac{a}{M^2}$ $b_S = \dfrac{b}{M}$	a_S	Nm^4/kg^2	specific constant of internal pressure
	a	Nm^4/mol^2	molar constant of internal pressure
	M	kg/mol	molar mass
	b_S	m^3/kg	specific internal volume
	b	m^3/mol	molar internal volume

In the technical literature, the specific constants are frequently denoted by a and b. The notation a_S and b_S (s standing for specific) used here serves only for a clear distinction.

5. Pressure of the Van der Waals interaction

$$p = \frac{n \cdot R \cdot T}{V - nb} - a\frac{n^2}{V^2}.$$

A graphic representation of the pressure as a function of the volume (at constant temperature) is given by the difference of a simple hyperbola and a quadratic hyperbola (**Fig. 19.20**).

➤   In general, the calculation of the volume from the pressure is no longer unique.

▲   For high temperatures and low densities, the Van der Waals equation approaches the equation for the ideal gas.

Isotherm, a curve for constant temperature.

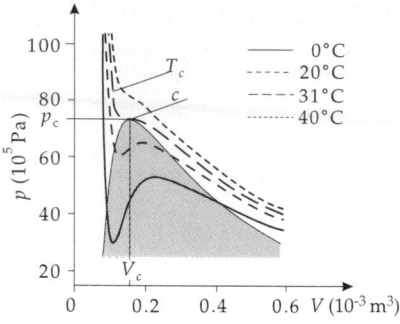

Figure 19.20: Van der Waals isotherms for various temperatures in the pV-diagram. Grey: phase coexistence region, c: critical point (saddle point), T_c: critical isotherm, p_c, V_c: pressure and volume at the critical point.

19.6.2.3   Region of phase coexistence

For low temperatures and certain volumes, the pressure becomes negative according to the Van der Waals equation. Furthermore, for positive pressure values there are also regions in which the pressure decreases with decreasing volume. In these regions, the system cannot be stable, but will contract on its own to a smaller volume. These unstable regions describe the gas-liquid phase transition. The gaseous and the liquid phase occur simultaneously.

Phase coexistence region, a region in which two phases may coexist (see p. 722).

Maxwell construction, the prescription to replace the isotherms in the non-equilibrium region by horizontal lines through the phase coexistence region (**Fig. 19.21**).

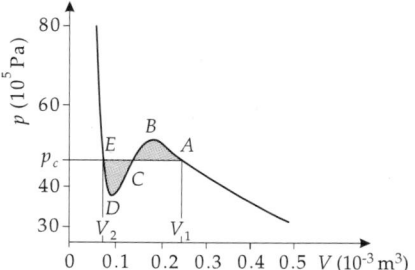

Figure 19.21: Maxwell construction for Van der Waals isotherms. The magnitudes of the areas between the curve and the straight substitution line must balance each other.

19.6.2.4   Critical point

Critical isotherm, curve for the temperature T_c at which the pressure as a function of the volume has a saddle point.

Critical temperature, T_c, the temperature corresponding to the critical isotherm.

critical temperature (Van der Waals equation)			Θ
	Symbol	Unit	Quantity
$T_c = \dfrac{8a}{27Rb}$ $R = 8.314 \text{ J}/(\text{K mol})$	T_c	K	critical temperature
	a	Nm^4/mol^2	molar coefficient of internal pressure
	b	m^3/mol	molar coefficient of internal volume
	R	J/(K mol)	universal gas constant

Critical point, the saddle point of the critical isotherm. Below the critical temperature one may always construct horizontal straight lines. Above the critical temperature the derivative $\mathrm{d}p/\mathrm{d}V$ is always negative.

Critical pressure, p_c, pressure at the critical point.

critical pressure (Van der Waals equation)			$\mathbf{ML^{-1}T^{-2}}$
$p_c = \dfrac{a}{27b^2}$	Symbol	Unit	Quantity
	p_c	Pa	critical pressure
	a	Nm^4/mol^2	molar coefficient of internal pressure
	b	m^3/mol	molar internal volume

Critical molar volume, v_c, the volume of one mole at the critical point.

critical molar volume $= 3 \cdot$ molar internal volume			$\mathbf{L^3N^{-1}}$
$v_c = 3b$	Symbol	Unit	Quantity
	v_c	m^3/mol	critical molar volume
	b	m^3/mol	molar internal volume

19.6.2.5   Law of coinciding states

Reduced variable, representation of a state variable in units of the value at the critical point,

$$\overline{p} = \frac{p}{p_c} \,, \quad \overline{v} = \frac{v}{v_c} \,, \quad \overline{T} = \frac{T}{T_c} \,.$$

Law of coinciding states, statement introduced by Van der Waals: all simple gases satisfy the same Van der Waals equation in the reduced variables.

  Simple gas, gas of particles having a small electric dipole moment and whose atoms or molecules are not strongly correlated even in the liquid phase.

■   Noble gases, N_2, O_2, H_2 or CO, CH_4 are simple gases.

Van der Waals equation in reduced variables:

$$\left(\overline{p} + \frac{3}{\overline{v}^2} \right) (3\overline{v} - 1) = 8\overline{T}.$$

19.6.2.6   Van der Waals equation as virial expansion

1. Approximation of the Van der Waals equation

An approximation to the Van der Waals equation is obtained by replacing the molar density n/V in the term corresponding to the internal pressure by the value of the ideal gas, $n/V \approx p/RT$:

$$\left(p + \frac{p^2}{(RT)^2} a \right) (V - nb) = nRT,$$

$$R = 8.314 \, \text{J/K mol} \,.$$

Representation as expansion:

$$pV = \frac{nRT}{1 + \dfrac{pa}{(RT)^2}} + pnb.$$

Using constants normalized to the particle number,

$$a' = \frac{a}{N_A^2}, \; b' = \frac{b}{N_A}, \; N_A = 6.022\,136\,7 \cdot 10^{23}\,\text{mol}^{-1},$$

the representation reads:

$$\left(p + \frac{p^2}{(kT)^2}\,a'\right)(V - Nb') = NkT,$$

$$k = 1.380\,66 \cdot 10^{-23}\,\text{J/K}.$$

Representation as expansion:

$$pV = \frac{NkT}{1 + \dfrac{pa}{(kT)^2}} + pNb.$$

2. Representation of the approximation using specific constants

If specific constants are used for technical thermodynamics, then:

conversion of molar into specific constants			
	Symbol	Unit	Quantity
$a_s = \dfrac{a}{M^2}$	a_s	Nm^4/kg^2	specific constant of internal pressure
$b_s = \dfrac{b}{M}$	a	Nm^4/mol^2	molar constant of internal pressure
	M	kg/mol	molar mass
$R_s = \dfrac{R}{M}$	b_s	m^3/kg	specific internal volume
	b	m^3/mol	molar internal volume
$R = 8.314\,\text{J/(K mol)}$	R	J/(K mol)	universal gas constant
	R_s	J/(K kg)	specific gas constant

Representation with specific quantities:

$$\left(p + \frac{p^2}{(R_sT)^2}\,a_s\right)(V - mb_s) = mR_sT.$$

Expansion for low pressure and high temperatures:

$$pV = nRT + n\left(b - \frac{a}{RT}\right)p + \cdots.$$

19.6.3   Equation of states for liquids and solids

Liquids and solids expand in all directions under heating—as do gases. One should notice, however, the **anomaly of water** between 0 °C and 4 °C. Microscopically, the changes of the macroscopic dimensions of a body with temperature originate from changes of the potential and kinetic energies, hence from the variations of the interatomic and intermolecular distances.

1. Equation of state for solids and liquids,

describes the variation of the volume with temperature and pressure.

▲ The change of the volume of a solid or a liquid is to first approximation linearly related to the change of temperature and pressure.

This formulation yields a good description over a wide range of the variables.

equation of state of solid or liquid			
	Symbol	Unit	Quantity
$V(T, p) = V_0 \{1 + \gamma (T - T_0)$	V	m^3	volume
	T	K	temperature
$- \kappa (p - p_0)\}$	p	Pa	pressure
	γ	K^{-1}	volume-expansion coeff.
	κ	Pa^{-1}	compressibility

$V_0 = V(T_0, p_0)$ is an arbitrary initial state. Temperature differences $T - T_0$ may be given also in °C instead of kelvin.

2. Special coefficients of the equation of state

Volume-expansion coefficient, γ, SI unit 1/kelvin, describes the temperature-dependent volume expansion at constant pressure.

Representation as partial derivative:

$$\gamma = \lim_{\Delta T \to 0} \frac{\Delta V}{V_0 \Delta T} = \frac{1}{V_0} \frac{\partial V}{\partial T}\bigg|_{p=p_0} .$$

Compressibility, κ, describes the pressure-dependent change of volume at constant temperature.

Representation as partial derivative:

$$\kappa = -\frac{1}{V_0} \frac{\partial V}{\partial p}\bigg|_{T=T_0} .$$

Compression modulus, K, the reciprocal value of the compressibility,

$$K = \frac{1}{\kappa} .$$

➤ In ultrasound technology, the compression modulus K is also denoted by C_B.

| M | The compressibility may be determined statically (directly) by measuring the change of volume for a known force and surface, and dynamically by ultrasonic experiments. Strictly speaking, in the latter approach it is the compression modulus that is measured.

Expansion coefficients of numerous materials may be found in the **Tab. 22.3**.

For numerous materials, the expansion coefficient lies:

- for solids in the range $\gamma \approx 10^{-5}$ K^{-1},
- for liquids about 1–2 orders of magnitude above this value ($10^{-3} - 10^{-4}$ K^{-1}).

Values of compressibility may also be found in **Tab. 22.3**. For solids and liquids, they are of the order of magnitude $\kappa \approx 10^{-6}$ bar^{-1}.

➤ The compressibility of liquids and solids is far lower than that of gases.

Small changes in temperature cause changes of volume similar to those caused by large changes of pressure. The consequence is that even small changes in temperature at constant volume may cause very high pressures.

■ If water were not compressible, the water level of the oceans would rise by about 30 m and large coastal regions would be submerged!

3. *Linear expansion coefficient,*

α, describes the variation of a length with temperature:

$$L_2 = L_1 + \Delta L = L_1 + \alpha L_1 \Delta T$$
$$= L_1(1 + \alpha \Delta T).$$

Representation as partial derivative:

$$\alpha = \frac{1}{L} \frac{\partial L}{\partial T}\bigg|_{p=p_0}.$$

➤ The linear expansion of bodies must be taken into consideration for constructions underlying fluctuations of temperature.

■ The space between lengths of railroad tracks is placed there to allow for the thermal expansion of the steel.

Bridges have a fixed bearing at one end and a roller bearing at the other end.

| M | **Dilatometer,** measures the linear expansion of a sample by the capacitance of a cell into which the sample is mounted.

The linear expansion under temperature changes may serve for measuring the temperature.

■ **Mercury thermometer.**

Bimetal, distinct expansion of two metallic strips. Application: control rods in muffle furnaces.

4. *Surface expansion coefficient,*

β, describes the change of a surface with temperature:

$$A_2 = A_1 + \Delta A = A_1 + \beta A_1 \Delta T$$
$$= A_1(1 + \beta \Delta T).$$

If the linear expansion is small compared with the total length, the linear expansion coefficient α, the surface expansion coefficient β, and the volume expansion coefficient γ are related as follows:

$$\beta = 2\alpha, \qquad \gamma = 3\alpha.$$

19.6.3.1 Anomaly of water

Nearly all substances have a positive expansion coefficient over the entire range of temperatures, i.e., the volume increases with increasing temperature, independent of the temperature range.

Water anomaly, the peculiar property of water not to have a positive expansion coefficient at every temperature.

▲   The expansion coefficient of water between $0\,°C$ and $4\,°C$ is **negative**. At $4\,°C$, $\gamma = 0$.

▲   Water has maximum density at $4\,°C$.

■   A liter of water at $4\,°C$ is heavier than a liter of water at the freezing point. Moreover, there is a step-like increase of volume at freezing. Hence, ice floats on water.

Fig. 19.22 shows the volume expansion of 1 kg of water between $-10\,°C$ and $50\,°C$. Two striking properties emerge:

•   At low temperature, the expansion coefficient is negative.

•   At high temperature, the rise is not linear. The expansion coefficient is not constant, but temperature-dependent.

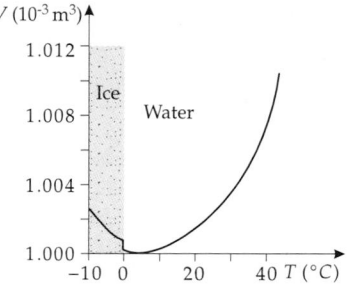

Figure 19.22: Thermal expansion of water. The minimum of the curve is at $4\,°C$.

The behavior of water under pressure is analogous to the temperature dependence.

▲   Under pressure, ice melts to water.

The statement that water cannot freeze to ice when under pressure is equivalent to the statement above.

■   Lakes do not freeze from the bottom.

This anomaly of water is of great importance for many biological processes.

20
Heat, conversion of energy and changes of state

20.1 Energy forms

The total energy E of a system is a macroscopic quantity that plays an important role in thermodynamics. The total energy is the product of the mean energy of the particles times the particle number. The energy of a particular particle, as well as the distribution of the total energy E over the individual particles, is of minor importance.

▲ **First law of thermodynamics**: the total internal energy of a system is a conserved quantity. Energy cannot be created or destroyed, but only transferred from one system to another.

Energy may occur in various forms, and energy transport may proceed in different ways. Various forms of energy may be partly converted from one to another.

■ Heat appears when braking a moving body by friction.
Generators convert mechanical work into electric energy.

Efficiency of conversion, the ratio of the converted energy to the input energy (see p. 71). The remaining fraction of energy is not lost, but occurs in another form of energy.

■ In a combustion engine, chemical energy is partly converted into mechanical work and partly into heat.

20.1.1 Energy units

The following energy units are used preferentially:

- **Newton meter**, Nm, used for mechanical work.
- **Joule**, J, used for heat.
- **Watt second**, Ws, used for electric work.
- ▲ The units of energy are equivalent to each other:

$$1\,\text{Nm} \;=\; 1\,\text{J} \;=\; 1\,\text{Ws} \;=\; 1\,\text{VAs} \;=\; 1\,\frac{\text{kg\,m}^2}{\text{s}^2}.$$

675

■   A current of 1 ampere flowing for one second through a voltage of 6 volts needs just as much energy as lifting a weight of 6 newtons by 1 meter.

20.1.1.1   Non-SI units

Erg, 10^{-7} joule.

$$1\,\text{J} = 10^7\,\text{erg}.$$

Calorie, cal, an older, no longer recognized unit, the amount of heat that is needed to heat 1 g of water at 14.5 °C by one degree:

$$1\,\text{cal} = 4.187\,\text{J} \qquad 1\,\text{J} = 0.239\,\text{cal}.$$

British Thermal Unit or **BTU**, a no longer recognized unit that is still used in Anglo-Saxon countries:

$$1\,\text{BTU} = 1055.06\,\text{J}.$$

Electron volt, a quantity used in atomic and nuclear physics, representing the work done if an elementary charge is accelerated by a potential difference of 1 V.

➤   By setting $\hbar = c = 1$ instead of $\hbar c \approx 197.32\,\text{MeV fm}$, the energy in quantum mechanics may also be represented by an inverse length in fm ($= 10^{-15}$ m):

$$1\,\text{eV} = 1.602 \cdot 10^{-19}\,\text{J} = 5.063 \cdot 10^{-9}\,\text{fm}^{-1}\hbar c\,, \qquad 1\,\text{J} = 6.242 \cdot 10^{18}\,\text{eV}\,.$$

20.1.2   Work

1. Work in thermodynamic systems

Work, corresponds in thermodynamics to the mechanical definition of work: The **work performed on the system** is counted **positive** and the **work extracted from the system** is **negative**.

Work, W, SI unit newton meter (Nm), the product of the force acting along a path times the distance covered:

work = force · path		$\mathbf{ML^2T^{-2}}$	
$\Delta W = -\vec{\mathbf{F}} \cdot \Delta \vec{\mathbf{s}},$	Symbol	Unit	Quantity
	W	Nm	work
$W_{1.2} = -\int_{s_1}^{s_2} \vec{\mathbf{F}} \cdot \mathrm{d}\vec{\mathbf{s}}$	$\vec{\mathbf{F}}$	N	force
	$\vec{\mathbf{s}}$	m	distance

The work is a scalar product of two vectors.

▲   Forces acting perpendicular to the displacement do not do work.

2. Compression work,

is done when a gas is compressed against the internal pressure (**Fig. 20.1**).

■   The volume of a cylinder filled with gas is reduced.

Work is the product of pressure and change in volume. The change in volume may proceed by displacing the boundary surface of a volume.

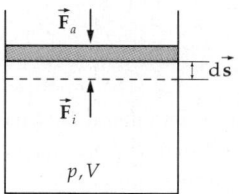

Figure 20.1: Work upon compression. The work to be done corresponds to the product of internal force times displacement, or pressure times volume difference.

work = pressure · surface · displacement	ML^2T^{-2}		
	Symbol	Unit	Quantity
$\Delta W = p\,A\Delta s$	W	Nm	work
	p	Pa	pressure
	A	m^2	surface area
	Δs	m	displacement

▲   The displacement is counted positively if the volume is diminished. As a result of this definition of displacement, Δs and ΔV have different signs,

$$\Delta V = -A\Delta s\,.$$

Therefore, the work must be viewed as the negative value of the product of pressure and change in volume. The change in volume is positive for enlarging, and negative for diminishing.

work = −pressure · volume change	ML^2T^{-2}		
	Symbol	Unit	Quantity
$\Delta W = -p\Delta V$	ΔW	Nm	work
	p	Pa	pressure
	ΔV	m^3	volume change

▲   The mechanical work done ΔW not only depends on the limits of integration, i.e., on the initial and final state of the system, but also on the path from the initial to the final state. Mathematically, this means that there is no total differential $dW = F\,ds$.

20.1.3 Chemical potential

Chemical potential, μ, SI unit joule, quantity of work to be done in order to account for a change in particle number so that the system remains in equilibrium.

chemical potential = $\dfrac{\text{input work}}{\text{change of particle number}}$	ML^2T^{-2}		
	Symbol	Unit	Quantity
$\mu = \dfrac{\Delta W}{\Delta N}$	μ	J	chemical potential
	W	J	work
	N	1	particle number

Hence, the work received or needed when ΔN additional particles are added to the system is

$$\Delta W = \mu \cdot \Delta N .$$

Energy is needed because the particles added cannot be introduced cold into the system without thermal changes. In order to be in thermal equilibrium with the system the particles must carry the mean energy of the particles already present.

20.1.4 Heat

Heat is a particular form of energy connected with the temperature increase of a substance. An input of heat causes a temperature increase. The relation for the input of heat and temperature increase is determined by a material property, the **heat capacity** C.

➤ In a phase transition, there may be an input or output of heat (e.g., melting heat or evaporation heat) without a change in temperature. However, in this case the heat capacity tends to infinity, and the definition given here can no longer be applied.

1. Quantity of heat

Heat, ΔQ, SI unit joule, the energy absorbed in a temperature increase ΔT:

quantity of heat = heat capacity · temperature difference		$\mathbf{ML^2T^{-2}}$	
$\Delta Q = C \Delta T$	Symbol	Unit	Quantity
$Q_{1.2} = \displaystyle\int_{T_1}^{T_2} C \mathrm{d}T$	Q	J	quantity of heat
	C	J/K	heat capacity
	T	K	temperature
$\mathrm{d}Q = C\mathrm{d}T$			

The differential representation holds in a mathematically strict sense only if no additional mechanical or chemical work is expended. Otherwise, $\mathrm{d}Q$ is not a total differential.

2. Measurement of heat

Heat is measured in calorimeters by determining the change in temperature for known heat capacity C_K of the calorimeter construction. Possible losses of heat must be taken into account:

$$\Delta Q = C_K \cdot \Delta T + \text{heat loss}.$$

Calorimeters are used for measuring quantities of heat. The most common types are:

Liquid calorimeter, most common construction: The reaction vessel is placed in a container with liquid, insulated against the surroundings.

Metallic calorimeter, particularly suited for wide temperature ranges: A block of metal (silver, copper, aluminum) confines the reaction zone.

Combustion calorimeters, are used for fast combustion reactions. Examples:
- **Bomb calorimeter** after Berthelot (for solids and liquids),
- **Exchange calorimeter** (also called wet calorimeter, for gases),
- **Mixture calorimeter** (dry calorimeter, also for gases).

➤ The heat exchange in chemical reactions may be determined with an accuracy of picodegrees. Principle of measurement: a 0.4 μm thick aluminum film is deposited on one side of a 0.4 mm long and 1.5 μm thick silicon strip. The system responds

to heating like a bimetallic strip does. The magnitude of bending is determined from the reflection angle of a laser beam.
■   A block of copper (mass 200 g) with a heat capacity of 76.6 J/K is heated from 17 to 23 °C. Assuming no heat loss, the quantity of heat absorbed is

$$\Delta Q = C \cdot \Delta T = 76.6 \text{ J/K} \cdot (23 - 17)\,°C = 459.6 \text{ J}.$$

20.2   Energy conversion

Different forms of energy may be converted into each other.
■   A weight can be lifted by electric energy.
    A generator converts mechanical work to electric energy.
In principle, one might assume that these conversions proceed completely.
▲   A real energy converter always shows losses.
But these losses do not mean that energy is lost; rather that only a fraction of the energy has been converted into the desired form.
■   In the conversion of mechanical energy, waste heat may occur.
▲   The total energy is a conserved quantity. Energy is not lost.
It turns out, however, that not all forms of energy can be converted completely into each other.
▲   Heat cannot be converted completely into mechanical or electric energy.
➤   This is the statement of the **second law of thermodynamics**.
By contrast, mechanical and electric energy may be converted completely into heat.

20.2.1   Conversion of equivalent energies into heat

Thermal energy may be produced in different ways. Possible ways include the conversion of mechanical energy (e.g., by friction) or of electric energy.

20.2.1.1   Electric energy

Electric energy may be converted loss-free into heat by the ohmic resistance of a conductor. But heat energy cannot be converted completely into electric energy.

heat = voltage · current · time			ML^2T^{-2}
	Symbol	Unit	Quantity
$Q = V \cdot I \cdot t$	Q	J	quantity of heat
	V	V	electric voltage
$Q = P \cdot t$	I	A	electric current
	t	s	time
	P	W	power

■   Assume an immersion heater (220 V nominal voltage, 4.5 A current input) heats water for 1 minute. The electric energy is converted completely into heat. The quantity of heat obtained is

$$Q = W_{el} = P_{el} \cdot t = V \cdot I \cdot t = 220 \text{ V} \cdot 4.5 \text{ A} \cdot 60 \text{ s} = 59400 \text{ Ws} = 59.4 \text{ kJ}.$$

The quantity of heat is sufficient to heat a glass of water (200 ml) by 75 °C.

Using Ohm's law, one obtains for the heat released at a resistor:

$\text{heat} = \dfrac{(\text{voltage})^2}{\text{resistance}} \cdot \text{time}$			$\mathbf{ML^2T^{-2}}$
	Symbol	Unit	Quantity
$Q = \dfrac{V^2 t}{R}$	Q	J	produced heat
	V	V	voltage
	t	s	time interval
$Q = I^2 Rt$	R	Ω	electric resistance
	I	A	current

■ A voltage of 5 volts is applied across a resistor ($R = 4.7$ kΩ). The waste heat of the resistor per hour is

$$Q = \frac{V^2}{R}t = \frac{(5 \text{ V})^2}{4.7 \text{ k}\Omega} \cdot 3600 \text{ s} = 19.15 \text{ J}.$$

20.2.1.2 Mechanical energy

Mechanical energy, like electric energy, may be converted completely into heat. By contrast, thermal energy cannot be converted completely into mechanical energy. Here, mechanical energy may occur as kinetic energy or as potential energy (e.g., tension of a spring).

$\text{heat} = \text{kinetic energy} + \text{potential energy}$			$\mathbf{ML^2T^{-2}}$
	Symbol	Unit	Quantity
$\Delta Q = \Delta W_{\text{kin}} + \Delta W_{\text{pot}}$	ΔQ	J	heat produced
	ΔW_{kin}	Nm	input of kinetic energy
	ΔW_{pot}	Nm	input of potential energy

■ A sphere of mass 5 g with a velocity of 150 m/s is stopped by a sandbag. The kinetic energy is completely converted into heat. The heat released is

$$Q = W_{\text{kin}} = \frac{1}{2}mv^2 = \frac{1}{2} \cdot 0.005 \text{ kg} \cdot \left(150 \, \frac{\text{m}}{\text{s}}\right)^2 = 56.25 \text{ Nm} = 56.25 \text{ J}.$$

20.2.1.3 Combustion energy

Combustion energy, most important form of conversion of chemical energy into heat. Here, predominantly materials containing carbon and hydrogen are oxidized.
■ Oil and natural gas consist mainly of hydrocarbon chains (predominantly alkanes) of various lengths. In their combustion, predominantly carbon dioxide (CO_2) and water (H_2O) are released, but also other substances, for example sulphur dioxide (SO_2) and nitrogen oxides (NO_x) due to contaminations.

1. Specific caloric value,

H, of a liquid or solid material, SI unit joule per kilogram, the thermal energy released per mass unit by combustion if the water vapor produced in the reaction is not condensed.

specific caloric value = $\dfrac{\text{quantity of heat}}{\text{mass of substance}}$			L^2T^{-2}
	Symbol	Unit	Quantity
$H = \dfrac{Q}{m}$	H	J/kg	specific caloric value
	Q	J	heat produced
	m	kg	mass of substance burned

One might also define the specific caloric value for gaseous substances in a similar way. However, the volume of a gas is more easily determined than its mass.

➤ Since the volume depends on temperature and pressure, the standard volume under standard conditions ($p = 101.325$ kPa, $T = 273.15$ K $= 0$ °C) is being used.

2. Specific caloric value of gases,

H_g, SI unit joule per cubic meter, the heat released by gaseous substances per unit of volume under standard conditions.

specific caloric value of gases = $\dfrac{\text{quantity of heat}}{\text{volume}}$			$ML^{-1}T^{-2}$
	Symbol	Unit	Quantity
$H_g = \dfrac{Q}{V_n}$	H_g	J/m^3	specific caloric value of gases
	Q	J	quantity of heat
	V_n	m^3	volume at standard conditions

The specific caloric values of selected substances are listed in **Tab. 22.9**.

■ Most solid (dry) fuels have a caloric value of approximately $20 - 50$ MJ/kg, oil of approximately $40 - 50$ MJ/kg, gases, approximately $10 - 130$ MJ/m^3.

3. Gross caloric value,

H_o, of a substance, the energy per unit mass produced directly by combustion.

However, part of this energy is needed to vaporize the water produced in the combustion of hydrogen. This energy may be used again in the condensation of the water vapor.

➤ Caloric value and gross caloric value differ by the heat of evaporation of the water produced.

4. Upper and lower heating value

Upper heating value H_o, formerly used notation for the gross caloric value or combustion heat.

Lower heating value H_l, formerly used notation, nowadays called caloric value.

Heating power boiler: In older technical systems only, the (lower) heating value is of importance for the useful thermal energy. More recent systems are operated such that the temperature of the waste gases lies below the dew-point so that the condensation heat of the evaporated water is regained. Hence, the full heating value may be exploited, which means, e.g., for gas heating about 10 % additional utilization.

For the quantity of heat produced by combustion:

quantity of heat = mass · caloric value			$\mathbf{ML^2T^{-2}}$
	Symbol	Unit	Quantity
$Q = m \cdot H$	Q	J	useful quantity of heat
	m	kg	mass (solid/liquid substances)
$Q = V_n \cdot H_g$	H	J/kg	specific caloric value (solid/liquid)
	V_n	m^3	gas volume, standard conditions
	H_g	J/m^3	specific caloric value (gas)

■ 300 g of charcoal is burned. The quantity of heat released is

$$Q = m \cdot H = 0.3 \text{ kg} \cdot 31 \, \frac{\text{MJ}}{\text{kg}} = 9.3 \text{ MJ} \,.$$

20.2.1.4   Solar energy

The irradiation of Earth by the Sun represents heat transport by radiation. The radiation may, for example, be converted into heat. Here the absorptance of the irradiated substance, as well as the angle between the insolation and the normal of the irradiated area, have to be taken into account.

quantity of heat ~ irradiated area · absorption factor · cos(angle)			$\mathbf{ML^2T^{-2}}$
	Symbol	Unit	Quantity
	Q	J	quantity of heat
	q_S	W/m^2	solar constant
$Q = q_S \cdot A \cdot \alpha \cdot t \cdot \cos \varphi$	A	m^2	irradiated area
	α	1	absorption factor
	t	s	time
	φ	1	incidence angle

Solar constant, annual mean of the power of the insolation on Earth per unit area,

$$q_S = 1.37 \, \frac{\text{kW}}{\text{m}^2} \,.$$

The solar constant is only a nominal value, ignoring the influence of clouds, dust, etc.
    About half of the energy flow radiated to Earth is absorbed in the atmosphere.
■ A plate of size 50 cm × 50 cm is irradiated for one hour by the Sun at an angle of 30° with respect to the normal. Assuming an absorption rate of 35 % (including absorption in air), the heat absorption is

$$Q = q_S \cdot A \cdot \alpha \cdot t \cdot \cos \varphi$$

$$= 1.37 \, \frac{\text{kW}}{\text{m}^2} \cdot 0.25 \text{ m}^2 \cdot 0.35 \cdot 3600 \text{ s} \cdot \cos 30° \approx 374 \text{ kJ} \,.$$

20.2.2 Conversion of heat into other forms of energy

The conversion of thermal energy into other forms of energy proceeds in general by means of heat engines, which operate according to the principle of the Carnot cycle (see p. 702).

The basic principle is to induce various changes of state in a system of substances by alternately bringing the system into contact with a cold and a hot thermal reservoir or "heat bath." The system thereby transports heat from the hot bath to the cold bath and does mechanical work that may be converted into other forms of energy.

Efficiency η of energy conversion, a nondimensional quantity, the ratio of the gained mechanical work to the total energy conversion.

The efficiency is always smaller than unity,

$$\eta < 1 .$$

▲   Thermal energy cannot be converted completely into other forms of energy.

The efficiency of a heat engine depends sensitively on the temperatures of the hot and cold bath between which heat is being exchanged.

ideal efficiency = 1 − temperature of cold reservoir / temperature of warm reservoir			**1**

$\eta_C = 1 - \dfrac{T_c}{T_h}$	Symbol	Unit	Quantity
	η_C	1	ideal efficiency
	T_c	K	temperature of cold reservoir
	T_h	K	temperature of hot reservoir

20.2.3 Exergy and anergy

There are forms of energy that can be converted completely into other forms of energy and forms for which this is not true.

■   Mechanical energy can be converted (almost) completely into electric energy, and vice versa. Mechanical energy and electric energy can be converted completely into thermal energy. On the other hand, thermal energy cannot be converted completely into electric or mechanical energy.

1. Classification of forms of energy

Energy forms may be classified as follows:

- **Exergy**, E_x, SI unit joule, fraction of the energy that can be converted **without limit** into other forms of energy.
- Energy forms that can be converted into exergy only in a limited way.
- **Anergy**, B, SI unit joule, fraction of energy that **cannot** be converted at all.
- ■ Unlimited convertible forms (exergy) include mechanical and electric energy. Limited convertible forms are heat, internal energy and enthalpy. They contain fractions of anergy.

2. Separation of the total energy

The total energy may be separated into two parts: mechanically usable energy and mechanically unusable energy.

▲   The total energy consists of exergy and anergy,

$$W_{\text{tot}} = E_x + B .$$

➤   Of course, one of the two parts may vanish.

3. Energy conversion principles

For the conversion of energy:
- Exergy may be converted into anergy.
- Anergy cannot be converted into exergy.

This is directly related to the second law of thermodynamics.

▲ Processes in which exergy is transformed into anergy are irreversible.

▲ In reversible processes, there is no conversion of exergy into anergy.

20.3 Heat capacity

20.3.1 Total heat capacity

1. Heat capacity,

C, SI unit joule per kelvin, sometimes also called total heat capacity, a material property of a body to be able to change its temperature under a certain input of heat. It depends on the amount of substance.

heat capacity $= \dfrac{\text{quantity of heat}}{\text{temperature difference}}$			$\mathbf{ML^2T^{-2}\Theta^{-1}}$
$C = \dfrac{\Delta Q}{\Delta T}$	Symbol	Unit	Quantity
	Q	J	quantity of heat
$C = \dfrac{\mathrm{d}Q}{\mathrm{d}T}$	C	J/K	heat capacity
	T	K	temperature

Temperature differences may also be measured in degrees Celsius instead of Kelvin, without recalculating the formulae.

In a phase transition, the heat capacity of a substance may become formally infinite, since heat is incorporated without leading to a change of temperature.

2. Measurement of heat capacity

The heat capacity of an unknown substance may be determined by measuring the change of the temperature for a known input of heat. The heat influx by conversion of electric energy may be determined to a high precision by measuring current, voltage and time of heating. But one must take into account the efficiency of the heating and the heat capacity of the heating material or heat container (water equivalent of calorimeter),

$$C = \frac{\eta \Delta Q}{\Delta T} - C_K, \qquad \eta: \text{efficiency}, \quad C_K: \text{water equivalent}.$$

■ A liquid is heated by an immersion heater (1000 W) for 15 s and shows a temperature increase of 7.18 K. The heat capacity is

$$C = \frac{\Delta Q}{\Delta T} = \frac{15 \text{ kJ}}{7.18 \text{ K}} = 2.09 \text{ kJ/K}.$$

3. Product representation of heat capacity

The heat capacity may be written as the product of the specific (molar) heat capacity and the total mass (total number of moles). So, this property of the substance used may be factorized into a (general) material property, and the quantity of substance that can be measured easily.

heat capacity = specific heat capacity · total mass			$\mathbf{ML^2T^{-2}\Theta^{-1}}$
	Symbol	Unit	Quantity
$C = n\,c_{mol}$	C	J/K	heat capacity
	n	mol	quantity of substance
$C = m\,c$	m	kg	total mass
	c_{mol}	J/(K mol)	molar heat capacity
	c	J/(K kg)	specific heat capacity

■ Half a liter (500 g) of water with a specific heat capacity of $c = 4.182$ kJ/(kg K) has the heat capacity

$$C = m \cdot c = 0.5 \text{ kg} \cdot 4.182 \text{ kJ/(K kg)} = 2.091 \text{ kJ/K}.$$

20.3.1.1 Heat capacity of mixtures of substances

▲ The total heat capacity of a mixture of different substances is the sum of the individual heat capacities:

$$C = C_1 + C_2 + C_3 + \cdots.$$

20.3.1.2 Water equivalent

In the evaluation of the temperature change of liquids (also solids or gases), the heat capacities of the surrounding vessels, as well as of the measuring device (e.g., thermo probes), must be taken into account. This heat capacity is called **water equivalent** and is denoted C_K or W.

$$W = C_K = m_k \cdot c_k.$$

The total heat capacity of the system is:

total heat capacity = heat capacity + water equivalent			$\mathbf{ML^2T^{-2}\Theta^{-1}}$
	Symbol	Unit	Quantity
$C_{tot} = C + W$	C_{tot}	J/K	total heat capacity
	C	J/K	heat capacity of substance
	W	J/K	water equivalent

M In order to determine the water equivalent, the calorimeter is filled with a definite quantity of water. A definite quantity of heat is absorbed, and the increase in temperature is measured.

20.3.2 Molar heat capacity

Molar heat capacity, c_{mol}, SI unit per kelvin and per mole, the heat capacity of one mole of a certain substance.

It may be defined analogously to the specific heat capacity.

1. Representation of molar heat capacity

Molar heat capacity, the quantity of heat absorbed by one mole of a substance per unit of temperature change.

molar heat capacity = $\dfrac{\text{quantity of heat}}{\text{quantity of substance} \cdot \text{temperature}}$			$\mathbf{ML^2T^{-2}\Theta^{-1}N^{-1}}$
$c_{mol} = \dfrac{\Delta Q}{n \Delta T}$ $\Delta Q = c_{mol} n \Delta T$	**Symbol**	**Unit**	**Quantity**
	c_{mol}	J/(K mol)	molar heat capacity
	ΔQ	J	quantity of heat
	ΔT	K	change of temperature
	n	mol	quantity of substance

Temperature differences can be measured also in degrees Celsius instead of Kelvin, without rewriting formulae.

2. Molar heat capacity as material property

The molar heat capacity is a material property that is defined by the quotient of heat capacity and the number of moles.

The molar heat capacity is the heat capacity per mole of a substance.

➤   In some books on thermodynamics, the molar heat capacity defined here is denoted specific heat capacity or specific heat. This may cause confusion.

molar heat capacity = $\dfrac{\text{heat capacity}}{\text{quantity of substance}}$			$\mathbf{ML^2T^{-2}\Theta^{-1}N^{-1}}$
$c_{mol} = \dfrac{C}{n}$ $c_{mol} = \dfrac{C \cdot N_A}{N}$ $N_A = 6.022\,136\,7 \cdot 10^{23} \text{ mol}^{-1}$	**Symbol**	**Unit**	**Quantity**
	c_{mol}	J/(K mol)	molar heat capacity
	C	J/K	heat capacity
	n	mol	quantity of substance
	N	1	particle number
	N_A	mol^{-1}	Avogadro number

For temperatures above 200 K, the molar heat capacity of solids is $3R = 24.9$ J/(K mol). This follows from the Dulong-Petit rule (see p. 691).

3. Representation by specific heat capacity

The definition of the molar heat capacity has the disadvantage that one must first determine the molar quantity of the substance considered. It is related to the specific heat capacity (easier to handle) via the molar mass.

molar heat capacity = specific heat capacity · molar mass	$ML^2T^{-2}\Theta^{-1}N^{-1}$		
	Symbol	Unit	Quantity
$c_{\mathrm{mol}} = c \cdot M$	c_{mol} c M	J/(K mol) J/(K kg) kg/mol	molar heat capacity specific heat capacity molar mass

■   Water has a molar mass of 18 g/mol and a specific heat capacity of 4.182 kJ/(kg K). The molar heat capacity is

$$c_{\mathrm{mol}} = c \cdot M = 4.182 \text{ kJ/(kg K)} \cdot 0.018 \text{ kg/mol} = 75.28 \text{ J/(K mol)} .$$

20.3.3   Specific heat capacity

1. Specific heat capacity,

c, SI unit joule per kelvin and per kilogram, the quantity of heat to be transferred to one kilogram of substance per degree of temperature increase.

specific heat capacity = $\dfrac{\textbf{quantity of heat}}{\textbf{temperature difference} \cdot \textbf{mass}}$	$L^2T^{-2}\Theta^{-1}$		
	Symbol	Unit	Quantity
$c = \dfrac{\Delta Q}{m \Delta T}$ $\Delta Q = cm\Delta T$	c ΔQ ΔT m	J/(K kg) J K kg	specific heat capacity quantity of heat change of temperature total mass

Temperature differences can be measured in degrees Celsius instead of Kelvin without rewriting formulae.

The specific heat capacity is often referred to as the specific heat; in some books, however, the term specific heat is used for the heat content per unit mass.

2. Representation as quotient

The specific heat capacity corresponds to the quotient of heat capacity and mass, or of molar heat capacity and molar mass.

➤   In some books on thermodynamics, the notion specific heat capacity is used for the molar heat capacity. Furthermore, the specific heat capacity is sometimes simply denoted specific heat. This may cause confusion.

specific heat capacity = $\dfrac{\textbf{(molar) heat capacity}}{\textbf{(molar) mass}}$	$L^2T^{-2}\Theta^{-1}$		
	Symbol	Unit	Quantity
$c = \dfrac{C}{m}$ $c = \dfrac{C}{n \cdot M}$ $c = \dfrac{c_{\mathrm{mol}}}{M}$	c c_{mol} C n M m	J/(K kg) J/(K mol) J/K mol kg/mol kg	specific heat capacity molar heat capacity heat capacity quantity of substance molar mass total mass

M | The specific heat capacity is determined by measuring the heat capacity and the mass of the substance considered.

▲ | The specific heat capacity depends on the material.

For specific heat capacities of important substances see **Tab. 22.3**.

■ | The values of c lie in the range 0.1–3 kJ/(kg K), for water approx. 4.2 kJ/(kg K).

■ | A metallic block of 250 g has a heat capacity of 224 J/K. What about the metal?

$$c = \frac{C}{m} = \frac{224 \text{ J/K}}{0.25 \text{ kg}} = 896 \text{ J/(kg K)}.$$

This is the specific heat capacity of aluminum.

20.3.3.1 Additional properties of specific heat capacity

- In general, the specific heat capacity depends on the temperature.
- The specific heat capacity tends to infinity for a first-order phase transition or a λ-transition. Therefore, in these cases one quotes the latent heat of melting or evaporation.
- For a second-order phase transition, the specific heat capacity has an anomaly at the critical point.
- The specific heat capacity of all substances tends to zero at absolute zero $T = 0$ K: $c_{T \to 0} = 0$.

20.3.3.2 Specific heat capacity of mixtures of substances

The specific heat capacity of a mixture of substances is equal to the sum of the individual heat capacities divided by the total mass:

$$c = \frac{C}{m} = \frac{m_1 c_1 + m_2 c_2 + m_3 c_3 + \cdots}{m_1 + m_2 + m_3 + \cdots}.$$

■ | A mixture of 30 g NaCl ($c = 867$ J/(K kg)) and 5 g KCl ($c = 682$ J/(kg K)) has a specific heat capacity of

$$c = \frac{m_1 c_1 + m_2 c_2}{m_1 + m_2}$$

$$= \frac{0.03 \text{ kg} \cdot 867 \text{ J/(kg K)} + 0.005 \text{ kg} \cdot 682 \text{ J/(kg K)}}{0.03 \text{ kg} + 0.005 \text{ kg}} \approx 841 \text{ J/(kg K)}.$$

20.3.3.3 Specific heat capacity of gases

The specific heat capacity may be measured either at constant pressure (volume varies with temperature), or at constant volume (pressure varies with temperature).

Notation:

c_V volume remains constant, pressure varies;

c_p pressure remains constant, volume varies.

Analogously, total (C_V, C_p) and molar heat capacities ($c_{V \text{ mol}}$, $c_{p \text{ mol}}$) for constant volume and constant pressure may be defined.

The specific heat capacity at constant pressure is larger than the specific heat capacity at constant volume.

$$c_p > c_V .$$

The quantity of heat supplied at constant pressure, ΔQ, will not only heat the system, but also expand it, and thus do volume work against the external pressure (atmospheric pressure).

▲   The quantity of heat supplied is not only used for heating, but is also needed to do work against the external pressure.

heat exchange at constant pressure			$\mathbf{ML^2T^{-2}}$
	Symbol	Unit	Quantity
$c_p m \Delta T = c_V m \Delta T + p \Delta V$	c_p	J/(K kg)	specific heat cap. at const. pressure
	c_V	J/(K kg)	specific heat cap. at const. volume
	m	kg	total mass
	ΔT	K	change of temperature
	p	Pa	pressure
	ΔV	m^3	change of volume

20.3.3.4   Specific heat capacity of ideal gas

1. Representation of specific heat capacities

For a gas with f degrees of freedom, the molar or specific heat capacity at constant volume is:

molar and specific heat capacity of ideal gas			
	Symbol	Unit	Quantity
$c_{V\,\mathrm{mol}} = R \cdot \dfrac{f}{2}$	$c_{V\,\mathrm{mol}}$	J/(K mol)	molar heat cap. const. vol.
$c_V = R_s \cdot \dfrac{f}{2}$	c_V	J/(K kg)	specific heat cap. const. vol.
	f	1	number of degrees of freedom
$R = 8.314\,\mathrm{J/(K\,mol)}$	R	J/(K mol)	universal gas constant
	R_s	J/(K kg)	specific gas constant

For an ideal gas $pV = nRT \implies p\Delta V = nR\Delta T$ at constant pressure. Inserting for $p\Delta V$ yields

$$c_p m \Delta T = c_V m \Delta T + nR\Delta T .$$

2. Difference of specific heat capacities

▲   The difference between the specific heat capacities at constant pressure and constant volume, respectively, is a material-dependent constant, the specific gas constant R_s.

difference of specific heat capacities			$\mathbf{L^2T^{-2}\Theta^{-1}}$
	Symbol	Unit	Quantity
$c_p - c_V = \dfrac{n}{m} R$ $= \dfrac{R}{M} = R_s$ $R = 8.314 \text{ J/(K mol)}$	c_p	J/(K kg)	spec. heat capacity at const. pressure
	c_V	J/(K kg)	spec. heat capacity at const. volume
	n	mol	quantity of substance
	m	kg	total mass
	M	kg/mol	molar mass
	R	J/(K mol)	universal gas constant
	R_s	J/(K kg)	specific gas constant

▲   For the molar heat capacity in the ideal gas:

$$c_{p\text{ mol}} - c_{V\text{ mol}} = R\,.$$

Specific gas constant, **individual gas constant** R_s, the material-dependent proportionality factor used in technical thermodynamics in the equation of state of an ideal gas.

Universal gas constant R, the material-independent proportionality factor appearing in the equation of state of an ideal gas (see p. 650),

$$R = 8.3145 \; \frac{\text{J}}{\text{K mol}}\,.$$

The expansion work may be described by compressibility and coefficients of expansion.

3. Relation between specific heat capacities

relation between specific heat capacities			$\mathbf{L^2T^{-2}\Theta^{-1}}$
	Symbol	Unit	Quantity
$c_p = c_v + T\dfrac{\alpha^2}{\kappa\rho}$	c_p	J/(K kg)	specific heat cap. at const. pressure
	c_V	J/(K kg)	specific heat cap. at const. volume
	T	K	temperature
	ρ	kg/m^3	density
	α	K^{-1}	coefficient of expansion
	κ	Pa^{-1}	compressibility

20.3.3.5   Adiabatic index

Adiabatic index, κ, dimensionless quantity, quotient of the specific heat capacities of an ideal gas,

$$\frac{c_p}{c_V} = \kappa\,.$$

➤   There is a risk of confusion with the compressibility κ. However, the latter has a dimension, contrary to the adiabatic index.

Isentropic index, alternative denotation of the adiabatic index.

For an ideal gas:

adiabatic index $= 1 + \dfrac{2}{\text{number of degrees of freedom}}$			1
$\kappa = 1 + \dfrac{2}{f}$	Symbol	Unit	Quantity
	κ	1	adiabatic index
	f	1	number of degrees of freedom

20.3.3.6   Specific heat capacity of liquids and solids

Nearly exclusively the value of c_p (easier to measure) is given.

➤ Liquids exhibit rather different dependences on pressure and temperature.

Dulong-Petit rule, a simple rule for the specific heat capacity of metals:

▲ All metals have the **constant molar heat capacity** of $c_p \approx 25$ JK^{-1}mol^{-1} over a wide range of temperatures.

For the specific heat capacity the following holds:

specific heat capacity (const. pressure) $\approx \dfrac{25 \text{ J}/(\text{K mol})}{\text{molar mass}}$			$L^2 T^{-2} \Theta^{-1}$
$c_p \approx \dfrac{1}{M} \cdot 25 \dfrac{\text{J}}{\text{K mol}}$	Symbol	Unit	Quantity
	c_p	J/(K kg)	spec. heat cap. at const. pressure
	M	kg/mol	molar mass

➤ This no longer holds for temperatures markedly below 200 K. For $T \to 0$, one finds $c \to 0$ because of the third law of thermodynamics.

20.4   Changes of state

20.4.1   Reversible and irreversible processes

1. Equilibrium state,

a state in which the macroscopic parameters no longer vary.

▲ According to general experience, in an isolated system processes proceed on their own until an equilibrium is reached.

2. Irreversible process,

a process that cannot proceed on its own in the reverse sequence (**Fig. 20.2 (b)**).

➤ All transitions from non-equilibrium to equilibrium are irreversible.

■ Two metallic plates at different temperatures brought in contact equilibrate their temperatures.

Irreversible processes proceed via **non-equilibrium states**.

▲ Irreversible processes increase the microscopic disorder (entropy) of the system.

3. Reversible process,

a process proceeding only through equilibrium states (**Fig. 20.2 (a)**).

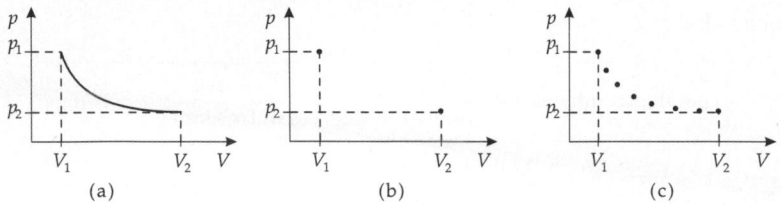

Figure 20.2: Changes of state. (a): reversible process, (b): irreversible process, (c): quasi-reversible process.

Reversible processes are an idealization, which, strictly speaking, does not exist. If a system is in equilibrium, the state variables have time-independent values, and nothing changes macroscopically.

Reversible changes of state may, however, be simulated approximately by small (infinitesimal) changes of the state variables which means that the equilibrium is disturbed only slightly. If these changes proceed slowly enough, the system always has sufficient time to equilibrate.

Quasi-reversible process, a process involving only very small changes of state (**Fig. 20.2 (c)**).

4. Particular importance of reversible processes

The importance of reversible changes of state lies in the fact that at all stages of the process one has an equilibrium state with definite values of the state properties, so that the total changes of state properties may be obtained by integration over the infinitesimal reversible steps.

➤ This is not possible for irreversible processes.

■ Isothermal expansion, e.g., expansion of a gas in a heating bath.
Reversible performance is achieved by slowly pulling back the piston, irreversible performance by a jerky motion of the piston.

20.4.2   Isothermal processes

1. Characteristics of isothermal processes

Isothermal process, a process in which the **temperature remains constant**.
Isotherms for the ideal gas are **hyperbolic sections** in the $p-V$-plane,

$$p \cdot V = \text{const.}, \; T = \text{const.}$$

Hence, the pressure decreases in an isothermal expansion and increases in an isothermal compression like $1/V$.

This is just the **law of Boyle–Mariotte (Fig. 20.3)**.

For $T = \text{const.}$ the change in internal energy is equal to zero.

internal energy = constant			$\mathbf{ML^2T^{-2}}$
	Symbol	Unit	Quantity
$\Delta U = C_V \Delta T = 0$	U	J	internal energy
$\Delta U = \Delta Q + \Delta W$	C_V	J/K	heat capacity
	T	K	temperature
$\Delta Q = -\Delta W$	ΔQ	J	heat transferred
	ΔW	J	work

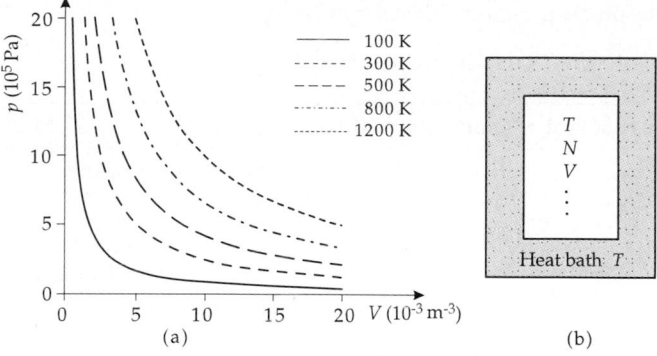

Figure 20.3: Isotherms. (a): ideal gas, (b): system in a heat bath.

▲ In isothermal processes, the heat supplied is equal to the volume work of the gas.
➤ This follows from the first law of thermodynamics. The minus sign indicates that the system does work if heat is absorbed.

2. Isothermal process: work done and change of entropy

Work done by the gas during a change of state at $T = $ const.:

$$W_{12} = p_1 V_1 \ln\left(\frac{V_2}{V_1}\right) = p_2 V_2 \ln\left(\frac{V_2}{V_1}\right),$$

$$= nRT \ln\left(\frac{V_2}{V_1}\right) = m R_s T \ln\left(\frac{V_2}{V_1}\right).$$

In terms of the pressure:

$$W_{12} = p_1 V_1 \ln\left(\frac{p_1}{p_2}\right) = p_2 V_2 \ln\left(\frac{p_1}{p_2}\right),$$

$$= nRT \ln\left(\frac{p_1}{p_2}\right) = m R_s T \ln\left(\frac{p_1}{p_2}\right).$$

The change of entropy is

$$\Delta S = (C_p - C_V) \ln\left(\frac{V_2}{V_1}\right) = C_p \ln\left(\frac{V_2}{V_1}\right) + C_V \ln\left(\frac{p_2}{p_1}\right).$$

20.4.3   Isobaric processes

1. Characteristics of isobaric processes

Isobaric process, a process in which the **pressure remains constant**.

Isobars are **horizontal straight lines** ($p = $ const.) in the p–V-diagram (**Fig. 20.4 (a)**): The volume increases with increasing temperature—the system changes from a lower isotherm to a higher isotherm.

➤ The linear relation between volume and temperature just corresponds to the **law of Gay-Lussac**.

The **work due to change of volume** in an isobaric process is

$$W_{12} = p(V_1 - V_2).$$

2. Isobaric process: change of heat and entropy

For $p = $ const., the absorbed heat Q_{12} is given by:

change of heat ∼ temperature difference			$\mathrm{ML^2T^{-2}}$
	Symbol	Unit	Quantity
$Q_{12} = m c_p (T_2 - T_1)$	Q	J	quantity of heat
	m	kg	mass of gas
$= C_p (T_2 - T_1)$	n	mol	molar quantity
	c_p	J/(K kg)	spec. heat capacity
$= n c_{p\,\mathrm{mol}} (T_2 - T_1)$			at const. pressure
	C_p	J/K	heat capacity
	$c_{p\,\mathrm{mol}}$	J/(K mol)	molar heat capacity
	T	K	temperature

For $p = $ const., the change of entropy is given by

$$\Delta S = C_p \ln\left(\frac{T_2}{T_1}\right) = C_p \ln\left(\frac{V_2}{V_1}\right).$$

20.4.4 Isochoric processes

1. Characteristics of isochoric processes

Isochoric process, a process in which the **volume remains constant (Fig. 20.4 (b))**.

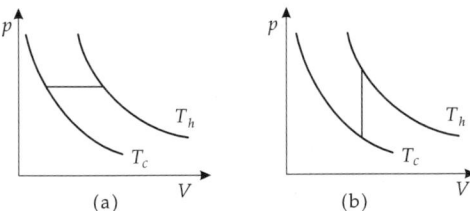

Figure 20.4: Changes of state. (a): isobaric process, (b): isochoric process. For isobaric expansion or isochoric increase of pressure, $T_1 = T_c$ (cold) and $T_2 = T_h$ (hot).

Isochors are **vertical straight lines** ($V = $ const.) in the p–V-diagram.
The pressure increases with increasing temperature, the system changes from a lower isotherm to a higher isotherm.
➤ The linear relation between pressure and temperature just corresponds to the **law of Gay-Lussac**.
Because $V = $ const., the volume work vanishes,

$$\Delta W = p \Delta V = 0.$$

2. Isochoric process: change of heat and entropy

For $V = $ const., the change of heat corresponds to the change of the internal energy.

change of heat $\sim$ temperature difference			$\mathbf{ML^2T^{-2}}$
	Symbol	Unit	Quantity
$Q_{12} = mc_V(T_2 - T_1)$	Q	J	quantity of heat
	m	kg	mass of gas
$= C_V(T_2 - T_1)$	n	mol	molar quantity
	c_V	J/(K kg)	spec. heat capacity at
$= nc_{V\,\text{mol}}(T_2 - T_1)$			const. volume
	C_V	J/K	heat capacity
$= \Delta U$	$c_{V\,\text{mol}}$	J/(K mol)	molar heat capacity
	T	K	temperature
	U	J	internal energy

For $V = $ const., the change of entropy is given by

$$\Delta S = C_V \ln\left(\frac{T_2}{T_1}\right) = C_V \ln\left(\frac{p_2}{p_1}\right).$$

20.4.5 Adiabatic (isentropic) processes

1. Characteristics of adiabatic and isentropic processes

Isentropic process, a process in which the **entropy remains constant**.
  Adiabatic process, a process in which **no heat** is exchanged with the environment.
■   Reactions in closed systems (e.g., Dewar flasks) are adiabatic.
▲   For reversible processes, the notions adiabatic and isentropic may be used synonymously.
➤   In regions of low temperature, however, in the demagnetization of crystals, adiabatic and isentropic processes may proceed differently.
Isentrops and **adiabats** in the p–V-diagram are **steeper than isotherms (Fig. 20.5)**,

$$pV^\kappa = \text{const.}, \quad \kappa > 1.$$

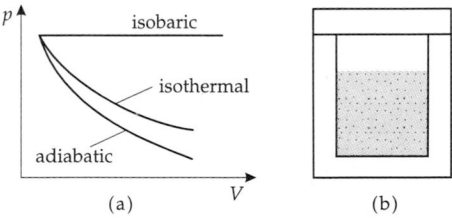

Figure 20.5: Changes of state. (a) isobaric, isothermal and adiabatic process, (b): closed system.

2. Adiabatic index,

κ, dimensionless quantity, the exponent of the volume in the adiabatic equation. The adiabatic index is equal to the ratio of the (specific) heat capacities for constant pressure and constant volume,

$$\kappa = \frac{C_p}{C_V} = \frac{c_p}{c_V} = \frac{c_{p\,\text{mol}}}{c_{V\,\text{mol}}}.$$

▲   For an ideal monatomic gas, $\kappa = 5/3$.

▲   The specific heat capacities c_p and c_V differ by the specific gas constant R_s,

$$c_p - c_V = R_s .$$

▲   Analogously, the molar heat capacities differ by the universal gas constant R,

$$c_{p\,\text{mol}} - c_{V\,\text{mol}} = R .$$

In an adiabatic process, the change of entropy and heat is equal to zero,

$$\Delta Q = 0, \qquad \Delta S = 0 .$$

3. Adiabatic process: change of internal energy

The work due to a change of volume is equal to the change of internal energy.

work ∼ temperature difference			ML^{2}T^{-2}
	Symbol	Unit	Quantity
$W_{12} = m c_V (T_2 - T_1)$ $= C_V (T_2 - T_1)$ $= \Delta U$	W_{12}	J	work
	m	kg	mass of gas
	c_V	J/(K kg)	spec. heat capacity at const. volume
	C_V	J/K	heat capacity
	T	K	temperature
	U	J	internal energy

20.4.5.1   Polytropic processes

1. Characteristics of polytropic processes

Polytropic process, a process in which the **product pV^n remains constant**.

polytropic equation			
	Symbol	Unit	Quantity
$p \cdot V^n = \text{const.}$ $T \cdot V^{n-1} = \text{const.}$	p	Pa	pressure
	V	m^3	volume
	n	1	polytropic coefficient
	T	K	temperature

Polytropic index, n, nondimensional quantity, exponent of the volume variable in the polytropic equation.

The polytrope may be understood as a generalization of the processes discussed so far:

special cases of the polytropic process		
$n = 0$	$p = \text{const.}$	isobaric process
$n = 1$	$pV = nRT = \text{const.}$	isothermal process
$n = \kappa$	$pV^\kappa = \text{const.}$	adiabatic process
$n \to \infty$	$p^{1/\infty} V = \text{const.}$	isochoric process

Mostly, one restricts oneself to the cases $1 < n < \kappa$ describing systems in which heat is exchanged with the environment, but no complete balance of heat is reached.

■   Processes proceeding very fast in noninsulated systems belong to this group.
In the p–V diagram, the course of polytropes belonging to $1 < n < \kappa$ is steeper than that of isotherms, but is flatter than that of isentrops, a specific example being $pV^n = \text{const.}$

2. Polytropic process: change of state variables

Work due to the change of volume:

$$W_{12} = \frac{p_2 V_2 - p_1 V_1}{n - 1}.$$

The absorbed heat is given by:

absorbed heat ~ temperature difference			$\mathbf{ML^2T^{-2}}$
	Symbol	Unit	Quantity
$Q_{12} = mc_V(T_2 - T_1)\dfrac{n - \kappa}{n - 1}$	Q	J	quantity of heat
	m	kg	mass of gas
	c_V	J/(K kg)	spec. heat capacity at
			const. volume
$= C_V(T_2 - T_1)\dfrac{n - \kappa}{n - 1}$	C_V	J/K	heat capacity
	T	K	temperature
	n	1	polytropic index
	κ	1	adiabatic index

Change of entropy:

$$\Delta S = C_V \frac{n - \kappa}{n - 1} \ln\left(\frac{T_2}{T_1}\right) = C_V(\kappa - n) \ln\left(\frac{V_2}{V_1}\right).$$

20.4.6 Equilibrium states

Equilibrium, the state achieved in a system on its own after sufficient time.
Depending on the external conditions, the **equilibrium state** is characterized as follows:

Closed isochoric states: maximum of entropy S.
Isothermal-isobaric states: minimum of free enthalpy $G = U + pV - TS$.
Isothermal-isochoric states: minimum of free energy $F = U - TS$.
Adiabatic-isobaric states: minimum of enthalpy $H = U + pV$.
Adiabatic-isochoric states: minimum of internal energy U.

Differentials of the thermodynamic potentials			$\mathbf{ML^2T^{-2}}$
	Symbol	Unit	Quantity
$dU = -p\,dV + T\,dS$	U	J	internal energy
	F	J	free energy
$dF = -p\,dV - S\,dT$	H	J	enthalpy
	G	J	free enthalpy
$dH = V\,dp + T\,dS$	p	Pa	pressure
	V	m^3	volume
$dG = V\,dp - S\,dT$	T	K	temperature
	S	J/K	entropy

Survey of equilbrium conditions					
system is …	isothermal	isobaric	isochoric	adiabatic	closed
entropy S maximum			$\mathrm{d}V = 0$		$\mathrm{d}U = 0$
internal energy U minimum			$\mathrm{d}V = 0$	$\Delta Q = 0$	
free energy F minimum	$\mathrm{d}T = 0$		$\mathrm{d}V = 0$		
enthalpy H minimum		$\mathrm{d}p = 0$		$\Delta Q = 0$	
free enthalpy G minimum	$\mathrm{d}T = 0$	$\mathrm{d}p = 0$			

20.5   Laws of thermodynamics

Law of thermodynamics, a fundamental relation among properties of state that holds empirically for all known systems.

■   The first law of thermodynamics states essentially that no energy may be lost or created in any form.

20.5.1   Zeroth law of thermodynamics

Equilibrium state, that macroscopic state of a closed system taken after sufficient time on its own.

▲   In the equilibrium state, the macroscopic properties of a state no longer change with time.

If two systems are joined together, exchange processes will proceed until the **intensive** quantities (pressure, temperature, chemical potential) of the systems are balanced.

When approaching thermal equilibrium, an **exchange of heat** continues until the temperatures of both systems are equal.

Zeroth law of thermodynamics, describes the equivalence of thermal systems:

▲   All systems in thermal equilibrium with one system are also in mutual thermal equilibrium with each other.

➤   The operation of a thermometer is based on this law.

20.5.2   First law of thermodynamics

Conserved quantity, a property of state that does not change in the system. A conserved quantity may be used to characterize the macroscopic state.

■   The total energy E of the closed system (see p. 623) is a conserved quantity.

In physics, the principle of energy conservation is of fundamental importance.

▲   All experience confirms the assumption that this principle is correct both for macroscopic and for microscopic dimensions.

➤   Besides the work expended or received by a system, one must also take into account the heat exchanged with the environment.

Internal energy, U, the total energy present in the internal degrees of freedom of a gas. In a closed system, the internal energy is identical to the total energy of the system.

1. Formulation of the first law of thermodynamics

First law of thermodynamics: The total change of energy of a system includes exchange of work and heat.

▲ The change of the internal energy in any (reversible or irreversible) change of state is given by the sum of work ΔW and heat ΔQ exchanged with the environment:

internal energy = work + heat			$\mathbf{ML^2T^{-2}}$
	Symbol	Unit	Quantity
$\Delta U = \Delta W + \Delta Q$	U	J	internal energy
	W	J	work
	Q	J	quantity of heat

- $\Delta W < 0$: work expended by the system,
- $\Delta W > 0$: work done on the system.
➤ One can also find the inverse definition in the literature.

The work and the heat exchanged with the environment depend on the manner in which the process is carried out. This is of importance, e.g., in chemical reactions for the concept of the reaction device.

➤ Work and heat are not total differentials. Therefore, the change is designated here by a Δ for the sake of clarity.

2. Work for reversible processes

work = −pressure · change of volume			$\mathbf{ML^2T^{-2}}$
$\Delta W_{rev} = -p \, \Delta V$	Symbol	Unit	Quantity
	W	J	work
$W_{rev} = -\int_{V_1}^{V_2} p \, dV$	p	Pa	pressure
	V	m^3	volume

➤ In irreversible processes, one may have $\Delta W_{irr} = 0$.

3. Heat for reversible processes

heat = temperature · change of entropy			$\mathbf{ML^2T^{-2}}$
$\Delta Q_{rev} = T \, \Delta S$	Symbol	Unit	Quantity
	Q	J	quantity of heat
$Q_{rev} = \int_{S_1}^{S_2} T \, dS$	T	K	temperature
	S	J/K	entropy

➤ This holds only for the **reversible** case.

Representation in terms of the heat capacity C_V at constant volume holds only for the reversible case:

heat = heat capacity · change of temperature			$\mathbf{ML^2T^{-2}}$
$\Delta Q_{\mathrm{rev}} = C_V \, \Delta T$	Symbol	Unit	Quantity
$Q_{\mathrm{rev}} = \displaystyle\int_{T_1}^{T_2} C_V \, \mathrm{d}T$	Q C_V T	J J/K K	quantity of heat heat capacity at const. volume temperature

▲   While the formulas for the partial contributions above are valid only for reversible processes, the first law of thermodynamics always holds.

20.5.2.1   Equivalent formulations of the first law of thermodynamics

Selection of various formulations of the first law of thermodynamics, which are all equivalent:

▲   **In the energy balance of a system, the sum of exchanged work and heat yields the total change of energy of the system.**
   This knowledge is due to Robert Mayer (1814–1878) and J.P. Joule (1818–1889), who proved by precise experiments that heat is a special form of energy.

▲   **The internal energy U of a system is a state function.** This means that the total energy content of a system is always the same no matter what process was used to reach the macrostate.

▲   **There is no *perpetuum mobile* of the first kind.**
   The term *perpetuum mobile* of the first kind denotes a machine that operates in a cycle and generates energy without extracting it from its environment.

▲   **The change of the internal energy in an arbitrary, infinitesimal change of state is a total differential.**
   The change of the internal energy depends only on the initial and final state, not on the path.

20.5.2.2   Microscopic aspects of the first law of thermodynamics

If neither heat nor work is added to the system, then the mean kinetic energy of the molecules $\frac{1}{2}m\overline{v^2}$ does not change.

If the system is heated through the walls of the cylinder without doing work, the kinetic energy of the molecules is increased by collisions with the wall (**Fig. 20.6 (a)**). In the collisions, energy is transferred from the wall to the particles. The system is heated, the walls are cooled.

If the system does expansion work, i.e., the piston is displaced outward, then the molecules lose kinetic energy by collisions with the piston moving away. The particles slow down, and the system cools (**Fig. 20.6 (c)**).

■   A camping gas cartridge or can of shaving cream cools during the outflow of the gas. If the piston moves inward, i.e., compression work is performed on the system, the particles colliding with the piston get an additional momentum from the motion of the piston, which also increases the kinetic energy (**Fig. 20.6 (b)**).

➤   In an irreversible expansion of real gases, the Joule–Thomson effect (see gas liquefaction – Joule–Thomson effect) causes heating or cooling, depending on the inversion temperature.

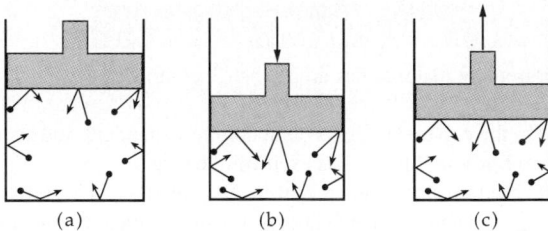

Figure 20.6: Change of the mean molecular velocity under compression (b) and expansion (c) of the system (a).

20.5.3 Second law of thermodynamics

▲ All experience confirms that the entropy takes a maximum value in the equilibrium state,

$$S = S_{\text{max}} \quad \text{in equilibrium.}$$

1. Formulation of the second law of thermodynamics

There are no processes in nature in which the total entropy decreases.

All irreversible processes in a closed system are connected with an increase of entropy. After a change of state, the system must move again to equilibrium, whereby the entropy increases,

$$\Delta S \geq 0.$$

In subsystems, $\Delta S < 0$ may be valid. But this is possible only by input of work. The system delivering this work correspondingly increases its entropy.

Reversible processes: the entropy remains constant,

$$\mathrm{d}S = 0.$$

Irreversible processes: the entropy increases,

$$\mathrm{d}S > 0.$$

2. Equivalent formulations of the second law of thermodynamics

▲ **There is no *perpetuum mobile* of the second kind.**

A *perpetuum mobile* of second kind is a machine that does nothing but performs work by cooling down a heat reservoir, that is, it would transform heat completely into work.

One always needs a second reservoir to be heated up.

▲ **There is no process that converts anergy into exergy.**

Heat cannot be converted completely into mechanical work, only the exergetic fraction of heat is convertible into work.

▲ **Any closed macroscopic system tends towards the most probable state.**

This is the state characterized by the largest number of microscopic possibilities, i.e., by the highest entropy (disorder).

20.5.4   Third law of thermodynamics

At finite temperature, any material has an intrinsic excitation energy corresponding to the quantity of heat.

■ Oscillations in the crystal lattice are temperature-dependent, intrinsic excitations.

At absolute zero, a body no longer has excitation energy.

■ All lattice oscillations in a solid are frozen. Nevertheless, at $T = 0$ the kinetic energy is not zero, since the atoms carry out quantum-mechanical **zero-point oscillations**.

1. Third law of thermodynamics,

defines the absolute value of the entropy at absolute zero.

▲ **At absolute zero, every body has zero entropy,**

$$S = 0 \text{ for } T = 0 \text{ K}.$$

2. Equivalent formulations of the third law of thermodynamics

▲ The specific heat capacity of all substances vanishes at absolute zero. The specific heat of all materials disappears at absolute zero.

$$c_{T=0} = 0.$$

▲ Absolute zero can never be reached experimentally; it is a theoretical reference only.

➤ Any quantity of heat (energy), however small, causes a finite increase of temperature.

20.6   Carnot cycle

20.6.1   Principle and application

1. Cycle,

a periodic process that, after a certain number of changes of state, reaches the initial state again (**Fig. 20.7 (a)**).

Carnot cycle, a cycle introduced by Carnot in 1824 with the ideal gas as working medium (**Fig. 20.7 (b)**).

The Carnot cycle allows production of work by heat exchange between cold and hot media.

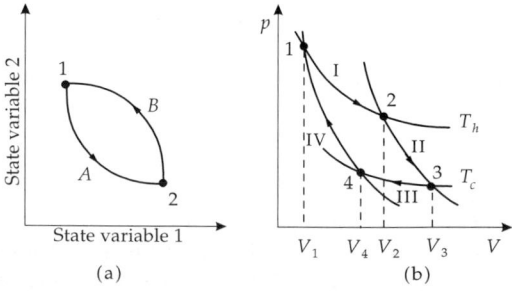

Figure 20.7: Cycles (schematic representation). (a): general cycle, (b): Carnot cycle.

2. Heat engine and refrigerator

Heat engine, a machine that does work by heat exchange.
- ■ Combustion motors, steam engines, turbines.
- ➤ The inverse process, the heating of a hot body by a cold body with the expenditure of work, is also possible.

Refrigerator or **heat pump**, a machine that heats up a hot system or cools down a cold system by input of work.
- ■ Refrigerator, air conditioning, heat pump.

The designation refrigerator or heat pump depends on whether one is referring to the heating of a hot system or the cooling of a cold system.
- ■ Machines based on a Carnot cycle may be used for continuous generation of low temperatures, also for air liquefaction in small quantities (see gas liquefaction).
- ▲ The ideal Carnot cycle can be realized technologically to a good approximation.

20.6.1.1 Stages of the Carnot cycle

The Carnot cycle involves four successive **reversible** partial processes (**Fig. 20.8**):
- isothermal expansion at high temperature T_h (I),
- adiabatic expansion with cooling to T_c (II),
- isothermal compression at low temperature T_c (III),
- adiabatic compression with heating to T_h (IV).

The working media are at temperatures T_h (hot) and T_c (cold), respectively.

1. First step: isothermal expansion

from volume V_1 to the volume V_2 at constant temperature T_h. For the isotherm,

$$\frac{V_2}{V_1} = \frac{p_1}{p_2}.$$

The energy of an ideal gas cannot change at constant temperature,

$$\Delta U_I = \Delta W_I + \Delta Q_I = 0.$$

The exchanged quantity of heat is

$$\Delta Q_I = -\Delta W_I = NkT_h \ln \frac{V_2}{V_1}.$$

2. Second step: adiabatic expansion

of the isolated working medium from V_2 to V_3 with cooling to the temperature of the cold medium. For the ideal gas,

$$\frac{V_3}{V_2} = \left(\frac{T_h}{T_c}\right)^{3/2}.$$

The work done by the gas is

$$\Delta W_{II} = \Delta U_{II} = C_V (T_c - T_h).$$

Because $\Delta Q_{II} = 0$ (adiabatic process), the work done during expansion is taken from the internal energy.

3. Third step: isothermal compression

of the system at temperature T_c from V_3 to V_4.

Analogous to step 1, the exchanged quantity of heat is

$$\Delta Q_{III} = -\Delta W_{III} = NkT_c \ln \frac{V_4}{V_3}.$$

The gas releases this amount of heat.

4. Fourth step: adiabatic compression

from V_4 to V_1 with heating to the temperature T_h.
The system returns to the initial state.

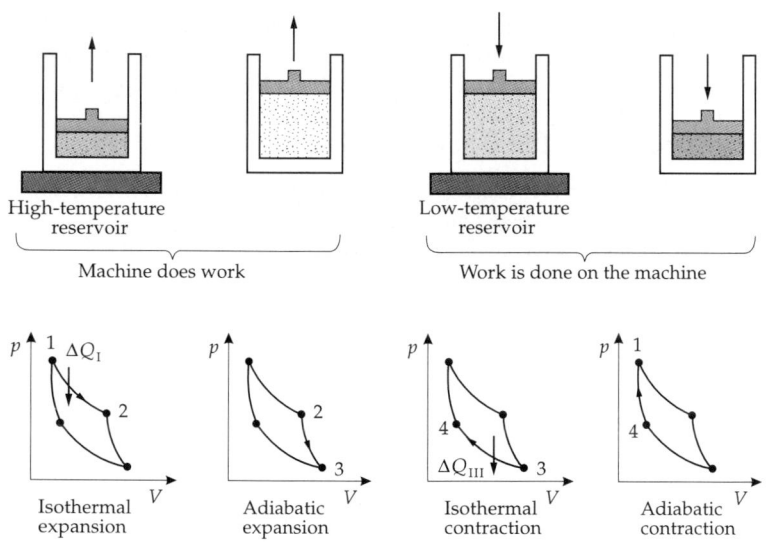

Figure 20.8: Partial steps in the Carnot cycle.

Work done on the gas:

$$\Delta W_{IV} = \Delta U_{IV} = C_V (T_h - T_c).$$

In the T-S diagram, the Carnot cycle is given by a rectangle defined by the straight lines $T = $ const. (isotherms) in steps I and III, and the straight lines $S = $ const. (adiabats) in steps II and IV (**Fig. 20.9**).

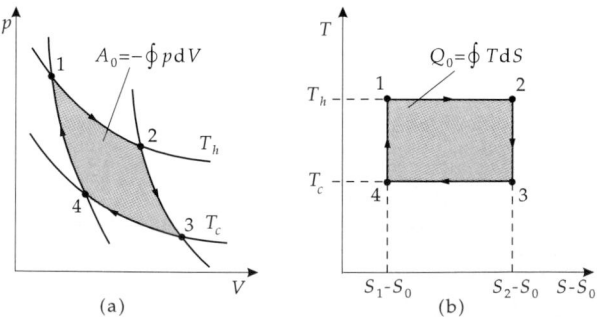

Figure 20.9: Carnot cycle in the p–V- and T-S-diagrams.

20.6.1.2   Energy balance and efficiency of the Carnot cycle

The total change of internal energy is

$$\Delta U_{\text{total}} = \underbrace{\Delta Q_I + \Delta W_I}_{I} + \underbrace{\Delta W_{II}}_{II} + \underbrace{\Delta Q_{III} + \Delta W_{III}}_{III} + \underbrace{\Delta W_{IV}}_{IV} = 0.$$

▲   The internal energy does not change (first law of thermodynamics).
Work generated in the process:

$$\Delta W = -Nk\,(T_h - T_c)\ln\frac{V_2}{V_1} = -\Delta Q\,.$$

Correspondingly, the converted quantity of heat is equal but of opposite sign to the work.
   Efficiency, the ratio of the generated work and the heat loss of the hot medium.

efficiency $= 1 - \dfrac{\text{low temperature}}{\text{high temperature}}$			1
	Symbol	Unit	Quantity
$\eta = 1 - \dfrac{T_c}{T_h} = \dfrac{T_h - T_c}{T_h}$	η	1	efficiency
	T_c	K	low temperature
	T_h	K	high temperature

The remaining part is nonconvertible heat (see p. 683).

20.6.2   *Reduced heat*

Reduced heat, quotient of heat and temperature.
➤   This definition leads directly to the concept of entropy.
In the Carnot cycle, the sum of the reduced heats is equal to zero,

$$\frac{\Delta Q_I}{T_h} + \frac{\Delta Q_{III}}{T_c} = 0.$$

The reduced heats of the processes II and IV are zero (adiabats).
▲   In a closed reversible process in arbitrarily small cycles, it follows that the reduced heat is conserved.
▲   Any closed process may be decomposed into Carnot cycles (**Fig. 20.10**).

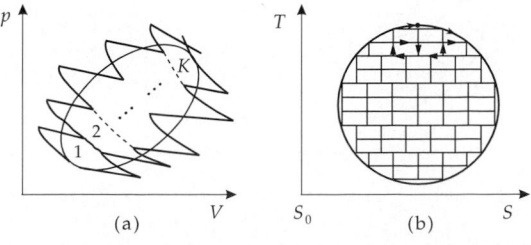

Figure 20.10: Decomposition of a cycle. (a): p–V-diagram, (b): T-S-diagram.

From the conservation of the reduced heat in the cycle, it follows that the reduced heat of a process is independent of the path,

$$\oint \frac{\Delta Q_{\text{rev}}}{T} = 0.$$

This is the second law of thermodynamics.

➤   The reduced heat $\Delta Q_{\text{rev}}/T$ forms a total differential.

The reduced heat directly implies the entropy:

$$\Delta S = \frac{\Delta Q_{\text{rev}}}{T}, \quad S_1 - S_0 = \int_0^1 \frac{\Delta Q_{\text{rev}}}{T}.$$

20.7   Thermodynamic machines

20.7.1   Right-handed and left-handed processes

1. Right-handed processes,

cycles running in the p–V-diagram to the right, i.e., clockwise (see **Fig. 20.9**).

■   The description of the Carnot cycle in the preceding section corresponds to a right-handed process.

▲   In right-handed processes, heat is taken from the hot system to do work.

The sum of the quantities of heat supplied to and withdrawn from the system during the process steps is negative, the total work done is positive:

$$\Delta Q < 0, \quad \Delta W > 0.$$

■   Heat engines are based on right-handed cycles.

2. Left-handed processes,

cycles running in the p–V-diagram to the left, i.e., counterclockwise (**Fig. 20.11 (a)**).

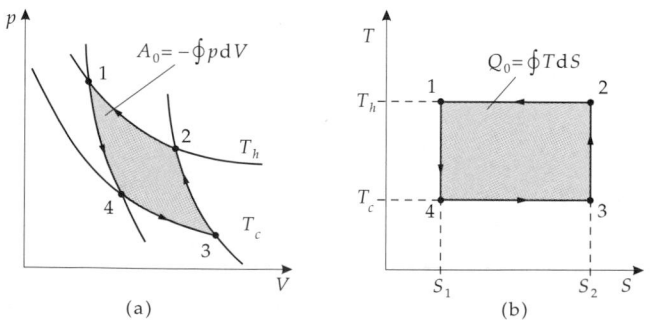

Figure 20.11: Left-handed process. (a): p–V-diagram, (b): T–S-diagram.

▲   In left-handed processes, work is expended in order to supply heat to the hot system. The sum of the quantities of heat supplied to and withdrawn from the system during the process steps is positive, the total work done is negative:

$$\Delta Q > 0, \quad \Delta W < 0.$$

■   Heat pumps and refrigerators are based on left-handed cycles.

20.7.2 Heat pump and refrigerator

1. Heat pump,

a thermodynamic machine operating according to the principle of a left-handed cycle that, with expenditure of work, pumps heat from the colder system to the warmer system. It may be used as a refrigerating machine to generate low temperatures (see generation of low temperatures), but also as a heater for heating a room from an environment at lower temperature.

■   Heat pumps installed in houses may be used in winter as heaters, and in summer as air conditioners. Both applications require input energy.

2. Coefficient of performance of a heat pump,

ε_W, dimensionless quantity, the ratio of the quantity of heat transferred to the hot system and the work expended.

coefficient of performance of a heat pump			1
$$\varepsilon_W = \frac{|Q|}{W}$$ $$\varepsilon_{W,C} = \frac{T_h}{T_h - T_c}$$ $$= \frac{1}{\eta_C}$$	Symbol	Unit	Quantity
	ε_W	1	coeff. of performance heat pump
	Q	J	released quantity of heat
	W	J	expended work
	$\varepsilon_{W,C}$	1	coeff. of performance Carnot cycle
	T_h	K	high temperature
	T_c	K	low temperature
	η_C	1	efficiency Carnot cycle

▲   The coefficient of performance ε_W in the Carnot cycle is always larger than unity.
▲   The coefficient of performance ε_W is largest for small temperature differences.

3. Refrigerating machine and its coefficient of performance

Refrigerating machine, a machine operating according to the same principle as the heat pump, taking heat from the colder system and pumping it into the warmer system.

➤   Heat pumps and refrigerating machines are distinguished only in the technical application. For the heat pump, the interest lies on the hot system to be heated, for the refrigerating machine on the cold system to be cooled.

Coefficient of performance of a refrigerating machine ε_K, dimensionless quantity, the ratio of the quantity of heat withdrawn from the cold system and the work expended.

coefficient of performance of a refrigerating machine			1
	Symbol	Unit	Quantity
$\varepsilon_K = \dfrac{|Q|}{W}$ $\varepsilon_{K,C} = \dfrac{T_c}{T_h - T_c}$	ε_K	1	coefficient of performance refrigerating machine
	Q	J	quantity of heat withdrawn
	W	J	expended work
	$\varepsilon_{K,C}$	1	coefficient of performance Carnot cycle
	T_h	K	high temperature
	T_c	K	low temperature

▲   The coefficient of performance ε_K of the Carnot cycle is always larger than unity.
▲   The coefficient of performance ε_K increases with decreasing temperature difference.

20.7.3   Stirling cycle

1. Stirling cycle,

cycle represented in **Fig. 20.12**, consisting of two isothermal and two isochoric partial processes.
▲   The efficiency of the Stirling cycle is equal to the efficiency of the Carnot cycle.
Efficiency of the Stirling cycle is given by:

efficiency of the Stirling cycle			1
	Symbol	Unit	Quantity
$\eta = 1 - \dfrac{T_c}{T_h}$	η	1	efficiency
	T_c	K	low temperature
	T_h	K	high temperature

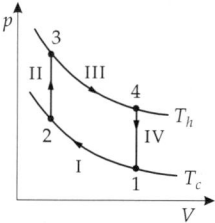

Figure 20.12: Stirling cycle. Working steps of the Stirling cycle: (I) isothermal compression, (II) isochoric heating, (III) isothermal expansion, (IV) isochoric cooling.

2. Stirling engine,

also **hot-air engine**, using a cycle with a fixed quantity of gas moved between two heat reservoirs (**Fig. 20.13**).
    Isothermal compression and expansion:
• working piston is shifted,
• displacer piston is not shifted.
Isochoric heating and cooling:
• working piston is not shifted,
• displacer piston is shifted.

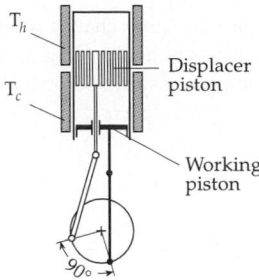

Figure 20.13: Stirling engine.

The Stirling engine has two pistons, the displacer piston and the working piston moving with a 90° phase shift relative to each other.

Working steps of the Stirling engine:

Isothermal compression: The displacer piston remains at its top end point and prevents contact to the hot heat bath while the working piston compresses the gas.

Isochoric heating: The displacer piston moves downwards while the working piston stands at its top end point. The gas is displaced upwards and contacts the hot heat bath.

Isothermal expansion: While the displacer piston remains at the bottom end point, the working piston moves downwards. The gas expands.

Isochoric cooling: The working piston remains at the bottom end point and the displacer piston moves upwards. The gas is displaced from the hot to the cold temperature reservoir.

The practical use of the Stirling engine suffers from incomplete heat transfer during the displacement.

| M | The efficiency is improved by inserting **regenerators** consisting of metal chips into the displacer, which support cooling and heating of the air flowing through. |

20.7.4 Steam engine

Clausius–Rankine cycle, cycle in the region of phase coexistence between liquid and gaseous phases (**Fig. 20.14**). It consists of two isentropic and two isobaric partial processes:
- isentropic (adiabatic) compression (I),
- isobaric supply of heat (II),
- isentropic expansion (III),
- isobaric extraction of heat (IV).

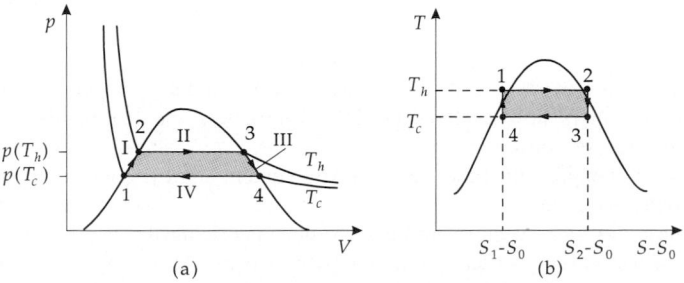

Figure 20.14: Clausius-Rankine cycle. (a): p–V-diagram, (b): T–S-diagram.

➤   The isobaric supply or extraction of heat does not lead to a change of temperature, but, as condensation heat, to a change of the fractions of liquid and gaseous phase.

▲   The efficiency η depends sensitively on the enthalpies of steam before (H_2) and after (H_3) expansion.

The indices refer to the points plotted in **Fig. 20.14**.

efficiency of a steam engine			1
$\eta = \dfrac{H_2 - H_3}{H_2 - H_4} \approx 1 - \dfrac{H_3}{H_2}$	Symbol	Unit	Quantity
	η	1	efficiency
	H	J	enthalpy

Steam engine, a machine based on the Clausius–Rankine cycle (**Fig. 20.15**).

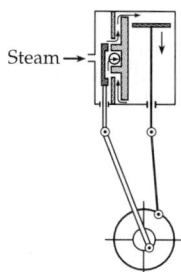

Steam →

Figure 20.15: Steam engine, schematic.

The high-pressure steam enters through the inlet (left), the low-pressure steam is expelled through the exhaust (small circle, mid-left). Piston and valve operate with phase shifts.

20.7.5   Open systems

1. Closed system,

a system with a fixed quantity of matter involved in the working process.

■   The Stirling engine is a closed system.

➤   Combustion engines in a closed system cannot use the combustion gas as a working medium. However, the combustion gas is used in open systems as a working medium.

2. Open system,

a system for which a certain number of particles leaves the system per unit time, and a certain number enters the system (see p. 624). Nevertheless, the total number of particles in the system may be conserved.

■   In the Otto engine, the gasoline–air mixture enters and the combusted gas escapes through the exhaust.

As a substitute, one usually considers a system including the particles crossing the boundary of the working system during the process time. At the beginning, this substitute system contains all particles that will enter the working system during the process and, at the end, all particles that have left the system during the process. This substitute system may have different pressures, volumes and temperatures before and after the process.

The balance of enthalpy is:

balance of enthalpy in open system			ML^2T^{-2}
$\Delta H = \Delta W_{ext} + \Delta Q$	Symbol	Unit	Quantity
	H	J	enthalpy
	W_{ext}	J	external work
	Q	J	supplied quantity of heat

If the flow velocities and potential energies of the entering and exiting particles are different, the corresponding energy differences must be summed:

$$\Delta H + \Delta W_{kin}^{flow} + \Delta W_{pot} = \Delta W_{ext} + Q.$$

3. Technical work,

also **operation work**, the total work done by a machine (theoretically) during a process step. It includes:
- injection of particles,
- change of volume,
- ejection of particles.

It can be defined as an integral,

$$W_t = \int_{p_1}^{p_2} V\,dp.$$

20.7.6   Otto and Diesel engines

20.7.6.1   Otto cycle

1. Otto cycle,

a cycle in an open system, consisting of two isentropic and two isochoric partial processes (**Fig. 20.16**):
- isentropic (adiabatic) compression,
- isochoric heating,
- isentropic expansion,
- isochoric cooling.

Efficiency η, depends on the volumes in the compressed and expanded state:

efficiency of the Otto engine			1
$\eta = 1 - \dfrac{1}{\varepsilon^{\kappa-1}}$	Symbol	Unit	Quantity
	η	1	efficiency
	ε	1	compression ratio
$\varepsilon = \dfrac{V_1}{V_2}$	κ	1	adiabatic index
	V	m^3	volume

2. Otto engine,

combustion engine operating in the Otto cycle. A homogeneous air–fuel mixture cyclically undergoes a fast combustion reaction by external ignition (spark plug).

Work steps of the Otto engine (**Fig. 20.16**):
- ab: intake of the fuel-air mixture,
- bc: compression stroke,
- cd: ignition of the fuel mixture, heating of the combustion gases,
- de: power stroke,
- e: opening of the exit valve,
- ba: exhaust stroke.

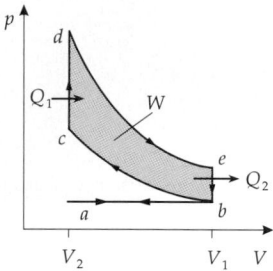

Figure 20.16: Otto cycle.

➤   Anti-knock compounds in the gasoline help prevent self-ignition.

20.7.6.2   Diesel cycle

1. Diesel cycle and Diesel engine

Diesel cycle, a cycle in an open system, consisting of two isentropic steps, one isochoric step and one isobaric partial step:
- isentropic (adiabatic) compression,
- isobaric heating,
- isentropic expansion,
- isochoric cooling.

Work steps of the Diesel engine (see **Fig. 20.17**):
- ab: intake of air,
- bc: compression stroke,
- cd: injection of fuel and combustion,
- de: power stroke,
- e: opening of the exit valve,
- ba: exhaust stroke.

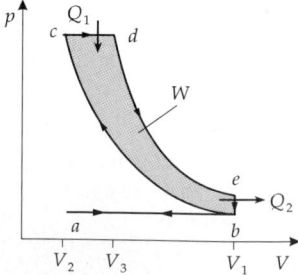

Figure 20.17: Diesel cycle.

2. Efficiency of the Diesel engine

Efficiency η, depends on the volumes in the compressed ($V_3 > V_2$) and expanded (V_1) state:

efficiency of the Diesel engine			**1**
$$\eta = 1 - \dfrac{\left(\dfrac{V_3}{V_2}\right)^{\kappa} - 1}{\kappa\left(\dfrac{V_3}{V_2} - 1\right)\left(\dfrac{V_1}{V_2}\right)^{\kappa-1}}$$	Symbol	Unit	Quantity
	η	1	efficiency
	κ	1	adiabatic index
	V	m³	volume

Diesel engine, a combustion engine operating in the Diesel cycle. The fuel is injected into compressed air. The combustion proceeds cyclically by self-ignition.

➤ Although at equal compression ratios the Diesel engine has a lower efficiency than the Otto engine, it can achieve higher compression ratios, meaning the overall efficiency of the Diesel engine is better than that of the Otto engine.

20.7.7 Gas turbines

1. Joule cycle,

an open cycle used for example in jet engines of airplanes. It consists of two isentropic and two isobaric partial steps (**Fig. 20.18 (a)**):

- isentropic (adiabatic) compression (I),
- isobaric heating (II),
- isentropic expansion (III),
- isobaric cooling (IV).

Efficiency η, depends on the temperatures before (T_1) and after (T_2) compression, or on the pressures, respectively:

$$\eta = 1 - \frac{T_1}{T_2} = 1 - \left(\frac{p_1}{p_2}\right)^{\frac{\kappa-1}{\kappa}}.$$

2. Ericsson cycle,

a closed cycle consisting of two isothermal and two isobaric partial processes (**Fig. 20.18 (b)**):

- isothermal compression (I),
- isobaric heating (II),
- isothermal expansion (III),
- isobaric cooling (IV).

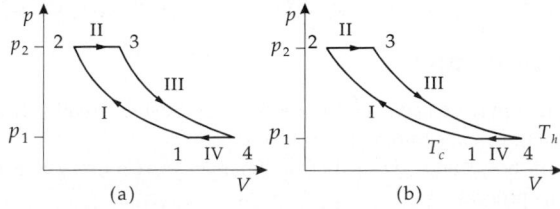

Figure 20.18: Cycles. (a): Joule cycle, (b): Ericsson cycle.

Efficiency η, depends on the temperatures:

$$\eta = 1 - \frac{T_c}{T_h} = \eta_C .$$

▲   Under ideal conditions, the efficiency may reach that of the Carnot cycle.

20.8   Gas liquefaction

Gas liquefaction at temperatures below the critical point may occur directly by compression.

■   Ammonia (NH_3), sulphur dioxide (SO_2) and chlorine (Cl_2) are gases whose critical temperatures lie above room temperature.

Otherwise, the gas must first be cooled below the critical temperature.

20.8.1   Generation of low temperatures

Low temperatures may be generated by:
• heat exchange with freezing mixtures,
• withdrawal of heat by dissolving of substances,
• cooling by a heat pump,
• employing the Joule–Thomson effect.

20.8.1.1   Freezing mixtures

Freezing mixtures, in general solid-liquid mixture systems used as reservoir for generating constant low temperatures. These mixtures must be brought first to this temperature by other means.

One uses systems at the melting point because, here, heat fluctuations do not lead to temperature changes, but, as latent heat, cause only fluctuations in the relative mass ratio of solid and liquid phases.

Low-temperature mixtures, see **Tab. 22.6**.

20.8.1.2   Heat of dissolution

Heat of dissolution, quantity of heat needed to dissolve a quantity of solid substance in a liquid substance.

If a substance is dissolved in a liquid, then heat is withdrawn from the liquid.

▲   Thereby, the temperature may drop below the melting point of the pure solvent without solidification of the system (freezing-point depression, see p. 737).

■   Casting salt on streets to prevent ice formation exploits this principle.

➤   Freezing mixtures of solutions consist of the solvent in solid phase (e.g., ice = frozen water) and the liquid phase with the dissolved substance (e.g., the salt solution).

20.8.1.3   Heat pump

Cooling of a system may be achieved by a left-handed cycle. Hereby, heat is withdrawn from the cold system by expenditure of work (see p. 707).

▲   According to the second law of thermodynamics, a second system must always be heated in this process.

■   Production of small quantities of liquid air or liquid helium.

➤ A system may also be cooled by a right-handed cycle, but only as long as the other system is colder than the system to be cooled.

20.8.2 Joule–Thomson effect

Gases kept in a vessel under increased pressure cool down when flowing out of a nozzle if the gas temperature lies below the **inversion temperature**. The process is an irreversible, adiabatic expansion of a real gas, since no external work is expended in the expansion ($\Delta W = 0$), and the expansion proceeds so rapidly that no heat can be exchanged with the environment ($\Delta Q = 0$).

The change of temperature occurs only for real (Van der Waals) gases, but not for an ideal gas. For control of the adiabatic expansion, the outflowing gas is slowed by a **throttle** (**Fig. 20.19**).

Joule–Thomson coefficient, δ, determines the inversion curve:

Joule–Thomson coefficient			$\mathbf{M^{-1}LT^2\Theta}$
	Symbol	Unit	Quantity
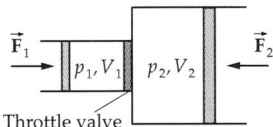	δ	K/Pa	Joule–Thomson coefficient
	V	m^3	volume
	C_p	J/K	heat capacity at const. pressure
	T	K	temperature
	α	1/K	isobaric-expansion coefficient
	c_p	J/(K kg)	spec. heat capacity at const. pressure
	ρ	kg/m^3	density

$$\delta = \frac{V}{C_p}(T\alpha - 1) = \frac{T\alpha - 1}{c_p \rho}$$

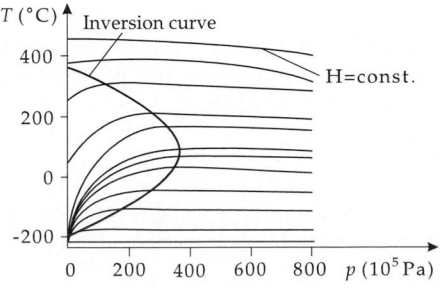

Throttle valve

Figure 20.19: Joule–Thomson effect.

Inversion curve, temperature-pressure curve for which δ just vanishes (**Fig. 20.20**).

Inversion temperature, temperature below which an irreversible expansion leads to a cooling of the gas.

T (°C)

Inversion curve

H=const.

400

200

0

-200

0   200   400   600   800 p (10^5 Pa)

Figure 20.20: Inversion curve.

➤   An irreversible expansion above the inversion temperature leads to a heating.
Inversion temperature of the Van der Waals gas:

inversion temperature = 6.75 · critical temperature			ϴ
	Symbol	Unit	Quantity
$T_i = \dfrac{2a}{Rb} = 6.75\,T_c$	T_i	K	inversion temperature
	T_c	K	critical temperature
	a	Nm^4/mol^2	internal-pressure constant
	b	m^3/mol	internal volume
	R	J/(K mol)	universal gas constant

This value is the maximum value. The inversion temperature T_i depends on the pressure.
➤   Instead of the molar constants a, b and R, the specific constants may also be used
here.

20.8.2.1   Linde process

Linde process, method of liquefaction of air according to the Joule–Thomson principle
(**Fig. 20.21**).

In order to lower the temperature of the high-pressure gas, **countercurrent heat ex-
changers** are used for liquefaction of air. The expanded cooled gas is brought in thermal
contact with the high-pressure gas in a system of pipes in which the high-pressure gas and
the cooled gas flow in opposite directions.

▲   This method operates only for gases whose inversion temperature for given pressure
of the compressor lies above room temperature.

■   Air, CO_2, N_2, ... may be liquefied in this manner.

➤   For hydrogen and helium, pre-cooling is needed because their inversion temperatures
(hydrogen $T_i \approx -80°C$) lie below room temperature.

Liquid hydrogen may be used for pre-cooling of helium. But this is no longer done
because of explosion hazard and expense.

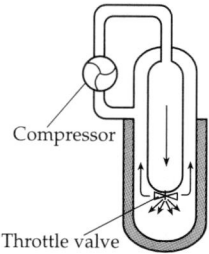

Compressor

Throttle valve

Figure 20.21: Linde process
(schematic).

In a **reversible** expansion of real gases, the temperature always decreases because the
gas must also do external work. The process of adiabatic expansion (see p. 707) has the
advantage of higher efficiency and is therefore used to liquefy helium.

20.8.2.2   Claude process

Claude process, an air liquefaction process wherein throttling is partially replaced by an
adiabatic expansion. The yield of liquid air is increased by the expansion. Furthermore,
part of the work expended is regained.

21
Phase transitions, reactions and equalizing of heat

21.1 Phase and state of aggregation

21.1.1 Phase

Homogeneous system, a system with properties that are uniform throughout.

■ A vessel of dry air under standard conditions.

Heterogeneous system, a system with properties that may change discontinuously at boundary surfaces.

■ A vessel containing water, steam and air.

Phase, homogeneous part of a heterogeneous system.

 Phase boundary, a separating interface between two phases.

■ A closed pot containing water and steam.
 The surface of water is a phase boundary. There is a gaseous phase (steam) and a liquid phase (water).

Phase transition, change of a substance in its intrinsic structure that affects the order of the system. This change in the order of the system causes a change of the temperature dependence of its properties.

■ If water is heated, it starts to boil when the boiling temperature is reached. An additional supply of heat does not lead to an increase of temperature, but only to further evaporation of water (**Fig. 21.1**).

21.1.2 Aggregation states

Aggregation state of a substance, a phase of a substance determined by certain properties and by the intrinsic structure.

 Four states of aggregation exist:

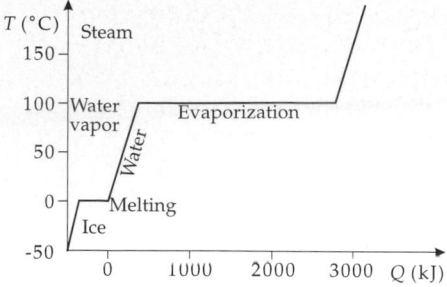

Figure 21.1: Temperature increase by supply of heat.

Solid: The body has a fixed internal order, e.g., a crystal lattice, with very strong internal interactions. It has a fixed shape with a defined surface. It takes a fixed volume, which changes only under high pressure.

Liquid: A liquid has no tightly fixed internal order, but is governed by strong internal interactions. A liquid does not take a specific shape, but still has a definite surface. It occupies a fixed volume, which changes only under high pressure.

Gas: A gas has no internal order and is governed only by weak internal interactions. A gas has neither a fixed shape nor a surface. It rather adapts to any volume, filling it entirely. Its volume may be changed by pressure.

Plasma: The plasma state occurs at very high energies. The atoms are ionized and decomposed into charged constituents. A plasma has no fixed intrinsic structure, but is governed by electromagnetic interactions (see p. 573).

▲   By supplying energy, a body may be converted from a solid state to a liquid or gaseous state, and a liquid may turn into a gaseous state.

21.1.3 Conversions of aggregation states

1. Boiling and condensation

Boiling of a substance, conversion of a liquid into a gas. Boiling of a substance occurs if the vapor pressure of the substance becomes higher than the environmental pressure.

Latent heat of vaporization, the amount of energy needed to vaporize a liquid.

Specific latent heat of vaporization, the amount of heat needed to vaporize 1 kg of a material. The specific latent heat depends on the pressure and the temperature.

specific latent heat $= \dfrac{\textbf{latent heat}}{\textbf{mass}}$			$L^2 T^{-2}$
$l_v = \dfrac{\Delta Q}{m}$	Symbol	Unit	Quantity
	l_v	J/kg	specific latent heat of vaporization
	ΔQ	J	latent heat of vaporization
	m	kg	mass of vaporized substance

$\boxed{\text{M}}$ In order to determine latent heats of vaporization (or condensation), vapor is condensed in specially designed calorimeters. The heat transfer to the calorimeter is measured.

Boiling point, the temperature at which the substance boils. The boiling point depends on the external pressure.

Boiling points of many substances are given in **Tab. 22.1.2**.

■   Boiling points of several elements (in ° C): aluminum 2467, lead 1740, mercury 356.58, oxygen (O_2) −182.96, hydrogen (H_2) −252.8, nitrogen (N_2) −195.8.

Condensation, conversion of a gas into a liquid.

Condensation occurs at the same temperature as boiling. Under specific conditions, a material may boil at temperatures slightly above the boiling point (see p. 727), or condense at temperatures slightly below that point (see p. 727).

Latent heat of condensation, the heat released in the condensation of a gas. Its numerical value equals that of the heat of vaporization.

2. Melting and solidification

Melting, the conversion of a solid into a liquid. Melting occurs if the sublimation pressure of the solid becomes lower than the vapor pressure of the liquid.

Solidification, conversion of a liquid into a solid. Solidification occurs at the same temperature as melting.

Melting point, the temperature at which a solid substance melts or a liquid substance solidifies. The melting point depends on the external pressure.

For the melting points of many substances, see **Tab. 22.1.2**.

■ Melting points of several elements (in ° C): aluminum 660.4, lead 327.5, iron 1535, gold 1064.4, mercury -38.87, oxygen (O_2) -218.4, hydrogen (H_2) -259.34, nitrogen (N_2) -209.86.

Latent heat of fusion, the quantity of heat that is released when a liquid freezes.

Specific latent heat of fusion, the amount of energy needed to melt 1 kg of a material:

specific latent heat of fusion = $\dfrac{\text{melting heat}}{\text{mass}}$			$\mathbf{L^2 T^{-2}}$
$l_f = \dfrac{\Delta Q}{m}$	Symbol	Unit	Quantity
	l_f	J/kg	specific latent heat of fusion
	ΔQ	J	melting heat
	m	kg	mass of molten substance

Latent heat of melting, the heat released during solidification of a liquid. Its numerical value equals that of the heat of fusion.

Sublimation, conversion of a solid into a gas.

Desublimation, the inverse process.

Heat of sublimation, quantity of heat to be supplied to sublimate a body.

➤ The sublimation heat is equal to the sum of melting heat and evaporation heat.

21.1.4 Vapor

Wet steam, **saturated vapor**, occurs in the coexistence of liquid and gaseous states in equilibrium.

Saturated vapor pressure, p_D, SI unit Pascal, vapor pressure of the saturated gas. The value depends exponentially on the temperature.

Vapor pressure curve, curve $p_D(T)$ representing the saturated vapor pressure of a two-phase system as a function of temperature (**Fig. 21.2**).

Nonsaturated vapor, vapor that is not in equilibrium with the liquid.

➤ In the course of time, the liquid vaporizes until equilibrium is reached or all the liquid is vaporized.

Triple point, point in which solid, liquid and gaseous phase are in mutual equilibrium (**Fig. 21.3**). At the triple point both the pressure p_{tr} and the temperature T_{tr} are fixed (see p. 734).

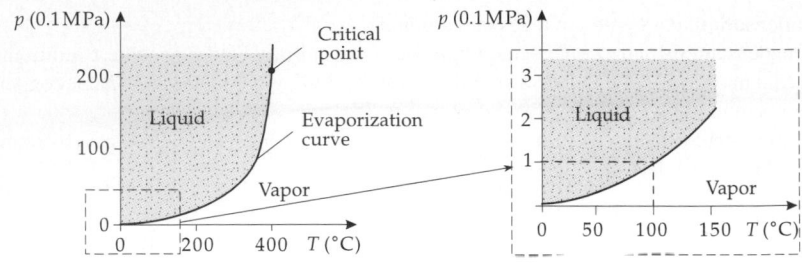

Figure 21.2: Vapor pressure curve.

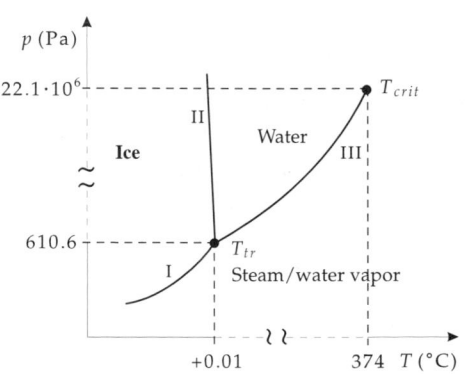

Figure 21.3: Phase diagram of water with triple point T_{tr} and critical temperature T_{crit}. I: sublimation pressure curve, II: melting pressure curve, III: vapor pressure curve.

■ For water, the temperature and the pressure at the triple point are 273.16 K and 610.6 Pa, respectively.

➤ Triple points are particularly well suited as fixed points for establishing the temperature scale.

21.2   Order of phase transitions

Change of entropy, due to the heat supplied or carried off in a phase transition, the entropy (disorder) of the system (which differs in the phases) is changed.

change of entropy = $\dfrac{\text{heat supplied (or released)}}{\text{temperature}}$			$\mathbf{ML^2T^{-2}\Theta^{-1}}$
	Symbol	Unit	Quantity
$\Delta S = \dfrac{\Delta Q}{T}$	S	J/K	entropy
	Q	J	quantity of heat
	T	K	temperature

21.2.1   First-order phase transition

Phase transition of first order, characterized by an additional heat supply (release) during the phase transition. Consequence:

▲ The **entropy jump** in the S-T-diagram (**Fig. 21.4 (a)**) is ascribed to the additional heat supply at the phase transition point.

■   Transitions between the different aggregation states are first-order transitions, except for the transition at the critical point.

Relation between the supplied (released) quantity of heat ΔQ and the change in temperature ΔT:

$$\Delta Q = C \, \Delta T \, , \quad C : \text{ heat capacity.}$$

▲   Since the temperature remains constant at the phase transition, the heat capacity tends to infinity at the phase transition of first order:

$$C \longrightarrow \infty \, .$$

The volume also displays a step-like behavior in the p–V-diagram.

▲   The compressibility of the substance diverges at the first-order phase transition:

$$\kappa = \left. \frac{1}{V} \frac{\partial V}{\partial p} \right|_{T = \text{const.}} \longrightarrow \infty \, .$$

Characterization of the first-order phase transitions:
- a jump in entropy,
- the heat capacity approaches infinity,
- the compressibility approaches infinity.

21.2.2 Second-order phase transition

Second-order phase transition, characterized by a kink in the temperature (or entropy) curve (**Fig. 21.4 (b)**) in the T-S-diagram.

➤   The entropy curve $T(S)$ has no jump, $\Delta S = 0$, but the derivative of the entropy with respect to the temperature changes discontinuously at the transition point.

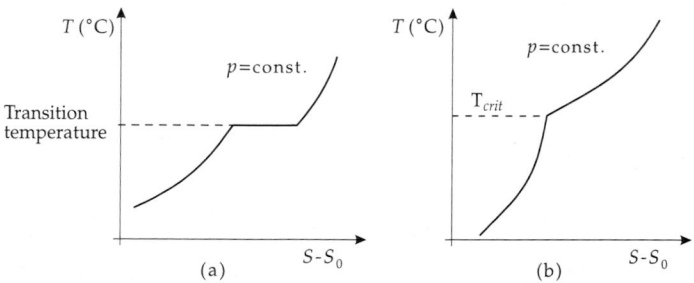

Figure 21.4: Phase transitions. (a): phase transition of first order, (b): phase transition of second order.

■   Phase transitions at the critical point are second-order phase transitions.

Second-order phase transitions are characterized by:
- a continuous break point in the entropy,
- a finite jump of heat capacity,
- the compressibility diverges.

21.2.3 Lambda transitions

Lambda transition, **λ-transition**, characterized by:

- The entropy as function of the temperature T exhibits no kink, but has a vertical tangent at a temperature T_λ.

- The derivative of the entropy with respect to the temperature diverges, $\dfrac{dS}{dT} \to \infty$.

- The heat capacity diverges, $C \to \infty$.

The heat-capacity curve shows a characteristic λ-shape (**Fig. 21.5**)

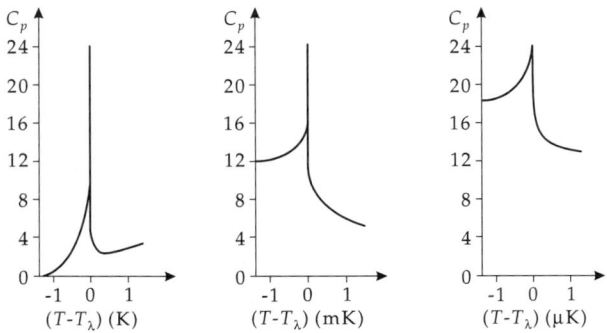

Figure 21.5: λ-transition. From left to right, the width of the interval about the transition temperature becomes smaller and smaller.

■   The transitions to superfluidity in ^{3}He and ^{4}He, as well as some conversions of crystal structures in binary alloys, are λ-transitions.

21.2.4 Phase-coexistence region

Coexistence region, two phases may exist simultaneously.

In the coexistence region, the temperature is constant for isobars.

The coexistence region is characterized by an entropy jump in the T-S-diagram, and by a jump in volume in the p–V-diagram. The coexistence region of two phases becomes smaller with increasing pressure and increasing temperature (**Fig. 21.6**), and finally vanishes at the critical point.

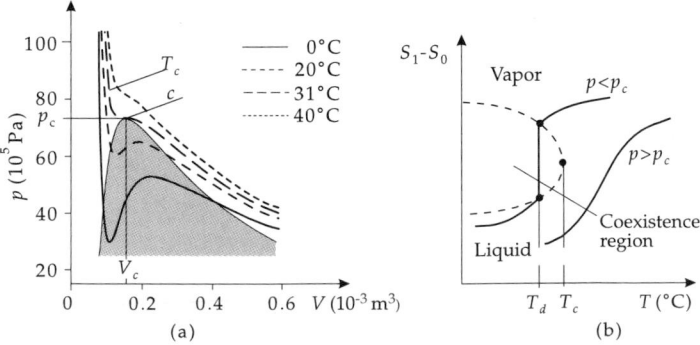

Figure 21.6: Coexistence region of two phases (schematically). (a): p–V-diagram, (b): S-T-diagram.

Critical point, the position in the phase diagram at which the region of coexistence shrinks to a point.

▲   There is no longer a phase transition above the critical point.

➤   It makes no sense to talk of distinct phases above the critical point.

21.2.5 Critical indices

At the critical point, there are no longer phase boundaries.

➤   Large density fluctuations may occur, manifested, e.g., by critical opalescence, where an extremely strong scattering of light is observed.

■   A transparent substance suddenly becomes impermeable to light. A tiny fog forms.

▲   At the critical point, several thermodynamics quantities become infinite.

For describing the behavior of diverging quantities near the critical point, power-series expansions are used.

Critical indices, the exponents of these expansions.

For the liquid–gas phase transition one needs six critical indices for which the standard notations $\alpha, \alpha', \beta, \gamma, \gamma', \delta$ have become customary.

Density difference, the difference of the densities of liquid and gas, $z = \rho_{fl} - \rho_g$. For $T \to T_C$, it approaches zero as

$$z = \rho_{fl} - \rho_g \sim \left(1 - \frac{T}{T_c}\right)^{\beta}.$$

The **specific heat capacity** at the critical volume $C_{V=V_c}$ may diverge for $T \to T_c$ when approaching the critical temperature from above or from below:

$$C_{V=V_c} \sim \begin{cases} \left(\dfrac{T}{T_c} - 1\right)^{-\alpha} & \text{if} \quad T|_{\rho \approx \rho_c} \geq T_c \,, \\[4mm] \left(1 - \dfrac{T}{T_c}\right)^{-\alpha'} & \text{if} \quad T|_{\rho \approx \rho_c} \leq T_c \,. \end{cases}$$

Compressibility, displays an analogous trend:

$$\kappa \sim \begin{cases} \left(\dfrac{T}{T_c} - 1\right)^{-\gamma} & \text{if} \quad T \geq T_c \,, \\[4mm] \left(1 - \dfrac{T}{T_c}\right)^{-\gamma'} & \text{if} \quad T \leq T_c \,. \end{cases}$$

Critical isotherm:

$$p - p_c \sim |\rho - \rho_c|^{\delta} \quad \text{for} \quad T = T_c \,.$$

Simple gases show a similar behavior with respect to the critical indices.

21.3   Phase transition and Van der Waals gas

21.3.1   Phase equilibrium

Vapor-pressure curve, curve representing the saturated vapor pressure of a two-phase system as a function of temperature (**Fig. 21.7**).

▲   The vapor pressure $p_g(T)$ is a pure function of temperature and does not depend on the vapor volume V. A change in the vapor volume affects only the quantity of vapor.

➤   Excess vapor is condensed again to liquid. If the quantity of vapor is too low, the liquid continues to vaporize until saturation is reached.

In equilibrium between vapor and liquid, a certain vapor pressure p_g is established that can be calculated with the Clausius–Clapeyron equation.

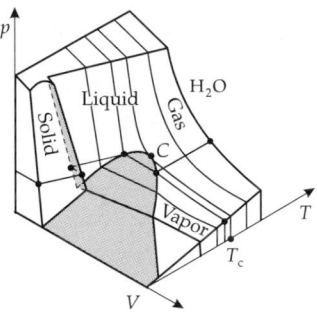

Figure 21.7: Three-dimensional $p(V, T)$ phase diagram. The vapor pressure curve $p(T)$ and the p–V diagram are projections onto the p–T and p–V planes, respectively.

The equilibrium conditions are:

$p_{fl} = p_g$	mechanical stability,
$T_{fl} = T_g$	thermal stability,
$\mu_{fl}(p, T) = \mu_g(p, T)$	chemical stability.

21.3.2   Maxwell construction

1. Equation of state according to Van der Waals

The Van der Waals equation of state for real gases takes into account the internal volume of the molecules and the (attractive) forces between the molecules. It allows for a simple approximation for real gases, but is inaccurate for liquids.

Van der Waals equation			$\mathbf{ML^2T^{-2}}$
	Symbol	Unit	Quantity
	p	Pa	pressure
	n	mol	number of moles
$\left(p + \left(\dfrac{n}{V}\right)^2 a\right)(V - nb) = nRT$	V	m^3	volume
	a	Nm^4/mol^2	internal-pressure constant
$R = 8.314 \, J/(K\,mol)$	b	m^3/mol	internal-volume constant
	R	$J/(K\,mol)$	universal gas constant
	T	K	temperature

This equation of state allows for **metastable** and **unstable** regions.
- Metastable regions display a negative derivative of the pressure with respect to the volume, and thus a positive compressibility. Metastable states may be reached during changes of state (see p. 727).
- Unstable regions show a positive derivative of the pressure with respect to the volume and negative compressibility. They cannot be reached by reversible changes of state.

2. Maxwell construction,

a prescription to replace a section of the curve $p(V)$ by a horizontal line.

In the region between the outer intersection points of the horizontal line with the Van der Waals curve (**Fig. 21.8**):

▲   The area in the p–V-diagram under the horizontal line must be equal to the area under the Van der Waals curve.

Thus, the intersection points between the curve and the horizontal line are chosen such that the area enclosed by the horizontal line, the x-axis and the vertical lines through the (outer) intersection points equalizes the corresponding area obtained by using the Van der Waals curve as upper boundary.

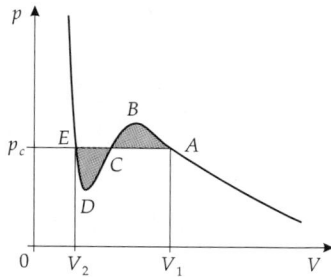

Figure 21.8: Maxwell construction.

In other words:

▲   The area enclosed by the horizontal line and the curve above the horizontal line is equal to the area enclosed by the horizontal line and the curve below the horizontal line.

▲   The outer intersection points of the horizontal line with the curve define the phase coexistence region.

In this interval, the Van der Waals curve describes the metastable and unstable regions which, however, cannot be reached in equilibrium. If the state of the system is changing, the metastable and unstable regions may be reached (see p. 727).

The length of the horizontal distance decreases continuously with increasing temperature. Hence, with increasing temperature, the phase coexistence region also shrinks. The length of the horizontal section approaches zero at the critical point.

3. Critical point and critical temperature in the Maxwell construction

Critical point, the state at which the phase coexistence region shrinks to a point.

Critical isotherm, isotherm passing through the critical point.

Temperature, pressure and molar volume at the critical point may be calculated:

▲   The critical point must be a saddle point on a Van der Waals isotherm.

➤   The critical isotherm is the only Van der Waals isotherm that has a saddle point.

Critical temperature, T_c, the temperature belonging to the critical isotherm.

critical temperature (Van der Waals equation)			Θ
	Symbol	Unit	Quantity
$T_c = \dfrac{8\,a}{27\,Rb}$	T_c	K	critical temperature
	a	Nm^4/mol^2	internal-pressure coefficient
$R = 8.314 \text{ J}/(\text{K mol})$	b	m^3/mol	internal volume
	R	$J/(\text{K mol})$	universal gas constant

4. Critical pressure in the Maxwell construction,

p_c, pressure at the critical point.

critical pressure (Van der Waals equation)			$ML^{-1}T^{-2}$
	Symbol	Unit	Quantity
$p_c = \dfrac{a}{27\,b^2}$	p_c	Pa	critical point
	a	Nm^4/mol^2	internal-pressure coefficient
	b	m^3/mol	internal volume

Above the critical temperature, no Maxwell construction is possible. Liquid and gas then can no longer be distinguished.

▲   In processes that do not intersect the range of coexistence in the phase diagram, two phases may be converted into each other without a phase transition. To carry out such a process one must go beyond the critical point.

■   The isothermal compression of a gas below the critical temperature leads to a phase transition. Heating a liquid to a temperature above the critical one, and subsequent isothermal expansion followed by isochoric cooling (**Fig. 21.9**), converts the liquid into a gas without a perceptible phase transition.

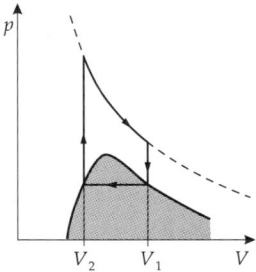

Figure 21.9: Scheme of a closed cycle with only one perceptible phase transition.

21.3.3   Delayed boiling and delayed condensation

The metastable regions (with negative derivative of the pressure with respect to the volume) of the Van der Waals isotherm may be realized in experiment in a non-equilibrium state.

■   If a gas is compressed isothermally with great care (avoiding shaking and presence of condensation centers), then one can follow the Van der Waals isotherm even beyond the intersection point with the horizontal Maxwell line, nearly up to the maximum of the curve.

Delayed condensation, vapor does not condense, although the temperature drops below the condensation temperature.

Delayed boiling, liquid does not boil, although its temperature exceeds the boiling temperature.

Superheated liquid, liquid brought to the metastable range by isochoric heating.

Supercooled vapor, gas brought to the metastable range by isochoric cooling.

▲ Even for slight disturbances, the metastable system changes in a shock-like manner into the stable state of phase coexistence.

➤ In practice, these unstable ranges are avoided by adding beads (condensation nuclei) or stirring the boiling liquid.

There are analogous phenomena in the solid–liquid phase transition.

21.3.4 Theorem of corresponding states

Reduced variable, the representation of a state variable in units of its value at the critical point:

$$\overline{p} = \frac{p}{p_c}, \quad \overline{v} = \frac{v}{v_c}, \quad \overline{T} = \frac{T}{T_c}.$$

▲ The reduced variables $\overline{p}, \overline{v}, \overline{T}$ are dimensionless.

Simple gas, gas of particles having no large electric dipole moments and whose atoms and molecules are not strongly correlated even in the liquid phase.

■ Noble gases, N_2, O_2, H_2 or CO, CH_4 are simple gases.

Theorem of corresponding states, a statement introduced by Van der Waals:

▲ All simple gases satisfy the same Van der Waals equation in the reduced variables.

Van der Waals equation in reduced variables:

$$\left(\overline{p} + \frac{3}{\overline{v}}\right)(3\overline{v} - 1) = 8\overline{T}.$$

21.4 Examples of phase transitions

21.4.1 Magnetic phase transitions

Paramagnets require higher field strengths than **ferromagnets** to reach **saturation magnetization**.

1. Curie temperature

When an external magnetic field is removed, a solid in a ferromagnetic state retains a permanent magnetic polarization. The magnitude of the magnetization depends strongly on the mechanical and thermal history of the material.

➤ Ferromagnetism is found mostly in solids with a well defined crystal structure. Amorphous ferromagnets represent exceptions.

Curie temperature, conversion temperature of the transition from ferromagnetism to paramagnetism. Ferromagnetism is established only below the Curie temperature.

■ The elements iron, cobalt and nickel exhibit ferromagnetic properties below the Curie temperature.

The Curie temperatures of various metals are listed in **Tab. 22.1.3**.

2. Weiss domains and Bloch walls

In a non-magnetic ferromagnetic, the atomic magnetic dipoles are not oriented randomly but are aligned parallel within larger regions of extension of several tenths of millimeters. These regions have a macroscopic magnetic dipole moment.

Weiss domains, designation of regions with parallel alignment of the magnetic dipoles. In a nonmagnetized ferromagnet, the dipole moments of the individual magnetized Weiss domains are randomly oriented (**Fig. 21.10**). Therefore, the material as a whole appears to be nonmagnetic.

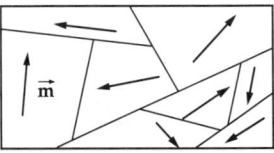

Figure 21.10: Weiss domains (schematic). $\vec{m}$: magnetic dipole moment.

Spontaneous magnetization, varies continuously between the individual Weiss domains over a range of ca. 300 atoms.

Bloch walls, interfaces between the Weiss domains.

▲   An external magnetic field causes the Weiss domains with similar alignment to enlarge until all domains are equally aligned: **saturation magnetization**.

21.4.2   Order–disorder phase transitions

In phase transitions of this kind, the low-temperature phase exhibits a certain order of atoms or molecules which get lost above the transition temperature.

➤   In principle, the solid–liquid and solid–gaseous transitions may also be understood as order–disorder transitions. But it is a convention to include only solid–solid phase transitions in this category.

Positional order, the arrangement of atoms or molecules in a crystal lattice.

➤   Order–disorder phase transitions also include conversions in the arrangement of atoms on the lattice sites.

■   Phase transition in β-brass (CuZn) at $T = 465\ °C$: In the low-temperature phase, brass displays a structure in which copper and zinc are well ordered in distinct sublattices. At higher temperatures, the elemental atoms are statistically distributed.

Orientational order describes the relative orientation of certain molecules with respect to each other.

■   Ammonium halides NH_4Cl, NH_4Br and NH_4J. Here the NH_4^+–tetrahedrons may take two distinct orientations in the crystal lattice (**Fig. 21.11**). Above the critical temperature, both orientations occur statistically distributed. Below $T_c = 256\ K$ all tetrahedrons in NH_4Cl have the same orientation, while in NH_4Br below T_c the tetrahedrons take an alternating orientation.

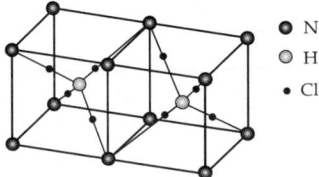

● N
○ H
• Cl

Figure 21.11: Possible orientation of the NH_4^+–tetrahedron in NH_4Cl.

21.4.3   Change in the crystal structure

1. Phase transitions of the type solid-solid

The solid phases of many substances may take different crystal structure, depending on pressure and temperature (for alloys also depending on the composition).
- For ice at pressures up to 8000 bar, five distinct modifications are known (ice I, II, III, IV, V). Ordinary ice at $p \approx 1$ bar is only one of them.
- ➤ At extremely high pressures, several nonmetals may even be transformed to a **metallic phase**.
- Carbon and hydrogen have this property.

If no appropriate catalyst is present, the solid–solid phase transitions sometimes may be delayed appreciably.
- Actually, a diamond is not stable at atmospheric pressure (see **Fig. 21.12**). However, the phase transition is appreciably delayed: a diamond is practically stable.

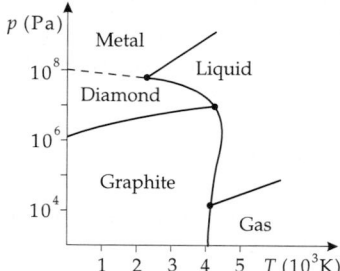

Figure 21.12: Phase diagram of ^{12}C (without fullerenes).

2. Structures of carbon

So far, three important stable forms of solid carbon are known:

a) Graphite, the most stable phase; planar, honeycomb-like structures, metallic conductivity. Three valence electrons are used for bonds with the neighbors in the plane. The fourth electron may be displaced freely in this plane (**sp^2-hybrid**) generating the conductivity of graphite. The individual planes are not linked among each other by chemical interactions, and thus may be shifted against each other (graphite is suitable as lubricant).

b) Diamond, very hard, at standard atmospheric pressure metastable (but in fact stable) phase, tetrahedron-like structures, insulator. Chemically resistant material with low frictional coefficient and high thermal conductivity. All four valence electrons are used for simple bonds with four neighbors each. Application as material for coating of tools, anticorrosion layers, wear-resistant surface coating and as passive material in microelectronics. Diamond layers can be produced synthetically as polycrystals, and with high purity by precipitation from the gaseous phase.

c) Fullerene, spherical closed structures of carbon. Three valence electrons are used for bonds with the neighbors, the fourth points to the outer surface of the spherical shell. Semiconducting material, soluble by several organic solvents, similar softness as graphite. Production by vaporization of graphite in an electric arc in a low-pressure noble-gas atmosphere (**Huffmann–Kraetschmer method**). Possible use for batteries (take-in of electrons), superconductivity (fullerene–alkali mixtures), photochemistry (photosensitizer), microconductors, optical switching elements.

d) Most important kinds of fullerenes:
- **Buckminster fullerene** C_{60}, most famous and most stable modification with the shape of a soccer ball, consisting of 12 pentagons and 20 hexagons. Main product of the Huffmann-Krætschmer method.
- C_{70}, second-most-frequent modification, with the shape of an American football.
- **Buckybabies**, unstable structures C_{32}, C_{44}, C_{50} and C_{58}, nearly spherical.
- **Buckygiants**, unstable, large structures C_{240}, C_{540}, C_{960}.

e) Fullerene-like structures:
- **Buckytubes**, tube-like macromolecules, similar to graphite, with a length of micrometers and microscopic diameter (some nanometers). Application in electric engineering (molecular wires).
- **Buckyonions**, onion-like arrangement of spherical fullerenes. Use not yet known, high resistance to compression expected.

21.4.4   Liquid crystals

In some organic substances with high molecular weight and a stretched form of the molecules, the long-range order remains in melting. Such molecules also have an alignment in the liquid phase, and hence are not isotropic.

Liquid crystals may occur in a variety of structures, e.g., in smectic or nematic phases (see p. 730).

➤ With increasing temperature, some substances may form various types of liquid crystals. Then they have several conversion temperatures.

Usually, liquid crystals are formed by complicated organic substances. Many of them have conversion temperatures and melting points in the range of 100 °C.

➤ Liquid crystals only received technical interest after substances with lower conversion temperatures were discovered.

Optic anisotropy of nematic liquid crystals leads to a strong scattering of light.

➤ In the phase transition to an isotropic liquid, the scattering disappears.

In liquid crystals of sufficiently large electric dipole moment, the permeability of light, the reflection, and the optical activity may be controlled in a simple manner by applying an electric field almost without power supply.

■ LCDs (**liquid crystal displays**) are based on this principle.

21.4.5   Superconductivity

Superconductors, electric conductors for which the direct-current resistance drops to an extremely small value when the temperature falls below a critical temperature T_c. The charge carriers are not single electrons, but **Cooper pairs**. For most of the metallic superconductors, the **transition temperature** lies at about 1–10 kelvin.

High-temperature superconductors, **HTCS**, ceramic superconductors based on copper oxide with high **critical temperature**.

➤ For high-temperature superconductors, liquid air is sufficient as coolant. For cooling of the common metallic superconductors, the more expensive liquid helium is necessary. But because of the thermal motion of the magnetic flux lines, the high-temperature superconductors have a relatively high electric resistance, which decreases only continuously with decreasing temperature. Besides the material instabilities, the electric resistance represents an appreciable limitation for technical applications.

Meissner–Ochsenfeld effect, shielding of external magnetic fields by the superconductor up to a critical magnetic field strength for which the superconductivity breaks down (see solid-state physics, superconductors). For intensities below the critical magnetic field strength H_c, no magnetic field can penetrate into the superconductor.

■ Superconductors are mainly used for loss-free circuit lines and for generation of high magnetic flux densities by superconducting magnetc coils.

21.4.6 Superfluidity

Superfluidity, the ability of a liquid to creep up at the walls of vessels and to overcome potential barriers.

■ If a beaker is immersed in a superfluid such that the bottom is below the surface of liquid but the rims project beyond the surface, the superfluid creeps up to the rim of the beaker and flows down the inner wall until the liquid level inside the beaker equals the level outside.

In superfluid liquids:

• the viscosity is zero, $\eta \to 0$,
• the thermal conductivity is infinite: $\lambda \to \infty$.

No temperature gradients arise, since all heat fluctuations are compensated immediately.

Helium II, a superfluid phase with maximum order. Below a pressure of 25 bar, there is no conversion of helium II into solid helium for arbitrary low temperature.

For standard atmospheric pressure, liquid helium converts into helium II at 2.2 K.

• Helium II has an extremely high thermal conductivity.
• Helium II does not boil—as other liquids do—with formation of vapor bubbles in the liquid volume. Helium II instead evaporates from the surface.

The viscosity of helium II may take extremely low values.

■ Helium II can still flow through the smallest capillaries that may not allow even gaseous helium to flow through.

21.5 Multicomponent gases

Multicomponent gas, a gas with more than one distinguishable type of particle (**component**).

Mole fraction, x_i, dimensionless quantity, the fraction of one kind of particles of the total quantity:

mole fraction = $\dfrac{\text{number of particles of one type}}{\text{total number of particles}}$			1
	Symbol	Unit	Quantity
$x_i = \dfrac{N_i}{N}$	x_i	1	mole fraction, type i
	N_i	1	particle number, type i
	N	1	total number of particles

The sum over all mole fractions yields unity:

$$\sum_{i=1}^{K} x_i = \sum_{i=1}^{K} \frac{N_i}{N} = \frac{N}{N} = 1 \,.$$

Concentration, c_i, the quantity of a substance i per unit volume or unit mass.

There are several definitions for the description of concentration (see p. 742). The notion of concentration used here for the solutions involves the molarity, the ratio of the quantity of substance of a dissolved substance to the volume of the solvent,

$$c_i = \frac{n_i}{V}.$$

Mass fraction, ξ_i, the ratio of the total mass of a given type of particle to the total mass of all particles. The mass fraction is equal to the product of the mole fraction and the ratio of the molar mass of the i type of particles and the molar mass of the total system.

mass fraction = $\dfrac{\text{total mass of type } i}{\text{total mass of all particles}}$			1
	Symbol	Unit	Quantity
$\xi_i = \dfrac{m_i}{m_{\text{tot}}}$	ξ_i	1	mass fraction type i
	m_i	kg	total mass type i
	m_{tot}	kg	total mass all particles
$= x_i \dfrac{M_i}{M_{\text{tot}}}$	x_i	1	mole fraction
	M_i	kg/mol	molar mass type i
	M_{tot}	kg/mol	molar mass total mixture

➤   In some books, the mass fraction is labeled x_i, and the mole fraction κ_i.

21.5.1 Partial pressure and Dalton's law

Total pressure p of a mixture of gases, SI unit pascal, the sum over all forces F exerted by momentum transfer onto an area A:

$$p = \frac{F}{A}.$$

Partial pressure p_i of one type of particle, the sum over all forces F exerted by momentum transfer by the specific type of particle onto an area A:

partial pressure = $\dfrac{\text{fraction of force}}{\text{area}}$			$\text{ML}^{-1}\text{T}^{-2}$
	Symbol	Unit	Quantity
$p_i = \dfrac{F_i}{A}$	p_i	Pa	partial pressure particles, type i
	F_i	N	fraction of force perp. area, type i
	A	m^2	area

Dalton's law:

The sum over all partial pressures of a gas consisting of various components yields the total pressure:

$$\sum_{i=1}^{K} p_i = \sum_{i=1}^{K} \frac{F_i}{A} = \frac{F}{A} = p.$$

The components of a gas are distributed independently of each other over the entire volume. Every component behaves as if there were no other components.

▲   Any component occupies the volume uniformly.

▲   In equilibrium, the partial pressure of a component is the same everywhere.

The quotient of partial pressure p_i and total pressure p is equal to the mole fraction x_i of the gas,

$$\frac{p_i}{p} = \frac{N_i}{N} = x_i \, .$$

21.5.2 Euler equation and Gibbs–Duhem relation

Euler equation, representation of the internal energy $U(T, S, p, V, \mu_i, N_i)$ as a function of the other variables for an isolated system in equilibrium:

Euler equation			$\mathbf{ML^2T^{-2}}$
$U = TS - pV + \sum_i \mu_i N_i$	Symbol	Unit	Quantity
	U	J	internal energy
	T	K	temperature
	S	J/K	entropy
	p	Pa	pressure
	V	m^3	volume
	μ_i	J	chemical potential, type i
	N_i	1	particle number, type i

Gibbs–Duhem relation, differential relation: The intensive variables $T, p, \mu_1, \ldots, \mu_K$ conjugated to the extensive variables $S, V, N_1, \ldots, N_K$ cannot all be independent of each other,

$$0 = S \, dT - V \, dp + \sum_{i=1}^{K} N_i \, d\mu_i \, .$$

Differential representation of the internal energy:

$$dU = T \, dS - p \, dV + \sum_{i=1}^{K} \mu_i \, dN_i \, .$$

➤   This representation is connected with the Gibbs–Duhem relation if the total differential of the Euler equation is formed:

$$dU = T \, dS - p \, dV + \sum_{i=1}^{K} \mu_i \, dN_i + S \, dT - V \, dp + \sum_{i=1}^{K} N_i \, d\mu_i \, .$$

Temperature, pressure and chemical potential (intensive variables) are the derivatives of the internal energy with respect to the extensive variables entropy, volume and particle number:

$$\frac{\partial U}{\partial S} = T, \qquad \frac{\partial U}{\partial V} = -p, \qquad \frac{\partial U}{\partial N_l} = \mu_l, \quad l = 1, \ldots, K.$$

21.6 Multiphase systems

Heterogeneous system, the properties of the system change discontinuously at certain interfaces.

■   A vessel with water, water vapor and air.

Phase, a homogeneous part of a heterogeneous system.

  Phase boundary, the separating interface between two phases.

■   A pot with water and water vapor has the water surface as phase boundary. There is a gaseous phase (water vapor) and a liquid phase (water).

21.6.1 Phase equilibrium

In a system with P phases $(i) = 1, 2, \ldots, P$ and K components $l = 1, 2, \ldots, K$, every phase obeys the equation:

$$\mathrm{d}U^{(i)} = T^{(i)}\mathrm{d}S^{(i)} - p^{(i)}\mathrm{d}V^{(i)} + \sum_{l=1}^{K} \mu_l^{(i)}\mathrm{d}N_l^{(i)}, \quad i = 1, 2, \ldots, P.$$

➤   For a complete description of the system $K + 2$ extensive properties of state are sufficient.

If the total system is in thermodynamic equilibrium, the intensive properties of the P phases and K components satisfy:

$$
\begin{array}{ll}
T^{(1)} = T^{(2)} = \cdots = T^{(P)} & \text{thermal equilibrium,} \\
p^{(1)} = p^{(2)} = \cdots = p^{(P)} & \text{mechanical equilibrium,} \\
\mu_l^{(1)} = \mu_l^{(2)} = \cdots = \mu_l^{(P)}, \quad l = 1, \ldots, K & \text{chemical equilibrium.}
\end{array}
$$

■   For the liquid-gas system in equilibrium:

$$T_{fl} = T_g, \qquad p_{fl} = p_g, \qquad \mu_{fl} = \mu_g.$$

If the system is not in thermal equilibrium, then an energy exchange proceeds until the temperature T is equalized. Analogously, in the absence of chemical equilibrium, an exchange of particles proceeds until the chemical potentials μ_l of every type of particle l are equal to each other. In the absence of mechanical equilibrium, a redistribution of volumes occurs until the pressure is equalized.

■   In a closed pot, water vaporizes until the saturated vapor pressure is reached. In an open pot, the environment must be included into the system. If the air is nonsaturated, water vaporizes completely before equilibrium can be achieved.

21.6.2 Gibbs phase rule

Gibbs phase rule, specifies the number F of intensive variables (degrees of freedom) needed for a complete description of the system.

number of degrees of freedom = number of components + two − number of phases			
	Symbol	Unit	Quantity
$F = K + 2 - P$	F	1	number of degrees of freedom
	K	1	number of components
	P	1	number of phases

➤ The notion of the degree of freedom used here should not be confused with the microscopic number of degrees of freedom f of molecules that may receive kinetic energy.

■ In a closed pot with steam ($K = 1$), three extensive variables are needed to describe the system completely, e.g., S, V, N. One of them (e.g., V) fixes only the size of the system. The intensive properties are already described completely by $F = 1+2-1 = 2$ intensive variables, e.g., pressure and temperature.

Vapor-pressure curve, curve $p_D(T)$, the vapor pressure of a two-phase system as a function of the temperature.

According to $F = K + 2 - P = 1+2-2 = 1$ in a (one-component) two-phase system, there is only one degree of freedom. Pressure and temperature of the system are dependent on each other.

Triple point of a one-component system, the point at which three phases are all in equilibrium. Here, $F = 1 + 2 - 3 = 0$.

▲ At the triple point, all intensive variables are fixed.

■ For water $T_{tr} = 273.16$ K and $p_{tr} = 610.6$ Pa.

➤ Triple points are particularly well suited as fixed points in establishing a temperature scale.

At the triple point, only the relative ratio of quantities of the various phases is variable.

The triple-point values of numerous substances are listed in the **Tab. 22.1/1**.

21.6.3   Clausius–Clapeyron equation

Clausius–Clapeyron equation, a differential equation for the vapor pressure $p(T)$ if entropy and volume per particle are known functions of T and p.

Clausius–Clapeyron equation			$\mathbf{ML^{-1}T^{-2}\Theta^{-1}}$
	Symbol	Unit	Quantity
$\dfrac{dp}{dT} = \dfrac{s_g - s_{fl}}{v_g - v_{fl}}$	p	Pa	pressure
	T	K	temperature
$= \dfrac{S_g - S_{fl}}{V_g - V_{fl}}$	S_{fl}, S_g	J/K	entropy liquid, gaseous phase
	V_{fl}, V_g	m^3	volume liquid, gaseous phase
$= \dfrac{Q}{(V_g - V_{fl})T}$	s_{fl}, s_g	J/(K kg)	specific entropy, liquid, gas
	v_{fl}, v_g	m^3/kg	specific volume, liquid, gas
	Q	J	evaporation heat

➤ V_g and V_{fl} do not mean the volumes of the entire liquid and gaseous phase, but rather the volumes taken by the same quantity of substance as liquid and as gas, respectively.

➤ Instead of the specific quantities, one may also use molar quantities or the entropy and volume per particle.

In most cases $V_g \gg V_{fl}$, one then obtains the approximation:

$$\frac{\mathrm{d}p}{\mathrm{d}T} \approx \frac{Q}{V_g T} \, .$$

➤   Of course, this approximation no longer holds near the critical point.
Representation in terms of specific quantities:

pressure difference =	specific latent heat · density / temperature			$\mathrm{ML^{-1}T^{-2}\Theta^{-1}}$
	Symbol	Unit	Quantity	
$\dfrac{\mathrm{d}p}{\mathrm{d}T} \approx \dfrac{l_v \rho_g}{T}$	p	Pa	pressure	
	T	K	temperature	
	l_v	J/kg	specific latent heat of vaporization	
	ρ_g	kg/m^3	density of gas	

21.7   Vapor pressure of solutions

Vapor pressure curve, the curve $p_D(T)$ representing the vapor pressure of a two-phase system as a function of the temperature.
➤   Here the saturated vapor pressure in equilibrium is used.
Boiling of a substance, the vapor pressure equals the environmental pressure.
    Solidification of a substance, the sublimation pressure is lower than the vapor pressure.

21.7.1   Raoult's law

Raoult's law, describes the reduction of vapor pressure of a solvent on dissolving a slow-evaporating substance.
▲   The relative lowering of vapor pressure is proportional to the **molar fraction** of the dissolved substance.

vapor pressure reduction / original vapor pressure = mole fraction (dissolved substance)				1
	Symbol	Unit	Quantity	
$\dfrac{\Delta p}{p(T)} = x_{\mathrm{st}}$	Δp	Pa	vapor pressure reduction	
	$p(T)$	Pa	original vapor pressure	
	x_{st}	1	mole fraction of dissolved substance	

➤   This law is valid only for very low concentrations (**Fig. 21.13**). Raoult's law has an appreciably wider range of application if the chemical activity is used in place of the mole fraction.

21.7.2   Boiling-point elevation and freezing-point depression

On dissolving a substance, the vapor pressure of the solvent is lowered. Therefore, the system reaches the vapor pressure corresponding to the environmental pressure only at higher temperatures.

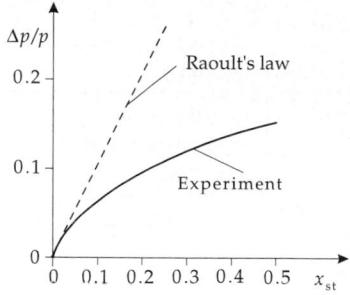

Figure 21.13: Comparison of Raoult's law with experiment.

1. Boiling-point elevation

▲   On dissolving a substance, the boiling temperature is raised in proportion to the quantity of the dissolved substance.

Ebullioscopic constant, E, SI unit kelvin, proportionality factor between rise in boiling point and mole fraction of the dissolved substance.

boiling-point elevation ~ mole fraction			Θ
	Symbol	Unit	Quantity
$\Delta T = E \cdot x_{\text{diss. subst.}}$	ΔT	K	rise in boiling point
	E	K	ebullioscopic constant
	$x_{\text{diss. subst.}}$	1	mole fraction

For ebullioscopic constants, see **Tab. 22.8/2**.

2. Freezing-point depression,

lowering of the freezing temperature because the vapor-pressure curve intersects the sublimation curve only at lower temperature (**Fig. 21.14**).

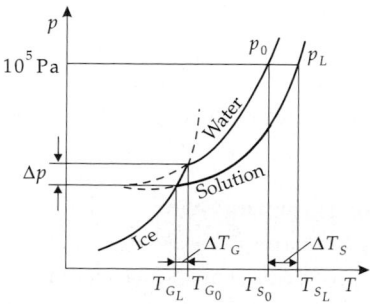

Figure 21.14: Rise in boiling point ΔT_S and fall in freezing point ΔT_G.

■   The spreading of salt in winter serves to lower the freezing point of water, in order to prevent formation of ice.

▲   On dissolving a substance, the melting point is lowered in proportion to the quantity of the dissolved substance.

Cryoscopic constant, K, SI unit kelvin, proportionality factor between freezing-point depression and mole fraction of the dissolved substance.

freezing-point depression $\sim$ mole fraction			Θ
	Symbol	Unit	Quantity
$\Delta T = K \cdot x_{\text{diss. subst.}}$	ΔT	K	freezing-point depression
	K	K	cryoscopic constant
	$x_{\text{diss. subst.}}$	1	mole fraction

For cryoscopic constants, see **Tab. 22.8/2**.

➤   For electrolytic solvents, dissociation must be taken into account. Dissociation modifies the mole fraction.

21.7.3   Henry–Dalton law

Henry–Dalton law:
The pressure of a gas above a solvent is proportional to the concentration x of the dissolved gas; for known reference points p_0, x_0:

$$\frac{x}{x_0} = \frac{p}{p_0} .$$

The law is also valid to a good approximation for the partial pressures of several gases.

■   In a closed bottle of mineral water an equilibrium is established between dissolved CO_2 (forming carbonic acid) and gas.

●   The Henry–Dalton law describes the vapor pressure of a gas dissolved in a liquid.

●   Raoult's law refers to the solution of a slowly evaporating substance, and the solvent produces the vapor pressure (see **Fig. 21.15**).

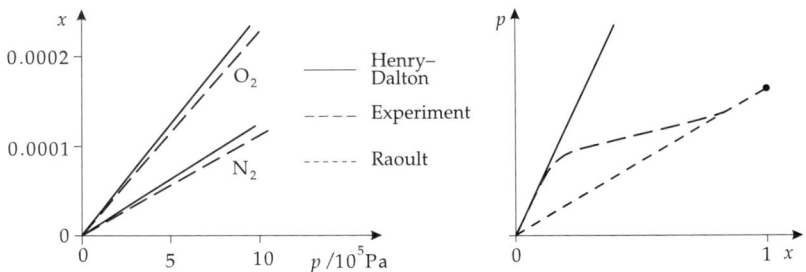

Figure 21.15: Comparison of Raoult's law and Henry–Dalton law with experiment.

21.7.4   Steam–air mixtures (humid air)

Steam–air mixtures are of great importance, e.g., for energy production and air conditioning.

■   Gasoline-vapor–air mixtures in combustion engines, or steam–air mixtures in air conditioning.

Dehydration of a gas, by withdrawal of water by chemicals, molecular sieves, freeze-out, heating or mixing with dry air. Generally, dehydrators are used.

■   Drying agents include: silica gel, phosphorus pentoxide and sulfuric acid.

Humidification of a gas, spraying of water, cooling, or mixing with humid air.

1. Atmospheric humidity

Absolute humidity, f, the quotient of the **mass of water** present, m_W (gaseous and liquid), and the volume of air, V_L:

$$f = \frac{m_W}{V_L}, \quad [f] = \text{kgm}^{-3}.$$

Water content, degree of moisture, x, the ratio of the mass of water, m_W, to the mass of air, m_L:

$$x = \frac{m_W}{m_L}, \quad [x] = 1.$$

Relative atmospheric humidity, relative moisture, φ, dimensionless quantity, the ratio of partial vapor pressure of water, p_D, to the saturated vapor pressure, p_S, at the corresponding temperature:

relative moisture = $\dfrac{\text{absolute moisture}}{\text{maximum moisture}}$			1
$\varphi = \dfrac{p_D}{p_S}$	Symbol	Unit	Quantity
	φ	1	relative moisture
	p_D	Pa	partial pressure
	p_S	Pa	saturated vapor pressure

The relative moisture is generally given as a percentage:
- **nonsaturated vapor** $\varphi < 100\,\%$,
- **saturated vapor** $\varphi = 100\,\%$.
- ➤ For humans at **room temperature**, a relative moisture of 50 % is considered to be comfortable.

M **Hygrometers** are devices for measuring the relative atmospheric moisture.
Hair hygrometers are based on the variation of length of animal hair with humidity.
Dew-point hygrometers are based on the determinantion of the dew-point.
Aspiration hygrometers measure the temperature depression resulting from the evaporation of water.
Psychrometry measures the moisture by comparing a temperature measurement with a thermometer kept at $\varphi = 100\,\%$ and another thermometer kept at room moisture, **see Tab. 22.8/3**.
Electronic hygrometers measure, e.g., the modified capacitance of a capacitor.
▲ The relative atmospheric moisture increases on cooling the steam–air mixture. This is caused by the lowering of the vapor pressure of water with the temperature (**Fig. 21.16**).
If the temperature decreases to the **dew-point**, then **condensation water** is observed:

$$p_D = p_S.$$

2. Saturated steam

Saturated steam, dry-saturated steam with $\varphi = 100\,\%$ exactly. Saturated steam is extremely unstable; a small withdrawal of heat may lead to formation of fog.
  Supersaturated steam, occurs at temperatures below the dew point. Small water droplets are formed that precipitate as fog.

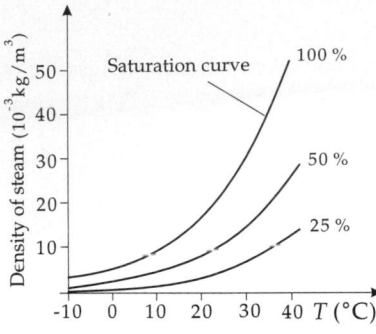

Figure 21.16: Density of steam as a function of the temperature.

Condensation nuclei, small solid particles on which small **water droplets** are formed so that condensation is enhanced.

Fog arises by formation of water droplets at condensation nuclei.

Clouds arise by the rise of moist masses of air that cool at high altitude.

Desublimation leads to formation of solid water (ice crystals, snow) at crystallization nuclei in the atmosphere at very low temperatures.

Hail arises when liquid water (rain drops) is cooled in cold air to temperatures below $0\,°C$.

Wet steam, two-phase mixture of saturated steam and liquid at boiling temperature. Rising vapor bubbles from the boiling liquid may transport small quantities of water.

The mass of the wet steam includes the mass of the saturated steam and the mass of water,

$$m_{\text{wetsteam}} = m_{\text{saturatedsteam}} + m_{\text{water}}.$$

Steam content, x_{steam}, ratio of masses of saturated steam and wet steam, respectively.

Water content, x_{water}, ratio of mass of water to mass of wet steam,

$$x_{\text{steam}} = \frac{m_{\text{saturatedsteam}}}{m_{\text{wetsteam}}}, \qquad x_{\text{water}} = \frac{m_{\text{water}}}{m_{\text{wetsteam}}}.$$

Superheated steam, steam at a temperature above the temperature corresponding to the saturated state.

▲   Superheated steam is nonsaturated.

3. Density of moist air

Density of moist air, sum of the specific density of dry air and the specific density of the steam fraction.

density of moist air			ML^{-3}
	Symbol	Unit	Quantity
$\rho_{\text{moist}} = \dfrac{1}{T}\left(\dfrac{p_{\text{dry}}}{R_{\text{dry}}} + \dfrac{p_{\text{steam}}}{R_{\text{steam}}}\right)$ $p_{\text{dry}} = p_{\text{moist}} - p_{\text{steam}}$ $R_{\text{dry}} = 287\ \text{J/(kg K)}$ $R_{\text{steam}} = 462\ \text{J/(kg K)}$	ρ T p_{moist} p_{dry} p_{steam} R_{dry} R_{steam}	kg/m^3 K Pa Pa Pa J/(kg K) J/(kg K)	density temperature pressure of moist air pressure of dry air pressure of steam spec. gas constant of dry air spec. gas constant of steam

The conditional equation for dry air (second row) follows from Dalton's law (see p. 732).
▲   Moist air is lighter than dry air.

4. Specific enthalpy of moist air

▲   The enthalpy of moist air is equal to the sum of the enthalpy of dry air and the enthalpy of steam.

The specific enthalpy of moist air is the sum of the specific enthalpy of dry air and the specific enthalpy of steam multiplied by the degree of moisture:

specific enthalpy of moist air			L^2T^{-2}
	Symbol	Unit	Quantity
$h_{moist} = h_{dry} + x h_{steam}$	h_{moist}	J/kg	spec. enthalpy of moist air
	h_{dry}	J/kg	spec. enthalpy of dry air
	h_{steam}	J/kg	spec. enthalpy of steam
	x	1	degree of moisture

The change of the specific enthalpy is determined by the change in temperature and by the specific heat capacity at constant pressure. For the steam, the specific evaporation enthalpy $\Delta h_{evapor.}$ must be added (see **Tab. 22.8/6**).

specific change of enthalpy			L^2T^{-2}
	Symbol	Unit	Quantity
$\Delta h_{dry} = c_{p\,dry} \Delta T$	h_{dry}	J/kg	spec. enthalpy dry air
	h_{steam}	J/kg	spec. enthalpy steam
$\Delta h_{steam} = c_{p\,steam} \Delta T$	$h_{evaporation}$	J/kg	spec. evaporation enthalpy
$+ \Delta h_{evapor.}$	T	K	temperature
	$c_{p\,dry}$	J/(kg K)	spec. heat capacity dry air
	$c_{p\,steam}$	J/(kg K)	spec. heat capacity steam

5. Mollier diagram

Mollier diagram, graphic representation of the relation between degree of moisture, relative atmospheric moisture, temperature and specific enthalpy.

h, x-diagram, exact denotation of this special type of diagram from which one may deduce the dependence of the specific enthalpy h on the degree of moisture x (**Fig. 21.17**).
➤   Usually, the degree of moisture x is plotted on the abscissa and the temperature T on the ordinate. Points referring to equal specific enthalpy correspond to the inclined straight lines, points referring to equal relative atmospheric moisture correspond to the rising lines curving to the right.
Saturation line, the line referring to the relative moisture $\varphi = 100\,\%$, lower boundary of the diagram.
▲   An h, x-diagram is valid only for a fixed **fixed total pressure** (range).
The partial pressure of the steam is variable and proportional to the degree of moisture. Therefore, there are also other representations in which the specific enthalpy depends on the vapor pressure, or an assignment vapor pressure—moisture is implemented in the picture.

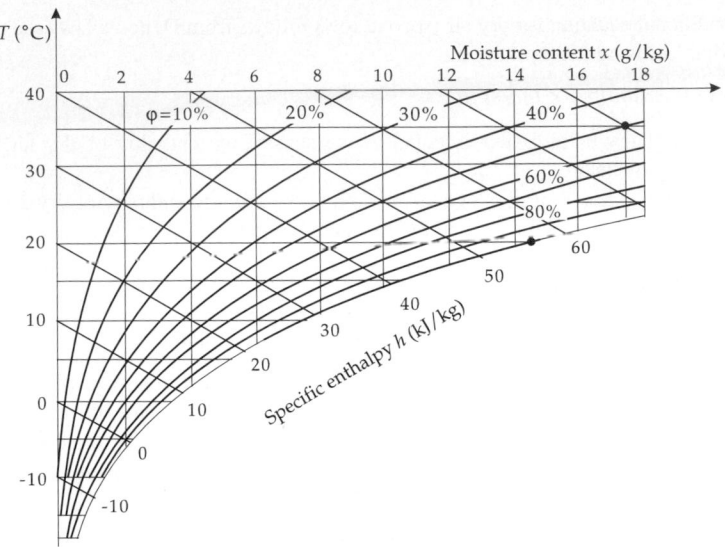

Figure 21.17: Representation of a h, x-diagram following Mollier. The horizontal lines correspond to equal temperature, the vertical lines to equal degree of moisture, the slanting lines to equal enthalpy and the curves inclined to the right correspond to equal relative moisture.

21.8 Chemical reactions

Chemical reaction, a process in which particles of one type react with particles of another type to yield new particles. These conversions are written with reaction equations.

■   Two molecules of hydrogen and one molecule of oxygen react to form two water molecules,

$$2\,H_2 + O_2 \rightleftharpoons 2\,H_2O.$$

Reaction equation, describes the initial substances and the final products of a reaction and their quantitative contributions.

Reactants, initial substances of a reaction.

Products, substances produced in the reaction.

Notation of a reaction equation in which the substances $A_1, A_2, \ldots$ are converted into the substances $B_1, B_2, \ldots$:

reaction equation			
	Symbol	Unit	Quantity
$a_1 A_1 + a_2 A_2 + \cdots$	a_1, a_2	1	stoichiometric coefficient reactant 1, 2
$\rightleftharpoons b_1 B_1 + b_2 B_2 + \cdots$	b_1, b_2	1	stoichiometric coefficient product 1, 2

Stoichiometric coefficients, a_i, b_i, specify how many particles of one type participate in a reaction process.

■ In the formula given above, a_1 particles of type A_1 react with a_2 particles of type A_2, forming b_1 particles of type B_1, and so on.

➤ As is indicated by the notation $\rightleftharpoons$, the backward reaction is also determined. The relation between both reactions is determined by the law of mass action (see. p. 745).

21.8.1 Stoichiometry

Stoichiometry, quantitative evaluation of chemical reactions.

1. Relations between the mass ratios

In order to guarantee that the reaction

$$a_1 A_1 + a_2 A_2 + \cdots \rightleftharpoons b_1 B_1 + b_2 B_2 + \cdots$$

can proceed to completion, the mass ratios must obey the relation:

mass ratio $\sim$ molar mass ratio			
	Symbol	Unit	Quantity
$\dfrac{m_1}{m_2} = \dfrac{a_1 M_1}{a_2 M_2}$	m_1, m_2	kg	total mass substance 1, 2
	a_1, a_2	1	stoichiometric coefficient substance 1, 2
	M_1, M_2	kg/mol	molar mass substance 1, 2

2. Solutions

Concentration, c_i, quantity of a substance i per unit volume or unit mass. The following terms are used:

- **Mass fraction**, fraction of the mass of the specific substance related to the total mass,

$$\xi_i = \frac{m_i}{m_{\text{tot}}} = x_i \frac{M_i}{M_{\text{tot}}} .$$

- **Mass percentage**, specification of the mass fraction as a percentage,

$$\text{mass-\%} = \xi_i \cdot 100\,\% = \frac{m_i}{m_{\text{tot}}} \cdot 100\,\% .$$

- ratio of mass of substance i to the remaining mass $m_{\text{remain.}}$,

$$\frac{m_i}{m_{\text{remain.}}} = \frac{m_i}{m_{\text{tot}} - m_i} = \frac{\xi_i}{1 - \xi_i} .$$

This ratio is particularly suitable for preparing solutions.
- **Molarity** c, the number of moles of a substance per liter of solvent.
- Mass of a substance per unit of volume of a solvent. This specification is also useful for preparing solutions.
- **Volume percentage**, the ratio of the volume of a substance to the total volume,

$$\text{vol.-\%} = \frac{V_i}{V_{\text{tot}}} \cdot 100\,\% .$$

This definition is meaningful only for mixing liquids.

The concept of concentration used here for solutions employs the molarity,

$$c_i = \frac{n_i}{V} = \frac{\text{mole quantity}}{\text{volume solvent}}.$$

■   A 0.5-molar NaCl solution contains 0.5 mol of NaCl per liter of water.
For the preparation of a solution with the desired molarity c, one has:

mass of substance to be dissolved			M
	Symbol	Unit	Quantity
$m = cVM$	m	g	mass of substance to be dissolved
	c	mol/ℓ	molarity
	V	ℓ	volume of solvent
	M	g/mol	molar mass of substance to be dissolved

■   To prepare 250 mℓ 0.1-molar NaF solution, one needs

$$m = 0.1\,\text{mol}/\ell \cdot 0.25\,\ell \cdot 42\,\text{g}/\ell = 1.05\,\text{g}.$$

Normality, the number of reactive monovalent reaction groups per liter of solvent.
➤   The equal charge of the monovalent reaction groups is of importance. So, for NaOH
one does not count both groups Na^+ and OH^-, but only one group.
■   A 1-molar HCl solution forms $H_3O^+\,Cl^-$ and is also 1-normal.
A 1-molar H_2SO_4 solution forms $2H_3O^+\,SO_4^{2-}$ and is 2-normal.

21.8.2   Phase rule for chemical reactions

Extended Gibbs phase rule, describes the number of degrees of freedom including chemical reactions:

number of degrees of freedom = components + two − number of phases − number of reactions			
	Symbol	Unit	Quantity
$F = K + 2 - P - R$	F	1	number of degrees of freedom
	K	1	number of components
	P	1	number of phases
	R	1	number of reaction equations

The total number of extensive variables is $(K - R + 2)$.

21.8.3   Law of mass action

Notation for a reaction equation:

$$a_1 A_1 + a_2 A_2 + \cdots \rightleftharpoons b_1 B_1 + b_2 B_2 + \cdots.$$

1. Law of mass action,

describes the equilibrium concentrations of the initial and final substances of a chemical reaction:

law of mass action			
	Symbol	Unit	Quantity
$$\frac{x_{B_1}^{b_1} x_{B_2}^{b_2} \cdots}{x_{A_1}^{a_1} x_{A_2}^{a_2} \cdots} = K(p, T)$$	a_1, a_2	1	stoichiometric coefficient reactant 1, 2
	b_1, b_2	1	stoichiometric coefficient product 1, 2
	x_{A_1}, x_{A_2}	1	mole fraction reactant 1, 2
	x_{B_1}, x_{B_2}	1	mole fraction product 1, 2
	K	1	equilibrium constant

➤ Instead of the mole fraction, the absolute concentration of the substance also may be used. But then the equilibrium constant has to be adapted correspondingly,

$$\frac{c_{B_1}^{b_1} c_{B_2}^{b_2} \cdots}{c_{A_1}^{a_1} c_{A_2}^{a_2} \cdots} = K' .$$

▲ In the law of mass action, the initial substances (reactants) raised to the power of their multiplicity occur in the denominator, and the produced substances (products) raised to the power of their multiplicity occur in the numerator.

2. Statements of the law of mass action

The equilibrium constant $K(p, T)$ describes the point of equilibrium and, therefore, the dominance of a forward reaction or backward reaction.

$K > 1$: The equilibrium lies on the side of the products. The concentration of the products dominates in equilibrium.
For an equal concentration of products and reactants the forward reaction dominates.

$K < 1$: The equilibrium lies on the side of the reactants.
The concentration of reactants dominates in equilibrium.
For equal concentration of products and reactants the backward reaction dominates.

▲ For the course of the reaction, the **products of concentrations** in the numerator and denominator are of importance, but not the individual concentration.

■ If the concentrations are changed such that the product of the concentrations remains constant, then the final concentration does not change, e.g.,

$$\frac{x_C}{x_A \cdot x_B} = \frac{x_C}{0.5 x_A \cdot 2 x_B} = K .$$

This is of particular importance if the equilibrium constant K is small.

■ A substance P has to be produced with an expensive raw material T and an inexpensive raw material B. The scheme of the reaction is

$$T + B \rightleftharpoons P , \quad \frac{x_P}{x_T \cdot x_B} = K .$$

For an incomplete reaction (K very small), an optimal use of the expensive substance can be achieved if the inexpensive substance is added in excess. For twice the concentration of the inexpensive material, only half the expensive material is needed to produce the same quantity of the product.

3. Equilibrium constant,

the equilibrium constant is related to the chemical potentials μ_j of the reaction partners,

$$K(p, T) = e^{\left[-\frac{1}{kT} \left(\sum_j b_j \mu_j(p, T) - \sum_i a_i \mu_i(p, T) \right) \right]}.$$

The equilibration constant depends on pressure and temperature.
Description in terms of the balance of free enthalpy:

equilibrium constant			
	Symbol	Unit	Quantity
$K(p, T) = e^{\left(-\frac{\Delta G(p,T)}{kT} \right)}$	K	1	equilibrium constant
	G	J	free enthalpy
	k	J/K	Boltzmann constant
	T	K	temperature

For the most important reactions, acid-base reactions, dissociations, the equilibrium constants are listed in tables on chemistry.
➤ Frequently, concentrations remaining constant, e.g., solvents (e.g., H_2O) are included in the constant.

21.8.4 pH-value and solubility product

For the dissociation (decomposition) of water,

$$2\,H_2O \rightleftharpoons H_3O^+ + OH^-$$

according to the law of mass action:

$$\frac{x_{H_3O^+} \cdot x_{OH^-}}{x_{H_2O}^2} = K, \qquad \frac{c_{H_3O^+} \cdot c_{OH^-}}{c_{H_2O}^2} = K'.$$

The concentration of water, remaining constant, is included in the constant.
■ For water:

$$c_{H_3O^+} \cdot c_{OH^-} = 10^{-14}\,\text{mol}^2/\ell^2 \text{ (at } T = 22\,°\text{C)} .$$

One may determine the OH^- concentration from the H_3O^+ concentration in water.

1. pH-value and pOH-value

pH-value, negative logarithm (base 10) of the H_3O^+ concentration,

$$\text{pH} = -\log\left(\frac{c_{H_3O^+}}{1\,\text{mol}/\ell} \right) .$$

pOH-value, negative logarithm (base 10) of the OH^- concentration,

$$pOH = -\log\left(\frac{c_{OH^-}}{1\ \text{mol}/\ell}\right).$$

➤ The sum of both values is equal to 14,

$$pH + pOH = 14.$$

- Acidic solutions have high H_3O^+ concentrations and low pH-values.
- Basic solutions have lower H_3O^+ concentrations and high pH-values.
- Neutral solutions have equal concentrations of H_3O^+ and OH^-,

$$pH = pOH = 7.$$

➤ There is no exact thermodynamic relation between the activity of the hydrogen ions and the pH-value. The conventional pH-scales are realized by a series of buffer solutions.

The H_3O^+ or OH^- concentrations may be varied by adding acids or bases.

◼ In fact, hydrochloric acid may dissociate completely into water,

$$HCl + H_2O \quad \rightarrow \quad H_3O^+ + Cl^-.$$

The resulting H_3O^+ excess leads to a lower pH-value determined by

$$pH = -\log\left(\frac{c_{H_3O^+}}{1\ \text{mol}/\ell}\right) = -\log\left(\frac{c_{\text{acid}} + c_{\text{diss.}}}{1\ \text{mol}/\ell}\right),$$

where for the concentrations:

$$(c_{\text{acid}} + c_{\text{diss.}}) \cdot c_{\text{diss.}} = 10^{-14} \text{mol}^2/\ell^2.$$

2. Acid and base constants

Acid constant, K_S, describes the dissociation of acids.
Base constant, K_B, describes the dissociation of bases.

➤ The degree of dissociation of acids and bases is higher for diluted acids and bases than for concentrated acids and bases.

p_{K_S} and p_{K_B} give the negative logarithm of the acid constants and base constants, respectively:

$$p_{K_S} = -\log K_S, \qquad p_{K_B} = -\log K_B.$$

3. Solubility product,

L, law of mass action for dissolved salts. It describes the ionic concentration of a saturated solution.

➤ The salt not dissolved is deposited at the bottom and can be ignored. So, only the terms in the numerator remain.

■   Solubility product of AgCl. On dissolving in water, the dissociation of silver chloride

$$AgCl \text{ (deposit)} \rightleftharpoons Ag^+ + Cl^- \text{ (saturated solution)}$$

is determined by the solubility product,

$$c_{Ag^+} \cdot c_{Cl^-} = L = 1.6 \cdot 10^{-10} \frac{mol^2}{\ell^2} \,.$$

4. Effective concentrations,

also **activity concentrations**, a, take into account the interaction between ions. They are used in the law of mass action instead of the concentrations of substances, which can be fixed analytically.

| M | Only the mean activity coefficient f, the geometric mean of the activity coefficients of anions and cations, can be measured.

Activity, a, effective concentration of the solvent,

$$a = f \cdot c \,.$$

21.9   Equalization of temperature

Heat can flow spontaneously only from the warmer system to the colder system. In doing so, the warmer system cools and the colder system is heated.

In this process, the total entropy increases.

Heat exchange, occurs by a direct contact of two substances of different temperature.

Final temperature, T_f, temperature of the total system after the heat exchange has terminated.

21.9.1   Mixing temperature of two systems

1. Richmann's mixing rule,

in mixing two systems, the final temperature is determined by the total heat capacities of the substances.

Richmann's mixing rule			
	Symbol	Unit	Quantity
$C_A(T_f - T_A) = C_B(T_B - T_f)$	C_A, C_B	J/K	heat capacity substance A, B
	T_A, T_B	K	initial temperature of substance A, B
	T_f	K	final temperature of system

It is assumed that no mechanical work or heat is released into the environment. The process is irreversible: $\Delta S > 0$.

The internal energy U of the total system remains constant. The total balance of the quantity of heat also remains constant, since the quantity of heat released by one system is absorbed by the other system. However, the entropy increases.

mixing temperature $= \dfrac{\text{sum (heat capacities} \cdot \text{temperature)}}{\text{sum heat capacities}}$				Θ
	Symbol	Unit	Quantity	
$T_f = \dfrac{C_A T_A + C_B T_B}{C_A + C_B}$ $= \dfrac{c_A m_A T_A + c_B m_B T_B}{c_A m_A + c_B m_B}$	T_f	K	final temperature of the system	
	C_A, C_B	J/K	heat capacity substance A, B	
	T_A, T_B	K	initial temperature substance A, B	
	c_A, c_B	J/(K kg)	specific heat capacity substance A, B	
	m_A, m_B	kg	mass substance A, B	

2. Systems of equal specific heat capacity

▲   For equal specific heat capacity $c_A = c_B$, the mixing temperature T_f depends only on the masses of the systems m_A and m_B,

$$T_f = \frac{m_A T_A + m_B T_B}{m_A + m_B} \quad \text{for} \quad c_A = c_B .$$

Systems with equal masses, $m_A = m_B$, reach the mean value of the temperatures as mixing temperature,

$$T_f = \frac{T_A + T_B}{2} \quad \text{for} \quad C_A = C_B.$$

If one system is **much larger** than the other one ($m_B \gg m_A$ for $c_A = c_B$ or $C_B \gg C_A$), the mixing temperature is nearly the temperature of the larger system,

$$T_f \approx \frac{C_A}{C_B} T_A + T_B \approx T_B \quad \text{for} \quad C_B \gg C_A .$$

A **heat bath** with fixed temperature must have a much higher heat capacity than the system being heated in the bath. Water is particularly well suited as a carrier of the heat bath because of its high specific heat capacity.

3. Several systems with distinct specific heat capacities

If several systems are put together, the final temperature is

$$T_f = \frac{C_1 T_1 + C_2 T_2 + C_3 T_3 + \cdots}{C_1 + C_2 + C_3 + \cdots} = \frac{c_1 m_1 T_1 + c_2 m_2 T_2 + c_3 m_3 T_3 + \cdots}{c_1 m_1 + c_2 m_2 + c_3 m_3 + \cdots} .$$

21.9.2 Reversible and irreversible processes

In the irreversible case (direct contact), the final temperature is

$$T_f = \frac{C_A T_A + C_B T_B}{C_A + C_B} .$$

For an irreversible process (direct contact) for the total system, $\Delta U = \Delta Q = 0$. If a heat engine is fitted between A and B, then the process is reversible, and

$$\Delta S = \Delta S_A + \Delta S_B = 0.$$

The final temperature is given by

$$T_f = \left(T_A^{C_A} T_B^{C_B} \right)^{\frac{1}{C_A + C_B}}.$$

- A **reversible** process yields the **geometric mean** of T_A and T_B, weighted by C_A, C_B.
- An **irreversible** process yields the **arithmetic mean** weighted by C_A, C_B.

Reversible case with dimensionless quantities in base and exponent:

$$T_f = T_A \cdot \left(1 + \frac{T_B - T_A}{T_A} \right)^{\frac{C_B}{C_A + C_B}}.$$

For very small temperature differences, the power expression for the reversible case may be expanded. In first order, the formula for the irreversible case is the result.

▲   The final temperature in the reversible case is lower than the final temperature in the irreversible case,

$$T_f^{\text{rev}} < T_f^{\text{irr}}.$$

In the reversible case, the work done by the heat engine is given by

$$\Delta W = \Delta U = C_A (T_f - T_A) + C_B (T_f - T_B).$$

21.10   Heat transfer

Heat transfer, **heat transport**, may proceed by three distinct mechanisms.

a) Convection, the heat energy is transported by the **flow** of a **fluid** (liquid or gaseous material).

■   The supply of warm seawater from the tropics to the Northern Hemisphere by the Gulf Stream; the cooling of engines by fans and ventilators.

b) Heat radiation, emission or absorption of electromagnetic radiation. Each body with finite temperature emits heat radiation.

■   Irradiation of Earth by the Sun; infrared lamps.

c) Heat conduction, requires direct contact between two bodies. The particles of one body transfer energy to the particles of the other body by collisions.

■   Heat transfer through windows and walls; saucepan on an electric range.
     Heat conduction is of importance for many physical and chemical processes.

■   In exothermic chemical reactions, heat must be removed in order to guarantee the safety of the reaction device, and to prevent the chemical reaction from running out of control. In endothermic chemical reactions, heat must be supplied in order to sustain the reaction.

21.10.1 Heat flow

Heat flow, heat flux, Φ, SI unit watt (= joule per second), the quantity of heat exchanged per unit of time. The differential notation is obtained by the limit of the time interval equal to zero.

heat flow = $\dfrac{\text{quantity of heat}}{\text{time interval}}$			$\mathbf{ML^2T^{-3}}$
$\Phi = \dfrac{Q}{t}$	Symbol	Unit	Quantity
	Φ	J/s = W	heat flow
$\Phi = \lim\limits_{\Delta t \to 0} \dfrac{\Delta Q}{\Delta t} = \dfrac{dQ}{dt}$	ΔQ	J	quantity of heat
	Δt	s	time interval

■ A body cooling slowly (and uniformly) within 15 seconds releases a quantity of heat of 90 J to the environment. The heat flow is

$$\Phi = \frac{\Delta Q}{\Delta t} = \frac{90 \text{ J}}{15 \text{ s}} = 6 \text{ W}.$$

M The heat flow can be determined by the law of heat transmission (see p. 759) by attaching a small mat of known heat conductivity coefficient (see p. 753) equipped with thermo-probes to a thermal contact spot and measuring the difference between the temperatures on both sides of the mat. The advantage of this method is that no precise information on the material at the contact spot is needed. The disadvantage is that the heat flow is affected by the measurement, and therefore can be determined only with limited accuracy.

21.10.2 Heat transfer

1. Heat transfer,

the heat transport between two substances of different temperature through an interface. Heat conduction, convection and heat radiation occur at the same time (**Fig. 21.18**).

▲ The exchanged heat is proportional to the product of the surface area, the temperature difference, and the duration of time.

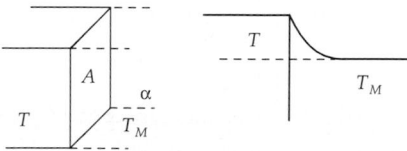

Figure 21.18: Heat transfer.

2. Heat transmission coefficient,

also **heat transfer coefficient,** α, SI unit watt per kelvin and per square meter. Proportionality factor determining the intensity of the heat transfer.

The heat transfer coefficient specifies the ability of a medium (gas or liquid) to remove heat from a substance.

The heat transfer coefficient depends on properties of the heat-removing medium (specific heat, density, heat conductivity coefficient), and on the surface of the heated or cooled substance.

quantity of heat ~ area · temperature difference · time			ML^2T^{-2}
	Symbol	Unit	Quantity
$\Delta Q = \alpha \cdot A$ $\cdot (T - T_M) \cdot \Delta t$	ΔQ	J	released quantity of heat
	α	$W/(K\,m^2)$	heat transfer coefficient
	A	m^2	contact surface
	T	K	temperature of substance
	T_M	K	temperature of medium
	Δt	s	time interval

The time interval Δt should be not too large, since the exchange of heat also causes a change of temperature.

Because only temperature differences are considered, these may be given also in degrees Celsius.

■ A cube of iron with an edge length of 30 cm and a temperature of 70 °C is cooling in air (20 °C). Let the cooling at the bottom face be negligible, so that only five faces contribute to the heat exchange:

$$A = 5 \cdot (30 \text{ cm})^2 = 0.45 \text{ m}^2.$$

The loss of heat during half a minute is

$$Q = \alpha At(T_A - T_B) = 5.8 \ \frac{W}{m^2\,K} \cdot 0.45 \text{ m}^2 \cdot 30 \text{ s} \cdot 50 \text{ °C} = 3.915 \text{ kJ}.$$

▲ The flow velocity of the heat-removing medium is very important for cooling processes.

For heat transfer coefficients, see **Tab. 22.4/3**.

Their order of magnitude covers a wide range depending on whether one considers gases at rest (about 10 $W/(m^2\,K)$), rapidly moving gases (about 100 $W/(m^2\,K)$), water (several 100 up to several thousands $W/(m^2\,K)$) or even condensing water vapor (above 10000 $W/(m^2\,K)$).

3. Heat flow

Heat flow in heat transition:

heat flow ~ area · temperature difference			ML^2T^{-3}
	Symbol	Unit	Quantity
$\Phi = \dfrac{dQ}{dt} = \alpha$ $\cdot A \cdot (T - T_M)$	Φ	J/s	heat flow
	α	$W/(K\,m^2)$	heat transfer coefficient
	A	m^2	contact surface
	T	K	temperature of substance
	T_M	K	temperature of medium

4. Time dependence of cooling by heat transition

In the steady state, the temperature curve follows an exponential function. The speed of cooling is influenced by the magnitude of the contact area and the heat capacity of the cooled or heated substance.

Since the heat flow, the rate of change of the quantity of heat released, is proportional to the difference in temperatures between the substances and the temperature change by heat removal, the change of the temperature difference $T_d = T_{\text{substance}} - T_{\text{medium}}$ is determined by a differential equation, the solution to which is an exponential function:

temperature dependence of cooling			Θ
	Symbol	Unit	Quantity
$\dfrac{dT_d}{dt} = \dfrac{-\alpha \cdot A}{C} \cdot T_d$ $T(t) = (T_0 - T_M)\, e^{-\frac{\alpha A}{C} t} + T_M$	T	K	temperature of substance
	T_0	K	initial temperature
	T_M	K	temperature of medium
	α	W/(K m^2)	heat transfer coefficient
	A	m^2	contact area
	C	J/K	heat capacity of substance
	t	s	time passed

▲   The heat capacity of the heat-removing medium should be much larger than the heat capacity of the cooled or heated substance.

21.10.3   Heat conduction

1. Heat conduction,

heat transfer in a medium as energy transport caused by collision processes between neighboring molecules. Heat conduction in a medium at rest implies no convection. In a moving medium, there is heat transfer by convection and conduction.

quantity of heat $\sim \dfrac{\text{area}}{\text{thickness}} \cdot$ temperature difference $\cdot$ time			$\mathbf{ML^2T^{-2}}$
	Symbol	Unit	Quantity
$\Delta Q = \lambda \cdot \dfrac{A}{s} \cdot (T_A - T_B) \cdot \Delta t$	ΔQ	J	transferred quantity of heat
	λ	W/(K m)	heat transfer coefficient
	A	m^2	contact area
	s	m	thickness of wall
	Δt	s	time interval
	T_A, T_B	K	temperatures

Definition in terms of heat flow:

heat flow $\sim \dfrac{\text{area}}{\text{thickness}} \cdot$ temperature difference			$\mathbf{ML^2T^{-3}}$
	Symbol	Unit	Quantity
$\Phi = \dfrac{dQ}{dt} = \lambda \cdot \dfrac{A}{s} \cdot (T_A - T_B)$	Φ	J/s	heat flow
	λ	W/(K m)	heat transfer coefficient
	A	m^2	contact area
	s	m	thickness of wall
	T_A, T_B	K	temperatures

■   For a temperature difference of 20 K, the heat loss per second through a wall of glass
    (1 m^2) with a thickness of 5 mm is

$$\Phi = \lambda \frac{A}{s}(T_A - T_B) = 1 \ \frac{W}{m \ K} \cdot \frac{1 \ m^2}{0.005 \ m} \cdot 20 \ K = 4 \ kW.$$

2. Heat conductivity,

thermal conductivity, **heat conductivity coefficient**, λ, SI unit watt per kelvin and per
meter, describes the property of a material to conduct heat.

The heat conductivity is determined by intrinsic properties of the material. The density
of the substance, the specific heat, the mean velocity and the mean free path of the particles
involved in the heat transport are of importance.

For heat conductivity coefficients, see **Tab. 22.3**.

■   The heat conductivity coefficient amounts to several hundreds W/(m K) for metals,
    0.1 to 1 W/(m K) for liquids, and about 0.02 W/(m K) for gases.

3. Microscopic description of heat conduction

In gases, the gas particles collide with each other. They exchange momentum and energy
and continue to travel with altered velocities. These collision processes are of primary
importance for the transport of energy and matter.

Mean free path, l, specifies the length of free flight of a particle (atom, molecule or
metal electron) between two successive collisions with other particles.

Mean velocity, **average velocity**, $\bar{v}$, the arithmetic mean of the velocities (without taking into account the directions).

For a Maxwell–Boltzmann distribution (see p. 655), one has

$$\bar{v} = \sqrt{\frac{8kT}{\pi m_N}} = \sqrt{\frac{8}{3\pi}}\sqrt{\overline{v^2}}.$$

Heat conductivity, λ, ability of the system to transport heat.

heat conductivity (microscopic)			$\mathbf{MLT^{-3}\Theta^{-1}}$
	Symbol	Unit	Quantity
$\lambda = \frac{1}{3}\bar{v}l\rho c_V$	λ	W/(K m)	heat conductivity
	$\bar{v}$	m/s	mean velocity
	l	m	mean free path
	ρ	kg/m^3	density
	c_V	J/(K kg)	spec. heat at const. volume

➤   Instead of the product of density and specific heat capacity, one may also use the
    product of molar density and molar heat capacity, or the product of particle density
    and specific heat per particle.

For monatomic gases ($f = 3$ degrees of freedom):

heat conductivity (monatomic gas)			$\mathbf{MLT^{-3}\Theta^{-1}}$
$\lambda = \dfrac{1}{2} k \bar{v} l \rho_N$ $k = 1.380\,66 \cdot 10^{-23}$ J/K	Symbol	Unit	Quantity
	λ	W/(K m)	heat conductivity
	k	J/K	Boltzmann constant
	$\bar{v}$	m/s	mean velocity
	l	m	mean free path
	ρ_N	1/m^3	particle density

4. Heat conductivity of gas mixtures

For gas mixtures, the thermal conductivity of the mixture may be calculated approximately by adding the thermal conductivities, weighted by the relative concentration, of the gases:

total heat conductivity of mixture			$\mathbf{MLT^{-3}\Theta^{-1}}$
$\lambda = x_1\lambda_1 + x_2\lambda_2 + \cdots$ $x_1 = \dfrac{n_1}{n_1 + n_2 + \cdots}$	Symbol	Unit	Quantity
	λ	W/(m K)	total heat conductivity
	x_1	1	mole fraction gas 1
	λ_1	W/(m K)	heat conductivity gas 1
	n_1	mol	quantity of particles gas 1

➤ The measurement of the thermal conductivity of gases is an important method to analyze gases, in particular to investigate the admixture of impurities in gases (gas chromatography).

M The measurement of the thermal conductivity of gases for the purpose of analyzing impurities is made by a comparative measurement with a control gas without impurities (**Fig. 21.19**). The gases (M) to be measured and the gases (V) for comparison are heated in chambers. The heating wires are connected in a type of Wheatstone bridge (see Wheatstone bridge) set to zero by adjusting the resistors until the current in **a** vanishes when the same gas flows through all four arms. For a change of concentration in the gas to be analyzed, its thermal conductivity is also changed, and therefore the heat emission by the heating wire. So, the temperature and the electric resistance of the heating wire are changed. The measuring instrument, calibrated to show Vol.-%, shows a drop of electric voltage proportional to the perturbation of the thermal conductivity, and hence proportional to the concentration of the impurity gas.

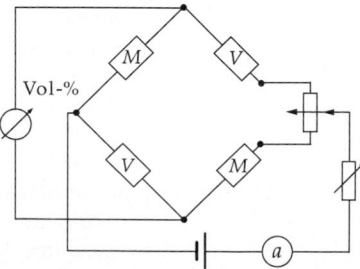

Figure 21.19: Measurement of heat conductivity: Wheatstone bridge circuit with the gas (*M*) to be measured and the gas (*V*) for comparison.

5. Heat conduction through several walls with the same surface

Heat conduction through several walls of thicknesses $s_1, s_2, \ldots$ for equal areas of surface and coefficients of heat conductivity:

$$\Phi = \frac{\lambda \cdot A}{s_1 + s_2 + \cdots}(T_A - T_B).$$

Heat conduction for different coefficients of heat conductivity and equal thickness:

$$\Phi = \frac{1}{\dfrac{1}{\lambda_1} + \dfrac{1}{\lambda_2} + \dfrac{1}{\lambda_3} + \cdots + \dfrac{1}{\lambda_n}}\frac{A}{s}(T_A - T_B).$$

Heat conduction for different coefficients of heat conductivity and different values of thickness:

$$\Phi = \frac{1}{\dfrac{s_1}{\lambda_1} + \dfrac{s_2}{\lambda_2} + \dfrac{s_3}{\lambda_3} + \cdots}A(T_A - T_B).$$

■ A wooden wall of 2 cm thickness is placed behind a glass wall of 5 mm thickness and an area of 1 m² ($\lambda = 0.2\,\text{W}/(\text{m K})$). For a temperature difference of 20 °C, the loss of heat per second is

$$\Phi = \frac{A(T_A - T_B)}{s_1/\lambda_1 + s_2/\lambda_2},$$

$$= \frac{1\,\text{m}^2 \cdot 20\,°\text{C}}{(0.005\,\text{m}/(1\,\text{W}/(\text{m K}))) + (0.02\,\text{m}/(0.2\,\text{W}/(\text{m K})))} = 190.5\,\text{W}.$$

6. Heat flow through a single layer tube wall

See **Fig. 21.20** for reference.

heat flow $\sim \dfrac{2\pi\ \text{tube length}}{\ln\ (\text{ratio of diameters})}$			$\mathbf{ML^2T^{-3}}$
	Symbol	Unit	Quantity
$\Phi = \dfrac{2\pi l}{\ln\left(\dfrac{d_A}{d_I}\right)}\lambda(T_A - T_B)$	Φ	W	heat flow
	d_I	m	inner diameter of tube
	d_A	m	outer diameter of tube
$= \dfrac{2\pi l}{\ln\left(\dfrac{d_I + 2s}{d_I}\right)}\lambda(T_A - T_B)$	s	m	thickness of tube wall
	l	m	tube length
	λ	W/(K m)	heat conductivity coefficient
	T_A, T_B	K	temperatures

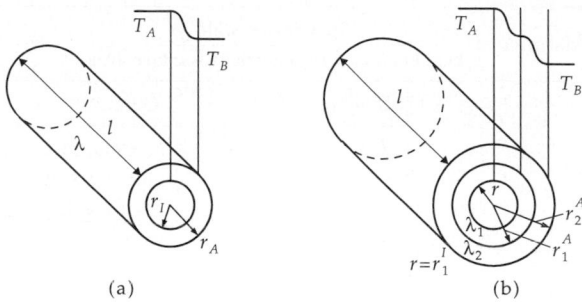

Figure 21.20: Heat conduction in a tube. (a): tube without peripheral layer, (b): tube with peripheral layer.

■ For a temperature difference of 25 °C, the heat flow through a tube of concrete (inner diameter 40 cm, length 3 m, thickness 4 cm) is

$$\Phi = \frac{2\pi l}{\ln\left(\dfrac{d_I + 2s}{d_I}\right)} \lambda (T_A - T_B) = \frac{2\pi \cdot 3\text{ m}}{\ln\left(\dfrac{0.4\text{ m} + 0.08\text{ m}}{0.4\text{ m}}\right)} \cdot 1 \, \frac{\text{W}}{\text{m K}} \cdot 25\,^\circ\text{C} = 2.6\text{ kW}.$$

Heat flow through a tube wall consisting of several layers:

$$\Phi = \left[\frac{1}{2\pi l \lambda_1} \ln\left(\frac{d_1^A}{d_1^I}\right) + \frac{1}{2\pi l \lambda_2} \ln\left(\frac{d_2^A}{d_2^I}\right) + \cdots \right]^{-1} (T_A - T_B)$$

$$= \left[\frac{1}{2\pi l \lambda_1} \ln\left(\frac{d_1^I + 2s_1}{d_1^I}\right) + \frac{1}{2\pi l \lambda_2} \ln\left(\frac{d_1^I + 2s_1 + 2s_2}{d_1^I + 2s_1}\right) + \cdots \right]^{-1} (T_A - T_B).$$

The tubes must fit directly into each other without a gap, i.e., the inner radius of tube 2 must be equal to the outer radius of tube 1, $d_1^A = d_2^I$.

▲ Air gaps must be treated as separate tubes with the heat conductivity coefficient of air.

21.10.4 Thermal resistance

1. Definition of thermal resistance

Thermal resistance, R_{th}, SI unit kelvin per watt, proportionality factor between heat flow and temperature difference.

thermal resistance =	temperature difference / heat flow	$\mathbf{M^{-1}L^{-2}T^3\Theta}$	
	Symbol	Unit	Quantity
$R_{\text{th}} = \dfrac{T_A - T_B}{\Phi}$	R_{th}	K/W	thermal resistance
	T_A, T_B	K	temperatures
	Φ	W	heat flow

▲ The thermal resistance depends on the heat conductivity coefficient, the thickness of the wall and the cross-sectional area.

thermal resistance = $\dfrac{\text{thickness of wall}}{\text{heat conductivity coefficient} \cdot \text{surface area}}$			$\mathbf{M^{-1}L^{-2}T^3\Theta}$
	Symbol	Unit	Quantity
$R_{\text{th}} = \dfrac{s}{\lambda A}$	R_{th}	K/W	thermal resistance
	s	m	thickness of wall
	λ	W/(K m)	heat conduction coefficient
	A	m^2	surface area

2. Analogies to the theory of electricity

The (electric) resistance influences the (electric) current for given temperature (or voltage) difference.

Analogies between quantities of thermodynamics and theory of electricity:

temperature difference	ΔT	corresponds to	potential difference (= voltage)	V
heat flow	Φ	corresponds to	current	I
thermal resistance	R_{th}	corresponds to	electric resistance	R
heat conductivity	λ	corresponds to	electric conductance	κ
series of walls		corresponds to	electric resistors in series	

▲   Like the electric resistance, the thermal resistance depends on the surface and the length of the resistor (thickness of wall), and on the (specific) conductivity.

3. Ohm's law of thermodynamics

The relation between temperature, heat flow and thermal resistance can be written formally as Ohm's law:

heat flow = $\dfrac{\text{temperature difference}}{\text{thermal resistance}}$			$\mathbf{ML^2T^{-3}}$
	Symbol	Unit	Quantity
$\Phi = \dfrac{T_A - T_B}{R_{\text{th}}}$	Φ	W	heat flow
	T_A, T_B	K	temperatures
	R_{th}	K/W	thermal resistance

4. Series connection of several thermal resistors

If several walls (heat resistors) are placed one behind the other, then this arrangement is treated analogously to the series connection of electric resistors (**Fig. 21.21**).

total resistance = sum of individual resistances			$\mathbf{M^{-1}L^{-2}T^3\Theta}$
	Symbol	Unit	Quantity
$R_{\text{tot}} = R_1 + R_2 + R_3 + \cdots$	R_{tot}	K/W	total resistance
$R_1 = \dfrac{s_1}{\lambda_1 A_1}$,	$R_1, R_2, \ldots$	K/W	resistance wall 1, 2, ...
	$s_1, s_2, \ldots$	m	thickness wall 1, 2, ...
$R_2 = \dfrac{s_2}{\lambda_2 A_2}$	$\lambda_1, \lambda_2, \ldots$	W/(K m)	heat conductivity coefficient wall 1, 2, ...
	$A_1, A_2, \ldots$	m^2	surface wall 1, 2, ...

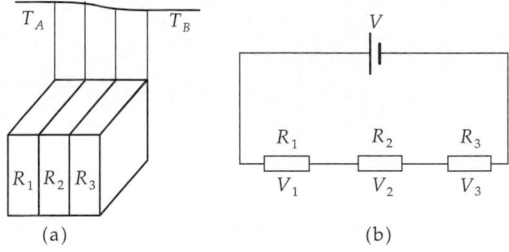

Figure 21.21: Thermal resistance. (a): series connection of thermal resistors, (b): analogy to electricity.

total heat current = $\dfrac{\text{temperature difference}}{\text{sum of individual resistances}}$			$\mathbf{ML^2T^{-3}}$
	Symbol	Unit	Quantity
$\Phi = \dfrac{T_A - T_B}{R_1 + R_2 + \cdots}$	Φ	W	heat flow
	T_A, T_B	K	temperatures
	R_1, R_2	K/W	resistance wall 1, 2

21.10.5 Heat transmission

1. Heat transmission,

the heat transfer between two liquid or gaseous substances A and B through one or several walls (**Fig. 21.22**).

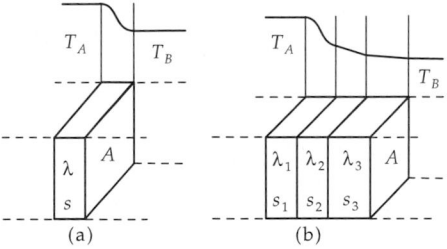

Figure 21.22: Heat conduction (a): through one wall, (b): through several walls.

The heat transfer proceeds by the following steps (**Fig. 21.23**):
- Heat transfer from substance A to the first wall: heat transmission coefficient α_1.
- Heat conduction through wall 1 of thickness s_1 and surface area A: heat conductivity coefficient λ_1.
- Heat conduction through subsequent walls.
- Heat transfer from the last wall (surface area A) to substance B: heat transmission coefficient α_2.
- ■ Thermopane windows are designed to minimize heat loss from the interior of a building into the environment. They consist of two panes of glass with a gap between them. The gap is filled with either air or a specially chosen gas mixture.

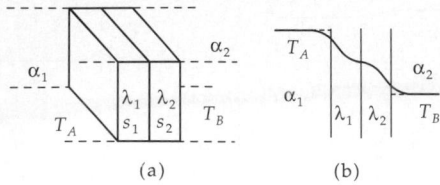

Figure 21.23: Heat transmission through several layers. (a): arrangement of layers, (b): trend of temperature.

2. Heat flow and thermal resistance

Description of the heat flow in terms of the thermal resistance:

heat flow = $\dfrac{\textbf{temperature difference}}{\textbf{sum of individual resistances}}$			$\mathbf{ML^2T^{-3}}$
	Symbol	Unit	Quantity
$\Phi = \dfrac{T_A - T_B}{R_A + R_B + R_1 + R_2 + \cdots}$	Φ	W	heat flow
	T_A, T_B	K	temperatures
	R_A, R_B	K/W	thermal resistance medium A, B
	R_1, R_2	K/W	thermal resistance wall 1, 2

The thermal resistances of the media in front of and behind the walls are:

thermal resistance = $\dfrac{1}{\textbf{heat transmission coefficient} \cdot \textbf{area}}$			$\mathbf{M^{-1}L^{-2}T^3\Theta}$
	Symbol	Unit	Quantity
$R_A = \dfrac{1}{\alpha_1 A}$	R_A, R_B	K/W	thermal resistance of medium A, B
$R_B = \dfrac{1}{\alpha_2 A}$	α_1, α_2	W/(K m^2)	heat transmission coefficient medium A, B
	A	m^2	contact surface

The thermal resistances of the walls are:

thermal resistance = $\dfrac{\textbf{thickness of wall}}{\textbf{thermal conduction coefficient} \cdot \textbf{surface area}}$			$\mathbf{M^{-1}L^{-2}T^3\Theta}$
	Symbol	Unit	Quantity
$R_1 = \dfrac{s_1}{\lambda_1 A}$	R_1, R_2	K/W	resistance of wall 1, 2
	s_1, s_2	m	thickness of wall 1, 2
$R_2 = \dfrac{s_2}{\lambda_2 A} \cdots$	λ_1, λ_2	W/(K m)	thermal conductivity of wall 1, 2
	A	m^2	surface area of wall 1, 2

Description of the heat flow after direct substitution:

$$\Phi = \frac{1}{\dfrac{1}{\alpha_1} + \dfrac{1}{\alpha_2} + \dfrac{s_1}{\lambda_1} + \dfrac{s_2}{\lambda_2} + \cdots} \cdot A(T_A - T_B).$$

■ Let the wall of a room consist of two rows of bricks with a thickness of 9 cm and an air gap of 5 cm in between them. For a temperature difference of 15 °C, the loss of heat per second and square meter is

$$Q = \frac{A t (T_A - T_B)}{1/\alpha_1 + 1/\alpha_2 + s_1/\lambda_1 + s_2/\lambda_2 + s_3/\lambda_3}$$

$$= \frac{1 \text{ m}^2 \cdot 1 \text{ s} \cdot 15 \text{ °C}}{2 \cdot (1/(8.1 \text{ W}/(\text{m}^2 \text{ K}))) + 0.05 \text{ m}/(0.026 \text{ W}/(\text{m K})) + 2 \cdot 0.09 \text{ m}/(0.6 \text{ W}/(\text{m K}))}$$

$$= 6.07 \text{ J}.$$

3. Heat transfer coefficient

Heat transfer coefficient, k-value, k, SI unit watt per kelvin and per square meter, describes the total heat transfer between two media separated by walls. For many systems (e.g., walls of buildings with fixed thickness), the k-value has been tabulated (see **Tab. 22.4/1** and **Tab. 22.4/2**).

heat flow $\sim$ surface area · temperature difference			$\mathbf{ML^2T^{-3}}$
	Symbol	Unit	Quantity
$\Phi = k \cdot A \cdot (T_A - T_B)$	Φ	W	heat flow
	k	W/(K m²)	heat transfer coefficient
	A	m²	cross-sectional area
	T_A, T_B	K	temperatures

Calculation of the heat transfer coefficient:

$\dfrac{1}{\textbf{heat transfer coefficient}} = \dfrac{1}{\textbf{heat transmission coefficient}} + \dfrac{1}{\textbf{heat conductivity coefficient}}$			$\mathbf{M^{-1}T^3\Theta}$
	Symbol	Unit	Quantity
$\dfrac{1}{k} = \dfrac{1}{\alpha_1} + \dfrac{1}{\alpha_2}$ $+ \dfrac{s_1}{\lambda_1} + \dfrac{s_2}{\lambda_2}$ $+ \cdots$	k	W/(K m²)	heat transfer coefficient
	α_1, α_2	W/(K m²)	heat transmission coefficient medium A, B
	s_1, s_2	m	thickness of wall 1, 2
	λ_1, λ_2	W/(K m)	heat conductivity coefficient wall 1, 2

Connection with the total resistance:

thermal resistance $= \dfrac{1}{\textbf{heat transfer coefficient · area}}$			$\mathbf{M^{-1}L^{-2}T^3\Theta}$
$R_{\text{tot}} = \dfrac{1}{kA}$ $= R_A + R_B + R_1$ $+ R_2 + \cdots$	Symbol	Unit	Quantity
	R	K/W	thermal resistance
	k	W/(K m²)	heat transfer coefficient
	A	m²	cross-sectional area

4. Heat transmission through an encased tube

Description of the heat flow in terms of the thermal resistance:

heat flow = $\dfrac{\textbf{temperature difference}}{\textbf{sum of individual resistances}}$			$\mathbf{ML^2T^{-3}}$
	Symbol	Unit	Quantity
$\Phi = \dfrac{T_A - T_B}{R_A + R_B + R_1 + R_2 + \cdots}$	Φ	W	heat flow
	T_A, T_B	K	temperatures
	R_A, R_B	K/W	thermal resistance of medium A, B
	R_1, R_2	K/W	thermal resistance of tube $1, 2 \ldots$

The thermal resistances of the media are:

thermal resistance = $\dfrac{1}{\textbf{heat transfer coefficient} \cdot \textbf{area}}$			$\mathbf{M^{-1}L^{-2}T^3\Theta}$
	Symbol	Unit	Quantity
$R_A = \dfrac{1}{l \cdot \pi \, d_1^I \cdot \alpha_1}$	R_A, R_B	K/W	thermal resistance of medium A, B
	α_1, α_2	W/(K m^2)	heat transfer coefficient medium A, B
$R_B = \dfrac{1}{l \cdot \pi \, d^A \, \alpha_2}$	d_1^I	m	inner diameter of tube 1
	l	m	tube length
	d^A	m	outer diameter of outer-most tube

For the thermal resistances of the tube walls (**Fig. 21.24**):

thermal resistance = $\dfrac{\ln(\text{ratio of diameters})}{\textbf{heat conductivity coefficient} \cdot \textbf{tube length}}$			$\mathbf{M^{-1}L^{-2}T^3\Theta}$
	Symbol	Unit	Quantity
$R_1 = \dfrac{1}{2\pi l \lambda_1} \ln\left(\dfrac{d_1^A}{d_1^I}\right)$	R_1, R_2	K/W	thermal resistance tube 1, 2
	d_1^I	m	inner diameter tube 1
	d_1^A	m	outer diameter tube 1
$R_2 = \dfrac{1}{2\pi l \lambda_2} \ln\left(\dfrac{d_2^A}{d_2^I}\right)$	λ_1	W/(K m)	heat conductivity coefficient tube 1
	l	m	tube length

The outer diameter of the inner tube is always equal to the inner diameter of the outer tube:

$$d_1^A = d_2^I, \quad d_2^A = d_3^I, \quad \ldots$$

➤   If there is an air gap between two tubes, then this air gap must be treated like a tube with the thermal conductivity of air.

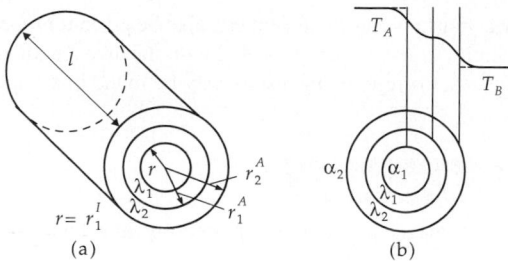

Figure 21.24: Heat transmission through several tube-like layers. (a): construction of tube, (b): the trend of radial dependence temperature.

Heat transfer resistances represented in terms of thickness of tubes:

thermal resistance = $\dfrac{1}{\text{heat transfer coefficient} \cdot \text{area}}$			$\mathbf{M^{-1}L^{-2}T^{3}\Theta}$
	Symbol	Unit	Quantity
	R_A, R_B	K/W	thermal resistance of medium A, B
$R_A = \dfrac{1}{l \cdot \pi \cdot d \cdot \alpha_1}$	α_1, α_2	W/(K m^2)	heat transfer coefficient 1, 2
$R_B = \dfrac{1}{l \cdot \pi \cdot (d + 2s_1 + 2s_2) \cdot \alpha_2}$	s_1, s_2	m	thickness of tube wall 1, 2
	d	m	inner diameter of tube 1
	l	m	tube length

Thermal resistances of the tube walls:

thermal resistance = $\dfrac{\ln(\text{ratio of diameters})}{\text{heat conductivity coefficient} \cdot \text{tube length}}$			$\mathbf{M^{-1}L^{-2}T^{3}\Theta}$
	Symbol	Unit	Quantity
	R_1, R_2	K/W	thermal resistance of tube 1, 2
$R_1 = \dfrac{1}{2\pi l \lambda_1} \ln\left(\dfrac{d + 2s_1}{d}\right)$	λ_1	W/(K m)	heat conductivity coefficient of tube 1
$R_2 = \dfrac{1}{2\pi l \lambda_2} \ln\left(\dfrac{d + 2s_1 + 2s_2}{d + 2s_1}\right)$	s_1	m	thickness of wall tube 1
	d	m	inner diameter of tube 1
	l	m	tube length

Total thermal resistance:

$$R_{\text{tot}} = \frac{1}{l\pi d\alpha_1} + \frac{1}{l\pi(d + 2s_1 + 2s_2)\alpha_2} + \frac{1}{2\pi l\lambda_1} \ln\left(\frac{d + 2s_1}{d}\right)$$

$$+ \frac{1}{2\pi l\lambda_2} \ln\left(\frac{d + 2s_1 + 2s_2}{d + 2s_1}\right) + \cdots .$$

➤   Formally, a heat transmission coefficient can also be given for tubes. However, here it is more meaningful to give a quantity scaled with the tube length instead of a quantity scaled with an area. Corresponding results may be found in the specialized literature.

21.10.6 Heat radiation

Heat radiation, electromagnetic radiation emitted by any body of finite temperature $T \neq 0$ K.
    Stefan–Boltzmann law, relation between the thermal energy emitted by an area A at temperature T per unit time, and the temperature.
▲   The radiated energy increases with the fourth power of the temperature.
Stefan–Boltzmann constant, radiation constant of a black body, σ,

$$\sigma = 5.67 \cdot 10^{-8} \text{ W/m}^2 \cdot \text{K}^4 \,.$$

Emittance, $\varepsilon \leq 1$, a dimensionless quantity depending strongly on the material and the surface conditions, as well as on the temperature of the radiating body.
    The **frequency dependence** of the heat radiation is described by **Planck's radiation law** (see p. 818).
    The Earth receives thermal energy from the Sun by heat radiation. The magnitude of the thermal energy received per unit of time and area is called the **solar constant**. According to ISO standards, its magnitude is

$$q_S = 1.37 \text{ kW/m}^2 \,.$$

According to CIE standards, $q_s = 1.35 \text{ kW/m}^2$.

21.10.7 Deposition of radiation

If radiation impinges upon the surface of a substance, the following processes may occur (**Fig. 21.25**):
- **Absorption**: the radiant power is deposited and converted to another type of energy.
- **Transmission**: the radiation passes through the substance unhindered.
- **Reflection**: the radiation is reflected.

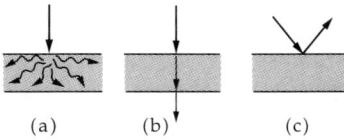

(a)   (b)   (c)

Figure 21.25: Deposition of radiation. (a): absorption, (b): transmission, (c): reflection.

    The three effects mentioned do not occur separately but, in general, simultaneously.
    The radiation is partly absorbed, partly transmitted and partly reflected. The fractions of the total radiation are specified by the associated coefficients.

1. Absorptance,

α, dimensionless, ratio of the absorbed radiant power to the total radiant power.

absorptance = $\dfrac{\text{absorbed radiant power}}{\text{total radiant power}}$			1
$\alpha = \dfrac{\Phi_a}{\Phi_0}$	Symbol	Unit	Quantity
	α	1	absorptance
	Φ_a	W	absorbed radiant power
	Φ_0	W	total radiant power

The absorptance depends on the wavelength of the radiation, and on the temperature.
■   Red glass absorbs radiation of wavelengths of colors other than red.
     Leaves appear as green, since they preferably absorb the red range of white light.
➤   Recording of absorption spectra of a substance in an ultraviolet (UV) spectrometer
     may be used for analyzing materials.

2. Black body radiator,

a substance with the absorptance $\alpha = 1$.
This property cannot be technically realized in full.
■   Solar panels are black in order to absorb as much of the incident light as possible.

3. Kirchhoff's law,

The absorptance is equal to the emittance (see p. 818).

4. Transmittance,

τ, dimensionless, ratio of the total radiant power transmitted through the substance and the
total radiant power.

transmittance = $\dfrac{\text{transmitted radiant power}}{\text{total radiant power}}$			1
$\tau = \dfrac{\Phi_t}{\Phi_0}$	Symbol	Unit	Quantity
	τ	1	transmittance
	Φ_t	W	transmitted radiant power
	Φ_0	W	total radiant power

Similar to the absorptance, the transmittance also depends on wavelength and temperature.

5. Reflectance,

ρ, dimensionless, ratio of the radiant power reflected by a substance and the total radiant
power.

reflectance = $\dfrac{\text{reflected radiant power}}{\text{total radiant power}}$			1
$\rho = \dfrac{\Phi_r}{\Phi_0}$	Symbol	Unit	Quantity
	ρ	1	reflectance
	Φ_r	W	reflected radiant power
	Φ_0	W	total radiant power

The reflectance depends on the wavelength and temperature.

▲ No radiant power is lost (energy conservation). The fractions of absorbed, transmitted, and reflected radiation together must yield the total radiant power.

absorption + transmission + reflection = 1			
	Symbol	Unit	Quantity
$\alpha + \tau + \rho = 1$	α	1	absorptance
	τ	1	transmittance
	ρ	1	reflectance

21.11   Transport of heat and mass

Heat flow density, q_{th}, SI unit watt per square meter, limit of the quantity of heat flowing per unit time Δt through an area element ΔA:

$$q_{th} = \lim_{\Delta t \to 0} \lim_{\Delta A \to 0} \frac{\Delta Q}{\Delta t \Delta A} = \frac{d^2 Q}{dt\, dA} .$$

The vector of heat flow density $\vec{q}_{th}$ has the magnitude of the heat flow density q_{th} and points along the direction of heat transport. Hence, the vector points along the steepest decrease of temperature.

21.11.1   Fourier's law

Fourier's law, the heat flow proceeds along the steepest decrease of temperature.

$$\vec{q}_{th} = -\lambda \cdot \operatorname{grad} T .$$

Heat conductivity λ, material-dependent proportionality constant in Fourier's law.
➤ The quantity λ is identical to the constant in the law of heat conduction (see p. 753, **Tab. 22.3/3–10**).
The total heat flow is obtained as an integral of the heat flow density vector, normal to the surface, over the surface,

$$\Phi = \frac{dQ}{dt} = - \int_{\text{surface}} \vec{q}_{th} \cdot \vec{n}\, dA = - \int (q_x n_x + q_y n_y + q_z n_z)\, dA .$$

$\vec{n}$ is the unit vector of the surface normal. The minus sign indicates that the heat flows from the warmer to the cooler region.

21.11.2   Continuity equation

1. Heat flow density vector,

description of the heat flow by means of the heat flow density vector as integral over the surface passed,

$$\Phi = - \int \vec{q}_{th} \cdot \vec{n}\, dA = - \int \left(\frac{\partial q_x}{\partial x} + \frac{\partial q_y}{\partial y} + \frac{\partial q_z}{\partial z} \right) dV .$$

Rewritten form with vector differential operators:

$$\Phi = \frac{\partial Q}{\partial t} = -\int \text{div } \vec{q}_{\text{th}} \, dV.$$

The conversion is carried out by means of the Gaussian integral theorem.

Specific quantity of heat per volume, e, thermal analog to the electric charge density,

$$e = \frac{dQ}{dV}, \qquad Q = \int e \, dV.$$

This expression may be included in the representation of heat flow,

$$\Phi = \frac{d}{dt} \int e \, dV = -\int \text{div } \vec{q}_{\text{th}} \, dV.$$

2. *Continuity equation of thermodynamics,*

equation expressing the conservation of the 'heat density.'

▲   The specific quantity of heat may be changed only by an in-flow or out-flow of heat, as expressed in terms of a heat flow,

$$\frac{\partial e}{\partial t} + \text{div } \vec{q}_{\text{th}} = 0.$$

This equation is derived from the definition of heat flow under the assumption of an arbitrary volume in the integral.

▲   The continuity equation, as well as its derivation, holds only if heat is conducted by equalization of temperature and **no** work is done on or by the system.

If work W is done, then according to the first law of thermodynamics

$$\frac{\partial e}{\partial t} + \text{div } \vec{q}_{\text{th}} = -\frac{d^2 W}{dV dt} = \frac{dp}{dt}.$$

Here it has been assumed that W is given by the integral of the pressure p during a change of volume dV,

$$\Delta W = -\int p \, dV.$$

If the change in pressure is considered as given, then it may be viewed as a source term for heat (loss). But other changes in energy may also contribute to a change of Q.

21.11.3   Heat conduction equation

1. *Laws of heat conduction*

quantity of heat = heat capacity · temperature difference			$\mathbf{ML^2T^{-2}}$
	Symbol	Unit	Quantity
$\Delta Q = C_V \Delta T = c \cdot m \cdot \Delta T$	ΔQ	J	quantity of heat
	C_V	J/K	heat capacity at const. volume
	ΔT	K	temperature difference
	c	J/(K kg)	specific heat capacity
	m	kg	mass

Heat conduction equation, describes the quantity of heat transported per unit time and unit volume,

$$c\rho \frac{\partial T}{\partial t} - \lambda \, \text{div grad} \, T = \frac{dp}{dt}, \quad \frac{\partial T}{\partial t} - \frac{\lambda}{c\rho} \left(\frac{\partial^2 T}{\partial x^2} + \frac{\partial^2 T}{\partial y^2} + \frac{\partial^2 T}{\partial z^2} \right) = \frac{1}{c\rho} \frac{dp}{dt}.$$

This equation is derived by applying Fourier's law, the continuity equation and the definition of the quantity of heat in terms of the heat capacity.

2. Thermal diffusivity,

κ, SI unit square meter per second, proportionality constant specifying how rapidly a spatial temperature difference equalizes:

thermal diffusivity = $\dfrac{\textbf{heat conductivity}}{\textbf{specific heat capacity} \cdot \textbf{density}}$			$\mathbf{L^2 T^{-1}}$
	Symbol	Unit	Quantity
$\kappa = \dfrac{\lambda}{c\rho}$	κ	m^2/s	thermal diffusivity
	λ	$W/(m\,K)$	heat conductivity
	c	$J/(K\,kg)$	specific heat capacity
	ρ	kg/m^3	density

➤ Instead of density times specific heat capacity, one may also insert the product of molar density and molar heat capacity, or the product of particle density and specific heat per particle.

Heat conduction equation without source term in short notation:

$$\frac{\partial T}{\partial t} - \kappa \Delta T = 0,$$

Δ being the Laplace operator.

21.11.4 Fick's law and diffusion equation

1. Basic laws for mass transport

Concentration differences may be described by analogy to heat differences.

Vector of particle flux density, $\vec{j}$, vector pointing along the steepest decrease of the particle density, and hence of the concentration ρ_N. Its magnitude represents the change of the particle number per unit time.

▲ **Fick's law**, describes the connection between the particle flux density vector and the particle density:

$$\vec{j} = -D \, \text{grad} \, \rho_N \, .$$

Diffusion constant, D, specifies how rapidly the system follows the gradient of concentration.

▲ **Continuity equation**, relation between the particle flux and the particle density:

$$\frac{\partial \rho_N}{\partial t} + \text{div} \, \vec{j} = w \, .$$

The expression w on the right-hand side involves the change of the total particle number, which may be caused, e.g., by a change in the chemical potential.

If $w = 0$, the particle density may change only where the incoming and outgoing fluxes are not balanced against each other.

▲   **Diffusion equation**, equation for the time variation of the particle density:

$$\frac{\partial \rho_N}{\partial t} - D\Delta\rho_N = w,$$

where Δ is the Laplace operator.

The diffusion equation is obtained from Fick's law and the continuity equation.

➤   The diffusion equation may also be established for the molar densities or mass densities, instead of the particle densities.

2. Microscopic description

Mean free path, l, the mean distance of free flight of a particle between two successive collisions with other particles.

Average velocity, **mean velocity**, $\bar{v}$, arithmetic mean of velocities (without taking into account the directions of motion).

For a Maxwell–Boltzmann distribution (see p. 655):

$$\bar{v} = \sqrt{\frac{8kT}{\pi m_N}} = \sqrt{\frac{8}{3\pi}}\sqrt{\overline{v^2}}.$$

Diffusion constant, D, describes the transport of matter:

diffusion constant (microscopically)			$\mathbf{L^2 T^{-1}}$
	Symbol	Unit	Quantity
$D = \dfrac{1}{3}\bar{v}l$	D	m^2/s	diffusion constant
	$\bar{v}$	m/s	mean velocity
	l	m	mean free path

21.11.5   Solution of the equation of heat conduction and diffusion

The solution of the diffusion equation in three-dimensional space is

$$\rho(x, y, z, t) = \sqrt{\left(\frac{1}{4\pi Dt}\right)^3}\, e^{-\frac{x^2+y^2+z^2}{4Dt}}.$$

In the position coordinates x, y, z, the function describes the density distribution or concentration distribution ρ as a Gaussian with the center at the origin of the coordinate system. The width of the curve is determined by the denominator $4Dt$ in the exponent of the exponential function. The width of the function increases over time. Simultaneously, the value of the function in the center decreases over time due to the negative power $t^{-3/2}$ in the normalization factor.

▲   However, the description of diffusion processes is valid only if there are no additional flows (or vortices). Otherwise, additional macroscopic flows (e.g., the stirring of the dissolved substance in the solvent) may dominate the whole process.

If one considers a point at which the density is c times $(0 < c < 1)$ the density at the center,

$$\rho_1 = c \cdot \rho_0 \qquad \rho_1 = \rho(x_1, 0, 0, t) \qquad \rho_0 = \rho(0, 0, 0, t),$$

the time evolution of the distance of this point from the center is:

$$x = \sqrt{-(\ln c) \cdot 4Dt}.$$

The extension velocity of the density cloud decreases in time,

$$v = \frac{dx}{dt} = \sqrt{-\frac{(\ln c)D}{t}}.$$

The path-time law involving the root of time, $x \sim \sqrt{Dt}$, is typical for diffusion processes. The edges of the distribution (c very small) are moving outward faster than the regions with large c.

Illustrative interpretation:

The initially high concentration (e.g., of a drop of color in a liquid) is diminished by spreading (the drop of color fades), but the total number of particles remains constant.

➤   The space integral over ρ does not depend on the time,

$$\int \rho dV = \int \sqrt{\left(\frac{1}{4\pi Dt}\right)^3} e^{-\frac{x^2+y^2+z^2}{4Dt}} \, dxdydz = 1.$$

Formula symbols used in thermodynamics

Symbol	Unit	Designation
α	1	absorptance
α	$W/(K\,m^2)$	heat-transmission coefficient
α	$1/K$	linear-expansion coefficient
β	$1/K$	surface-expansion coefficient
γ	$1/K$	volume-expansion coefficient
ε	1	compression ratio
ε_{kin}	J	mean kinetic energy
η	$1/(Pa\,s)$	viscosity
η	1	efficiency
η_C	1	efficiency of Carnot cycle
ϑ	$°C$	degree Celsius
κ	m^2/s	thermal diffusivity
κ	1	adiabatic exponent
κ	Pa^{-1}	compressibility
λ	$W/(K\,m)$	heat-conductivity coefficient
μ	J	chemical potential
ξ_i	1	mass fraction
ρ	1	reflectance
ρ	kg/m^3	density
ρ_m	mol/m^3	molar density
ρ_N	$1/m^3$	particle density
σ	m^2	cross-section
σ	$W/(m^2 K^4)$	Stefan-Boltzmann constant
τ	s	collision time
τ	1	transmittance
φ	1	relative moisture
Φ	$J/s = W$	heat flow
a	Nm^4/mol^2	molar internal-pressure constant
a_s	Nm^4/kg^2	specific internal-pressure constant
A	m^2	area
b	m^3/mol	molar internal volume
b_s	m^3/kg	specific internal volume
B	J	anergy
$B(T)$	mol/m^3	second virial coefficient
c	mol/ℓ	molarity
c	$J/(K\,kg)$	specific heat capacity
c_{mol}	$J/(K\,mol)$	molar heat capacity
c_p	$J/(K\,kg)$	specific heat capacity, constant pressure
c_V	$J/(K\,kg)$	specific heat capacity, constant volume
C	J/K	heat capacity
$C(T)$	mol^2/m^6	third virial coefficient
d	m	diameter

(*continued*)

Symbol	Unit	Designation
D	m^2/s	diffusion constant
E	K	ebullioscopic constant
E	J	total energy
E_x	J	exergy
f	1	number of degrees of freedom
f	kg/m^3	absolute moisture
f	Hz	collision frequency
f_{max}	kg/m^3	maximum moisture
F	J	free energy
F	N	force
G	J	free enthalpy
h	m	altitude above sea level
h	J/kg	specific enthalpy
H	J	enthalpy
H	J/kg	specific calorific value
H_g	J/m^3	specific calorific value of gases
H_o	J/kg	specific gross-calorific value
j	$1/(m^2\,s)$	particle flux density
k	J/K	Boltzmann constant
k	$W/(K\,m^2)$	heat transmittance
K	K	cryoscopic constant
K	1	equilibrium constant (law of mass action)
l	m	tube length
l	m	mean free path
l	J/kg	specific latent heat
m	kg	total mass
m_N	kg	particle mass
M	kg/mol	molar mass
n	1	polytrope exponent
n	mol	quantity of substance
N	1	particle number
N_A	mol^{-1}	Avogadro's number
N_L	m^{-3}	Loschmidt constant
p	Pa	pressure
p_c	Pa	critical pressure
p_D	Pa	partial pressure
p_n	Pa	standard pressure
p_S	Pa	saturated-vapor pressure
P	W	power
q_{th}	W/m^2	heat-flow density
q_S	W/m^2	solar constant
Q	J	quantity of heat
R	$J/(K\,mol)$	universal gas constant
R_s	$J/(K\,kg)$	specific gas constant
R_{th}	K/W	thermal resistance
s	m	wall thickness
s	$J/(K\,kg)$	specific entropy

(*continued*)

Symbol	Unit	Designation
S	J/K	entropy
t	s	time
T	K	temperature
T_c	K	critical temperature
T_c	K	temperature, cold bath
T_h	K	temperature, heat bath
T_i	K	inversion temperature, Joule-Thomson
T_n	K	standard temperature
U	J	internal energy
v	m^3/kg	specific volume
v_c	m^3/mol	critical molar volume
v_{rms}	m/s	root-mean-square velocity
v_w	m/s	most-probable velocity
$\overline{v}$	m/s	mean velocity
V	m^3	volume
V_m	m^3/mol	molar volume
V_n	m^3	standard volume
W_{kin}	J	total kinetic energy
$\overline{W}$	J	mean energy
x	1	degree of moisture
x_i	1	mole fraction sort i
z	m	altitude

Universal constants and their values		
k	$1.38066 \cdot 10^{-23}$ J/K	Boltzmann constant
N_A	$6.0221367 \cdot 10^{23}$ mol^{-1}	Avogadro's number
N_L	$2.68675 \cdot 10^{25}$ m^{-3}	Loschmidt constant
p_n	101 325 Pa	standard pressure
q_S	1.37 kW/m^2	solar constant
R	8.314 J/(K mol)	universal gas constant
T_n	273.15 K	standard temperature
σ	$5.67 \cdot 10^{-8}$ W/(m^2 K^4)	Stefan-Boltzmann constant

22
Tables on thermodynamics

22.1/1 Calibration points of temperature scales

Substance			Temperature	
Formula	Name	Point	T/K	$\vartheta/°C$
^{4}He	helium-4	λ-point	2.18	-270.97
^{4}He	helium-4	boiling point	4.21	-268.94
$p - H_2$	parahydrogen	triple point	13.81	-259.34
$n - H_2$	hydrogen (normal)	triple point	13.97	-259.18
$p - H_2$	parahydrogen	boiling point	20.27	-252.88
$n - H_2$	hydrogen (normal)	boiling point	20.39	-252.76
Ne	neon	triple point	24.56	-248.59
Ne	neon	boiling point	27.07	-246.08
N_2	nitrogen	phase transition	35.5	-237.65
O_2	oxygen	phase transition	43.7	-229.79
O_2	oxygen	triple point	54.36	-218.79
N_2	nitrogen	triple point	63.14	-210.01
N_2	nitrogen	boiling point	77.35	-195.80
O_2	oxygen	boiling point	90.18	-182.97
C_5H_{12}	isopentane	melting point	113.5	-159.65
C_7H_{14}	methyl cyclohexane	melting point	146.85	-126.30
$C_4H_{10}O$	diethyl ether	melting point	156.85	-116.30
CS_2	carbon disulphide	melting point	161.55	-111.60

(*continued*)

22.1/1   Calibration points of temperature scales (*continued*)

Substance			Temperature	
Formula	Name	Point	T/K	$\vartheta/°C$
C_7H_8	toluene	melting point	178.05	−95.10
CO_2	carbon dioxide	melting point	194.65	−78.50
$CHCl_3$	trichloromethane	melting point	209.65	−63.50
Hg	mercury	melting point	234.28	−38.87
H_2O	water	melting point	273.15	0.00
H_2O	water	triple point	273.16	0.0100
$C_{12}H_{10}O$	diphenyl ether	triple point	300.03	26.88
$Na_2SO_4 \cdot 10\,H_2O$	sodium sulphate	phase transition	305.43	32.38
H_2O	water	boiling point	373.15	100.00
$C_7H_6O_2$	benzoic acid	triple point	395.51	122.36
In	indium	melting point	429.76	156.61
$C_{10}H_8$	naphthalene	boiling point	491.15	218.0
Sn	tin	melting point	505.05	231.9
$C_{14}H_{10}O$	benzophenone	boiling point	579.05	305.9
Cd	cadmium	melting point	594.05	320.9
Pb	lead	melting point	600.65	327.50
Hg	mercury	boiling point	629.73	356.58
Zn	zinc	melting point	692.73	419.58
S	sulphur	boiling point	717.82	444.67
Sb	antimony	melting point	903.65	630.5
Al	aluminum	melting point	934	660.37
Ag	silver	melting point	1235	961.93
Au	gold	melting point	1338	1064.43
Cu	copper	melting point	1356	1083
Ni	nickel	melting point	1728	1455
Co	cobalt	melting point	1768	1495
Pd	palladium	melting point	1827	1554
Pt	platinum	melting point	2045	1772
Rh	rhodium	melting point	2239	1966
Ir	iridium	melting point	2683	2410
W	tungsten	melting point	3683	3410

22.1.2   Melting and boiling points

22.1/2   Melting and boiling points of elements

Element	Melting point $\vartheta/°C$	Boiling point $\vartheta/°C$
actinium	1050	3200 ± 300
aluminum	660.37	2467
americium	994 ± 4	2607
antimony	630.5	1750
arsenic	817 at 2.8 MPa	613 sublimation
barium	725	1640
beryllium	1275 ± 5	2970
bismuth	271.3	1560 ± 5
boron	2300	2550
bromine (Br_2)	-7.2	58.78
cadmium	320.9	765
calcium	839 ± 2	1484
carbon	sublimation at 3652	
cerium	798	3443
cesium	28.40 ± 0.01	669.3
chlorine(Cl_2)	-100.98	-34.6
chromium	1857 ± 20	2672
cobalt	1495	2870
copper	1083.4 ± 0.2	2567
dysprosium	1412	2467
europium	822	1527
fluorine	-219.62	-188.14
gadolinium	1313	3273
gallium	29.78	2403
germanium	93704	2830
gold	1064.43	2808 ± 2
hafnium	2227 ± 20	4602
holmium	1474	2700
hydrogen (H_2)	-259.34	-252.8
indium	156.61	2080
iodine (J_2)	113.5	184.35
iridium	2410	4130
iron	1535	2750
krypton	-156.6	-152.30 ± 0.10
lanthanum	918	3464
lead	327.502	1740
lithium	180.54	1342
magnesium	648.8	1107
manganese	1244 ± 3	1962
mercury	-38.87	356.58
molybdenum	2610	5560
neodymium	1021	3074

(continued)

22.1/2   Melting and boiling points of elements (*continued*)

Element	Melting point $\vartheta/°C$	Boiling point $\vartheta/°C$
neptunium	630 ± 1	
nickel	1455	2730
niobium	2468 ± 10	5127
nitrogen (N_2)	-209.86	-195.8
osmium	2700	>5300
oxygen (O_2)	-218.4	-182.962
palladium	1554	2970
phosphorus (red)	590 bei 4.3 MPa	
phosphorus (yellow)	44.1	280
platinum	1772	3827 ± 100
plutonium	641	3232
polonium	254	962
potassium	63.25	760
praseodymium	931	3520
promethium	1042	(3000)
protactinium	< 1600	
radium	700	< 1140
radon	-71	-61.8
rhenium	3180	5627
rhodium	1966 ± 3	3727 ± 100
rubidium	38.89	686
ruthenium	2310	3900
samarium	1074	1794
scandium	1541	2836
selenium	217	684 ± 1.0
silicon	1410	2355
silver	961.93	2212
sodium	91.81 ± 0.03	882.9
strontium	769	1384
sulphur (mcl.)	119.0	
sulphur (rh.)	112.8	444.674
tantalum	2996	5425 ± 100
tellurium (a.)	449.5 ± 0.3	989.8 ± 3.8
tellurium (rh.)	452	1390
terbium	1356	3230
thallium	303.5	1457 ± 10
thorium		
thulium	1545	1950
tin (cub.)	231.9681	2270
titanium	1660 ± 10	3287
tungsten	3410 ± 20	5660
uranium	1132.3 ± 0.8	3818
vanadium	1890 ± 10	3380
xenon	-111.9	-107.1 ± 3
ytterbium	819	1196
zinc	419.58	907
zirconium	1852 ± 2	4377

22.1/3   Conversion temperatures of inorganic compounds

Substance	Melting point $\vartheta/°C$	Boiling point $\vartheta/°C$
aluminum carbonate	stable up to 1400	diss. 2200
aluminum oxide	2072	2980
aluminum phosphate	> 1500	
aluminum sulphite	1100	subl. 1500
ammonia	−77.7	−33.35
ammonium chloride	subl. 340	520
ammonium nitrate	169.6	210
ammonium thiocyanate	149.6	diss. 170
antimony bromide	96.6	280
antimony chloride	2.8	79
antimony oxide	656	1550
antimony trihydride	−88	17.1
barium oxide	1918	ca. 2000
barium permanganate	3.77	diss. 200
beryllium bromide	490 ± 10	520
beryllium chloride	405	520
beryllium iodide	510 ± 10	590
beryllium oxide	2530 ± 30	ca. 3900
bismuth bromide	218	453
bismuth selenide	710	diss.
bismuth sulphide	diss. 685	
boric acid	236 ± 1	
boron carbide	2350	> 3500
boron oxide	45 ± 2	ca. 1860
cadmium bromide	567	863
cadmium chloride	568	960
cadmium fluoride	1100	1758
cadmium iodide	387	796
cadmium oxide	> 1500	subl. 1559
cadmium telluride	1121	1091
calcium bromide	742	1815
calcium carbide	stab. 25 - 447	2300
calcium carbonate	1339	diss. 898.6
calcium chloride	782	> 1600
calcium fluoride	1423	ca. 2500
calcium iodide	784	ca. 1100
calcium oxide	2614	2850
cesium bromide	636	1300
cesium chloride	645	1290
cesium fluoride	682	1251
cesium iodide	626	1280
chromium carbide	1980	3800
chromium oxide	2266 ± 25	4000
cobalt fluoride	ca. 1200	1400
copper chloride	620	993
copper iodide	605	1290

(continued)

22.1/3   Conversion temperatures of inorganic compounds (*continued*)

Substance	Melting point $\vartheta/°C$	Boiling point $\vartheta/°C$
dysprosium bromide	881	1480
dysprosium chloride	718	1500
dysprosium fluoride	1360	> 2200
dysprosium iodide	955	1320
erbium fluoride	1350	2200
erbium iodide	1020	1280
europium bromide	677	1880
europium chloride	727	> 2000
europium fluoride (EuF$_2$),	1380	> 2400
(EuF$_3$)	1390	2280
europium iodide (EuI$_2$)	527	1580
fluorine dioxide	−223.8	−144.8
gallium arsenide	1238	
gallium dichloride	164	535
gallium trichloride	77.9 ± 0.2	201.3
heavy water	3.82	101.42
holmium bromide	914	1470
holmium chloride	718	1500
holmium fluoride	1143	> 2200
holmium iodide	989	1300
hydrogen disulphide	−89.6	70.7
hydrogen fluoride	−83.1	19.54
hydrogen peroxide	−0.41	150.2
hydrogen sulphide	−85.5	−60.7
indium antimonide	535	
indium arsenide	943	
indium phosphide	1070	
indium telluride	667	
iron oxide	1594 ± 5	
lead bromide	373	916
lead fluoride	855	1290
lead iodide	402	954
lithium oxide	> 1700	
magnesium chloride	714	1412
magnesium fluoride	1261	2239
magnesium oxide	2852	3600
mercury bromide	236	322
mercury chloride	276	302
mercury iodide	259	354
orthophosphoric acid	73.6	diss. 200
ozone	−192.7 ± 2	−111.9
potassium bromide	734	1435
potassium chlorate	356	diss. 400
potassium hydroxide	360.4 ± 0.7	1320 - 1324
potassium perchlorate	610 ± 10	diss. 400
radium bromide	728	subl. 900

(*continued*)

22.1/3   Conversion temperatures of inorganic compounds (*continued*)

Substance	Melting point $\vartheta/°C$	Boiling point $\vartheta/°C$
rubidium bromide	693	1340
rubidium chloride	718	1390
rubidium fluoride	795	1410
silicon dioxide (quartz)	1610	2230
silicon tetrahydride (silane)	−185	−111.8
sodium amide	210	400
sodium bromide	747	1390
sodium chloride	801	1413
sodium cyanide	563.7	1496
sodium fluoride	993	1695
sodium hydroxide	318.4	1390
sodium iodide	661	1304
sodium metaborate	966	1434
strontium chloride	875	1250
strontium fluoride	1473	2489
strontium oxide	2430	≈ 3000
sulphuric acid (100 %)	10.36	330 ± 0.5
tellurium bromide	210	339
terbium bromide	827	1490
terbium fluoride	1172	2280(?)
terbium iodide	946	> 1300
tetrachlorosilane	−70	57.57
tetrafluorosilane	−90.2	−86
thallium bromide	480	815
thallium chloride	430	720
thorium carbide	2655 ± 25	ca. 5000(?)
thorium oxide	3220 ± 50	4400
thorium tetraiodide	566	839
thulium bromide	952	1440
thulium fluoride	1158	> 2200
thulium iodide	1015	1260
titanium carbide	3140 ± 90	4820
titanium di-iodide	600	1000
titanium dioxide	1830 - 1850	2500 - 3000
titanium fluoride	1200	1400
titanium monoxide	1750	> 3000
titanium nitride	2930	
tungsten carbide	2870 ± 50	6000
tungsten dicarbide	2860	6000
vanadium carbide	2810	3900
vanadium dioxide	1967	
vanadium (V) oxide	690	diss. 1750
vanadium (III) oxide	1970	
ytterbium bromide	677	1800
ytterbium chloride	702	1900
ytterbium fluoride	1052	2380

22.1/4   Melting and boiling points of organic compounds

Compound	Molecule formula	Melting point $\vartheta/°C$	Boiling point $\vartheta/°C$
alkanes			
methane	CH_4	−182.48	−161.49
ethane	CH_3CH_3	−183.27	−88.62
propane	$CH_3CH_2CH_3$	−187.69	−42.07
butane	$CH_3(CH_2)_2CH_3$	−138.35	−0.5
pentane	$CH_3(CH_2)_3CH_3$	−129.72	36.07
hexane	$CH_3(CH_2)_4CH_3$	−95.35	68.74
heptane	$CH_3(CH_2)_5CH_3$	−90.61	98.42
octane	$CH_3(CH_2)_6CH_3$	−56.8	125.66
nonane	$CH_3(CH_2)_7CH_3$	−53.52	150.79
decane	$CH_3(CH_2)_8CH_3$	−29.66	174.12
isobutane	$(CH_3)_2CHCH_3$	−159.6	−11.73
isopentane	$(CH_3)_2CHCH_2CH_3$	−159.9	27.85
alkenes (olefins)			
ethene	$CH_2 = CH_2$	−169.15	−103.71
propene	$CH_2 = CHCH_3$	−185.25	−47.7
cyclohexene	$CH_2(CH_2)_3CH = CH$	−103.7	83.2
alkynes			
ethyne	$CH \equiv CH$	−80	−83.4
propyne	$CH_3C \equiv CH$	−102.7	−23.22
aromatic hydrocarbons			
benzene	C_6H_6	5.53	80.1
naphthaline	$C_{10}H_8$	80.29	217.95
toluene	$C_6H_5CH_3$	−94.99	110.63
ethyl benzene	$C_6H_5CH_2CH_3$	−94.98	136.19
propyl benzene	$C_6H_5(CH_2)_2CH_3$	−99.5	159.22
o-xylene	$C_6H_4(CH_3)_2$	−25.18	144.41
styrene	$C_6H_5CH = CH_2$	−30.63	145.2
amines			
methyl amine	CH_3NH_2	−93.49	−6.33
dimethyl amine	$(CH_3)_2NH$	−92.19	6.88
trimethyl amine	$(CH_3)_3N$	−117.3	2.87
ethyl amine	$CH_3CH_2NH_2$	−81	16.58
propyl amine	$CH_3CH_2CH_2NH_2$	−83	48.5
aniline	$C_6H_5NH_2$	−63	184.13
organic halogen compounds			
chloromethane	CH_3Cl	−97.72	−24.22
bromomethane	CH_3Br	−93.6	3.56
iodomethane	CH_3I	−66.45	42.43
dichloromethane	CH_2Cl_2	−95.14	39.75
trichloromethane	$CHCl_3$	−63.49	61.73
tetrachloromethane	CCl_4	−23.02	76.54
tetrabromomethane	CBr_4	92	190
tetraiodomethane	CI_4	171	135

(continued)

22.1/4 Melting and boiling points of organic compounds (*continued*)

Compound	Molecule formula	Melting point $\vartheta/°C$	Boiling point $\vartheta/°C$
organic halogen compounds (*continued*)			
chloroethane	CH_3CH_2Cl	−136.4	12.27
bromoethane	CH_3CH_2Br	−117.6	38.35
chlorobenzene	C_6H_5Cl	−45.58	131.7
bromobenzene	C_6H_5Br	−30.82	156.06
iodobenzene	C_6H_5I	−30.63	145.2
alcohols			
methanol	CH_3OH	−97.68	64.51
ethanol	CH_3CH_2OH	−114.1	78.32
1-propanol	$CH_3CH_2CH_2OH$	−126.2	97.2
1-butanol	$CH_3(CH_2)_2CH_2OH$	−89.3	117.73
2-propanol	$CH_3CHOHCH_3$	−88.5	82.5
1-pentanol	$CH_3(CH_2)_3CH_2OH$	−78.2	138.35
ethene glycol	CH_2OHCH_2OH	−13.56	197.3
glycerol	$CH_2OHCHOHCH_2CH_3$	18.6	290
cyclohexanol	$CH_2(CH_2)_4CHOH$	25.15	161.5
ethers			
dimethyl ether	CH_3OCH_3	−141.49	−24.84
diethyl ether	$CH_3CH_2OCH_2CH_3$	−116.3	34.55
methyl phenyl ether	$C_6H_5OCH_3$	−37.3	154
aldehydes			
formaldehyde	$HCHO$	−92	−19.1
acetaldehyde	CH_3CHO	−123	20.4
propionaldehyde	$CH_3CH_2CH_2CHO$	−80	48
butanal	$CH_3CH_2CH_2CHO$	−96.4	74.8
isobutyraldehyde	$(CH_3)_2CHCHO$	−65	64.1
benzaldehyde	C_6H_5CHO	−26	178
ketones			
acetone	CH_3COCH	−94.7	56.29
ethyl methyl ketone	$CH_3CH_2COCH_3$	−86.69	79.64
acetophenone	$C_6H_5COCH_3$	19.65	202
carboxylic acids			
formic acid	$HCOOH$	8.4	100.56
acetic acid	CH_3COOH	16.63	117.9
propionic acid	CH_3CH_2COOH	−20.8	140.99
butyric acid	$CH_3CH_2CH_2COOH$	−4.26	163.53
chloroacetic acid	$ClCH_2COOH$	63	189.5
dichloroacetic acid	$Cl_2CHCOOH$	10.8	192.5
trichloroacetic acid	Cl_3CCOOH	56.3	197.55
glycine	NH_2CH_2COOH	234	286
lactic acid	$CH_3CHOHCOOH$	18	56
oxalic acid	CO_2HCOOH	157	189.5
adipic acid	$CO_2H(CH_2)_4COOH$	152	267
benzoic acid	$C_6H_5CO_2H$	121.7	249

(*continued*)

22.1/4   Melting and boiling points of organic compounds (*continued*)

Compound	Molecule formula	Melting point $\vartheta/°C$	Boiling point $\vartheta/°C$
carboxylic acidic derivates			
acetyl chloride	CH_3COCl	-112	51
acetyl bromide	CH_3COBr	-96	76.7
acetyl iodide	CH_3COJ	0	108
acetamide	CH_3CONH_2	82.15	221.1
methyl formate	HCO_2CH_3	-99	32
methyl acetate	$CH_3CO_2CH_3$	-98.05	56.9
ethyl acetate	$CH_3CO_2CH_2CH_3$	-39.5	77.1
acetic anhydride	$(CH_3CO)_2O$	-73.05	140
others			
urea	NH_2CONH_2	132.75	decays

22.1/5   Melting point T_f and boiling point T_s of oils

Substance	$T_f/°C$	$T_s/°C$
diesel fuel	-5	$60\ldots300$
heating oil	-5	$200\ldots350$
machine oil	-5	$380\ldots400$
tar	-15	300
transformer oil	-5	175
gas oil	-30	$200\ldots300$
linseed oil	-15	316
kerosene	-70	$150\ldots300$
turpentine oil	-10	160
gasoline	$-30\ldots-50$	$67\ldots100$

22.1/6   Melting temperatures of high-temperature ceramics

Substance	$\vartheta_{melt}/°C$	Substance	$\vartheta_{melt}/°C$
HfC	3890 ± 150	NbB_4	2900
TaC	3880 ± 150	VC	2810
ZrC	3530	HfO_2	2790
NbC	3480	W_2B	2770 ± 80
HfB_2	3250 ± 100	W_2C	2730 ± 15
TiN	3205	UO_2	2730
TiC	3147	WC	2720
TaB_2	3100	MoC	2700
TaN	3087 ± 50	ZrO_2	2700
NbB_2	3000	ZrB_{12}	2680
HfN	2982	YN	2670
ZrN	2982	ThC_2	2656 ± 75
TiB_2	2980	ScN	2650
ThO_2	2950	UN	2650 ± 100
ThN	2630 ± 50	BeO	2440
CoO	2603	Cr_2O_2	2400

(*continued*)

22.1/6   Melting temperatures of high-temperature ceramics (*continued*)

Substance	ϑ_{melt} /°C	Substance	ϑ_{melt} /°C
NbB_6	2540	Nb_5Si_3	2440
SmB_6	2540	TaB	2430
LaB_6	2530	ThS	2425
Ta_4Si	2510	TaS	2425
MgO	2500	Nb_2N	2420
Ta_5Si_3	2500	Y_2O_3	2410
UB_4	2495	AlN	2400
SrO	2460	U_2C	2400
CeS	2450	VB_2	2400 ± 50
$WB(\alpha)$	2400 ± 100	Be_3N_5	2205
UB_2	2385	BaS	2205
VN	2360	Be_3N_2	2200
MoB	2350	Ti_2B	2200
UC	2315	CrB_2	2200 ± 50
La_2O_3	2310	$TaSi_2$	2200
YC_2	2300 ± 50	Nd_2S_3	2200
W_2B_5	2300 ± 50	GeB_6	2190
BeB_6	2300	WSi_2	2165
YB_6	2300	ThB_6	2150
CaC_2	2300	ZrSi	2150
Th_2S	2300	Mo_2B	2140
Th_4S_7	2300	NdS	2140
NbB	2280	Ti_5Si_3	2120
ScB_2	22500	GdB_6	2100
Mo_3B_4	2250	Th_3N_4	2100
VB	2250	MoB_2	2100
Zr_5Si_3	2250	La_2S_3	2100
UC_2	2250	V_3B_2	2070
SrB_6	2235	Al_2O_3	2050
UB_{12}	2235	CrB	2050
CaB_6	2230	Ce_3S_4	2050 ± 75
BaB_6	2230	$MoSi_2$	2030
Ba_3N_2	2220	TiO	2020
ThB_4	2210	$Al_2O_3 \cdot BaO$	2000

22.1/7   Melting temperatures of low-melting alloys at the eutectic point

Substance	ϑ_{melt} /°C	Substance	ϑ_{melt} /°C
(92.2 % Hg; 2.8 % Na)	−48.2	(90 % K; 10 % Na)	17.5
(94.5 % Cs; 5.5 % Na)	−30	(56 % Na; 44 % K)	19
(93 % Cs; 7 % N)	−28	(85.2 % Na; 14.8 % Hg)	21.4
(78 % K; 22 % Na)	−11.4	(60 % Na; 40 % K)	26
(80 % K; 20 % Na)	−10	(70 % Na; 30 % K)	41
(91.8 % Rb; 8.2 % Na)	−4.5	(50 % Na; 50 % Hg)	45
(70 % K; 30 % Na)	−3.5	(70 % Hg; 30 % Na)	55
(60 % K; 40 % Na)	5	(80 % Na; 20 % K)	58
(50 % K; 50 % Na)	11	(60 % Na; 40 % Hg)	60

22.1.3 Curie and Néel temperatures

22.1/8   Ferromagnetic phase transitions—Curie temperature

Substance	T_C/K	Substance	T_C/K
Co	1400.15	CrO_2	380.15
Dy	105.15	UH_3	180.15
Er	29.15	silicon-iron (4 Si)	963.15
Fe	1033.15	alperm (16 Al)	673.15
Gd	289.15	permalloy (78.5 Ni)	873.15
Ho	29.15	super permalloy (78.5 Ni)	673.15
MnSb	587.15	hipernik (50 Ni)	773.15
Ni	627.15	permendur (50 Co)	1253.15
Tb	221	perminvar (25 Co, 45 Ni)	988.15
Tm	22(?)	perminvar (7 Co, 70 Ni)	923.15
FeRh	675		

22.1/9   Antiferromagnetic phase transitions—Néel temperature

Substance	T_N/K	Substance	T_N/K
Ce	125	Ho	131.55
$CoCl_2$	521.45	Mn	103.15
CoO	274.93	MnF_2	66.45
Cr	473.15	MnO	122.15
Cr_2O_3	305.95	Nd	7.5
Dy	178.5	$NiCl_2$	49.55
Er	85	NiO	523.15
Eu	87	Pr	< 1.5
$FeCO_3$	57.15	Sm	15
$FeCl_2$	23.45	Tb	229
FeF_2	78.35	$TiCl_2$	103.15
FeO	198.15	Tm	51–60
FeRh	350		

22.1/10   Ferro- and antiferroelectric transitions—Curie temperature

Substance	Type of transition	T_C/K	Substance	Type of transition	T_C/K
$BaTiO_3$	F	193.15	$(NH_4)H_2PO_4$	AF	148.15
	F	278.15	$NaNbO_3$	AF	793.15
	F	393.15	$NaTaO_3$	AF	748.15
CsH_2PO_4	F	160.15	$PbTiO_3$	F	763.15
KD_2PO_4	F	216.15	$PbZrO_3$	AF	506.15
KH_2PO_4	F	123.15	RbH_2PO_4	F	147.15
$KNbO_3$	F	708.15	WO_3	AF	1010.15
$KTaO_3$	F	13.15			

22.2   Characteristics of real gases

22.2/1   Values of temperature, pressure and density at the critical point

Gas	T_c/K	p_c/MPa	$\rho_c/(10^2\ kg\ m^{-3})$
oxygen	155	5.06	4.1
nitrogen	126	3.39	3.11
hydrogen	33	1.29	0.31
helium	5	0.23	0.69
neon	44	2.72	4.84
argon	151	4.85	5.31
chlorine	417	7.69	5.73
carbon monoxide	133	3.48	3.01
carbon dioxide	304	7.36	4.68
sulphur dioxide	431	7.86	5.24
methane	191	4.62	1.62
air	132	3.77	
ethane	305	4.88	2.03
propane	370	4.24	2.20
butane	425	3.78	2.28
isobutane	408	3.64	2.21
ammonia	405	11.2	2.35
hydrogen sulphide	374	8.98	3.49
ethene	283	5.10	2.27
ethyne	309	6.22	2.31
dinitrogen oxide	310	7.24	4.59
nitrogen monoxide	180	6.56	5.20
dichlorodifluoromethane	385	4.10	5.55
trifluoromethane	471	4.36	5.54

22.2/2   Molar mass, specific gas constant and density of gases

The density refers to standard conditions $T = 273.15$ K, $p = 101325$ Pa

Gas	$M/(g\ mol^{-1})$	$R_s/(J\ K^{-1}kg^{-1})$	$\rho/(kg\ m^{-3})$
air	28.96	286.91	1.293
chlorine	70.91	117.19	3.214
methane	16.04	517.97	0.717
ethane	30.07	276.35	1.357
ethene	28.05	296.21	1.260
ethyne	26.04	319.14	1.175
propane	44.10	188.45	2.010
propene	42.08	197.48	1.915
ammonia	17.03	487.9	0.771
carbon monoxide	28.01	296.67	1.250
carbon dioxide	44.01	188.81	1.977
oxygen	32.00	259.69	1.429
nitrogen	28.02	296.61	1.250
nitrogen monoxide	30.01	276.93	1.340
hydrogen	2.02	4122.0	0.0899
steam	18.02	461.25	0.804

22.2/3   Van der Waals constants

Gas	$a/(N\,m^4\,mol^{-2})$	$b/(10^{-6}\,m^3\,mol^{-1})$
acetone	1.58	98.5
ammonia	0.422	37.2
argon	0.136	32.3
ethanol	1.22	84
helium	0.0035	23.8
krypton	0.234	39.9
methane	0.228	27.1
methanol	0.95	67
neon	0.21	17.1
propane	0.92	84.5
1-propanol	1.5	101
oxygen	0.138	31.8
nitrogen	0.141	39.2
water	0.555	30.5
hydrogen	0.0245	26.6
xenon	0.415	51

22.2/4   Pressure and temperature at the triple point

Substance	T_t/K	p_t/hPa	Substance	T_t/K	p_t/hPa
ammonia	195.5	60.6	neon	24.56	431
carbon dioxide	216.56	5180	parahydrogen	13.81	70.4
oxygen	543.6	1.5	water	273.16	6.1
nitrogen	63.14	12.53			

22.3   Thermal properties of substances

22.3.1   Viscosity

The viscosity is given for the temperature 0 °C or 20 °C and standard pressure.

22.3/1   Dynamic viscosity of gases

Gas	$\eta(0\,°C)/(10^{-6}\,Pa\,s)$	$\eta(20\,°C)/(10^{-6}\,Pa\,s)$
ammonia	9.3	10.2
chlorine	12.3	13.5
ethene	9.4	10.3
ethyne	9.5	10.4
carbon monoxide	16.6	18.0
carbon dioxide	13.7	15.0
air	17.2	18.4

(continued)

22.3/1   Dynamic viscosity of gases (*continued*)

Gas	$\eta(0\ {}^\circ C)/(10^{-6}\ Pa\,s)$	$\eta(20\ {}^\circ C)/(10^{-6}\ Pa\,s)$
methane	10.2	11.0
sulphur dioxide	11.6	12.8
oxygen	19.2	20.7
nitrogen	16.5	17.8
hydrogen	8.4	9.0

22.3/2   Dynamic viscosity of liquids

Substance	$\eta(0\ {}^\circ C)/(10^{-6}\ Pa\,s)$	$\eta(20\ {}^\circ C)/(10^{-6}\ Pa\,s)$
acetone	395	322
benzene	910	648
trichloromethane	700	570
ethanol	1780	1200
heptane	517	409
methanol	820	587
pentane	282	232
mercury	1685	1554
toluene	768	585
water	1792	1002

22.3.2   Expansion, heat capacity and thermal conductivity

The tables below list the following thermal quantities:
- linear expansion coefficient α at 25 °C,
- specific heat capacity c_p at constant pressure at 25 °C,
- thermal conductivity λ at 27 °C.

22.3/3   Thermal properties of pure metals

Metal	$\alpha\ /(10^{-6}\ K^{-1})$	$c_p\ /(J \cdot g^{-1} \cdot K^{-1})$	$\lambda\ /(W \cdot cm^{-1} \cdot K^{-1})$
aluminum	23.1	0.897	2.37
antimony	11.0	0.207	0.243
barium	20.6	0.204	0.184
beryllium	11.3	1.825	2.00
bismuth	13.4	0.122	0.0787
cadmium	30.8	0.232	0.968
calcium	22.3	0.647	2.00
cerium	5.2	0.192	0.113
cesium		0.242	0.359
chromium	4.9	0.449	0.937
cobalt	13.0	0.421	1.00
copper	16.5	0.385	4.01

(*continued*)

22.3/3   Thermal properties of pure metals (*continued*)

Metal	$\alpha /(10^{-6} \, \text{K}^{-1})$	$c_p /(\text{J} \cdot \text{g}^{-1} \cdot \text{K}^{-1})$	$\lambda /(\text{W} \cdot \text{cm}^{-1} \cdot \text{K}^{-1})$
dysprosium	9.9	0.173	0.107
erbium	12.2	0.168	0.145
europium	35	0.182	0.140
gadolinium	9	0.236	0.105
gallium		0.371	0.406
gold	14.2	0.129	3.17
hafnium	5.9	0.144	0.230
holmium	11.2	0.165	0.162
indium	32.1	0.233	0.816
iridium	6.4	0.131	1.47
iron	11.8	0.449	0.802
lanthanum	12.1	0.195	0.134
lead	28.9	0.129	0.353
lithium	46	3.582	0.847
lutetium	9.9	0.154	0.164
magnesium	24.8	1.023	1.56
manganese	21.7	0.479	0.0782
mercury		0.140	0.0834
molybdenum	4.8	0.251	1.38
neodymium	9.6	0.190	0.165
neptunium			0.063
nickel	13.4	0.444	0.907
niobium	7.3	0.265	0.537
osmium	5.1	0.130	0.876
palladium	11.8	0.244	0.718
platinum	8.8	0.133	0.716
plutonium	46.7		0.0674
polonium			0.200
potassium		0.757	1.024
praseodymium	6.7	0.193	0.125
promethium	11		0.15
rhenium	6.2	0.137	0.479
rhodium	8.2	0.243	1.500
rubidium		0.363	0.582
ruthenium	6.4	0.238	1.17
samarium	12.7	0.197	0.133
scandium	10.2	0.568	0.158
silver	18.9	0.235	4.29
sodium	71	1.228	1.41
strontium	22.5	0.301	0.353
tantalum	6.3	0.140	0.575
technetium			0.506
terbium	10.3	0.182	0.111
thallium	29.9	0.129	0.461
thorium	11.0	0.113	0.540
thulium	13.3	0.160	0.169

(*continued*)

22.3/3   Thermal properties of pure metals (*continued*)

Metal	$\alpha\,/(10^{-6}\,\mathrm{K}^{-1})$	$c_p\,/(\mathrm{J}\cdot\mathrm{g}^{-1}\cdot\mathrm{K}^{-1})$	$\lambda\,/(\mathrm{W}\cdot\mathrm{cm}^{-1}\cdot\mathrm{K}^{-1})$
tin	22.0	0.228	0.666
titanium	8.6	0.523	0.219
tungsten	4.5	0.132	1.74
uranium	13.9	0.116	0.276
vanadium	8.4	0.489	0.307
ytterbium	26.3	0.155	0.385
yttrium	10.6	0.298	0.172
zinc	30.2	0.388	1.16
zirconium	5.7	0.278	0.227

22.3/4   Thermal properties of construction and building materials

Material	$\alpha\,/(10^{-6}\,\mathrm{K}^{-1})$	$c_p\,/(\mathrm{J}\cdot\mathrm{g}^{-1}\cdot\mathrm{K}^{-1})$	$\lambda\,/(\mathrm{W}\cdot\mathrm{m}^{-1}\cdot\mathrm{K}^{-1})$
metals			
steel, V2A	16.0	0.51	14
steel, unalloyed	11 ... 13	0.49	47 ... 58
cast iron	10.5	0.532	58
aluminum bronze	24	0.435	128
bronze	17.5	0.37	64
constantan	15	0.410	23.3
brass	18	0.385	113
monel	14	0.43	19.7
nickel brass	18.36	0.398	48
phosphorus bronze	18.9	0.36	110
concrete			
standard concrete (1:2:4)	12	0.88	1.4 ... 1.5
ferroconcrete	10 ... 15	0.88	0.39 ... 1.6
wood			
oak	≈ 3	2.4	0.17
maple	≈ 3	1.6	0.16
birch	≈ 3	1.9	0.142
beech	≈ 3	2.1	0.17
alder	≈ 3	1.4	0.17
ash	≈ 3	1.6	0.16
pine	≈ 3	1.4	0.14
spruce	≈ 3	2.1	0.14

(*continued*)

22.3/4   Thermal properties of construction and building materials (*continued*)

Material	$\alpha\,/(10^{-6}\,\mathrm{K}^{-1})$	$c_p\,/(\mathrm{J}\cdot\mathrm{g}^{-1}\cdot\mathrm{K}^{-1})$	$\lambda\,/(\mathrm{W}\cdot\mathrm{m}^{-1}\cdot\mathrm{K}^{-1})$
building bricks			
brick	6	0.92	1
sandstone	7 ... 12	0.71	2.3
chamotte	5	0.8	≈ 1.2
slate		0.76	≈ 0.5
marble	≈ 11	0.84	2.8
glass			
window glass	7.9	0.84	0.81
quartz glass	0.6	0.73	0.81
glass wool		0.84	≈ 0.04

22.3/5   Thermal properties of gases

Substance	$c_p/$ $(\mathrm{J}\cdot\mathrm{g}^{-1}\cdot\mathrm{K}^{-1})$	$c_v/$	$\lambda/$ $(\mathrm{W}\cdot\mathrm{m}^{-1}\cdot\mathrm{K}^{-1})$
ethene	1.47	1.173	0.017
ammonia	2.056	1.568	0.022
argon	0.52	0.312	0.016
acetylene	1.616	1.300	0.018
chlorine	0.473	0.36	0.0081
hydrogen chloride	0.795	0.567	0.013
furnace gas	1.05	0.75	0.02
helium	5.20	3.121	0.143
carbon dioxide	0.816	0.627	0.015
carbon monoxide	1.038	0.741	0.023
krypton	0.25	0.151	0.0088
city gas	2.14	1.59	
air, dry	1.005	0.718	0.02454
methane	2.19	1.672	0.030
neon	1.03	0.618	0.046
propane	1.549	1.360	0.015
oxygen	0.909	0.649	0.024
carbon disulphide	0.582	0.473	0.0069
sulphur dioxide	0.586	0.456	0.0086
hydrogen sulphide	0.992	0.748	0.013
nitrogen	1.038	0.741	0.024
hydrogen	14.05	9.934	0.171
steam (100 °C)	1.842	1.381	0.016
xenon	0.16	0.097	0.0051

22.3/6   Thermal properties of liquid substances

Substance	$c_p/$ $(J \cdot g^{-1} \cdot K^{-1})$	$\lambda/$ $(W \cdot m^{-1} \cdot K^{-1})$
diethyl ether	2.298	0.13
ethyl alcohol	2.38	
acetone	2.22	0.16
petrol	2.02	0.13
benzene	1.70	0.15
diesel fuel	2.05	0.15
glycerol	2.37	0.29
heating oil	2.07	0.14
linseed oil	1.88	0.17
petroleum ether	1.76	0.14
mercury	0.138	10
colza oil	1.97	0.17
nitric acid, conc.	1.72	0.26
lubricating oil	2.09	0.13
sulphuric acid, conc.	1.42	0.47
transformer oil	1.88	0.13
trichloroethylene	0.93	0.12
toluene	1.67	0.14
water	4.187	0.60

22.3/7   Thermal properties of plastic materials

Material	$\alpha/$ $(10^{-6}\ K^{-1})$	$c_p/$ $(J \cdot g^{-1} \cdot K^{-1})$	$\lambda/$ $(W \cdot m^{-1} \cdot K^{-1})$
acryl	90	1.5	
polyvinyl chloride (PVC); flexible	240	1...2	0.16
polyvinyl chloride (PVC); stiff	50	0.9	0.16
polyethylene		2.3	
polystyrene	70	1.3	
polyester	80	2.1	0.23
polyester, 70 % fiberglass	12		0.17
bakelite (with wood flour)	50	1.5	0.34
bakelite (with asbestos)	30	1.3	0.6
rubber (slightly vulcanized)	220	2.1	0.15
rubber (with soot)	160	1.6	0.17

22.3/8   Heat conductivity and specific heat capacity of solid materials

Material	$c_p/$ $(J \cdot g^{-1} \cdot K^{-1})$	$\lambda/$ $(W \cdot m^{-1} \cdot K^{-1})$
asbestos	0.816	
basalt	0.86	1.67
ice	2.09	2.33
gypsum	1.1	0.81
mica	0.87	0.35

(continued)

22.3/8   Heat conductivity and specific heat capacity of solid materials (*continued*)

Material	$c_p/$ $(\mathrm{J \cdot g^{-1} \cdot K^{-1}})$	$\lambda/$ $(\mathrm{W \cdot m^{-1} \cdot K^{-1}})$
graphite	0.71	168
hard rubber	1.42	0.17
charcoal	0.84	0.084
limestone	0.909	2.2
boiler scale	0.80	1.2...3
colophony	1.30	0.317
cork	≈ 2.0	≈ 0.05
chalk	0.84	0.92
leather, dry	≈ 1.5	0.15
paper	1.336	0.14
paraffin	3.26	0.26
pitch		0.13
porcelain	≈ 1	≈ 1
quartz	0.80	9.9
black, soot	0.84	0.07
sand, dry	0.80	0.58
emery	0.96	11.6
snow	4.187	
silicon carbide	0.67	15.2
hard coal	1.02	0.24
beef fat	0.88	
tombac	0.381	159
clay, dry	0.88	≈ 1
peat dust, dry	1.9	0.08
vulcanized fiber	1.26	0.21
wax	3.34	0.084

22.3/9   Heat conduction of thermal insulators

Material	$\lambda /(\mathrm{W \cdot m^{-1} \cdot K^{-1}})$
flexible material in layers	
hair felt	0.038
balsam wool	0.039
felt, 75 % hair, 25 % flax	0.039
felt, 50 % hair, 50 % flax	0.038
flax fiber between paper	0.04
thermofelt (flax + asbestos)	0.053
loose material	
rock wool	0.037...0.042
glass wool	0.042
granular cork	0.043...0.045
gypsum powder	0.075...0.086
sawdust	0.059...0.061
charcoal	0.052...0.056

22.3/10   Heat conduction at various temperatures

Substance	$\lambda(0\,^\circ\text{C})/$ $(\text{W}\,\text{m}^{-1}\,\text{K}^{-1})$	$\lambda(50\,^\circ\text{C})/$ $(\text{W}\,\text{m}^{-1}\,\text{K}^{-1})$	$\lambda(100\,^\circ\text{C})/$ $(\text{W}\,\text{m}^{-1}\,\text{K}^{-1})$
asbestos	0.15	0.18	0.195
acetone	0.17	0.16	0.15
aniline	0.19	0.177	0.167
ethanol	0.188	0.177	—
castor oil	0.184	0.177	0.172
foamed concrete	0.11	0.11	0.13
water	0.551	0.648	0.683

22.3/11   Volume expansion of water at various temperatures

$\vartheta/\,^\circ\text{C}$	$\gamma/(10^{-4}\text{K}^{-1})$	$\vartheta/\,^\circ\text{C}$	$\gamma/(10^{-4}\text{K}^{-1})$
5 – 10	0.53	20 – 40	3.02
10 – 20	1.50	40 – 60	4.58
20	2.07	60 – 80	5.87

22.3/12   Volume expansion of liquids

The values of volume expansion of liquids are given for a temperature of 18 °C.

Substance	$\gamma/(10^{-4}\text{K}^{-1})$	Substance	$\gamma/(10^{-4}\text{K}^{-1})$
acetone	14.3	kerosene	10.0
aniline	8.5	methanol	11.9
trichloromethane	12.8	1-propanol	9.8
diethyl ether	16.3	mercury	1.8
ethanol	11.0	nitric acid	12.4
rock oil	9.2	turpentine oil	9.4
glycerol	5.0	toluene	10.8

22.4   Heat transmission

22.4/1   Thermal transmittance k in $\dfrac{\text{W}}{\text{m}^2 \cdot \text{K}}$ (approximate values)

Material	wall thickness/cm								
	0.3	1	2	5	10	12	15	20	25
glass	5.8	5.6							
wooden wall			3.8	2.4		1.7			
gravel concrete			4.1	3.5		3.1	2.8		
slag brick						2.7			1.7
ferroconcrete				4.2	3.7		3.3	2.9	

22.4/2   Thermal transmittance k **in** $\dfrac{\text{W}}{\text{m}^2 \cdot \text{K}}$ **for building bricks**

| | wall thickness/cm | | | | | | |
| | inner wall | | | outer wall | | | |
Material	9	19	24	24	30	39	49
full brick	2.56	1.94	1.73	2.00	1.78	1.45	1.22
horizontally perforated brick	2.00	1.63	1.36	1.50	1.28	1.10	0.87
vertically perforated brick	2.36	1.69	1.49	1.69	1.48	1.19	1.00
clinker brick	2.73		1.99	2.35			
sandy limestone							
perforated bricks	2.24	1.88	1.62	1.85	1.57	1.37	1.10
full bricks	2.52	2.19	1.94	2.28	1.97	1.74	1.43
hard bricks	2.56	2.23	2.02	2.35	2.04	1.80	1.49
metallurgic bricks	2.24	1.88	1.60	1.81	1.57	1.37	1.10
gas concrete							
$600 \text{ kg} \cdot \text{m}^{-3}$	1.64	1.28	1.04	1.12	0.94	0.80	0.62
$800 \text{ kg} \cdot \text{m}^{-3}$	1.77	1.41	1.15	1.26	1.06	0.91	0.71
$1000 \text{ kg} \cdot \text{m}^{-3}$	1.90	1.52	1.26	1.38	1.17	1.01	0.79
lightweight concrete solid bricks							
$1200 \text{ kg} \cdot \text{m}^{-3}$	2.00	1.63	1.36	1.50	1.30	1.10	0.87
$1400 \text{ kg} \cdot \text{m}^{-3}$	2.17	1.81	1.52	1.72	1.48	1.29	1.02
$1600 \text{ kg} \cdot \text{m}^{-3}$	2.36	1.99	1.71	1.97	1.71	1.50	1.21
lightweight concrete hollow-block bricks							
$1400 \text{ kg} \cdot \text{m}^{-3}$			1.30	1.45	1.27		
$1600 \text{ kg} \cdot \text{m}^{-3}$			1.42	1.59	1.38		

22.4/3 Heat transmission coefficients α (guide values)

Conditions	$\alpha/(\mathrm{W}\cdot\mathrm{m}^{-2}\cdot\mathrm{K}^{-1})$
air along plane polished surface	
$\quad$ speed of air $v \le 5\mathrm{m}\cdot\mathrm{s}^{-1}$	$5.6 + \dfrac{4v}{\mathrm{m}\cdot\mathrm{s}^{-1}}$
$\quad$ speed of air $v > 5\mathrm{m}\cdot\mathrm{s}^{-1}$	$7.12 \cdot \left(\dfrac{v}{\mathrm{m}\cdot\mathrm{s}^{-1}}\right)^{0.78}$
air along plane iron wall	
$\quad$ speed of air $v \le 5\mathrm{m}\cdot\mathrm{s}^{-1}$	$5.8 + \dfrac{4v}{\mathrm{m}\cdot\mathrm{s}^{-1}}$
$\quad$ speed of air $v > 5\mathrm{m}\cdot\mathrm{s}^{-1}$	$7.14 \cdot \left(\dfrac{v}{\mathrm{m}\cdot\mathrm{s}^{-1}}\right)^{0.78}$
air along plane brickwork	
$\quad$ speed of air $v \le 5\mathrm{m}\cdot\mathrm{s}^{-1}$	$6.2 + \dfrac{4.2v}{\mathrm{m}\cdot\mathrm{s}^{-1}}$
$\quad$ speed of air $v > 5\mathrm{m}\cdot\mathrm{s}^{-1}$	$7.52 \cdot \left(\dfrac{v}{\mathrm{m}\cdot\mathrm{s}^{-1}}\right)^{0.78}$
air perpendicular to metallic wall	
$\quad$ at rest	$3.5\ldots35$
$\quad$ moderate motion	$23\ldots70$
$\quad$ rapid motion	$58\ldots290$
water around pipes	
$\quad$ at rest	$350\ldots580$
$\quad$ flowing	$350 + 2100\sqrt{\dfrac{v}{\mathrm{m}\cdot\mathrm{s}^{-1}}}$
water in vessels	$580\ldots2300$
water in vessels, stirred	$2300\ldots4700$
boiling water in pipes	$4700\ldots7000$
boiling water at metallic surfaces	$3500\ldots5800$
condensing steam	$11\,600$

22.5   Practical correction data

22.5.1   Pressure measurement

p_0 and ρ_0 denote the pressure and density of air at sea level and $\vartheta = 15\,°C$.

22.5/1   Standard atmosphere in relative units

Altitude/m	p/p_0	ρ/ρ_0	$\vartheta/°C$	Altitude/m	p/p_0	ρ/ρ_0	$\vartheta/°C$
0	1	1	15	5000	0.533	0.601	−17.5
1000	0.887	0.907	8.5	6000	0.465	0.538	−24
2000	0.784	0.822	2	7000	0.405	0.481	−30.5
3000	0.692	0.742	−4.5	8000	0.351	0.428	−37
4000	0.608	0.669	−11	10000	0.261	0.337	−50

22.5/2   Air pressure p as a function of altitude h, absolute units

h/m	p/hPa	h/m	p/hPa	h/m	p/hPa	h/m	p/hPa
0	1013.25	700	931.9	2000	795.0	6000	471.8
100	1001.3	800	920.8	2400	756.3	8000	356.0
200	989.5	900	909.7	2800	719.1	10000	264.4
300	977.7	1000	898.8	3200	683.4	12000	193.3
400	966.1	1200	877.2	3600	649.2	15000	120.4
500	954.6	1400	856.0	4000	616.4	17500	81.2
600	943.2	1600	835.3	5000	540.2	20000	54.75

22.5.1.1   Conversion to sea level

As a rule, pressure data refer to sea level. For that reason, the reference data must be corrected. The altitude of the place of measurement above sea level and the temperature difference between the place of measurement and sea level must be taken into account. The influence of the geographic latitude is, as a rule, masked by the inaccuracies in the temperature of the air column. The correction is made as follows. From the first table, one picks the factor that accounts for the altitude and the air temperature. With this value, one then uses the second table and takes, for this factor, the quantity of correction that must be added to the measured quantity. The unit adopted is the non-SI unit: torr = 1 mm Hg.

22.5/3   Temperature and altitude factors

Altitude/m	Temperature of air column/°C			
	−16	0	16	28
2000	1.2	1.1	1.0	1.0
2100	11.5	10.8	10.2	9.7
2200	23.0	21.6	20.3	19.5
2300	34.5	32.5	30.5	29.2
2400	46.0	43.3	40.7	38.9
2500	57.5	54.1	50.9	48.6
2600	69.0	64.9	61.0	58.3
2700	80.6	75.7	71.2	68.1
2800	92.1	86.5	81.4	77.8
2900	103.6	97.4	91.5	87.5
3000	115.1	108.2	101.7	97.3
3100	126.6	119.0	111.9	107.0
3200	138.1	129.8	122.0	116.7
3300	149.6	140.6	132.2	126.4
3400	161.1	151.4	142.4	136.2
3500	172.6	162.3	152.5	145.9
3600	184.1	173.1	162.7	155.6
3700	195.6	183.9	172.9	165.3
3800	207.1	194.7	183.1	175.0
3900	218.6	205.5	193.2	184.8
4000	230.1	216.3	203.4	194.5
4100	241.6	227.1	213.5	204.2
4200	253.1	237.9	223.7	213.9
4300	264.6	248.8	233.9	223.6
4400	276.1	259.6	244.0	233.4
4500	287.6	270.4	254.2	243.1
4600	299.1	281.2	264.4	252.8
4700	310.6	292.0	274.5	262.5
4800	322.1	302.8	284.7	272.2
4900	333.6	313.6	294.9	282.0

22.5/4   Additive correction term for pressure measurement

Temp. altitude factor	Barometer measured value in mm Hg					
	780	760	740	720	700	
1	0.9	0.9	0.9	0.8	0.8	—
5	4.5	4.4	4.3	4.2	4.0	—
10	9.0	8.8	8.6	8.3	8.1	—
15	13.6	13.2	12.9	12.5	12.2	—
20	18.2	17.7	17.2	16.8	16.3	—
25	22.8	22.2	21.6	21.0	20.4	—
30	27.4	26.7	26.0	25.3	24.6	—
35	—	31.2	30.4	29.6	28.8	—
	760	740	720	700	680	660
40	35.8	34.9	33.9	33.0	32.0	31.1
45	40.4	39.3	38.3	37.2	36.2	35.1
50	45.0	43.8	42.7	41.5	40.3	39.1
55	49.7	48.4	47.1	45.8	44.5	43.1
60	—	52.9	51.5	50.1	48.6	47.2
65	—	57.5	55.9	54.4	52.8	51.3
70	—	62.1	60.4	58.7	57.1	55.4
75	—	66.7	64.9	63.1	61.3	59.5
	720	700	680	660	640	
80	69.5	67.5	65.6	63.7	61.7	—
85	74.0	72.0	69.9	67.9	65.8	—
90	78.6	76.4	74.2	72.1	69.9	—
95	83.2	80.9	78.6	76.3	74.0	—
100	87.9	85.4	83.0	80.5	78.1	—
105	—	89.9	87.4	84.8	82.2	—
110	—	94.5	91.8	89.1	86.4	—
115	—	99.1	96.3	93.4	90.6	—
120	—	103.7	100.7	97.8	94.8	—
125	—	108.3	105.3	102.2	99.1	—
	680	660	640	620	600	
125	105.3	102.2	99.1	96.0	92.9	—
130	109.8	106.6	103.3	100.1	96.9	—
135	114.3	111.0	107.6	104.3	100.9	—
140	118.9	115.4	111.9	108.4	104.9	—
145	123.5	119.9	116.3	112.6	109.0	—
150	128.2	124.4	120.6	116.9	113.1	—
155	—	128.9	125.0	121.1	117.2	—
160	—	133.5	129.4	125.4	121.4	—
165	—	138.1	133.9	129.7	125.5	—
170	—	142.7	138.4	134.0	129.7	—

(*continued*)

22.5/4 Additive correction term for pressure measurement (*continued*)

Temp. altitude factor	Barometer measured value in mm Hg				
	640	620	600	580	560
170	138.4	134.0	129.7	125.4	121.1
175	142.9	138.4	133.9	129.5	125.0
180	147.4	142.8	138.2	133.6	129.0
185	151.9	147.2	142.4	137.7	132.9
190	153.5	151.6	146.7	141.8	136.9
195	161.1	156.1	151.0	146.0	141.0
200	165.7	160.5	155.4	150.2	145.0
205	170.4	165.0	159.7	154.4	149.1
210	—	169.6	164.1	158.6	153.2
215	—	174.1	168.5	162.9	157.3
	620	600	580	560	540
215	174.1	168.5	162.9	157.3	151.7
220	178.7	172.9	167.2	161.4	155.7
225	183.3	177.4	171.5	165.6	159.7
230	188.0	181.9	175.8	169.8	163.7
235	192.6	186.4	180.2	174.0	167.8
240	—	191.0	184.6	178.2	171.9
245	—	195.5	189.0	182.5	176.0
250	—	200.1	193.4	186.8	180.1
255	—	204.7	197.9	191.1	184.3
260	—	209.4	202.4	195.4	188.4
	580	560	540	520	
260	202.4	195.4	188.4	181.5	—
265	206.9	199.8	188.4	181.5	—
270	211.5	204.2	196.9	189.6	—
275	216.0	208.6	201.1	193.7	—
280	220.6	213.0	205.4	197.8	—
285	225.2	217.5	209.7	201.9	—
290	229.9	222.0	214.0	206.1	—
295	—	226.5	218.4	210.3	—
300	—	231.0	222.8	214.5	—
	560	540	520	500	480
305	235.6	227.2	218.8	210.3	201.9
310	240.2	231.6	223.0	214.4	205.9
315	244.8	236.0	227.3	218.6	209.8
320	249.4	240.5	231.6	222.7	213.8
325	254.1	245.0	236.0	226.9	217.8
330	—	249.6	240.3	231.1	221.8
335	—	254.1	244.7	235.3	225.9
340	—	258.7	249.1	239.6	230.0
345	—	263.3	253.6	243.8	234.1

22.5.1.2   Mercury barometer measurements (temperature correction)

This correction has its origin in the thermal expansion of mercury, as well as of the measuring scale.

22.5/5   Barometric measurements with a brass scale

The quantities given in the table must be subtracted from the measuring value. Intermediate values may be estimated by linear interpolation.

Temperature $\vartheta/°C$	Measuring value in mm								
	620	630	640	650	660	670	680	690	700
0	0	0	0	0	0	0	0	0	0
5	0.51	0.51	0.52	0.53	0.54	0.55	0.56	0.56	0.57
10	1.01	1.03	1.04	1.06	1.08	1.09	1.11	1.13	1.14
15	1.52	1.54	1.56	1.59	1.61	1.64	1.66	1.69	1.71
20	2.02	2.05	2.08	2.12	2.15	2.18	2.21	2.25	2.28
25	2.52	2.56	2.60	2.64	2.68	2.72	2.77	2.81	2.85
30	3.02	3.07	3.12	3.17	3.22	3.27	3.32	3.36	3.41
35	3.52	3.58	3.64	3.69	3.75	3.81	3.86	3.92	3.98

Temperature $\vartheta/°C$	Measuring value in mm								
	710	720	730	740	750	760	770	780	790
0	0	0	0	0	0	0	0	0	0
5	0.58	0.59	0.60	0.60	0.61	0.62	0.63	0.64	0.64
10	1.16	1.17	1.19	1.21	1.22	1.24	1.26	1.27	1.29
15	1.74	1.76	1.78	1.81	1.83	1.86	1.88	1.91	1.93
20	2.31	2.34	2.38	2.41	2.44	2.47	2.51	2.54	2.57
25	2.89	2.93	2.97	3.01	3.05	3.09	3.13	3.17	3.21
30	3.46	3.51	3.56	3.61	3.66	3.71	3.75	3.80	3.85
35	4.03	4.09	4.15	4.21	4.26	4.32	4.38	4.43	4.49

22.5/6   Barometric measurements with a glass scale

The quantities given in the table must be subtracted from the measuring value. Intermediate values may be estimated by linear interpolation.

Temperature $\vartheta/°C$	Measuring value in mm								
	700	710	720	730	740	750	760	770	780
0	0	0	0	0	0	0	0	0	0
5	0.060	0.061	0.062	0.063	0.064	0.064	0.065	0.066	0.067
10	0.121	0.122	0.124	0.126	0.127	0.129	0.130	0.132	0.134
15	0.181	0.184	0.186	0.189	0.191	0.193	0.196	0.198	0.201
20	0.242	0.245	0.248	0.252	0.255	0.258	0.261	0.264	0.268
25	0.303	0.307	0.311	0.315	0.319	0.323	0.327	0.331	0.335
30	0.363	0.368	0.373	0.378	0.383	0.387	0.392	0.397	0.402

22.5.2   Volume measurements—conversion to standard temperature

22.5/7   Temperature correction for aqueous solutions

Frequently, the values for aqueous solutions refer to the standard temperature of 20 °C. The volume measurement is made, however, at another temperature. The following table lists an additive correction to the volume to be measured, by reference to the standard volume at 20 °C. The volume-expansion coefficient of glass is assumed to be 0.000025 per degree.

Temperature $\vartheta/°C$	Volume at 20 °C						
	2000	1000	500	400	300	250	150
15	−1.54	−0.77	−0.38	−0.31	−0.23	−0.19	−0.12
16	−1.28	−0.64	−0.32	−0.26	−0.19	−0.16	−0.10
17	−0.99	−0.50	−0.25	−0.20	−0.15	−0.12	−0.07
18	−0.68	−0.34	−0.17	−0.14	−0.10	−0.08	−0.05
19	−0.35	−0.18	−0.09	−0.07	−0.05	−0.04	−0.03
21	0.37	0.18	0.09	0.07	0.06	0.05	0.03
22	0.77	0.38	0.19	0.15	0.12	0.10	0.06
23	1.18	0.59	0.30	0.24	0.18	0.15	0.09
24	1.61	0.81	0.40	0.32	0.24	0.20	0.12
25	2.07	1.03	0.52	0.41	0.31	0.26	0.15
26	2.54	1.27	0.64	0.51	0.38	0.32	0.19
27	3.03	4.52	0.76	0.61	0.46	0.38	0.23
28	3.55	1.77	0.89	0.71	0.53	0.44	0.27
29	4.08	2.04	1.02	0.82	0.61	0.51	0.31
30	4.62	2.31	1.16	0.92	0.69	0.58	0.35

22.5.2.1   Measurements with a glass constant-volume thermometer

22.5/8   Temperature correction for a glass constant-volume thermometer

The following table lists the additive correction due to the thermal expansion of glass, with respect to 20 °C.

Temperature $\vartheta/°C$	Measured volume in milliliters					
	2000	1000	500	400	300	250
15	−0.25	−1.12	−0.06	−0.05	−0.04	−0.031
16	−0.20	−0.10	−0.05	−0.04	−0.03	−0.025
17	−0.10	−0.08	−0.04	−0.03	−0.02	−0.019
18	−0.10	−0.05	−0.02	−0.02	−0.02	−0.12
19	−0.05	−0.02	−0.01	−0.01	−0.01	−0.006
21	0.05	0.02	0.01	0.01	0.01	0.006
22	0.10	0.05	0.02	0.02	0.02	0.012
23	0.15	0.08	0.04	0.03	0.02	0.019
24	0.20	0.10	0.05	0.04	0.03	0.025
25	0.25	0.12	0.06	0.05	0.04	0.031
26	0.30	0.15	0.08	0.06	0.04	0.038
27	0.35	0.18	0.09	0.07	0.05	0.044
28	0.40	0.20	0.10	0.08	0.06	0.050
29	0.45	0.22	0.11	0.09	0.07	0.056
30	0.50	0.25	0.12	0.10	0.08	0.062

22.6   Generation of liquid low-temperature baths

To generate constant low temperatures, one may use solid-liquid mixtures at the melting point. This bath must be stirred. Cooling is undertaken, depending on the required temperature, with dry ice (−78 °C) or liquid air (−190 °C). The substances listed in the subsequent table may be used as temperature baths (T_K = melting point, T_S = boiling point).

22.6/1   Liquid baths at low temperatures

Substance	T_K /°C	T_S /°C
isopentane	−159.9	27.85
methyl cyclopentane	−142.4	71.8
allyl chloride	−134.5	45
n-pentane	−129.7	36.1
allyl alcohol	−129	97
ethyl alcohol	−117.3	78.5
carbon disulphide	−110.8	46.3
isobutyl alcohol	−108	108.1
acetone	−95.4	56.2
toluene	−95	110.6
ethyl acetate	−84	77

(continued)

22.6/1   Liquid baths at low temperatures (*continued*)

Substance	T_K /°C	T_S /°C
dry ice + acetone	−78	—
p-cymen	−67.9	177.1
trichloromethane	−63.5	61.7
N-methyl aniline	−57	196
chlorobenzene	−45.6	132
anisole	−37.5	155
bromobenzene	−30.8	156
tetrachloromethane	−23	76.5
benzonitrile	−13	205

22.7   Dehydrators

Dehydration of gases may be achieved by absorption (chemical effect) or by adsorption (physical effect).

22.7/1   Efficiency of chemical dehydration

Substance	Residual water in $mg/(10^{-3}$ m$^3)$ of dry air	Substance	Residual water in $mg/(10^{-3}$ m$^3)$ of dry air
P_2O_5	< 1 mg in 40 m^3	NaOH molten	0.16
$Mg(ClO_4)_2$ anhyd.	—	$CaBr_2$	0.18
BaO	0.00065	$CaCl_2$ molten	0.34
KOH molten	0.002	$Ba(ClO_4)_2$	0.82
CaO	0.003	$ZnCl_2$	0.85
H_2SO_4	0.003	$ZnBr_2$	1.16
$CaSO_4$ anhyd.	0.005	$CaCl_2$ granular	1.5
Al_2O_3	0.005	$CuSO_4$ anhyd.	2.8

22.7/2   Efficiency of physical dehydration

The dehydrators are ordered according to increasing efficiency.

argil (fired at low temperatures)
asbestos
charcoal
clay
porcelain (fired at low temperatures)
glass wool
diatomite
silica gel
freezing-through

22.8   Vapor pressure

22.8.1   Solutions

22.8/1   Saturated vapor pressure at 20 °C

Substance	p_D/hPa	Substance	p_D/hPa
acetone	240	methanol	129
benzene	100	pentane	565
trichloromethane	213	tetrachloromethane	121
diethyl ether	584	toluene	29.3
ethanol	587	water	23.4

22.8/2   Cryoscopic and ebullioscopic constants

Substance	K/K	E/K	Substance	K/K	E/K
ammonia	1320	340	acetic acid	3900	3070
benzene	5070	2640	ethanol	—	1070
diethyl ether	1790	1830	carbon tetrachloride	29800	4880
tetrachloromethane	4900	3800	water	1860	520

22.8.2   Relative humidity

22.8/3   Psychrometry

Determination of the relative humidity through the temperature difference $\Delta\vartheta$ of two thermometers, one of which measures at 100 % humidity (ϑ_f), while the other measures at the local value (ϑ_R). For $\Delta\vartheta = 0$, $\varphi = 100\%$.

ϑ_R /°C	φ in % at $\Delta\vartheta$ /°C										
	0	1	2	3	4	5	6	7	8	9	10
0	100	81	63	45	28	11	—	—	—	—	—
2	100	84	68	51	35	20	—	—	—	—	—
4	100	85	70	56	42	28	14	—	—	—	—
6	100	86	73	60	47	35	23	10	—	—	—
8	100	87	75	63	51	40	28	18	7	—	—
10	100	88	76	65	54	44	34	24	14	4	—
12	100	89	78	68	57	48	38	29	20	11	—
14	100	90	79	70	60	51	42	33	25	17	9
16	100	90	81	71	62	54	45	37	30	22	15
18	100	91	82	73	64	56	48	41	34	26	20
20	100	91	83	74	66	59	51	44	37	30	24
22	100	92	83	76	68	61	54	47	40	34	28
24	100	92	84	77	69	62	56	49	43	37	31
26	100	92	85	78	71	64	58	50	45	40	34
28	100	93	85	78	72	65	59	53	48	42	37
30	100	93	86	79	73	67	61	55	50	44	39

22.8.3 Vapor pressure of water

22.8/4   Vapor pressure of water at low temperatures

$\vartheta/°C$	p_D/hPa	$\vartheta/°C$	p_D/hPa	$\vartheta/°C$	p_D/hPa	$\vartheta/°C$	p_D/hPa
0	6.0	10	12.1	20	23.4	30	43.2
2	7.0	12	13.8	22	27.0	32	48.6
4	8.0	14	15.8	24	30.5	34	54.3
6	9.2	16	18.6	26	34.3	36	60.6
8	10.5	18	21.1	28	38.6	38	67.6

22.8/5   Vapor pressure and specific enthalpy of water

Temperature $\vartheta/°C$	Density $\varrho/(kg\,m^{-3})$	Specific volume $v/(10^{-3}m^3\,kg^{-1})$	Vapor pressure p_D/bar	Specific enthalpy $h/(kJ\,kg^{-1})$
5	1000	1.0	0.0087	21, 0
10	1000	1.0	0.0123	42.0
15	999	1.001	0.0170	62.9
20	998	1.002	0.0234	83.9
25	997	1.003	0.0317	104.8
30	996	1.004	0.0424	125.7
40	992	1.008	0.0738	167.4
50	988	1.012	0.1234	209
60	983	1.017	0.1992	251
70	978	1.023	0.3116	293
80	972	1.029	0.4736	335
90	965	1.036	0.7011	377
100	958	1.044	1.013	419
120	943	1.061	1.985	504
140	926	1.080	3.614	589
160	907	1.102	6.181	675
180	887	1.128	10.03	763
200	864	1.157	15.55	852
250	799	1.251	39.78	1085
300	712	1.404	85.93	1345
350	574	1.741	165.35	1672

The following table is of importance for the description of saturated steam and wet steam. The indices D and W denote steam and (boiling) water, respectively; p – vapor pressure, ϑ – temperature, v – specific volume and h – specific enthalpy.

22.8/6   Specific volume and specific enthalpy of steam

p/bar	$\vartheta/\,^\circ\mathrm{C}$	$v_W/$ $(10^{-3}\mathrm{m^3\,kg^{-1}})$	$v_D/$ $(\mathrm{m^3\,kg^{-1}})$	$h_W/$ $(\mathrm{kJ\,kg^{-1}})$	$h_D/$ $(\mathrm{kJ\,kg^{-1}})$
0.01	6.98	1.0001	129.2	29.34	2514
0.02	17.53	1.0012	67.01	73.46	2533.6
0.04	28.98	1.0040	34.80	121.41	2554.5
0.06	36.18	1.0064	23.74	151.50	2567.5
0.08	41.53	1.0084	18.10	173.86	2577.1
0.1	45.83	1.0102	14.67	191.83	2584.4
0.2	60.09	1.0172	7.650	251.45	2609.9
0.4	75.88	1.0265	3.993	317.65	2636.9
0.6	85.95	1.0333	2.732	359.93	2653.6
0.8	93.51	1.0387	2.087	391.72	2665.8
1.0	99.63	1.0434	1.694	417.51	2675.4
1.4	109.3	1.0513	1.236	458.42	2690.3
2.0	120.23	1.0608	0.8854	504.70	2706.3
3.0	133.54	1.0735	0.6056	561.43	2724.7
4.0	143.62	1.0839	0.4622	604.67	2737.6
5.0	151.84	1.0928	0.3747	640.12	2747.5
6.0	158.84	1.1009	0.3155	670.42	2755.5
8.0	170.41	1.1150	0.2403	720.94	2767.5
10.0	179.88	1.1274	0.1943	762.61	2776.2
12.0	187.96	1.1386	0.1632	798.43	2782.7
15.0	198.29	1.1539	0.1317	844.67	2789.9
20.0	212.37	1.1766	0.09954	908.59	2797.2
30.0	233.84	1.2163	0.06663	1008.4	2802.3
40.0	250.33	1.2521	0.04975	1087.4	2800.3
50.0	263.91	1.2858	0.03943	1154.5	2794.2

22.9   Specific enthalpies

22.9/1   Specific calorific value H_u (mean values)

Solid substances	$H_u/(MJ \cdot kg^{-1})$
anthracite	33.4
lignitic coal	9.6
lignitic coal, hard	17
lignitic coal, briquette	20
fat coal	31.0
coke	29.2
wood, dry	13.3
lean coal	31.0
peat, dry	14.6
furnace coke	30.1
charcoal	31

Liquid substances	$H_u/(MJ \cdot kg^{-1})$
ethyl alcohol	26.9
benzene	40.2
diethyl ether	34
petroleum	41
diesel fuel	42.1
heating oil	41.8
petrol	42.5
methyl alcohol	19.5
kerosene	40.8
spirit (95 %)	25.0
coal tar	34

Gaseous substances	$H_u/(MJ \cdot m^{-3})$
ethyne	85.99
butane	124
natural gas, wet	29
ethane	64.5
natural gas, dry	43.9
furnace gas	5
methane	35.9
propane	93.4
city gas	20
hydrogen	10.8
propene	88.0

22.9/2   Specific melting and evaporation enthalpies of pure metals

Metal	Spec. melting enthalpy Δh_s /(kJ · kg^{-1})	Spec. evapor. enthalpy Δh_v /(kJ · kg^{-1})
aluminum	397	10 900
antimony	167	1050
barium	56	1100
beryllium	1390	32 600
bismuth	52.2	725
cadmium	56	890
calcium	216	3750
cerium	39	2242
cesium	16.4	496
chromium	280	6700
cobalt	275	6503
copper	205	4790
dysprosium	68.1	—
erbium	119	—
europium	60.6	—
gadolinium	63.6	—
gallium	80.8	3640
gold	65.7	1650
hafnium	146	3703
holmium	103	—
indium	28.5	1970
iridium	117	3900
iron	277	6340
lanthanum	81.3	2880
lead	23.0	8600
lithium	603	20 500
lutetium	126	—
magnesium	368	5420
manganese	266	4190
mercury	11.8	285
molybdenum	290	5610
neodymium	49.5	—
nickel	303	6480
niobium	334	7492
osmium	289	—
palladium	157	—
platinum	111	2290
potassium	59.6	1980
praseodymium	48.9	—
rhenium	178	3797
rhodium	218	—
rubidium	25.7	880
ruthenium	193	—

(*continued*)

22.9/2 Specific melting and evaporation enthalpies of pure metals (*continued*)

Metal	Spec. melting enthalpy Δh_s /(kJ·kg^{-1})	Spec. evapor. enthalpy Δh_v /(kJ·kg^{-1})
samarium	57.3	—
scandium	314	6785
silver	105	2350
sodium	113	390
strontium	94	1585
tantalum	199	4162
terbium	67.9	—
thallium	20.6	794.6
thorium	59.5	2344
thulium	99	—
tin	59.6	2450
titanium	324	8990
tungsten	192	4350
uranium	36.6	1731
vanadium	452	8998
ytterbium	44.3	—
yttrium	128	4421
zinc	111	1755
zirconium	219	6382

22.9/3 Relative volume change in melting

Substance	$\Delta V/V$	Substance	$\Delta V/V$
aluminum	0.066	magnesium	0.042
antimony	−0.0094	mercury	0.036
cadmium	0.047	potassium	0.024
gallium	−0.03	silver	0.05
gold	0.0519	sodium	0.025
indium	0.025	tin	0.026
lead	0.036	water (ice)	−0.083
lithium	0.015	zinc	0.069

22.9/4 Temperature dependence of evaporation heat

Substance	0 °C	20 °C	60 °C	100 °C	140 °C	180 °C	220 °C
methanol	1220	1190	1130	1030	906	743	472
ethanol	927	925	894	827	717	584	370
1-propanol	—	—	—	688	598	488	358
diethyl ether	388	367	329	287	234	134	—
acetic acid	—	352	376	387	385	368	344

22.9/5 **Specific melting enthalpies** ΔH_s **and evaporation enthalpies** ΔH_v

Substance	$\Delta h_s/$ $(kJ \cdot kg^{-1})$	$\Delta h_v/$ $(kJ \cdot kg^{-1})$
1-pentanol	—	502
1-propanol	86.5	750
acetic acid	192	406
acetone	98	525
aluminum oxide	1069	4730
ammonia	—	1370
argon	29.44	163
benzene	128	394
boron	2055	50000
bromine	67.8	183
butane	80.34	385
butyl alcohol	121.35	616
cane sugar	56	—
carbon dioxide	180.7	136.8
carbon disulphide	57.8	352
carbon monoxide	29.86	216
chlorine	90.48	290
deuterium	98.5	304
diethyl ether	—	384
ethane	95.23	489
ethanol	108	840
ethene	119.68	483
ethyl chloride	—	382
fluorine	81.9	172
formic acid	276	432
frigen 11 (CCl_3F)	50.24	182
frigen 12 (CCl_2F_2)	34.27	162
frigen 21 ($CHCl_2F$)	—	242
frigen 22 ($CHClF_2$)	47.68	234
glycerol	201	—
heptane	141	318
hexane	152	332
hydrogen chloride	—	443
hydrogen fluoride	—	375
iodine	124	172
krypton	19.52	108
methane	58.62	510
methanol	92	1100
methyl acetate	—	406
naphthalene	148	314
neon	16.58	91.2
nitrobenzene	94.2	397
nitrogen	25.74	198

(*continued*)

22.9/5   Specific melting enthalpies ΔH_s and evaporation enthalpies ΔH_v (*continued*)

Substance	$\Delta h_s /$ $(\text{kJ} \cdot \text{kg}^{-1})$	$\Delta h_v /$ $(\text{kJ} \cdot \text{kg}^{-1})$
octane	181	299
oxygen	13.87	213
ozone	—	316
pentane	116	360
phenol	122	510
phosphorus, white	21.0	400
phosphorus trihydride	33.33	430
potassium chloride	342	2160
potassium nitrate	107	—
propane	57.36	426
propene	71.48	438
pyridine	105	450
selenium	68.6	1200
silicon	164	14 050
sodium chloride	500	2900
sulphur	42	290
sulphur dioxide	115.64	390
sulphuric acid	109	—
toluene	—	364
trichloromethane	75	279
water	334	2265
water, heavy	318	2072
xenon	—	99.2
xylene	109	343

Part V
Quantum physics

23
Photons, electromagnetic radiation and light quanta

The particle character of light was first demonstrated in three phenomena: thermal radiation, photo effect and Compton scattering.

23.1 Planck's radiation law

A significant hint of the failure of classical physics arose from investigations of thermal radiation (Planck, 1900). According to Einstein (1905) electromagnetic radiation is quantized in **photons**.

1. Photons and Planck's quantum of action

Photons, symbol γ, the energy quanta of the electromagnetic field.

Photon energy, E_{Ph}, proportional to the frequency f or the angular frequency $\omega = 2\pi f$. Usually it is given in electron volts (eV),

$$E_{Ph} = hf = \hbar\omega.$$

Photon momentum, $\vec{p}_{Ph}$, proportional to the wave number vector $\vec{k}$ (with $|\vec{k}| = \frac{2\pi}{\lambda}$, λ is the wavelength of the electromagnetic radiation),

$$\vec{p}_{Ph} = \hbar\vec{k}, \quad |\vec{p}_{Ph}| = \hbar k = h/\lambda.$$

The vector $\vec{k}$ points along the propagation direction of the electromagnetic radiation.

Planck's quantum of action, a universal constant,

$$h = 6.626\,075\,5(40) \cdot 10^{-34}\ \text{J s},$$

$$\hbar = \frac{h}{2\pi} = 1.054\,572\,66(63) \cdot 10^{-34}\ \text{J s} = 6.582\,122(20) \cdot 10^{-22}\ \text{MeV s}.$$

2. *Thermal radiation and the blackbody radiator*

Thermal radiation, temperature radiation, the electromagnetic radiation of a body at finite temperature. The body also absorbs a fraction of the thermal radiation from its environment. There is a permanent exchange of energy between the body and its environment. In the end, this process leads to temperature equilibrium.

Blackbody radiator, a body with the reflectance zero. A blackbody absorbs any incident radiation completely.

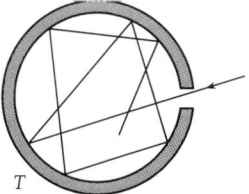

Figure 23.1: Model of the blackbody.

Cavity radiator, model of a blackbody radiator (**Fig. 23.1**): a box with a small aperture in the wall. The wall is impenetrable for radiation from inside (ideally reflecting) and has a definite temperature. The probability that a photon enters the cavity through the aperture and, after multiple reflection by the inner walls, leaves the cavity through the aperture again, is negligible (absorptance $\alpha = 1$). The aperture appears absolutely black.

Cavity radiation, the thermal radiation leaving the aperture of a cavity radiator. The spectral distribution of the radiation energy density of the cavity radiation depends on the temperature of the cavity radiator.

➤ According to Kirchhoff's law (see p. 765), the spectral radiance $L_{e,f}$ of an arbitrary thermal radiator may be reduced to that of a black body.

For the radiation field in the interior of the cavity, one defines:

radiant energy density			$\mathbf{ML^{-1}T^{-2}}$
	Symbol	Unit	Quantity
$u = \dfrac{Q}{V}$	u	J/m^3	radiant energy density
	Q	J	radiant energy
	V	m^3	volume

3. *Planck's radiation law*

This law describes the frequency and temperature dependence of the radiant energy density of the cavity radiation:

spectral radiant energy density			$\mathbf{ML^{-1}T^{-1}}$
	Symbol	Unit	Quantity
$u_f(f, T) = \dfrac{8\pi f^2}{c^3} \cdot \dfrac{hf}{e^{hf/(kT)} - 1}$	$u_f(f, T)$	J s m^{-3}	spectral radiant energy density
	c	m s^{-1}	speed of light
	f	s^{-1}	frequency
	h	J s	quantum of action
	k	J K^{-1}	Boltzmann constant
	T	K	temperature

This law connects the classical picture of continuous emission and absorption of electromagnetic waves with the photon picture of quantized electromagnetic radiation.

➤ Conversion of the radiant density (radiant power density) $L_{e,f}$ into the energy density u_f of unpolarized uniform and isotropic cavity radiation:

$$u_f = 2 \int d\Omega \, \frac{L_{e,f}}{c} = 8\pi \, \frac{L_{e,f}}{c} \, .$$

4. Connection between radiant energy density and frequency

The dependence of the spectral radiant energy density of the cavity radiation on the angular frequency ω or wavelength λ reads as follows:

$$u_\omega(\omega, T) = u_f(f, T) \cdot \frac{df}{d\omega} = \frac{1}{2\pi} \cdot u_f(f, T) \, ,$$

$$u_\omega(\omega, T) = \frac{\hbar \omega^3}{\pi^2 c^3} \frac{1}{e^{\hbar\omega/(kT)} - 1} \, ,$$

$$u_\lambda(\lambda, T) = u_f(f, T) \cdot \left| \frac{df}{d\lambda} \right| = \frac{f^2}{c} \cdot u_f(f, T) \, ,$$

$$u_\lambda(\lambda, T) = \frac{8\pi hc}{\lambda^5} \frac{1}{e^{hc/(k\lambda T)} - 1} \, .$$

| M | Planck's radiation law is the basis of optical pyrometry for measuring high temperatures.

5. Wien's displacement law and limiting cases of Planck's formula

▲ **Wien's law** for $hf \gg kT$:

$$u_f(f, T) = \frac{8\pi f^3 h}{c^3} e^{-\frac{hf}{kT}} \, .$$

▲ **Rayleigh-Jeans law** for $hf \ll kT$:

$$u_f(f, T) = \frac{8\pi f^2}{c^3} kT \, .$$

▲ **Wien's displacement law:** With increasing temperature, the maximum of the spectral radiant energy density $u_f(f, T)$ is shifted to higher photon energy, i.e., to higher frequencies (shorter wavelengths) (**Fig. 23.2**):

Wien's displacement law			L
$\lambda_{\max} = \dfrac{b}{T}$	Symbol	Unit	Quantity
	$\lambda_{\max}$	m	wavelength at max. $u_f(f, T)$
$b = 2.8978 \cdot 10^{-3}$ m $\cdot$ K	b	m $\cdot$ K	Wien's constant
	T	K	temperature

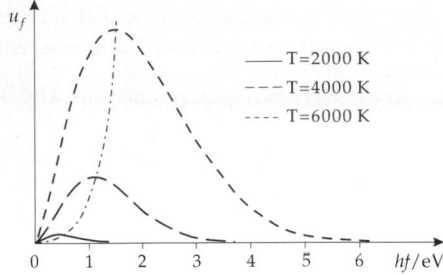

Figure 23.2: Radiant energy density $u_f(f, T)$ for various temperatures according to Planck's radiation law. Dashed-dotted line: Rayleigh-Jeans law.

6. Stefan-Boltzmann law

Integration of the spectral radiant energy density over all frequencies yields the total radiant flux Φ_{tot} of a radiation emitted by an area A. The total radiant flux Φ_{tot} is proportional to the fourth power of the temperature T.

total radiant flux $\sim$ temperature^4			$\mathbf{ML^2T^{-3}}$
	Symbol	Unit	Quantity
$\Phi_{tot} = \sigma \cdot A \cdot T^4$	Φ_{tot}	W	total radiant flux
$\sigma = 5.67051(19)$	A	m^2	area
$\cdot 10^{-8}$ W/(m^{2}K^4)	σ	W/(m^{2}K^4)	Stefan-Boltzmann constant
	T	K	temperature

23.2 Photoelectric effect

Photoeffect, photons eject electrons from a material.

1. Properties of photoelectrons

Photoelectrons, electrons ejected out of a material in the photoeffect.

Photoelectric current, photocurrent, arises if there is an appropriate potential difference between the irradiated body and an anode. The ejected electrons move to the anode.

Photoelectric Einstein equation, describes the kinetic energy E_{kin} of electrons ejected from the body by the incident radiation:

kinetic energy of photoelectrons			$\mathbf{ML^2T^{-2}}$
	Symbol	Unit	Quantity
$E_{kin} = hf - W_A$	E_{kin}	J	kinetic energy
	h	J s	quantum of action
	f	s^{-1}	frequency
	W_A	J	work function

The kinetic energy of the photoelectrons depends on the frequency of the incident radiation, but not on the radiation intensity (**Fig. 23.3**). The radiation intensity determines only the intensity of the photocurrent (**Fig. 23.4**).

2. Work function

Work function, W_A, the minimum energy required for the ejection of an electron from a material. The work function typically amounts to several electron volts (see **Tab. 29.3**).

■   Work function W_A of several elements (in eV): K 2.30, Na 2.75, Hg 4.49, Ge 5.0. For any material there is a threshold frequency for the photoeffect (**red limit**). Below this threshold frequency f_0, no photoeffect occurs (**Fig. 23.3**):

$$f_0 = \frac{W_A}{h} \, .$$

The chemical structure and surface properties determine the work function W_A, and hence the threshold frequency f_0. The photoeffect may be explained only in the framework of the photon model of electromagnetic radiation.

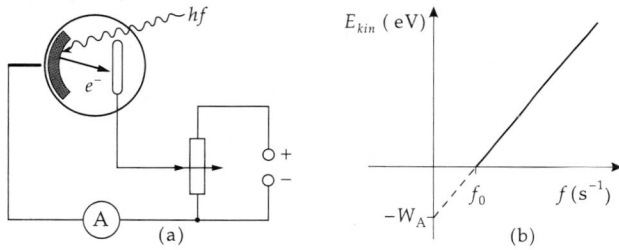

Figure 23.3: Left: experimental set-up for measuring the photoeffect. Right: dependence of the kinetic energy of photoelectrons on the frequency f of the incident radiation.

3. Use of the photoeffect for measurements

| M | When a suppression voltage is applied, the photocurrent vanishes at a threshold voltage V_G, which is related to the maximum velocity v_{max} of the photoelectrons by $eV_G = mv_{max}^2/2$. The quantum of action h can be determined by measuring the incident frequency f and the threshold voltage V_G. The measurement yields a linear relation between the suppression voltage at which the photocurrent vanishes, and the frequency (**Fig. 23.3**). The slope of the straight line yields Planck's constant, or quantum of action, $h = e\,\mathrm{d}V_G/\mathrm{d}f$.

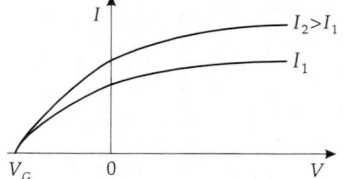

Figure 23.4: Photocurrent I as function of the applied voltage V for different intensities I of the incident radiation.

| M | **Internal photoeffect**, leads to a change of the electric conductance of semiconductors. The phenomenon is used for light-intensity measurement with semiconductor diodes.

23.3   Compton effect

1. Scattering of photons by electrons

Compton effect, a shift of the wavelength (and hence the frequency) in the elastic scattering of photons by free electrons. The shift increases with the scattering angle, but does not depend on the wavelength of the incident radiation (**Fig. 23.5**):

Compton shift of wavelength			L
	Symbol	Unit	Quantity
$\Delta\lambda = \dfrac{h}{m_e c}(1 - \cos\varphi)$	$\Delta\lambda$	m	shift of wavelength
	h	J s	quantum of action
	m_e	kg	electron mass
	c	m s^{-1}	speed of light
	φ	1	scattering angle of photon

2. Conservation laws for scattering

Momentum and energy conservation for the scattering process (relativistically):

$$m_e c^2 + hf = \frac{m_e c^2}{\sqrt{1 - \beta^2}} + hf',$$

$$\hbar\vec{k} = \frac{m_e \vec{v}'_e}{\sqrt{1 - \beta^2}} + \hbar\vec{k}',$$

with $|\vec{k}| = \dfrac{2\pi}{\lambda}$ and $\beta = \dfrac{v}{c}$.

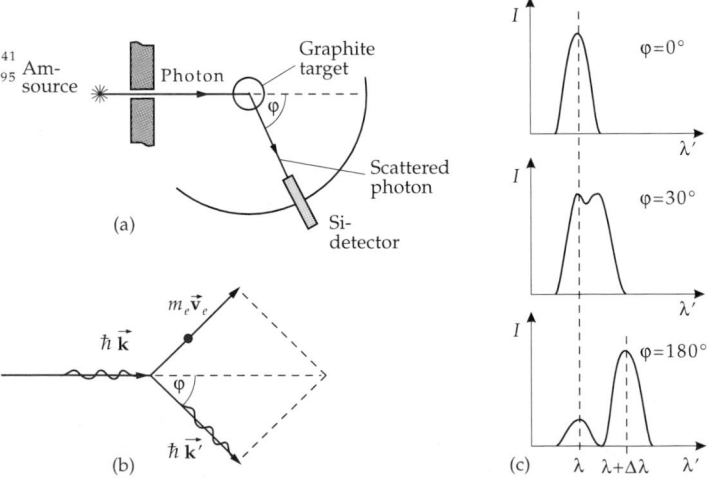

Figure 23.5: Compton effect. (a): experimental set-up, (b): kinematics of photon-electron scattering, (c): intensity I of the scattered radiation as a function of the wavelength λ' of the scattered radiation for various scattering angles φ. λ: wavelength of the incident radiation.

- The electron is at rest before the reaction.
- $\vec{k}$ is the wave number vector pointing along the direction of photon propagation.
- The primed quantities refer to the situation after the collision.
➤ The non-shifted line in the spectrum of the scattered radiation corresponds to photon scattering by strongly bound electrons. The momentum transfer, which in this case is imparted to the atom as a whole, is very low, and therefore the wavelength of the radiation remains almost unchanged in the scattering (Thomson scattering).

3. Compton wavelength

of the electron, λ_C, the proportionality factor in the formula for Compton scattering:

$$\lambda_C = \frac{h}{m_e c} = 2.426\,310\,58(22) \cdot 10^{-12} \text{ m}.$$

Frequently,

$$\lambdabar_C = \frac{\hbar}{m_e c} = 3.861\,593\,23(35) \cdot 10^{-13} \text{ m}$$

is denoted the Compton wavelength.
➤ The Compton effect may occur in photon scattering by any electrically charged particles. Then the corresponding mass of these particles must be inserted in the formula to obtain their Compton wavelength.
■ $\lambda_C^{\text{proton}} = 1.321\,41 \cdot 10^{-15} \text{ m} \approx 1 \text{ fm}$.

4. Radiation pressure

Also called **light pressure**, momentum transfer in the reflection of electromagnetic radiation by a body (change of photon momentum in the reflection). The radiation pressure of sunlight on a mirror is of the order of magnitude of 10^{-11} bar, which is unobservably small. Since the radiation pressure for small particles may reach the magnitude of the gravitational attraction, it affects astrophysical processes. For example, the tail of a comet always points away from the Sun as a result of radiation pressure.
➤ During a period of 28 months, the spherical satellite Vanguard 1 (diameter 16 cm) was displaced from its orbit by 1600 m due to radiation pressure.
➤ Intense laser light may reach intensities of 10^{18} W/cm^2. Such radiation may generate a pressure of ca. 100 Mbar on the outer surface of a plasma, which may lead to plasma compression. In this way, access to new ranges of pressure and temperature in plasma physics are obtained.

24

Matter waves—wave mechanics of particles

Quantum mechanics, the theory of the laws of motion in the atomic range (spatial extension $< 10^{-8}$ m).

➤ For particle velocities $v \approx c$, with c being the speed of light in a vacuum, relativistic quantum mechanics must be used for describing the phenomena.

24.1 Wave character of particles

24.1.1.1 Basic assumptions of quantum mechanics

Quantum mechanics is based on two hypotheses:

1. Planck's quantum hypothesis

In emission and absorption of electromagnetic radiation by atoms, the energy may be exchanged only in definite amounts (**quanta**).

energy of photon			**ML^{2}T^{-2}**
	Symbol	Unit	Quantity
$E = \hbar \cdot \omega, \quad \omega = 2\pi f,$	E	J	energy
	ω	rad s^{-1}	angular frequency
$\hbar = \dfrac{h}{2\pi}$	f	s^{-1}	frequency
	h	J s	quantum of action

In atomic physics, the quantity **electron volt** is frequently used as an energy unit. One electron volt corresponds to an energy of $1.602\,177\,33(49) \cdot 10^{-19}$ J.

Wave-number vector, $\vec{\mathbf{k}}$, a vector along the propagation direction of the electromagnetic wave, with magnitude $|\vec{\mathbf{k}}| = \frac{2\pi}{\lambda}$. λ is the wavelength of the electromagnetic radiation.

Photon momentum, proportional to the wave-number vector $\vec{\mathbf{k}}$:

momentum of photon			MLT^{-1}
$\vec{p} = \hbar \cdot \vec{k}, \quad k = \dfrac{2\pi}{\lambda},$	Symbol	Unit	Quantity
	$\vec{k}$	m^{-1}	wave-number vector
$\hbar = \dfrac{h}{2\pi}$	h	J s	quantum of action
	$\vec{p}$	kg m/s	momentum vector

2. Matter waves

A **de Broglie wavelength** may be assigned to any free particle; it is inversely proportional to the particle momentum.

de Broglie wavelength			L
	Symbol	Unit	Quantity
$\lambda = \dfrac{h}{p}$	λ	m	wavelength
	h	J s	quantum of action
	p	kg m/s	momentum

■ After being accelerated by a voltage V, an electron with

$$m = 9.109\,389\,7(54) \cdot 10^{-31} \text{ kg} \quad \text{(electron mass)}$$

$$e = -1.602\,177\,33(49) \cdot 10^{-19} \text{ C} \quad \text{(electron charge)}$$

has the wavelength

$$\lambda = \frac{h}{\sqrt{2m|e|V}} = \sqrt{\frac{150.5}{V}} \cdot 10^{-10} \text{ m} \quad (V \text{ in volts}).$$

■ de Broglie wavelength (in m): electron (1 eV) $1.23 \cdot 10^{-9}$, electron (10^2 eV) $0.12 \cdot 10^{-9}$, α particle (10^2 eV) $1.4 \cdot 10^{-12}$, thermal neutrons (0.025 eV) $0.18 \cdot 10^{-9}$, golf ball ($v = 25$ m/s) $5.8 \cdot 10^{-34}$.

24.1.1.2   Wave-particle duality

Wave-particle duality, the property of atomic particles (photons, electrons, nucleons, atoms, molecules) to behave either as particles with defined values of energy and momentum (in emission and absorption processes or collisions), or as a wave (in propagation, diffraction and interference).

| M | **Electron diffraction**, coherent diffraction of electron beams by periodic structures so that an interference pattern arises behind the sample. This is a demonstration of the wave property of electrons.

Electron diffraction is used for investigating the structure of surfaces or thin layers (for the principle of measurement see **Fig. 24.1**).

Electron microscope (E. Ruska, Nobel Prize, 1986), exploits the short wavelength of accelerated electrons. The resolving power is better by a factor of 1000 than that of the light microscope.

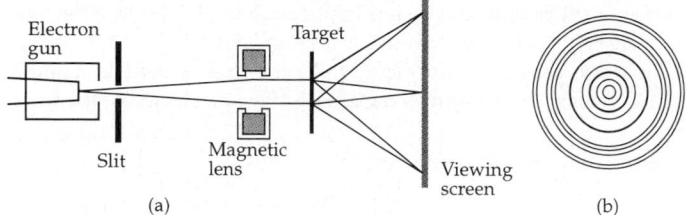

Figure 24.1: Electron diffraction. (a): basic experimental set-up, (b): interference pattern.

24.2 Heisenberg's uncertainty principle

The notion of a particle trajectory, i.e., the specification of the particle coordinates as a function of time, loses its meaning in quantum mechanics. It is no longer possible to assign simultaneously a defined spatial position and an exact momentum value to the particle. A plane wave with a defined wave number $\vec{k}$ corresponds to a free particle with constant momentum $\vec{p}$ and is infinitely extended: the particle is not localized in space.

Heisenberg's uncertainty principle, connects the uncertainty Δp_x in the determination of the component p_x of the momentum with the uncertainty Δx in a simultaneous determination of the position coordinate x.

uncertainty in position · uncertainty in momentum $\geq$ Planck's constant/2			$\mathbf{ML^2T^{-1}}$
	Symbol	Unit	Quantity
$\Delta x \cdot \Delta p_x \geq \dfrac{\hbar}{2}$	Δx	m	uncertainty in position
	Δp_x	kg m/s	uncertainty in momentum
	$\hbar (= h/(2\pi))$	J s	quantum of action

For objects on the atomic scale, the **measuring process** is inevitably connected with some influence on the quantity to be measured. Any reduction of the fluctuation of the measured values of the particle position increases the fluctuation of the measured momentum values. This is not due to inaccuracy of the methods of measurement adopted, but is instead a basic principle of nature.

➤   The momentum component p_y and the position coordinate x can be measured simultaneously without fluctuation.

The uncertainty principle holds also for other canonically conjugated quantities, the products of which have the dimension of action. For the angle ϕ and the angular momentum l, $\Delta\phi \cdot \Delta l \geq \hbar/2$. For the energy E and the time t, there also exists an uncertainty principle, $\Delta E \cdot \Delta t \geq \hbar/2$.

24.3 Wave function and observable

1. Wave function and probability of finding a particle

Wave function, $\psi(x, y, z, t)$, a complex function describing the state of a particle quantum-mechanically completely. It serves as a mathematical tool and cannot be determined experimentally.

▲ The wave function contains the full information on the results of measurements of physical quantities of a quantum-mechanical system.

Probability density: The probability $dw(x, y, z, t)$ of finding a particle at time t at position $\vec{r} = (x, y, z)$ in volume dV is given by the square of the magnitude of the wave function:

probability density = |wave function|^2				1
	Symbol	Unit	Quantity	
$dw(x, y, z, t) = \|\psi(x, y, z, t)\|^2 \, dV$	w	1	probability density	
	ψ	m$^{-3/2}$	wave function	
	dV	m^3	volume element	

▲ The wave function has the meaning of a probability amplitude.

Normalization of the wave function, the integration of the probability density over the entire space must yield unity, since the probability of finding the particle somewhere must be unity,

$$\int |\psi(x, y, z, t)|^2 \, dV = 1 \, .$$

▲ The wave function must be normalizable.

2. Wave function of a free particle

Free particles, described by plane waves:

wave function of free particles				L$^{-3/2}$
	Symbol	Unit	Quantity	
	a	m$^{-3/2}$	amplitude	
	j	1	imaginary unit	
$\psi(\vec{r}, t) = a \cdot e^{j[(\vec{k}\cdot\vec{r})-\omega t]}$	ω	rad s^{-1}	angular frequency	
	t	s	time	
$= a \cdot e^{\frac{j}{\hbar}[(\vec{p}\cdot\vec{r})-E(p)t]}$	$\vec{k}$	m^{-1}	wave-number vector	
	$\vec{r}$	m	position vector	
	$\psi(\vec{r}, t)$	m$^{-3/2}$	wave function	
	$\vec{p}$	kg m/s	momentum	
	$E(p)$	F	energy	

3. Wave packet,

the superposition of many plane waves of neighboring frequencies. For one-dimensional motion along the x-direction, a wave packet has the form (**Fig. 24.2**):

$$\psi(x, t) = \frac{1}{2\pi} \int\limits_{-\infty}^{+\infty} f(k) \, e^{j[k\cdot x-\omega(k)t]} \, dk \, .$$

The amplitude function (spectral function) $f(k)$ determines the weight distribution of the plane waves of various frequencies. In most cases, the probability of finding the particle differs from zero only in a limited space region: the particle is localized. In this case,

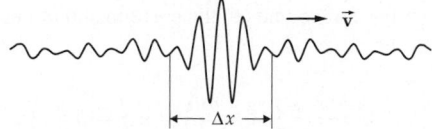

Figure 24.2: Schematic
illustration of a wave packet.

the appearance of many distinct frequencies in the wave function also means a large momentum width: a reduction of the fluctuation of the measured position values increases the uncertainty of the momentum.

The amplitude function $f(k)$ also determines the uncertainty in position and momentum at the initial time. Over the course of time, the center of gravity of the wave packet moves with a mean velocity given by $f(k)$. The momentum uncertainty Δp is conserved while the uncertainty in position Δx increases: the wave packet blurs (**Fig. 24.3**).

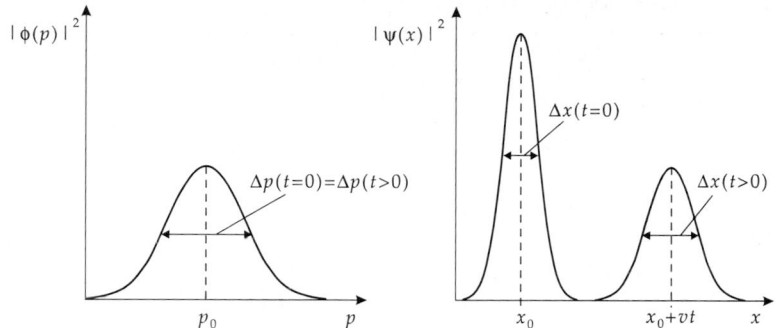

Figure 24.3: Blurring of a wave packet. $|\phi(p,t)|^2$, $|\psi(x,t)|^2$: probability density for momentum p and position x; p_0, x_0: mean values of momentum and position at time $t = 0$; v: mean velocity (group velocity); $\Delta p(t)$, $\Delta x(t)$: uncertainty in momentum and position at time t.

4. Observable,

O, a physical quantity that may be observed, i.e., may be defined by a prescription for the measuring procedure.

■   Energy, position, momentum, orbital angular momentum, spin.

In quantum mechanics, one assigns an operator $\hat{O}$ to any observable O operating on the wave function ψ.

▲   Time is not an operator in quantum mechanics; it is a parameter of the wave function. When transforming quantities to quantum mechanics, the operators are constructed according to the structure of the classical quantities.

■   In quantum mechanics, the Cartesian components of the classical orbital angular momentum $\vec{l} = \vec{r} \times \vec{p}$:

$$l_x = yp_z - zp_y, \quad l_y = zp_x - xp_z, \quad l_z = xp_y - yp_x$$

are replaced by the operators $\hat{l}_x, \hat{l}_y, \hat{l}_z$ constructed in the same manner from the components of the position operator $\hat{\vec{r}}$ and the momentum operator $\hat{\vec{p}}$:

$$\hat{l}_x = \hat{y}\hat{p}_z - \hat{z}\hat{p}_y, \quad \hat{l}_y = \hat{z}\hat{p}_x - \hat{x}\hat{p}_z, \quad \hat{l}_z = \hat{x}\hat{p}_y - \hat{y}\hat{p}_x .$$

Inserting the Cartesian representation for the components of the position and momentum operators, one obtains:

$$\hat{l}_x = y\left(-j\hbar\frac{\partial}{\partial z}\right) - z\left(-j\hbar\frac{\partial}{\partial y}\right), \qquad \hat{l}_y = z\left(-j\hbar\frac{\partial}{\partial x}\right) - x\left(-j\hbar\frac{\partial}{\partial z}\right),$$

$$\hat{l}_z = x\left(-j\hbar\frac{\partial}{\partial y}\right) - y\left(-j\hbar\frac{\partial}{\partial x}\right).$$

The operator of orbital angular momentum is a vector operator, with

$$\hat{\mathbf{l}} = (\hat{l}_x, \hat{l}_y, \hat{l}_z), \quad \hat{\mathbf{l}}^2 = \hat{l}_x^2 + \hat{l}_y^2 + \hat{l}_z^2.$$

5. Survey of important observables

Physical quantity	Symbol	Operator
position component i	$\hat{x}_i$	$x_i, \quad i = 1, 2, 3$
momentum component i	$\hat{p}_{x_i}$	$-j\cdot\hbar\dfrac{\partial}{\partial x_i}, \quad i = 1, 2, 3$
orbital angular momentum components:		
$\quad$ x-direction	$\hat{l}_x$	$j\cdot\hbar\left(\sin\varphi\cdot\dfrac{\partial}{\partial\vartheta} + \cot\vartheta\cdot\cos\varphi\cdot\dfrac{\partial}{\partial\varphi}\right)$
$\quad$ y-direction	$\hat{l}_y$	$-j\cdot\hbar\left(\cos\varphi\cdot\dfrac{\partial}{\partial\vartheta} - \cot\vartheta\cdot\sin\varphi\cdot\dfrac{\partial}{\partial\varphi}\right)$
$\quad$ z-direction	$\hat{l}_z$	$-j\cdot\hbar\dfrac{\partial}{\partial\varphi}$
square of orbital angular momentum	$\hat{l}^2$	$-\hbar^2\,\Delta_{\vartheta,\varphi}$
energy	$\hat{H}$	$-\dfrac{\hbar^2}{2m}\Delta + V$

The coordinates in a Cartesian frame are designated by the indices $i = 1, 2, 3$. The components of the orbital angular momentum operator are given in spherical coordinates.
Angular component of the Laplace operator Δ in spherical coordinates:

$$\Delta_{\vartheta,\varphi} = \frac{1}{r^2\sin\vartheta}\frac{\partial}{\partial\vartheta}\left(\sin\vartheta\frac{\partial}{\partial\vartheta}\right) + \frac{1}{r^2\sin^2\vartheta}\frac{\partial^2}{\partial\varphi^2}.$$

■ Applying the position operator $\hat{x}$ on the wave function ψ means multiplication of the wave function by the position coordinate x. Applying the momentum operator $\hat{p}_x$ on the wave function ψ means partial derivation of the wave function with respect to x and multiplication by the number $-j\hbar$.

6. Eigenfunction,

ψ_n of the operator $\hat{O}$: application of the operator $\hat{O}$ on the function ψ_n reproduces the function multiplied by the **eigenvalue** a_n; the index n labels the various eigenfunctions and the related eigenvalues,

$$\hat{O}\,\psi_n = a_n\,\psi_n, \quad n = 1, 2, 3, \ldots.$$

■ The one-dimensional motion of a free particle of momentum p along the x-direction is described by a plane wave. The spatial part of the wave function is given by

$$\varphi(x) = e^{jkx} = e^{\frac{j}{\hbar} px}.$$

Application of the momentum operator $\hat{p}_x$ on the wave function φ yields

$$\hat{p}_x \, \varphi(x) = \frac{\hbar}{j} \frac{\partial}{\partial x} \varphi(x) = p \, \varphi(x).$$

A plane wave is an eigenfunction of the momentum operator with the eigenvalue p.

➤ A plane wave is also an eigenfunction of the energy operator (Hamiltonian) $\hat{H} = \hat{p}^2/(2m)$ with the eigenvalue $E = p^2/(2m)$.

Degeneracy, there are several eigenfunctions $\psi_{n1}, \psi_{n2}, \dots$ with a given eigenvalue a_n:

$$\hat{O}\,\psi_{n1} = a_n\,\psi_{n1},\dots, \quad \hat{O}\,\psi_{nN} = a_n\,\psi_{nN}, \quad N\text{-fold degeneracy}.$$

Parity, π, of a wave function, characterizes the behavior of the wave function $\psi(\vec{r})$ under reflection at the coordinate origin, $\vec{r} \longrightarrow -\vec{r}$ (**Fig. 24.4**),

$$\psi(-\vec{r}) = +\psi(\vec{r}), \quad \pi = +1, \quad \text{even parity,}$$
$$\psi(-\vec{r}) = -\psi(\vec{r}), \quad \pi = -1, \quad \text{odd parity.}$$

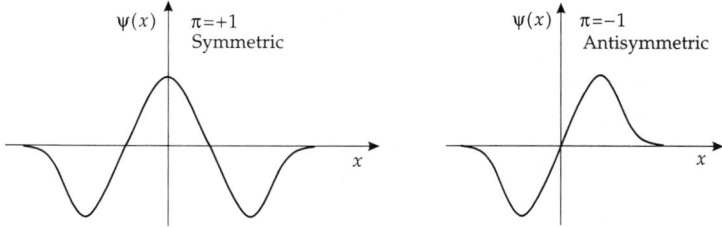

Figure 24.4: Parity of a wave function $\psi(x)$. $\pi = +1$: even parity, symmetric function, $\pi = -1$: odd parity, antisymmetric function.

7. Simultaneous eigenfunction,

a function ψ is a simultaneous eigenfunction of a set of operators $\hat{O}_1, \dots, \hat{O}_k$,

$$\hat{O}_1\,\psi = a_1\,\psi, \dots, \quad \hat{O}_k\,\psi = a_k\,\psi.$$

■ Simultaneous eigenfunctions of the operators $\hat{\mathbf{l}}^2, \hat{l}_z$ are the spherical harmonics (spherical surface harmonics) $Y_l^m(\vartheta, \varphi)$:

$$\hat{\mathbf{l}}^2\, Y_l^m(\vartheta, \varphi) = \hbar^2\, l(l+1)\, Y_l^m(\vartheta, \varphi),$$
$$\hat{l}_z\, Y_l^m(\vartheta, \varphi) = \hbar\, m\, Y_l^m(\vartheta, \varphi).$$

Possible quantum numbers of the orbital angular momentum are $l = 0, 1, 2, \dots$. In the illustrative vector model, they specify the magnitude of the orbital angular momentum vector, $|\vec{\mathbf{l}}| = \hbar\sqrt{l(l+1)}$. For a given value of l, there are $2l + 1$ values of the **magnetic quantum number** m that specify the possible orientations (projections) of the orbital angular momentum vector with respect to the z-axis as quantization axis (**directional quantization**), $m = -l, -l+1, \ldots, 0, \ldots l - 1, l$ (**Fig. 24.5**). The angle α between the quantization axis and the angular momentum vector satisfies $\cos \alpha = m/\sqrt{l(l+1)}$.

➤   There is no function that would be a simultaneous eigenfunction of the operator of an additional orbital angular momentum component l_x or l_y. In an angular momentum state characterized by the quantum numbers l, m these orbital angular momentum components do not take fixed values; their expectation values vanish.

■   Angular momentum quantum numbers: $l = 2$, $m = -2, -1, 0, 1, 2$.

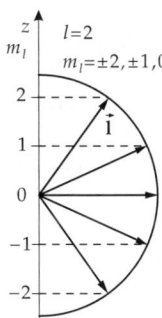

Figure 24.5: Vector model of the orbital angular momentum $\vec{\mathbf{l}}$. Directional quantization for $l = 2$.

8. Eigenvalues of operators and their meaning

▲   The eigenvalues of an operator representing an observable in quantum mechanics are real,

$$a_n^* = a_n \, .$$

▲   The eigenvalues of an operator $\hat{O}$ are the possible measured values of the observable O. After a measurement of an observable O that gave the value a_n, let the system be in the eigenstate ψ_n:

$$\text{state } \psi \quad \xrightarrow{\text{measurement } O} \quad \text{measured value } a_n, \text{ state } \psi_n \, .$$

▲   Any wave function ψ may be expanded in terms of the complete set of normalized eigenfunctions ψ_n of the operator $\hat{O}$,

$$\psi = \sum_n c_n \, \psi_n \, .$$

The wave function ψ is normalized if

$$\sum_n |c_n|^2 = 1 \, .$$

The expansion coefficient c_n yields the probability $|c_n|^2$ of obtaining the value a_n when measuring the observable O of a system in the state ψ.

Repeated measurements of the observable O of a system in the eigenstate ψ_n always lead to the same measured value a_n, without any fluctuation of the measured values of the individual measurements. If the observable O is repeatedly measured for a system which is in an arbitrary state ψ that is not an eigenstate of $\hat{O}$, the measured values fluctuate about the expectation value.

9. Expectation values of observables

Expectation value $\overline{O}$ of the observable O in the state ψ, the mean value of the measured values of the observable O measured on a system in the state ψ,

$$\overline{O} = \int \psi^* \, \hat{O} \, \psi \, dV = \sum_n |c_n|^2 \, a_n \, .$$

The expectation value is in general time-dependent.

- The possible measured values of the position of a particle moving in x-direction cover the interval $[-\infty, +\infty]$, i.e., the position operator $\hat{x}$ has a continuous spectrum of eigenvalues. If the particle is in the state ψ, the weight function for averaging the possible measured values to get the expectation value is given by the probability density of finding the particle at the position x in the element dx,

$$dw(x, t) = |\psi(x, t)|^2 \, dx \, .$$

The expectation value of the position is

$$\bar{x} = \int_{-\infty}^{+\infty} x \, dw(x, t) = \int_{-\infty}^{+\infty} x |\psi(x, t)|^2 \, dx = \int_{-\infty}^{+\infty} \psi(x, t)^* \, x \, \psi(x, t) \, dx \, .$$

- The expectation value of the momentum component p_x in the state ψ is

$$\overline{p_x} = \int_{-\infty}^{+\infty} \psi^* \cdot \frac{\hbar}{j} \frac{d}{dx} \psi \, .$$

10. Matrix representation of operators

Matrix representation of the operator $\hat{O}$ in the basis of the functions φ_i, $i = 1, \ldots, N$:

$$O_{ik} = \int \varphi_i^* \, \hat{O} \, \varphi_k \, dV \, , \quad i, k = 1, \ldots, N \, .$$

▲ Observables are represented in quantum mechanics by **Hermitean matrices**:

$$O_{ik}^* = O_{ki} \, .$$

The quadratic matrix becomes diagonal if the orthonormalized eigenfunctions ψ_n are used as a basis:

$$O_{nm} = \int \psi_n^* \, \hat{O} \, \psi_m \, dV = a_m \int \psi_n^* \, \psi_m \, dV = a_m \, \delta_{nm} \, .$$

The diagonal elements are the eigenvalues a_m, i.e., the possible measured values.

If two observables O_1, O_2, the operators of which have eigenfunctions $\psi_n^{(1)}$, $\psi_m^{(2)}$, are measured successively, the state generated by the first measurement is in general disturbed by the second measurement:

$$\text{state } \psi \quad \xrightarrow{\text{measurement } O_1} \quad a_n, \psi_n^{(1)} \quad \xrightarrow{\text{measurement } O_2} \quad b_m, \psi_m^{(2)} \, .$$

11. Commutators of operators

Commutator, $\hat{C}$, of the operators $\hat{O}_1$ and $\hat{O}_2$, an operator defined by

$$\hat{C} = [\hat{O}_1, \hat{O}_2] = \hat{O}_1\,\hat{O}_2 - \hat{O}_2\,\hat{O}_1\,.$$

Two operators are said to commute if their commutator vanishes,

$$\hat{C} = [\hat{O}_1, \hat{O}_2] = 0\,.$$

Then,

$$\hat{O}_1\,(\hat{O}_2\,\psi) = \hat{O}_2\,(\hat{O}_1\,\psi)\,.$$

Commuting operators $\hat{O}_1$, $\hat{O}_2$ have a simultaneous system of eigenfunctions ψ_{nm}, with the eigenvalues a_n, b_m,

$$\hat{O}_1\,\psi_{nm} = a_n\,\psi_{nm}\,, \qquad \hat{O}_2\,\psi_{nm} = b_m\,\psi_{nm}\,.$$

▲   Commuting operators represent compatible measurements:

$$\text{state } \psi \xrightarrow{\text{measurement } O_1} \{a_n, \psi_{nm}\} \xrightarrow{\text{measurement } O_2} \{b_n, \psi_{nm}\}\,.$$

The state n generated by the measurement of O_1 is not disturbed by the measurement of O_2. The second measurement merely specifies the state m.

a) Commutation relations for position and momentum operators: relations between the products of position and momentum operators ($i = 1, 2, 3$):

$$[\hat{x}_i, \hat{p}_k] = \hat{x}_i \cdot \hat{p}_k - \hat{p}_k \cdot \hat{x}_i = j \cdot \hbar \cdot \delta_{ik} \quad \text{with} \quad \delta_{ik} = \begin{cases} 1: & k = i \\ 0: & k \neq i \end{cases}$$

These commutation relations establish the validity of Heisenberg's uncertainty relation for position and momentum (see p. 827).

b) Commutation relations for orbital angular momentum operators:

$$[\hat{l}_x, \hat{l}_y] = \hat{l}_x \cdot \hat{l}_y - \hat{l}_y \cdot \hat{l}_x = j\hbar\hat{l}_z\,,$$

$$[\hat{l}_y, \hat{l}_z] = \hat{l}_y \cdot \hat{l}_z - \hat{l}_z \cdot \hat{l}_y = j\hbar\hat{l}_x\,,$$

$$[\hat{l}_z, \hat{l}_x] = \hat{l}_z \cdot \hat{l}_x - \hat{l}_x \cdot \hat{l}_z = j\hbar\hat{l}_y\,.$$

The operator of the square of the orbital angular momentum commutes with the operators of all components of the orbital angular momentum,

$$[\hat{\mathbf{l}}^2, \hat{l}_x] = [\hat{\mathbf{l}}^2, \hat{l}_y] = [\hat{\mathbf{l}}^2, \hat{l}_z] = 0\,.$$

➤   Any set of operators whose components satisfy commutation relations of this kind represents an angular momentum.

12. Hamiltonian and time evolution

Hamiltonian, $\hat{H}$, operator of the total energy of a quantum-mechanical system. The Hamiltonian determines the time evolution of the state function ψ.

- Free particle of mass m: $\hat{H} = \dfrac{\hat{p}^2}{2m}$.

  Particle of mass m in a potential V: $\hat{H} = \dfrac{\hat{p}^2}{2m} + V(\hat{\mathbf{r}})$.

  Particle of mass m in a one-dimensional oscillator potential: $\hat{H} = \dfrac{\hat{p}_x^2}{2m} + \dfrac{m}{2}\omega^2 \hat{x}^2$.

  Electron in hydrogen atom: $\hat{H} = \dfrac{\hat{p}^2}{2m} - \dfrac{e^2}{r}$.

Time evolution operator, $\hat{U}(t, t_0)$, describes the time evolution of a state ψ from the time t_0 to the time t,

$$\psi(t) = \hat{U}(t, t_0)\,\psi(t_0)\,, \quad \hat{U}(t_0, t_0) = 1\,, \quad \hat{U}(t, t_0) = e^{-\frac{j}{\hbar} H(t - t_0)}\,.$$

13. Schrödinger and Heisenberg pictures

Schrödinger picture, formulation of quantum mechanics with time-independent operators $\hat{O}^S$ and time-dependent states ψ^S,

$$\frac{\partial \hat{O}^S}{\partial t} = 0\,, \quad \frac{\partial \psi^S(t)}{\partial t} = -\frac{j}{\hbar} H \psi^S(t) \quad \textbf{Schrödinger equation.}$$

Heisenberg picture, formulation of quantum mechanics with time-dependent operators $\hat{O}^H$ and time-independent states ψ^H,

$$\frac{\partial \psi^H}{\partial t} = 0\,, \quad \frac{d\hat{O}^H(t)}{dt} = +\frac{j}{\hbar}[H, \hat{O}^H(t)] \quad \textbf{Heisenberg equation.}$$

Connection between the two representations: the quantities coincide at time $t = t_0$,

$$\psi^S(t) = \hat{U}(t, t_0)\,\psi^H\,, \quad \hat{O}^H(t) = \hat{U}^\dagger(t, t_0)\,\hat{O}^S\,\hat{U}(t, t_0)\,.$$

- ▲ Schrödinger picture and Heisenberg picture are equivalent formulations of quantum mechanics. They provide identical physical statements (expectation values of observables).

24.4   Schrödinger equation

▲ Electromagnetic waves in a vacuum (speed of light c) and matter waves for free particles obey different **dispersion relations** $\omega = \omega(k)$.

Electromagnetic waves: $\omega(k) = c \cdot k$, matter waves: $\omega(k) = \dfrac{m_0 c^2}{\hbar} + \dfrac{\hbar k^2}{2m_0}$.

The various dispersion relations correspond to different differential equations for wave propagation.

1. Differential equation for the wave function (Schrödinger equation)

Schrödinger equation, a differential equation for wave functions governing the behavior of atomic particles in the nonrelativistic limit, similar to Newton's equation of motion that determines the motion of a classical point mass. The Schrödinger equation is a linear and homogeneous partial differential equation, of first order in the time, and of second order in the space, variable. The solutions of the Schrödinger equation are complex functions.

The time-dependent Schrödinger equation for a particle of mass m in a potential $V(\vec{r})$ reads:

time-dependent Schrödinger equation			$\mathbf{ML^{1/2}T^{-2}}$
$-\dfrac{\hbar}{j}\dfrac{\partial\psi(\vec{r},t)}{\partial t} = \hat{H}\,\psi(\vec{r},t)$ $\hat{H} = \dfrac{\hat{p}^2}{2m} + V(\hat{\vec{r}})$ $-\dfrac{\hbar}{j}\dfrac{\partial\psi(\vec{r},t)}{\partial t} = -\dfrac{\hbar^2}{2m}\Delta\psi(\vec{r},t) + V(\vec{r})\psi(\vec{r},t)$ $\Delta = \dfrac{\partial^2}{\partial x^2} + \dfrac{\partial^2}{\partial y^2} + \dfrac{\partial^2}{\partial z^2}$	Symbol	Unit	Quantity
	ψ	$m^{-3/2}$	wave function
	j	1	imaginary unit
	m	kg	particle mass
	Δ	m^{-2}	Laplace operator
	$V(\vec{r})$	J	potential
	$\hbar$	J s	quantum of action

▲ The Hamiltonian of a quantum-mechanical system determines the time evolution of this system.

➤ The time evolution of a state according to the time-dependent Schrödinger equation has to be distinguished from the changes of the state caused by the interference of a measuring apparatus. After measuring an observable $\hat{O}$ resulting in the measured value a_n, the system is in the eigenstate ψ_n.

■ A free particle is represented by a plane wave. Inserting this wave function into the Schrödinger equation, differentiation with respect to time yields the factor hf, and application of the Laplace operator yields $\dfrac{h^2k^2}{8\pi^2m}$. The common factor $ae^{j[2\pi ft-(\vec{k}\cdot\vec{r})]}$ in the Schrödinger equation cancels out:

$$hf = \frac{h^2k^2}{8\pi^2m} + V(r),$$

$$= \frac{p^2}{2m} + V(r).$$

This is the law of energy conservation, with hf being the energy of a quantum of frequency f.

2. Normalization of the wave function,

corresponds to the requirement that the probability of finding the particle *anywhere* be equal to unity for any point in time t:

$\int |\text{wave function}|^2 \cdot \text{volume element} \equiv 1$			1
$\displaystyle\int_V dw(x,y,z,t) = \int_V |\psi(x,y,z,t)|^2\,dV$ $\equiv 1$	Symbol	Unit	Quantity
	ψ	$m^{-3/2}$	wave function
	dV	m^3	volume element

▲   A solution to the Schrödinger equation can only then be interpreted as a probability amplitude if the function is normalizable.

➤   A plane wave is not normalizable. The normalized wave function of a free particle is a wave packet.

3. Stationary states

Stationary state, a state for which the probability density of finding the particle is time-independent. The wave function of a stationary state is given by

$$\psi(\vec{r}, t) = e^{\frac{j}{\hbar} Et} \cdot \varphi(\vec{r}), \quad \hat{H}\varphi(\vec{r}) = E\varphi(\vec{r}), \quad |\psi(\vec{r}, t)|^2 = |\varphi(\vec{r})|^2.$$

Stationary Schrödinger equation, equation of motion of a particle with a spatial probability density that does not depend on time:

$$\hat{H}\varphi = E\varphi, \quad \frac{h^2}{8\pi^2 m}\Delta\varphi + (E - V(\vec{r}))\varphi = 0.$$

The normalization condition for $\psi(\vec{r}, t)$ requires:

$$\int_0^\infty |\varphi(\vec{r})|^2 \, dV = 1.$$

Energy eigenfunctions, the solutions to the stationary (time-independent) Schrödinger equation. These solutions exist only for certain **eigenvalues** of the energy E.

   Energy eigenvalues, energies for which solutions to the stationary Schrödinger equation exist.

   Energy spectrum of the particle (or particle system), the set of all eigenvalues E.

   If the potential $V(r)$ is a monotonously rising function and if $\lim_{r\to\infty} V(r) = 0$, then the energy eigenvalues form a **discrete** spectrum in the range $E < 0$. For $E \geq 0$, the energy eigenvalues form a **continuum**.

24.4.1 Piecewise constant potentials

Piecewise constant potential, one-dimensional potential of constant value, interrupted by **finite** potential steps.

   General formulation for the solutions of the time-independent Schrödinger equation for a particle of mass m and energy E in a constant potential V:

$V = 0$ $\qquad \varphi(x) = A \cdot e^{\pm jk_1 x}, \quad k_1 = \sqrt{\frac{2m}{\hbar^2} E} = \frac{p_1}{m}.$

Plane wave propagating to the left or right,
wave number k_1, particle momentum p_1.

$V = V_0 > 0 \qquad E > V_0$:

$$\varphi(x) = A \cdot e^{+jk_2 x} + B \cdot e^{-jk_2 x}, \quad k_2 = \sqrt{\frac{2m}{\hbar^2}(E - V_0)} = \frac{p_2}{m}.$$

Plane wave propagating with the amplitude A to the right
and with amplitude B to the left, wave number k_2,
particle momentum p_2.
$E < V_0$:

$$\varphi(x) = A \cdot e^{+k_3 x} + B \cdot e^{-k_3 x}, \quad k_3 = \sqrt{\frac{2m}{\hbar^2}(V_0 - E)}.$$

Increasing or decreasing exponential function, classically
forbidden motion.

➤   For reasons of normalization, wave functions that do not vanish asymptotically ($x \to \pm\infty$) are excluded.

In general, the particle is reflected with some probability and transmitted with some other probability by a potential step, even if the total energy is larger than the step in the potential.

Transmission coefficient, T, the ratio of transmitted particle flux to incident particle flux.

Reflection coefficient, R, the ratio of reflected particle flux to incident particle flux. Since the total particle number is conserved, $R = 1 - T$.

1. Potential step

Potential formulation (**Fig. 24.6**):

$$V(x) = \begin{cases} 0 & \text{for} \quad x < 0, \\ V_0 > 0 & \text{for} \quad x \geq 0. \end{cases}$$

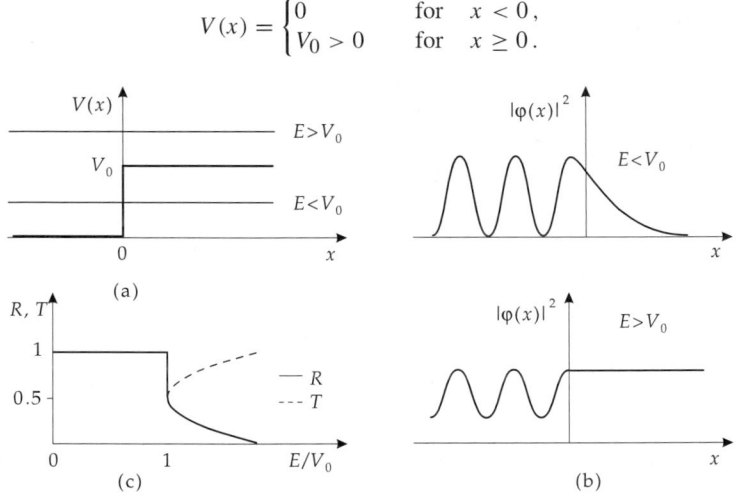

Figure 24.6: Potential step. (a): potential, (b): probability density $|\varphi(x)|^2$ for $E < V_0$ and $E > V_0$, (c): reflection coefficient R and transmission coefficient T as a function of the ratio E/V_0.

Total energy $E < V_0$: $R = 1, \ T = 0$.

Total energy $E > V_0$: $R = \left(\dfrac{k_1 - k_2}{k_1 + k_2}\right)^2$, $T = \dfrac{4k_1 k_2}{(k_1 + k_2)^2}$.

➤   According to classical mechanics, there is no particle motion with $E < V_0$ and $x > 0$ because the kinetic energy then would have to be negative. In quantum mechanics, however, the probability density of the particle in this range differs from zero, since its localization at the classical turning point $x = 0$ causes a momentum uncertainty that leads to energies above the potential step. According to the uncertainty principle, for energy and time this energy uncertainty ΔE may be maintained only for a finite time interval Δt, hence a particle incident from the left may not be observed at $x \to +\infty$. The probability density decreases exponentially in the classically forbidden range, i.e., the particle is reflected.

2. Potential barrier

Potential formulation:

$$V(x) = \begin{cases} 0 & \text{for} \quad |x| > a, \\ V_0 > 0 & \text{for} \quad |x| \leq a. \end{cases}$$

$$E < V_0: T = \left(1 + \frac{V_0{}^2}{V_0{}^2 - (2E - V_0)^2} \sinh^2 (2ak_3)\right)^{-1}.$$

$$E > V_0: T = \left(1 + \frac{V_0{}^2}{V_0{}^2 - (2E - V_0)^2} \sin^2 (2ak_2)\right)^{-1}.$$

Reflection coefficient: $R = 1 - T$.

Approximation for $E < V_0$ and $2a \cdot k_3 \gg 1$:

$$T \approx \left(\frac{2k_1 k_3}{k_1^2 + k_3^2}\right)^2 e^{-4ak_3} \approx e^{-4ak_3}, \quad a \cdot k_3 \gg 1.$$

➤ A potential step may be viewed as a potential barrier of infinite width. In this case, no tunnel effect (see below) may occur ($T = 0$, $R = 1$).

▲ For $E < V_0$ the transmission coefficient increases monotonously with increasing incident energy E; the reflection coefficient correspondingly decreases. For fixed energy $E < V_0$, the transmission coefficient increases with decreasing width $2a$ of the potential barrier.

▲ For $E > V_0$ there is no reflection by the potential barrier ($R = 0$, $T = 1$) if the energy E coincides with a **resonance energy**, given by:

$$2ak_2 = n\pi, \quad n = 1, 2, \ldots.$$

■ In α-decay of heavy atomic nuclei, α-particles are emitted with kinetic energies that are far below the maximum value of the potential barrier that arises as sum of a repulsive Coulomb potential and an attractive nuclear potential. For ^{212}Po, the height of the potential barrier is about 30 MeV, the decay energy of the α-particles is 8.9 MeV. These energetic relations and their connection with the lifetime of the decaying nucleus can be understood on the basis of the tunnel effect.

Tunnel effect, a potential barrier of height V_0 and width $2a$ is traversed by a particle of energy $E < V_0$. Such a process is forbidden according to classical mechanics. When localizing the particle at the classical turning point, the wave function involves momentum components that correspond to energies above the potential barrier. The uncertainty principle between energy and time allows the maintenance of this uncertainty in energy ΔE over a time interval Δt, which is sufficiently long to observe the particle behind a potential barrier of finite width (**Fig. 24.7**).

Tunnel microscope: A metallic pin is moved at a distance of several nm over a species surface to be studied such that the tunnel current is kept constant by varying the distance between the pin and the species by means of a piezocrystal. The crystal driving voltage provides a mapping of the surface structure.

3. Potential well

Potential formulation:

$$V(x) = \begin{cases} 0 & \text{for} \quad |x| \le a, \\ V_0 > 0 & \text{for} \quad |x| > a. \end{cases}$$

$E < V_0$: discrete spectrum, bound states.
$E > V_0$: continuous spectrum, scattering states, reflection and transmission.

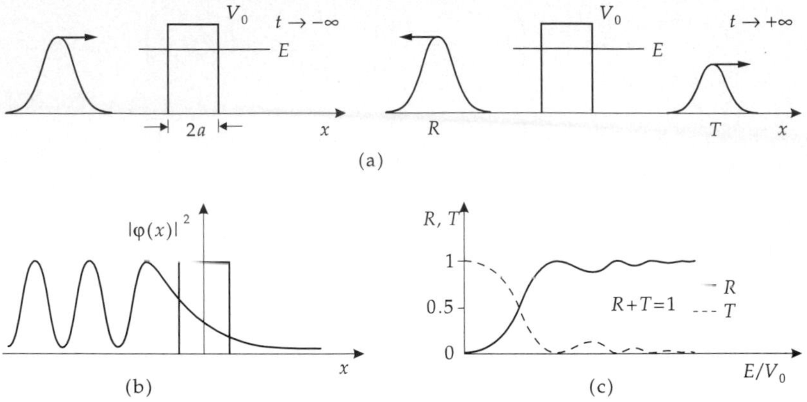

(a)

(b)

(c)

Figure 24.7: Tunnel effect. (a): separation of the incident wave packet into a reflected and a transmitted fraction (solution to the time-dependent Schrödinger equation), (b): probability density $|\varphi(x)|^2$ (stationary solution to the Schrödinger equation), (c): variation of the transmission coefficient T and reflection coefficient R with the ratio E/V_0.

Constraints equation for bound states:

$$K^2 - k^2 + 2kK \cot 2ka = 0 , \quad k^2 = \frac{2m}{\hbar^2} E , \quad K^2 = \frac{2m}{\hbar^2} (V_0 - E) .$$

➤ The equation for determining the energy eigenvalues may be solved graphically.

▲ The number and position of the bound energy levels depend on $V_0 a^2$. For $V_0 a^2 < \pi^2 \hbar^2/(8m)$ there exists only one bound state.

▲ The distance between successive energy eigenvalues increases with the excitation energy.

▲ The bound particle may be found with some probability beyond the turning points of classical motion.

▲ The wave function of the ground state has positive parity.

▲ Successive (as function of energy) eigenfunctions of the spectrum have opposite parities.

4. Infinitely high potential well

Potential formulation:

$$V(x) = \begin{cases} 0 & \text{for} \quad |x| \le a/2 , \\ \infty & \text{for} \quad |x| > a/2 . \end{cases}$$

▲ The wave function vanishes for $|x| \ge a/2$. It obeys the boundary condition

$$\varphi(-a/2) = \varphi(a/2) = 0 .$$

The wave function has a kink at these points (discontinuous derivative).

▲ In an infinitely high potential well, there are only bound states.

▲ The energy between successive energy eigenvalues increases with the excitation energy.

▲ The wave function of the ground state has positive parity.

▲ Alternate energy eigenfunctions of the spectrum have opposite parities.

Energy eigenvalues (**Fig. 24.8a**):

$$E_n = \frac{\pi^2 \hbar^2}{2ma^2} n^2, \quad n = 1, 2, 3, 4, \ldots .$$

Ground-state energy (zero-point energy):

$$E_1 = \frac{\pi^2 \hbar^2}{2ma^2}.$$

Eigenfunctions of positive parity (**Fig. 24.8b**):

$$\varphi_n(x) = \sqrt{\frac{2}{a}} \cos \frac{n\pi x}{a}, \quad n = 1, 3, 5, \ldots .$$

Eigenfunctions of negative parity (**Fig. 24.8b**):

$$\varphi_n(x) = \sqrt{\frac{1}{a}} \sin \frac{n\pi x}{a}, \quad n = 2, 4, 6, \ldots .$$

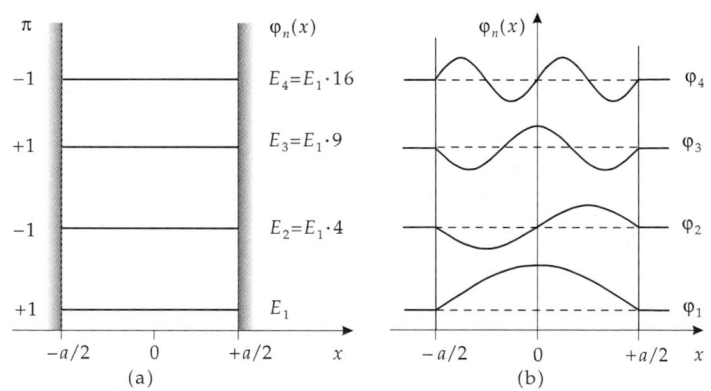

Figure 24.8: Infinitely high potential well. (a): schematic spectrum of energy eigenvalues $E_n = \frac{\pi^2 \hbar^2}{2ma^2} \cdot n^2$, π: parity of eigenfunctions, (b): eigenfunctions $\varphi_n(x)$.

24.4.2 Harmonic oscillator

Harmonic oscillator, a particle of mass m under the influence of a restoring force proportional to the displacement from the rest position and producing vibrations of a certain eigenfrequency along one or several spatial directions.

1. Time-independent Schrödinger equation

of the one-dimensional harmonic oscillator with angular frequency ω:

$$\frac{d^2}{dx^2} \varphi(x) + \frac{8\pi^2 m}{h^2} (E - \frac{m\omega^2}{2} x^2) \varphi(x) = 0.$$

▲   The energy states of the harmonic oscillator are quantized (**Fig. 24.9**),

$$E_n = \hbar\omega(n + \frac{1}{2}), \quad n = 0, 1, 2, 3\ldots,$$

$E_0 = \hbar\omega/2$ is the zero-point energy. There are no eigenstates with the asymptotics of scattering states.

▲   The energy levels of the harmonic oscillator are equally spaced,

$$\Delta E = E_{n+1} - E_n = \hbar\omega.$$

▲   The particle may be found with a non-zero probability beyond the turning points of classical motion.

▲   The wave function of the ground state has positive parity.

▲   Alternate energy eigenfunctions of the spectrum have opposite parity.

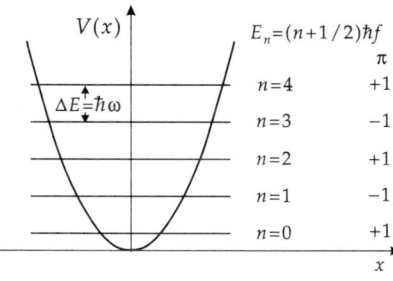

Figure 24.9: Harmonic oscillator. Spectrum of energy eigenvalues, $\Delta E = \hbar\omega$.

2. Eigenfunctions of the harmonic oscillator

The eigenfunctions of the harmonic oscillator are given by

$$\varphi_n(x) = (r_0)^{1/4} \sqrt{\frac{1}{2^n\, n!\sqrt{\pi}}}\, e^{-r_0 x^2/2}\, H_n(\sqrt{r_0}x).$$

$r_0 = \sqrt{m\omega/\hbar}$ is the **oscillator parameter**, H_n are the **Hermite polynomials (Fig. 24.10)**:

$$H_0(z) = 1, \quad H_1(z) = 2z, \quad H_2(z) = 4z^2 - 2, \quad H_3(z) = 8z^3 - 12z, \ldots.$$

The probability densities $|\varphi_n(x)|^2$ for the first few states are shown in **Fig. 24.11**.

▲   The momentum (hence the energy) of a particle localized about the minimum of the potential differs from zero because of Heisenberg's uncertainty relation. The ground-state energy of the harmonic oscillators does not coincide with the minimum value of the potential energy function.

Zero-point energy, ground-state energy, lowest energy of the harmonic oscillator:

$$E_0 = \frac{1}{2}\hbar\omega.$$

➤   The harmonic oscillator serves as a model for many kinds of excitations, among them:
•   vibrations in molecules and atomic nuclei,
•   lattice vibrations in a crystalline solid.

Phonon, frequently used name for an energy quantum of the harmonic oscillator with $E = hf = \hbar\omega$. If this amount of energy is transferred to the harmonic oscillator, then it is excited into the next higher energy state.

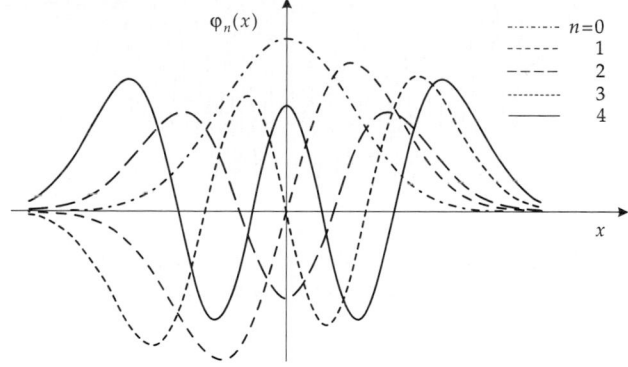

Figure 24.10: Harmonic oscillator: energy eigenfunctions $\varphi_n(x)$, $n = 0, 1, 2, 3, 4$.

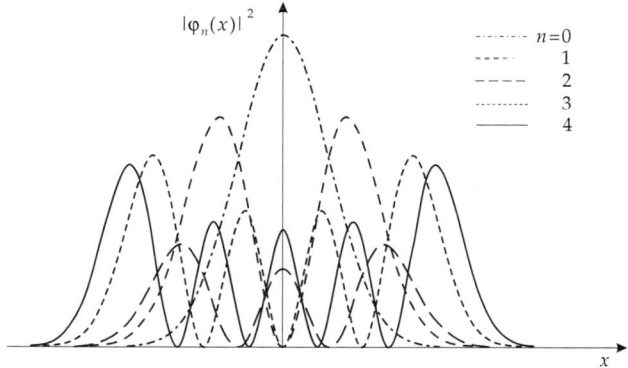

Figure 24.11: Harmonic oscillator: probability density $|\varphi_n(x)|^2$.

3. Bohr's correspondence principle

Correspondence principle of Bohr: The classical description of a mechanical system must correspond to the quantum-mechanical description in the limit of large quantum numbers (**Fig. 24.12**).

■ For large quantum numbers n, the trend of the probability density of a quantum-mechanical particle in a one-dimensional oscillator potential corresponds to the probability of finding a classical particle: maximum in the vicinity of the classical turning points (the particle velocity has a minimum value), and minimum at the equilibrium position (the particle velocity has a maximum value).

24.4.3 Pauli principle

Fermions, particles with half-integer spin.

■ Electrons and nucleons (neutrons, protons) are fermions with spin $s = 1/2$.

▲ Fermions obey the **Pauli principle**. The wave function of a system of indistinguishable fermions must be antisymmetric with respect to permutation of any two particles.

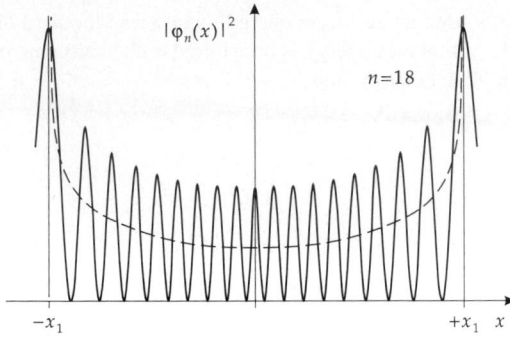

Figure 24.12: Probability density of a particle in an eigenstate of the harmonic oscillator with large quantum number. $\pm x_1$: turning points of classical motion. Dashed line: classical probability of finding the particle.

Antisymmetric wave function for two particles:

$$\Psi_a(1, 2) = \frac{1}{\sqrt{2}}(\psi_n(1) \cdot \psi_m(2) - \psi_n(2) \cdot \psi_m(1)) \,.$$

n and m denote arbitrary complete sets of quantum numbers. The function $\Psi_a(1, 2)$ is normalized. It changes sign under permutation of the particles 1 and 2,

$$\Psi_a(2, 1) = -\Psi_a(1, 2) \,.$$

Pauli principle: If two fermions are indistinguishable and the quantum numbers n and m coincide, $\Psi_a \equiv 0$, i.e., the probability of finding two particles in the same state equals zero. Two indistinguishable fermions must not occupy the same state (**exclusion principle**).

➤   The Pauli principle provides an understanding of the structure of the electron shell of atoms, as well as of atomic nuclei.

24.5   Spin and magnetic moments

24.5.1   Spin

Spin, intrinsic angular momentum (eigen angular momentum) of elementary particles. The spin has a definite fixed value for any kind of elementary particles. Unlike the orbital angular momentum, the spin quantum number may also take half-integer values.

1. Experimental demonstration of spin

Stern–Gerlach experiment (1921): A beam of silver atoms is sent through an inhomogeneous magnetic field. The individual electron, which according to the shell structure of the Ag atom determines the total angular momentum of the atom, carries no orbital angular momentum. Hence, an atomic magnetic moment can only be due to the spin of this electron. According to classical theory, one would expect a broad distribution of the outgoing beam, since any orientation of the magnetic moment connected with the spin against the magnetic field should be allowed. One observes, however, a splitting of the beam into two components, which demonstrates the value of the electron spin to be $s = 1/2$, with two possible orientations $m_s = \pm 1/2$ with respect to the direction of the magnetic field (see **Fig. 24.13**).

Rabi experiment (1938), permits the measurement of the much smaller nuclear moments by means of successive magnetic fields of different orientation.

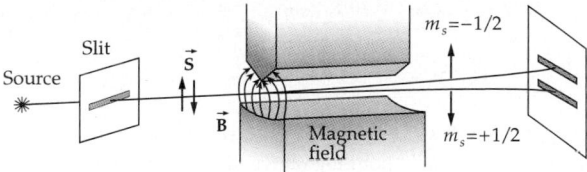

Figure 24.13: Stern–Gerlach experiment demonstrating the electron spin.

2. Spin operators and their properties

Spin operator for particles with the spin (intrinsic angular momentum) 1/2, vector operator with Cartesian components $\hat{s}_x$, $\hat{s}_y$, $\hat{s}_z$ (**Fig. 24.14**),

$$\hat{\mathbf{s}} = (s_x, s_y, s_z), \quad \hat{\mathbf{s}}^2 = s_x^2 + s_y^2 + s_z^2.$$

Commutation relations for the spin operator, correspond to the commutation relations of an angular momentum operator,

$$[\hat{s}_x, \hat{s}_y] = \hat{s}_x \cdot \hat{s}_y - \hat{s}_y \cdot \hat{s}_x = j\hbar \hat{s}_z,$$

$$[\hat{s}_y, \hat{s}_z] = \hat{s}_y \cdot \hat{s}_z - \hat{s}_z \cdot \hat{s}_y = j\hbar \hat{s}_x,$$

$$[\hat{s}_z, \hat{s}_x] = \hat{s}_z \cdot \hat{s}_x - \hat{s}_x \cdot \hat{s}_z = j\hbar \hat{s}_y.$$

and

$$[\hat{\mathbf{s}}^2, \hat{s}_x] = [\hat{\mathbf{s}}^2, \hat{s}_y] = [\hat{\mathbf{s}}^2, \hat{s}_z] = 0.$$

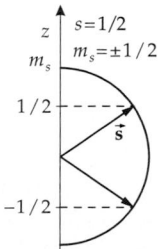

Figure 24.14: Vector model of electron spin $\vec{\mathbf{s}}$ ($s = 1/2$).

3. Pauli spin matrices,

$\hat{\sigma}_x$, $\hat{\sigma}_y$, $\hat{\sigma}_z$, representation of the operators corresponding to the spin components by 2×2-matrices,

$$\hat{\mathbf{s}} = \frac{\hbar}{2}\hat{\sigma}, \quad \hat{\sigma}_x = \begin{pmatrix} 0 & 1 \\ 1 & 0 \end{pmatrix}, \quad \hat{\sigma}_y = \begin{pmatrix} 0 & -j \\ j & 0 \end{pmatrix}, \quad \hat{\sigma}_z = \begin{pmatrix} 1 & 0 \\ 0 & -1 \end{pmatrix}.$$

Spin eigenfunction, χ_{sm_s}, simultaneous eigenfunction of the operator of the z-component of spin with the eigenvalue $\pm \hbar m_s$, and of the operator of the square of spin with the eigenvalue $s(s + 1)\hbar^2 = \frac{3}{4}\hbar^2$,

$$\hat{s}_z \, \chi_{sm_s} = \hbar m_s \, \chi_{sm_s}, \quad m_s = \pm\frac{1}{2},$$

$$\hat{\mathbf{s}}^2 \, \chi_{sm_s} = \hbar^2 s(s + 1)\chi_{sm_s}, \quad s = \frac{1}{2}.$$

Eigenstate with $m_s = +1/2$: spin pointing in positive z-direction.
Eigenstate with $m_s = -1/2$: spin pointing in negative z-direction.
  Representation of the spin eigenfunctions by column matrices:

$$\chi_{sm_s=\frac{1}{2}} = \begin{pmatrix} 1 \\ 0 \end{pmatrix}, \quad \chi_{sm_s=-\frac{1}{2}} = \begin{pmatrix} 0 \\ 1 \end{pmatrix}.$$

Arbitrary normalized spin state:

$$\chi = a \, \chi_{sm_s=\frac{1}{2}} + b \, \chi_{sm_s=-\frac{1}{2}} = \begin{pmatrix} a \\ b \end{pmatrix}, \quad |a|^2 + |b|^2 = 1.$$

$|a|^2$, $(|b|^2)$ is the probability for measuring the spin component $m_s = +1/2, (-1/2)$ in z-direction.
  Spin orientation along the direction ϑ, φ:

$$a = \cos(\vartheta/2) \, e^{-j\frac{\varphi}{2}}, \quad b = \sin(\vartheta/2) \, e^{j\frac{\varphi}{2}}.$$

The general wave function of a particle with spin 1/2 has two components,

$$\psi(\vec{\mathbf{r}}, s, t) = \begin{pmatrix} \psi_+(\vec{\mathbf{r}}, t) \\ \psi_-(\vec{\mathbf{r}}, t) \end{pmatrix}.$$

$|\psi_+(\vec{\mathbf{r}}, t)|^2 \, dV$, $(|\psi_-(\vec{\mathbf{r}}, t)|^2 \, dV)$ is the probability for finding the particle at time t in the volume element dV about the position $\vec{\mathbf{r}}$, with the spin pointing along the positive, (negative) z-direction.

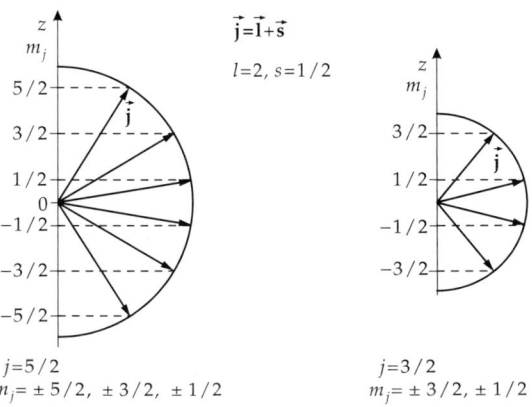

Figure 24.15: States of the total angular momentum $\vec{\mathbf{j}} = \vec{\mathbf{l}} + \vec{\mathbf{s}}$ for $l = 2$. m_j: magnetic quantum number for the z-component of total angular momentum $\vec{\mathbf{j}}$.

4. Total angular momentum,

$\hat{\vec{j}}$ of an electron, obtained by vector addition of the orbital angular momentum $\hat{\vec{l}}$ and spin $\hat{\vec{s}}$,

$$\hat{\vec{j}} = \hat{\vec{l}} + \hat{\vec{s}}, \quad \hat{j}_z = \hat{l}_z + \hat{s}_z.$$

The possible values of the quantum numbers for $\hat{\vec{j}}^2$ and $\hat{j}_z$ are (**Fig. 24.15**):

$$j = l + 1/2, \, j = l - 1/2, \quad m_j = -j, \ldots, +j.$$

An angular momentum state j has only $2j + 1$ possible orientations with respect to the quantization axis.

24.5.2 Magnetic moments

1. Magnetic moment of orbital motion,

$\hat{\vec{\mu}}_l$, expressed by the orbital angular momentum operator $\hat{\vec{l}}$:

operator of orbital magnetic moment			
	Symbol	Unit	Quantity
$\hat{\vec{\mu}}_l = -g_l \dfrac{e_0}{2m_e} \cdot \hat{\vec{l}}$	$\hat{\vec{\mu}}_l$	JT^{-1}	operator of orbital magnetic moment
	g_l	1	g-factor of orbital angular momentum
$= -g_l \, \mu_B \cdot \dfrac{\hat{\vec{l}}}{\hbar}$	e_0	C	elementary charge
	m_e	kg	electron mass
$g_l = 1$	$\hat{\vec{l}}$	J	orbital angular momentum operator

Bohr magneton, μ_B, universal constant:

$$\mu_B = -\frac{e_0 \cdot \hbar}{2 \cdot m_e} = 5.788\,382\,63(52) \cdot 10^{-11} \text{ MeV/T} = 9.274\,015\,4 \cdot 10^{-24} \text{ J/T}.$$

2. Magnetic spin moment,

$\hat{\vec{\mu}}_s$, expressed by the spin operator $\hat{\vec{s}}$:

operator of magnetic spin moment			
	Symbol	Unit	Quantity
$\hat{\vec{\mu}}_s = -g_s \dfrac{e_0}{2m_e} \cdot \hat{\vec{s}}$	$\hat{\vec{\mu}}_s$	JT^{-1}	operator of magnetic spin moment
	g_s	1	g-factor of spin
$= -g_s \, \mu_B \cdot \dfrac{\hat{\vec{s}}}{\hbar}$	e_0	C	elementary charge
	m_e	kg	electron mass
$g_s = 2.0023$	$\hat{\vec{s}}$	J	spin operator

Gyromagnetic factor, g, proportionality coefficient between angular momentum and magnetic moment of the electron:

$$g_l = 1, \quad g_s \approx 2.$$

➤ Relativistic quantum theory shows that the gyromagnetic spin factor does not exactly equal the value 2,

$$\frac{g_s - 2}{2} = (1\,159.652\,193 \pm 0.000\,010) \cdot 10^{-6}.$$

➤ The magnetic spin moment of the electron nearly corresponds to the magnetic moment of an orbital motion with angular momentum $l = 1$.

▲ Magnetic moment and angular momentum of the electron have opposite orientation both for orbital and spin magnetism, respectively.

3. Total magnetic moment,

$\hat{\vec{\mu}}$, of the electron in the atom, sum of the magnetic moments of spin and orbital motion (**Fig. 24.16**),

$$\hat{\vec{\mu}} = \hat{\vec{\mu}}_l + \hat{\vec{\mu}}_s = -\frac{e_0}{2m_e}(\vec{l} + 2 \cdot \vec{s}).$$

▲ In the vector model, the total magnetic moment $\hat{\vec{\mu}}$ of the electron is not antiparallel to the total angular momentum $\vec{j} = \vec{l} + \vec{s}$.

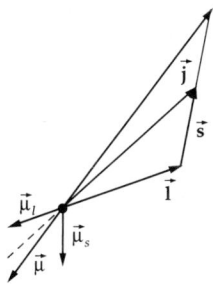

Figure 24.16: Coupling of orbital angular momentum $\vec{l}$ and spin $\vec{s}$ to the total angular momentum $\vec{j}$, and magnetic moments $\vec{\mu}_l$, $\vec{\mu}_s$, $\vec{\mu} = \vec{\mu}_l + \vec{\mu}_s$.

4. Potential energy in the magnetic field

The potential energy E_{pot} of an unbound electron in a uniform magnetic field $\vec{B} = (0, 0, B_z)$ along the z-direction is given by

$$E_{\text{pot}} = -\vec{\mu}_s \cdot \vec{B} = g_s \frac{e_0}{2m_e} s_z \cdot B_z.$$

For an eigenstate of the z-component of the spin operator with the projection quantum number $m_s = \pm 1/2$:

$$E_{\text{pot}} = g_s \frac{e_0 \hbar}{2m_e} m_s \cdot B_z = g_s \mu_B \cdot m_s \cdot B_z.$$

Similarly, for an electron with orbital angular momentum (l, m):

$$E_{\text{pot}} = g_l \frac{e_0 \hbar}{2m_e} m \cdot B_z = g_l \mu_B \cdot m \cdot B_z.$$

The potential energy of an atom in a state with the quantum numbers J, M_J for the total angular momentum and its projection onto the z-axis, L for the total orbital angular momentum, and S for the total spin, in a uniform magnetic field $\vec{\mathbf{B}} = (0, 0, B_z)$ along the z-axis is given by:

potential energy in a magnetic field			$\mathbf{ML^2T^{-2}}$
	Symbol	Unit	Quantity
$E_{\mathrm{pot}} = -\vec{\mu} \cdot \vec{\mathbf{B}}$ $= g(L, S, J) \cdot \mu_{\mathrm{B}} \cdot M_J \cdot B$	E_{pot}	J	potential energy
	M_J	1	quantum number of projection of total angular momentum
	$g(L, S, J)$	1	Landé factor
	μ_{B}	J/T	Bohr magneton
	B	T	magnetic flux density

Landé factor $g(L, S, J)$, describes the dependence of the gyromagnetic ratio on the quantum numbers of the term:

$$g(L, S, J) = 1 + \frac{J(J+1) - L(L+1) + S(S+1)}{2J(J+1)} .$$

Larmor precession, precession of the vector of the magnetic moment of an atomic system in an external magnetic field $\vec{\mathbf{B}}$ with constant angular velocity about the field direction.

Larmor frequency, frequency of Larmor precession, for the orbital magnetism given by

$$\omega_L = g_l \, \mu_{\mathrm{B}} \cdot \frac{B}{\hbar} .$$

Nuclear magnetic spin, generated by the magnetic moment of the atomic nucleus as a consequence of the nuclear spin.

|M| Nuclear magnetic spin is used to cool bodies to temperatures of the order of μK. An external magnetic field aligns the magnetic moments of the atomic nuclei of a pre-cooled material. After switching off the magnetic field, the atomic nuclei again approach a statistically disordered state. This process is carried out adiabatically ($\Delta Q = 0$). A lowering of the degree of order, which would correspond to an increase of entropy, is therefore connected with a decrease of temperature. The lowest temperature of a test sample reached so far is about $5 \cdot 10^{-6}$ K. To date, the lowest temperature measured for a system of Cu nuclei is about $50 \cdot 10^{-9}$ K.

25
Atomic and molecular physics

Atoms, the smallest particles of a chemical element possessing its chemical properties. An electrically neutral atom consists of a Z-fold positively charged **nucleus**, and Z negatively charged **electrons** (shell) moving in the Coulomb field of the nucleus.

Nuclear charge number, **atomic number**, Z, number of protons forming the atomic nucleus.

▲ Atoms are electrically neutral. The sum of the electrons of an atom equals the number of protons of the atomic nucleus.

Atomic radius, R_A, of the order of magnitude 10^{-10} m (formerly used: 1 angstrom = 1 Å = 10^{-10} m). The radius of the atomic nucleus, in contrast, is of the order of magnitude 1 fm = 10^{-15} m.

Atomic and ionic radii of the elements are compiled in **Tab. 29.2**. The values depend on the method of measurement and should be considered only as a guide. The trend of atomic radii with atomic number is plotted in **Fig. 25.1**.

■ Atomic radii of some elements (in nm): He 0.122, Li 0.155, O 0.056, Fe 0.126, Rb 0.248, U 0.153.

Ions, electrically charged particles that occur when an atom releases or accepts electrons (see p. 552).

The ionic charge is given by a superscript to the right of the atomic symbol: H^+ (single positively charged hydrogen ion), Cl^- (single negatively charged chlorine ion).

Ionization energy, E_i, or **ionization work**, W_i, the energy expended to remove an electron from a stationary bound atomic state (**Fig. 25.2**).

25.1 Fundamentals of spectroscopy

Energy levels, stationary states of the atom with a definite energy. The energy levels are specified by additional quantum numbers, such as total orbital angular momentum L, total spin S and total angular momentum J.

Ground state, the stationary state with lowest energy.

Excited state, a state with an energy above the ground-state energy.

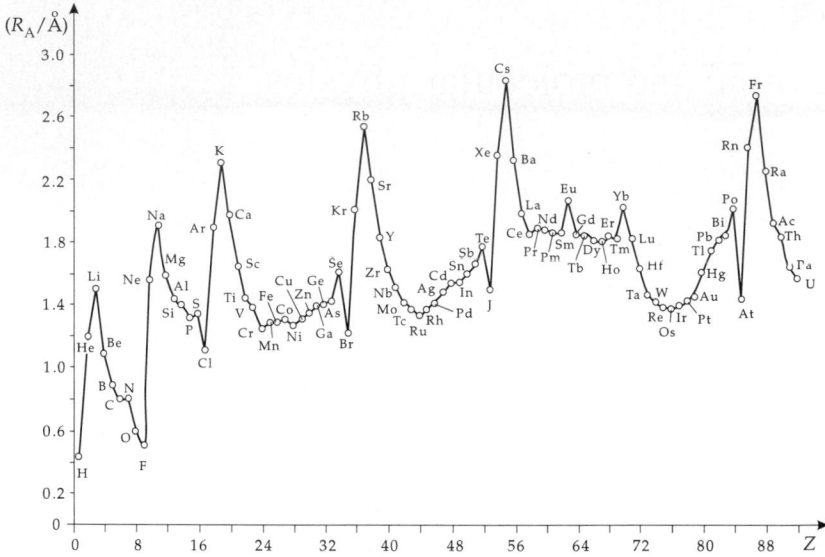

Figure 25.1: Atomic radius R_A plotted against atomic number Z.

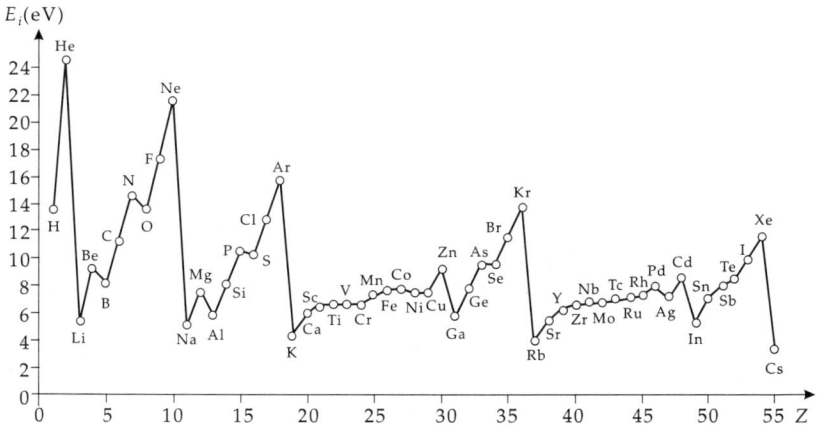

Figure 25.2: Ionization energy E_i plotted against atomic number Z.

Level scheme, a graphical representation of the energies of the stationary states of an atom.

Spectroscopy, measurement and analysis of the radiation emitted or absorbed by atoms (or molecules, atomic nuclei, etc.).

Spectrum, the dependence of the intensity of radiation emitted or absorbed by atoms, molecules, nuclei, etc., on the frequency or wavelength of the radiation.

1. Emission spectrum,

the frequency distribution of the radiation emitted by a substance. Emission spectra are measured for transitions from an excited atomic state to the ground state, or to another energetically lower atomic state.

| M | The sample may be stimulated to emit radiation through electron collisions in gas discharges, in a high-frequency plasma or by spark discharge, in an electric arc, and by thermal excitation. Emission spectra are measured by decomposing the radiation emitted by excited atoms into components of different wavelengths by a spectrograph.

▲   The emission spectrum of the hydrogen atom is a **line spectrum**.

2. Line shape of spectral lines

Line shape, profile of the intensity $I(\omega)$ in a small frequency range about a spectral line ω_0 that corresponds to a spontaneous transition from the stationary state i into the stationary state f,

$$I(\omega) \sim \frac{(\Delta\omega)/2}{(\omega - \omega_0)^2 + (\Delta\omega)^2/4} \, .$$

Natural line width, $\Delta\omega$, difference between the frequency values at which the intensity curve drops to half the peak value I_{max} (see **Fig. 25.3**).

▲   The line width $\Delta\omega$ corresponds to an energy uncertainty of the initial state, $\Delta E = \hbar\Delta\omega$, which according to Heisenberg's uncertainty relation is related to the **mean lifetime** τ of the initial state i as follows:

$$\Delta E \sim \hbar/\tau \, .$$

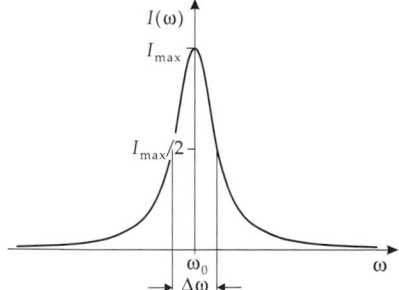

Figure 25.3: Width $\Delta\omega$ of a spectral line.

   Line broadening, enlargement of the width of an experimentally observed spectral line versus the natural line width. Line broadening can be caused by the Doppler effect, by atomic collisions depending on the pressure, and by interaction with radiation fields.

■   The mean lifetime of excited atomic states lies in general between 10^{-7} s and 10^{-8} s. Hence, the frequency uncertainties range up to $\Delta\omega \approx 10^8$ Hz.

➤   Transitions from metastable states of long lifetime ($\tau \approx 10^{-3}$ s) have a small line width ($\Delta\omega \approx 10^3$ Hz).

▲   The emission or absorption spectra of molecules consist of sequences of lines that, at low resolution of the spectral apparatus, appear as structureless bands (**band spectrum**).

▲   The thermal radiation emitted by bodies is electromagnetic radiation with a **continuous spectrum**.

3. Absorption spectrum,

the frequency distribution of the incident radiation intensity that is attenuated in the sample. Absorption spectra usually correspond to transitions of atoms from the ground state to excited states. An example is shown in **Fig. 25.4**.

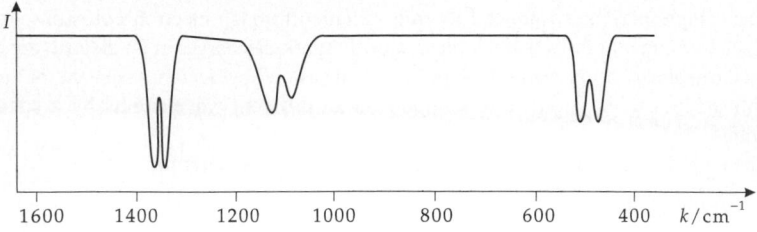

Figure 25.4: Absorption spectrum of SiO_2 in the infrared. k: wave number, I: intensity of radiation.

M   Absorption spectra are observed when white light passes through "cold" vapor or "cold" gases. The wavelengths corresponding to the absorbed frequencies appear as black lines in the transmission spectrum.

Resonance spectroscopy, measurement of the absorption of an incident radiation of fixed wavelength by a sample as a function of an external parameter (temperature, pressure, magnetic field).

25.2   Hydrogen atom

A hydrogen atom is an electrically neutral object consisting of an electron and a proton, and bound by the electrostatic interaction. The binding energy in the ground state amounts to about 13.6 eV, the atomic radius is about 0.5 Å.

Electron, elementary particle with a negative charge $-e$ (e: elementary charge) and a rest mass m_e,

$$e = 1.602\,177\,33(49) \cdot 10^{-19}\,\text{C},$$

$$m_e = 9.109\,389\,7(54) \cdot 10^{-31}\,\text{kg}.$$

Proton, elementary particle with a positive charge e and a rest mass $m_p \approx 1836\,m_e$,

$$m_p = 1.672\,623\,1(10) \cdot 10^{-27}\,\text{kg}.$$

The numbers in brackets give the uncertainty of the last digits.

Deuteron, an atomic nucleus consisting of a proton and a neutron. The **neutron** is electrically neutral and about 2.5 electron masses heavier than the proton.

Deuterium, heavy hydrogen. The nucleus of the deuterium atom is a deuteron.

Hydrogen-like systems, systems whose energetic behavior is determined by a single electron. Hydrogen-like systems are the ions He^+, Li^{2+}, Be^{3+}, ..., U^{91+}.

■   In 1993 the hyperfine splitting of $^{209}Bi^{82+}$ was measured for the first time at the heavy-ion storage ring ESR at GSI (Darmstadt, Germany).

Alkali atoms Li, Na, K, Rb, Cs, Fr, show similarity to the hydrogen atom: the nucleus and the inner electrons represent a positively charged center, the weakly bound **valence electron** moves about this center.

Rydberg atoms, highly excited hydrogen atoms or hydrogen-like atoms (principal quantum number $n > 100$). Their radii range up to $\approx 5 \cdot 10^{-7}$ m; this corresponds to the size of a virus.

25.2.1   Bohr's postulates

1. Formulation of Bohr's postulates

a) Bohr postulate (**postulate of stationary states**):

Atoms may exist in certain stationary states in which they do not radiate energy. These stationary states correspond to the "orbits" in the classical picture along which the electrons move in a planetary motion. They do **not** emit electromagnetic radiation when in these orbits, despite the radial acceleration.

b) Bohr postulate (**postulate of quantization of orbits**):

The orbital angular momentum of an electron in a stationary orbit is equal to an integer multiple of $\hbar$:

$$l_n = r_n \cdot m_e v_n = n \cdot \hbar, \quad \hbar = \frac{h}{2\pi}, \quad n = 1, 2, 3, \ldots.$$

r_n is the radius of the nth orbit; n is a natural number, $n > 0$.

In the stationary state n, the hydrogen atom has an energy

$$E_n = -\frac{Z^2 e^4 m_e}{8 h^2 \varepsilon_0^2} \cdot \frac{1}{n^2} \qquad \varepsilon_0\text{: electrical permittivity of free space.}$$

c) Bohr postulate (**Bohr frequency condition**):

An atom emits a quantum of electromagnetic radiation (**photon**) when an electron changes from an orbit with number m to an orbit with a smaller number n.

Energy of photon, difference of the energies of the electron in the orbits before and after the transition:

$$E = \hbar\omega = hf = E_m - E_n .$$

Bohr's postulates cannot be derived from classical physics and are explained only by quantum mechanics. The concept of electron orbit proposed in Bohr's atomic model is successful because of the wave nature of the electron and because of Heisenberg's uncertainty relation.

➤   Bohr's postulates can be used to explain the line spectrum of the hydrogen atom.

2. Bohr radii

Bohr orbital radius, r_n, follows from the equilibrium condition for the centrifugal force and the Coulomb force on a classical circular orbit, and from the second Bohr postulate:

Bohr orbital radius				L
	Symbol	Unit	Quantity	
$\dfrac{Ze^2}{4\pi\varepsilon_0 r_n^2} = m_e \cdot \dfrac{v_n^2}{r_n}$	Z	1	atomic number	
	e	C	elementary charge	
$r_n \cdot m_e v_n = n \cdot \hbar$	ε_0	$\mathrm{C\,V^{-1}m^{-1}}$	electric permittivity of free space	
$r_n = 4\pi\varepsilon_0 \dfrac{n^2 \hbar^2}{m_e Z e^2}$	r_n	m	Bohr orbital radius	
	m_e	kg	electron mass	
	v_n	m/s	orbital velocity	

Bohr radius, r_1, frequently denoted by a_0 or a_∞, radius of the orbit with $n = 1$,

$$r_1 = 0.529\,177\,249(24) \cdot 10^{-10} \text{ m} \approx 0.5 \text{ Å} .$$

M   The **Franck–Hertz experiment** (1913) confirmed Bohr's postulates by the demonstration of a discrete energy transfer by accelerated electrons to mercury atoms in a triode-like vacuum tube.

3. Frequencies in the hydrogen spectrum

Hydrogen spectrum, a line spectrum consisting of several series:

frequencies in the hydrogen spectrum			T^{-1}
	Symbol	Unit	Quantity
$f_{mn} = cR_H \left(\dfrac{1}{n^2} - \dfrac{1}{m^2} \right) \quad n < m$ $R_H = 1.0967\,758\,10 \cdot 10^7 \text{ m}^{-1}$	f_{mn} c R_H n, m	s^{-1} $m\,s^{-1}$ m^{-1} 1	frequency speed of light Rydberg constant for H-atom natural numbers

wavelengths in hydrogen spectrum			**L**
	Symbol	Unit	Quantity
$\lambda_{mn} = \dfrac{1}{R_H} \left(\dfrac{n^2 \cdot m^2}{m^2 - n^2} \right) \quad n < m$ $R_H = 1.0967\,758\,10 \cdot 10^7 \text{ m}^{-1}$	λ_{mn} R_H n, m	m m^{-1} 1	wavelength Rydberg constant for H-atom natural numbers

Principal quantum numbers, n, discrete values of the series $n = 1, 2, \ldots$, describe the energy spectrum of the hydrogen atom.

4. Series and series formulas of the hydrogen spectrum,

for a scheme of the series see **Fig. 25.5**.

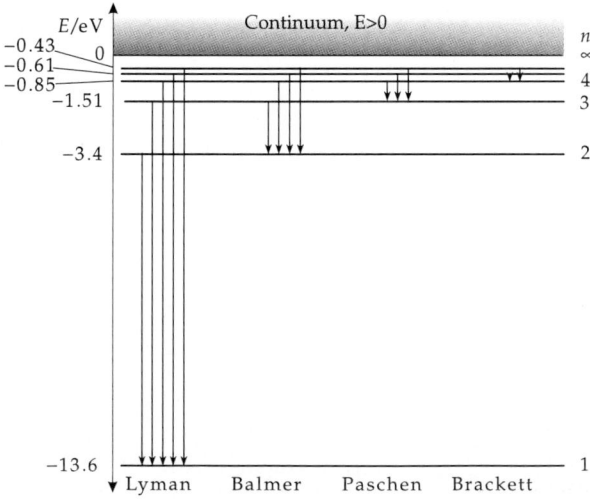

Figure 25.5: Series in the line spectrum of the hydrogen atom. Term scheme and transitions, n: principal quantum number.

▲   For the principal quantum numbers in the series formulas $m > n$.

Lyman series ($n = 1$) in the ultraviolet, **Balmer series** ($n = 2$) in the visible range, **Paschen series** ($n = 3$) in the near infrared, **Brackett series** ($n = 4$) and **Pfund series** ($n = 5$) in the far infrared frequency range (**Fig. 25.6**).

Term, T_n, given by

$$T_n = \frac{cR_H}{n^2}.$$

▲   The lines of the hydrogen spectrum may be represented as differences of terms.

The **Rydberg constant** R_H for the hydrogen atom determined from the spectra differs slightly from the calculated Rydberg constant R_∞:

Rydberg constant, R_∞ (assuming an infinitely heavy center of force),

$$R_\infty = \frac{m_e e^4}{8\varepsilon_0^2 h^3 \cdot c} = 1.097\,373\,156\,83(4) \cdot 10^7 \text{ m}^{-1}.$$

When calculating R_H, one has to take into account that the proton has a finite mass m_p as compared with the electron mass m_e (reduced mass $\mu = m_p m_e / (m_p + m_e)$):

$$R_H = \frac{R_\infty}{1 + m_e/m_p}.$$

Series limit, the maximum value of the frequency of a line in a series. For $m \to \infty$, the **energy of the limit frequency** $f_{\lim} = f_{\infty n}$ in the hydrogen atom is:

$$E_n = hf_{\lim} = \frac{hR_H c}{n^2}.$$

▲   The ground state of the hydrogen atom lies at $E_1 = -13.595$ eV.

▲   There are additional frequencies, also above the frequency limit, corresponding to transitions between continuum states and discrete atomic states.

5. Degeneracy in the hydrogen spectrum

The total angular momentum $\vec{j}$ of the electron in the hydrogen atom is calculated by vector addition of the orbital angular momentum $\vec{l}$ and the spin $\vec{s}$, $\vec{j} = \vec{l} + \vec{s}$. For $l > 0$, the possible values of the quantum number j that determines the magnitude of the total angular momentum are given by $j = l \pm 1/2$.

Accidental degeneracy in the hydrogen atom, a degeneracy of the energy levels specific to the Coulomb potential ($\sim 1/r$), see **Fig. 25.7**.

▲   The energy of the stationary states of a hydrogen atom depends almost entirely on the principal quantum number n. To an energy state E_n belong wave functions with orbital angular momentum quantum numbers $l = 0, 1, 2, \ldots, n - 1$.

Degeneracy in the energy spectrum of the hydrogen atom:

E_1	$n = 1$	$l = 0$	ground state
E_2	$n = 2$	$l = 0, 1$	first excited state
E_3	$n = 3$	$l = 0, 1, 2$	second excited state
E_4	$n = 4$	$l = 0, 1, 2, 3$	third excited state
$\ldots$	$\ldots\ldots$	$\ldots\ldots\ldots$	$\ldots\ldots\ldots\ldots$

➤   In the ground state, $n = 1$, the orbital angular momentum vanishes: $l_{n=1} \equiv 0$.

➤   The exclusive dependence of the energy eigenvalues on the principal quantum number holds for all hydrogen-like systems, provided the magnetic interaction between the orbital motion and the electron spin is ignored.

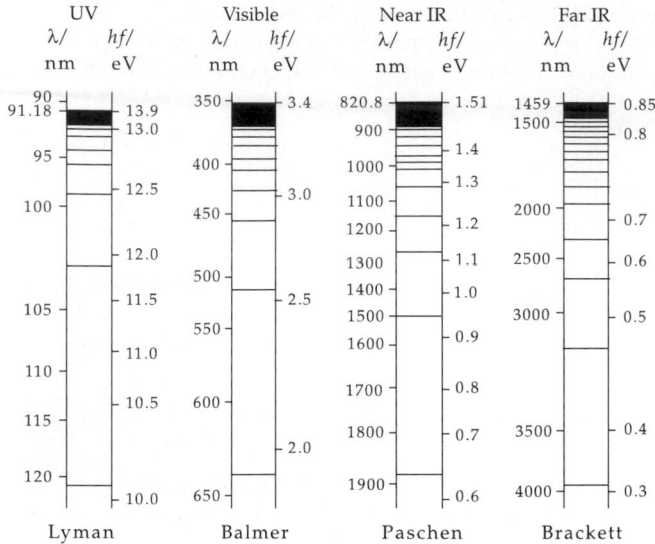

Figure 25.6: Series in the line spectrum of the hydrogen atom: wavelengths λ and energies hf.

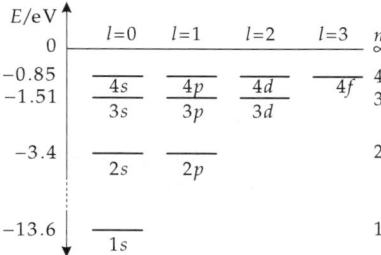

Figure 25.7: Accidental degeneracy of the states of the hydrogen atom with respect to the quantum number l of the orbital angular momentum.

6. Fine structure of the hydrogen spectrum

If **spin–orbit coupling** is taken into account, there arises a fine-structure splitting of the levels in hydrogen and hydrogen-like systems. The energy of the stationary states in the hydrogen atom then depends only on the quantum number j of the total angular momentum. The states remain partly degenerate with respect to the orbital angular momentum quantum number l: the levels with $l = j - 1/2$ and $l = j + 1/2$ have the same energy.

Fine structure splitting of levels in hydrogen and hydrogen-like systems caused by relativistic effects like electron spin (**Fig. 25.8**).

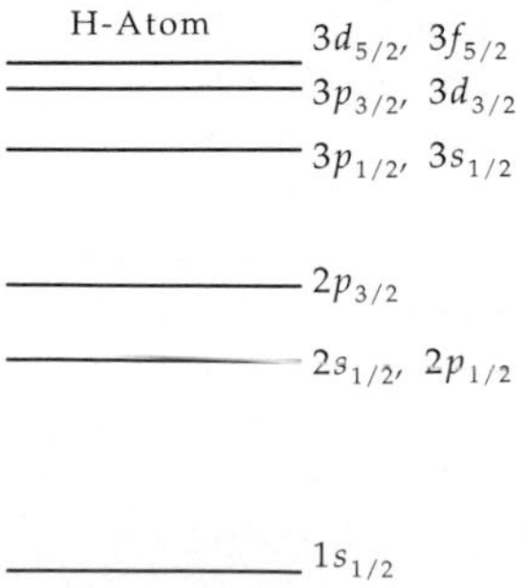

Figure 25.8: Fine structure of the hydrogen spectrum. Classification of the states by nl_j, n: principal quantum number, l: orbital angular momentum quantum number, j: quantum number of the total angular momentum.

fine-structure formula by Sommerfeld			$\mathbf{ML^2T^{-2}}$
	Symbol	Unit	Quantity
$$E_{nj} = -\frac{R_\infty \cdot h \cdot Z^2}{n^2}$$ $$\cdot \left[1 + \frac{Z^2 \cdot \alpha^2}{n^2} \cdot \left(\frac{n}{j + \frac{1}{2}} - \frac{3}{4} \right) \right]$$ $$\alpha = 1/137.035\,989\,5(61)$$	E_{nj} j α R_∞ Z h n	J J s 1 m^{-1} 1 J s 1	energy eigenvalue total angular momentum fine-structure constant Rydberg constant nuclear charge number quantum of action principal quantum number

Fine-structure constant α, ratio of the "orbital velocity" of the first Bohr orbit (orbit radius $r_1 = \dfrac{\varepsilon_0 \cdot h^2}{2\pi \cdot m_e \cdot e^2}$) to the speed of light c,

$$\alpha = \frac{2\pi e^2}{h \cdot c} = 1/137.035\,989\,5(61) \,.$$

➤ The splitting of the $l = j \pm 1/2$ levels, which is only $4.375 \cdot 10^{-6}$ eV for the terms $2s_{1/2}$ and $2p_{1/2}$, (**Lamb shift**) can be explained by quantum electrodynamics.

25.3 Stationary states and quantum numbers in the central field

Potential energy of an electron in the field of the atomic nucleus, which for hydrogen-like systems takes into account the screening of the nuclear Coulomb field by the inner-shell electrons via the introduction of an effective atomic number $Z^* < Z$,

$$V_C(r) = -\frac{1}{4\pi\varepsilon_0} \frac{Z^* e^2}{r} \,.$$

r is the distance of the electron from the center of the nucleus.

Application of the operator $\hat{\mathbf{l}}^2$ on a wave function ψ_{nl} specified by the orbital angular momentum quantum number l ($l = 0, 1, 2, \ldots$) yields:

$$\hat{\mathbf{l}}^2 \psi_{nl} = \hbar^2 \, l(l+1) \, \psi_{nl} \,.$$

Centrifugal potential, additional potential for electrons in a state with $l \neq 0$:

$$V_Z^{(l)}(r) = \frac{\hbar^2}{2m} \cdot \frac{l(l+1)}{r^2} \,.$$

The centrifugal potential causes the electrons in states with larger angular momentum to be pushed farther outward, similar to planetary motion.

1. Effective central potential in the many-electron atom

Effective potential, $V_{\text{eff}}^{(l)}(r)$, a central potential consisting of the sum of the screened Coulomb potential of the atomic nucleus and the centrifugal potential (**Fig. 25.9**),

$$V_{\text{eff}}^{(l)}(r) = V_{\text{C}}(r) + V_{\text{Z}}^{(l)}(r) .$$

effective central potential			$\mathbf{ML^2T^{-2}}$
	Symbol	Unit	Quantity
	V_{eff}	J	potential energy
	Z^*	1	effective atomic number
$V_{\text{eff}}^{(l)}(r) = -\dfrac{1}{4\pi\varepsilon_0}\dfrac{Z^*e^2}{r}$	ε_0	C V^{-1}m^{-1}	electric permittivity of free space
$+\dfrac{\hbar^2}{2m_{\text{e}}}\cdot\dfrac{l(l+1)}{r^2}$	r	m	distance electron – atomic center
	l	J s	quantum number orb. ang. momentum
	m_{e}	kg	electron mass
	$\hbar$	J s	quantum of action

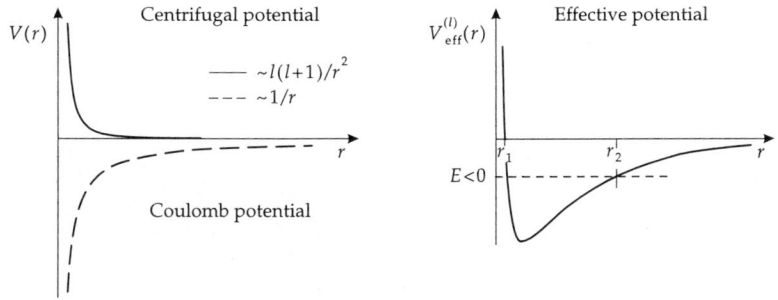

Figure 25.9: Effective potential $V_{\text{eff}}^{(l)}(r)$ (schematic). (a): Coulomb potential and centrifugal potential, (b): total potential. r_1, r_2: inversion points of classical motion of a particle of energy $E < 0$.

2. Wave function of a particle and radial quantum number

Wave function of a particle in a central potential, in spherical coordinates (r, ϑ, φ) separable into radial and angular components:

$$\psi_{n_r lm} = \frac{u_{n_r l}(r)}{r} Y_l^m(\vartheta, \varphi), \qquad \int_0^\infty |u_{n_r l}(r)|^2\, dr = 1 .$$

The angular component is represented by the spherical harmonics (spherical surface harmonics) Y_l^m.

➤ This formulation for the wave function holds not only for the Coulomb potential, but for any central potential $V(r)$, independent of the detailed radial variation.

Radial quantum number, n_r, number of zeros of the radial wave function $u_{n_r l}(r)$, without counting the trivial zeros at $r = 0$ and $r = \infty$ (**Fig. 25.10**). Possible values for n_r: $n_r = 0, 1, 2, \ldots$.

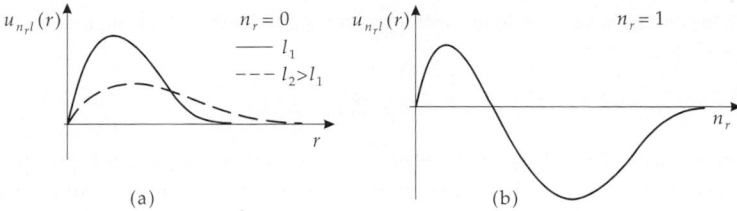

Figure 25.10: Radial wave function $u_{n_r l}(r)$ (schematic). (a): wave function without node, various angular momentum quantum numbers l, (b): wave function with one node. The trivial zeros of $u(r)$ at $r = 0$ and $r \to \infty$ are not counted.

3. Orbital angular momentum and magnetic quantum numbers

Orbital angular momentum quantum number, l, integer quantity specifying the orbital angular momentum state of a particle. Possible values for l: $l = 0, 1, 2, \ldots$.

In spectroscopic notation, different values of the orbital angular momentum quantum number l are expressed by letters:

l	0	1	2	3	4	...
name	s	p	d	f	g	...

Magnetic quantum number m, integer quantum number, specifies the component of the orbital angular momentum along the quantization axis (z-axis). Possible values of m for given value of l: $m = -l, -l + 1, \ldots, 0, \ldots, l - 1, l$.

▲   An angular momentum l has $2l + 1$ possible orientations along the quantization axis.
Parity, π, of the wave function $\psi_{n_r l m}$, determined by the behavior of the spherical harmonic under reflection at the coordinate origin $\vartheta \longrightarrow \pi - \vartheta, \varphi \longrightarrow \varphi + \pi$,

$$Y_l^m(\vartheta, \varphi) \longrightarrow Y_l^m(\pi - \vartheta, \varphi + \pi) = (-1)^l \cdot Y_l^m(\vartheta, \varphi), \quad \pi = (-1)^l.$$

Orbital angular momenta $l = 0, 2, 4, \ldots$: states of positive parity, $\pi = +1$.
   Orbital angular momenta $l = 1, 3, 5, \ldots$: states of negative parity, $\pi = -1$.
▲   The energy eigenvalues of a particle in the central potential depend only on the radial quantum number n_r and the orbital angular momentum quantum number l, $E = E_{n_r l}$.

4. Level degeneracy in the central potential

Degeneracy of a level, the feature that several quantum-mechanical states with different quantum numbers belong to a given energy value.
▲   The stationary state of a particle in a central potential with the energy eigenvalue $E_{n_r l}$ shows a natural $(2l + 1)$-fold degeneracy with respect to the magnetic quantum number m.
Accidental degeneracy in the hydrogen atom, a degeneracy of the energy levels specific for the Coulomb potential ($\sim 1/r$). The energy of the states of the hydrogen atom depends only on the **principal quantum number** n,

$$n = n_r + l + 1.$$

▲   To the energy state E_n belong wave functions ψ_{nl} with the orbital angular momentum quantum number,

$$l = 0, 1, 2, \ldots, n - 1.$$

➤   The exclusive dependence of the energy eigenvalues on the principal quantum number holds for all hydrogen-like systems provided the spin–orbit interaction is ignored.

5. States of positive energy: scattering states

Scattering states, solutions to the Schrödinger equation for positive energy values $E = p^2/2m$. The spectrum of eigenvalues is continuous. For large distances from a spherically symmetric scattering potential that decreases faster than $1/r$ at large distances, the wave function consists of an incident plane wave with the wave vector $\vec{\mathbf{k}}$, and an outgoing spherical wave with scattering amplitude $f_{\vec{\mathbf{k}}}(\vartheta)$:

$$\psi(\vec{\mathbf{r}}) \longrightarrow e^{j\vec{\mathbf{k}}\vec{\mathbf{r}}} + f_{\vec{\mathbf{k}}}(\vartheta) \cdot \frac{e^{jkr}}{r} \quad \text{for} \quad r \to \infty.$$

The absolute square of the scattering amplitude $|f_{\vec{\mathbf{k}}}(\vartheta)|^2$ determines the probability for scattering of the particle through an angle ϑ relative to the incident wave, which is given by the orientation of $\vec{\mathbf{k}}$.

6. Probability density for electrons

Electron density $w(\vec{\mathbf{r}})$ in an atom, determined by the quantity

$$w(\vec{\mathbf{r}}) = |\psi(\vec{\mathbf{r}})|^2.$$

Radial probability density $W(r)\,\mathrm{d}r = 4\pi|\psi|^2 r^2 \mathrm{d}r$, probability of finding the electron within a spherical shell of radii r and $r + \mathrm{d}r$ about the nucleus (**Fig. 25.11**). The position of the peak of the function $W(r)$ determines the most likely distance of the electron from the nucleus.

Only the **s-electrons** ($l = 0$) have a nonvanishing probability $w(\vec{\mathbf{r}})$ to be found at the position of the atomic nucleus ($r \to 0$).

Angular distribution of the electron density, determined by the quantum numbers of the orbital angular momentum and its projection onto a given z-axis (**Fig. 25.12**).

Selection rules, conditions for the transition of an atomic electron from some energy level to another one with the emission or absorption of a photon. In electric dipole transitions, the orbital angular momentum quantum numbers may change as follows:

$$\Delta l = \pm 1 \quad \text{and} \quad \Delta m = 0, \pm 1.$$

The principal quantum numbers of the levels involved in the transition affect the radiation intensity.

7. Shell structure of the electron shell

Electron shells, the set of atomic electrons occupying states with the same principal quantum number n forms a shell.

Spectroscopic classification by the principal quantum numbers:

n	1	2	3	4	...
shell	K	L	M	N	...

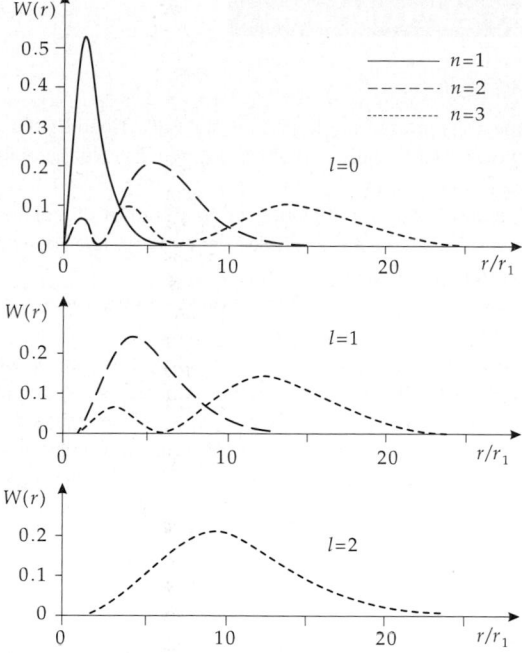

Figure 25.11: Radial probability density of electrons for s-, p- and d-states in the hydrogen atom. r_1 is the Bohr radius.

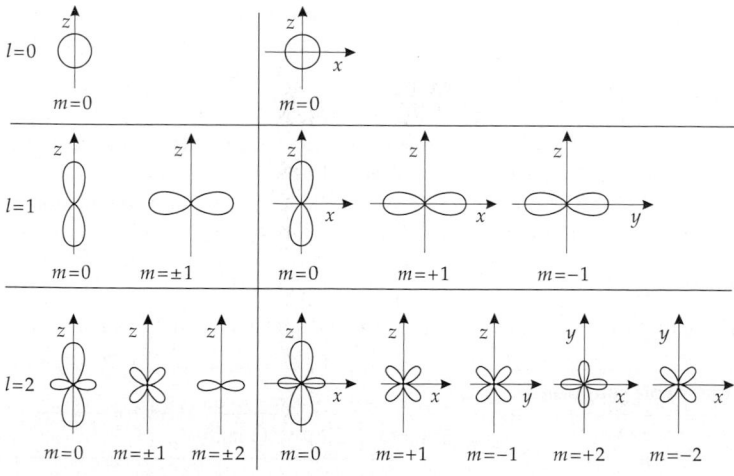

Figure 25.12: Angular dependence of the electron density for s-, p- and d-states. Quantization axis is the z-axis.

25.4 Many-electron atoms

1. Vector model of atom

▲ **Pauli principle**: Any atomic single-particle state described by the quantum numbers n, l, m can be occupied by only two electrons, which have the possible spin orientations $m_s = \pm 1/2$ (see p. 844).

Vector model of the atom, the orbital angular momentum of any electron is represented by the vector $\vec{l}$, the spin by the vector $\vec{s}$. These vectors may take only certain orientations relative to the z-axis (directional quantization).

The vector model serves to systematize the complex spectra of many-electron atoms and in studies of the fine structure of spectra.

Directional quantization, the feature of the angular momenta of an electron that the projections of the vectors $\vec{l}$ and $\vec{s}$ onto a selected direction in space (e.g., an external magnetic field in z-direction) may take only discrete values. The selected direction is called **quantization axis**. The component of the vector $\vec{l}$ along the quantization axis may take only the $(2l + 1)$ integer values $l, l - 1, \ldots, 0, \ldots, -l + 1, -l$ (in units of $\hbar$). The vector $\vec{s}$, on the contrary, has only the components $+1/2$ and $-1/2$ along the quantization axis (in units of $\hbar$).

2. Total angular momentum in the vector model

total angular momentum vector			$\mathbf{ML^2T^{-1}}$
	Symbol	Unit	Quantity
$\vec{j} = \vec{l} + \vec{s}$ $j_z = l_z + s_z$	$\vec{j}$	J s	total angular momentum
	$\vec{l}$	J s	orbital angular momentum
	$\vec{s}$	J s	spin
	j_z	J s	z-component total angular momentum
	l_z	J s	z-component orbital angular momentum
	s_z	J s	z-component of spin

According to quantum-mechanical vector addition, the total angular momentum of an electron of orbital angular momentum l may take only the values $j = 1/2$ (for $l = 0$), or $j = l + 1/2, l - 1/2$ (for $l > 0$). Thus, the vector $\vec{j}$ has $2j + 1$ possible orientations relative to the z-axis. The projections of spin and orbital vectors add up, $m_j = m_l + m_s$.

3. Spin–orbit coupling

Spin–orbit coupling, interaction between magnetic spin and orbital moment, given by:

spin–orbit coupling			
	Symbol	Unit	Quantity
$V_{ls} = -\dfrac{Ze^2}{2m_e^2 c^2} \dfrac{1}{r^3} \vec{l} \cdot \vec{s}$	Z	1	atomic number
	e	C	electron charge
	m_e	kg	electron mass
	c	m/s	speed of light
	$\vec{l}$	J s	orbital angular momentum
	$\vec{s}$	J s	spin

The energy of an atomic electron depends on the relative orientation of spin and orbital angular momentum because of the magnetic interaction between spin moment and orbital moment. States with orbital angular momentum $\vec{l}$ and spin $\vec{s}$ aligned parallel or antiparallel, respectively, differ in energy. A level with the quantum number l splits into two levels with the quantum numbers $j = l + 1/2$ and $j = l - 1/2$, hence a **fine structure** of the spectral lines is observed (**Fig. 25.13**).

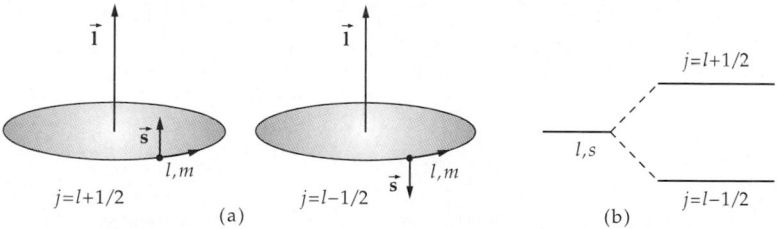

Figure 25.13: Spin–orbit coupling. (a): Illustration of the magnetic interaction between spin and orbital motion, (b): spin–orbit splitting of a level with orbital angular momentum l.

4. LS-coupling,

a coupling scheme for weak spin–orbit interaction. One first couples the orbital angular moments of the atomic electrons considered to a total orbital angular momentum $\vec{L}$,

$$\vec{L} = \sum_{i=1}^{N} \vec{l}_i \quad \text{with} \quad |\vec{L}| = \hbar\sqrt{L(L+1)}\,;$$

then the atomic electron spins are coupled to a total spin $\vec{S}$:

$$\vec{S} = \sum_{i=1}^{N} \vec{s}_i \quad \text{with} \quad |\vec{S}| = \hbar\sqrt{S(S+1)}\,.$$

The total angular momentum $\vec{J}$ of the atom is given by the **vector sum** of the total orbital angular momentum $\vec{L}$ and the total spin $\vec{S}$:

$$\vec{J} = \vec{L} + \vec{S} \quad \text{with} \quad |\vec{J}| = \hbar\sqrt{J(J+1)}\,.$$

The quantum number J may take the values

$$J = L + S, \ L + S - 1, \ldots, \ |L - S| + 1, \ |L - S|\,.$$

J takes $2S + 1$ values if $L \geq S$, and $2L + 1$ values if $L \leq S$ (**Fig. 25.14**).

▲   The LS-coupling scheme is an appropriate starting point for approximate solutions if the spin–orbit interaction causes only a weak perturbation of the electron motion. It is used preferably in spectral analyses of light atoms.

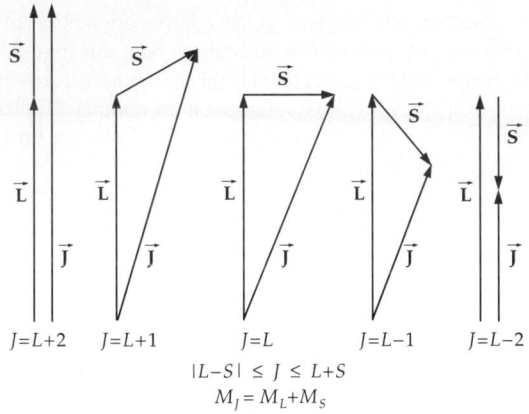

$$J=L+2 \qquad J=L+1 \qquad J=L \qquad J=L-1 \qquad J=L-2$$

$$|L-S| \le J \le L+S$$
$$M_J = M_L + M_S$$

Figure 25.14: LS-coupling for $L > S$ and $S = 2$ (schematic).

5. jj-coupling,

a coupling scheme for strong spin–orbit interaction. Here the orbital angular momentum $\vec{l}_i$ and the spin $\vec{s}_i$ of an atomic electron are first coupled to an individual total angular momentum of this electron:

$$\vec{j}_i = \vec{l}_i + \vec{s}_i .$$

The total angular momentum $\vec{J}$ of the atom is then obtained by summing the total angular momenta of the individual electrons:

$$\vec{J} = \sum_{i=1}^{N} \vec{j}_i \quad \text{with} \quad |\vec{J}| = \hbar\sqrt{J(J+1)} .$$

▲   The jj-coupling scheme is an appropriate starting point for approximate solutions when the spin–orbit interaction is strong. Its use is preferable in spectrum analysis of heavy atoms.

➤   In the analytical treatment based on the Schrödinger equation, the angular-momentum vectors are replaced by the corresponding operators.

6. Multiplets in the term structure

Multiplet, a group of energy levels (terms) belonging to different values of the quantum number J of the atomic total angular momentum.

   Multiplicity, the number of terms belonging to a multiplet (L, S, J) of energy levels:

$$S \le L: \quad \text{multiplicity } 2S+1, \qquad S > L: \quad \text{multiplicity } 2L+1 .$$

■ $S = 0$:   multiplicity 1, singlet system,

   $S = \frac{1}{2}$:   multiplicity 2, doublet system,

   $S = 1$:   multiplicity 3, triplet system.

▲   The following spectroscopic notation is used for characterizing the terms of a many-electron system:

multiplicity total orbital angular momentum total angular momentum			
	Symbol	Unit	Quantity
$^{2S+1}L_J$	S	1	quantum number of total spin
	L	1	quantum number of total orbital angular momentum
	J	1	quantum number of total angular momentum

7. Hund rules

In accordance with the **Pauli principle**, the electrons fill the quantum states such that:
1. the maximum total spin S,
2. the maximum total orbital angular momentum L,
3. the total angular momentum $J = L - S$ for less-than-half-filled shells,
   the total angular momentum $J = L + S$ for more-than-half-filled shells.

   Selection rules, relations among the quantum numbers of two stationary atomic states that must be satisfied in order to get an allowed radiative dipole transition:

$$\Delta S = 0, \quad \Delta L = \pm 1, \quad \Delta J = 0, \pm 1 \text{ (but not } 0 \longrightarrow 0), \quad \Delta M_J = 0, \pm 1.$$

8. Example: Helium atom

In the helium atom (nuclear charge number $Z = 2$), the spins of the two electrons couple to $S = 0$ (singlet) or $S = 1$ (triplet). Two term systems arise: **parahelium** ($S = 0$) and **orthohelium** ($S = 1$). The spin function in the singlet state behaves **antisymmetrically** under permutation of the particle coordinates; the spin function in the triplet is **symmetric** under permutation of the particles. The low-lying energy states correspond to the occupation of the lowest single-electron states in the Coulomb potential (**Fig. 25.15**):

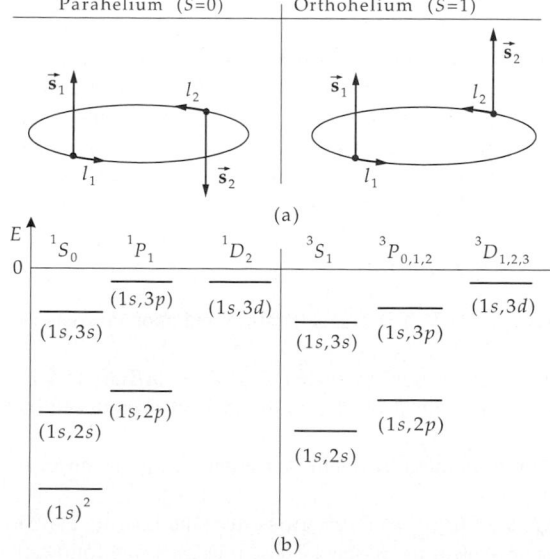

Figure 25.15: Helium atom. (a): parahelium (spin singlet, $S = 0$) and orthohelium (spin triplet, $S = 1$), (b): structure of terms (schematic). $(nl, n'l')$: electron configuration (n principal quantum number, l orbital angular momentum), $^{2S+1}L_J$: spectroscopic classification of the terms by the total spin (S), total orbital angular momentum (L) and total angular momentum (J).

electron 1	electron 2	configuration	orbital angular momentum
$1s$	$1s$	$(1s)^2$	$L = 0$ (S)
$1s$	$2s$	$(1s, 2s)$	$L = 0$ (S)
$1s$	$3s$	$(1s, 3s)$	$L = 0$ (S)
$1s$	$2p$	$(1s, 2p)$	$L = 1$ (P)
$1s$	$3p$	$(1s, 3p)$	$L = 1$ (P)
$1s$	$3d$	$(1s, 3d)$	$L = 2$ (D)

Terms in parahelium: $^1S_{J=0}$, $^1P_{J=1}$, $^1D_{J=2}$.
Terms in orthohelium: $^3S_{J=1}$, $^3P_{J=0,1,2}$, $^3D_{J=1,2,3}$.
The fine-structure splitting of the terms ^{2S+1}L in the helium atom with respect to the allowed J-values is very small.

➤ According to the Pauli principle, the electron configuration $(1s)^2$ cannot occur in orthohelium, since both the spatial function and the spin function would be symmetric under permutation of the particles. But the total function must be antisymmetric under permutation of all variables (position, spin).
The comparable states in orthohelium are more tightly bound than those in parahelium (positive exchange energy in symmetric spatial states).

9. Isotopic shift,

mass-dependent shift of the hyperfine structure multiplets observed in isotopic mixtures. It is caused by
• different values of the Rydberg constant of the isotopes due to nuclear drag (different nuclear masses of the isotopes),
• different deviations of the nuclear Coulomb field from the field of a point charge (different nuclear quadrupole moments for different isotopes).

25.5   X-rays

1. Characteristic x-rays,

arise in electron transitions from outer shells to atomic inner-shell states with small principal quantum number n. In the excitation of characteristic x-radiation by bombarding a metallic electrode by accelerated electrons, the electron collisions create holes in the inner electron shells that are subsequently filled by electrons from electron shells with a higher principal quantum number m (see **Fig. 25.16**). Such a transition is accompanied by emission of an **x-ray quantum** (photon) of energy,

$$hf_{mn} = E_m - E_n .$$

▲ X-ray quanta are in the energy range keV. The characteristic x-radiation consists of individual sharp lines.
Principal lines of characteristic x-ray spectra of several elements are listed in **Tab. 29.4/1**.
Primary radiation, characteristic x-radiation generated in the ionization produced by electron collisions.
Fluorescence radiation, x-radiation generated by photo ionization, i.e., in the absorption of x-ray photons by atoms.
If an electron is removed from the K-shell ($n = 1$), electrons from the L ($n = 2$)-, M ($n = 3$)-shell, etc., may go to the free places in the K-shell. These transitions are followed

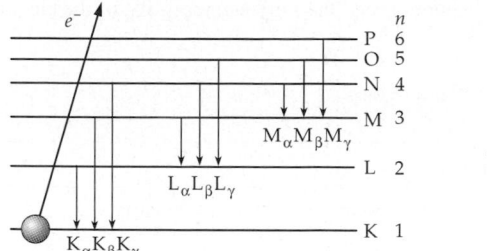
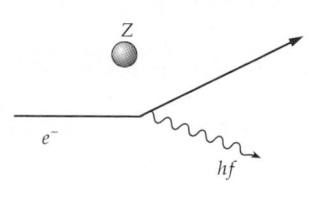

Figure 25.16: Characteristic x-radiation and bremsstrahlung from fast electrons ($v/c \leq 1$) deflected by an atomic nucleus.

by transitions to the secondary free places in the higher shells. The process terminates only when all atomic states are occupied again by electrons after **electron capture**, i.e., the atom is electrically neutral again.

K-series, spectral lines corresponding to transitions of electrons from outer shells to the K-shell. Analogously there are L-, M-series, etc.

The lines of a series are distinguished by a Greek letter as index (K_α, K_β, K_γ, ...).

■ K_{α_1} denotes an x-ray transition from the $2p_{3/2}$-state of the L-shell to the $1s_{1/2}$-state of the K-shell. K_β corresponds to a transition from the M- to the K-shell. K_γ corresponds to a transition from the N- to the K-shell, etc.

2. Moseley's law for characteristic frequencies

Moseley's law			1
	Symbol	Unit	Quantity
$\sqrt{\dfrac{f_{mn}}{cR}} = a(Z - \sigma)$	a	1	constant
	f_{mn}	s^{-1}	frequency
	R	m^{-1}	Rydberg constant
	c	$m\,s^{-1}$	speed of light
	Z	1	atomic number
	σ	1	screening constant

The constant a depends on the quantum numbers of the shells involved in the transition.

Screening constant, σ, a quantity that takes into account the screening of the valence electrons by the inner electrons, i.e., the valence electrons do not feel the full nuclear charge, but a smaller effective charge. According to Moseley's law, the frequencies of the α-lines of an element with atomic number Z are:

$$f_{K_\alpha} = \frac{3}{4} cR(Z - 1)^2,$$

$$f_{L_\alpha} \approx \frac{5}{36} cR(Z - 7.4)^2.$$

3. Bremsstrahlung,

a continuous x-ray spectrum generated by the deflection of electrons in the Coulomb field of the nucleus. The bremsstrahlung spectrum terminates at a certain lowest wavelength λ_{min}.

▲   The energy of the x-ray quanta cannot exceed the kinetic energy W_k of the electrons generating them:

$$W_k = e V_0 = h f_{max} = hc/\lambda_{min} .$$

wavelength limit of bremsstrahlung			L
	Symbol	Unit	Quantity
$\lambda_{min} = \dfrac{c\,h}{e\,V_0}$ $\approx 1.24\ \text{Å} \quad \text{for} \quad V_0 = 10^4\ \text{V}$	λ_{min} c h e V_0	m m/s J s C V	wavelength limit speed of light quantum of action elementary charge acceleration voltage

M   Measurements of the short-wave limit of a bremsstrahlung spectrum yield precise values of h.

25.5.1 Applications of x-rays

M   X-rays have a significant penetration depth in materials due to their high energy. This fact is used for **measurement of thickness**, **materials testing** and **quality control**.

1. Absorption of x-rays

Absorption coefficient, linear-attenuation coefficient, μ, reciprocal value of the penetration depth at which the radiation intensity drops to the fraction $1/e$ (e ≈ 2.718). The coefficient of x-ray absorption by matter decreases with increasing acceleration voltage V_0 (frequency of x-ray quanta). For mass-attenuation coefficient μ/ρ for x-rays, see **Tab. 29.6/1**.

   Absorption edges break the monotonous dependence of the absorption coefficient at frequencies at which the energy of the x-ray quanta is sufficiently high to release electrons from the K-, L-, M-, ... shell of the atom. At these energies, the absorption coefficient increases discontinuously (**Fig. 25.17**).

2. Auger effect,

a two-step process. The atom is excited by absorption of an x-ray quantum releasing an electron from a deeper shell (preferably the K-shell). The hole is then filled by an electron from a higher (L-, M-, ...) shell. The released energy ΔE is used to separate an additional electron (Auger electron) from an outer shell. The process is a **radiationless transition**.

M   X-ray quanta are measured by exploiting their ability to **ionize** or **dissociate** atoms or molecules. X-ray quanta may ionize atoms or molecules in a gas volume which, after acceleration by an electric field, produce a current pulse (**counter**). They may also be registered photographically via blackening of an x-ray film.

M   **X-ray computer tomography**, method of generating cross-sectional images of bodies. The principle rests on the dependence of the absorption coefficient on the transmission direction. The tomogram reflects the inhomogeneity of the irradiated body (**Fig. 25.18**). The distribution of inhomogeneity (mostly inhomogeneity of density) is calculated in three dimensions by a mathematical deconvolution of the intensity attenuation measured in various directions.

M   **Positron-emission tomographs** (PET), internal γ-source (positron emitter) may visualize dynamical processes in a body. The principle of measurement is similar to that of the x-ray computer tomograph.

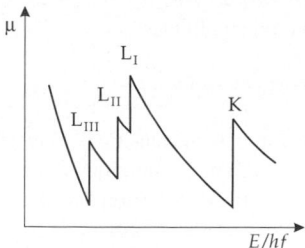

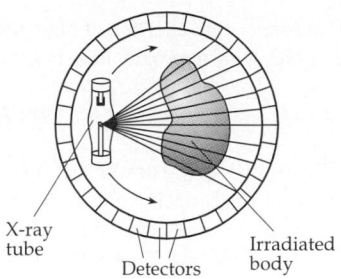

X-ray
tube   Irradiated
            Detectors   body

Figure 25.17: Absorption edges in the x-ray   Figure 25.18: Principle of computer
attenuation coefficient.   tomography.

25.6 Molecular spectra

Molecular spectra, consist of sequences of lines, bands and band systems. They originate from:

- electronic transitions, radiation in the infrared, visible and ultraviolet frequency range,
- vibrational transitions, radiation in the infrared spectral range,
- rotational transitions, radiation in the far infrared spectral range.

1. Vibrational spectra

Vibrational excitations, correspond to oscillations of the atoms of a molecule against each other along their connecting line. The center of mass of the molecule remains at rest, the electronic state remains unchanged.

Lennard–Jones potential, a model potential for diatomic molecules (**Fig. 25.19**):

$$V(r) = \left(-\frac{a}{r^6} + \frac{b}{r^{12}} \right).$$

The constants a and b are material parameters and are to a large extent independent of temperature. The high power of the second term means that the repulsive force becomes effective only when the distance between the particles is small.

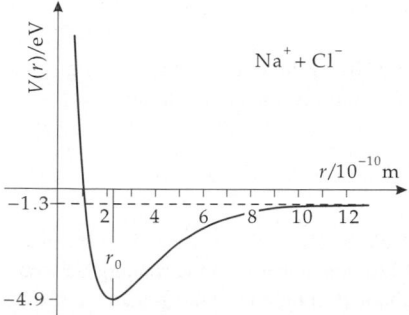

Figure 25.19: Ion binding in the NaCl molecule. Potential energy V as a function of the ionic distance r, equilibrium distance: $r_0 \approx 2.5 \cdot 10^{-10}$ m.

For dissociation energies of diatomic molecules, see **Tab. 29.1/6**.
The harmonic approximation holds for small oscillation amplitudes:

$$V(r) = \text{const} \cdot (r - r_0)^2 \,.$$

Vibrational spectrum of diatomic molecules, corresponds to transitions between the vibrational states of the molecule. In the approximation of small vibrational amplitudes, the atoms of the molecule vibrate against each other about the equilibrium distance r_0 and form a harmonic oscillator with equally spaced energy levels (**Fig. 25.20**):

quantum-mechanical vibrator			ML^2T^{-2}
	Symbol	Unit	Quantity
$E_{\text{vib}} = hf\left(v + \dfrac{1}{2}\right)$	E_{vib}	J	energy
	h	J s	quantum of action
$f = \dfrac{1}{2}\pi\left(\dfrac{k}{\mu}\right)^{1/2}$	f	s^{-1}	frequency
	v	1	vibrational quantum number
$v = 0, 1, 2, \dots$	k	kg/s^2	force constant
	μ	kg	reduced mass

▲   Vibrational spectra are characterized by **equally spaced** energy levels.

➤   About 20 vibrational levels are known in the NaCl molecule. The level spacing is about 0.04 eV.

2. Rotational spectra

Rotational spectra of diatomic molecules correspond to electromagnetic transitions between the rotational states of the molecule. The molecule rotates as a whole about an axis perpendicular to the molecular axis, without change of the atomic distance, or individual parts of the molecule rotate relative to each other (inner rotation).

Rigid rotator, the distance between the atoms of a diatomic molecule does not change under rotation (dumbbell model). The energy of a rigid rotator is determined only by its moment of inertia I and its angular momentum J (**Fig. 25.21**):

quantum-mechanical rotator			ML^2T^{-2}
	Symbol	Unit	Quantity
$E_{\text{rot}}(J) = \dfrac{\hbar^2}{2I}J(J+1)$	E_{rot}	J	energy
	h	J s	quantum of action
	J	1	rotational quantum number
	I	$kg\,m^2$	moment of inertia at r_0

▲   Rotational spectra are characterized by a **linear increase** in the spacing between neighboring energy levels with increasing rotational quantum number,

$$\Delta E = E_{\text{rot}}(J) - E_{\text{rot}}(J - 1) = \frac{\hbar^2}{I}J \,.$$

➤   The quantity $\hbar^2/I$ has a value of 10^{-4} eV to 10^{-2} eV for typical molecules. The spacing between neighboring rotational levels is lower than the spacing between vibrational levels. In the NaCl molecule, about 40 rotational states have been observed.

➤   There exists a series of rotational states for each given vibrational state.

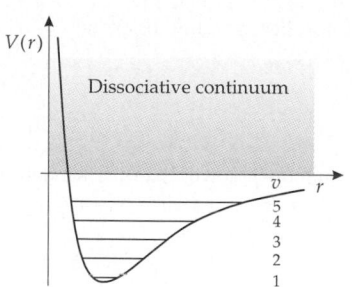

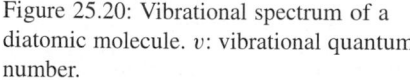

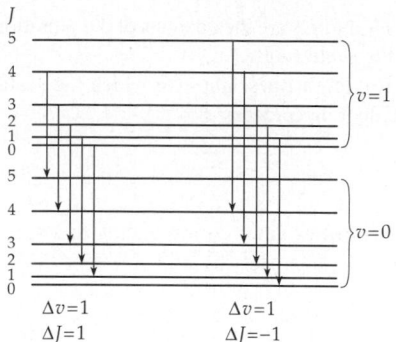

Figure 25.20: Vibrational spectrum of a diatomic molecule. v: vibrational quantum number.

Figure 25.21: Rotational-vibrational states of a diatomic molecule and allowed transitions. v: vibrational quantum number, J: rotational quantum number.

Selection rule for transitions between vibrational states:

$$\Delta v = \pm 1 \, .$$

Selection rule for transitions between rotational states:

$$\Delta J = \pm 1 \, .$$

   Rotational-vibrational band, a group of spectral lines related to transitions between rotational states built upon different vibrational states.
   Dissociation continuum, continuum limit joining a band towards the short wavelength spectral range. The continuum corresponds to dissociation of the molecule into free states of its constituents.

➤   The vibrational and rotational spectra of diatomic molecules are a result of the motion of the nuclei only. Moreover, there are also transitions between different electron configurations of the molecule, with energies between 1 eV and 10 eV. If the electron state changes, the binding potential between the ions or atoms of the molecule also changes. Hence, the equilibrium distance, the moment of inertia, the vibrational frequency and therefore also the excitation energies of the vibrational and rotational states are modified.

Electronic band spectrum, a complex spectrum with band structure generated by a manifold of transitions that involve a simultaneous change of the electronic, vibrational and rotational state of a molecule.

3. Raman spectra,

are produced by inelastic scattering of photons by molecules. In Raman scattering, in addition to the spectral lines of the primary light source, there also arise lines shifted symmetrically from these lines, weak lines of lower and higher frequency (see **Fig. 25.22**):

$$hf_0 + E_1 \rightarrow hf_r + E_1 \quad (a)$$

$$hf_0 + E_1 \rightarrow hf_s + E_2 \quad (b)$$

$$hf_0 + E_2 \rightarrow hf_a + E_1 \quad (c).$$

E_1 and E_2 are the energies of the vibrational or rotational states of the molecule involved in the scattering.

Rayleigh lines, lines for which the scattered photon has the same frequency f_r as the incident f_0 (process a):

$$f_r = f_0 .$$

Stokes lines, lines corresponding to a scattered frequency f_s lower than the incident frequency f_0 (process b):

$$f_s = f_0 - \frac{E_2 - E_1}{h} .$$

The photon transfers energy to the molecule.

Anti-Stokes lines, lines corresponding to a scattered frequency f_a higher than the incident frequency f_0 (process c):

$$f_a = f_0 + \frac{E_2 - E_1}{h} .$$

The photon carries off vibrational or rotational energy from an excited molecule.

M   Raman spectra yield information on the eigenfrequencies, the moments of inertia and the shape of molecules.

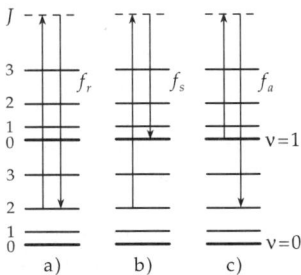

Figure 25.22: Raman spectra. Excitation of a virtual intermediate level (dashed line). (a): Rayleigh lines, (b): Stokes lines, (c): anti-Stokes lines.

25.7   Atoms in external fields

1. Electron in a magnetic field

Hamiltonian of an electron in a magnetic field $B_z \perp x, y$-plane:

$$\hat{H} = \frac{1}{2m}(p_x^2 + (p_y + m\omega_c x)^2), \quad \text{with} \quad \omega_c = \frac{eB_z}{m} .$$

Changing from classical momenta to **momentum operators**:

$$p_x \to -j \cdot \hbar \frac{\partial}{\partial x} \qquad p_y \to -j \cdot \hbar \frac{\partial}{\partial y} .$$

Substitution: $q = (x - x_0) = x + \dfrac{\hbar k_y}{m\omega_c} .$

Schrödinger equation for x-direction:

$$\left(-\frac{\hbar^2}{2m} \cdot \frac{\partial^2}{\partial x^2} + \frac{1}{2}m\omega_c^2 q^2\right)\psi(x, y) = E\psi(x, y).$$

➤   This differential equation is similar to that of a harmonic oscillator (see p. 90). The energy levels form an equally spaced spectrum:

$$E_n = \hbar\omega_c\left(n + \frac{1}{2}\right).$$

Cyclotron frequency, the angular frequency ω_c:

$$\omega_c = \frac{eB_z}{m}.$$

➤   These equations also hold for free electrons in solids, and for nucleons in the nucleus. Because of the modification of free motion by the potential in the medium, the particle mass must be replaced by the so-called "effective mass" m^*.

2. Zeeman effect,

a splitting of spectral lines in the magnetic field, caused by the shift of atomic energy levels due to the interaction of the atomic magnetic moment with the external magnetic field. The splitting is proportional to the magnetic flux density B.

Transverse Zeeman effect, the emitted light is observed perpendicular to the orientation of the magnetic field lines.

Longitudinal Zeeman effect, the emitted light is observed along the direction of the magnetic field lines.

Normal Zeeman effect in transverse observation: splitting of a line of frequency f into a triplet. The triplet consists of the non-shifted line and of two lines shifted symmetrically to higher and lower frequencies $f \pm \Delta f$ (see **Fig. 25.23**). The normal Zeeman effect occurs only in **singlet systems** ($S = 0$, $J = L$). The magnetic moment of the atom is then determined by the orbital moment. The term L splits into $2L + 1$ terms, which are separated by $\Delta E = \mu_B \cdot B$, μ_B being the Bohr magneton. The selection rule $\Delta M = 0, \pm 1$ provides a splitting into three lines, independent of L.

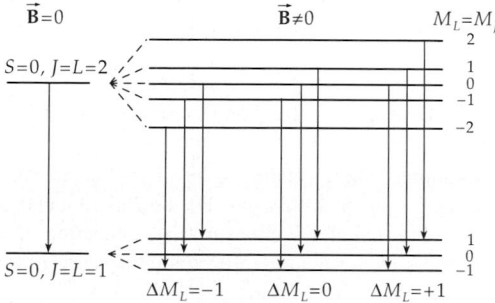

Figure 25.23: Normal Zeeman effect.

Anomalous Zeeman effect, complicated splitting of the spectral lines in a magnetic field. It occurs when the terms involved in the transitions are not spin singlets (**Fig. 25.24**).

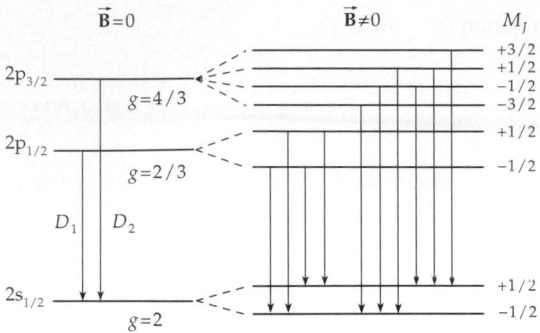

Figure 25.24: Anomalous Zeeman effect. Splitting of the ground state ($2s_{1/2}$) and of the first two excited states ($2p_{1/2}$, $2p_{3/2}$) of the Na atom in a magnetic field $\vec{B}$. g: Landé factor. The arrows mark the allowed transitions (selection rule $\Delta M_J = 0, \pm 1$).

3. Energy splitting in a magnetic field

energy splitting in magnetic field			$\mathbf{ML^2T^{-2}}$
	Symbol	Unit	Quantity
	ΔE	J	energy splitting
	m_j	1	quantum number
$\Delta E = g(L, S, J) \cdot m_J \cdot \mu_B \cdot B$			projection of tot. ang. mom.
	$g(L, S, J)$	1	Landé factor
	μ_B	J/T	Bohr magneton
	B	T	magnetic flux density

Landé factor, $g(L, S, J)$, describes the dependence of the gyromagnetic ratio on the quantum numbers of the term:

$$g(L, S, J) = 1 + \frac{J(J+1) - L(L+1) + S(S+1)}{2J(J+1)}.$$

■   Anomalous Zeeman effect in the sodium spectrum: the D_1-line (transition $2p_{1/2} \longrightarrow 2s_{1/2}$) and the D_2-line (transition $2p_{3/2} \longrightarrow 2s_{1/2}$) split into 4 and 6 lines, respectively.

Paramagnetic electron resonance, selective absorption of electromagnetic radiation by atoms of a substance. The frequencies correspond to transitions between Zeeman levels in an external magnetic field.

M	**Electron spin resonance**: The substance to be studied is placed in a magnetic field. The spin degeneracy is removed. A weak HF field is radiated onto the specimen, and the damping of the oscillator is measured as a function of the frequency. The damping reaches a maximum when the radio frequency coincides with the frequency of a transition between the Zeeman levels.

4. Stark effect,

the splitting of spectral lines under the influence of an electric field. This splitting is very small even in strong fields of 10^3 to 10^6 V/cm. In order to observe the effect, high-resolution spectrometers are needed.

Quadratic Stark effect, the splitting varies with the square of the electric field strength. The quadratic Stark effect occurs in atoms that have no permanent electric dipole moment in the ground state. In an external electric field $\vec{E}$ the atoms are polarized. The induced electric dipole moment $\vec{d}$ is proportional to $\vec{E}$. In the field $\vec{E}$ it has the potential energy $-\vec{d} \cdot \vec{E} \sim \vec{E}^2$. Thus, the quadratic Stark effect is connected with the **electric polarizability** of atoms.

Linear Stark effect, occurs in hydrogen and in hydrogen-like atoms, in which a degeneracy of states of equal principal number n with respect to the orbital angular momentum arises, i.e., states of different parity (e.g., $l = 0$ and $l = 1$) are mixed (**Fig. 25.25**).

▲   Hydrogen in the ground state ($n = 1$, $l = 0$) does **not** display the linear Stark effect.

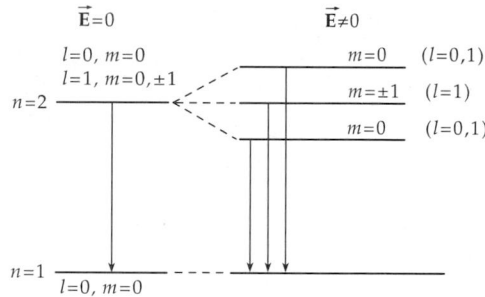

Figure 25.25: Linear Stark effect in the hydrogen atom.

25.8 Periodic Table of elements

1. Basic assumptions for explaining the Periodic Table

a) Model of independent particles, every electron of an atom moves independently of the other electrons in an effective potential. The mutual repulsion of the electrons yields only a weak residual interaction. This model combined with the Pauli principle explains the Periodic Table of elements.

▲   **Pauli principle**: In a system of indistinguishable particles with half-integer spin (see p. 889), no more than one particle can occupy a given one-particle state ($n\,l\,m_l m_s$).
Applied to the atom, this means: every electron in an atom has its own set of quantum numbers n, l, m_l and m_s that differs from the quantum numbers for any other electron.

b) Electron shells, the set of electrons occupying states with the same principal quantum number n.

Subshell, in the general case of a central potential deviating from the Coulomb potential, the degeneracy with respect to the orbital angular momentum quantum number l observed in the hydrogen atom is removed. The energy levels of a shell characterized by the quantum number l then form a subshell each. Formation of shells in the atom means a grouping of the levels according to the energy: the energy spacing between the subshells remains smaller than the energy separation of the main shells.

▲   Pauli principle: In a many-electron system, at most $2n^2$ electrons may occupy a shell with the principal quantum number n.
The electrons of the first ten elements occupy the following single-electron states (the arrow indicates the spin orientation):

atomic number	element	shell state	K 1s	2s	M 2p	ionization energy /eV
1	H		↑			13.6
2	He		↑↓			24.6
3	Li		↑↓	↑		5.4
4	Be		↑↓	↑↓		9.32
5	B		↑↓	↑↓	↑	8.296
6	C		↑↓	↑↓	↑↑	11.256
7	N		↑↓	↑↓	↑↑↑	14.545
8	O		↑↓	↑↓	↑↑↑↓	13.614
9	F		↑↓	↑↓	↑↑↑↓↓	17.418
10	Ne		↑↓	↑↓	↑↑↑↓↓↓	21.559

2. Filling of the electron states

Sequence of filling the electron states in the shells and in the subshells, distinguished by the orbital angular momentum l, corresponds to the sequence of the energy levels with given n and l:

Within a shell the state with $l = 0$ is occupied first, and then the states with larger l up to $l = n - 1$ follow.

Within a subshell, the successive filling is such that a maximum value of the total angular momentum is achieved.

Orbital, a state defined by n and l.

Valence electrons, determine the chemical and optical properties of the atoms. They belong to the s- and p-subgroups of the shell with the highest value of n of a given atom.

a) Inert-gas atoms, atoms with completely filled shells. For this reason, they respond inertly in chemical processes. Their ionization energy is very large.

b) Transition elements, elements with a modified sequence of shell occupation. It is energetically advantageous to fill first electron states with the next higher principal quantum number $n + 1$, but lower orbital angular momentum quantum number l before closure of the shell n. This refers to the orbitals $(n + 1)$s and $(n + 1)$p as compared to the orbitals nd and nf.

c) Transuranic elements, elements with atomic numbers above $Z = 92$. The atomic nuclei of these elements are not stable. They do not occur in nature.

The naming of the transuranic elements with atomic numbers 104 – 109 was controversial for a long time. The final names selected are: Rutherfordium (104), Dubnium (105), Seaborgium (106), Bohrium (107), Hassium (108), Meitnerium (109).

The heaviest **artificially produced elements** have been made in heavy-ion-induced nuclear reactions. The velocity filter Ship at Gesellschaft für Schwerionenforschung (GSI) in Darmstadt, Germany detected:

Bohrium, $_{107}$Bh (named after Niels Bohr): production reaction ^{209}Bi $+ ^{54}$Cr. It belongs to the 6d-transition metals. The 5f-, 6s-, 6p- and 7s-shells are occupied. The 6d-shell is half filled with five electrons. The element should have chemical properties like manganese and rhenium.

Hassium, $_{108}$Hs (named after the German federal state of Hesse, location of GSI): production reaction ^{208}Pb $+ ^{58}$Fe. This element belongs to the 6d-transition metals, with similar chemical properties as iron, osmium and ruthenium. The 5f-, 6s-, 6p- and 7s-shells are occupied. The 6d-shell carries six electrons. Up until 1993, four atoms of this element have been detected.

Meitnerium, $_{109}$Mt (named after Lise Meitner): production reaction ^{209}Bi $+ ^{58}$Fe. This element is a 6d-transition metal with properties similar to cobalt, rhodium and iridium. The 5f-, 6s-, 6p- and 7s-shells are occupied. The 6d-shell carries seven electrons.

d) Superheavy elements, elements with $Z \geq 110$. $Z = 112$ were observed for the first time in 1996. In 1999 elements $Z = 114, 116, 118$ were reported by a Lawrence Berkeley Laboratory team.

3. Magnetic moment of the atom,

is determined by the contribution of the spin moment and the contribution from incompletely filled subshells.

▲   In an occupied s-state, the magnetic spin moments of the electrons compensate each other.

▲   In occuped p-, d-, f-subgroups, besides the magnetic spin moments, the magnetic orbital moments are compensated too. The magnetic moment of these atoms equals zero.

Diamagnetism occurs in all elements with completed subshells.

Paramagnetism occurs in elements with incompletely filled subshells. These atoms have a nonvanishing magnetic moment.

4. Ionization potentials and atomic radii

Ionization potentials: **Tab. 29.1/1**, atomic and ionic radii: **Tab. 29.2**.

25.9   Interaction of photons with atoms and molecules

25.9.1   Spontaneous and induced emission

Absorption, a photon is absorbed by an atom. The atom thereby changes to a higher energy state (**Fig. 25.26**).

1. Spontaneous and induced emission

Spontaneous emission, the emission of photons by excited atoms (molecules) without a fixed phase relation between photons emitted by different atoms (molecules) in identical excited states.

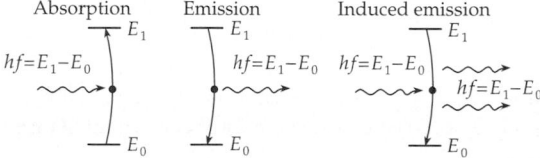

Figure 25.26: Schematic pictures of absorption and emission of photons.

Induced emission, the emission of photons of energy hf from excited atomic or molecular states under the action of an electromagnetic field of the same frequency. In this case the incident and the emitted photon have the same phase. After the process, the number of photons with frequency f in the radiation field has increased by 1.

➤   The property of coherence of photons in the induced emission is used in the **quantum generators**, the **lasers** and **masers** (light/microwave amplification by stimulated emission of radiation).

Occupation number N_1, number of atoms in a certain energy state E_1. It is temperature-dependent. The occupation numbers are changed by emission and absorption.

Occupation number ratio, determined by the **Boltzmann distribution** at a definite temperature in thermal equilibrium (see p. 576):

Boltzmann distribution			1
	Symbol	Unit	Quantity
$\dfrac{N_1}{N_2} = e^{-\frac{E_2-E_1}{kT}}$	N_1, N_2	1	occupation numbers
	E_1, E_2	J	energy of states
	k	J/K	Boltzmann constant
	T	K	temperature

In thermal equilibrium, the occupation of the lower-lying level dominates.

2. Occupation inversion,

inversion, **occupation inversion**, a process leading to a higher occupation of the upper energy level than that of the lower level, by input of energy.

■   **Three-level laser**: There exists a metastable state of relatively long lifetime ($\tau \approx 10^{-3}$ s) (**Fig. 25.27**). Normally, the lifetime of an excited atom is about $\tau \approx 10^{-8}$ s. A further level lying above the metastable one is excited, e.g., by intense illumination of short-wave light ($E_1 \to E_3$). The laser medium is selected such that spontaneous transitions $E_3 \to E_2$ are favored above $E_3 \to E_1$. The different holding times in the various states and the different transition probabilities between the states cause the occupation number of level 2 to dominate over that of level 1 ($N_2 > N_1$).

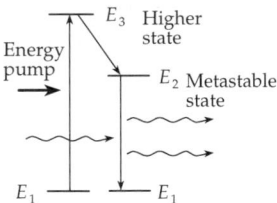

Figure 25.27: Three-level laser.

Optical resonators, the induced emitted light quanta are forced by a mirror system to stay in a limited space region. Thereby the number of coherent light quanta increases in an avalanche-like manner.

| M | **Helium-neon laser**, belongs to the group of gas lasers (**Fig. 25.28**).

The excitation is carried out by electron collisions in a gas-discharge tube. The laser-active part is formed by a capillary. In the gas mixture (He-Ne; $p_{He} : p_{Ne} = (5 \dots 10) : 1$), the helium atoms are excited by electron collisions via an intermediate level at 25 eV into the metastable levels 2^3s and 2^1s. The excited He atoms transfer their energy by collisions completely to the metastable 2s and 3s levels of neon atoms, thereby generating an occupation inversion. Different laser radiation may be extracted by stimulated transitions into the 2p and 3p levels. As a rule, the spectral lines lying in the IR range are suppressed in favor of the line with $\lambda = 632.8$ nm by an appropriate choice of the resonator mirrors.

➤ Laser power of 10 GW is possible, though only in short pulses of 10^{-9} s duration.

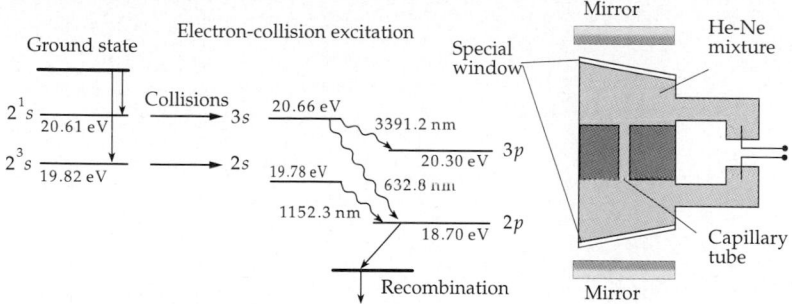

Figure 25.28: Helium-neon laser. (a): Operation mode, (b): constructive design (schematic).

26
Elementary particle physics—standard model

26.1 Unification of interactions

26.1.1 Standard model

Standard model, model of fundamental particles and their interactions based on the **electroweak theory** and **quantum chromodynamics** (theory of color interaction).

1. Fundamental particles

12 fermions (spin-1/2 particles):
- six **quarks** and
- six **leptons**,

each group being subdivided into three **generations** according to increasing mass. All matter is composed of these particles and their antiparticles:

	Quarks	Q/e	Leptons	Q/e
1st generation	d (down)	$-1/3$	electron neutrino ν_e	0
	u (up)	$+2/3$	electron e^-	-1
2nd generation	s (strange)	$-1/3$	muon neutrino ν_μ	0
	c (charme)	$+2/3$	muon μ	-1
3rd generation	b (bottom)	$-1/3$	tau neutrino ν_τ	0
	t (top)	$+2/3$	tau τ	-1

2. Fundamental interactions

Universality, the observation that the particle families differ only in their mass, but not in their interaction.

Four fundamental interactions describe the known physical world completely (**Fig. 26.1**):

- gravitation,
- electromagnetism,
- strong nuclear force,
- weak nuclear force.

Interaction type	Strength (relative)	Range (m)	Interaction between	Field quanta (gauge bosons)
strong	1	$\approx 10^{-15}$	color charges and quarks	gluons g
electromagnetic	10^{-2}	∞	electric charges	photons γ
weak	10^{-14}	$\approx 2 \cdot 10^{-18}$	leptons and hadrons	$W^{\pm}, Z^0$ bosons
gravitation	10^{-38}	∞	all particles	gravitons

The four interactions are mediated by exchange particles, the vector bosons:

- graviton,
- photon γ,
- gluon g,
- $W^{\pm}, Z^0$.

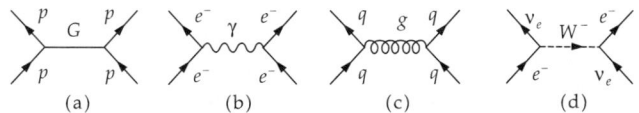

| (a) | (b) | (c) | (d) |

Figure 26.1: Elementary graphs of (a): gravitational interaction, (b): electromagnetic interaction, (c): strong interaction, (d): weak interaction.

26.1.1.1 Gravitational interaction

Gravitational interaction, the attractive interaction between masses. It is described in terms of the exchange of hypothetically massless **gravitons** with spin 2 as field quanta (see p. 887).

gravitational force = constant $\cdot \dfrac{\text{mass}_1 \cdot \text{mass}_2}{\text{distance}^2}$			$\mathbf{MLT^{-2}}$
$\vec{F}_G = -G \cdot \dfrac{M_1 \cdot M_2}{r^2} \cdot \dfrac{\vec{r}}{r}$ $G = 6.672\,59(85)$ $\cdot 10^{-11}\,\mathrm{N\,m^2\,kg^{-2}}$	Symbol	Unit	Quantity
	$\vec{F}_G$	N	gravitational force
	G	$\mathrm{N\,m^2\,kg^{-2}}$	gravitational constant
	M_1, M_2	kg	masses
	$\vec{r}$	m	distance vector between masses

The gravitational interaction has an infinite range and cannot be shielded.

➤   The hypothesis of a fifth force, which may be introduced as an additional Yukawa-type term to the gravitation potential Φ, with a strength parameter α and a range parameter λ,

$$\Phi(r) = -G\,\frac{M}{r}\,(1 + \alpha e^{-r/\lambda}),$$

leads to a gravitational constant that would depend on the distance r of the test body from the gravitating mass M. This hypothesis has not been substantiated by experiment so far.

26.1.1.2   Electromagnetic interaction

Electromagnetic interaction, the interaction between electric charges, currents, and magnetic moments. It is explained by the exchange of massless **photons** with spin 1 as field quanta. The interaction between electric charges is described by Coulomb's law.

Coulomb force = constant $\cdot$ $\dfrac{\text{charge}_1 \cdot \text{charge}_2}{\text{distance}^2}$			$\mathbf{MLT^{-2}}$
$\vec{F}_{el} = \dfrac{1}{4\pi\varepsilon_0} \cdot \dfrac{Q_1 \cdot Q_2}{r^2}\,\dfrac{\vec{r}}{r}$	Symbol	Unit	Quantity
	$\vec{F}_{el}$	N	Coulomb force
$\varepsilon_0 = 8.854\,187\,817$	Q_1, Q_2	C	charges
	ε_0	$CV^{-1}m^{-1}$	permittivity of free space
$\cdot\,10^{-12}\ \mathrm{C\ V^{-1}\ m^{-1}}$	r	m	distance between charges

➤   The ratio of the gravitational force and the Coulomb force between two protons is:

$$\frac{F_G}{F_{el}} = G \cdot 4\pi\varepsilon_0 \cdot \frac{m_p^2}{e^2} \approx 0.83 \cdot 10^{-36}.$$

For a given distance apart, the electrostatic interaction between protons is about 10^{36} times stronger than the gravitational force between them.

26.1.1.3   Weak interaction

1. Weak interaction

the interaction responsible for the **decay** of heavy leptons and quarks into the lighter ones. This decay is described as the exchange of $W^{\pm}$, Z^0 vector bosons with spin 1 and large mass. As a consequence, the weak interaction has a short range.

■   **Free neutrons** decay via the **weak interaction** into three particles: $n \rightarrow p + e^- + \bar{\nu}_e$: a proton, an electron and a neutral electron antineutrino (**Fig. 26.2**). The mean lifetime of the free neutron

$$\tau = (889.1 \pm 2.1)\ \mathrm{s}$$

is larger by a factor of 10^{27} than the time of 10^{-23} s that is characteristic for processes governed by the strong interaction.

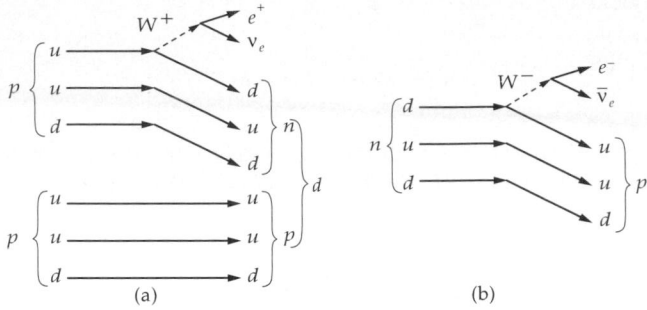

Figure 26.2: Quark line diagrams. (a): fusion of hydrogen, (b): neutron decay.

2. Properties of the weak interaction

- The interaction strength is **significantly smaller** than the strength of the strong interaction, and at low energies also smaller than that of the electromagnetic interaction. It is, however, larger than the strength of the gravitational interaction.
- The weak interaction has an extremely short range, smaller than 10^{-17} m.
- There are **no bound states**.

The **short range of the weak interaction** has its origin in the large rest mass of the $W^\pm$ and Z^0 bosons. According to the uncertainty relation, the exchange of virtual particles must satisfy the condition $\Delta E \cdot \Delta t \geq \hbar$, $\Delta E = m_W c^2$ being the rest energy of the boson. These bosons may propagate only a distance of the order of:

$$R_0 \approx \frac{\hbar}{m_W c} \approx \frac{200 \text{ MeV} \cdot \text{fm}}{100 \cdot 10^3 \text{ MeV}} = 2 \cdot 10^{-18} \text{ m} \,,$$

even if they were to move with the speed of light in a vacuum. This corresponds to about one one-thousandth the range of the nuclear force.

➤ For photons $m_{ph} = 0$, and hence $R_0 = \infty$: the electromagnetic field has an infinite range.

Virtual particle, a particle with energy and momentum that do not satisfy the relativistic energy–momentum relation for free particles:

$$\frac{E^2}{c^2} - p^2 \neq m_0^2 c^2 \,.$$

▲ Virtual particles exist only for short times (mediators of interactions).

26.1.1.4   Strong interaction

Strong interaction, the interaction responsible for the binding of nuclear constituents, and therefore for the existence of stable nuclei. The strong interaction between **quarks** as constituents of hadrons and mesons is attributed to the exchange of massless **gluons**, which are field quanta with spin 1.

▲ Properties of the strong interaction:

- The strong interaction is **attractive** at distances of $r \approx 2 \cdot 10^{-15}$ m, and **repulsive** for $r < 10^{-15}$ m.
- It has a **short range** ($\approx 10^{-15}$ m).

- Within this range, it is about 100 to 1000 times **stronger** than the **electromagnetic interaction**.
- A **saturation** of the binding energy per nucleon with increasing nuclear mass number occurs.
- The strong interaction is **state-dependent**. The interaction between two nucleons depends on the relative orientation of the nucleon spins, the isospin T of the two-nucleon system, and the orbital momentum of the relative motion.
- The strong interaction is **charge-independent**. In the nucleon–nucleon system with isospin $T = 1$, the same force $V(n-n) = V(p-p) = V(p-n)$ acts independently of the charge state of the nucleon pair (assuming otherwise identical quantum numbers of the nucleon-pair state). The interaction $V(p-n)$ in the isospin state $T = 0$ differs from the interaction in the $T = 1$-state.

Mechanism of the strong interaction is based on the **color force**, i.e., the exchange of massless colored gluons with spin 1 as field quanta.

At sufficiently large distances, the strong interaction may also be described effectively by the exchange of mesons between the nucleons.

Yukawa potential, an approximation for the attractive part of the potential between two nucleons:

Yukawa potential			$\mathbf{MLT^{-2}}$
	Symbol	Unit	Quantity
$V_K = V_0 \cdot \dfrac{1}{r} \, \mathrm{e}^{-\frac{r}{r_0}}$	V_K	J/m	Yukawa potential
	V_0	J	interaction strength
	r_0	m	range
	r	m	nucleon distance

26.1.2 Field quanta or gauge bosons

1. Gauge bosons

or **field quanta**, mediators of the interactions (bosons, integer spin values).

Graviton, spin 2, gauge boson (mediator) of the gravitational interaction. The graviton is expected to be massless and uncharged. The graviton has not yet been detected experimentally.

Photon, spin 1, gauge boson of the electromagnetic interaction in quantum electrodynamics (QED). This theory takes into account the quantum nature of the electromagnetic field and correctly describes the experimental deviations from the description based on potentials (Coulomb force, Maxwell equations).

▲ The photon has rest mass $m_\gamma = 0$ and charge $q_\gamma = 0$.

Free photons are the energy quanta of light, **virtual photons** are the mediators of the electromagnetic interaction.

$W^\pm, Z^0$ **bosons**, spin 1, field quanta of the weak interaction:
- $W^\pm$ with mass $m = 80.22 \pm 0.26$ GeV,
- Z^0 with mass $m = 91.173 \pm 0.020$ GeV.

Z-bosons are the reason neutrinos are repelled by electrons and quarks.

2. Electroweak interaction

(*Salam* and *Weinberg*), unified theory of electromagnetic and weak interactions. The existence of the Z^0 particle was predicted by this theory.

➤ Another prediction of the unified theory, the Higgs particle with $m_H \approx 300$ GeV, has not yet be observed.

3. Field quanta of the strong interaction

Gluons (the Greek root for glue), spin 1, field quanta of the strong or color interaction (**quantum chromodynamics**). The gluons bind the quarks together. Eight distinct gluons should exist in total, which differ in their color. Like photons, they are quanta without rest mass. Unlike photons, which may propagate over unlimited distances, gluons are restricted to a finite space region of about 10^{-15} m diameter, since they carry color charge and therefore strongly interact among each other. See **Fig. 26.3** for a comparison of size scales of matter and their various binding forces. Presumably, free quarks or gluons will never be observed, since the force between two quarks increases with their separation r (assuming a linear quark-quark potential $V(r) = Br$, $B > 0$).

Glueballs, particles consisting exclusively of gluons. There are experimental hints of such bound systems of field quanta.

Type of Matter	Interaction	Mediator
Quark	strong	Gluon
Hadron	strong	Gluon
Nuclei	electromagnetic	Photon
Atoms	electromagnetic	Photon
Molecules	gravitational	Graviton
Solar system		

Figure 26.3: Scales of matter and corresponding forces.

M Gluons arise e.g. in the annihilation of a high-energy positron-electron pair. A quark and an antiquark are generated in these annihilations. If the energy of the electrons and positrons is sufficiently large, one or several gluons may be formed in the separation of the quark-antiquark pair.

Quark, antiquark, and gluon cannot propagate as individual particles beyond distances of 10^{-15} m, since they immediately produce additional particles. Characteristic **hadron jets** are formed in this way.

4. Theoretical approaches of elementary-particle physics

Gauge theory, mathematical formulation of the interactions, derived from a symmetry principle: The basic equation is invariant under certain transformations of the wave function.

■ Electroweak theory and quantum chromodynamics are gauge theories. One hopes to formulate a unification of the interaction, possibly including a gravitation theory based on a gauge theory.

Coupling constants g_1, g_2, g_3, parameters of electromagnetic, weak and strong interaction. They determine the relative strength of the corresponding forces between the particles. The coupling constants depend on the momentum and energy transferred in the interaction process.

Asymptotic freedom: The coupling constant g_3 of the strong interaction becomes small for large momentum transfer or at small distances. The quarks then behave like quasi-free particles. Perturbation theory can be applied.

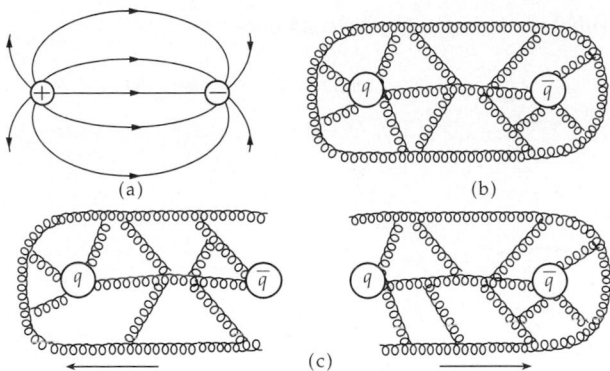

Figure 26.4: Quark confinement. (a): field lines of an electric dipole, (b): field configuration of gluons between the quark q and the antiquark $\bar{q}$, (c): formation of quark-antiquark pairs (mesons) in breaking the flow tube (arrows show directions of motion).

Quark confinement, expected, but not yet strictly proven, consequence of quantum chromodynamics that quarks cannot be observed as free particles. It follows from the property of the quark–quark interaction via the exchange of self-interacting gluons: for larger distances of quark and antiquark as constituents of a meson, the energy of the quark-antiquark pair increases proportional to the distance, hence new quark-antiquark pairs are formed that combine to become colorless mesons (**Fig. 26.4**).

26.1.3 Fermions and bosons

The elementary particles are grouped in two classes according to their spin values: fermions and bosons.

1. Fermions

All elementary particles with half-integer spin $(1/2, 3/2, 5/2, \ldots)$. They obey Fermi statistics and are governed by the Pauli principle (see p. 844).

Fermi–Dirac statistics, quantum statistics for a system consisting of fermions in equilibrium.

Fermi distribution, gives the mean number n_i of noninteracting fermions in the state i with energy E_i (**Fig. 26.5**):

Fermi distribution			1
	Symbol	Unit	Quantity
$$n_i = \dfrac{g}{e^{\frac{E_i-\mu}{kT}} + 1}$$ $$g = 2s + 1$$	n_i	1	particle number
	g	1	weight factor
	E_i	J	energy of state i
	μ	J	chemical potential
	k	J/K	Boltzmann constant
	T	K	temperature
	s	1	particle spin

Chemical potential, μ, determined by the condition

$$\sum_i n_i = N \quad \text{(total number of fermions)}.$$

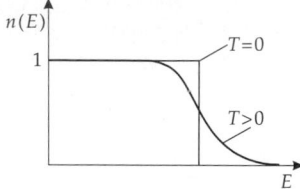

Figure 26.5: Fermi distribution.

2. Bosons

Elementary particles with integer spin, they obey Bose–Einstein statistics and are **not** governed by the **Pauli principle**.

Bose–Einstein statistics, describe the statistical distribution according to the quantum mechanics of indistinguishable particles with integer spin (0, 1, 2, ...).

➤   A state i may be occupied by an arbitrary number of bosons.

Bose–Einstein distribution, describes the mean particle number n_i of noninteracting particles with integer spin in a state i with energy E_i:

Bose–Einstein distribution				1
	Symbol	Unit	Quantity	
$n_i = \dfrac{g}{e^{\frac{E_i - \mu}{kT}} - 1}$ $g = 2s + 1$	n_i	1	particle number	
	g	1	weight factor	
	E_i	J	energy of state i	
	μ	J	chemical potential	
	k	J/K	Boltzmann constant	
	T	K	temperature	
	s	1	particle spin	

The weight factor g is equal to 1 for bosons with spin $s = 0$, and $g = 3$ for bosons with spin $s = 1$. For fermions with spin $s = 1/2$, $g = 2$. In general: $g = 2s + 1$.

▲   All fundamental particles have a non-zero spin value.

▲   Gauge bosons, the field quanta of the fundamental interactions, have the following spin values: spin 1 for photons, $W^\pm$, Z^0 and gluons, spin 2 for the hypothetical graviton.

➤   The boson nature of photons is of importance for the laser principle: there can be an arbitrary number of photons with identical phase at a given position in the same energy state.

3. Bose–Einstein condensation

transition of a non-interacting particle system obeying Bose–Einstein statistics into a state in which all particles occupy the lowest energy state. Bose–Einstein condensation is expected at high particle number densities n or low temperatures T if the distance between the particles (mass m) becomes comparable to the de Broglie wavelength λ of the particles in thermal motion,

$$n\lambda^3 \geq 2.612\,, \quad \lambda = \sqrt{\hbar^2/(2\pi mkT)}\,,$$

(k Boltzmann constant). Bose–Einstein condensation is disturbed by the interatomic interactions: if there are strong forces between the molecules, a normal liquid arises rather than a Bose–Einstein condensate.

The Bose–Einstein condensation of a weakly interacting system of bosons was demonstrated in 1995 for a gas of rubidium atoms by combining the methods of laser cooling and evaporation cooling of a gas confined in a magnetic trap. By laser pre-cooling in a magneto-optical trap a cold, dense cloud of rubidium atoms is produced which then is brought into a magnetic trap. Here it is cooled down by evaporative cooling to a temperature of 170 nK at a density of $3 \cdot 10^{12}$ cm^3. Bose–Einstein condensation of about 2000 atoms has been observed in the center of the trap, manifesting itself as a drastic change of the position and momentum distribution of the particles. A second, non-condensed component was observed in the vicinity of the condensate.

Evaporative cooling, selective removal of energy-rich particles from the system. After thermalization, the remaining system has a lower mean energy.

➤ Bose–Einstein condensation has also been demonstrated for lithium atoms which attract each other by the weak Van der Waals forces.

26.2 Leptons, quarks, and vector bosons

26.2.1 Leptons

Leptons, a class of particles governed by the electroweak interaction, but not by the strong interaction.

▲ Leptons have spin 1/2, they are fermions.
▲ There are six types of leptons and their corresponding antiparticles.
▲ All leptons are structureless, point-like particles.

Properties of leptons:

Name	Mass $m/(\mathrm{MeV}/c^2)$	Charge Q/e
electron e	$0.510\,999\,06 \pm 0.000\,000\,15$	-1
electron neutrino ν_e	$< 7.3 \cdot 10^{-6}$	0
muon μ	$105.658\,389 \pm 0.000\,034$	-1
muon neutrino ν_μ	< 0.27	0
tau lepton τ	1776.3 ± 2.4	-1
tau neutrino ν_τ	< 31	0

Name	Magnetic dipole moment μ/μ_B	Electric $d/(e \cdot \mathrm{cm})$	Lifetime
electron e	$1.001\,159\,652\,193$ $\pm 0.000\,000\,000\,010$	$(-0.3$ $\pm 0.8) \cdot 10^{-26}$	$\tau > 3.8 \cdot 10^{23}$ a
electron neutrino ν_e	$< 1.08 \cdot 10^{-9}$		$\tau/m_{\nu_e} > 300$ s/eV

Name	Magnetic dipole moment μ/μ_B	Electric dipole moment $d/(e \cdot \text{cm})$	Lifetime
muon μ	$(1.001\,165\,923$ $\pm 0.000\,000\,008)\dfrac{m_e}{m_\mu}$	$(+3.7$ $\pm 3.4) \cdot 10^{-19}$	$\tau = (2.197\,03$ $\pm 0.000\,05) \cdot 10^{-6}$ s
muon neutrino ν_μ	$< 7.4 \cdot 10^{-6}$		$\tau/m_{\nu_\mu} > 15.4$ s/eV
tau lepton τ			$\tau = (0.305$ $\pm 0.006) \cdot 10^{-12}$ s
tau neutrino ν_τ	$< 4 \cdot 10^{-6}$		

Lepton charge, lepton number L, a charge-like quantum number similar to baryon number B.

▲ For a system of elementary particles, the baryon and lepton charges are summed separately.

▲ **Lepton charge** is conserved in all nuclear reactions.

▲ All leptons have lepton number $L = \pm 1$.

▲ All leptons have baryon number $B = 0$.

■ The electron has lepton charge $+1$, the positron lepton charge -1.

■ The **photon** γ has both baryon charge $B = 0$ and lepton charge $L = 0$.

Positron, the antiparticle of the electron.

26.2.2 Quarks

Hadrons, all particles governed by the strong interaction. They have an intrinsic structure. Baryons and mesons are hadrons. Any baryon is composed of three quarks. Any meson consists of a quark-antiquark pair.

▲ All hadrons have lepton number $L = 0$.

Quarks, particles invented hypothetically in order to explain the similarity of baryon and meson multiplets.

▲ Quarks are structureless and point-like. There are six kinds of quarks and six kinds of antiquarks, just as there are six kinds of leptons and six kinds of antileptons. Quarks (q) and antiquarks ($\bar{q}$) have baryon numbers $\frac{1}{3}$ and $-\frac{1}{3}$, respectively.

Properties of quarks:

Name		$m/(\text{MeV}/c^2)$	Q/e	I	I_z	s	π	S	Charm	Bottom	Top
down	d	$5\ldots 15$	$-\frac{1}{3}$	$\frac{1}{2}$	$-\frac{1}{2}$	$\frac{1}{2}$	$+$	$-$	$-$	$-$	$-$
up	u	$2\ldots 8$	$+\frac{2}{3}$	$\frac{1}{2}$	$\frac{1}{2}$	$\frac{1}{2}$	$+$	$-$	$-$	$-$	$-$
strange	s	$100\ldots 300$	$-\frac{1}{3}$	0	0	$\frac{1}{2}$	$+$	-1	$-$	$-$	$-$
charm	c	$1300\ldots 1700$	$+\frac{2}{3}$	0	0	$\frac{1}{2}$	$+$	$-$	$+1$	$-$	$-$
bottom	b	$4700\ldots 5300$	$-\frac{1}{3}$	0	0	$\frac{1}{2}$	$+$	$-$	$-$	-1	$-$
top	t	174000 ± 17000	$+\frac{2}{3}$	0	0	$\frac{1}{2}$	$+$	$-$	$-$	$-$	$+1$

Q: charge, I: isospin, I_z: isospin projection, s: spin, π: parity, S: strangeness.

▲   Because of the relatively long lifetime of the mesons and hadrons, which are composed of the $c/\bar{c}$-quarks, $b/\bar{b}$-quarks and $t/\bar{t}$-quarks, new quantum numbers are assigned to them: to the c-quark: **charm**; to the b-quark: **bottom**; to the t-quark: **top**.

Top-quark. The top-quark t was detected in 1994 in proton-antiproton collisions with a center-of-mass energy of 1.8 TeV. A light quark within the proton collides with a light antiquark within the antiproton, forming a $t\bar{t}$-pair. The top-quark t decays almost exclusively into a b-quark and a W^+-meson, which in turn decays either (67 %) hadronically into two quarks (u (or c) and $\bar{d}$ (or $\bar{s}$)), or (33 %) leptonically into $e^+ + \nu_e$ or $\mu^+ + \nu_\mu$. Similarly, $\bar{t}$ decays into $\bar{b}$ and W^-, with a subsequent hadronic or leptonic decay of the W^--meson. The neutrinos manifest themselves as a missing value in the energy balance for the reconstructed event. The quarks and antiquarks hadronize with formation of hadronic showers (jets), whereby the particle tracks arising in the hadronization of the b-quarks are characterized by a displacement of the vertex away from the interaction point. The $t\bar{t}$-pair thus can be detected by events characterized by the appearance of two charged leptons and at least two jets.

The mass of the top-quark is $(174 \pm 17)\,\text{GeV}/c^2$, and thus 35 times larger than the mass of the b-quark. The t-quark is the heaviest elementary particle known so far.

Baryon charge, **baryon number** B, a charge-like quantum number assigned to the elementary particles (charge-like means: it is an additive scalar quantity, like the electric charge).

▲   All quarks have baryon number $B = \pm\frac{1}{3}$ and lepton number $L = 0$.

▲   All baryons have lepton number $L = 0$.

▲   The **baryon number** is conserved in all particle conversions.

➤   This conservation law guarantees that the number of particles and antiparticles belonging to a family remains unchanged. Protons and neutrons have baryon charge $+1$. Electrons and positrons have baryon number 0.

1. Flavors: strangeness, charm, bottom, and top

▲   Hadrons are composed of quarks. The six quark types are called **flavors**.

▲   Mesons consist of one quark and one antiquark each. Their baryon number therefore equals 0.

▲   Baryons consist of three quarks, their baryon number is 1.

▲   The strong interaction is flavor-blind, i.e., it does not distinguish between the kinds of quarks.

▲   Flavor changes are mediated by the weak interaction.

All baryons in the baryon decuplet and the baryon octet may be constructed from the down-, up- and strange-quarks.

The three edges in the baryon decuplet violate the Pauli principle if there is no new degree of freedom characterizing the quarks.

2. Color

▲   **Color**, a new degree of freedom ascribed to quarks and gluons. It has the character of a (color) charge responsible for the color interaction.

▲   Any quark occurs in three colors. Convention: **red** r, **blue** b, **green** g.

▲   Antiquarks carry the complementary colors (anti-red $\bar{r}$, anti-blue $\bar{b}$, anti-green $\bar{g}$).

▲   All hadrons are color-neutral (white).

   Baryons: the three quarks have distinct colors adding up to zero (white).

   Mesons are formed by $q\bar{q}$-pairs with complementary colors $r\bar{r}$, $b\bar{b}$, $g\bar{g}$.

▲   Gluons, the mediators of the strong color interaction, are themselves colored:

Unlike the photon, which is electrically neutral, the gluons have a "color charge". There are eight distinct color combinations of gluons: $r\bar{b}$, $r\bar{g}$, $b\bar{g}$, $g\bar{b}$, $b\bar{r}$, $g\bar{r}$, $(r\bar{r} + g\bar{g} - 2b\bar{b})/\sqrt{6}$, $(r\bar{r} - g\bar{g})/\sqrt{2}$.

26.2.3   Hadrons

Hadrons, spatially extended elementary particles governed by the strong interaction (see p. 892). **Mesons** and **baryons** are distinguished by their spin values.

Mesons, elementary particles composed of a quark–antiquark pair. They are governed by the strong interaction and have integer spin values.

▲   Mesons have baryon number $B = 0$.

Baryons, elementary particles composed of three quarks. They are governed by the strong interaction and have half-integer spin. Baryons are fermions.

▲   Baryons have baryon number $B = \pm 1$.

Fig. 26.6 illustrates the quark composition of various baryons and mesons.

1. Strangeness and heavy baryons

Strangeness, S, property of certain elementary particles to be generated by the strong interaction and to decay via the weak interaction. A new quantum number is assigned to these particles that describes this property and is mediated by a quark, the strange quark.

▲   The strangeness quantum number is conserved in strong and electromagnetic interactions; it is not conserved in weak interaction.

▲   If the decay of a particle implies the violation of a conservation law, the process is suppressed, and this corresponds to a prolongation of the lifetime of the particle.

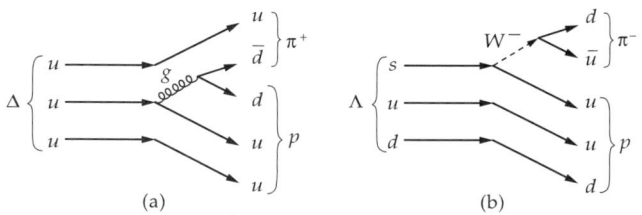

Figure 26.6: Quark line diagrams. (a): delta decay, (b): lambda decay.

2. Hyperons and kaons

Hyperons, particles with half-integer spin value s and masses above the nucleon mass. They belong to the family of baryons and carry strangeness ($S \neq 0$).

K-meson, kaon, an unstable elementary particle from the meson family. It carries strangeness $S = \pm 1$ and decays after a mean lifetime τ of 10^{-8}–10^{-10} s. This time is very large as compared with the characteristic time for processes governed by strong interaction. A time of 10^{-10} s is typical for processes proceeding via the weak interaction. There are four K-mesons: K^+, K^0, $\overline{K}^0$, K^-.

▲   Frequently, hyperons and K-mesons are produced in pairs.

M   **Fig. 26.7** schematically shows the reaction:

$$\pi^+ + p \rightarrow \Lambda^0 + K^0 + \pi^+ + \pi^+$$
$$\Lambda^0 \rightarrow \pi^- + p$$
$$K^0 \rightarrow \pi^+ + \pi^-.$$

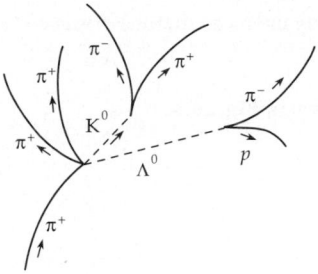

Figure 26.7: Schematic view of a bubble-chamber exposure: A π^+ strikes an (invisible) proton at rest.

Λ^0-**hyperon**, a neutral elementary particle invisible in a bubble-chamber, as is any neutral particle. It decays into a proton and a π^--meson with a mean lifetime of $2.6 \cdot 10^{-10}$ s.

K^0-**meson**, a neutral K-meson, also invisible in bubble-chamber records. The K^0-meson decays with a mean lifetime of 10^{-10} s into a π^+-meson and a π^--meson.

3. Table of mesons with spin 1 (vector mesons)

Name	Symbol	$m/(\text{MeV}/c^2)$	Q/e	S	Γ/MeV	Quark content
rho-meson	ρ^+	768.1 ± 0.5	1	0	151.5 ± 1.2	$u\bar{d}$
	ρ^0	768.1 ± 0.5	0	0	151.5 ± 1.2	$(u\bar{u} - d\bar{d})/\sqrt{2}$
	ρ^-	768.1 ± 0.5	-1	0	151.5 ± 1.2	$d\bar{u}$
omega-meson	ω	781.95 ± 0.14	0	0	8.43 ± 0.10	$(u\bar{u} + d\bar{d})/\sqrt{2}$
phi-meson	ϕ	1019.413 ± 0.008	0	0	4.43 ± 0.06	$s\bar{s}$
kaon	K^{*+}	891.59 ± 0.24	1	1	49.8 ± 0.8	$u\bar{s}$
	K^{*0}	896.10 ± 0.28	0	1	50.5 ± 0.6	$d\bar{s}$
	K^{*-}	891.59 ± 0.24	-1	-1	49.8 ± 0.8	$\bar{u}s$
	$\bar{K}^{*0}$	896.10 ± 0.28	0	-1	50.5 ± 0.6	$\bar{d}s$

4. Table of mesons with spin 0 (pseudoscalar mesons)

Name	Symbol	$m/(\text{MeV}/c^2)$	Q/e	S	τ/s	Quark content
charged pion	$\pi^\pm$	139.5679 ± 0.0007	± 1	0	$(2.6030$ $\pm 0.0024) \cdot 10^{-8}$	$u\bar{d}$ $d\bar{u}$
neutral pion	π^0	134.9743 ± 0.0008	0	0	$(8.4$ $\pm 0.6) \cdot 10^{-17}$	$(u\bar{u} - d\bar{d})$ $/\sqrt{2}$
eta-meson	η	547.45 ± 0.19	0	0	$\approx 0.55 \cdot 10^{-18}$	$(u\bar{u} + d\bar{d}$ $-2s\bar{s})/\sqrt{6}$
	η'	957.75 ± 0.14	0	0	$\approx 0.33 \cdot 10^{-20}$	$(u\bar{u} + d\bar{d}$ $+s\bar{s})/\sqrt{3}$
kaon	$K^\pm$	493.646 ± 0.009	± 1	± 1	$(1.2371$ $\pm 0.0029) \cdot 10^{-8}$	$u\bar{s}$ $\bar{u}s$
neutral kaon	K^0	497.671 ± 0.031	0	1	$50\% K_S^0,$ $50\% K_L^0$	$d\bar{s}$
	$\bar{K}^0$	497.671 ± 0.031	0	-1	$50\% K_S^0,$ $50\% K_L^0$	$\bar{d}s$
K short	K_S^0	497.671 ± 0.031	0	—	$(0.8922$ $\pm 0.0020) \cdot 10^{-10}$	—
K long	K_L^0	497.671 ± 0.031	0	—	$(5.17$ $\pm 0.04) \cdot 10^{-8}$	—

Notation in the table: m mass; Q electric charge; S strangeness; τ mean lifetime.

▲   Giving the decay width is equivalent to giving the mean lifetime τ, since $\Gamma = \hbar/\tau$.

5. Ordering scheme of the meson family

The spin O meson family ordering scheme is shown in **Fig. 26.8**.

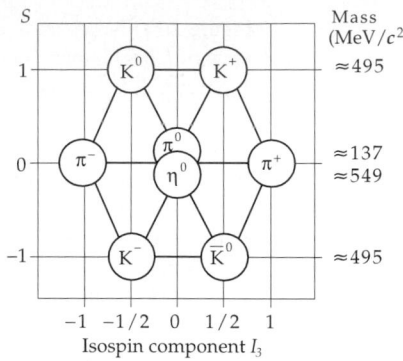

Figure 26.8: Ordering scheme of the meson family with spin 0 (pseudoscalar mesons).

Additional mesons, also composed of c-($\bar{c}$-) and b-($\bar{b}$-)quarks:

- D- and D^*-mesons with charm $C = \pm 1$,
- D_S- and D_S^*-mesons with charm and strangeness $C = S = \pm 1$,
- B- and B^*-mesons with bottom $B = \pm 1$.

6. Quarkonium

Quarkonium, a quark-antiquark state (=meson) involving the heavy quarks,

- charmonium ($c\bar{c}$): e.g. the J/ψ with $m = 3096.93 \pm 0.09$ MeV/c^2,
- bottonium ($b\bar{b}$): e.g. the Υ with $m = 9460.32 \pm 0.22$ MeV/c^2.

➤   The names are formed in analogy to positronium, the bound $e^+ - e^-$ state.

The excited states are similar to the level series found in atomic physics (**Fig. 26.9**).

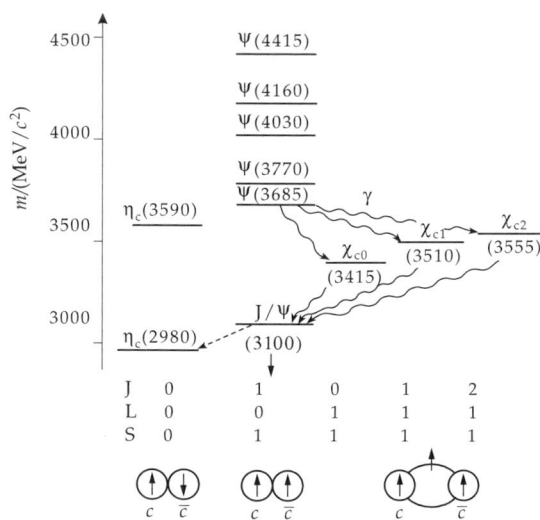

Figure 26.9: Mass spectrum of the charmonium states with spin S, orbital angular momentum L, and total angular momentum J.

7. Baryons with spin 1/2

Spin-1/2 octet, ordering scheme of the baryons with spin $\frac{1}{2}$ (see **Fig. 26.10**).

Survey of mass m, charge Q, mean lifetime τ, magnetic dipole moment μ, electric dipole moment d, and strangeness S:

Name	Symbol	$m/\mathrm{MeV}/c^2$	Q/e	τ/s
proton	p	$938.272\,31 \pm 0.000\,28$	1	$> 10^{31}$ yr
neutron	n	$939.565\,63 \pm 0.000\,28$	0	889.1 ± 2.1
lambda	Λ	1115.63 ± 0.05	0	$(2.632 \pm 0.020) \cdot 10^{-10}$
sigma	Σ^+	1189.37 ± 0.07	1	$(0.799 \pm 0.004) \cdot 10^{-10}$
	Σ^0	1192.55 ± 1.10	0	$(7.4 \pm 0.7) \cdot 10^{-20}$
	Σ^-	1197.43 ± 0.06	-1	$(1.479 \pm 0.011) \cdot 10^{-10}$
xi	Ξ^0	1314.90 ± 0.6	0	$(2.90 \pm 0.09) \cdot 10^{-10}$
	Ξ^-	1321.32 ± 0.13	-1	$(1.639 \pm 0.015) \cdot 10^{-10}$

Name	Symbol	μ/μ_N	$d/(e \cdot \mathrm{cm})$	S	Quark content
proton	p	$2.792\,847\,39 \pm 6 \cdot 10^{-8}$	$(-4 \pm 6) \cdot 10^{-23}$	0	uud
neutron	n	$-1.913\,042\,7 \pm 5 \cdot 10^{-7}$	$< 12 \cdot 10^{-26}$	0	udd
lambda	Λ	-0.613 ± 0.004	$< 1.5 \cdot 10^{-16}$	-1	sdu
sigma	Σ^+	2.42 ± 0.05		-1	suu
	Σ^0			-1	sdu
	Σ^-	-1.160 ± 0.025		-1	sdd
xi	Ξ^0	-1.250 ± 0.014		-2	ssu
	Ξ^-	-0.6507 ± 0.015		-2	ssd

8. Baryons with spin 3/2

Spin-3/2 decuplet, ordering scheme for baryons with spin $\frac{3}{2}$. Baryon multiplet of three quarks (see **Fig. 26.11**):

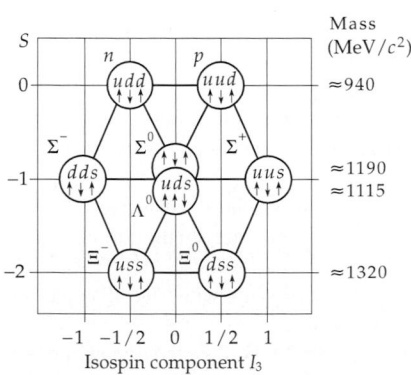

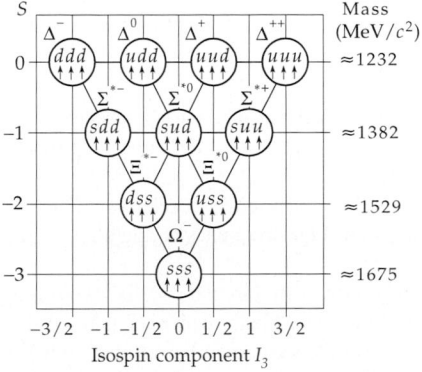

Figure 26.10: Ordering scheme of baryons with spin 1/2.

Figure 26.11: Spin-3/2 baryon decuplet of three quarks.

The antibaryons have similar multiplets.

➤ For the isospin component $I_3 = 0$ and strangeness -1, there exist two states distinguished by the isospin quantum number I: $I = 0$: Λ^0, $I = 1$: Σ^0.

Baryon family with spin 3/2

Name	Symbol	$m/(\text{MeV}/c^2)$	Q/e	τ/s	S	Quark content
omega	Ω^-	1672.43 ± 0.32	-1	$(0.822 \pm 0.012) \cdot 10^{-10}$	-3	sss
xi	Ξ^{*0}	1531.80 ± 0.32	0	$\Gamma = 9.1 \pm 0.5$ MeV	-2	ssu
	Ξ^{*-}	1535.0 ± 0.6	-1	$\Gamma = 9.9 \pm 1.8$ MeV	-2	ssd
sigma	Σ^{*+}	1382.8 ± 0.4	1	$\Gamma = 35.8 \pm 0.8$ MeV	-1	suu
	Σ^{*-}	1387.2 ± 0.5	-1	$\Gamma = 39.4 \pm 2.1$ MeV	-1	sdd
	Σ^{*0}	1383.7 ± 1.0	0	$\Gamma = 36 \pm 5$ MeV	-1	sdu
delta	Δ^{++}	1232	2	$\Gamma = 115 - 125$ MeV	0	uuu
	Δ^+	1232	1	$\Gamma = 115 - 125$ MeV	0	uud
	Δ^0	1232	0	$\Gamma = 115 - 125$ MeV	0	udd
	Δ^-	1232	-1	$\Gamma = 115 - 125$ MeV	0	ddd

26.2.4   Accelerators and detectors

Subatomic structures may be studied only by highly energetic projectiles (incident particles). According to

$$\lambda = \frac{h}{p},$$

(see p. 826) the wavelength λ of the beam of matter shortens with increasing momentum, i.e., finer and finer details may be resolved.

In order to produce new pairs of particles of mass m, a specific threshold energy is required:

$$E = 2mc^2 .$$

Any increase in energy achieved in accelerators may thus provide new knowledge.

1. Accelerators

Linear accelerator, a particle accelerator with a linear succession of high-frequency acceleration segments. The beam of projectiles passes them only once before hitting the target.

Cyclotron, circular accelerator. The particles follow orbits in a transverse magnetic field. The high-frequency acceleration voltage of fixed frequency acts on the particles many times.

Synchrotron, circular accelerator with a magnetic field varying with time. The particle orbit is a closed path, which the particles traverse many times.

Collider, accelerator based on the synchrotron principle. Two beams moving in opposite direction are made to collide with a small angle between them. For the same beam energy, a much higher energy in the center-of-mass system of the colliding particles is reached than in fixed-target accelerators.

■ Examples of colliders are the electron–proton storage ring HERA at DESY in Hamburg, with a circumference of 6.3 km (30-GeV electrons colliding with 820-GeV protons) and the electron-positron collider LEP at CERN in Geneva, with 26.7 km circumference and $60 + 60$ GeV beam energy. The Z- and W-bosons, the carriers of the weak interaction, were demonstrated unambiguously for the first time in 1983 at the proton-antiproton collider at CERN (1984 Nobel Prize to C. Rubbia and S. van der Meer).

Currently, the proton–proton collider LHC at CERN is under construction with 8 TeV available energy.

Luminosity, $L^* = N_S/\sigma$, unit $s^{-1}cm^{-2}$, important characteristic for storage rings, gives the number of reactions of a certain type N_S per second, divided by the reaction cross section σ.

Linear collider, arrangement of two oppositely directed linear accelerators. The particles traverse the acceleration sections only once before collision. But since the particles travel in straight lines, the large radiation losses due to deflections in storage rings are avoided. Currently, several 0.5 TeV e^+e^- collision machines are being planned: TESLA (20 km length, 1.3 GHz frequency) and S-Band (25 km, 3 GHz), both at DESY, CLIC (6.25 km, 30 GHz) at CERN, and several others.

2. Detectors

- **Nuclear plates**, photographic **emulsions** blackened along the tracks of the detected particles.
- **Bubble chamber**, formerly used for measuring elementary particles. A liquid is kept under pressure close to its boiling point in a large chamber. By a sudden lowering of pressure, the liquid is brought to a superheated state. Highly energetic charged particles passing this region generate an ionization track along which the surrounding liquid starts to boil. This causes a change in the refractive index, and the track can be observed in transmission or reflection of light. The bubble chamber is sensitive for about 10 ms after lowering of pressure. The charged particles are deflected by magnetic fields (Lorentz force). The charge and velocity of the particles can be extracted from the track curvature. The energy of the particles is determined from the ionization density. Liquid hydrogen or propane have been used as detector liquids.
- **Streamer chamber**, detector in which the passage of particles is made to produce luminous discharges along the track by the application of pulses of high voltage. The tracks are photographed for later analysis.
- **Ionization chamber**, detector that measures the primary ionization generated by the particle. The detector works with a counter gas in an electric field.
- **Čerenkov counter**, a detector in which the particles move with a speed above the phase velocity of light through an optically strongly refractive material, and thus generate a cone-like electromagnetic wave front. The particle velocity can be determined from the angle of the light cone. Recent application is the **R**ing **I**maging **Ch**erenkov counter (RICH).
- **Semiconductor detectors**, determine the ionization $\Delta E/\Delta x$, and possibly also the deposited energy E.
- **Silicon-strip detector**, strips of boron on a silicon monocrystal. The p-n junction is operated with back-bias. The electrons produced by a charged particle passing through the detector are collected on the anodes of the stripes.
- **Scintillation counter**, particle detection by fluorescent light quanta in the passage of a charged particle through a scintillator. Amplification of the light signal by secondary-electron multipliers. The high time resolution enables high counting rates. The spatial resolution is low.

- **Proportional chamber**, consists of planes of parallel anode wires (thickness about 50 μm, distance about 1 mm) between metallic cathode planes. Argon-alcohol mixture as filling gas. High precision in spatial localization of particle track.
- **TPC** (time projection chamber), track detector allowing the reconstruction of the particle trajectory, taking into account the drift times of electrons produced by the ionization processes. The position and time coordinates of the particles are determined by hundreds of anode wires or pads.

26.3 Symmetries and conservation laws

Homogeneity of time, properties of the laws of nature do not change in time. Physical quantities of a system homogeneous in time do not depend on time t, but only on time differences Δt. This is a deeper explanation of energy conservation.

Homogeneity of space, properties of the laws of nature do not vary in space. Physical quantities of a system homogeneous in space do not change in displacements (translations) $\vec{r} \rightarrow \vec{r} + \Delta \vec{r}$. This is a deeper explanation of momentum conservation.

Isotropy of space, the equivalence of all directions in space. The properties of a system do not change under rotations. A consequence of the isotropy of space is the conservation of angular momentum.

Noether theorem: the correspondence of fundamental symmetries and conservation laws. The invariance of the field-theoretical action integral with respect to an n-parametric continuous transformation group implies the existence of n conservation laws.

26.3.1 Parity conservation and the weak interaction

Mirror symmetry of the world means that the mirror object of any object may also exist as a real object.

Parity conservation, mirror symmetry of the world in quantum mechanics. It always holds when the strong or electromagnetic interaction is responsible for the reaction.

■   Excited atoms in a field-free space radiate electromagnetic waves isotropically. If the atom is put into the magnetic field of a pair of coils, the atomic levels of different angular momentum projections relative to the field direction are split (Zeeman effect). The radiation pattern is mirror-symmetric with respect to the plane of the circular current. It does not change if the current-flow direction is reversed (**Fig. 26.12**).

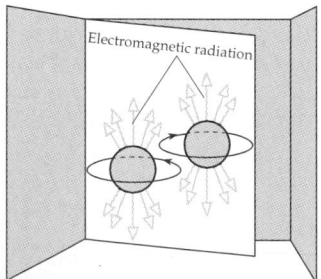

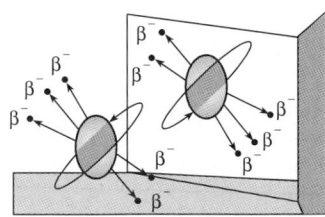

Figure 26.12: Electromagnetic radiation in the Zeeman effect, parity conservation.

Figure 26.13: Schematic diagram of parity violation.

Parity operator, $\hat{P}$, generates a spatial reflection of the wave function: $\hat{P}\psi(\vec{r}) = \psi(-\vec{r})$.

Parity violation, the non-conservation of parity during a nuclear, atomic, or elementary-particle reaction. An example of a parity-violating reaction is nuclear β-decay, as is schematically shown in **Fig. 26.13**.

■ A β-emitter (e.g. a ^{60}Co-source) is put into a uniform magnetic field at low temperature. Let the magnetic moments of the ^{60}Co-nuclei be fully polarized. The counting rate of a β-sensitive detector (e.g. an anthracene scintillator) is measured as a function of the heating time of the sample. Simultaneously, the γ-radiation emitted by the source is recorded. With increasing temperature, the polarization is gradually removed by thermal motion, and the β-asymmetry disappears. In a second measurement, the polarity of the magnetic field is reversed. **Fig. 26.14** shows the results.

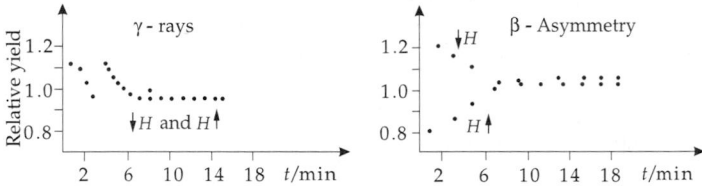

Figure 26.14: Experimental results on β-decay of ^{60}Co. H: magnetic field.

The asymmetry of β-emission depends on the orientation of the magnetic field H and is therefore not mirror-symmetric. The asymmetry of the γ-quanta, on the contrary, is independent of the magnetic field direction.

▲ Parity is **not** conserved in **weak interactions**.

Consequence of parity violation:

▲ The spins of **electron** and **neutrino** from weak decays always point **opposite to the propagation direction** (negative helicity). The spins of the corresponding antiparticles, **positron** and **antineutrino**, on the contrary, always point **along the propagation direction** (positive helicity).

▲ An intrinsic parity $\pi_n = \pi_p = +1$ is assigned to protons and neutrons. Intrinsic parity $\pi_e = +1$ is also ascribed to the electron. A system of two particles A and B has parity

$$\pi = (-1)^l \pi_A \cdot \pi_B,$$

where l is the quantum number of the orbital angular momentum of the relative motion. Parity is a multiplicative quantum number.

26.3.2 Charge conservation and pair production

▲ Elementary-particle and nuclear reactions always proceed in such a way that the total charge does not change; electric charge, baryon charge, and lepton charge are added separately and are conserved in all reactions.

■ An example of the conservation of electric charge is alpha decay:

$$\underbrace{^{14}_{7}\text{N} + ^{4}_{2}\text{He}}_{9} \rightarrow \underbrace{^{1}_{1}\text{H} + ^{17}_{8}\text{O}}_{9}.$$

Pair production, a reaction in which electromagnetic radiation (γ-quantum) is converted into a particle and the corresponding antiparticle, e.g. production of an electron–positron pair:

$$\gamma \rightarrow e^+ + e^- .$$

Because of energy and momentum conservation, the e^+e^- pair formation may proceed only in the external field of a third particle (e.g. an atomic nucleus). Pair production is a threshold reaction. Because of the finite rest mass of the electron and the positron ($m_e \cdot c^2 \approx 511$ keV), this reaction occurs only at γ-energies above 1.022 MeV.

Pair annihilation, a process in which a particle and its antiparticle (with a vanishing total momentum) annihilate to form electromagnetic radiation. Because of momentum conservation, at least two photons must emerge:

$$e^+ + e^- \rightarrow 2\gamma .$$

Antiparticles, elementary particles the charge-like quantum numbers have the opposite signs (but the same magnitude) with respect to the conjugated particles.

➤ The conservation law of electric charge would allow the conversion of a γ-quantum into an electron and a proton. This reaction is **not observed**; both the baryon number as well as the lepton number would then not be conserved.

Antiproton, the antiparticle of the proton. It has electric charge $q_{\bar{p}} = -1e$, baryon number $B_{\bar{p}} = -1$, and parity $\pi_{\bar{p}} = -1$.

According to the abovementioned conservation laws, conversion of a γ-quantum into a proton and an antiproton is possible. The threshold energy for this reaction is

$$Q_{\text{thr}} \geq 2 \cdot m_p \cdot c^2 = 2 \cdot 938.2796 \text{ MeV.}$$

The charge-like quantum numbers of various elementary particles are listed below.

Elementary particle	Electric charge	Baryon charge	Lepton charge
proton	+1	+1	0
neutron	0	+1	0
electron	−1	0	+1
positron	+1	0	−1
π^+, π^0, π^--mesons	+1, 0, −1	0	0
photon	0	0	0
neutrino	0	0	+1
antiproton	−1	−1	0
antineutron	0	−1	0
antineutrino	0	0	−1

26.3.3 Charge conjugation and antiparticles

Charge conjugation, C, symmetry operation connecting particles and antiparticles. Charge conjugation is connected with a discontinuous transformation. Under this transformation, a particle is substituted for its antiparticle.

▲   For any particle there exists an antiparticle. It has the same mass and lifetime as the particle, but opposite charge-like quantum numbers.

➤   For symmetry with respect to charge conjugation, the universe should not only be electrically neutral, but there should be as many particles as antiparticles. All present observations, however, indicate an asymmetry of the universe.

$\hat{C}$**-operator**, operator carrying out the transformation particle→ antiparticle. Applying this operator twice successively leads back to the original particle.

26.3.4   Time-reversal invariance and inverse reactions

Time-reversal invariance, the symmetry of physical phenomena with respect to time reversal.

$\hat{T}$**-operator**, operator causing time reversal, i.e., replacement of t by $-t$.

■   In an inelastic collision between two particles A and B, particles C and D are produced in the final state. The probability for the transition of the system from the initial state i to the final state f is denoted w_{fi}. The probability for the inverse process (initial state f^* goes into the final state i^*) is $w_{i^* f^*}$. Time-reversal invariance requires:

$$w_{fi} = w_{i^* f^*} .$$

The following table shows the behavior of several physical quantities with respect to time reversal $\hat{T}$, charge conjugation $\hat{C}$, and space inversion $\hat{P}$.

Quantity	Symmetry operation		
	$\hat{T}$	$\hat{C}$	$\hat{P}$
momentum $\vec{\mathbf{p}}$	$-\vec{\mathbf{p}}$	$\vec{\mathbf{p}}$	$-\vec{\mathbf{p}}$
spin $\vec{\mathbf{J}}$	$-\vec{\mathbf{J}}$	$\vec{\mathbf{J}}$	$\vec{\mathbf{J}}$
electric field $\vec{\mathbf{E}}$	$\vec{\mathbf{E}}$	$\vec{\mathbf{E}}$	$-\vec{\mathbf{E}}$
magnetic field $\vec{\mathbf{H}}$	$-\vec{\mathbf{H}}$	$\vec{\mathbf{H}}$	$\vec{\mathbf{H}}$
dipole moment (electric) $\vec{\mathbf{J}} \cdot \vec{\mathbf{E}}$	$-\vec{\mathbf{J}} \cdot \vec{\mathbf{E}}$	$\vec{\mathbf{J}} \cdot \vec{\mathbf{E}}$	$-\vec{\mathbf{J}} \cdot \vec{\mathbf{E}}$

▲   **Time-reversal invariance** has been confirmed for reactions governed by strong or electromagnetic interaction.

▲   Symmetry of the interaction under separate $\hat{C}$-, $\hat{P}$- or $\hat{T}$-transformation is not a universal law of nature.

▲   The electromagnetic, weak, and strong interactions are invariant under the application of all three operations in any order.

➤   A consequence of the $\hat{C}\hat{P}\hat{T}$-invariance that may be confirmed by experiment is the equality of the mean lifetimes, the masses and the magnitudes of magnetic moments of particles and antiparticles. Up to the present no experiments are known that violate the $\hat{C}\hat{P}\hat{T}$-invariance.

26.3.5   Conservation laws

Conservation laws and interaction symmetries are closely related:

▲   If a symmetry is broken, then a conservation law is violated.

Universal conservation laws and their validity for the various interactions:

Conservation law/ quantum number	Interaction			
	strong	electromagnetic	weak	gravitational
energy E	+	+	+	+
momentum $\vec{p}$	+	+	+	+
angular momentum $\vec{J}$	+	+	+	+
charge-like:				
electr. charge Q	+	+	+	+
baryon charge B	+	+	+	+
lepton charge L	+	+	+	+
spin-like:				
spin $\hat{s}$	+	+	+	+
isospin $\hat{I}$	+	—	—	—
isospin component I_z	+	+	—	—
strangeness S	+	+	—	—

Conservation law	Physical origin	Type of conservation law
energy	homogeneity of time	geometric
momentum	homogeneity of space	geometric
angular momentum	isotropy of space	geometric
$\hat{C}\hat{P}$-invariance	right-left symmetry of space	geometric
$\hat{T}$-invariance	symmetry of time $(t, -t)$	geometric
electric charge	unknown	charge
baryon charge	unknown	charge
lepton charge	unknown	charge
strangeness	unknown	

26.3.6 Beyond the standard model

M **Lifetime of proton**, should be $\tau = 4.5 \cdot 10^{29 \pm 1.7}$ yr according to the prediction of the **Grand Unified Theory (GUT)** of Georgi and Glashow (i.e., by many orders of magnitude larger than the age of the universe). Various experiments have yielded lower limits of 10^{31} up to $5 \cdot 10^{32}$ years. The experiments were carried out in salt mines, gold mines, and mountain tunnels in order to shield them from cosmic radiation.

The energy-dependent coupling parameters for the interactions, including a GUT, are shown in **Fig. 26.15**.

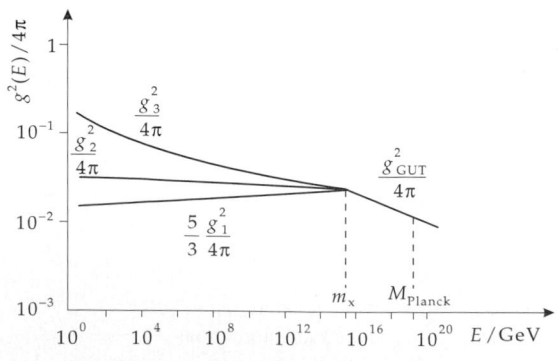

Figure 26.15: Energy-dependent coupling parameters g_1 (electromagnetic interaction), g_2 (weak interaction), g_3 (strong interaction), and g_{GUT} (Grand Unified Theory).

Supersymmetry model (SUSY), model of unification postulating a series of new elementary particles associated with the known ones:

■   Neutralino, chargino, sneutrino, selectron, smuon, squark and gluino.

A mass $m > 15$ GeV/c^2 has been predicted for the lightest supersymmetric particle. None of these particles has been detected up to the present. The mass spectrum could be measured at future proton storage rings like LHC, or at future electron-positron linear accelerators.

Magnetic monopole, isolated occurrence of magnetic elementary charges as required by unified theories. The existence of magnetic monopoles would violate time-reversal invariance. This would not constitute a basic problem, since violation of this symmetry has already been observed for neutral kaons.

| M | Magnetic monopoles have not yet been observed, despite intense search: detectors have been exposed to cosmic rays in balloon ascents, and lunar rocks have been investigated. In 1975 the discovery of a magnetic monopole was announced, but it is now believed that the event was caused by an extremely heavy nucleus. Magnetic monopoles might be 10^{16} times heavier than a proton.

Majorana neutrinos, massive neutrinos, neutrinos with $m_\nu \neq 0$. According to standard **electroweak theory** $m_\nu = 0$. Finite neutrino rest masses would have far-reaching consequences for the theory: for example, the **lepton number would not be conserved**.

| M | The experimental upper limit for the electron-neutrino mass is about 7 eV/c^2. These experiments actually measure the value of m_ν^2, which turned out to be partly negative.

An upper limit for the neutrino mass is provided by the explosion of a star 165,000 years ago: the difference in the arrival time of the neutrinos and of the light from the supernova (SN1987A) observed in 1987 leads to a rest mass of $m < 7$ eV/c^2.

Planck mass, $M = \sqrt{\hbar c/G} \approx 1.2 \cdot 10^{19}$ GeV/c^2 (G is the gravitational constant), mass or energy beyond which gravitation, according to the **general theory of relativity**, essentially determines the physics of elementary particles.

27
Nuclear physics

27.1 Constituents of the atomic nucleus

Atomic nucleus, bound system of A **nucleons**.
    Nucleon, generic term for proton and neutron.
    Proton, positively charged elementary particle with spin 1/2. The magnitude of the electric charge of the proton corresponds to the elementary charge.
    Neutron, neutral elementary particle with spin 1/2.

1. Basic characteristics of the atomic nucleus

Atomic number, **proton number**, Z, number of protons in the atomic nucleus, hence number of electrons in the neutral atom.
    Neutron number, N, number of neutrons in the atomic nucleus.
    Mass number, A of the atomic nucleus, the total number of nucleons in the nucleus,

$$A = Z + N.$$

Notation: The atomic number Z is given as left subscript of the atomic symbol X, the mass number A as left superscript, the neutron number N as right subscript:

$$^A_Z X_N.$$

    even-even nuclei, even proton number Z, even neutron number N,
    even-odd nuclei, even proton number Z, odd neutron number N,
    odd-even nuclei, odd proton number Z, even neutron number N,
    odd-odd nuclei, odd proton number Z, odd neutron number N.

2. Isotopes, isobars and isotones

Isotopes, atomic nuclei with the same atomic number Z, but different neutron numbers N.
■ $^A_Z X_N$ and $^{A+1}_Z X_{N+1}$ are isotopes. Example: the carbon isotopes ^{12}C, ^{13}C and ^{14}C.

➤   Basically, isotopes are chemically equivalent. Only processes that depend on mass
exhibit a slightly different behavior for different isotopes (differences in the physical-
chemical equilibria, differences in diffusion velocity, isotopic shifts in atomic spectra,
resonance frequencies in molecules, critical temperature of superconductors). These
phenomena are called **isotope effects**.

Isobars, atomic nuclei with equal mass number A, but different proton numbers Z. Isobars
belong to different chemical elements.

■ ${}_{Z}^{A}X_N$ and ${}_{Z+1}^{A}Y_{N-1}$ are isobars. Example: ${}^{14}C$ and ${}^{14}N$.

Isotones, atomic nuclei with equal neutron number N, but different atomic numbers Z.
Isotones belong to different chemical elements.

■ ${}_{Z}^{A}X_N$ and ${}_{Z+1}^{A+1}Y_N$ are isotones.

3. Isospin and generalized Pauli principle

Isospin, $\vec{t}$, operator of isospin, has all the mathematical properties of the spin operator
$\hat{s} = \hat{\sigma}/2$ (in units of $\hbar$),

$$\vec{t} = (\hat{t}_x, \hat{t}_y, \hat{t}_z), \quad \vec{t} = \vec{\tau}/2, \quad t = 1/2, \quad m_t = \pm 1/2.$$

Proton and neutron may be considered to be two states of the nucleon with different isospin
orientation m_t (third component of isospin):

$$m_t = +1/2: \quad \text{proton}, \qquad m_t = -1/2: \quad \text{neutron}.$$

Charge operator $\hat{q}$ of the nucleon, has eigenvalues 0 (neutron) and e (proton),

$$\hat{q} = \frac{e}{2}(1 + \hat{\tau}_z), \qquad e: \text{elementary charge}.$$

In a vector model, the isospins $\vec{t}_1, \vec{t}_2$ of two nucleons couple to the total isospin $\vec{T}$, with
quantum numbers T, $M_T = m_{t_1} + m_{t_2}$:

isospin singlet:	$T = 0$,	$M_T = 0$	neutron-proton system,
isospin triplet:	$T = 1$,	$M_T = 1$	proton-proton system,
		$M_T = 0$	neutron-proton system,
		$M_T = -1$	neutron-neutron system.

Symmetry of the isospin function of the two-nucleon system under permutation of the
isospin coordinates of the two nucleons:

$T = 0$: antisymmetric isospin function,
$T = 1$: symmetric isospin function.

Charge independence of nuclear forces, the two-nucleon force in the isospin triplet
state of a pair of nucleons (pp, pn or nn) is independent of its charge if the electromagnetic
interaction is ignored. The np-force in the isospin singlet state differs from that in the
isospin triplet state.

Generalized Pauli principle, the wave function of a many-nucleon system must be
antisymmetric under simultaneous permutation of the spin, isospin and space coordinates
of any two nucleons.

■   The ground state of the deuteron is an isospin-singlet state ($T = 0$, $M_T = 0$, antisymmetric isospin function) and a spin-triplet state ($S = 1$, symmetric spin function)
of the neutron-proton system. According to the generalized Pauli principle, the spatial function must be symmetric under permutation of the particle coordinates, i.e., the
quantum number L of the orbital angular momentum of the relative motion may take
only even values: $L = 0, 2, 4, \ldots$.

4. Table of fundamental properties of nucleons

Property	Proton	Neutron
mass	$1.672\,623\,1(10) \cdot 10^{-27}$ kg	$1.674\,928\,6(10) \cdot 10^{-27}$ kg
charge	$+1.602\,177\,33 \pm 0.000\,004\,65 \cdot 10^{-19}$ C	0
lifetime	$\geq 10^{31}$ yr	(889 ± 2.1) s
spin ($\hbar$)	$1/2$	$1/2$
magnetic moment	$(+2.792\,847\,39$ $\pm 0.000\,000\,06) \cdot \mu_K$	$(-1.913\,042\,7$ $\pm 0.000\,000\,5) \cdot \mu_K$
gyromagnetic ratio	$5.585\,692$	$-3.826\,3$
isospin projection	$+1/2$	$-1/2$

5. Nuclear spin resonance

Nuclear magneton, unit of the magnetic moment of atomic nuclei,

$$\mu_K = e\,\hbar/(2m_{proton}) = 3.152\,451\,66(28) \cdot 10^{-14} \text{ MeV T}^{-1}.$$

M   The proton spin is measured by means of the **paramagnetic nuclear spin resonance** (**NMR**, **N**uclear **M**agnetic **R**esonance): The magnetic moment $\vec{\mu}_p$ of the proton may take only definite orientations in a magnetic field $\vec{B}$ (directional quantization). These directions correspond to different energies. If a sample (e.g., water) is put in a magnetic field, a spin polarization of the protons of hydrogen arises. A high-frequency field is applied by a coil and the frequency is varied continuously. If the frequency f reaches a value corresponding to a transition from one spin state to another, the RF circuit embedding the coil is damped (**Fig. 27.1**).

➤   NMR is used to analyze the structure of organic molecules; it also has applications in medicine (nuclear spin tomograph).

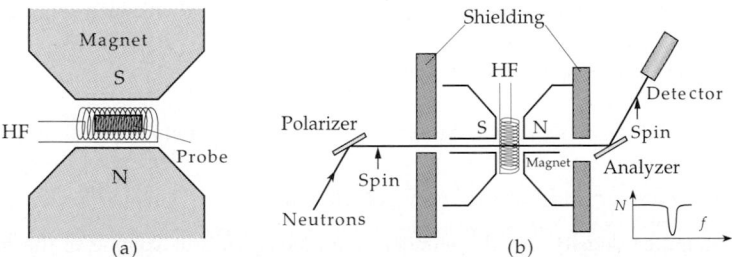

Figure 27.1: (a): Principle of nuclear magnetic resonance measurement (NMR). (b): NMR measurement of the neutron spin via the counting rate N as a function of the frequency f of the HF field.

6. Magnetic moment of nucleons

Both the neutron and proton have a non-zero magnetic moment.

Magnetic moment of neutron:

$$\mu_n = -(1.913\,042\,7 \pm 0.000\,000\,5) \cdot \mu_K .$$

$\boxed{\text{M}}$ The most precise method of measuring the magnetic moment of the neutron is the **nuclear spin resonance technique**: A neutron beam ($E_n \approx 25$ meV) is polarized by means of a polarizer and passes through a uniform magnetic field superimposed on an HF field. After passing the magnetic field, the polarization of the neutron is determined by an analyzer. Here, the magnetic scattering by magnetically saturated surfaces of a ferromagnetic material due to the magnetic moment of the neutron is employed. If the polarization coincides with the direction of magnetization of the analyzer, the scattering reaches a maximum. The frequency of the HF field at which the magnetic moment of the neutron flips is determined with such an analyzer.

Magnetic moment of the proton:

$$\mu_p = +(2.792\,847\,39 \pm 0.000\,000\,06) \cdot \mu_K .$$

The Bohr magneton and the nuclear magneton are based on a point-like, structureless, charged particle. Since the neutron is electrically neutral, one might expect a vanishing magnetic moment. But the magnetic moment of the proton, as well as the magnetic moment of the neutron, deviate appreciably from the expected values. The measured values therefore indicate that nucleons are not point-like particles.

▲   Nucleons are spatially extended objects with an intrinsic structure. Protons and neutrons consist of three constituent quarks, gluons and virtual quark-antiquark pairs.

➤   To date, attempts to measure an electric dipole moment of the neutron have failed. Recent experiments based on magnetic resonance techniques have shown that the electric dipole moment of the neutron must be less than $4 \cdot 10^{-25}$ $e \cdot$ cm, if it exists at all.

27.2   Basic quantities of the atomic nucleus

Shape of atomic nuclei, mostly deformed in an axially symmetric manner, spherical near closed nucleon shells.

   Nuclear radius R, may be estimated by the formula

$$R = r_0 \cdot A^{1/3} , \qquad r_0 \approx 1.2 \text{ fm} = 1.2 \cdot 10^{-15} \text{ m}, \, A = \text{atomic mass number}.$$

1. Nucleon number and mass-density distribution

Nucleon-number density, ρ_0, the number of nucleons per unit volume in the nuclear interior is almost constant for all nuclei:

$$\rho_0 = 0.17 \cdot 10^{45} \text{ nucleons/m}^3 = 0.17 \text{ nucleons/fm}^3 .$$

This value corresponds to a mass density of atomic nuclei of about $2.7 \cdot 10^{17}$ kg/m^3. The highest density of a macroscopic solid is $\rho = 22\,570$ kg/m^3 for the metal osmium. Hence, the nuclear density exceeds the density of solids under standard conditions by **13 orders of magnitude**.

Mass-density distribution, $\rho(r)$, density of the atomic nucleus as function of the distance r from the center of the nucleus (**Fig. 27.2**), empirically determined as

$$\rho(r) = \frac{\rho_0}{1 + e^{(r-R)/a}} .$$

The parameter a measures the thickness b of the surface layer within which the nuclear density drops from 90 % to 10 % of the central density: $b = 4.4\,a$, $a \approx 0.6$ fm.

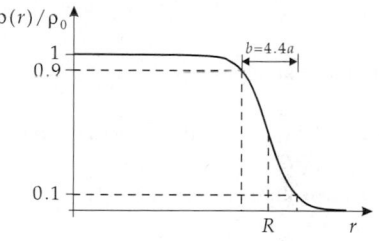

Figure 27.2: Mass-density distribution in the atomic nucleus. R: nuclear radius, a: surface parameter, b: thickness of surface layer, ρ_0: central nucleon-number density.

M　The charge distribution in the nucleus is measured by scattering of charged particles (e^-, p, α-particles) (**Rutherford scattering, Fig. 27.3 (a)**). The mass-density distribution of heavy nuclei may deviate slightly from the charge-density distribution, due to the neutron excess. The nuclear radius R and the radius parameter r_0 can be derived from scattering data, assuming an appropriate form factor for the charge distribution.

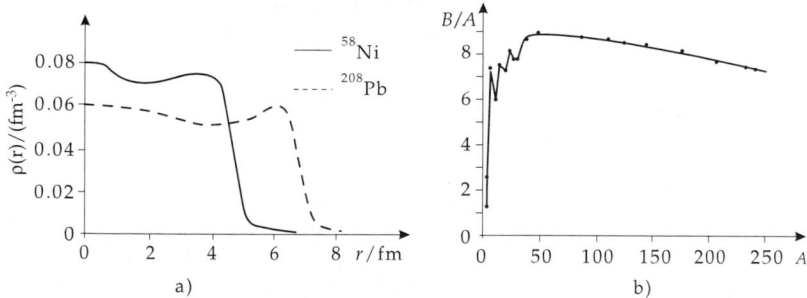

Figure 27.3: (a): Measured charge distribution in the ^{58}Ni- and ^{208}Pb-nuclei. r is the distance from the nuclear center. (b): Binding energy per nucleon B/A in MeV as function of the atomic mass number A.

2. Binding energy and mass defect

Binding energy, B, the energy released when free nucleons are bound together to form a nucleus. SI unit is the joule, J. Usually the binding energy is given in MeV:

$$1 \text{ MeV} = 1.6022 \cdot 10^{-13} \text{ J} .$$

▲　The mass of a stable atomic nucleus is smaller than the sum of the masses of the constituent nucleons.

Mass defect, $\Delta W(A, Z)$, the difference between the sum of masses of all nucleons and the nuclear mass $m_K(A, Z)$,

$$\Delta W(A, Z) = Z \cdot m_p + (A - Z) \cdot m_n - m_K(A, Z) .$$

According to mass-energy equivalence, the mass defect ΔW is related to the binding energy B,

$$B = \Delta W(A, Z) \cdot c^2, \qquad 1\ \text{MeV}/c^2 = 1.7827 \cdot 10^{-30}\ \text{kg}.$$

| **M** | Atomic masses may be determined by mass spectrometers from the deflection of ions in electric and magnetic fields. The binding-energy difference of atomic nuclei can also be determined from the decay energy in β-decay, or the Q-value of nuclear reactions.

3. Atomic mass unit,

u, equal to $1/12$ of the mass of a neutral atom of the carbon isotope ^{12}C:

$$u = \frac{1}{12} m_{^{12}\text{C}} = \frac{1\ \text{g}}{N_\text{A}} = 1.660\,540\,2(10) \cdot 10^{-27}\ \text{kg} \quad (N_\text{A}\text{: Avogadro's number}).$$

This unit is convenient in nuclear physics, since the masses of all atomic nuclei may be given by almost-integer multiples of u.

Quantity	Symbol	Value	Error /ppm
atomic mass unit	u	$931.494\,32$ MeV$/c^2$	0.30
electron mass	m_e	$0.510\,999\,06$ MeV$/c^2$	0.30
muon mass	m_μ	$105.658\,389$ MeV$/c^2$	0.32
proton mass	m_p	$938.272\,31$ MeV$/c^2$	0.30
neutron mass	m_n	$939.565\,63$ MeV$/c^2$	0.30
Planck's constant	$\hbar$	$6.582\,122\,0 \cdot 10^{-22}$ MeV $\cdot$ s	0.30

4. Binding energy per nucleon

Binding energy per nucleon B/A, a measure for the stability of an atomic nucleus. Mean experimental value: $B/A \approx 8$ MeV.

▲ Nuclei are bound with ca. 1 % of their mass.

For light nuclei, the binding energy per nucleon increases with the mass number. The most stable atomic nucleus is iron (^{56}Fe) with a binding energy per nucleon of ≈ 8.8 MeV. For $A > 56$, the binding energy per nucleon decreases with increasing nucleon number (**Fig. 27.3 (b)**). Therefore, nuclear energy may be released either by fusion of light nuclei, or by fission of heavy nuclei.

➤ The local maxima of the binding energy in the range of light nuclei (e.g., for $^4_2$He) are caused by closure of neutron and/or proton shells (see p. 917), analogous to the strong binding of the electron shell in inert-gas atoms.

Saturation of nuclear forces, the binding energy per nucleon is approximately constant at about 8 MeV.

▲ The magnitude of the binding energy of a nucleus determines its stability against decay.

27.3 Nucleon-nucleon interaction

27.3.1 Phenomenologic nucleon-nucleon potentials

The potential V_{12} of the interaction between two nucleons may be determined up to energies of about 300 MeV from the elastic nucleon-nucleon scattering by a **phase-shift**

analysis. One measures the differential cross-section in **single-scattering experiments,** and spin-dependent quantities (polarization, depolarization) in **multiple-scattering experiments,** or in experiments with **polarized particle beams** or/and **polarized targets.**

General formulation:

$$V_{12} = V_W(r) + V_B(r)\,(\vec{\sigma}_1 \cdot \vec{\sigma}_2) + V_H(r)\,(\vec{\tau}_1 \cdot \vec{\tau}_2) + V_M(r)\,(\vec{\sigma}_1 \cdot \vec{\sigma}_2)(\vec{\tau}_1 \cdot \vec{\tau}_2)$$

$$+ V_T\,S_{12} + V_{LS}\,(\vec{L} \cdot \vec{S})\,.$$

Wigner force, V_W, central force depending only on the nucleon distance r.

1. Exchange forces

Exchange force, a central force depending on the state of the nucleon-nucleon system: magnitude and sign (attraction or repulsion) depend on the symmetry of the spin function (total spin $S = 0$ or $S = 1$), the isospin function (total isospin $T = 0$ or $T = 1$) or the spatial function (orbital angular momentum $L = 0, 2, 4, \ldots$ or $L = 1, 3, 5, \ldots$).

Bartlett force, $\sim \vec{\sigma}_1 \cdot \vec{\sigma}_2$, exchange force that distinguishes between the spin states $S = 0$ and $S = 1$.

Heisenberg force, $\sim \vec{\tau}_1 \cdot \vec{\tau}_2$, exchange force that distinguishes between the isospin states $T = 0$ and $T = 1$.

Majorana force, $\sim (\vec{\sigma}_1 \cdot \vec{\sigma}_2)(\vec{\tau}_1 \cdot \vec{\tau}_2)$, exchange force that distinguishes between states with even and odd orbital angular momentum.

■ For an interaction consisting of Wigner and Bartlett forces, the total potential is

$$V_{12} = V_W - 3 \cdot V_B \quad \text{for } S = 0\,,$$

$$V_{12} = V_W + 1 \cdot V_B \quad \text{for } S = 1\,.$$

2. Tensor forces and spin-orbit coupling

Tensor force, S_{12} a static noncentral force depending on the relative orientation of the nucleon spins $\vec{s}_1, \vec{s}_2$ with respect to the distance vector $\vec{r}$ of the two nucleons (**Figs. 27.4** and **27.5**, $\vec{s} = \hbar\vec{\sigma}/2$),

$$S_{12} = 3\,\frac{(\vec{\sigma}_1 \cdot \vec{r})(\vec{\sigma}_2 \cdot \vec{r})}{r^2} - \vec{\sigma}_1 \cdot \vec{\sigma}_2\,.$$

The electric quadrupole moment of the deuteron originates from the tensor term of the nucleon-nucleon force.

Spin-orbit coupling, $\sim \vec{L} \cdot \vec{S}$, a velocity-dependent noncentral force that depends on the relative orientation of the total spin $\vec{S}$ and the orbital angular momentum $\vec{L}$ of the relative motion of the nucleons.

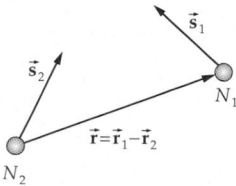

Figure 27.4: Tensor force S_{12} between two nucleons N_1, N_2. $\vec{s}_1, \vec{s}_2$: nucleon spins.

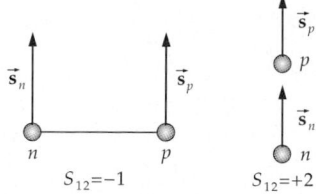

Figure 27.5: Tensor force S_{12} in special configurations of the neutron(n)-proton(p) system.

3. Hard-core

Hard-core potential, infinite repulsive potential in the form factor of the nucleon-nucleon potential. Two nucleons may not approach each other to distances below the hard-core radius r_c, $r_c \approx 0.6$ fm (**Fig. 27.6**). The hard-core potential contributes to the saturation of nuclear binding.

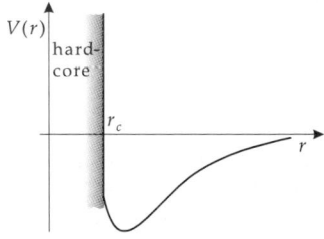

Figure 27.6: Hard-core potential with hard-core radius r_c. r: distance between the nucleons.

27.3.2 Meson exchange potentials

Meson exchange: The emission of a virtual meson of finite mass by a nucleon and absorption of this meson by a second nucleon modifies the momentum states of the nucleons. This effect may be interpreted as the action of a force. The range of this force R is inversely proportional to the mass m of the exchanged meson,

$$R \approx \hbar/(m\,c)\,.$$

1. Yukawa potential,

a nucleon-nucleon potential caused by the exchange of a single pion ($m_\pi c^2 \approx 140$ MeV) (**one-pion-exchange potential OPEP, Fig. 27.7 (a)**). The Yukawa potential includes central forces with exchange character and the long-range tensor force. The r-dependence is given by

$$V_Y = \mathrm{e}^{-\mu r}/(\mu r)\,, \quad \mu = m_\pi c/\hbar\,.$$

The one-pion-exchange potential provides a satisfactory description of the nucleon-nucleon interaction at nucleon separations $r \geq 2$ fm.

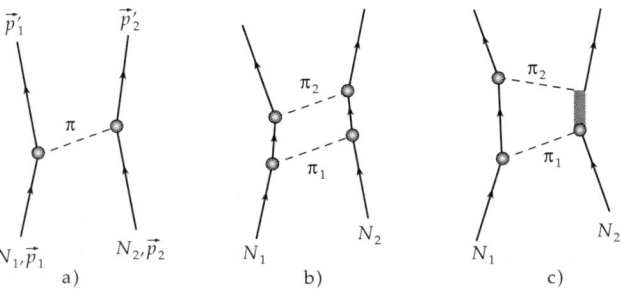

Figure 27.7: Exchange of virtual mesons between two nucleons N_1, N_2. (a): one-pion exchange, (b): 2π-exchange, (c): 2π-exchange with virtual excitation of the $\Delta(1232)$-resonance in the nucleon.

➤ The uncorrelated exchange of two pions may be simulated to a good approximation by the exchange of a fictitious scalar meson, the σ-meson with a mass of ≈ 400 MeV. The σ-meson mediates the attractive component of the nucleon-nucleon force at medium distances (**Fig. 27.7 (b)**).

2. Boson-exchange potential,

nucleon-nucleon potential corresponding to a correlated multi-pion exchange involving heavy mesons with integer spin (**Fig. 27.7 (c)**).

2π-channel: isovector ρ-meson (spin $I = 1$, isospin $T = 1$),
3π-channel: isoscalar ω-meson (spin $I = 1$, isospin $T = 0$).

The boson-exchange model describes the nucleon-nucleon interaction at short distances (but $r > r_C$).

➤ The spin-orbit coupling in the nucleon-nucleon potential is due to the exchange of vector mesons. It is a short-range force.

27.4   Nuclear models

27.4.1   Fermi-gas model

Fermi-gas model, considers the nucleus an ensemble of A nucleons moving without mutual interaction in a limited space region that corresponds to the nuclear volume. In the ground state, the nucleons occupy discrete momentum states of increasing energy up to the **Fermi momentum** p_F, which is determined by the nuclear density ρ,

$$p_F = \hbar k_F, \quad k_F = \left(\frac{3}{2}\pi^2\rho\right)^{1/3} \approx 1.36 \text{ fm}^{-1}.$$

Fermi energy, maximum kinetic energy of a nucleon in the Fermi gas,

$$\varepsilon_F = \frac{\hbar^2}{2m} k_F^2 \approx 37 \text{ MeV}.$$

27.4.2   Nuclear matter

Nuclear matter, a nuclear model that treats the atomic nucleus as an infinite system of nucleons (nucleon number $A \to \infty$, volume $V \to \infty$) with a fixed particle-number density ρ at temperature $T = 0$,

$$\lim_{A,V \to \infty} \frac{A}{V} = \rho = \text{const.}$$

The mass difference between neutron and proton, and the Coulomb interaction between the protons, is ignored. The nucleons interact through a two-particle force represented by a realistic potential derived from the free nucleon-nucleon scattering. The binding energy per nucleon B/A is calculated in the approximation of independent pairs as a function of the particle number density ρ. For low densities, the kinetic energy of the nucleons dominates. With increasing density, the influence of the attractive components of the nucleon-nucleon interaction, which leads to binding, is however more and more counteracted by the repulsive short-range components. This interplay yields a minimum of the binding energy per

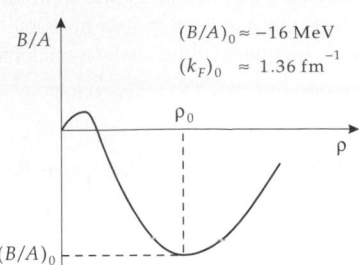

Figure 27.8: Nuclear matter. Binding energy per nucleon B/A versus particle-number density ρ (schematic). ρ_0: saturation density, $(B/A)_0$: binding energy per nucleon at saturation density, $(k_F)_0$: wave number corresponding to the Fermi momentum at saturation density.

nucleon as a function of the nuclear density. The minimum of the curve corresponds to the saturation values for density and binding energy in the nucleus; the value found for B/A may be compared with the volume term of the **Bethe–Weizsäcker formula** (see below).

27.4.3 Droplet model

Droplet model, treats the nucleons as molecules of an incompressible, charged liquid drop.
   Ground state, the energetically lowest state of the nucleus.

1. Bethe–Weizsäcker formula,

based on the droplet model, yields the binding energies of nuclei in the ground state:

binding energy = volume- + surface- + Coulomb- +symmetry- + pairing energy			$\mathbf{ML^2T^{-2}}$
	Symbol	Unit	Quantity
$E_B = a_V \cdot A - a_O \cdot A^{2/3}$ $- a_C \cdot \dfrac{Z^2}{A^{1/3}}$ $- a_S \cdot \dfrac{(A-2Z)^2}{A} + \varepsilon_P$	a_V	MeV	volume energy per nucleon
	a_O	MeV	coefficient of surface energy
	a_C	MeV	coefficient of Coulomb energy
	a_S	MeV	coefficient of symmetry energy
	ε_P	MeV	pairing energy
	A	1	mass number
	Z	1	atomic number

Values of the constants:

Constant	a_V	a_O	a_C	a_S	ε_P
E /MeV	15.85	18.34	0.71	23.22	0 oder $\pm 11.46/\sqrt{A}$

2. Properties of the components in the binding energy

Volume energy ($E_V \sim R^3 \sim A$), a consequence of the short range of nuclear forces. Only the next neighbors of a nucleon are reached by the nuclear force. The volume energy corresponds to the binding energy in the limit of large mass numbers A for $N = Z$ and ignoring the Coulomb interaction between the protons. The linear dependence of the volume energy on A expresses the **saturation** property of nuclear forces.

Surface energy ($E_O \sim R^2 \sim A^{\frac{2}{3}}$), a consequence of the fact that the nucleons at the surface of a finite nucleus cannot saturate their interactions with neighboring nucleons. The surface energy reduces the nuclear binding.

Coulomb energy ($E_C \sim R^{-1} \sim A^{-\frac{1}{3}}$), corresponding to the electric repulsion between protons. The Coulomb energy reduces the nuclear binding.

Symmetry energy ($E_S \sim (N - Z)^2/A$), expresses the trend to particular stability of nuclei with $N = Z$ for small A. Light nuclei become less stable if $|N - Z|$ increases.

Pairing energy, the energy gain δ when two neutrons or protons form a pair with total spin $S = 0$. The pairing energy is an empirical correction to the pure droplet model (compare Cooper pairing, p. 1044), which results in a stronger binding of nuclei with even neutron and/or proton number:

$$\varepsilon = \begin{cases} \delta: & N \text{ even, } Z \text{ even}, \\ 0: & N \text{ odd, } Z \text{ even, or vice versa}, \\ -\delta: & N \text{ odd, } Z \text{ odd}. \end{cases} \quad \delta = 11.46/\sqrt{A} \text{ MeV}$$

3. Line of beta-stability,

the line in the N-Z plane about which the stable nuclei are arranged (**Fig. 27.9**).

➤　Light nuclei are particularly stable for $Z = N$. The doubly-magic tin isotope with $Z = N = 50$ is the heaviest nucleus with equal number of neutrons and protons accessible to experiment. Heavier nuclei with $N = Z$ decay by spontaneous proton emission.

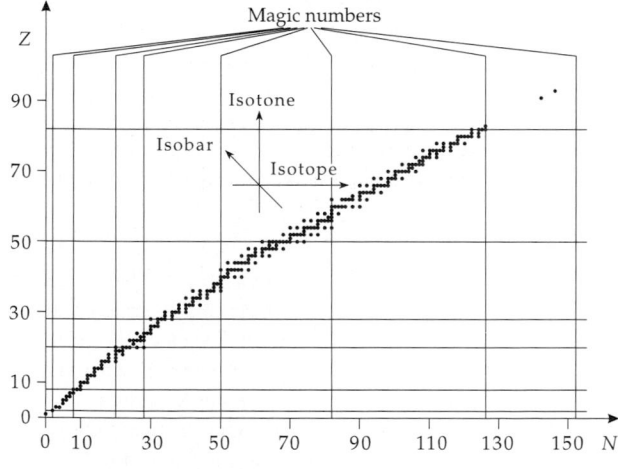

Figure 27.9: Line of β-stability in the N-Z plot. The arrows indicate the directions along which particular nuclei are arranged. The magic numbers (full lines) mark the shell closure for protons and neutrons, respectively.

27.4.4　Shell model

Shell model, a description of the motion of nucleons in terms of noninteracting particles in a **mean nuclear potential** generated by the nucleons themselves.

This description of nucleonic motion in the nucleus corresponds to the treatment of electronic motion in the electron shell of the atomic nucleus. But while the electrons are moving in a given external field, the Coulomb potential of the nucleus, the nuclear shell model is based on a replacement of the two-particle forces between the nucleons by an effective mean nuclear potential. The remaining two-particle residual interaction between the nucleons is assumed to be weak.

▲   The nuclear shell model describes the energy spectrum of light nuclei and of heavier nuclei near shell closure (magic nucleon numbers) rather well if the two-nucleon residual interaction is taken into account.

The mean potential is frequently approximated by an **oscillator potential**, or by a potential with a radial dependence suggested by the mass-density distribution of the nucleus. In mass regions in which the nuclear shape deviates from the spherical shape, a **deformed mean potential** must be used. The mean nuclear potential field is characterized by the presence of a strong **spin-orbit coupling** term $V_{ls}(r)(\hat{\vec{l}} \cdot \hat{\vec{s}})$, which causes an energy difference between single-particle states with parallel and antiparallel orientation of the nucleonic spin $\vec{s}$ and the orbital angular momentum $\vec{l}$.

1. Single-particle states in the shell model

The mean potential is used for calculating the single-particle states (energy levels) of the nucleons in the nucleus. The **quantum numbers** of the single-particle states are:

- $n = 0, 1, 2, \ldots$
  radial quantum number, number of zeros of the radial wave function,
- $l = 0, 1, 2, \ldots$
  orbital angular momentum quantum number,
- $j = l \pm 1/2$
  quantum number of the total angular momentum $\vec{j} = \vec{l} + \vec{s}$,
- $m_j = m_l + m_s$, $m_j = -j, \ldots, j$
  quantum number of the projection of the total angular momentum $j_z = l_z + s_z$. The quantities m_l and m_s are the projection quantum numbers for orbital angular momentum and spin of the nucleon, respectively.

Conventional spectroscopic classification of the single-particle states: $(n+1)l_j$.

The **single-particle energies** ε depend only on the quantum numbers n, l, j: $\varepsilon = \varepsilon_{nlj}$.

2. Shell structure of the energy states

The single-particle states in the mean nuclear potential are energetically grouped in shells: the energy separation between the levels within a shell is much smaller than the energy separation between the shells (**Fig. 27.10**).

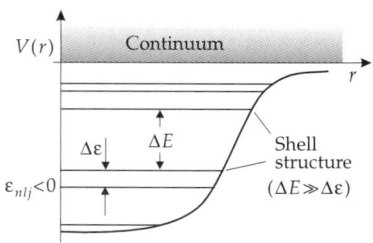

Figure 27.10: Shell structure of the single-particle states in the mean shell model potential $V(r)$ (schematic). ε_{nlj}: single-particle energies, n: radial node number, l: quantum number of orbital angular momentum, j: quantum number of the total angular momentum.

3. Nucleon configuration

Nucleon configuration, a specific occupation of the single-particle states

$$(n_1 l_1 j_1), \ (n_2 l_2 j_2), \ldots (n_f l_f j_f)$$

by the A nucleons of the nucleus,

$$(n_1 l_1 j_1)^{N_1} \ (n_2 l_2 j_2)^{N_2} \cdots (n_f l_f j_f)^{N_f}, \quad N_1 + N_2 + \cdots + N_f = A.$$

A single-particle state (nlj) can be occupied by at most $2j + 1$ neutrons and protons. Configuration: $(nlj)^{2j+1}$.

4. Magic nuclei

Magic numbers, numbers of protons or neutrons for which the nucleus is particularly stable, as compared with neighboring nuclei:

$$N, Z: \ 2, 8, 20, 28, 50, 82, 126 \quad \text{and} \quad N = 184.$$

▲   In magic nuclei the shells are completely filled.
▲   Particularly many stable elements exist with magic neutron numbers.
Doubly-magic nuclei, nuclei for which the neutron number **and** the proton number are equal to a magic number.
■ $^4_2\text{He}_2$, $^{16}_8\text{O}_8$, $^{40}_{20}\text{Ca}_{20}$, $^{208}_{82}\text{Pb}_{126}$.
▲   Doubly-magic nuclei are particularly stable. Their abundance in nature is higher than that of their neighbors.

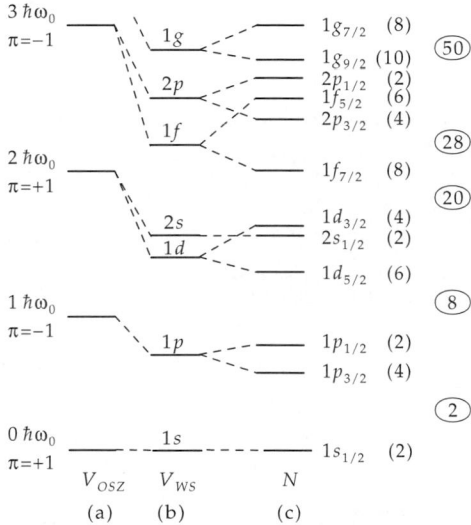

Figure 27.11: Single-particle states in the mean shell model potential. Spectroscopic classification: $(n + 1)lj$, n: node number of the radial function, l: orbital angular momentum, j: total angular momentum. (a): oscillator potential, (b): central potential of finite depth with Woods-Saxon radial shape, (c): central potential of finite depth with spin-orbit coupling (Nilsson). Numbers in brackets: maximum occupation numbers for one kind of nucleons, numbers in circles: magic numbers.

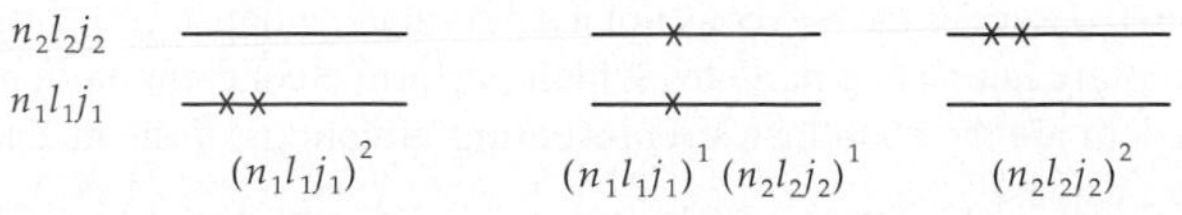

$n_2 l_2 j_2$

$n_1 l_1 j_1$

$(n_1 l_1 j_1)^2 \qquad (n_1 l_1 j_1)^1 (n_2 l_2 j_2)^1 \qquad (n_2 l_2 j_2)^2$

Figure 27.12: Two-particle configurations for two single-particle states $(n_1 l_1 j_1)$, $(n_2 l_2 j_2)$.

5. Role of residual interaction

Configuration mixing, a state in which the wave functions of different nucleon configurations are superposed coherently due to the residual interaction between the nucleons.

➤   If the residual interaction is a two-body force, then it may connect only such configurations that differ in the single-particle states of at most two particles.

6. Excited states in the shell model

Single-particle excitation, transition of a single nucleon from a single-particle state (nlj) to an energetically higher single-particle state $(n'l'j')$.

Particle-hole excitation, excitation of a single nucleon from a fully occupied shell. Transition from the configuration $(n_h l_h j_h)^{2j+1}$ to the configuration $(n_h l_h j_h)^{-1}(n_p l_p j_p)^1$.

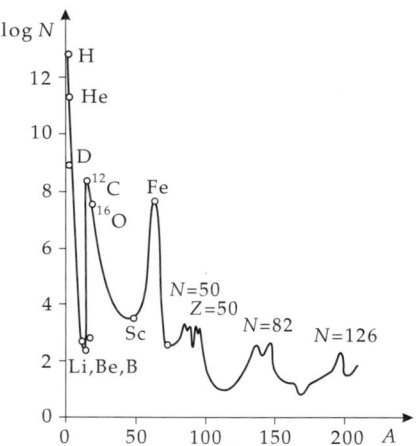

Figure 27.13: Cosmic abundance of elements. N, Z: magic numbers.

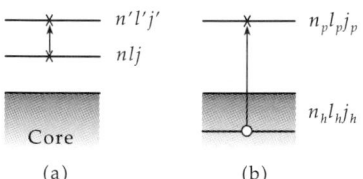

(a)   (b)

Figure 27.14: Elementary excitation in the shell model. (a): single-particle excitation $(nlj) \longrightarrow (n'l'j')$, (b): particle-hole excitation $(n_h l_h j_h)^{-1}(n_p l_p j_p)^1$.

27.4.5 Collective model

Collective model, describes the nucleons not as individual, independent particles, but as an ensemble of strongly interacting particles which perform a coherent motion. The relevant degrees of freedom are the coordinates representing vibrations of the nuclear surface and rotations of the nucleus.

Rotational and vibrational excitations occur the same way they do in molecules.

1. Vibrations of the nuclear surface

Vibrational excitations, harmonic oscillations of the nuclear surface about the equilibrium shape of the nucleus with angular frequency ω_I. The vibration is characterized by the angular momentum I (multipolarity) and by the number n_I of excitation quanta (phonons). In harmonic approximation, an equally spaced spectrum of excited states is produced, and $E_{I\,n_I+1} - E_{I\,n_I} = \hbar\omega_I$ occurs for any value of the angular momentum I:

vibrational excitation			$\mathbf{ML^2T^{-2}}$
	Symbol	Unit	Quantity
$E_{I\,n_I} = \left(n_I + \dfrac{1}{2}\right) \cdot \hbar \cdot \omega_I$	$E_{I\,n_I}$	J	excitation energy
	$\hbar$	J s	quantum of action/(2π)
	ω_I	rad s^{-1}	angular frequency
	I	1	angular momentum quantum number
	n_I	1	vibrational quantum number

Quadrupole vibrations ($I = 2$) occur as the lowest vibrational excitations in nuclei with $N = Z$. If two quadrupole vibrational quanta are excited ($n_2 = 2$), three degenerated states occur with total angular momenta (nuclear spins) $J = 0, 2, 4$. In real nuclei, this degeneracy is removed by the interaction between the phonons: one actually observes a trio of states that are closely grouped about the energy of the two-phonon state at $2 \cdot \hbar\omega_2$ (**Fig. 27.15**).

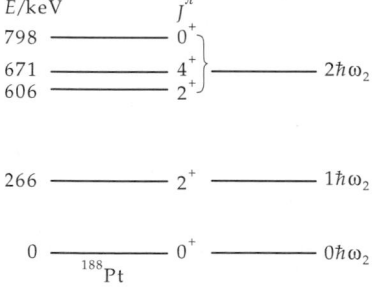

Figure 27.15: Excitation of quadrupole vibrations ($I = 2$) in ^{188}Pt. E: excitation energy, $\hbar\omega_2$: excitation energy of a quadrupole phonon, J^π: spin and parity of the level.

2. Electric quadrupole moment,

Q_0, characterizes nuclei with a deformed charge distribution in the ground state:

$$Q_0 = \frac{2}{5} Ze(b^2 - a^2).$$

b and a are the half-axes of the ellipsoid, Z is the charge number of the nucleus.

3. Nuclear rotations

Rotational excitations, rotation of a nucleus with a permanent deformation in the ground state, with angular momentum J about an axis perpendicular to the symmetry axis, without excitation of intrinsic nucleonic motion. The excitation energy of the rotational states is determined by the moment of inertia Θ of the nucleus. The separation between subsequent states in the rotational spectrum increases with the angular momentum of rotation.

rotational excitation			$\mathbf{ML^2T^{-2}}$
	Symbol	Unit	Quantity
$E_J = \dfrac{\hbar^2}{2\Theta} J(J+1)$	E_J	J	excitation energy
	$\hbar$	J s	quantum of action/2π
	J	1	angular momentum quantum number
	Θ	kg m^2	moment of inertia

For axially symmetric nuclei with shapes that are invariant against a rotation through an angle π about an axis perpendicular to the symmetry axis, for reasons of symmetry the rotational quantum number J is restricted to even values $J = 0, 2, 4 \ldots$.

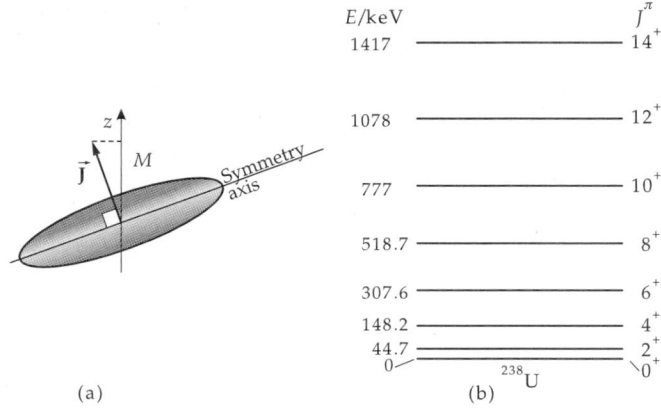

(a)   (b)

Figure 27.16: Excitation of rotational states in atomic nuclei. (a): angular momentum $\vec{\mathbf{J}}$ of the rotation about an axis perpendicular to the symmetry axis. M: angular momentum projection to the z-axis (quantization axis), (b): rotational band in ^{238}U.

➤ The moment of inertia of nucleus is about a factor of two smaller than the moment of inertia of a solid body of the same shape and mass density.

27.5 Nuclear reactions

27.5.1 Reaction channels and cross-sections

Nuclear reaction, conversion of a nucleus by interaction (collision) with another nucleus, a hadron, a lepton or a gamma quantum. Reaction equation:

$$a + A \longrightarrow b + B, \qquad A(a, b)B.$$

a: incident particle (projectile),   A: target nucleus,
b: outgoing particle (ejectile),   B: remaining nucleus.

Types of nuclear reactions:

elastic scattering:	$a + A \longrightarrow a + A$,
inelastic scattering:	$a + A \longrightarrow a' + A^*$,
radiative capture:	$a + A \longrightarrow B + \gamma$,
rearrangement reaction:	$a + A \longrightarrow b + B$, $a \neq b$,
multi-particle reaction:	$a + A \longrightarrow B + b_1 + b_2 + \cdots$,
fusion:	$a + A \longrightarrow C^*$,
induced nuclear fission:	$a + A \longrightarrow B_1 + B_2$.

1. Characteristics of reaction channels

Reaction channel, α, subdivision λ of a number N of nucleons into two groups $N_1, N_2, N_1 + N_2 = N$ that are spatially separated from each other and have intrinsic states specified by excitation energy, spin I_1, I_2, parity π_1, π_2 and possibly other quantum numbers κ_1, κ_2:

channel index: $\alpha = \{\lambda, I_1, I_2, \pi_1, \pi_2, \kappa_1, \kappa_2\}$, $\lambda = (N_1, N_2)$, $N = N_1 + N_2$.

Channel radius, R_α, minimum distance between the nucleon groups N_1, N_2 at which there is not yet a strong interaction between the two nuclei.

Interaction region, part of the configuration space in which the mass centers of both nuclei are separated by a distance $R < R_\alpha$ for all partitions λ.

Entrance channel, reaction channel in which the system is found at time $t \to -\infty$ (initial state).

Exit channel, reaction channel in which the system is found at time $t \to +\infty$ (final state).

Open channel, reaction channel allowed by energy conservation.

Closed channel, reaction channel forbidden by energy conservation.

2. Channel spin and total angular momentum

Channel spin $\vec{S}_i$ in the entrance channel, vector addition of the spins $\vec{I}_a$ and $\vec{I}_A$ of incident particle a and target nucleus A to a total spin $\vec{S}_i$,

channel spin = spin$_a$ + spin$_A$			ML^2T^{-1}
	Symbol	Unit	Quantity
$\vec{S}_i = \vec{I}_a + \vec{I}_A$	$\vec{S}_i$	J s	channel spin
$\|I_a - I_A\| \le S_i \le I_a + I_A$	$\vec{I}_a$	J s	spin of projectile a
	$\vec{I}_A$	J s	spin of target A

Analogously, for the channel spin in the exit channel S_f:

$$\vec{I}_b + \vec{I}_B = \vec{S}_f, \quad |I_b - I_B| \le S_f \le I_b + I_B.$$

The vector addition of the channel spin $\vec{S}$ and the orbital angular momentum of relative motion $\vec{L}$ yields the total angular momentum $\vec{J}$ of the corresponding channel,

$$\vec{L} + \vec{S} = \vec{J}, \quad |L - S| \le J \le S + L.$$

total angular mom. = channel spin + orbital angular mom.			$\mathbf{ML^2T^{-1}}$
	Symbol	Unit	Quantity
$\vec{J} = \vec{S} + \vec{L}$	$\vec{J}$	J s	total angular momentum
	$\vec{S}$	J s	channel spin
$\|L - S\| \le J \le S + L$	$\vec{L}$	J s	orbital angular momentum of relative motion

3. Example: Nuclear reactions on lithium

Proton-induced nuclear reactions on $_3^7\text{Li}$ at an incidence energy of several MeV:

entrance channel: $\text{p} + _3^7\text{Li}$,

exit channels: $\text{p} + _3^7\text{Li}$,

$\text{p}' + _3^7\text{Li}^*$,

$\text{n} + _4^7\text{Be}$,

$\alpha + \alpha$,

$\alpha + \alpha + \gamma$,

$\alpha + \text{t} + \text{p}$.

4. Reference frames

Laboratory system, the reference frame in which the target nucleus is at rest in the initial state.

Center-of-mass system, the reference frame in which the center of mass of projectile and target nucleus is at rest.

▲   If the mass of the scattering center is very large compared with the mass of the incident particle; the laboratory and center-of-mass coordinates coincide.

5. Energy transfers in nuclear reactions

Q-value, energy change Q of a nuclear reaction, difference of the kinetic energies in the exit channel f (after the reaction) and the entrance channel i (before the reaction) E_f and E_i in the center-of-mass system:

$$Q = E_f - E_i.$$

The Q-value of a reaction in which a light particle a (mass m_a) with the kinetic energy E_a hits a target nucleus A (mass M_A) at rest, generating a final nucleus B (mass M_B) with kinetic energy E_B and a light particle b (mass m_b) with kinetic energy E_b under the reaction angle θ, is given by

$$\begin{aligned}
Q &= E_B + E_b - E_a \\
&= (m_a + M_A - M_B - m_b) \cdot c^2 \\
&= E_b \left(1 + \frac{m_b}{M_B}\right) - E_a \left(1 - \frac{m_a}{M_B}\right) - \frac{2}{M_B} \sqrt{E_a E_b m_a m_b} \cos\theta.
\end{aligned}$$

Exothermal reactions, reactions with positive Q-value, $Q > 0$: energy is released.

Endothermal reactions, reactions with negative Q-value, $Q < 0$: energy is needed. The reaction is observed only above a threshold energy.

■
$$\begin{array}{r} {}^3_2\text{He} + n \rightarrow {}^4_2\text{He} + Q \\ m_{{}^3\text{He}} = 3.0392471 \text{ u} \\ +m_{\text{n}} = 1.00866497 \text{ u} \\ \hline \sum = 4.047912 \text{ u} \qquad m_{{}^4\text{He}} = 4.002603256 \text{ u} \end{array}$$

The mass of ^{4}He is smaller than the first sum. The Q-value of the reaction is positive.

■
$${}^{10}\text{B} + n \rightarrow {}^7\text{Li} + {}^4\text{He} + Q$$

$$\begin{array}{ll} m_{{}^{10}\text{B}} = 10.01293800 \text{ u} & m_{{}^7\text{Li}} = 7.01600450 \text{ u} \\ +m_{\text{n}} = 1.00866497 \text{ u} & m_{{}^4\text{He}} = 4.002603256 \text{ u} \\ \hline \sum = 11.02160297 \text{ u} & \sum = 11.01860775 \text{ u} \end{array}$$

The second sum is smaller than the first sum. The Q-value of the reaction is positive. In this reaction, energy is released.

6. Cross-sections of nuclear reactions

Cross-section, σ, dimension of an area, a measure of the probability that the system changes from the entrance channel to a definite exit channel.

$$\sigma = \frac{\text{number of reactions/unit time}}{\text{number of incident particles/(unit time} \cdot \text{unit area)}} .$$

Unit of the cross-section in atomic and nuclear physics: **barn** b ($1 \text{ b} = 10^{-28} \text{ m}^2$).

The cross-section depends on the projectile-target combination, and on the incident energy.

Differential cross-section $d\sigma/d\Omega$, cross-section for a reaction with an outgoing particle observed in the solid angle element $d\Omega = \sin\theta \, d\theta \, d\phi$.

Doubly differential cross-section $d^2\sigma/(d\Omega \, dE)$, cross-section for a reaction with an outgoing particle observed in the solid angle element $d\Omega$ and the energy interval dE.

Total cross-section, σ_{tot}, the integral of the differential cross-section over the full solid angle,

$$\sigma_{\text{tot}}(E) = \int \left(\frac{d\sigma(E, \theta, \phi)}{d\Omega} \right) \cdot d\Omega .$$

Total cross-section, also the sum of the total interaction cross-sections $\sigma_{\alpha\alpha'}$ over all open reaction channels α',

$$\sigma_{\text{tot}} = \sum_{\alpha'} \sigma_{\alpha\alpha'} .$$

Nomenclature for the cross-sections according to the type of reaction:
- **Elastic scattering cross-section**, σ_{s}, cross-section for elastic scattering of an incident particle by a target nucleus.
- **Inelastic scattering cross-section**, σ_{in}, cross-section for inelastic scattering of an incident particle by a target nucleus.
- **Reaction cross-section**, σ_{ab}, cross-section for the transition from the entrance channel a into the exit channel b.
- **Absorption cross-section**, σ_{c}, cross-section for absorption of an incident particle by the sample. For neutrons, this quantity is frequently called the **capture cross-section**.

27.5.2 Conservation laws in nuclear reactions

▲ In nuclear reactions, the baryon number (number of nucleons) and the electric charge are conserved as well as the energy, momentum and angular momentum.

▲ In processes governed by strong interaction, the parity π, and for special two-particle interactions the isospin $\vec{T}$, are also conserved:

$$\pi_a \cdot \pi_A \cdot (-1)^{L_i} = \pi_b \cdot \pi_B \cdot (-1)^{L_f},$$

$$\vec{T}_a + \vec{T}_A = \vec{T}_b + \vec{T}_B.$$

27.5.2.1 Energy and momentum conservation

The **kinematics** of nuclear reactions is determined by energy and momentum conservation (**Fig. 27.17**). Both conservation laws hold generally, i.e., for all interactions. They are the starting point for calculating the kinematics of collision processes.

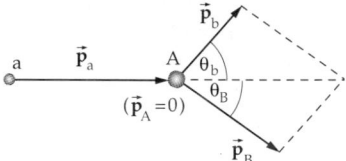

Figure 27.17: Momentum conservation in elastic collisions (laboratory system). $\vec{p}$: momentum before collision, $\vec{p}_b + \vec{p}_B$: momentum after collision.

If a particle with kinetic energy $E_{kin}(a)$ hits a target nucleus A at rest ($E_{kin}(A) = 0$), for a reaction A(a,b)B with Q-value Q at the reaction angles θ_b, θ_B:

$$E_{kin}(a) = E_{kin}(b) + E_{kin}(B) - Q,$$

$$\frac{p_a^2}{2m_a} = \frac{p_b^2}{2m_b} + \frac{p_B^2}{2m_B} - Q,$$

$$\vec{p}_a = \vec{p}_b + \vec{p}_B.$$

This system of equations yields for particle b:

$$E_{kin}(b) = E_{kin}(a) - E_{kin}(B) + Q, \quad p_b = \sqrt{2m_b \cdot E_{kin}(b)}, \quad \sin\theta_b = \frac{p_B}{p_b} \cdot \sin\theta_B,$$

$$p_b = \frac{\sqrt{2m_a \cdot E_{kin}(a)} \cdot \cos\theta_B}{(1 + \frac{m_b}{m_B})}$$

$$\pm \sqrt{\left(\frac{\sqrt{2m_a \cdot E_{kin}(a)} \cdot \cos\theta_B}{(1 + \frac{m_b}{m_B})}\right)^2 + \frac{2E_{kin}(a)(m_b - m_a) + 2Q \cdot m_b}{(1 + \frac{m_b}{m_B})}}.$$

Threshold energy, the energy needed to start a certain reaction. This threshold energy arises in **endothermal reactions** ($Q < 0$),

$$E_{kin}(a, \text{threshold}) = -\frac{m_a + m_A}{m_A} Q \quad \text{with} \quad Q < 0.$$

■ $p + p \rightarrow p + p + \pi$

In this reaction, a π-meson is generated. The Q-value of this reaction therefore equals the mass of the π-meson multiplied by the square of the speed of light in a vacuum c:

$$Q = -m_\pi \cdot c^2 \approx -140 \text{ MeV}.$$

Threshold energy: $\dfrac{m_p + m_p}{m_p} \cdot m_\pi \cdot c^2 \approx 2 \cdot 140 \text{ MeV}.$

27.5.2.2   Angular momentum conservation

Impact parameter, b, perpendicular distance between the path of the incident particle from the target nucleus before a collision process. For a given incident energy $E_{\text{kin}}(a) = p_a^2/(2m_a)$, the impact parameter determines the orbital angular momentum L of the relative motion of the two reaction partners, $L = p_a \cdot b$ (**Fig. 27.18**).

▲   Because of the finite range R of nuclear forces, the energy of the incident particles determines the possible values of angular momenta involved in the reaction (**Fig. 27.19**),

$$L_{\text{max}} = p_a \cdot R.$$

s-wave scattering, scattering of particles by atomic nuclei where only particles with orbital angular momentum $L = 0$ (**central collisions**) contribute to the cross section.

➤   In low-energy nucleon-nucleon scattering, angular momenta $L \geq 1$ may be ignored. s-wave scattering dominates the scattering of slow neutrons ($E \approx 1$ eV) by nuclei.

p-wave scattering, scattering with angular momentum $L = 1$, contributes significantly to the neutron-nucleus scattering cross-section already at neutron energies of about 1 MeV.

➤   In the calculation of scattering cross-sections and angular distributions of 14 MeV neutrons by nuclei, one has to take into account orbital angular momenta up to $L \approx 14$.

▲   Angular momentum conservation: The total angular momentum in the entrance channel i equals the total angular momentum in the exit channel f:

$$\vec{J}_i = \vec{S}_i + \vec{L}_i = \vec{S}_f + \vec{L}_f = \vec{J}_f.$$

▲   Conservation of the total angular momentum permits the conversion of orbital angular momentum in the initial state into nuclear spin in the final state.

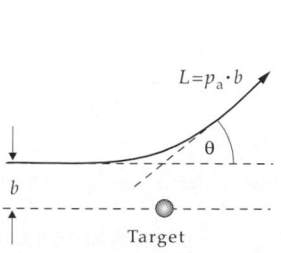

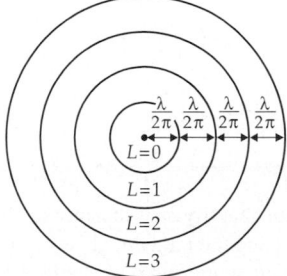

Figure 27.18: Impact parameter b and scattering angle θ of a trajectory with orbital angular momentum $L = p_a \cdot b$, p_a: momentum of the incidence particle.

Figure 27.19: Probabilities for finding the incidence particle versus distance between particle and scattering center for various orbital angular momenta L of partial waves. λ: De Broglie wavelength.

➤ High orbital angular momenta ($L \approx 100\ \hbar$) are reached in heavy-ion-induced nuclear reactions with a specific energy of about 10 MeV/nucleon. In this way, excitation states with high spins (**high-spin states**) may be reached.

27.5.3 Elastic scattering

1. Rutherford scattering,

the scattering of charged particles in the Coulomb field of the nucleus.

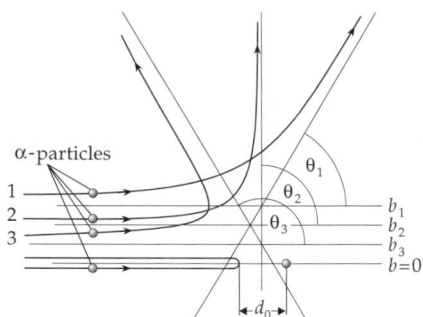

Figure 27.20: Rutherford scattering of α-particles by nuclei. d_0: minimum distance in a central collision.

2. Rutherford scattering formula

Differential cross-section of Rutherford scattering in the center-of-mass frame:

Rutherford scattering formula			$\mathbf{L}^2$
	Symbol	Unit	Quantity
$$\frac{d\sigma_R}{d\Omega} = \left(\frac{Z \cdot Z' \cdot e^2}{4E_0}\right)^2$$ $$\cdot \frac{1}{\sin^4(\theta/2)}$$ $$\cdot \left(\frac{1}{4\pi\varepsilon_0}\right)^2$$	$\dfrac{d\sigma_R}{d\Omega}$	b/sr	differential cross-section
	Z	1	charge number of projectile
	Z'	1	charge number of target nucleus
	E_0	J	kinetic energy of projectile
	θ	rad	scattering angle
	e	C	elementary charge
	ε_0	$C\,V^{-1}\,m^{-1}$	electric permittivity of free space

▲ The quantity d_0 is distance of closest approach between the incident particle of energy E_0 and the target nucleus in a central collision.

■ In the scattering of α-particles by heavy nuclei at a kinetic energy of 15.8 MeV, d_0 is about $1.2 \cdot 10^{-15}$ m.

3. Mott scattering,

the scattering of very energetic particles (velocity v close to the speed of light c). The theory of Mott scattering takes into account the influence of the spin of the interacting particles and yields the relativistic correction to the Rutherford scattering cross section $\frac{d\sigma_R}{d\Omega}$:

$$\frac{d\sigma_M}{d\Omega} = \frac{d\sigma_R}{d\Omega} \frac{\cos^2(\theta/2)}{1 + 2 \cdot (v/c)^2 \cdot \sin^2(\theta/2)} \,.$$

27.5.4 Compound-nuclear reactions

Compound-nuclear reaction, a reaction model based on the idea of the nucleus as a drop of a nuclear liquid (see p. 916). The kinetic energy of the incident particle and the binding energy released in its capture by the target nucleus are statistically distributed over all nucleonic degrees of freedom—as in the transfer of thermal energy to a liquid. A highly heated **compound nucleus** C is generated with an excitation energy given by the sum of the incidence energy $E_{kin}(a)$ and the binding energy $E_B(a)$ of the particle a in the nucleus B,

$$a + A \longrightarrow C^* \,, \quad E^*(C) = E_{kin}(a) + E_B(a) \,.$$

1. Probability of formation and decay of compound nuclei

The probability of formation of a compound nucleus is large when this excitation energy coincides with the energy of a compound-nuclear level. On the other hand, the compound nucleus has a long lifetime, since it decays only when an amount of energy above the binding energy is concentrated into a nucleon or a group of nucleons by collisions between the nucleons,

$$C^* \longrightarrow b + B \,.$$

■ In the capture of slow neutrons with an incident energy of only 1 eV, about 8 MeV is released in nuclei of medium mass number due to the binding energy of a neutron.

▲ Formation and decay of the compound nucleus are independent processes. The cross-sections of nuclear reactions proceeding through highly excited long-living compound-nuclear states show narrow, closely spaced resonances as a function of the incidence energy (**Fig. 27.21**).

➤ The lifetime of a compound-nuclear state is about 10^{-18} s. It is thus several orders of magnitude larger than the transit time of the incident particle across the nucleus. In heavy nuclei, the width of neutron resonances is about 10^{-2} eV; the mean separation of the resonances is about 50 keV.

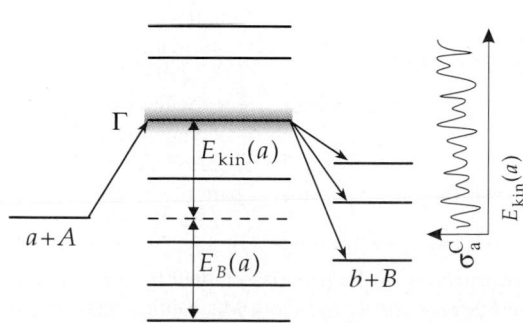

Figure 27.21: Compound-nuclear reaction $a + A \longrightarrow C^* \longrightarrow b + B$ (schematic). Γ: total width of resonance, σ_a^C: cross-section for formation of the compound nucleus versus kinetic energy of the incidence particle $E_{kin}(a)$, with resonances at quasi-stationary states of the compound nucleus C.

2. Cross-section of the compound-nuclear reaction A(a,b)B:

$$\sigma_{ab} = \sigma_a^C \cdot P_b, \quad P_b = \frac{\Gamma_b}{\Gamma}, \quad \Gamma = \sum_i \Gamma_i, \quad i = a, b, c, \dots.$$

σ_a^C: cross-section for compound-nucleus formation,
P_b: probability for decay of the compound nucleus with emission of particle b,
Γ_b: partial width for the decay $C^* \longrightarrow b + B$,
Γ: total width of compound-nuclear level.
▲   The separation between neighboring resonances decreases with increasing excitation energy of the compound nucleus, the resonance width increases, i.e., the resonances begin to overlap.

$1/v$-**law** for the capture cross-section of slow neutrons of energy E:

$$\sigma^C \sim \frac{1}{\sqrt{E}} \sim \frac{1}{v}, \quad v: \text{ neutron velocity}.$$

■   Several formation and decay channels of the compound nucleus ^{51}Cr* are shown in **Fig. 27.22**.

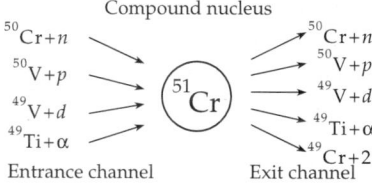

Compound nucleus

Entrance channel   Exit channel

Figure 27.22: Reactions with formation of the compound nucleus ^{51}Cr through various entrance channels decay through various exit channels.

3. Breit–Wigner formula,

describes the energy variation of the cross-section of the compound-nuclear reaction A(a,b)B near a resonance (**Fig. 27.23**):

Breit–Wigner formula			$\mathbf{L^2}$
	Symbol	Unit	Quantity
	$\sigma(a, b, E)$	m^2	cross-section of reaction $a \rightarrow b$
	E_r	MeV	resonance energy
$\sigma(a, b, E) = \sigma(a, E_r)$	E	MeV	particle energy
	$\sigma(a, E_r)$	m^2	compound-nuclear formation cross-section
$\cdot \dfrac{\Gamma \cdot \Gamma_b}{(E - E_r)^2 + (\frac{1}{2}\Gamma)^2}$	Γ	MeV	total width of compound-nuclear resonance
	Γ_b	MeV	partial width for exit channel b

Evaporation spectrum, the energy distribution of the particles emitted by a highly excited compound nucleus. The spectrum largely corresponds to a Maxwellian distribution

(**Fig. 27.24**). The number $N(E)\,dE$ of particles emitted in the energy interval between E and $E + dE$ is

$$N(E)\,dE \sim E\,e^{-E/(kT)}\,dE\,, \quad T: \text{ nuclear temperature}.$$

▲ The angular distribution of the reaction products of a compound-nuclear reaction is in general **isotropic**.

M Resonance reactions of neutrons are of practical importance in the operation of nuclear reactors. They affect the neutron transport and lead to unwanted neutron losses.

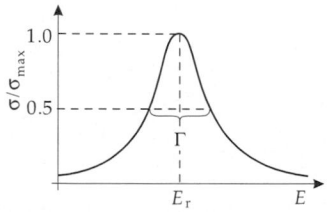

Figure 27.23: Breit–Wigner resonance with half-width Γ. E_r: resonance energy.

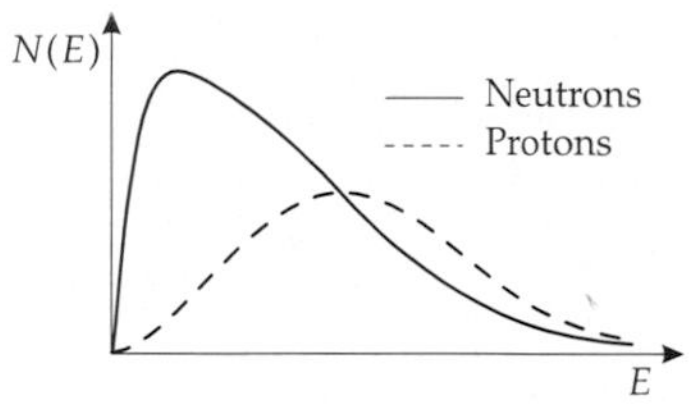

Figure 27.24: Evaporation spectrum for neutrons and protons (schematic).

27.5.5 Optical model

Optical model, considers the atomic nucleus as a refracting and absorbing medium. It provides cross-sections for elastic scattering and absorption of the incident particle. The optical model may be applied to the interaction of neutrons, protons, complex light particles (deuterons, α-particles), heavy ions and mesons with nuclei.

Optical potential, $U(r)$, function of the distance r of the incident particle from the center of the target nucleus, consists of a complex spherical potential and a spin-orbit coupling term:

$$U(r) = -V\,f(r) - jW\,g(r) + W_{ls}(r)\,(\vec{\sigma} \cdot \vec{l})\,.$$

Frequently used form factors:

$$f(r) = \frac{1}{1 + e^{(r-R)/a}}\,, \qquad g(r) = e^{-(r-R)^2/b^2}\,.$$

R: nuclear radius, a, b: surface parameters.

The form factor $f(r)$ of the real part follows the radial mass density distribution in the nucleus (**Woods-Saxon potential, Fig. 27.25**). The form factor $g(r)$ of the imaginary part of the optical potential simulates particle absorption at the nuclear surface. The strength parameters V and W depend on the incidence energy (**Fig. 27.26**).

▲ The cross-sections as functions of incidence energy calculated with the optical model exhibit **giant resonances** with resonance widths of several MeV.

27.5.6 Direct reactions

Direct reactions differ from compound-nuclear reactions in the following ways:

• the reaction time ($\approx 10^{-22}$ s) corresponds to about the transit time of the incident particle across the target nucleus,

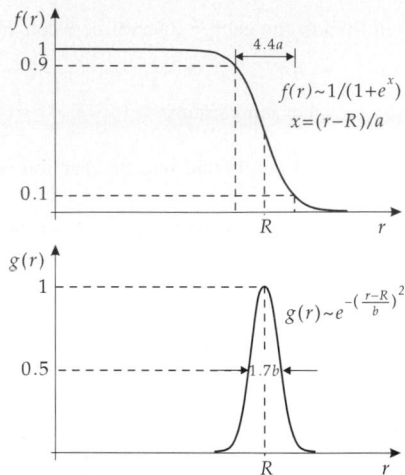

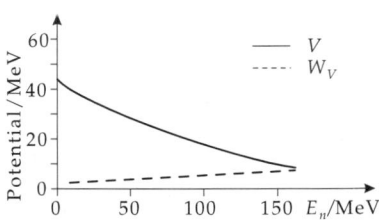

Figure 27.25: Form factors of the optical potential. $f(r)$: real part (Woods-Saxon potential), $g(r)$: imaginary part (Gaussian potential).

Figure 27.26: Dependence of the strength parameters of the optical potential on the incidence energy E.

- a direct transition proceeds from the entrance channel to the exit channel without formation of a quasi-stationary intermediate state of the total system,
- only few nucleonic degrees of freedom are involved in the reaction,
- the reaction proceeds preferably at the nuclear surface,
- the energy dependence of the cross-section displays broad giant resonances.

Stripping reaction, a direct reaction in which a particle is stripped from the projectile when it passes the target nucleus, and is captured into a single-particle state in the mean nuclear potential of the target nucleus. The process is mediated by a peripheral interaction of the projectile with the target nucleus.

Pick-up reaction, a direct reaction in which the projectile passing the target nucleus picks a particle from a single-particle state in the mean potential of the target nucleus. The process is mediated by a peripheral interaction of the projectile with the target nucleus.

➤   Direct reactions of this type are used to determine single-particle states in nuclei.

Direct inelastic scattering, a collision process in which preferably collective vibrational and rotational states of the target nucleus are excited.

Intermediate processes, reactions in which the formation of an intermediate state of the total system begins but the decay into the exit channel proceeds before a complete equilibrium state is established. The spectra and angular distributions of the reaction products show features of both compound-nuclear and direct reactions.

27.5.7   Heavy-ion reactions

Heavy-ion reactions, reactions in which nuclei with relatively high atomic number $Z > 2$, $A > 4$ are used as incident particles.

1. Coulomb barrier and kinetic energy per nucleon

Coulomb barrier, T_C, the minimum value of the kinetic energy of the incident particle needed to reach the range of nuclear forces:

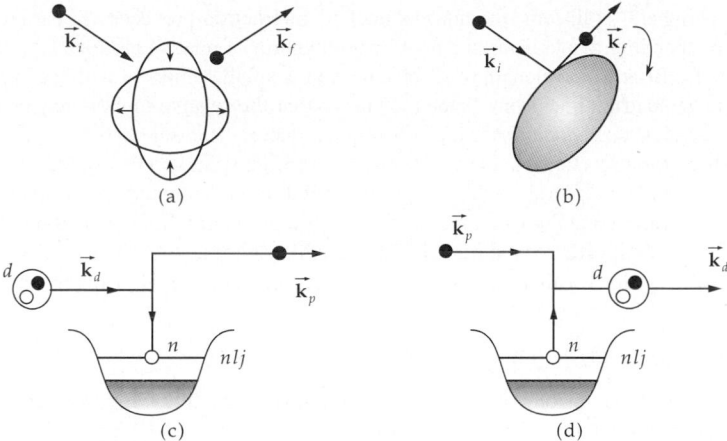

Figure 27.27: Direct reactions (schematic figure). $\vec{k}$: wave vectors. (a): vibrational excitation, (b): rotational excitation, (c): stripping reaction A(d,p)B, capture of the neutron into the single-particle state (nlj) of the target nucleus, (d): pick-up reaction A(p,d)B.

Coulomb barrier			$\mathbf{ML^2T^{-2}}$
	Symbol	Unit	Quantity
$T_C = \dfrac{Z_1 \cdot Z_2 \cdot e^2}{(R_1 + R_2)} \cdot \dfrac{1}{4\pi\varepsilon_0}$	T_C	J	Coulomb barrier
	Z_1, Z_2	1	atomic numbers
	R_1, R_2	m	nuclear radii
	e	C	elementary charge
	ε_0	$CV^{-1}m^{-1}$	electric-permittivity constant

■ For the reaction $^{40}_{20}Ca_{20}$ on $^{208}_{82}Pb_{126}$, the Coulomb barrier is 211 MeV, i.e., 5.3 MeV/nucleon.

Specific energy ε, kinetic energy per nucleon,

$$\varepsilon = \frac{E_{kin}}{A} .$$

Classification of heavy-ion reactions by the specific energy ε:

$\varepsilon < 10$ MeV/A:	low-energy heavy-ion reactions,
10 MeV/$A < \varepsilon < 100$ MeV/A:	heavy-ion reactions at medium energies,
100 MeV/$A < \varepsilon < 10$ GeV/A:	relativistic heavy-ion reactions,
$\varepsilon > 10$ GeV/A:	ultra-relativistic heavy-ion reactions .

2. Features particular to heavy-ion reactions

- Because the incident particle often has a mass comparable to that of the target, a large fraction of the kinetic energy goes into center-of-mass motion.
- Both reaction partners have a high charge, hence Coulomb effects become significant and many phenomena result from the interplay of Coulomb and nuclear forces.
- In the interaction region, intermediate states with as many as 300 to 400 nucleons are formed. Therefore, in the description of the system, macroscopic aspects may be taken into account to a larger extent than in light-body induced reactions.

- In peripheral collisions, the nucleus-nucleus interaction proceeds via partial waves corresponding to a large orbital angular momentum of relative motion ($L \geq 100\,\hbar$).
- The De Broglie wavelength of relative motion is small compared with the characteristic geometric dimensions of the system, so that the relative motion may be treated by classical considerations using collision parameters and trajectories.
- ▲ In heavy-ion reactions, nuclear states with very high spins can be excited.
- ■ In the reaction $^{40}_{20}\mathrm{Ca}_{20} \rightarrow {}^{208}_{82}\mathrm{Pb}_{126}$, an orbital angular momentum of about $140\,\hbar$ may be reached at the Coulomb barrier. Such high angular momenta allow the production of **superdeformed nuclei** with cigar-like shapes.
- ■ For $^{40}_{20}\mathrm{Ca}_{20}$-ions with an energy of 10 MeV per nucleon, the De Broglie wavelength is $\lambda = 0.5$ fm.

3. Reaction types in heavy-ion reactions

Depending on the collision parameter, one distinguishes the following reaction types in low energy, heavy-ion reactions (**Fig. 27.28**):

- ▲ **Coulomb processes**, elastic Rutherford scattering and Coulomb excitation of collective states of the target nucleus and/or the projectile for large values of the collision parameter at which nuclear forces are not yet effective ($L \gg L_{\mathrm{gr}}$, L_{gr}—angular momentum at grazing incidence).
- ▲ **Quasi-elastic reactions**, direct reactions for collision parameters corresponding to grazing incidence of the projectile ($L \approx L_{\mathrm{gr}}$). The small reaction time of $\approx 10^{-22}$ s allows an excitation of only few nuclear degrees of freedom. The exchange of energy and nucleons between projectile and target nucleus is still weak.
- ▲ **Deep-inelastic reactions**, reactions at medium values of the collision parameter ($L_{\mathrm{crit}} < L < L_{\mathrm{gr}}$), which proceed via formation of a relatively long-living two-nuclei system with a lifetime of $\approx 10^{-21}$ s. In this system, many degrees of freedom are excited without reaching a compound-nuclear state. A strong exchange of energy and of nucleons between projectile and target nucleus is observed.

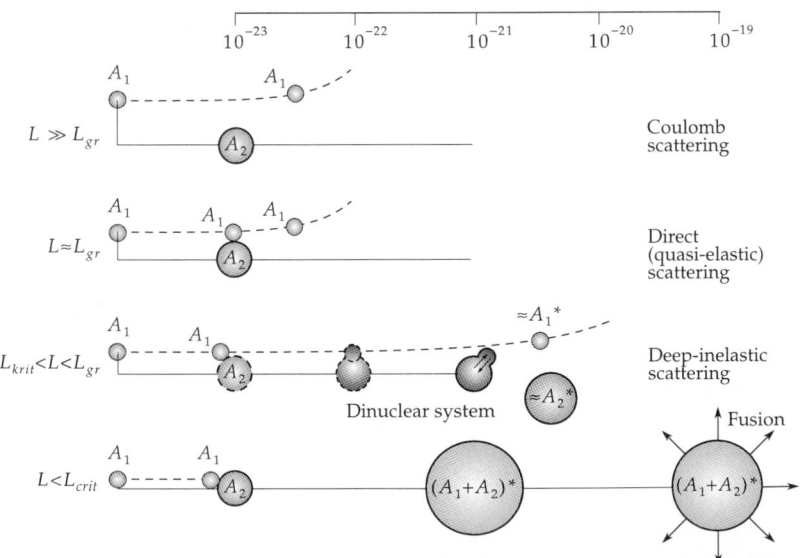

Figure 27.28: Classification of low-energetic heavy-ion reactions A_1+A_2 by the collision parameter (orbital angular momentum L). L_{gr}: orbital angular momentum for grazing nucleus-nucleus interaction, L_{crit}: orbital angular momentum at which fusion begins.

▲   **Fusion reactions**, formation of a highly excited compound nucleus with a lifetime of $\approx 10^{-18}$ s for small collision parameters ($L < L_{\mathrm{crit}}$). The compound nucleus decays by emission of particles and γ-rays, or by fission.

■   The cross-section of the reaction ^{40}Ar(379 MeV) $+^{232}$Th shows that, besides the quasi-elastic peak near the incidence energy, there is a second relative maximum at an energy loss of ≈ 160 MeV that corresponds to a deep-inelastic process.

■   In the deep-inelastic reaction $^{86}_{36}$Kr(515 MeV)$+^{166}$Er, one observes reaction products similar to the projectile with nuclear charge numbers between $Z = 28$ and $Z = 45$.

➤   Nuclei far from the line of stability are produced in heavy-ion reactions.

Islands of stability, regions in the Z-N plane, stabilized by magic proton numbers Z and neutron numbers N. These nuclei should have very long lifetimes compared with neighboring nuclides in the Z-N diagram. According to model calculations, islands of stability are expected around $Z = 114$ and $N = 184$.

Superheavy elements, elements with $Z \geq 110$.

➤   The heaviest **transuranium elements Bohrium** ($_{107}$Bh), **Hassium** ($_{108}$Hs) and **Meitnerium** ($_{109}$Mt) as well as the elements with $Z = 110 - 112$, 114, 116, 118 have surprisingly long lifetimes ($\tau \approx$ ms). The long lifetime suggests a new shell structure in this region.

4. Higher-energy heavy-ion collisions

Multifragmentation, decay of the highly-excited compressed nucleon system formed in heavy-ion collisions of intermediate energy into numerous fragments with a broad distribution of charge and mass numbers. A nuclear phase transition liquid-gas is expected to play a role.

Relativistic heavy-ion collisions, heavy-ion reactions with extremely high incidence energies produced at CERN (Geneva) and at AGS (Brookhaven). These reactions may generate new states of matter:

•   **Resonance matter**, enrichment of normal nuclear matter by excited unstable nucleonic states (Δ- and N^*-resonances).

•   **Antimatter**, formed from the antiparticles of nucleons: $\bar{p}, \bar{n}, \bar{d}$ (antideuteron), $\bar{\alpha}$ . . .

•   **Hypernuclei**, and **multi-hyperon matter**, consisting of nucleons and hyperons (Λ-, Σ^-- and Ξ^--particles).

•   **Quark-gluon plasma**, phase of nuclear matter in which quarks and gluons move almost freely, instead of being bound in baryons and mesons. This **deconfinement** is expected to occur only at very high baryonic and energy densities (1–3 GeV/fm^3).

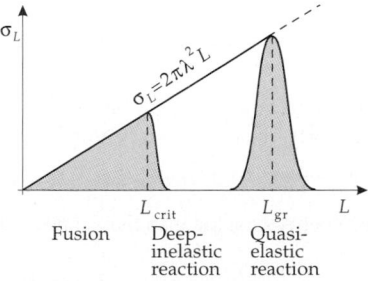

Figure 27.29: Schematic subdivision of the cross-section σ_L of a low-energetic, heavy-ion reaction. L_{gr}: orbital angular momentum for a grazing nucleus-nucleus interaction, L_{crit}: orbital angular momentum for beginning fusion.

27.5.8   Nuclear fission

Nuclear fission, the process of decomposition of a heavy nucleus into two fractions (**fission products**) of almost equal size, and several neutrons (**fission neutrons**). Nuclear fission can be induced by capture of neutrons or photons by the nucleus.

■ $^{235}U + n \rightarrow X + Y + \nu\,n + 200$ MeV, ν : number of fission neutrons.

On average, $\nu = 2.43 \pm 0.07$ neutrons with a mean energy of 2 MeV are emitted per fission event.

1. Cause of nuclear fission

➤   Nuclear fission may be explained by the **droplet model** and by the **shell model**. At low excitation energy, the nucleus carries out surface vibrations of small amplitude about the equilibrium shape in the ground state. The surface tension thereby creates a potential barrier causing stability of the nucleus against large deformations. If the excitation energy increases, this **fission barrier** may be overcome: the nuclear deformation increases until the nucleus forms a neck and finally breaks into two fractions, which then separate under the influence of the repulsive Coulomb potential.

Fission barrier, potential energy barrier that prevents fission.

nucleus	binding energy of neutron	fission barrier
^{235}U	6.5 MeV	^{236}U: 6 MeV
^{238}U	6 MeV	^{239}U: 7 MeV

➤   Since the binding energy of a neutron in ^{235}U exceeds the fission barrier, ^{235}U is usually chosen as the main fuel material in thermal nuclear reactors.

2. Spontaneous fission and fission isomerism

Spontaneous fission, fission from the ground state of nuclei with $Z^2/A > 17$ by tunneling the fission barrier. The half-life for spontaneous fission is larger than the half-life for α-decay.

■ ^{235}U: α-decay: $T_{\frac{1}{2}} = 7.1 \cdot 10^8$ yr,   spontaneous fission: $T_{\frac{1}{2}} = 1.8 \cdot 10^{17}$ yr.

Fission isomerism, appearance of a second minimum in the nuclear potential as a function of the separation between the fission products, caused by shell effects. In neutron-induced fission, the nucleus first passes to an excited state belonging to the first potential minimum, which couples to states belonging to the second minimum. Fission finally proceeds by decay from the states in the second minimum by the tunnel effect.

■   Example:

$$^{16}O + ^{238}U \rightarrow {}^{251}Fm^* + 3\,n$$

The excited Fermium nucleus $^{251}Fm^*$ decays by fission with a half-life of $T_{\frac{1}{2}} \approx 0.014$ s.

➤   Ternary fission (three pieces) of heavy nuclei occurs with low probability.

▲   The kinetic energy of the fission products nearly equals the total energy released in fission.

▲   As a rule, fission products are radioactive.

▲   Fission products decay preferably by neutron emission, but also by γ- and β-decay.

3. Fission neutrons and mass distribution

Prompt neutrons, neutrons emitted simultaneously with fission.

  Delayed neutrons, neutrons emitted by the fission products after the primary fission process. This emission is delayed typically between 0.2 s and 60 s.

➤   Delayed neutrons play a fundamental role in the operation of controlled chain reaction devices.

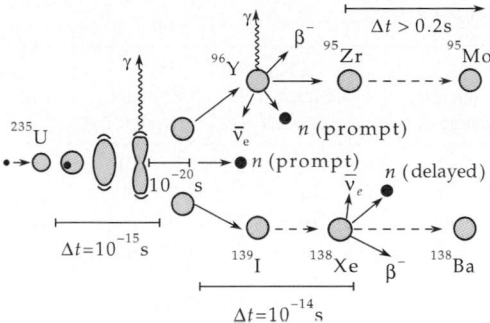

Figure 27.30: Time evolution of the fission of a uranium nucleus.

Mass distribution in fission, abundance distribution of fission products.

▲   As a rule, the mass distribution is asymmetric (mass ratio of fission products ≈ 3 : 2).

■   For ^{235}U, symmetric fission is 600 times less likely than asymmetric fission.

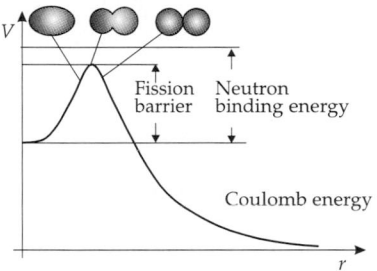

Figure 27.31: Nuclear fission. Potential energy V and nuclear shape as functions of the separation r of the fission products.

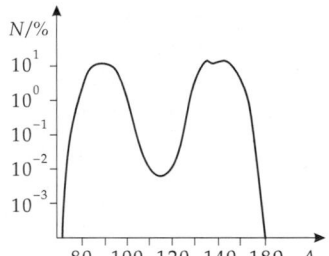

Figure 27.32: Mass distribution of the fission products in the fission of uranium.

27.6 Nuclear decay

Radioactive decay, spontaneous decay of unstable nuclides with the emission of particles or photons. The decays proceed via radioactive decay series into stable nuclides.

Radionuclide, nuclide undergoing radioactive decay.

Radioactive isotopes, particular species of radionuclide.

Radioactivity, the property of nuclides or macroscopic quantities of matter (atmosphere, waters, rocks, building materials) to emit radioactive radiation.

Natural radioactivity, radioactivity of nuclides occurring in nature.

Artificial radioactivity, radioactivity occurring in nuclides produced artificially, e.g., in nuclear reactions.

Modes of radioactivity:

Decay mode	Change of		
	nuclear charge ΔZ	neutron number ΔN	mass number ΔA
α-decay (emission of a He nucleus)	-2	-2	-4
β-decay (e^+- or e^--emission)	± 1	∓ 1	0
γ-decay (emission of a photon)	0	0	0
electron capture	-1	$+1$	0
proton emission	-1	0	1
neutron emission	0	-1	-1
cluster radioactivity	$-Z_{cluster}$	$-N_{cluster}$	$-(Z_{cluster} + N_{cluster})$
spontaneous fission	$\approx \frac{1}{2}Z$	$\approx \frac{1}{2}N$	$\approx \frac{1}{2}A$

▲   **Radioactive decay** is a statistical process.

27.6.1   Decay law

1. Decay constant

Decay constant, λ, specifies the probability of a certain radioactive decay mode. It is independent of space and time, but is specific to the particular nucleus.
▲   Every radionuclide has a unique decay constant.
The decay constant gives the fraction of nuclei decaying per second.

number of decays = −decay constant · number of nuclei · time			**1**
	Symbol	Unit	Quantity
$dN = -\lambda \cdot N \cdot dt$	dN	1	number of decays
	λ	s^{-1}	decay constant
	N	1	number of radioactive nuclei
	dt	s	time interval

Radioactive decay follows the exponential decay law (**Fig. 27.33**):

decay law			**1**
	Symbol	Unit	Quantity
$N(t) = N_0 e^{-\lambda \cdot t}$	$N(t)$	1	number of radioactive nuclei at time t
	N_0	1	number of nuclei at time $t = 0$
	λ	s^{-1}	decay constant
	t	s	time variable

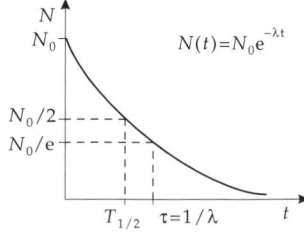

Figure 27.33: Exponential decay law. λ: decay constant, τ: mean lifetime, $T_{\frac{1}{2}}$: half-life.

Mean lifetime τ (SI unit: second s) of radioactive nuclei, reciprocal value of the decay constant:

$$\tau = \frac{1}{\lambda}.$$

2. Half-life,

$T_{1/2}$ (SI unit: second s), the time interval after which the number of radioactive nuclei drops to half of the initial number:

$$T_{\frac{1}{2}} = \frac{\ln 2}{\lambda} = \ln 2 \cdot \tau.$$

Partial decay constant, λ_k, the probability for a particular decay mode k. For radioactive isotopes that may decay via several modes:

$$\lambda = \sum_k \lambda_k.$$

3. Activity,

A, the number of decays per unit time,

$$A = -\frac{dN}{dt}.$$

activity			$\mathbf{T^{-1}}$
	Symbol	Unit	Quantity
$A = \lambda \cdot N = \lambda \cdot N_0 e^{-\lambda \cdot t}$	M	kg/mol	molar mass of the substance
	m	kg	mass of the substance
$= \lambda \cdot \dfrac{m \cdot N_A}{M}$	N	1	number of radioactive nuclei
	N_A	mol^{-1}	Avogadro's number
	λ	s^{-1}	decay constant

Becquerel (Bq), SI unit of activity,

$$1\ \text{Bq} = \frac{1\ \text{decay}}{\text{s}}.$$

Specific activity, A_s, the activity per unit mass of the substance,

$$A_s = \frac{A}{m}, \quad m:\ \text{mass}.$$

4. Radionuclides in the environment

Typical concentration of several radionuclides in the environment:

Substance	Radionuclide	Half-life $T_{\frac{1}{2}}$/yr	Concentration 10^{-3}Bq/l
ground water	^{3}H	12.232	20 – 100
	^{40}K	$1.26 \cdot 10^9$	4 – 400
	^{238}U	$4.51 \cdot 10^9$	1 – 200
surface water	^{3}H	12.232	40 – 400
	^{40}K	$1.26 \cdot 10^9$	40 – 2000
	^{238}U	$4.51 \cdot 10^9$	– 40
drinking water	^{3}H	12.232	20 – 70
	^{40}K	$1.26 \cdot 10^9$	200
	^{238}U	$4.51 \cdot 10^9$	– 40

5. Decay chains,

arise when a nuclide produced in a radioactive decay may again be radioactive.

▲   For the number of radioactive **parent** and **daughter nuclides** present at time t, the following decay law holds:

change of daughter nuclei / time unit = production rate – decay rate			$\mathbf{T^{-1}}$
	Symbol	Unit	Quantity
$\dfrac{dN_D}{dt} = \lambda_P \cdot N_P - \lambda_D \cdot N_D$	N_D	1	number of daughter nuclides
	N_P	1	number of parent nuclides
	t	s	time variable
	λ_D	s^{-1}	decay constant of daughter nucleus
	λ_P	s^{-1}	decay constant of parent nucleus

decay law for daughter nuclide			1
	Symbol	Unit	Quantity
$N_D(t) = N_P(0)\dfrac{\lambda_P}{\lambda_D - \lambda_P}$ $\cdot \left(e^{-\lambda_P \cdot t} - e^{-\lambda_D \cdot t} \right)$	N_D	1	number of daughter nuclides
	$N_P(0)$	1	number of parent nuclides at time $t = 0$
	t	s	time variable
	λ_D	s^{-1}	decay constant of daughter nucleus
	λ_P	s^{-1}	decay constant of parent nucleus

Radioactive equilibrium, stationary state of a daughter isotope with an equal number of production- and decay reactions in a certain time interval:

$$\frac{dN_D}{dt} = 0 \,.$$

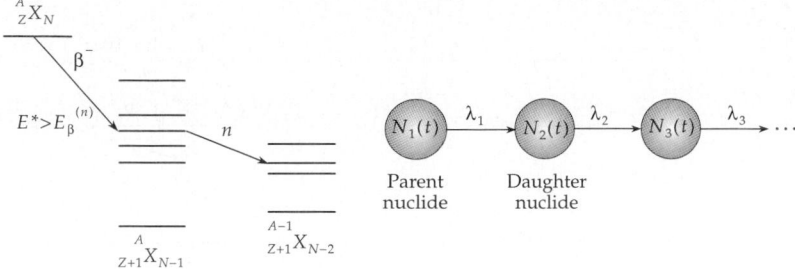

Figure 27.34: Decay chain (schematic).

In equilibrium:

$$N_P \cdot \lambda_P = N_D \cdot \lambda_D , \qquad \frac{N_P}{N_D} = \frac{T_{\frac{1}{2}P}}{T_{\frac{1}{2}D}} .$$

N_P: number of parent nuclides, N_D: number of daughter nuclides,
$T_{\frac{1}{2}P}$:   half-life of parent nuclide, $T_{\frac{1}{2}D}$:   half-life of daughter nuclide.

6. *Example: uranium-radium decay chain,*

(**Fig. 27.35**), in the uranium series

$$\frac{N_{Ra}}{N_U} = 0.36 \cdot 10^{-6} .$$

Hence, one has to process tons of uranium in order to get one gram of radium.

27.6.2   α-decay

α-**decay**, the emission of a He nucleus of mass number $A = 4$ and nuclear charge number $Z = 2$ (**Fig. 27.36**).
    Decay equation:

$$^{A}_{Z}X_N \rightarrow {}^{A-4}_{Z-2}X_{N-2} + {}^{4}_{2}\alpha_2 .$$

$$^{212}_{84}Po_{128} \longrightarrow {}^{208}_{82}Pb_{126} + {}^{4}_{2}\alpha_2 .$$

▲   The kinetic energies E_α of the particles emitted in α-decay form a line spectrum. Typical energies of α-particles are between 4 MeV and 9 MeV.
■ ^{212}Po: $E_\alpha = 8.9$ MeV, ^{232}Th: $E_\alpha = 4.1$ MeV.
▲   The half-lifes of many α-radioactive nuclei are relatively large, since the α-decay proceeds by the **tunnel effect**. The potential wall at the nuclear surface resulting from the overlay of the attractive nuclear potential and the repulsive Coulomb potential is higher than the kinetic energy of the emitted α-particles. In order to leave the nucleus the α-particles have to tunnel through the potential wall (see p. 839, **Fig. 27.37**).

Geiger-Nutall relation, empirical connection between the decay constant λ and the kinetic energy E_α of the α-particles:

$$\ln \lambda = k_1 + k_2 \cdot \ln E_\alpha .$$

The constants k_1 and k_2 characterize the different decay chains.

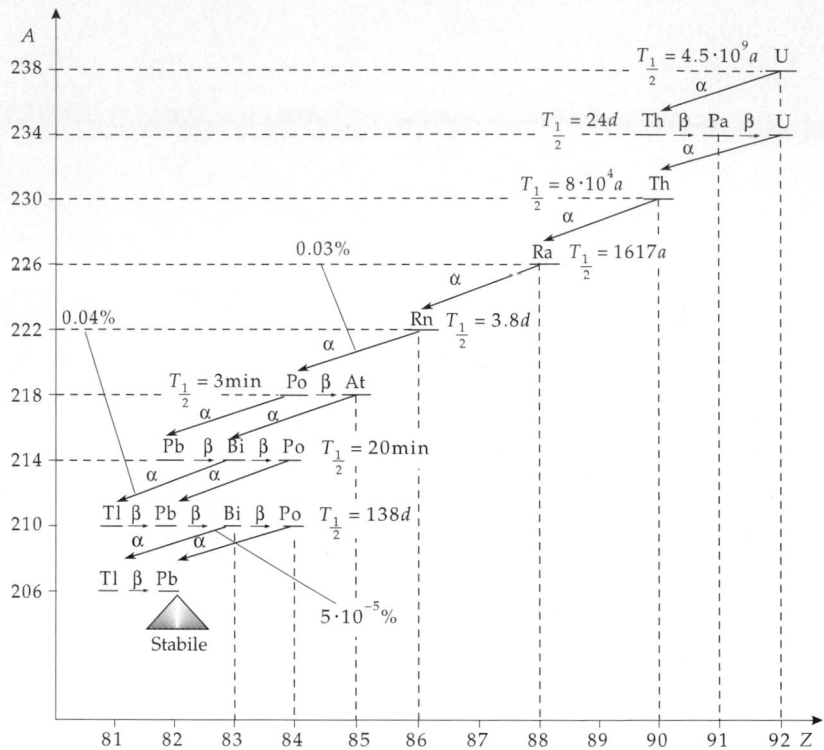

Figure 27.35: Uranium-radium decay chain.

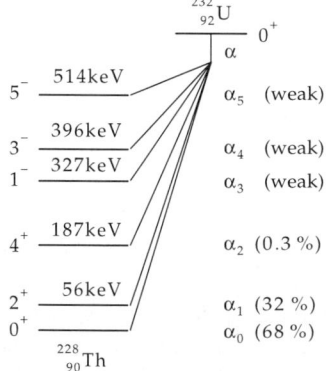

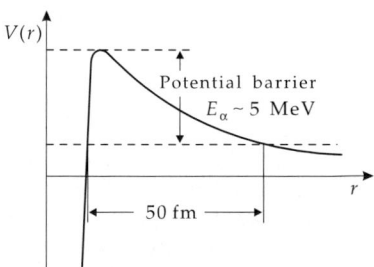

Figure 27.36: α-decay. In the decay of $^{232}_{92}U_{140}$ into $^{282}_{90}Th_{138}$, six α-groups of different kinetic energy and intensity are observed corresponding to different excited states of the final nucleus.

Figure 27.37: α-decay as tunnel effect through the Coulomb barrier.

Penetrability, D, of the Coulomb potential wall:

penetrability of a potential wall				**1**
		Symbol	Unit	Quantity
$$D = e^{-\frac{4\pi \cdot R}{\lambda_B}} \cdot \gamma$$		D	1	penetrability
		R	m	nuclear radius
		λ_B	m	De Broglie wavelength
$$\gamma = \sqrt{\frac{B}{E}} \arccos\sqrt{\frac{E}{B}} - \sqrt{1 - \frac{E}{B}}$$		B	J	height of potential wall
		e	C	elementary charge
		ε_0	C/(Vm)	electric permittivity constant
$$B = \frac{Z \cdot z \cdot e^2}{4\pi \varepsilon_0 R}$$		E	J	kinetic energy of particle
		Z	1	charge number of nucleus
$$\lambda_B = \frac{h}{\sqrt{2mB}}$$		z	1	charge number of emitted particle
		m	kg	mass
		h	J s	Planck's constant

This relation holds for all charged particles.

27.6.3 β-decay

β-**decay**, includes three modes of nuclear conversions caused by weak interactions:
- β^--**decay**, instability of an atomic nucleus against emission of an electron,
- β^+-**decay**, instability of an atomic nucleus against emission of a positron,
- **electron capture**, capture of an atomic electron by the nucleus.

In $\beta^{\pm}$-decay, there are three particles in the final state:

$$n \longrightarrow p + e^- + \bar{\nu}_e, \quad p \longrightarrow n + e^+ + \nu_e.$$

Neutrino, ν, a particle invented by Pauli (1931), at first hypothetically, in order to preserve the validity of energy and angular momentum conservation in β-decay. The neutrino carries no electric charge and presumably also no rest mass, but has spin $s = 1/2$ and lepton number ± 1. Recent experiments have given an indication of a very small, but non-zero, mass for the neutrino.

Electrons, positrons and neutrinos do not exist in the nucleus as constituents. They are generated just at the moment of decay by the weak interaction between the nucleons.

Equation for radioactive decay of a nucleus X:

$$_Z^A X_N \rightarrow {}_{Z\pm 1}^A X_{N\mp 1} + e^{\mp} + \begin{pmatrix} \bar{\nu}_e \\ \nu_e \end{pmatrix}.$$

Electron capture, e-capture, capture of an atomic electron by the nucleus with conversion of a proton into a neutron.

Equation for decay:

$$e^- + {}_Z^A X_N \longrightarrow {}_{Z-1}^A X_{N+1} + \nu_e.$$

K-capture, capture of an electron from the K-shell, the most intense transition, since the probability of finding an electron within the nuclear range is a maximum for the K-shell.

➤ The electron hole remaining in the K-shell is filled by an electron transition in the shell with the emission of characteristic X-rays or an Auger electron.

1. β-stability

β-stability, the property of isotopes to be stable against β-decay.

▲ All nuclides occuring in nature lie in the "valley of stability" of the Z-N diagram. Nuclides on the left side of the energy-Z diagram of isobars show β^--decay. Nuclides on the right side show β^+-decay.

■ β-decays of the isobars with $A = 41$ (**Fig. 27.38 (a)**).

▲ The **energy spectrum** of electrons emitted in β-decay is continuous up to a **maximum energy** E_0 (**Fig. 27.38 (b)**).

➤ A two-body decay to an isobar and a β-particle would display a discrete energy spectrum because of energy and momentum conservation.

If the neutrino had a rest mass differing from zero (Majorana neutrino), the energy distribution in the above figure would diverge from the solid line just below the maximum energy and follow the trend indicated by the dashed line, with a vertical tangent at the endpoint.

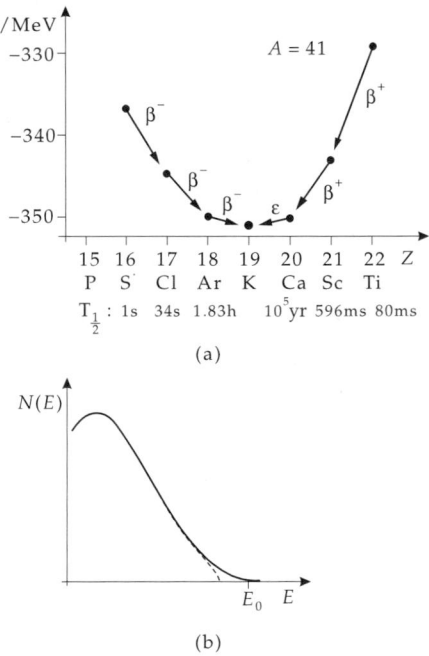

Figure 27.38: (a): β-decays of the isobars with $A = 41$. The binding energy B, the decay mode (β^+, β^- or electron capture ε) and the half-life $T_{\frac{1}{2}}$ are given. (b): energy spectrum in β-decay. E_0: maximum energy. Dashed: trend for a finite neutrino mass.

2. Fermi plot,

also **Curie plot**, representation of the measured β-energy distribution in a diagram of the form:

Curie representation of the β-spectrum			**1**
	Symbol	Unit	Quantity
$C(\varepsilon) = \sqrt{\dfrac{N(\eta)}{F(Z, \eta)\eta^2}}$	$C(\varepsilon)$	1	Curie function
	$N(\eta)$	1	number of electrons
	$F(Z, \eta)$	1	Fermi function
$\eta = \dfrac{p}{m_0 c}$	p	kg m s^{-1}	momentum
	η	1	momentum/$(m_0 c)$
	E	kg m s^{-2}	energy
$\varepsilon = \dfrac{E}{m_0 c^2}$	ε	1	energy/$(m_0 c^2)$
	m_0	kg	electron mass
	c	m s^{-1}	speed of light

3. Fermi function,

$F(Z, \eta)$, takes into account the distortion of the electron and positron wave function ψ at the position of the nucleus by the Coulomb field of the nucleus:

$$F(Z, \eta) = \frac{|\psi(0)_{\text{Coulomb}}|^2}{|\psi(0)_{\text{free}}|^2} .$$

The Fermi function depends strongly on the element (Z).

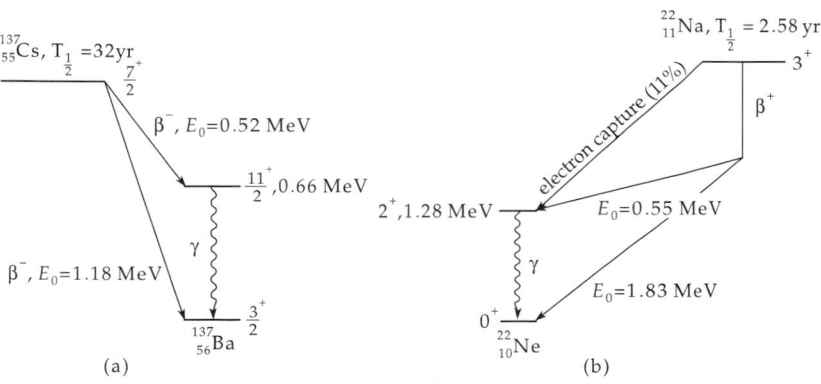

Figure 27.39: β-decay and electron capture. E_0: maximum energy in the β-spectrum. (a): decay scheme for the β^--decay of $^{137}_{55}$Cs, (b): decay scheme for the β^+-decay of $^{22}_{11}$Na.

4. Selection rules for β-transitions

The β-transitions between nuclear states obey selection rules in spin and parity.
    Allowed transitions, the Fermi plot of the β-spectrum is a straight line.
    Forbidden transitions, the Fermi plot of the β-spectrum deviates from a straight line.
    ft-value, a method of classifying β-decays, connected with the measured half-life $T_{\frac{1}{2}}$:

ft-value ~ half-life				1
$$ft = T_{1/2} \int_{1}^{\varepsilon_0} F(Z, \varepsilon)\varepsilon\sqrt{\varepsilon^2 - 1}$$ $$\cdot (\varepsilon_0 - \varepsilon)^2 d\varepsilon$$	Symbol	Unit	Quantity	
	$F(Z, \varepsilon)$	1	Fermi function	
	ε	1	energy/$(m_0 c^2)$	
	ε_0	1	maximum energy/$(m_0 c^2)$	
	$T_{1/2}$	s	half-life	

Superallowed transitions: $\log ft \approx 3.5$.
Allowed transitions: $\log ft \approx 5$.
Forbidden transitions: $\log ft = 9 \ldots 18$.

27.6.4 γ-decay

γ-**decay**, emission of a photon by an excited nucleus. The excitation may be preceded by α- or β-decay, by a nuclear reaction, or an inelastic collision with another nucleus. Similar to electrons in the atomic shell, the atomic nuclei have discrete energy levels and can emit electromagnetic radiation with characteristic line spectra.

Equation of decay:

$$_Z^A X_N^* \longrightarrow _Z^A X_N + \gamma \,.$$

■ $_{27}^{60}$Co-sample as γ-source:

The β-decay of $_{27}^{60}$Co ($T_{1/2} = 5.2$ yr) populates the excited states $E^* = 2.505$ MeV, $J^\pi = 4^+$ (99.9 %) and $E^* = 1.332$ MeV, $J^\pi = 2^+$ (0.1 %) of the nucleus $_{28}^{60}$Ni. The corresponding endpoint energies in the β-spectrum are 314 keV and 1480 keV, respectively. In the transitions $4^+ \longrightarrow 2^+$ and $2^+ \longrightarrow 0^+$ (ground state), the Ni-nucleus emits γ-radiation of 1.173 MeV and 1.332 MeV, respectively (**Fig. 27.40**).

Nuclear isomerism, occurrence of long-lived excited states in nuclei, caused by large differences in the spins of the levels involved in possible transitions.

Nuclear resonance fluorescence, the re-absorption of a γ-rays after emission by a nucleus of the same species. Resonance absorption is suppressed by the recoil-energy loss and by the Doppler effect: the photon energy available for a new excitation of a nucleus is smaller than the de-excitation energy ΔE of the isotope. The thermal motion of nuclei causes a broadening of the line, both in the emission and absorption spectrum.

M   **Mössbauer effect** (Rudolf Mössbauer, Nobel Prize, 1961), amplification of resonance absorption in crystals at low temperatures, since the recoil momentum must then be transferred to the crystal as a whole. The resonance width is then so small that energy spectra can be measured with a resolution up to 10^{-9} eV.

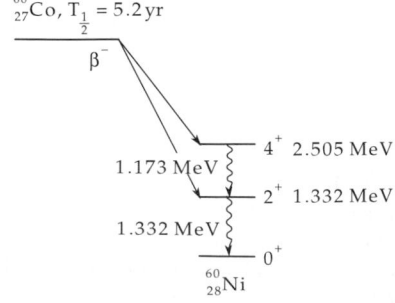

Figure 27.40: Decay scheme of $_{27}^{60}$Co.

27.6.5 Emission of nucleons and nucleon clusters

Delayed nucleon emission, emission of nucleons following a radioactive decay (e.g., β-decay) populating excited states in the daughter nucleus with excitation energy E^* above the nucleon binding energy $E_B^{(N)}$ (**Fig. 27.41**).

➤ Delayed emission of α-particles has also been observed.

Spontaneous nucleon emission, decay of nuclides generated in nuclear reactions beyond the limit of nuclear stability (vanishing binding energy for nucleons at sufficient distance from the line of stability) by spontaneous nucleon emission (proton emission at high proton excess, neutron emission at high neutron excess).

Cluster decay, the decay of nuclei by emission of **clusters** (^{12}C, ^{14}C and other nuclei). This decay mode suggests the importance of shell closure for the stability of atomic nuclei.

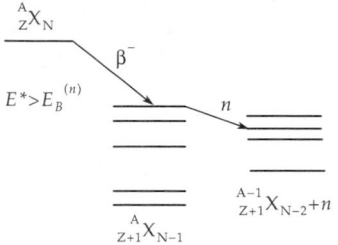

Figure 27.41: Decay scheme for delayed nucleon emission. E^*: excitation energy, $E_B^{(n)}$: neutron binding energy.

27.7 Nuclear reactor

Chain reaction, nuclear fission reactions that become self-sustaining by the release of a sufficient number of neutrons per fission event at a controlled constant rate (reactor), or suddenly (atomic bomb) (**Fig. 27.42**).

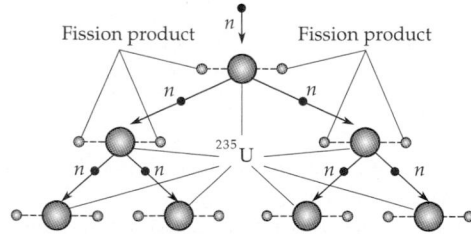

Figure 27.42: Scheme for a chain reaction.

1. Characteristics of the chain reaction

Multiplication factor, k, the number of neutrons released in a chain reaction available for an additional fission process.

▲ Condition for a chain reaction is $k \geq 1$.

Subcritical assembly, a device for nuclear fission in which the multiplication factor is less than unity. In order to maintain nuclear fission, an external neutron source is necessary.

Critical assembly, **controlled chain reaction**, a device for nuclear fission in which the multiplication factor is set to unity.

Supercritical assembly, the multiplication factor is larger than unity. The chain reaction then increases in an uncontrolled manner. The consequence is an explosion.

Mean fission-neutron number, ν, the number of neutrons released on the average per fission event. In real assemblies, this number is reduced by radiative capture in the fuel and external nuclei, as well as by the escape of neutrons from the active zone.

Fast-fission factor, ε, the factor by which the number of fission neutrons is modified due to the release of additional neutrons from fission of ^{238}U and ^{235}U by fast neutrons.

Resonance factor, ψ, a measure of the neutron loss due to neutron absorption in the energy range in which the resonance absorption cross-sections of uranium are particularly high.

Resonance-escape probability, p, the probability of avoiding resonance absorption:

$$p = 1 - \psi .$$

Fission probability, f, the ratio of the fission cross-section to the total absorption cross-section.

Leakage rate, L, the probability that neutrons will escape from the surface of the reactor.

2. Neutron balance and reactivity excess

neutron balance in the reactor			1
	Symbol	Unit	Quantity
	k	1	multiplication factor
	ν	1	mean neutron number per ^{235}U-fission
$k = \nu \cdot \varepsilon \cdot p \cdot f \cdot L$	ε	1	fast-fission factor by fission of ^{238}U
	p	1	resonance-escape probability
	f	1	fission probability
	L	1	leakage rate

▲ **Reactivity excess**:

$$\delta = k - 1 > 0 .$$

The condition must be fulfilled in order to compensate for the fuel consumption and "poisoning" of the fuel by fission fragments that capture neutrons.

Control rods, rods of strongly neutron-absorbing material used to reduce the reactivity excess to zero.

Delayed neutrons, neutrons emitted by fission products. They allow a response time for controlling the reaction in second range.

3. Moderators and neutron spectrum

Moderators, substances with small mass number (H,D,B,C,O) and low neutron absorption cross-section used to thermalize fast-fission neutrons (mean energy ≈ 2 MeV). The moderation proceeds mainly by elastic collisions with the moderator nuclei, which reduces their kinetic energy to the thermal energy region in which the fission cross-section is high.

■   Water is frequently used as moderator in thermal reactors.

Neutron spectrum, the energy spectrum of neutrons. **Fig. 27.43** shows the spectrum of neutrons produced in a fission event for a reactor with moderator.

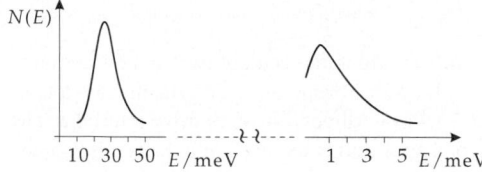

Figure 27.43: Neutron spectrum of a thermal reactor.

Thermal neutrons, are in thermal equilibrium with the moderator. Their velocity distribution is well described by a Maxwellian distribution. Most-probable values for velocity and kinetic energy: $v = 2200$ ms^{-1}, $E = 0.0253$ eV.

27.7.1 Types of reactors

The various types of reactors are distinguished by the following characteristics:
- energy of neutrons triggering the fission, and kind of fissionable material,
- kind of coolant,
- kind of moderator.

Thermal reactors, the fission proceeds mainly by capture of thermal neutrons ($E_n \approx$ 0.025 eV).

Fast reactors, the fission proceeds mainly by means of fast neutrons ($E_n > 0.1$ MeV).

As **fissionable material**, U^{235} (frequently weakly enriched), U^{233} (bred from Th^{232}) and Pu^{239} (bred from U^{238}), as well as mixtures of these, are used.

Moderators: usually water, heavy water or graphite. **Coolants**: water, gases (CO_2, He); in fast breeders (see below): liquid sodium.

1. Pressurized-water reactors,

thermal reactors using **enriched uranium** with about 5 % ^{235}U. Water is used as moderator **and** coolant. An increased pressure (15.8 MPa) leads to a shift of the boiling point.

➤   The natural abundance of ^{235}U in the uranium isotopes is 0.72 %.

First cycle, **coolant cycle**, passing directly through the active zone of the reactor. This coolant cycle is closed.

Active zone, the region of the reactor in which the fuel is located and nuclear fission proceeds.

Second cycle, is used to cool the first cycle and drives the generators directly.

Spent-fuel elements, fuel elements in which the fraction of ^{235}U is no longer sufficient to sustain a chain reaction (< 0.8 % ^{235}U).

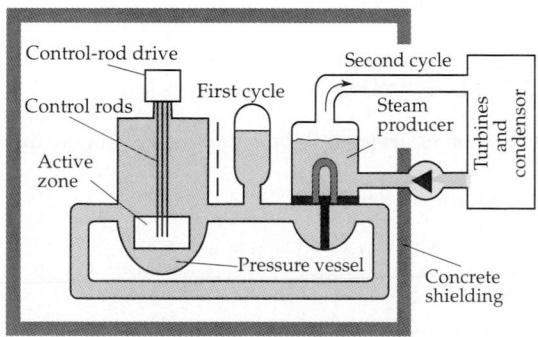

Figure 27.44: Scheme of a pressurized-water reactor.

2. Boiling-water reactors,

thermal reactors with enriched uranium as fuel in which the coolant (water) flows through the active zone from bottom to top. Part of the water evaporates. The steam (steam temperature about 286 °C; pressure of about 7 MPa) is directly used to drive a turbine. The steam leaving the turbine is liquified in a condenser and is fed back into the active zone by pumps.

3. Breeding process and breeder reactors

Breeding of nuclear fuel, production of thermally fissionable fuel nuclides $^{233}_{92}$U and $^{239}_{90}$Pu in reactors by neutron capture into $^{232}_{90}$Th and $^{238}_{92}$U.

■

$$n + {}^{232}_{90}\text{Th} \longrightarrow {}^{233}_{90}\text{Th} \longrightarrow {}^{233}_{91}\text{Pa} \longrightarrow {}^{233}_{42}\text{U}.$$

Breeding rate, the ratio of the number of fissionable nuclei formed by neutron capture to the number of nuclei consumed by fission events.
▲   If the breeding rate is larger than unity, the reactor produces more fuel than it consumes.
Breeder reactors, reactors with breeding rates larger than unity.

  Fast breeders, use uranium in natural isotopic abundance and plutonium (about 80 % UO_2; 20 % PuO_2) for the fuel elements. In the breeding blanket there is UO_2 with a **depletion** of ^{235}U. **Breeding** proceeds via the following process:

$$^{238}_{92}\text{U} + n \rightarrow {}^{239}_{92}\text{U} \begin{array}{c} \nearrow \gamma \\ \longrightarrow \\ \beta^-; 23.5 \text{ min} \end{array} {}^{239}_{93}\text{Np} \begin{array}{c} \longrightarrow \\ \beta^-; 2.36 \text{ d} \end{array} {}^{239}_{94}\text{Pu}.$$

Liquid sodium is used as coolant. A moderator is not appropriate. The $^{24}_{11}$Na produced in the active zone remains in the first cycle in the safety zone of the reactor.

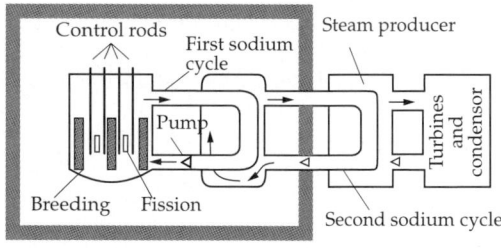

Figure 27.45: Scheme of a fast breeder.

27.8 Nuclear fusion

Nuclear fusion, the fusion of light nuclei. Energy is released in the fusion of light nuclei (see p. 911).
■   Several fusion reactions of light nuclei:

$$D + D \longrightarrow T + p + 4.04 \text{ MeV},$$

$$D + T \longrightarrow {}^4\text{He} + n + 17.6 \text{ MeV},$$

$$T + T \longrightarrow {}^4\text{He} + 2n + 11.3 \text{ MeV}.$$

For other possible fusion reactions, see **Tab. 29.5/2**.

➤   The Sun and the stars get their energy from such fusion reactions.

Hydrogen burning, fusion of four protons via several intermediate reactions to a stable α-particle, the energy release is 26.7 MeV.

■   Fusion of 1 g hydrogen yields about $6 \cdot 10^{11}$ J.

Helium burning fusion of three α-particles to a ^{12}C-nucleus.

1. Proton-proton processes,

Hydrogen cycle, hydrogen burning in which the light nuclei Li, Be and B are also involved as nuclear catalysts in the reaction chain. The reaction chains I, II, III are mainly distinguished by the energy fraction going into neutrinos (**Fig. 27.46**). Reaction chain I is denoted as the **deuteron cycle**.

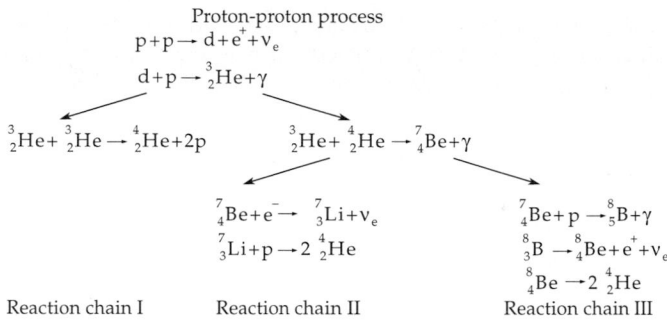

Figure 27.46: Reaction chains of the proton-proton process.

2. CNO cycles,

the hydrogen burning occurring in the Sun. The reaction chains involve the light nuclei C, N and O as nuclear catalysts (**Fig. 27.47**).

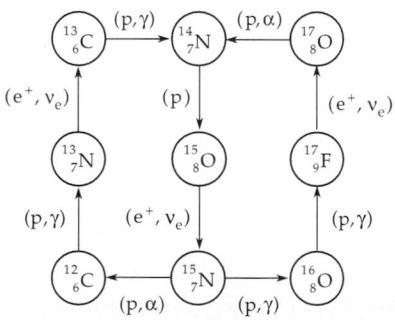

Figure 27.47: CNO cycle. Double-cycle, determined by the branching ratio of the reactions ^{15}N(p,α)^{12}C and ^{15}N(p,γ)^{16}O.

3. Carbon-nitrogen cycle,

CN cycle, a reaction chain proposed by Bethe to explain the Sun's energy (**Fig. 27.48**).

Salpeter process, fusion of three α-particles to form a ^{12}C-nucleus in a two-step process:

$$^{4}\text{He} +^{4}\text{He} + 95 \text{ keV} \longrightarrow {}^{8}\text{Be} + \gamma \,, \quad {}^{8}\text{Be} +^{4}\text{He} \longrightarrow {}^{12}\text{C} + \gamma + 7.4 \text{ MeV} \,.$$

Carbon-Nitrogen Cycle

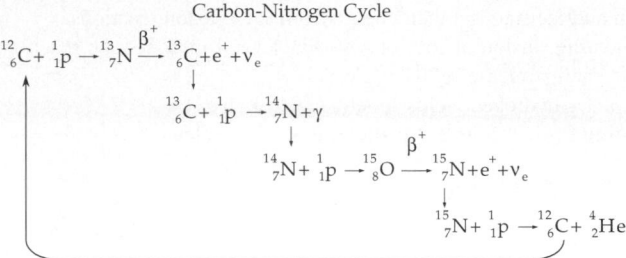

Figure 27.48: CN cycle.

The Coulomb barrier has to be overcome in the fusion of two nuclei. For the hydrogen cycle, the required energy is 0.5 MeV. This corresponds to a temperature of about $5.8 \cdot 10^9$ K. For the carbon–nitrogen cycle, a temperature about four times higher than for the deuteron cycle is needed.

4. Fusion reactor,

a nuclear reactor in which a controlled fusion reaction takes place. The fuel is in the plasma state. The necessary kinetic energy of the reaction partners corresponds to a plasma temperature of about 10^8 K.

Plasma, a gaseous mixture of free electrons, ions and electrically neutral particles.

Confinement, inclusion of a plasma in a limited volume. This confinement may not consist of conventional materials because of the high temperature. Moreover, in order to gain energy in a fusion reactor, the high-temperature plasma must be kept together for a sufficient time interval.

Magnetic confinement, a plasma at low fuel density is kept together for a longer time by a magnetic field of special configuration.

Inertial confinement, the fuel is compressed by energy supplied by laser, electron or heavy-ion beams. It is kept together for a short time at high density by its own inertia.

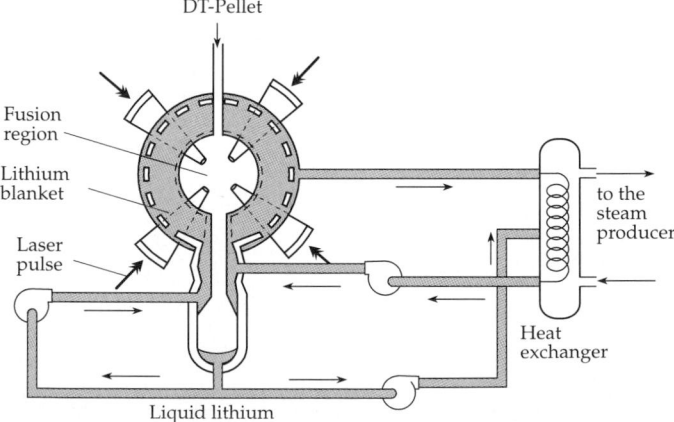

Figure 27.49: Scheme of a fusion reactor with inertial confinement and ignition of the fuel pellet by laser pulses.

5. Lawson criterion,

balance equation for maintaining the fusion process in a plasma (break-even condition):

Lawson criterion			$\mathbf{ML^2T^{-2}}$
	Symbol	Unit	Quantity
$(E_F + E_P + E_\gamma) \cdot (\eta + \varepsilon)$ $= E_P + E_\gamma$	E_F	J	fusion energy
	E_P	J	thermal plasma energy
	E_γ	J	bremsstrahlung energy
	η	1	efficiency of energy conversion
	ε	1	efficiency of energy supply

In 1993 the Lawson criterion was approached to within one order of magnitude at the Joint European Torus (JET).

27.9 Interaction of radiation with matter

27.9.1 Ionizing particles

Ionizing particles, all charged particles; they produce positive ions and electrons by collisions with electrons in the atomic shells.

    Ionization, the production of a secondary electrons and reduce the kinetic energy of the incident particle.

1. Ionization losses,

decrease of the kinetic energy of the incident particle by ionization processes.

    Bremsstrahlung, the energy radiation caused by the acceleration of charged particles in the Coulomb field of the atomic nucleus.

    Radiation losses, decrease of the kinetic energy of the incident particle by production of bremsstrahlung through the electromagnetic interaction with the atomic nucleus.

▲   The radiation losses of heavy charged particles are negligible compared with the ionization losses. Energy losses by bremsstrahlung become important only at energies $> m_0 c^2$ (for protons $> 10^3$ MeV).

▲   For electrons, the stopping power rapidly increases at energies > 1 MeV due to the bremsstrahlung losses (relativistic rise).

▲   Heavy charged particles have a material-dependent finite range R in matter.

2. Range and Bragg peak

Mean range, $\bar{R}$, the penetration depth at which the incoming particle flux is reduced to half of the initial value (**Fig. 27.50**).

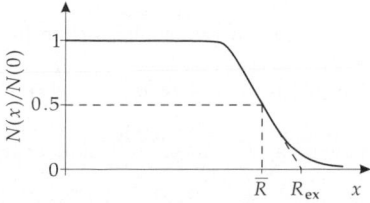

Figure 27.50: Range of heavy charged particles in matter. x: penetration depth, $\bar{R}$: mean range, R_{ex}: extrapolated range.

Extrapolated range R_{ex}, the intersection point of the tangent to the relative flux density (as function of the penetration depth) at the inflexion point, and the x-axis.

Bragg maximum, Bragg peak, the ionizing power of heavy charged particles (including protons) takes a maximum at the end of their trajectory in the target material (**Fig. 27.51**).

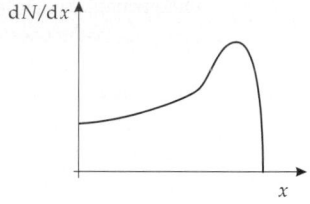

Figure 27.51: Specific ionizing power dN/dx of heavy charged particles versus penetration depth x.

M   Application of heavy-ion beams and proton beams in technology and medicine: Due to the Bragg peak, the depth of penetration in solids (ion implantation, doping) or organic tissue (tumor therapy) may be controlled precisely (± 1 mm) via the bombarding energy.

3. Energy-range relation,

connection between kinetic energy E_{kin} of the incident particles (charge Z) and their range R in a medium,

$$R \sim E_{\text{kin}}^2/Z^2, \quad v \ll c,$$

$$R \sim E_{\text{kin}}/Z^2, \quad v \approx c.$$

■ α-particles of energy $E = 5$ MeV have a range in air of 3.5 cm. The range of these α-particles in aluminum is only 23 μm.

For the range of α-particles see **Tab. 29.6/3**.

▲   Unlike the trajectories of heavy charged particles, the trajectories of electrons are not straight lines in the target. Hence, there is no unique range for electrons.

➤   Photons also do not have a defined range in matter.

4. Stopping power,

S, differential energy loss dE along the path element dx,

$$S = -\frac{dE}{dx}.$$

▲   The stopping power depends on the square of the charge number of the incident particle.

➤   In **dosimetry**, the quantity S is also called **linear energy transfer power (LET,** linear energy transfer) L_∞.

The stopping power for heavy charged particles of energy $E \ll m_0 c^2$ is well described by the **Bethe-Bloch equation**:

Bethe-Bloch equation			$\mathbf{MLT^{-2}}$
	Symbol	Unit	Quantity
$$S = \frac{Z \cdot z^2 \cdot e^4 \cdot N_A \cdot m_i}{8\pi \varepsilon_0^2 \cdot m_e \cdot E_{kin} \cdot M_A}$$ $$\cdot \rho \cdot \ln\left(\frac{4m_e \cdot E_{kin}}{\bar{I} \cdot m_i}\right)$$	S Z z N_A m_i ε_0 m_e E_{kin} M_A $\bar{I}$ ρ e	MeV/cm 1 1 mol^{-1} kg C V^{-1} m^{-1} kg J g/mol J kg/m^3 A s	stopping power atomic number of target atom charge number of projectile Avogadro's number mass of projectile permittivity of free space rest mass of electron kinetic energy of projectile molar mass of target material mean ionization energy density elementary charge

5. Stopping power for electrons

stopping power for electrons			$\mathbf{MLT^{-2}}$
	Symbol	Unit	Quantity
$$S = \frac{Ze^4 N_A}{8\pi \varepsilon_0^2 m_e v^2 M_A}$$ $$\cdot \rho \cdot \ln\left(\frac{m_e v^2 E_{kin}}{2\bar{I}^2(1-\beta^2)}\right)$$ $$+ f(\beta)$$	S Z N_A m_i ε_0 m_e E_{kin} M_A $\bar{I}$ v β $f(\beta)$ ρ e	MeV/cm 1 mol^{-1} kg C V^{-1} m^{-1} kg J g/mol J m/s 1 J/m kg/m^3 A s	stopping power atomic number of target atom Avogadro constant mass of projectile electric permittivity of free space rest mass of electron kinetic energy of projectile molar mass of target mean ionization energy electron velocity v/c relativistic correction density elementary charge

■ The differential ionization power of electrons is about 1000 times smaller than that of α-particles.

6. *Mass stopping power and specific ionization*

Mass stopping power, S_m, the ratio of stopping power S and density ρ of the target material,

$$S_m = -\frac{1}{\rho}\frac{\mathrm{d}E}{\mathrm{d}x}\,.$$

| **M** | From this quantity, the mass stopping power of heterogeneous materials may be determined by weighting with the mass fractions of the corresponding components.

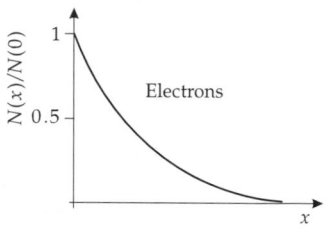

Figure 27.52: Penetration of electrons in matter. x: penetration depth, $N(x)$: particle number at depth x.

Specific ionization, j, ratio of mass stopping power S_m and mean ionization energy $\bar{I}$,

$$j = S_m/\bar{I}\,.$$

The number of ion pairs $\mathrm{d}N$ produced along a path element $\mathrm{d}x$ is given by

$$\mathrm{d}N = j \cdot \mathrm{d}x\,.$$

➤ Particles of equal charge and energy but different mass may be distinguished from each other by their specific ionization.

■ An electron of energy $E_{\mathrm{kin}} = 10^5$ eV produces about 200 ion pairs per 1 cm path in air. A proton of the same energy produces about 10^4 ion pairs along the same path length.

27.9.2   *γ-radiation*

Attenuation of γ-radiation by a layer of matter of thickness d and density ρ is described by an exponential attenuation law:

attenuation law for γ-radiation			$\mathbf{L^{-2}T^{-1}}$
	Symbol	Unit	Quantity
$\varphi = \varphi_0\,e^{-\mu d}$	φ	$\mathrm{m}^{-2}\mathrm{s}^{-1}$	particle flux density behind absorber
	φ_0	$\mathrm{m}^{-2}\mathrm{s}^{-1}$	particle flux in front of absorber
	μ	m^{-1}	linear attenuation coefficient
	d	m	thickness of layer

Mass-attenuation coefficient $\mu_{\mathrm{M}} = \mu/\rho$ (SI unit m^2/kg), linear-attenuation coefficient referred to the density.

1. Photoelectric effect,

production of secondary electrons by the interaction of photons with bound electrons.
For secondary electron emission, see **Tab. 29.3/5**.
▲   The photoelectric effect is the dominant interaction for $E_\gamma < 0.5$ MeV.
Mass-attenuation coefficient for photons τ/ρ (SI unit m^2/kg), increases rapidly with Z and decreases with increasing photon energy:

$$\frac{\tau}{\rho} \sim \frac{Z^4}{(hf)^3}.$$

2. Compton effect,

describes the elastic collision of photons by free electrons.
Compton mass-attenuation coefficient σ/ρ (SI unit m^2/kg), nearly independent of the atomic number Z and inversely proportional to the γ-energy:

$$\frac{\sigma}{\rho} \sim \frac{1}{hf}.$$

▲   The Compton effect dominates for medium photon energies
(H_2O: 30 keV $< hf < 25$ MeV;   Pb: 500 keV $< hf < 5$ MeV).

3. Pair production,

the creation of an electron-positron pair in the Coulomb field of the atomic nucleus. The reaction threshold is $hf = 2m_ec^2 = 1.022$ MeV (see p. 902).
Mass-attenuation coefficient for pair production κ/ρ (SI unit m^2/kg), proportional to Z and increasing logarithmically with increasing γ-energy:

$$\frac{\kappa}{\rho} \sim Z \ln(hf).$$

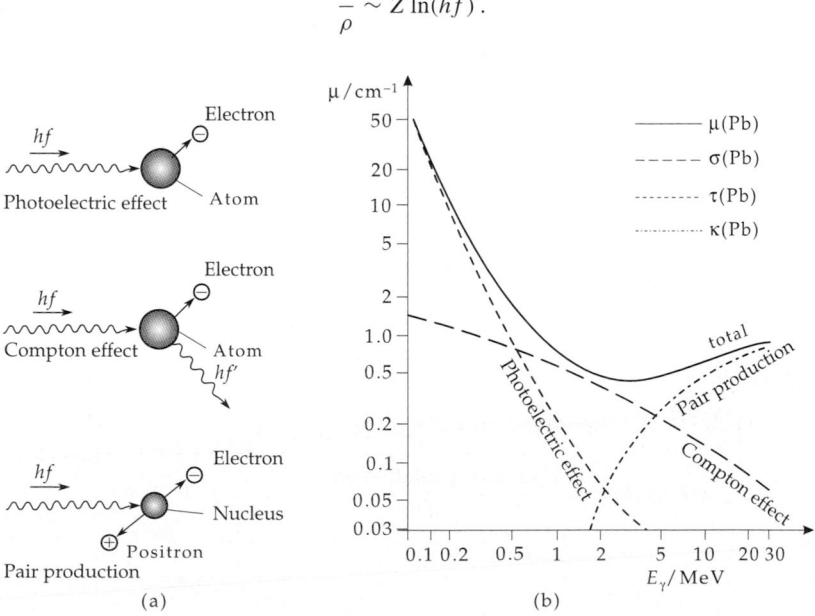

Figure 27.53: (a): Interaction of γ-radiation with matter. (b): Linear-attenuation coefficients of γ-radiation in lead.

4. Total-attenuation coefficient,

μ (SI unit m^2/kg), composed additively from the photo-absorption coefficient τ, the Compton attenuation coefficient σ, and the pair-production coefficient κ:

$$\mu = \tau + \sigma + \kappa.$$

For the mass-attenuation coefficient of X-rays, see **Tab. 29.6/1**.

Linear-attenuation coefficient, μ', product of mass-attenuation coefficient and density (SI unit m^{-1}),

$$\mu' = \mu \cdot \rho, \qquad \rho: \text{density}.$$

27.10 Dosimetry

Dosimetry, measurement techniques for ionizing radiation, x-rays, γ-radiation and neutrons.

1. Definition of activity

Activity, A, a measure of the decay rate of a radionuclide. It does not take into account the different biological efficiency of the radiation modes.

activity = $\dfrac{\text{number of decays}}{\text{time}}$			T^{-1}
	Symbol	Unit	Quantity
$A = \dfrac{\mathrm{d}N}{\mathrm{d}t}$	A	Bq	activity
	N	1	number of decays
	t	s	time

Becquerel, SI unit of activity,

$$[A] = \text{Bq} = \frac{1 \text{ decay}}{\text{s}}.$$

➤   The formerly used unit of 1 curie = 1 Ci is of historical origin and corresponds to the number of decays of 1 g ^{226}Ra per second:

$$1 \text{ Ci} = 3.7 \cdot 10^{10} \text{ Bq}.$$

2. Energy dose

(short form: dose), D, a measure of the physical radiation impact:

energy dose = $\dfrac{\text{absorbed radiation energy}}{\text{mass}}$			L^{2}T^{-2}
	Symbol	Unit	Quantity
$D = \dfrac{\Delta W}{\Delta m}$	D	Gy	energy dose
	ΔW	J	absorbed radiation energy
	Δm	kg	mass

Gray, SI unit of energy dose,

$$[D] = \text{Gy} = \frac{\text{J}}{\text{kg}}.$$

➤ The unit "rad" was used until 1985,

$$1 \text{ rad} = 10^{-2} \text{ Gy}.$$

➤ In organic tissue and water, the dose of 1 Gy corresponds to an increase of temperature of 0.00024 K. But the energy release proceeds in a very narrow region. Therefore,
vital molecules may be destroyed.

▲ When evaluating a radiation dose, the biological efficiency of the various radiation
modes has to be taken into account.

3. Equivalent dose,

H, takes into account the efficiency of the different radiation modes:

equivalent dose = evaluation factor · energy dose			$\mathbf{L^2 T^{-2}}$
	Symbol	Unit	Quantity
$H = q \cdot D$	H	Sv	equivalent dose
	D	Gy	energy dose
	q	1	evaluation factor

Sievert, SI unit of equivalent dose,

$$[H] = \text{Sv} = \frac{\text{J}}{\text{kg}}.$$

➤ Until 1979 the "rem" served as the unit of equivalent dose,

$$1 \text{ rem} = 10^{-2} \text{ Sv}.$$

4. Evaluation factor,

q, factor to evaluate the biological effect of a certain dose of radiation. It is composed of a
quality factor *Q* that takes into account the radiation mode, and a factor *N* that takes into
account the distribution of the radiation in space and time:

$$q = QN$$

For irradiation of a body from outside, $N = 1$.

Quality factor, *Q*, connected with the linear energy transfer (LET) capacity of charged
particles for unlimited energy transfer.

Mean quality factors $\bar{Q}$ for various radiation modes:

radiation mode	$\bar{Q}$
x-rays, γ	1
electrons, positrons	1
thermal neutrons	2.3
fast neutrons	10
α-particles	20
heavy ions	20

Dose rate, the equivalent dose per unit time.

■ $\dfrac{\text{Sv}}{\text{h}}, \dfrac{\text{Sv}}{\text{min}}, \dfrac{\text{Sv}}{\text{s}}$

5. Particle and energy flux densities

Spectral particle radiance, p_E, particle flux density in relation to the solid angle and energy:

$$p_E(\vec{r}) = \phi_E(\vec{r}, t, E, \Omega) \text{, unit: } s^{-1}J^{-1}sr^{-1}m^{-2}.$$

Spectral particle flux density, ϕ_E, the integral of the spectral particle radiance over the solid angle:

$$\phi_E(\vec{r}, t, E) = \int p_E(\vec{r})d\Omega.$$

Particle flux, Φ, is obtained from the spectral particle flux density by integration over kinetic energy and time:

particle flux			L^{-2}
	Symbol	Unit	Quantity
$\Phi(\vec{r}) = \displaystyle\int_{t_1}^{t_2}\int_0^\infty\int_0^{4\pi} p_E(\vec{r})d\Omega dEdt$ $= \dfrac{dN}{dA_\perp}$	$\Phi(\vec{r})$	m^{-2}	particle flux
	$p_E(\vec{r})$	$1/(s\,J\,sr\,m^2)$	spectral particle radiance
	Ω	sr	solid angle
	E	J	energy
	t	s	time
	N	1	particle number
	$A_\perp$	m^2	area

▲   The particle flux is the number of particles flowing in a per unit time interval in the normal direction through an area element of a spherical surface about the source.

Particle flux density, ϕ, the particle fluence per unit time.

particle flux density = particle density · velocity			$L^{-2}T^{-1}$
	Symbol	Unit	Quantity
$\phi(\vec{r}, t) = \dfrac{\Phi(\vec{r})}{t} = n \cdot v$	$\phi(\vec{r}, t)$	$m^{-2}\,s^{-1}$	particle flux density
	$\Phi(\vec{r})$	m^{-2}	particle flux
	t	s	time
	n	m^{-3}	particle density
	v	$m\,s^{-1}$	particle velocity

Spectral energy flux density, ψ, the product of particle flux density and energy:

$$\psi = E \cdot \phi_E(\vec{r}, t, E).$$

Energy flux density, I_E, the integral of the product of particle flux density and energy integrated over energy:

$$I_E = \int E \cdot \phi_E(\vec{r}, t, E)dE.$$

Energy flux, the time integral of the energy flux density.

6. *Attenuation law*,

determines the attenuation of a beam by a certain material of thickness dz:

attenuation			$\mathbf{L^{-2}T^{-1}}$
	Symbol	Unit	Quantity
$d\psi = -\psi \cdot \mu \cdot dz$	$d\psi$	$m^{-2}\,s^{-1}$	attenuation of spectral energy flux density
	ψ	$m^{-2}\,s^{-1}$	spectral energy flux density
	μ	m^{-1}	linear mass attenuation coefficient
	dz	m	thickness of material

Integration of the above relation yields the attenuation law:

$$\psi(z) = \psi_0 e^{-\mu \cdot z}.$$

This law holds only for a narrow beam and, because of the sensitive energy dependence of the mass-attenuation coefficient, only for mono-energetic radiation.

Half-value depth, s, the thickness of the material at which half of the incident radiation quanta have interacted with the material:

$$s = \frac{\ln 2}{\mu}.$$

7. *Energy-transfer coefficient*

Energy-conversion coefficient, **linear energy-transfer coefficient**, μ_{tr}, determines the energy transfer from the radiation to the attenuating layer:

linear energy-transfer coefficient			$\mathbf{L^{-1}}$
	Symbol	Unit	Quantity
$\mu_{tr} = \dfrac{1}{W} \cdot \dfrac{dW_{kin}}{dz}$	μ_{tr}	m^{-1}	linear energy-transfer coefficient
	W	J	total radiant energy
	dW_{kin}	J	kinetic energy of secondary electrons
	dz	m	thickness of layer

8. Kerma

(**k**inetic **e**nergy **r**eleased per unit **ma**ss), K, describes the first stage of the interaction of indirectly ionizing radiation (e.g., neutrons):

indirectly ionizing radiation			$\mathbf{L^2 T^{-2}}$
	Symbol	Unit	Quantity
$K = \dfrac{1}{\rho}\dfrac{\mathrm{d}E_{\mathrm{tr}}}{\mathrm{d}V}$	K	Gy	kerma
	ρ	kg/m^3	material density
	E_{tr}	J	kinetic energy of released charged particles
	V	m^3	volume

➤   When giving a kerma, one must refer to the specific material.

9. Relative biological efficiency

(RBE) of a kind of radiation x for a biological endpoint a (e.g., a given value of the survival probability of some kinds of cells), is determined by comparison with a reference dose:

$$RBE_a = \left(\frac{D_{\mathrm{ref}}}{D_x}\right)_a .$$

The reference dose causes the same biological effect as the dose D_x.
    Frequently, ^{60}Co-γ-radiation or a 250 keV-X-radiation is used as reference dose.

27.10.1 Methods of dosage measurements

Personal dosimetry, the measurement of the dose at a place near the surface of the body representative for radiant exposure.

M **Ionization chamber**, gas counter with gas amplification 1, used in the range of dose of μGy up to 10^3 Gy. The discharge of a cylindrical capacitor is measured. The residual charge is a measure of the dose (**Fig. 27.54 (b)**).

➤   Ionization chambers are used for personal dosimetry. They provide quick and sufficiently accurate information. The ionization chamber is an integrating dosimeter.

Gas amplification, the increase of free charge carriers by secondary ionization of the primarily produced ions accelerated in the electric field.

M **Proportional counter**, gas counter with a gas amplification up to 10^4.

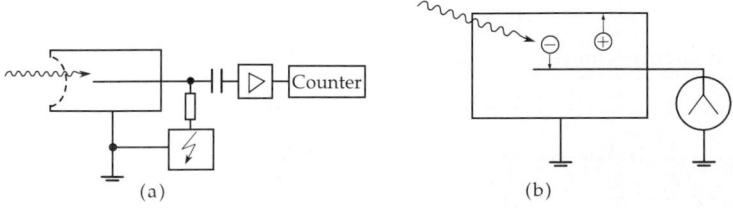

Figure 27.54: (a): Sketch of a proportional counter, (b): sketch of an ionization chamber.

The pulse height of the current pulse produced is proportional to the energy of the incident radiation. The number of pulses is a measure of the number of incident radiation quanta (**Fig. 27.54 (a)**).

| M | **Geiger-Müller counters**, trigger counters operating with a gas amplification of about 10^8. The proportionality of pulse height and energy of the incident radiation is lost. These dosimeters are used in local dose and dose rate measurements.

Local dose, the equivalent dose for soft tissue at a certain place in the radiation field within a certain time interval.

| M | **Film dosimeters**, detectors that exploit the blackening of photographic material by incident radiation. They are used in the dose range between 0.1 mSv and 1 Sv and are suitable for photon energies between 20 keV and 3 MeV. Film dosimeters are applied in personal dosimetry, in particular to keep track of the dose received by radiation-exposed persons. By means of radiation converters (e.g., Cd plates for neutrons in γ-radiation), this dosimeter may be used universally. It integrates the total dose.

| M | **Thermoluminescence dosimeter**, converts the energy of ionizing radiation stored in a solid into light via heating. This storage of energy is a solid-state effect (see p. 1063).

Radiotoxicity, the toxicity of radionuclides for the human body due to the emitted radiation.

Biological half-life, the time over which an activity present in the body is reduced by excretion to half of the initial value.

Nuclide	Physical half-life	Biological half-life	Critical organ
Class 1 of radiotoxicity: maximum 3.7 kBq			
^{90}Sr	28.1 yr	11 yr	bones
^{210}Pb	22 yr	730 d	bones
^{210}Po	138 d	40 d	spleen
^{233}U	$1.63 \cdot 10^5$ yr	300 d	bones

Nuclide	Physical half-life	Biological half-life	Critical organ
Class 2 of radiotoxicity: maximum 37 kBq			
^{22}Na	2.58 yr	19 d	whole body
^{137}Cs	26.6 yr	100 d	muscle
^{144}Ce	285 d	330 d	bones
^{131}I	8.0 d	180 d	thyroid gland
Class 3 of radiotoxicity: maximum 370 kBq			
^{14}C	5570 yr	35 yr	fatty tissue
^{24}Na	15 h	19 d	whole body
^{105}Rh	1.54 d	28 d	kidney
^{109}Cd	1.3 yr	100 d	liver
Class 4 of radiotoxicity: maximum 3.7 MBq			
^{3}H	12.6 yr	19 d	whole body
^{238}U	$4.5 \cdot 10^9$ yr	300 d	kidney

27.10.2   Environmental radioactivity

Cosmic radiation, radiation incident from outer space on Earth. The primary cosmic radiation consists mainly of protons and α-particles interacting with the nuclei of molecules of air ($^{14}_{7}$N, $^{16}_{8}$O). Components of the secondary radiation: p, n, π, μ, K, e, γ, ν.

The neutrino flux has no influence on the radiation exposure of humans, since neutrinos are governed only by the weak interaction.

➤ The mean dose rate of cosmic radiation at sea level is about $3 \cdot 10^{-4}$ Sv/yr.

➤ Singular events have been observed in cosmic rays from which one may conclude the existence of particles of energy $> 10^{20}$ eV. The nature (new exotic particles, energetic photons or atomic nuclei) and origin (extragalactic sources, neutron stars, collision fronts in the halo of our galaxy) of these particles is not yet clear.

Terrestrial radiation, radiation of the natural radioactive nuclides with very long half-life, and of their products.

Cosmic radiation produces the radioactive isotopes tritium $^3_1$H and $^{14}_6$C.

Terrestrial doses for various places, and some extreme values:

Place/country	Equivalent dose (10^{-5} Sv/yr)
Nile Delta, Egypt	350
Paris, France	350
Grand Central Station, New York	525
Katzenbuckel/Baden-Württemberg, Germany	630
India/Kerala state	≤ 2700
Brazil/Atlantic coast	≤ 8700

Nuclei produced in the upper atmosphere fall to Earth's surface by sedimentation, rainfall or convection.

	Until 1963	1963/1964	1979
		/ (Bq/kg(H$_2$O))	
rainfall, central Europe (annual average)	740	222000	9250
rainfall, European west coast (annual average)	296	92500	2960
ground water, central Europe	444	166500	7400
surface water, North Atlantic	22.2	1850	555

■ Besides tritium, radioactive hydrogen and radioactive carbon, the air contains mainly radon and its decay products. Radon escapes from clefts in the Earth's crust and is swept to the surface by spring water.

Fall-out, increase of radioactivity, in particular the tritium abundance, in the earth surface, as a consequence of the **above-ground atomic explosions** in the 1960s.

Self-radiation of the human body, originates from radioactive isotopes ingested in food, and by breathing.

▲ The natural self-radiation level is about $3 \cdot 10^{-4}$ Sv/yr.

Natural exposure, the sum of all three components: cosmic, terrestrial and self-radiation.

▲ Currently, the natural exposure is about $1.1 \cdot 10^{-3}$ Sv/yr.

Some parts of the body are exposed much more by the inhalation of radioactive decay products.

■ For example, exposure of the lungs is about $1.2 \cdot 10^{-2}$ Sv/yr.

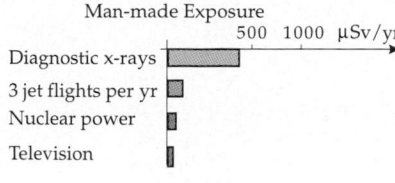

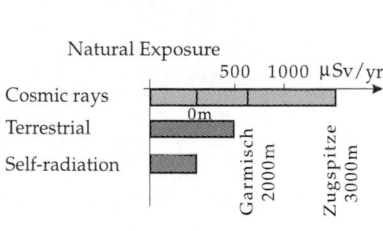

Figure 27.55: Comparison of man-made and natural exposures. Two relatively high altitude locations in Germany are compared with sea level.

Man-made or artificial exposure, the exposure produced by man. Among these are:
- nuclear power plants,
- medical diagnostics,
- building materials.

Activity of various building materials:

Building material	^{226}Ra (α-emitter)	^{232}Th (α-emitter)	^{40}K (β-emitter)
		/ (Bq/kg)	
building bricks	52.5	49.2	652
sandstone	11.5	4.1	273
concrete	26.3	21.8	437

▲   X-ray diagnostics and radiation therapy are the main exposure factors.

Dose rate of various x-ray sources:

Device	Dose rate Sv/h (distance 10 cm)
color TV	$0.6 \cdot 10^{-6}$
monitor screens	$5 \cdot 10^{-6}$
oscilloscores	$1 \cdot 10^{-6}$
radar control screens	$4 \cdot 10^{-6}$

28
Solid-state physics

28.1 Structure of solid bodies

28.1.1 Basic concepts of solid-state physics

Solid, matter in the solid physical state. Solids may be classified according to the state of order of their structural constituents (atoms, ions, molecules):

- **Crystalline solid** (**crystal**), a solid with periodic order of its structural constituents. Regular, periodically repeating configurations of structural elements occur in all three dimensions.
- **Amorphous solid**, a solid without long-range order of the structural elements. There are no periodically repeating configurations of structural elements.
- ■ Alkali metals have a crystalline structure. Diamond is crystalline carbon. Common salt (sodium chloride, NaCl) exhibits a crystalline structure.
- ■ Alloys and gels are amorphous solids.

Many solid materials (e.g., **glasses** or **polymers**) cannot be included in this scheme. Polymers have a partly periodic order. Solids exist with micro-crystalline structure.

Solids are distinguished from each other by their response to a physical influence:

- **Isotropic solid**, no space direction is preferred over the others. The response of the solid is direction-independent.
- ▲ Frequently, amorphous solids are isotropic.
- **Anisotropic solid**, certain space directions are different from the others. The response of the solid is direction-dependent.
- ▲ The periodic structures in crystals define preferred spatial orientations.

Monocrystal, idealized solid with a periodically repeating atomic structure that extends over the entire volume. The crystal axes have about the same orientation relative to a body-fixed coordinate frame in all regions of the body.

- ■ Salts crystallizing out of **solutions** are often monocrystals.
- M **Monocrystal growing** from melts (one component), from solutions (several components) or from the gaseous phase.

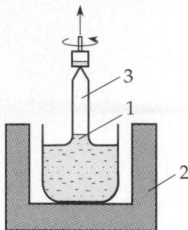

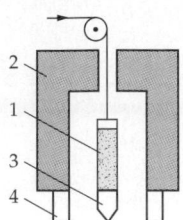

Figure 28.1: Schematic view of the
Czochralski method. 1 – melt, 2 – heater,
3 – growing monocrystal.

Figure 28.2: Schematic view of the
Bridgeman method. 1 – melt, 2 – heater,
3 – growing crystal, 4 – cooling.

Czochralski method: The crystal is drawn directly from the melt (**Fig. 28.1**).

Bridgeman method: The crystal grows in a crucible, which descends at a constant speed from the hot zone into the cold zone (**Fig. 28.2**).

The methods mentioned above have the disadvantage that the crystal is polluted by oxygen incorporated from the crucible walls.

Zone melting method: The impure material is melted by a narrow melting device moving slowly along the sample. A monocrystal forms behind the heating zone. Impurities prefer the liquid phase and are removed.

Lattice defect, deviation from the ideal structure of strict spatial periodicity by lattice defects (dislocations, vacancies, stacking disorders, etc.).

▲   The type and abundance of lattice defects essentially determines the physical properties of a solid.

Polycrystalline solids, the monocrystalline regions (crystallites) frequently extend over few micrometers only, the crystallite orientations vary randomly.

■   Metals crystallized after melting are usually polycrystalline.

Grain, monocrystalline region in a solid.

Grain boundaries, separate the monocrystalline regions of a polycrystalline solid.

Texture, distribution of the orientation of grains in a polycrystalline solid.

28.1.2   Structure of crystals

Crystal lattice, periodic, three-dimensional arrangement of atoms, molecules or ions; their type and geometric structure determines the outward appearance and the physical properties of the crystal.

Space lattice, **point lattice**, mathematical abstraction of the crystal lattice to a spatially periodic arrangement of points corresponding to the lattice sites. The kind of atoms or molecules at the lattice sites is thereby ignored.

Base, a group of atoms or molecules ascribed to any lattice point or any elementary parallelepiped.

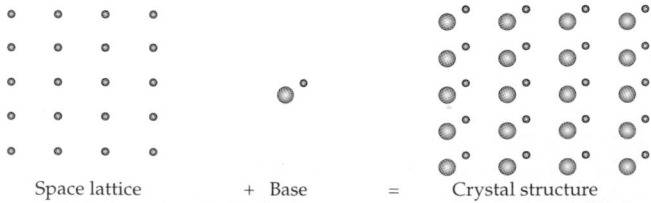

Figure 28.3: On the concept of crystalline structure.

1. Crystal structure,

determined by the symmetry of the structure, the lattice parameters (lengths and angles), and the specification of the center-of-mass positions in the asymmetric unit of the elementary cell.

Elementary cell, element of the crystal lattice from which the complete lattice may be reproduced by translation.

Asymmetric unit, smallest spatial fraction of an elementary cell from which the entire elementary cell may be obtained by symmetry operations.

Translation, displacement of an elementary cell in space by the translation vector $\vec{T}$.

2. Lattice vectors and crystal axes

Fundamental translation vectors, **lattice vectors** $\vec{a}$, $\vec{b}$, $\vec{c}$, displacements $\vec{T} = \vec{a}n_1 + \vec{b}n_2 + \vec{c}n_3$ along integral multiples of these vectors map a crystal lattice onto itself.

■   Let $\vec{r}$ be an arbitrary point in space. The lattice at the point

$$\vec{r}\,' = \vec{r} + n_1\vec{a} + n_2\vec{b} + n_3\vec{c} \qquad (n_1, n_2, n_3 \text{ are integers})$$

is identical to the lattice at the point $\vec{r}$. The lattice vectors $\vec{a}$, $\vec{b}$, $\vec{c}$ span a parallelepiped.

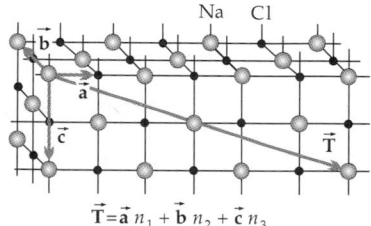

$$\vec{T} = \vec{a}\,n_1 + \vec{b}\,n_2 + \vec{c}\,n_3$$

Figure 28.4: On the concept of the translation vector.

▲   A point lattice is uniquely defined by the fundamental translation vectors (lattice vectors) $\vec{a}$, $\vec{b}$, $\vec{c}$.

Crystal axes, directions defined by the fundamental lattice vectors $\vec{a}$, $\vec{b}$ and $\vec{c}$.

Lattice constants, magnitudes of the fundamental lattice vectors $\vec{a}$, $\vec{b}$ and $\vec{c}$, specify the distances of the bases along the crystal axes.

3. Primitive elementary cell,

elementary cell with the minimum volume for a given lattice structure. The primitive elementary cell contains only one lattice point.

➤   Although the primitive parallelepiped has one lattice point on each of its eight edges, these have to be shared over the eight elementary cells contacting each other there.

The lattice vectors shown in **Fig. 28.5** each span a primitive elementary cell.

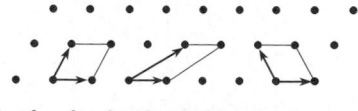

Figure 28.5: Primitive elementary cells.

➤   It is not always suitable, or customary, to choose the elementary cell to be as small as possible. The elementary cells of tungsten and copper shown in the following figure exhibit the cubic symmetry of these metals better.

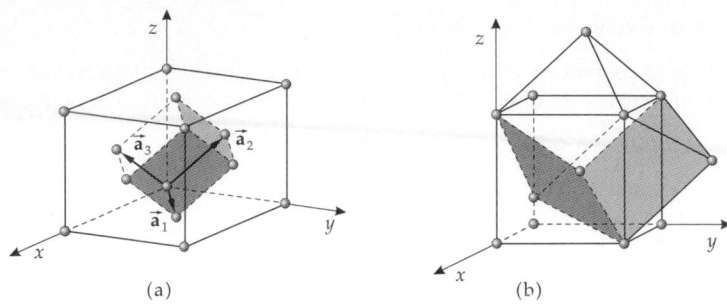

Figure 28.6: Elementary cell. (a): copper (face-centered cubic lattice), (b): tungsten (body-centered cubic lattice).

4. Crystal system and lattice types

Crystal system, subdivision of crystals into seven crystal systems according to the following criteria:
- lattice constants are equal or different,
- angles between the crystal axes.

Lattice types:
- **Primitive lattice**: all lattice points are on the edges of the elementary cell.
- **Face-centered lattice**: additional atoms occur at the intersection points of the face diagonals of the elementary cell.
- **Base-centered lattice**: besides the atoms at the edges, there is one atom at each intersection point of the face diagonals of two opposite faces.
- **Body-centered lattice**: besides the atoms at the edges, there is one atom at the intersection point of the space diagonals of the elementary cell.

28.1.3   Bravais lattices

1. Types of Bravais lattices

Bravais lattice, notation for individual lattice types. In 3D space, there are 14 distinct Bravais lattices:

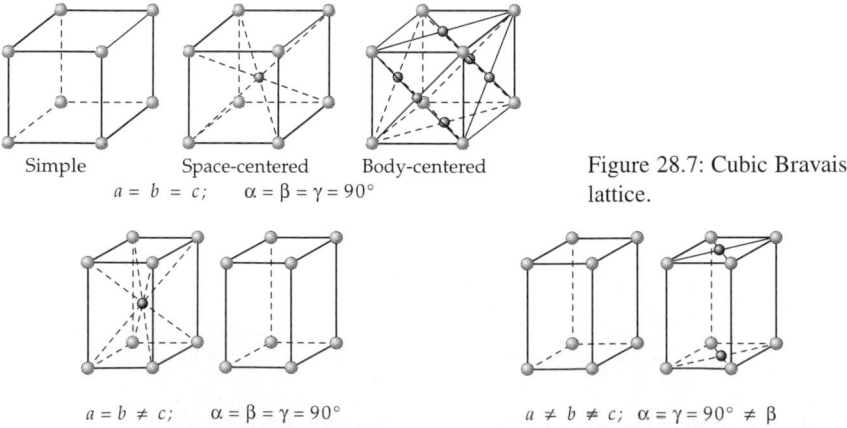

| Simple | Space-centered | Body-centered | Figure 28.7: Cubic Bravais lattice. |

$$a = b = c; \quad \alpha = \beta = \gamma = 90°$$

$$a = b \neq c; \quad \alpha = \beta = \gamma = 90°$$

Figure 28.8: Tetragonal Bravais lattice.

$$a \neq b \neq c; \quad \alpha = \gamma = 90° \neq \beta$$

Figure 28.9: Monoclinic Bravais lattice.

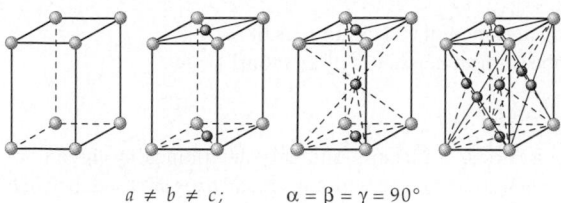

Figure 28.10: Orthorhombic
Bravais lattice.

$a \neq b \neq c;$ $\alpha = \beta = \gamma = 90°$

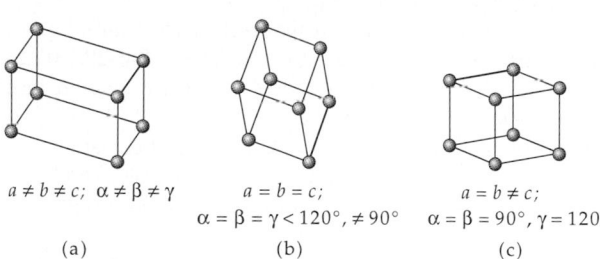

$a \neq b \neq c;\ \alpha \neq \beta \neq \gamma$ $a = b = c;$ $a = b \neq c;$
                                $\alpha = \beta = \gamma < 120°, \neq 90°$ $\alpha = \beta = 90°, \gamma = 120°$
(a)   (b)   (c)

Figure 28.11: Bravais lattices. (a): triclinic, (b): rhombohedral, (c): hexagonal.

The following structures are important for metals:
- the face-centered cubic lattice (fcc),
- the body-centered cubic lattice (bcc),
- the hexagonal compact packing of spheres (hcp).
➤ The lattice types of important elemental crystals are given in the **Periodic Table of elements**.

2. Packing density of the elementary cell

Compact packing of spheres, regular arrangement of spheres of equal size with a minimum of empty space between them. One distinguishes between **hexagonal** and **face-centered** compact packing of spheres.

Fig. 28.12 shows a layer of compact packing of spheres with the centers at A.

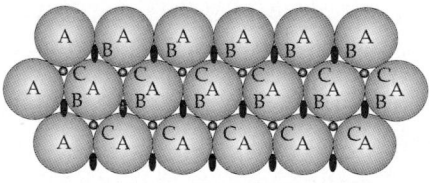

Figure 28.12: Compact
packing of spheres.

If the second layer is arranged in the positions B (or in the equivalent positions C), there are two possibilities for arranging the third layer:
- The spheres of the third layer may be placed above the positions A. The result is a sequence $ABABA\ldots$ (**hexagonal** structure).
- The spheres of the third layer occupy the positions above C. The result is a sequence of planes $ABCABC\ldots$ (**face-centered cubic**).
▲ In the compact packing of spheres, every sphere of a plane touches six other spheres of the same plane and three spheres in each of the two neighboring planes.

Packing density, fraction of space occupied in the elementary cell by the volume of the spheres.
- In both types of structures for the compact packing of spheres, the packing density amounts to 74 %.

- For comparison: the packing density of the bcc-lattice is 68 %.

Coordination number, the number of next neighbors of an atom.

3. Lattice planes and Miller indices

Lattice plane, arbitrary plane in a lattice. A plane is uniquely determined by three non-collinear points. The intersection points of the plane with the crystal axes are used to define the lattice plane. **Miller indices**, abbreviation for the specification of lattice planes for given crystal axes. They are fixed as follows:

- The intersection points of the plane with the crystal axes defined by the lattice vectors $\vec{a}$, $\vec{b}$, $\vec{c}$ are determined in units of the lattice constants (**Fig. 28.14**).
- The reciprocal values of the numbers obtained in this way are reduced to the least common denominator.
- The numerators of the fractions are the Miller indices of the lattice plane.
- The planes are identified by Miller indices given in brackets: (hkl).
- If an intersection point is at infinity, the corresponding index is zero.

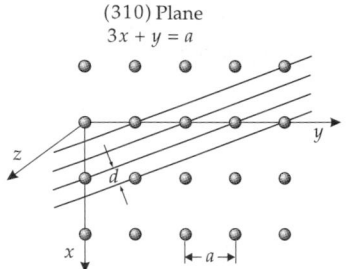

Figure 28.13: Lattice plane perpendicular to the z-axis, lattice sites in the x-y-plane.

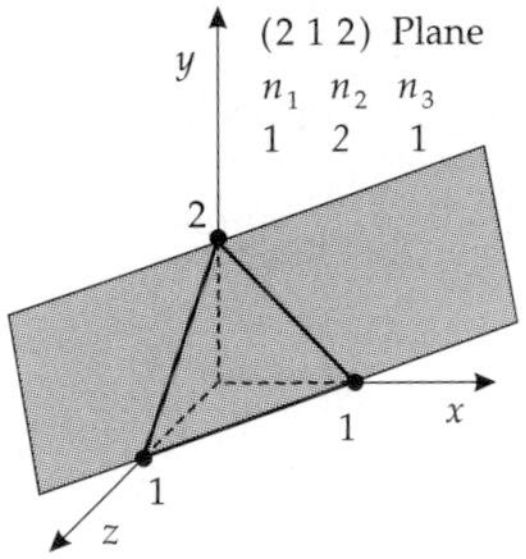

Figure 28.14: Construction scheme of the Miller indices: example $(h, k, l) = (2, 1, 2)$.

- ■ For the plane with intersection points 6, 2, 3, the reciprocal values are 1/6, 1/2, 1/3 → 1/6, 3/6, 2/6.

  Hence, the Miller indices are (132).
- If the plane intersects one or several crystal axes on the negative side of the origin, the index is specified by an upper horizontal bar.
- ■ $(h\bar{k}l)$ means that the plane intersects the $\vec{b}$-axis in the negative range.

Crystal direction, direction of a vector in the basis of the fundamental lattice vectors; its components are integer numbers (**Fig. 28.16**).

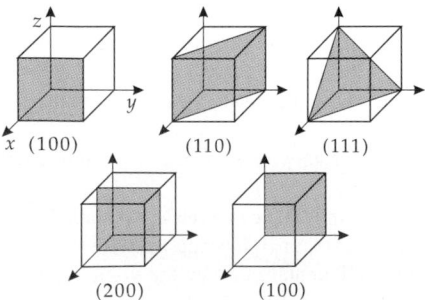

Figure 28.15: Several crystal planes in a cubic lattice.

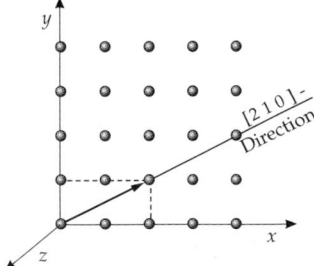

Figure 28.16: Crystal direction.

These integer numbers are put into **square** brackets: $[hkl]$.

➤   In **cubic** crystals the **direction** $[hkl]$ is always **perpendicular to the plane** (hkl) with the same indices. In other crystal systems this does not hold in general.

Atomic coordinates u, v, w, determine the positions of lattice points in an elementary cell. They are given as fractions of the lattice constants a, b, c along the crystal axes.

28.1.3.1   Simple crystal structures

1. NaCl

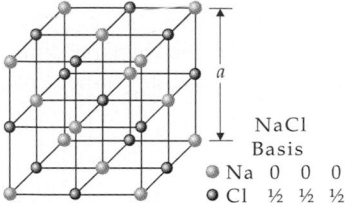

Crystal	a/nm	Crystal	a/nm
LiH	0.408	AgBr	0.577
NaCl	0.563	MgO	0.420
KCl	0.629	MnO	0.443
PbS	0.592	UO	0.492

Figure 28.17: NaCl-structure and representative crystals of the NaCl-structure (a: lattice constant).

Bravais lattice: fcc, base: 1 sodium- and 1 chlorine atom (separation: $\frac{1}{2}$ of space diagonal), number of base units per elementary cell: 4, coordination number: 6.

Atomic coordinates:

$$\text{Na:} \quad 000; \quad \frac{1}{2}\frac{1}{2}0; \quad \frac{1}{2}0\frac{1}{2}; \quad 0\frac{1}{2}\frac{1}{2} \qquad \text{Cl:} \quad \frac{1}{2}\frac{1}{2}\frac{1}{2}; \quad 00\frac{1}{2}; \quad 0\frac{1}{2}0; \quad \frac{1}{2}00$$

2. CsCl

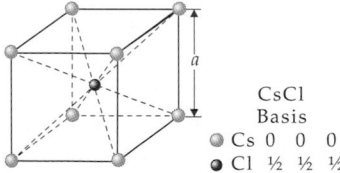

Crystal	a/nm	Crystal	a/nm
CsCl	0.411	AgMg	0.328
TlBr	0.397	LiHg	0.329
TlI	0.420	AlNi	0.288
NH_4Cl	0.387	BeCu	0.270

Figure 28.18: CsCl-structure and representative crystals of the CsCl-structure (a: lattice constant).

Bravais lattice: simple cubic, base: 1 cesium- and 1 chlorine atom (separation $\frac{1}{2}$ of space diagonal), number of base units per elementary cell: 1, coordination number: 8.

Atomic coordinates:

$$\text{Cs:} \quad 000 \qquad \text{Cl:} \quad \frac{1}{2}\frac{1}{2}\frac{1}{2}$$

28.1.4 Methods for structure investigation

1. X-ray diffraction,

most common method for structure investigation. It is based on the diffraction of x-rays by the lattice atoms. Wavelength of photon (energy E_γ):

$$\lambda_\gamma = \frac{1.24}{E_\gamma/\text{keV}} \text{ nm}.$$

Diffraction occurs on the atomic electrons. Hence, the intensity of diffraction depends strongly on atomic number Z.

X-ray diffraction is not very sensitive for elements of low atomic number. The positions of oxygen atoms or hydrogen atoms may barely be determined by x-ray diffraction. Moreover, elements of neighboring atomic numbers may barely be distinguished from each other.

2. Electron diffraction,

diffraction of electrons by atomic nuclei, therefore sensitively dependent on the atomic number. The wavelength of an electron with energy E_e is

$$\lambda_e = \frac{1.2}{\sqrt{E_e/\text{eV}}} \text{ nm}.$$

Electrons are charged particles and about 2000 times lighter than neutrons. They interact very intensely with matter electromagnetically, hence do not penetrate deeply into the crystal. Electron diffraction is therefore of particular importance for structure investigations of surfaces and thin layers.

3. Neutron diffraction,

exploits the wave property of the neutron for diffraction by periodic structures. Neutron diffraction by a crystal lattice occurs if the de Broglie wavelength of the neutrons (energy E_n) is similar to the separation of the lattice planes in the crystal. The wavelength of the neutron is

$$\lambda_n = \frac{0.028}{\sqrt{E_n/\text{eV}}} \text{ nm}.$$

Coherent scattering of neutrons occurs at the atomic nuclei of the structure components. The intensity of the diffraction depends on the neutron scattering cross-section of the nucleus. Structure analysis may be done with experiments with thermal neutrons ($E_n \approx 0.025$ eV).

Neutron diffraction allows both the determination of the position of elements of low atomic number, and also the discrimination between neighboring elements of the Periodic Table.

Magnetic scattering of neutrons, scattering by the magnetic moments of the atoms due to the interaction with the magnetic moment of the neutron.

4. Bragg condition,

premise for constructive interference in the reflection of incident radiation by the lattice planes of the crystal. If the condition is not fulfilled, the radiation interferes destructively.

Bragg condition			L
	Symbol	Unit	Quantity
$n\lambda = 2d \cdot \sin \Theta$	n	1	integer number
	λ	m	wavelength
	d	m	separation of lattice planes
	Θ	rad	glancing angle

➤   The wavelength must be within the range given by the structure of the crystal in order to produce measurable Bragg reflections.

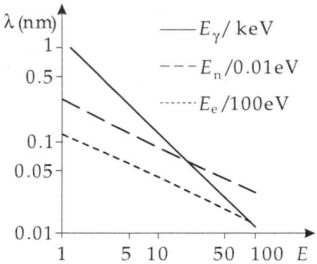

Figure 28.19: Wavelengths of x-ray photons, neutrons and electrons as a function of their energy.

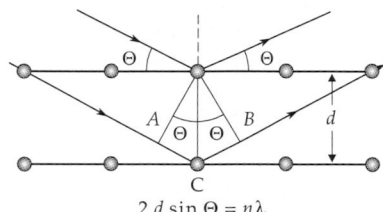

Figure 28.20: Bragg condition. Θ: glancing angle. The angle of incidence relative to the normal to the lattice planes is $\pi/2 - \Theta$. A, B: wave fronts, ABC: path difference $2d \sin \Theta$.

5. Methods of x-ray and neutron scattering

a) Laue method: In this method, a **fixed monocrystal** is irradiated by x-ray or neutron beams with a **continuous, "white" spectrum**. The Bragg condition is fulfilled only for certain wavelengths. Constructive interferences arise at certain angles, producing point-like reflections. The pattern of reflections is determined by the structure of the crystal. This method is particularly convenient for a rapid determination of crystal orientations and crystal symmetries. It is rarely used for structure investigations.

b) Rotating-crystal method: A **monocrystal** in a **mono-energetic** x-ray or neutron beam is **rotated** about a fixed axis. The Bragg condition is fulfilled at certain rotation angles at which point-like constructive interferences occur.

c) Debye-Scherrer method: This method is applied for the investigation of powders. The **powder specimen is irradiated** by a **mono-energetic beam**. The crystallites in the powder sample are statistically oriented. Diffracted beams emerge from crystallites that are randomly oriented in such a way that the primary beam hits several lattice planes at an angle that fulfils the Bragg condition.

The Debye-Scherrer method is applied for measurements of the variation of the lattice constants with the temperature, or the variation of the composition of an alloy. A practical advantage of the method is that monocrystals are not needed.

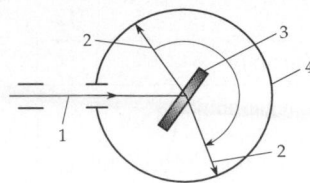

Figure 28.21: Rotating-crystal method.
1 - primary beam, 2 – scattered beam,
3 – rotating monocrystal, 4 – film.

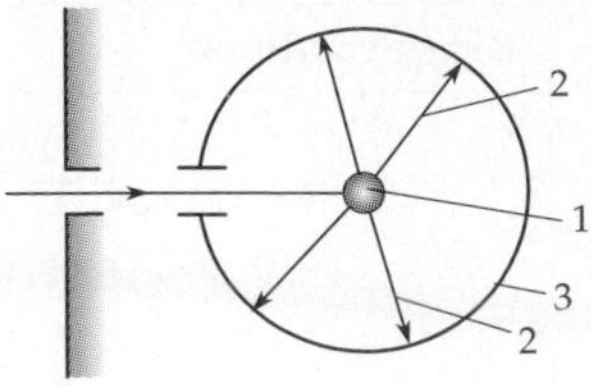

Figure 28.22: Debye-Scherrer method.
1 – polycrystal, 2 – scattered beam, 3 – film.

28.1.5   Bond relations in crystals

1. Survey of the types of bond in crystals

Type of bond	Ionic (heteropolar)	Covalent (unipolar)
properties	insulator at low temperatures, ionic conduction at high temperatures, plastically deformable	insulator, semiconductor, brittle, high melting point
interaction	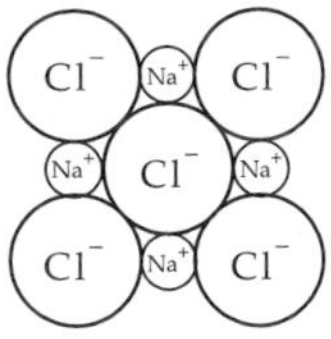	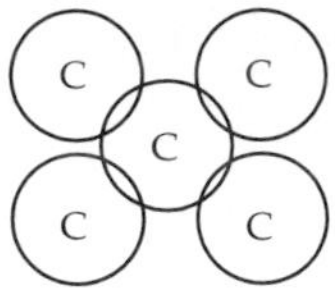
examples	alkali halogenides	organic molecules; C; Si; InSb
binding energy (eV/atom)	6 – 20	1 – 7

Type of bond	Metallic	Van der Waals
properties	electric conductor, good thermal conductor, plastic, high reflectance in IR and visible spectrum	insulator, low melting point, easily compressible, transparent in the far UV
interaction	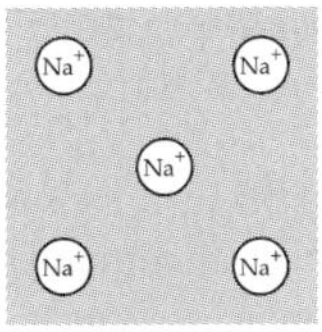	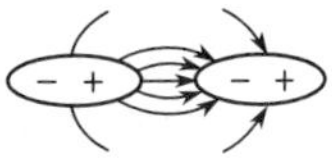
examples	metals, alloys	noble-gas crystals, H_2, O_2, polymers, molecular crystals
binding energy (eV/atom)	1 – 5	$10^{-2} – 10^{-1}$

Lattice energy, difference of energy between the free atoms and the crystal.

▲   A crystal is only stable if its total energy is lower than the total energy of the free atoms or molecules of which it is composed.

2. Ionic bond,

caused by the attracting Coulomb force between different charged ions.

■   Common salt, Na^+Cl^-, is a typical ionic crystal.

binding energy in ionic bond			ML^2T^{-2}
	Symbol	Unit	Quantity
$E_B = \dfrac{Q^2}{4\pi\varepsilon_0} \cdot \dfrac{\alpha}{r}$	E_B	J	binding energy
	Q	A s	charge
	ε_0	A s/(V m)	permittivity of free space
	r	m	distance
	α	1	Madelung constant

Ionic binding forces have a long range. Frequently, the effect of not only the next but also of even more distant neighbors has to be taken into account.

Madelung constant, α, determines the strength of the ionic bond by taking into account the more distant ionic charges:

Madelung constant			1
	Symbol	Unit	Quantity
$\alpha = \sum\limits_{j} \dfrac{\pm R}{r_j}$	R	m	distance of next neighbors
	r_j	m	distance between ion j and reference ion

For a negative reference ion, positive ions get a sign $+$, and negative ions a sign $-$.

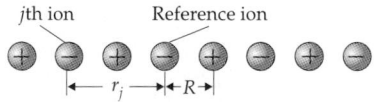

Figure 28.23: On calculating the Madelung constant.

Table of typical values of the Madelung constant α:

Structure	NaCl	CsCl	ZnS (cubic)
α	1.747558	1.747558	1.6381

Repulsive interaction, occurs because of the Coulomb force and the Pauli principle (see p. 844) if two atoms approach each other closely and their electron shells overlap.

• At low temperature, ionic crystals are insulators.

• At high temperature, ionic conduction occurs. Ionic crystals are plastically deformable.

3. Metallic bond,

originates from the electrostatic interaction of the valence electrons released by the atoms with all positive atomic cores of the crystal. The binding partners are not rigidly coupled; the free valence electrons have a high mobility and are not localized.

■   Sodium, aluminum, iron.

Transition metals, metals with an incomplete d-shell (3d-, 4d-, 5d-metals), i.e., all metals beyond the eight main groups of the Periodic Table of elements (see p. 877). They are characterized by a high binding energy. Additional binding forces are generated by the interaction between the inner d-shells.

■   Copper, silver, gold.

The metallic bond is weaker than the ionic bond. Hence, the lattice energy of an alkali metal crystal is significantly lower than that of an ionic alkali halogenide crystal. Example: NaCl: 8.1 eV/atom, Na: 1.1 eV/atom.

| **M** | Crystals with a metallic bond are electric conductors and good thermal conductors. They are plastically deformable. They are strongly reflecting in the IR and the visible spectral range. |

4. Covalent bond,

unipolar bond, electron pair binding via the exchange interaction. This type of bond is dominant in the elements of the third through the fifth main group of the Periodic Table. The unfilled valence electron shells may organize a closed, noble-gas-like electron configuration that involves the valence electrons of the next neighbors.

■   Many carbon compounds are covalently bound, in particular diamond and organic molecules.

Electron exchange, an affiliation of an electron pair to two neighboring atoms.

   Exchange interaction of covalent bond, a force mediated by the exchange of electrons between atoms. The spins of the electrons are oriented antiparallel (singlet state), so that (due to the Pauli principle) the spatial wave function of the two electrons is symmetric. For a spatially symmetric electron wave function (antiparallel spins: antisymmetric spin wave function), the probability density of finding the particle at the center between the binding partners is larger than in the case of a spatially antisymmetric wave function for electrons with parallel spins (triplet state – symmetric spin function). The singlet configuration of electrons yields an energy contribution—as compared with separated atoms—which leads to a binding of the two atoms.

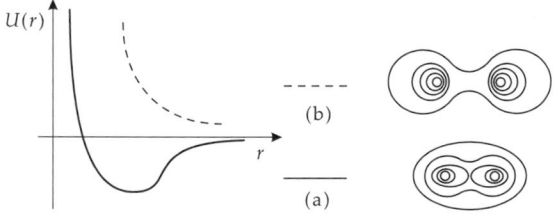

Figure 28.24: Binding potentials as a function of the interatomic distance r for electron pairs with (a) antiparallel spins (bound state), and (b) parallel spins (scattering state). The right side of the figure sketches the contour lines of the electron density distribution: despite the exchange force, the electrons remain closely to the atoms.

▲   Covalent bonds are bonds between neutral atoms. A configuration with parallel orientation of the spins of electrons involved in the exchange does not lead to a binding of the atoms.

▲   Important examples are covalently bound semiconductors.

▲   Besides crystals with ionic or covalent bonds, there are also crystals with a mixed bond.

5. Van der Waals bond,

weakly attracting dipole-dipole interaction occurring when instantaneous dipole moments are mutually induced in the crystal atoms or molecules. The interaction from these induced dipole moments (dipole-dipole interaction) results in a weak attractive electric force.

Van der Waals binding energy			ML^2T^{-2}
	Symbol	Unit	Quantity
$U(r) \approx -\dfrac{C}{r^6}$	$U(r)$	J	binding potential
	C	J m^6	interaction constant
	r	m	distance

▲ C is of the order of magnitude of 10^{-77} J m^6.

▲ The Van der Waals potential is the most important attractive interaction in noble-gas crystals and between organic molecules.

➤ For a correct description of the experimental data, an additional weak-repulsive potential of the hard-core type $\sim r^{-12}$ is needed.

The **Lennard-Jones potential** results from combining the hard-core repulsion with the Van der Waals potential

Lennard-Jones potential			ML^2T^{-2}
	Symbol	Unit	Quantity
$U(r) = 4\varepsilon\left[\left(\dfrac{\sigma}{r}\right)^{12} - \left(\dfrac{\sigma}{r}\right)^6\right]$	$U(r)$	J	binding potential
	r	m	distance
$C = 4\varepsilon\sigma^6$	ε	J	parameter
	σ	m	parameter

with new parameters ε and σ, where $C = 4\varepsilon\sigma^6$.

Table of ε, σ and C for the noble gases:

Noble gas	He	Ne	Ar	Kr	Xe
$\varepsilon/10^{-23}$ J	14	50	167	225	320
$\sigma/10^{-10}$ m	2.56	2.74	3.40	3.65	3.98
$C = 4\varepsilon\sigma^6/(10^{-77}\text{J m}^6)$	0.016	0.085	1.032	2.128	5.088

28.2 Lattice defects

Lattice defect, deviation from the ideal structure of strict spatial periodicity by construction faults (vacancies, dislocations, stacking disorders, etc.).

▲ Type and abundance of lattice defects modify the mechanical, electric, magnetic and optical properties of solids in a characteristic manner.

28.2.1 Point defects

1. Vacancies

Vacancies, atoms missing on regular lattice sites.
  Divacancies, neighboring vacancies.

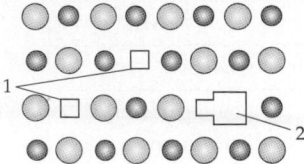

Figure 28.25: Lattice plane of a two-atomic lattice with vacancies (1 – vacancies, 2 – divacancy).

Vacancy-formation energy, E_V, energy expended to remove an atom from the lattice compound and lift it to the crystal surface.

vacancy density in equilibrium			$\mathbf{L^{-3}}$
	Symbol	Unit	Quantity
$n = N \cdot e^{-\frac{E_V}{k_B T}}$	n	m^{-3}	vacancy density
	N	m^{-3}	particle density
	E_V	J	vacancy-formation energy
	k_B	$J \cdot K^{-1}$	Boltzmann constant
	T	K	temperature

■   At room temperature $\frac{n}{N} \approx 10^{-17}$.

At 1000 K the vacancy concentration increases to $\frac{n}{N} \approx 10^{-5}$.

▲   In ionic crystals it is more advantageous energetically to produce the same number of cation and anion vacancies.

| M | **Measurement of vacancy concentrations**: The vacancy concentration may be calculated from the difference between the relative linear expansion $\Delta L/L$ in heating and the relative lattice change $\Delta a/a$ determined by means of x-ray diffraction. A vacancy affects the diffraction only weakly, but the length of the sample increases if atoms migrate from the crystal interior to the surface.

Vacancy concentrations have been determined for about two decades by means of positron annihilation spectroscopy (PAS). Positrons from a positron source (e.g., ^{22}Na) are thermalized in a solid by collisions with the lattice atoms and then captured into the vacancies. The vacancies represent a negatively charged sink relative to their environment. The positrons captured into such sinks produce annihilation photons that have different characteristics from those produced by freely moving positrons.

2. Frenkel defects, lattice impurities and color centers

Interstitial atoms, additional atoms built into the lattice between the regular lattice sites.

Frenkel defect, consists of a vacancy and an atom at an interstitial position in the vicinity of the vacancy where the atom would fit in. There is an attractive interaction between the interstitial atom and the vacancy.

▲   Frenkel defects are the most abundant point defects in silver halogenides.

Lattice impurities, impurity atoms built in:

•   at regular lattice sites (**substitutional**), or

•   between the lattice sites (**interstitial**).

▲   Lattice impurities in semiconductors play a dominant role as donors or acceptors.

Color centers, lattice impurities that absorb visible light.

▲   Color centers occur in ionic crystals. They cause a coloring of these crystals which—as a rule—are transparent in the optical range of the spectrum.

F-center, simplest color center consisting of an anion vacancy and an excess electron bound to this vacancy.

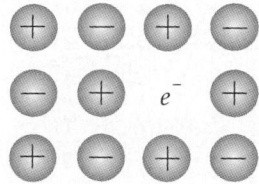

Figure 28.26: F-center.

28.2.2   One-dimensional defects

Dislocation, a linear arrangement of point defects.
▲   Dislocations generate a stress field in their vicinity.
Edge dislocation, a lattice plane terminates in the crystal, like a wedge does.
   Low external stresses may move dislocations if the binding forces have no preferred orientation.
   Glide plane, a crystal plane along which two parts of the crystal glide over each other.
▲   The gliding direction is perpendicular to the dislocation line (symbol ⊥) for edge dislocations.

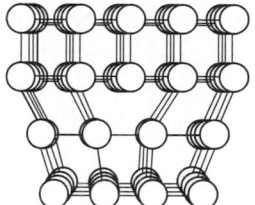

Figure 28.27: Edge dislocation.

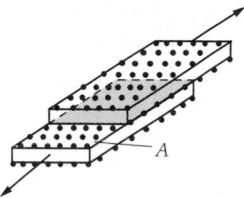

Figure 28.28: Glide plane.

■   Forces of 1N/cm² are sometimes sufficient to move a dislocation.
Screw dislocation, can be visualized as follows: A crystal is cut across the middle. Then a shear stress is applied parallel to the cut edge one atomic distance away.
▲   The crystal lattice is displaced parallel to the dislocation line by one atomic plane.
Burgers vector, $\vec{b}$, together with the direction of the **dislocation line** $\vec{s}$ characterize the geometric properties of a dislocation. The Burgers vector $\vec{b}$ is always a lattice vector.

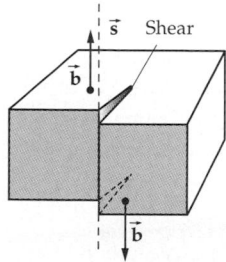

Figure 28.29: Schematic figure on the generation of a screw dislocation. $\vec{s}$: dislocation line, $\vec{b}$: Burgers vector.

•   A closed loop is made from atom to atom about a dislocation line that lies completely in the non-disturbed crystal region.
•   This loop, starting from the same atom, is transferred into the corresponding ideal crystal without dislocation. The loop is then no longer closed.
•   The missing vector required to complete the loop is the Burgers vector $\vec{b}$.
▲   For **edge dislocations**, the Burgers vector is **perpendicular** to the dislocation line.

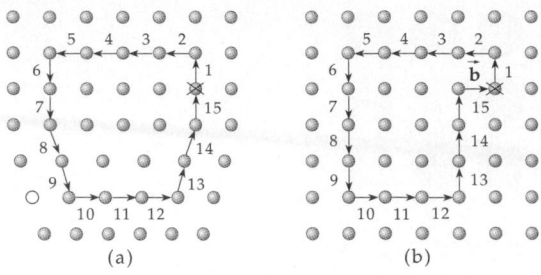

Figure 28.30: Burgers vector $\vec{b}$ of an edge dislocation. (a): circulation in the distorted crystal region, (b): circulation in the non-distorted crystal region. The step sequence is indicated.

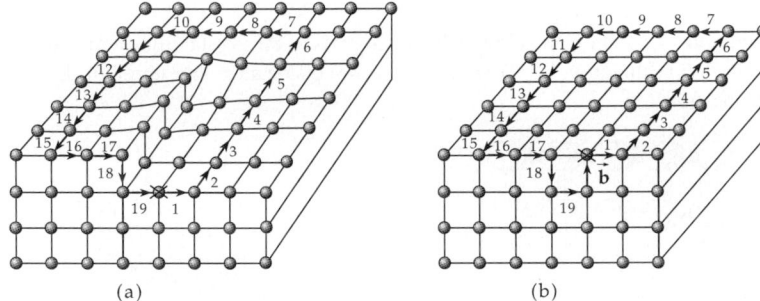

Figure 28.31: Burgers vector $\vec{b}$ of a screw dislocation. (a): circulation in the distorted crystal region, (b): circulation in the non-distorted crystal region. The step sequence is indicated.

▲   For **screw dislocations**, the Burgers vector and the dislocation line are **parallel** to each other.

Dislocation density, number of dislocation lines per unit area.

■   In strongly deformed metal crystals one observes dislocation densities of $10^{11} - 10^{12}$ cm^{-2}.

Plasticity, a measure of irreversible shape variability of solids under external deforming forces.

▲   The more dislocations exist in a crystal, the higher is its plasticity.

M   Dislocations may be etched by appropriate bases or acids. The etching speed in the region distorted by the dislocation is higher than in the non-disturbed crystal. The resulting etch pits may be counted by a microscope or an electron microscope.

28.2.3   Two-dimensional lattice defects

Grain boundaries, boundaries between monocrystalline regions (grains).

Small-angle grain boundaries, boundaries of crystallites the grain boundaries of which enclose angles of only few degrees because of twisting of the crystallites forming the grain boundary. **Fig. 28.32** sketches a small-angle grain boundary formed by successive edge dislocations.

Stacking disorder, two atomic planes are displaced in their plane by a vector that is **not** a lattice vector.

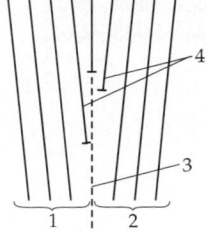

Figure 28.32: Schematic representation of a small-angle grain boundary.
1 – crystal 1, 2 – crystal 2, 3 – grain boundary, 4 – edge dislocations.

28.2.4 Amorphous solids

Amorphous solids, solids without long-range order. A certain short-range order may exist in the vicinity of individual atoms.

▲   Amorphous solids are always produced by the freezing of a disorder.

▲   The amorphous state is a metastable state, i.e., after extended preservation (sometimes years) the substance recrystallizes.

| M | Thermal treatment converts the amorphous solid into a crystalline state.

Metallic glasses, amorphous alloys displaying the properties of metals:

●   elastic at high mechanical stress,

●   magnetic,

●   good thermal conductivity,

●   electrically conducting,

and properties of glasses,

●   mechanically hard,

●   corrosion-resistive.

Cooling speeds of 10^6 K/s and more are required in order to produce amorphous metals. Simple metals may not be able to be produced as stable amorphous materials. Besides the metal, a so-called glass-former (boron or phosphorus) must be added to an alloy. Metallic glasses occur only for a thickness up to 50 μm. The cooling speed is too low for higher thicknesses.

| M | **Melt-spinning** is the most common method for the production of metallic glasses (**Fig. 28.33**).

Metallic glasses are used for:

●   transformer sheets, because of low eddy current losses,

●   hard recording-head material, fast remagnetization capability,

●   magnetic memories.

Nanocrystalline materials, solids consisting up to about 50 % of lattice defects (**Fig. 28.34**).

Nanocrystalline materials are generated by local energy supply, i.e., by producing a high density of lattice defects.

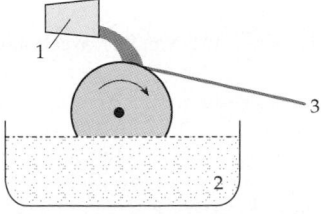

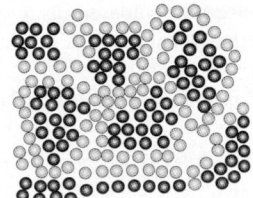

Figure 28.33: Scheme of the melt-spinning method. 1 – melting pot, 2 – coolant liquid, 3 – amorphous tape.

Figure 28.34: Scheme of a nanocrystalline material.

28.3   Mechanical properties of materials

Mechanical stress, σ, force referring to the cross-sectional area generated by a solid to prevent deformation.

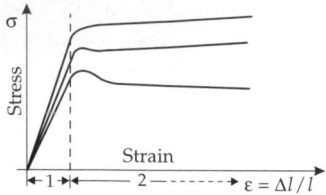

Figure 28.35: Stress-strain diagram. 1 – Hooke's region, 2 – plastic region.

Hooke's law, a linear relation between stress and strain (see p. 156).
Elastic region, interval in which Hooke's law is valid.

Hooke's law: stress $\sim$ strain			$ML^{-1}T^{-2}$
	Symbol	Unit	Quantity
$\sigma = E \cdot \varepsilon$	σ	$N\,m^{-2}$	stress
	E	$N\,m^{-2}$	elasticity modulus
$\varepsilon = \Delta l / l$	ε	1	strain
	l	m	length
	Δl	m	change of length

Newton's law: The viscous or plastic behavior of a material is proportional to the expansion velocity.

Newton's law: strain $\sim$ expansion velocity			$ML^{-1}T^{-2}$
	Symbol	Unit	Quantity
	σ	$N\,m^{-2}$	stress
$\sigma = \eta_0 \cdot \dfrac{d\varepsilon}{dt}$ $\varepsilon = \dfrac{\Delta l}{l}$	η_0	$N\,m^{-2}\,s$	dynamical viscosity
	$d\varepsilon/dt$	s^{-1}	expansion velocity
	l	m	length
	Δl	m	change of length

Creeping, a typical property of polymers, which also occurs for other materials. It refers to the compliance of a substance under an applied mechanical stress.

28.3.1   Macromolecular solids

Macromolecular solids, solids formed from very long molecules.
▲   Macromolecular solids are held together by covalent and Van der Waals binding forces.
▲   Macromolecular solids may be either amorphous or crystalline.

28.3.1.1   Polymers

Monomers, molecules that form the basic reactive units of polymers.
   Polymers, macromolecules formed from monomers via chemical reactions (conversion of the monomer into a reactive state by breaking of bonds, growth of chains by attachment

of reactive monomers, termination of chains by attachment of a molecule). The process of bonding of monomers to long chains is called as **polymerization**.

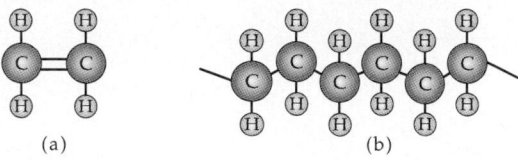

Figure 28.36: Scheme for the polymerization of polyethylene. (a): monomer (ethene), (b): polymer (polyethylene).

1. Characteristics of polymers

Molecular mass distribution, variation of the molecular mass due to different lengths of chains.

▲   The molecular mass distribution determines the performance of the material.

▲   The broader the molecular mass distribution, i.e., the larger the range of variation of molecular masses, the wider the temperature range over which the polymers soften.

Mean relative molecular mass, degree of polymerization, M_r, measure of the length of a macromolecule.

mean relative molecular mass M_r			1
	Symbol	Unit	Quantity
$M_r = \dfrac{m_M}{u}$	M_r	1	mean relative molecular mass
	m_M	kg	molecular mass
	u	kg	mass of monomer

➤   Mean relative molecular masses range from 10^3 to 10^6.

▲   The mean relative molecular mass is a measure of the viscosity of the material. The viscosity increases with molecular mass.

▲   Polymers do not exist in the gaseous phase.

The order of polymers may be:

•   **statistical** (ball structure), or

•   **paracrystalline** (chain molecules aligned with each other in a certain order).

▲   The tensile strength of polymer materials is strongly dependent on the temperature.

▲   Solid polymers are visco-elastic substances.

➤   The order of polymers may be described theoretically by field-theoretical methods developed originally for treating magnetic systems (Pierre-Gilles de Gennes, Nobel Prize, 1991).

2. Elasticity and plasticity of polymers

Elasticity, deformations that ocurred in the past are no longer present; the deformations are fully reversible.

   Plasticity, the deformations are irreversible and are preserved in the future.

■   Rubber is largely **elastic**, plasticine is **plastic**.

Visco-elasticity, after applying a constant strain, there occurs at first a small elastic extension, followed by a plastic deformation. After removing the strain, the elastic extension disappears, but the plastic deformation is retained (**Fig. 28.37**).

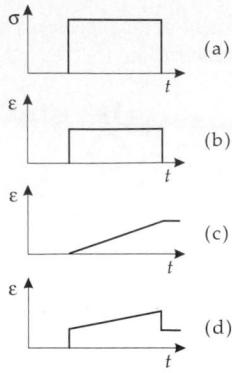

Figure 28.37: Visco-elastic behavior. (a): applied stress, (b): elastic behavior, (c): plastic behavior, (d): visco-elastic behavior.

▲ Visco-elastic behavior is caused by the shearing of macromolecules (chain molecules) against each other.

Loading speed, $d\sigma/dt$, speed of change of stress on the sample.

Deformation speed, $d\varepsilon/dt$, speed of response of the body against a load by strain.

Maxwell model of the visco-elastic behavior:

Maxwell model			$\mathbf{T^{-1}}$
	Symbol	Unit	Quantity
$\dfrac{d\varepsilon}{dt} = \dfrac{1}{G}\dfrac{d\sigma}{dt} + \dfrac{\sigma}{\eta}$	$d\varepsilon/dt$	s^{-1}	deformation speed
	G	$N\,m^{-2}$	shear modulus
	σ	$N\,m^{-2}$	stress
	η	$N\,m^{-2}\,s$	dynamic viscosity

▲ At very low shear velocity, a polymer behaves like a viscous liquid.

▲ At extremely high shear velocities (e.g., by a stroke), a polymer behaves like an elastic solid.

■ Silly-putty toy.

28.3.1.2 Thermoplasts

Thermoplasts, easily melting and swelling polymer materials of high solubility. Recycling is possible with low supplied energy.

■ Polyethylene (PE), polyvinyl chloride (PVC), polystyrene (PS), polyamide (nylon, perlon), polyester, polyacrylonitrile, polycarbonates.

28.3.1.3 Elastomers

Elastomers, almost fully elastic polymers.

▲ Elastomers swell readily, do not melt and are not soluble.

▲ The elastic behavior results because of the **wide-meshed** cross-linkage of the macro-molecules.

Vulcanization, process of cross-linking of the macromolecules after shaping. The degree of cross-linking of the molecules is essential for the elasticity of the material.

■ Elastomers: synthetic rubber, neoprene, polyurethane, silicon rubber.

Relaxation, behavior of a polymer, the strain of which returns exponentially to zero after removing the shear stress.

Voigt-Kelvin model of relaxation			1
$\varepsilon(t) = \dfrac{\sigma_0}{E}\left(1 - e^{-\frac{t}{\tau}}\right)$ after applying $\varepsilon(t) = \dfrac{\sigma_0}{E} e^{-\frac{t}{\tau}}$ after removing	Symbol	Unit	Quantity
	ε	1	extension
	E	N m^{-2}	elasticity modulus
	σ_0	N m^{-2}	strain
	τ	s	relaxation time
	t	s	time

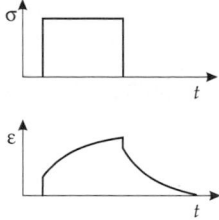

Figure 28.38: Relaxing polymer (schematic).

28.3.1.4 Duromers

Duromers (duroplasts), very **close-meshed**, interlaced, very hard, **inelastic** polymers.
▲ Duromers are neither meltable nor soluble.
■ Duromers: bakelites, formaldehyde resins and epoxide resins.

28.3.2 Compound materials

Compound materials, various substances joined to another substance—the compound material.
■ Reinforced concrete, fiberglass-stabilized polyester and laminated fabric.
Layer-compound materials, compound materials produced by layer-on-layer stacking of individual material components.
■ **Bimetal**, compound material consisting of two materials (metals) of different thermal expansion and used as a temperature-controlled switch.
Particle-compound materials, substances consisting of a matrix with small particles deposited into the structure.
  Dispersion hardening, deposition of hard particles, e.g., carbides, oxides and silicides, in a soft matrix. Thereby the resistivity increases because of suppression of the dislocation motions.
■ Dispersion-hardened alloys are used in turbine blades.
▲ Metallic particles dispersed in a matrix of elastomers may lead to electric conduction: **conducting elastomers**.

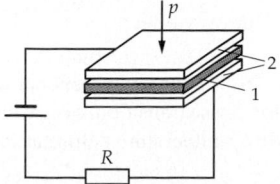

Figure 28.39: Principle scheme of a pressure sensor. 1 – conducting elastomer, 2 – conducting plates.

Fiber-reinforced compound materials, materials with very long (endless fibers) or short (short fibers) metallic or nonmetallic fibers embedded into a (metallic or nonmetallic) matrix.

▲   The hard fibers take a part of the forces.

Whiskers, monocrystalline fibers with extremely high values of rupture resistance.

■   Fiber-reinforced compound materials are used in light-weight construction of cars and airplanes.

28.3.3   Alloys

1. Main properties of alloys

Alloys, mixtures of several metals to a coherent body.

Limiting cases:

●   **Heterogeneous mixture**, the components are not miscible. The alloy then always consists of distinct crystal types.

■   Copper-lead.

●   **Mixed crystals**, the components are miscible in any mixing ratio. A homogeneous alloy results that contains only one crystal type.

■   Copper-nickel.

Intermetallic compounds, for certain compositions the components form compounds characterized by a crystal lattice.

■ Fe_3Al.

2. Temperature-dependent shape variation of alloys

Shape-memory alloy, **memory alloy**, an alloy that shows a temperature-dependent shape variation.

▲   Shape memory is caused by a **martensitic phase transition**, a diffusionless and reversible phase transformation characterized by coupled atomic displacements by magnitudes that are small compared with the atomic separation. A visible shape variation arises.

▲   Shape-memory alloys have different coefficients of thermal expansion in different directions, both in **magnitude** as well as in **sign**. They are by 3 to 4 orders of magnitude greater than those of an ordinary metal.

▲   The volume of a sample increases in heating.

Properties of memory alloys:

●   **superelastic performance**,

●   **high damping capability**.

One-way effects, memory effects in which the state before deformation is reached again after heating and is preserved in cooling.

Original shape   Distortion   Heating   Cooling

Figure 28.40: Memory alloy. One-way effect.

Two-way effects, are produced irreversibly in a deformation by additional dislocation motions. When heating to above the temperature of the phase transition, a high-temperature deformation arises, and in cooling a corresponding low-temperature deformation.

▲   This conversion may be repeated many times.

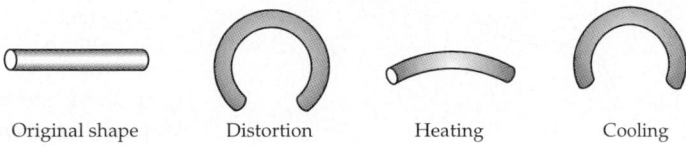

Figure 28.41: Memory alloy. Two-way effect.

All-round effects, occur in certain Ni-Ti alloys. The initial material is deformed and then undergoes a thermal treatment at 400 °C − 500 °C (**tempering**). The result is a complete shape inversion under temperature change.

▲   This conversion may be repeated many times.

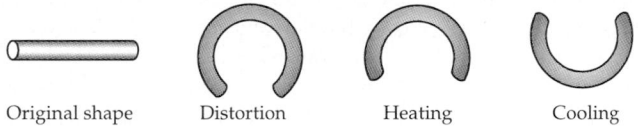

Figure 28.42: Memory alloy. All-round effect.

3. Application of memory effects

- Antennas for space vehicles may consist of a compact winding of thin wire. They widen by the heat of sun to a circular shape with a diameter of several kilometers.
- **Cold welding**, connection of tubes. A sleeve of memory alloy is produced with an inner diameter several percent smaller than the outer diameter of the parts to be connected. At the temperature of liquid nitrogen, the sleeve widens to fit over the outer diameter of the tubes to be joined. At room temperature, the sleeve shrinks in diameter and stretches in the axial direction. A solid, hermetically tight joint results.

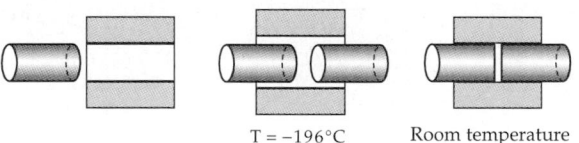

Figure 28.43: Cold welding.

- Surgery for bone fractures: A spring clamp of given size and shape is stretched at low temperature. The ends are fixed by screws at both sides of the fracture. The alloy is chosen so that the spring clamp at body temperature remembers its initial shape and simultaneously turns over into the **superelastic state**. On knitting of the bones, the residual deformation gradually reduces, nevertheless a constant pressure stress is maintained.

4. Superelasticity,

property of certain alloys to maintain the capability of elastic extension beyond the Hooke region. When relaxing after reaching the 10 %-extension, the relaxation line runs somewhat less steeply than, but nearly parallel to, the load line. No permanent deformation remains.

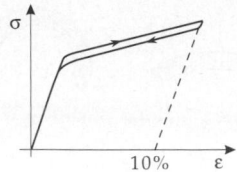

Figure 28.44: Strain-extension diagram in the superelastic case.

28.3.4   Liquid crystals

1. Types of liquid crystals

Liquid crystals display both the properties of a liquid and those of a crystalline medium in a certain range of temperature or concentration of a solvent. Liquid crystals are formed by stretched molecules, mostly of aromatic compounds.

Nematic phases, liquid crystals. On average, the longitudinal axes of the molecules are aligned parallel within larger or smaller regions. The molecules may, however, be shifted arbitrarily along these axes and twisted against each other about the axes.

Smectic phases, liquid crystals in which the molecules also occur with parallel longitudinal axes, but in layers.

In the time and spatial average, a parallel alignment of the longitudinal axes occurs only over small ranges.

▲   External fields may generate the ideal case – parallel ordering of all molecules over a larger region.

Cholesteric phase, special case of the nematic phase. Nematically ordered regions are ordered in layers whereby the orientations of the longitudinal axes are twisted from layer to layer.

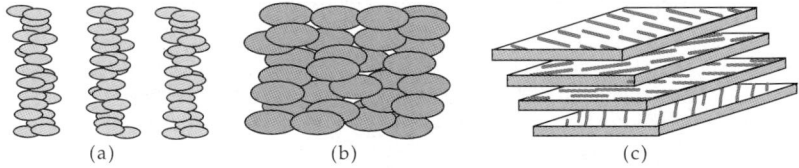

Figure 28.45: (a): smectic phase, (b): nematic phase, (c): cholesteric phase.

2. Properties of liquid crystals

Orientation elasticity, property of orientation of the longitudinal molecular axes under the influence of an external perturbation. After removing the perturbation, the initial state is restored.

Optical birefringence, optical anisotropy, displayed in particular by liquid crystals in the cholesteric phase.

Selective total reflection, only certain wavelengths are reflected. Selective total reflection is a feature of cholesteric liquid crystals built of **twisted nematic structures**. It depends on pressure and temperature variations, as well as on electric and magnetic fields.

▲   The selectively reflected wavelength depends on the pitch of the helix, and on the mean refractive index of the liquid crystal.

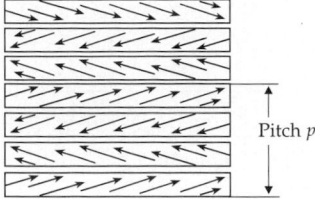

Pitch p

Figure 28.46: Twisted
nematic structure.

3. Applications of liquid crystals

- Liquid crystals are used for surface-covering measurement of temperature in medical diagnostics. Incident white light appears as colored when reflected, and the color corresponds to the body's surface temperature.
- Liquid crystal display element (**LCD**): A $10 - 20\ \mu$m thick layer of a nematic liquid is put between two electrodes. The molecules of the liquid crystal are deposited at the electrodes with preferred directions twisted against each other by $\pi/2$. The electrodes are transparent. If light that is linearly polarized parallel to a preferred direction is projected onto this twisted nematic phase, the polarization direction is also rotated by $\pi/2$ when it traverses the cell. When the individual segments of the electrode are triggered by a voltage of $10 - 20\,$V, the original orientation of the molecules of the liquid is disturbed, since the molecules now align with respect to the applied electric field. An analyzer placed behind the cell with a rotated transmission direction of $\pi/2$ with respect to the polarizer distinguishes whether an electrode segment has been activated or not; an activated electrode element appears as dark.
- Despite the high voltage, power consumption of an LCD practically vanishes, since the alignment of molecules requires very low energy.

28.4 Phonons and lattice vibrations

28.4.1 Elastic waves

1. Lattice vibrations,

vibrations of the lattice elements $n, n + 1$, etc. about their equilibrium positions.

▲ For small displacements, Hooke's law applies (harmonic lattice vibrations).

Elastic constant, C_n, interaction constant between planes with a separation of $n \cdot a$, a being the lattice constant.

equation of motion for one atom per elementary cell			MLT^{-2}
	Symbol	Unit	Quantity
	C_n	kg s^{-2}	elastic constant
	u_s	m	displacement of plane s
$M\dfrac{\mathrm{d}^2 u_s}{\mathrm{d}t^2} = F_s$			
$= \sum_n C_n \cdot (u_{s+n} - u_s)$	u_{s+n}	m	displacement of plane at d $= n \cdot a$
	M	kg	atomic mass
	F_s	kg m s^{-2}	force
	t	s	time

2. Elastic waves,

solutions u_s of the equation of motion:

elastic wave			L
	Symbol	Unit	Quantity
$u_s(\vec{r}, t) \sim e^{j(\vec{k}\vec{r} - \omega t)}$	u_s	m	displacement
	$\vec{k}$	m^{-1}	wave vector
	ω	rad/s	angular frequency
	$\vec{r}$	m	position vector
	t	s	time

3. Dispersion of elastic waves

Dispersion, $\omega(\vec{k})$, dependence of the angular frequency ω of elastic waves on the wave vector $\vec{k}$.

For a monoatomic cubic lattice in which only nearest neighbors interact ($n = 1$), for the propagation directions parallel to the [100]-, [110]- and [111]-direction (reduction to the one-dimensional problem of a one-dimensional linear wave):

dispersion			T^{-1}
	Symbol	Unit	Quantity
$\omega = \sqrt{\dfrac{4C_1}{M}} \left| \sin\left(\dfrac{ka}{2}\right) \right|$	ω	rad/s	angular frequency
	k	m^{-1}	wave number
	a	m	lattice constant
	C_1	$kg\,s^{-2}$	elastic constant
	M	kg	mass of atom

4. Phonons

First Brillouin zone, range of the physically meaningful values of the wave vector $\vec{k}$ (see **Fig. 28.47**). The range of $-\pi \cdots +\pi$ for the phase ka includes **all** independent values of ω. The statement that two neighboring atoms are out of phase by more than π is physically meaningless, since there exists a physically identical phase with a value within the range $-\pi \cdots +\pi$.

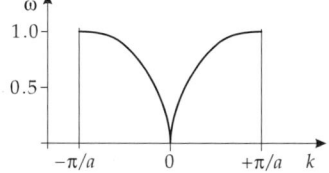

Figure 28.47: First Brillouin zone.

▲   The wave number k may be restricted to the range $-\pi/a \leq k \leq +\pi/a$.

Phonon, energy quantum of an elastic wave. The denotation is analogous to that of the **photon**, the energy quantum of an electromagnetic wave.

▲   The elastic energy of a lattice is quantized.

▲   The propagation of phonons is described by their wave vector $\vec{k}$ and the dispersion relation $\omega(\vec{k})$.

▲   Phonons interact with particles, or with fields, as if they had a quasi-momentum $\hbar\vec{k}$.

Quasi-momentum of a phonon $\hbar\vec{\mathbf{k}}$, a quantity with the dimension of a momentum, which does not actually exist in the crystal, but which obeys selection rules for allowed transitions between quantum states similar to momentum conservation.

5. Measurement methods for phonons

Phonon spectrum, the energy distribution of the elastic waves in a solid.

| M | **Inelastic neutron scattering**, most important method of measurement of the phonon spectrum of a solid. Because of their zero charge, the neutrons are not affected by the Coulomb field of nuclei. They interact directly with the nuclei of a solid lattice.

The **kinematics of neutron scattering** is determined by the conservation laws for energy and momentum.

energy and momentum conservation in neutron scattering			
	Symbol	Unit	Quantity
$E_f = E_i \pm \hbar \cdot \omega$ $\vec{\mathbf{p}}_f = \vec{\mathbf{p}}_i \pm \hbar\vec{\mathbf{k}}$	ω $\vec{\mathbf{k}}$ E_i, E_f	rad/s m^{-1} J	phonon frequency wave vector energy of incoming and outgoing neutron
	$\vec{\mathbf{p}}_i, \vec{\mathbf{p}}_f$	$\mathrm{kg\ m\ s}^{-1}$	momentum of incoming and outgoing neutron
	$\hbar$	J s	quantum of action

The $(+)$ signs apply to scattering processes in which a phonon is annihilated, the $(-)$ signs to processes creating a phonon. The quantity v_s denotes the velocity of sound, $\omega = v_s \cdot k$.

| M | In order to determine the dispersion relation, and hence the elastic constants, the energy loss or energy gain of the scattered neutrons must be measured as a function of the direction of scattering $\vec{\mathbf{p}}_f - \vec{\mathbf{p}}_i$. Typical neutron energies for such measurements are in the range of several meV (milli electron volts).

6. Types of phonons

Longitudinal phonons, correspond to vibrations of the medium along approximately the direction of propagation of the elastic wave.

Transverse phonons, energy quanta of vibrations of the medium approximately perpendicular to the direction of propagation of the wave. Exact parallelness, or orthogonality, occurs only for certain symmetry directions of the lattice or in the limit of isotropic media.

Acoustic phonons: The atoms of a primitive elementary cell vibrate along the same direction (analogous to an in-phase vibration of coupled oscillators). There are always three acoustic branches. For low wave numbers, an approximately linear relation exists, $\omega \approx ck$, and hence a sound velocity.

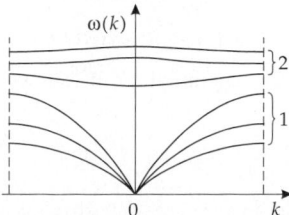

Figure 28.48: Schematic trend of the dispersion relation $\omega(k)$ in the long-wave limit. (1): acoustic phonons, (2): optical phonons.

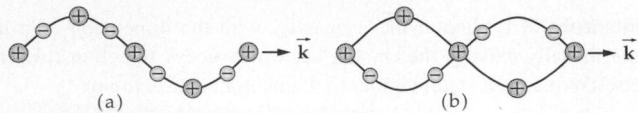

Figure 28.49: Vibrational states of a transverse-phonon wave. (a): acoustic branch, (b): optical branch.

Optical phonons: If the primitive elementary cell contains $N > 1$ atoms then—besides the acoustic phonons—$3N - 1$ additional "optical" branches occur that correspond to opposite relative vibrations of the various atoms of the elementary cell (analogous to the opposite-phase vibrations of coupled oscillators). The natural frequencies of the optical phonons are higher than those of the acoustic ones.

■   In a two-atomic lattice (e.g., NaCl) the atoms vibrate against each other.

7. Equations of motion of elastic waves

Equation of motion for elastic waves in crystals with two atoms per elementary cell, assuming an interaction between next neighbors only (for propagation directions of waves coinciding with symmetry directions where the lattice planes each contain only one type of atom):

equation of motion, two atoms per elementary cell			$\mathbf{MLT^{-2}}$
	Symbol	Unit	Quantity
$M_1 \dfrac{d^2 u_{2i+1}}{dt^2} = C_1 \cdot (u_{2i+2} + u_{2i} - 2u_{2i+1})$	u_i	m	displacement of lattice plane i
$M_2 \dfrac{d^2 u_{2i}}{dt^2} = C_1 \cdot (u_{2i+1} + u_{2i-1} - 2u_{2i})$	C_1	kg s^{-2}	elastic constant
	M_1, M_2	kg	atomic masses

▲   The coupled system of differential equations has only then a solution if the following dispersion relation holds:

$$\omega^2 = C_1 \left(\frac{1}{M_1} + \frac{1}{M_2} \right) \pm C_1 \sqrt{ \left(\frac{1}{M_1} + \frac{1}{M_2} \right)^2 - \frac{4 \sin^2 (k \cdot a)}{M_1 \cdot M_2} } .$$

▲   For small k, i.e., for very long waves ($\lambda \gg a$):

$$\omega^2 \approx 2C_1 \left(\frac{1}{M_1} + \frac{1}{M_2} \right) \quad \textbf{optical branch,}$$

$$\omega^2 \approx \frac{2C_1}{M_1 + M_2} k^2 a^2 \quad \textbf{acoustic branch.}$$

8. Phonon velocity

Group velocity, $v_{gr} = \dfrac{d\omega}{d\vec{k}}$ of the elastic wave, velocity of the phonons.

For mono-atomic lattices (atomic mass M, lattice separation a), it follows from the dispersion relation that

$$v_{gr} = \sqrt{ \frac{C_1 a^2}{M} } \cos \frac{ka}{2} .$$

▲ **At the boundary of the Brillouin zone** ($ka = \pm\pi$), the **group velocity always vanishes**. These elastic waves are therefore standing waves.

▲ **Elastic constant** C_1 and **elasticity modulus** E are proportional to each other:

$$C_1 = a \cdot E, \quad a: \text{lattice constant.}$$

➤ In **ionic crystals**, optical phonons give rise to a strong electric polarization so that this type of vibration may be excited very efficiently by **photons**, i.e., by electromagnetic fields.

Gap, the frequency range between the acoustic and optical branches that is **not** included in the phonon spectrum. Crystals do not display natural vibrations in this frequency range, so that electromagnetic waves may propagate only with strong damping: the reflectance in this frequency range is therefore very high.

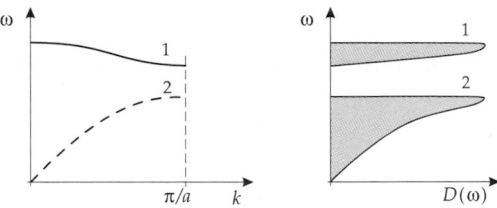

Figure 28.50: Schematic representation of the gap in the state density $D(\omega)$ of the phonon spectrum. (1): optical frequency, (2): acoustic frequency.

| M | The dispersion in prisms of ionic crystals is employed in infrared spectroscopy.

28.4.2   Phonons and specific heat capacity

According to classical mechanics, any vibrational lattice component of a solid has three translational degrees of freedom. An equivalent statement says that at finite temperature $T > 0$ phonons are excited in the lattice. The temperature dependence of the excitation of degrees of freedom manifests itself as a thermodynamically measurable quantity related to the specific heat $C(T)$.

Heat capacity, C_V, derivative of the internal energy with respect to the temperature at constant volume:

$$C_V = \left(\frac{\partial U}{\partial T}\right)_V.$$

Specific heat capacity, c_V, ratio of the heat capacity C_V to the mass m of the substance:

$$c_V = \frac{C_V}{m}.$$

Molar heat capacity, C_{mol}, ratio of the heat capacity C_V to the quantity of substance $n = m/M$, with M denoting the molar mass:

$$C_{mol} = \frac{C_V}{n}.$$

Dulong-Petit law: the molar heat capacity is a constant.

At room temperature, this law holds for nearly all solids.

Dulong-Petit law			$\mathbf{ML^2T^{-2}\Theta^{-1}}$
$C_{mol} = 3N_A k_B$	Symbol	Unit	Quantity
$= 24.9\dfrac{J}{mol \cdot K}$	C_{mol}	$J\,K^{-1}\,mol^{-1}$	molar heat capacity
	N_A	mol^{-1}	Avogadro constant
	k_B	$J\,K^{-1}$	Boltzmann constant

Low temperatures $(T \to 0)$: the specific heat capacity for insulators varies as T^3, and for metals as T as T goes to zero:

$$c_V \sim \begin{cases} T^3 & \text{insulators} \\ T & \text{metals} \end{cases} \quad \text{for} \quad T \to 0.$$

Bose-Einstein distribution, probability distribution $n(\omega, T)$, of finding a state of energy $\hbar\omega$ in thermal equilibrium at temperature T,

$$n(\omega, T) = \frac{1}{e^{\frac{\hbar\omega}{k_B T}} - 1}.$$

State density $D(\omega)$, distribution of the vibrational states over the range of frequencies. $D(\omega)d\omega$ is the number of natural vibrations in the frequency range between ω and $\omega + d\omega$.
Internal energy U **of the crystal**:

internal energy of a crystal with the state density $D(\omega)$			$\mathbf{ML^2T^{-2}}$
	Symbol	Unit	Quantity
	U	J	internal energy
	ω	rad/s	angular frequency of an oscillator
$U = \displaystyle\int_0^\infty \hbar\omega\, n(\omega, T) D(\omega) d\omega$	$D(\omega)$	s/rad	state density
	$n(\omega, T)$	1	Bose-Einstein distribution function
	T	K	temperature
	$\hbar$	J s	quantum of action

28.4.3 Einstein model

All N lattice atoms oscillate harmonically and isotropically, independent of each other, with the same angular frequency ω_E about their equilibrium positions.
State density in the Einstein model:

$$D(\omega) = N \cdot \delta(\omega - \omega_E).$$

Here, $\delta(\omega - \omega_E)$ is the delta function,

$$\delta(\omega - \omega_E) = \begin{cases} 0 & \text{for} \quad \omega \neq \omega_E \\ \to \infty & \text{for} \quad \omega = \omega_E \end{cases}, \quad \int_{-\infty}^{\infty} \delta(\omega - \omega_E)d\omega = 1.$$

internal energy of N oscillators in the Einstein model	ML^2T^{-2}		
	Symbol	Unit	Quantity
$U = \dfrac{f \cdot N\hbar\omega}{e^{\frac{\hbar\omega}{k_B T}} - 1}$	U	J	internal energy
	N	1	number of oscillators
	ω	rad/s	angular frequency of oscillator
	k_B	J K^{-1}	Boltzmann constant
	T	K	temperature
	f	1	number of degrees of freedom

Heat capacity:

$$C_V = f \cdot N k_B \left(\frac{\hbar\omega}{k_B T} \right)^2 \cdot \frac{e^{\frac{\hbar\omega}{k_B T}}}{\left(e^{\frac{\hbar\omega}{k_B T}} - 1 \right)^2} \cdot$$

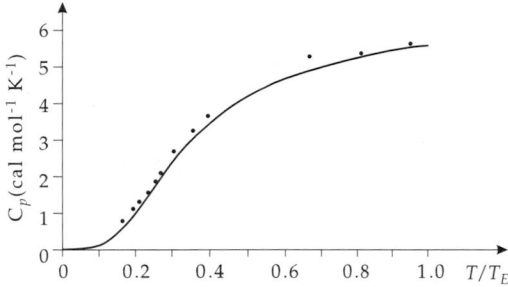

Figure 28.51: Comparison of the measured molar quantity of heat of diamond with the theoretical curve calculated in the Einstein model for a parameter value $T_E = \frac{\hbar\omega}{k_B} = 1320$ K.

In the limit of high temperature, the Einstein model yields the law of Dulong-Petit. At very low temperatures, it provides too low a value for C_V.

28.4.4 Debye model

Debye model, the state density increases as the square of ω up to the frequency limit ω_D. At this **Debye frequency** ω_D, the state density drops suddenly to zero.

state density in the Debye model	T		
	Symbol	Unit	Quantity
$D(\omega) = \begin{cases} \omega^2/\omega_D^3 & \text{for } \omega \leq \omega_D \\ 0 & \text{for } \omega > \omega_D \end{cases}$	$D(\omega)$	s rad^{-1}	state density
	ω	rad s^{-1}	angular frequency
	ω_D	rad s^{-1}	Debye frequency
$\omega_D^3 = 6\pi^2 v_s^3 N/V$,	v_s	m s^{-1}	sound velocity
$\omega = v_s \cdot k$	k	m^{-1}	wave number
	N	1	number of oscillators
	V	m^3	volume

The sound velocity v_s is a constant with $\omega = v_s \cdot k$. In the Debye model, the group velocities are replaced by the mean sound velocities.

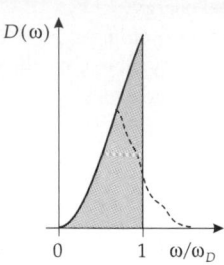

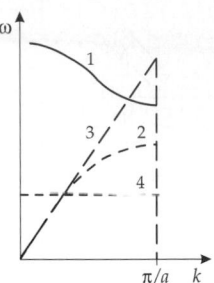

Figure 28.52: State density in the Debye model for a simple cubic lattice. Shadowed area: integration over the Debye sphere; dashed line: integration over the first Brillouin zone.

Figure 28.53: Dispersion in the Einstein and Debye models. 1 – optical branch, 2 – acoustic branch, 3 – Debye model, 4 – Einstein model.

Debye temperature, T_D, determined from the Debye frequency ω_D:

Debye temperature			Θ
	Symbol	Unit	Quantity
$T_D = \dfrac{\hbar \omega_D}{k_B} = \dfrac{\hbar v_s}{k_B} \cdot \left(\dfrac{6\pi^2 N}{V} \right)^{1/3}$	ω_D	rad s^{-1}	Debye frequency
	v_s	m s^{-1}	sound velocity
	N	1	number of oscillators
	V	m^3	volume
	k_B	J K^{-1}	Boltzmann constant
	$\hbar$	J s	quantum of action
	T_D	K	Debye temperature

N: total number of particles in the volume V.

Internal energy for very low temperatures $T \ll T_D$ in any direction of lattice:

internal energy in the Debye model			$\mathbf{ML^2T^{-2}}$
	Symbol	Unit	Quantity
$U = \dfrac{3}{5}\pi^4 N k_B T \left(\dfrac{T}{T_D} \right)^3$	U	J	internal energy
	N	1	number of oscillators
	k_B	J K^{-1}	Boltzmann constant
	T	K	temperature
	T_D	K	Debye temperature

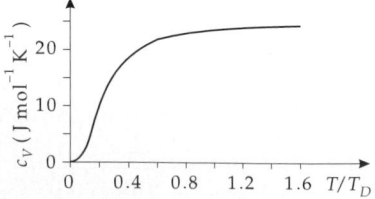

Figure 28.54: Specific heat capacity c_V of a solid according to the Debye model. The T^3-law corresponds to the range $T/T_D < 0.1$.

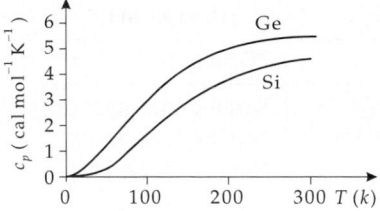

Figure 28.55: Specific heat capacity c_p of silicon and germanium.

Debye's T³-law for low temperatures $T \ll T_D$:

Debye's T³-law for $T \ll T_D$			$\mathbf{L^2T^{-2}\Theta^{-1}}$
	Symbol	Unit	Quantity
$C_V \approx \dfrac{12}{5}\pi^4 N \cdot k_B \left(\dfrac{T}{T_D}\right)^3$	C_V	J K^{-1} kg^{-1}	heat capacity
	N	1	number of oscillators
	k_B	J K^{-1}	Boltzmann constant
	T	K	temperature
	T_D	K	Debye temperature

28.4.5 Heat conduction

1. Insulators

Heat conduction in insulators, energy transport mediated by the motion of phonons in a solid.

    Free-phonon gas, model according to which the phonons move freely and independently of each other, like the molecules of a gas.

▲ Phonons propagate in a solid with velocity of sound. The heat transport mediated by them proceeds much slower, however, since the phonons collide with each other and with impurities, thereby permanently changing their direction of motion.

Mean free path of phonons, Λ_{ph}, the average distance traveled by a phonon between two collisions.

▲ The heat conduction in an insulator may be modeled by the phonon gas.

heat conductivity λ in insulators			$\mathbf{MLT^{-3}\Theta^{-1}}$
	Symbol	Unit	Quantity
$\lambda = \dfrac{1}{3}v\Lambda_{Ph}C_{Ph}\rho_{Ph}$	λ	W/(m K)	heat conductivity
	v	m/s	mean phonon velocity
	Λ_{Ph}	m	mean free path of phonon
	C_{Ph}	J K^{-1}	heat capacitance of phonon gas
	ρ_{Ph}	m^{-3}	phonon density

➤ The mean group velocity and the specific heat capacitance may be estimated with the Debye model. The mean free path cannot be derived from the Debye model, since it would yield an infinitely large mean free path.

▲ At low temperature, the mean free path is determined essentially by the scattering of phonons on lattice defects.

Heat flow density, j_q, the heat transported per unit area and unit time caused by a temperature difference.

heat flow density j_q in insulators			MT^{-3}
	Symbol	Unit	Quantity
$j_q = \lambda \cdot \dfrac{\Delta T}{\Delta x}$	j_q	$W\,m^{-2}$	heat flow density
	λ	$W/(m\,K)$	heat conductivity
	$\Delta T/\Delta x$	$K\,m^{-1}$	temperature gradient

➤ Heat conduction is a non-stationary process. A very small volume element may, however, be considered to be in thermodynamic equilibrium.

2. Metals

Heat conduction in metals, differs from heat conduction in insulators by the additional heat transport due to the free electrons.

electronic heat conductivity λ_{el} in metals			$MLT^{-3}\Theta^{-1}$
	Symbol	Unit	Quantity
$\lambda_{el} = \dfrac{1}{3} v_{el} \Lambda_{el} C_{el} \rho_{el}$	λ_{el}	$W/(m\,K)$	heat conductivity of electrons
	v_{el}	m/s	mean velocity of electrons
	Λ_{el}	m	mean free path of electrons
	C_{el}	J/K	heat capacitance of electron gas
	ρ_{el}	m^{-3}	density of electron gas

➤ The heat capacity of the electron gas is significantly lower than the heat capacity of the phonon system. On the contrary, the mean velocity of electrons is much higher than the mean group velocity (sound velocity) of phonons. The mean free path of electrons also exceeds the mean free path of phonons.

▲ In metals, heat is mainly transported by the electron gas.

Wiedemann-Franz law: The heat conductivity of metals is directly proportional to the electric conductivity κ.

Wiedemann-Franz law			$MLT^{-3}\Theta^{-1}$
	Symbol	Unit	Quantity
$\lambda_{el} = \dfrac{\pi^2}{3}\left(\dfrac{k_B}{e}\right)^2 T\kappa$	λ	$W\,m^{-1}\,K^{-1}$	heat conductivity
	k_B	$J\,K^{-1}$	Boltzmann constant
	e	C	elementary charge
	κ	$\Omega^{-1}\,m^{-1}$	electric conductivity
	T	K	temperature

28.5 Electrons in solids

Electrical conductivity, κ, of a metal, ratio of current density and electric field strength. It is inversely proportional to the specific electrical resistivity ρ,

$$\kappa = \frac{1}{\rho}.$$

➤ The SI unit of electric conductivity is $(\Omega\,m)^{-1}$.

■ The specific electrical resistivity ρ of solids varies from $10^{-8}\,\Omega\,m$ to $10^{13}\,\Omega\,m$.

Classification of substances according to their specific electrical resistivity:

- **conductors:** $\rho < 10^{-5}\,\Omega\,m \Longleftrightarrow \kappa > 10^5(\Omega\,m)^{-1}$ (z.B. Cu $5.88\cdot10^7$, Ag $6.21\cdot10^7$, Au $4.55\cdot10^7$)
- **semiconductors:** $10^{-5}\,\Omega\,m < \rho < 10^7\,\Omega\,m \Longleftrightarrow 10^{-7}(\Omega\,m)^{-1} < \kappa < 10^5(\Omega\,m)^{-1}$
- **insulators:** $\rho > 10^7\,\Omega\,m \Longleftrightarrow \kappa < 10^{-7}(\Omega\,m)^{-1}$

28.5.1 Free-electron gas

Ideal **Fermi gas**, many-body state of free, non-interacting particles that obey the Pauli principle.

1. Eigenfunctions and eigenvalues of free electrons

The **wave function of the free electron** in the stationary state is a **plane wave**:

$$\varphi = \frac{1}{\sqrt{2\pi}}e^{j\vec{k}\vec{r}} \quad \text{normalization to } \delta\text{-function}.$$

Since the electrons are confined in the solid, their probability density at the boundary must vanish. If the solid is approximated by a cube of edge length L with periodic boundary conditions, the components of the wave number vector along the cube edges are integer multiples of $2\pi/L$:

components of wave number vector			L^{-1}
$k_x = \dfrac{2\pi}{L}\cdot n_x,$	Symbol	Unit	Quantity
	k_x, k_y, k_z	m^{-1}	components of wave number vector
$k_y = \dfrac{2\pi}{L}\cdot n_y,$	n_x, n_y, n_z	1	integer numbers
	L	m	edge length of normalization volume
$k_z = \dfrac{2\pi}{L}\cdot n_z$			

▲ Free electrons in a solid may take only discrete energy values:

energy values of free electrons in a solid			ML^2T^{-2}
$E = \dfrac{\hbar^2}{2m}\cdot\vec{k}^2$	Symbol	Unit	Quantity
	E	J	energy of electron
	m	kg	electron mass
$= \dfrac{2\pi^2\hbar^2}{mL^2}(n_x^2 + n_y^2 + n_z^2)$	L	m	edge length of cube
	n_x, n_y, n_z	1	integer numbers

▲ The Pauli principle prevents all electrons from occupying the lowest energy state ($n_x = n_y = n_z = 1$). Each energy state may be occupied by at most two electrons with opposite spins.

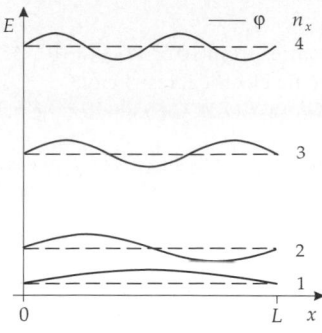

Figure 28.56: Energy levels
(- - -) and wave functions
(φ) of an electron gas in a
cube of edge length L.

2. Characteristics of a Fermi gas

Position space, configuration space, a space spanned by the position vectors $\vec{r}$. A point in the position space has Cartesian coordinates (x, y, z).

Momentum space, a space spanned by the momentum vectors $\vec{p}$. A point in the momentum space has Cartesian coordinates p_x, p_y, p_z.

k-space, a space spanned by the wave vectors $\vec{k}$. A point in the k-space has Cartesian coordinates k_x, k_y, k_z.

A particle with the momentum $\vec{p} = \hbar\vec{k}$ has coordinates $(k_x, k_y, k_z) = \hbar^{-1}(p_x, p_y, p_z)$ in k-space.

Ground state, the state with lowest energy. The ground state of an N-particle system is constructed by successively putting the particles into the lowest possible one-particle state—beginning with the lowest one—until all N particles are placed.

Fermi level, the highest occupied energy level in the ground state of a system of fermions.

Fermi sphere, volume in momentum space occupied by electrons of a non-interacting electron gas (Fermi gas) in the ground state.

Fermi momentum, p_F, radius of the Fermi sphere. The Fermi momentum is the maximum magnitude of a particle of mass m in a Fermi gas, $p_F = \hbar k_F = \sqrt{2mE_F}$.

Fermi velocity, v_F, velocity of the particles (electrons) of mass m at the surface of the Fermi sphere:

$$v_F = \hbar k_F / m \, .$$

Fermi energy, E_F, energy of the Fermi level, surface of the Fermi sphere.

relation between Fermi energy and momentum			$\mathbf{ML^2T^{-2}}$
	Symbol	Unit	Quantity
	E_F	J	Fermi energy
$E_F = \dfrac{p_F^2}{2m} = \dfrac{\hbar^2 k_F^2}{2m}$	p_F	kg m/s	Fermi momentum
	k_F	m^{-1}	Fermi wave number
	m	kg	mass of particle
	$\hbar$	J s	quantum of action/(2π)

An electron gas is in the ground state only for $T = 0$. For finite temperature, some of the electrons will attain a momentum above $\hbar k_F$ due to the thermal energy and will leave the Fermi sphere: the surface of the Fermi sphere becomes diffuse.

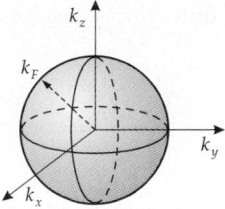

Figure 28.57: Fermi sphere.

3. Electron number density in a Fermi gas

Elementary volume in k-space:

$$V_k = \left(\frac{2\pi}{L}\right)^3 .$$

▲ Only two electrons with opposite spin may be placed in the elementary volume. For a three-dimensional electron gas, the Fermi sphere has a volume

$$V_F = \frac{4\pi}{3} k_F^3 .$$

Number of particles in the Fermi sphere of radius k_F,

$$N = 2 \cdot \frac{V_F}{V_k} = \frac{L^3}{3\pi^2} k_F^3 = \frac{V k_F^3}{3\pi^2} ,$$

where the factor 2 accounts for the spin. $V = L^3$ is the volume in position space.

Fermi wave number and Fermi energy of an N-electron system			
$k_F = \left(\dfrac{3\pi^2 N}{L^3}\right)^{1/3}$	Symbol	Unit	Quantity
	k_F	m^{-1}	Fermi wave number
	E_F	J	Fermi energy
$E_F = \dfrac{\hbar^2}{2m}\left(\dfrac{3\pi^2 N}{L^3}\right)^{2/3}$	L	m	width of potential well
	m	kg	electron mass
	N	1	number of electrons

▲ The electron number density n determines the position of the Fermi level, i.e., the magnitude of the Fermi momentum,

$$n = \frac{N}{L^3} = \frac{N}{V} .$$

The Fermi momentum increases if, for a constant particle number N, the volume V confining the Fermi gas is reduced.

4. Experimental determination of the electron number density

Electron number densities are determined experimentally by means of the **Hall effect**. A current of density $j_x = n \cdot e \cdot v_x$ flows in x-direction through a conducting slab of width b

and thickness d; n denotes the electron density, v_x the drift velocity, and e the elementary charge.

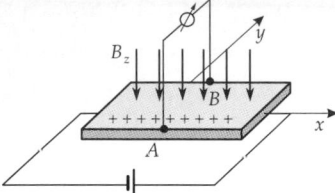

Figure 28.58: Hall effect.

The electrons moving in the magnetic field $\vec{B}_z$ perpendicular to the conductor plane are affected by the Lorentz force,

$$F_L = -e \cdot v_x \cdot B_z .$$

This force displaces the electrons perpendicular to the original current direction $\vec{e}_x$ and perpendicular to the orientation of the transverse magnetic flux density. A potential difference arises between the points A and B (**Hall voltage**):

$$V_H = B_z v_x b = \frac{1}{n \cdot e} j_x B_z b = R_H j_x B_z b .$$

Hall coefficient $R_H = \dfrac{1}{n \cdot e}$ (see **Tab. 29.7/1**).

5. Quantum Hall effect,

at very low temperatures (liquid helium, $T \approx 4\,\mathrm{K}$) and very strong magnetic fields (superconducting coil), the Hall resistance $R_{\mathrm{H}} = V_H / I_x$ of an extremely thin ("two dimensional") layer is quantized in a way related to the quantum of action h and the elementary charge e via

$$R_{\mathrm{H}} = \frac{h}{e^2} = 25812.807 \ \Omega .$$

When varying the magnetic field or the current only, the Hall resistances are measured,

$$R_{\mathrm{Hall}} = \frac{1}{n} \frac{h}{e^2} , \quad n \text{ integer.}$$

This effect was observed for the first time in 1977 by **Klaus von Klitzing** in studies of the Hall effect in silicon field-effect transistors (Nobel Prize, 1985).

➤   Because of the high precision in the determination of R_{Hall}, the quantized Hall effect serves as definition of a **standard resistance**.

| M | The fine-structure constant α may be measured with very high precision via the quantized Hall effect:

$$\alpha = \frac{1}{2\varepsilon_0 c} \frac{e^2}{h} = \frac{1}{2\varepsilon_0 c} / R_{\mathrm{Hall}} .$$

6. Table of several parameters of the Fermi level of various metals

	Alkali metals			Transition metals		
	Li	Na	K	Cu	Ag	Au
electron concentration n in 10^{22} cm^{-3}	4.6	2.5	1.34	8.5	5.76	5.9
Fermi energy E_F in eV	4.7	3.1	2.1	7.0	5.5	5.5
Fermi wave number k_F in 10^{10} m^{-1}	1.1	0.9	0.73	1.35	1.19	1.20
Fermi velocity v_F in 10^6 m/s	1.3	1.1	0.85	1.56	1.38	1.39

7. State density in Fermi systems

State density, $D(E)$, the number of energy states per unit volume and energy interval dE.

state density per unit of volume and energy			$\mathbf{M^{-1}L^{-5}T^2}$
	Symbol	Unit	Quantity
$D(E) = \dfrac{1}{V}\dfrac{dN}{dE}$	$D(E)$	m^{-3} J^{-1}	state density
	dE	J	considered energy interval
	dN	1	number of states in the energy interval dE
	V	m^3	volume

state density in the ground state for $T = 0$			$\mathbf{M^{-1}L^{-5}T^2}$
	Symbol	Unit	Quantity
$D_0(E) = \dfrac{1}{2\pi^2}\left(\dfrac{2m}{\hbar^2}\right)^{3/2}\cdot\sqrt{E}$	$D_0(E)$	m^{-3} J^{-1}	state density for $T = 0$
	m	kg	electron mass
	$\hbar$	J s	quantum of action $/(2\pi)$
	E	J	energy of electron gas

8. Fermi–Dirac distribution function,

$f(E, T)$, the probability distribution in a free electron gas of temperature T for occupation of a quantum state with the energy E,

$$f(E, T) = \frac{1}{e^{\frac{E - E_F}{k_B T}} + 1}.$$

▲ For $T > 0$, the state density D_0 has to be multiplied with the Fermi–Dirac distribution $f(E, T)$ to obtain the state density $D(E, T)$ (**Fig. 28.59**).

state density for $T > 0$			$\mathbf{M^{-1}L^{-5}T^2}$
	Symbol	Unit	Quantity
$D(E,T) = f(E,T)D_0(E)$	$D(E,T)$	$m^{-3}\,J^{-1}$	state density for $T > 0$
	$D_0(E)$	$m^{-3}\,J^{-1}$	state density for $T = 0$
$= \dfrac{1}{2\pi^2}\left(\dfrac{2m}{\hbar^2}\right)^{3/2}$	$f(E,T)$	1	Fermi–Dirac distribution
	m	kg	electron mass
	$\hbar$	J s	quantum of action$/(2\pi)$
$\cdot\; \dfrac{\sqrt{E}}{e^{\frac{E-E_F}{k_B T}} + 1}$	k_B	$J\,K^{-1}$	Boltzmann constant
	T	K	temperature
	E_F	J	Fermi energy
	E	J	energy of electron

➤ When increasing the temperature from 0 to T, electrons from below the Fermi energy are thermally excited to above the Fermi energy. In a solid, the electrons in the vicinity of the Fermi surface may receive energy from the phonons.

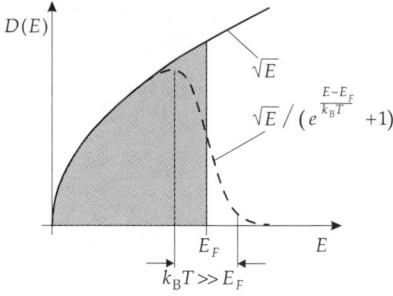

Figure 28.59: State density D of a Fermi gas as a function of the energy E. Dashed line: density of the occupied states for a finite temperature T ($k_B T \ll E_F$). Shadowed area: density of the occupied states for $T = 0$.

9. Fermi temperature and heat capacity

Fermi temperature, T_F, corresponding to the Fermi energy E_F:

$$T_F = E_F/k_B.$$

➤ The Fermi temperature T_F is not the physical temperature of the system, but rather a quantity that compares the Fermi energy with the temperature.

▲ Only electrons at the surface of the Fermi sphere are mobile and contribute to the specific heat. They correspond to a fraction T/T_F of all electrons.

Heat capacity of the electron gas, C_e, depends linearly on the temperature.

internal energy and heat capacity of an electron gas			
	Symbol	Unit	Quantity
$U \approx N(k_B T)\dfrac{T}{T_F}$	U	J	internal energy
	C_e	$J\,K^{-1}$	heat capacity of electron gas
	N	1	number of electrons
$C_e \approx 2k_B N \dfrac{T}{T_F}$	T	K	temperature
	T_F	K	Fermi temperature
	k_B	$J\,K^{-1}$	Boltzmann constant

28.5.2 Band model

1. Bloch theorem and the model of almost free electrons

Bloch theorem: The solutions to the Schrödinger equation $\psi_k(\vec{r})$ for a **periodic potential** $V(\vec{r}) = V(\vec{r} + \vec{T})$ are always of the form:

Bloch function			$\mathbf{L^{-3/2}}$
$\psi_k(\vec{r}) = u_k(\vec{r})\,e^{j\vec{k}\vec{r}}$	Symbol	Unit	Quantity
	$\psi_k(\vec{r})$	$m^{-3/2}$	state function
$u_k(\vec{r} + \vec{T}) = u_k(\vec{r})$	$u_k(\vec{r})$	$m^{-3/2}$	periodic function
	$\vec{r}$	m	position vector
	$\vec{k}$	m^{-1}	wave vector

▲ $\vec{T}$ is a **fundamental translation vector** (see p. 969) in the crystal lattice.
Kronig–Penney model, a δ-potential is assumed at the positions of the atoms.
▲   **Energy gaps** occur in the Kronig–Penney model.
   Almost-free electrons, a model for describing conduction mechanisms in metals based on the assumption that the electrons are only weakly disturbed by the periodic lattice potential, but may be scattered at the lattice sites according to the Bragg condition.

2. Bragg reflection condition and standing electron waves

Bragg reflection condition, condition for the reflection of a wave by a crystal lattice. Given wavelengths may be reflected only at certain glancing angles θ (incidence angles $\pi/2 - \theta$).

Bragg reflection condition			L
$2a\sin\theta = n\lambda$	Symbol	Unit	Quantity
	a	m	lattice constant
	θ	rad	glancing angle
	λ	m	wavelength
	n	1	integer number

Bragg condition in one dimension			
$\lambda_n = \dfrac{2a}{n}$	Symbol	Unit	Quantity
	a	m	lattice constant
with $\quad k_n = \pm\dfrac{2\pi}{\lambda_n} = \pm\dfrac{n\pi}{a}$	λ_n	m	wavelength
	k_n	m^{-1}	wave number
	n	1	integer number

Standing electron waves in the crystal, generated by constructive interference of electron waves scattered at the lattice sites.
   If Bragg reflection occurs, standing waves are formed ($n = 1$):

$$\psi(+) = e^{jk_1 x} + e^{-jk_1 x} = 2\cos\left(\frac{\pi x}{a}\right),$$

$$\psi(-) = e^{jk_1 x} - e^{-jk_1 x} = 2j\sin\left(\frac{\pi x}{a}\right).$$

probability of presence of standing electron waves			
$\rho(+) = |\psi(+)|^2 \sim \cos^2 \dfrac{\pi x}{a}$ $\rho(-) = |\psi(-)|^2 \sim \sin^2 \dfrac{\pi x}{a}$	Symbol	Unit	Quantity
	$\rho(+), \rho(-)$	m^{-3}	probability densities
	x	m	position
	a	m	lattice constant

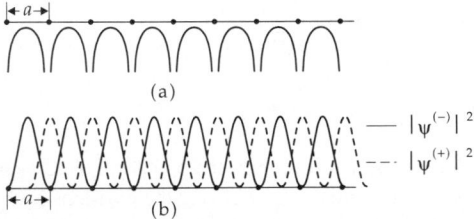

(a)

$$— \; |\psi^{(-)}|^2$$
$$--- \; |\psi^{(+)}|^2$$

(b)

Figure 28.60: Schematic representation of the potential energy (a) and the probability densities of standing waves (b).

Depending on the type of interference, the electrons may mainly be found:
- close to the atomic centers ($x = 0, a, 2a, \ldots$, maxima of $\rho(+)$), or
- removed from the atomic centers ($x = a/2, 3a/2, \ldots$, maxima of $\rho(-)$).

The two states have *different* energies.

➤ The expectation value of the potential energy of a traveling wave not obeying the Bragg condition is larger than that in the state $\psi(+)$, but smaller than that in the state $\psi(-)$. According to the model, energies between these levels may not occur for traveling waves.

3. Energy bands and energy gaps

Energy band, synonym for a limited but continuous energy range.

Energy gap, E_g, forbidden energy interval between allowed energy bands.

If the Fermi energy falls within an allowed energy band, then at $T > 0$ electrons may occupy higher energy states *without* crossing an energy barrier, i.e., even at very low temperature. If the Fermi edge falls within a forbidden band, then the electrons need at least the gap energy (energy barrier) in order to change to an excited state.

- **Valence band**, allowed energy band in which all electron states are occupied at $T = 0$.
- **Conduction band**, allowed energy band of energy higher than that of the valence band.
- ▲ Electrons in the conduction band contribute to electric conduction.
- ▲ In the ground state ($T = 0$), the conduction band is *not* fully occupied.

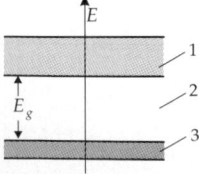

Figure 28.61: Band scheme with valence band, conduction band and energy gap. 1 – conduction band empty, 2 – energy gap E_g, 3 – valence band occupied.

4. Metals, insulators and semiconductors

Metals, substances with the Fermi energy about in the middle of an allowed band. The energy band is therefore not fully occupied, and hence is a conduction band. There are nearly as many unoccupied states as occupied states, so many electrons may move in the conduction band even at low temperature.

Insulators, dielectrics, substances for which the Fermi energy falls in the forbidden range between two bands. The thermal energy is not sufficiently high to lift enough electrons from the fully occupied valence band into the empty conduction band.

Semi-metals, poorly conducting metals for which the Fermi energy lies close to the top, or at the bottom, of an allowed band. If the Fermi level lies near the bottom of the band, then only few electrons are available to take energy from the electric field and to participate in the process of conduction. On the other hand, if the Fermi energy lies close to the upper edge of the band, then a sufficient number of electrons is available, but the number of allowed free states is low.

Semiconductors, have a narrow forbidden range ($E_g \approx 1\text{eV}$) within which the Fermi energy falls. Electrons from the fully occupied valence band may overcome the small-energy gap and reach the free-conduction band by thermal excitations at temperatures $T > 0$.

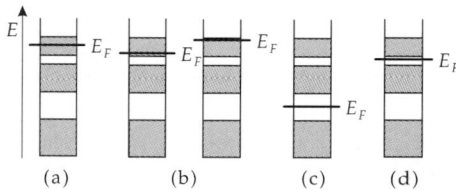

Figure 28.62: Band scheme for various substances. (a): metal, (b): semi-metal, (c): insulator, (d): semiconductor.

5. Fermi energy and optical properties

The **optical properties of solids** are sensitively determined by the position of the Fermi energy. Visible light covers the energy range 1.6 eV $< E <$ 3.2 eV. The gap between bands in dielectrics (insulators) amounts to about 4 eV. The energy of the visible light is not sufficient to lift electrons from the valence band into the next higher band.

▲ All ideal **dielectrics** are transparent in the visible spectrum of light. The impermeability of many dielectric minerals is connected to their impurity.

▲ **Metals** contain sufficiently many free electrons and free allowed energy states in order to absorb light quanta. Therefore, metals are **opaque for light**. On the other hand, an electron may lose energy by creating a photon of corresponding energy. Both processes have equal probabilities. Therefore, metals are good reflectors.

   A prerequisite for high **reflectance** and **absorptance** is a clean surface. Oxidation frequently leads to formation of dielectric surface layers.

■ Ordinary mirrors reflect light by a metallic layer (e.g., silver) evaporated behind the glass.

▲ Semiconductors with band gaps of 1 eV may absorb light quanta. An electron may overcome the energy gap between valence band and conduction band at the cost of the energy of an absorbed photon (**photo current**).

6. Occupation numbers and equation of motion

Occupation number, the number of electrons occupying an energy band. For isolated atoms, the occupation number of energy states that are classified by the principal quantum number n and the orbital angular momentum quantum number l is given by $2(2l + 1)$.

▲   Energy bands are described by the same quantum numbers as in the isolated atom.

■   The lithium atom has three electrons. Two electrons occupy the energetically lowest level (1s-level), which is thereby completely filled. The excess electron populates the 2s-state at slightly higher energy. If lithium atoms form a crystal, there arises a localized core state of 1s-type and an energy band of 2s-type above it. Every lithium atom contributes two electrons to the 1s-core state, which then is fully occupied. The third electron populates the 2s-band. This band is only half-filled. So a lithium crystal is a metal.

The other alkali metals Na, K, Rb, Cs and Fr behave analogously.

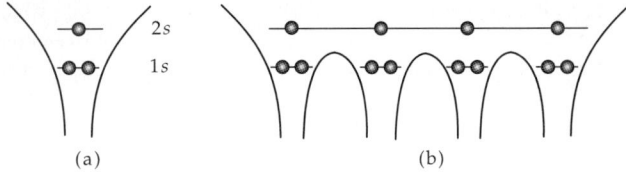

Figure 28.63: (a): Energy levels in the Li atom, (b): energy band (2s) and localized 1s-core states in a Li crystal.

Equation of motion of an electron in a solid under the influence of the forces of the crystal lattice:

equation of motion of an electron			$\mathbf{MLT^{-2}}$
	Symbol	Unit	Quantity
$\hbar \dfrac{d\vec{k}}{dt} = m^* \cdot \dfrac{d\vec{v}_{gr}}{dt}$	F	kg m s^{-2}	force
	k	m^{-1}	wave number of electron
$= \vec{F}$	m^*	kg	effective electron mass
	v_{gr}	m/s	group velocity of electron wave
$v_{gr} = \dfrac{1}{\hbar} \cdot \dfrac{d\varepsilon}{dk}$	$\varepsilon(k)$	J	dispersion of electron
	$\hbar$	J s	quantum of action/(2π)

Effective mass, m^*, takes into account the dependence of the electron energy on the wave number (dispersion).

effective electron mass in a solid			$\mathbf{M}$
	Symbol	Unit	Quantity
$m^* = \dfrac{\hbar^2}{\dfrac{d^2\varepsilon}{dk^2}}$	m^*	kg	effective mass
	$\hbar$	J s	quantum of action/(2π)
	ε	J	electron energy
	k	m^{-1}	wave number

▲   Narrow energy bands correspond to a large effective mass.

■ Na: In sodium the 3s-band is half-filled. The motion of the electrons is almost free:

$$\frac{m^*}{m} \approx 1 .$$

Fe, Co, Pt: 3d-transition metals. Here, the 4s-band is filled first.
All s-bands are very narrow, i.e., m^* is large:

$$\frac{m^*}{m} \approx 10 .$$

28.6 Semiconductors

Semiconductors, dielectric with a small gap distance (energy gap between the conduction band and the valence band).

Elemental semiconductors, elements of the IV-th group of the Periodic Table with four valence electrons.

■ Element semiconductors: C, Si, Ge, Sn (for their properties see **Tab. 29.9/1**).

Compound semiconductor, chemical compound with the properties of a semiconductor (see **Tab. 29.9/2**).

Intrinsic conduction of a semiconductor, arises when electrons from the valence band reach the empty conduction band by thermal excitation, or by incident light.

Defect electrons, **holes**, electrons missing from complete occupation of the valence band. The holes behave like positively charged particles in a sea of electrons.

▲ In intrinsically conducting semiconductors, free electrons and holes are always produced pairwise.

1. Electron density and conductance in semiconductors

density of free electrons = density of holes			$\mathbf{L^{-3}}$
	Symbol	Unit	Quantity
$n = p$	n	m^{-3}	density of free electrons
	p	m^{-3}	density of holes

The conductance κ is determined by the product of the mobility μ and the number of free charge carriers n, p.

conductivity of a semiconductor			$\mathbf{I^2 T^3 M^{-1} L^{-3}}$
	Symbol	Unit	Quantity
	κ	$\Omega^{-1}\,m^{-1}$	conductance
	e	C	elementary charge
$\kappa = e(\mu_n \cdot n + \mu_p \cdot p)$	μ_n	$m^2/(V\,s)$	mobility of electrons
	μ_p	$m^2/(V\,s)$	mobility of holes
	n	m^{-3}	density of free electrons
	p	m^{-3}	density of holes

electron density in the conduction band			L^{-3}
	Symbol	Unit	Quantity
	n	m^{-3}	density of free electrons
	E_L	J	bottom of the conduction band
$n = n_L \cdot e^{-\frac{E_L - E_F}{k_B T}}$	E_F	J	Fermi energy
	n_L	m^{-3}	effective electron density in the conduction band
	k_B	J K^{-1}	Boltzmann constant
	T	K	temperature

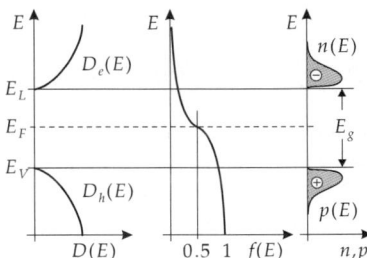

Figure 28.64: State density D, distribution function f and densities of charge carriers n, p of a semiconductor. E_V: top edge of valence band, E_L: bottom of conduction band, E_F: Fermi energy, E_g: energy gap.

density of holes in the valence band			L^{-3}
	Symbol	Unit	Quantity
	p	m^{-3}	density of holes
	E_V	J	top edge of valence band
$p = n_V \cdot e^{-\frac{E_F - E_V}{k_B T}}$	E_F	J	Fermi energy
	n_V	1	effective density of holes in the valence band
	k_B	J K^{-1}	Boltzmann constant
	T	K	temperature

➤ The mobilities of electrons μ_n and holes μ_p are strongly dependent on the semiconducting material.

▲ The electron mobilities of pure semiconductors are only weakly dependent on the temperature,

$$\mu(T) = \mu_0 \left(\frac{T}{T_0}\right)^{3/2}.$$

Intrinsic charge-carrier density, n_i, density of free charge carriers for intrinsically conducting semiconductors.

intrinsic charge-carrier density n_i			$\mathbf{L^{-3}}$
	Symbol	Unit	Quantity
$n_i = \sqrt{n_L n_V} \cdot e^{-\frac{E_g}{2k_B T}}$	n_i	$\mathrm{m^{-3}}$	intrinsic charge-carrier density
	n_L, n_V	$\mathrm{m^{-3}}$	effective state densities in conduction band and valence band
	E_g	J	energy gap
	T	K	temperature
	k_B	J/K	Boltzmann constant

➤ The intrinsic conductivity σ is very small. At room temperature,

$$k_B T \approx \frac{1}{40} \ \mathrm{eV} \,.$$

For an energy gap of $E_g \approx 1$ eV

$$\sigma \approx 10^{-8} \ \Omega^{-1} \ \mathrm{m^{-1}} \,.$$

M The resistance of a semiconductor $R(T)$ may be used as a temperature sensor for low temperatures according to the relation

$$R(T) \approx R_0 \cdot e^{\frac{-E_g}{2k_B T}} \,.$$

Here, R_0 is a material-dependent constant.

2. *Properties of important elemental semiconductors Ge, Si*

	Ge	Si
Data on crystal structure		
structure	diamond	diamond
lattice constant a	0.564613 nm	0.543095 nm
atomic density n	$4.42 \cdot 10^{22} \ \mathrm{cm^{-3}}$	$0.5 \cdot 10^{22} \ \mathrm{cm^{-3}}$
Electrical properties		
energy gap E_g	0.66 eV	1.11 eV
intrinsic carrier density n_i	$2.24 \cdot 10^{13} \ \mathrm{cm^{-3}}$	$1.14 \cdot 10^{10} \ \mathrm{cm^{-3}}$
relative permittivity constant ε_r	16	11.8
mobility μ_n	$3900 \ \mathrm{cm^2 \ V^{-1} \ s^{-1}}$	$1350 \ \mathrm{cm^2 \ V^{-1} \ s^{-1}}$
mobility μ_p	$1900 \ \mathrm{cm^2 \ V^{-1} \ s^{-1}}$	$480 \ \mathrm{cm^2 \ V^{-1} \ s^{-1}}$
Effective state density		
conduction band n_L	$1.04 \cdot 10^{19} \ \mathrm{cm^{-3}}$	$3.22 \cdot 10^{19} \ \mathrm{cm^{-3}}$
valence band n_V	$6.03 \cdot 10^{18} \ \mathrm{cm^{-3}}$	$1.83 \cdot 10^{19} \ \mathrm{cm^{-3}}$

28.6.1   Extrinsic conduction

Impurity atoms deposited in pure semiconductors modify the resisitivity appreciably.

➤   An addition of 1 ppm (= 10^{-6}) of impurity atoms may increase the conductivity by a factor of more than 100.

1. Donor,

impurity atom with a larger number of valence electrons than that of the atoms of the pure semiconductor lattice. The excess electrons are not needed for the lattice binding and may be separated from the atomic site without much energy expenditure.

▲   In terms of the band model, these electrons form localized levels just below the conduction band.

■   For elemental semiconductors of the IV-th group (e.g., Ge), the elements of the V-th group (e.g., P) are donors.

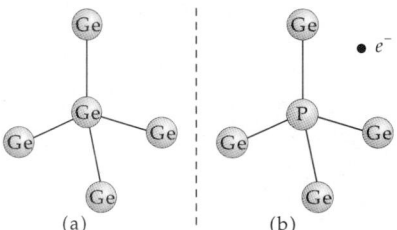

Figure 28.65: Doping of a germanium crystal with phosphorus atoms (schematic). (a): non-doped germanium crystal, (b): germanium crystal doped with phosphorus atoms.

Doping of a germanium crystal with phosphorus atoms: The non-saturated electron of the pentavalent phosphorus atom makes a bond with the positive ion, which leads to a hydrogen-like state. The binding energy of this system is only 0.01 eV for germanium, and 0.03 eV for silicon.

2. Acceptor,

impurity atom with fewer valence electrons than the lattice atoms. It offers a low-lying energy level in the crystal compound to another electron. Since in filling a vacancy another vacancy arises, i.e., the hole arises at another position, the phenomenon is called hole conduction.

▲   In terms of the band model, these electrons form **localized levels** just above the valence band.

■   For the elemental semiconductors of the IV-th group, the elements of the III-rd group are acceptors.

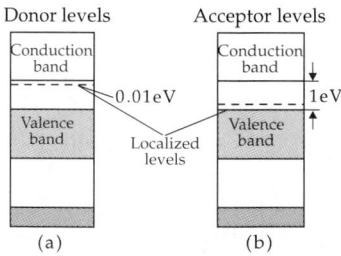

Figure 28.66: Band scheme with localized electron levels. (a): donor levels, (b): acceptor levels.

3. Doping of semiconductors

Doping, process in which impurity atoms (donors, acceptors) with a different number of valence electrons are implemented into a pure semiconductor lattice.

For localized levels in various semiconductors, see **Tab. 29.9/3 and 29.9/4**.

Majority-charge carriers, charge carriers participating predominantly in electrical conduction.

n-doping, doping with donors; electron conduction predominates.

p-doping, doping with acceptors; hole conduction predominates.

n-conducting semiconductor, semiconductor with $n > p$; electron conduction predominates.

p-conducting semiconductor, semiconductor with $p > n$; hole conduction predominates.

➤ Without an applied voltage, electrons diffuse from the n-region to the p-region where—despite charge neutrality—free lattice sites are nevertheless available. This is due to the electron excess in the n-region and the electron deficit in the p-region.

Space charge regions, at the interface a positive space charge forms in the n-region and a negative space charge in the p-region.

Junctions of p- and n-semiconducting regions: A p–n junction arises in a monocrystal containing two oppositely doped regions. The region with implanted acceptor atoms is p-conducting. In the region with implanted donor atoms, the electrons are majority-charge carrier.

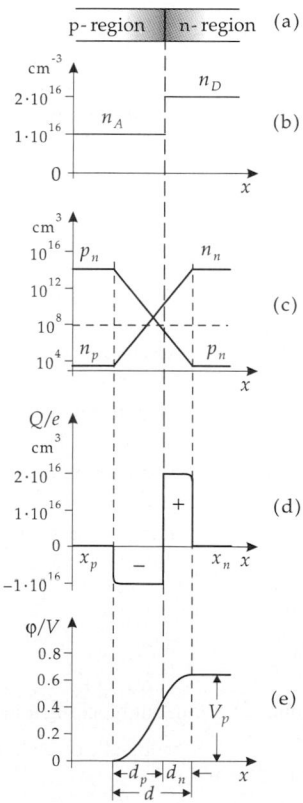

Figure 28.67: Properties of doped semiconductors. (a): p-n interface, (b): acceptor and donor concentration, (c): charge carrier density, (d): space charge regions of widths d_n (negative) and d_p (positive), (e): potential difference between the n- and p-region.

Widths of the negative and positive space charge regions d_n and d_p because of charge neutrality are given by:

widths of space charge regions			**L**
	Symbol	Unit	Quantity
$d_n \cdot n_D = d_p \cdot n_A$	d_n, d_p	m	width of the negative and positive space charge region
	n_D, n_A	m^{-3}	majority-charge carrier density

➤ Similar to parallel plate capacitors, the space charge regions generate a potential gradient, the diffusion voltage.

Diffusion voltage, V_D, the potential difference between the n- and p-conducting regions:

diffusion voltage across the pn-junction			$\mathbf{L^2 T^{-3} M I^{-1}}$
	Symbol	Unit	Quantity
	V_D	V	diffusion voltage
	n_A	m^{-3}	acceptor concentration
$V_D = \dfrac{k_B T}{e} \ln \dfrac{n_A n_D}{n_i^2}$	n_D	m^{-3}	donor concentration
	n_i	m^{-3}	intrinsic charge carrier density
	e	C	elementary charge
	k_B	$J\,K^{-1}$	Boltzmann constant
	T	K	temperature

The width of the space charge region is given by:

width of the space charge region			**L**
	Symbol	Unit	Quantity
	d	m	width of space charge region
	ε_r	1	relative permittivity constant
$d = \sqrt{\dfrac{2\varepsilon_r \varepsilon_0 V_D}{e} \cdot \dfrac{n_A + n_D}{n_A \cdot n_D}}$	ε_0	C/(V m)	permittivity constant of free space
	V_D	V	diffusion voltage
	n_A	m^{-3}	acceptor concentration
	n_D	m^{-3}	donor concentration
	e	C	elementary charge

28.6.2 Semiconductor diode

Diode, a circuit element conducting the current in one direction, but blocking it in the other direction.

Semiconductor diode, a circuit element with a pn-junction.

1. Main characteristics of semiconductor diodes

Anode, the electrode at the p-region of the diode.

Cathode, the electrode at the n-region of the diode.

Reverse voltage, V_{Sp}, negative voltage across the p- and the n-regions, causes a broadening of the space charge region: the charge carriers are pushed out by the electric field,

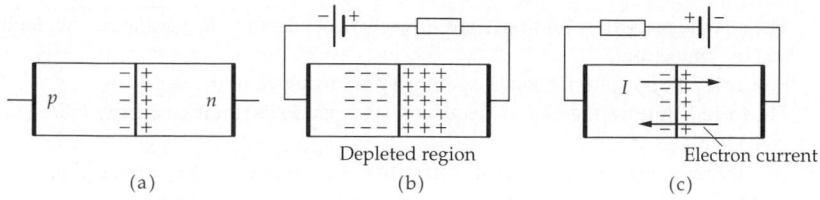

Figure 28.68: pn-junction, external voltage (a): zero, (b): negative (reverse direction), (c): positive (forward direction).

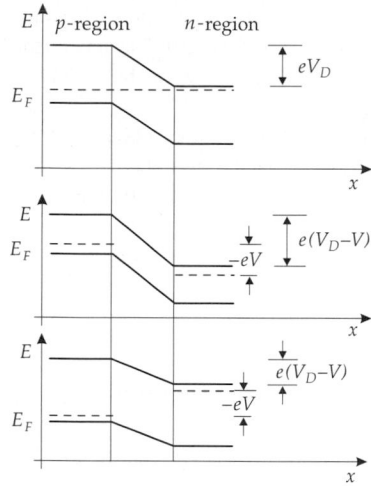

Figure 28.69: Energy levels at the pn-junction according to the band model.

and the current is interrupted to a large extent; the space charge region acts as a **depletion layer**.

Avalanche breakdown, a steep increase of the diode current at maximum negative voltage, usually far above 6 V.

Zener effect, similar to the avalanche breakdown, but causing a rapid increase at a lower voltage (below 6 V).

Breakdown voltage, **Zener voltage**, V_Z, negative voltage at which the avalanche breakdown or Zener breakdown sets in.

➤ When the breakdown voltage is exceeded, the component may be destroyed.

A positive voltage between the p- and n-regions enhances the diffusion process from the n-region to the p-region: the electrons are accelerated by the electric field against the field direction. The current increases exponentially with the voltage (**Shockley diode formula**).

Shockley diode formula			**I**
	Symbol	Unit	Quantity
$I = I_{\mathrm{Sp}}\left(e^{V/V_T} - 1\right)$	I_{Sp}	A	reverse current
	E_g	eV	energy gap
	V_T	V	temperature voltage
$I_{\mathrm{Sp}} \sim e^{-E_g/k_{\mathrm{B}}T}$	V	V	pn-voltage
	I	A	current in pn-direction
$V_T = k_{\mathrm{B}}T/e$	k_{B}	J K^{-1}	Boltzmann constant
	T	K	temperature
	e	C	elementary charge

▲   The electric properties of diodes are strongly dependent on the geometry, the doping
    and the temperature.
➤   Properties of the material and the geometry are included in the factor I_{Sp}.
➤   The temperature voltage V_T is frequently set equal to the thermal energy $k_B T$ and is
    given in eV.

Reverse direction, the anode potential is **negative** with respect to the cathode potential.

Reverse current, I_{Sp}, the leakage current of a pn-junction operated in reverse direction.
The leakage current is caused by electrons from the p-region and holes from the n-region,
i.e., by minority-charge carriers driven through the depletion layer by the electric field.

Forward direction, anode potential **positive** with respect to cathode potential.

Threshold voltage, V_S, positive voltage; when exceeding it, the diode resistance becomes low, meaning it conducts current. V_S cannot be fixed exactly because of the steep
but smooth increase of current with increasing voltage. In practice, the transition from the
blocking to the conducting state may occur suddenly.

Reverse recovery time, τ, the time needed by a pn-junction after polarity changing of
the voltage in order to change from the blocking to the conducting state.

Characteristic curves, graphical description of the current versus voltage dependence
of a circuit element.

There are various types of diodes, distinguished mainly by the magnitude of doping of
the two regions. This affects both the values of characteristics as well as the characteristic
curves.

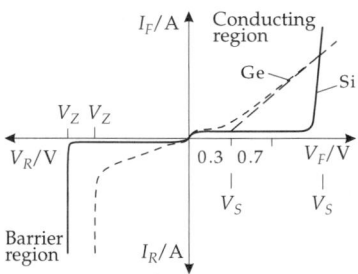

Figure 28.70: Characteristic curves of typical germanium and silicon diodes V_F: voltage in forward direction, V_R: voltage in reverse direction, V_Z: Zener voltage, V_S: threshold voltage, I_F: continuous forward current, I_R: reverse current.

➤   As a rule, the cathode of a diode is indicated by a ring on the component itself, and
    in a circuit the symbol for a diode has a vertical line at the cathode.

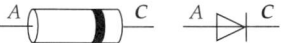   

Figure 28.71: Schematic drawing (left) of a diode and symbol (right) for a diode in a circuit diagram; A: anode, C: cathode.

Figure 28.72: Switching diode: circuit symbol and typical characteristic values V_S: low (Si: 0.7 V, Ge: 0.3 V), V_Z: 50...100 V, I_F: 50...200 mA, I_R: ≈ 1 nA, τ: 2...20 ns.

2. Switching diode,

fast diode. In the forward direction, the diode conducts with a low forward resistance; in
the reverse direction, it blocks the current down to a very low leakage value. Switching
diodes are produced very economically. Because of their versatility, they are also called
universal diodes.

| M | Universal diode for switching, limiting, decoupling, and for logic circuits.

3. Schottky diode,

very fast diode suitable for high frequencies. It does not have a pn-junction, but rather a metal-semiconductor junction, with the result that only majority carriers contribute to the current conduction. The Schottky diode responds very quickly to voltage variations so that currents may be switched reliably even in the GHz-range. Its characteristic curve is similar to that of a switching diode, but increases less steeply in the forward direction.

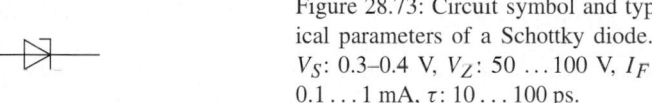

Figure 28.73: Circuit symbol and typical parameters of a Schottky diode. V_S: 0.3–0.4 V, V_Z: 50 … 100 V, I_F: 0.1 … 1 mA, τ: 10 … 100 ps.

■   Application in high-frequency circuits (up to about 40 GHz).

4. Rectifier diode,

allows a high power loss and current-pulse capability in contrast to switching diodes. Current-pulse capability is of particular importance in rectifier circuits directly connected to the power supply, since very high currents (> 10 A) may occur in the forward direction. Owing to the high voltages acting on line rectifiers, the reverse current should be very low because, otherwise, additional losses arise. The characteristic curve corresponds to that of a switching diode.

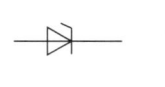

Figure 28.74: Circuit symbol and typical parameters of a rectifier diode. V_F: ≤ 1 V, V_Z: up to 500 V, I_R: ≈ 50 μA, τ: about μs (very short in high-frequency rectifiers).

■   **Bridge rectifier**:
     If V_E is positive, current flows via the diodes D_1 and D_2 through the load resistor R_L. The diodes D_3 and D_4 are non-conducting. During the next half-wave, D_4 and D_3 are conducting while D_1 and D_2 are non-conducting. The current through R_L has the same direction in both half-waves. The advantage of this circuit over rectifiers with only one diode is that the current flows through the load resistor even during the negative half-wave. The voltage level still strongly fluctuates. This fluctuation may be reduced by connecting a smoothing capacitor C parallel to R_L.

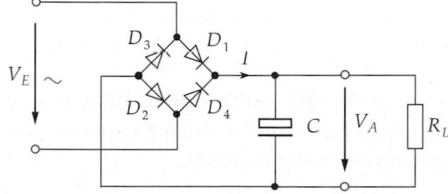

Figure 28.75: Circuit of a bridge rectifier.

5. Z-diode,

highly doped reversed-biased diode. It behaves as a switching diode does in the forward and reverse directions, but has a much lower Zener voltage V_Z, which is very precisely specified by type (by the high doping, the field strength in the interface becomes very high, causing additional breaking of electron-hole bonds, and hence additional charge carriers, which contribute to the current flow). Unlike switching diodes, the breakdown in the Z-diode is intended and does *not* result in a damage of the diode.

Figure 28.76: Circuit symbol
of a Z-diode.

■   Z-diodes are used for the limitation and stabilization of voltages.

6. DIAC trigger diode,

DIode **A**lternating **C**urrent switch. Unlike all other types of diodes, it consists of two pn-junctions and becomes conducting above a defined voltage.

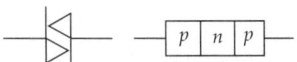

Figure 28.77: Circuit symbol
and layer structure of a
DIAC.

In principle, a DIAC represents a pair of diodes connected in series and reversed. When a voltage is applied, one diode is forward-biased, the other is reversed-biased. Thereby only a low residual current $I \le 100\ \mu$A flows as long as the voltage is not beyond the break-down voltage V_Z of a pn-junction. Then the DIAC suddenly becomes low-resistive, and the current steeply increases while the voltage is falling. If the applied voltage is lowered again, the DIAC becomes currentless when the voltage falls below a holding voltage V_H. Because of the symmetry of the layers, the polarity of the DIAC does not play any role.

Figure 28.78: Characteristic
curve of a DIAC.

DIACs are applied in situations in which short, well defined current pulses are required to trigger an (electronic) switch safely at a precisely defined voltage.

7. Photodiode,

varies its forward resistance depending on the luminous intensity incident on the diode. The photodiode is operated *reversed biased*.

➤   Photodiodes are operated in reverse direction below the breakdown voltage (low depleted-layer capacitance for short response times). The reverse current over a broad range depends essentially on the illuminance ($\approx 0.1\ \mu$A/lx), and only weakly but linearly on the reverse voltage.

The charge carriers bound in the doped crystal of the photodiode may be lifted from the valence band to the conduction band by energy supply through incident light (photoelectric effect, creation of electron-hole pairs). The energy of the light quanta,

$$E_{ph} = hf,$$

must be higher than the binding energy of the charge carriers at the lattice sites, h being the quantum of action and f the photon frequency (see **Tab. 29.9/5**).

➤   If the frequency becomes too low, i.e., the wavelength too large, no charge carriers are released despite the high light intensity (spectral range: Si-diodes $0.6 \ldots 1\ \mu$m, Ge-diodes $0.5 \ldots 1.7\ \mu$m).

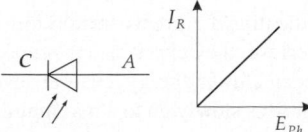

Figure 28.79: Circuit symbol and characteristic curve of a photodiode.

In principle, the photo effect also occurs for common pn-junctions. For the photodiode, however, the effect is optimized by the composition and doping.

8. PIN diode,

an ohmic resistor, current-dependent for high-frequency signals, operated in forward direction.

For the PIN diode, an undoped insulating layer (i-layer) is included between the p-region and the n-region. The intrinsic layer contains almost no free charge carriers, therefore it is insulating in the reverse direction. But in the forward direction, charge carriers from the doped layers may flow into the insulating layer, which then becomes conducting.

▲   A PIN diode is a current-dependent resistor for high-frequency alternating currents. A control direct current I_d, which fixes the resistance value, is superimposed on a high-frequency alternating current for which the PIN diode represents an ohmic resistance R.

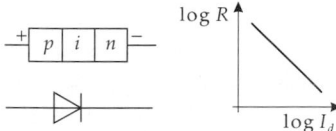

Figure 28.80: Structure, circuit symbol and characteristic line of a PIN diode. i: undoped insulating layer.

■   Application: current-controlled switch for high-frequency signals.

9. Step-recovery diode,

(SRD), the current flow in the depletion layer is terminated *suddenly*, not continuously, when changing from forward to reverse direction.

➤   In principle, all diodes exhibit this effect. For the step-recovery diode, it is particularly pronounced due to the doping.

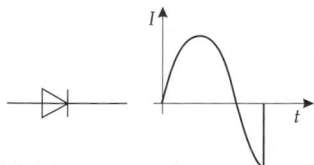

Figure 28.81: Circuit symbol and characteristic behavior of the step-recovery diode.

■   Application: generation of steep pulses, frequency multipliers up to the GHz range.

10. Tunnel diode

Tunnel effect, quantum-mechanical effect that allows a particle to overcome a high, but not too wide, potential barrier with a certain probability depending on the height and width of the barrier, although classically the motion is forbidden.

Tunnel diode, a highly doped germanium diode. The doping is so strong that the interface between the layers, the depletion layer, becomes very thin. The wave nature of the electrons enables them to overcome this thin potential barrier (tunnel effect), although the field strengths are not sufficiently high. For low positive pn-voltage, the effect causes a linear current rise that is non-typical for diodes (the tunnel currents from p to n and vice versa just compensate for $V = 0$; for a voltage increase, the current rise is proportional to the voltage). For a further increase of voltage fewer and fewer energy levels are available to which the electrons might tunnel, hence the current rises slowly up to a maximum and then drops again, exhibited in the declining part of the characteristic curve. At high voltage, the common diffusion current becomes dominant again. For negative pn-voltage, the tunnel diode becomes conducting at once. This behavior is exploited in the *backward diode*. The tunnel diode is characterized by its very fast switching time of about 100 ps, which enables application in high-frequency technology.

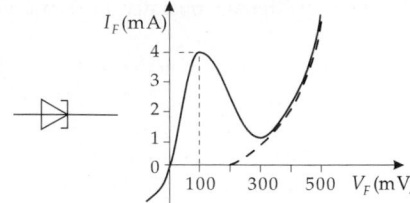

Figure 28.82: Circuit symbol and characteristic curve of a tunnel diode.

■   Application: very fast trigger diode, ultrahigh-frequency oscillator, reduction of damping of oscillating circuits.

11. Backward diode,

has a lower doping than the tunnel diode, and therefore a significantly reduced current peak at positive voltage, but keeps the property of conductance at negative voltage. This causes behavior in the backward diode just opposite to that of a common diode. It is *reversed biased* and then is conducting at negative voltage without threshold voltage, and has a high resistance at positive voltage up to a relatively low reverse voltage.

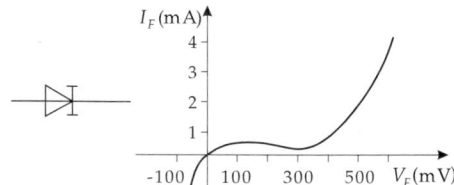

Figure 28.83: Circuit symbol and characteristic curve of the backward diode.

■   Application: high-frequency rectifier for low voltages.

12. Capacitance diode (varactor),

a voltage-dependent capacitance, reversed-biased.

   In the capacitance diode, the depletion layer acts as a capacitor the plate area of which remains constant while the distance of the plates, and thus the capacitance, is varied by the applied control voltage. This effect arises in all diodes. Capacitance diodes are distinguished by a large ratio of highest ($5 \ldots 300$ pF) to lowest ($1 \ldots 5$ pF) capacitance to be reached, and by a very low internal resistance, and thus high quality.

■   Application: tuning of radio receivers and TV sets.

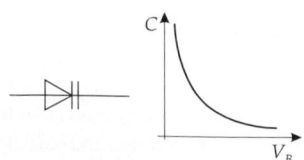

Figure 28.84: Circuit symbol and characteristic curve of the capacitance diode.

13. Light-emitting diode (LED),

a source of light with material-dependent frequency. The light intensity is controlled by the current through the forward-biased pn-layer.

If the n-layer is very highly doped as compared with the p-layer, then the conduction current consists mainly of electrons and depends only to a very small extent on hole conduction. The electrons reaching the p-layer in the forward direction recombine with the holes present there. Energy is released, which is emitted as light in the infrared or visible spectrum, depending on the material. If the radiation, which occurs more or less intensely in any diode, is guided to the outside, the result is a LED.

➤ LEDs are produced not from silicon or germanium, but from GaAsP (III-V-compounds). Their efficiency amounts to several percent in the infrared range and less than 0.1 percent otherwise.

Figure 28.85: Light-emitting diode. red, yellow: GaAsP (gallium arsenide phosphide), green: GaP (gallium phosphide), blue: SiC (silicon carbide), infrared: GaAs (gallium arsenide), GaAlAs (gallium aluminum arsenide).

▲ The frequency of the emitted light depends on the energy gain in recombination.

■ Application as signal lamps, entertainment electronics, opto-electronic couplers, fiber systems.

28.6.3 Transistor

Transistor, a semiconductor component with at least two pn-junctions, mainly for control and amplification of signals, but also used as an electronic switch.

One distinguishes between bipolar and unipolar (field effect) transistors. Bipolar transistors are current-controlled, whereas unipolar transistors are voltage-controlled. This means that unipolar transistors consume significantly less power than bipolar ones. Therefore, in present day applications, bipolar transistors are increasingly being replaced by unipolar ones, in particular in the microelectronics of large-scale integrated circuits.

28.6.3.1   Bipolar transistors

Bipolar transistor, consists essentially of two pn-junctions. The sequence of layers defines the type of the transistor (**npn**- or **pnp**-transistor).

npn-transistor, bipolar transistor with a layer sequence npn.

pnp-transistor, bipolar transistor with a layer sequence pnp, frequently replaced by a npn-transistor.

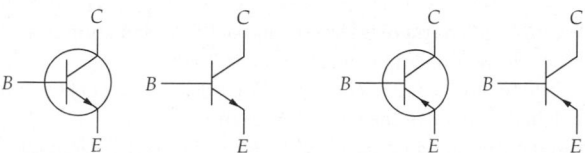

Figure 28.86: Circuit symbol for npn- and pnp-transistors, each with the old (with a circle) and new notation.

Base, B, the electrode at the central layer. The control signals are applied to the base.

Collector, C, the electrode at one of the outer layers. In general, the collector is at positive potential for npn-transistors and at negative potential for pnp-transistors, with respect to the emitter.

Emitter, E, the electrode at the second outer layer.

▲ As a rule, transistors are *not* configurated symmetrically. Collector terminal and emitter terminal must not be interchanged.

➤ Mnemonic rule: The collector collects majority charge carriers of the central layer and emits them again at the emitter. So, the current flow of the majority carriers of the base always goes from the collector to the emitter.

➤ Because of its more frequent use, only the npn-type will be treated below. The pnp-transistor is equivalent in the function and inverse in circuit technology. In most cases, it may be replaced by a npn-transistor.

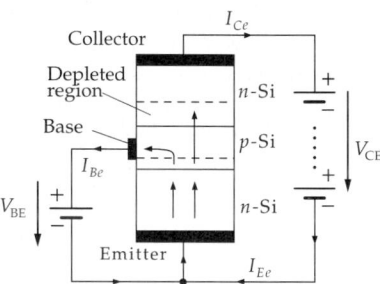

Figure 28.87: Configuration and functional principle of a bipolar transistor, I_{Be}, I_{Ce}, I_{Ee} electron currents.

Let there be a positive voltage V_{CE} across collector C and emitter E. If B is negative with respect to E, then no current may flow to C, since both the BC-diode as well as the EB-diode are reversed-biased. If, however, B is positive with respect to E, the BE-diode is forward-biased, and electrons from the n-zone may reach the p-zone. If the mean free path of the electrons for recombination with a lattice hole is large enough and the p-layer is narrow enough, the electrons may diffuse to the BC-junction, where they are extracted towards the collector due to the positive V_{CE} voltage: a current flows.

Notations for transistor circuits:

I_C	collector current
I_B	base current
I_E	emitter current
V_{CE}	collector-emitter voltage
V_{BE}	base-emitter voltage
V_{BC}	base-collector voltage.

➤ For pnp-transistors, the base has to be negative-biased relative to the emitter.

▲ The transistor acts as current-amplifier: a low base current causes a large collector current.

Four-quadrant family of characteristics, a compact representation of the dependence of all input and output currents and input and output voltages.

Input characteristic, the relation $I_B = I_B(V_{BE})$ at V_{CE} = const. (third quadrant). In principle it is the characteristic of the base-emitter diode.

Output characteristic, the relation $I_C = I_C(V_{CE})$ with the parameter I_B (first quadrant).

Saturation region, the region of the output characteristics in which I_C strongly increases with V_{CE} (V_{CE} small).

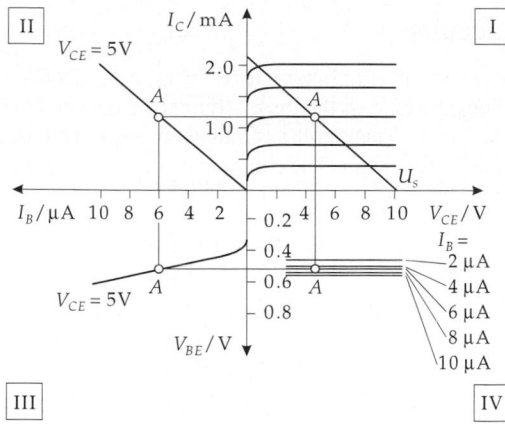

Figure 28.88: Four-quadrant family of characteristics of a npn-transistor in emitter connection. The points A mark operating points in the linear parts of the characteristic curves.

Active region, the part of the output characteristic in which I_C depends very little on V_{CE}, but strongly depends on I_B. Transistors in amplifier circuits are operating in this region.

Current-gain characteristic or **transfer characteristic,** the relation $I_C = I_C(I_B)$ with $V_{CE} = $ const. (second quadrant).

Reaction characteristic, back reaction of the output voltage V_{CE} on the input voltage $V_{BE} = V_{BE}(V_{CE}, I_B)$ (fourth quadrant). In the active region, the back reaction ≈ 0, i.e., V_{BE} is independent of V_{CE}.

Control characteristic, a combination of input characteristic and current-gain characteristic $I_C = I_C(V_{BE})$ at $V_{CE} = $ const.

Absolute maximum ratings, maximum values for the connection of a transistor. If these values are exceeded, the transistor may be destroyed. Transistors are particularly sensitive to base voltages or base currents that are too high, since then the very narrow depleted layer is affected. A high power consumption in the output circuit that is too high may also lead to damage. The maximum ratings can be found in the data sheets for the corresponding type of component.

Operating point, determines the region in the family of characteristics in which the transistor operates. In **analog technology,** the transistor is frequently used for amplification of time varying currents or voltages. In order not to distort the signals, these have to fall in the linear range of the characteristics. But since the characteristics are extremely nonlinear about the origin, the signal has to be raised to a linear section, the operating point (points A in the family of characteristics). This is done by means of an external connection in which a direct voltage is superimposed on the alternating signal.

Collector resistor, resistor before the collector. The **emitter resistor** and the **base resistor** are similarly defined.

Resistance load line, serves to determine the operating point in the family of characteristics and is fixed by the collector resistance R_C (in the common emitter circuit). The collector resistance provides a dependence between I_C and V_{CE} according to Ohm's law,

$$I_C = \frac{V_S - V_{CE}}{R_C},$$

which has to be fulfilled in addition to the relation $I_C = I_C(V_{CE})$ given by the transistor. Hence, the operating point is fixed for given I_B.

➤ Setting of the operating point is of central importance for any transistor circuit and is crucial for its correct operation. The maximum ratings of the transistor always have to be taken into account.

28.6.3.2   Basic transistor circuits

Basic transistor circuits, fundamental circuits of a transistor. There are three distinct basic circuits for bipolar transistors, depending on which of the three terminals is the common reference point for the input and output signals. There is the common emitter circuit, the common base circuit, and the common collector circuit. The common emitter circuit is the most suitable circuit for voltage amplification.

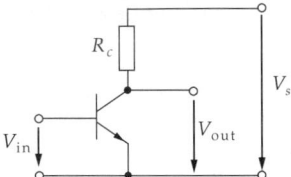

Figure 28.89: Principle of the common emitter circuit.
$V_{in} = V_{BE}$: input voltage,
$V_{out} = V_{CE}$: output voltage.

Common emitter circuit, the emitter is the common reference point for the input and output signals.

parameters of the transistor for common emitter circuit			
$R_{BE} = \dfrac{\partial V_{BE}}{\partial I_B}$	Symbol	Unit	Quantity
	R_{BE}	Ω	differential input resistance
$v_r = \dfrac{\partial V_{BE}}{\partial V_{CE}}$	v_r	1	voltage reaction
	β	1	small-signal current gain
	R_{CE}	Ω	differential output resistance
$\beta = \dfrac{\partial I_C}{\partial I_B}$	V_{BE}	V	base-emitter voltage
	V_{CE}	V	collector-emitter voltage
$R_{CE} = \dfrac{\partial V_{CE}}{\partial I_C}$	I_B	A	base current
	I_C	A	collector current

Two-port network, a vector group the internal structure and performance of which is ignored; only the functional relation between input and output quantities is known.

Two-port equations, the conditional equations of a two-port network. They link the input and output quantities of the network.

A transistor may be regarded as a two-port network. One electrode is common for the input and output of the two-port network (E for the emitter circuit). The transfer of the input quantities V_{BE} and I_B through the transistor may be calculated by means of the two-port equations. Similar relations may be given for the collector and base circuits, respectively.

two-port equations of transistor in common emitter circuit			
	Symbol	Unit	Quantity
	ΔV_{BE}	V	change of base voltage
	ΔI_B	A	change of base current
$\Delta V_{BE} = R_{BE}\Delta I_B + v_r \Delta V_{CE}$	ΔV_{CE}	V	change of output voltage
	ΔI_C	A	change of collector current
$\Delta I_C = \beta \Delta I_B + \dfrac{1}{R_{CE}}\Delta V_{CE}$	R_{BE}	Ω	differential input resistance
	R_{CE}	Ω	differential output resistance
	β	1	small-signal current gain

▲   The quantities

$$R_{\text{in}} = \frac{\Delta V_{\text{BE}}}{\Delta I_{\text{B}}} \qquad \text{input resistance}$$

$$B = \frac{\Delta I_{\text{C}}}{\Delta I_{\text{B}}} \qquad \text{current gain}$$

$$R_{\text{out}} = \frac{\Delta V_{\text{CE}}}{\Delta I_{\text{C}}} \qquad \text{output resistance}$$

may be assigned to the differential quantities. In the active region, the differential and integrated values agree reasonably well.

characteristic quantities of the common emitter circuit			
voltage gain: $v_v = \dfrac{\Delta V_{\text{CE}}}{\Delta V_{\text{BE}}} = -\dfrac{\beta R_{\text{C}}}{R_{\text{BE}}}$ $v_v \approx -100 \ldots -200$	Symbol	Unit	Quantity
	R_{in}	Ω	input resistance
	R_{out}	Ω	output resistance
	R_{C}	Ω	collector resistance
	β	1	small-signal current gain
input resistance: $R_{\text{in}} \approx R_{\text{BE}} = \dfrac{V_T}{I_{\text{B}}} \approx \dfrac{40\ \text{mV}}{I_{\text{B}}}$	R_{BE}	Ω	input resistance at operating point
	V_T	V	temperature voltage
	$\parallel$		parallel connection
output resistance: $R_{\text{out}} = R_{\text{CE}} \parallel R_{\text{C}}$	v_v	1	small-signal voltage gain

➤   The negative sign of v_v means a 180°-phase shift of the output signal with respect to the input signal. Of course, v_v is limited by the voltage reaction and by R_{CE} and cannot be enlarged arbitrarily by simply increasing R_{C}.

Negative feedback in an amplifier circuit, a method to feed the output signal in opposition, i.e., with opposite phase, back to the input signal. The gain of the circuit is always lowered, but the operating point is stabilized, since the circuit is readjusting itself. The characteristic curve is thereby linearized.

Negative voltage feedback, negative feedback in which the *output voltage* is fed back to the input via a voltage divider. The gain of the transistor stage becomes independent of the transistor parameters and is determined almost entirely by the external wiring.

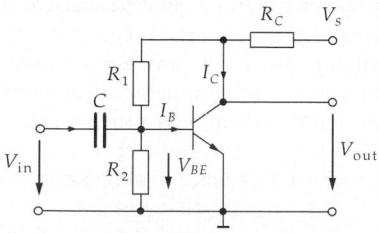

Figure 28.90: Common emitter circuit with negative voltage feedback.

Voltage gain:

$$v \approx -\frac{R_1 + R_2}{R_2}.$$

Input resistance:

$$\frac{1}{R_{\text{in}}} = \frac{1}{R_1} + \frac{1}{R_{\text{BE}}} + \frac{v_v}{R_2} \Rightarrow R_{\text{in}} \ll R_{\text{BE}}.$$

Output resistance:

$$R_{\text{out}} = (R_C \| R_{CE}) \cdot \frac{v}{v_v}.$$

The lower gain also causes a stabilization of the operating point according to $\Delta V_{CE} = v_D \Delta V_{\text{BE}}$, with the **drift gain**

$$v_D = 1 + \frac{R_1}{R_2}.$$

Negative current feedback, negative feedback in which the voltage generated by the *output current* is coupled back to the input with opposite phase.

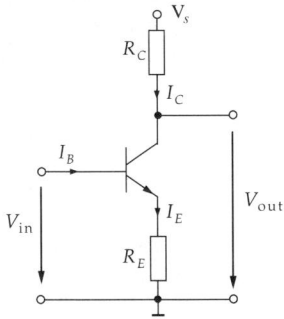

Figure 28.91: Common emitter circuit with negative current feedback.

Voltage gain	Input resistance	Output resistance	Current gain
$v \approx -\dfrac{R_C}{R_E}$	$R_{\text{in}} = R_{\text{BE}} + \beta R_E \gg R_{\text{BE}}$	$R_{\text{out}} \approx R_C$	$v_i \approx \beta$

➤ A large input resistance is an efficient way for amplifiers to keep the load of the signal source low.

▲ Owing to the high output resistance, the common emitter circuit with negative current feedback is suitable as *constant current source*.

■ **Amplifier stage in common emitter circuit:** The input capacitor C_1 prevents a short-circuit of the bias voltage of the base by the signal generator. The output capacitor C_2 DC-decouples the load resistance from the collector voltage. The emitter capacitor C_E AC-connects R_E.

Common collector circuit, a basic circuit in which the collector is the common reference point for the input and output signals.

➤ A transistor stage in common collector circuit is frequently called **emitter follower**.

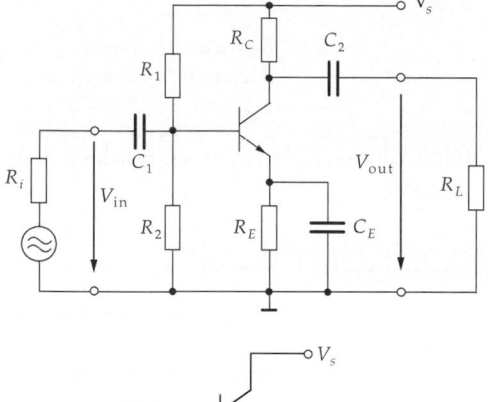

Figure 28.92: Amplifier stage in common emitter circuit.

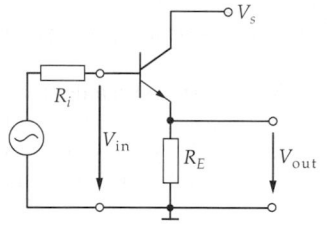

Figure 28.93: Common collector circuit.

Voltage gain	Current gain	Input resistance	Output resistance
$v_v \approx 1$	$v_i \approx \beta$	$R_{\mathrm{in}} \approx \beta \cdot R_{\mathrm{BE}}$	$R_{\mathrm{out}} \approx \dfrac{R_{\mathrm{BE}} + R_{\mathrm{E}}}{\beta} \ll R_{\mathrm{BE}}$

■   Owing to the high input resistance and low output resistance, common collector circuits are often used as **impedance converters**, i.e., matching pads between high-resistance signal sources and low-resistance loads.

Common base circuit, a basic circuit in which the base is the common reference point for the input and output signal.

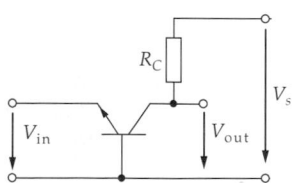

Figure 28.94: Common base circuit.

Voltage gain	Current gain	Input resistance	Output resistance
$v_v = \dfrac{\beta R_{\mathrm{C}}}{R_{\mathrm{BE}}}$	$v_i \approx 1$	$R_{\mathrm{in}} = \dfrac{R_{\mathrm{BE}}}{\beta}$	$R_{\mathrm{out}} \approx R_{\mathrm{C}}$

The voltage gain is the same as in a common emitter circuit. However, in the base circuit the output signal is in phase with the input signal, hence a negative voltage feedback is prevented. Input and output are completely decoupled by the constant base potential.

■   The common base circuit has a very high critical frequency, hence a much larger bandwidth than an emitter stage.

28.6.3.3   Darlington transistor

Darlington transistor, a series connection of two transistors. The total current gain corresponds to the product of the individual current gain factors and is connected like a single transistor with a high amplification.

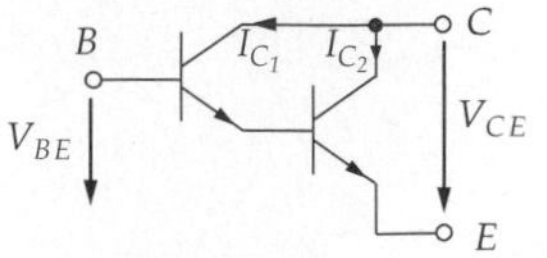

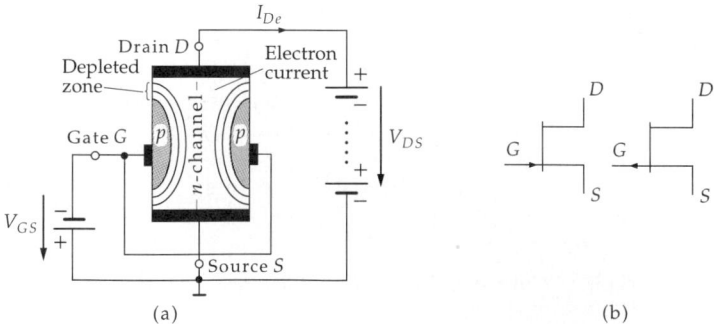

Figure 28.95: Circuit diagram of a Darlington transistor.

■ Such a high amplification ($\beta > 1000$) may be necessary when matching high-resistive voltage sources to low-resistive loads.

28.6.4   Unipolar (field effect) transistors

Field effect transistor (FET), voltage-free and, hence, nearly zero control power transistor that in most cases may replace a bipolar transistor.

Substrate, doped semiconductor block into which the pn-junctions needed for the function of the FET have been diffused.

Whereas in the bipolar transistor two kinds of charge carriers, electrons and holes, are involved in current conduction, the unipolar transistor consists of a substrate in which only the majority carriers are conducting: *either* electrons *or* holes. The charge carriers are influenced by an applied external field that controls the current flow. Hence, the control draws no power.

28.6.4.1   Junction field effect transistor (JFET)

Junction FET, (JFET), consists of a doped Si-crystal as substrate into which a channel-like zone (thickness $\approx 1\ \mu$m) with inverse doping is embedded. Depending on the doping of the channel, one distinguishes n- and p-channel FET. The figure displays an n-channel FET.

Figure 28.96: (a): Configuration and operation mode of an n-channel junction FET, I_{De} electron current. (b): circuit symbols for n-channel and p-channel junction FET.

Drain D and **source** S, the terminals of a FET connected to the conduction channel. The signal to be controlled is applied to D and S.

Gate G, the terminal at a thin p-zone diffused into the n-channel to which the control voltage is applied.

Bulk B, the terminal attached to the substrate, existing only for MOSFETs. In many cases, B is internally connected with the source.

If a voltage V_{DS} is applied across D and S, then an electron current flows through the n-channel as it does through an ohmic resistor. If G becomes negative with respect to S ($V_{GS} < 0$), then the pn-junction between S and G is reversed-biased.

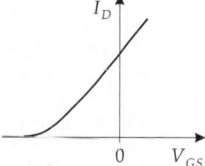

Figure 28.97: Characteristic
of a junction-FET.

At the interface, a region free of charge carriers is formed which extends more and more into the n-channel as the reverse voltage V_{GS} increases. Thereby, the channel cross-section is diminished, and the resistance is increased: the channel resistance may be controlled via the gate voltage.

Depletion-type FET, a FET that conducts the DS-current without a gate voltage applied.

Enhancement-type FET, reversed-biased without a gate voltage applied.

➤   The junction-FET is a depletion-type FET.

Contrary to bipolar transistors, junction-FETs are in many cases configured symmetrically, hence D and S may be interchanged.

➤   For a negative gate, a gate current of only 1 pA to 1 μA flows. If, however, the gate voltage V_{GS} becomes positive, then the pn-junction between gate and n-channel becomes conducting. In this case, the FET consumes power.

28.6.4.2   Insulated Gate FET (IGFET, MOSFET)

MOS technology (MOS: metal-oxide-silicon), manufacturing principle for FETs according to which the gate is separated from a pn-junction by a thin but sophisticated insulating layer (usually metal oxide).

MOSFET, a FET manufactured in MOS technology that has the advantage of also remaining currentless for a positive gate voltage.

Enhancement or enrichment mode, an enhancement-type MOSFET. One distinguishes p- and n-channel MOSFETs. For the n-channel, MOSFET two n-doped islands, the source S and the drain D, are implemented into a p-doped substrate. No current may flow between D and S if a voltage V_{DS} is applied, since one pn-junction is reversed-biased independent of the sign of the voltage. The surface is coated with a thin insulating layer onto which a metallic layer, the gate G, is evaporated as a terminal. The substrate itself may get a separate terminal, the bulk B, or may be internally connected with the source. This terminal becomes important for the power-FET. If the gate is positive with respect to the source, then the minority carriers of the p-region, the electrons, are electrostatically pulled close to the insulating layer so that an n-conducting channel arises between S and D.

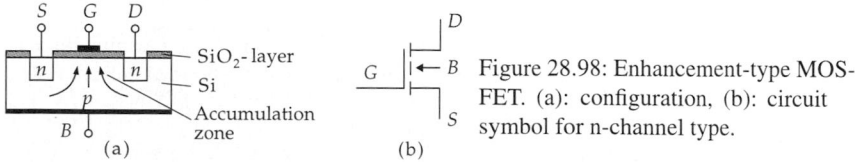

Figure 28.98: Enhancement-type MOSFET. (a): configuration, (b): circuit symbol for n-channel type.

➤   In the enhancement mode, the minority carriers in the substrate are accumulated between the n-conducting islands and constitute the conduction electrons that contribute to the current flow.

The higher the gate voltage, the higher the number of electrons in the DS-channel and the lower the conduction resistance.

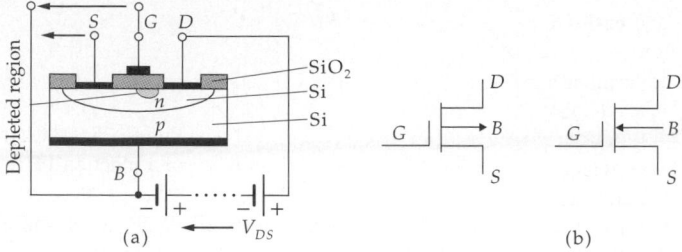

Figure 28.99: Depletion-type MOSFET. (a): configuration, (b): circuit symbols for p-channel and n-channel types.

Depletion mode, a depletion-type field effect transistor, analogous to the junction-FET. Here a thin, e.g., n-conducting channel is effected between the islands of enhancement-type, which admits a current flow without applying a gate voltage: the FET is depletion-type. If the gate voltage becomes negative, the majority carriers in the n-channel are pushed out of the channel, and a smaller number of conduction electrons remain there: the resistance increases. The particular feature of this FET is that, for a positive gate voltage, the conduction electrons in the n-channel are enriched by minority carriers of the substrate, and thus the drain current may increase.

➤   If bulk and source are internally connected, this property is indicated by the circuit symbol:

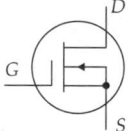

Figure 28.100: Circuit symbol of a depletion type n-channel MOSFET with source and bulk terminals connected internally.

■ . The FET became indispensable for large-scale integrated circuits due to its zero power control and because of the possibility to produce it with shorter switching times on a smaller and smaller area of substrate.

Dual-gate MOSFET, corresponds to a normal MOSFET, but has two gate terminals G_1 and G_2 arranged one behind the other above the conduction channel. The independent wiring of the gates allows the control of the current flow independently as long as the current is not turned off completely by one of the gates.

■   Application: adjustable amplifier in high-frequency circuits. One gate controls the desired signal, the other one controls the transconductance of the MOSFET.

28.6.5 Thyristor

Thyristor or **four-layer diode**, a semiconductor with a pnpn-structure, meaning three depletion layers. Like a common diode, the thyristor may conduct the load current only in one direction.

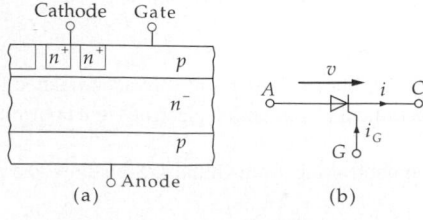

Figure 28.101: Thyristor. (a): layer structure, (b): circuit symbol.

Anode and **cathode**, as for a common diode the outermost of the p- and n-layers, respectively.

Gate, the terminal at the inner p-layer that, for a positive voltage with respect to the cathode, makes the thyristor conduct.

Forward blocking region of the thyristor, range of voltages up to a maximum positive voltage V_{DRM}, which must not bc exceeded. In this voltage region, the thyristor is reverse-biased via the depletion layer.

➤   If the thyristor is loaded with a higher voltage, a **forward breakover ignition** occurs in which the thyristor suddenly becomes transmitting. Caution: This may destroy the thyristor!

Forward leakage current, the residual current flowing in a thyristor operated in the forward blocking region.

Forward conduction region, that part of the characteristic of a thyristor into which the bias point of the thyristor is shifted from the forward blocking region by a positive gate voltage.

Triggering current, i_G, the current at the gate that floods the central depletion layer with charge carriers and *triggers* the thyristor, i.e., makes it conducting.

Reverse blocking region, the region of negative voltage between anode and cathode. In the reverse blocking region, the thyristor cannot become conducting because both outer depletion layers are inverse-biased.

Reverse breakdown voltage, the maximum negative voltage V_{RRM} that may be connected to the thyristor.

Reverse blocking current, residual current i_R of several μA through the thyristor operated in the reverse blocking region.

➤   When the reverse breakdown voltage V_{RRM} is exceeded, the reverse blocking current increases avalanche-like, and the thyristor is destroyed.

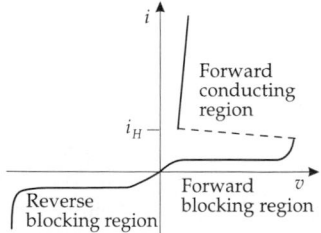

Figure 28.102: Characteristics of the thyristor.

Holding current i_H, the current (usually between 10 and 100 mA) above which a triggered thyristor remains conducting despite missing gate voltage. If the current is made sufficiently high, a short trigger pulse at the gate is sufficient to make the thyristor permanently conducting.

Trigger pulse, voltage pulse at the gate, switches the thyristor to the conducting state as long as a sufficiently high holding current flows in the triggered state.

Trigger time, the time interval needed by the thyristor to switch from the blocking to the conducting state. It depends on the steepness of the trigger pulse.

■   **Phase angle control**: By short periodic current pulses at the gate of a thyristor, certain phases of the alternating signal may be reduced by an appropriate phase relation of the pulses with respect to a control alternating voltage. This works only in the positive half-wave, since for negative voltages the thyristor always blocks. However, if the thyristor is triggered by the current pulse during the positive half-wave, then the voltage drop across it remains zero until the alternating voltage falls below the holding voltage.

➤ There are thyristors that can be used up to blocking voltages of several kV and currents up to several kA. Their range of application is restricted to the kHz-region.

28.6.5.1   Triac

Triac (**TRI**ode **A**lternating **C**urrent switch), acts as two inverse-parallel connected thyristors and is frequently denoted **bidirectional thyristor**. It may control both positive and negative half-waves of an alternating voltage.

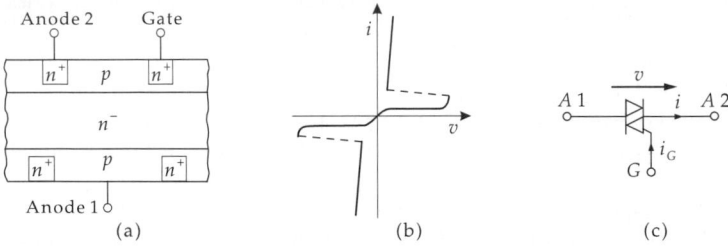

Figure 28.103: Triac. (a): configuration, (b): characteristic, (c): circuit symbol.

28.6.5.2   Gate turn-off thyristor (GTO)

Gate turn-off thyristor, (**GTO**, **G**ate **T**urn **O**ff thyristor), may be triggered by a positive gate pulse and switched off again by a negative one. There are GTOs both with symmetric as well as with asymmetric blocking capability.

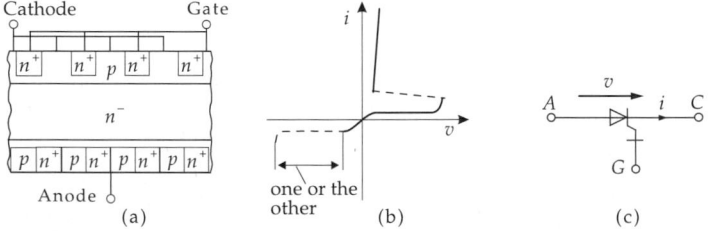

Figure 28.104: GTO. (a): configuration, (b): characteristic, (c): circuit symbol.

■ Generation of a sinusoidal output voltage from a direct voltage by means of pulse DC-AC inverters.

28.6.5.3   Insulated-gate bipolar thyristor (IGBT)

IGBT, a combination of MOS technology and technology of bipolar transistors. For switching on and switching off, only low control power is needed. The transmission resistance is very low.

28.6.6   Integrated circuits (IC)

Integrated circuit (IC), a circuit consisting of several transistor functions integrated on a single semiconducting substrate of small size.

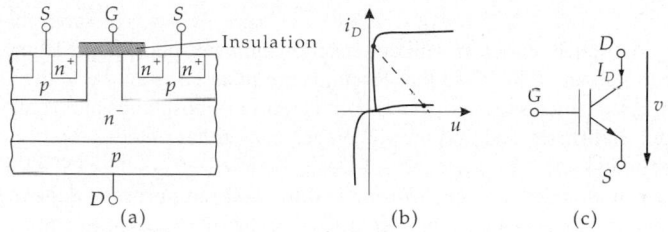

Figure 28.105: IGBT. (a): configuration, (b): characteristic, (c): circuit symbol.

28.6.6.1 Production of ICs

Wafer, silicon substrate on which the structures needed for producing an IC are deposited.

M **Vapor-phase epitaxy**, method for the deposition of Si-layers on a wafer. In an oven, single Si atoms obtained by chemical reactions of Si-containing gases are deposited on the wafer.

■ At 1250 °C, $SiCl_4$ reacts with H_2 to Si and HCl. The HCl is extracted while the silicon is deposited.

➤ The layers may be **doped**, whereby the H_2 is guided first through gases containing boron (p-doping) or phosphorus (n-doping).

Oxidation, the deposition of a SiO_2-layer on a wafer for
- insulation,
- protection against impurities in the pn-junctions,
- generation of circuit structures.

28.6.6.2 Generation of circuit structures

General procedure (see **Fig. 28.106**):
(a) Deposition of a SiO_2-layer on the Si-wafer.
(b) Upon it, a layer of photosensitive material is deposited.
(c) **Photolithography**: masking (covering) of the regions where the SiO_2 is to be re-moved, and radiant exposure with UV-light (modifies the chemical properties of the irradiated and non-irradiated areas).
(d) **Development** in a suitable chemical solution uncovers the SiO_2 in the non-irradiated areas.
(e) Etching of the SiO_2 at the uncovered areas.
(f) Removal of the photosensitive material.

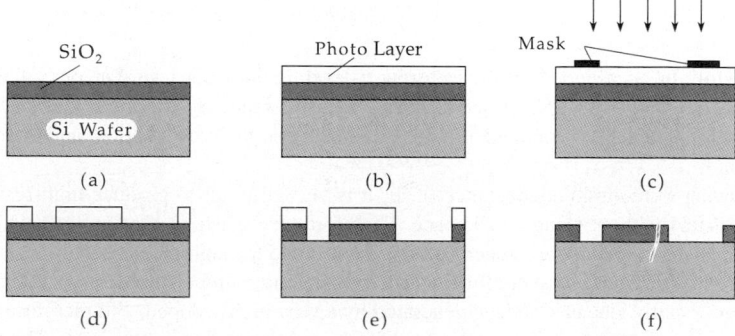

Figure 28.106: Photolithographic production of an IC. For legend see text.

Doping

| M | In an atmosphere enriched with either boron atoms or phosphorus atoms, silicon is heated to about 1000 °C so that Si-atoms are released from the lattice compound, leaving free lattice sites into which boron atoms or phosphorus atoms may be incorporated (**indiffuse**), and thus the silicon becomes either p-doped (boron) or n-doped (phosphorus). |

▲   The penetration depth of the diffusion is **time- and temperature-dependent**.

■   Phosphorus atoms penetrate into Si to 1 μm if the substrate is heated to 1000 °C for one hour.

➤   The diffusion rate in SiO_2 is significantly lower than that in pure silicon. The structures generated by **photolithography** determine which regions are doped

Production of electronic components

Transistor and **diode** (see **Fig. 28.107**):

(a) Deposition of an n-doped layer on a p-doped substrate. Part of this layer becomes the collector.

(b) By oxidation and photolithography, the state in (b) is generated.

(c) Indiffusion of acceptor atoms into the exposed part of the n-layer: this region corresponds to the base.

(d) After further oxidation and photolithography, another n-layer is indiffused into a part of the p-layer. This region is the emitter.

(e) Once again, oxidation and cutting-out three windows above the collector, base and emitter, and evaporation of an Al-layer generates the pad electrodes.

➤   For the production of diodes, the steps (d) and (e) are dropped.

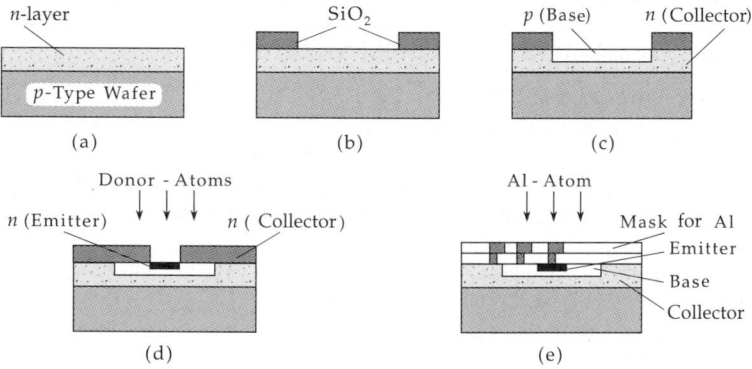

Figure 28.107: Generation of a transistor function. For legend, see text.

Resistor: In an n-doped layer, a narrow p-layer is embedded so that one of the pn-junctions is operated reverse-biased and thereby a resistance is generated. The magnitude of the resistance depends on the length of the p-channel, the cross-section, and the doping strength.

➤   Owing to the high conductance of Si, it is very difficult to produce high-resistance resistors without using much space. Therefore, the resistor is frequently replaced by a transistor, and the resistance value is determined through the base current.

Capacitor: A capacitor essentially consists of two conducting electrodes separated by an insulator. Usually, one electrode is generated by a very highly doped, and therefore highly conducting, p- or n-region. An insulating SiO_2-layer is deposited on this layer. The second electrode is produced by evaporating a thin aluminum film on this oxide layer.

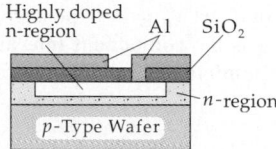

Figure 28.108: Capacitor on a silicon chip.

▲   In most cases, integrated circuits are realized in **MOS technology** because of the low power drain of the individual transistor functions, in order to avoid a too-strong heating of the component.

➤   Nevertheless, for extremely large integrated circuits, there may arise problems with heat extraction. Therefore, the components must be equipped with cooling facilities.

M   In practice, cooling of ICs is frequently done through good thermal contact with a medium of high thermal conductivity, as a rule copper. Most recent findings indicate that **diamond** (98.9 % ^{12}C and 1.1 % ^{13}C), by reducing the ^{13}C-component (to 0.001 %) and cooling to 80 K (liquid nitrogen), has a thermal conductivity $\lambda > 2000$ W cm^{-1} K^{-1} (for comparison: copper: $\lambda = 4.01$ W cm^{-1} K^{-1}). Thus, a power density **500 times higher** might be achieved.

28.6.7 Operational amplifiers

Operational amplifier, a multi-stage amplifier with a high gain that may get a definite fixed gain value by **external wiring**, or may carry out **mathematical operations**.

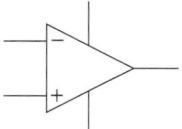

Figure 28.109: Circuit symbols of an operational amplifier. "$-$" denotes the inverting, "$+$" the non-inverting input.

➤   The connections shown as vertical lines indicate the (symmetric) voltage supply of the operational amplifier; as a rule, they are not plotted.

Inverting input terminal, the output signal is inverted (opposite phase) to the input signal.

Non-inverting input terminal, the output signal is non-inverted (in phase) to the input signal.

Difference amplifier, the basic component of an operational amplifier. It consists of two—possibly identical—transistors:

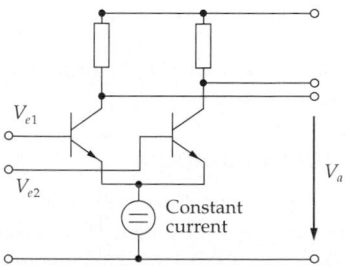

Figure 28.110: Difference amplifier.

M   If both inputs have **equal** voltage, then one should get $V_a = 0$. In practice, however, $V_a \neq 0$ **always**. The reason for this behavior is the **component tolerance** of the transistors and resistors, which results in an **asymmetry** of the difference amplifier.

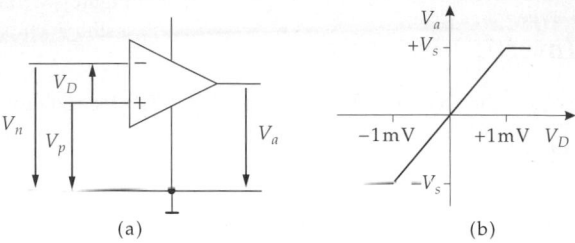

(a)   (b)

Figure 28.111: Operational amplifier. (a): connection, (b): characteristic.

▲   An operational amplifier always amplifies the **difference** of the voltages at the inputs.

output voltage of an operational amplifier			
	Symbol	Unit	Quantity
$V_a = A(V_p - V_n)$	V_a	V	output voltage
	V_p	V	input voltage of the non-inverting terminal
$= AV_D$	V_n	V	input voltage of the inverting terminal
	V_D	V	difference voltage
	A	1	gain

▲   An operational amplifier must be operated only with very **small** voltage differences (order of millivolt).

Linear region, the range of voltage differences V_D in which the operational amplifier acts as a **voltage amplifier** (to about ± 1 mV).

Saturation region, the difference voltages are beyond the linear region. The output voltage no longer changes when V_D is increased; it remains constant at the supply voltage $\approx \pm V_s$.

Ideal operational amplifier, an operational amplifier with the following properties:

	Ideal	Real
open-circuit voltage gain A	∞	$10^3 \ldots 10^6$
input resistance R_e (at both inputs)	∞	≈ 1 MΩ
output resistance R_a	0	≈ 100 Ω

➤   All statements about the operational amplifier always refer to the **ideal** operational amplifier. In practice, minor deviations will always occur.

28.6.7.1   Negative-feedback operational amplifier

▲   For the operation of amplifiers, a stable bias point in the **linear range** of the operational amplifier has to be adjusted so that the amplifier does not run into a saturation state. This is done by **negative feedback**, as is done for the transistor amplifier.

Negative feedback, the output signal V_a of the operational amplifier is fed back to the **inverting input** (with opposite phase). Hence, deviations from the bias point will be fed back with inverse sign, and therefore are weakened.

28.6.7.2   Inverting amplifier

gain of inverting amplifier			1
$\dfrac{V_a(j\omega)}{V_e(j\omega)} = -\dfrac{Z'(j\omega)}{Z(j\omega)} \cdot \dfrac{A(j\omega)}{1 + A(j\omega)}$ $\approx -\dfrac{Z'(j\omega)}{Z(j\omega)}$ $\beta(j\omega) \approx \dfrac{Z(j\omega)}{Z(j\omega) + Z'(j\omega)}$	Symbol	Unit	Quantity
	V_a	V	output voltage
	V_e	V	input voltage
	Z, Z'	Ω	resistances
	A	1	open-circuit voltage gain

▲   The gain of the inverting amplifier is, for a sufficiently large open-circuit voltage gain, **independent** of the architecture of the operational amplifier and determined **only** by the external connection.

▲   An inverting amplifier **multiplies** $V_e(j\omega)$ **by a constant factor** $-Z'(j\omega)/Z(j\omega)$.

➤   One may also make a **non-inverting** amplifier using an operational amplifier (**Fig. 28.112 (b)**) with a gain

$$\frac{V_a}{V_e} = 1 + \frac{R_0}{R_1}.$$

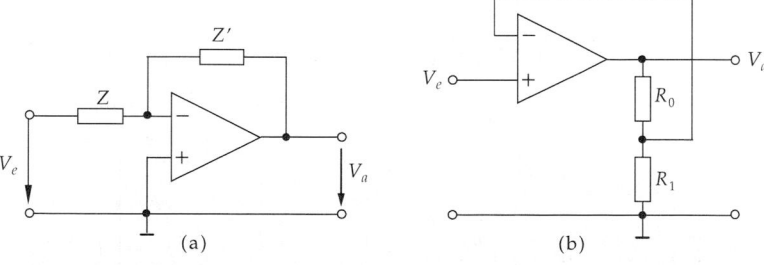

Figure 28.112: (a): Inverting amplifier, (b): non-inverting amplifier. Z and Z' denote (real or complex) resistances.

28.6.7.3   Summing amplifiers

▲   The resistances $R_1, \ldots, R_n$ determine the weighting factors of the input voltages $V_1, \ldots, V_n$. The output voltage corresponds to the sum of the weighted input voltages, multiplied by a factor determined by the coupling resistance R_0.

characteristic data of the summing amplifier

	Symbol	Unit	Quantity
$V_a = -R_0 \left(\dfrac{V_1}{R_1} + \cdots + \dfrac{V_n}{R_n} \right)$			
	V_a	V	output voltage
	$V_1, \ldots, V_n$	V	input voltages
$V_a = -\dfrac{R_0}{R} (V_1 + \cdots + V_n)$	R_0	Ω	coupling resistance
	$R_1, \ldots, R_n$	Ω	weighting factors
for $R = R_1 = \cdots = R_n$			

Subtractor, analogous to the adder, the **non-inverting** input is set to the voltage level to be subtracted.

➤   Addition and subtraction may be done **simultaneously** with a single operational amplifier.

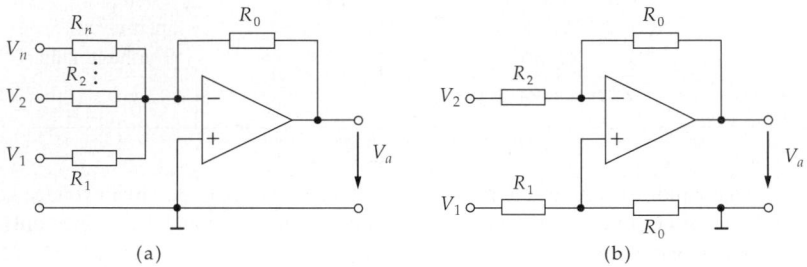

Figure 28.113: (a): Summing amplifier, (b): adder and subtractor.

28.6.7.4 Integrator

For sinusoidal signals $V_e(j\omega)$ of angular frequency ω, the impedance $Z_C(j\omega)$ of a capacitor with the capacitance C is

$$Z_C(j\omega) = \frac{1}{j\omega C}.$$

With $Z(j\omega) = R$ and $Z'(j\omega) = Z_C(j\omega)$, one obtains an inverting amplifier.

performance of the integrator

	Symbol	Unit	Quantity
$V_a(j\omega) = -\dfrac{V_e(j\omega)}{j\omega RC}$			
	V_a	V	output voltage
	V_e	V	input voltage
$V_a(t) = -\dfrac{1}{RC} \displaystyle\int V_e(t)\, dt$	R	Ω	resistance
	C	F	capacitance

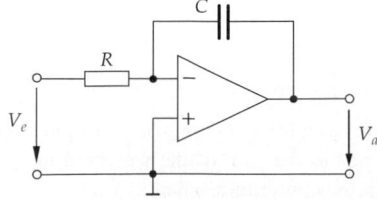

Figure 28.114: Circuit diagram of an integrator.

Summing integrator, an integrator in which the charging current is supplied via separate resistors $R_1, \ldots, R_n$, analogous to the summing amplifier:

$$V_a(t) = -\frac{1}{C} \int \left(\frac{V_1(t)}{R_1} + \cdots + \frac{V_n(t)}{R_n} \right) dt \, .$$

28.6.7.5 Differentiator

With $Z(j\omega) = 1/(j\omega C)$ and $Z'(j\omega) = R$ one obtains for the inverting amplifier:

performance of the differentiator			
	Symbol	Unit	Quantity
$V_a(j\omega) = -j\omega RC \cdot V_e(j\omega)$	V_a	V	output voltage
	V_e	V	input voltage
$V_a(t) = -RC\dfrac{dV_e(t)}{dt}$	R	Ω	resistance
	C	F	capacitance

➤ In practice, the differentiation property is not as effective as the integrating one:

- For high frequencies ω, the approximation to the ideal operational amplifier is not as good, since the open-circuit voltage gain $A \to A/(j\omega RC)$ is lowered, and hence $A \to \infty$ is no longer fulfilled.
- High-frequency noise components at the amplifier input are amplified particularly well.
- For large ω, and therefore small value of $1/(j\omega C)$, the internal resistance R_i of the signal generator becomes noticeable.

M Application: analog computers.

Mathematical problems, e.g., integration of differential equations, may be carried out by means of operational amplifiers.

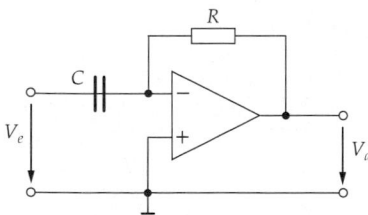

Figure 28.115: Circuit diagram of a differentiator.

28.6.7.6 Voltage followers

Voltage follower, the full output signal is fed back to the inverting input (100 % negative feedback): the output signal exactly follows the input signal,

$$\frac{V_a}{V_e} \approx 1 \, .$$

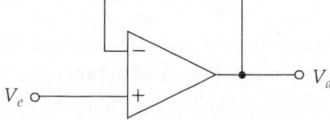

Figure 28.116: Voltage follower.

▲   The output resistance is very low, whereas the input resistance is very high.
■   The voltage follower is frequently used as **impedance converter**.

28.6.7.7   Positive-feedback operational amplifier

Positive feedback, the output signal is fed back to the **non-inverting** input. Owing to the amplifying effect, the operational amplifier is driven into the **saturation state**.

   Flip-flop circuit, a circuit with two stable output states. A flip-flop circuit produces **square-wave signals**.

28.6.7.8   Schmitt trigger

Schmitt trigger, a flip-flop circuit that jumps to the alternative state if one of the two definite input signal levels is exceeded. Switching between the stable states proceeds very fast.

   Falling below V_e^{on} the circuit trips to the "on" state, exceeding V_e^{off} it trips back to the "off" state. One has:

$$V_e^{on} = \frac{R_1}{R_1 + R_2} V_a^{min},$$

$$V_e^{off} = \frac{R_1}{R_1 + R_2} V_a^{max}.$$

Switching hysteresis, the difference $V_e^{off} - V_e^{on}$. It is connection-dependent and cannot be made arbitrarily small.

$$V_e^{off} - V_e^{on} = (V_a^{max} - V_a^{min})(\frac{R_1}{R_1 + R_2} - \frac{1}{A})$$

$$\approx (V_a^{max} - V_a^{min})\frac{R_1}{R_1 + R_2}.$$

▲ $R_1/(R_1 + R_2)$ must **always** be larger than A^{-1}.

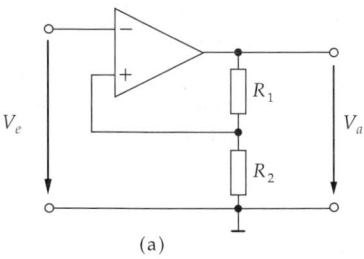
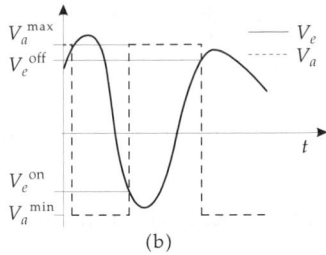

Figure 28.117: Schmitt trigger, (a): circuit, (b): operational mode.

28.7   Superconductivity

Superconductivity, a state of order of matter occurring in many metals and compounds with metallic conductivity. Superconducting properties are destroyed by magnetic correlations (**Fig. 28.118**).

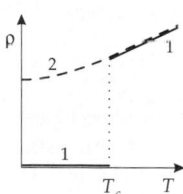

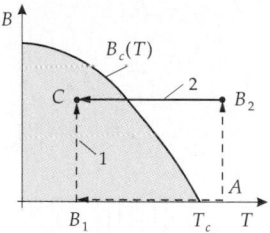

Figure 28.118: Temperature dependence of the electric resistance of a superconductor (1) and a normal conductor (2).

Figure 28.119: Alternative cycles in the B-T phase diagram for superconductor and ideal conductor. For the ideal conductor, the final state (C) is path-dependent. In a superconductor, the final state C is path-independent (thermodynamically stable state).

For superconductors, two effects are of particular importance:
- When cooling a sample below a characteristic temperature T_c, the specific electric resistivity $\rho(T)$ drops to a value that cannot be distinguished experimentally from $\rho = 0$.
- For temperatures $T < T_c$ and magnetic fluxes $B < B_{c1}$, the substances are ideal diamagnets (Meissner-Ochsenfeld effect).

Characteristic physical quantities of several superconductors may be found in the tables.

28.7.1 Fundamental properties of superconductivity

1. Meissner-Ochsenfeld effect,

also **Meissner effect**, the ideal diamagnetic behavior of a superconductor in a weak magnetic field. If a superconductor in a magnetic field ($B < B_{c1}$) is cooled below its critical temperature T_c, the magnetic field lines are expelled from the interior of the superconductor. Thereby an induced persistent screening current flows in a thin surface layer of the sample, its magnetic field just compensating the external flux density. The ideal diamagnetism cannot be traced back to the ideal conductance.

➤ Below T_c (in the superconducting state) the thermodynamic quantities, and several physical transport quantities of most superconductors, exhibit an exponential temperature dependence. This behavior suggests the formation of an energy gap at the Fermi energy in the superconducting state.

▲ The magnetic susceptibility of an ideal type-I superconductor is

$$\chi = -\frac{1}{4\pi} \quad \text{(cgs)}, \qquad \chi = -1 \quad \text{(SI)}.$$

▲ The specific heat has a λ-anomaly at T_c .
   For $T < T_c$ it displays an exponential temperature dependence.
▲ The ultrasonic attenuation in the superconducting state behaves like the specific heat.
M The temperature dependence of the ultrasonic attenuation (which is proportional to the number of normally conducting electrons) was one of the first experimental confirmations of the BCS theory.

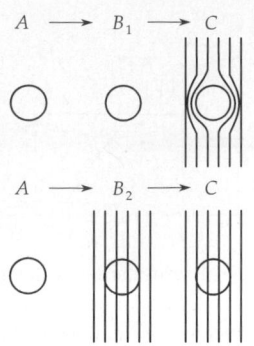
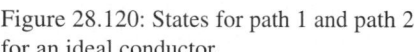
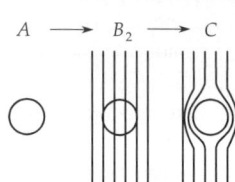

Figure 28.120: States for path 1 and path 2 for an ideal conductor.

Figure 28.121: States for path 2 for a type-I superconductor.

2. Theory of superconductivity

BCS theory (after Bardeen, Cooper and Schrieffer), a fundamental microscopic theory of superconductivity. It describes the coupling of two electrons with opposite spins and momenta by means of a phonon.

The attractive Coulomb force between an electron and the ion cores generates a local and instantaneous deformation of the lattice. Owing to the large mass of the lattice atoms and the associated inertia of the lattice, this deformation is not immediately canceled by the thermal motion. A second electron then may find itself in a force field of positive charge, and may be attracted. Therefore, an attractive interaction between two electrons arises via a lattice deformation. This coupling is energetically favorable if both the spins and momenta of the two electrons are aligned antiparallel to each other.

So, by means of a phonon, a new quasi-particle arises from two electrons, which is denoted a Cooper pair. Each electron gains an amount of energy of $E_G/2$ by the pair formation. Moreover, an energy gap of width E_G occurs in the electron distribution at the Fermi energy. This energy gap determines the physical properties of the BCS superconductor. The width of the gap varies exponentially with decreasing temperature. Therefore, all physical properties of a solid that are related to the conduction electrons exhibit an exponential temperature dependence.

Cooper pair, quasi-particle of the BCS theory. Its spin is an integer, hence the Pauli principle does not apply to Cooper pairs. Cooper pairs are governed by Bose-Einstein statistics.

▲   All Cooper pairs may occupy the lowest energy state (Bose-Einstein condensation). Therefore, they all have a fixed phase relation that may lead to formation of macroscopic quantum states.

▲   There is no inelastic scattering in the motion of Cooper pairs as long as the energy loss is less than the energy gap.

3. Isotope effect and Josephson effect

Isotope effect, dependence of the critical temperature T_c on the mass M of the isotope of the superconductor,

$$M^\alpha \cdot T_c = \text{const.}, \qquad \alpha \approx 0.5 .$$

| M | The parameter α depends on the series of isotopes. The most frequent experimental value is about $1/2$. Such a value is expected according to the BCS theory. The isotope |

effect is interpreted as experimental confirmation of the BCS theory and the role of lattice vibrations in the formation of Cooper pairs.

Josephson effect, the tunneling of Cooper pairs through a thin insulating layer between two superconductors. It is based on the fixed phase relation among the Cooper pairs (phase-coherence effect, macroscopic quantum states). A tunnel current flows without an external potential difference. A phase change arises in the tunneling of the Cooper pairs between the two superconductors.

| **M** | The phase-coherence effects in superconductors are of great importance for measuring very small magnetic fields. Such measurement systems are called **SQUID** (superconducting **qu**antum **i**nterferometer **d**evice). They are employed in solid-state physics, geophysics, biophysics and medicine.

4. Critical current density,

the current density at which the superconducting state converts to normal conduction. The reason is a possible energy loss in the inelastic scattering of Cooper pairs that is higher than the energy gap.

■ The current density is $j = 2env$. n is the number of Cooper pairs, and v is the drift velocity of Cooper pairs. Consider a crystal lattice of mass M that contains a defect. The lattice moves with the velocity v relative to the electron gas. If an excitation energy ε is transferred to the lattice by collision, then both energy and momentum must be conserved:

$$\frac{1}{2}Mv^2 = \frac{1}{2}Mv'^2 + \varepsilon, \quad M\vec{v} = M\vec{v}' + \hbar\vec{k}.$$

Hence:

$$0 = \hbar\vec{k}\vec{v} + \frac{\hbar^2 k^2}{2M} + \varepsilon.$$

If the mass of the crystal is very large ($M \to \infty$), then

$$v_c = \frac{\varepsilon}{\hbar k},$$

v_c being the velocity for the energy $\varepsilon = E_g$. The existence of an energy gap E_g prevents inelastic scattering for velocities $v < v_c$. For higher velocities, inelastic scattering may occur.

5. Critical magnetic flux density,

B_C, a consequence of the existence of a critical current. The superconducting state breaks down above a critical magnetic field strength.

■ Besides the SQUID systems, the technical application of superconductors lies mainly in the construction of high-flux magnets. Here, the critical current density of the materials used is the crucial quantity. Presently, wires of Nb-compounds are produced which are embedded in a Cu-matrix. The maximum flux density of such magnets is about 20 tesla.

Pinning, the fixing of magnetic **flux tubes** in a type-II superconductor at a definite position in the superconductor. The creation of pinning centers occurs because the Lorentz force between the magnetic flux tubes during a current flow causes a motion of the tubes resulting in a release of heat. Materials with pinned flux tubes are called hard superconductors. They have a higher critical current density and are used for construction of magnets.

Pinning centers, places at which the magnetic flux tubes in type-II superconductors may be fixed. Such pinning centers may be dislocations, grain boundaries or segregations, i.e., defects in the crystal lattice.

6. Type-I and type-II superconductors

Type-I and type-II superconductors, superconductors of the first and second kind. A sufficiently strong magnetic field destroys the superconductivity and the diamagnetic behavior of the sample. Type-I and type-II superconductors behave differently in a magnetic field.

- **Type-I** (also **soft superconductors**): for increasing magnetic flux density, a sudden transition from superconductivity to normal conduction occurs at $H = H_c$. The persistent screening currents flow in a thin surface layer of thickness λ (**London penetration depth**). The values of H_c are too low for type-I superconductors to be used in superconducting magnetic coils.

- **Type-II** (frequently alloys or transition metals with high electric resistance in the normal state, i.e., small mean free path of electrons in the normal state). The transition from the superconducting state to the normally conducting state does not occur discontinuously, but extends over an interval of magnetic field strengths between H_{c1} and H_{c2}. At $H_{c1} < H_c$, the field begins to penetrate into the sample, forming normally conducting flux tubes (**vortices**). The exit points of the flux tubes may be made visible in the electron microscope by means of small ferromagnetic particles; they again form ordered structures. The magnetic moment of the vortices is quantized. Superconductivity disappears completely only for field intensities $> H_{c2}$.

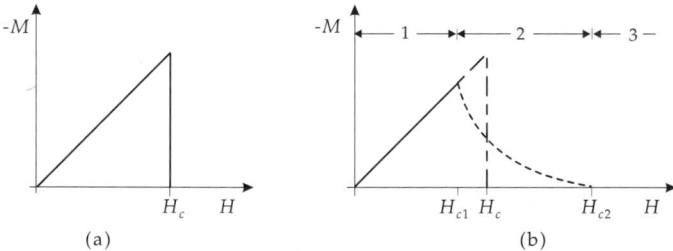

Figure 28.122: Magnetization curves $M(H)$ of superconductors. (a): type-I superconductor, (b): type-II superconductors. 1 – superconducting state, 2 – mixed state, 3 – normally-conducting state. The negative sign of M corresponds to diamagnetic behavior.

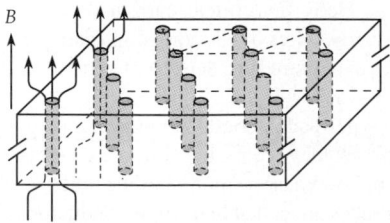

Figure 28.123: Vortex lattice of flux tubes in type-II superconductors.

Flux quantum, the elementary quantity of magnetic flux. In flux tubes, it is equal to

$$\Phi_0 = \frac{h}{2e} = 2 \cdot 10^{-15} \text{ Vs}.$$

➤ The number 2 in the denominator is a consequence of the double charge of a Cooper pair.

7. London penetration depth and Ginsburg-Landau parameter

London penetration depth, usually denoted λ. It determines the penetration depth of a magnetic field into a superconductor.

Coherence length, usually denoted ξ. It corresponds to the spatial extension of a Cooper pair. The ratio of λ to ξ, the Ginsburg-Landau parameter κ, distinguishes between type-I and type-II superconductors.

Ginsburg-Landau parameter,

$$\kappa = \frac{\lambda}{\xi}.$$

▲ Superconductor of the first kind: $\kappa < \dfrac{1}{\sqrt{2}}$.

▲ Superconductor of the second kind: $\kappa > \dfrac{1}{\sqrt{2}}$.

| M | Because of the complete expulsion of magnetic fields from the interior of a superconductor, superconducting materials are used to shield unwanted electromagnetic fields.

28.7.2 High-temperature superconductors

High-temperature superconductor (**HTSC**), superconducting copper-oxide compounds with critical temperatures $T_c \geq 80$ K. They are crystallized in the tetragonal **perovskite structure**. This leads to an anisotropy of the superconducting properties.

➤ HTSC exhibit an appreciable residual resistance due to the thermal motion of the magnetic flux lines.

➤ HTSC may achieve great importance in the technical applications of superconductivity. In order to reach the superconducting state, it is no longer necessary to use expensive liquid helium; the temperature of liquid nitrogen is sufficient.

The HTSC most investigated at present is $YBa_2Cu_3O_7$. Depending on the oxygen content of the sample, the critical temperature is $60 - 93$ K.

➤ These superconductors are ceramic and exhibit a relatively low critical current density at $T = 77$ K in zero-field ($B = 0$ T).

1. Families of high-temperature superconductors and material-specific properties

The following table gives the most important families of high-temperature superconductors:

Denotation	Chemical formula	Maximum T_c
123-HTSC	$(Y, Eu, Gd, \cdot)Ba_2CU_3O_7$	92 (YBCO)
bismuth-22$(n-1)n$	$Bi_2Sr_2Ca_{n-1}Cu_nO_{2n+4}$	90 (Bi-2212) 122 (Bi-2223) 90 (Bi-2234)
thallium-22$(n-1)n$	$Tl_2Ba_2Ca_{n-1}Cu_nO_{2n+4}$	110 (Tl-2212) 127 (Tl-2223) 119 (Tl-2234)
thallium-12$(n-1)n$	$Tl(Sr, Ba)_2Ca_{n-1}Cu_nO_{2n+3}$	90 (Tl-1212) 122 (Tl-1223) 122 (Tl-1234) 110 (Tl-1245)

▲   In all HTSC a certain number of CuO layers with interpolated layers of Y or Ca ions are arranged to a pack. The conducting CuO layers are separated by insulating layers (BaO, SrO or TcO layers).

▲   In HTSC the superconducting properties are strongly anisotropic (j_c, $H_{c1,2}$ ∥ to the CuO layer 5 to 10 times stronger than j_c, $H_{c1,2}$ ⊥ to the CuO layer).

▲   The many grain boundaries in the ceramic HTSC become barriers for the Cooper pairs and reduce the critical current.

2. Methods of producton of HTSC layers

Epitaxial HTSC films are obtained by growth of films on monocrystalline substrates. The anisotropy of the HTSC is utilized in the production of these monocrystalline layers, j_c increases. $SrTiO_3$, $LaHCO_3$ and also Al_2O_3 are used as substrates.

   Texturization, another method to increase the critical current density. The random distribution of crystallites is converted by controlled crystallization to a more or less oriented distribution of the crystal axes about a given direction.

➤   This method of texturization is applied to compact HTSC ceramics.

■   Superconducting resonators: because of their energy gap, HTSC exhibit significantly lower HF losses in the frequency range up to 100 GHz than normal conductors (**Fig. 28.125**).

■   Miniaturization of antennas in the lower GHz-range and for millimeter wave antennas. They exhibit significantly lower losses than normal conductors (**Fig. 28.124**).

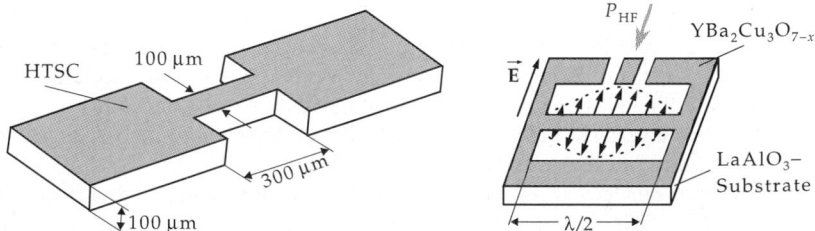

Figure 28.124: Model of an antenna made of HTSC layers.

Figure 28.125: Model of a resonator using $YBa_2Cu_3O_7$.

■   High-current conductors in low magnetic fields: the Bi-2223 phase is used as superconductor. The ceramicized powder is filled into Ag-tubes. These tubes are cast

or rolled and treated by heat. Critical current densities (at $T = 77$ K and 0 T) of 13000 A/cm^2 have been reached.
➤ HTSC ceramics have some undesirable material properties, e.g.:
• high brittleness,
• high instability against extraction of oxygen.

28.8   Magnetic properties

Magnetism, a quantum-mechanical phenomenon, state of order of matter occurring in conductors and insulators in several forms. Metals at low temperatures are ordered either as superconductors or magnetically aligned systems.

Magnetization, M, defined as the quotient of the magnetic moment and the volume of the sample. M depends on the strength of the external magnetic field, and on the temperature.

Definitions:
cgs system: $\vec{B}' = \vec{B}_a + 4\pi\vec{M}$,
SI system: $\vec{B} = \vec{B}_a + \mu_0\vec{M}$.
$\vec{B}_a$: external magnetic flux density.

Magnetic susceptibility, χ_m, quotient of the magnitude of magnetization $|\vec{M}|$ and the magnitude of the magnetic field strength $|\vec{H}|$,

$$\chi_m = \frac{M}{H}, \quad \text{or} \quad \chi_m = \frac{\partial M}{\partial H}.$$

Dimension: the magnetic susceptibility is dimensionless, according to $\vec{B} = \mu_0\vec{H} + \vec{I} = \mu_0(\vec{H} + \vec{M})$, the definition of the magnetic polarization $\vec{I} = \mu_0\chi_m\vec{H}$, and the constitutive equation $\vec{B} = \mu_r\mu_0\vec{H}$ (μ_0: magnetic field constant, μ_r: relative permeability, $\mu = \mu_r\mu_0$: permeability). The magnetic susceptibility is related to the relative permability in an isotropic medium by

$$\chi_m = \mu_r - 1.$$

The numerical measures of the susceptibility in the cgs system and in the SI system differ by a factor 4π.

1. Kinds of magnetism

For **paramagnetic substances** $\chi_m > 0$ ($\vec{M}$, $\vec{H}$ parallel),
    for **diamagnetic substances** $\chi_m < 0$ ($\vec{M}$, $\vec{H}$ antiparallel),
        for **ferromagnets** χ_m depends on magnetization history.

| M | Magnetic susceptibilities are measured by means of a magnetic balance via the force $\vec{F}$ on a sample in an inhomogeneous magnetic field $\vec{H}$,

$$F_x \sim V \cdot \vec{H} \cdot \frac{d\vec{H}}{dx}.$$

It has to be assumed that the sample is small enough that both $\vec{H}$ and $\frac{d\vec{H}}{dx}$ effectively do not vary through the volume of the sample. This method allows the measurement of changes in the susceptibility down to 10^{-10}.

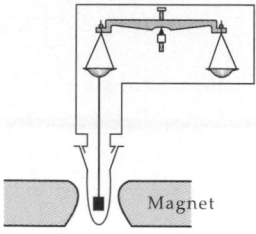

Magnet

Figure 28.126: Magnetic
balance.

2. Diamagnetism,

is connected with the tendency of electric charges to shield the interior of a body against
an external magnetic field.

➤   This is analagous to Lenz's law in electrodynamics.

Diamagnetic molar susceptibility, after Langevin, generated by the electrons of the individual atoms:

diamagnetic molar susceptibility			$L^3 mol^{-1}$
	Symbol	Unit	Quantity
$\chi_d = -\mu_e \dfrac{N_A Z e^2}{6m} \langle r^2 \rangle$ (SI)	χ_d	1	diamagnetic susceptibility
	μ_e	A m^2	magnetic moment of electron
$\chi_d = -\dfrac{N_A Z e^2}{6mc^2} \langle r^2 \rangle$	Z	1	atomic number
	e	C	elementary charge
	N_A	mol^{-1}	Avogadro's number
	m	kg	electron mass
	c	m s^{-1}	speed of light

Here, $\langle r^2 \rangle$ is the mean-squared distance of the electrons from the atomic nucleus.
Typical values of the diamagnetic molar susceptibility are

	He	Ne	Ar	Kr	Xe
χ_{d_μ} (in 10^{-12} m^3/mol)	-1.9	-7.2	-19.4	-28.0	-43.0.

➤   The formula given above presupposes that the field direction and the symmetry axis
of the system coincide. In many molecules this is not the case, however.

▲   Superconductors of the first kind also behave like ideal diamagnets.

3. Paramagnetism,

occurs in:
- atoms, molecules and lattice defects with an odd number of electrons. The total spin
cannot be zero in this case;
- free atoms and ions with a partially filled inner shell, e.g., in transition metals, rare
earths and actinides;

➤   inclusion of these atoms into a crystal lattice is not necessarily related to the paramagnetic behavior of the entire solid.

- several substances with an even number of electrons;
- metals.

4. Langevin equation and Curie's law

Magnetization of a mole of a substance with an atomic magnetic moment μ is described by the Langevin equation:

$$M = N_A \cdot \mu \cdot L(x), \quad x = \frac{\mu \cdot H}{k_B \cdot T}.$$

The Langevin function $L(x)$ is given by

$$L(x) = \coth x - \frac{1}{x}.$$

For high temperatures $T \gg \dfrac{\mu H}{k_B}$, $x \ll 1$, expansion of the coth-function yields

$$L(x) \approx \frac{x}{3}.$$

The dependence of the magnetic susceptibility on the temperature in this approximation is given by **Curie's law**:

$$\chi_M = \frac{M}{H} = \frac{N_A \mu^2}{3k_B T} = \frac{C_p}{T}.$$

The quantity $C_p = N_A \mu^2/(3k_B)$ depends on the substance.

M Owing to the $\dfrac{1}{T}$-behavior of the magnetic susceptibility, Curie's law allows the use of paramagnetic salts for measuring low temperatures ($T < 1$ K). The **paramagnetism of the conduction electrons** arises from the spin moment of the electrons. For $\dfrac{\mu_B H}{k_B T} \ll 1$,

$$\chi_M = \frac{N_A \mu_B^2}{k_B T}, \quad \mu_B\text{: Bohr magneton.}$$

The Bohr magneton μ_B is defined in the cgs system as $e\hbar/(2mc)$, in the SI system as $e\hbar/(2m)$. It corresponds essentially to the magnetic spin moment of a free electron.

Only conduction electrons in the vicinity of the Fermi energy may contribute to the paramagnetic susceptibility. This fraction is given by T/T_F. The contribution of the conduction electrons to the susceptibility is

$$\chi_{el} = \chi_m \frac{T}{T_F} = \frac{N_A \mu_B^2}{k_B T_F}.$$

▲ The conduction electrons yield a **temperature-independent** contribution to the susceptibility at high temperatures.

▲ At low temperatures, all electron spins are aligned parallel to the field.

28.8.1   Ferromagnetism

1. Generation of ferromagnetism

Ferromagnets contain spontaneously aligned domains with equal orientation of the magnetization. These domains are denoted Weiss domains. Ferromagnetism is caused by unoccupied inner electron shells.

Exchange integral, I, determines the interaction energy E_{int} of neighboring atoms via the magnetic dipole-dipole interaction of the electron spins $\vec{s}_i, \vec{s}_{i+1}$ ($i, i+1$: neighboring sites in a linear spin chain):

$$E_{\text{int}} = -\frac{2I}{\hbar^2} (\vec{s}_i \cdot \vec{s}_{i+1}).$$

The exchange integral I depends on the overlap of the probability densities of the electrons in both atoms. The interaction is therefore limited to immediately neighboring atoms.

▲   Electrons with antiparallel spins attract each other ($I > 0$) if the electrostatic repulsion is ignored.

▲   A purely magnetic dipole-dipole interaction cannot be the origin of the alignment of the domains.

▲   The spin state of neighboring atomic electrons is influenced by the conduction electrons.

▲   Ferromagnetism is related to conduction electrons. Therefore, ferromagnetism arises only in metals.

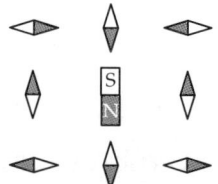

Figure 28.127: Orientation of atomic dipoles under the influence of a central dipole.

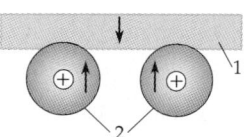

Figure 28.128: Exchange interaction between adjacent atoms by means of conduction electrons. 1 – conduction electrons, 2 – atoms.

2. Langevin equation of ferromagnetism

Molecular field, a model field generated by spontaneous magnetization:

$$\vec{H}_{\text{molecular field}} = \lambda \cdot \vec{M}.$$

The atomic magnetic moments are subject to the external field $\vec{H}$ and to this molecular field. The magnetization is given by

$$M = N_A \mu_B \tanh \frac{\mu_B (H + \lambda M)}{k_B T}.$$

In the absence of an external magnetic field,

$$M = N_A \mu_B \tanh \frac{\lambda \mu_B M}{k_B T} = f(M, T).$$

Fig. 28.129 shows graphical solutions of this equation and their temperature dependence.

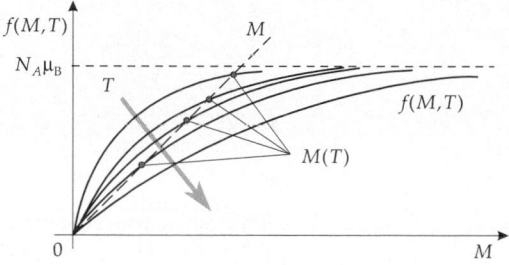

Figure 28.129: Graphical solution of the Langevin equation.

There is no solution if the slope of the function $f(M, T)$ is smaller than or equal to 1. Then, magnetization breaks down. This phenomenon occurs at temperatures above the Curie temperature T_C,

$$T > T_C = \frac{N_A \mu_B^2 \lambda}{k_B}.$$

Curie-Weiss law, describes the magnetization for $T > T_C$:

$$M = \frac{T_C \cdot H}{\lambda(T - T_C)}, \qquad \chi_m = \frac{C}{T - T_C}.$$

3. Magnetic hysteresis

Hysteresis, the dependence of a physical state in a solid on the former states.

Magnetic hysteresis, dependence of the magnetic flux density on the magnetic field strength. The phenomenon occurs in all ferromagnetic and ferrimagnetic substances.

Initial magnetization curve, the path of magnetization of a sample not previously subjected to an external field as a function of the applied magnetic field.

Saturation magnetization, M_S, is reached if all atomic magnetic dipoles are aligned parallel. The entire sample then consists of only one domain.

Remanence (residual magnetism), B_R, the residual magnetization remaining when the magnetic field H drops to zero after having reached the saturation magnetization.

Coercive field strength, H_c, the field strength that has to be applied opposite to the original direction of the magnetic field in order to reduce the magnetization M to zero.

▲　The area enclosed by the hysteresis curve represents the energy loss, i.e., the absorption of magnetic energy in the material by remagnetization.

▲　For small variations of the field intensity, the domains are displaced again reversibly.

Barkhausen effect, irreversible displacements and rotations of domain walls at higher field strengths. **Fig. 28.131** below shows a section of the hysteresis curve with high resolution.

Soft magnets, magnets with a narrow and flat hysteresis. They have low coercive field strengths and low remanence.

Hard magnets, magnets with an almost rectangular hysteresis with high remanence and large coercive field strength.

■　Ferromagnets are of great technical importance. Soft-magnetic materials are used in transformers, in electromagnets and for magnetic shielding. Hard magnets are used as permanent magnets in generators and machines. Most important use is in storage media (e.g., for recorder tapes, video tapes, hard disks).

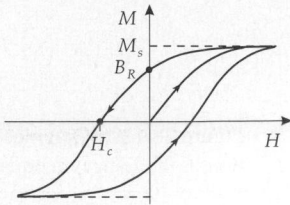

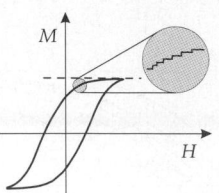

Figure 28.130: Ferromagnetic hysteresis. M_s: saturation magnetization, B_R: remanence, H_C: coercive field strength.

Figure 28.131: Barkhausen discontinuities.

28.8.2 Antiferromagnetism and ferrimagnetism

Antiferromagnetism and ferrimagnetism, there exist sublattices with opposite magnetization.

Antiferromagnetism, the magnetization of the sublattices is compensating, since the antiparallel-aligned magnetic moments of the structure components are of equal magnitude. The resultant magnetization is zero, no domains occur. The substance behaves diamagnetically.

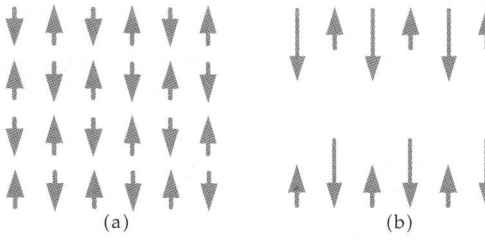

Figure 28.132: Antiferromagnet (a) and ferrimagnet (b).

Néel temperature, T_N, the temperature above which all atomic moments are statistically disordered due to thermal motion. The substance is then paramagnetic. For $T \geq T_N$, the susceptibility is given by

$$\chi_m = \frac{C}{T + T_N},$$

T_N representing paramagnetic Néel temperature.

■ Manganese oxide (MnO) is antiferromagnetic.

Ferrimagnetism, the magnetic moments of the sublattices are only partly compensating, since the antiparallel-aligned magnetic moments of adjacent structure components have different magnitudes. The substance behaves like a weak ferromagnet.

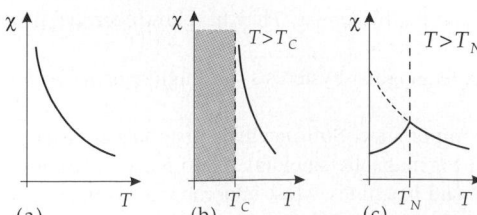

Figure 28.133: Susceptibility versus temperature of: (a) paramagnet, (b) ferromagnet (with complex behavior in the shadowed region), and (c) antiferromagnet. T_C: Curie temperature, T_N: Néel temperature.

- Iron oxide Fe_2O_3 behaves ferrimagnetically. In this compound, the iron atom occurs in two-valued and three-valued forms. Correspondingly, there are two atomic moments of different magnitude.

➤ The theoretical description of antiferromagnetism and ferrimagnetism is, similar to that of ferromagnetism, based on the molecular field approximation. The molecular fields of the two sublattices receive different signs.

28.9 Dielectric properties

Dielectric, a crystal with a conductance by about 20 orders of magnitude smaller than that of a metal. The capacitance of a capacitor increases if a dielectric is placed between the capacitor plates.

Polarization, $\vec{P}$, electric dipole moment of a solid per unit volume.

Orientation polarization, the alignment of a polar molecule in an electric field. The charge distribution in the molecule remains unchanged.

Displacement polarization, displacement of electric charges in an dielectric under the influence of an electric field $\vec{E}$. Neutral molecules change to dipoles.

➤ In both cases, the polarization results in a **separation of charge**.

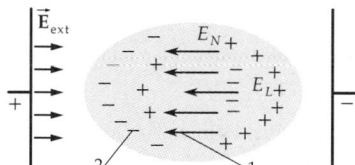

Figure 28.134: Displacement polarization. 1 – dipoles generated by the local field E_L, 2 – charges generated by the de-electrification field E_N.

The induced or permanent dipoles are aligned by the electric field.

1. Electric displacement density in the dielectric

Electric displacement density, $\vec{D}$, characterizes the electric field in a dielectric:

electric displacement density $\vec{D}$			ITL^{-2}
	Symbol	Unit	Quantity
$\vec{D} = \varepsilon_0\vec{E} + \vec{P}$	$\vec{D}$	$C\,m^{-2}$	electric displacement density
	$\vec{E}$	$V\,m^{-1}$	electric field strength
	$\vec{P}$	$C\,m^{-2}$	electric polarization
	ε_0	$C\,V^{-1}\,m^{-1}$	permittivity constant of free space

2. Charge separation in a dielectric

Electric susceptibility, χ, the amount of charge separation in a dielectric. χ describes the macroscopic dielectric property of the material.

➤ For low electric field strengths, the electric polarization is proportional to the electric field intensity:

$$\vec{P} = \varepsilon_0\chi\vec{E},$$

where χ is the electric susceptibility, $\vec{E}$ is the electric field strength and ε_0 is the permittivity constant of free space.

In a few exceptional cases, a constant term appears in the formula (e.g., Seignette salts).

For low electric field strengths, it holds:

displacement density $\vec{D}$ for low electric field strengths			ITL^{-2}
	Symbol	Unit	Quantity
$\vec{D} = \varepsilon_0\vec{E} + \varepsilon_0\chi\vec{E}$	$\vec{D}$	C m^{-2}	electric displacement density
$\quad = \varepsilon_0\varepsilon_r\vec{E}$	$\vec{E}$	V m^{-1}	electric field strength
	χ	1	electric susceptibility
$\varepsilon_r = 1 + \chi$	ε_0	C V^{-1} m^{-1}	permittivity constant of free space
	ε_r	1	relative permittivity

➤ Laser light may generate such high field intensities that the approximation of a linear relation between polarization and electric field intensity is no longer valid. The polarization then has to be expanded into a power series,

$$\vec{P} = \varepsilon \left(A + \chi E + \chi' E^2 + \cdots \right) \frac{\vec{E}}{E}.$$

▲ In anisotropic materials, the relative permittivity is a tensor.
▲ The relative permittivity is frequency-dependent.

3. Polarizability and local field

Polarizability, α_i, determines the magnitude of the dipole moment $\vec{p}_i$ generated under the influence of an electric field at the position of a dipole,

$$\vec{p}_i = \alpha_i \cdot \vec{E}_{Li},$$

where $\vec{E}_{Li}$ is the local field intensity at position i. Polarizability is an atomic quantity and depends on the structure of the crystal.

Local field, $\vec{E}_L$, superposition of the external field $\vec{E}_{ext}$ with the field $\vec{E}_{sample}$ of the dipoles of the sample,

$$\vec{E}_L = \vec{E}_{ext} + \vec{E}_{sample}.$$

➤ As a rule, one restricts oneself to geometrically simple test bodies such as ellipsoids, spheres or disks.

De-electrification field, $\vec{E}_N$, the field generated by the charges on the surface of a test body (e.g., ellipsoid) directed opposite to the external field and depends on the geometry of the sample. Inside the sample,

$$\vec{E} = \vec{E}_{ext} + \vec{E}_N$$

with

$$\vec{E}_N = -\frac{1}{\varepsilon_0} N\vec{P}; \quad N = \begin{cases} 1 & \text{ellipsoid} \\ \frac{1}{3} & \text{sphere} \\ 1 & \text{disk area} \perp \vec{E}_{ext} \\ 0 & \text{disk area} \parallel \vec{E}_{ext} \end{cases}.$$

Lorentz field, $\vec{\mathbf{E}}_i$, electric field inside a fictitious cavity in the interior of a polarized dielectric,

$$\vec{\mathbf{E}}_i = -\vec{\mathbf{E}}_N = -\frac{N}{\varepsilon_0} \cdot \vec{\mathbf{P}}.$$

N is determined by the geometrical shape of the cavity.

4. Dipole field in the crystal lattice

Dipole field, $\vec{\mathbf{E}}_D(\vec{\mathbf{r}})$, electric field at the distance $\vec{\mathbf{r}}$ from a **point dipole** at the position $\vec{\mathbf{r}} = \vec{\mathbf{0}}$, with the dipole moment $\vec{\mathbf{p}}$:

electric field of a dipole			$\mathbf{LT^{-3}MI^{-1}}$
	Symbol	Unit	Quantity
$\vec{\mathbf{E}}_D(\vec{\mathbf{r}}) = \dfrac{3(\vec{\mathbf{p}} \cdot \vec{\mathbf{r}})\vec{\mathbf{r}} - r^2\vec{\mathbf{p}}}{4\pi \varepsilon_0 r^5}$	$\vec{\mathbf{E}}_D(\vec{\mathbf{r}})$	V/m	dipole field
	$\vec{\mathbf{r}}$	m	distance vector to the dipole
	$\vec{\mathbf{p}}$	C m	dipole moment
	ε_0	C/(V m)	permittivity constant of free space

Dipole field in a crystal lattice:

$$\vec{\mathbf{E}}_D = \sum_i \vec{\mathbf{E}}_D(\vec{\mathbf{r}}_i).$$

➤ The dipole field $\vec{\mathbf{E}}_D$ depends on the lattice structure.
➤ For all lattices with cubic symmetry, the sum over the lattice yields zero, i.e., the dipole field vanishes, $\vec{\mathbf{E}}_D = 0$. For lattices with tetragonal perovskite structure ($\longrightarrow$ high-temperature superconductor), this is not so.
■ The local field for **cubic lattice types** with a sphere as test body is

$$\vec{\mathbf{E}}_L = \vec{\mathbf{E}}_{ext} - \frac{1}{\varepsilon_0} \cdot \vec{\mathbf{P}} + \frac{1}{3\varepsilon_0} \cdot \vec{\mathbf{P}}.$$

This local field generates the local polarization of a lattice atom.
▲ For N_V lattice atoms of equal kind per unit of volume the polarization of the test body is

$$\vec{\mathbf{P}} = \varepsilon_0 N_V \alpha \vec{\mathbf{E}}_L = \varepsilon_0 N_V \alpha \left(\vec{\mathbf{E}} + \frac{1}{3\varepsilon_0} \cdot \vec{\mathbf{P}}\right).$$

polarization of a spherical test body			$\mathbf{ITL^{-2}}$
	Symbol	Unit	Quantity
$\vec{\mathbf{P}} = \varepsilon_0 \chi \vec{\mathbf{E}}$	$\vec{\mathbf{P}}$	C m^{-2}	polarization
	χ	1	electric susceptibility
$\chi = \dfrac{N_V \alpha}{1 - \frac{1}{3}N_V \alpha}$	$\vec{\mathbf{E}}$	V m^{-1}	electric field strength
	N_V	1	atomic density in the lattice
	α	1	polarizability

➤   If the crystal is composed of different species of atoms, and if the atoms have different polarizability, then one has to sum over the atoms.

electric susceptibility			1
$$\chi = \dfrac{\displaystyle\sum_i N_i\alpha_i}{1 - \dfrac{1}{3}\displaystyle\sum_i N_i\alpha_i}$$	Symbol	Unit	Quantity
	χ	1	electric susceptibility
	N_i	1	number of atoms i
	α_i	1	polarizability of atoms i

5. Electronic and ionic polarization

Electronic polarization, deformation and displacement of the electron cloud of an atom relative to the practically point-like positively charged atomic nucleus (**Fig. 28.135**).

➤   Electronic polarization may always occur.

➤   In the field of an electromagnetic radiation, the electronic polarization is not a static quantity. It will oscillate in the rhythm of the electromagnetic waves. But accelerated charges radiate energy: the forced oscillation of the electronic charge cloud is damped. Therefore, the polarizability α_i, and thus the susceptibility χ, are complex numbers. The relative permittivity ε_r also becomes complex.

▲   For a dielectric in an alternating electromagnetic field, the optical quantities refractive index n and absorption coefficient κ and the electric susceptibility χ are related as follows:

relative permittivity ε_r			1
	Symbol	Unit	Quantity
$\varepsilon_r = 1 + \chi = (n + j\kappa)^2$	ε_r	1	relative permittivity
	χ	1	electric susceptibility
	n	1	refractive index
	κ	1	absorption coefficient
	j	–	imaginary unit

Ionic polarization, occurs in ionic crystals. The positive and negative ions are deflected differently by an electric field.

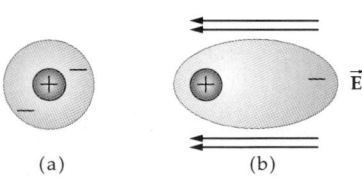

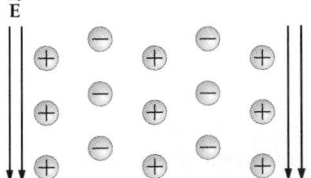

Figure 28.135: Electronic polarization in an electric field $\vec{E}$. Shadowed area: electronic cloud. (a): charge distribution in an atom without external field, (b): charge distribution in an atom in the field.

Figure 28.136: Ionic polarization in an electric field $\vec{E}$.

Total polarization, the sum of ionic and electronic polarizations.

28.9.1   Para-electric materials

Para-electric materials, substances containing electric dipoles even absent an external electric field that are, however, disordered due to thermal motion.

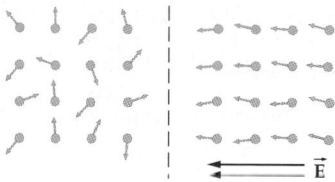

Figure 28.137: Orientation polarization in para-electric substances.

Orientation polarizability, α_{orient}, a function of frequency, and complex because of damping,

$$\alpha_{\text{orient}} = \frac{\alpha_0}{1 - j\omega\tau} \, .$$

τ is a characteristic time constant—the relaxation time. α_0 is the static polarizability when applying a field that is constant in time.

■   Orientation polarization occurs in liquid crystals.

➤   The relative permittivity $\varepsilon_r = 1 + \chi$ for water at room temperature is 81 under an applied static field ($\omega = 0$).

In the range of visible light, the corresponding value is only 1.77. Therefore, water is transparent for light. The difference of the relative permittivity for a static field and for visible light is due to the orientation polarization, which is essentially completely suppressed at high frequencies because of damping.

Dielectric losses, w, arise when applying an electric field because of the resistance against a polarization,

$$w = \text{Im}\,(\chi) \cdot E^2 \omega \, ,$$

where $\text{Im}\,(\chi)$ is the imaginary part of the complex electric susceptibility.

28.9.2   Ferroelectrics

1. Electrets

Ferroelectric crystals exhibit a spontaneous polarization absent even an external electric field.

Electrets, ferroelectric crystals with a permanent dipole moment. Their polarization cannot be influenced by an external field.

➤   Electrets are analogous to permanent magnets.

■   Examples of electrets: nylon and wax.

▲   As a rule, ferroelectric crystals show a hysteresis similar to ferromagnetic materials.

➤   The hysteresis of electrets is almost a rectangle.

Ferroelectric Curie temperature, T_C, the temperature above which the crystal is no longer in a ferroelectric state.

| M | **Production of electrets**: in a thermal or photoelectric method. A sample is heated beyond the Curie temperature, and in this state is exposed to a strong electric field. The dipoles aligned by the field are then frozen by cooling. Thermally, this is a non-

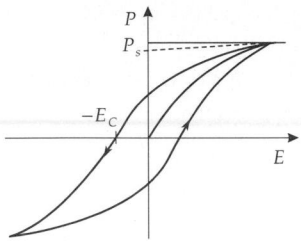

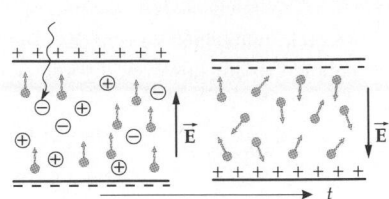

Figure 28.138: Ferroelectric hysteresis. P_s: spontaneous polarization, E_c: coercive field strength.

Figure 28.139: Influence of ionizing radiation on the charge distribution in electrets.

equilibrium state. It will pass over to the equilibrium state with a relaxation time τ. For electrets, this relaxation time is in the range of years.

▲   Ionizing radiation generates free charge carriers in an electret. As a result, the surface charge changes. The internal field is inverted.

■   Electrets are used in radiation detection.

2. *Piezoelectricity,*

property of a dielectric to become polarized under the influence of a mechanical deformation and, conversely, to become deformed under the influence of an electric field (electrostriction). The origin of piezoelectricity is the difference between the elasticity moduli for the two sublattices of positive and negative ions.

▲   Ionic crystals may exhibit piezoelectricity. The lack of a symmetry center is a necessary condition.

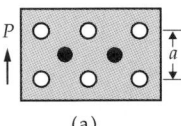 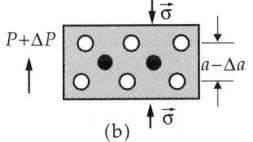

Figure 28.140: Piezoelectricity (schematic). (a): crystal without mechanical stress, (b): crystal with mechanical stress σ. ΔP: piezoelectric polarization induced by stress.

(a)   (b)

■   **Conversion of pressure to electric voltage**:
•   piezoelectric gas lighter,
•   piezoelectric microphone.
      Conversion of electric voltage to deformation and vice versa:
•   oscillating quartz.
➤   Piezoelectric crystals are not always ferroelectric. Example: quartz.
Domains, regions in ferroelectrics over which the polarization has equal orientation for all structural components. In adjacent domains, other orientations are prevalent.
▲   Domains have a size of several micrometers.
➤   So far, no satisfactory microscopic explanation of ferroelectricity has been found.

28.10   Optical properties of crystals

▲   Crystals that are not electrically conducting at room temperature are usually transparent.

▲   Colorless crystals do not have the possibility of exciting electron states or vibrational states in the visible spectral range.

➤   The wavelengths in the visible spectral range are between 360 nm and 740 nm. This range of wavelengths corresponds to energies between 3.4 eV and 1.7 eV.

28.10.1 Excitons and their properties

Exciton, bound electron-hole pair. In the creation of an exciton, the binding energy E_B is released. Therefore, at least the energy E_g is needed for the generation of an unbound particle-hole pair, whereas for generating a bound particle-hole pair only the smaller energy $E_g - E_B$ is needed.

➤   Excitons may move through the crystal. They transport excitation energy, but no charge.

Recombination, decay of the exciton. The electron falls back into the unoccupied state (hole). The released excitation energy leaves the crystal as radiation.

➤   The electron-hole pair may be considered analoguous to the positronium atom (bound $e^+ e^-$-system).

Energy level of an exciton. The energy level of weakly bound excitons (Mott-Wannier excitons) relative to the top of the valence band is described by the following formula:

energy level of the Mott-Wannier exciton			$\mathbf{ML^2T^{-2}}$
	Symbol	Unit	Quantity
	E_n	J	exciton energy
	E_g	J	energy gap
	μ	kg	reduced mass of electron-hole system
$E_n = E_g - \dfrac{\mu e^4}{8h^2 \varepsilon_0^2 \varepsilon_r^2 n^2}$	m_e^*	kg	effective mass of electron
$\dfrac{1}{\mu} = \dfrac{1}{m_e^*} + \dfrac{1}{m_h^*}$	m_h^*	kg	effective mass of hole
	e	C	elementary charge
	h	J s	quantum of action
	ε_r	$C^2\,N^{-1}\,m^{-2}$	relative permittivity of crystal
	n	1	principal quantum number
	ε_0	A s/(V m)	permittivity constant of free space

■ Cu_2O is a crystal; its absorption spectrum at low temperature due to exciton excitations is described by the above equation.

[M]   Absorption spectra are measured by means of a set-up sketched in **Fig. 28.142**.

Frenkel exciton, bound electron-hole pair localized at a lattice atom of the crystal. An ideal Frenkel exciton travels as a wave through the entire crystal, but the electron and hole always remain close to each other.

▲   In alkali-halide crystals, the excitons of lowest energy are localized at the negative halogen ions.

▲   Pure alkali-halide crystals are transparent in the visible range of the spectrum. The absorption in the ultraviolet range exhibits considerable structure.

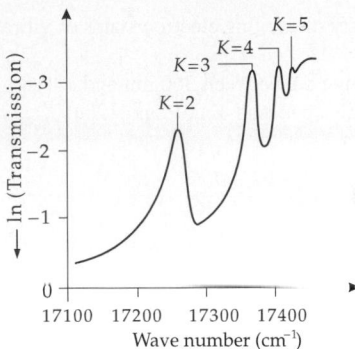

Figure 28.141: Absorption spectrum of Cu$_2$O.

Figure 28.142: Optical spectrometer. 1 – tungsten incandescent filament, 2 – lens, 3 – sample, 4 – Dewar vessel, 5 – entrance slit, 6 – photomultiplier, 7 – Rowland circle, 8 – concave grating.

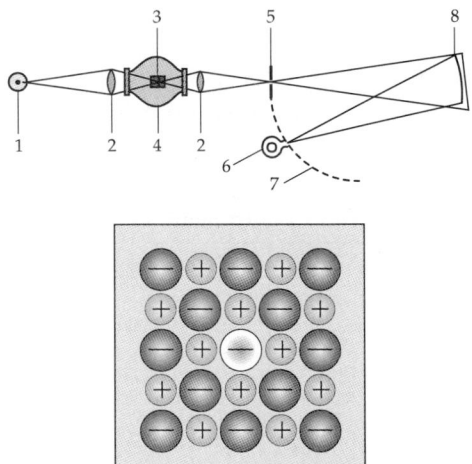

Figure 28.143: Schematic representation of a Frenkel exciton, localized at an atom of an alkali-halide crystal.

28.10.2   Photoconductivity

Photoconductivity, the increase of the electric conductivity of an electrically insulating crystal under the influence of radiation. In the elementary process of photoabsorption, an electron from the valence band is lifted up to the conduction band (thereby leaving a hole in the valence band).

▲   Both the holes and the electrons may contribute to the conductivity.

Time variation of the electron concentration n in the frame of a simple model (electron-hole pairs are created uniformly over the entire crystal; the recombination proceeds via the direct annihilation of electron-hole pairs) follows from a balance equation:

time variation of electron concentration			$\mathbf{L^{-3}T^{-1}}$
	Symbol	Unit	Quantity
$\dfrac{dn}{dt} = L - An^2$	n	m^{-3}	electron concentration
	L	$s^{-1}\,m^{-3}$	absorption probability
	A	$m^3\,s^{-1}$	measure for recombination probability

In the steady state, $\dfrac{dn}{dt} = 0$ and

$$n_0 = \sqrt{\frac{L}{A}}.$$

Time constant, t_0, characterizes the speed of decrease of charge carriers after switching off the source of light:

$$n = \frac{n_0}{1 + \dfrac{t}{t_0}}.$$

During the time t_0, the concentration of charge carriers drops to $n_0/2$.

Sensitivity, G, ratio of photon flow I to absorption probability,

$$G = \frac{I}{L \cdot d \cdot e} \qquad d\text{: thickness of sample.}$$

Traps, defects in the crystal that offer energy levels in the range between the conduction band and the valence band, and thus may "hold back" an electron or a hole between the energy bands.

■ Traps crucially affect the time behavior of the photoconducting cell in an **exposure meter**, or in the **luminescence layer** of a TV tube.

28.10.3 Luminescence

Luminescence, absorption of energy by matter and subsequent re-emission in the visible spectral range, or in adjacent spectral regions.

➤ The type of excitation does not matter.

Luminophors, crystalline solids capable of luminescence.

Fluorescence, emission of light during the excitation, or within a very short time delay of 10^{-8} s after the excitation.

➤ The time interval of 10^{-8} s is of the order of the lifetime of an atomic energy state for an allowed electric dipole transition in the visible spectral range.

Phosphorescence, afterglow during a finite time after switching off the excitation.

➤ The delay time may vary over a broad range: alkaline earth, zinc sulphide and zinc silicate luminophors have afterglow times between μs (**TV screens**) and several hours (**luminous dials**).

➤ Many solids have a low efficiency for conversion of other forms of energy to radiation.

Activators, substances that in weak admixtures may cause an appreciable increase of the efficiency.

28.10.4 Optoelectronic properties

Opto-electronics, deals with the phenomena involved in the conversion of electric energy into optical energy and vice versa.

▲ The most important component is the semiconductor pn-junction.

Light-emitting diode (LED) or **luminescence diode**, consists of a pn-junction.

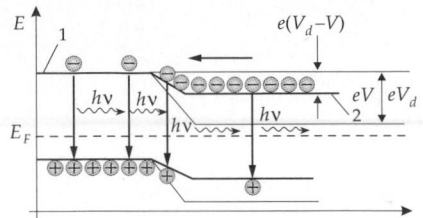

Figure 28.144: Schematic picture of a pn-junction of an LED. 1 – p-region, 2 – n-region.

The band deformation is weakened by a voltage in the flow direction. The electrons then must spend only the energy $e(V_d - V)$ in order to pass from the n-region to the p-region. Conversely, this also holds for the holes. In the vicinity of the junction, the electrons and holes recombine and thereby release the energy of the band gap E_g in the form of photons.

▲   LEDs produce almost monochromatic, but in general **incoherent**, light of wavelength

$$\lambda \approx \frac{1.24}{E_g(\text{eV})} \, \mu\text{m}$$

(E_g in electron volts). The color of the LED is therefore determined by the width of the forbidden zone.

▲   The radiant power released is proportional to the current.

▲   LEDs have very long lifetimes.

Laser diode, LD, pn-junction with very high doping $n_D \approx 10^{19}$ cm^{-3} (**degenerated semiconductor**).

▲   Laser diodes produce **coherent** radiation.

▲   Electrons occupy the conduction band in the n-region. Conversely, the holes occupy the valence band.

Population inversion for laser diodes: Energetically high-lying states in the conduction band are occupied by electrons while low-lying states are empty (occurs in the junction region of the active zone).

▲   Hence, the basic condition for stimulated emission of the laser is fulfilled.

Resonator mirrors, necessary for feedback, they form the boundary surfaces of the semiconductor crystal. The reflecting-end faces are cleavage faces of the crystal that are perfectly planar and parallel. Because of the high refractive index of semiconductors, the reflection is very strong.

Spontaneous emission ($\rightarrow$ atomic physics), occurs for low current intensities.

Threshold current, I_{th}, current intensity above which stimulated emission occurs.

Longitudinal vibrational modes of laser, standing waves constituting the laser spectrum. Owing to the finite length L of the laser diode (distance between the reflecting planes) the only standing waves occurring have wavelengths

$$\lambda = \frac{m}{n} \frac{L}{2}; \quad m = 1, 2, 3, \ldots,$$

n being the refractive index of the crystal.

Formula symbols used in quantum physics

Symbol	Unit	Designation
α	$\mathrm{C\ m^2 V^{-1}}$	polarizability
α	1	fine-structure constant
α	1	Madelung constant
β	1	small-signal current amplification
γ	$\mathrm{N\ m^2/kg^2}$	gravitational constant
Γ	MeV	decay width
Δ	$1/\mathrm{m}^2$	Laplace operator
ε	1	energy/$(m_0 c^2)$
ε	1	stretching
ε	1	efficiency of energy supply
ε	1	fast-fission factor
ε	J	electron energy
ε	J	Lennard-Jones parameter
ε_P	J	pairing energy
ε_0	$\mathrm{A\ s/V\ m}$	permittivity constant of free space
$\mathrm{d}\varepsilon/\mathrm{d}t$	s^{-1}	stretching velocity
η	1	momentum/$(m_0 c)$
η	1	efficiency
η_0	$\mathrm{N\,m^{-2}s}$	dynamical viscosity
θ, Θ	rad	angle
Θ_D	K	Debye temperature
κ/ρ	m^2/kg	mass attenuation coefficient of pair production
κ	1	absorption coefficient
κ	$\Omega^{-1}\,\mathrm{m}^{-1}$	electric conductivity
λ	$1/\mathrm{s}$	decay constant
λ	$\mathrm{W/(m\ K)}$	thermal conductivity
λ	m	wavelength
Λ	m	mean free path
μ	$1/\mathrm{m}$	linear-attenuation coefficient
μ	kg	reduced mass
μ	J/T	magnetic moment
μ	J	chemical potential
μ_B	J/T	Bohr magneton
μ_K	J/T	nuclear magneton
$\hat{\mu}_l, \hat{\mu}_s$	J/T	operator of magnetic moment
μ_n	$\mathrm{m}^2/(\mathrm{V\ s})$	mobility of electrons
μ_p	$\mathrm{m}^2/(\mathrm{V\ s})$	mobility of holes
ν	1	mean neutron number
π	1	parity
ρ	m^{-3}	particle density

(continued)

Symbol	Unit	Designation
σ/ρ	m^2/kg	mass-attenuation coefficient of Compton scattering
σ	b	cross-section
σ	$J/m^2\,K^4$	Stefan-Boltzmann constant
σ	$N\,m^{-2}$	tension
σ	1	screening constant
σ	m	Lennard-Jones parameter
τ/ρ	m^2/kg	mass-attenuation coefficient of photo effect
τ	s	mean lifetime
τ	s	relaxation time
$\Phi(\vec{r})$	$1/m^2$	particle flux
$\phi(\vec{r}, t)$	$1/(m^2 s)$	particle flux density
Φ_{tot}	W	total radiant flux
φ	$1/m^2 s$	particle flux density behind absorber
φ	rad	scattering angle
χ	1	electric susceptibility
χ_d	1	diamagnetic susceptibility
χ_μ	m^{-1}	molar susceptibility
χ_m	1	magnetic susceptibility
ψ	$m^{-3/2}$	wave function
$\psi_k(\vec{r})$	$m^{-3/2}$	state function
ω_D	$rad\,s^{-1}$	Debye frequency
ω	$rad\,s^{-1}$	angular frequency
Ω	sr	solid angle
a	m	lattice constant
a_C	MeV	coefficient of Coulomb energy
a_O	MeV	coefficient of surface energy
a_S	MeV	coefficient of symmetry energy
a_V	MeV	volume energy per nucleon
A	Bq	activity
A	$m^3\,s^{-1}$	recombination probability
A	1	mass number
A	1	amplification
b	m K	Wien's constant
B	1	baryon number
B	1	bottom quantum number
B	J	binding energy
B	T	magnetic flux density
c	m/s	speed of light
c_V	$J/(kg\,K)$	heat capacity
C	F	capacitance of capacitor
C	$J\,m^6$	Van der Waals interaction constant
C	1	charm quantum number
$\hat{C}$	1	charge conjugation operator
C_e	$J\,K^{-1}$	heat capacity of electron gas
C_{el}	J/K	heat capacity of electron gas
C_n	$kg\,s^{-2}$	elastic constant
C_{Ph}	$J\,K^{-1}$	heat capacity of phonon gas

(continued)

Symbol	Unit	Designation
d	m	interplanar crystal spacing
d_n	m	width of negative space charge region
d_p	m	width of positive space charge region
D	Gy	energy dose
$\vec{\mathbf{D}}$	$A\,s\,m^{-2}$	electric displacement density
$D(\omega)$	s	state density
e	A s	elementary charge
E	$N\,m^{-2}$	elasticity modulus
E	J	energy
dE	J	energy interval
$\vec{\mathbf{E}}$	$V\,m^{-1}$	electric field strength
E_B	J	binding energy
$E_D(\vec{\mathbf{r}})$	V/m	dipole field
E_F	J	Fermi energy
E_g	J	energy gap
E_I	J	ionization energy
E_{kin}	J	kinetic energy
E_L	J	lower edge of conduction band
E_N	J	exciton energy
E_V	J	energy of vacancy formation
E_V	J	lower edge of valence band
f	1/s	frequency
f	1	degrees of freedom
f	1	fission probability
$f(E, T)$	1	Fermi distribution
$\vec{\mathbf{F}}$	N	force
$F(Z, \eta)$	1	Fermi function
F_S	$kg\,m\,s^{-2}$	deforming force
G	$N\,m^{-2}$	shear modulus
g	1	Landé factor
g_i	1	weight factor
g_s, g_l	1	g-factor
h	J s	quantum of action
$\hbar$	J s	quantum of action $(h/2\pi)$
H	Sv	dose equivalent
$\hat{H}$	J	Hamiltonian
I	$kg\,m^2$	moment of inertia
$\bar{I}$	J	mean ionization energy
I	1	isospin quantum number
$\vec{\mathbf{I}}, \vec{\mathbf{j}}, \vec{\mathbf{J}}$	J s	total angular momentum
I_{Sp}	A	diode reverse current
ΔI_B	A	change of base current
ΔI_C	A	change of collector current
j	1	imaginary unit
j_q	$W\,m^{-2}$	heat-flow density
J	1	rotational quantum number
J, j	1	angular momentum quantum number

(*continued*)

Symbol	Unit	Designation
k	J/K	Boltzmann constant
k	1	multiplication factor
k	m^{-1}	wave vector (magnitude)
$\vec{k}, \vec{K}$	1/m	wave vector
k_B	$J\,K^{-1}$	Boltzmann constant
k_F	m^{-1}	Fermi momentum
K	m^{-1}	wave number
K	Gy	kerma
$\vec{l}, \vec{L}$	J s	orbital angular momentum
L	1	leakage rate
L	1	lepton number
L	$s^{-1}\,m^{-3}$	absorption probability
l, L	1	orbital angular momentum quantum number
$L_{e,\nu}(T)$	$W\,s/(m^2\,sr)$	spectral radiant density
m	kg	particle mass
m^*	kg	effective mass
m_e	kg	electron mass
m_j	1	magnetic quantum number
m_M	kg	molecular mass
M	kg	atomic mass
M	kg/mol	molar mass
M_r	1	mean relative molecular mass
n	m^{-3}	vacancy density
n	m^{-3}	density of free electrons
n, m	1	principal quantum number
$n(\omega, T)$	1	Bose-Einstein distribution function
n_A	m^{-3}	acceptor concentration
n_D	m^{-3}	donor concentration
n_i	m^{-3}	intrinsic charge carrier density
n_L	m^{-3}	effective electron density in conduction band
n_V	1	effective hole density
N	m^{-3}	particle density
N_1, N_2	1	occupation numbers
N_A	1	Avogadro's constant
p	m^{-3}	density of holes
p	1	resonance escape probability
$\vec{p}$	kg m/s	momentum
$\vec{p}, d$	C m	electric dipole moment
$\vec{P}$	$A\,s\,m^{-2}$	electric polarization
$p_E(\vec{r})$	$1/(Js\,sr\,m^2)$	spectral particle radiance
$\hat{P}$	1	reflection operator
Q	A s	charge
Q	J	radiant energy
Q	J	heat change
R_{BE}	Ω	differential input resistance

(*continued*)

Symbol	Unit	Designation
R_C	Ω	collector resistance
R_CE	Ω	differential output resistance
R_H	$1/\text{m}$	Rydberg constant hydrogen
R_∞	$1/\text{m}$	Rydberg constant
r_n	m	Bohr radius
$\vec{s}, \vec{S}$	Js	spin
S	MeV/cm	stopping power
S	1	strangeness quantum number
S, s	1	spin quantum number
T	K	temperature
$\hat{T}$	1	time-reversal operator
T_F	K	Fermi temperature
$T_{1/2}$	s	half-life
T_C	J	Coulomb barrier
u	J/m^3	radiant-energy density
u	kg	atomic mass unit
u_k	m	displacement of kth lattice plane
$u_k(\vec{r})$	$\text{m}^{-3/2}$	periodic function
$u_\nu(\nu, T)$	J s/m^3	spectral radiant energy density
u_s	m	displacement of plane
u_{s+n}	m	displacement of plane with distance $n \cdot a$
U	J	internal energy
$U(R)$	J	binding energy
V_0	V	acceleration voltage
V_D	V	difference voltage
V_D	V	diffusion voltage
V_in	V	input voltage
V_n	V	voltage at the inverted input terminal
V_out	V	output voltage
V_p	V	voltage at the non-inverted input terminal
V_T	V	temperature voltage
ΔV_BE	V	change of base voltage
ΔV_CE	V	change of output voltage
V	m^{-3}	volume
$V(r)$	J	potential
v	1	vibrational quantum number
v	m/s	mean phonon velocity
v_el	m/s	mean electron velocity
v_gr	m/s	group velocity of electron wave
v_r	1	reverse voltage transfer
w	1	probability density
W_A	J	work function
W_I	J	ionization energy
Z	1	atomic number
Z	Ω	complex resistance
Z^*	1	effective atomic number

29
Tables in quantum physics

29.1 Ionization potentials

29.1/1 Ionization energies of elements

The following table lists the ionization energies E_i in eV for the elements, and for various charge states.

Z	Charge state											
	1+	2+	3+	4+	5+	6+	7+	8+	9+	10+	11+	12+
1 H	13.598											
2 He	24.587	54.416										
3 Li	5.392	75.638	122.451									
4 Be	9.322	18.211	153.893	217.713								
5 B	8.298	25.154	37.930	259.368	340.217							
6 C	11.260	24.383	47.887	64.492	392.077	489.981						
7 N	14.534	29.601	47.448	77.472	97.888	552.057	667.029					
8 O	13.618	35.116	54.934	77.412	113.896	138.116	739.315	871.387				
9 F	17.422	34.970	62.707	87.138	117.240	157.161	185.182	953.886	1103.89			
10 Ne	21.564	40.962	63.45	97.11	126.21	157.93	207.27	239.09	1195.797	1362.164		
11 Na	5.139	47.286	71.64	98.91	138.39	172.15	208.47	264.18	299.87	1465.091	1648.659	
12 Mg	7.646	15.035	80.143	109.24	141.26	186.50	224.94	265.90	327.95	367.53	1761.802	1962.613
13 Al	5.986	18.828	28.447	119.99	153.71	190.47	241.43	284.59	330.21	398.57	442.07	2085.983
14 Si	8.151	16.345	33.492	45.141	166.77	205.05	246.52	303.17	251.10	401.43	476.06	523.50
15 P	10.486	19.725	30.18	51.37	65.023	230.43	263.22	309.41	371.73	424.50	479.57	560.41
16 S	10.360	23.33	34.83	47.30	72.68	88.049	280.93	328.23	279.10	447.09	504.78	564.65
17 Cl	12.967	23.81	39.61	53.46	67.8	98.03	114.193	348.28	400.05	455.62	529.26	591.97
18 Ar	15.759	27.629	40.74	59.81	75.02	91.007	124.319	143.456	422.44	478.68	538.95	618.24
19 K	4.341	31.625	45.72	60.91	82.66	100.00	117.56	154.86	175.814	503.44	564.13	629.09
20 Ca	6.113	11.871	50.908	67.10	84.41	108.78	127.70	147.24	188.54	211.270	591.25	656.39
21 Sc	6.54	12.80	24.76	73.47	91.66	111.1	138.0	158.7	180.02	225.32	249.832	685.89
22 Ti	6.82	13.58	27.491	43.266	99.22	119.36	140.8	168.5	193.2	215.91	265.23	291.497

	I	II	III	IV	V	VI	VII	VIII	IX	X	XI	XII
23 V	6.74	14.65	29.310	46.707	65.23	128.12	150.17	173.7	205.8	230.5	255.04	308.25
24 Cr	6.766	16.50	30.96	49.1	69.3	90.56	161.1	184.7	209.3	244.4	270.8	298.0
25 Mn	7.435	15.640	33.667	51.2	72.4	95	119.27	196.46	221.8	248.3	286.0	314.4
26 Fe	7.870	16.18	30.651	54.8	75.0	99	125	151.06	235.04	262.1	290.4	330.8
27 Co	7.86	17.06	33.50	51.3	79.5	102	129	157	186.13	276	305	336
28 Ni	7.635	18.168	35.17	54.9	75.5	108	133	162	193	224.5	321.2	352
29 Cu	7.726	20.292	36.83	55.2	79.9	103	139	166	199	232	266	368.8
30 Zn	9.394	17.964	39.722	59.4	82.6	108	134	174	203	238	274	310.8
31 Ga	5.999	20.51	30.71	64								
32 Ge	7.899	15.934	34.22	45.71	93.5							
33 As	9.81	18.633	28.351	50.13	62.63	127.6						
34 Se	9.752	21.19	30.820	42.944	68.3	81.70	155.4					
35 Br	11.814	21.8	36	47.3	59.7	88.6	103.0	192.8				
36 Kr	13.99	24.359	36.95	52.5	64.7	78.5	111.0	126	230.39			
37 Rb	4.177	27.28	40	52.6	71.0	84.4	99.2	136	150	277.1		
38 Sr	5.695	11.030	43.6	57	71.6	90.8	106	122.3	162	177	324.1	374.0
39 Y	6.38	12.24	20.52	61.8	77.0	93.0	116	129	146.52	191	206	
40 Zr	6.84	13.13	22.99	34.84	84.5							
41 Nb	6.88	14.32	25.04	38.3	5055	102.6	125	153	183			
42 Mo	7.099	16.15	27.16	46.4	61.2	68	126.8	161	192			
43 Tc	7.28	15.26	29.54	43	59	76	94	119	147	185		
44 Ru	7.37	16.76	28.47	46.5	63	81	100	126	155	216		
45 Rh	7.46	18.08	31.06	45.6	67	85	105	132		225		
46 Pd	8.34	19.43	32.93	48.8	66	90	110			178		

(continued)

Z	Charge state											
	1+	2+	3+	4+	5+	6+	7+	8+	9+	10+	11+	12+
47 Ag	7.576	21.49	34.83	52	70	89	116	139	162	187		
48 Cd	8.991	16.904	4405	550	73	94	115	146	170	185		
49 In	5.785	19.86	28.0	58	77	98	120	144	178	204		
50 Sn	4.332	14.63	30.7	46.4	81.1	103	126	150	176	213		
51 Sb	8.64	16.7	24.8	44.1	63.8	107.6	132	157	184	211		
52 Te	9.01	18.8	30.6	37.9	66	83	137.1	164	192	220		
53 I	10.44	19.0	31.4	41.7	71	83	104	169.9	200	229		
54 Xe	12.127	21.2	32.1	45.5	57	89	102	126	204.3	238		
55 Cs	3.893	25.1	34.6	45.5	62	74	108	122	150	256		
56 Ba	5.210	10.01	37	48.8	62	80	93	106	144	158		
57 La	5.61	11.43	19.17	52	66	80	100	114	151	165		
58 Ce	6.91	12.3	19.5	36.7	70	85	100	122	137	172		
59 Pr	5.76	10.55	21.62	39.95	57.45							
60 Nd	5.49	10.72										
61 Pm	5.55	10.90										
62 Sm	5.63	11.07										
63 Eu	5.67	11.25										
64 Gd	6.14	12.1										
65 Tb	5.85	11.52										
66 Dy	5.93	11.67										
67 Ho	6.02	11.80										
68 Er	6.10	11.93										

69 Tm	6.18	12.05	23.71							
70 Yb	6.254	12.17	25.2							
71 Lu	5.426	13.9	19							
72 Hf	7.0	14.9	23.3	33.3						
73 Ta	7.89	16.2	22.3	33.1	45					
74 W	7.98	17.7	24.1	35.4	48	61				
75 Re	7.88	16.6	26	37.7	51	64	79			
76 Os	8.7	17	25	40	54	68	83	99		
77 Ir	9.1	17.0	27	39	57	72	88	104	121	
78 Pt	8.96	18.54	28.5	41.1	55	75	92	109	127	146
79 Au	9.223	20.5	30.5	43.5	58	73	96	114	133	153
80 Hg	10.434	18.761	34.21	46	61	77	94	120	139	159
81 Tl	3.106	20.42	29.8	50.7	64	81	98	116	145	166
82 Pb	7.415	15.03	31.93	42.3	69.73	84	103	122	142	173
83 Bi	7.287	19.3	25.6	45.3	56	94.42	107	127	148	169
84 Po	8.2	19.4	27.3	38	61	73	112	132	154	176
85 At	9.2	20.1	29.3	41	51	78	91	138	160	183
86 Rn	10.745	21.4	29.4	43.8	55	67	97	111	166	190
87 Fr	3.98	22.5	33.5	43	59	71	84	117	133	197
88 Ra	5.277	10.144	34.3	46.4	58.5	76	89	103	140	156
89 Ac	6.89	11.5	—	49	62	76	95	109	123	164
90 Th	6.95	11.5	20.0	28.7	65	8,	94	115	130	145
91 Pa	—	—	—	—	—	84	100	115	138	154
92 U	6.2	—	—	—	—	—	104	121	137	126

29.1/2 Ionization energies of nitrogen compounds

Molecule	E_i (eV)	Molecule	E_i (eV)	Molecule	E_i (eV)
NH	13.10	C_3HN	11.6	$CH_3N - NH_2$	5.07
NH_2	11.4	CH_3CHCN	9.76	C_2N_2	13.8
NH_3	10.15	$C_3H_5NH_2$	9.6	$(CH_3)_2N - NH_2$	8.12
ND_3	11.52	$n - C_3H_7NH_2$	9.17	$(CH_3)_3N_2$	4.95
CN	15.13	$(CH_3)_3N$	8.32	$NCC \equiv CCN$	11.4
HCN	13.86	C_4N	12.3	NH_3	10.3
CH_3NH_2	9.41	$(CH_3)_2CCN$	9.15	CH_3N_3	9.5
CH_5N	8.97	$n - C_4H_9NH_2$	9.19	NF	12.0
C_2N	12.8	$(C_6H_5)_2NH$	8.44	NF_2	12.0
CH_2CN	10.87	C_5N	12.0	NF_3	13.2
CH_3CN	11.96	C_6H_7N	7.70	CH_2FCN	13.0
C_2H_5N	9.94	C_7H_9N	7.34	N_2F_4	12.04
$(CH_3)_2NH$	8.4	N_2	15.51	CNCl	12.49
$C_2H_5NH_2$	9.32	N_2^+	50	CH_2ClCN	12.2
C_2H_3CN	10.75	N_2H_2	9.85	CNBr	11.95
C_2H_5CN	11.85	N_2H_3	7.88	CNI	10.98
C_3N	14.3	N_2H_4	9.56		

29.1/3 Ionization energies of hydrocarbon compounds

Molecule	E_i (eV)	Molecule	E_i (eV)
H_2	15.427	$C_5H_2 = C(CH_3) - CH = CH_2$	8.85
graphite	3.8	$CH_3CH_2CCH_3 = CH_2$	9.12
CH_2	11.82	$CH_3CH_2CH_2CH = CH_2$	9.50
CH	9.86	C_5H_{12}	10.37
CD_3	9.95	C_6H_4	10.23
CH_4	12.99	C_6H_6	9.245
CD_4	13.25	$CH_2 = C(CH_3 - C)CH_3 = CH_2$	8.72
C_2H_2	11.41	C_6H_{10}	8.945
C_2H_3	9.45	$C_4H_9CH = CH_2$	9.46
C_2H_4	10.516	$(CH_3)_2CHCH = CHCH_3$	8.30
$(C_4H_8)_4$	9.23	C_6H_{12}	9.08
C_2H_5	8.80	C_6H_{14}	10.17
C_2H_6	11.65	C_7H_7	7.73
C_3H_3	8.25	C_7H_8	8.820
$C_3HC \equiv CH$	10.34	$CH_3C_6H_{11}$	9.86
$CH_3CH \equiv CH_2$	9.73	C_7H_{16}	10.06
C_3H_8	11.08	$C_6H_5CH = CH_2$	8.86
$CH \equiv C - C \equiv CH$	10.73	$(CH_3)_2C_6H_4$	8.56
$CH_2 = CH - CH = CH_2$	9.07	$C_6H_5CH_2CH_3$	8.76
$CH_3C \equiv CCH_3$	11.46	$C_6H_{13}CH = CH_2$	9.52
$CH_3CH_2CH = CH_2$	9.58	C_8H_{18}	10.24
$(CH_3)_2C = CH_2$	9.23	$C_6H_5C_3H_7$	8.72
$CH_3C_3H_5$	9.88	C_9H_{20}	1021
C_4H_{10}	9.08	$C_{10}H_8$	8.12
C_5H_6	8.58	$C_{14}H_{10}$	7.38

29.1/4   Ionization energies of halogen compounds

Molecule	E_i (eV)	Molecule	E_i (eV)	Molecule	E_i (eV)
HF	15.77	CH_2Cl	9.70	CBr	10.11
F_2	15.83	CCl_2	8.78	CH_2Br	8.34
CF	13.81	CH_3Cl	11.28	$CHBr_2$	8.13
CF_2	13.30	CF_3Cl	12.92	CH_3Br	10.54
CHF_2	9.45	CClF	13.13	$CHBrF_2$	12.1
CF_3	10.10	CCl_2F	8.96	CF_3Br	12.3
CH_3F	12.85	CCl_3	7.92	CH_2Br_2	10.8
CF_7	17.8	CCl_4	11.1	C_2H_3Br	9.80
C_2H_3F	10.37	CH_2Cl_2	11.4	$cycl-BrHC=CHBr$	9.69
$H_2C=CF_2$	10.30	CF_2Cl_2	11.8	C_2HBr_3	9.27
C_2HF_3	10.14	$CHCl_3$	11.42	C_2H_5Br	10.29
C_2H_4	10.12	C_2HCl_3	9.47	$CH_3-C\equiv CBr$	10.1
C_2H_5F	12.00	C_2H_3Cl	9.995	C_6H_5Br	9.41
$CH_2=CHCF_3$	10.9	$Cl_2C=CH_2$	9.79	HI	10.38
C_6H_4F	10.86	$cycl-ClHC-CHCl$	9.67	IF_5	13.5
C_6H_5F	9.197	$C_2F_2Cl_2$	10.0	ICl	10.4
C_6ClF_5	10.4	C_2F_3Cl	10.4	IBr	10.3
C_6BrF_5	9.6	C_2Cl_4	9.5	I_2	9.28
$C_6F_5CH_3$	9.6	C_2H_5Cl	10.97	CH_3I	9.51
HCl	12.74	$CH_3C\equiv CCl$	9.9	CF_3I	10.0
ClF_3	13.0	HBr	11.62	C_2H_5I	9.33
Cl_2	11.48	C_6H_5Cl	9.07	$n-C_3H_7I$	9.41
CCl	12.9	Br_2	10.55	$CH_2-C_4H_9$	9.19
CCl_2	13.10	BrCl	11.1	C_6H_5I	9.10

29.1/5   Ionization energies of oxygen compounds

Molecule	E_i (eV)	Molecule	E_i (eV)	Molecule	E_i (eV)
OH	13.18	O_2	14.01	O_3	11.7
H_2O	12.60	O_2^+	50	FO	13.0
CO	14.01	HO_2	11.53	F_2O	13.7
CO^+	43	H_2O_2	1092	$(CF_3)_2C=O$	11.82
CH_2O	10.90	CO_2	13.79	ClO	10.4
CH_3O	9.2	HCOO	9.0	$COCl_2$	11.77
CH_2OH	8.2	COOH	8.7	$CH_2ClCOCH_3$	9.91
CH_3OH	10.95	HCOOH	11.05	$CHCl_2COCH_3$	10.12
$CH_2=C=O$	9.60	$HFC=O$	11.4	ClO_2	11.1
C_2H_5O	9.2	CHOCHO	9.48	ClO_3	11.7
C_2H_4OH	7.0	$(H_2CO)_2$	10.51	ClO_3F	13.6
C_2H_5OH	10.25	CH_3COOH	10.38	NO	9.25
$(CH_3)_2O$	10.00	$HCOOCH_3$	10.82	$NH_2HC=O$	10.16
$n-C_3H_7OH$	10.42	CH_3COCHO	9.60	N_2O	12.63
$n-C_4H_9OH$	10.30	C_2H_5COOH	10.47	NO_2	9.78...12.3
$(C_2H_5)_2O$	9.53	CH_3COOCH_3	10.27	CH_5ONO	10.7
C_6H_5OH	8.50	$n-C_3H_7COOH$	10.2	CH_3NO_2	11.34
$(C_6H_5)HC=O$	9.51				

29.1/6   Dissociation energies of diatomic molecules

Molecule	E_d (eV)	Molecule	E_d (eV)	Molecule	E_d (eV)	Molecule	E_d (eV)
Ag_2	1.8	BeF	7.0	H_2	4.48	MnO	3.4
AgBr	3.1	BrCl	2.23	HD	4.51	Na_2	0.7
AgCl	3.4	BrF	2.4	HT	4.52	NaBr	3.8
AgH	2.36	BrO	2.4	HBr	3.75	NaCl	4.2
AgI	2.6	CaBr	2.9	HCl	4.43	NaF	5.0
AgO	2.5	CaCl	2.8	HF	5.9	NaH	2.1
AgSn	2.55	CaF	3.1	HI	3.05	NaI	3.1
AuCu	2.4	CsBr	4.3	Hg_2	0.06	NaK	0.61
AlBr	4.6	CaH	1.7	HgBr	0.7	N_2	9.76
AlC	1.9	CaI	2.8	HgCl	1.0	NBr	2.9
AlCl	5.1	CaO	5.0	HgF	1.8	NF	2.6
AlF	7.65	CaS	3.0	HgH	0.38	NH	3.6
AlH	2.9	C_2	6.2	HgI	0.36	NO	3.5
AlI	3.84	CCl	2.8	HgS	2.8	NS	5.0
AlO	5.0	CF	4.7	I_2	1.54	O_2	5.1
AlS	3.5	CH	3.47	IBr	1.82	OH	4.4
AsN	6.6	CN	8.4	ICl	2.15	P_2	5.0
AsO	5.0	CO	11.1	IF	2.9	Rb_2	0.48
Au_2	2.28	Cl_2	2.48	IO	1.9	RbBr	4.0
AuAl	3.1	ClF	2.6	K_2	0.51	RbCl	4.4
AuCl	3.1	ClO	2.8	KBr	3.95	RbF	5.4
AuCr	2.2	Cs_2	0.45	KCl	4.4	RbH	1.8
AuH	3.1	CsCl	4.4	KF	5.1	RbI	3.3
AuMg	2.7	CsF	5.0	KH	1.86	S_2	4.3
AuSn	2.55	CsH	1.9	KI	3.33	SF	2.8
BBr	4.5	CsI	3.6	Li_2	1.1	SH	3.5
BCl	5.2	Cu_2	0.2	LiBr	4.4	SO	5.3
BF	8.1	CuBr	3.4	LiCl	4.8	Tl_2	4.59
BH	3.0	CuCl	3.7	LiF	6.0	TlBr	3.4
BO	7.45	CuF	3.0	LiH	2.4	TlCl	3.8
BaBr	2.8	CuH	2.9	LiI	3.6	TlF	4.7
BaCl	2.7	CuI	3.0	LiO	3.43	TlH	2.0
BaF	3.8	CuO	4.8	MnBr	3.2	TlI	2.8
BaH	1.8	D_2	4.55	MnCl	3.9	ZnCl	2.6
BaO	4.7	F_2	1.6	MnF	5.0	ZnH	0.85
BaS	2.4	FO	1.9	MnH	2.2	ZnI	1.4
BeCl	4.8						

29.2   Atomic and ionic radii of elements

The values for atomic and ionic radii of elements depend on the method of measurement. Therefore, the data on atomic and ionic radii compiled in this table must be considered approximate values only.

29.2/1   Atomic and ionic radii of elements

Atomic number	Element	Charge	Radius (nm)	Atomic number	Element	Charge	Radius (nm)
1	H	−1	0.154	16	S	−2	0.184
		0	0.46			0	0.095
2	He	0	0.122			+2	0.219
3	Li	0	0.155			+4	0.037
		+1	0.068			+6	0.030
4	Be	0	0.113	17	Cl	−1	0.181
		+1	0.044			0	0.089
		+2	0.035			+5	0.034
5	B	0	0.091			+7	0.027
		+1	0.035	18	Ar	0	0.192
		+3	0.023			+1	0.154
6	C	−4	0.260	19	K	0	0.236
		0	0.077			+1	0.133
		+4	0.016	20	Ca	0	0.197
7	N	−3	0.171			+1	0.118
		0	0.071			+2	0.099
		+3	0.016	21	Sc	0	0.164
		+5	0.013			+3	0.073
8	O	−2	0.132	22	Ti	0	0.146
		−1	0.176			+1	0.096
		0	0.056			+2	0.094
		+1	0.022			+3	0.076
		+6	0.009			+4	0.068
9	F	−1	0.133	23	V	0	0.134
		0	0.053			+2	0.088
		+7	0.007			+3	0.074
10	Ne	0	0.160			+4	0.063
		+1	0.112			+5	0.059
11	Na	0	0.189	24	Cr	0	0.127
		+1	0.097			+1	0.081
12	Mg	0	0.160			+2	0.089
		+1	0.082			+3	0.063
		+2	0.066			+6	0.052
13	Al	0	0.143	25	Mn	0	0.130
		+3	0.051			+2	0.080
14	Si	−4	0.271			+3	0.066
		−1	0.384			+4	0.060
		0	0.134			+7	0.046
		+1	0.065	26	Fe	0	0.126
		+4	0.042			+2	0.074
15	P	−3	0.212			+3	0.064
		0	0.130	27	Co	0	0.125
		+3	0.044			+2	0.072
		+5	0.035			+3	0.063

(*continued*)

29.2/1 Atomic and ionic radii of elements (*continued*)

Atomic number	Ele-ment	Charge	Radius (nm)	Atomic number	Ele-ment	Charge	Radius (nm)
28	Ni	0	0.121	41	Nb	0	0.145
		+2	0.069			+1	0.100
		+3	0.035			+4	0.074
29	Cu	0	0.128			+5	0.069
		+1	0.096	42	Mo	0	0.139
		+2	0.072			+1	0.093
30	Zn	0	0.139			+4	0.070
		+1	0.088			+6	0.062
		+2	0.074	43	Tc	0	0.136
31	Ga	0	0.139			+7	0.098
		+1	0.081	44	Ru	0	0.134
		+3	0.062			+4	0.067
32	Ge	−4	0.272	45	Rh	0	0.134
		0	0.139			+3	0.068
		+2	0.073			+4	0.065
		+4	0.053	46	Pd	0	0.137
33	As	−3	0.222			+2	0.080
		0	0.148			+4	0.065
		+3	0.058	47	Ag	0	0.144
		+5	0.046			+1	0.126
34	Se	−2	0.191			+2	0.089
		−1	0.232	48	Cd	0	0.156
		0	0.160			+1	0.114
		+1	0.066			+2	0.097
		+4	0.050	49	In	0	0.166
		+6	0.042			+1	0.130
35	Br	−1	0.196			+3	0.081
		0	0.105	50	Sn	−4	0.294
		+5	0.047			−1	0.370
		+7	0.039			0	0.158
36	Kr	0	0.198			+2	0.093
37	Rb	0	0.248			+4	0.071
		+1	0.147	51	Sb	−3	0.245
38	Sr	0	0.215			0	0.161
		+2	0.112			+3	0.076
39	Y	0	0.181			+5	0.062
		+3	0.089	52	Te	−2	0.211
40	Zr	0	0.160			−1	0.250
		+1	0.109			0	0.170
		+2	0.074			+1	0.082
						+4	0.070
						+6	0.056

(*continued*)

29.2/1   Atomic and ionic radii of elements (*continued*)

Atomic number	Element	Charge	Radius (nm)	Atomic number	Element	Charge	Radius (nm)
53	I	−1	0.220	70	Yb	0	0.193
		0	0.124			+3	0.081
		+5	0.062	71	Lu	0	0.174
		+7	0.050			+3	0.085
54	Xe	0	0.218	72	Hf	0	0.159
55	Cs	0	0.268			+4	0.078
		+1	0.167	73	Ta	0	0.146
56	Ba	0	0.221			+5	0.068
		+1	0.153	74	W	0	0.140
		+2	0.134			+4	0.070
57	La	0	0.187			+6	0.062
		+1	0.139	75	Re	0	0.137
		+3	0.106			+4	0.072
		+4	0.090			+7	0.056
58	Ce	0	0.183	76	Os	0	0.135
		+1	0.127			+4	0.088
		+3	0.103			+6	0.069
		+4	0.092	77	Ir	0	0.135
59	Pr	0	0.182			+4	0.068
		+3	0.101	78	Pt	0	0.138
		+4	0.090			+2	0.080
60	Nd	0	0.182			+4	0.065
		+3	0.099	79	Au	0	0.144
61	Pm	0	—			+1	0.137
		+3	0.098			+3	0.085
62	Sm	0	0.181	80	Hg	0	0.160
		+3	0.096			+1	0.127
63	Eu	0	0.202			+2	0.110
		+2	0.109	81	Tl	0	0.171
		+3	0.095			+1	0.147
64	Gd	0	0.179			+3	0.095
		+3	0.094	82	Pb	0	0.175
65	Tb	0	0.177			+2	0.080
		+3	0.092			+4	0.065
		+4	0.084	83	Bi	−4	0.213
66	Dy	0	0.177			0	0.182
		+3	0.091			+1	0.098
67	Ho	0	0.176			+3	0.096
		+3	0.089			+5	0.071
68	Er	0	0.175	84	Po	+6	0.067
		+3	0.088	85	At	+7	0.062
69	Tm	0	0.174	87	Fr	0	0.280
		+3	0.087			+1	0.180

(*continued*)

29.2/1   Atomic and ionic radii of elements (*continued*)

Atomic number	Ele-ment	Charge	Radius (nm)	Atomic number	Ele-ment	Charge	Radius (nm)
88	Ra	0	0.235			+4	0.097
		+2	0.143			+6	0.080
89	Ac	0	0.203	93	Np	0	0.150
		+3	0.118			+3	0.110
90	Th	0	0.180			+4	0.095
		+4	0.102			+7	0.071
91	Pa	0	0.162	94	Pu	0	0.162
		+3	0.113			+3	0.108
		+4	0.098			+4	0.093
		+5	0.089	95	Am	+3	0.107
92	U	0	0.153			+4	0.092

29.3   Electron emission

29.3/1   Work function W_A of electrons from pure elements

The table lists the values for various methods of measurement. The following abbreviations have been used for these methods: T: thermal ionization; P: photoemission; CPD: contact potential difference; F: field emission. For monocrystalline samples, the crystallographic directions whose work function has been measured are given. Data in italics are relatively uncertain (method of measurement not clear, preparation of sample not clear).

Element	W_A /eV	Crystal direction	Method	Element	W_A /eV	Crystal direction	Method
Ag	*4.26*		P	Ca	2.87		P
	4.64	(100)	P	Cd	*4.22*		CPD
	4.52	(110)	P	Ce	2.9		P
	4.74	(111)	P	Co	5.0		P
Al	4.28		P	Cr	4.5		P
	4.41	(100)	P	Cs	2.14		P
	4.06	(110)	P	Cu	4.65		P
	4.24	(111)	P		4.59	(100)	P
As	*3.75*		P		4.48	(110)	P
Au	5.1		P		4.94	(111)	P
	5.47	(100)	P		4.53	(112)	P
	5.37	(110)		Eu	2.5		P
	5.31	(111)		Fe	4.5		P
B	*4.45*		T		4.67	(100)	P
Ba	2.7		T		4.81α	(111)	P
Be	4.98		P		4.70α		P
Bi	*4.22*		P		*4.62β*		P
C	*5.0*		CPD		*4.68γ*		P

(continued)

29.3/1   Work function W_A of electrons from pure elements (*continued*)

Element	W_A /eV	Crystal direction	Method	Element	W_A /eV	Crystal direction	Method
Ga	4.2		CPD	Pt	5.65		P
Ge	5.0		CPD		5.7	(111)	P
	4.80	(111)	P	Rb	2.16		P
Gd	3.1		P	Re	4.96		T
Hf	3.9		P		5.75	(1011)	F
Hg	4.49		P	Rh	4.98		P
In	4.12		P	Ru	4.71		P
Ir	5.27		T	Sb	4.55		—
	5.42	(110)	F		(amorphous)		
	5.76	(111)			4.7	(100)	—
	5.67	(100)	F	Sc	3.5		P
	5.00	(210)	F	Se	5.9		P
K	2.30		P	Si (n)	4.85		CPD
La	3.5		P	Si (p)	4.91	(100)	CPD
Li	2.9		F		4.60	(111)	P
Lu	3.3		CPD	Sm	2.7		P
Mg	3.66		P	Sn	4.42		CPD
Mn	4.1		P	Sr	2.59		T
Mo	4.6		P	Ta	4.25		T
	4.53	(100)	P		4.15	(100)	T
	4.95	(110)	P		4.80	(110)	T
	4.55	(111)	P		4.00	(111)	T
	4.36	(112)	P	Tb	3.0		P
	4.50	(114)	P	Te	4.95		P
	4.55	(332)	P	Th	3.4		T
Na	2.75		P	Ti	4.33		P
Nb	4.3		P	Tl	3.84		CPD
	4.02	(001)	P	U	3.63		P&CPD
	4.87	(110)	P		3.73	(100)	P&CPD
	4.36	(111)	T		3.90	(110)	P&CPD
	4.63	(112)	T		3.67	(113)	P&CPD
	4.29	(113)	T	V	4.3		P
	3.95	(116)	T	W	4.55		CPD
	4.18	(310)	T		4.63	(100)	F
Nd	3.2		P		5.25	(110)	F
Ni	5.15		P		4.47	(111)	F
	5.22	(100)	P		4.18	(113)	CPD
	5.04	(110)	P		4.30	(116)	T
	5.35	(111)	P	Y	3.1		P
Os	4.83		T	Zn	4.33		P
Pb	4.25		P		4.9	(0001)	CPD
Pd	5.12		P	Zr	4.05		P
	5.6	(111)	P				

29.3/2   Work function for adsorbed surfaces

Adsorbent	Adsorbate	W_A /eV	Adsorbent	Adsorbate	W_A /eV
Be	Cs	1.94	Pt	O	6.55
C	Cs	1.37	Pt	Na	2.10
Ti	Cs	1.32	Pt	K	1.62
Cr	Cs	1.71	Pt	Rb	1.57
Fe	Cs	1.82	Pt	Cs	1.38
Ni	Cs	1.37	Pt	Ba	1.9
Cu	Ba	3.35	Pt	Ba	3.28
Ge	Ba	2.2	Au	O	6.46
Zr	Cs	3.93	Au	O	5.66
Mo	Cs	1.54	Au	Ba	2.3
Mo	Th	2.58	Au	Ba	3.35
Ag	Ba	1.56	WO	Na	1.72
Hf	Cs	3.62	WO	K	1.76
Ta	Cs	1.1	steel	Cs	1.41
Ta	Cs	1.6	steel (304)	Cs	1.52
W	Li	2.18	Ag_2O	Cs	0.75
W	O	6.20	NbC	Cs	1.2
W	Ba	1.75	ZrC	Cs	1.60
W	La	2.2	Mo_2C	Cs	1.45
W	Th	2.63	Ta_2C	Cs	1.4
Re	Cs	1.45	$MoSi_2$	Cs	1.75
Re	Th	2.58	WSi_2	Cs	1.47

29.3/3   Thermoelectric emission properties of a tungsten cathode

Basic properties of a thermocathode are: current density of thermoemission j_T; evaporation speed v_V of the activated surface material. From these quantities, the efficiency of the thermocathode may be evaluated: $\eta = j_T/v_V$.

T /K	j_T (A/cm^2)	v_V (g/(cm^2 s))	T /K	j_T (A/cm^2)	v_V (g/(cm^2 s))
2100	$3.9 \cdot 10^{-3}$	$2.0 \cdot 10^{-13}$	2600	$7.0 \cdot 10^{-1}$	$3.9 \cdot 10^{-9}$
2200	$1.3 \cdot 10^{-2}$	$2.1 \cdot 10^{-12}$	2700	1.6	$1.8 \cdot 10^{-8}$
2300	$4.1 \cdot 10^{-2}$	$1.8 \cdot 10^{-11}$	2800	3.5	$7.4 \cdot 10^{-8}$
2400	$1.2 \cdot 10^{-1}$	$1.2 \cdot 10^{-10}$	2900	7.3	$2.8 \cdot 10^{-7}$
2500	$3.0 \cdot 10^{-1}$	$7.6 \cdot 10^{-10}$	3000	14.0	$9.5 \cdot 10^{-7}$

29.3/4   Photo cathodes of alkali antimonides

Photo cathode	Quantum yield $\dfrac{\text{electrons}}{\text{photon}}$	Limit wavelength λ_0 (nm)	Sensitivity $(\mu A/lm)$	Energy gap (eV)	Type	Thermal noise (A/cm^2)
K_3Sb	0.07	550	12	1.4	p	—
K_2CsSb	0.3	660	100	1.0	p	10^{-17}
$K_2CsSb(O)$	0.35	780	130	1.0	p	10^{-16}
Na_3Sb	0.02	330	?	1.1	n	—
Na_2KSb	0.30	600	60	1.0	p	10^{-16}
Rb_3Sb	0.10	580	25	1.0	p	—
Cs_3Sb	0.15	580	25	1.6	p	10^{-16}
Cs_3Sb on MgO	0.20	650	80	1.6	p	10^{-15}
$(Cs)Na_2KSb$	0.30	870	300	1.0	p	10^{-15}

29.3/5   Basic properties of secondary-electron emission

The secondary-electron yield δ is the number of emitted electrons per incident electron. The maximum value δ_{max} and the corresponding energy of the primary electron E_{max} are compiled for various elements in the table below. The energies of primary electrons leading to a yield of 1 are also given.

Element	δ_{max}	E_{max} (eV)	E_I (eV)	E_{II} (eV)
Ag	1.5	800	200	$>$ 2000
Al	1.0	300	300	300
Au	1.4	800	150	$>$ 2000
B	1.2	150	50	600
Ba	0.8	400	—	—
Bi	1.2	550		
Be	0.5	200	—	—
C (diamond)	2.8	750		$>$ 5000
C (graphite)	1.0	300	300	300
C (black)	0.45	500	—	—
Cd	1.1	450	300	700
Co	1.2	600	200	
Cs	0.7	400	—	—
Cu	1.3	600	200	1500
Fe	1.3	400	120	1400
Ga	1.55	500	75	
Ge	1.15	500	150	900
K	0.7	200	—	—
Li	0.5	85	—	—
Mg	0.95	300	—	—
Mo	1.25	375	150	1200
Na	0.82	300	—	—
Nb	1.2	375	150	1050
Ni	1.3	550	150	$>$ 1500
Pb	1.1	500	250	1000

(*continued*)

29.3/5   Basic properties of secondary-electron emission (*continued*)

Element	δ_{max}	E_{max} (eV)	E_I (eV)	E_{II} (eV)
Pd	> 1.3	> 250	120	
Pt	1.8	700	350	3000
Rb	0.9	350	—	—
Sb	1.3	600	250	2000
Si	1.1	250	125	500
Sn	1.35	500		
Ta	1.3	600	250	> 2000
Th	1.1	800		
Ti	0.9	280	—	—
Tl	1.7	650	70	> 1500
W	1.4	650	250	> 1500
Zr	1.1	350		

29.4   X-rays

29.4/1   Main lines of the characteristic x-ray spectrum of various elements (K-series)

Element	Wavelength λ (m^{-12})		
	α_2	α_1	β
lead	17.0	16.5	14.6
chromium	229.4	229.0	208.5
iron	194.0	193.6	175.7
germanium	125.8	125.4	112.9
gold	18.5	18.0	15.9
cobalt	179.3	178.9	162.1
copper	154.4	154.1	139.2
manganese	210.6	210.2	191.0
nickel	166.2	165.8	150.0
selenium	110.9	110.5	99.2
silicon	712.8	712.5	676.8
uranium	13.1	12.6	11.1
tungsten	21.4	20.9	18.4
zinc	143.9	143.5	129.5

29.5   Nuclear reactions

29.5/1   Cross-section for scattering of neutrons by various elements

Element	Fast neutrons σ_{tot} (b)	Thermal neutrons		
		σ_S (b)	σ_{Ab} (b)	σ_A (b)
H	0.9	38 (H_2)	0.33	
He	1.4	0.8		
Al	1.7	1.4	0.23	0.23
Fe	3.0	11.4	2.53	0.003
Ni	3.2	17.5	4.6	0.03
Cu	3.2	7.8	3.7	0.64; 2.9
Ge	3.4	9	2.4	0.002; 0.02; 0.2; 0.6
Cd	4.3	7	2600	0.1; 0.3; 0.04
Hg	4.8	21	380	0.025; 1.0
Pb	4.7	11.4	0.17	0.0003
^{232}Th	7.2	12.6	7.4	7.4
^{238}U	5.2	8.3	7.68	2.73; 0.76
^{238}U	1.3		687	107; 580 (fission)
^{239}Pu	2.0		1065	315; 750 (fission)

29.5/2   Nuclear-fusion reactions

Reaction	Reaction energy Q(MeV)
$^2_1H + ^3_1H \rightarrow ^4_2He + ^1_0n$	17.61
$^2_1H + ^2_1H \rightarrow ^3_2He + ^1_0n$	3.27
$^2_1H + ^2_1H \rightarrow ^3_1H + ^1_1p$	4.03
$^2_1H + ^3_2He \rightarrow ^4_2He + ^1_1p$	18.35
$^1_1p + ^{11}_5B \rightarrow 3 \cdot ^4_2He$	8.7
$^{12}_6C + ^1_1H \rightarrow ^{13}_7N + \gamma$	1.9
$^{13}_7N \rightarrow ^{13}_6C + e^+$	1.2
$^{13}_6C + ^1_1H \rightarrow ^{14}_7N$	1.9
$^{14}_7N + ^1_1H \rightarrow ^{15}_8O + \gamma$	7.3
$^{15}_8O \rightarrow ^{15}_7N + e^+$	1.7
$^{15}_7N + ^1_1H \rightarrow ^{12}_6C + ^4_2He$	4.9
$^2_1H + ^1_1H \rightarrow ^3_2He + \gamma$	5.4
$^2_1H + ^2_1H \rightarrow ^4_2He + \gamma$	23.8
$^3_2He + ^1_1H \rightarrow ^4_2He + e^+$	18.7
$^3_2He + ^3_1H \rightarrow ^4_2He ^2_1H$	14.3

29.6   Interaction of radiation with matter

29.6/1   Mass-attenuation coefficient μ/ρ in $10^{-1}\,\mathrm{m}^2/\mathrm{kg}$ for x-rays

Element	Wavelength λ(nm)									
	0.02	0.04	0.06	0.08	0.10	0.12	0.14	0.16	0.18	0.2
Ag	5.4	37	17	39	71	120	174	250	354	436
Al	0.27	1.05	3.3	7.3	14.0	24	36	55	79	106
C	0.167	0.243	0.40	0.80	1.40	2.5	3.9	5.8	7.9	10.0
Cu	1.45	10	32	71	134	218	42	60	85	119
Fe	1.06	7.1	23.5	50.7	95	170	270	390	61	78
N	0.177	0.34	0.73	1.51	2.6					
O	0.183	0.336	0.730	1.53						
Pb	4.6	33	77	147	77	128	180	258	360	

29.6/2   Mass-attenuation coefficient for electrons in aluminum

Energy E (keV)	μ/ρ (m^2 kg^{-1})	Energy E (keV)	μ/ρ (m^2 kg^{-1})
0.9	$2.5 \cdot 10^5$	100.0	13
5.8	$1.5 \cdot 10^4$	200.0	2.9
10.5	$3.5 \cdot 10^3$	460.0	0.9
46.6	$7.4 \cdot 10^1$	660.0	0.6

29.6/3   Range of α-particles in air, biological tissue and aluminum

Energy E (MeV)	Air, R (cm)	Tissue, R (μm)	Aluminum, R (μm)
4.0	2.5	31	16
5.0	3.5	43	23
6.0	4.6	56	30
7.0	5.9	72	38
8.0	7.4	91	48
9.0	8.9	110	58
10.0	10.6	130	69

29.7   Hall effect

29.7/1   Hall coefficient for metals

The Hall coefficient is given for temperatures between 0 °C and 30 °C.

Metal	$R_H/(10^{-10}\ m^3C^{-1})$
Li	−1.7
Be (99.5 %)	+7.7
Na	−2.1
Mg	−0.83
Al (99.5 %)	0.33
K	−4.2
Ca (99 %)	−1.78
Ti (99.91 %)	−0.26
Ti (99.87 %)	+0.10
V	+0.82
V (99.63 %)	+0.79
Cr (99.9 %)	+3.63
Mn (99.99 %)	+0.84
Cu	−0.536
Zn (technical)	+1.04
Ga	−0.63
Rb	−4.2
Y (99.2 %)	−0.770
Y (monocr. $\frac{\rho_{273K}}{\rho_{4.2K}} = 10.4$)	
H ∥ c	−1.72
H ⊥ c	−0.47
Y ($\frac{\rho_{273K}}{\rho_{4.2K}} = 16$)	
H ∥ c	+1.5
H ⊥ c	+0.4
Zr (97.3 % Zr; 2.4 % Hf)	+1.385
Zr ($\frac{\rho_{273K}}{\rho_{4.2K}} = 38$)	+2.15
Nb	+0.88
Mo	+1.80
Ru	+2.2
Rh (99.5 %)	+0.505
Pd	−0.845

(*continued*)

29.7/1   Hall coefficient for metals (*continued*)

Metal	$R_H/(10^{-10}\ m^3C^{-1})$
Ag (technical)	−0.897
Ag (99.9 %)	−0.909
Cd (99.9 %)	+0.531
In	−0.073
Sn	−0.022
Cs	−7.8
La (99.8 %)	−0.8
Ce (99.88 %)	+1.81
Pr (99.9 %)	+0.709
Nd (99.98 %)	+0.971
Sm	−0.21
Sm ($\frac{\rho_{273K}}{\rho_{4.2K}} = 17.3$)	−0.5
Tm	−1.5
Yb	+3.7
Lu	−0.53
Lu (monocr. $\frac{\rho_{273K}}{\rho_{4.2K}} = 25$)	
H ∥ c	−2.6
H ⊥ c	+0.4
Hf (99.4 %)	+0.42
Ta (99.8 %)	+0.971
W	+1.18
Re	+3.15
Re ($\frac{\rho_{273K}}{\rho_{4.2K}} = 27$)	+1.6
Ir	+0.402
Pt	−1.27
Pt (99.9 %)	−0.214
Au	−0.705
Hg	< 0.02
Tl	+0.24
Th	−1.2
U	+0.34

29.8 Superconductors

29.8/1 Selected properties of superconducting elements

Essential properties of superconductors are the critical temperature T_c and the critical field strength H_c.

Element	T_c (K)	H_c (A/m)
W	0.0154 ± 0.0005	91.51 ± 2.39
Be	0.026	
Lu	0.1 ± 0.03	27852.115 ± 3978.87
Ir	0.1125 ± 0.001	1273.24 ± 3.97
Hf	0.128	1010.63
U	0.2	
Ti	0.40 ± 0.04	4456.34
Ru	0.49 ± 0.015	5490.85 ± 159.15
Cd	0.517 ± 0.002	2228.17 ± 79.58
Zr	0.61 ± 0.15	3740.14
Zr(ω)	0.65; 0.95	
Os	0.66 ± 0.03	5570.42
Zn	0.85 ± 0.01	4297.18 ± 23.87
Mo	0.915 ± 0.005	7639.44 ± 238.73
Gd	1.083 ± 0.0001	4639.37 ± 15.92
Al	1.175 ± 0.002	8347.68 ± 23.87
Th	1.38 ± 0.02	127.32 ± 238.73
Pa	1.4	
Re	1.697 ± 0.006	15915.49 ± 397.89
Tl	2.38 ± 0.02	14164.79 ± 159.15
In	3.408 ± 0.001	22401.06 ± 159.15
Sn	3.722 ± 0.001	24271.13 ± 159.15
Hg(β)	3.949	26 976.76
Hg(α)	4.154 ± 0.001	32706.34 ± 159.15
Ta	4.47 ± 0.04	65969.72 ± 477.46
La(α)	4.88 ± 0.02	63661.98 ± 795.77
V	5.40 ± 0.05	112 045.08
Gd(β)	5.9; 6.2	44 563.38
La(β)	6.00 ± 0.1	87 216.91; 127 323.95
Gd(γ)	7	75 598.60
Pb	7.196 ± 0.006	63900.71 ± 79.57
Tc	7.8 ± 0.1	112 204.23
Gd(Δ)	7.85	64 855.63
Nb	9.25 ± 0.02	163 929.59

29.8/2   Superconducting compounds and alloys with $T_c > 10$ K

Substance	T_c (K)	Substance	T_c (K)
Al_2CMo_3	10.0	$Nb_{0.3}SiV_{2.7}$	12.8
CW	10	$BaBi_{0.2}O_3Pb_{0.8}$	13.2
$Nb_{0.18}Re_{0.82}$	10	$SiV_{2.7}Zr_{0.3}$	13.2
B_2LuRu	10	LiO_4Ti_2	13.7
$Ir_{0.4}Nb_{0.6}$	10	$Br_2Mo_6S_6$	13.8
$RhTa_3$	10	$N_{0.93}Nb_{0.85}Zr_{0.15}$	13.8
CMO_xNb_{1-x}	10.2(max)	InV_3	13.9
CTa	10.3	$Mo_{0.57}Re_{0.43}$	14.0
$NbTc_3$	10.5	$Ge_{0.1}Si_{0.9}V_3$	14.0
$Mo_{\approx 0.60}Re_{0.395}$	10.6	CMo	14.3
Mo_3Ru	10.6	$GaNB_3$	14.5
NZr	10.7	$Al_{0.1}Si_{0.9}V_3$	14.5
$Cu_{1.8}Mo_6S_8$	10.8	Mo_3Tc	15
$NbSnTa_2$	10.8	$Mo_6Pb_{0.9}S_{7.5}$	15.2
$Nb_{0.75}Zr_{0.25}$	10.8	$B_{0.1}Si_{0.9}V_3$	15.8
$Nb_{0.66}Zr_{0.33}$	10.8	$MoTc_3$	15.8
Nb_3Pt	10.9	$C_{0.1}Si_{0.9}V_3$	16.4
$SiTi_{0.3}V_{2.7}$	10.9	Nb_2SnTa	16.4
C_3La	11.0	Nb_3Sn_2	16.6
GeV_3	11	GaV_3	16.8
$Mo_{0.52}Re_{0.48}$	11.1	$C_{0.66}Th_{0.13}Y_{0.21}$	17
B_4Rh_4Y	11.3	$PbTa_3$	17
$Cr_{0.3}SiV_{2.7}$	11.3	SiV_3	17.1
$Ge_{0.5}Nb_3Sn_{0.5}$	11.3	$Nb_{2.5}SnTa_{0.5}$	17.6
$LaMo_6Se_8$	11.4	$Nb_{2.75}SnTa_{0.25}$	17.8
$AuNb_3$	11.5	$AlNb_3$	18.0
CNb	11.5	$(Ca, La)_2CuO_4$	18
C_3Y_2	11.5	Nb_3Sn	18.05
B_4LuRh_4	11.7	Nb_3Si	19
$Mo_{0.3}SiV_{2.7}$	11.7	$Al_{\approx 0.8}Ge_{\approx 0.2}Nb_3$	20.7
AlV_3	11.8	$GeNb_3$	23.2
$Mo_{0.3}Tc_{0.7}$	12.0	$(Ba, La)_2CuO_4$	36
CMo_2	12.2	$Cu(La, Sr)_2O_4$	39
Mo_6Se_8Tl	12.2	$Ba_2Cu_3LaO_6$	80
$Nb_2SnTa_{0.5}V_{0.5}$	12.2	$Ba_2Cu_3O_7Y$	90
$B_{0.03}C_{0.51}Mo_{0.47}$	12.5	$Ba_2Cu_3O_7Tm$	101
Mn_3Si	12.5	$Bi_2CaCu_2O_8Sr_2$	110
$Al_{0.5}Ge_{0.5}Nb$	12.6	$Ba_2CaCu_2O_8Tl_2$	120
Mo_3Os	12.7		

29.9   Semiconductors

29.9.1   Thermal, magnetic and electric properties of semiconductors

29.9/1   Element semiconductors

The given values refer to standard conditions.

Substance	Formation enthalpy ($kJ \cdot mol^{-1}$)	Relative permittivity ε_r	Refractive index n	Energy gap E_g (eV)	Mobility $\mu(cm^2\ V^{-1}\ s^{-1})$	
					electrons	holes
C	714.4	5.7	2.419	5.4	1800	1400
Si	324	11.8	3.99	1.107	1900	500
Ge	791	16	3.99	0.67	3800	1820
α-Sn	267.5			0.08	2500	2400

29.9/2   Compound semiconductors

Substance	Formation enthalpy ($kJ \cdot mol^{-1}$)	Relative permittivity ε	Refractive index n	Energy gap E_g(eV)	Mobility $\mu(cm^2\ V^{-1}\ s^{-1})$		Application
					electrons	holes	
ZnS	477	8.9	2.356	3.54	180		luminous compound
ZnSe	422	9.2	2.89	2.58	540	28	
ZnTe	376	10.4	3.56	2.26	340	100	
CdTe	339	7.2	2.5	1.44	1200	50	
HgSe	247			2.12	20000		
AlAs	627	10.9		2.16	1200	420	
AlSb	585	11	3.2	1.60	200...400	550	
GaP	635	11.1	3.2	2.24	300	150	LED (green); IR-diodes
GaAs	535	13.2	3.30	1.35	8800	400	LED; FET; IR-diodes
GaSb	493	15.7	3.8	0.67	4000	1400	
InP	560	12.4	3.1	1.27	4600	150	Gunn elements
InAs	477	14.6	3.5	0.36	33000	460	Hall generator, $R_H = 100\ cm^3/A\ s$
InSb	447	17.7	3.96	0.163	78000	750	Hall generator, $R_H = 400\ cm^3/A\ s$
Bi_2Te_3	—	—	—	0.15	800	400	electrical coolant
PbTe	393	280	—	0.21	1600	750	IR-detector
PbS	435	—	—	0.37	800	1000	photo resistor, IR-detector

29.9/3 Properties of doping in Si

The energy E_i of the donor levels D gives the distance from the bottom of the conduction band; the energy E_i of the acceptor levels A is the distance from the edge of the valence band.

	Al	As	Au	B	Bi	Cu	Fe	Ga
type	A	D	A	A	D	A	A	A
E_i(eV)	0.057	0.049	0.35; 0.67	0.046	0.069	0.24; 0.72	0.4; 0.66	0.065

	In	Li	O	P	S	Sb	Tl	Zn
type	A	D	D	D	D	D	A	A
E_i (eV)	0.16	0.033	0.03–0.06	0.044	0.18; 0.37	0.039	0.26	0.31; 0.66

29.9/4 Properties of doping in Ge

The energy E_i of the donor levels D gives the distance from the bottom of the conduction band; the energy E_i of the acceptor levels A is the distance from the edge of the valence band.

	Al	Ag	As	Au	B	Be	Bi
type	A	D	D	A	A	A	D
E_i (eV)	0.0102	0.13; 0.5; 0.7	0.0127	0.16; 0.59; 0.75	0.0104	0.07	0.012

	Cd	Co	Cr	Cu	Fe	Ga	In
type	A	A	A	A	A	A	A
E_i (eV)	0.05; 0.15	0.09; 0.25; 0.48	0.07; 0.12	0.4; 0.33; 0.53	0.35; 0.52	0.0108	0.0112

	Li	Mn	Ni	O	P	Pt	S
type	D	A	D	D	D	A	D
E_i (eV)	0.0093	0.16; 0.42	0.22; 0.49	0.01	0.012	0.04; 0.20; 0.67	0.18

	Sb	Se	Te	Tl	Zn		
type	D	D	D	A	A		
E_i (eV)	0.0096	0.014; 0.28	0.11; 0.30	0.01	0.03; 0.09		

29.9/5 Effect of ionizing radiation on semiconducting materials

This table lists the ionization energies for electron-hole pair formation and the pair densities g_0 produced per 10^{-2} J/kg.

Material	E_{ion}(eV)	$g_0 (cm^{-3})$
silicon	3.6	$10 \cdot 10^{13}$
silicon dioxide	≈ 18	$\approx 8 \cdot 10^{12}$
gallium arsenide	≈ 4.8	$\approx 7 \cdot 10^{13}$
germanium	2.8	$1.2 \cdot 10^{14}$

Part VI
Appendix

30
Measurements and measurement errors

Statistics offers a number of methods that, under certain conditions, permit the specification of the **expectation value** (mean value) and the **variance** (deviations from the mean value) of the random quantity considered (e.g., a random sample or a measurement/run) or on the **correlation** between random quantities. Hence, an **error estimation** relative to the actual value becomes feasible.

30.1 Description of measurements

Measurement, a quantitative determination of a physical quantity in an experiment by comparison with its basic unit.

Measured quantity, **measured variable**, nomenclature for the property to be determined by a measurement, a statistical survey, a sampling, or by carrying out a random experiment.

Discrete measured quantities
- number of dots on a die 1 to 6, faces of a coin (heads or tails).

Continuous measured quantities
- measured values for the capacitance of a capacitor or the value of a resistance.

30.1.1 Quantities and SI units

Physical phenomena may be described by mathematical objects (numbers, vectors, functions, etc.) and relations between them (equations). The goal of physics is the experimental determination and the possible precision description of natural phenomena by means of the underlying laws.

Physical quantity, serves for the description of physical states and processes. A physical quantity must be measurable in a way based on a **measuring prescription** using **measuring equipment**, i.e., it must be convertible by a physical process into a phenomenon (e.g., deflection of a pointer) that is directly accessible to human experience.

Unit, a convention allowing the quantification of the observation of a physical unit. For example: the unit mass is the mass of the international prototype of the kilogram, i.e., all masses are measured in terms of multiples and fractions of this unit mass. The definition of a unit involves fixing the physical phenomenon that will be considered the measure (or a definite quantity) of the physical quantity (e.g., mass of the kilogram prototype; distance traversed by light in a vacuum during a definite time; absolute temperature of the triple point of water, etc.). The unit gets a name (e.g., kilogram), which is denoted in formulas by a standard **abbreviation** (e.g., kg).

▲   Any physical quantity is specified by its **numerical value (numerical measure)** $\{G\}$ and its **unit** $[G]$:

$$G = \{G\} \cdot [G] \,.$$

System of units, a set of units that enables the quantification of all measurable physical quantities. **Fundamental quantities** or **basic quantities** of a system of units with their **basic units** are chosen in such a way that the units of all measurable quantities may be derived from the basic units.

SI units, defined in the **Système International d'Unités** (International System of Units). For the set of SI basic quantities, see p. 1125; for the list of SI units see p. 1126.

➤   The SI was established by the *Conférence Générale des Poids et Mesures* (General Conference on Weights and Measures), which was founded on May 20, 1875, by the Meter Convention and currently comprises 47 member countries. It is represented and managed by the *Bureau International des Poids et Mesures* (International Board on Weights and Measures) in Sèvres, France. The *International Standardization Organization* **ISO** and the *International Union of Pure and Applied Physics* **IUPAP** promote international recommendations for use of the system.

Besides the SI, several other units still exist. Their use is accepted in selected fields (e.g., carat as a weight unit for precious stones, diopter as unit of refractive power) (see p. 1128).

➤   Quantities not established in the SI or elsewhere should not be used. This concerns in particular the former technical systems of measures based on the kilopond and centimeter-gram-second (cgs) systems.

The various systems of units differ not only by the choice of the basic units, but also by the definition of basic units and derived units. For example, in the SI the mass is a basic unit and the force a derived unit, whereas in the kilopond-system the mass is derived from the basic unit of force.

➤   Units should be denoted only such as defined in SI. Examples: K for kelvin, not °K, but °C for degrees Celsius; kilometer per hour (km/h), but not hour kilometer or kilometer/hour. The unit should always be separated from the numerical value by a thin space, e.g., 35 mm film, but not 35mm film or 35-mm-film (exception: the symbols °, ′ and ″ for degree, minute and second).

Derived units, compound units, defined by equations relating physical quantities. Derived units may be given by multiplication or division of basic units. For example, the SI unit of velocity, meter per second (m/s), is obtained by division of the basic units meter (m) and second (s). Powers may also be used:

$$1 \, m \cdot m = 1 \, m^2 \,.$$

For clarity, negative exponents may be written instead of division slashes; brackets should be used where confusion may arise:

$$1 \, kg/(m \cdot s^2) = 1 \, kg \, m^{-1} s^{-2} \,.$$

▲   Independently of the selected system of units, any derived unit is specified by citing which basic units are involved.

Dimension, for any physical quantity a specification of the combination of basic units which compose it. In this book, the dimension of the quantity under discussion is given in all formula tables in a mini-box in the upper right-hand corner.

■   The unit of the dynamic viscosity is

$$1 \text{ Pa} \cdot \text{s} = 1 \text{ N}/(\text{m}^2 \cdot \text{s}) = (1 \text{ kg} \cdot \text{m}/\text{s}^2)/(\text{m}^2 \cdot \text{s})$$
$$= 1 \text{ kg}/(\text{m} \cdot \text{s}) = 1 \text{ kg} \cdot \text{m}^{-1} \cdot \text{s}^{-1}.$$

Its dimension is written in a system-independent form as

$$\text{ML}^{-1}\text{T}^{-1}.$$

➤   Compound units are pronounced as follows: units multiplied by each other are simply put in a row, units divided by each other are connected by "per." Example:

$$\text{kg m/s}^2 = \text{"kilogram meter per square second."}$$

The pronunciation km/h = "kilometer per hour" is correct.

➤   Some compound units have been given **special names** that are used instead of the compound name, such as hertz (1/s), newton (kg · m/s²) and others.

Nondimensional units, quantities with the unit 1, i.e., their numerical value is independent of the selected system of units. These are in particular percentages, statements relative to another quantity, and angles.

Conversions of units serve for the determination of comparable quantities expressed in different units. They are made by replacing a unit in a formula by a conversion factor and another unit. For example, to convert the old unit kilopond to the new unit newton, one adopts the conversion formula

$$1 \text{ kp} = 9.80665 \text{ N}.$$

One ounce per cubic inch (oz/in³) is converted to metric units as

$$1 \text{ oz/in}^3 = \frac{1 \text{ oz}}{(1 \text{ in})^3} = \frac{0.02835 \text{ kg}}{(0.0254 \text{ m})^3} = \frac{0.02835}{0.0254^3} \frac{\text{kg}}{\text{m}^3} = 1730 \text{ kg/m}^3 .$$

Decimal prefixes, prefixes, are used for denoting decimal multiples and fractions of basic units. Prefixes above 10^6 are represented by capital letters, all remaining prefixes by lower-case letters (see p. 1126). Example:

$$1 \text{ km} = 1 \text{ kilometer} = 10^3 \text{ m} = 1000 \text{ m}.$$

➤   Only **one** decimal prefix is admitted in front of a unit.

➤   Exception: For historical reasons, the units derived from the basic unit kilogram (kg) are the gram ($= 10^{-3}$ kg), the milligram ($= 10^{-6}$ kg), etc.

➤   Powers also refer to the decimal prefix:

$$1 \text{ cm}^2 = 1 \text{ square centimeter} = 1 \text{ (cm)}^2 = 1 \cdot (10^{-2} \text{ m})^2 = 10^{-4} \text{ m}^2 .$$

Natural constants, the characteristic quantities of certain natural phenomena that—to our knowledge—in all physical processes have a fixed value, e.g., the gravitational constant or the speed of light in a vacuum. Some of them are used for fixing the basic quantities, since they may be measured independently; their values in the system of units are then exact.

➤   The values of the natural constants are fixed by measurements. A **balancing calculation** (**regression**) yields those values for which the measurements are the least contradictory; the most recent data are compiled periodically by CODATA.

➤   Several constants are fixed by standards for technical use.

Material constants, characterizing specific properties of materials. They may depend on the composition of the material and on external parameters such as pressure, tension, etc.

Natural constants, on the contrary, have an arbitrarily precisely ascertainable value limited only by the measurement accuracy of the apparatus.

➤   The numerical values of the natural constants depend on the selected system of units. Inversely, the system of units is determined by the specification of these numerical values. Several natural constants have the dimension 1 (such as the fine-structure constant (see p. 859) $\alpha \approx 1/137$) and therefore have the same numerical value in all systems of units.

30.2   Error theory and statistics

30.2.1   Types of errors

Measured values of physical quantities are always subject to errors, i.e., they deviate from the true value.

30.2.1.1   Measured result

Measured result, **measured value**, **actual value**, the value of one or several measured variables after a measurement, in general not exactly reproducible but fluctuating about a **mean value** or **true value** in repeated measurements.

■   This may be, e.g., the length of a screw from industrial production, the result of a numerical random generator, the energy of a particle in a real gas, or the amount of rainfall during 24 hours.

Run, compilation of several measured results. A **primary list** of data is generated from the measurements.

30.2.1.2   Measurement error

Measurement error, deviation of a measured value from the true value. Depending on their origin, one distinguishes so-called **systematic** errors from **statistical** errors.

Systematic errors, errors characteristic of the method of measurement. They are due to the experimental arrangement or the measuring process (e.g., wrong calibration of the measuring device) and may be avoided only partly by variation of the experimental set-up.

Statistical errors, **random errors**, deviations caused by the experimentalist (e.g., reading errors), by uncontrollable perturbations (e.g., influence of temperature, variations of atmospheric pressure, etc.) or by the random nature of the events considered (e.g., radioactive decay).

Accuracy of measurement, in an experiment determined by systematic errors and statistical errors.

True error, δx_{iw}, deviation of the ith measurement with the measured result x_i from the "true value" x_w. Mostly unknown, since x_w is unknown,

$$\delta x_{iw} = x_i - x_w .$$

Absolute error, measurement error referring to the individual measurement.

Apparent error, deviation of the measured value x_i from the arithmetic mean $\bar{x}$ as approximate value of the true value,

$$v_i = x_i - \bar{x} .$$

Average error, **linear variance**, the mean value of the magnitude of the apparent error for n individual measurements,

$$d_x = \bar{v}_i = \frac{1}{n} \sum_{i=1}^{n} |x_i - \bar{x}| .$$

Relative error, v_{rel}, the absolute error divided by the mean value, a dimensionless quantity,

$$v_{\text{rel}} = \frac{v_i}{\bar{x}} = \frac{x_i - \bar{x}}{\bar{x}} .$$

Percentage error, $v_\%$, the relative error given as a percentage, $v_\% = v_{\text{rel}} \cdot 100\,\%$.

Absolute maximum error, $\delta z_{\max}$, upper error margin of a quantity $z = f(x, y)$ depending on parameters x and y that are subject to errors,

$$\delta z_{\max} = \left| \frac{\partial}{\partial x} f(\bar{x}, \bar{y}) \delta x \right| + \left| \frac{\partial}{\partial y} f(\bar{x}, \bar{y}) \delta y \right| .$$

Relative maximum error, $\delta z_{\max}/\bar{z}$, absolute maximum error divided by the mean value.

■ A wire (length L, radius R) is extended by a force F (tension σ) by ΔL. The elasticity modulus E of the wire can be determined by measuring L, R, F and ΔL. According to Hooke's law,

$$\frac{\Delta L}{L} = \frac{1}{E} \cdot \sigma , \quad \sigma = \frac{F}{A} , \quad A = \pi R^2 .$$

Because of

$$E = \frac{F}{\pi R^2} \cdot \frac{L}{\Delta L},$$

the relative maximum error of the statement on E may then be calculated from the errors δL, δR, δF, $\delta(\Delta L)$ of the individual measurement:

$$\left| \frac{\delta E}{E} \right|_{\max} = \left| \frac{\delta F}{F} \right| + 2 \left| \frac{\delta R}{R} \right| + \left| \frac{\delta L}{L} \right| + \left| \frac{\delta(\Delta L)}{\Delta L} \right| .$$

The error of the measurement of the radius enters the relative maximum error of the elasticity modulus with the factor 2.

Mean error of an individual measurement, $\overline{\delta x}$:

$$\sigma_n = \overline{\delta x} = \sqrt{\frac{1}{(n-1)} \sum_{i=1}^{n} (x_i - \bar{x})^2}, \quad \bar{x} \text{ is the arithmetic mean.}$$

Mean error of the mean value, $\overline{\delta \bar{x}}$:

$$\bar{\sigma}_n = \overline{\delta \bar{x}} = \sqrt{\frac{1}{n(n-1)} \sum_{i=1}^{n} (x_i - \bar{x})^2}, \quad \bar{x} \text{ is the arithmetic mean.}$$

▲ The mean error $\overline{\delta \bar{x}}$ of the mean value $\bar{x}$ equals the mean error $\overline{\delta x}$ of an individual measurement x_i divided by the square root of the number of measurements:

$$\overline{\delta \bar{x}} = \frac{\overline{\delta x}}{\sqrt{n}}.$$

30.2.1.3 Error propagation

Error propagation, the error of a physical quantity $f(x_0, y_0, \ldots)$ composed of directly measured partial quantities $x_0, y_0, \ldots$ may be calculated from the errors of the partial quantities.

Error propagation in an individual measurement,

$$\overline{\delta f(x_0, y_0)} = \frac{\partial f(x, y)}{\partial x}\bigg|_{x_0, y_0} \overline{\delta x} + \frac{\partial f(x, y)}{\partial y}\bigg|_{x_0, y_0} \overline{\delta y}.$$

Gauss' law of error propagation, propagation of the errors of mean values,

$$\overline{\overline{\delta f(x_0, y_0)}} = \sqrt{\left(\frac{\partial f(x, y)}{\partial x}\bigg|_{x_0, y_0} \overline{\delta x}\right)^2 + \left(\frac{\partial f(x, y)}{\partial y}\bigg|_{x_0, y_0} \overline{\delta y}\right)^2}.$$

■ The density ρ of a spherical body is determined indirectly by measuring the mass m and the radius R of the sphere, $\rho = \rho(m, R)$. The error of the measurement of the density follows from the errors of mass and radius.

30.2.2 Mean values of runs

Arithmetic mean, empirical expectation value, approximate value of the true value of a run of n individual measurements. Frequently, the equally weighted mean of the n measured values that are subject to error is given:

$$\bar{x} = \frac{1}{n} \sum_{i=1}^{n} x_i = \frac{1}{n} \sum_{j=1}^{k} H_j \cdot x_j = \sum_{j=1}^{k} h_j \cdot x_j,$$

i.e., the n measured values are distributed over $k \leq n$ distinct x_j-values with the rate H_j.

▲ Center-of-gravity property, the sum of the deviations of the measured values of the primary data list from the arithmetic mean is by definition identically zero,

$$\sum_{i}^{n} (x_i - \bar{x}) \equiv 0.$$

▲ Linearity of the arithmetic mean,

$$\overline{(ax + b)} = a\bar{x} + b.$$

▲ a, b constants, x measured variable.

▲   Quadratic minimum property, the sum of the **squares** of the deviations of all measured values x_i from the average value $\bar{x}$ takes a minimum:

$$\sum_i^n (x_i - \bar{x})^2 = \text{minimum}.$$

➤   This property is a basic ingredient of a **balancing calculation**.

▲   Combination of measurements, the mean of a total measurement involving n measured values equals the sum of the mean values of the partial measurements, weighted
by the relative fractions of measured points $n_i / \sum n_i = n_i / n$,

$$\bar{x} = \sum \bar{x}_i \cdot \frac{n_i}{n} = \sum \bar{x}_i n_i / \sum n_i .$$

▲   If the results of the run are given by a rate distribution, then

$$\bar{x} = \frac{1}{\sum_i^k H_i} \sum_{i=1}^k x_i H_i .$$

▲   Here, x_i are the class means of the classes K_i $(i = 1, \ldots, k)$.

Quantile, percentile of order p, a measured value that is **not** below a fraction p of all
measured values of the primary data list, and **not** above a fraction $1 - p$, a characteristic
quantity for describing the relative position of the individual measured values among each
other.

   Median, central value, $\tilde{x}$, special case of a percentile, defined as the value **bisecting** the
series of the n measured values of the primary data list when ordered by the magnitude.

   Median for even number of measured values:

$$\tilde{x} = \frac{x_{\frac{n}{2}} + x_{\frac{n}{2}+1}}{2} .$$

Median for odd number of measured values:

$$\tilde{x} = x_{\frac{n+1}{2}} .$$

➤   The median is applied mainly in the following situations:

a)   classes at the boundaries of the ordered primary list are missing;

b)   extreme measured values occur that would falsify the result;

c)   variations of the measured values above and below the mean value do not affect this
value.

▲   The sum of the absolute magnitudes of the deviations of all measured values x_i from
the median $\tilde{x}$ is smaller than the sum of the deviations from any other value a:

$$\sum_{i=1}^n |x_i - \tilde{x}| < \sum_{i=1}^n |x_i - a|, \quad \begin{array}{l} \text{for all} \quad a \neq \tilde{x}, \\ \text{for all} \quad x_{\frac{n}{2}} \leq a \leq x_{\frac{n}{2}+1}, \end{array} \quad \begin{array}{l} \text{if } n \text{ odd,} \\ \text{if } n \text{ even.} \end{array}$$

Quadratic mean:

$$x_{\text{quad}} = \sqrt{\frac{1}{n} \sum_{i=1}^n x_i^2} .$$

Geometric mean:

$$\hat{x} = \sqrt[n]{\prod_{i=1}^{n} x_i} = (x_1 \cdot x_2 \cdot \cdots \cdot x_n)^{1/n} \, .$$

➤ The geometric mean is used in particular for quantities governed by laws that lead to geometric sequences.

◼ Mean average **growth velocity** or **rate of increase** of time-dependent processes (radioactive decay, lifetime of components),

$$\hat{x} = (x_1 \cdot x_2 \cdot \cdots \cdot x_n)^{1/n}, \quad x_i > 0 \, .$$

▲ The logarithm of the geometric mean is equal to the arithmetic mean of the logarithms of all measured values,

$$\ln \hat{x} = \frac{1}{n}(\ln x_1 + \cdots + \ln x_n) \, .$$

Growth velocity, the average percentage development from x_n to x_{n+1} (specifications in percent fractions of a total set A),

$$\overline{W} = \sqrt[n-1]{\frac{x_n}{x_1}} \cdot 100 \, \% \, .$$

Rate of increase, the average percentage evolution by $\bar{R}$ percent,

$$\bar{R} = \left(\sqrt[n-1]{\frac{x_n}{x_1}} - 1 \right) \cdot 100 \, \% \, .$$

➤ If there is no percentage evolution, then the absolute values $a_1 = x_1 \cdot A$, $a_n = x_n \cdot A$ may be inserted instead of x_1, x_n.

Harmonic mean:

$$x_h = \frac{n}{\displaystyle\sum_{i=1}^{n} \frac{1}{x_i}} \, .$$

▲ **Theorem of Cauchy**: There exists the following hierarchy of mean values x_{quad}, x_h, $\hat{x}$ and $\bar{x}$:

$$x_{\min} \leq x_h \leq \hat{x} \leq \bar{x} \leq x_{\text{quad}} \leq x_{\max} \, .$$

30.2.3 Variance

Variance, **mean square deviation**, **standard deviation**, measure of the variance caused by measurement errors, fluctuation of the measured values about the true value.

Span, **variation width**, difference between the largest and smallest measured value,

$$\delta x_{\max} = x_{\max} - x_{\min} \, .$$

➤ The span is mostly used for a small number of measured values. Application in statistical quality controls.

Mean absolute deviation about the value C,

$$\overline{|\delta x|_C} = \frac{1}{n} \sum_{i=1}^{n} |x_i - C| .$$

➤ Normally, $C = \tilde{x}$ (median) or $C = \bar{x}$ (arithmetic mean) are used.
➤ If a rate table ordered by classes is given, then the centers of the classes are inserted as measured values x_i.

Root-mean-square deviation, standard deviation, empirical variance:

$$\sigma_n = \sqrt{\overline{(\delta x)^2}} = \sqrt{\frac{1}{n-1} \sum_{i=1}^{n} (x_i - \bar{x})^2} .$$

▲ If the run data are given in terms of a rate distribution, then

$$\sigma_n = \sqrt{\overline{(\delta x)^2}} = \sqrt{\frac{1}{n-1} \sum_{i=1}^{k} (x_i - \bar{x})^2 H(x_i)}, \quad n = \sum_{i} H(x_i) .$$

➤ In the case of subdivision into classes, the class centers are often inserted instead of the unknown measured values.

Empirical variance, σ_n^2, square of the standard deviation, also denoted as **variance**.

The empirical variance σ_n is an **unbiased** estimate for the variance of an underlying probability function on the parent population.

Relative variance measure, variation coefficient, percentage value of the variance measure related to the arithmetic mean,

$$\overline{(\delta x)^2}_{\text{rel}} = \frac{\overline{(\delta x)^2}}{\bar{x}} \cdot 100\% .$$

30.2.4 Correlation

Covariance of two measured quantities x, y, $\text{cov}(x, y)$, the expectation value of the product of the deviations of the corresponding quantities from their mean values,

$$\text{cov}(x, y) = \overline{(x - \bar{x})(y - \bar{y})} .$$

Correlation coefficient of x, y, ρ_{xy}, covariance of x, y, divided by the product of the root-mean-square deviations σ_x, σ_y,

$$\rho_{xy} = \frac{\text{cov}(x, y)}{\sigma_x \cdot \sigma_y}, \quad -1 \leq \rho_{xy} \leq 1 .$$

• If x and y are **statistically independent random variables**, then $\rho_{xy} = 0$; x and y are **not correlated**.
• x and y are linearly dependent, $y = ax + b$ (a, b : real numbers), if and only if $\rho_{xy} = \pm 1$.
• The sign of the correlation coefficient indicates whether a positive or negative correlation exists:

positive correlation, an increase (decrease) of x causes an increase (decrease) of y,

negative correlation, an increase (decrease) of x causes a decrease (increase) of y.

30.2.5 Regression analysis

Regression, the optimal adjustment of a properly selected parameter-dependent regression fit $y = f(x, a, b, ...)$ to n given data points $(x_1, y_1), (x_2, y_2), ..., (x_n, y_n)$ of two correlated random variables.

Sum of error squares, sum of the squares of the differences between the measured values y_i and the function values of the regression fit f at the points x_i,

$$\sum_{i=1}^{n} [y_i - f(x_i, a, b, ...)]^2 .$$

Principle of least squares, allows calculation of the parameter set $a, b, ...$ that provides the optimal adjustment of the regression fit to the given data points by the condition that the sum of the errors squared takes a minimum (Gauss' minimum principle),

$$\sum_{i=1}^{n} [y_i - f(x_i)]^2 = \min.$$

Linear regression, regression fit with a straight line as formulation,

$$y = ax + b.$$

The formulation is appropriate if the two random variables are almost linearly correlated.

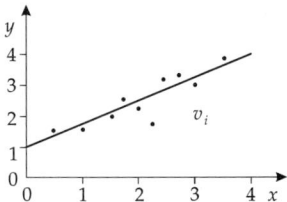

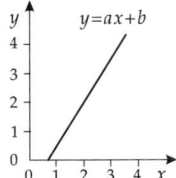

Figure 30.1: Adjustment of a curve to given data points by the principle of least squares.

Figure 30.2: Linear regression.

30.2.6 Rate distributions

Primary list, a list of all measured values of a run. Identical measured results may occur repeatedly.

■   When producing n capacitors with a capacitance of $C = 100 \ \mu F$, the value for the individual component is in general not exactly $100 \ \mu F$, but fluctuates about this value. The data follow a characteristic distribution about the desired value $C = 100 \ \mu F$. To get a deeper understanding of the type of distribution, and of the nature of the underlying probability process, one determines the so-called relative rate distribution and compares it with special probability functions that may be derived from known probability structures. For example, the hypergeometric distribution may be traced back to the very simple and clear **jar model**.

In our example, the individual measured quantity is the capacitance of each capacitor. The data constitute the so-called **primary list**:

Capacitor no.	1	2	3	4	5	6	$\cdots$	n
Capacitance in μF	101.1	99.6	101.4	103.3	98.0	99.5	$\cdots$	C_n

Class K_i, a set of several elements (measured values) of a primary list with defined properties which are combined under the index i.

■ The daily production output of n capacitors of a given capacitance C may be classified, e.g., by subdividing the capacitances into $N = 8$ intervals ($N = 8$ classes).

Class	Limits of interval		Class	Limits of interval	
K_1		$C < 92.5$	K_5	$100.0 \leq$	$C < 102.5$
K_2	$92.5 \leq$	$C < 95.0$	K_6	$102.5 \leq$	$C < 105.0$
K_3	$95.0 \leq$	$C < 97.5$	K_7	$105.0 \leq$	$C < 107.5$
K_4	$97.5 \leq$	$C < 100.0$	K_8	$107.5 \leq$	C

➤ Classes need not always be defined. For discrete measured values $x = X_i$ repeating in the primary list, the coinciding values may of course be considered a class of its own, $K_i = X_i$.

Class center, **interval center**, arithmetic mean of the interval limits of a class.

➤ It is more suitable to form the arithmetic mean of all measured values of the corresponding class. But sometimes the individual measured values are not known, or are discarded in data-taking for reasons of time (computational effort in very extensive surveys). Therefore, the interval center is in general an approximate quantity.

Rate $H_i = H(K_i)$, number of measured values from the primary list falling into the class K_i.

➤ If measured values occur repeatedly in the primary list, a discrete measured value may also be taken as a class.

Rate table, tabular mapping of each class onto the corresponding number (**rate**) of measured values.

■ The rate table of a daily production output, related to the capacitance of the capacitors, might look as follows:

K_i	K_1	K_2	K_3	K_4	K_5	K_6	K_7	K_8	sum
$H(K_i)$	133	43789	189345	281321	255128	206989	26923	155	1003783

Rate distribution, **rate histogram**, graphic representation of a rate table.

■ The rate table given above is represented by the **bar graph** shown in **Fig. 30.3**. For visual representation, one also uses other diagrams, e.g., the **pie chart**.

Relative rate, the relative rate h_i of the class K_i for n measured values in total,

$$h_i = \frac{H_i}{n}.$$

Relative rate distribution, **normalized rate distribution** h_i,

$$\sum_{i=1}^{N} h_i = 1.$$

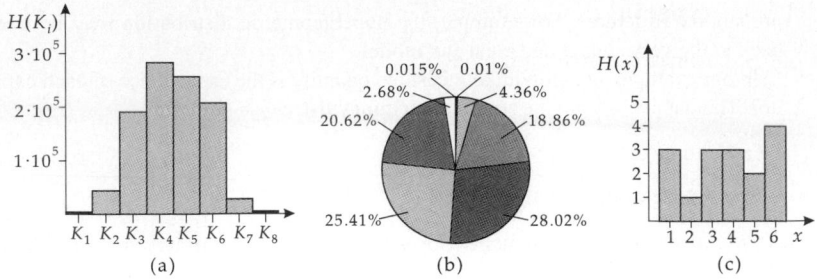

(a)   (b)   (c)

Figure 30.3: Representation of a rate table. (a): bar graph, (b): pie chart, (c): distribution with three cluster points.

The relative rate may also be represented graphically by a histogram.

▲   When dividing the (relative) rate by a constant factor c, the arithmetic mean remains unchanged,

$$\frac{\sum_{i}^{N} x_i \cdot H(x_i)/c}{\sum_{i}^{N} H(x_i)/c} \equiv \bar{x} .$$

Modal value, density mean, x_m, the most frequently measured value in a series of measured values.

➤   For runs with several cluster points, there are also several density means. Each cluster point has to be considered separately.

Urn model or jar model, n marbles are picked out of a vessel (urn or jar) containing N marbles, M of them being black and $N - M$ being white. If p is the probability to pick a black sphere, then the probability to pick a white sphere is $1 - p$. We are looking for the probability of finding k marbles of a definite color among the n marbles picked out (when repeating the experiment n times, a certain event happens exactly k times).

Selection **with retjar** (with return), every sphere is returned after selection.

Selection **without retjar** (without return), the marbles picked are not returned to the jar.

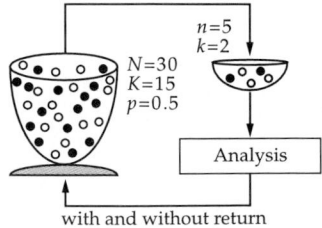

with and without return

Figure 30.4: Urn model.

Single probability, $P(k)$, the probability that a discrete random variable takes the value k in a single measurement.

30.2.6.1   Special discrete distributions

- **Hypergeometric distribution**:

$$P(k) = \frac{\left(\begin{array}{c} pN \\ k \end{array}\right)\left(\begin{array}{c} N(1-p) \\ n-k \end{array}\right)}{\left(\begin{array}{c} N \\ n \end{array}\right)}, \quad p \cdot N : \text{integer}.$$

Expectation value: $n \cdot p$.
Variance: $\sigma^2 = n \cdot p(1-p)[(N-n)/(N-1)]$.

- **Binomial distribution**:

$$P(k) = \left(\begin{array}{c} n \\ k \end{array}\right) p^k(1-p)^{n-k}.$$

Expectation value: $n \cdot p$.
Variance: $\sigma^2 = n \cdot p(1-p)$.

- **Poisson distribution**:

$$P(k) = \frac{c^k}{k!} \cdot e^{-c}, \quad k = 0, 1, 2, \ldots; \quad c > 0.$$

Expectation value: c.
Variance: $\sigma^2 = c$.

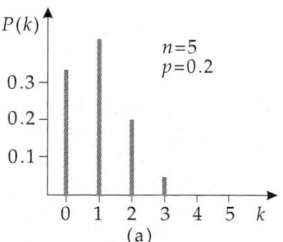
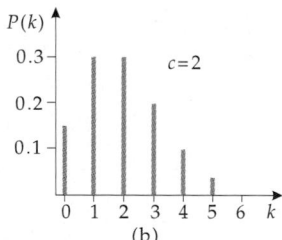

Figure 30.5: (a): binomial distribution, (b): Poisson distribution.

▲ The hypergeometric distribution corresponds to the jar model without returning the marbles picked. The binomial distribution corresponds to the jar model with return of the marbles picked.

▲ The binomial distribution follows from the hypergeometric distribution if the number of marbles in a jar model becomes very large ($N \to \infty$) and the number of random samples n remains small.

▲ The Poisson distribution follows from the binomial distribution if the number of random samples n in the jar model becomes very large and the marked fraction p is very small but finite, $n \to \infty$, $p \to 0$.

Probability density, $f(x)$, density of the distribution of a continuous random variable, or idealized analytic function for the probability density of discrete random variables.

30.2.6.2   Special continuous distributions

- **Gauss' distribution, normal distribution:**

$$f(x) = \frac{1}{\sigma\sqrt{2\pi}} e^{-(x-m)^2/(2\sigma^2)} .$$

Expectation value: m.
Variance: σ^2.

- **Standard normal distribution, Gauss' normal distribution**, special case of the normal distribution with $m = 0$ and $\sigma = 1$.
- **Exponential distribution:**

$$f(x) = \lambda e^{-\lambda x} , \quad \lambda > 0, \quad x \geq 0 .$$

Expectation value: $1/\lambda$.
Variance: $\sigma^2 = 1/\lambda^2$.

- **Weibull distribution:**

$$f(x) = \frac{\gamma}{\beta} \left(\frac{x-\alpha}{\beta}\right)^{\gamma-1} e^{-((x-\alpha)/\beta)^\gamma} , \quad x \geq \alpha .$$

Expectation value: $\beta\,\Gamma(1 + 1/\gamma) + \alpha$.
Variance: $\sigma^2 = \beta^2 \{\Gamma(1 + 2/\gamma) - [\Gamma(1 + 1/\gamma)]^2\}$, $\quad \Gamma(k)$: gamma function.

- χ^2**-distribution** with the degree of freedom n: a distribution resulting for the measured quantity $\chi^2 = Y_n = x_1^2 + x_2^2 + \cdots + x_n^2$ if the individual measured values x_i, ($i = 1, \ldots, n$) follow a standard normal distribution,

$$f_\chi(Y_n; n) = \frac{1}{2^{n/2}\Gamma(n/2)} Y_n^{(n/2)-1} e^{-Y_n/2} .$$

Expectation value: n.
Variance: $\sigma^2 = 2n$.

- t**-distribution, student's distribution**, distribution of the measured quantity $T_n = x/\sqrt{Y_n/n}$ if x obeys a standard normal distribution and Y_n obeys a $f_\chi(Y_n; n)$-distribution,

$$f_t(T_n; n) = \frac{\Gamma((n+1)/2)}{\sqrt{n\pi}\,\Gamma(n/2)} \left(1 + \frac{T_n^2}{n}\right)^{-(n+1)/2} .$$

Expectation value: 0.
Variance: $\sigma^2 = n/(n-2)$.

The **normal distribution** is symmetric about its maximum at $x = m$. The maximum value of the function $f(x)$ is $1/(\sigma\sqrt{2\pi})$. The normal distribution has inflexion points at $x = m \pm \sigma$. About 99.7 % of the measured values fall into the interval $x = m \pm 3\sigma$, about 95.5 % into the interval $x = m \pm 2\sigma$, and about 68 % into the interval $x = m \pm \sigma$. The variance σ^2 may be extracted from the **half-width** b of the curve, i.e., the width of the curve at half maximum, $\sigma^2 = 0.18 \cdot b^2$. For a finite number n of measurements, the arithmetic mean $\bar{x}$ of the measured values is the best estimate for the expectation value m.

The normal distribution is normalized to 1,

$$\int_{-\infty}^{\infty} f(x)\,dx = 1 .$$

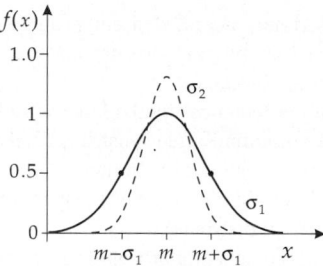

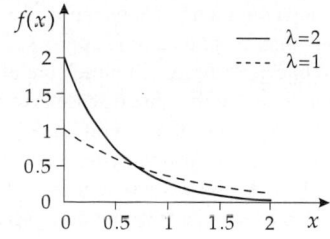

Figure 30.6: Normal distribution.
Maximum: $M = 1/(\sigma\sqrt{2\pi})$, inflexion
points: $(m \pm \sigma)$, half-width: b.

Figure 30.7: Exponential distribution.

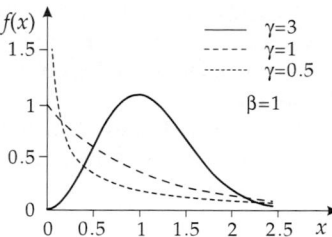

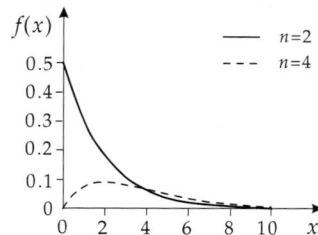

Figure 30.8: Weibull distribution.

Figure 30.9: χ^2-distribution.

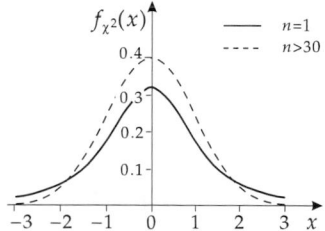

Figure 30.10: t-distribution.

Central value limit theorem, with increasing n, the sum of n independent random variables obeying the same distribution always converges towards the normal distribution.

▲ Owing to the multiple superposition of error sources, measuring errors are in general normally distributed.

30.2.7 Reliability

Events depending on time (e.g., radioactive decay, failure of an electric component) may be reasonably described several special quantities.

Lifetime, time between failures of objects. The distribution of the failures over time may be purely incidental (non-aging objects), or may be altered by external influences (aging objects).

Non-aging objects, objects with a finite **lifetime**, the failure is purely random and follows a distribution based on a purely combinatoric random principle (**jar model, Poisson distribution, exponential distribution**). They are not influenced by aging processes, as, e.g., external wearing phenomena.

■ Electronic components such as resistors, capacitors, integrated circuits (under specified conditions of application, i.e., no excessive load by too high currents or too high voltages) are to a good approximation non-aging objects.

■ Objects of finite "lifetime" are also found in non-technical fields. For example, the infection with a rare disease is to a good approximation Poisson-distributed, the time intervals between several infections follow an exponential distribution.

▲ The failures of non-aging objects follow a Poisson distribution with respect to time. The times between the failures obey an exponential distribution.

Aging objects, objects with a finite **lifetime** that undergo an aging process. The aging may affect the purely random decay process, hence may modify the distribution of failures (see Weibull distribution).

■ Typical examples for aging objects are engines, tires, tools.

▲ The failure of aging objects is no longer Poisson-distributed. In order to describe the time distances between the failures, a more sophisticated form of the distribution has to be used. Frequently, the time distance between failures may be represented by a superposition of several exponential distributions. The lifetime of aging objects may in some cases be represented by a **Weibull distribution**.

The exponential distribution and the Weibull distribution are special cases of **reliability**.

Reliability, $Z(t)$, the average number of parts $N(t)$ still functioning after the time t, related to an initial set N_0. General set-up for describing aging processes as a function of time:

$$Z(t) = \frac{N(t)}{N_0} = e^{-\int_0^t \lambda(t')dt'} .$$

$Z(t)$ is the probability that a part did **not** yet fail after the time t.

Failure probability, $F(t)$, average number of parts $N_0 - N(t)$ that failed after the time t, relative to the initial quantity N_0,

$$F(t) = 1 - Z(t) .$$

$F(t)$ is the probability that a part failed after the time t.

Failure density, ρ, the average number of failures per unit time at the moment t relative to the initial set N_0,

$$\rho(t) = \frac{dF(t)}{dt} = -\frac{dZ(t)}{dt} = \lambda(t)Z(t) .$$

➤ The integral over the failure density is just the quantity of failures relative to the initial quantity N_0,

$$\int_0^t \rho(t')dt' = -\int_0^t \frac{dZ(t')}{dt'}dt' = -(Z(t) - Z(0)) = 1 - Z(t) = F(t) .$$

Failure rate, the average number of failures per time unit, relative to the number of still-functioning parts $N(t)$,

$$\lambda(t) = -\frac{1}{N(t)}\frac{dN(t)}{dt} = -\frac{1}{Z(t)}\frac{dZ(t)}{dt} = \frac{\rho(t)}{Z(t)} .$$

Mean time between failures (MTBF):

$$MTBF = \int_0^\infty Z(t)dt .$$

▲   The probability that the total system is still functioning after the time t is equal to the product of the reliabilities of the individual systems,

$$Z_{\text{total}} = Z_1 Z_2 \ldots Z_n .$$

Non-aging objects:

$$\lambda_{\text{total}} = \lambda_1 + \lambda_2 + \cdots + \lambda_n .$$

▲   If the rate λ and the operating time t are small, the failure rate may be approximated by the number of failures per initial quantity and operating time,

$$\lambda \approx \frac{1 - N(t)}{N_0 \cdot t} = \frac{\text{failures}}{\text{initial quantity} \cdot \text{operating time}} .$$

▲   For non-aging objects, $Z(t)$ is the exponential distribution ($\lambda =$ const.), and the failure time is thus $1/\lambda$.

■   Some failure rates (λ in fit = failure/10^9 h):

wrap connections	0.0025
mica capacitor	1
HF-coil	1
metal-layer resistor	1
paper capacitor	2
transistor	200
light-emitting diode (50 % loss of luminosity)	500

31
Vector calculus

Vector, a quantity characterized by a **magnitude** and an **orientation**. A vector is represented graphically by an arrow whose length represents the magnitude of the vector.

■ Velocity, momentum, electric field intensity are vectors, like the position vector pointing from the origin of the coordinate frame to a defined position.

Vectors are distinguished by their behavior under rotations of the coordinate frame. Since they have a direction measured relative to a coordinate frame, their components (not their magnitude) change under rotation of the reference frame. On the contrary, **scalars** do not change their value under rotation of the reference frame; they are real or complex numbers.

■ Time, mass, charge and temperature are scalars.

When they represent physical quantities, both scalars and vectors have a **unit** that has to be specified in addition. In the case of vectors, the unit refers to the magnitude of the vector.

➤ Although the magnitude of the vector is shown by the length of the arrow, it may have an arbitrary unit. For example, the unit of a force vector is the newton.

Component representation, representation of the vector in a Cartesian coordinate frame. In order to represent an arbitrary vector, the base of the vector arrow is put at the origin of a Cartesian coordinate frame, and the coordinates of its end point can be specified by a column vector:

$$\vec{a} = \begin{pmatrix} a_x \\ a_y \\ a_z \end{pmatrix} \quad \longleftrightarrow \quad \vec{a} = a_x \vec{e}_x + a_y \vec{e}_y + a_z \vec{e}_z \,,$$

$\vec{e}_x, \vec{e}_y, \vec{e}_z$ being unit vectors pointing along the positive coordinate axes. The components of the vector have the same unit as the vector itself,

$$[a_x] = [a_y] = [a_z] = [\vec{a}] \,.$$

1115

Magnitude of a vector, the length of the vector arrow. In a component representation, it is given by

$$|\vec{a}| = \sqrt{a_x^2 + a_y^2 + a_z^2}\,.$$

For the unit,

$$[|\vec{a}|] = [\vec{a}]\,.$$

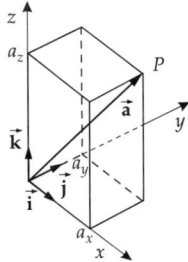

Figure 31.1: Component representation of a vector $\vec{a}$ in a three-dimensional Cartesian reference frame.

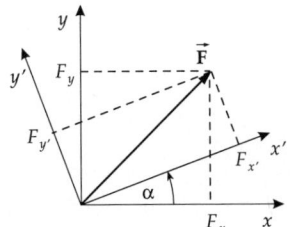

Figure 31.2: Behavior of a vector under rotation of the coordinate frame. (F_x, F_y) and $(F_{x'}, F_{y'})$ are the components of the vector $\vec{F}$ in two frames rotated with respect to each other by the angle α.

31.1.2 Multiplication by a scalar

A vector may be multiplied by a real or complex number (scalar).

Multiplication by a scalar, every component is multiplied by the real or complex number α:

$$\alpha\vec{a} = \begin{pmatrix} \alpha a_x \\ \alpha a_y \\ \alpha a_z \end{pmatrix}.$$

The length of the vector is changed by the factor $|\alpha|$: $|\alpha\vec{a}| = |\alpha|\,|\vec{a}|$. If $\alpha < 0$, the resulting vector points opposite to the original vector.

Inverse vector, opposite vector, the vector obtained by multiplication by -1. It has the same length as the original vector, but points in the opposite direction.

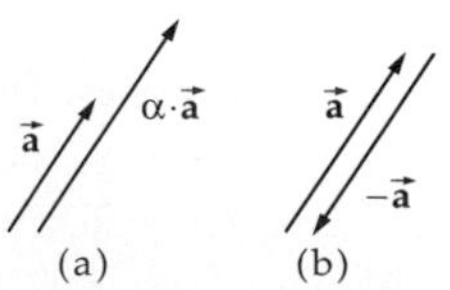

Figure 31.3: Vector multiplication. (a): multiplication of a vector $\vec{a}$ by a scalar α, (b): opposite vector $-\vec{a}$.

31.1.3   Addition and subtraction of vectors

Vectors may be added and subtracted if they have the same units.

Vector addition, the individual components are added:

$$\vec{a} + \vec{b} = \begin{pmatrix} a_x + b_x \\ a_y + b_y \\ a_z + b_z \end{pmatrix},$$

where $\vec{a}$ and $\vec{b}$ are arbitrary vectors having identical units. $\vec{a}$ and $\vec{b}$ form a parallelogram; the resulting vector is the diagonal.

The same result is obtained by putting one vector at the end of the other vector; the resulting vector points from the initial point of the first vector (origin) to the endpoint of the second vector.

Vector subtraction, achieved by adding the opposite vector:

$$\vec{a} - \vec{b} = \vec{a} + (-1) \cdot \vec{b}.$$

The vector $\vec{a} - \vec{b}$ is also called the "difference" vector; it points from the endpoint of vector $\vec{b}$ to the endpoint of vector $\vec{a}$.

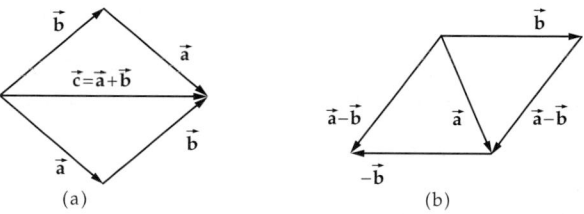

(a)   (b)

Figure 31.4: Vector addition. (a): addition, (b): subtraction of the vectors $\vec{a}$ and $\vec{b}$.

Unit vector along $\vec{a}$, a vector of length 1 pointing along the vector $\vec{a}$. It is obtained by dividing the vector $\vec{a}$ by its length,

$$\vec{e} = \frac{\vec{a}}{|\vec{a}|}.$$

Unit vectors are used to specify a direction.

31.1.4   Multiplication of vectors

There are two kinds of vector multiplication.

1. Scalar product,

$\vec{a} \cdot \vec{b}$, its value is a real number (scalar). The scalar product is given by the length of the normal projection of one vector onto the second vector multiplied by the magnitude of the second vector. If the angle α between the two vectors is larger than $90°$, then the scalar product is negative.

scalar product			
$\vec{a} \cdot \vec{b} = |\vec{a}| \, |\vec{b}| \cos\alpha$	Symbol	Unit	Quantity
	$\vec{a}, \vec{b}$	arbitrary	vectors
$= a_x b_x + a_y b_y + a_z b_z$	$a_x, b_x, \ldots$	arbitrary	components
	α	rad	angle between $\vec{a}$ and $\vec{b}$

The scalar product is commutative, i.e.,

$$\vec{a} \cdot \vec{b} = \vec{b} \cdot \vec{a}.$$

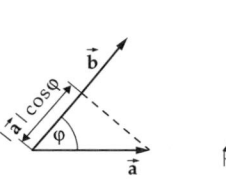
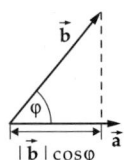
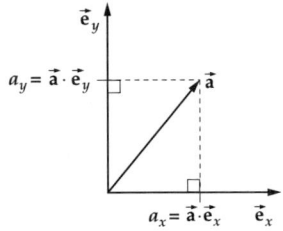

Figure 31.5: Scalar product of two vectors $\vec{a}$ and $\vec{b}$.

Figure 31.6: Components (a_x, a_y) of a vector $\vec{a}$ along the axes given by $\vec{e}_x, \vec{e}_y$.

The scalar product is used to form the **projection** of a vector onto the direction of another vector. In particular, one may decompose a given vector into its Cartesian components:

$$\vec{a} = \begin{pmatrix} a_x \\ a_y \\ a_z \end{pmatrix} = a_x \vec{e}_x + a_y \vec{e}_y + a_z \vec{e}_z,$$

$$a_x = \vec{a} \cdot \vec{e}_x, \quad a_y = \vec{a} \cdot \vec{e}_y, \quad a_z = \vec{a} \cdot \vec{e}_z,$$

where $\vec{e}_x$, $\vec{e}_y$ and $\vec{e}_z$ are unit vectors along the axes of a Cartesian coordinate frame.

Using the scalar product, one may check whether two vectors are perpendicular to each other.

▲   The scalar product of two vectors that are perpendicular to each other vanishes.

The length of a vector is equal to the root of the scalar product of the vector with itself:

$$|\vec{a}| = \sqrt{\vec{a} \cdot \vec{a}}.$$

It is always larger than or equal to zero.

Finally, one may calculate the angle α between two vectors $\vec{a}$ and $\vec{b}$ by the scalar product:

$$\cos\alpha = \frac{\vec{a} \cdot \vec{b}}{|\vec{a}| \, |\vec{b}|}.$$

2. *Vector product,*

cross-product, $\vec{a} \times \vec{b}$, a vector assigned to two vectors $\vec{a}$ and $\vec{b}$ that points perpendicular to $\vec{a}$ and $\vec{b}$. Its length is equal to the product of the lengths of the two vectors, and of the sine of the angle enclosed:

vector product			
$\lvert \vec{a} \times \vec{b} \rvert = \lvert \vec{a} \rvert \, \lvert \vec{b} \rvert \sin\alpha$	Symbol	Unit	Quantity
$\vec{a} \times \vec{b} = \begin{pmatrix} a_y b_z - b_y a_z \\ a_z b_x - b_z a_x \\ a_x b_y - b_x a_y \end{pmatrix}$	$\vec{a}, \vec{b}$	arbitrary	vectors
	$a_x, b_x, \ldots$	arbitrary	components
	α	rad	angle between $\vec{a}$ and $\vec{b}$

The vector product is used to construct a vector perpendicular to two given vectors. The vectors $\vec{a}$, $\vec{b}$ and $\vec{a} \times \vec{b}$ in this order of sequence form a **right-handed system**, like the thumb, forefinger and middle finger of the right hand.

➤ Distinctions between scalar and vector product: The vector product is a vector, the scalar product is a real number. The scalar product has a maximum value when the two vectors are parallel to each other; the magnitude of the vector product has a maximum when the vectors are perpendicular to each other.

The most important properties of the vector product are:

▲ $\vec{a} \times \vec{a} = 0$: the vector product of a vector with itself vanishes.

▲ $\vec{a} \times \vec{b} = -\vec{b} \times \vec{a}$: the vector product is anti-commutative.

▲ The unit vectors of a Cartesian coordinate frame are related as follows:

$$\vec{e}_x \times \vec{e}_y = \vec{e}_z \,; \quad \vec{e}_y \times \vec{e}_z = \vec{e}_x \,; \quad \vec{e}_z \times \vec{e}_x = \vec{e}_y \,.$$

The cross-products between identical unit vectors vanish:

$$\vec{e}_x \times \vec{e}_x = \vec{e}_y \times \vec{e}_y = \vec{e}_z \times \vec{e}_z = 0 \,.$$

▲ **Triple scalar product**, the scalar product of a vector $\vec{c}$ with the vector product of the vectors $\vec{a}$ and $\vec{b}$:

$$(\vec{a} \times \vec{b}) \cdot \vec{c}.$$

The triple scalar product is defined only in a three-dimensional space. The triple scalar product is a scalar; its absolute value is equal to the volume of the parallelepiped described by the vectors $\vec{a}, \vec{b}, \vec{c}$.

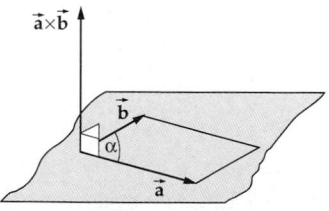

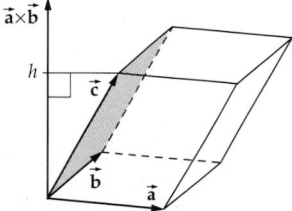

Figure 31.7: Vector product of two vectors $\vec{a}$ and $\vec{b}$. $\vec{a} \times \vec{b}$ points perpendicular to the vectors $\vec{a}$ and $\vec{b}$.

Figure 31.8: Triple scalar product.

The two-fold **cross-product** $\vec{a} \times (\vec{b} \times \vec{c})$ is a vector in the plane spanned by the vectors $\vec{b}$ and $\vec{c}$:

$$\vec{a} \times (\vec{b} \times \vec{c}) = \vec{b}(\vec{a} \cdot \vec{c}) - \vec{c}(\vec{a} \cdot \vec{b}) \,.$$

32
Differential and integral calculus

32.1 Differential calculus

Derivative of a function $y = f(x)$ at the point x, defined as the slope of the tangent to the function at the point x.

  Difference quotient, slope of the secant through the points $P(x, y)$ and $P_0(x_0, y_0)$,

$$\frac{\Delta y}{\Delta x} = \frac{\Delta f(x)}{\Delta x} = \frac{f(x) - f(x_0)}{x - x_0}.$$

Differential quotient $f'(x)$, limit value of the difference quotient for $P \to P_0$, $\Delta x \to 0$,

$$\frac{dy}{dx} = f'(x) = \lim_{\Delta x \to 0} \frac{\Delta y}{\Delta x} = \lim_{\Delta x \to 0} \frac{f(x + \Delta x) - f(x)}{\Delta x}.$$

The derivative of a function at the point P_0 corresponds to the gradient of its graph at the point P_0, $f'(x_0) = \tan \alpha$.

32.1.1 Differentiation rules

Constants rule, the derivative of a constant c is equal to zero,

$$c' = 0.$$

Factor rule, a constant factor c remains unchanged when the derivative is taken,

$$(c \cdot f(x))' = c \cdot f'(x).$$

Power rule, when carrying out the derivative of a power function, the exponent is lowered by unity, and the old exponent enters as a factor,

$$\frac{d}{dx} x^n = n \cdot x^{n-1}.$$

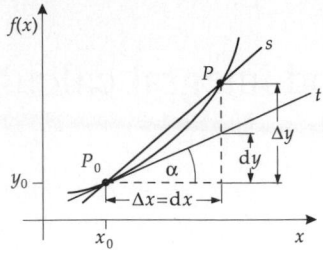

Figure 32.1: Derivative of a function $f(x)$. t: tangent, s: secant.

Sum rule, the derivative of a sum (difference) is equal to the sum (difference) of the derivatives,

$$(f(x) \pm g(x))' = f'(x) \pm g'(x).$$

Product rule:

$$(f(x) \cdot g(x))' = f(x) \cdot g'(x) + f'(x) \cdot g(x),$$

$$(f(x) \cdot g(x) \cdot h(x))' = f(x) \cdot g(x) \cdot h'(x) + f(x) \cdot g'(x) \cdot h(x) + f'(x) \cdot g(x) \cdot h(x).$$

Quotient rule:

$$\left(\frac{f(x)}{g(x)}\right)' = \frac{g(x) \cdot f'(x) - f(x) \cdot g'(x)}{g^2(x)},$$

$$\left(\frac{1}{g(x)}\right)' = \frac{-g'(x)}{g^2(x)}.$$

Chain rule:

$$(f(g(x))' = g'(x) \cdot f'(g(x)), \qquad \frac{df}{dx} = \frac{dg}{dx} \cdot \frac{df}{dg}.$$

$\dfrac{df}{dg}$: exterior derivative, $\dfrac{dg}{dx}$: interior derivative.

Logarithmic derivative, derivative of the logarithm $\ln y$ of the function y for $y > 0$,

$$(\ln y)' = \frac{y'}{y}.$$

32.2 Integral calculus

Integration, inverse of differentiation.

Antiderivative function, **integral function** $F(x)$ of a function $f(x)$. The derivative $F'(x)$ of the integral function is equal to $f(x)$. The function $F(x)$ is defined over the same interval as $f(x)$.

Integration of a function $f(x)$, determination of the integral function $F(x)$ of $f(x)$, the derivative of which is again the original function $f(x)$.

▲ To any integrable function, there exist infinitely many integral functions $F(x) + C$ that differ only by an additive **integration constant** C. All integral functions have the same slope at a fixed value x.

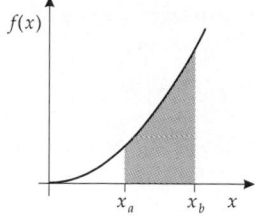

Figure 32.2: Definite integral A of the function $f(x)$.

Indefinite integral I, the integration constant C is not fixed,

$$I = \int f(x)\, \mathrm{d}x = F(x) + C\,.$$

Definite integral, upper and lower boundary of integration are fixed. The definite integral is a number,

$$A = \int_a^b f(x)\, \mathrm{d}x = F(b) - F(a)\,.$$

▲   The definite integral A corresponds to the area between the function $f(x)$ and the x-axis between $x = a$ and $x = b$. If $f(x)$ becomes also negative in the integration interval, then the definite integral is equal to the difference of the areas above and below the x-axis.

32.2.1   Integration rules

Constant rule, a constant factor may be pulled out of the integral,

$$\int c \cdot f(x)\, \mathrm{d}x = c \cdot \int f(x)\, \mathrm{d}x\,.$$

Sum rule, the integral over a sum of terms is equal to the sum of the integrals over the terms,

$$\int (f(x) + g(x))\, \mathrm{d}x = \int f(x)\, \mathrm{d}x + \int g(x)\, \mathrm{d}x\,.$$

Power rule:

$$\int x^n\, \mathrm{d}x = \frac{x^{n+1}}{n+1}\,, \quad n \neq -1\,.$$

Inversion rule, inversion of the sign of the definite integral under inversion of the integration boundaries,

$$\int_a^b f(x)\, \mathrm{d}x = -\int_b^a f(x)\, \mathrm{d}x\,.$$

Equality of upper and lower boundary, the integral vanishes,

$$\int_a^a f(x)\, \mathrm{d}x = 0\,.$$

Interval rule, definite integrals may be decomposed into integrals over parts of the interval,

$$\int_a^b f(x)\,dx = \int_a^c f(x)\,dx + \int_c^b f(x)\,dx\,.$$

Partial integration, inversion of the product rule of differentiation,

$$\int f(x)\cdot g'(x)\,dx = f(x)\cdot g(x) - \int f'(x)\cdot g(x)\,dx\,.$$

Substitution rule:

$$\int f(g(x))\cdot g'(x)\,dx = \int f(z)\,dz\,,\quad z = g(x)\,.$$

Logarithmic integration:

$$\int \frac{f'(x)}{f(x)}\,dx = \ln|f(x)| + C\,.$$

32.3   Derivatives and integrals of elementary functions

Given are the original function $f(x)$, its derivative $f'(x) = \dfrac{df}{dx}$ and the integral function $\int f(x)\,dx = F(x) + C.$

$f(x)$	$f'(x)$	$F(x)$	$f(x)$	$f'(x)$	$F(x)$
c	0	cx	e^x	e^x	e^x
x	1	$\frac{1}{2}x^2$	a^x	$a^x \ln(a)$	$\frac{a^x}{\ln(a)}$
x^a	ax^{a-1}	$\frac{x^{a+1}}{a+1}$	$\ln(x)$	$\frac{1}{x}$	$x\ln x - x$
$\frac{1}{x}$	$-\frac{1}{x^2}$	$\ln|x|$	$\log_a(x)$	$\frac{1}{x\ln(a)}$	$\frac{x\ln x - x}{\ln(a)}$
$\sin(x)$	$\cos(x)$	$-\cos(x)$	$\arcsin(x)$	$\frac{1}{\sqrt{1-x^2}}$	$x\arcsin(x) + \sqrt{1-x^2}$
$\cos(x)$	$-\sin(x)$	$\sin(x)$	$\arccos(x)$	$\frac{-1}{\sqrt{1-x^2}}$	$x\arccos(x) - \sqrt{1-x^2}$
$\tan(x)$	$\frac{1}{\cos^2(x)}$	$-\ln|\cos(x)|$	$\arctan(x)$	$\frac{1}{1+x^2}$	$x\arctan(x) - \frac{1}{2}\ln(1+x^2)$
$\cot(x)$	$\frac{-1}{\sin^2(x)}$	$\ln|\sin(x)|$	$\text{arccot}\,(x)$	$\frac{-1}{1+x^2}$	$x\,\text{arccot}\,(x) + \frac{1}{2}\ln(1+x^2)$
$\sinh(x)$	$\cosh(x)$	$\cosh(x)$	$\text{Arsinh}(x)$	$\frac{1}{\sqrt{x^2+1}}$	$x\text{Arsinh}(x) - \sqrt{x^2+1}$
$\cosh(x)$	$\sinh(x)$	$\sinh(x)$	$\text{Arcosh}(x)$	$\frac{1}{\sqrt{x^2-1}}$	$x\text{Arcosh}(x) - \sqrt{x^2-1}$
$\tanh(x)$	$\frac{1}{\cosh^2(x)}$	$\ln(\cosh(x))$	$\text{Artanh}(x)$	$\frac{1}{1-x^2}$	$x\text{Artanh}(x) + \frac{1}{2}\ln(1-x^2)$
$\coth(x)$	$\frac{-1}{\sinh^2(x)}$	$\ln|\sinh(x)|$	$\text{Arcoth}(x)$	$\frac{1}{1-x^2}$	$x\text{Arcoth}(x) + \frac{1}{2}\ln(x^2-1)$

33
Tables on the SI

33.0/1 International system of units (SI): Basic quantities

Denotation	Abbr.	Definition	Dim.
meter	m	The meter is the length of path traversed by light in a vacuum during the 1/299 792 458th fraction of a second.	**L**
kilogram	kg	The kilogram is the mass of an international prototype of the kilogram. It is a platinum-iridium cylinder deposited at the BIPM in Sèvres, near Paris.	**M**
second	s	The second is the duration of 9 192 631 770 vibrational periods of the radiation corresponding to the transition between the two hyperfine structure levels of the ground state of the Cs^{133} atom.	**T**
ampere	A	The ampere is the constant current that, when flowing through two infinitely extended conductors of negligible cross-sectional area positioned 1 meter apart in a vacuum, generates a force of $2 \cdot 10^{-7}$ N per meter of length.	**I**
kelvin	K	The kelvin is the 1/273.16th fraction of the thermodynamic temperature of the triple point of water.	**Θ**
mole	mol	The mole is the amount of substance that contains as many elementary constituents as there are atoms in 0.012 kg of carbon 12.	**N**
candela	cd	The candela is the luminosity in a given direction of a monochromatic source of radiation of frequency of $540 \cdot 10^{12}$ hertz and a radiant intensity in that direction of (1/683) watt per steradian.	**J**

33.0/2   Decimal prefixes

Prefix	Value	Abbreviation	Prefix	Value	Abbreviation
yocto	10^{-24}	y	deca	10^1	da
zepto	10^{-21}	z	hecto	10^2	h
atto	10^{-18}	a	kilo	10^3	k
femto	10^{-15}	f	mega	10^6	M
pico	10^{-12}	p	giga	10^9	G
nano	10^{-9}	n	tera	10^{12}	T
micro	10^{-6}	μ	peta	10^{15}	P
milli	10^{-3}	m	exa	10^{18}	E
centi	10^{-2}	c	zetta	10^{21}	Z
deci	10^{-1}	d	yotta	10^{24}	Y

33.0/3   Derived SI units

Denotation	Symbol	Defining equation	Unit	Name of unit
1. length				
angle	$\alpha, \varphi, \ldots$		rad	radian
solid angle	Ω		sr	steradian
length	$s, l, \ldots$		m	meter
area	A	$A = s^2$	m^2	
volume	V	$V = s^3$	m^3	
2. time and velocity				
time	t		s	second
vibrational period	T	$T = \dfrac{\text{time}}{\text{vibrations}}$	s	
frequency	f	$f = 1/T$	$\text{Hz} = 1/\text{s}$	hertz
velocity	$\vec{v}$	$v = \mathrm{d}s/\mathrm{d}t$	m s^{-1}	
angular velocity	$\vec{\omega}$	$\omega = \mathrm{d}\alpha/\mathrm{d}t$	rad s^{-1}	
acceleration	$\vec{a}$	$a = \mathrm{d}^2 s/\mathrm{d}t^2$	m s^{-2}	
angular acceleration	$\vec{\alpha}$	$\alpha = \mathrm{d}^2\varphi/\mathrm{d}t^2$	rad s^{-2}	
3. mechanics				
mass	m		kg	kilogram
density	ρ	$\rho = m/V$	kg m^{-3}	
force	$\vec{F}$	$F = m \cdot a$	$\text{N} = \text{kg m s}^{-2}$	newton
moment of inertia	J	$J = \sum_i m_i r_i^2$	kg m^2	
torque	τ	$\tau = r \times F$	N m	
momentum	$\vec{p}$	$p = m \cdot v$	kg m s^{-1}	
pressure	p	$p = F/A$	$\text{Pa} = \text{N m}^{-2}$	pascal
work, energy	W	$W = \int \vec{F} \cdot \mathrm{d}\vec{s}$	$\text{J} = \text{N m}$	joule
power	P	$P = \mathrm{d}W/\mathrm{d}t$	$\text{W} = \text{N m s}^{-1}$	watt
surface tension	σ	$\sigma = \mathrm{d}W/\mathrm{d}A$	N m^{-1}	
elasticity modulus	E	$E = \sigma/\varepsilon$	N m^{-2}	
compression modulus	K	$K = -V\mathrm{d}p/\mathrm{d}V$	N m^{-2}	
dynamic viscosity	η	$\eta = (F_R/A) \cdot \mathrm{d}d/\mathrm{d}v$	Pa s	
kinematic viscosity	ν	$\nu = \eta/\rho$	$\text{m}^2\,\text{s}^{-1}$	
efficiency	η	$\eta = P_{\text{eff}}/P_{ein}$	1	

(continued)

33.0/3   Derived SI units (*continued*)

Denotation	Symbol	Defining equation	Unit	Name of unit
4. electricity and magnetism				
electric charge	Q	$Q = I \cdot t$	$C = A\,s$	coulomb
electric voltage	V	$V = W/Q$	$V = J\,C^{-1}$	volt
electric field strength	$\vec{E}$	$\vec{E} = \vec{F}/Q$	$N\,C^{-1} = V\,m^{-1}$	
electric resistance	R	$R = V/I$	$\Omega = V\,A^{-1}$	ohm
electric conductance	G	$G = 1/R$	$S = \Omega^{-1}$	siemens
spec. el. resistance	ρ	$\rho = RA/l$	$\Omega\,m$	
spec. el. conductance	κ	$\kappa = 1/\rho$	$\Omega^{-1}m^{-1}$	
electric capacitance	C	$C = Q/V$	$F = C\,V^{-1}$	farad
permittivity	ε	$\varepsilon = D/E$	$F\,m^{-1}$	
magnetic flux	Φ	$\Phi = \int V dt$	$Wb = V\,s$	weber
inductance	L	$L = \Phi/I$	$H = V\,s\,A^{-1}$	henry
magn. flux density	$\vec{B}$	$B = d\Phi/dA$	$T = Wb\,m^{-2}$	tesla
magn. field strength	$\vec{H}$	$H = dI/ds$	$A\,m^{-1}$	
permeability	μ	$\mu = B/H$	$H\,m^{-1}$	
5. thermodynamics				
temperature	T		K	kelvin
quantity of heat	Q	(= form of energy)	J	joule
heat capacitance	C	$C = \Delta Q/\Delta T$	$J\,K^{-1}$	
spec. heat capacitance	c	$c = C/m$	$J\,K^{-1}\,kg^{-1}$	
heat conductivity	λ	$\lambda = l\,dQ/At\,dT$	$W\,K^{-1}\,m^{-1}$	
entropy	S	$S = Q/T$	$J\,K^{-1}$	
spec. caloric power	H	$H = Q/m$	$J\,kg^{-1}$	
internal energy	U	$U = \dfrac{f}{2}n_{mol}RT$	J	
free energy	F	$F = U - TS$	J	
enthalpy	H	$H = U + pV$	J	
free enthalpy	G	$G = U + pV - TS$	J	
6. physical chemistry				
particle number	N		1	
particle number density	n	$n = N/V$	m^{-3}	
amount of substance	n	$n = N/N_{A}$	mol	mole
7. light				
light intensity	I		cd	candela
light flow	Φ	$\Phi = \int I d\Omega$	$lm = cd\,sr$	lumen
amount of light	Q	$Q = \int \Phi dt$	$lm\,s$	
luminance	L	$L = dI/(dA\cos\theta)$	$cd\,m^{-2}$	
illuminance	E	$E = (d\Phi/dA)\cos\theta$	$lx = lm\,m^{-2}$	lux
exposure	H	$H = \int E dt$	$lx\,s$	
radiant flux	Φ_e	$\Phi_e = dW/dt$	W	
radiant intensity	I_e	$I_e = d\Phi_e/d\Omega$	$W\,sr^{-1}$	
radiant density	B_e	$B_e = dI_e/(dA\cos\theta)$	$W\,m^{-2}\,sr^{-1}$	
irradiance	E_e	$E_e = (d\Phi_e/dA)\cos\theta$	$W\,m^{-2}$	
irradiation	H_e	$H_e = \int E_e dt$	$J\,m^{-2}$	
focal length	f	$1/f = 1/a + 1/b$	m	

(*continued*)

33.0/3   Derived SI units (*continued*)

Denotation	Symbol	Defining equation	Unit	Name of unit
8. nuclear reactions				
decay constant	λ	$\lambda = -dN/(N\,dt)$	s^{-1}	
half-life	$T_{1/2}$	$T_{1/2} = \ln 2/\lambda$	s	
activity	A	$A = \dfrac{\text{decays}}{\text{time}}$	$Bq = s^{-1}$	becquerel
spec. activity	a	$a = A/m$	$Bq\,kg^{-1}$	
energy dose	D	$D = W/m$	$Gy - J\,kg^{-1}$	gray
energy dose rate	$\dot{D}$	$\dot{D} = dD/dt$	$Gy\,s^{-1}$	
equivalent dose	D_q	$D_q = q \cdot N \cdot D$ (1)	$Sv = J\,kg^{-1}$	sievert
cross-section	σ	$\sigma = \dfrac{-dN}{nN ds}$	m^2	
9. acoustics				
sound pressure	p		Pa	
sound pressure level	L_p	$L_p = 20\log_{10}(p/p_0)$	db	decibel
volume level	L_N	$L_N = 20\log_{10}(p/p_0)$	phon	phon

(1) q is a quality factor for the different types of radiation. N is the product of several factors that are defined in more detail by the ICRP (International Commission on Radiological Protection). They are related to biological efficiency.

33.0/4   Accepted non-SI units

This table surveys other accepted units and their conversion to SI units.

Quantity	Unit	Abbreviation	Relation to SI unit
generally valid			
plane angle	second	″	$1'' = (1/60)'$
	minute	′	$1' = (1/60)°$
	degree	°	$1° = (\pi/180)\,rad$
volume	liter	l	$1\,l = 10^{-3}\,m^3$
time	minute	min	$1\,min = 60\,s$
	hour	h	$1\,h = 60\,min = 3600\,s$
	day	d	$1\,d = 24\,h = 86400\,s$
	common year	a, yr	$1\,a = 365\,d = 8760\,h$
mass	ton	t	$1\,t = 10^3\,kg$
pressure	bar	bar	$1\,bar = 10^5\,Pa$
valid in special fields			
length in	light year	ly	$1\,ly = 9.4605 \cdot 10^{15}\,m$
astronomy	parsec	pc	$1\,pc = 3.0857 \cdot 10^{16}\,m = 3.26\,ly$
	astronomic unit	AU	$1\,AU = 1.4959787 \cdot 10^{11}\,m$
length in navigation	nautical mile	sm	$1\,sm = 1852\,m$
length in atomic physics	angstrom unit	Å	$1\,Å = 10^{-10}\,m$
velocity in air and sea navigation	knot	kn	$kn = 1\,sm\,h^{-1} = 0.514444\,m\,s^{-1}$

(continued)

33.0/4 Accepted non-SI units (*continued*)

Quantity	Unit	Abbreviation	Relation to SI unit
valid in special fields (*continued*)			
refractive power of lenses	dioptric	dpt	$1\ \mathrm{dpt} = \mathrm{m}^{-1}$
area of land	hectare	ha	$1\ \mathrm{ha} = 10^4\ \mathrm{m}^2$
	acre	a	$1\ \mathrm{a} = 10^2\ \mathrm{m}^2$
liquids	liter	l	$1\ \mathrm{l} = 1\ \mathrm{dm}^3 = 10^{-3}\ \mathrm{m}^3$
plane angle in geodesy	gon	gon	$1\ \mathrm{gon} = (\pi/200)\ \mathrm{rad}$
size of textile threads	tex	tex	$1\ \mathrm{tex} = 10^{-6}\ \mathrm{kg\,m}^{-1}$
mass of precious stones	carat	Kt	$1\ \mathrm{Kt} = 0.2\ \mathrm{g}$
mass in atomic physics	atomic mass unit	u	$1\ \mathrm{u} = 1.660\,540\,2 \cdot 10^{-27}\ \mathrm{kg}$
energy in atomic physics	electron volt	eV	$1\ \mathrm{eV} = 1.602\,177\,33 \cdot 10^{-19}\ \mathrm{J}$

33.0/5 Conversion table of energy units

	erg	J	kWh
1 erg	1	10^{-7}	$2.7778 \cdot 10^{-14}$
1 J	10^7	1	$2.7778 \cdot 10^{-7}$
1 kWh	$3.6 \cdot 10^{13}$	$3.6 \cdot 10^6$	1
1 kpm	$9.8066 \cdot 10^7$	9.8066	$2.72 \cdot 10^{-6}$
1 kcal	$4.1868 \cdot 10^{10}$	$4.1868 \cdot 10^3$	$1.16 \cdot 10^{-3}$
1 eV	$1.6021 \cdot 10^{-12}$	$1.6 \cdot 10^{-19}$	$4.45 \cdot 10^{-26}$

	kpm	kcal	eV
1 erg	$1.0197 \cdot 10^{-8}$	$2.3884 \cdot 10^{-11}$	$6.2419 \cdot 10^{11}$
1 J	$1.10197 \cdot 10{-1}$	$2.3884 \cdot 10^{-4}$	$6.2419 \cdot 10^{18}$
1 kWh	$3.6709 \cdot 10^5$	$8.6001 \cdot 10^2$	$2.25 \cdot 10^{25}$
1 kpm	1	$2.3427 \cdot 10^{-3}$	$2.6126 \cdot 10^{19}$
1 kcal	$4.2685 \cdot 10^2$	1	$2.6126 \cdot 10^{22}$
1 eV	$1.634 \cdot 10^{-20}$	$3.8276 \cdot 10^{-23}$	1

33.0/6 Wind forces

(as measured 10 m above ground)

Beaufort degree	Velocity	Dynamic pressure	Name / indication
3	3.4 to 5.3 m/s	ca. $0.017\ \mathrm{kN/m}^2$	wind / moves leaves
6	9.9 to 12.4 m/s	ca. $0.08\ \mathrm{kN/m}^2$	strong wind / moves strong boughs, howls
9	18.3 to 21.5 m/s	ca. $0.25\ \mathrm{kN/m}^2$	storm / moves loose stones
12	beyond 30 m/s	beyond $0.5\ \mathrm{kN/m}^2$	hurricane / moves heavy objects

33.0/7   Anglo-American units

Quantity	Unit	Abbreviation	Conversion to SI units
length	inch	in	1 in = 0.0254 m
	foot	ft	1 ft = 12 in = 0.3048 m
	yard	yd	1 yd = 3 ft = 0.9144 m
	statute mile	mile	1 mile = 1760 yd = 1609.34 m
	nautical mile	n mile	1 n mile = 1852 m
area	square inch	in^2	1 in^2 = 6.452 · 10^{-4} m^2
	square foot	ft^2	1 ft^2 = 144 in^2 = 0.0929 m^2
	square yard	yd^2	1 yd^2 = 9 ft^2 = 0.8361 m^2
	square mile	mile^2	1 mile^2 = 2.59 · 10^6 m^2
	acre	a	1 a = 4046.86 m^2
volume	cubic inch	in^3	1 in^3 = 1.63871 · 10^{-5} m^3
	cubic foot	ft^3	1 ft^3 = 0.02832 m^3
	cubic yard	yd^3	1 yd^3 = 0.76456 m^3
	gallon	gal	1 gal = 3.78541 · 10^{-3} m^3
	registerton	RT	1 RT = 100 ft^3 = 2.832 m^3
velocity	mile per hour	mph	1 mph = 1.609 km/h = 0.447 m/s
mass	grain	gr	1 gr = 6.4799 · 10^{-5} kg
	dram	dram	1 dram = 1.77184 · 10^{-3} kg
	ounce	oz	1 oz = 2.83495 · 10^{-2} kg
	pound	lb	1 lb = 0.45359 kg
	long hundredweight	long cwt	1 long cwt = 50.8023 kg
	short hundredweight	sh cwt	1 sh cwt = 45.3592 kg
	long ton	long tn	1 long tn = 1016.05 kg
	short ton	sh tn	1 sh tn = 907.185 kg
pressure	pound-force per square inch	lbf/in^2	1 lbf in^{-2} = 6.8947 · 10^3 Pa
	pound-force per square foot	lbf/ft^2	1 lbf ft^{-2} = 47.88 Pa
	ton-force per square foot	tonf/ft^2	1 tonf ft^{-2} = 107.252 · 10^3 Pa
energy	foot pound-force	ft lbf	1 ft lbf = 1.3558 J
	British thermal unit	Btu	1 Btu = 1055.06 J
power	horsepower	hp	1 hp = 745.7 W

Index

Thermodynamic formulas

Change of state	Isothermal process $\Delta T = 0$, T_{Pr}	Isobaric process $\Delta p = 0$, p_{Pr}	Isochoric process V_{Pr} (Isovolume) $\Delta V = 0$	Adiabatic process (Isentropic) S_{Pr} $\Delta Q = 0$, $S = $ const.	Polytropic process
constant	pV, $T = $ const.	V/T, $p = $ const.	p/T, $V = $ const.	$\Delta Q = 0$, $S = $ const.	$p \cdot V^n = $ const.
law	$\dfrac{p_1}{p_2} = \dfrac{V_2}{V_1}$	$\dfrac{V_1}{V_2} = \dfrac{T_1}{T_2}$	$\dfrac{p_1}{p_2} = \dfrac{T_1}{T_2}$	$\dfrac{p_1}{p_2} = \left(\dfrac{V_2}{V_1}\right)^{\kappa}$	$\dfrac{p_1}{p_2} = \left(\dfrac{V_2}{V_1}\right)^{n}$
polytropic coefficient	$n = 1$	$n = 0$	$n = \infty$	$n = \kappa$	$n = n$
internal energy $\Delta U = \Delta Q + \Delta W$	$\Delta U = 0$	$C_V(T_2 - T_1)$	$C_V(T_2 - T_1) = \Delta Q$	$C_V(T_2 - T_1) = \Delta W$	$C_V(T_2 - T_1)$
absorbed heat $\Delta Q = C\Delta T$	$\Delta Q = -\Delta W$	$C_p(T_2 - T_1)$	$C_V(T_2 - T_1) = \Delta U$	$\Delta Q = 0$	$\dfrac{n-\kappa}{\kappa-1}W =$ $C_V \dfrac{n-\kappa}{n-1}(T_2 - T_1)$
compression work $\Delta W = -\int p\,dV$	$p_1 V_1 \ln\left(\dfrac{V_2}{V_1}\right) = p_1 V_1 \ln\left(\dfrac{p_1}{p_2}\right)$	$p(V_1 - V_2)$	$\Delta W = 0$	$\Delta W = \Delta U =$ $\dfrac{p_2 V_2 - p_1 V_1}{\kappa - 1}$	$\dfrac{p_2 V_2 - p_1 V_1}{n-1} =$ $C_V \dfrac{\kappa-1}{n-1}(T_2 - T_1)$
technical work $\Delta W_t = \int V\,dp$	$\Delta W_t = \Delta W$	$\Delta W_t = 0$	$V(p_2 - p_1)$	$\Delta W_t = \kappa \Delta W$	$\Delta W_t = n\Delta W$
change of entropy $\Delta S = \dfrac{\Delta Q}{T}$	$mR_s \ln\left(\dfrac{V_2}{V_1}\right) = \dfrac{W}{T}$, $\dfrac{V_2}{V_1} = \dfrac{p_1}{p_2}$, $mR_s = \dfrac{pV}{T}$	$C_p \ln\left(\dfrac{T_2}{T_1}\right) = C_p\left(\dfrac{V_2}{V_1}\right)$	$C_V \ln\left(\dfrac{T_2}{T_1}\right) = C_V\left(\dfrac{p_2}{p_1}\right)$	$\Delta S = 0$	$C_V \dfrac{n-\kappa}{n-1} \ln\left(\dfrac{T_2}{T_1}\right)$